TABLE OF ATOMIC WEIGHTS AND NUMBERS

Based on the 1977 Report of the Commission on Atomic Weights of the International Union of Pure and Applied Chemistry. Scaled to the relative atomic mass of carbon-12.

Element	Symbol	Atomic Number	Atomic Weight	
Actinium	Ac	89	227.0278	(e)
Aluminum	Al	13	26.98154	
Americium	Am	95	(243)	(f)
Antimony	Sb	51	121.75	(a)
Argon	Ar	18	39.948	(a, b, c)
Arsenic	As	33	74.9216	
Astatine	At	85	(210)	(f)
Barium	Ba	56	137.33	(c)
Berkelium	Bk	97	(247)	(f)
Beryllium	Be	4	9.01218	
Bismuth	Bi	83	208.9804	
Boron	B	5	10.81	(b, d)
Bromine	Br	35	79.904	
Cadmium	Cd	48	112.41	(c)
Calcium	Ca	20	40.08	(c)
Californium	Cf	98	(251)	(f)
Carbon	C	6	12.011	(b)
Cerium	Ce	58	140.12	(c)
Cesium	Cs	55	132.9054	
Chlorine	Cl	17	35.453	
Chromium	Cr	24	51.996	
Cobalt	Co	27	58.9332	
Copper	Cu	29	63.546	(a, b)
Curium	Cm	96	(247)	(f)
Dysprosium	Dy	66	162.50	(a)
Einsteinium	Es	99	(252)	(f)
Erbium	Er	68	167.26	(a)
Europium	Eu	63	151.96	(c)
Fermium	Fm	100	(257)	(f)
Fluorine	F	9	18.998403	
Francium	Fr	87	(223)	(f)
Gadolinium	Gd	64	157.25	(a, c)
Gallium	Ga	31	69.72	
Germanium	Ge	32	72.59	(a)
Gold	Au	79	196.9665	
Hafnium	Hf	72	178.49	(a)
Helium	He	2	4.00260	(c)
Holmium	Ho	67	164.9304	
Hydrogen	H	1	1.0079	(b)
Indium	In	49	114.82	(c)
Iodine	I	53	126.9045	
Iridium	Ir	77	192.22	(a)
Iron	Fe	26	55.847	(a)
Krypton	Kr	36	83.80	(c, d)
Lanthanum	La	57	138.9055	(a, c)
Lawrencium	Lr	103	(260)	(f)
Lead	Pb	82	207.2	(b, c)
Lithium	Li	3	6.941	(a, b, c, d)
Lutetium	Lu	71	174.967	(a)
Magnesium	Mg	12	24.305	(c)
Manganese	Mn	25	54.9380	
Mendelevium	Md	101	(258)	(f)
Mercury	Hg	80	200.59	(a)
Molybdenum	Mo	42	95.94	
Neodymium	Nd	60	144.24	(a, c)
Neon	Ne	10	20.179	(a, d)
Neptunium	Np	93	237.0482	(e)
Nickel	Ni	28	58.70	
Niobium	Nb	41	92.9064	
Nitrogen	N	7	14.0067	
Nobelium	No	102	(259)	(f)
Osmium	Os	76	190.2	(c)
Oxygen	O	8	15.9994	(a, b)
Palladium	Pd	46	106.4	(c)
Phosphorus	P	15	30.97376	
Platinum	Pt	78	195.09	(a)
Plutonium	Pu	94	(244)	(f)
Polonium	Po	84	(209)	(f)
Potassium	K	19	39.0983	(a)
Praseodymium	Pr	59	140.9077	
Promethium	Pm	61	(145)	(f)
Protactinium	Pa	91	231.0359	(e)
Radium	Ra	88	226.0254	(c, e)
Radon	Rn	86	(222)	(f)
Rhenium	Re	75	186.207	
Rhodium	Rh	45	102.9055	
Rubidium	Rb	37	85.4678	(a, c)
Ruthenium	Ru	44	101.07	(a, c)
Samarium	Sm	62	150.4	(c)
Scandium	Sc	21	44.9559	
Selenium	Se	34	78.96	(a)
Silicon	Si	14	28.0855	(a)
Silver	Ag	47	107.868	(c)
Sodium	Na	11	22.98977	
Strontium	Sr	38	87.62	(c)
Sulfur	S	16	32.06	(b)
Tantalum	Ta	73	180.9479	(a)
Technetium	Tc	43	(98)	(f)
Tellurium	Te	52	127.60	(a, c)
Terbium	Tb	65	158.9254	
Thallium	Tl	81	204.37	(a)
Thorium	Th	90	232.0381	(c, e)
Thulium	Tm	69	168.9342	
Tin	Sn	50	118.69	(a)
Titanium	Ti	22	47.90	(a)
Tungsten	W	74	183.85	(a)
(Unnilhexium)	(Unh)	106	(263)	(f, g)
(Unnilpentium)	(Unp)	105	(262)	(f, g)
(Unnilquadium)	(Unq)	104	(261)	(f, g)
Uranium	U	92	238.029	(c, d)
Vanadium	V	23	50.9415	(a)
Xenon	Xe	54	131.30	(c, d)
Ytterbium	Yb	70	173.04	(a)
Yttrium	Y	39	88.9059	
Zinc	Zn	30	65.38	
Zirconium	Zr	40	91.22	(c)

Except as noted in the footnotes that follow, the atomic weight values are good to ±1 unit in the last place.

(a) Precise to ±3 units in the last place.

(b) Atomic weight cannot be expressed more precisely because among normal terrestrial materials there are known variations in isotopic compositions.

(c) Geological samples of this element have been found with anomalous isotopic composition and atomic weights different from this value.

(d) Considerable variations from this atomic weight value can occur in commercial samples because of changes in isotopic composition.

(e) The atomic weight of the radioisotope of longest half-life.

(f) The mass number of the radioisotope of longest half-life.

(g) The official name and symbol has not been agreed to. Element 105 is unofficially called hahnium. Element 104 is called rutherfordium by American scientists and kurchatovium by Russian scientists.

FUNDAMENTALS
OF CHEMISTRY

FUNDAMENTALS OF CHEMISTRY

JAMES E. BRADY
St. John's University
New York

JOHN R. HOLUM
Augsburg College
Minnesota

JOHN WILEY & SONS
New York   Chichester   Brisbane   Toronto

Photo Editor: Kathy Bendo

This book designed by Raphael Hernandez
Production Supervised by Linda Indig
Illustrations by John Balbalis

Library of Congress Cataloging in Publication Data:

Brady, James E
  Fundamentals of chemistry.

  Includes indexes.
  1. Chemistry. I. Holum, John R., joint author.
II. Title.
QD31.2.B69   540   80-21079
ISBN 0-471-05816-5

Printed in the United States of America

10 9 8 7 6 5 4 3 2 1

PREFACE

When we were students, those of us who teach chemistry saw the subject as something exciting and fascinating. Now, as teachers, we try to impart this same sense of excitement to our own students. Over the years this has become more difficult as the introductory chemistry course for science majors has gradually become largely a service course. Motivating our students—turning them on to chemistry—is often a major hurdle that we must overcome. Another problem is that many of our students are, to a certain degree, less well prepared by their high-school training than students of, say ten years ago. Nevertheless, the chemical principles and facts that they must learn and understand are about the same as before.

In writing this book, we considered both of these problems. To heighten students' interest and to make them aware of the importance of chemistry in their lives, the discussions of principles continually draw on students' everyday experience and nontechnical background for illustrative examples. This includes chemistry and chemicals encountered in day-to-day living as well as topics that have received widespread attention in the news media. Our discussions and examples are further complemented by the liberal use of photographs. In addition, we consulted faculty members in disciplines serviced by the general chemistry course in order to identify which topics are important to students majoring in those disciplines. These topics are often used as illustrations in the development of concepts and in example problems. Enrichment material and special applications of chemical principles to disciplines other than chemistry, especially the life sciences, also appear in "Special Topics" placed strategically throughout the book. In this way, students will understand how chemistry applies to their chosen specialty.

To help less well-prepared students, the level of mathematical sophistication has been limited to basic algebraic operations; lengthy derivations have been avoided. This, along with the informal writing style, was deliberately chosen to present the material in a nonthreatening way. Extreme care has been taken throughout to provide clear, concise explanations of topics that use a language level appropriate for today's college freshman. Particular attention has been given to the development of concepts, recognizing that first-year students often have as much difficulty with theory as they do with applications and problem solving.

Instructors will notice the teaching aids that we have introduced to improve student understanding and performance. Each section title appears in a short form to identify section content; then there is a brief introductory statement that summarizes the section. This will give students a brief preview of the section content and aid them later during review. Another review aid is chapter summaries that provide an overview of important concepts. Numerous marginal comments and figures are included to supplement the textual material. Within the text itself, new terms are set in blue boldface type when they are first introduced and defined. They also appear in the glossary at the end of the book.

To clarify problem solving and improve student performance, we have included a large number of worked-out examples, each identified by type so that the student will recognize the lesson to be learned or the principle demonstrated. The factor-label method is emphasized. Exercises immediately follow worked-out examples to help build confidence, reinforce principles, and allow students to test their understanding. The answers to all of these exercises are given in Appendix C.

End-of-chapter review exercises are divided into questions (nonmathematical) and problems. The problems are presented in pairs of the same or similar type. Answers to one member of each pair also appear in Appendix C. In each chapter there is an index to exercises, questions, and problems to help students structure their studies and to help instructors assign homework.

The order of topics also has been chosen with today's typical science major in mind. In the introductory chapter, groundwork is laid for the laboratory by the introduction of systems of units and the notion of significant figures. We have chosen to use a mix of SI and traditional metric units in this text for several reasons: (1) there are many scientists who have not accepted the full use of SI units; (2) students must be prepared to encounter the traditional units that appear in all but the most recent literature; and (3) the traditional units are still widely used in the "real world." Therefore, we use torr and atmospheres in our calculations involving gases; energy tables are given in both joules and calories; and volumes are given in milliliters or liters.

We discuss gases early in the text because it allows students to see how the ideas of science are applied to a familiar state of matter and how the study of matter leads to the development of theoretical models.

Stoichiometry is divided into two chapters. The first one deals with the mole concept and empirical and molecular formulas. The ideal gas law is discussed here because of its application to molecular weights of gases. The second stoichiometry chapter concentrates on chemical reactions and includes a treatment of solution stoichiometry so that quantitative experiments can be performed early in the laboratory.

Energy is the next topic (Chapter 5). The importance of the subject is obvious to the student and the goal of the chapter is to provide a sound understanding of the meaning of energy and how it relates to chemical systems. Some basic thermodynamic principles are presented here.

Before studying electronic structure and the principles of ionic and covalent bonding in Chapters 7 and 8, students are introduced in Chapter 6 to some basic chemical and physical properties of the elements and their compounds through a discussion of the periodic table and some periodic trends. The purpose here is to provide the student with some chemical background and to illustrate the need to develop a theoretical understanding of chemical structure. Although a conscious effort has been made to weave descriptive inorganic chemistry into discussions and examples throughout the book, Chapter 6 is the first of four chapters (6, 12, 13, and 20) that deal explicitly with this subject. In all of these chapters, extra care has been taken to relate chemistry to the students' daily experiences.

After presenting the basic concepts of covalent bonding in Chapter 8, modern theories of bonding are developed in Chapter 9 along with an overview of organic chemistry.

Chapters 10 and 11 discuss further the states of matter and solutions, emphasizing their physical behavior. Next, Chapters 12 and 13 provide an interlude of descriptive chemistry of the representative nonmetals. Chapter 12 focuses on simple species, and Chapter 13 examines macromolecular species, both natural and synthetic.

Chapters 14 to 17 examine factors that control the outcome of chemical reactions. Rates of reaction are discussed first, followed by the thermodynamic criteria for reaction spontaneity. Emphasis in the thermodynamics chapter (Chapter 15) emphasizes the development of concepts such as entropy and Gibbs free energy. Chapter 16 studies equilibrium and related factors Chapter 17 deals with acid-base equilibria in aqueous solutions.

Placing the electrochemistry chapter after the chapters on thermodynamics and equilibrium permits us to introduce the Nernst equation and the calculation of equilibrium constants from cell potentials.

The final three chapters are largely descriptive. After a survey of biochemistry (Chapter 19), we consider the properties and preparation of the metals, both representative and transition. The final chapter returns to nuclear reactions, the basic elements of which are presented in Chapter 5. In Chapter 21 the emphasis is on the applications of nuclear reactions to chemistry and to the fields of specialization of the students.

JAMES E. BRADY
JOHN R. HOLUM

ACKNOWLEDGMENTS

Doris Berg and June Brady were our typists, and we are happy to acknowledge with thanks their patient and careful work. To our wives, Mary Holum and June Brady, and to our children, Liz, Ann, and Kathryn Holum and Mark and Karen Brady, go special thanks for lifetimes of encouragement and support. We express special appreciation for the encouragement, good humor, and fine work provided by the staff at Wiley, particularly our editor, Gary Carlson; his assistant, Wendy Wanger; our picture editor, Kathy Bendo; our illustrator, John Balbalis; our designer, Raphael Hernandez; and our production supervisor, Linda Indig. We also appreciate the editorial assistance of Stephen Perine, who helped smooth out our stylistic differences. Finally, we are especially grateful to the following colleagues who have contributed so much to this book by their thoughtful reviews and criticisms of the manuscript and their many valuable suggestions.

David Becker
Oakland Community College

Jo A. Beran
Texas A and I University

Robert Coley
Montgomery County Community College

Gordon J. Ewing
New Mexico State University

Larry L. Funck
Wheaton College, Illinois

Helen Hauer
Delaware Technical and Community College

I. C. Hisatsune
The Pennsylvania State University

Delwin D. Johnson
Forest Park Community College

Dane Jones
California Polytechnic State University

Neil R. Kestner
Louisiana State University

Lavier J. Lokke
San Diego Mesa College

David Neher
Texas A and I University

Jack E. Powell
Iowa State University of Science and Technology

Muriel Ramsden
Union College

Dr. Vincent J. Sollimo
Burlington County College

J.E.B.
J.R.H.

SUPPLEMENTARY MATERIAL FOR STUDENTS AND TEACHERS

A complete package of supplements has been assembled to aid the teacher in presenting the course and to help students accomplish problem solving and other study assignments.

Study Guide for Fundamentals of Chemistry by James E. Brady and John R. Holum. This softcover book has been carefully structured to assist students in mastering the important subjects in the text. For each section there is a statement of objectives, followed by a brief review of major topics, sometimes with additional worked-out examples. Most sections have a brief "Self-Test," with answers provided, that supplement the text exercises. Each section concludes with a list of new terms, and each chapter has its own glossary.

Laboratory Manual for Fundamentals of Chemistry by Jo A. Beran. This manual features a thorough "techniques" section, with photographs of important manipulations, and 46 experiments sequenced to follow the topical development of the text. For the teacher, an instructor's manual accompanies the laboratory manual.

Solutions Manual for Fundamentals of Chemistry by Ernest R. Birnbaum. This softcover supplement provides detailed solutions to all of the numerical in-chapter exercises and end-of-chapter review problems.

Teacher's Manual for Fundamentals of Chemistry. This manual, available to teachers only, provides all of the usual services.

Transparency Masters. Instructors who adopt this book may obtain from Wiley, without charge, a set of 8½ × 11-inch, black-and-white line drawings that duplicate key figures and tables in the text. They can be used to prepare transparencies.

J.E.B.
J.R.H.

CONTENTS

CHAPTER ONE

INTRODUCTION

1.1 WHY STUDY CHEMISTRY?

Chemistry is needed in most programs in the other sciences because it describes and explains the composition and behavior of so many different substances

For those with a curious eye, the world about us is filled with fascinating happenings. Some of them—such as earthquakes, volcanos, and lightning—occur on a grand and fearsome scale. Others—the germination of a seed, for example—take place under more serene conditions. Around us each day fires burn, iron rusts, grass grows and the human race goes on, through growth, death, reproduction, and birth.

Throughout history, thoughtful people have marveled at the wonders of nature and have looked for an understanding of natural phenomena. The curiosity of prehistoric humans led them to attempt to understand fire, so that they could put it to work for their benefit. Attempts to explain fire continued until 1789 when the French scientist, Antoine Lavoisier, finally discovered that combustion occurs when oxygen in the air unites with certain substances in the fuel. Today, human curiosity continues as we seek, among many things, to unravel the secrets of life and harness the energy source that fuels the sun. Thus science, with all its specialties, has gradually evolved.

Today, we have come to realize that no branch of science can be entirely independent of the others. To be a scientist in the modern world requires basic knowledge in areas that overlap our own specialities. Indeed, the successes that have resulted from the cross-fertilization of ideas among the various scientific disciplines are apparent almost anywhere we look. They include everything from electronic calculators to nuclear power plants, from insecticides to plastics to miracle drugs. Therefore, it is easy to see that scientists must be able to communicate with one another. Biologists must be able to talk to chemists, and chemists must be able to talk to physicists. This, in fact, is probably the reason you are reading this book. It is likely that you will need a knowledge of chemistry in your planned specialty.

Chemistry is the study of the composition of substances and the way the behavior of substances relates to their composition. Its extreme importance among the sciences arises because so many disciplines must draw on certain

If all of the creations of scientific research were suddenly to disappear, civilization would come to a grinding halt.

chemical principles and concepts. Biologists must know some chemistry so that they can understand such processes as metabolism and energy conversions in organisms. People in the field of medical technology must understand the chemical basis of the tests and analyses that they perform. Pharmacists must understand the chemistry of drug reactions and interactions, and nutritionists must have some understanding of the way the body functions chemically so that they can provide the proper nutrients.

The list of specialties that require a knowledge of chemistry could go on for several pages. But the fact that you plan to go into one of these fields need not be the only reason for you to study chemistry. Throughout this book, you will see that chemistry touches your personal life every day in many, many ways. Learning chemistry, then, can also be a way of gaining a better understanding of your world.

1.2 THE SCIENTIFIC METHOD

Explanations that are broadened into theories often begin with puzzling facts and strange observations

Scientists are really very much like other people. They experience the same emotions, desires, fears, and prejudices that afflict all human beings. Perhaps the one thing that makes them a little different from others, however, is that they have learned to be very aware of things that take place around them and they have learned to analyze what they see in an orderly, logical way. As a student of science, one of your goals should be to sharpen your own awareness and mental abilities so that you can take your place as a member of the scientific community.

Scientists study nature by methods that are not very different from the way that small children explore the world around them. When a child touches a hot stove and gets burned, the idea is planted that there's some relationship between the stove and pain. If brave enough, perhaps the child will test his or her idea by touching the stove again. The logic that we apply to our work as scientists follows a sequence of steps that is only slightly more formalized than that of the child testing the stove. The sequence usually starts when we ask ourselves some spe-

Figure 1.1
A modern chemical research laboratory.

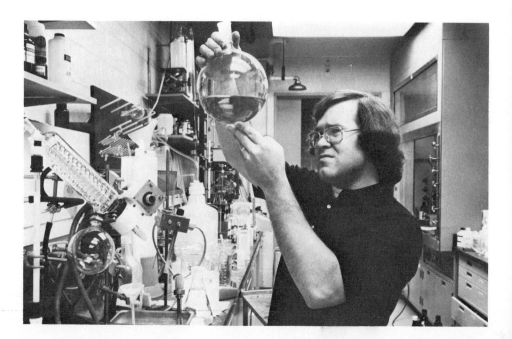

cific question about the behavior of nature. We begin to search for an answer by making observations. Most often this is done in a laboratory (Figure 1.1). We use laboratories because they allow us to control our experiments so that the observations we make are reproducible; that is, we can be sure that if we again do the same things in the same way, we will get the same results. From our observations, we learn certain **empirical facts,** so named because we learn them by observing the behavior of some physical, chemical, or biological system. For example, suppose we wish to study the behavior of gases, such as the air we breathe. We would soon discover that the volume of a gas depends on a number of factors, including the weight of the gas, its temperature, and its pressure. The existence of these factors and the various relationships among them will be our empirical facts.

Webster's defines *empirical* as ''pertaining to, or founded upon, experiment or experience.''

The empirical facts we gather from our observations are referred to as **data,** and we carefully record them. Then, looking at our data, we may be able to summarize some of the information we have discovered. Such summaries of data are called **generalizations.** For instance, one generalization we would probably make from our observation of gases is that when the temperature of a gas rises, the gas tends to expand and occupy a larger volume.

A generalization based on the results of many experiments is called a **law.** As useful as they may be in summarizing the results of experiments, laws can only state what happens. They do not explain *why* substances behave the way they do. Human beings are curious creatures though, and we seek explanations. Therefore, based on the results of our experiments, we formulate a **hypothesis**—a tentative explanation of why nature behaves in certain ways. We use a hypothesis to make predictions of new behavior and then design experiments to test our predictions. If the results of our experiments prove that our hypothesis was incorrect, we *must* discard it and seek a new one. However, if after repeated testing the explanation appears sound, it is raised to the status of a theory. A **theory** is a tested explanation of why nature behaves as it does. Because theories are usually very broad and have many subtle implications, performing every conceivable test that might disprove a theory is virtually impossible. Therefore, we can rarely prove that a theory is correct. Generally all we can do is find additional experimental support for it.

The sequence of steps just described—observation, formulation of a theory, and testing of a theory by additional observation—is called the **scientific method.** Despite its name, the use of this method isn't restricted to those who label themselves scientists. The child touching the hot stove learns by using these steps too, as does an automobile mechanic who listens to your engine and makes various tests to determine why it doesn't run well. The mechanic replaces parts and then starts the engine again to see if the problem has been found. A doctor certainly uses the scientific method all the time in diagnosing illness. Temperature, pulse, blood pressure, and ''where it hurts'' are all pieces of data from which the physician forms a hypothesis about what's causing your pain. Treatment is given based on this hypothesis, and your progress is observed to test it. In short, we all use the scientific method, usually more by instinct than design.

The scientific method is merely a formal description of the way humans approach the experiences of life.

From the preceding discussion you may get the impression that scientific progress always proceeds in a planned and orderly fashion. This isn't true—luck sometimes plays an important role. For example, in 1828 Frederick Wöhler, a German chemist, was heating a substance called ammonium cyanate in an attempt to add support to one of his theories. His experiment, however, produced an unexpected substance that he analyzed and found to be urea (a constitutent of urine). This was the first time that anyone had ever made a substance found in living creatures from a chemical having a nonorganic origin. The fact that this could be done was an important discovery. Nevertheless, if it had not been for Wöhler's application of the scientific method to his unexpected results, the significance of his experiment may have gone unnoticed.

As a final note here, it is significant that the most spectacular and dramatic changes in science occur when major theories are proven incorrect. This happens only rarely, but when it occurs, scientists are set scrambling and exciting new frontiers are opened.

1.3 THE METRIC SYSTEM

In the metric system we simply move the decimal point to switch to larger or smaller units

The process of measurement is a common and necessary activity in our lives. We measure the speed of our car to avoid getting a traffic ticket. We buy much of our food according to how much it weighs. Then we get on a scale to see whether we can afford to eat the amount of food we buy.

A necessary aspect to all measurements is a uniform and generally accepted system of units. In the United States we normally use the English system of units for everyday purposes. We measure distance in units of inches, feet, or miles; volume in ounces, quarts, or gallons; and weight in ounces, pounds, or tons. Sometimes it is necessary to convert from one unit to another, and no doubt you've become adept at performing simple conversions such as 6 ft = 72 in. However, using the English system, these conversions can involve tedious calculations.

In the sciences and in all of the other industrialized countries of the world, another set of units is used. Originally called the **metric system,** this set of units uses a decimal system in which conversion from one size unit to another involves simply moving the decimal point to the left or right an appropriate number of spaces. This is a clear advantage over our English system and accounts for its widespread use. In fact, the United States is in the process of a voluntary conversion to the metric system, and many metric units are finally beginning to appear on consumer products (Figure 1.2). Soft drinks, for example, are now available in 2-liter bottles and alcoholic beverages are now bottled in 1.75-liter containers. Road signs often display distances in both miles and kilometers, and packaged goods on grocery shelves give their contents in grams as well as in ounces or pounds.

How long would it take you to calculate the number of inches in a mile?

Only four countries do not use the metric system: the United States, Burma, South Yemen, and Brunei.

The Metric Conversion Act (1975) calls for a board to coordinate the voluntary switch to metric units in the United States.

Figure 1.2
Metric units are becoming a common sight on many highways and consumer products.

1.4 THE SI SYSTEM

Figure 1.3
The international standard for the kilogram, carefully preserved within multiple bell jars.

The SI system is the modern version of the metric system

The metric system has gradually evolved since it was first formulated by the French Academy of Sciences during the 1790's. In 1960, the General Conference on Weights and Measures adopted a modified version of the metric system called the **International System of Units,** abbreviated **SI** (from the French name, *Le Système International d'Unités*).

There are differences between some of the units of the old metric system and those of the SI. For the most part, scientists have accepted the new SI units readily but, as happens when any change is made, there are a few holdouts. In addition, much of the existing scientific literature was written before the SI was adopted. It is therefore a good idea to be familiar with the units of both systems. Although the primary emphasis here will usually be on the SI, the important differences will be pointed out to you as we encounter them.

The SI begins with the set of **base units** given in Table 1.1. Not all of these units will have meaning to you now, but we'll encounter most of them later in this book and their meanings will be made clear when they are needed. The size of each base unit has been very precisely defined. For example, the international standard for the SI unit of mass,[1] the kilogram, is a carefully preserved platinum-iridium alloy block (Figure 1.3) stored at the International Bureau of Weights and Measures in France. This metal block is defined as having a mass of *exactly* one kilogram.

Table 1.1
The SI Base Units

Quantity	Unit	Symbol
Length	Meter	m
Mass	Kilogram	kg
Time	Second	s
Electric current	Ampere	A
Temperature	Kelvin	K
Amount of substance	Mole	mol
Luminous intensity	Candela	cd

Objects created by human hands, of course, can be destroyed or lost and therefore are not the most desirable choices for standards. In the SI system, only the kilogram is defined by such an object. The other base units are established in terms of reproducible physical phenomena. The meter, for example, is defined as 1,650,763.73 wavelengths[2] of the orange-red light given off by a certain kind of atom of the element krypton. Small amounts of krypton are present in the air; so it is a standard available to everyone.

1.5 DERIVED UNITS

Other necessary units are formed by combining the SI base units in appropriate ways

The base units, which form the foundation of the SI system, are used to define additional **derived units.** For example, we know that if we wish to calculate the area

---

[1] Mass and weight are not identical. Later in this chapter we will take a closer look at the difference between them.
[2] The color of light is determined by the distance between crests of the light wave. This distance is called the wavelength. We'll discuss this in detail later in the book.

of a rectangular room, we multiply its length by its width. Similarly, the *unit* for area is derived by multiplying the *unit* for length by the *unit* for width.

$$\text{length} \times \text{width} = \text{area}$$
$$(\text{meter}) \times (\text{meter}) = (\text{meter})^2$$
$$m \times m = m^2$$

In the English system we measure area in square feet (ft²) or square yards (yd²). Carpeting, for example, is priced by the square yard.

The SI unit for area is therefore m^2 (meter squared, or square meter). In deriving this unit, we see a very important concept that we will use repeatedly throughout this book when we perform calculations. *Units undergo the same kinds of mathematical operations that numbers do.* Thus $m \times m = m^2$ just as $3 \times 3 = 3^2$. We will say more about how we use this concept in Section 1.10.

EXERCISE 1.1   What would be the SI derived unit for (a) volume (of a room, for example) and (b) speed? (*Hint.* In your car you measure speed in miles/hour.)

1.6 DECIMAL MULTIPLIERS

Larger and smaller metric units are created by multiplying the basic units by appropriate decimal factors

When making measurements, the basic units are not always handy. For instance, the meter (approximately 39 in.) is not convenient for expressing the dimensions of very small objects such as bacteria. Nor is it convenient for expressing very large distances such as those between planets or stars. In the SI system we can modify the basic units so that they more closely suit our needs by using **decimal multipliers.** These factors and the prefixes that we use to identify them are given in Table 1.2. Those that we will encounter most frequently are given in boldface.

When the name of a unit is preceded by one of these prefixes, the size of the basic unit is modified by that decimal multiplier. Notice also that each prefix has

Table 1.2
Decimal Multipliers that Serve as SI Unit Prefixes

Prefix	Pronunciation	Symbol	Multiplication Factor[a]
exa	*Texa*co	E	$1,000,000,000,000,000,000 = 10^{18}$
peta	*peta*l	P	$1,000,000,000,000,000 = 10^{15}$
tera	*terra*ce	T	$1,000,000,000,000 = 10^{12}$
giga	jig-a (*a* as in *a*bove)	g	$1,000,000,000 = 10^9$
mega	*mega*phone	M	$1,000,000 = 10^6$
kilo	kill-o (*o* as in *o*ver)	k	$1000 = 10^3$
hecto	heck-to (*to* as in *to*tal)	h	$100 = 10^2$
deka	deck-a (as in *a*bove)	da	$10 = 10^1$
deci	*deci*mal	d	$0.1 = 10^{-1}$
centi	*centi*grade	c	$0.01 = 10^{-2}$
milli	*milli*tant	m	$0.001 = 10^{-3}$
micro	*micro*phone	μ	$0.000\,001 = 10^{-6}$
nano	nan-oh (*a* as in *na*nny goat)	n	$0.000\,000\,001 = 10^{-9}$
pico	peek-oh	p	$0.000\,000\,000\,001 = 10^{-12}$
femto	fem-toe (*fem* as in *fem*inine)	f	$0.000\,000\,000\,000\,001 = 10^{-15}$
atto	*atto*mize	a	$0.000\,000\,000\,000\,000\,001 = 10^{-18}$

[a] Exponential forms such as 10^6 or 10^{-15} will be explained in Section 1.9.

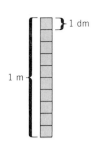

1 dm

1 m

its own abbreviation which is added to the abbreviation of the unit that it modifies. For instance, the prefix *kilo* implies 1000. Therefore, a kilometer is a unit that equals 1000 m.

$$1 \text{ km} = 1000 \text{ m}$$

Similarly, the prefix *deci-* signifies 1/10th. A decimeter, then, is 1/10th of a meter.

$$1 \text{ dm} = 0.1 \text{ m}$$

Of course, this also means that 1 m = 10 dm.

EXERCISE 1.2   Write out the full name of each of these abbreviations: (a) nm; (b) cm; (c) km; (d) pm; and (e) mm.

1.7 LABORATORY MEASUREMENTS

The most common measurements you will make in the laboratory are those of mass, volume, temperature, and length

You will routinely carry out many measurements in the course of your scientific studies. Some of the most common measurements you will make are those of length, volume, mass, and temperature. Probably, you already have a good idea of how much a foot, a quart, or a pound is. The SI units, however, may be another story. Table 1.3 gives some of the common conversions between the English system and the SI.[3] You should become familiar with the metric units and begin to develop a feel for their size.

Length

The SI base unit for length is the **meter (m).** Unfortunately, the meter is awkward to use for most laboratory purposes. More convenient units of length are the **cen-**

Table 1.3
Some Useful Conversions

	English to Metric	Metric to English
Length	1 in. = 2.540 cm	1 m = 39.37 in.
	1 yd = 0.9144 m	1 km = 0.6215 mile
	1 mile = 1.609 km	
Mass	1 lb = 453.6 g	1 kg = 2.205 lb
	1 oz = 28.35 g	
Volume	1 gal = 3.786 liter	1 liter = 1.057 qt
	1 qt = 946.4 ml	
	1 oz (liquid) = 29.6 ml	

[3] Originally, these conversions were established by measurement. For example, if a metric ruler is used to measure the length of an inch, it is found that 1 in. equals 2.54 cm. Later, to avoid confusion about the accuracy of such measurements, it was agreed that these relationships would be taken to be exact. For instance, 1 in. is now defined as *exactly* 2.540 cm. Exact conversions also exist for the other quantities in Table 1.2, but for simplicity many have been rounded off. For example, 1 lb = 453.59237 g exactly.

timeter (cm) and the **millimeter (mm).** They are related to the meter in the following way.

$$1 \text{ m} = 100 \text{ cm} = 1000 \text{ mm}$$

Many 12-in. rules have one side that is marked in centimeters and millimeters (Figure 1.4). Note that there are 10 mm in each centimeter. Also, notice that the line marking the fifth millimeter division is made slightly longer to locate the midpoint between the larger centimeter lines. This is a common practice when dividing a scale into 10 parts.

It is useful to remember that 1 in. equals approximately 2.5 cm (1 in. = 2.540 cm, exactly).

Figure 1.4
Centimeters and millimeters are conveniently sized units for most laboratory measurements. Here is a common ruler that is marked in both English and metric units.

Volume

To calculate the volume of a room you would multiply its length by its width, and then multiply that answer by the height of the room. If each of these dimensions were measured in meters, the unit for volume would be the cubic meter (m^3). That is somewhat more than a cubic yard—obviously much too large to use for laboratory purposes.

The traditional unit of volume used for the measurement of liquids in the metric system is the **liter.** In SI terms, a liter is defined as one cubic decimeter.

$$1 \text{ liter} = 1 \text{ dm}^3$$

However, the liter is even too large for most laboratory purposes and the glassware we use is usually marked in **milliliters (ml).**

$$1 \text{ liter} = 1000 \text{ ml}$$

Since 1 dm = 10 cm, $1 \text{ dm}^3 = (10 \text{ cm})^3 = 1000 \text{ cm}^3$. Therefore, 1 ml is exactly 1 cm^3 (cubic centimeter, often abbreviated cc).

Some typical laboratory glassware used for volume measurements is shown in Figure 1.5.

Mass

The terms *mass* and *weight* are often used interchangeably. Actually, the two terms refer to different things. **Mass** is a measure of how much stuff there is in a given sample. **Weight** is a measure of the force with which this stuff is attracted to the earth by gravity. A golf ball, for instance, has the same mass regardless of its location. However, on the moon where the gravitational attraction is only about 1/6 of that on earth, the weight of the golf ball would only be 1/6 of its weight on earth.

In the SI, the base unit for mass is the **kilogram (kg).** However, the **gram (g),** which is $\frac{1}{1000}$ of a kilogram, is a more convenient unit for most laboratory measurements. Determining the mass of a sample is accomplished by a proce-

This is the answer to Exercise 1.1(a).

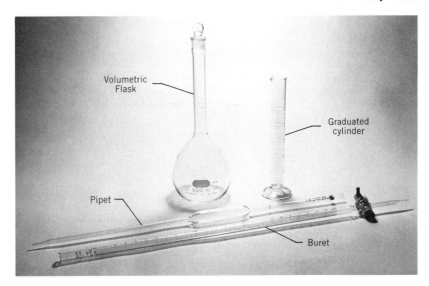

Figure 1.5
Some typical apparatus for measuring the volumes of liquids
in the laboratory.

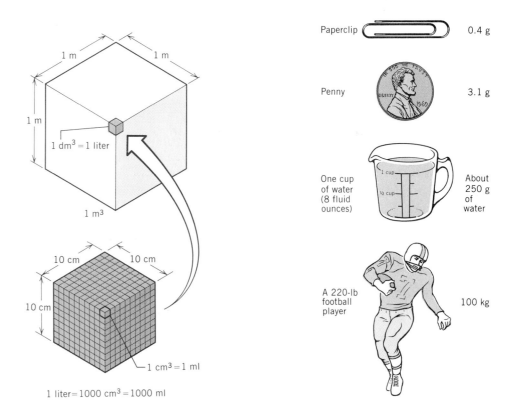

dure called **weighing,** even though we do not really measure the sample's
weight as we would that of a piece of fish or some vegetables on a spring scale
in a grocery store or supermarket. Instead, we *compare* the sample's weight (the
force of its attraction by gravity) with the weights of standard masses. The apparatus that we use is called a **balance.** Examples of typical laboratory balances

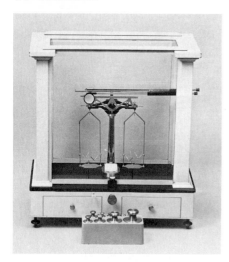

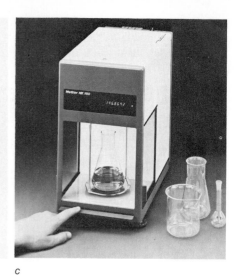

a   *b*   *c*

Figure 1.6
Typical laboratory balances. (*a*) A traditional two-pan balance. (*b*) A modern top-loading balance. (*c*) A modern analytical balance capable of measurements to the nearest 0.0001 g.

are shown in Figure 1.6. To find out the mass of our sample, we place it on one pan and the standard masses on the other. When the weight of the sample and the weights of the standards are in balance (that is, when they match), the masses of both are the same.

Notice that two of the balances in Figure 1.6 appear to have only one pan. This is true of most modern balances. The standard masses supplied with these balances by the manufacturer are located within the balance case, where they are protected from dust (and the probing fingers of curious people).

Temperature

Temperature is something with which we are all familiar, and *hot, warm, cool,* and *cold* are terms we all use quite frequently. To actually measure the temperature, we use a **thermometer**—a long glass tube with a very thin bore connected to a reservoir containing a liquid, usually mercury (Figure 1.7). As the temperature rises, the liquid in the reservoir expands and its length in the column increases.

Figure 1.7
A typical laboratory thermometer.

Thermometers are graduated, or marked, in degrees according to one of two temperature scales. These scales both use the temperature at which water freezes[4] and the temperature at which it boils as reference points. On the **Fahrenheit** scale, water freezes at 32 °F and boils at 212 °F. This is probably the scale with which you are most familiar. In recent years, however, you have probably noticed an increased use of the **Celsius** scale (formerly called the centi-

---

[4] Water freezes and ice melts at the *same* temperature. A mixture of ice and water has a temperature of 32 °F or 0 °C. If heat is removed, some water freezes and more ice is formed. If heat is added, some ice melts and more water is formed. The temperature, however, remains constant.

Biologists who study microorganisms carry out many of their experiments at 37 °C because that is normal human body temperature.

grade scale), especially in weather forecasts. This is the scale we employ in the sciences. On the Celsius scale, water freezes at 0 °C and boils at 100 °C (Figure 1.8).

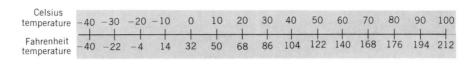

| Celsius temperature | −40 | −30 | −20 | −10 | 0 | 10 | 20 | 30 | 40 | 50 | 60 | 70 | 80 | 90 | 100 |
| Fahrenheit temperature | −40 | −22 | −4 | 14 | 32 | 50 | 68 | 86 | 104 | 122 | 140 | 168 | 176 | 194 | 212 |

Figure 1.8
Comparison of the Celsius and Fahrenheit temperature scales.

As you can see, on the Celsius scale there are 100 degree units between the freezing point of water and its boiling point, while on the Fahrenheit scale there are 180 degree units between these reference points. Each Celsius degree, therefore, is nearly twice as large as a Fahrenheit degree (actually, 5 Celsius degrees are equal to 9 Fahrenheit degrees). To convert between °C and °F we can use either of the relationships:[5]

$$°F = \frac{9}{5} °C + 32 \qquad (1.1)$$

$$°C = \frac{5}{9} (°F - 32) \qquad (1.2)$$

EXAMPLE 1.1 **Converting from °C to °F**

Problem: Thermal pollution is a serious problem near power plants. For example, trout will die if the water temperature rises above approximately 25 °C. What is this temperature in °F?

Solution: We can use Equation 1.1

$$°F = \frac{9}{5} °C + 32$$

Substituting, we get

$$°F = \frac{9}{5} (25) + 32$$
$$= 45 + 32$$
$$= 77$$

Trout die at temperatures above 77 °F.

EXAMPLE 1.2 **Converting from °F to °C**

Problem: A student, writing a letter to a friend in Europe, wanted to impress on him how warm the recent weather had been. That day the temperature had climbed to 95 °F. What Celsius temperature should the student have reported to his friend?

Solution: To convert to °C, we use Equation 1.2.

$$°C = \frac{5}{9} (°F - 32)$$

[5] Most equations will be numbered to make it easy to refer to them either later on in the book or in class discussions. However, the really important equations will be put in "boxes."

Substituting, we get

$$°C = \frac{5}{9}(95 - 32)$$

$$= \frac{5}{9}(63)$$

$$= 35$$

The temperature had been 35 °C that day.

EXERCISE 1.3 What Celsius temperature corresponds to a room temperature of 86 °F? What Fahrenheit temperature corresponds to −17.8 °C?

EXERCISE 1.4 Hypothermia is a condition caused by the loss of deep-body heat. When body temperature falls below 90 °F, unconsciousness can occur. Death from heart failure can occur when the body temperature drops below about 85 °F. What are these temperatures in °C?

The SI uses the Kelvin temperature scale which, not surprisingly, has a degree unit called the **kelvin (K).** Notice that the symbol for this unit is K, not °K. Kelvin temperatures must be used in many equations where the temperature enters directly into the calculations. We will come across this many times throughout the book.

Figure 1.9 shows how the Fahrenheit, the Celsius, and the Kelvin temperature scales relate to each other. You can see that the kelvin is exactly the same size as the Celsius degree. The only difference between these two scales is the zero point. The zero point on the Kelvin scale—called absolute zero—corresponds to the lowest temperature that is possible. It is 273.15 degree units lower than the zero point on the Celsius scale. This means that 0 K equals −273.15 °C and 0 °C equals 273.15 K. Thermometers are never marked with the Kelvin scale; so if we need to express temperature in kelvins, we have to do some arithmetic. The relationship between kelvins and degrees Celsius is

$$K = °C + 273.15$$

In our calculations, we'll round this to three digits and use the equation,

$$\boxed{K = °C + 273}$$

(1.3)

Figure 1.9
Comparison among Kelvin, Celsius, and Fahrenheit scales.

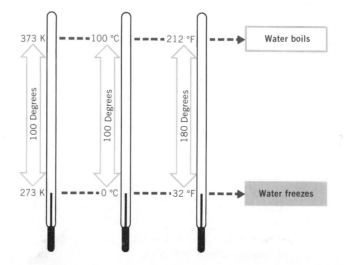

EXAMPLE 1.3   **Converting Between °C and Kelvins**

Problem: Liquid oxygen (sometimes abbreviated LOX) is used in liquid-fuel rockets. Its boiling point is −183 °C. What is this temperature in kelvins?

Solution: Using Equation 1.3,

$$K = -183 + 273$$
$$= 90$$

The kelvin temperature at which liquid oxygen boils is 90 K.

EXERCISE 1.5   A substance is heated from 300 K to 315 K. What is the change in temperature expressed in °C?

1.8 SIGNIFICANT FIGURES

The number of significant figures indicates the precision or reliability of a measurement

When a scientist writes down a number that has been obtained from a measurement, two kinds of information are given at the same time. One of these, of course, is the magnitude of the measurement. *The other is the extent of its reliability.*

Let's look at an example. Suppose you hired two different people to measure the width of your room. The first person reports that the room is 11.2 ft wide. The second tells you that he measured the width at the same place and obtained 11.13 ft. Obviously the room can have only one true width at any given place. What, then, do these two numbers tell us?

The first number, 11.2 ft, implies that the width of the room was measured with a tape measure that required the person using it to estimate the tenths place. Figure 1.10a illustrates how a tape measure such as this would be marked. Because the tenths place must be estimated, different people measuring the width of the room might report distances that differ by ±0.1 ft. Therefore, we must view the reported value of 11.2 ft as uncertain by ±0.1 ft.

Figure 1.10
(a) Measurement of length with a tape measure marked only every 1 ft. An estimate must be made of the tenths place. (b) A portion of the scale of another tape measure that is marked every 0.1 ft.

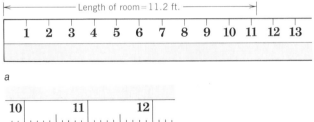

The second measurement, 11.13 ft, suggests that the tape measure used to obtain it was more finely divided, as in Figure 1.10b. The markings on this scale allow the user to be certain of the tenths place, but require that the hundredths place be estimated. Therefore, we can expect the second measurement to be uncertain by about ±0.01 ft.

We would certainly expect the second measurement to be more reliable than the first because it has a smaller amount of uncertainty. We are told this by the way the two measurements are recorded. *The reliability of a piece of data is indicated by the number of digits used to represent it.* Digits that result from measurement, such that only the digit furthest to the right is not known with certainty are called **significant figures.** The *number* of significant figures in a number is equal to the number of digits known for sure, *plus* one that is uncertain. The first measurement, 11.2 ft, has only three significant figures. The second, more reliable measurement, 11.13 ft, has four significant figures.

Accuracy and precision

Accuracy and precision are frequently used terms with which you should be familiar. **Accuracy** refers to freedom from error, mistake or misfunctioning of the measuring instrument. The closer a measurement corresponds to the actual, true value, the more accurate the measurement. Therefore, accurate measurements depend on careful calibration to be sure that your thermometer, meter stick, etc., are in proper working order. Accuracy also depends on your skill in using the instrument.

Precision refers to how *reproducible* measurements are. The first value for the width of the room mentioned earlier (11.2 ft) is reproducible to the nearest 0.1 ft. By this we mean that we can be reasonably sure that anyone else measuring the width of the room with this same tape measure will obtain a value that differs from 11.2 ft by no more than 0.1 ft. The second value (11.13 ft) is reproducible to the nearest 0.01 ft and is therefore considered more precise. In general, the more significant figures there are in a reported value, the greater its precision. We usually assume (and sometimes we are painfully wrong) that a very precise measurement is also highly accurate.

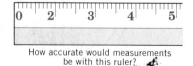

How accurate would measurements be with this ruler?

Combining numbers in calculations

Usually when we perform several different measurements, the results must be combined through arithmetic calculations to give some desired final answer. For instance, calculating the area of a rectangular room requires two measurements—the length and the width—which are multiplied together to give the answer we want. If one of these measurements is very precise but the other is not, we can't expect too much precision in the calculated area. To have an idea of how reliable the calculated area is, we need to have a way of being sure that the answer reflects the precision of the original measurements. To make sure this happens we follow certain rules to avoid answers that may contain too many or two few significant figures.

For multiplication and division, the number of significant figures in the answer should not be greater than the number of significant figures in the least precise factor. Let's look at an illustration that could typically involve some measured numbers.

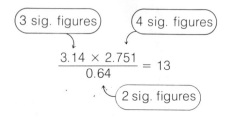

$$\frac{3.14 \times 2.751}{0.64} = 13$$

The answer to this problem that is displayed on a calculator[6] is 13.497093. However, the rule says that the answer should have only as many significant figures

---

[6] Calculators usually give answers with too many significant figures. An exception is when the answer has zeros at the right which are significant digits. An answer of 1.200 would be displayed simply as 1.2; if the zeros belong in the answer, don't forget to write them down.

Most calculations in this text will be carried to three significant figures.

as the least precise factor. Since the least precise factor, 0.64, has only two significant figures, the answer should have only two. This means that we must round off the calculator answer to 13.

For addition or subtraction, the answer should have the same number of decimal places as the quantity having the least number of decimal places. As an example, consider the following problem involving the addition of measured values.

$$
\begin{array}{r}
3.247 \\
41.36 \\
+\ \underline{125.2} \leftarrow \text{(This number has only 1 decimal place.)} \\
169.8 \leftarrow \text{(Answer should be rounded to 1 decimal place.)}
\end{array}
$$

The reason for this rule is fairly obvious. The digits listed beneath the 7 in 3.247 are unspecified. We know nothing about them. Therefore, when 7 is added to these unknown digits, the result is completely uncertain and we are not justified in writing the 7 in the answer. The same argument applies to the hundredths column. Beneath the 4 and 6 is another unknown digit. When this column is added the answer will be unknown. We are not justified in writing a digit in the answer in the hundredths place.

To obtain the sum in this problem we can round off the terms that are being added.

$$3.2 + 41.4 + 125.2 = 169.8$$

Many people prefer to add the original numbers and then round the answer. If we enter the original numbers into a calculator, we obtain the sum, 169.807. Rounding to the first decimal place again gives 169.8.

Not all the numbers we use come from measurement. Sometimes in performing calculations, we use numbers that come from a direct count of objects or that result from definitions. Such numbers are called **exact numbers** and are considered to possess an infinite number of significant figures. An example is the number of inches in one foot. By definition, there are exactly 12 in. in 1 ft, no more or less. When we write 1 ft = 12 in., there is absolutely no uncertainty in either quantity; so we can express them to any desired number of digits with complete confidence. Therefore, when using exact numbers in calculations, we can forget about them as far as significant figures are concerned. We determine the number of significant figures in the answer in the usual way, but we take into account only those numbers that arise from measurement.

EXAMPLE 1.4 **Calculations Using Exact Numbers**

Problem: A student measured her height to be 64.25 in. What is her height measured in feet?

Solution: We know that to obtain feet from inches we must divide the number of inches by 12. The arithmetic involved is 64.25 ÷ 12. We can consider the 12 to contain as many significant figures as we wish. The answer, then, should contain 4 significant figures because that is the number of significant figures in 64.25.

$$\frac{64.25}{12} = 5.354 \text{ ft}$$

EXERCISE 1.6 Perform the following calculations, making sure that each answer has the proper number of significant figures. Assume that *each* number is the result of a measurement: (a) 4.8 × 392; (b) 7.255 ÷ 81.334; (c) 0.2983 + 1.52; and (d) 14.5403 − 0.022.

EXERCISE 1.7   Calculate the height of the student in Example 1.4 in units of (a) yards and (b) miles. Be sure to express the answer to the correct number of significant figures.

1.9 SCIENTIFIC NOTATION

Very large and very small numbers are easier to work with when they are put into exponential forms

Some numbers that we deal with—such as the number of red blood cells in the body—are very large. Others—such as the mass of a single blood cell—are very tiny. To avoid the problems of doing arithmetic with such cumbersome quantities, it is often convenient to express them in **scientific notation** (also often called **exponential notation**). In this system numbers are written as the product of two factors. The first is a decimal number that usually ranges between 1 and 10, and the second is 10 raised to an appropriate power. For example, 500 can be written as

$$500 = 5 \times 100 = 5 \times 10 \times 10$$
$$= 5 \times 10^2$$

Notice that the exponent on the 10 is equal to the number of places that the decimal has been moved to the left to obtain the 5. For instance, we can write 6253 as

$$6253 = 6.253 \times 10^3$$

decimal is moved 3 places to the left

When writing small numbers in scientific notation, the decimal point must be moved to the *right* to give a number between 1 and 10. The exponent on the 10 is therefore negative. For example,

$$0.00063 = 6.3 \times 10^{-4}$$

decimal is moved 4 places to the right

EXAMPLE 1.5   **Writing Numbers in Scientific Notation**

Problem: There are approximately 25,000,000,000,000 red blood cells in the average human.[7] Write this number in scientific notation.

Solution: To write this number in scientific notation we must move the decimal point 13 places to the left to locate it between the 2 and the 5. Therefore, we write the number as 2.5×10^{13}.

EXAMPLE 1.6   **Converting from Scientific Notation to Decimal Notation**

Problem: A typical bacteria cell has a length of about 5×10^{-6} m. Write this number in standard decimal notation.

Solution: The exponential part of this number, 10^{-6}, tells us we must move the decimal 6 places to the left to write the number in the usual decimal form. The answer is 0.000005 m.

---

[7] Outside the United States, a comma is sometimes used as a decimal marker (i.e., decimal point). To avoid confusion in international dealings, numbers are written using thin spaces instead of commas to separate groups of digits; for example, 25 000 000 000 000. Small numbers such as 0.00063 are written 0.000 63.

EXERCISE 1.8   Write these numbers in scientific notation: (a) 23,000; (b) 21,700,000; (c) 0.0015; and (d) 0.000027.

EXERCISE 1.9   Write these numbers in standard decimal notation: (a) 2.7×10^3; (b) 3.5×10^{28}; and (c) 2×10^{-12}.

Besides being a compact way of writing very large or very small numbers, scientific notation allows us to avoid the confusion that can result when zeros may be significant figures in a number. Suppose, for instance, that you are told that the value of a measurement is 200 cm. Just how precise is this measurement? Is it 200 ± 100 cm, 200 ± 10 cm, or 200 ± 1 cm? When the number such as 200 is written in standard notation, there is no indication of its precision. However, by presenting the measurement in scientific notation, we can use the decimal portion to indicate the correct number of significant figures.

$$\begin{aligned} 200 \text{ cm} &= 2 \times 10^2 \text{ cm} & \text{(1 significant figure)} \\ &= 2.0 \times 10^2 \text{ cm} & \text{(2 significant figures)} \\ &= 2.00 \times 10^2 \text{ cm} & \text{(3 significant figures)} \end{aligned}$$

Throughout the remainder of this book we will assume that numbers such as 200, 760, or 450 have three significant figures unless we specify otherwise. Only if we wish to indicate fewer than three significant figures will we write such numbers in scientific notation. This conforms to the practice used by most scientists.

It would be a good idea for you to review the procedures for handling arithmetic computations involving numbers in exponential form provided in Appendix A. *You should do this even if you have an electronic calculator that can deal with numbers expressed in scientific notation.*

1.10   THE FACTOR-LABEL METHOD

Units, like numbers, can be multiplied or divided (canceled)

You are undoubtedly accustomed to performing simple conversions such as changing feet to inches or pounds to ounces. Probably, you view such problem solving as routine because you are very familiar with the units; you've developed a "feel" for them. In chemistry, however, some of the units will be new to you and setting up a problem to obtain the correct answer may be a challenge. The **factor-label method** is a system that we use to aid us in performing the proper arithmetic. We use it to make sure we have set up a problem correctly. The method consists of using *conversion factors* to change the units given to us in the data to the units asked for in the answer.

given quantity $\times$ conversion factor(s) = desired quantity

A **conversion factor** is a fraction obtained from a *valid relationship between units.* For example, suppose we wish to convert 3.00 in. to centimeters. The relationship between inches and centimeters is

1.00 in. = 2.54 cm

If we divide both sides of this equation by 1.00 in., we obtain a conversion factor.

$$\frac{1.00 \text{ in.}}{1.00 \text{ in.}} = \frac{2.54 \text{ cm}}{1.00 \text{ in.}} = 1$$

Notice that we have canceled the units inches from the numerator and denominator. Units behave just as numbers do in mathematical operations. This fact is a key part of the factor-label method. Also notice that the conversion factor 2.54 cm/1.00 in. is equal to 1. This means that if we multiply some quantity by 2.54 cm/1.00 in., which is the same as multiplying by 1, we will not change its magnitude. We can use this conversion factor to convert 3.00 in. to centimeters as follows:

$$3.00 \; \cancel{in.} \times \frac{2.54 \; cm}{1.00 \; \cancel{in.}} = 7.62 \; cm$$

Again, we cancel the units inches. The only units left are centimeters, which are the units we are looking for.

We can also construct a conversion factor, 1.00 in./2.54 cm, from the relationship between inches and centimeters. What would have happened if we had used this factor in our calculation?

$$3.00 \; in. \times \frac{1.00 \; in.}{2.54 \; cm} = 1.18 \; in.^2/cm$$

These units are nonsense! Who ever heard of square inches per centimeter?

In this case, inches do not cancel! We get units of in.²/cm because in. × in. = in.². Even though our calculator is very good at arithmetic, the quantity 1.18 is not the number of centimeters in 3.00 in. *The factor-label method lets us know the answer is wrong because the units are wrong!*

All chemists and physicists use the factor-label method. There must be a good reason why.

We will use the factor-label method extensively throughout this book to aid us in setting up problems. Below are several examples that further illustrate how the method is used. Even though this method may seem strange to you now, you should practice using it because it is a powerful tool that can help you solve problems correctly.

EXAMPLE 1.7   **Calculation Using the Factor-Label Method**

Problem: Convert 3.25 m to millimeters.

Solution: The problem can be stated in mathematical form as

$$3.25 \; m = ? \; mm$$

To solve the problem we must have a relationship between meters and millimeters. Using the information in Table 1.2, we can write

$$1 \; m = 1000 \; mm$$

We could also use the relationship, 1 mm = 10^{-3} m.

We can use this to construct two different conversion factors,

$$\frac{1 \; m}{1000 \; mm} \quad \text{and} \quad \frac{1000 \; mm}{1 \; m}$$

Which one shall we use? We know that we must eliminate meters and this requires using the second factor so that meters cancel.

$$3.25 \; \cancel{m} \times \frac{1000 \; mm}{1 \; \cancel{m}} = 3250 \; mm$$

EXAMPLE 1.8   **Calculation Using the Factor-Label Method**

Problem: A liter, which is slightly more than a quart, is defined as 1 cubic decimeter (1 dm³). How many liters are there in 1 cubic meter (1 m³)?

Solution: This problem is really very easy if we solve it step by step. It is worded in terms of liters, cubic decimeters, and cubic meters. We know that we will have to get rid of the cubic meters and cubic decimeters to end up with liters, the unit asked for. This means that we will need two conversion factors, one based on the relationship given in the problem

$$1 \text{ liter} = 1 \text{ dm}^3$$

and the other based on the relationship between meters and decimeters

$$1 \text{ m} = 10 \text{ dm}$$

But we need the relationship between *cubic* meters and *cubic* decimeters. Since we know that units undergo the same mathematical operations as numbers, this relationship is easy to find.

$$1 \text{ m}^3 = 1 \text{ m} \times 1 \text{ m} \times 1 \text{ m}$$
$$= 10 \text{ dm} \times 10 \text{ dm} \times 10 \text{ dm}$$
$$= 1000 \text{ dm}^3$$

Now we can solve the problem.

$$1 \text{ m}^3 \times \frac{1000 \text{ dm}^3}{1 \text{ m}^3} \times \frac{1 \text{ liter}}{1 \text{ dm}^3} = 1000 \text{ liters}$$

Notice that we have "strung together" two conversion factors. The first had to have m³ in the denominator, the second then had to have dm³ in the denominator so that the units of the answer could be liters.

EXAMPLE 1.9 Calculation Using the Factor-Label Method

Problem: In 1975 the world record for the long jump was 29.21 ft. Use the factor-label method to convert this distance to meters.

Solution: One set of relationships that we can use to construct the necessary conversion factors is

$$1 \text{ ft} = 12 \text{ in.}$$
$$1 \text{ in.} = 2.540 \text{ cm}$$
$$100 \text{ cm} = 1 \text{ m}$$

The given value of 29.21 ft is converted to meters by setting up successive conversion factors in such a way that all the units except meters will cancel.

$$29.21 \text{ ft} \times \frac{12 \text{ in.}}{1 \text{ ft}} \times \frac{2.540 \text{ cm}}{1 \text{ in.}} \times \frac{1 \text{ m}}{100 \text{ cm}} = 8.903 \text{ m}$$

ft to in.
ft to cm
ft to m

Keep an eye on the units that you're trying to get rid of. Set up the units to cancel, and the arithmetic will take care of itself.

Notice that if we were to stop after the first conversion factor, our answer would have the units, inches. After the second factor the units would be centimeters. After the third factor we can stop because the units are finally meters—the units we want. Also notice that the answer has been rounded to four significant figures because that is how many there were in the measured distance.

This is not the only way to solve the problem. Other sets of conversion factors could have been chosen. For example, we could also have used

$$3 \text{ ft} = 1 \text{ yd}$$
$$1 \text{ yd} = 0.9144 \text{ m}$$

Then the problem would be set up as follows:

$$29.21 \text{ ft} \times \frac{1 \text{ yd}}{3 \text{ ft}} \times \frac{0.9144 \text{ m}}{1 \text{ yd}} = 8.903 \text{ m}$$

Many problems that you meet have more than one path to the answer. There isn't any *one* correct way to set up the solution. The important thing is for you to be able to reason your way through a problem and find some set of relationships that can take you from the given information to the answer. The key is to keep in mind the units that must be gotten rid of by cancellation.

EXERCISE 1.10 Use the factor-label method to perform the following conversions: (a) 3.00 yd to inches; (b) 1.25 km to centimeters; (c) 3.27 mm to feet; and (d) 20.2 miles/gal to liters/km.

1.11 MATTER AND ENERGY

Matter has mass, occupies space, and is able to do work if it possesses energy

Some examples of matter.

Energy is so important that we will devote an entire chapter to it later.

Matter is anything that has mass and occupies space. It is the stuff our universe is made of—air, water, rocks, even people. **Energy** is something that matter possesses if the matter is able to do work. An object has energy if it is able to push, through some distance, a force that pushes back.

Energy and matter are closely related. That, in fact, is the message in Albert Einstein's famous equation, $E = mc^2$, which we will discuss in Chapter Five. In chemistry we are concerned about matter and energy because the chemicals that we work with are forms of matter and because when the chemicals interact with each other they gain or lose energy. To discuss chemistry further we must learn more about both energy and matter. Let's begin by discussing energy.

Nearly everyone is familiar with the word energy. Yet few people really understand what it is (even though they think they do). This is because energy is quite different from matter. We can't touch it, smell it, or see it; all we can do is observe its effects on matter.

In defining energy we said that an object having it is able to do work. For example, you possess energy because you are able to move your limbs in opposition to the force of gravity. Coal and oil possess energy because they can be burned and the heat released can be harnessed to run machinery. But how do objects possess energy?

Basically there are two ways that a substance can have energy. One of these is as kinetic energy. **Kinetic energy** is the energy that an object has because of its motion. The amount of kinetic energy an object has depends on both its mass and its speed. We all know that a heavy sledgehammer swung rapidly will drive a stake into the ground faster than will gentle taps with a light hammer! If we know the mass and speed of an object, we can calculate its kinetic energy from the equation

$$\text{K.E.} = \tfrac{1}{2} mv^2 \tag{1.4}$$

where m is the object's mass and v is its velocity (speed).

The second way for a substance to possess energy is to store it as **potential energy.** When you wind a watch or an alarm clock, the energy you expend winding it is stored in a spring as potential energy. This potential energy is gradually converted back to kinetic energy as the timepiece operates. The foods that we eat possess a different kind of potential energy sometimes called chemical energy. That's the energy stored in chemicals that is released when they change to substances with less potential (chemical) energy. The chemical energy in foods is released by the process of metabolism, and it ultimately appears as movements of muscles and as body heat. Nuclear energy is still another kind of potential energy that can be released in nuclear power plants or atomic and hydrogen bombs.

Energy can appear in a variety of forms. It can appear entirely as mechanical energy; that is, energy of motion or kinetic energy. It can also occur as electrical energy, light energy, or sound energy. Eventually, however, all energy ends up as heat. The concept of heat immediately brings to mind the notion of temperature. It is important to remember that heat and temperature are quite different. Heat is a form of energy; temperature is a measure of the *intensity* of heat. Heat and temperature are related by the fact that heat always flows, without any outside help, from warm objects to cool ones. A hot cup of coffee, for example, becomes cool because heat flows from the coffee into the surroundings.

Both fires could be at the same temperature, but the forest fire has more heat in it than the flame of the match.

1.12 BEHAVIOR OF MATTER

To understand the way matter behaves we must study its various characteristics

If we are to study and understand the behavior of matter, we must first learn something about its characteristics or **properties.** These allow us to distinguish among the many substances[8] we encounter. To help us organize our thinking we classify the properties of matter into different types.

Physical properties are characteristics that can be specified without referring to other substances. For example, at 25 °C, water in an open container exists as a liquid. This liquid form is a physical property that we observe for all samples of pure water under these conditions. Solid, liquid, and gaseous forms are called the **states** of matter. The state of a particular kind of matter depends on conditions such as temperature and pressure. For instance, water in an open vessel exists in the liquid state at 25 °C, but below 0 °C it is a solid (ice) and above 100 °C it becomes a gas (steam).

Some other physical properties are mass, volume, melting point, color, and electrical conductivity. If we want to specify the amount of matter in something,

---

[8] A substance is matter that has a particular set of properties that identify it. It is the chemical material of which things are composed.

Special Topic 1.1 Using Colors in Medical Diagnostic Tests

Even though color may sometimes be a deceiving physical property (see footnote 9), it is a property easily identified by nearly everyone. Many important medical diagnostic tests rely on the identification of the colors of chemicals produced during the test. For example, diabetics regularly test their urine using special, chemically treated strips of papers that develop colors which are related to the amount of sugar in the urine. See Color Plate 1A.

we give its mass. If we want to specify the amount of space something occupies, we give its volume. Solid water (ice) melts at one particular temperature (32 °F or 0 °C), and copper is a reddish-colored metal that is a good conductor of electricity.

Chemical properties relate to how substances change into other substances, usually by interacting (reacting) with other chemicals. An iron nail, when exposed to moisture and oxygen, undergoes a chemical change, or **chemical reaction,** as it slowly forms a new substance called iron oxide—rust. This is a chemical property of iron that distinguishes it from another metal such as gold, which is unaffected by oxygen and water. It is also a property that makes iron jewelry less desirable than gold jewelry! Sometimes, under proper conditions, one substance will change into others all by itself. An example is nitroglycerine—when it explodes it changes into a mixture of nitrogen, oxygen, carbon dioxide, and water. This transformation is a chemical property of nitroglycerine.

The properties of matter allow us to distinguish among the many substances we encounter, just as we routinely use taste, color, odor, and physical appearance to identify foods. It is doubtful whether anyone reading this book would deliberately bite into a rock instead of a snack when they are hungry. *For the purposes of identification, the most desirable properties are those that do not depend on the size of the sample.* Color, odor, melting point and electrical conductivity are some examples.[9] Mass and volume are examples of properties that do depend on sample size.

To be really useful, properties must be able to be measured quantitatively; that is, we must be able to assign numbers to them. One useful property is **density,** which is simply the ratio of an object's mass to its volume. We can express this mathematically as

$$\boxed{d = \frac{m}{V}} \tag{1.5}$$

To calculate an object's density we must make two measurements; one of them is the object's mass and the other is its volume.

EXAMPLE 1.10 Calculating Density

Problem: A student measured the volume of an iron nail to be 0.880 cm³ (or 0.880 ml). She found that its mass was 6.92 g. What is the density of iron?

Solution: To calculate density we have to take the ratio of mass to volume.

$$\text{density} = \frac{6.92\ \text{g}}{0.880\ \text{cm}^3}$$

$$= 7.86\ \text{g/cm}^3$$

[9] Color can often be a useful property for identification, but there are instances where you can be fooled, particularly when particle size is very small. For example, you know that silver is a white metal with a high luster. However, in a very finely divided state, as in the image on black and white photographic film or paper, metallic silver appears black.

This could also be written as

$$\text{density} = 7.86 \text{ g/ml}$$

The density of iron calculated in Example 1.10 tells us how much of this kind of matter is packed into each cubic centimeter (or milliliter). Each pure substance has its own characteristic density. Gold, for example, is much more dense than iron. Each cubic centimeter of gold has a mass of 19.3 g; its density is 19.3 g/cm³. By comparison, the density of water is 1.00 g/cm³, and the density of air at room temperature is only about 0.0012 g/cm³.

There is more mass in 1 cm³ of gold than there is in 1 cm³ of iron.

Density is useful because it provides a conversion factor relating mass and volume.

EXAMPLE 1.11 **Calculations Using Density**

Problem: The density of a typical piece of Spanish mahogany is 0.86 g/cm³. What is the volume of a piece of mahogany weighing 75 g?

Solution: The density tells us that 1 cm³ = 0.86 g for this wood. We can use this relationship to construct a conversion factor.

$$75 \text{ g mahogany} \times \frac{1 \text{ cm}^3 \text{ mahogany}}{0.86 \text{ g mahogany}} = 87 \text{ cm}^3 \text{ mahogany}$$

EXERCISE 1.11 A bar of aluminum has a volume of 1.45 ml. Its mass is 3.92 g. What is its density?

✷ EXERCISE 1.12 What volume would 2.86 g of silver occupy? The density of silver is 10.5 g/cm³. What is the mass of 16.3 cm³ of silver?

1.13 ELEMENTS, COMPOUNDS, AND MIXTURES

The three classifications of matter are elements, compounds, and mixtures

When we change matter by chemical reactions, the simplest substances that we ever obtain are called **elements**. No matter how hard we try, we are unable by chemical means to decompose elements into still simpler substances that we can keep, store, and look at. Elements are, therefore, the simplest forms of matter that we encounter in the laboratory. Some examples of elements are silver, gold, iron, hydrogen, and oxygen. So far, scientists have discovered 105 different elements, many of them relatively rare. The names of all of the elements are listed on the inside front cover of this book, but we will be most interested in only a few of them.

Elements combine in various proportions to give all of the more complex substances that we find in nature. Hydrogen and oxygen combine to form water; iron and oxygen combine to form rust. The word **compound** is used to describe a substance formed from two or more elements in which the elements are always combined in the same fixed ratio. For example, water is a compound composed of hydrogen and oxygen. In water, regardless of its source, we always find these elements in a ratio of 1 g of hydrogen to 8 g of oxygen. Hydrogen peroxide (a bleach) is another compound of hydrogen and oxygen in which these two elements are combined in a different fixed ratio, 1 g of hydrogen to 16 g of oxygen.

Hydrogen peroxide is also an antiseptic.

Elements and compounds are examples of **pure substances.** The composi-

Mixtures containing sugar can have variable composition.

tion of a pure substance is always the same, regardless of its source. All samples of table salt, for instance, contain the same proportions of the elements, sodium and chlorine; all samples of water contain the same proportions of hydrogen and oxygen. Pure substances are rare, however. Usually we encounter mixtures of compounds or elements. Unlike pure substances, **mixtures** have variable composition. Sugar and water are examples of compounds. When the sugar is dissolved in the water, a mixture is formed whose composition depends on the amount of sugar that is added to the water. (Some people like their coffee sweeter than others!)

We can further classify mixtures as being either homogeneous or heterogeneous. A **homogeneous** mixture has the same properties (e.g., density and composition) throughout. Examples are thoroughly stirred mixtures of sugar and water or salt and water. We call a homogeneous mixture a **solution.**

A **heterogeneous** mixture consists of two or more regions or phases that differ in properties. A mixture of gasoline and water is an example of a two-phase mixture in which the gasoline floats on the water as a separate layer (Figure 1.11). Another example is a mixture of ice and water. Even though both phases have the same composition, many of their other properties are different.

The relationships among elements, compounds, and mixtures are summarized in Figure 1.12.

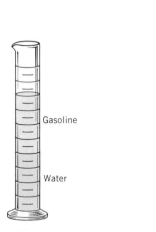

Figure 1.11
A two-phase heterogeneous mixture.

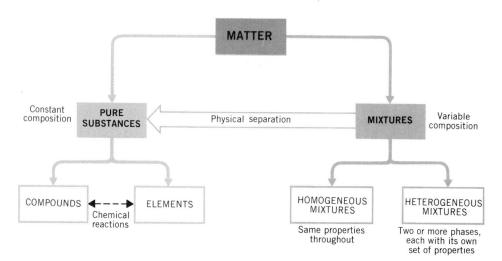

Figure 1.12
Classification of matter.

1.14 CHEMICAL SYMBOLS

Chemical symbols are shorthand notations for the names of elements

Each of the 105 elements has been assigned its own unique symbol, which we will find useful for writing chemical formulas and chemical equations. In most cases the symbols are formed from one or two letters of the English name for the element. The symbol for carbon is C, for bromine it is Br, and for barium it is Ba. In some instances the symbols are derived from the Latin names given to those elements that were discovered long ago by alchemists. Table 1.4 contains a list of elements whose symbols come to us in this way.[10]

Latin was the universal language of science in the early days of chemistry.

---

[10] The symbol for tungsten is W, from the name wolfram. This is the only element having a symbol unrelated to its English name that is not derived from a Latin name.

Table 1.4
Elements Having Symbols Derived from Their Latin Names

Element	Symbol	Latin Name
Sodium	Na	Natrium
Potassium	K	Kalium
Iron	Fe	Ferrum
Copper	Cu	Cuprum
Silver	Ag	Argentum
Gold	Au	Aurum
Mercury	Hg	Hydrargyrum
Antimony	Sb	Stibium
Tin	Sn	Stannum
Lead	Pb	Plumbum

Regardless of the origin of the symbol, the first letter is always capitalized and the second letter is always written in lower case. For example, the symbol for copper is Cu, not CU. Be very careful to follow this rule so that you can avoid confusion between such symbols as Co (cobalt) and CO (carbon monoxide).

1.15 CHEMICAL FORMULAS

Chemical formulas give the composition of compounds

Chemical formulas convey several kinds of information. In one sense, the **formula** for a compound is a shorthand way of writing the name for a compound, in the same way that chemical symbols are shorthand notations for the names of elements. The most important characteristic of a formula, however, is that it specifies the composition of a complex chemical substance.

The smallest bits of an element that retain their chemical identity are atoms. Scientists have reached this conclusion based on many observations and measurements, which we will discuss in Chapter Three. When elements combine to form compounds, their atoms come together in definite proportions. Often, these atoms join to produce larger stable particles called **molecules.** Sometimes the combining atoms acquire electrical charges and cling together in the compound because of the attraction between opposite charges. Later in the book, when we wish to refer to individual atoms or molecules, in general, we will simply use the term *particle*.

In a chemical formula for a compound, the constituent elements are identified by their chemical symbols. In addition, the subscripts in the formula tell us the number of atoms of each of the elements that are present. Water, as you probably know, has the formula H_2O. This tells us that water is composed of hydrogen (H) and oxygen (O). It also tells us that there are two H atoms for every one O atom. (When no subscript is written, we assume it to be 1.) Similarly, the formula for chloroform, $CHCl_3$, tells us that there are one atom of carbon, one atom of hydrogen, and three atoms of chlorine. Some of the elements themselves occur as simple molecules. Examples are oxygen and nitrogen, the major gases found in the atmosphere. When these elements are not combined with other elements in compounds, we always find them as molecules having the formulas O_2 and N_2.

For more complicated compounds, we sometimes find formulas containing parentheses. An example is the formula for urea, $CO(NH_2)_2$. This formula tells us that the atoms within the parentheses, NH_2, are repeated twice. The formula for urea could be written as CON_2H_4. There are good reasons for writing certain formulas with parentheses, but we will discuss them later.

EXAMPLE 1.12   Counting Atoms in Formulas

Problem: How many atoms of each element are expressed in the formula for sodium bicarbonate (baking soda), $NaHCO_3$?

Solution: We simply look at the subscripts following the symbols for the elements. Remember that the subscript 1 is implied when no subscript is written. This formula tells us that the smallest unit of the compound contains:

1 Na
1 H
1 C
3 O

EXAMPLE 1.13   Counting Atoms in Formulas

Problem: How many atoms of each element are represented in the formula for calcium bicarbonate, $Ca(HCO_3)_2$?

Solution: The parentheses tell us that the subscript 2 applies to everything in parentheses. We could, it we wished, rewrite the formula $CaH_2C_2O_6$. Then we can see that the formula has:

1 Ca
2 H
2 C
6 O

EXERCISE 1.13   How many atoms of each element are expressed by the formulas: (a) $NiCl_2$, (b) $FeSO_4$, and (c) $Ca_3(PO_4)_2$?

1.16   CHEMICAL REACTIONS

Chemical reactions produce amazing changes in the properties of the reacting chemicals

In the first section of this chapter, chemistry was defined as the study of the composition of substances and the way the behavior of substances relates to their composition. One important way that substances behave is that most of them undergo chemical reactions that transform them into new substances. A fascinating thing about these chemical reactions is that when they occur tremendous changes in properties are often observed. This is because the composition of the individual chemical components changes.

A typical example of a chemical reaction occurs between substances that we call acids and bases. Many common substances fall into these categories. Muriatic acid (which can be purchased in most hardware stores) and stomach acid both contain hydrochloric acid. Lemon juice contains citric acid, and vinegar contains acetic acid. In general, acids taste sour (although you should never taste any chemical unless your instructor tells you to, and even then be sure he or she likes you!). Acids also react with many metals (Figure 1.13) and, on prolonged contact, can cause serious injury to the skin.

Bases are also found among common substances. Drain cleaners and oven cleaners often contain the base sodium hydroxide, sometimes called lye. Household ammonia is also a base. Bases are bitter to the taste and they feel soapy

Figure 1.13
Some metals are attacked by acids. Here an ordinary steel paperclip dissolves in concentrated hydrochloric acid. The bubbles contain gaseous hydrogen that is formed as one of the products of the reaction.

when you get them on your fingers. In fact, they actually react with the oils and fats in your skin to make soap! Bases are said to be *caustic,* meaning that they are able to burn and eat away your skin. It's wise to avoid direct contact with them as much as possible, particularly in concentrated form, because they can produce serious injury.

It is a chemical property of acids and bases that when they are mixed they react with each other. Their acidic and basic properties disappear, and we say that they have *neutralized* each other. In their place we observe the properties of the products of the chemical reaction. For example, if muriatic acid and lye are mixed in *just* the right proportions, a solution containing only ordinary table salt is formed. That's pretty amazing when you think about it. Two solutions, which by themselves can cause serious injury if swallowed, are mixed and the solution that results contains only salt. A few hundred years ago you could have been quite a magician with that trick!

Neutralization reactions similar to the one above are quite common. We swallow milk of magnesia, a white insoluble base, to neutralize excess stomach acid. Sodium bicarbonate does the same thing, but it produces carbon dioxide gas as one of the products and that makes us burp.

As you study chemistry and observe chemical reactions in the laboratory, keep it in mind that very dramatic changes in properties are occurring. Look for them—it's one of the things that makes chemistry exciting.

1.17 CHEMICAL EQUATIONS

A chemical equation gives a "before-and-after" picture of a chemical reaction

A **chemical equation** describes what happens when a chemical reaction occurs. It uses chemical formulas to provide us with a before-and-after picture of the chemical substances involved. For example, the reaction between muriatic acid (hydrochloric acid) and lye (sodium hydroxide) discussed in the previous section can be represented by the equation,

$$HCl + NaOH \longrightarrow NaCl + H_2O$$

HCl is the formula for hydrochloric acid, and NaOH is the formula for sodium hydroxide. These two substances are called the **reactants** and are written on the left side of the arrow. The **products** of the reaction—the substances present after the reaction is over—are written on the right of the arrow. For this reaction, the products are sodium chloride, NaCl, and water. The arrow means "reacts to yield." Thus, this equation tells us that *hydrochloric acid and sodium hydroxide react to yield sodium chloride and water.*

Many reactions are more complex than the one we've just examined. An example is the burning of butane, C_4H_{10}, the fluid used in disposable cigarette lighters (Figure 1.14). The equation for this chemical reaction is

$$2C_4H_{10} + 13O_2 \longrightarrow 8CO_2 + 10H_2O$$

In this equation there are numbers, called **coefficients,** appearing in front of each of the formulas. These are present to balance the equation. An equation is **balanced** if there are the same number of atoms of each element indicated on both sides of the arrow. The 2 before the C_4H_{10} tells us that two molecules of butane react. This involves a total of 8 carbon atoms and 20 hydrogen atoms. On the right we find 8 molecules of CO_2, which contain a total of 8 carbon atoms. Similarly, 10 water molecules contain 20 hydrogen atoms. Finally, we can count 26 oxygen atoms on both sides of the equation. We will see how to balance equations in Chapter Four.

Figure 1.14
The combustion of butane, C_4H_{10}.

EXAMPLE 1.14 Reading Chemical Equations

Problem: The chemical equation for the combustion of acetylene, C_2H_2, is

$$2C_2H_2 + 5O_2 \longrightarrow 4CO_2 + 2H_2O$$

How many atoms of oxygen are shown on each side of this equation?

Solution: On the left we see $5O_2$, which represents 10 oxygen atoms (5 molecules, each with 2 oxygens). On the right we have $4CO_2$, which contains 8 oxygens (4×2), plus $2H_2O$ which contains 2 oxygens (2×1). The total on the right is $8 + 2 = 10$ oxygen atoms.

EXERCISE 1.14 How many atoms of each element appear on each side of the equation

$$Mg(OH)_2 + 2HCl \rightarrow MgCl_2 + 2H_2O$$

SUMMARY

Scientific Method. Chemistry is important because you will need it in your chosen field and because chemistry affects your life directly. Like other scientists, chemists employ the scientific method. Observations are made and the empirical data that are collected from many experiments are often summarized in scientific laws. Hypotheses and theories that explain the data are tested in other experiments. For making measurements, a modified metric system of units has been adopted called the SI. This system defines a set of base units (Table 1.1) that may be used to derive units for other quantities, such as area or volume. The size of units can be modified with decimal multipliers (Table 1.2), which make them more closely suit our needs in particular circumstances.

Laboratory Measurements. In the laboratory we routinely measure mass (which is different from weight), volume, temperature and length. Convenient units for these purposes are grams, milliliters, degrees Celsius or kelvins, and centimeters or millimeters.

Significant Figures. The reliability of a measured number is expressed by the number of significant figures that it contains. These are the number of digits known with certainty, plus the one that possesses some uncertainty. A measured number is precise if it contains many significant figures. A measurement is accurate if its value is very close to the true value. When combining measurements by multiplication or division, the answer should not contain more significant figures than the least precise factor. When addition or subtraction is used, the answer is rounded to have the same number of decimal places as the quantity having the fewest decimal places. Exact numbers do not enter into determining the number of significant figures in a calculated quantity. Scientific notation is useful for writing large or small numbers in compact form and for expressing unambiguously the number of significant figures in a number.

Factor-Label Method. The factor-label method is based on the ability of units to undergo the same mathematical operations as numbers. Conversion factors are constructed from valid relationships between units. Unit cancellation serves as a guide to the use of conversion factors and aids us in correctly setting up the arithmetic for a problem.

Matter and Energy. Matter and energy are both of fundamental concern in studying chemistry. A substance can have energy as kinetic energy (due to motion) or potential energy (stored energy). Eventually all forms of energy are converted to heat.

Properties. When studying matter we are concerned about its physical properties and chemical properties. Physical properties can be described without reference to other substances. Chemical properties relate to how substances change into others. Properties are most useful if they are independent of sample size. An example is the physical property, density—the ratio of an object's mass to its volume.

Elements, Compounds, and Mixtures. The simplest forms of matter that chemists are concerned with are elements. Elements combine in fixed proportions to form compounds. Elements and compounds are pure substances that may be combined in varying proportions to give mixtures. A one-phase mixture is homogeneous; if it consists of two or more phases, it is heterogeneous.

Atoms, Molecules, and Reactions. Elements are assigned chemical symbols. These are used to write chemical formulas whose subscripts specify the numbers of atoms of each element that are present. Sometimes atoms combine to produce larger, stable particles called molecules. When elements and compounds react, the properties of the reaction mixture change as the products are formed. Chemical equations present before-and-after descriptions of chemical reactions. When balanced, the equation contains coefficients that make the numbers of atoms of each kind the same among the reactants and products.

REVIEW QUESTIONS

1.15. You are probably not a chemistry major. Therefore, after some thought, give two reasons why a course in chemistry will be beneficial to *you* in the pursuit of your particular major.

1.16. With what is the science of chemistry concerned?

1.17. After you read this question, look around and then list ten items you can see that are composed of synthetic materials not found in nature.

1.18. Are there any objects in your room composed *entirely* of natural materials? What are they?

1.19. What steps are involved in the Scientific Method?

1.20. What is the function of a laboratory?

1.21. Give definitions for (a) *data,* (b) *hypothesis,* (c) *law,* and (d) *theory.*

1.22. What advantages does the metric system of units have over the English system?

1.23. Several commercial products that use metric units are pictured in Figure 1.2. Can you find any others among the items with which you come in contact each day?

1.24. What does the abbreviation SI stand for?

1.25. Which SI base unit is defined in terms of a physical object?

1.26. On page 20 we saw that kinetic energy can be calculated by using the equation K.E. = $\frac{1}{2}mv^2$. The SI unit for mass is the kilogram (kg), and the derived unit for velocity or speed is meter/second (m/s). What is the derived unit for energy?

1.27. What is the meaning of these prefixes: (a) centi, (b) milli, (c) kilo, (d) micro, (e) nano, (f) pico, and (g) mega?

1.28. What abbreviation is used for each of the above prefixes?

1.29. What units are most useful in the laboratory for measuring (a) length, (b) volume, and (c) mass?

1.30. How do mass and weight differ?

1.31. How is mass measured? What is the difference between a balance and a scale?

1.32. What do we mean by significant figures?

1.33. What is accuracy? What is precision?

1.34. What is a conversion factor? We *cannot* use the equation 1 yd = 2 ft to construct a proper conversion factor relating yard to feet. Why? Can we construct a conversion factor relating cm to m from the equation 1 cm = 1000 m?

1.35. Define (a) matter, (b) property of matter, (c) physical property, and (d) chemical property.

1.36. Choose an object on your desk and list as many of its physical properties as you can. Which properties are independent of the size of the object?

1.37. Suppose there are two glasses in front of you, one containing water and one containing gasoline. What physical properties could you use to differentiate between them? What chemical properties could you use?

1.38. If you were given only the masses of the water and gasoline in the preceding question, or if you were given only the volume of each liquid, you could not tell them apart. However, if you were given both their masses *and* their volumes, you could determine which was the gasoline and which was the water. How?

1.39. Define (a) element, (b) compound, (c) mixture, (d) homogeneous, and (e) heterogeneous.

1.40. What is the chemical symbol for each of the following elements? (a) chlorine, (b) sulfur, (c) iron, (d) silver, (e) sodium, (f) phosphorus, (g) iodine, (h) copper, (i) mercury, and (j) calcium.

1.41. What is the name of each of the following elements? (a) K, (b) Zn, (c) Si, (d) Sn, (e) Mn, (f) Mg, (g) Ni, (h) Al, (i) C, and (j) N.

1.42. What is the difference between an atom and a molecule?

1.43. How many atoms of hydrogen are represented in each of the following formulas? (a) $NaHCO_3$, (b) H_2SO_4, (c) C_3H_8, (d) $HC_2H_3O_2$, (e) $(NH_4)_2SO_4$, and (f) $(CH_3)_3COH$.

1.44. How many atoms of each kind are represented in the

following formulas? (a) Na_3PO_4, (b) $Ca(H_2PO_4)_2$, (c) C_4H_{10}, (d) $Fe_3(AsO_4)_2$, and (e) $Cu(NO_3)_2$.

1.45. Asbestos, a known cancer-causing agent, has as a typical formula, $Ca_3Mg_5(SI_4O_{11})_2(OH)_2$. How many atoms of each element are given in this formula?

1.46. Consider the balanced equation:
$2Fe(NO_3)_3 + 3Na_2CO_3 \rightarrow Fe_2(CO_3)_3 + 6NaNO_3$.
(a) How many atoms of Na are on each side of the equation?
(b) How many atoms of C are on each side of the equation?
(c) How many atoms of O are on each side of the equation?

1.47. What are the numbers that are written in front of the formulas in a balanced chemical equation called?

1.48. Define (a) *kinetic energy*, (b) *potential energy*. What two things determine how much kinetic energy a body possesses?

1.49. How do "temperature" and "heat" differ? How do we usually measure temperature?

1.50. What reference points do we use in calibrating the scale of a thermometer? What temperature on the Celsius scale do we assign to each of these reference points?

REVIEW PROBLEMS

1.51. What number should replace the question mark in each of the following?
(a) 1 cm = ? m
(b) 1 km = ? m
(c) 1 m = ? pm
(d) 1 dm = ? m
(e) 1 g = ? kg
(f) 1 mg = ? g

1.52. What numbers should replace the question marks below?
(a) 1 nm = ? m
(b) 1 μg = ? g
(c) 1 kg = ? g
(d) 1 Mg = ? g
(e) 1 mg = ? g
(f) 1 dg = ? g

1.53. Perform the following conversions.
(a) 10.0 cm to km
(b) 5.3 g to mg
(c) 5.3 mg to kg
(d) 37.5 ml to liters
(e) 0.125 liter to ml
(f) 342 nm to mm

1.54. Perform the following conversions; express your answers in scientific notation.
(a) 1.83 nm to cm
(b) 3.55 g to mg
(c) 8.44 km to cm
(d) 33 m to mm
(e) 0.55 dm to km
(f) 53.8 kg to mg

1.55. Perform the following conversions.
(a) 36 in. to cm
(b) 5.0 lb to kg
(c) 3.0 qt to ml
(d) 1 cup (8 oz) to ml
(e) 55 mi/hr to km/hr
(f) 50.0 mi to km

1.56. Perform these conversions:
(a) 250 ml to qt
(b) 2.0 ft to m
(c) 1.33 kg to lb
(d) 1.75 liters to fluid oz
(e) 75 km/hr to mi/hr
(f) 80.0 km to mi

1.57. Cola is often sold in 12-ounce cans. How many milliliters is this?

1.58. How many fluid ounces are in a 2-liter bottle of cola?

1.59. A metric ton is 1000 kg. How many pounds is this?

1.60. In the United States, a "long ton" is 2240 lb. What is this weight expressed in metric tons (1 metric ton = 1000 kg)?

1.61. Perform the conversions:
(a) 8.0 yd² to m²
(b) 3.4 in.² to cm²
(c) 1.5 ft³ to liters

1.62. Perform these conversions:
(a) 85 cm² to in.²
(b) 3.3 m³ to ft³
(c) 144 in.² to m²

1.63. The strange Ydarb tribe of a former French colony thrives on cabbage and canned potatoes. Cabbage costs 31 francs per head and potatoes costs 17 francs per can. The average Ydarb consumes 3 cans of potatoes per head of cabbage. If a Ydarb spends 124 francs on cabbage, how much money will then be spent on potatoes?

1.64. A projectile is fired at a speed of 2155 ft/sec. What is this speed expressed in km/hr?

1.65. A clinical thermometer registers a patient's temperature to be 37.13 °C. What is this temperature in °F?

1.66. The coldest permanently inhabited place on earth is the Siberian village of Oymyakon in the U.S.S.R. In 1964 the tempreature reached a shivering −96 °F! What is this temperature in °C?

1.67. Helium has the lowest boiling point of any liquid. It boils at 4 K. What is its boiling point in °C and °F?

1.68. When an object is heated to high temperature, it glows and gives off light. The color balance of this light depends on the temperature of the glowing object. Photographic lighting is described, in terms of its color balance, as a temperature in kelvins. For example, a certain electronic flash gives a color balance (called color temperature) rated at 5800 K. What is this temperature expressed in °C?

1.69. Express the following numbers in scientific notation *without using a calculator.*
(a) 245
(b) 31,000
(c) 0.00287
(d) 45,000,000
(e) 0.00000004
(f) 324,400

1.70. Express the following numbers in scientific notation *without using a calculator.*
(a) 3389
(b) 0.000025
(c) 81,300,000
(d) 0.0225
(e) 2.33
(f) 18,300

1.71. Write the following numbers in standard, nonexponential form.
(a) 2.1×10^3
(b) 3.35×10^{-4}
(c) 3.8×10^6
(d) 4.6×10^{-10}
(e) 34.6×10^{-2}
(f) 8.5×10^4

1.72. Write these numbers in standard, nonexponential form:
(a) 4.27×10^{-4}
(b) 7.11×10^7
(c) 33.5×10^{-6}
(d) 2.85×10^{-3}
(e) 5.0000×10^4
(f) 17.2×10^5

1.73. How many significant figures do the following measured quantities have?
(a) 2.75 cm
(b) 39.24 mm
(c) 12.0 g
(d) 0.0021 kg
(e) 0.0006080 m
(f) 0.002 ml

1.74. How many significant figures do the following measured quantities have?
(a) 0.240 g
(b) 11.303 m
(c) 0.0008 kg
(d) 615.0 mg
(e) 1.00005 liter
(f) 3.505 mm

1.75. Perform the following arithmetic and round off the answers to the correct number of significant figures. Assume that all of the numbers were obtained by measurement.
(a) 0.022×315
(b) $83.25 - 0.1075$
(c) $(84.4 \times 0.02)/(31.22 \times 9.8)$
(d) $(33.4 + 112.7 + 0.002)/(6.488)$
(e) $(315.44 - 208.1) \times 8.8175$

1.76. Perform the following arithmetic and round off the answers to the correct number of significant figures. Assume that all of the numbers were obtained by measurement.
(a) $3.58/1.739$
(b) $4.02 + 0.001$
(c) $(22.4 \times 8.3)/(1.142 \times 0.002)$
(d) $(1.345 + 0.022)/(13.36 \times 8.4115)$
(e) $(74.335 - 74.332)/(4.75 \times 1.114)$

1.77. Gasoline has a density of about 0.65 g/ml. How much do 25 gallons weigh in kilograms? In pounds?

1.78. A sample of kerosene weighs 25.3 g. Its volume was measured to be 31.7 ml. What is the density of the kerosene?

1.79. Acetone, the solvent in nail polish remover, has a density of 0.791 g/ml. What is the volume of 10.0 g of acetone?

1.80. A glass apparatus contains 21.335 g of water when filled at 25 °C. At this temperature, water has a density of 0.99704 g/ml. What is the volume of this apparatus?

CHAPTER TWO

PHYSICAL PROPERTIES OF GASES

2.1 QUALITATIVE FACTS ABOUT GASES

The four important measurable properties of gases are volume, temperature, pressure, and mass

We begin our study of matter with gases because their physical properties are fairly easy to understand and they offer clues to the structure of all matter. Gases are also important substances in their own right. Some, such as oxygen, are essential to our lives. Carbon dioxide is a gas we have to remove from our bodies as we breathe. Hydrogen may one day be pumped around the nation through pipelines and used as a fuel to replace natural gas. Ammonia is an important fertilizer.

You are probably already familiar with many of the ways in which gases behave physically. If you've ever punctured a tire you've seen air rush out, never in. Air always loses pressure when given a larger volume—like the entire atmosphere. When the tire is patched, you can make air rush in, not out of the tire by using a tire pump. Air pressure always rises when more air is squeezed into what is essentially a fixed volume. You have seen warning labels on aerosol cans: Do Not Incinerate. If a *confined* gas—the aerosol propellant—is heated its pressure will always rise, often by enough to burst the container.

Pressure and heat are two of the agents of physical change among gases. To study these and other agents we need to learn how four physical properties of gases are related. These properties are volume, mass, temperature, and pressure.

The volume of a gas is always the same as the volume of its container. Gases always spread out to fill whatever space they have available, which is one of their important physical properties. The volume of a gas is usually given in the unit of liter or milliliter, although sometimes the SI unit—the cubic meter, m^3—is used.

A tire pump can be used to increase the pressure in a bicycle tire.

The mass of a gas sample is normally given in grams or milligrams. For the most part, we assume that the gases we work with are confined. Therefore, once we specify the mass, we can usually assume it doesn't change. This simplifies our study because with mass a constant we can limit our work to the relations among the other three properties—pressure, volume, and temperature. Later, in Chapter Three, we'll return to the effects of mass and its interrelations with the other properties.

$$K = {}^\circ C + 273.15$$

The most convenient unit for gas temperature is the kelvin. Temperature in other units must first be changed to kelvins in calculations involving gases.

Gas pressures are reported in a variety of names, according to the profession and the country. The underlying concept is the same for all, however, and before venturing into the units we'll study that concept.

2.2 PRESSURE

Pressure is defined as force per unit area

The first physical property that we'll discuss is pressure. Pressure is created anytime a force is applied to some surface. If you've used your thumb and finger to crack open a peanut, you probably found it easier when you put the force on the narrow ridge of the shell at one end than somewhere along a wider side. Instinctively you did exactly the right thing. You concentrated the force on a very small area, the ridge, and that created a substantial pressure. Evidently an important relation exists among force, area, and pressure. To develop it we need to define force.

A **force** is anything that will cause something else to change its motion or direction. Force is the name we give to whatever makes something move, stop, veer to one side, come, go, rise, or fall, or crack apart. A force is also anything that might be acting to counterbalance any of these changes. Whenever *nothing* is happening you can be sure that two or more forces are exactly balanced. One of the very important forces in nature is gravity. The earth exerts this force of attraction on anything else having mass, ranging from the moon to a falling leaf, from a soaring bird to the air itself. In fact, the force of gravity pulling on the air creates the opposing effect of the air pushing on the earth, a force that creates the atmospheric pressure. We have to be very careful now, however. Force and pressure are not the same things.

Pressure is force per unit area, the result of dividing the force by the area on which it acts.

$$\text{pressure} = \frac{\text{force}}{\text{area}}$$

You no doubt know from experience how important the distinction between force and pressure is. The force of gravity pulling on a heavy backpack filled with books or camping gear hurts the shoulder far less if it acts on large straps rather than on narrow ropes. The force is the same in both cases, but when it's distributed over wide, large-area straps the force per unit area or pressure is much less than when it's concentrated on narrow, small-area ropes. Thus, the mass of an object can be an origin of force, but how much *pressure* that mass can exert depends on the area, as illustrated in Example 2.1.

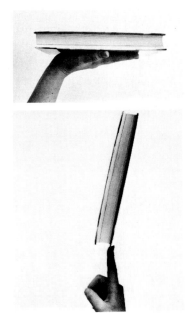

With the force distributed over a large area the net pressure is small, (top): when the force acts on a small area the net pressure is large (bottom).

EXAMPLE 2.1 Calculating Pressure

Problem: Let's assume that this book has a weight of 3.25 lb, and that the palm of your hand has an area of 19.5 in.². If you rest the book, cover side down, on your palm, what is the resulting pressure? If you balance the book on an edge so that it presses on an area of a finger $\frac{1}{16}$ in. by $\frac{1}{2}$ in., what is the pressure now? (Calculate to three significant figures.)

Solution: Pressure is force divided by area. In the first experience

$$\text{pressure} = \frac{3.25 \text{ lb}}{19.5 \text{ in.}^2} = 0.167 \text{ lb/in.}^2$$

In the second experience we'll assume that "$\frac{1}{16}$ inch" means 0.0625 in. and that "$\frac{1}{2}$ inch" means 0.500 in. That makes the area 0.0313 in.² (rounding from 0.3125). Therefore,

$$\text{pressure} = \frac{3.25 \text{ lb}}{0.0313 \text{ in.}^2} = 104 \text{ lb/in.}^2$$

The weight of the book hasn't changed, but the pressure has become over 600 times as great. Ouch!

EXERCISE 2.1 Butter is often sold in a 1-pound container measuring 5.00 × 4.72 × 1.22 in. What pressure does this much butter cause when it's in this container and lying with its largest side down? What is its pressure when lying with its long narrow side down? Calculate your answer to three significant figures in lb/in.²

Measuring atmospheric pressure

We can detect the existence of atmospheric pressure by a very simple device known as a Torricelli barometer, named after Evangelista Torricelli (1608–1647), an Italian physicist. As seen in Figure 2.1, a Torricelli barometer consists of a long narrow tube (80 cm or more in length) that is sealed at one end, filled with mercury, and then inverted into a dish of mercury. When inverted, some mercury flows out, but not all of it. The pressure of the atmosphere bearing down on the surface of the mercury in the dish keeps some of the mercury in the tube, and the height of the mercury column is a direct measure of the pressure. At the top of the tube is a space once filled with mercury but now has nothing. It is literally empty space. Nothing up there presses down on the mercury inside the tube. In that space is a **vacuum,** a region empty of all matter. (It's not actually a perfect vacuum. We describe it as a partial vacuum because a very low concentration of mercury vapor is present. Its effect is so small that we may ignore it.)

Mercury—a shiny, metallic element—is a liquid above −39 °C. Handle with care! Its vapor is poisonous.

Figure 2.1
The mercurial or Torricelli barometer.

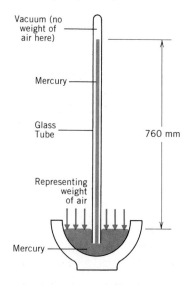

How high the column of mercury in a Torricelli barometer will stand varies with several factors. One is the location of the barometer on the planet. On a mountain where the air is less dense than at sea level, the air pressure is less and the mercury column is shorter. Mercury itself expands when heated and contracts when cooled (the principle of the thermometer). Therefore, the column's height will vary with temperature. Corrections for temperature, however, have been worked out, and barometers used in scientific work come with correction tables. Finally, the height of the mercury column varies with the weather. (Who hasn't heard a weather forecaster speak of changing pressures!) At ordinary temperatures and at sea level the mercury column will be about 760 mm high.

Before studying the specific units for the pressure of a gas we might point out one important property of the pressure of all fluids. A **fluid** is any material whose shape adjusts spontaneously and rather quickly to the shape of its container, which means that both liquids and gases are fluids. The pressure exerted by a fluid acts equally on every side of its container and on any object suspended in the fluid. Gas pressure, for example, is felt equally by all the interior surfaces of its container.

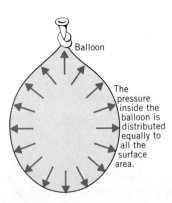
The pressure inside the balloon is distributed equally to all the surface area.

2.3 UNITS OF PRESSURE

The standard atmosphere, the torr, the millimeter of mercury, and the pascal are widely used units of pressure in various scientific fields

Evangelista Torricelli.

The relationship between the height of a column of mercury in a Torricelli barometer and air pressure gave us one of our oldest units of pressure, the **millimeter of mercury,** abbreviated **mm Hg.** One millimeter of mercury (1 mm Hg) is the pressure caused by a column of mercury 1 mm in height. Since air pressure can vary with both temperature and geographic location, scientists use the concept of a *standard* atmosphere of pressure for reference purposes. It is defined not just in terms of the height of a mercury column, but also for a particular temperature of the mercury and a particular place. One standard **atmosphere** is the pressure exerted by a column of mercury that is 760 mm high at a temperature of 0 °C. (That's how high mercury will stand under those conditions in a Torricelli barometer.) The symbol for standard atmosphere is **atm.**

$$1 \text{ atm} = 760 \text{ mm Hg (0 °C), exactly} \qquad (2.1)$$

The **torr,** a smaller unit, honoring Torricelli, is 1/760 of a standard atmosphere. This is the unit we use throughout this book.

$$1 \text{ torr} = \frac{1}{760} \text{ atm, exactly} \qquad (2.2)$$

The *torr* and the *mm Hg* units are the same.

The SI unit of pressure is the **pascal,** which honors Blaise Pascal (1623–1662), a French scientist. We'll omit the formal definition of the pascal; it depends on a knowledge of physics we will not need. All we need is the relation of torr to pascal:

$$133.3224 \text{ Pa} = 1 \text{ torr} \qquad (2.3)$$

EXAMPLE 2.2   Converting Pressure Units

Problem: Show that 1 atm is roughly 101 kPa (kilopascals).
Solution: Use equations 2.1 and 2.3 as sources of conversion factors.

$$\frac{760 \text{ torr}}{1 \text{ atm}} \times \frac{133.3224 \text{ Pa}}{1 \text{ torr}} = 101{,}325.0 \text{ Pa/atm}$$

Next, change pascals/atm to kilopascals/atm by moving the decimal point three places to the left:

$$101{,}325.0 \text{ Pa/atm} = 101.3250 \text{ kPa/atm}$$

Rounding this we see that 1 atm is roughly equal to 101 kPa.

EXERCISE 2.2   Suppose one day you found that the pressure in the chemistry laboratory was 730 torr. What was the pressure in standard atmospheres, in kilopascals, and in mm Hg?

EXAMPLE 2.3   Using Units of Pressure

Problem: A suction pump works by creating a vacuum into which some fluid will rise. Suppose that you are working for a construction company and someone tells you to take a suction pump and remove the water from a pit. The water level is over 35 feet below the place where you must put the pump. Can any suction pump do the job? (The density of water is 1.00 g/ml, and the density of mercury is 13.6 g/ml.)

Solution: To answer the question we must first find out how high a column of *water* can be held by the pressure of one atmosphere. Let's begin by assuming that the pump is capable of pulling a perfect vacuum. We already know that a perfect vacuum can make a column of mercury stand only 760 mm high. Comparing the density of mercury with that of water, we can see that the mercury is 13.6 times as dense as water. Thus all other factors being equal, a perfect vacuum should make a column of water stand 13.6 times as high as it would a column of mercury.

Therefore, height of water = 760 mm × 13.6 = 10,366 mm (we'll round later). Now all we have to do is convert from millimeters to feet:

$$10{,}336 \text{ mm} \times \frac{1 \text{ cm}}{10 \text{ mm}} \times \frac{1 \text{ in.}}{2.54 \text{ cm}} \times \frac{1 \text{ ft}}{12 \text{ in.}} = 33.9 \text{ ft (rounded)}$$

mm to cm

to in.

to ft

In other words, even the best vacuum pump will be able to draw water up only 33.9 feet. Drawing it higher, such as 35 feet in our problem, can't be done by a vacuum pump. A pump creating pressure will be needed to *push* the water up.

EXERCISE 2.3 In the weather reports, atmospheric pressure is generally given in "inches," meaning "inches of *mercury.*" What is a pressure of 760 torr (1 standard atm) in "inches"?

EXERCISE 2.4 One of the oldest units for atmospheric pressure is pounds per square inch (lb/in.²). Find out what a standard atmosphere is in these units. Give the answer in 3 significant figures. *Hint.* Use the results of Example 2.3. Calculate the mass in pounds of a uniform column of water 33.9 feet high having an area of 1.00 in.² at its base. (The density of water is 1.00 g/ml; 1 ml = 1 cm³; 1 lb = 454 g; 1 in. = 2.54 cm.)

2.4 THE PRESSURE-VOLUME RELATIONSHIP

Robert Boyle.

The volume of a gas is inversely proportional to the pressure acting on it (at constant temperature and mass)

Now that we have the units for describing some important properties of gases we can begin a study of the ways that gases physically respond to changes in pressure, volume, and temperature. How, for example, does the volume of a gas change if you change the pressure on it? Over 300 years ago the English scientist Robert Boyle (1627–1691) used a simple J-tube, such as the one in Figure 2.2, to get data concerning the relationship of pressure to volume. The short end of the J-tube was sealed. Mercury was then poured into the open tube until it rose just short of the lower end of the sealed tube and trapped a quantity of gas at essentially 1 atm of pressure. See part (*a*) of Figure 2.2. When more mercury was

Figure 2.2
The J-tube experiment and the effect of pressure on volume. Boyle used a J-tube with a much longer open end to test the effects of still higher pressures.

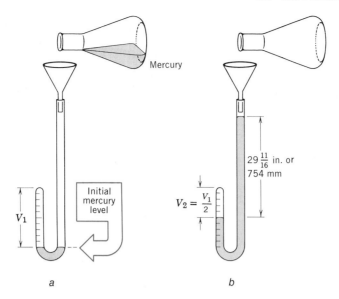

Mercury

$29\frac{11}{16}$ in. or 754 mm

Initial mercury level

V_1

$V_2 = \dfrac{V_1}{2}$

a   *b*

poured in, part (*b*), the extra weight squeezed the entrapped air and caused its volume to decrease. (This result was no surprise to Boyle or to anybody else of that time who had given the matter any serious thought. The difference that Boyle made was in getting accurate data.) When enough mercury was added to cut the volume of the trapped air in half, the difference in the heights of the mercury columns inside the sealed and open ends of the J-tube was $29\frac{11}{16}$ in. This is the same as 754 mm Hg, which is very close to 1 standard atm. That was the *extra* pressure over and above the original pressure, which was also very close to 1 standard atm. Thus the air initially at 1 atm now had an additional pressure of 1 atm for a total pressure on the gas of 2 atm. By doubling the pressure, Boyle had reduced the volume by one half. Of course, Boyle obtained much more data than this, and the curve of Figure 2.3 shows how pressure-volume data are related.

Figure 2.3
The relationship of volume to pressure—Boyle's law. The pressure unit for this curve is *inches of mercury,* the unit Boyle used. One atmosphere of pressure is 30 in. of mercury. When 48 units of volume at 30 in. of mercury pressure were subjected to twice as much pressure (60 in.), the volume was cut in half to 24 units. Doubling the pressure again, to 120 in., reduced the pressure by a factor of one-half again, to 12.5 units.

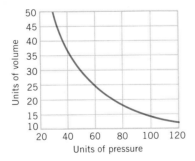

Boyle obtained the same kinds of results for different gases. As long as the temperature was held constant throughout each experiment, Boyle observed a reciprocal relation that we now call the **pressure-volume law** or **Boyle's law:**

Pressure-Volume Law (Boyle's Law). At a constant temperature the volume of a given sample of a gas is inversely proportional to the applied pressure.

$$V \propto \frac{1}{P} \quad (T \text{ and mass are constant}) \tag{2.4}$$

The symbol ∝ means "is proportional to."

We remove the proportionality sign, ∝, by introducing a proportionality constant, *C*:

$$V = \left(\frac{1}{P}\right)(C) \tag{2.5}$$

Using the pressure-volume law we can find what value one of the factors, pressure or volume, will become if we change the other, provided we keep the temperature and mass of the sample constant. Let's use the subscript 1 to indicate starting conditions. Thus P_1 and V_1 stand for initial values of pressure and volume. We'll use the subscript 2 for the new values after some change has occurred. By rearranging equation 2.5 we get

$$P_1 V_1 = C \qquad (C \text{ is the proportionality constant}) \tag{2.6}$$

But after any change at constant temperature on the same sample, the constant C must still be the same. (Otherwise we couldn't call it a "constant"!) Therefore, at the new values, P_2 and V_2:

$$P_2 V_2 = C \tag{2.7}$$

Since C is the same for both these equations, we can combine them:

$$P_1 V_1 = P_2 V_2 \tag{2.8}$$

When we know any three of the four values in this equation we can calculate the fourth, which typically will be either a final pressure, P_2, or a final volume, V_2. To find the new pressure (P_2) we multiply the old pressure (P_1) by a ratio of volumes (V_1/V_2).

$$P_2 = P_1 \times \left(\frac{V_1}{V_2}\right) \tag{2.9}$$
— a ratio of volumes

To find a new volume (V_2) we multiply the old volume (V_1) by a ratio of pressures (P_1/P_2).

$$V_2 = V_1 \times \left(\frac{P_1}{P_2}\right) \tag{2.10}$$
— a ratio of pressures

You should not memorize equations 2.9 or 2.10. All you need to do is apply the pressure-volume relation, which you should memorize, and the idea of multiplying some given value by a ratio.

EXAMPLE 2.4   Calculating with the Pressure-Volume Law

Problem: A sample of oxygen with a volume of 500 ml at a pressure of 760 torr is to be compressed to a volume of 450 ml. What pressure will be needed to do this? (The temperature is constant.)

Solution: Step 1. Assemble the data. You will always find it easier to work any gas law problem if you identify and assemble all the pertinent data first.

$$V_1 = 500 \text{ ml} \qquad P_1 = 760 \text{ torr}$$
$$V_2 = 450 \text{ ml} \qquad P_2 = ?$$

Step 2. Think about how the pressure must change. It must either increase or decrease. Here, since the volume is to decrease, the pressure *must increase*. That's what the pressure-volume law tells us.

Step 3. Form a ratio. In this example it will be a ratio of volumes. Two ratios can be formed from our data:

$$\frac{500 \text{ ml}}{450 \text{ ml}} \quad \text{and} \quad \frac{450 \text{ ml}}{500 \text{ ml}}$$

Which one do we use to multiply by P_1? Only the ratio of $\frac{500 \text{ ml}}{450 \text{ ml}}$ can make P increase. The other ratio is a fraction less than one, and using it would make P decrease. Therefore, we use the first ratio:

$$P_2 = 760 \text{ torr} \times \frac{500 \text{ ml}}{450 \text{ ml}}$$

$$= 844 \text{ torr (rounded from 844.444)}$$

EXERCISE 2.5 A sample of nitrogen has a volume of 880 ml and a pressure of 740 torr. If the volume is changed to 440 ml at the same temperature, what will be the pressure.

EXERCISE 2.6 If 200 ml of helium at 760 torr is given a pressure of 800 torr, what volume will the sample have, assuming there is no change in temperature?

No real gas obeys the pressure-volume relationship *exactly* over a wide range of conditions. However, the *P-V* data for most gases closely fit Boyle's equation, particularly when the pressures are low and the temperatures are relatively high. At high pressures and low temperatures—conditions that favor the change of a gas into its liquid form—gases depart from Boyle's law behavior more and more. The hypothetical gas that would obey this law exactly can still be imagined, however, and such a gas is called an **ideal gas.** We say that real gases approach ideal gas behavior as the pressure is lowered and the temperature is raised.

2.5 PARTIAL PRESSURE RELATIONSHIPS

Each gas in a mixture contributes its partial pressure to the total pressure

In some very important applications of the pressure-volume law the gas is not a pure substance but is instead a mixture of gases. A good example of this type of gas is the air we breathe and the air we exhale. What happens to the pressure, for example, when two or more gases that do not chemically react are mixed together? John Dalton (1766–1844), an English scientist, was interested in this question. His work led to the **law of partial pressures,** sometimes called **Dalton's law of partial pressures.**

Law of Partial Pressures (Dalton's Law). The total pressure exerted by a mixture of gases is the sum of their individual partial pressures.

$$P_{total} = P_a + P_b + P_c + \cdots \tag{2.11}$$

The elements oxygen and nitrogen occur as molecules with the formulas O_2 and N_2, not as atoms with the formulas O and N.

The **partial pressure** of a gas in a mixture of gases is the pressure that it would exert on the container if it were present all alone in that same volume. In equation 2.11 subscript letters were used to identify the individual gases, but in working with actual mixtures a particular partial pressure is often identified by using the chemical formula of the gas. For example, oxygen is O_2; its partial pressure in a mixture may be symbolized as P_{O_2} (or sometimes PO_2).

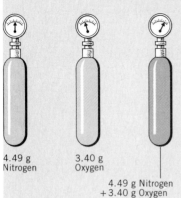

EXAMPLE 2.5 **Calculating Partial Pressure**

Problem: Imagine that we have three containers of exactly the same volume, each with a pressure gauge, and that all of them are completely empty (i.e., each contains a perfect vacuum). We put 4.49 g of nitrogen in the first tank and its pressure then reads 300 torr. We put 3.42 g of oxygen into the second tank and its pressure becomes 200 torr. What will be the final pressure in the third tank if we put 4.49 g of nitrogen *and* 3.42 g of oxygen into it?

Solution: The law of partial pressures tells us that each gas will act as if it alone were in the tank (assuming, of course, that no *chemical* reaction occurs). Therefore,

4.49 g Nitrogen

3.40 g Oxygen

4.49 g Nitrogen + 3.40 g Oxygen

$$P_{total} = P_{O_2} + P_{N_2}$$
$$= 300 \text{ torr} + 200 \text{ torr} = 500 \text{ torr}$$

EXERCISE 2.7 At sea level and 0 °C, the partial pressure of nitrogen in clean, dry air is 601 torr when the total pressure is 760 torr. If the only other constituent is oxygen, what is its partial pressure under these conditions?

When a gas is prepared in a laboratory, it is normally collected over water, as illustrated in Figure 2.4. The gas will therefore contain some water vapor. Actually, a space above any liquid will always contain some of the vapor of that liquid. (Remember that even the vacuum above the mercury in the Torricelli barometer contained some mercury vapor, but the amount was so small that we could ignore it.) Vapor exerts a pressure just like any other gas. This pressure is called the **vapor pressure** of the liquid, and its value depends *only* on the temperature of the liquid. Table 2.1 shows the vapor pressure for water over a wide range of temperatures.

The vapor pressure of mercury is only 0.0012 torr or 1.4×10^{-6} atm at 20 °C.

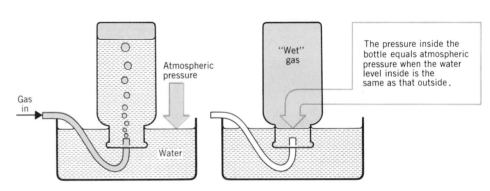

Figure 2.4
When a gas is collected over water, the gas picks up water vapor. The total pressure of the gas mixture inside the bottle therefore includes the partial pressure of the water vapor at the temperature of the water.

Notice that the vapor pressure for water is 47 torr at 37 °C, the temperature of the human body. Now look at Table 2.2, which gives the partial pressures of the respiratory gases at various stages in the breathing process. The partial pressure of water vapor for the air in the lungs (alveolar air) and for exhaled air is 47

Table 2.1
Vapor Pressure of Water

Temperature (°C)	Vapor Pressure (torr)
0	4.579
5	6.543
10	9.209
15	12.79
20	17.54
25	23.76
30	31.82
35	42.18
37 (body temperature)	47.07
40	55.32
45	71.88
50	92.51
55	118.0
60	149.4
65	187.5
70	233.7
75	289.1
80	355.1
85	433.6
90	527.8
95	633.9
100	760.0

Table 2.2
The Changing Composition of Air during Respiration

Gas	Partial Pressure (torr)		
	Inhaled Air	Exhaled Air	Alveolar Air[b]
Nitrogen	594.70	569	570
Oxygen	160.00	116	103
Carbon dioxide	0.30	28	40
Water vapor	5.00[a]	47	47
Totals	760.00	760	760

[a] The average value representing air of about 20 percent relative humidity, a familiar weather report term that we will not further define.

[b] Alveolar air is air within the alveoli, thin-walled air sacs emmeshed in beds of fine blood capillaries. Little more than bubbles of tissue, they are at the terminals of the successively branching tubes making up the lungs. We have about 300 million alveoli in our lungs.

torr. Also notice that the partial pressures of all the gases listed add up to 760 torr, the standard atmospheric pressure. It is a great illustration of both the law of partial pressures and the fact that water vapor acts just like any other gas.

Whenever a gas is collected over water, the *net* pressure of the gas (i.e., the pressure it would have if it were dry) will always be the total pressure less the vapor pressure of water, P_{water}, at the temperature of the water:

$$P_{dry\ gas} = P_{total} - P_{water} \qquad (2.12)$$

As you can see, this equation is simply an application of the equation for the law of partial pressures.

EXAMPLE 2.6 Correcting for the Water Vapor Present in a Gas Collected over Water

Problem: Suppose you prepare oxygen using an apparatus such as the one shown in Figure 2.4. You collect the sample over cold water (20 °C) and make sure that the water level inside the bottle is exactly the same as it is outside. (This ensures that the total pressure inside the gas-collecting bottle is the same as the atmospheric pressure outside.) Atmospheric pressure that day is 738 torr. The volume of the gas is 310 ml. Calculate the partial pressure of oxygen in the bottle and find what its volume would be if it were *dry* at 760 torr and 20 °C. (Work to three significant figures.)

Solution: Step 1. Correct for the partial pressure of water. Table 2.1 shows that at 20 °C the vapor pressure of water is 17.5 torr.

$$P_{O_2} = P_{total} - P_{water\ vapor}$$
$$= 738\ torr - 17.5\ torr$$
$$= 721\ torr\ (rounded)$$

Step 2. Find the dry volume at 760 torr. This is a straight pressure-volume problem where the new volume, V_2, is found by multiplying the original volume, V_1, by a ratio of pressures. Since P_2, the new pressure (760 torr), is greater than P_1, the corrected pressure (721 torr), the new volume must be *less* than the original. We therefore use $\frac{721\ torr}{760\ torr}$ as the correct ratio of pressures:

$$V_2 = 310\ ml \times \frac{721\ torr}{760\ torr} = 294\ ml$$

The answer means that if you could take the water vapor out of the sample of oxygen you collect, the oxygen would occupy a volume of 294 ml under a pressure of 760 torr.

EXERCISE 2.8 Suppose you prepared a sample of nitrogen and collected it over water at 15 °C when the atmospheric pressure was 745 torr. This sample occupied 310 ml. Find the partial pressure of nitrogen in the sample and its dry volume under a pressure of 760 torr and 15°C.

2.6 THE TEMPERATURE-VOLUME RELATIONSHIP

Gases expand when heated and contract when cooled in direct proportion to the change in temperature (at constant pressure and mass)

Both the pressure-volume law and the law of partial pressures resulted from studies of fixed quantities of gases kept at constant temperature throughout a given experiment. Suppose that the pressure instead of the temperature is kept constant on a weighed sample of gas. How does the volume respond to temperature? Jacques Alexander César Charles (1746–1823), a French physicist, became interested in this question after watching some early experiments with hot air balloons.

For the kinds of quantitative measurements that Charles used, the problem is in making sure that the pressure does remain constant throughout the experiment. Figure 2.5 is one solution. We know that the pressure on the mercury in the open bulb shown at the right in this figure is the same as that of the surrounding

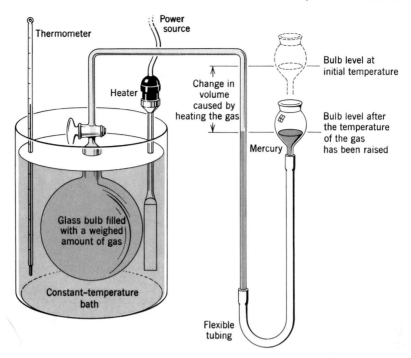

Figure 2.5
Charles' law. Shown here is an apparatus for studying how
the volume of a gas changes with temperature while the
pressure is kept constant.

atmosphere. The apparatus therefore is designed to use atmospheric pressure
as the constant pressure for the experiment. The bulb is connected to a vertical
glass tube by a flexible rubber hose that lets us move the mercury bulb up and
down. As long as the mercury level in the bulb is adjusted to be the same as the
mercury level in the glass tube, the pressure acting on the gas inside the apparatus will also be the same as the atmosphere's. If any heating or cooling of the
gas changes its volume, the mercury level in the glass arm will also change. Before that change in volume is measured, the bulb level has to be raised or lowered to even the two mercury levels, to make sure that the pressure inside is
again the same as that of the atmosphere outside.

The heater in Figure 2.5 is used to change the bath temperature. For low temperatures a refrigerating device could be used. At each new temperature setting,
the gas in the glass bulb is given enough time to come to the temperature of the
bath. Then, after adjusting the mercury levels, the change in volume is measured
by the change in the height of the mercury column in the glass tube.

Figure 2.6 gives some plots of data for temperature and volume for different
masses of the same gas. This gas happens to change to its liquid form at
−100 °C. However, because each line is a straight line it is easily extrapolated
(i.e., reasonably extended) to lower temperatures (dashed lines). The extrapola-
tions are actually reasonable *predictions* of how the gas would behave if it never
changed to a liquid. All the solid lines show a *direct* proportionality between volume and temperature. All the extensions point to −273 °C, absolute zero (zero on
the Kelvin temperature scale), the point where the gas volume has dropped to a
hypothetical value of zero as a result of being cooled. All other gases show similar behavior, which is now summarized in the **temperature-volume law** of
gases, often called **Charles' law:**

Figure 2.6
Charles' law. Each line shows
how volume changes with temperature for a different mass of the
same gas.

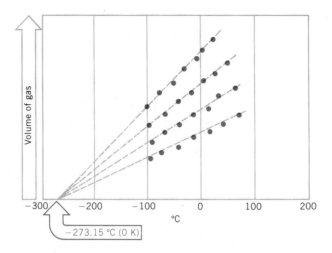

Volume of gas

−300   −200   −100   0   100   200

°C

−273.15 °C (0 K)

Temperature-Volume Law (Charles' Law). The volume of a given sample of gas kept at constant pressure is directly proportional to its Kelvin temperature.

$$V \propto T \qquad \text{(fixed mass and pressure)} \qquad (2.13)$$

Letting V be the volume and T be the temperature in kelvins, we may write 2.13 as an equation, where C' is the constant of proportionality:

$$V = C'T \qquad (2.14)$$

or

$$\frac{V}{T} = C' \qquad (2.15)$$

(The prime in C' simply tells us that this constant is not the same as the one in the pressure-volume law.) Using subscripts 1 and 2 in the usual way, we may express the law in a way that lets us do calculations without finding the value of C':

$$\frac{V_1}{T_1} = \frac{V_2}{T_2} \qquad (2.16)$$

This equation has to be true because both V_1/T_1 and V_2/T_2 have to be equal to the same constant, C', as long as the mass and pressure of the gas do not change. Since we normally want to find either V_2 or T_2 after some change, we'll use equation 2.16 in one of two ways:

$$V_2 = V_1 \left(\frac{T_2}{T_1} \right) \qquad (2.17)$$

⌞——— a ratio of Kelvin temperatures

Or,

$$T_2 = T_1 \left(\frac{V_2}{V_1} \right) \qquad (2.18)$$

⌞———a ratio of volumes

Temperature, T, must be in kelvins, but volume, V, may be in any unit provided the units for both V_1 and V_2 are the same.

EXAMPLE 2.7 Calculating with the Temperature-Volume Law

Problem: Anesthetic gas is normally given to a patient when the room temperature is 20.0 °C and the patient's body temperature is 37.0 °C. What would this temperature change do to 1.60 liter of gas if the pressure and mass stay constant? (Work to three significant figures.)

Solution: *Step 1.* Collect the data in one place and *change all Celsius temperatures to kelvins.*

$$V_1 = 1.60 \text{ liter} \quad T_1 = 293 \text{ K } (20.0 + 273)$$
$$V_2 = \quad ? \quad\quad T_2 = 310 \text{ K } (37.0 + 273)$$

Step 2. Use the temperature-volume law to decide how the volumes must change. Will it increase or decrease? In this example the volume must *increase* since the temperature increases.

Step 3. Form a ratio of temperatures that is greater than 1, and multiply it by V_1.

$$V_2 = V_1 \left(\frac{T_2}{T_1} \right)$$

$$V_2 = 1.60 \text{ liter} \times \frac{310 \text{ K}}{293 \text{ K}}$$

$$= 1.69 \text{ liter}$$

(The *change* in volume, 90 ml, is less than 6%, but it's a factor that anesthesiologists and inhalation therapists must consider.)

EXERCISE 2.9 To make 300 ml of oxygen at 20 °C change in volume to 250 ml, what must be done to the sample if its pressure and mass are kept constant?

EXERCISE 2.10 Why doesn't it make any sense to extrapolate the plots in Figure 2.6 below a temperature of −273 °C?

2.7 THE COMBINED GAS LAW

The pressure-volume and temperature-volume laws can be merged into a combined gas law

In most real situations in the laboratory or in industrial and medical applications, keeping the temperature or the pressure of a gas constant is either expensive or completely unnecessary. In the handling of any given mass of a gas, generally all three of the other properties—pressure, volume, and temperature—undergo some changes. The relationship connecting these three that has been found to be true for most gases under moderate conditions of pressure and temperature is that the volume of a gas is proportional to the ratio of the Kelvin temperature to the gas pressure.

$$V \propto \frac{T}{P} \quad \text{(at constant mass)}$$

This relationship can be expressed as an equation by introducing a new proportionality constant, which we will call C''.

$$V = \frac{T}{P} \times C'' \tag{2.19}$$

Rearranging this equation we have:

$$\frac{PV}{T} = C'' \qquad \text{(at constant mass)} \tag{2.20}$$

Using subscripts in our usual way, we can change equation 2.20 into a form that removes the need to calculate the value of C''.

$$\frac{P_1 V_1}{T_1} = \frac{P_2 V_2}{T_2} \qquad \text{(at constant mass)} \tag{2.21}$$

This relationship has to be true because $P_1 V_1 / T_1$ and $P_2 V_2 / T_2$ both equal the same constant, C''. As always the temperature must be in kelvins, but pressure and volume may be in any units. Equation 2.20 is the **combined gas law** in mathematical terms, and equation 2.21 is the form most useful for calculations.

Combined Gas Law. For a given mass of gas the product of its pressure and volume divided by its Kelvin temperature is a constant.

The combined gas law must contain the other gas laws as special cases, as we will discover in the next example.

EXAMPLE 2.8 **Relating the Combined Gas Laws to the Pressure-Volume Law**

Problem: Prove that the equation for the combined gas law (equation 2.21) reduces to the equation for the pressure-volume law under the right conditions.

Solution: The "right condition" for the pressure-volume law is *constant temperature* (assuming constant mass, also). That means that $T_1 = T_2$, and the two can be canceled:

$$\frac{P_1 V_1}{\cancel{T_1}} = \frac{P_2 V_2}{\cancel{T_2}}$$

Therefore,

$$P_1 V_1 = P_2 V_2$$

which is one way to state the pressure-volume law.

EXERCISE 2.11 Prove that the combined gas law as expressed in equation 2.21 reduces to the equation for the temperature-volume law, equation 2.16, under the right condition.

Equation 2.21 contains another gas law, the **pressure-temperature law.** This relationship was discovered by Joseph Gay-Lussac (1778–1850), a French scientist. Using Example 2.8 as a guide, solve Exercise 2.12 to discover how the pressure of a gas will change with the Kelvin temperature when the mass and the *volume* of the sample are fixed.

EXERCISE 2.12 What does equation 2.21 reduce to if the gas volume is a constant? Express your result in the form of an equation similar to equation 2.14. You will then have a mathematical expression for the pressure-temperature law, also known as **Gay-Lussac's law.** Write a statement of that law similar to the statements for the other gas laws. (Prepare both a written statement and an equation.)

EXERCISE 2.13 A steel cylinder contains a sample of neon, the gas in neon signs, at a pressure of 760 torr and a temperature of 25 °C. What will the pressure of the neon be if the cylinder is heated to 800 °C, as might happen if the cylinder were caught in a storeroom fire? Work this out using the same process of reasoning that you used to deduce the pressure temperature law in Exercise 2.12. Calculate your answer to three significant figures and express the result both in torr and in atm.

EXAMPLE 2.9 **Using the Combined Gas Law**

Problem: If a sample of argon, the gas in electric light bulbs, is at 760 torr when the volume is 100 ml and the temperature is 35.0 °C, what must its temperature be if its pressure becomes 720 torr and its volume 200 ml? (Calculate to three significant figures.)

Solution: Step 1. Assemble the data and change degrees Celsius to kelvins.

$$V_1 = 100 \text{ ml} \quad P_1 = 760 \text{ torr} \quad T_1 = 308 \text{ K} (35.0 + 273)$$
$$V_2 = 200 \text{ ml} \quad P_2 = 720 \text{ torr} \quad T_2 = ?$$

Step 2. Think about what the pressure change alone must mean for the new temperature. According to the pressure-temperature law, a drop in pressure means a drop in temperature. Therefore, we'll need a ratio of pressures that is a fraction *less* than 1 to act on T_1, specifically $\frac{720 \text{ torr}}{760 \text{ torr}}$ rather than $\frac{760 \text{ torr}}{720 \text{ torr}}$. Next, think how the change in volume must relate to the change in temperature. A rise in volume must mean a rise in temperature, according to the temperature-volume law. That requires a ratio of volumes *greater* than 1 to act on T_1, specifically $\frac{200 \text{ ml}}{100 \text{ ml}}$, not $\frac{100 \text{ ml}}{200 \text{ ml}}$.

Step 3. Calculate the answer using the ratios deduced by reasoning.

$$T_2 = (308 \text{ K}) \times \left(\frac{200 \text{ ml}}{100 \text{ ml}}\right) \times \left(\frac{720 \text{ torr}}{760 \text{ torr}}\right)$$

ratio of volumes raises T — ratio of pressures lowers T

$$= 584 \text{ K (rounded from 583.579)} \quad \text{or} \quad 311 \text{ °C} (584 - 273)$$

The net effect is a large increase in temperature. The small temperature-lowering effect of the change in pressure is overwhelmed by the large increase in volume, which requires a rise in temperature.

EXERCISE 2.14 Rearrange the equation for the combined gas law (2.21) to write equations that express the following relationships in terms of P, V, and T, using appropriate subscripts.
(a) $P_2 = P_1 \times$ (ratio of volumes) $\times$ (ratio of Kelvin temperatures)
(b) $V_2 = V_1 \times$ (ratio of pressures) $\times$ (ratio of Kelvin temperatures)
(c) $T_2 = T_1 \times$ (ratio of volumes) $\times$ (ratio of pressures), which is what we did in Example 2.9.

EXERCISE 2.15 What will be the final pressure of a sample of nitrogen with a volume of 950 m³ at 745 torr and 25 °C if it is heated to 60 °C and given a final volume of 1150 m³? (Calculate to three significant figures.)

2.8 LAW OF EFFUSION

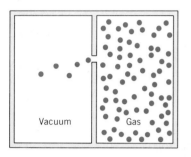

Figure 2.7
Effusion.

Less dense gases effuse more rapidly than more dense gases

The **effusion** of a gas is its movement through an extremely tiny opening into a region of lower pressure. (See Figure 2.7.) Diffusion, which is not the same as effusion, is an intermingling of one substance with another.

The rate at which a gas effuses depends on its density (its mass per unit volume). The English scientist Thomas Graham (1805–1869) found that the more dense the gas the slower it effuses. The exact relation is called the **law of gas effusion,** or sometimes **Graham's law:**

> **Law of Gas Effusion (Graham's Law).** The rate of effusion of a gas is inversely proportional to the square root of its density when the pressure and temperature of the gas are constant.
>
> $$\text{effusion rate} \propto \frac{1}{\sqrt{d}} \qquad \text{(constant } P \text{ and } T\text{)} \qquad (2.22)$$

Using d for gas density and C^* for a constant of proportionality, the relation of effusion rate to density can be written as an equation:

$$(\text{effusion rate}) \times \sqrt{d} = C^* \qquad \text{(constant } P \text{ and } T\text{)} \qquad (2.23)$$

The value of C^* is virtually the same for all gases. That important fact allows us to predict what will be the relative rates of effusion of two gases under the same conditions simply by comparing their densities. Letting subscripts $_A$ and $_B$ identify gases A and B, we can rewrite equation 2.23 as an expression that gives us this comparison:

$$(\text{effusion rate})_A \times (\sqrt{d_A}) = (\text{effusion rate})_B \times (\sqrt{d_B}) \qquad (2.24)$$

Now, we can rearrange this equation to show how the ratio of effusion rates relates to a ratio of gas densities:

$$\frac{(\text{effusion rate})_A}{(\text{effusion rate})_B} = \sqrt{\frac{d_B}{d_A}} \qquad (2.25)$$

EXAMPLE 2.10 **Using the Law of Gas Effusion**

Problem: A store receives a shipment of defective balloons. Each balloon has a tiny pinhole of the same size. If one balloon is filled with helium and another is filled with air to the same volume and pressure, which balloon will deflate faster and by how much? The density of helium at room temperature is 0.00016 g/ml and that of air is 0.0012 g/ml.

Solution: Use equation 2.25.

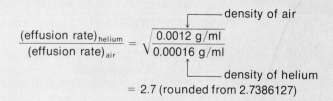

$$\frac{(\text{effusion rate})_{helium}}{(\text{effusion rate})_{air}} = \sqrt{\frac{0.0012 \text{ g/ml}}{0.00016 \text{ g/ml}}}$$

density of air

density of helium

= 2.7 (rounded from 2.7386127)

The answer means that helium will effuse 2.7 times as fast as air through the same sized pinhole. In fact, helium will effuse through the taut rubber material of a defect-free balloon. You may well remember from childhood the disappointment of finding that your helium-filled balloon went absolutely limp overnight as these photos show.

(Left) Two newly filled balloons—on the left, air-filled; on the right, helium-filled. (Right) The two balloons a day later. Helium has escaped by effusion from its balloon.

EXERCISE 2.16

Natural gas pipelines cannot be kept totally free of developing microscopic leaks. Natural gas is methane which has a density at room temperature and 760 torr of 0.654 g/liter. Under these conditions the density of hydrogen is 0.0818 g/liter. This difference in density, in the light of Graham's law, suggests a potential problem in using natural gas pipelines for sending hydrogen, if we should ever adopt hydrogen as an alternative to methane. What is this problem? *Hint.* Show how the rate of effusion of hydrogen compares to that of methane.

2.9 KINETIC THEORY OF GASES

The gas laws can be explained by assuming that gases consist of many small particles moving randomly according to the laws of physics

Have you ever taken something apart to see how it works—a clock, a radio, a car engine? Back in the nineteenth century that same kind of curiosity moved many scientists to ask how gases work. They were aware of the very predictable behavior of gases, and they had a few simple laws. They wondered what had to be true about the "structure" of a gas to explain its behavior. The **kinetic theory of gases** was the answer. This theory begins with a set of postulates that theorize what gases must be like, and it assumes that the laws of physics hold just as much with gases as with anything else.

The kinetic theory describes a "working model" of a gas.

Postulates of the kinetic theory of gases

1 A gas consists of an extremely large number of very tiny particles that are in constant random motion.
2 The gas particles are perfectly round, very hard, and have a total volume so small in relation to that of their container that it can be ignored.

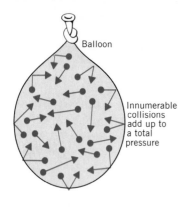

Balloon

Innumerable collisions add up to a total pressure

3 The particles often collide with each other and with the walls of the container, and they move in straight lines between collisions. (This means that it is assumed that particles exert no forces of attraction or repulsion on each other.)

Physics—the study of matter, energy, and physical changes—has a branch called mechanics that deals with the laws of behavior of moving objects. (Billiard players master at least some of the applications of these laws if not their mathematical forms.) In effect, the postulates of the kinetic theory treat gases as though they consist of super-small, constantly moving billiard balls that bounce off each other and the sides of their container.

Using the laws of motion and some insights from statistics, scientists deduced that the pressure of a gas is the net effect of the innumerable collisions made by the gas particles on the container's walls. They used the postulates to calculate the pressure at constant temperature, and the result agreed beautifully with the pressure-volume relation.

The biggest triumph of the kinetic theory came with its explanation for gas temperature. Each particle in a gas has a particular mass and a certain velocity. Some will momentarily be fully stopped and have zero velocity. Others, after being struck particularly hard, will have a very high velocity. With collisions happening at random all the time, the velocity of any one particle will change constantly. But remember that, according to the postulates, a sample of gas has many particles. Therefore, we may speak of an average velocity.

Any object with both mass and velocity has energy. As we learned in Section 1.11, energy of motion is called kinetic energy, K.E.

Particles in air have average velocities of about 400 km/sec (900 miles per hour)!

$$K.E. = \tfrac{1}{2}mv^2 \tag{2.26}$$

where m = mass and v = velocity. Since the particles have an average velocity, they must also have an average kinetic energy. *Individual* particles will have energies ranging from zero (when they are motionless for an instant) to extremely high values, but overall there is an average energy. Figure 2.8 shows the statistical distribution of kinetic energies among the huge number of particles in a gas

Figure 2.8
The distribution of kinetic energies among gas particles. The solid curve shows how at room temperature (about 300 K) the relative number of particles having a particular value of kinetic energy changes with changes in kinetic energy. The flatter curve indicates what happens to the plot at a higher temperature.

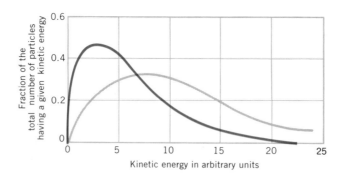

Kinetic energy in arbitrary units

sample. The peak of the curve corresponds approximately to the average energy. When scientists used the postulates to calculate pressure, they found that the product of the pressure and the volume of a gas is proportional to the average kinetic energy of the particles.:

$$PV \propto \tfrac{1}{2}m\overline{v^2} \tag{2.27}[1]$$

The combined gas law tells us that the product of the pressure and volume of a fixed mass of gas is also proportional to the Kelvin temperature; $PV \propto T$. If PV is

---

[1] The line over v^2 in equation 2.27 signifies that the velocity term is the *mean square* velocity, not the average velocity squared. The distinction is not important here because the two are nearly the same in a sample of innumerable gas particles.

proportional to both the average kinetic energy and to the Kelvin temperature, we must conclude that *the Kelvin temperature of a gas is directly proportional to the average kinetic energy of its particles:*

$$T \propto \tfrac{1}{2}m\overline{v^2} \tag{2.28}$$

EXAMPLE 2.11   **Reasoning with the Kinetic Theory**

Problem: If you cause the temperature of a gas to rise, what is actually happening to the gas particles?

Solution: The increase in temperature must mean an increase in the average kinetic energy (equation 2.28). Since m, the mass, can't change, v, the velocity, must increase. In other words, additional heat makes the particles move, on the average, at higher velocities.

EXERCISE 2.17   According to the results of the kinetic theory, what happens to the gas particles at 0 K, absolute zero?

EXERCISE 2.18   What impossible conditions would have to be met to lower the temperature of a gas below 0 K?

2.10   KINETIC THEORY AND THE GAS LAWS

The kinetic theory can be used to explain each gas law

The kinetic theory is our first encounter with both a scientific theory and a scientific model. A **scientific model** is a picture or mental construction derived from a set of ideas and assumptions that we imagine to be true because they enable us to explain certain observations and measurements. The gas laws are really the summaries of innumerable observations of real gases. Let us see how the kinetic theory accounts for each one.

At constant volume the pressure of a gas is proportional to its Kelvin temperature, according to the pressure-temperature law. Raising the temperature raises the average velocity of the gas particles (Example 2.11). At higher average velocities the gas particles hit the container's walls harder, creating the higher pressure (Figure 2.9 on the next page).

At constant pressure the volume of a gas is proportional to its Kelvin temperature, according to the temperature-volume law. When one "wall" of the container is the surface of a column of mercury and it can move in or out, a change in temperature no longer has to change the pressure. An increase in temperature, tending to make the pressure increase, instead makes the gas expand its volume (Figure 2.10). A decrease in temperature, which otherwise would reduce the gas's pressure, results in the outside pressure pushing the movable "wall" in until the pressure inside and out are again equal.

The volume of a gas is inversely proportional to its pressure (at constant temperature), according to the pressure-volume law. Suppose that the volume of a gas is allowed to increase at a constant temperature. The gas will thin out, and it will have fewer particles per unit volume. Therefore, fewer particles will be able to strike a unit area of the container's wall per second. This reduction in the number of hits per unit area per second will mean a reduction in gas pressure. If the gas volume decreases, then the gas will have more particles in each unit of volume. Now more particles will hit a unit area of the container's wall per second. This will mean a higher pressure (Figure 2.11).

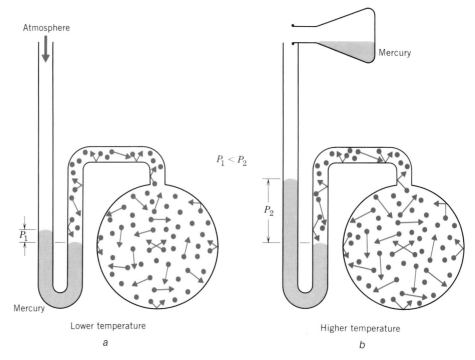

Figure 2.9
The kinetic theory and the pressure-temperature law (law of Gay-Lussac). (a) Here the gas is shown exerting a slight pressure, P_1, in excess of the atmospheric pressure. (b) At a higher temperature the gas particles move with greater kinetic energy and higher average velocity. This would cause the mercury column to be pushed out but, by adding more mercury to prevent that, the volume occupied by the gas is kept constant. The additional mercury, of course, is a measure of the increased pressure caused by the increased temperature.

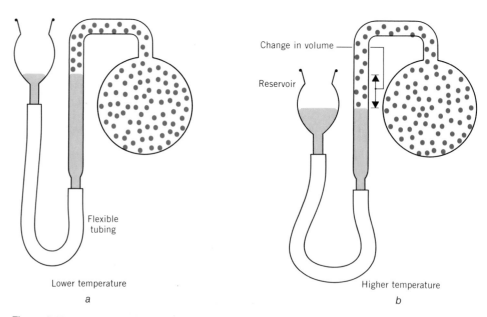

Figure 2.10
The kinetic theory and the temperature-volume law (Charles' law). In both (a) and (b) the pressure inside the gas container is the same because the level of mercury in the tube is the same as the level in the reservoir of mercury. (a) At the lower temperature the gas occupies a lower volume. (b) At the higher temperature the gas particles move with greater average energy and greater average velocity. If the level of mercury did not drop to allow for a greater volume, the pressure in (b) would be greater. By letting mercury flow into the reservoir, however, more volume is given to the gas.

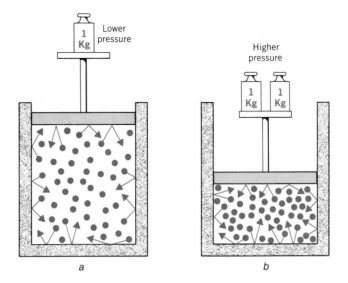

Figure 2.11
The kinetic theory and the pressure-volume law (Boyle's law).
In both (a) and (b) the total number of collisions occurring
at the cylinder's walls is the same. However, as shown here,
when the volume is made smaller the number of collisions
per unit area—which causes the pressure—increases.

The law of partial pressures is actually evidence for the postulate that the gas particles neither attract nor repel each other. Only if the particles of each gas act independently can the partial pressures of each gas add up in a simple way to total pressure (Figure 2.12).

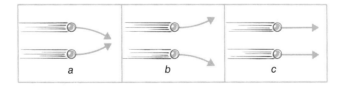

Figure 2.12
Gas particles neither attract nor repel each other in the model
of an ideal gas postulated by the kinetic theory. (a) If they
attracted each other their paths would curve together. (b) If they
repelled each other their paths would curve away. (c) When
there is neither attraction nor repulsion the paths of the particles
could be affected only by collisions, not by near misses. The
fact that partial pressures are additive according to Dalton's
law supports the view given in (c). If (a) were correct, the
total pressure would be less than the sum of the partial
pressures because particles about to collide with the wall would
be held back by attractions and cause less pressure. If (b)
were correct, the total pressure would be greater than the sum
of the partial pressures because particles about to collide
with the wall would be given extra "pushes" and cause more
pressure.

According to the law of effusion, when the rates of effusion of two gases are compared at the same pressure and *temperature,* the less dense effuses more rapidly. But one result of the kinetic theory was to show that the same temperature means the same average kinetic energy. Since K.E. $= \frac{1}{2}mv^2$, a particle of low mass can have the same kinetic energy as one of higher mass only if the velocity of the particle of lower mass is *higher* than that of the particle of greater mass. The higher velocity causes a higher rate of effusion (Figure 2.13). Mass, of course, is not the same as density, and so we have to ask if a gas whose particles are of low mass have a correspondingly low density. As found by experiments, the answer is yes. When gases are compared at identical temperatures and pressures, those with high densities have particles with higher masses than those with low densities. Therefore, a gas with high density will diffuse less rapidly than one of lower density at the same temperature and pressure. Because of the velocity-*squared* term in the equation for kinetic energy, a *square-root* term must occur in the equation for comparing rates of effusion (equation 2.25).

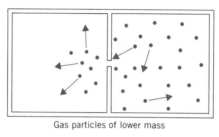
Gas particles of lower mass

a

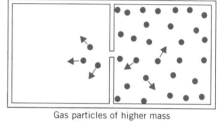
Gas particles of higher mass

b

Figure 2.13
The kinetic theory and the law of effusion (Graham's law). Each of the chambers has the same volume. The temperatures and pressures are the same in (*a*) and (*b*). These conditions ensure that the particles in (*a*) and (*b*) have the same average kinetic energy, K.E. $= \frac{1}{2}m\bar{v}^2$. In (*a*), however, the particles of lower mass must have higher average velocities so that their values of kinetic energy can be the same as the particles in (*b*), which have higher masses and therefore lower average velocities. The particles in (*a*) with higher velocities therefore effuse more rapidly than those in (*b*).

SUMMARY

Gas Properties. To describe the physical state of a gas we use as many of the following measurable properties as we need.

Property	Symbol	Usual Units
Pressure	P	torr
Volume	V	ml or liter
Temperature	T	kelvins (always)
Mass	m	g or mg
Partial pressure	P_a	torr; (a) stands for the formula of the gas
Density	$d = m/V$	g/liter or g/ml

Gas Laws. An ideal gas is a hypothetical gas that obeys the combined gas law exactly over all ranges of pressure or temperature. A real gas behaves very much like an ideal gas at normal pressures and temperatures.

Pressure-Volume Law (Boyle's law). Volume varies inversely with pressure at constant temperature and mass: $V \propto 1/P$.

Temperature-Volume Law (Charles' Law). Volume varies directly with Kelvin temperature at constant pressure and mass: $V \propto T$.

Temperature-Pressure Law (Gay-Lussac's Law). Pressure varies directly with Kelvin temperature at constant volume and mass: $P \propto T$.

Law of Partial Pressures (Dalton's Law). The total pressure for a mixture of gases is the sum of the partial pressures of the individual gases: $P_{total} = P_a + P_b + \cdots$.

Combined Gas Law. The product of P and V divided by the Kelvin temperature for a gas is a constant: $PV/T = $ constant.

Law of Effusion (Graham's Law). The rate of effusion of a gas is inversely proportional to the square root of its density (at constant P and T): (effusion rate) $\times \sqrt{d} = C^*$.

Kinetic Theory. An ideal gas consists of many particles each having essentially no volume and having neither attractive nor repulsive forces between them. They are in a state of utterly chaotic, random motion. When the laws of motion and statistics are applied to this model and the results are compared to the general gas law, the Kelvin temperature of a gas is directly proportional to the average kinetic energy of the gas particles. Pressure is the result of forces of collisions of the particles with the container's walls.

INDEX TO EXERCISES, QUESTIONS, AND PROBLEMS
(Those numbered above 36 are review problems.)

REVIEW QUESTIONS

2.19. What is the difference between force and pressure?

2.20. What causes the atmosphere to exert a force on the earth?

2.21. Why is the high density of mercury an advantage in a Torricelli barometer?

2.22. Mercury has a low vapor pressure. What advantage does this give to the use of mercury in a Torricelli barometer?

2.23. At 20 °C the density of mercury is 13.5 g/ml and that of water is 1.00 g/ml. Also at 20 °C, the vapor pressure of mercury is 0.0012 torr and that of water is 18 torr. Give and explain two reasons why water would be an inconvenient fluid to use in a Torricelli barometer.

2.24. How are the torr and the mm Hg related? _1 torr = 1 mm Hg_ are =

2.25. How are the mm Hg and the atm related? _760 = 1 atm_

2.26. State the pressure-volume law in both words and in the form of an equation. _V varies inversely t pressure, B.al. law $V \propto 1/p$_

2.27. Give the temperature-volume law in both words and in the form of an equation. _$V \propto T$_

2.28. What is the law of partial pressures? State it in words and in the form of an equation. _$P_{total} = P_a + P_b + P_c$_

2.29. In what way does the law of partial pressures provide evidence for a postulate of the kinetic theory?

2.30. Of the four important variables in the study of the physical properties of gases, which are assumed to be held at constant values in each of these laws?

(a) Boyle's law _mass & temp_
(b) Charles' law _mass + pressure_
(c) Law of partial pressures
(d) Gay-Lussac's law _mass & volume_
(e) Graham's law of effusion _pressure & temp_
(f) Combined gas law _$PV/T = $ constant_

2.31. Explain *how* heat makes a gas expand at constant pressure. (Describe a mechanical model that connects the rising temperature to the response of the gas—expansion.)

2.32. Explain *how* heat makes a confined gas have a higher pressure at constant volume.

2.33. Although mass is one of the four important variables in studying the physical properties of gases, it does not appear as a term in the combined gas law. (We assumed that the mass is a constant in that law.) However, where might mass be introduced into equation 2.20? (We will actually make that modification in the next chapter.)

2.34. Gases, if left to themselves, must migrate from a region of high pressure to a region of low pressure. In what specific way can the kinetic theory be used to explain this direction of migration?

2.35. What aspects of the kinetic theory of gases guarantees that two gases given the freedom to mix will always mix, entirely?

2.36. If the molecules of a confined gas are somehow given a lower average kinetic energy, what physical properties of the gas will change (and in what direction)?

REVIEW PROBLEMS

2.37. If a force of 200 lb acts on an area of 1 ft², what is the pressure in pounds per square inch? In atmospheres?

2.38. Which exerts the higher pressure, a force of 100 lb acting on 25 in.² or a force of 25 lb acting on 5 in.²?

2.39. A scientist observed that the atmospheric pressure in the laboratory was 744 mm Hg. What was the pressure in torr?

2.40. One day in the lab the atmospheric pressure was 755 torr. What was it in mm Hg?

2.41. A scientist observed that the atmospheric pressure was 750 torr. What fraction of one standard atmosphere is that?

2.42. What would be the pressure in torr if it is 0.850 atm?

2.43. A pressure of 0.445 atm would be how many torr?

2.44. A pressure of 755 torr would be how many kilopascals?

2.45. What are the partial pressures in kilopascals of nitrogen and oxygen in inhaled air? (Use the data in Table 2.2.)

2.46. In kilopascals, what are the partial pressures of nitrogen and oxygen in exhaled air? (Use the data in Table 2.2.).

2.47. To compress nitrogen at 760 torr from 750 ml to 500 ml, what must the new pressure be if the temperature is kept constant?

2.48. If oxygen at 950 torr is allowed to expand at constant temperature until its pressure is 760 torr, how much larger will the volume become?

2.49. Helium in a 100-ml container at a pressure of 500 torr is transferred to a container with a volume of 250 ml. What is the new pressure,
(a) if no change in temperature occurs?
(b) if its temperature changes from 20 °C to 15 °C?

2.50. What will have to happen to the temperature of a sample of methane if 1000 ml at 740 torr and 25 °C is given a pressure of 814 torr and a volume of 900 ml?

2.51. A sample of nitrogen at 760 torr with a volume of 100 ml is carefully compressed at constant temperature in successive changes of pressure, equaling 20 torr at a time, until the final pressure is 1000 torr. Calculate each new volume and prepare a plot of P versus V showing P on the horizontal axis.

2.52. A sample of helium is confined in a heavy-walled container with a volume of 450 ml fitted with a pressure gauge. The initial pressure is 760 torr. The helium is then heated in increments of 20 K starting at 293 K until the final temperature is 393 K. Prepare a plot of P versus T using the horizontal axis for T.

2.53. What volume of "wet" methane would you have to collect at 20 °C and 740 torr to be sure the sample contained 240 ml of dry methane at the same pressure?

2.54. What volume of "wet" oxygen would you have to collect if you needed the equivalent of 260 ml of dry oxygen at 760 torr and the atmospheric pressure in the lab that day was 746 torr? The oxygen is to be collected over water at a temperature of 15 °C.

2.55. Fuel is ignited in a diesel engine when it is injected into hot compressed air. The compression is what heats the air. In a typical high-speed diesel engine the chamber in the cylinder has a diameter of 10.8 cm and a length of 13.3 cm. (Find the volume by the equation, volume = 3.14 × (radius)² × length.) On compression, the length of the chamber is *reduced* by 12.7 cm (a "5-in. stroke"). The compression of the air changes its pressure from 1.0 atm to 34.0 atm. The initial temperature of the air is 363 K. What will be its final temperature, just before the fuel injection, as a result of the compression? (Calculate to three significant figures.)

2.56. Suppose early one cool morning (61 °F) you take your bike for a long bike hike. You check the tire pressure and find that it's 50.0 lb/in². (That means that the air pressure in the tire is actually 50.0 lb/in.² + 14.7 lb/in.² = 64.7 lb/in.². Tire gauges tell us how much *over* the atmospheric pressure the measured pressure is, and we may take atmospheric pressure to be 14.7 lb/in.².) By late afternoon the temperature has climbed to 98.5 °F. This rise in temperature plus the heat created by road friction makes the air temperature inside the tire 105 °F. What will the pressure gauge read now, assuming that the volume of air in the tire has not changed?

2.57. Under conditions for which the density of carbon dioxide is 1.96 g/liter and that of nitrogen is 1.25 g/liter, which gas will effuse more rapidly? What will be the ratio of the rates of effusion of nitrogen to carbon dioxide?

2.58. Uranium hexafluoride is a white solid that readily passes directly into the vapor state. At room temperature (20.0 °C) its vapor pressure is 120 torr. A trace amount (about 0.7%) of the uranium in this compound is of the type that can be used in nuclear power plants or atomic bombs. It's called uranium-235. Essentially all of the rest, called uranium-238, is useless for direct application in power plants or bombs. In fact, its presence interferes. During World War II a massive government effort was made to separate the two kinds of uranium. Gas effusion was used because the density of the hexafluoride made from uranium-235 is 2.2920 g/liter at 120 torr and 20.0 °C while under these same conditions the density of the hexafluoride made from uranium-238 is 2.3119 g/liter. (In the gaseous states, these compounds behave remarkably like ideal gases.) Calculate how much more rapidly the lower density compound will effuse compared to the higher density compound. (The very small difference that you will find was actually enough, but repeated effusions were necessary.)

CHAPTER THREE

ATOMS AND MOLECULES. A QUANTITATIVE APPROACH

3.1 LAWS OF CHEMICAL COMBINATION

Elements always combine in constant ratios by mass

We learned in the last chapter that the physical behavior of gases is predictable. But what about the physical behavior of other matter? And what about the *chemical* behavior of gases and other matter? Do chemical properties follow any consistent pattern? The science of chemistry couldn't exist if they didn't. Fundamental to any science is the notion that the objects studied always behave in predictable ways. What if steel could switch from being hard and strong to being soft and rubbery at unpredictable times? What if table salt were essential in our diets on some days but poisonous on others, and we had no way of telling when it was safe? Fortunately, we never have to deal with such unpredictability, but the question still remains *"In what ways* is matter predictable?"

Two of the consistently observed patterns of the behavior of matter are now summarized in two laws of chemical combination, the law of conservation of mass and the law of definite proportions. Evidence for these laws came from the work of many scientists in the eighteenth and early nineteenth century. The patterns these laws summarized were, like the gas laws, powerful clues to the fundamental structure of all matter. These two laws, for example, provided scientists with some of the earliest and most convincing evidence that elements consist of atoms.

> **Law of Conservation of Mass.** Mass is neither gained nor lost in a chemical change; mass is conserved.

> **Law of Definite Proportions.** In a given chemical compound, the elements are always combined in the same proportion by mass.

The law of conservation of mass means that if we placed chemical reactants in a sealed vessel that permitted no matter to escape, the mass of the vessel and its contents after a reaction would be identical to its mass before. While that may seem quite obvious to us now, it wasn't quite so obvious in the early history of modern chemistry. In early studies either one reactant or a product was sometimes a gas, and gases can easily enter or escape open reaction vessels. Not until scientists made sure that *all* reactants and all products were identified, collected quantitatively, and weighed could the law of conservation of mass rest on solid evidence.

The law of definite proportions may be illustrated by a beautiful, naturally occurring mineral called iron pyrite by geologists and ruefully dubbed "fools gold" by miners. (Color plate 1B is a picture of iron pyrite. It has a beautiful golden sheen, but it sells for about $5/lb.) Iron pyrite is a compound of two elements, iron and sulfur. When a 1.000-g sample of iron pyrite is broken down into its elements there is *always* obtained 0.4655 g of iron and 0.5345 g of sulfur. The ratio of sulfur to iron is $\dfrac{0.5345 \text{ g sulfur}}{0.4655 \text{ g iron}}$ = 1.148 g of sulfur/1.000 g of iron. Provided that the samples of iron pyrite are pure and are not contaminated by other minerals or by dirt or sand, they all contain sulfur and iron combined in the same ratio by mass—1.148 g of sulfur to 1.000 g of iron. The size of the pyrite sample doesn't matter for our purposes. A 20.000-g sample of iron pyrite contains 10.690 g of sulfur combined with 9.310 g of iron. The ratio of 10.690 g of sulfur to 9.310 g of iron is 1.148 g sulfur/1.000 g iron, the same as for the smaller sample. The elements sulfur and iron are always combined in pure iron pyrite, regardless of its source, in the same proportion by mass. Figure 3.1 gives a visualization of these relationships.

All other compounds have a constant composition regardless of source. For example, no matter where in the world a sample of pure water is taken, the elements hydrogen and oxygen always are combined in water in the ratio of 7.921 g of oxygen to 1.000 g of hydrogen. In table salt the elements sodium and chlorine are invariably combined in a ratio of 1.542 g of chlorine to 1.000 g of sodium.

$10.690 \div 9.310 = 1.148$

Figure 3.1

The iron and sulfur combined in iron pyrite are in a ratio of 1.148 g of sulfur to 1.000 g of iron.

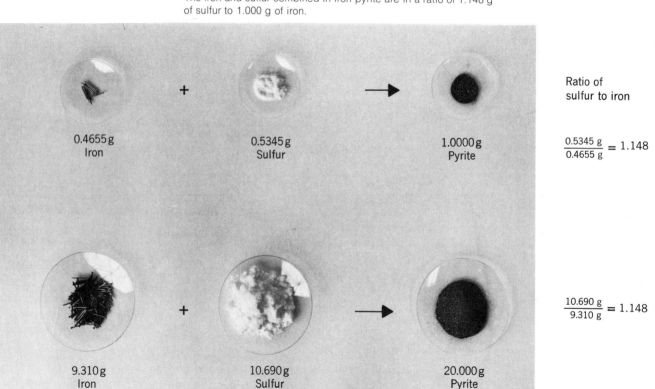

| 0.4655 g Iron | + | 0.5345 g Sulfur | → | 1.0000 g Pyrite | Ratio of sulfur to iron $\dfrac{0.5345 \text{ g}}{0.4655 \text{ g}} = 1.148$ |

| 9.310 g Iron | + | 10.690 g Sulfur | → | 20.000 g Pyrite | $\dfrac{10.690 \text{ g}}{9.310 \text{ g}} = 1.148$ |

Aspirin consists of the elements carbon, hydrogen, and oxygen always combined in the proportion: 13.392 g of carbon to 1.000 g of hydrogen to 7.931 g of oxygen. The consistency summarized by the law of definite proportions proved to be so complete that the law is used to define what is a chemical compound and what is not. Any substance consisting of two or more chemically combined elements always present in a definite proportion by mass is classified as a chemical compound, as we learned in Chapter 1.

3.2 DALTON'S ATOMIC THEORY

Dalton proposed that each element consists of indestructible particles called atoms that are identical in mass

Some Greek philosophers in the fifth century B.C., using speculation and reason (but not experimentation) reached the conclusion that if a piece of some substance, say gold, were cut in two, and each smaller piece were again cut in two, and if this process were continued (Figure 3.2), all of the successively smaller pieces would still be gold—but only up to a point. Ultimately, they reasoned, you would wind up with a piece of gold that could not be cut and still give two pieces of *gold*. In fact, they reasoned, you couldn't cut it at all, and that gold was made up of these very small uncuttable particles. The Greek word for "not cut" is *atomas*, and from it we get our word *atom*.

Figure 3.2
Cutting a piece of gold successively into equal parts gives smaller and smaller pieces of gold. Eventually a "magic knife" would be needed to work with invisible pieces. Yet the pieces would still be gold—but only up to a point, at an atom of gold. An atom of gold is the smallest sample of gold, and it can't be cut to give still smaller pieces that are still gold.

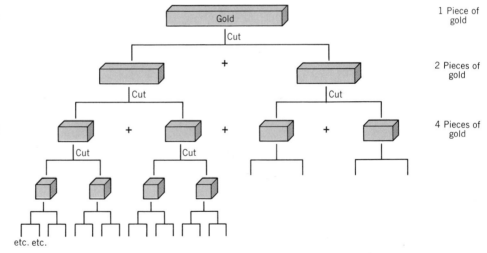

1 Piece of gold

2 Pieces of gold

4 Pieces of gold

etc. etc.

The Greek idea of an atom that could not be cut or broken bore no scientific fruit until John Dalton saw in this concept a way to make sense out of the law of definite proportions. Dalton believed that this law and the law of conservation of mass *compelled* a belief in atoms as the real particles of which all elements and compounds are made. The alternative view of matter could be that there is absolutely no limit to the number of times you could subdivide gold or any other element. As late as the latter part of the nineteenth century a few scientists believed this.

Dalton postulated that if the idea of atoms were to explain the laws of chemical combination, then these atoms must have certain properties. A list of these postulated properties constitutes **Dalton's atomic theory.**

1 Matter consists of definite particles called atoms.
2 Atoms are indestructible. In chemical reactions, the atoms rearrange, but they do not themselves break apart.

3 All atoms of a particular element are identical in mass (and in other properties).

4 Atoms of different elements are different in mass (and in other properties).

5 When the atoms of different elements combine to form compounds, new and more complex particles form. The elements that combine to make the compound provide their atoms in a definite ratio to make the more complex particles of the compound.

Now let us see how the laws of chemical combination support the postulated properties of atoms (Dalton's theory) and thereby support the idea that atoms exist. We will use the example of sodium chloride, table salt. We said earlier that the two elements are combined in this compound in a ratio of 1.542 g of chlorine to 1.000 g of sodium. Every analysis of sodium chloride ever done has given the same ratio, regardless of the way the sample was made or from what part of the world it came. The only way we can have such a definite proportion *by mass,* said Dalton, is for the elements to have combined in a definite proportion *by whole atoms* (corresponding to whole numbers), with each atom retaining its own particular mass that is characteristic for the element. Dalton was saying that in chemical compounds atoms do not exist as fractions; nor can atoms of the same element (taken from different sources) have different masses. If either were possible then the proportions by mass of the elements could not be so constant from sample to sample.

In noticing that the actual proportions by mass of the elements in a compound are not 1 to 1, Dalton concluded that the atoms of different elements must have different masses. For example, if we take the simplest way to imagine a ratio of atoms in table salt, 1 atom of sodium to 1 atom of chlorine, then a chlorine atom would need to have a mass 1.542 times the mass of a sodium atom. Figure 3.3 gives a visualization of these relationships.

If Dalton had thought (or if it were true) that all atoms of all elements had *identical* masses, then a mass ratio of 1.542 g of chlorine to 1.000 g of sodium in table salt would mean an atom ratio of 1542 atoms of chlorine to 1000 atoms of sodium. All other compounds would likewise have ratios by atoms equally complicated. But from the inception of Dalton's theory scientists found it much easier to believe in relatively simple ratios between atoms in their compounds. Subsequent research verified that belief.

Law of multiple proportions

One powerful piece of evidence for Dalton's theory came when Dalton and other scientists studied elements that could combine to give two (or more) compounds. Iron and sulfur, for example, form more than one compound. If we assume simple ratios, a few of the many theoretically possible combinations of iron and sulfur atoms are:

(Fe) is an iron atom. (S) is a sulfur atom.

Suppose that FeS and FeS_2 exist. If Dalton's theory is right, then making FeS_2 from a *fixed* mass of iron, say 1.000 g of iron, will require exactly twice the mass of sulfur as making FeS from 1.000 g of iron. That is because, for a given amount

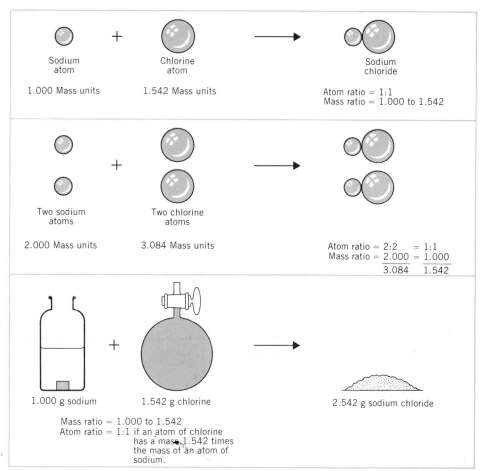

Figure 3.3
How ratios by atoms and ratios by masses can be different. (For color pictures of the substances given here, see Color Plate 1C.)

of iron, there are twice as many sulfur atoms in FeS_2 as in FeS. In other words, the ratio of the *mass* of sulfur in FeS_2 to its mass in FeS should be 2 to 1, a ratio of simple whole numbers, if each compound is made from the identical mass of iron. And that result has been repeatedly observed. Here are the mass ratios in two known compounds of iron and sulfur (where we use the names given by geologists for these minerals until we can take up the topic of formal chemical nomenclature):

Mineral	Mass Ratio of the Elements	
Pyrite (FeS_2)	1.000 g iron to 1.148 g sulfur	A mass ratio of 2 to 1 for sulfur
Troilite (FeS)	1.000 g iron to 0.574 g sulfur	

Compare the masses of sulfur that combine in different ways with 1.000 g of iron:

$$\frac{1.148 \text{ g sulfur}}{0.574 \text{ g sulfur}} = \frac{2}{1}$$

It would be extremely hard to imagine any other view of matter besides Dalton's atomic view that could account for a simple, whole-number ratio like this. A great many other examples of elements forming two or more compounds are known. One example involves tin and oxygen. Tin is the metal used to coat steel in "tin cans" to prevent rusting. In one compound the ratio by mass of the elements is

1.000 g of oxygen to 7.420 g of tin. In the other compound the mass ratio is 1.000 g of oxygen to 3.710 g of tin. Now compare the two masses of tin that combine with 1.000 g of oxygen:

$$\frac{7.420 \text{ g tin}}{3.710 \text{ g tin}} = \frac{2}{1}$$

The ratio is a ratio of simple whole numbers, 2 to 1. Out of these and many other examples came the third law of chemical combination, the law of multiple proportions:

> **Law of Multiple Proportions.** Whenever two elements form more than one compound, the different masses of one that combine with the same mass of the other are in the ratio of small whole numbers.

Some problems with Dalton's theory

Dalton turned out to be wrong in some ways, but luckily the errors did not affect the basic postulate that atoms do exist. Atoms are not, as Dalton postulated, indestructible. When given enough energy in "atom-smashing" machines, atoms split into a number of fragments including tiny subatomic particles such as electrons, protons, and neutrons. Dalton's theory works because in all *chemical* changes such fragmentation never occurs, not in any way that observably affects mass relationships. In many chemical reactions atoms pass some of their tiniest, least massive subatomic particles—electrons—back and forth. We will see, however, that when elements combine to form a compound the total number of these electrons remains the same, so the total mass of the compound is unchanged compared to the sum of the masses of the elements in it.

Isotopes

Another instance in which Dalton was incorrect was his postulate that all the atoms of an element have identical masses. Actually, most elements are mixtures of a small number of different kinds of atoms that have identical chemical properties, but slightly different masses. The atoms of the *same* element that have slightly different masses are called **isotopes.** An iron nail, for example, is made up of a mixture of four isotopes. One isotope makes up 91.66% of any iron sample and, if we arbitrarily assign it a mass of 1.000000 mass units, then the relationships by mass of the four isotopes of iron are:

Iron Isotope 1		Iron Isotope 2		Iron Isotope 3		Iron Isotope 4
(5.82% of all iron atoms)		(91.66% of all iron atoms)		(2.19% of all iron atoms)		(0.33% of all iron atoms)
0.964382 mass units per atom	to	1.000000 mass units per atom	to	1.017887 mass units per atom	to	1.035727 mass units per atom

The existence of isotopes could not have influenced the development of Dalton's theory for three very important reasons. First, the *average* mass of the isotopes of any element is a constant that holds with extremely few exceptions for that element regardless of where it is obtained anywhere in the earth or atmosphere. Second, whenever we take a sample of an element large enough to see, it contains an enormous number of atoms. We cannot help always having the same average mass if we take the same huge number of atoms. Third, all the isotopes of an element have identical *chemical* properties—all give the same kinds of chemical reactions. Thus, although the existence of isotopes makes Dalton's third postulate not literally true, any element behaves *chemically* as if it were true. All four isotopes of iron, for example, will combine with sulfur in the same ways to give the minerals iron pyrite or troilite, depending on the mass ratios used.

3.3 ATOMIC WEIGHTS

An atomic weight is the average relative mass of an element's atoms on a scale using atoms of carbon-12 as the reference

Dalton's theory spurred a great deal of research into finding the relative masses of the atoms of the elements. One reason why that research was vital to the future of chemistry and to all of its related fields was the need for a practical way to use a weighing balance for measuring samples of elements that would contain their atoms in the ratios needed for making compounds. For example, suppose an artist asked you to make a sample of the mineral troilite mentioned in the previous section. This substance can be used as a pigment in ceramics, but for best results it must be free of uncombined iron and sulfur. You might go to a chemical reference and learn that the formula of troilite is FeS. In this mineral iron atoms and sulfur atoms are combined in a ratio of 1 to 1. For every atom of iron there is one of sulfur. The practical problem is that atoms are too small to be picked out and counted like so many cups and saucers. Yet that 1 to 1 ratio by atoms is the crucial ratio regardless of how much the individual atoms of iron or sulfur weigh.

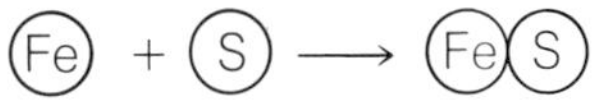

$$Fe + S \longrightarrow FeS$$

Since we can't pick and count individual atoms, we have to have a way of obtaining atoms in the ratios we want by some measurement we can actually do—such as measuring quantities by mass.

We found out in the last section that the mass ratio in FeS is 1.000 g of iron to 0.574 g of sulfur. Knowing now that this corresponds to a 1 to 1 ratio by atoms, we can see that iron atoms must be heavier than sulfur atoms, heavier by the ratio $\frac{1.000}{0.574} = 1.742$. An atom of iron has a mass 1.742 times the mass of an atom of sulfur. Therefore, any time we take masses of iron and sulfur in such a way that the quantity of iron weighs 1.742 times the quantity of sulfur we can be sure of getting iron and sulfur atoms in a 1 to 1 ratio.

This quantity of iron . . .	. . . has as many atoms of iron as there are sulfur atoms in . . .	. . . this quantity of sulfur.	The mass ratio of Fe to S
1.000 g Fe		0.574 g S	$\frac{1.000}{0.574} = 1.742$
1.742 g Fe		1.000 g S	$\frac{1.742}{1.000} = 1.742$
2.000 g Fe		1.148 g S	$\frac{2.000}{1.148} = 1.742$
3.484 g Fe		2.000 g S	$\frac{3.484}{2.000} = 1.742$

Each mass in the left column is 1.742 times the mass given opposite it in the right column.

What we need now is a reference against which to compare the relative masses of the elements. The decision is quite arbitrary; sulfur, for example, could be used. However, by worldwide agreement among scientists, the most abundant isotope of carbon is the reference standard. And in order to allow the lightest of all atoms, those of the chief isotope of hydrogen, to have a relative mass equal at least to one whole number, the relative mass of the atoms of that carbon isotope is set at 12.0000 exactly. This carbon isotope is therefore called carbon-12. The number 12 is called the mass number of this isotope. Each isotope has its own **mass number,** which we will for the present define as the relative mass of the isotope's atoms (based on carbon-12) rounded to the nearest whole number.

Once we have chosen a standard, the masses of the other elements can be determined on a relative basis. For example, relative to carbon-12 atoms, the average mass of the atoms of the iron isotopes discussed earlier (mass numbers 54, 56, 57, and 58) is 55.847. (In other words an average atom of iron has a mass that is $\dfrac{55.847}{12.0000}$ or 4.6539 times the mass of an atom of carbon-12.) If we had a pile of carbon-12 atoms with a mass of 12.0000 g and a pile of iron atoms with a mass of 55.847 g, the two piles would have identical numbers of atoms. See Figure 3.4.

Naturally occurring sulfur consists of a mixture of four isotopes, with mass numbers of 32, 33, 34, and 36. The average mass of the atoms in this mixture, relative to carbon-12, is 32.06, which means that an average atom of sulfur has a mass that is $\dfrac{32.06}{12.0000}$ or 2.672 times the mass of an atom of carbon-12.

The element hydrogen as it occurs in nature consists of atoms of two isotopes with mass numbers of 1 and 2. Their average mass, relative to carbon-12, is 1.0079. In other words, the average atom of hydrogen has a mass that is $\dfrac{1.0079}{12.0000}$ or 0.08399 times the mass of an atom of carbon-12. (For the actual mass of a carbon-12 atom see Special Topic 3.1.)

In 10,000 atoms of natural carbon there are 9889 atoms of carbon-12 and 111 atoms of carbon-13

The standard definition of mass number will be given in Chapter 5.

Sulfur Isotope	Percent Abundance
S-32	95.0%
S-33	0.76%
S-34	4.22%
S-36	0.014%

Hydrogen Isotope	Relative Abundance
H-1	99.985%
H-2	0.015%

Figure 3.4
Each quantity of these elements contains the same number of atoms.

32.1 g yellow sulfur   55.8 g iron   12.0 g carbon

The relative average mass of the naturally occurring atoms of an element, based on the carbon-12 reference, is called the **atomic weight** of that element. Atomic *weight* is the term still most widely used by scientists instead of atomic *mass,* which technically would be correct and which is now recommended by the SI.

Some isotopes are made in special laboratories, and do not occur in nature.

Figure 3.5
Atomic weight scale.

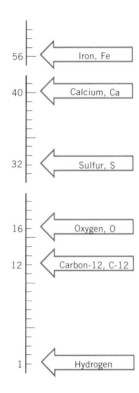

Figure 3.5 illustrates how we can understand atomic weights as numbers on a scale starting with the element having the lightest atoms, hydrogen, and moving to higher numbers through the reference isotope. For a complete list of the most precise values of the atomic weights, however, refer to the inside front cover.

The vital conclusion from this study of atomic weights is that if we take samples of any two elements in ratios according to their atomic weights, we are certain to obtain their atoms in a ratio of 1 to 1. Whenever ratios by atoms other than 1 to 1 are needed, the ratios by atomic weights can be adjusted correspondingly.

EXAMPLE 3.1   *Understanding Atomic Weights*

Problem: Calcium oxide (CaO) is used in making stucco, plaster, and mortar. It can be made directly from the elements calcium and oxygen, which are combined in this substance in a 1 to 1 ratio by atoms. Calculate the ratio by grams of calcium to oxygen (to three significant figures) that would give the purest sample of calcium oxide.

Solution: For a problem like this all you have to do is look up the atomic weights and round them to the specified number of significant figures.

calcium, 40.1 (from 40.08);   oxygen, 16.0 (from 15.9994)

To obtain a 1 to 1 ratio by atoms, merely take masses of these elements in the ratio of 40.1 to 16.0. You could actually take 40.1 g of calcium and 16.0 g of oxygen, or you could operate on a larger or smaller scale, for example, 4.01 g of calcium to 1.60 g of oxygen, or 401 g of calcium and 160 g of oxygen, or 200 g of calcium (rounded from 200.5) and 80.0 g of oxygen. Each of these mass ratios gives a 1 to 1 atom ratio, although the actual numbers of atoms varies with the scale.

EXERCISE 3.1   Beryllium oxide is used to make certain ceramics. The elements beryllium and oxygen occur in this compound in a ratio of 1 to 1 by atoms. In what ratio do they occur by mass?

3.4   FORMULA WEIGHTS

The sum of the atomic weights of the atoms in a chemical formula is the formula weight of the compound

The concept of a relative mass applies not only to atoms but also to molecules. And because of the law of conservation of mass we know that when atoms combine to make molecules their relative masses appear unchanged in the larger particles. The name **formula weight** is given to the relative mass of a particle consisting of all of the atoms specified in the formula.[1] The formula weight of a compound is the sum of the atomic weights of the atoms in its formula. The simplest way to learn how to calculate formula weights is to work some examples. Except where specified otherwise, round atomic weights to three significant figures in these calculations.

EXAMPLE 3.2   *Calculating a Formula Weight*

Problem: Calculate the formula weight of water, H_2O.
Solution: First, assemble the atomic weights, rounding the values given in the table (inside the front cover) to three significant figures. Multiplying the atom's atomic weight by the subscript (if any) appearing to its right in the formula and then add.

---

[1] Many references call this a *molecular weight,* instead, but *formula weight* is a much more general term. Not all compounds consist of molecules; some are made of particles called ions. Moreover, we can use *formula weight* for *atomic weight* whenever that might be useful. The formula of sodium, for example, is Na, and we could just as well say that the formula weight of sodium is 22.98977 as to call that number its atomic weight. Thus *formula weight* applies in general to anything having a chemical formula—compound or element—regardless of the kinds of particles that make it up.

$$H_2O = 2H + O$$
$$= (2 \times 1.01) + (1 \times 16.0)$$
$$= 18.0, \text{ the formula weight of water}$$

EXAMPLE 3.3   **Calculating a Formula Weight**

Problem: Calculate the formula weight of aspirin, $C_9H_8O_4$
Solution: Look up the atomic weights and round:

carbon, 12.0; hydrogen, 1.01; oxygen, 16.0

Multiply by the appropriate subscripts and add:

$$C_9H_8O_4 = 9C + 8H + 4O$$
$$= (9 \times 12.0) + (8 \times 1.01) + (4 \times 16.0)$$
$$= 108 + 8.08 + 64.0$$
$$= 180, \text{ the formula weight of aspirin (rounded)}$$

EXERCISE 3.2   Calculate the formula weight of each substance.
(a) Sodium chloride (table salt), $NaCl$
(b) Sucrose (table sugar), $C_{12}H_{22}O_{11}$

As we learned in Section 1.15, some formulas have parentheses. A subscript outside the parentheses is a multiplier for all atoms inside. Thus, there are 4 carbon, 6 hydrogen, and 4 oxygen atoms indicated in the formula $Ca(C_2H_3O_2)_2$.

EXAMPLE 3.4   **Calculating a Formula Weight**

Problem: Find the formula weight of calcium acetate, $Ca(C_2H_3O_2)_2$, the stiffening agent in "canned heat."
Solution: The atomic weights, rounded, are

Ca, 40.1; C, 12.0; H, 1.01; and O, 16.0

$$Ca(C_2H_3O_2)_2 = Ca + 4C + 6H + 4O$$
$$= (1 \times 40.1) + (4 \times 12.0) + (6 \times 1.01) + (4 \times 16.0)$$
$$= 158, \text{ the formula weight of calcium acetate}$$

EXERCISE 3.3   Calculate the formula weights of each of the following.
(a) Calcium propionate, $Ca(C_3H_5O_2)_2$, a food additive
(b) Ammonium ferrous sulfate, $(NH_4)_2Fe(SO_4)_2$, a substance used in photography

---

3.5 THE MOLE

The lab-sized unit of a substance is its formula weight taken in grams, a quantity called one mole of the substance

In Section 3.3 we learned that samples of elements weighed in ratios that correspond to their atomic weights contain their atoms in 1 to 1 ratios. Formula weights can be used in the same way as atomic weights. Samples of any compounds that are weighed in ratios corresponding to their formula weights contain their formula units in 1 to 1 ratios. By *formula unit* we mean whatever particle is represented by the chemical formula. It could be an atom, or a molecule, or some other kind of particle depending of the nature of the substance. For example, the formula weight of water is 18.0, and the atomic weight (or formula weight) of sodium is 23.0. Therefore, 18.0 g of water has as many formula units (which are molecules) as 23.0 g of sodium (which has atoms as formula units).

The formula weight of a substance taken in *grams* is one unit of that substance for experimental laboratory work, and the SI name of this unit is **mole** (abbreviated **mol**). For instance, the formula weight of aspirin is 180; therefore 1 mol of aspirin = 180 g of aspirin. Another term sometimes used for the quantity equaling one mole is **molar mass.** In some context, for example, it might be useful to refer to 180 g of aspirin as 1 molar mass of aspirin.

Avogadro's number

The number of formula units in one mole is so large that we really can't fully appreciate it, but one mole of any pure substance contains 6.02×10^{23} formula units. This number is called **Avogadro's number** in honor of Amadeo Avogadro (1776–1856), an Italian scientist. The large size of Avogadro's number points out that atoms and molecules must be extremely tiny. Only the most powerful, highly magnified electron microscope pictures have ever resolved an element into images of individual atoms. See Figure 3.6.

More precisely, 1 mol has 6.022045×10^{23} formula units.

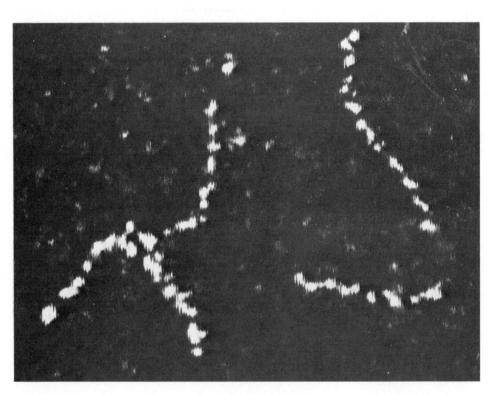

Figure 3.6
The white spots in this photograph were caused by individual atoms of thorium.

Actually, Avogadro's number is the second meaning of the word *mole. Mole* stands for a lab-sized unit of a chemical—one molar mass—and also for the number—Avogadro's number—just as other words represent numbers.

12	dozen
144	gross
6.02×10^{23}	Avogadro's number

For example, the expressions "a mole of virus particles" and "Avogadro's number of virus particles" mean exactly the same—a quantity of virus particles numbering 6.02×10^{23}. In all chemical applications of Avogadro's number, however, the fact of central importance is that *equal numbers of moles have equal numbers of formula units.* For example, 10.5 mol of sugar has as many formula units (molecules) as 10.5 mol of iron (whose formula units are atoms).

EXAMPLE 3.5 **Understanding Avogadro's Number**

Problem: How many carbon atoms are in 6.00 g of carbon?

Solution: Work from the basic meaning of Avogadro's number—the number of formula units per mole of substance. Since carbon's atomic weight is 12.0, one mole of carbon consists at the same time of 12.0 g of carbon and 6.02×10^{23} atoms of carbon. These facts give us a choice of two conversion factors:

$$\frac{6.02 \times 10^{23} \text{ atoms C}}{12.0 \text{ g C}} \quad \text{or} \quad \frac{12.0 \text{ g C}}{6.02 \times 10^{23} \text{ atoms C}}$$

If we multiply what was given, the 6.00 g of carbon, by the first conversion factor, the units "g C" will cancel and our answer will be in atoms of carbon:

$$6.00 \text{ g C} \times \frac{6.02 \times 10^{23} \text{ atoms C}}{12.0 \text{ g C}} = 3.01 \times 10^{23} \text{ atoms C}$$

EXERCISE 3.4 How many molecules of water, H_2O, are in 250 g of water, roughly 1 cup?

One of the advantages of the mole concept is that it allows us to think about formulas on two levels at the same time, the atom/molecule level of individual particles and the level of practical, lab-sized quantities. A dozen molecules of H_2O are made of two dozen atoms of H and one dozen of O. But if our minds are on practical, lab-sized quantities when we think about the formula H_2O, we can immediately visualize one *mole* of H_2O as consisting of two moles of H and one mole of O. The numbers are in the same ratio whether we deal with individual particles or with moles of them. And when we are at the mole level we can easily convert mole numbers into corresponding masses of substance according to the needs of an experiment.

H_2O	consists of	2H	+	O
1 molecule		2 atoms		1 atom
1 dozen molecules		2 dozen atoms		1 dozen atoms
6.02×10^{23} molecules		12.04×10^{23} atoms		6.02×10^{23} atoms
1 mol		2 mol		1 mol
18.0 g H_2O		2.0 g H		16.0 g O

The symbol *mol* stands for both the singular and plural.

EXAMPLE 3.6 **Using the Mole Concept**

Problem: How many moles of sulfur must be combined with 2.0 mol of iron to give iron sulfide, FeS?

Solution: The formula of iron sulfide, FeS, tells us that the atom ratio of iron to sulfur is 1 to 1. In lab-sized amounts, the ratio is 1 mol of iron to 1 mol of sulfur. If we write this ratio as a fraction, it can serve as a conversion factor in either of two ways.

$$\frac{1 \text{ mol Fe}}{1 \text{ mol S}} \quad \text{or} \quad \frac{1 \text{ mol S}}{1 \text{ mol Fe}}$$

To solve our problem we use the second factor to make "mol Fe" cancel.

$$2.0 \text{ mol Fe} \times \frac{1 \text{ mol S}}{1 \text{ mol Fe}} = 2.0 \text{ mol S}$$

We calculate that 2.0 mol S will combine with 2.0 mol Fe.

EXAMPLE 3.7   **Using the Mole Concept**

Problem: Methane (natural gas) has the formula CH_4. If a sample of methane contains 0.30 mol C, how many moles of H are present?

Solution: Because the formula says that 1 atom of carbon combines with 4 atoms of hydrogen, we automatically know that 1 mol of C combines with 4 mol of H. We therefore have two ratios to choose from as conversion factors.

$$\frac{1 \text{ mol C}}{4 \text{ mol H}} \quad \text{or} \quad \frac{4 \text{ mol H}}{1 \text{ mol C}}$$

To get "mol C" to cancel and leave our answer in "mol H" we have to use the second factor.

$$0.30 \text{ mol C} \times \frac{4 \text{ mol H}}{1 \text{ mol C}} = 1.2 \text{ mol H}$$

EXERCISE 3.5   How many moles of Fe combine with 0.22 mol of O to give rust, Fe_2O_3?

EXERCISE 3.6   In 0.50 mol of Fe_2O_3, how many mol of Fe and how many mol of O are present?

In laboratory work, there are two important kinds of calculations involving moles. One is to find the number of grams that will deliver a certain number of moles—a moles to grams conversion. The other is to find the number of moles in a given mass of a sample—a grams to moles conversion. Both calculations are worked from the basic definition of the mole used in its first meaning, that of molar mass.

EXAMPLE 3.8   **Calculating Grams from Moles**

Problem: Calcium arsenate, $Ca_3(AsO_4)_2$, is a poison sometimes used to kill insects on plants. What is the mass of 0.586 mol of calcium arsenate?

For $Ca_3(AsO_4)_2$:

3 Ca: 3 × 40.1 = 120
2 As: 2 × 74.9 = 150
8 O: 8 × 16.0 = 128
                  398

Solution: In just about any kind of problem involving moles, the first step should be to find the formula weights of the chemicals involved. The formula weight of calcium arsenate is 398. This number gives us a choice of two conversion factors.

$$\frac{398 \text{ g } Ca_3(AsO_4)_2}{1 \text{ mol } Ca_3(AsO_4)_2} \quad \text{or} \quad \frac{1 \text{ mol } Ca_3(AsO_4)_2}{398 \text{ g } Ca_3(AsO_4)_2}$$

The problem calls for an answer in units of *mass,* so we have to use the first factor to cancel our "mol $Ca_3(AsO_4)_2$."

$$0.586 \text{ mol Ca}_3\text{(AsO}_4\text{)}_2 \times \frac{398 \text{ g Ca}_3\text{(AsO}_4\text{)}_2}{1 \text{ mol Ca}_3\text{(AsO}_4\text{)}_2}$$

$$= 233 \text{ g Ca}_3\text{(AsO}_4\text{)}_2, \text{ the answer.}$$

Always double check your answer to see that it makes sense. Notice that the problem calls for about half a mole. To make sense, therefore, the answer should be about half the number of grams in a mole.

The mass of calcium arsenate (233 g) found in Example 3.8, which is the same as 0.586 mol of calcium arsenate, contains exactly the same number of formula units as there are in 0.586 mol of water, 0.586 mol of glucose, 0.586 mol of iron, or 0.586 mol of hydrogen.

EXERCISE 3.7 Calculate the number of grams in 0.586 mol of each of the following substances.
(a) Water, H_2O
(b) Glucose, $C_6H_{12}O_6$, a sugar in grape juice and honey
(c) Iron, Fe
(d) Methane, CH_4

EXAMPLE 3.9 **Calculating Moles from Grams**

Problem: Sodium bicarbonate, $NaHCO_3$, is one ingredient of baking powder.[2] In one experiment in a series of tests of different ratios of ingredients, a scientist used 21.0 g $NaHCO_3$. How many moles were in this sample?

For $NaHCO_3$:   *Solution:* First find the formula weight of $NaHCO_3$. It's 84.0, which gives us the choice of these two conversion factors.

1 Na	23.0
1 H	1.01
1 C	12.0
3 O	48.0
	84.0

$$\frac{84.0 \text{ g NaHCO}_3}{1 \text{ mol NaHCO}_3} \quad \text{or} \quad \frac{1 \text{ mol NaHCO}_3}{84.0 \text{ g NaHCO}_3}$$

To change grams to moles we have to pick the factor that lets us cancel grams—the second of these choices.

$$21.0 \text{ g NaHCO}_3 \times \frac{1 \text{ mol NaHCO}_3}{84.0 \text{ g NaHCO}_3} = 0.250 \text{ mol NaHCO}_3$$

Thus, the sample used, 21.0 g $NaHCO_3$, was a quarter of a mole of $NaHCO_3$.

EXERCISE 3.8 Calculate the number of moles in 100.0 g of each of the following samples:
(a) Water, H_2O
(b) Glucose, $C_6H_{12}O_6$
(c) Iron, Fe
(d) Methane, CH_4

EXERCISE 3.9 Why does 100 g of methane gas, CH_4, have so many more moles than 100 grams of glucose?

[2] Don't confuse baking *powder* with baking *soda* if you're an amateur baker. Baking soda is sodium bicarbonate, $NaHCO_3$. Baking powder is a special mixture of compounds.

EXERCISE 3.10 If 40 g of an element contains 12.04×10^{23} atoms, what is its atomic weight?

3.6 PERCENTAGE COMPOSITION

The percentage by weight of an element in a compound is the same as the number of grams of that element present in 100 grams of the compound

No intelligent experimental work involving chemical reactions can be done without using the mole concept and, to carry out mole calculations, formula weights have to be calculated from formulas. Thus chemical formulas are essential, and determining them has been one of the very important tasks of chemistry.

Determining a chemical formula begins with finding what elements are present. This work is called qualitative analysis. Then the ratios by *mass* of these elements are determined by measuring how much of each element can be obtained from a given mass of the compound. Such experimental work is called quantitative analysis. Using atomic weights as conversion factors, the ratios by mass are next mathematically converted into their corresponding ratios by *moles*. Of course, as we have just seen, the ratios by moles are numerically identical to ratios by *atoms*. Therefore we can now write a formula in which the subscripts show ratios by atoms, and the subscripts are expressed in whole numbers since atoms come in whole units. In short, the general strategy for determining a chemical formula is to identify the elements, find their ratios by mass, calculate ratios by atoms using atomic weights, and express these ratios in whole numbers as subscripts by the atomic symbols in the formula.

In scientific reports the ratios by mass of the elements in a compound are expressed as percentages by weight.[3] The clearest way to understand a **percentage by weight** is to view it as representing the number of grams of a particular element in 100 g of the compound. A complete list of the percentages by weight of a compound's elements is the **percentage composition** of the compound. Percentages by weight are usually expressed in four significant figures because the experimental data are that precise.

EXAMPLE 3.10   **Calculating Percentage Composition from Mass Data**

Problem: A sample of a liquid with a mass of 8.657 g was decomposed into its elements to give 5.217 g of carbon, 0.9620 g of hydrogen, and 2.478 g of oxygen. What is the percentage by weight of each element in this liquid?

Solution: Let's call the liquid "X." We were given the grams of each element in 8.657 g X. We're asked to find how many grams of carbon, how many of hydrogen, and how many of oxygen are in 100 g of X. We therefore do simple percents:

$$\frac{5.217 \text{ g C}}{8.657 \text{ g } X} \times 100 = 60.26 \ \%\text{C}$$

$$\frac{0.9620 \text{ g H}}{8.657 \text{ g } X} \times 100 = 11.11 \ \%\text{H}$$

---

[3] We continue to use what we believe is the most widely employed expression for this concept in chemistry—percentage by weight—instead of percentage by mass (which is the better term).

$$\frac{2.478 \text{ g O}}{8.657 \text{ g } X} \times 100 = 28.62 \text{ \%O}$$

$$99.99\%$$

(The percentages, of course, have to add up to 100—allowing for slight errors caused by rounding.) The percentages tell us that in 100 g of compound "X" there are 60.26 g of carbon, 11.11 g of hydrogen, and 28.62 g of oxygen.

EXERCISE 3.11   A sample having a mass of 0.4620 g of an unknown white powder was found to have these elements in the given quantities: 0.1945 g of carbon, 0.0297 g of hydrogen, and 0.2377 g of oxygen. Calculate the percentage composition of this substance.

It is often necessary to use a formula to calculate percentage composition in order to check experimental results. Suppose you suspect that an unknown compound is calcium oxide, CaO. You found that when a sample of the unknown was decomposed it contained 71.47% calcium and 28.53% oxygen. Are the data correct for the formula CaO, or would they correspond to some other formula for the oxide of calcium? To find out, you have to be able to calculate *theoretical percentages,* the percentage composition of a compound based on its formula. Let us calculate the theoretical percentage composition of CaO to see if the values check with the observed values.

EXAMPLE 3.11   **Calculating Percentage Compositions From Formulas**

Problem: What is the percentage by weight of each element in CaO?
Solution: First calculate the formula weight of CaO; do this to *four* significant figures.

$$\text{Ca} + \text{O} = \text{CaO}$$
$$40.08 + 16.00 = 56.08$$

Remember what these data mean:

$$56.08 \text{ g CaO contains } 40.08 \text{ g Ca}$$
$$56.08 \text{ g CaO contains } 16.00 \text{ g O}$$

Now find the percentages:

For Ca: $\frac{40.08 \text{ g Ca}}{56.08 \text{ g CaO}} \times 100 = 71.47\%$ Ca in CaO

For O: $\frac{16.00 \text{ g O}}{56.08 \text{ g CaO}} \times 100 = 28.53\%$ O in CaO

$$100.00\%$$

Notice that the theoretical percentages add up to 100%, as they must, and that they agree with the observed values. The unknown must be CaO.

EXERCISE 3.12   What is the percentage by weight of each element in butane, C_4H_{10}, a liquid fuel used in cigarette lighters.

As a general rule, if experimental and theoretical percentages differ by no more than 0.30% the data are said to agree. Thus, if the theoretical or calculated percentage of, say, carbon in a compound is 47.46%, experimental results anywhere between 47.16% and 47.76% are acceptable.

EXERCISE 3.13 Chromium carbonyl, $Cr(CO)_6$, has been tried as a gasoline additive to improve engine performance. When it was first prepared, it was considered an unusual compound. The reported analyses were 23.75% Cr, 32.69% C, and 43.55% O. Were these data consistent with the formula $Cr(CO)_6$? Do the calculations.

3.7 EMPIRICAL FORMULAS

A compound's empirical formula tells what elements are present and in what ratios by atoms (or by moles)

The next step in finding the formula of a compound is to use the percentages of the elements to calculate the ratios of the different atoms present. The chemical formula that gives these ratios by atoms in their smallest whole numbers is called the **empirical formula.** We will see how to use percentages by weight to calculate empirical formulas in this section, but as a quick preview of the next section we must mention that the empirical formula is often not a complete description of the composition of a substance. For example, the empirical formula of hydrogen peroxide, a bleaching agent, is simply HO—a 1 to 1 ratio of hydrogen atoms to oxygen atoms. Hydrogen peroxide, however, consists of individual molecules that are made of two hydrogen atoms and two oxygen atoms. The formula of the molecule—the *molecular* formula—is therefore H_2O_2. Similarly, the empirical formula of acetylene, a gas used in welding, is CH, but individual molecules of acetylene have the composition C_2H_2. The solvent benzene also has the empirical formula CH, but its individual molecules have the composition C_6H_6. For many substances, the empirical formula is the only kind of formula we can write, because they do not consist of individual molecules. Sodium chloride (table salt), NaCl, is a typical example. We will study the condition and arrangement of its atoms in Chapter 6. Regardless of the kind of substance, however, all compounds have empirical formulas, and let us now see how we can use atomic weights and percentages by weight to calculate them.

EXAMPLE 3.12   Calculating Empirical Formulas From Percentage Compositions

Problem: What is the empirical formula of a gas having 42.86% carbon and 57.14% oxygen?

Solution: Remember that the percentage data mean 42.86 *grams* of carbon and 57.14 *grams* of oxygen in 100 grams of the compound. But the ratio by grams can't be used in an empirical formula, only a ratio by moles or, the same thing, by atoms. We have to use atomic weights to change the grams to moles.

$$\text{For carbon: } 42.86 \text{ g C} \times \frac{1 \text{ mol C}}{12.01 \text{ g C}} = 3.569 \text{ mol C}$$

$$\text{For oxygen: } 57.14 \text{ g O} \times \frac{1 \text{ mol O}}{16.00 \text{ g O}} = 3.571 \text{ mol O}$$

Conversion factors based on atomic weights

We could write the formula as $C_{3.569}O_{3.571}$, but that violates the convention. To reduce these subscripts to the set of *whole* numbers that stand for the same ratio, we divide each of them by the smallest one present.

$$C_{\frac{3.569}{3.569}}O_{\frac{3.571}{3.569}} = C_{1.000}O_{1.001} = CO$$

In these calculations, if we are within ± 0.05 of a whole number we can round to that number. Therefore the empirical formula is CO. It's not that the formula $C_{3.569}O_{3.517}$ is wrong if we understand the subscripts as giving *mole* ratios. But we want to be able to interpret an empirical formula in terms of atoms, too, and they always combine in ratios of whole numbers. (The gas, CO, incidentally, is carbon monoxide, a poison.)

EXAMPLE 3.13 **Calculating Empirical Formulas From Percentage Compositions**

Problem: Barium carbonate, a white powder used in paints, enamels and ceramics, has this percentage composition: Ba, 69.58; C, 6.09; O, 24.32. What is its empirical formula?

Solution: Convert the grams (given by the percentages) to moles.

For barium: $69.58 \text{ g Ba} \times \dfrac{1 \text{ mol Ba}}{137.3 \text{ g Ba}} = 0.5068 \text{ mol Ba}$

For carbon: $6.09 \text{ g C} \times \dfrac{1 \text{ mol C}}{12.01 \text{ g C}} = 0.5071 \text{ mol C}$

For oxygen: $24.32 \text{ g O} \times \dfrac{1 \text{ mol O}}{16.00 \text{ g O}} = 1.520 \text{ mol O}$

Our preliminary empirical formula is therefore,

$$Ba_{0.5068}C_{0.5071}O_{1.520}$$

We divide each subscript by the smallest, 0.5068:

$$Ba_{\frac{0.5068}{0.5068}}C_{\frac{0.5071}{0.5068}}O_{\frac{1.520}{0.5068}} = Ba_1C_{1.001}O_{2.999}$$

The result is very close to $BaCO_3$, which is the empirical formula of barium carbonate.

EXAMPLE 3.14 **Calculating Empirical Formulas from Percentage Compositions**

Problem: Propane, a component of liquified petroleum gas (LPG) and a common fuel, has 81.83% carbon and 18.17% hydrogen. What is its empirical formula?

Solution: Convert the grams (given by the percentages) to moles:

For carbon: $81.83 \text{ g C} \times \dfrac{1 \text{ mol C}}{12.01 \text{ g C}} = 6.813 \text{ mol C}$

For hydrogen: $18.17 \text{ g H} \times \dfrac{1 \text{ mol H}}{1.001 \text{ g H}} = 18.15 \text{ mol H}$

We could now write the formula as $C_{6.813}H_{18.15}$. We again divide each subscript by the smallest:

$$C_{\frac{6.813}{6.813}}H_{\frac{18.15}{6.813}} = C_1H_{2.664}$$

This time the procedure didn't work. The number 2.664 is too far from a whole number to round off. The ratio of 1 to 2.664 is, in a *mole*-ratio sense,

The mole ratio in the gas is 1 mol C to 2.665 mol H, but we can't use that for the *atom* ratio.

correct; we've just not found the simplest whole numbers for this ratio. Let's multiply each subscript in $C_1H_{2.664}$ by a simple whole number, 2.

$C_{1\times2}H_{2.664\times2} = C_2H_{5.328}$. That didn't work either because 5.328 is still so far from a whole number to round it off. Let's try using 3 instead of 2 on the earlier ratio of 1 to 2.664.

$C_{1\times3}H_{2.664\times3} = C_3H_{7.992}$. Now we can safely round. The empirical formula of propane is C_3H_8. Let's review the approach. After getting the right mole ratio from the percentage data, divide each number of that ratio by the smallest number. Now at least one number in the ratio will be a whole number, 1. If the others aren't close enough to whole numbers to round, multiply the numbers in this ratio by 2. If that doesn't work, try 3. Keep trying until all the numbers in the ratio are whole numbers.

EXERCISE 3.14 "Calomel" is the common name of a white powder once used in the treatment of syphilis. Its percentage composition is 84.98% mercury and 15.02% chlorine. What is its empirical formula?

3.8 MOLECULAR FORMULAS

A molecular formula goes beyond the information in an empirical formula to give the actual composition of a molecule

If you did Exercise 3.14 you found an empirical formula of HgCl, a 1 to 1 ratio of mercury atoms to chlorine atoms. But that ratio could also be expressed in the formula Hg_2Cl_2, or Hg_3Cl_3, etc. The formula of calomel is, in fact, not HgCl but Hg_2Cl_2, and that kind of formula is called a molecular formula. A **molecular formula** is a chemical formula that gives the actual composition of a molecule.

To find out that calomel is Hg_2Cl_2 and not HgCl all that is needed is one more piece of data, the formula weight of the compound. The formula weight of calomel, as determined by experiment, is 472.09. The formula HgCl corresponds to a formula weight of 236.04. That's half of 427.09. The only way to double 236.04 and have the same basic ratio of atoms of mercury to atoms of chlorine is to double all the atoms in HgCl and write the formula as Hg_2Cl_2.

EXAMPLE 3.15 **Calculating a Molecular Formula from an Empirical Formula Using the Formula Weight**

Problem: Styrene, the raw material for polystyrene foam plastics, has the empirical formula CH. Its formula weight is 104. What is the molecular formula of styrene?

Solution: First we calculate the formula weight as if the true molecular formula were the same as the empirical formula. For CH, that comes to $12.01 + 1.01 = 13.02$. Since this is far from the formula weight, 104, we ask how many times 13.02 equals 104? The answer will be how many units of CH are in one molecule, since each unit's formula weight is 13.02.

$$\frac{104}{13.02} = 7.99$$

The answer can be rounded off to 8 and, therefore, the molecular formula is $8 \times$ CH, which we write C_8H_8.

Most methods for determining formula weights have precisions that allow no more than three significant figures in the results. But this is nearly always good enough for calculating a molecular formula from an empirical formula.

EXAMPLE 3.16 **Calculating a Molecular Formula from Percentage Composition and Formula Weight**

Problem: Isobutylene is a raw material for synthetic rubber. It consists of 85.71% carbon and 14.29% hydrogen. Its formula weight is found to be 57. What is the molecular formula of isobutylene?

Solution: First find the empirical formula. Change the percentages (masses per 100 g of compound) to moles:

Carbon: $85.71 \text{ g C} \times \dfrac{1 \text{ mol C}}{12.01 \text{ g C}} = 7.137 \text{ mol C}$

Hydrogen: $14.29 \text{ g H} \times \dfrac{1 \text{ mol H}}{1.001 \text{ g H}} = 14.28 \text{ mol H}$

The empirical formula could be written: $C_{7.317}H_{14.28}$. To reduce this to the lowest whole numbers, divide each number by the smallest, 7.137:

$$C_{\frac{7.137}{7.137}}H_{\frac{14.28}{7.137}} = C_1H_2 = CH_2.$$

Now, use the formula weight to find the molecular formula. The formula weight of CH_2 is 14.03 (12.01 + 2 × 1.01). How many units of CH_2 "fit" into a formula weight of 57?

$$\frac{57}{14.03} = 4 \text{ (rounded from 4.063)}$$

Therefore, the molecular formula of isobutylene is C_4H_8.

EXERCISE 3.15 A compound was found to contain 87.42% nitrogen and 12.58% hydrogen. Its formula weight was determined to be 32. What was its molecular formula?

3.9 FORMULA WEIGHTS OF GASES

Equal volumes of gases under standard conditions contain the same number of moles

The formula weights of gases are easy to measure, thanks to a relationship called Avogadro's law. Avogadro studied some interesting data observed earlier by Gay-Lussac about the relationships among the *volumes* of reacting gases and their gaseous products. Dalton had studied the mass relationships in many reactions, but he could find no explanation for the volume relationships of reactions of gases. *When measured at the same pressures and temperatures, the volumes of gaseous reactants and gaseous products are in ratios of simple whole numbers.*[4] Sometimes the sum of the volumes of reactants equals the total volume of products. For example,

hydrogen + chlorine ⟶ hydrogen chloride
[one volume]   [one volume] ⟶   [two volumes]

---

[4] This statement is sometimes called Gay-Lussac's law of combining volumes.

Often, however, the sum of the volumes of the reactants does not equal that of the products. For example,

hydrogen + oxygen ⟶ water vapor
[two volumes] [one volume] ⟶ [two volumes]

What intrigued Avogadro were the simple ratios of volumes, ratios expressed in whole numbers. He formulated two hypotheses to explain how they could be in whole numbers.

1 The fundamental particles of hydrogen, oxygen, and chlorine are not atoms but *diatomic molecules,* particles with two atoms bound together. (For example, according to Avogadro the molecular formula of hydrogen gas is H_2 not H; and the molecular formula of oxygen gas is O_2 not O; and of chlorine, Cl_2 not Cl.)
2 Equal volumes of gases contain the same numbers of molecules when measured under the same pressure and temperature. (For example, at 25 °C and 760 torr, according to Avogadro, one liter of hydrogen has the same number of molecules as one liter of oxygen or one liter of any gas.)

With these hypotheses, the reaction that forms water vapor could be illustrated as follows:

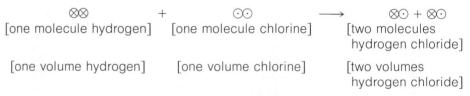

[two molecules hydrogen] [one molecule oxygen] [two molecules water]
water vapor]

Thus, if the reacting particles are actually diatomic molecules, then their ratio of 2 to 1 matches the volume ratio of 2 to 1. Moreover, if the ratio by volumes is the same as the ratio by molecules, a volume of any of these gases must have the same number of molecules as a volume of any other gas.

The reaction by which hydrogen chloride forms can be shown in a similar way. Assuming that hydrogen, chlorine, and hydrogen chloride are diatomic molecules gives a ratio by molecules that matches the ratio by volumes.

[one molecule hydrogen] [one molecule chlorine] [two molecules hydrogen chloride]

[one volume hydrogen] [one volume chlorine] [two volumes hydrogen chloride]

In the formation of hydrogen chloride, it's as if one pair of linked train engines and a pair of linked cabooses unlink and join to each other to make two trains made up of one engine and one caboose.[5]

The important implication of Avogadro's success in explaining the volume data for gas reactions is that gas volume is proportional to the *numbers* of particles, not to the masses of the particles. The only way we can change the *volume* of a gas without changing its pressure or without raising or lowering its temperature is to add more *particles* of the gas, not somehow make each particle heavier or lighter. Of course, if volume is proportional to molecules, it's also proportional to moles, *n*.

Gas volume ∝ number of moles of gas (at constant *T* and *P*)

$$V \propto n \text{ (constant } T \text{ and } P)$$

The diatomic elements are:

Hydrogen H_2
Nitrogen N_2
Oxygen O_2
Fluorine F_2
Chlorine Cl_2
Bromine Br_2
Iodine I_2

[5] The logic used by Avogadro would also hold if hydrogen were H_4 or H_6, and if oxygen and chlorine had formulas to match. Assuming them to be diatomic was the simplest hypothesis, and it turned out to be right. Their molecules in any case could not have an *odd* number of atoms and permit the 1 to 1 correspondence between molecules and volumes.

The modern form of Avogadro's second hypothesis above, now called **Avogadro's law,** is given in terms of moles.

> **Avogadro's Law.**   Under identical conditions of pressure and temperature, equal volumes of gases have equal numbers of moles.

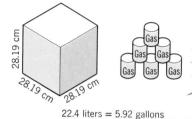

28.19 cm
28.19 cm   28.19 cm

22.4 liters = 5.92 gallons

Exactly how much volume one mole of gas occupies varies with the temperature and the pressure. Therefore, we need a set of reference conditions, called the **standard conditions of temperature and pressure,** or **STP** for short. Standard pressure is arbitrarily defined as 1 atmosphere or 760 torr. Standard temperature is 273.15 K (0 °C), which we will round to 273 K. The volume of one mole of a gas, any gas, under STP is 22.4 liters, or very close to that, as the experimental data of Table 3.1 show. This volume is called the **standard molar volume.**

$$1 \text{ molar volume} = 22.4 \frac{\text{liters}}{\text{mol}} \quad \text{(at STP)}$$

Table 3.1
Molar Volumes of Some Gases at STP

Gas	Formula	Molar Volume (liters)	Molar Mass (grams)
Helium	He	22.398	4.00
Argon	Ar	22.401	39.9
Hydrogen	H_2	22.410	2.02
Nitrogen	N_2	22.413	28.0
Oxygen	O_2	22.414	32.0
Carbon Dioxide	CO_2	22.414	44.0

Sealed glass bulbs are used to weigh gas samples.

Now we can see how the formula weight of a gas can be determined experimentally. This is done by measuring how much volume a weighed sample of the gas occupies at STP and converting that volume to moles. Then the ratio of grams to moles gives the formula weight. For example, suppose that 32 g of a gas occupies 22.4 liters at STP. Since 22.4 liters is the volume of *one* mole of any gas at STP, one mole of this gas must weigh 32 g. The formula weight is therefore 32. Operating at exactly STP, however, is seldom convenient, and we don't have to. We can now use Avogadro's law to rewrite the combined gas law into a form that lets us use a variety of experimental conditions.

3.10   IDEAL GAS LAW

For an ideal gas, PV/nT is a constant called the universal gas constant and symbolized by R

In Section 2.7 we learned the combined gas law, $PV/T = C''$. However, this holds only for a fixed mass or, the equivalent, a fixed number of moles of the gas. The constant C'' is actually not a constant, therefore, except in the special circumstance that the number of moles of gas in the sample is fixed. To make the combined gas law more general we take note of Avogadro's law. The value of the constant C'' must be directly proportional to the number of moles in the sample. For example, if by doubling the number of moles, n, at constant P and T we also double the volume, the combined gas law can hold only by also doubling the value of C''. Hence, we can write

$$C'' \propto n$$

Or

$$C'' = n \times \text{(a new constant)}$$

The new constant is given the symbol R and is called the **universal gas constant.** Now we can write the combined gas law in a more general form, called the **ideal gas law** (or, in some references, the universal gas law): $PV/T = nR$. This is usually rearranged and written as

$$\boxed{PV = nRT} \qquad (3.1)$$

The value of R depends on the units we pick for P and V. (T is always in kelvins.) For one mole of an ideal gas for which $V = 22.4$ liter, $T = 273$ K and $P = 1.00$ atm, we can find R as follows.

$$R = \frac{PV}{nT} = \frac{(1.00 \text{ atm})(22.4 \text{ liter})}{(1 \text{ mol})(273 \text{ K})} = 0.0821 \frac{\text{liter atm}}{\text{mol K}}$$

EXAMPLE 3.17 **Calculating with the Universal Gas Constant, $R = \dfrac{0.0821 \text{ liter atm}}{\text{mol K}}$**

Problem: If a sample of oxygen at 21.0 °C and 740 torr weighs 16.0 g, what volume would it occupy?

Solution: We must first convert the given data into the units used in the universal gas constant. Thus we have to convert degrees Celsius to kelvins by adding 273 (20.0 °C = 293 K). Then convert grams of oxygen to moles. Since the correct formula of oxygen is O_2, not O, the formula weight of oxygen is 32.0.

$$16.0 \text{ g} \times \frac{1 \text{ mol}}{32.0 \text{ g}} = 0.500 \text{ mol } O_2$$

Finally, convert 740 torr to atm by using the fact that 1.00 atm = 760 torr:

$$740 \text{ torr} \times \frac{1.00 \text{ atm}}{760 \text{ torr}} = 0.974 \text{ atm}$$

Collect the data:

$$n = 0.500 \text{ mol} \qquad\qquad T = 293 \text{ K}$$

$$R = 0.0821 \frac{\text{liter atm}}{\text{mol K}} \qquad P = 0.974 \text{ atm}$$

Now use the ideal gas law equation, 3.1, to find V, canceling units where possible:

$$V = \frac{nRT}{P} = \frac{(0.500 \text{ mol}) \left(0.0821 \dfrac{\text{liter atm}}{\text{mol K}}\right)(293 \text{ K})}{(0.974 \text{ atm})}$$

$$= 12.3 \text{ liters (rounded from 12.3487)}$$

EXERCISE 3.16 What volume will a sample of methane, CH_4, weighing 10.2 g occupy at 25.0 °C and 755 torr?

Now let's see how we can use the ideal gas law to determine formula weights of gases. We do this by using the ideal gas equation to find the number of moles, n, of a gas in a sample weighed under measured values of P, V, and T. A formula weight, of course, is the ratio of grams to moles. When we know both the grams of a sample and the moles in it, finding the formula weight is easy. All we have to do is divide the grams by moles.

EXAMPLE 3.18   Calculating the Formula Weight of a Gas

Problem: Water can be decomposed electrically into two gases. In one experiment, one of the gases that was collected had a mass of 1.090 g and a volume of 850 ml at 745 torr and 25 °C. What was its formula weight?

Solution: First collect the data.

n = ? (But we know that the sample has a mass of 1.090 g.)

V = 850 ml = 0.850 liter

P = 745 torr or 0.980 atm $\left(\text{since } 745 \text{ torr} \times \dfrac{1.00 \text{ atm}}{760 \text{ torr}} = 0.980 \text{ atm} \right)$

T = 298 K (25 + 273)

Now use the ideal gas equation to find n, the number of moles.

$$PV = nRT$$

$$(0.980 \text{ atm}) \times (0.850 \text{ liter}) = (n) \times \left(0.0821 \dfrac{\text{liter atm}}{\text{mol K}} \right) \times (298 \text{ K})$$

$$n = 0.0340 \text{ mol (rounded)}$$

The formula weight, the ratio of grams to moles, is then

$$\dfrac{1.090 \text{ g}}{0.0340 \text{ mol}} = 32.1 \text{ g/mol (rounded)}$$

The formula weight of the gas is 32.1. (The gas is oxygen.)

EXERCISE 3.17   A sample of the other gas that can be obtained by the electrical decomposition of water (Example 3.18) had a mass of 0.0682 g and a volume of 817 ml at 745 torr over Hg at 25 °C. What was its formula weight? What was its likely formula?

EXERCISE 3.18   What volume of nitrogen, N_2, would have to be collected at 744 torr at 28.0 °C to have a sample containing 0.015 mol?

SUMMARY

Laws of Chemical Combination. When accurate masses of all of the reactants and products are measured and compared, no observable changes in mass accompany chemical reactions (law of conservation of mass). The mass ratios of the elements in any compound are constant regardless of geologic sources (law of definite proportions). Whenever two elements form more than one compound, then the dif-

ferent masses of one element that combine with a fixed mass of the second are in a ratio of simple whole numbers (law of multiple proportions).

Dalton's Atomic Theory. John Dalton used the laws of chemical combination as the chief evidence for his theory that all matter consists of very small, indestructible particles called atoms; that all atoms of one element must be of identical mass; that atoms of different elements are of different masses; and that chemical changes are nothing more than the regrouping of atoms. We now know that elements generally consist of isotopes—atoms of the same element having slightly different masses but the same chemical properties—but their *average* mass is constant.

Mole Concept. Atomic weights are the relative masses of the elements on a scale where the carbon-12 isotope is assigned a mass value of 12.0000 units. The sum of the atomic weights of all the elements appearing in a chemical formula gives the formula weight. A sample of any pure substance—element or compound—having a mass in *grams* equal to the formula weight is called one mole of that substance, with units of g/mol. That quantity consists of Avogadro's number (6.02×10^{23}) of formula units, a number that can also be called a mole. "Mole" therefore has two meanings—a quantity in grams equal to the formula weight and the number 6.02×10^{23} itself.

Chemical Formulas. A molecular formula gives the actual composition of a molecule. An empirical formula gives only the ratio of the atoms (expressed in the smallest whole numbers) present in the substance. The formula weight is needed to check if an empirical formula is identical with a molecular formula, and (if not) we calculate the latter from the former. Empirical formulas may be calculated from percentage compositions, which are determined by direct chemical analyses. The percentage of an element in a compound is the same as the number of grams of that element in 100 g of the compound.

Ideal Gas Equation. $PV = nRT$, the ideal gas equation, relates the four experimental variables for a gas—pressure, volume, moles, and temperature. Under standard conditions of pressure and temperature (STP), 0 °C and 1 atm, all gases have the same volume per mole—22.4 liters (standard molar volume). The general gas constant, R, is $0.0821 \frac{\text{liter atm}}{\text{mol K}}$ when P is in atm, V is in liters, and T is in kelvins. The number of moles, n, of a gas can be calculated using the ideal gas equation. When the mass of the sample is also known, then the ratio of mass to moles will give the formula weight of the gas.

INDEX TO EXERCISES, QUESTIONS, AND PROBLEMS
(Those numbered above 42 are review problems.)

REVIEW QUESTIONS

3.19. Name and state the three laws of chemical combination.

3.20. In your own words give the five postulates of Dalton's atomic theory.

3.21. Which postulate is based on the law of conservation of mass?

3.22. Which postulate of Dalton's theory is based on the law of definite proportions?

3.23. To carry out a chemical reaction as neatly and cleanly as possible, why is it important to measure substances in particular ratios by masses?

3.24. Using atomic weights from the table inside the cover, and rounding to their nearest whole numbers, how many times heavier is a magnesium atom than a carbon atom?

3.25. Which has the larger mass, a molecule of water or a mole of water?

3.26. In your own words what is the difference between a molecule and a mole?

3.27. What is the relationship between "formula weight" and molar mass?

3.28. What are the units of molar mass?

3.29. How many molecules of oxygen are in a sample having a mass of 32 g? (The formula weight of oxygen is 32.) What is the name of that number?

3.30. To determine the empirical formula of a new substance, what information must we obtain?

3.31. What is the difference between an empirical formula and a molecular formula?

3.32. What information must we obtain to get a molecular formula from an empirical formula?

3.33. Describe in your own words how we can use the percentage composition of a substance to figure out an empirical formula. How do we interpret percentage data?

3.34. What were Avogadro's two hypotheses for explaining the volume data for gas reactions?

3.35. What is the universal gas law?

3.36. What determines the units for the universal gas constant?

3.37. What are the standard conditions of temperature and pressure (STP)?

3.38. What is the difference between a mole of oxygen and a molar volume of oxygen?

3.39. Under what conditions is a value of 22.4 liters for the molar volume correct?

3.40. What makes it possible for the molar *volumes* of hydrogen and oxygen to be the same but their molar *masses* to be so different (2 g/mol vs. 32 g/mol)?

3.41. At STP, how many molecules of hydrogen are in 22.4 liters?

3.42. Explain how the gas constant can have different values and still be called a constant.

REVIEW PROBLEMS

3.43. (a) In one experiment 2.50 g of tin reacted with chlorine to make 3.99 g of tin chloride. How much chlorine was used (in grams)?

(b) In another experiment, under different conditions, 4.64 g of tin reacted with chlorine to make 10.18 g of tin chloride, only this was not the same tin chloride as that obtained in (a). How much chlorine was used up (in grams)?

(c) Using data from part (a), how much chlorine combined with 1 g of tin to make the first compound?

(d) Using data from part (b), how much chlorine combined with 1 g of tin to make the second compound?

(e) Do the two quantities of chlorine that will combine with 1 g of tin give a ratio of simple whole numbers? If so, what is that ratio?

3.44. (a) In one experiment 5.00 g of nitrogen combined with oxygen to give 10.71 g of one of the nitrogen oxides, nitric oxide, which is an air pollutant. How much oxygen was used up (in grams)?

(b) In another experiment under different conditions, 3.00 g of nitrogen combined with oxygen to make 9.86 g of another oxide of nitrogen, also an air pollutant. How much oxygen was used up (in grams)?

(c) Using the data from parts (a) and (b) how much oxygen combined with 1 g of nitrogen to make each oxide of nitrogen?

(d) Do the two quantities of oxygen that will combine with 1 g of nitrogen give a ratio of simple whole numbers? If so, what is that ratio?

3.45. What is the formula weight of each of the following?
(a) Sodium carbonate, Na_2CO_3
(b) Cholesterol, $C_{27}H_{46}O$
(c) Calcium cyclamate, $Ca(C_6H_{12}NSO_3)_2$ (sweetening agent)

3.46. Calculate the formula weight of each of the following.
(a) Calcium nitrate, $Ca(NO_3)_2$ (used in matches, firecrackers, and explosives)
(b) Magnesium phosphate, $Mg_3(PO_4)_2$ (used in antacid)
(c) Ascorbic acid, $C_6H_8O_6$ (Vitamin C)

3.47. What is the mass of 0.100 mol of each of the substances given in problem 3.45?

3.48. What is the mass of 0.250 mol of each of the substances in problem 3.46?

3.49. How many moles of water are in 9.00 g of water, H_2O?

3.50. How many moles of sodium nitrate are in 1.70 g of sodium nitrate, $NaNO_3$, a substance used as a fertilizer or to make gunpowder.

3.51. Ammonium sulfate, $(NH_4)_2SO_4$, is a fertilizer used to supply both nitrogen and sulfur. How many grams of ammonium sulfate are in a 35.8 mol?

3.52. A 0.500 mol sample of table sugar, $C_{12}H_{22}O_{11}$, weighs how many grams.

3.53. A solution of zinc chloride, $ZnCl_2$, in water is used to soak the ends of wooden fence posts to preserve them in the ground. One ratio used is 840 g $ZnCl_2$ to 4 liters water. How many moles of $ZnCl_2$ are in 840 g of $ZnCl_2$?

3.54. Thallium sulfate, Tl_2SO_4, a powerful poison, was illegally used in 1970–1971 in poison baits to control predators (e.g., coyotes) on western range lands. Hundreds of bald and golden eagles died from taking the baits. A 1.00 kg can of Tl_2SO_4 contains how many moles Tl_2SO_4?

3.55. Borazon, one crystalline form of boron nitride, BN, is very likely the hardest of all substances. If one sample contained 3.02×10^{23} atoms of boron, how many atoms and how many grams of nitrogen were also combined in the sample?

3.56. When water is dropped onto calcium carbide, CaC_2, acetylene (C_2H_2) forms, which can be ignited to supply light. Calcium carbide has therefore been used to make signal flares for use on lakes or the ocean. If one sample of CaC_2 contained 12.04×10^{23} atoms of carbon, how many atoms and how many grams of calcium were also combined in this sample? (The other product of this reaction is calcium hydroxide, $Ca(OH)_2$.)

3.57. Iodine must be present in the diet to prevent a thyroid condition called goiter (a very disfiguring enlargement of the throat area). "Iodized" salt is available in all supermarkets and, when used, keeps the incidence of goiter at nearly zero. Sodium iodide, NaI, could be used, but it is less stable than calcium iodate, $Ca(IO_3)_2$. How many atoms of iodine are in 0.500 mol of $Ca(IO_3)_2$? How many grams of $Ca(IO_3)_2$ are needed to supply that much iodine?

3.58. Ammonium carbonate, $(NH_4)_2CO_3$, is present in a 1 to 1 ratio by weight with powdered soap in "ammoniated washing powder." How many atoms of nitrogen are in 0.750 mol of $(NH_4)_2CO_3$? How many grams of this compound would supply that much nitrogen?

3.59. "Diammonium phosphate," $(NH_4)_2HPO_4$, is made from phosphate rock and ammonia (by an indirect process) and is widely used as a fertilizer to supply both nitrogen and phosphorus. How many grams are in 6.26 mol $(NH_4)_2HPO_4$?

3.60. Calcium dihydrogen phosphate, $Ca(H_2PO_4)_2$, is a component of both "superphosphate" and "triple superphosphate" fertilizers. How many grams are present in 4.34 mol $Ca(H_2PO_4)_2$?

3.61. A 100-pound sack of "diammonium phosphate" (question 3.59) contains how many moles of this compound?

3.62. A hopper containing 500 lb of calcium dihydrogen phosphate (question 3.60) contains how many moles of $Ca(H_2PO_4)_2$?

3.63. Sodium perborate, $NaBO_3$, is present in "oxygen bleach"; it acts by releasing oxygen, which has the bleaching action. How many grams are in 4.65 mol of $NaBO_3$)?

3.64. Barium sulfate, $BaSO_4$, is given as a thick slurry in flavored water before X rays are taken of the intestinal tract. The barium blocks X rays and the tract therefore becomes outlined. How many grams are in 0.568 mol of barium sulfate?

3.65. One recipe for making soap calls for mixing 13 oz of sodium hydroxide, NaOH, dissolved in 5 cups water, with 6 lb of tallow. Assume that the formula of tallow is $C_{57}H_{110}O_6$. The mole ratio theoretically should be 3 mol of NaOH to 1 mol of tallow. (NaOH is commonly known as "lye.") If excess NaOH remains the soap will be extra harsh on the skin. If excess tallow remains, the soap will be greasy and will have poorer cleansing power. Calculate the actual mole ratio used in this recipe. (Round at the end to one significant figure.)

3.66. One recipe for hydroponic plant food calls for 1.50 oz KNO_3, 1.00 oz $CaSO_4$, 0.750 oz $MgSO_4$, 0.500 oz $CaHPO_4$, and 0.250 oz $(NH_4)_2SO_4$ dissolved in 10 gal of water. How many grams and how many moles of each compound are present in this volume of solution? (Hydroponics is a method of raising plants using a balanced nutrient medium with or without any mechanical support by soil, sand, or gravel.)

3.67. In the late 1970s, the United States manufacture of ammonia, NH_3, an important fertilizer and a raw material for the chemical industry, was about 32 billion lb per year. How many moles per year was that?

3.68. In the late 1970s, about 35 million tons of sulfuric acid were made in the United States each year. How many moles is that? (Each ton is 2000 lb.)

3.69. The following are molecular formulas. Write their empirical formulas.
(a) C_2H_6 (ethane) (b) $C_6H_{12}O_6$ (glucose) (c) $C_9H_8O_3$ (aspirin)

3.70. Write the empirical formula for each of the following.
(a) S_8 (sulfur) (b) H_2O_2 (hydrogen peroxide) (c) H_2O (water)

3.71. Calculate the percentage of nitrogen in these two important nitrogen fertilizers—ammonia (NH_3) and urea (N_2H_4CO).

3.72. Calculate the percentage of nitrogen and the percentage of phosphorus in each of these "ammonium phosphate" fertilizers—$NH_4H_2PO_4$ and $(NH_4)_2HPO_4$.

3.73. Phencyclidine ("angel dust") is $C_{17}H_{25}N$. A sample suspected of being this drug was analyzed and found to have 83.71% C, 10.42% H, and 5.61% N. Are these data consistent for phencyclidin within the limits of error that permit a $\pm 0.30\%$ deviation from the theoretical values? Calculate to four significant figures the percentages of the elements in this compound.

3.74. Lysergic acid diethylamide, LSD, is $C_{20}H_{25}N_3O$. One suspected sample was found to contain 74.07% C, 7.95% H, and 9.99% N. Are these data consistent for LSD within the allowed limits of error (see problem 3.73)? Calculate the theoretical percentages to four significant figures.

3.75. A sample of an unknown gas having a mass of 3.620 g was made to decompose into 2.172 g of oxygen and 1.448 g of sulfur. Prior to the decomposition, this sample occupied a volume of 1120 ml when its pressure was 750 torr and its temperature 25 °C.
(a) What is the percentage composition of the elements in this gas?
(b) What is the empirical formula of the gas?
(c) What is its molecular formula?

3.76. A sample of an unknown gas with a mass of 1.620 g occupied a volume of 941 ml when its pressure was 748 torr and its temperature was 20 °C. When made to decompose into its elements, 1.389 g of carbon and 0.2314 g of hydrogen were obtained.
(a) What is the percentage composition of the elements in this gas?
(b) Determine its empirical formula.
(c) Determine its molecular formula.

3.77. An unknown gas consisting of molecules composed of carbon and hydrogen atoms reacted with oxygen to give two gases, carbon dioxide and water vapor. The proportions by their volumes, measured at the same temperature and pressure, are: 1 vol unknown + 2 vol oxygen gave 2 vol water and 1 vol carbon dioxide. Use the symbol $\oplus$ for an atom of carbon, $\mathbb{O}$ for an atom of oxygen, and $\otimes$ for an atom of hydrogen.
(a) Write symbols for the molecules of oxygen, carbon dioxide, and water. (Do not be concerned about the se-

quence in which the atoms of these molecules are joined nor with geometric arrangements within the molecules.)

(b) Using these symbols and the volume data, deduce what the simplest composition of one molecule of the gas would be.

3.78. Pictured below is the balance arm of a simple weighing balance. The small vertical lines correspond to 1 g units of mass, and the numbers identify every fifth line. Below this drawing are three just like it, except the numbers are omitted. The lines still correspond to 1 g intervals. The total capacity of each balance is 100 g.

(a) Using drawing a, mark the balance for its potential use in measuring water (H_2O) in units of *moles* water. Identify the lines corresponding to 1 mol, 1.5 mol, 2 mol, 2.5 mol and so on up in 0.5 mol increments as far as one can go. This scale could now be used for obtaining samples of water directly by moles. The indirect method requiring one to convert moles to grams could be avoided.

(b) Mark drawing b for use in measuring moles of methyl alcohol (CH_4O) directly. Do this as you were instructed in part (a).

(c) Mark drawing c for measuring moles of aluminum in 0.5 mole units as instructed in part (a).

(d) If the research of a particular chemist routinely required the use of 500 chemicals, each with different formula weights, how many balances would be needed if they were marked to measure moles, not grams. If you were funding that research, what approach would you have the chemist use, the direct (in moles) or the indirect (converting moles to grams)?

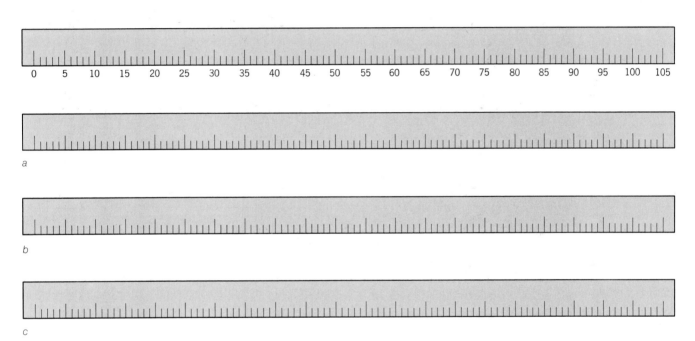

0 5 10 15 20 25 30 35 40 45 50 55 60 65 70 75 80 85 90 95 100 105

a

b

c

CHAPTER FOUR

QUANTITATIVE ASPECTS OF CHEMICAL REACTIONS

4.1 STOICHIOMETRY

The stoichiometry of a reaction is the quantitative description of the relative amounts of the substances involved

"Stoke-ee-ah-meh-tree" from the Greek *stoicheion,* meaning "element," and *metron,* meaning "measure."

When performing experiments it is both unscientific and possibly quite danger-ous to approach reactions haphazardly. Therefore, when we go into the labora-tory to work with chemicals we almost always use a balanced equation, whether the equation is of an actual or a predicted reaction. A balanced equation helps us because its coefficients give us the reaction's stoichiometry. The **stoichi-ometry** of a reaction is the description of the relative quantities *by moles* of the reactants and products as given by the coefficients of the balanced equation for the reaction. In a sense, stoichiometry is the molar bookkeeping of chemistry, and the books have to balance. No verifiable, quantitative data concerning chemical reactions is possible without stoichiometry. This is true of chemical ap-plications in such different fields as agriculture, clinical analyses, drug reac-tions, food chemistry, inhalation therapy, nutrition, crime lab analyses, geochemistry—the list could really go on and on. Stoichiometry begins with writ-ing and balancing chemical equations, because these operations give the needed coefficients.

4.2 BALANCING EQUATIONS

In a balanced equation all atoms appearing among reactants must appear somewhere among the products

In Section 1.17, we learned that a balanced equation is a shorthand descrip-tion of a chemical reaction. For example, we can express the fact that "hy-drogen reacts with oxygen to give water" much more briefly in a balanced equation and in the same place specify the coefficients that give us the re-action's all-important stoichiometry. As the first step in writing the equation for this reaction we set down the chemical symbols for the reactants, H_2 (hydrogen) and O_2 (oxygen), separated by a plus sign to stand for "reacts with." Then we write an arrow that points to the chemical symbol for the product, H_2O (water).

$$H_2 + O_2 \longrightarrow H_2O \quad \text{(unbalanced)}$$

The reaction can't happen as written because one of the two atoms of oxygen on the left doesn't show up in the product. The procedure for bringing this equation into balance is to find the coefficients for each chemical symbol that insure that all atoms are accounted for, that all atoms of each element on one side of the arrow show up on the other side, too. If we write "2" next to H_2O, then we will show two atoms of oxygen on both the left and right sides:

$$H_2 + O_2 \longrightarrow 2H_2O \quad \text{(unbalanced)}$$

Since "$2H_2O$" also means $2 \times 2 = 4$ atoms of hydrogen, we now have four atoms of hydrogen on the right but only two on the left. To make the numbers of hydrogen atoms equal on both sides we write a coefficient of "2" for H_2 on the left side:

$$2H_2 + O_2 \longrightarrow 2H_2O \quad \text{(balanced)}$$

Warning. Do not change subscripts within a formula to get an equation to balance.

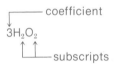

For example, writing H_2O_2 instead of $2H_2O$ in the first effort to get the above equation to balance would have been a drastic and altogether wrong move. The formula H_2O_2 stands for hydrogen peroxide, which is an entirely different substance from H_2O. Once we knew we had to write the formula of water and wrote it correctly as H_2O, then the only way we could double the oxygen atoms on the right was to write $2H_2O$ (which, of course, also doubled the hydrogen atoms on the right). Let's try this on another example.

EXAMPLE 4.1   **Writing a Balanced Equation**

Problem: Write the balanced equation for the formation of table salt, NaCl (sodium chloride), from sodium (Na) and gaseous chlorine (Cl_2).

Solution: First, we write the correct formulas in the style for an equation:

$$Na + Cl_2 \longrightarrow NaCl \quad \text{(unbalanced)}$$

There are two chlorine atoms on the left and only one on the right. That subscript of "2" in Cl_2 suggests that "2" should be the coefficient for NaCl:

$$Na + Cl_2 \longrightarrow 2NaCl \quad \text{(unbalanced)}$$

But now there are two sodium atoms on the right and only one on the left, so we write a coefficient of "2" for Na:

$$2Na + Cl_2 \longrightarrow 2NaCl \quad \text{(balanced)}$$

Now there are two sodium atoms on the left and two on the right; two chlorine atoms on the left and two on the right.

Example 4.1 illustrated a strategy for judging what coefficients to pick— letting subscripts suggest coefficients. For example, in the presence of electrical sparks or high-energy ultraviolet light, oxygen (O_2) will change to another form, ozone (O_3). To symbolize this change we begin with the formulas:

$$O_2 \longrightarrow O_3 \quad \text{(unbalanced)}$$

The subscript in O_3 (namely, 3) suggests a coefficient for O_2 just as the subscript in O_2 (2) suggests a coefficient for O_3 because $2 \times 3 = 3 \times 2$, a balance. Therefore we can write:

$$3O_2 \longrightarrow 2O_3$$

Now there are six oxygen atoms on the left (3×2) and six on the right (2×3). Let's try this strategy in a similar example.

EXAMPLE 4.2 **Balancing a Chemical Equation. Using Subscripts to Suggest Coefficients**

Problem: Although bright and shiny, aluminum objects are covered with a tight, invisible coating of aluminum oxide (Al_2O_3) that forms when freshly exposed aluminum (Al) reacts with oxygen. Write the balanced equation for this reaction.

Solution: First, the chemical symbols are set up in the pattern of an equation:

$$Al + O_2 \longrightarrow Al_2O_3 \quad \text{(unbalanced)}$$

Let's start with the imbalance in oxygen atoms. The subscript "3" in Al_2O_3 suggests that it be the coefficient for O_2 just as the "2" in O_2 suggests that it be the coefficient for Al_2O_3:

$$Al + 3O_2 \longrightarrow 2Al_2O_3 \quad \text{(unbalanced)}$$

Of course, doubling Al_2O_3 doubles the aluminum atoms on the right, from 2 to 4. But that's easily solved by writing a coefficient of "4" in front of Al:

$$4Al + 3O_2 \longrightarrow 2Al_2O_3 \quad \text{(balanced)}$$

EXERCISE 4.1 Balance These Equations by Letting Subscripts Suggest Coefficients

(a)[1] $P + O_2 \rightarrow P_4O_{10}$
(b)[2] $N_2 + H_2 \rightarrow NH_3$

Many reactions involve chemicals whose formulas include groups of atoms, sometimes in parentheses. *If these groups do not change in the reactions, treat them as units when balancing the equation.*

EXAMPLE 4.3 **Balancing Equations When Formulas Include Groups of Atoms**

Problem: Water that contains dissolved calcium compounds is called "hard" water, meaning that it is hard to use soap in such water. The soap forms a scum

---

[1] Some white flares work this way. Phosphorus burns intensely and one product is tetraphosphorus decaoxide, P_4O_{10}, a white powder.

[2] This is the way ammonia, NH_3, is made. Fritz Haber won the Nobel Prize in 1918 for working out the engineering and scientific details—which says as much about the importance of ammonia as it does about the problems of making it.

with the calcium compound. One way to "soften" such water is to add sodium carbonate, Na_2CO_3. It removes the calcium by forming calcium carbonate, which is insoluble in water. Balance the equation that illustrates this reaction:

(*aq*) aqueous solution
(*s*) solid

$$Ca(NO_3)_2(aq) + Na_2CO_3(aq) \longrightarrow CaCO_3(s) + NaNO_3(aq) \quad \text{(unbalanced)}$$

| calcium | sodium | calcium | sodium |
| nitrate | carbonate | carbonate | nitrate |

Solution: Neither the nitrate group (NO_3) nor the carbonate group (CO_3) change from one side to the other in this equation. Therefore, treat them as units. There are two units of (NO_3) on the left but only one on the right, so put a 2 in front of $NaNO_3$.

$$Ca(NO_3)_2 + Na_2CO_3 \longrightarrow CaCO_3 + 2NaNO_3 \quad \text{(balanced)}$$

That brings everything into balance. As a check, notice that there is 1 Ca on each side, 2 NO_3 on each side, 2 Na on each side, and 1 CO_3 on each side.

EXERCISE 4.2 Balance each of the following equations.
(a) $Mg + O_2 \longrightarrow MgO$
(b) $CH_4 + Cl_2 \longrightarrow CCl_4 + HCl$
(c) $NO + O_2 \longrightarrow NO_2$
(d) $NaOH + H_2SO_4 \longrightarrow Na_2SO_4 + H_2O$
(e) $CH_4 + O_2 \longrightarrow CO_2 + H_2O$
(f) $C_2H_6 + O_2 \longrightarrow CO_2 + H_2O$
(g) $Al(OH)_3 + H_2SO_4 \longrightarrow Al_2(SO_4)_3 + H_2O$

4.3 STOICHIOMETRIC CALCULATIONS

Coefficients give relative quantities of reactants and products by moles as well as by molecules

As we balance an equation, our thinking is in terms of the small particles represented by the formulas. However, when we use the equation to plan an experiment we must shift our thinking to huge collections of them—to *moles*. It is simply impossible in experimental work to use just one or a few molecules or atoms; they are too small to pick and count individually. It is all right during the balancing of an equation to view the ratios of coefficients as ratios of molecules or atoms: for example, in the reaction of hydrogen and oxygen that gives water:

$$2 \text{ molecules } H_2 + 1 \text{ molecule } O_2 \longrightarrow 2 \text{ molecules } H_2O$$

However, for laboratory work we somehow have to get large collections of each substance in the same ratios by molecules. Multiplying each coefficient by any constant number, including Avogadro's number (6.02×10^{23}), does not alter the ratios, but it gives us a lab-sized unit:

$$2 \times (6.02 \times 10^{23}) \text{ molecules } H_2 + 1 \times (6.02 \times 10^{23}) \text{ molecules } O_2$$
$$\longrightarrow 2 \times (6.02 \times 10^{23}) \text{ molecules } H_2O$$

The essential 2:1:2 ratio has not been changed, but the scale of the reaction has shifted to the mole level, because 6.02×10^{23} molecules of any substance is 1 mole of it. Now we can interpret the coefficients as follows:

$$2 \text{ mol } H_2 + 1 \text{ mol } O_2 \longrightarrow 2 \text{ mol } H_2O$$

The ratio of *moles* is identical to the ratio of *molecules;* it has to be since equal numbers of moles have equal numbers of molecules.

The ratio of the coefficients for any given chemical reaction is set by nature; we cannot change that ratio. The decision left for us is the *scale* of the reaction— how much do we want to use up or make? The number of options is infinite. We could have

$$0.02 \text{ mol } H_2 + 0.01 \text{ mol } O_2 \longrightarrow 0.02 \text{ mol } H_2O$$

or

$$1.36 \text{ mol } H_2 + 0.68 \text{ mol } O_2 \longrightarrow 1.36 \text{ mol } H_2O$$

or

$$88 \text{ mol } H_2 + 44 \text{ mol } O_2 \longrightarrow 88 \text{ mol } H_2O$$

In every case, the relative mole quantities of H_2 to O_2 to H_2O are 2 : 1 : 2. We could say that 2 mol H_2, 1 mol O_2 and 2 mol H_2O are *equivalent* to each other *in this reaction.* This does not mean that one chemical can actually substitute for any of the other chemicals. It means that a specific mole quantity of one substance *requires* the presence of a specific mole quantity of each of the other substances in accordance with the ratios of coefficients.

Figure 4.1 shows five different scales for the reaction of iron with sulfur to make iron sulfide, FeS. Notice that the *mole* ratios are the same regardless of the scale of the reaction.

EXAMPLE 4.4 **Working with Mole Relationships**

Problem: Two atoms of sulfur (S) will combine with three molecules of oxygen (O_2) to give two molecules of sulfur trioxide (SO_3), an air pollutant:

$$2S + 3O_2 \longrightarrow 2SO_3$$

How many *moles* of sulfur will combine with 9 moles of oxygen?

Solution: For this kind of problem the critically important step is to *use ratios of coefficients as conversion factors* to calculate how many moles of one chemical in the equation are equivalent to any given number of moles of any other substance shown. The conversion factor we need here involves the coefficients of oxygen and sulfur. These coefficients are 2 and 3—2 atoms of sulfur to 3 molecules of oxygen or 2 moles of sulfur to 3 moles of oxygen. *Remember that ratios of moles and ratios of molecules are identical.* This relationship can be stated in conversion factor form either as:

$$\frac{2 \text{ mol S}}{3 \text{ mol } O_2} \quad \text{or as} \quad \frac{3 \text{ mol } O_2}{2 \text{ mol S}}$$

What these say is that *in this equation* 2 mol of sulfur requires 3 mol of oxygen or, in other words, 2 mol of sulfur are chemically equivalent to 3 mol of oxygen. Now we need to find how much sulfur is equivalent to 9 mol of oxygen. We always pick the conversion factor according to what must cancel and what must remain. Here we want the answer in moles of sulfur (mol S), so we must get moles of oxygen (mol O_2) to cancel. Therefore, we multiply "9 mol O_2" by the *first* conversion factor:

$$9 \text{ mol } O_2 \times \frac{2 \text{ mol S}}{3 \text{ mol } O_2} = 6 \text{ mol S, our answer}$$

Figure 4.1
Atom/molecule ratios, mole ratios, and the scale of a reaction. Iron and sulfur, when heated, combine as follows: Fe + S → FeS. Sulfur atoms have smaller masses than iron atoms—by a factor of 32.1/55.8, the ratio of their atomic weights. Therefore, the piles of sulfur have lower masses than the piles of iron, but the *atoms* of each pile are always in a 1 to 1 ratio, the ratio required by nature for this reaction. The overall scale of the reaction, however, is picked up by the experimentalist.

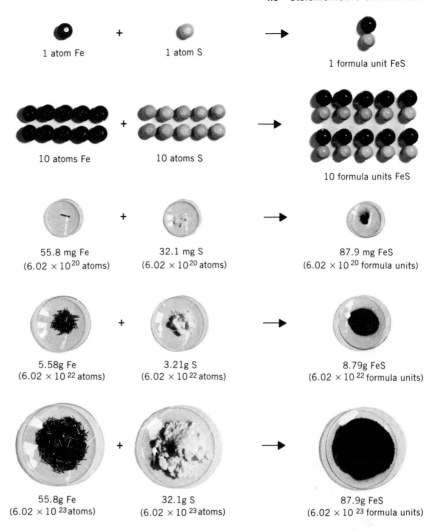

1 atom Fe + 1 atom S → 1 formula unit FeS

10 atoms Fe + 10 atoms S → 10 formula units FeS

55.8 mg Fe (6.02×10^{20} atoms) + 32.1 mg S (6.02×10^{20} atoms) → 87.9 mg FeS (6.02×10^{20} formula units)

5.58g Fe (6.02×10^{22} atoms) + 3.21g S (6.02×10^{22} atoms) → 8.79g FeS (6.02×10^{22} formula units)

55.8g Fe (6.02×10^{23} atoms) + 32.1g S (6.02×10^{23} atoms) → 87.9g FeS (6.02×10^{23} formula units)

EXERCISE 4.3   Referring to Example 4.4, how many moles of SO_3 can be made from 9 mol O_2?

EXERCISE 4.4   Referring to Example 4.4, how many moles of O_2 are needed to change 30 mol of S into SO_3?

EXERCISE 4.5   Solutions of ferric chloride, $FeCl_3$, are used in photoengraving and to make ink. This compound can be made by the reaction of 2 atoms of iron (Fe) and 3 molecules of chlorine (Cl_2):

$$2\ Fe + 3Cl_2 \longrightarrow 2FeCl_3$$

(a) Set up the three possible pairs of conversion factors based on the coefficients of this equation. One pair will involve Fe and Cl_2; another, Fe and $FeCl_3$; the third pair, Cl_2 and $FeCl_3$. Select from these six conversion factors to solve the next problems.
(b) How many moles of $FeCl_3$ form from 24 mol Cl_2?
(c) How many moles of Fe are needed to combine with 24 mol Cl_2 by this reaction?
(d) If 0.500 mol Fe is to be used by this reaction, how much chlorine is needed and how much $FeCl_3$ will form (answering both in moles)?

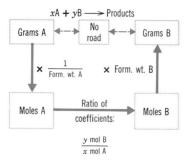

The reaction in Exercise 4.5 was described and problems related to it were solved in terms of moles. However, laboratory weighing balances are not marked for moles. Problem 3.78 in the last chapter gave a strong hint as to the reason why. The balances are marked in grams; therefore, reactions given in terms of moles have to be worked out in terms of grams before the reaction can be carried out. The only way available for relating *grams* of one compound to *grams* of another in a particular reaction is to go through their respective moles during the calculations.

EXAMPLE 4.5   **Working Mole Problems**

Problem: Aluminum oxide (Al_2O_3), a buffing powder, is to be made by combining 5.00 g of aluminum (Al) with oxygen (O_2). How much oxygen will be needed in moles? In grams?

Solution: All mole problems require a balanced equation, so the first step is to write the equation for the reaction. We did this particular equation in Example 4.2.

$$4Al + 3O_2 \longrightarrow 2Al_2O_3$$

Since the calculation route from moles to grams or grams to moles requires formula weights, these should be calculated early and set off to one side for reference as needed. The formula weights needed in this example are:

$$Al, 27.0; O_2, 32.0$$

Now comes the critical step — changing the grams of the given substance into moles. *Always get the problem to the mole level and solve it there.* Afterward, moles can be converted to grams. We were given 5.00 g of Al, and from the atomic weight we have a conversion factor of $\dfrac{1 \text{ mol Al}}{27.0 \text{ g Al}}$.

$$5.00 \text{ g Al} \times \frac{1 \text{ mol Al}}{27.0 \text{ g Al}} = 0.185 \text{ mol Al}$$

To find the oxygen needed to react with 0.185 mol of Al, we use the coefficients of Al and O_2 for a conversion factor to see how many moles of O_2 are equivalent to 0.185 mol of Al:

$$\frac{4 \text{ mol Al}}{3 \text{ mol O}_2} \quad \text{or} \quad \frac{3 \text{ mol O}_2}{4 \text{ mol Al}}$$

Since "mol O_2" is the unit of the answer we want, we pick the second factor so that "mol Al" will cancel out.

$$0.185 \text{ mol Al} \times \frac{3 \text{ mol O}_2}{4 \text{ mol Al}} = 0.139 \text{ mol O}_2$$

This is the answer at the mole level; 0.139 mol of O_2 is needed to combine with 0.185 mol of Al to make Al_2O_3 by our equation. The question also calls for the grams of O_2, and we know from the formula weight of O_2 that there are 32.0 g O_2/mol O_2. Therefore:

$$0.139 \text{ mol O}_2 \times \frac{32.0 \text{ g O}_2}{1 \text{ mol O}_2} = 4.45 \text{ g O}_2$$

That's the second answer; it takes 4.45 g of O_2 (0.139 mol) to combine with 5.00 g of Al (0.185 mol) to make Al_2O_3.

EXERCISE 4.6   Calculate how much Al_2O_3 in moles and in grams would form in the experiment of Example 4.5.

EXERCISE 4.7   Referring to Example 4.5, if 9.48 g of O_2 combined with aluminum to make Al_2O_3, how much aluminum in moles and in grams would be needed?

EXERCISE 4.8   Referring to Example 4.4, a sample of 8.00 g of S is changed to sulfur trioxide by the action of oxygen. How much oxygen in moles and grams is needed for this reaction?

EXAMPLE 4.6   **Working Mole Problems**

Problem: During its combustion, ethane (C_2H_6) combines with oxygen (O_2) to give carbon dioxide (CO_2) and water (H_2O). A sample of ethane was burned completely and the water that formed had a mass of 1.61 g. How much ethane was in the sample?

Solution: First, set up the equation using the formulas given:

$$C_2H_6 + O_2 \longrightarrow CO_2 + H_2O$$

Then balance it:

$$2C_2H_6 + 7O_2 \longrightarrow 4CO_2 + 6H_2O$$

Then compute the needed formula weights:

$$H_2O, 18.0; \qquad C_2H_6, 30.0$$

For calculating how many grams of one substance correspond to a certain number of grams of another we have to take the path that goes through moles. Therefore, we change the given, 1.61 g of H_2O, to moles using the formula weight of H_2O.

$$1.61 \text{ g } H_2O \times \frac{1 \text{ mol } H_2O}{18.0 \text{ g } H_2O} = 0.0894 \text{ mol } H_2O$$

Now we must calculate how many moles of ethane are needed to make 0.0894 mol H_2O by this reaction. We have these two possible conversion factors relating the two substances:

$$\frac{6 \text{ mol } H_2O}{2 \text{ mol } C_2H_6} \qquad \text{or} \qquad \frac{2 \text{ mol } C_2H_6}{6 \text{ mol } H_2O}$$

We pick the second conversion factor to multiply by 0.0894 mol H_2O to find moles C_2H_6.

$$0.0894 \text{ mol } H_2O \times \frac{2 \text{ mol } C_2H_6}{6 \text{ mol } H_2O} = 0.0298 \text{ mol } C_2H_6$$

This is the answer at the mole level—0.0298 mol of C_2H_6 must be burned to give 1.61 g (0.0894 mol) of H_2O. The question also calls for the answer in grams of C_2H_6. We use the formula weight of C_2H_6 to calculate the number of grams:

$$0.0298 \text{ mol } C_2H_6 \times \frac{30.0 \text{ g } C_2H_6}{1 \text{ mol } C_2H_6} = 0.894 \text{ g } C_2H_6$$

This is the other answer—0.894 g of C_2H_6 will combine with oxygen to make 1.61 g of H_2O.

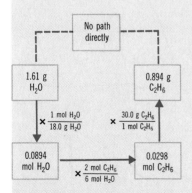

EXERCISE 4.9   Calculate the number of moles and grams of oxygen needed in Example 4.6.

EXERCISE 4.10   How much carbon dioxide in moles and in grams are also produced in Example 4.6?

EXERCISE 4.11   Taking the results of Example 4.6 and Exercises 4.9 and 4.10, calculate the sum of the masses of the reactants and the sum of the masses of the products. Do the results conform to the law of conservation of mass?

EXAMPLE 4.7   **Working Mole Problems by Accumulating Conversion Factors**

Problem: How might conversion factors be accumulated to shorten the solution to a problem such as that solved in Example 4.6?

Solution: Go back to Example 4.6 and review the three *separate* calculations. In the first, the result was 0.894 mol H_2O. In the second this result was converted to 0.0298 mol C_2H_6. The final answer came from the third conversion — of 0.0298 mol C_2H_6 to 0.894 g C_2H_6. Making separate calculations can be avoided and time saved by accumulating the conversion factors as follows:

$$1.61 \; \cancel{g\;H_2O} \times \frac{1 \; \cancel{mol\;H_2O}}{18.0 \; \cancel{g\;H_2O}} \times \frac{2 \; \cancel{mol\;C_2H_6}}{6 \; \cancel{mol\;H_2O}} \times \frac{30.0 \text{ g } C_2H_6}{1 \; \cancel{mol\;C_2H_6}} = 0.894 \text{ g } C_2H_6$$

0.0894 mol H_2O

0.0298 mol C_2H_6

0.894 g C_2H_6

Notice that all units cancel except the unit for the final answer. If these cancellations had not led to that result, then we would have known with certainty that the solution had been improperly set up. The first stage, of course, changes g H_2O into mol H_2O. Then mol H_2O is converted to the equivalent amount of C_2H_6—mol C_2H_6. Finally there is a shift back to the gram level in converting mol C_2H_6 into g C_2H_6. In using the method of accumulating conversion factors, the operations are rolled into one long setup. Then the arithmetic is done. In brief, we move from a mass of one substance to the mole level, stay there to find moles of any other substance in the reaction, and then move back to the gram level of that other compound. *The calculation path from grams of one substance to grams of anything else in the reaction always goes through moles.* There is no direct calculation path from grams of one compound to grams of another.

EXERCISE 4.12   Use the strategy of Example 4.7 to work Exercise 4.9 again.

EXERCISE 4.13   Work Exercise 4.10 again using the strategy of Example 4.7.

EXERCISE 4.14   Rework Exercises 4.6 and 4.7 by the method of accumulating conversion factors.

EXERCISE 4.15 Use the accumulation of conversion factors to work Exercise 4.8.

EXAMPLE 4.8 **Working Mole Problems Involving Gases**

Problem: The industrial synthesis of ammonia (NH_3) by the Haber process involves the following equation:

$$N_2(g) + 3H_2(g) \longrightarrow 2NH_3(g) \quad (g = \text{gaseous state})$$

If 725 liters of nitrogen, initially at 740 torr and 25 °C (298 K) are to be used to make ammonia, how much hydrogen will be needed in moles and in grams?

Solution: Instead of being given a certain *mass* of one substance, we are to start with a certain volume. We have to go from a gas volume to the mole level instead of going from grams to the mole level. The easiest way to do this is to use the ideal gas equation ($PV = nRT$) to find the number of moles of nitrogen from its given volume.

Since the value of R we are accustomed to using contains pressure in *atm* units $\left(R = 0.0821 \dfrac{\text{liter atm}}{\text{K mol}}\right)$ we have to convert 740 torr into atm to get the units to cancel properly.

$$740 \text{ torr} \times \frac{1.00 \text{ atm}}{760 \text{ torr}} = 0.974 \text{ atm}$$

Now we can solve for n, the number of moles of nitrogen:

$$n = \frac{PV}{RT} = \frac{(0.974 \text{ atm})(725 \text{ liter})}{\left(0.0821 \dfrac{\text{liter atm}}{\text{K mol}}\right)(298 \text{ K})}$$

$$= 28.9 \text{ mol nitrogen}$$

Using the coefficients of the balanced equation, we can devise a conversion factor for converting moles of nitrogen into the equivalent number of moles of hydrogen needed for the reaction:

$$28.9 \text{ mol } N_2 \times \frac{3 \text{ mol } H_2}{1 \text{ mol } N_2} = 86.7 \text{ mol } H_2$$

This gives us one answer: 86.7 mol of hydrogen are needed to combine with 725 liters of nitrogen. Convert this to grams of hydrogen in the usual way using the formula weight of hydrogen, 2.02.

$$86.7 \text{ mol } H_2 \times \frac{2.02 \text{ g } H_2}{1 \text{ mol } H_2} = 175 \text{ g } H_2$$

EXERCISE 4.16 How many moles and grams of ammonia would form given the data of Example 4.8? How many liters of NH_3 (at STP) does this represent?

4.4 PERCENTAGE YIELD AND LIMITING REACTANTS

Quite often one of the reactants limits the yield of a product even when the other reactants are present in sufficient quantities

The **percentage yield** of a product is the ratio, calculated as a percentage, of the actual quantity of product obtained from a reaction to the quantity predicted by the reaction's stoichiometry. For example, suppose we actually obtained

The actual quantity is called the "yield."

9.44 g of aluminum oxide, Al_2O_3, from 5.00 g of aluminum according to the equation

$$4Al + 3O_2 \longrightarrow 2Al_2O_3$$

(This was worked as Example 4.5 and Exercise 4.6.) The percentage yield would be 100%, since 9.44 g of Al_2O_3 is the maximum one can get from 5.00 g of Al under perfect conditions. However, if we obtain 4.72 g of Al_2O_3 as the actual yield in a particular experiment, the percentage yield would be

$$\frac{4.72 \text{ g } Al_2O_3}{9.44 \text{ g } Al_2O_3} \times 100 = 50.0\% \text{ yield}$$

In this situation, not all of the aluminum reacted with oxygen.

Sometimes in experimental work one of two (or more) reactants is deliberately used in a mole quantity that is greater than what actually corresponds to the reaction's stoichiometry. An excess of one reactant might help to insure that all of another reactant is used up. Often an excess of one reactant helps to speed up a reaction. (A large excess of oxygen, for example, makes things burn faster.) Whatever the reason for using an excess of a reactant, we can't make *more* product from it because the yield will be limited by the other reactant that is present in a deficient mole quantity. The deficient reactant is called the **limiting reactant.** The maximum or theoretical yield of a product is always based on the limiting reactant. The steps to use to find the limiting reactant and for then calculating the maximum yield are these.

Steps for Finding the Limiting Reactant and Maximum Yield
Step 1 Be sure you are working with a balanced equation.
Step 2 Calculate formula weights.
Step 3 Calculate the moles of all reactants from the masses given. (As always in stoichiometric problems, get the calculations to the mole level.)
Step 4 Use the coefficients of the balanced equation to compare the ratio of the *moles* of reactants taken with the ratio of the moles required by the stoichiometry. This comparison will give you the identity of the limiting reactant.
Step 5 Use the moles of the limiting reactant to calculate the yields of products.

Now let's work a problem applying these steps.

EXAMPLE 4.9   **Calculating Limiting Reactants and Maximum Yields**

$CHCl_3$ is a solvent that also works as an anesthetic

CCl_4 once a dry-cleaning solvent, is seldom used because it damages the liver.

Problem: Chloroform ($CHCl_3$) reacts with chlorine to give carbon tetrachloride (CCl_4) and hydrogen chloride (HCl) by the following equation:

$$CHCl_3 + Cl_2 \longrightarrow CCl_4 + HCl$$

In one experiment the reactants were taken in a ratio of 1 to 1 *by mass* instead of 1 to 1 by moles: 25.0 g of $CHCl_3$ were mixed with 25.0 g of Cl_2. What is the limiting reactant? What is the maximum yield of CCl_4 in moles and grams?

Solution: Step 1. The balanced equation has been given.
Step 2. Formula weights: $CHCl_3$, 119; Cl_2, 70.9; CCl_4, 154.
Step 3. Move to the mole level for both reactants. Only at this level can we use the coefficients of the reactants to see which one limits the yield of products.

Chloroform: $25.0 \text{ g CHCl}_3 \times \dfrac{1 \text{ mol CHCl}_3}{119 \text{ g CHCl}_3} = 0.210 \text{ mol CHCl}_3$ taken

Chlorine: $25.0 \text{ g Cl}_2 \times \dfrac{1 \text{ mol Cl}_2}{70.9 \text{ g Cl}_2} = 0.353 \text{ mol Cl}_2$ taken

Step 4. The coefficients of the reactants are 1 to 1 in the equation — 1 mol of $CHCl_3$ to 1 mol of Cl_2. Therefore, 0.210 mol of $CHCl_3$ would need only 0.210 mol of Cl_2, *which is less than the 0.353 mol of Cl_2 actually taken.* There is plenty of chlorine. The limiting reactant, then, is $CHCl_3$ because when it is gone no more products can be formed.

Step 5. Since there is a 1 to 1 ratio between $CHCl_3$, the limiting reactant, and CCl_4 in this reaction, the maximum or theoretical yield of CCl_4 is 0.210 mol. The theoretical yield is always based on the limiting reactant, and 0.210 mol of that reactant was taken. The theoretical yield of CCl_4 in grams is found by:

$$0.210 \text{ mol CCl}_4 \times \dfrac{154 \text{ g CCl}_4}{1 \text{ mol CCl}_4} = 32.3 \text{ g CCl}_4$$

The maximum yield of CCl_4 in this reaction is 32.3 g (0.210 mol).

EXAMPLE 4.10 **Calculating Limiting Reactants and Maximum Yields**

Problem: Aluminum chloride, Al_2Cl_6, is used to make the active ingredient in most antiperspirants. It can be made by the reaction of aluminum with chlorine according to the equation:

$$2Al + 3Cl_2 \longrightarrow Al_2Cl_6$$

What is the limiting reactant if 20.0 g of Al and 30.0 g of Cl_2 are used, and how much Al_2Cl_6 can form?

Solution: Step 1. We use the balanced equation given above.

Step 2. Formula weights: Al, 27.0; Cl_2, 70.9; Al_2Cl_6, 267.

Step 3. Convert what is given into moles:

Aluminum: $20.0 \text{ g Al} \times \dfrac{1 \text{ mol Al}}{27.0 \text{ g Al}} = 0.741 \text{ mol Al}$

Chlorine: $30.0 \text{ g Cl}_2 \times \dfrac{1 \text{ mol Cl}_2}{70.9 \text{ g Cl}_2} = 0.423 \text{ mol Cl}_2$

Step 4. The coefficients of the equation tell us that 2 mol of Al are needed for 3 mol of Cl_2. We have to use this ratio to compare the actual moles of reactants with each other in Step 3 until we can decide which reactant is the limiting reactant. Let's see how much Cl_2 is actually needed for 0.741 mol of Al (20.0 g):

$$0.741 \text{ mol Al} \times \dfrac{3 \text{ mol Cl}_2}{2 \text{ mol Al}} = 1.11 \text{ mol Cl}_2$$

In other words, 0.741 mol of Al would require 1.11 mol of Cl_2 to react completely. Obviously, the 0.423 mol of Cl_2 actually supplied falls far short of this. Therefore, chlorine is the limiting reactant. Let's check this by finding out how much aluminum would be needed for 0.473 mol of Cl_2 (30.0 g):

$$0.423 \text{ mol Cl}_2 \times \dfrac{2 \text{ mol Al}}{3 \text{ mol Cl}_2} = 0.282 \text{ mol Al}$$

In other words, the chlorine we have available would need only 0.282 mol of Al, far less than the 0.741 mol of Al available. This verifies that Cl_2 is the limiting reactant.

Step 5. Calculating the yield. Since chlorine is the limiting reagent, we calculate the yield of Al_2Cl_6 using the moles of chlorine consumed. The balanced equation tells us that 1 mol of Al_2Cl_6 can be produced from 3 mol of Cl_2; therefore:

$$0.423 \text{ mol } Cl_2 \times \frac{1 \text{ mol } Al_2Cl_6}{3 \text{ mol } Cl_2} = 0.141 \text{ mol } Al_2Cl_6$$

We convert the answer to grams:

$$0.141 \text{ mol } Al_2Cl_3 \times \frac{267 \text{ g } Al_2Cl_6}{1 \text{ mol } Al_2Cl_6} = 37.6 \text{ g } Al_2Cl_6 \qquad \text{(theoretical yield)}$$

The maximum yield of Al_2Cl_6 is 37.6 g.

EXERCISE 4.17 What is the limiting reactant in an experiment in which 10.0 g of sodium and 20.0 g of chlorine, Cl_2, are mixed to produce sodium chloride, NaCl?

4.5 REACTIONS IN SOLUTION

Whenever possible, reactions are carried out with all of the reactants in the same fluid phase

Dalton's theory and the kinetic model imply that the particles of reacting substances can react with each other only if they collide. For collisions to occur, the particles must obviously be free to move around. Only in the liquid and gaseous states is such free motion possible. When working with solids, therefore, we normally try to find a solvent for the solid reactants and put them into solution before mixing them. This makes it necessary for us to learn at this time some of the terms used in describing solutions. (We will discuss properties of solutions in far greater detail in Chapter Eleven.)

A **solution** is a uniform mixture of particles of the atomic or molecular size. A minimum of two substances are present. One is called the solvent and all others are called the solutes. The **solvent** is the medium into which the other substances are mixed or dissolved. The solvent is usually a liquid such as water. A **solute** is any substance dissolved by the solvent. It may originally have been a solid, a liquid, or a gas. We'll confine our attention in this section to **aqueous solutions,** solutions for which the solvent is water. In an aqueous solution of sugar, sugar is the solute and water is the solvent. Club soda is a solution of a gas, carbon dioxide (the solute), in water (the solvent). Antifreeze is mostly a solution of the liquid ethylene glycol (the solute) in water. (If we mix two liquids, which one is called the solute and which the solvent isn't important. Usually the liquid present in the greater amount is arbitrarily called the solvent.)

Several terms are used to describe solutions. A **dilute solution** is one in which the ratio of solute to solvent is very small, for example, a few crystals of sugar in a glass of water. In a **concentrated solution** the ratio of solute to solvent is large. Molasses is a concentrated solution of sugar and water. A **saturated solution** is one in which no more solute will dissolve at a particular temperature. It will just sit at the bottom of the container. In an **unsaturated solution,** the ratio of solute to solvent is lower than that of the saturated solution of the same solute. If more solute is added, some of it will dissolve. It's not easy but sometimes a **supersaturated solution** can be made. This is an unstable system in which the ratio of dissolved solute to solvent is higher than that of a saturated solution.

We can sometimes make a supersaturated solution by carefully cooling a saturated solution. Usually, the cooler a solution, the lower is the maximum possible ratio of solute to solvent. If we are very careful—using dust- and lint-free solvent and avoiding any shaking or stirring of the solution—the solute that should

a   b   c   d

Figure 4.2
Supersaturation. (*a*) A supersaturated solution. (*b*) A small seed crystal is added. (*c* and *d*) Excess solute quickly precipitates with its crystals growing away from the seed.

come out of solution as we cool it stays dissolved. However, if we tap the container, scratch its inner wall with a stirring rod, or add a "seed" crystal of the pure solute, the excess solute usually separates from the solution immediately. This event can be very dramatic and pretty to observe (see Figure 4.2). The formation of a solid from a solution is called **precipitation** and the solid itself is called the **precipitate.**

The ratio of solute to solvent necessary to prepare a saturated solution at a given temperature is called the **solubility** of that solute in that solvent. Table 4.1 gives some examples that show how widely solubilities can vary. (Notice that a saturated solution can still be relatively dilute. For example, only a very small amount of lead sulfate is dissolved in 100 g of water in a saturated solution.)

Table 4.1
Solubilities of Some Compounds in Water

Substance	Formula	Solubility (g/100 g water)[a]
Ammonium chloride	NH_4Cl	29.7 (0 °C)
Boric acid	H_3BO_3	6.35 (30 °C)
Calcium chloride	$CaCl_2$	74.5 (20 °C)
Copper sulfide	CuS	3.3×10^{-5} (18 °C)
Lead sulfate	$PbSO_4$	4.3×10^{-3} (25 °C)
Sodium hydroxide	$NaOH$	42 (0 °C)
		347 (100 °C)

[a] At the temperature given in parentheses.

4.6 MOLAR CONCENTRATION

The unit of moles per liter is one of the most useful for describing concentrations

Sometimes a particular reactant is available from the stockroom only as a solution. Assuming that we have calculated how many grams or how many moles we want, how do we proceed if the substance is already dissolved in the solvent we want to use? It would be nice to have a way to get the amount of *solute* we need simply by pouring out a certain volume of the solution.

Our problem is solved if we know the concentration of the solution. The **concentration** of a solution is a quantitative description of the relative amounts of solute and solution, and concentration is always given as a ratio. For stoichiometric calculations, the best way to describe the concentration of a solution is by the ratio of the moles of solute per liter of solution. The special name for this ratio is **molar concentration,** or **molarity,** abbreviated M.

$$M = \frac{\text{mol solute}}{\text{liter solution}}$$

Thus a bottle with the label "0.10 M NaCl" contains a solution of sodium chloride with a concentration of 0.10 mol of sodium chloride per liter of solution.

EXAMPLE 4.11   Using Molar Concentrations in Mole Problems

Problem: A student needs 0.25 mol of NaCl and all that is available is a solution labeled "0.40 M NaCl." What volume of the solution should be used? (Give the answer in milliliters.)

Solution: The best approach is to use the information on the label of the bottle, the molar concentration, to construct ratios that can be used in a factor-label approach. The label tells us we could use either of these two ratios:

$$\frac{0.40 \text{ mol NaCl}}{1 \text{ liter solution}} \quad \text{or} \quad \frac{1 \text{ liter solution}}{0.40 \text{ mol NaCl}}$$

Apply the factor-label method. Since we want our answer in a unit of volume, we must use the second ratio.

$$0.25 \text{ mol NaCl} \times \frac{1 \text{ liter solution}}{0.40 \text{ mol NaCl}} = 0.63 \text{ liter solution}$$

Each liter has 1000 ml, therefore:

$$0.63 \text{ liter solution} \times \frac{1000 \text{ ml}}{1 \text{ liter}} = 630 \text{ ml solution}$$

EXERCISE 4.18   A glucose solution with a molar concentration of 0.20 M is available. What volume of this solution must be measured to obtain 0.0010 mol of glucose. Give your answer in milliliters.

Another common problem involving molar concentrations is to calculate the grams of solute needed to make a given volume of solution having a specified molarity.

EXAMPLE 4.12   Preparing Solutions with Specific Molar Concentrations

Problem: How would you prepare 500 ml of 0.150 M Na_2CO_3 solution?

Solution: Although the problem is stated in a way that it would normally arise in the lab, what we really need to know is how many *grams* of Na_2CO_3 (sodium carbonate) are in 500 ml of 0.150 M Na_2CO_3 solution. The label may read in M (moles/liter) but the balance reads in grams. And to make the solution we have to weigh out grams first. Before we can calculate grams we have to calculate moles—in this example, how many moles of Na_2CO_3 are present in 500 ml of 0.150 M Na_2CO_3 solution? The key step in the calculation is to "translate the label," which means to use the concentration stated on the bottle's label to set up conversion factors. (It is also a good idea to substitute "1000 ml" for "1 liter" whenever the desired volume is in milliliters instead of liters.) For example, "0.150 M Na$_2$CO$_3$" lets us write:

$$\frac{0.150 \text{ mol Na}_2\text{CO}_3}{1000 \text{ ml Na}_2\text{CO}_3(aq)} \quad \text{or} \quad \frac{1000 \text{ ml Na}_2\text{CO}_3(aq)}{0.150 \text{ mol Na}_2\text{CO}_3}$$

To find the *moles* of Na$_2$CO$_3$ we need for 500 ml of Na$_2$CO$_3$ solution we use the factor-label method:

$$500 \text{ ml Na}_2\text{CO}_3(aq) \times \frac{0.150 \text{ mol Na}_2\text{CO}_3}{1000 \text{ ml Na}_2\text{CO}_3(aq)} = 0.0750 \text{ mol Na}_2\text{CO}_3$$

Now we convert moles of Na$_2$CO$_3$ to grams of Na$_2$CO$_3$ using the formula weight of Na$_2$CO$_3$, which is 106.

$$0.0750 \text{ mol Na}_2\text{CO}_3 \times \frac{106 \text{ g Na}_2\text{CO}_3}{1 \text{ mol Na}_2\text{CO}_3} = 7.95 \text{ g Na}_2\text{CO}_3$$

The answer, then, is that we need 7.95 g Na$_2$CO$_3$ to make 500 ml of 0.150 M Na$_2$CO$_3$.

EXERCISE 4.19   How would you prepare 250 ml of 0.200 M NaHCO$_3$?

Special pieces of glassware called volumetric flasks are available for preparing solutions of known concentrations. One was pictured in Chapter One, Figure 1.5. A line etched on the narrow neck of a volumetric flask shows where the liquid level should be for the flask to contain exactly the volume stated on its side. Since liquids expand or contract slightly with changes in temperature, the given capacity of the flask is correct only for the temperature stated by an etching on the flask. In the most careful work, the capacity of a flask is checked against precise standards. Any laboratory operation in which a stated capacity or a stated mass is checked against precisely known standards is called **calibration.** Figure 4.3 shows how a volumetric flask is used to make a solution of known molarity.

Figure 4.3
Preparation of 1 liter of a 1 M solution. The flask used is a volumetric flask and has an etched line on its neck marking a volume of 1 liter.

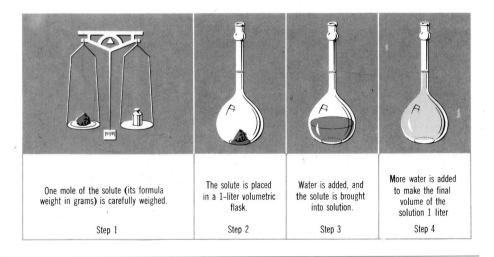

One mole of the solute (its formula weight in grams) is carefully weighed.	The solute is placed in a 1-liter volumetric flask.	Water is added, and the solute is brought into solution.	More water is added to make the final volume of the solution 1 liter
Step 1	Step 2	Step 3	Step 4

EXERCISE 4.20   How would you prepare 250 ml of a solution of glucose, C$_6$H$_{12}$O$_6$, that has a concentration of 0.0500 M?

Now that we know what molarity means and how to make solutions of known molar concentrations, let's learn how to do mole calculations when solutions are used, not the pure substances.

EXAMPLE 4.13 **Stoichiometric Calculations When Reactants Are in Solution**

Problem: What volume of 0.556 *M* HCl has enough hydrochloric acid (HCl) to combine exactly with 25.4 ml of aqueous sodium hydroxide (NaOH) with a concentration of 0.458 *M*. The equation for the reaction is:

$$HCl(aq) + NaOH(aq) \longrightarrow NaCl(aq) + H_2O(\ell)$$

Solution: This kind of a problem is nothing more than a variation of the problem of matching the moles given of one chemical to the moles of another according to the ratio of their coefficients in the balanced equation. What is different is that the chemicals are provided already in solution, and after finding the *moles* needed we have to calculate what *volume* contains these moles. So we first find the moles of NaOH in 25.4 ml of 0.458 *M* NaOH. The molarity of the solution gives us these conversion factors:

$$\frac{0.458 \text{ mol NaOH}}{1000 \text{ ml NaOH}(aq)} \quad \text{or} \quad \frac{1000 \text{ ml NaOH}(aq)}{0.458 \text{ mol NaOH}}$$

To see how much NaOH is in 25.4 ml NaOH(*aq*) we use the first:

$$25.4 \text{ ml NaOH}(aq) \times \frac{0.458 \text{ mol NaOH}}{1000 \text{ ml NaOH}(aq)} = 0.0116 \text{ mol NaOH}$$

Now we use a ratio of coefficients from the balanced equation to calculate how much 0.0116 mol NaOH is equivalent to in moles of HCl.

$$0.0116 \text{ mol NaOH} \times \frac{1 \text{ mol HCl}}{1 \text{ mol NaOH}} = 0.0116 \text{ mol HCl}$$

a ratio of the coefficients in the chemical equation .

But this 0.0116 mol HCl comes only in a solution with a concentration of 0.556 *M*. This molarity gives us the following conversion factors:

$$\frac{0.556 \text{ mol HCl}}{1000 \text{ ml HCl}(aq)} \quad \text{and} \quad \frac{1000 \text{ ml HCl}(aq)}{0.556 \text{ mol HCl}}$$

Therefore, to find the volume of HCl that holds 0.0116 mol HCl we do the following:

$$0.0116 \text{ mol HCl} \times \frac{1000 \text{ ml HCl}(aq)}{0.556 \text{ mol HCl}} = 20.9 \text{ ml HCl}(aq)$$

EXAMPLE 4.14 **Stoichiometric Calculations When Reactants Are in Solution**

Problem: How many milliliters of 0.114 *M* H_2SO_4 solution provide the sulfuric acid required to react with the sodium hydroxide in 32.2 ml of 0.122 *M* NaOH according to the following equation?

(*ℓ*) liquid

$$H_2SO_4(aq) + 2NaOH(aq) \longrightarrow Na_2SO_4(aq) + 2H_2O(\ell)$$

Solution: The molarity of the NaOH solution gives us these conversion factors:

$$\frac{0.122 \text{ mol NaOH}}{1000 \text{ ml NaOH}(aq)} \quad \text{and} \quad \frac{1000 \text{ ml NaOH}(aq)}{0.122 \text{ mol NaOH}}$$

Therefore, in 32.2 ml of 0.122 *M* NaOH there are:

$$32.2 \text{ ml NaOH}(aq) \times \frac{0.122 \text{ mol NaOH}}{1000 \text{ ml NaOH}(aq)} = 0.00393 \text{ mol NaOH}$$

How many moles of sulfuric acid are chemically equivalent to this much NaOH *in the reaction given?* To find out we multiply the moles of NaOH by a conversion factor based on the coefficients in the equation:

$$0.00393 \text{ mol NaOH} \times \frac{1 \text{ mol } H_2SO_4}{2 \text{ mol NaOH}} = 0.00197 \text{ mol } H_2SO_4$$

Now we calculate what volume of $H_2SO_4(aq)$ contains 0.00197 mol of H_2SO_4; and we use the given concentration of the $H_2SO_4(aq)$ to set up the correct conversion factor. A concentration of 0.114 *M* H_2SO_4 means either;

$$\frac{0.114 \text{ mol } H_2SO_4}{1000 \text{ ml } H_2SO_4(aq)} \quad \text{or} \quad \frac{1000 \text{ ml } H_2SO_4(aq)}{0.114 \text{ mol } H_2SO_4}$$

Therefore:

$$0.00197 \text{ mol } H_2SO_4 \times \frac{1000 \text{ ml } H_2SO_4(aq)}{0.114 \text{ mol } H_2SO_4} = 17.3 \text{ ml } H_2SO_4(aq)$$

Thus 17.3 ml of 0.114 *M* H_2SO_4 will exactly neutralize 32.2 ml of 0.122 *M* NaOH according to the equation given.

EXERCISE 4.21   What volume of 0.337 *M* KOH (potassium hydroxide) provides enough solute to combine with the sulfuric acid (H_2SO_4) in 18.6 ml of 0.156 *M* H_2SO_4? The reaction is:

$$2KOH(aq) + H_2SO_4(aq) \longrightarrow K_2SO_4(aq) + 2H_2O(\ell)$$

The reactions used in Examples 4.13 and 4.14 and in Exercise 4.21 are examples of a very important general reaction introduced in Chapter One, the neutralization of an acid by a base. Some of the common properties of acids and bases were given in Section 1.16, and any reaction that destroys either an acid or a base is called **neutralization.** In the next exercise, for example, sodium carbonate is used to destroy or neutralize an acid.

EXERCISE 4.22   Hydrochloric acid will react with sodium carbonate (Na_2CO_3) as follows:

$$2HCl(aq) + Na_2CO_3(s) \longrightarrow 2NaCl(aq) + CO_2(g) + H_2O(\ell)$$
$$\text{hydrochloric} \quad \text{sodium} \quad \text{sodium} \quad \text{carbon} \quad \text{water}$$
$$\text{acid} \quad \text{carbonate} \quad \text{chloride} \quad \text{dioxide}$$

What volume of 0.224 *M* HCl is needed to react with the solute in 24.2 ml of 0.284 *M* Na_2CO_3?

Another important kind of molarity calculation is done when a solution of very accurately known concentration—a **standard solution**—is used to determine the concentration of another solution. We might have, for example, a sodium hydroxide solution, but we don't know its molar concentration—it is the "unknown" solution. However, if we have a standard solution of sulfuric acid, we can use it to determine the molarity of the "unknown." The technique for this analysis, called

Special Topic 4.1 Titration

Titration is a procedure for adding a chemical dissolved in a solution to another substance, also in solution, until some visual effect such as a color change signals that the two compounds have reacted in a stoichiometric ratio. The apparatus is shown in the photograph. The long tube is a buret, and it is marked for volumes usually in increments of 0.10 ml. The stopcock at the bottom of the buret permits the analyst to control the amount of solution run out of the buret into the receiving flask. The substance being analyzed, the "unknown," is in this flask. For example, the 0.118 M H_2SO_4 solution of Example 4.15 would be placed in the buret. A carefully measured volume of the sodium hydroxide solution of unknown concentration would be in the flask. (Usually this solution is run into the flask from a second buret.) Then the analyst allows the sulfuric acid solution to flow slowly into the "unknown" until one last drop completes the neutralization of the sodium hydroxide. To tell when the stoichiometrically correct amount of acid has been added, the analyst adds a drop or two of an indicator solution to the unknown before starting the titration. An acid-base indicator is a dye that has one color in an acidic medium and a different color in a basic solution. Two common examples are litmus, which is red in acid and blue in base, and phenolphthalein, which is colorless in acid and pink in base. (The theory of acid-base indicators is discussed in Chapter Seven-

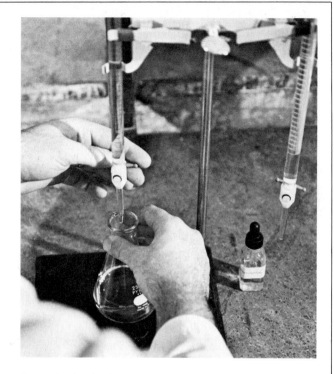

Apparatus for titration.

teen.) When the analyst observes the change from one color to the other, the titration is stopped. Usually that color transition takes place with the addition of only one (last) drop of acid.

titration, is described in Special Topic 4.1 (and in much greater detail in your lab manual). We will go much more into the theory of acid-base analyses in Chapter Seventeen, but we already have the background for doing the calculations.

EXAMPLE 4.15   Using Titration Data to Calculate Molarities

Problem: What is the molar concentration of a sodium hydroxide solution if 32.5 ml of this "unknown" require exactly 14.2 ml of 0.118 M H_2SO_4 to be neutralized according to the following equation?

$$2NaOH + H_2SO_4 \longrightarrow Na_2SO_4 + 2H_2O$$

Solution: We are asked to calculate a value for the molar concentration of the sodium hydroxide. Therefore, we must use the data to find the ratio of moles NaOH to 1000 ml of NaOH(aq). We know how many moles of H_2SO_4 were consumed:

$$14.2 \text{ ml } H_2SO_4(aq) \times \frac{0.118 \text{ mol } H_2SO_4}{1000 \text{ ml } H_2SO_4(aq)} = 0.00168 \text{ mol } H_2SO_4$$

Next, we will use a ratio of coefficients in the balanced equation to calculate how many moles of NaOH reacted with 0.00168 mol H_2SO_4.

$$0.00168 \text{ mol } H_2SO_4 \times \frac{2 \text{ mol NaOH}}{1 \text{ mol } H_2SO_4} = 0.00336 \text{ mol NaOH}$$

This much NaOH was dissolved in 32.5 ml of solution. This volume corresponds to 0.0325 liter of solution, so the ratio of moles to liters, the molarity of the NaOH solution, is

$$\frac{0.00336 \text{ mol NaOH}}{0.0325 \text{ liter}} = 0.103 \ M \text{ NaOH}(aq), \text{ our answer}$$

In other words the NaOH dissolved in 32.5 ml of 0.103 M NaOH(aq) is neutralized by the sulfuric acid dissolved in 14.2 ml of 0.118 M H$_2$SO$_4$(aq)

EXERCISE 4.23   What is the molar concentration of a dilute solution of nitric acid if 25.8 ml of this solution exactly neutralized 26.4 ml of 0.108 M KOH according to the following equation.

$$\text{HNO}_3(aq) + \quad \text{KOH}(aq) \longrightarrow \quad \text{KNO}_3(aq) \ + \text{H}_2\text{O}(\ell)$$

nitric	potassium	potassium
acid	hydroxide	nitrate

4.7 PERCENTAGE CONCENTRATION

A percentage concentration usually gives the number of grams of solute in 100 grams of solution, but other meanings are possible

Many times in the lab we might need just a few drops of a solution for a quick test tube test. Since there is no reason to do molarity calculations for such tests, there is no compelling need to take the trouble to prepare the test reagent according to molarity. Yet it's still a good idea to have some idea of its concentration, and knowing its percentage concentration both fills that need and allows for an easier job of making the test reagent.

Several varieties of "percent" concentrations are in use.[3] Only two, however, weight/weight percent and volume/volume percent are true percentage quantities. Others such as weight/volume percent and milligram percent will be defined here because of their widespread use, particularly in the biological sciences.

Weight/weight percent (w/w percent)
In most chemical laboratories "percent concentration" means **weight/weight percent**—the number of grams of solute per 100 grams of solution (not solvent, but *solution*). For example, a solution with the label "0.9% NaCl" means that there is a ratio of 0.9 g of NaCl per 100 g of solution.

EXAMPLE 4.16   **Using Weight/Weight Percents**

Problem: How much of a 4.00% (w/w) solution of salt is needed to obtain 0.500 g salt?

Solution: With all problems involving concentrations, begin by translating the label into its full units. The label given here means either one of two ratios, which may be used as needed in calculations as conversion factors:

$$\frac{4.00 \text{ g salt}}{100 \text{ g solution}} \quad \text{or} \quad \frac{100 \text{ g solution}}{4.00 \text{ g salt}}$$

---

[3] We will now yield to the most widespread practice among scientists in chemistry and biology of using "percent" when "percentage" is correct.

We want 0.500 g of salt, so we need to find out how many grams of the solution are needed. The second conversion factor applies because then "g salt" will cancel leaving units of "g solution":

$$0.500 \text{ g salt} \times \frac{100 \text{ g solution}}{4.00 \text{ g salt}} = 12.5 \text{ g solution}$$

The answer means that 12.5 g of the solution contains the desired 0.500 g of the salt when the concentration is 4.00% (w/w).

EXERCISE 4.24   Hydrochloric acid can be purchased from chemical supply houses as a solution that is 37% (w/w) HCl. What mass of that solution will contain 7.5 g HCl?

EXAMPLE 4.17   **Preparing Weight/Weight Percent Solutions**

Problem: "White" vinegar can be made by preparing a 5.0% (w/w) solution of acetic acid in water. How would you make 500 g of such a solution?

Solution: The correct mass of acetic acid has first to be calculated and then that quantity dissolved in water. Enough water is added until the final mass of the solution is 500 g. This kind of problem therefore comes down to calculating the needed mass of solute.

Translate the label; "5.0% (w/w) acetic acid" gives the two possible conversion factors:

$$\frac{5.0 \text{ g acetic acid}}{100 \text{ g solution}} \quad \text{or} \quad \frac{100 \text{ g solution}}{5.0 \text{ g acetic acid}}$$

Use the first factor of these two to get "g solution" to cancel and leave our answer in "g acetic acid":

$$500 \text{ g solution} \times \frac{5.0 \text{ g acetic acid}}{100 \text{ g solution}} = 25 \text{ g acetic acid}$$

If you dissolve 25 g of acetic acid in enough water to make the final mass of the solution 500 g, you may put "5.0% (w/w) acetic acid" on the label of the bottle.

EXERCISE 4.25   One common laboratory reagent is 10% (w/w) NaOH solution. How would you prepare 750 g of this solution?

Volume/volume percent (v/v percent)

When both solute and solvent are liquids, sometimes it is convenient to describe concentrations as **percents by volume** (v/v percent)—the number of volumes of the solute in 100 volumes of solution. ("Volume" may be any volume unit provided the same unit is used for both solute and solution.)

EXAMPLE 4.18   **Using Volume/Volume Percents**

Problem: A 40% (v/v) solution of ethylene glycol in water gives antifreeze protection to a car's cooling system to −24 °C (−12 °F). What volume of ethylene glycol would have to be used to make 5 quarts of this solution?

Solution: Translate the label; "40% (v/v) ethylene glycol" means:

$$\frac{40 \text{ vol ethylene glycol}}{100 \text{ vol solution}} \quad \text{or} \quad \frac{100 \text{ vol solution}}{40 \text{ vol ethylene glycol}}$$

In these factors "vol" may be any volume unit we want—ml, liter, quart, gallon, tank car or whatever we desire. The problem specifies "quarts." Since we want the answer in "quarts ethylene glycol," we have to use the first conversion factor:

$$5 \text{ quarts solution} \times \frac{40 \text{ quarts ethylene glycol}}{100 \text{ quarts solution}} = 2 \text{ quarts ethylene glucol}$$

If you dissolve 2 quarts ethylene glycol in enough water to make the final volume of the solution 5 quarts you can put "40% (v/v) ethylene glycol" on the label of the bottle.

EXERCISE 4.26 A 35% (v/v) solution of wood alcohol (methyl alcohol) in water gives substantially the same antifreeze protection as 40% ethylene glycol (referring to Example 4.17). How would you prepare 20 quarts of this solution? (Because wood alcohol boils away so much more readily than ethylene glycol, its solution in water is called "temporary antifreeze." It has to be frequently checked for strength.)

Other forms of percent concentration

In some scientific specialties scientists have developed hybrid expressions for percents. They are "hybrid" because their units don't actually cancel (as they ought to) when the percent is calculated. In some clinical areas, for example, you may see a **weight/volume percent,** meaning the number of grams of solute in 100 ml of solution (not solvent, *solution*). Thus 5% (w/v) glucose means 5 g glucose in 100 ml solution.

W/v percent is most common in the health sciences, but the unit of deciliter, dl, is used for 100 ml. Thus 5% (w/v) is written 5g/dl instead of 5g/100 ml to save space on clinical report sheets.

When the solute concentration is very low, the percent may be called -milligram-percent, meaning the number of milligrams of solute in 100 ml of solution. Since the solvent is usually water and since 1 ml of water is essentially the same as 1 gram of water, there is no meaningful distinction between w/w and w/v percents for very dilute solutions. Thus "50 mg-percent morphine sulfate (in water)" means either 50 mg of morphine sulfate per 100 ml of solution or 50 mg of morphine sulfate per 100 g of solution.

Problems involving these hybrid percents are worked by the same strategies used for w/w and v/v percents. The first step is to expand and translate the label until the units of volume are identified clearly.

4.8 MAKING DILUTE SOLUTIONS FROM CONCENTRATED SOLUTIONS

When a specific amount of a concentrated solution is diluted, the original quantity of solute becomes part of a larger volume

Stock solutions of common reagents are often prepared in large volumes of set concentrations. For example, the chief supply of hydrochloric acid in your stockroom may be 1.00 *M* HCl solution. Your experiment, however, may call for a less concentrated solution. Therefore, we need to know how to prepare a dilute solution from one that is more concentrated. Once the calculations are completed, you take the calculated volume of the concentrated solution, place it in a volumetric flask corresponding to the desired final volume, and then add solvent portion by portion, mixing the contents as you do, until the desired final volume of solution is reached.

To make the calculation, we begin with the final concentration and the desired final volume. We use these to find the moles of *solute* that will be present in the

These are the official abbreviations:

dil = dilute
concd = concentrated
concn = concentration

more dilute solution. If the concentration is in moles/liter, we can calculate the amount of solute by multiplying the volume by the molarity:

$$(\text{liters})_{dil} \times \left(\frac{\text{moles}}{\text{liter}}\right)_{dil} = \text{moles of solute to be in the final solution} \quad (4.1)$$

$$\uparrow \text{molarity}_{dil}$$

We get this quantity of solute by taking enough of the concentrated solution to deliver it:

$$(\text{liters})_{concd} \times \left(\frac{\text{moles}}{\text{liter}}\right)_{concd} = \text{moles of solute to be in the final solution} \quad (4.2)$$

$$\uparrow \text{molarity}_{concd}$$

Since the left sides of both Equations 4.1 and 4.2 equal the same thing—namely, the moles of solute that are to be present in the final solution—these two sides equal each other.

$$(\text{liters})_{concd} \times (\text{molarity})_{concd} = (\text{liters})_{dil} \times (\text{molarity})_{dil} \quad (4.3)$$

The units for volume and concentration are the same on both sides of this equation, so they cancel. Therefore, the unit of volume need not be liters, which often is awkward. The milliliter or any other unit we please may be used as long as the same unit is used on both sides of the equation. For example, we can rewrite Equation 4.3 as follows:

$$\boxed{\text{ml}_{concd} \times M_{concd} = \text{ml}_{dil} \times M_{dil}} \quad (4.4)$$

EXAMPLE 4.19   Doing the Calculations for Making Dilutions

Problem: How could you prepare 100 ml of 0.10 M NaCl from 0.50 M NaCl?
Solution: Step 1. Assemble the data.

$$\text{vol}_{concd} = ? \qquad \text{vol}_{dil} = 100 \text{ ml}$$

$$\text{concn}_{concd} = 0.50 \ M \qquad \text{concn}_{dil} = 0.1 \ M$$

Step 2. Use Equation 4.4

$$(\text{vol})_{concd}(0.50 \ M) = (100 \text{ ml})(0.10 \ M)$$

$$\text{vol}_{concd} = 20 \text{ ml}$$

This answer means that you would place 20 ml of 0.50 M NaCl into a graduated cylinder or into a 100 ml volumetric flask and then add water until the final volume was 100 ml. See Figure 4.4.

EXERCISE 4.27   Describe how you would make 500 ml of 0.20 M NaOH from 0.50 M NaOH.

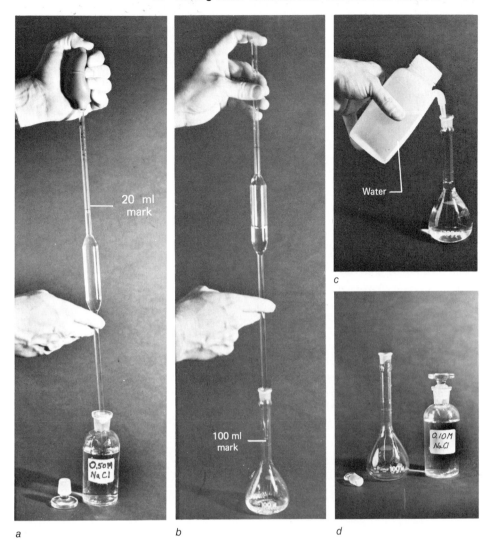

20 ml mark

0.50M NaCl

Water

c

100 ml mark

a

b

0.10M NaCl

d

Figure 4.4
Preparing solutions by dilution. (*a*) The calculated volume of the more concentrated solution is withdrawn from the stock solution. (*b*) It is placed in a volumetric flask of the correct capacity. (*c*) and (*d*) Water (or other solvent) is added, and the contents are frequently agitated to make them mix until the final volume (indicated by the etched line) is reached. Then the solution is transferred to a dry reagent bottle for storage.

EXERCISE 4.28   How many milliliters of 1.00 *M* HCl have to be diluted to make 250 ml of 0.125 *M* HCl?

EXERCISE 4.29   What is the molarity of a solution prepared by diluting 250 ml of 0.600 *M* NaOH to a volume of 600 ml?

EXERCISE 4.30   Very dilute solutions are often needed for measuring certain optical properties of solutes. If 1.00 ml of a solution of a dye having a concentration of 0.100 *M* is diluted to 50.0 ml, and then 1.00 ml of the resulting solution is taken and diluted to 100 ml, what is the concentration of the final solution?

SUMMARY

Stoichiometry. Because equal numbers of moles contain equal numbers of molecules, the coefficients in a balanced equation provide the key to calculating the quantities of chemicals involved in a reaction. At one level, these numbers give ratios in terms of molecules (or atoms or other kinds of formula units). For laboratory work, they also give the same ratios in terms of moles. When balancing an equation, only the coefficients may be adjusted; the subscripts within the formulas cannot be changed.

Mole Relationships. If "x" is the coefficient of substance "A" and "y" is the coefficient of substance "B" in some equation, for example,

$$xA + mW + \cdots \longrightarrow yB + nZ + \cdots$$

then the relationship of A to B in terms of moles is by either of two conversion factors:

$$\frac{x \text{ mol } A}{y \text{ mol } B} \quad \text{or} \quad \frac{y \text{ mol } B}{x \text{ mol } A}$$

Regardless of the units in which substance A are taken—atoms, molecules, liters (if a gas), grams, or volume of some solution—the quantity of substance B that is related to A by this equation can be found only by working the problem at the mole level. Any other units must first be converted to their equivalent number of moles. The many ways of going from any unit of A to any unit of B are outlined below.

Yields of Products. When one of two or more reactants is taken in a mole quantity that is less than the stoichiometry of the reaction requires, it is a limiting reactant and the calculation of yield is based on that chemical.

Stoichiometry When Reagents Are in Solution. Solutions provide a fluid medium that allows particles to collide and react. Several common qualitative terms are used to describe concentrations of solutions (the ratio of solute to solution volume)—dilute, concentrated, saturated, unsaturated, supersaturated—but for calculating the mole quantities of reactants, the molar concentration is the best description. It describes the moles of solute per liter of solution. In working mole problems that involve molar concentrations, always write out the units of molarity (including the formula of the solute). This is what "translate the label" means. The same advice holds for working with any of the percent concentrations. The general equation for calculating how to make a dilute solution from a concentrated solution is:

$$ml_{concd} \times M_{concd} = ml_{dil} + M_{dil}$$

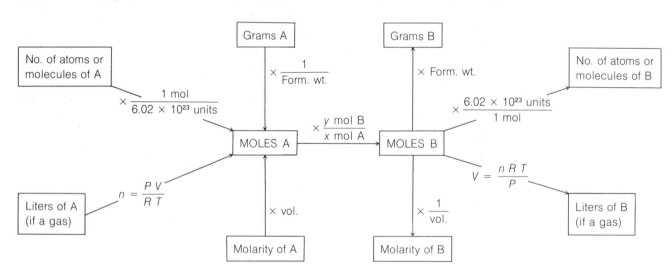

INDEX TO EXERCISES, QUESTIONS, AND PROBLEMS
(Those numbered above 52 are review problems.)

REVIEW QUESTIONS

4.31. Why is a balanced equation important in experimental work?

4.32. What important fact or principle makes it possible for us to think of coefficients either as representing the numbers of individual molecules (or atoms or other particles as the case may be) of substances or as representing numbers of *moles* of substances?

4.33. If an experimenter carelessly thinks that the coefficients of a balanced equation give proportions by masses, and carries out an experiment on that basis, there will usually be an insufficient amount of one reactant. What in general terms is that reactant called?

4.34. If two substances react completely on both a 1 to 1 ratio by mass and a 1 to 1 ratio by moles, what must be true about these substances?

4.35. What do we call the ratio, expressed as a percent, of the amount of product actually isolated from a reaction to the amount predicted by the stoichiometry of the reaction?

4.36. In a solution made by mixing sugar and water, which is the solute and which is the solvent.?

4.37. What is the solvent in an aqueous solution?

4.38. Can a saturated solution be dilute? Use data in Table 4.1 to cite an example.

4.39. Can a concentrated solution be unsaturated? Using Table 4.1, supply an illustration.

4.40. In what way is the concept of *concentration* different from that of *solubility*?

4.41. Write out in your own words all the information given on this label found on a reagent bottle: "0.500 *M* NaCl."

4.42. Does the label "0.500 *M* NaCl" tell how much solution is in the bottle?

4.43. Does the label "0.500 *M* NaCl" tell how much water was used to prepare, say, 1 liter of the solution?

4.44. Why is it unimportant from the standpoint of working mole problems to know exactly how much *solvent* was used to prepare a given quantity of a solution having some specific molar concentration?

4.45. What is the difference between "1 mol NaCl" and "1 *M* NaCl"?

4.46. What is the difference between a molecule, a mole, and a molar concentration?

4.47. Suppose you have prepared a 0.20 *M* salt solution in a 250 ml volumetric flask and then accidentally spill some of it. What happens to the concentration of the solution in the bottle?

4.48. Chemical handbooks have extensive tables giving to several significant figures the density of pure water at each degree Celsius between its freezing and boiling points. Well-equipped laboratories have constant-temperature baths, which are large vats of water that can be kept at a given temperature to within 0.1 degree. Given enough time, anything put into a constant-temperature bath comes to the bath temperature. These labs, of course, have high precision analytical balances. Describe in your own words how you could go into such a laboratory and check the calibration of a volumetric flask with the legend "100 ml at 20 °C" etched on its side.

4.49. How many millimoles make one mole?

4.50. Prove that the numerical value of a concentration given in mol/liter is the same as if it were stated in millimoles/milliliter.

4.51. If a solution is labeled "4.0% (w/w) NaCl," what specifically are the units for that concentration?

4.52. What are the specific units for each of the following concentrations?
(a) 10.0% (v/v) ethyl alcohol
(b) 9 mg% quinine sulfate
(c) 0.90% (w/v) NaCl

REVIEW PROBLEMS

4.53. A mixture of 0.1 mol of magnesium and 0.1 mol of chlorine reacted completely to form magnesium chloride according to the equation: $Mg + Cl_2 \rightarrow MgCl_2$.
(a) What information describes the *stoichiometry* of this reaction?
(b) What information gives the *scale* of the reaction?

4.54. When sulfur trioxide (SO_3), which is present in smoggy air in trace concentrations, reacts with water, sulfuric acid (H_2SO_4), a very corrosive acid, forms. Write a balanced equation and describe its stoichiometry in words.

4.55. Write the equation that expresses in acceptable chemical shorthand the information given in this statement: "Iron can be made to react with oxygen to give iron oxide having the formula Fe_2O_3."

4.56. A chemist describes a particular experiment this way: "0.0400 mol of H_2O_2 decomposed into 0.0400 mol of H_2O and 0.0200 mol O_2." Express the chemistry of this reaction by a conventional equation.

4.57. Balance each of the following equations.
(a) $Ca(OH)_2 + HCl \longrightarrow CaCl_2 + H_2O$

(b) $AgNO_3 + CaCl_2 \longrightarrow Ca(NO_3)_2 + AgCl$
(c) $Fe_2O_3 + C \longrightarrow Fe + CO_2$
(d) $NaHCO_3 + H_2SO_4 \longrightarrow Na_2SO_4 + H_2O + CO_2$
(e) $C_4H_{10} + O_2 \longrightarrow CO_2 + H_2O$

4.58. Balance each of the following equations.
(a) $SO_2 + O_2 \longrightarrow SO_3$
(b) $P_4O_{10} + H_2O \longrightarrow H_3PO_4$
(c) $Pb(NO_3)_2 + Na_2SO_4 \longrightarrow PbSO_4 + NaNO_3$
(d) $Fe_2O_3 + H_2 \longrightarrow Fe + H_2O$
(e) $Al + H_2SO_4 \longrightarrow Al_2(SO_4)_3 + H_2$

4.59. Balance each of the following equations.
(a) $Mg(OH)_2 + HBr \longrightarrow MgBr_2 + H_2O$
(b) $HCl + Ca(OH)_2 \longrightarrow CaCl_2 + H_2O$
(c) $Al_2O_3 + H_2SO_4 \longrightarrow Al_2(SO_4)_3 + H_2O$
(d) $KHCO_3 + H_3PO_4 \longrightarrow K_2HPO_4 + H_2O + CO_2$
(e) $C_9H_{20} + O_2 \longrightarrow CO_2 + H_2O$

4.60. Balance each of the following equations.
(a) $CaO + HNO_3 \longrightarrow Ca(NO_3)_2 + H_2O$
(b) $Na_2CO_3 + Mg(NO_3)_2 \longrightarrow MgCO_3 + NaNO_3$
(c) $Al(NO_3)_3 + H_2SO_4 \longrightarrow Al_2(SO_4)_3 + HNO_3$
(d) $Mg(HCO_3)_2 + HCl \longrightarrow MgCl_2 + H_2O + CO_2$
(e) $C_3H_{10}O + O_2 \longrightarrow CO_2 + H_2O$

4.61. The octane present in gasoline burns according to the following equation:

$$2C_8H_{18} + 25O_2 \longrightarrow 16CO_2 + 18H_2O$$
octane

(a) How many moles of O_2 are needed to react fully with 4 mol of octane?
(b) How many moles of CO_2 can form from 1 mol of octane?
(c) How many moles of water are produced by the combustion of 6 mol of octane?
(d) If this reaction is to be used to synthesize 8 mol of CO_2, how many moles of oxygen are needed? How many moles of octane?

4.62. The alcohol in "gasohol" burns according to the following equation:

$$C_2H_6O + 3O_2 \longrightarrow 2CO_2 + 3H_2O$$

(a) If 25 mol of ethyl alcohol burn this way, how many moles of oxygen are needed?
(b) If 30 mol of oxygen are consumed by this reaction, how many moles of alcohol are used up? How many moles of carbon dioxide form?
(c) In one test 23 mol of carbon dioxide were produced by this reaction. How many moles of oxygen were consumed to make that?
(d) In another test 41 mol of water were collected from this reaction. How many moles of alcohol had been burned. How many moles of oxygen were used up? How many moles of CO_2 also formed?

4.63. The combustion of a sample of butane (C_4H_{10}), lighter fluid, produced 2.46 g water:

$$2C_4H_{10} + 13O_2 \longrightarrow 8CO_2 + 10H_2O$$

(a) How many moles of water formed?
(b) How many moles of butane burned?
(c) How many grams of butane burned?
(d) How much oxygen was used up in moles? In grams?

4.64. One way to change iron ore, Fe_2O_3, into metallic iron is to heat it together with hydrogen:

$$Fe_2O_3 + 3H_2 \longrightarrow 2Fe + 3H_2O$$

(a) How many moles of iron are made from 25 mol Fe_2O_3?
(b) How many moles of hydrogen are needed to make 30 mol Fe?
(c) If 120 mol H_2O form, how many moles of Fe_2O_3 were used up?
(d) One metric ton has a mass of 1×10^6 g. If 20.0 metric tons of Fe_2O_3 are changed to iron by this reaction, how much hydrogen will be needed in moles? In kilograms?
(e) Theoretically, how much iron would form from 20.0 metric tons of iron ore by this method in moles? In grams? In metric tons?

4.65. Terephthalic acid, an important raw material for making Dacron, a synthetic fiber, is made from *para*-xylene by the following reaction:

$$C_8H_{10} + 3O_2 \xrightarrow[\text{(special conditions)}]{} C_8H_6O_4 + 2H_2O$$
para-
xylene   terephthalic
                    acid

(a) How much terephthalic acid could be made from 154 g of *para*-xylene in moles? In grams?
(b) At the end of the 1970s the annual U.S. production of terephthalic acid was 5.95×10^9 lbs (1 lb = 454 g). How much *para*-xylene was needed for this in moles? In grams? In metric tons? (*Note.* 1 metric ton = 10^6 g.)
(c) The annual production of *para*-xylene at the same time (part b) was 4.18×10^9 lb. What % of that was changed to terephthalic acid?

4.66. Adipic acid, a raw material for nylon, is made industrially by the oxidation of cyclohexane:

$$5O_2 + 2C_6H_{12} \xrightarrow[\text{(special conditions)}]{} 2C_6H_{10}O_4 + 2H_2O$$
            cyclohexane   adipic
                               acid

(a) How many moles oxygen would be needed to make 40 mol of adipic acid by this reaction?
(b) What is the theoretical yield of adipic acid if 164 g of cyclohexane are used in moles? In grams?
(c) The annual U.S. production of adipic acid was running at about 1.69×10^9 lb at the end of the 1970s (1 lb = 454 g). How much cyclohexane was needed for this, in moles? In grams? In metric tons (1 metric ton = 10^6 g)?
(d) At the same time (part c), the U.S. production of cyclohexane was 2.34×10^9 lb. What percent was converted to adipic acid?

4.67. The Solvay process is used to make sodium carbonate, Na_2CO_3, a chemical that ranked 11th among all chemicals in annual production in 1978. It begins with passing ammonia and carbon dioxide through a solution of sodium chloride. This makes sodium bicarbonate and ammonium chloride:

$$H_2O + NaCl + NH_3 + CO_2 \longrightarrow$$

sodium   ammonia   carbon
chloride   dioxide

$$NH_4Cl + NaHCO_3$$

ammonium   sodium
chloride   bicarbonate

In the next step, sodium bicarbonate is heated to make sodium carbonate:

$$2NaHCO_3 \longrightarrow Na_2CO_3 + CO_2 + H_2O$$

(a) How many moles of sodium carbonate could, in theory, be made from 100 mol of NaCl?
(b) What is the theoretical yield of sodium carbonate from 546 g of NaCl in moles? In grams?
(c) If all the 1978 production of sodium carbonate—15.2 × 10^9 lb (1 lb = 454 g)—had been made this way how much sodium chloride would have been needed, in theory. Give your answer in moles and in grams.

4.68. Rock phosphate, a mineral so important as a source of fertilizer for the world's food supply that it's called "white gold," is a mixture of calcium phosphate, $Ca_3(PO_4)_2$, calcium hydroxide, $Ca(OH)_2$, and calcium fluoride, CaF_2. Some references give $Ca_3(PO_4)_2 \cdot Ca(OH)_2$ as a sufficiently accurate formula. Where calcium fluoride is prominent, the formula is given as $Ca_3(PO_4)_2 \cdot CaF_2$. (The raised dot separating $Ca_3(PO_4)_2$ from CaF_2 is a standard technique for writing the formula of a complex substance when somewhat distinct chemicals are bound in a stoichiometric ratio.) Ordinary "superphosphate" fertilizer is made from phosphate rock by the action of sulfuric acid:

$$3H_2SO_4 + Ca_3(PO_4)_2 \cdot Ca(OH)_2 \longrightarrow$$

sulfuric   phosphate rock
acid   (approximate
          formula)

$$Ca(H_2PO_4)_2 + 3CaSO_4 + 2H_2O$$

calcium   calcium
dihydrogen sulfate
phosphate

(a) How much sulfuric acid (in moles and in metric tons) is needed to convert 100 metric tons of phosphate rock into "superphosphate" by this equation? (1 metric ton = 10^6 g.)
(b) The mixture of calcium dihydrogen phosphate and calcium sulfate produced by this equation (plus some water) is marketed as the "superphosphate." The 100 metric ton batch (part a) would produce how much calcium dihydrogen phosphate (in metric tons) and how much calcium sulfate (in the same unit)?

4.69. The chief process for converting iron ore, Fe_2O_3, into

iron is through the combined action of coal and oxygen. Partial combustion of the carbon in coal gives carbon monoxide, CO:

$$2C + O_2 \longrightarrow 2CO$$

In a series of steps, CO acts on Fe_2O_3 with the overall result:

$$Fe_2O_3 + 3CO \longrightarrow 2Fe + 3CO_2$$

(a) In an experiment done on a very small scale to test the efficiency of new approaches, a sample of 324 g of Fe_2O_3 was converted to iron. How much iron, in theory, could form from this in moles? In grams?
(b) The actual yield of iron in this test was 198 g. What was the percent yield?
(c) How much iron in theory could be made from a 100 metric ton load of iron ore (1 metric ton = 10^6 g)? Calculate your answer in moles and in metric tons.

4.70. The Synthane process developed by the U.S. Bureau of Mines for coal gasification converts the carbon in coal into methane, CH_4, by first heating the pulverized coal with steam and oxygen and then heating the resulting mixture of gases with carbon monoxide and hydrogen. A simple statement of the overall process is:

$$C + 2H_2 \longrightarrow CH_4$$

A feed of 1.38 × 10^4 metric tons of coal yields 580 metric tons of methane.
(a) If the coal were pure carbon, what would be the % yield?
(b) Recalculate the percent conversion if subbituminous coal with 44% (w/w) carbon is used.

4.71. One way to make phosphoric acid, H_3PO_4, a chemical needed to make "triple superphosphate" fertilizer from phosphate rock, is to boil a solution of tetraphosphorus decaoxide, P_4O_{10}, in water:

$$P_4O_{10} + 6H_2O \longrightarrow 4H_3PO_4$$

(a) In a small scale, laboratory experiment, 88.6 g of P_4O_{10} was changed to phosphoric acid this way. How much phosphoric acid formed in moles? In grams?
(b) If the experiment were done so that the final volume was 500 ml, what was the molarity of the phosphoric acid solution made?
(c) In 1978, 19.13 × 10^9 pounds (1 lb = 454 g) of phosphoric acid was made in the U.S. If all of it had been made by this reaction (which is untrue), how much tetraphosphorus decaoxide would have been needed in moles? In metric tons?

4.72. When fuels containing sulfur are burned, the gases leaving the furnace for the smokestack contain SO_2 from the reaction:

$$S + O_2 \longrightarrow SO_2$$

Unless emissions of SO_2 are controlled, bodies of water downwind will slowly become more acidic because SO_2

reacts with water to make sulfurous acid. One technique for removing SO_2 from smokestack gases is to let the gases interact with wet limestone, $CaCO_3$. The net effect is:

$SO_2 + CaCO_3$ + other chemicals
$$\longrightarrow CaSO_3 + \text{other products}$$

In the 1970s, about 400 million metric tons (400×10^6 metric tons) of coal were burned each year to make electricity in the U.S. If the average sulfur content was 2.5% (w/w) sulfur:
(a) How many metric tons of sulfur does this represent?
(b) How much sulfur dioxide would form from that sulfur in moles? In metric tons?
(c) How many metric tons of $CaSO_3$ would form?
(d) That much $CaSO_3$ represents a solid waste disposal problem. To get an idea of its magnitude, if all of the waste were loaded into railroad coal cars, each carrying 100 metric tons of $CaSO_3$, how many cars would be needed to haul it away?

4.73. A common laboratory preparation of hydrogen on a small scale uses the reaction of zinc with hydrochloric acid:

$$Zn + 2HCl \longrightarrow ZnCl_2 + H_2$$

(a) If 10.0 liters of hydrogen at 760 torr and 25 °C are wanted, how much zinc will be needed, in theory. Give your answer in moles and in grams.
(b) How many moles of HCl would be needed for part a?
(c) If the hydrochloric acid were available as a 6.00 M solution, how many milliliters of that solution would be sufficient?

4.74. Ammonia is converted to ammonium sulfate, an important fertilizer, by this reaction:

$$2NH_3 + H_2SO_4 \longrightarrow (NH_4)_2SO_4$$
ammonia   sulfuric   ammonium
             acid   sulfate

(a) If 100 kg of ammonium sulfate are to be made in one batch, how many liters of ammonia at STP would be needed?
(b) How many moles of H_2SO_4 would be required?
(c) If the H_2SO_4 were used in the form of a 6.00 M solution, what volume of that solution in liters would be needed?

4.75. One of the concerns about the world's dwindling supply of natural gas (methane, CH_4, mostly) is its need for making ammonia, an essential fertilizer, if enough food is to be raised to meet world needs. The overall equation for one process for making ammonia is described by the following equation, in which N_4O is used as an approximate formula for air, one of the raw materials:

$$7CH_4 + 10H_2O + 4N_4O \longrightarrow 16NH_3 + 7CO_2$$

Methane is sold in units of *tcf*, where 1 *tcf* = 1 trillion cubic feet = 10^{12} cubic feet = 28.3×10^{12} liters at STP.
(a) What volume of ammonia in liters at STP can be made by this process from 1 *tcf* of methane (also at STP)?
(b) How many moles of ammonia does this represent?

(c) One trillion cubic feet of methane therefore represents how many metric tons of ammonia (1 metric ton = 10^6 g)?

4.76. One industrial synthesis of acetylene, a gas used as a raw material for making countless synthetic drugs, dyes, and plastics, is the addition of water to calcium carbide:

$$CaC_2 + 2H_2O \longrightarrow Ca(OH)_2 + C_2H_2$$
calcium   calcium   acetylene
carbide   hydroxide

(a) In a small scale test to improve efficiency, 100 g of CaC_2 is converted to acetylene. What is the theoretical yield of acetylene in moles? In liters (at STP)?
(b) To make 1.00×10^6 liters of acetylene (at STP) by this method would require how much calcium carbide in moles? In kilograms?

4.77. A student tried to convert benzene (C_6H_6) into chlorobenzene (C_6H_5Cl). The other expected product was HCl. Starting with 16.2 g of benzene, 17.6 g of chlorobenzene was isolated. What was the % yield.

4.78. The reaction of bromine (Br_2) with ethane (C_2H_6) is used to make bromoethane (C_2H_5Br), a soil sterilant. The other product is hydrogen bromide (HBr). In one experiment, 12.0 g of ethane was mixed with 80.0 g of bromine. Bromoethane in the amount of 32.4 g was isolated. What was the limiting reactant and what was the % yield?

4.79. A research supervisor told a chemist to make 100 g of chlorobenzene from the reaction of benzene with chlorine and to expect a yield no higher than 65%. What would be the minimum mass of benzene that could give 100 g of chlorobenzene if the % yield was correct? The equation for the reaction is:

$$C_6H_5 + Cl_2 \xrightarrow[\text{conditions)}]{\text{(special}} C_6H_5Cl + HCl$$
benzene   chlorobenzene

4.80. Certain salts of benzoic acid have been used as food additives for decades. The potassium salt of benzoic acid, potassium benzoate, can be made by the action of potassium permanganate on toluene:

$$C_7H_8 + 2KMnO_4 \longrightarrow$$
toluene   potassium
            permanganate

$$KC_7H_5O_2 + 2MnO_2 + KOH + H_2O$$
potassium   manganese   potassium
benzoate   dioxide   hydroxide

If the yield of potassium benzoate cannot realistically be expected to be more than 68%, what is the minimum number of grams of toluene needed to achieve that yield while producing 10.0 g $KC_7H_5O_2$?

4.81. Calculate the number of grams of each solute that would have to be taken to make each of the following solutions.
(a) 250 ml of 0.100 M NaCl

(b) 100 ml of 0.440 M $C_6H_{12}O_6$ (glucose)

(c) 500 ml of 0.500 M H_2SO_4

4.82. How much solute in grams is needed to make each of the following solutions?

(a) 250 ml of 0.100 M Na_2SO_4

(b) 100 ml of 0.250 M Na_2CO_3

(c) 500 ml of 0.400 M NaOH

4.83. How much solute in grams or in milliliters, as the units call for, is needed to make each of these solutions?

(a) 100 g of 3.48% (w/w) NaCl

(b) 250 g of 1.98% (w/w) glucose

(c) 750 ml of 30.0% (v/v) ethylene glycol

4.84. How much solute is obtained from the following volumes of solutions? State your answer in grams or in milliliters according to the kind of % used.

(a) 75.0 g of 0.900% (w/w) sugar

(b) 150 g of 12.5% H_2SO_4 (w/w)

(c) 450 ml of 25.0% (v/v) methyl alcohol in water

4.85. Concentrated aqueous ammonia is 15 M NH_3. How would you prepare 500 ml of 1.0 M NH_3?

4.86. Concentrated nitric acid is 16 M HNO_3. How would you prepare 100 ml of 0.50 M HNO_3?

4.87. Concentrated acetic acid is 17.4 M $HC_2H_3O_2$. How would you prepare 1.00 liter of 1.00 M acetic acid?

4.88. Concentrated hydrochloric acid is 12 M HCl. How would you prepare 500 ml of 0.25 M HCl?

4.89. For an experiment, a student needed freshly made calcium carbonate. It can be made by mixing solutions of calcium chloride and potassium carbonate:

$$CaCl_2(aq) + K_2CO_3(aq) \longrightarrow CaCO_3(s) + 2KCl(aq)$$

| calcium | potassium | calcium | potassium |
| chloride | carbonate | carbonate | chloride |

The reactants were available as "0.500 M $CaCl_2$" and "0.750 M K_2CO_3."

(a) If the yield is 100%, what volumes of these solutions (in milliliters) should be mixed to produce 12.0 g of $CaCO_3$?

(b) If the yield is only 92%, what volumes have to be mixed to get 12.0 g of $CaCO_3$?

4.90. If some sulfuric acid is spilled on a lab bench, a safe way to neutralize it is to sprinkle solid sodium bicarbonate on it. The following reaction destroys the acid:

$$2NaHCO_3 + H_2SO_4 \longrightarrow Na_2SO_4 + 2CO_2 + 2H_2O$$

| sodium | sulfuric | sodium |
| bicarbonate | acid | sulfate |

Enough bicarbonate is added and stirred into the acid with a glass rod until the fizzing of CO_2 is over.

(a) If 40 ml of 6.0 M H_2SO_4 is spilled, will 25 g of $NaHCO_3$ be enough?

(b) If not, what percent of the acid will be neutralized?

4.91. What is the molar concentration of a sulfuric acid solution if 15.46 ml was neutralized by 33.48 ml of 0.1048 M NaOH as follows:

$$2NaOH(aq) + H_2SO_4(aq) \longrightarrow Na_2SO_4(aq) + 2H_2O$$

4.92. A 0.321 g sample of sodium carbonate, which was contaminated by sodium chloride, was dissolved in water, and then it required 35.4 ml of 0.144 M HCl to react completely with the sodium carbonate as follows:

$$2HCl(aq) + Na_2CO_3 \longrightarrow 2NaCl(aq) + CO_2(g) + H_2O(\ell)$$

(The impurity in the sample does not interfer with this analysis.) How much Na_2CO_3 was present in the sample in grams? What was the percentage purity of the sample?

CHAPTER FIVE

ENERGY. CHEMICAL, SOLAR, AND NUCLEAR

5.1 SOURCES OF ENERGY

The sun, chemicals that its energy helped to make, and unstable atomic nuclei are our chief sources of useful energy

Our principal sources of energy are outlined in Figure 5.1. Defined according to human needs, an **energy source** is any natural change in our environment that we can use to make it easier to do our own tasks, or it is any material that we can get to undergo such a change. For example, the wind is a natural change. It may blow whether we like it or not but, if we put a sail into the wind to let it push a boat, we can rest at the oars. Coal is one of nature's materials. If we let it burn to

Figure 5.1
The principal sources of energy on our planet (from J.R. Holum, *Topics and Terms in Environmental Problems,* 1978, Wiley-Interscience, New York, used by permission).

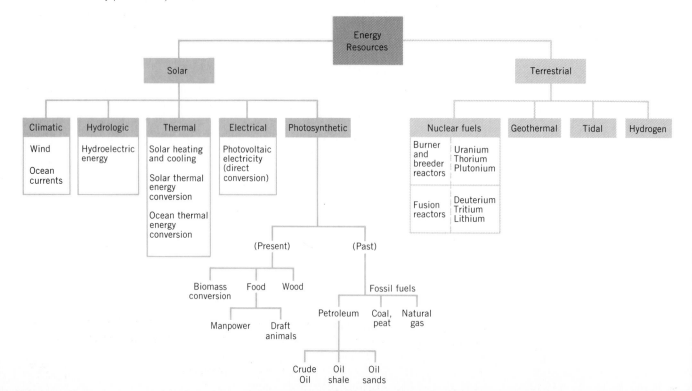

heat water that becomes steam under pressure, we can then generate electricity or drive locomotives.

In this chapter we will concentrate mostly on three sources of energy—chemical, solar, and nuclear. Our aim is not just to gain information about sources of energy, but also to learn about some of its forms—heat, work, electromagnetic, and nuclear. We seek a deeper understanding of the concept of energy, of some of the laws that operate when it is transferred and transformed, and of how information concerning energy gives us insights into the structure of matter. Important features of this structure—atomic nuclei, neutrons, protons, and electrons—will be introduced in this chapter.

We learned in Chapter One that energy is not a thing but an ability—the ability to do work. "Work" in science means more than "toting barges and lifting bales." The concept includes several kinds of work, but all come down to the pushing or pulling of something against an opposing force. For example, when hot gases expand in the cylinder of a gasoline engine, they push back the piston and ultimately move the car. This is mechanical work. A battery delivers its energy by forcing electricity through a wire. The current might run a small motor or operate a pocket calculator. Indirectly, then, the battery performs work that we may call electrical work. Regardless of the form of work, however, it can be converted completely and quantitatively into heat.

5.2 UNITS OF ENERGY

The calorie is the most used unit of energy in chemistry and biology at the present time, but the joule—the SI unit—is becoming more and more widely accepted

One of the important facts about the world we live in is that all forms of energy can be converted entirely into heat. Therefore, units for measuring heat energy are of particular importance. Many units have been developed, and we will study two, the calorie and the joule.

The **calorie (cal)** is the amount of energy that will raise the temperature of 1.00 g of water from 14.5 to 15.5 °C. This is a very small quantity of energy because both 1 g of water and 1 °C are small. Therefore, a multiple of the calorie, the kilocalorie (kcal), is very commonly used.

The Calorie used in nutrition is really the kilocalorie.

$$1 \text{ kcal} = 1000 \text{ cal (exactly)}$$

The SI unit of energy is the joule. Historically, it developed out of studies of kinetic energy, not heat energy, but it bears a definite relationship to the calorie because kinetic energy can be changed completely into heat. For example, when you apply the brakes to a moving car, the car's kinetic energy is converted to heat by friction, and the brakes get hot. The joule is defined in terms of the equation for kinetic energy, $KE = \frac{1}{2}mv^2$ (m = mass; v = velocity). One **joule (J)** is the kinetic energy possessed by a mass of 2 kg moving at a velocity of 1 m/sec.

$$1 \text{ J} = \frac{1}{2} (2 \text{ kg}) \left(\frac{1 \text{ m}}{\text{sec}}\right)^2$$

or

$$1 \text{ J} = 1 \text{ kg} \times \frac{\text{m}^2}{\text{sec}^2} \tag{5.1}$$

The energy you would deliver to your foot if you dropped 4.4 lb of butter on it from a height of 4 in. is about 1 J. The calorie is related to the joule as follows.

$$1 \text{ cal} = 4.184 \text{ J} \quad \text{(exactly)} \tag{5.2}$$

5.3 SPECIFIC HEAT AND HEAT CAPACITY

We can measure the energy transferring in an event if all of it can be used to make the temperature of water change.

The definition of the calorie is based on a thermal property of water called specific heat. The **specific heat** of any substance is the amount of heat needed to raise the temperature of 1 g of that substance by 1 °C. In general:

$$\text{specific heat} = \frac{\text{cal}}{\text{g °C}} \tag{5.3}$$

where °C stands for a *change* of temperature. The actual value of the specific heat of any substance varies slightly with the temperature of the substance. Table 5.1 gives some typical specific heats. Liquid water has one of the highest specific heats of all known substances. In the range from 0 to 100 °C it is (rounded) 1 cal/g °C. This is about ten times that of iron, which means that 1 cal will have ten times the effect on the temperature of iron as it will on the temperature of an equal mass of liquid water.

Table 5.1
Specific Heats

Substance	Specific Heat (25 °C)	
	$\dfrac{\text{cal}}{\text{g °C}}$	$\dfrac{\text{J}}{\text{g °C}}$
Ethyl alcohol	0.586	2.45
Gold	0.0308	0.129
Granite	0.192	0.803
Iron	0.1075	0.4498
Olive Oil	0.47	2.0
Water, liquid	0.99828	4.1796

The high specific heat of water is important to our lives. We have to maintain a very steady body temperature, 37 °C (98.6 °F). Changes of even a few degrees in either direction can lead to death. However, because our body contains so much water—about 60% of our weight is water—it has a very large heat capacity. The **heat capacity** of any object is the heat it will absorb divided by the rise in temperature.

Thus, the specific heat of some object is its heat capacity divided by its mass— its heat capacity per gram.

$$\text{heat capacity} = \frac{\text{heat}}{\text{°C}} \tag{5.4}$$

where "heat" may be in any energy unit we choose and "°C" means the number of degrees of *change* in temperature. Anything with a large heat capacity can take on or lose quite a bit of heat without undergoing very much change in temperature. Thus the water in our bodies acts something like a thermal "cushion," giving the body time to adjust to sudden large changes in outside temperature.

If an object of lower temperature is put in contact with one of a higher temperature, the lower temperature will change upward and the higher temperature will drop. The cause of these changes in temperature is the transfer of heat from the warmer to the cooler object. We sometimes speak of the "flow of heat." How much heat undergoes a transfer can be calculated from temperature changes, masses, and specific heat.

EXAMPLE 5.1 **Calculating Heat Transfers from Thermal Data and Masses**

Problem: A vat of 5.45 kg of water underwent a drop in temperature from 60.30 °C to 57.60 °C. How much energy left the water? Give your answer in calories, kilocalories, joules, and kilojoules. Find also the heat capacity of this vat of water in calories per degree celsius. In this range of temperatures the specific heat of water is $0.999 \frac{\text{cal}}{\text{g °C}}$.

Solution: We know that one of the final answers must be in calories, so we rewrite equation 5.3.

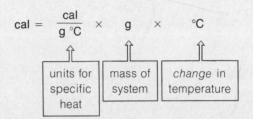

$$\text{cal} = \frac{\text{cal}}{\text{g °C}} \times \text{g} \times \text{°C}$$

units for specific heat   mass of system   *change* in temperature

The change in temperature = 60.30 °C − 57.60 °C = 2.70 °C. The mass is 5.45 kg = 5.45 × 10³ g. Therefore:

$$\text{cal} = 0.999 \frac{\text{cal}}{\text{g °C}} \times 5.45 \times 10^3 \text{ g} \times 2.70 \text{ °C} = 14.7 \times 10^3 \text{ cal}$$

Thus, when a mass of water of 5.45 × 10³ g undergoes a drop in temperature of 2.70 °C starting from 60.30 °C, 14.7 × 10³ cal of heat energy leave the water. Next, convert 14.7 × 10³ cal to kilocalories, joules, and kilojoules.

$$14.7 \times 10^3 \text{ cal} \times \frac{1 \text{ kcal}}{1000 \text{ cal}} = 14.7 \text{ kcal}$$

$$14.7 \times 10^3 \text{ cal} \times 4.184 \frac{\text{J}}{\text{cal}} = 61.5 \times 10^3 \text{ J}$$

$$61.5 \times 10^3 \text{ J} \times \frac{1 \text{ kJ}}{1000 \text{ J}} = 61.5 \text{ kJ}$$

To find the heat capacity of the vat, we take the ratio of the heat transferred out to the degrees of change in temperature accompanying that transfer. That gives us the *heat per degree,* or heat capacity.

$$\frac{14.7 \times 10^3 \text{ cal}}{2.70 \text{ °C}} = 5.44 \times 10^3 \text{ cal/°C} \quad \text{(rounded)}$$

The heat capacity of this vat, therefore, is 5.44 × 10³ cal/°C (22.8 × 10³ J/°C).

EXERCISE 5.1 How much heat will raise the temperature of 100 g water from 25.0 °C to 37.0 °C? $\left(\text{In this range, water's specific heat is } 1.00 \frac{\text{cal}}{\text{g °C}}. \right)$ Give your answer in calories, kilocalories, joules, and kilojoules.

5.4 ENERGY CHANGES IN CHEMICAL REACTIONS

**The heat of reaction measured in a calorimeter at
constant volume is the same as the change in energy, ΔE,
for the system**

One very important application of chemical reactions is to generate usable energy. We use the energy from burning gasoline to move us in a car or bus from home to work or to school. We use the energy from the breakdown of food in our bodies to provide foot power to walk to class. Power companies burn coal, oil, or gas to generate electricity to light our homes and power factories.

To study energy changes in chemical reactions, we must first carefully define some important terms. The word **system** refers to the particular part of the universe we wish to examine. The system may be the chemicals undergoing a reaction taking place in a beaker, the chemicals in a battery reacting to give electricity, or a living cell with all the complex changes going on inside it. The **surroundings** are whatever is entirely outside the system that we define. The surroundings of a system include everything else in the universe except the system itself. The system is separated from the surroundings by some boundary, which may be either real or imaginary. For instance, if our system is a solution in a beaker, the boundary is wherever the solution contacts the beaker or the air above it. If we wish to think in grand terms and view our entire planet as a system, we can isolate it in our minds from the rest of the universe by thinking of an imaginary boundary enveloping the earth at some arbitrary altitude.

The *total energy* possessed by any system, regardless of its composition, size, or shape, is the sum of its kinetic energy and its potential energy. We will use the symbol E to stand for the total energy.

$$E_{system} = (kinetic\ energy)_{system} + (potential\ energy)_{system} \qquad (5.5)$$

Unfortunately, we can never know what the total energy of a system is. Suppose we are working with a chemical reaction happening in a beaker resting on the laboratory bench. The system may *appear* to be standing still and thus have no kinetic energy. However, it is on the surface of a rotating planet that moves about the sun, which itself moves through a galaxy that moves through the universe. Obviously, the system's total kinetic energy cannot be calculated. We cannot measure its total potential energy either, because we have no way to measure the attractions between the beaker and all the other bodies in the universe. If we cannot evaluate E, what good is it?

The good news is that we really are not interested in knowing the total energy, E. We care only about what that system might do for us (or *to* us!) here on planet earth. For example, if we want to find out how much energy we can get out of gasoline when it burns, we really are not interested in how much total energy the combustion products have. What we do want to know is the *difference* between the energy of the products and the energy of the reactants. In other words, what we are interested in is how the total energy *changes* in going from reactants to products. Only that *change* in energy will be of help (or harm) to us.

We will use the symbol ΔE (read "delta E") to stand for this change in total energy.

$$\Delta E = E_{products} - E_{reactants} \qquad (5.6)$$

Any quantity in science described as "Δ" (something) is *always* a final value minus an initial value.

In many reactions the products have less energy than the reactants, and energy flows from the system to the surroundings when the reaction occurs. If the products have more energy than the reactants, the energy flows from the surroundings into the system. As we will see shortly, we can measure how much energy leaves the system and enters the surroundings, or how much energy moves in the other direction and, therefore, we can determine ΔE.

Notice that one thing we have strongly implied so far is that the quantity of energy that leaves a system is exactly the same quantity that moves into the surroundings. If the surroundings give energy to the system, the quantity that goes out of the surroundings equals what enters the system. *No energy is lost.* It just transfers from one place to another, and some of it changes form. The formal statement of this important fact is called the **law of conservation of energy.**

> **Law of Conservation of Energy.** The energy of the universe is constant; it can be neither created nor destroyed but only transferred and transformed.

To keep track of quantities of energy as they transfer from place to place, scientists have agreed on some rules for the algebraic signs of energy values. When a system gives off energy, the sign of ΔE is negative, to signify a *loss* of energy by the system. For example, suppose that the total energy of the products, $E_{products}$, of some reaction were 200 J and the total energy of the reactants, $E_{reactants}$, were 300 J; the energy given off would be 100 J, and ΔE would be computed as

Hereafter, we will assume that "ΔE" means "ΔE_{system}."

$$\Delta E = 200 \text{ J} - 300 \text{ J}$$
$$= -100 \text{ J}$$

"exo-" means "out"
"endo-" means "in"
"-thermic" means "heat"

A reaction like this, which gives off energy and has a negative value of ΔE, is said to be **exothermic.** A reaction that absorbs energy is said to be **endothermic.** The sign of ΔE for an endothermic reaction is positive.

When we carry out an exothermic reaction, some energy can appear as heat and some can appear as work. Suppose, for example, we let a battery discharge. Part of the energy is released as electrical work, but some also is given off as heat—the battery becomes warm. Interestingly, the slower we let the battery discharge, the more work and the less heat we will get. However, the total change in energy, the sum of the work plus heat energy, given off by a particular battery is always exactly the same, regardless of how we use the battery. This is true of any chemical change for a system; the total change in energy, work plus heat, is always the same, *provided we do the measurements under the same conditions.* By "the same conditions" we mean that each measurement begins with the reactants in the same physical form (solid, liquid, or gas); that they have the same composition, concentration, temperature, and pressure; and that the products at the end of each test have the same physical form, composition, concentration, temperature, and pressure. The values of all these variables—physical form, composition, concentration, temperature, and pressure—define the **state of a system.**

As the operation of the battery was described, the net change in energy, ΔE, is independent of how we carry out the change, provided that we begin at the same initial state and end at the same final state. For example, discharging a battery rapidly or slowly gives the same ΔE. Because the value of ΔE for any change depends only on what the initial and final states of the system are and not on how we go from the one to the other, ΔE is said to be a **state function.** One significance of a function of a system being a state function is that the number of details about intermediate stages and steps that we have to pay attention to when measuring a change in that function is greatly reduced.

This brings us to the question of how we can measure ΔE experimentally. If we burn a gallon of gasoline, a definite amount of energy, ΔE, is released by the system. As with the battery, this energy output can be divided into two parts, heat and work. How much heat or work we get depends on how we harness the energy. In an automobile some of the energy of combustion is used for work—the moving of the car from one place to another. The rest appears as heat, most of it leaving the car at the radiator but some with the exhaust. Even with identical weights and equipment, some cars go further than others on a gallon of gasoline because more of the available energy is harnessed as work. Of course, we could keep the car out of gear, burn a gallon of gasoline, and get no useful work at all: All the energy released appears as heat, just as if we burned the fuel in an open container. It is always possible to carry out a reaction so that no work is done. In other words, we can make all the energy appear as heat. (It would be nice if we could make *all* the energy appear as work, with no heat released and wasted, but that turns out to be impossible. We will say more about this subject in Chapter Fifteen.)

Nature has limits to our ability to get work out of heat.

One way to measure the heat released in a reaction is to let the reaction take place inside a sealed container that has been immersed in a vessel filled with a known amount of water. Such an apparatus is called a bomb calorimeter (Figure 5.2). The heat released by the reaction is absorbed by the water, causing the temperature of the water to rise. By measuring the change in temperature and by knowing how many calories are needed to raise the temperature of the calorimeter by 1 degree—the calorimeter's heat capacity—we can calculate the energy that is released when the reaction takes place. A bomb calorimeter is commonly used to determine the **heat of combustion** of a compound—the heat released when a particular quantity of the compound is burned in oxygen.

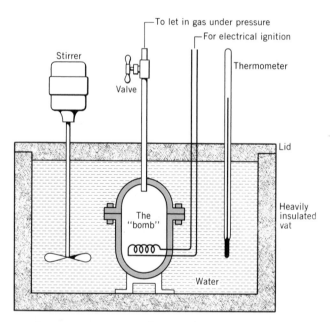

Figure 5.2
A calorimeter. The water bath is usually equipped with devices for adding or removing heat from the water in order to keep the temperature a constant.

EXAMPLE 5.2 **Determining the Heat of Combustion**

Problem: Many people are interested in the caloric content of foods—how many calories there are in a particular serving. Suppose 1.324 g of sucrose (table sugar) is burned in a bomb calorimeter, and the heat capacity of the calorimeter has earlier been found to be 2254 cal/°C. The temperature of the calorimeter increases by 2.314 °C (from 25.000 °C). Calculate the heat of combustion of sucrose in calories per gram and calories per mole. The formula of sucrose is $C_{12}H_{22}O_{11}$.

Solution: Using the calorimeter's heat capacity, we know that for each degree rise in temperature, 2254 cal are released by the reaction. Since the total rise is 2.314 °C, the total energy released is

$$2.314 \, ^\circ\cancel{C} \times 2254 \, \frac{cal}{^\circ\cancel{C}} = 5216 \, cal$$

This energy, 5216 cal, came from burning 1.324 g of sucrose. Therefore, for each gram of sucrose, the energy released is

$$\frac{5216 \, cal}{1.324 \, g \, sucrose} = 3940 \, cal/g \, sucrose$$

Thus the caloric content of sucrose is 3940 cal/g. To express this in calories per *mole*, we have to know the molar mass of sucrose, which is easily calculated to be $\frac{342.3 \, g \, sucrose}{1 \, mole \, sucrose}$. Therefore, the caloric content of a mole of sucrose is

$$\frac{342.3 \, \cancel{g \, sucrose}}{1 \, mol \, sucrose} \times \frac{3940 \, cal}{1 \, \cancel{g \, sucrose}} = \frac{1.349 \times 10^6 \, cal}{1 \, mol \, sucrose}$$

Since 1000 cal = 1 kcal, the molar heat of combustion of sucrose as determined by this experiment is 1.349×10^3 kcal/mol. This result may be compared to the generally accepted value of 1.3482×10^3 kcal/mol.

In the popular literature on nutrition and dieting the kilocalorie is always called the Calorie (with a capital C). Therefore, in such literature, the caloric content of sucrose (Example 5.2) would be reported as 1348 Calories per mole, 3.94 Calories per gram, or 112 Calories per ounce (based on the fact that there are 28.35 g/oz).

EXERCISE 5.2 Fats in food have much higher caloric values than sugar. One component of fat is stearic acid $C_{18}H_{36}O_2$. When a sample of 1.0122 g of stearic acid was burned completely in the bomb calorimeter of Example 5.2, the temperature of the calorimeter rose by 4.256 °C (from 25.000 °C). Calculate the heat of combustion of stearic acid in calories per gram and kilocalories per mole.

The bomb calorimeter is insulated. Ideally it is perfectly insulated, which means that no energy moves across the boundary between the system—the calorimeter—and the surroundings during the combustion. The law of conservation of energy tells us that if there is no energy flow between system and surroundings then *the total energy of the system cannot change during the combustion.* What, then, makes the calorimeter's temperature rise? Is energy created out of nothing? Not at all. What happens in the combustion is a conversion of potential energy into kinetic energy, and as we learned in Chapter Two, kinetic energy is associated with temperature. The molecules of reactants have more *potential* energy than those of the products. This potential energy is present because of the chemical bonds and the internal tensions that exist in the way the molecules are held together. Some of the reactant's potential energy changes to kinetic energy in the form of the increased kinetic energy of molecules of products. The decrease in the system's potential energy is balanced by an increase in its kinetic energy, which means an increase in heat energy. This is why the temperature rises—why, indeed, all insulated systems in which exothermic changes occur become warmer. Insulated systems in which endothermic changes take place become cooler.

5.5 ENTHALPY CHANGES IN CHEMICAL REACTIONS

The heat of reaction measured in a calorimeter at constant pressure is the same as the change in enthalpy, ΔH, for the system.

A reaction in a bomb calorimeter occurs at constant volume (provided the bomb doesn't explode!) This means that if a gas is produced or consumed during the reaction, there will be a change in pressure. Most of the reactions that we are interested in, however, occur at constant pressure. In the lab we routinely study reactions using open vessels that are exposed to a very nearly constant atmospheric pressure. If gases form, they automatically push out against the surrounding air and the pressure in the vessel stays the same. Reactions in living systems also occur at essentially constant pressure—that of the atmosphere—and they involve the exchange of the respiratory gases, oxygen and carbon dioxide. We have an important question, then. How does the heat released by a reaction done in an open vessel at constant pressure compare with the heat released when the same reaction occurs in a bomb calorimeter? The answer is that they are almost the same and sometimes identical.

Suppose we burn some octane (gasoline) at a temperature where *all* of the reactants and products are in their gaseous states.

$$2C_8H_{18}(g) + 25O_2(g) \longrightarrow 16CO_2(g) + 18H_2O(g)$$

octane

Letting the coefficients stand for *moles,* if we start with 2 mol of octane then, before the reaction, there is a total of 27 mol of gas. After the reaction there are 34 mol of gaseous products. If this reaction were done in a sealed container, the gases would exert a greater pressure after the reaction than before, simply because there are more molecules present afterward. However, if the same reaction occurs in a cylinder fitted with a movable piston, the gases could all remain confined as their behavior is studied, but the system could be kept at a constant pressure. With the movable piston the gases can push the piston out to make more room for their greater number of molecules. Pushing on the piston, of course, means that the expanding gases do measurable work—they make something move. It is the same work they would do simply to push away the surrounding atmosphere if the vessel were entirely open. Since there is a fixed amount of energy available, the energy that is used as work to push back the piston can't appear as heat energy. Therefore, when this reaction occurs at constant pressure, the energy that is available to appear as heat is less than it would be if the reaction occurred at constant volume in a sealed container.

One difference between running a reaction at constant pressure in a device with a piston and in an open vessel is that we have no way to harness the work energy released from the latter as the atmosphere is pushed away. This energy is lost into the surroundings, which is the price—a small price, actually—we pay for the convenience of using open vessels.

To discuss energy changes for reactions occurring at constant *volume,* we used the concept of total energy, E, and the change in energy ΔE. To describe energy changes at constant *pressure,* scientists have defined another term, **enthalpy,** or heat content, with the symbol H. Every substance is assumed to have some value of heat content or enthalpy but, like energy (E), H can't be known in any absolute sense. That doesn't matter, because we are interested only in *changes* in enthalpy, ΔH.

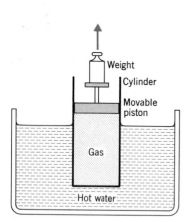

Weight
Cylinder
Movable piston
Gas
Hot water

H means hydrogen
H means enthalpy

$$\Delta H = H_{products} - H_{reactants} \qquad (5.7)$$

Any exothermic reaction at constant pressure has a negative ΔH, and an endothermic reaction has a positive ΔH.

How do ΔE and ΔH compare for the same reaction? They differ only in the pushing work associated with changing volumes. Therefore, for reactions in which the system experiences no change in volume or pressure, ΔE and ΔH are identical. Such conditions are common. They occur whenever all reactants and products are in the solid state, the liquid state, or in a liquid solution. These states vary in volume only slightly (if at all) as reactions occur or as temperatures fluctuate over small ranges. Only when a reaction produces a net change in the number of moles of gases can it cause any significant expansion work. Only then do ΔH and ΔE differ noticeably, but the difference is usually small. For example, the combustion of 1 mol of octane at 100 °C liberates about 1220 kcal of heat at constant pressure ($\Delta H = -1220.0$ kcal/mol). The identical reaction at constant volume liberates only about 2.6 kcal more ($\Delta E = -1222.6$ kcal/mol). Thus, for this reaction, ΔH and ΔE differ by roughly 0.2%.

Reactions at constant pressure will be our primary concern for the rest of our study. Therefore, the remainder of our discussion of energies of reactions in this chapter is devoted to enthalpy changes, ΔH, and we thus avoid any inaccuracies that might occur because of differences between ΔH and ΔE. Like ΔE, ΔH is a state function. The value of ΔH for a reaction depends only on the conditions defining the states of the reactants and products and not on whether the enthalpy appears entirely as heat or if some of this enthalpy changes into electrical energy or some other form of energy.

An enthalpy change, ΔH, can be measured in a vessel open to the atmosphere. In fact, fairly good rough estimates of enthalpy changes can be determined in an apparatus no more sophisticated than a couple of nested styrofoam cups. How much heat is released when a fixed quantity of something is dissolved in water—the **enthalpy of solution**—can be measured this way.

EXAMPLE 5.3   **Determining the Enthalpy of Solution**

Problem: As sulfuric acid dissolves in water, it generates considerable heat. A sample of 175 g of water was placed in a styrofoam cup and chilled to 10.0 °C. Then a 4.90-g sample of pure sulfuric acid (H_2SO_4), also at 10.0 °C, was added, and the contents of the cup were quickly stirred with a thermometer. The temperature rose rapidly to 14.9 °C. Since the solution is dilute, we can assume that its specific heat is the same as that of water in this range of temperature, $1.00 \dfrac{cal}{g \, °C}$. Ignore the fact that some of the heat released warms the glass of the thermometer itself. What is the change in enthalpy in kilocalories per mole for this change?

Solution: First, the total mass of the *solution* = 175 g + 4.90 g = 180 g (rounded). Second, the temperature change = 14.9 °C − 10.0 °C = 4.9 °C. Next, the number of calories that caused this rise (following the pattern of Example 5.1) is:

$$1.00 \; \frac{cal}{g \, °C} \; \times \; 180 \, g \; \times \; 4.9 \, °C \; = 880 \text{ cal (rounded to 2 significant figures)}$$

specific heat   mass of sample   temperature change

Thus 880 cal were liberated when 4.90 g of H_2SO_4 dissolved in 175 g of water. The question asks for the answer in terms of *moles,* not grams. Therefore, we need the molar mass of H_2SO_4, which is 98.1 g/mol. The sample in moles is

$$4.90 \text{ g } H_2SO_4 \times \frac{1 \text{ mol } H_2SO_4}{98.1 \text{ g } H_2SO_4} = 0.0500 \text{ mol } H_2SO_4.$$

Since the temperature *rises* the system will *release* heat to the surroundings. Therefore, the change is exothermic and we have to give a negative sign to the value of the heat we just calculated, −880 cal. Now we can calculate the value of $\Delta H_{solution}$, the ratio of heat exchanged to moles of solute involved, as follows, rounding to two significant figures.

$$\Delta H_{solution} = \frac{-880 \text{ cal}}{0.0500 \text{ mol } H_2SO_4} = -18 \times 10^3 \text{ cal/mol } H_2SO_4$$

or $-18 \text{ kcal/mol } H_2SO_4$

The enthalpy of solution for sulfuric acid in water at the concentration described is $-18 \text{ kcal/mol } H_2SO_4$ ($-75 \text{ kJ/mol } H_2SO_4$).

Values of $\Delta H_{solution}$ depend on how concentrated the solutions are made. The accepted value of ΔH solution for sulfuric acid is 17.7 kcal/mol H_2SO_4 for a solution made as dilute as described in Example 5.3.

EXERCISE 5.3 Sodium hydroxide, NaOH, is found in some drain cleaners. When it dissolves in water, considerable heat evolves. When 3.98 g of NaOH (molar mass, 40.0 g/mol) was dissolved in 175 g of water, both at the same temperature in a styrofoam cup, the temperature rose 5.5 °C. Calculate $\Delta H_{solution}$ in kilocalories per mole of NaOH.

For any reaction, ΔH is called the *heat of reaction.* Its value depends on the temperature and pressure at which we carry out the reaction because these are variables that help define the state of the system. So that we may compare the enthalpy change for one reaction with that of another, we use a standard set of reference conditions—1 atm of pressure (760 torr), 25 °C (298 K)—and the substance in its most stable form under these conditions. Carbon in the form of a diamond, for example, is not as stable as in the form of graphite at 1 atm and 25 °C. Therefore, enthalpy data for carbon are based on the graphite form. Similarly, the most stable state for hydrogen at 1 atm and 25 °C is H_2, not H. Any substance in its most stable chemical form at 25 °C and 1 atm is said to be in its **standard state.** All changes in enthalpy are measured and compared from this reference state. In a sense, we are assuming that the absolute value of the enthalpy of a substance, H, is zero when it is in its standard state and that all changes in enthalpy are measured from this arbitrary "ground zero."

When the enthalpy change of a reaction is determined with all the substances in their standard states and the scale of the reaction is in the mole quantities given by the coefficients of the balanced equation, ΔH is called the **standard heat of reaction,** symbolized as $\Delta H°$, where (°) signifies "standard." The units of $\Delta H°$ are in kilocalories or kilojoules. For example, a reaction between gaseous nitrogen and hydrogen produces ammonia according to the equation:

$$N_2(g) + 3H_2(g) \longrightarrow 2NH_3(g)$$

When specifically 1 mol of N_2 and 3 mol of H_2, both at 25 °C and 1 atm, change to 2 mol of NH_3, also at 25 °C and 1 atm, the reaction is accompanied by the release of 22.16 kcal (92.72 kJ). Therefore, for this reaction, $\Delta H° = -22.16$ kcal (-92.72 kJ).

Often you will find an equation followed by its value of $\Delta H°$. For example:

$$N_2(g) + 3H_2(g) \longrightarrow 2NH_3(g) \qquad \Delta H° = -22.16 \text{ kcal } (-92.72 \text{ kJ})$$

An equation that includes its value of $\Delta H°$ is called a **thermochemical equation.** With this kind of equation, it is very important to interpret the coefficients as standing for *moles,* not molecules.

If a reaction results in the formation of 1 mol of a compound *from its elements,* all in their standard states, the enthalpy change is given a special name. It is called the **standard heat of formation** of the compound. The symbol now used is $\Delta H_f°$, where "*f*" stands for "formation." The units for $\Delta H_f°$ are kcal/mol or kJ/mol, meaning kilocalories or kilojoules per mole of the *compound being formed.* When we write the thermochemical equation for the reaction, we sometimes break the rule requiring the lowest possible *whole* numbers for coefficients. To show just 1 mol of the designated product being formed, we take its normal coefficient and divide it into all the other coefficients, even when that gives simple fractions for other coefficients. For example, for the formation of ammonia we would divide the coefficients of the normal equation (just given) by 2.

$$\frac{1}{2} N_2(g) + \frac{3}{2} H_2(g) \longrightarrow NH_3(g) \qquad \Delta H_f° = -11.08 \text{ kcal/mol } (-46.36 \text{ kJ/mol})$$

Notice that $\Delta H°$ is divided by the same number as the coefficients, in this case 2, to give $\Delta H_f°$.

The standard heat of formation of each element in its most stable form at 25 °C and 1 atm is defined as zero. Because heats of formation are not absolute values but *changes* in values, they have to be relative to something. The question of what they should be relative to is most simply answered by making them relative to the elements.

Table 5.2 gives several standard heats of formation. The corresponding thermochemical equations are not included because we can write them whenever they are needed.

EXAMPLE 5.4 **Writing a Thermochemical Equation to Accompany $\Delta H_f°$**

Problem: Write the thermochemical equation for the formation of H_2O in its liquid state from the *elements* in their gaseous states.

Solution: First write the equation as you normally would:

$$2H_2(g) + O_2(g) \longrightarrow 2H_2O(\ell)$$

Since the coefficient of H_2O is 2, divide all the coefficients by 2:

$$H_2(g) + \frac{1}{2} O_2(g) \longrightarrow H_2O(\ell) \qquad \Delta H_f° = -68.32 \text{ kcal/mol } (-285.9 \text{ kJ/mol})$$

EXERCISE 5.4 Write the thermochemical equation to accompany the heat of formation of sodium bicarbonate, $NaHCO_3$, which is sometimes used as a home remedy for acid indigestion.

In writing thermochemical equations, it is very important that the physical state of each substance be specified. Notice, for example, that $\Delta H_f°$ for *gaseous* water is -57.80 kcal/mol (-241.8 kJ/mol), whereas $\Delta H_f°$ for *liquid* water is -68.32 kcal/mol (-285.9 kJ/mol). The heat of formation for liquid water is *more* than

Table 5.2

Standard Enthalpies of Formation of Typical Substances

Substance	ΔH_f° kcal/mol	kJ/mol	Substance	ΔH_f° kcal/mol	kJ/mol
Ag(s)	0.00	(0.00)	$H_2O_2(\ell)$	−44.84	(−187.6)
AgCl(s)	−30.36	(−127.0)	HCl(g)	−22.06	(−92.30)
Al(s)	0.00	(0.00)	HI(g)	6.35	(26.6)
$Al_2O_3(s)$	−399.09	(−1669.8)	$HNO_3(\ell)$	−41.40	(−173.2)
C(s, graphite)	0.00	(0.00)	$H_2SO_4(\ell)$	−193.91	(−811.32)
CO(g)	−26.42	(−110.5)	$HC_2H_3O_2(\ell)$	−116.4	(−487.0)
$CO_2(g)$	−94.05	(−393.5)	Hg(ℓ)	0.00	(0.00)
$CH_4(g)$	−17.889	(−74.848)	Hg(g)	14.54	(60.84)
$CH_3Cl(g)$	−19.6	(−82.0)	$I_2(s)$	0.00	(0.00)
$CH_3I(g)$	3.40	(14.2)	K(s)	0.00	(0.00)
$CH_3OH(\ell)$	−57.02	(−238.6)	KCl(s)	−104.18	(−435.89)
$CO(NH_2)_2(s)$ urea	−79.634	(−333.19)	$K_2SO_4(s)$	−342.66	(−1433.7)
$CO(NH_2)_2(aq)$	−76.30	(−319.2)	$N_2(g)$	0.00	(0.00)
$C_2H_2(g)$	54.194	(226.75)	$NH_3(g)$	−11.08	(−46.19)
$C_2H_4(g)$	12.496	(52.284)	$NH_4Cl(s)$	−75.38	(−315.4)
$C_2H_6(g)$	−20.236	(−84.667)	NO(g)	21.60	(90.37)
$C_2H_5OH(\ell)$	−66.356	(−277.63)	$NO_2(g)$	8.09	(33.8)
Ca(s)	0.00	(0.00)	$N_2O(g)$	19.49	(81.57)
$CaCO_3(s)$	−288.5	(−1207)	$N_2O_4(g)$	2.31	(9.67)
$CaCl_2(s)$	−190.0	(−795.0)	Na(s)	0.00	(0.00)
CaO(s)	−151.9	(−635.5)	$NaHCO_3(s)$	−226.5	(−947.7)
$Ca(OH)_2(s)$	−235.80	(−986.59)	$Na_2CO_3(s)$	−270.3	(−1131)
$CaSO_4(s)$	−342.42	(−1432.7)	NaCl(s)	−98.23	(−411.0)
$CaSO_4 \cdot \frac{1}{2}H_2O(s)$	−376.47	(−1575.2)	NaOH(s)	−102.0	(−426.8)
$CaSO_4 \cdot 2H_2O$	−483.06	(−2021.1)	$Na_2SO_4(s)$	−330.90	(−1384.5)
$Cl_2(g)$	0.00	(0.00)	$O_2(g)$	0.00	(0.00)
Fe(s)	0.00	(0.00)	Pb(s)	0.00	(0.00)
$Fe_2O_3(s)$	−196.5	(−822.2)	PbO(s)	−52.40	(−219.2)
$H_2(g)$	0.00	(0.00)	S(s)	0.00	(0.00)
$H_2O(g)$	−57.80	(−241.8)	$SO_2(g)$	−70.96	(−296.9)
$H_2O(\ell)$	−68.32	(−285.9)	$SO_3(g)$	−94.45	(−395.2)

that for gaseous water because, when gaseous water changes to a liquid, heat is released. These enthalpy relationships are most easily visualized by means of an *enthalpy diagram,* as given in Figure 5.3. Horizontal lines in such a diagram correspond to different values of absolute enthalpy, H, and higher values are given lines higher up than lower values. *Changes* in enthalpy, ΔH, are represented by the vertical distances separating these lines. For an exothermic reaction, the reactants will always appear on a higher line, as seen by the location of $H_2(g) + 1/2O_2(g)$ in Figure 5.3. Think of a line as representing the *sum* of the enthalpies of all substances on the line in the physical states specified. The

Figure 5.3
Enthalpy diagrams. (a) For the direct conversion of gaseous hydrogen and oxygen to *liquid* water. (b) For the stepwise conversion of these elements to liquid water going through the gaseous water stage. Overall, the enthalpy change is the same.

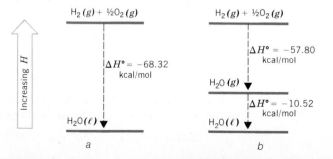

enthalpy diagram on the left in Figure 5.3 charts the enthalpy change for the direct conversion of gaseous hydrogen and oxygen to *liquid* water; 68.32 kcal/mol are released. The enthalpy diagram on the right shows the same change occurring in two steps. Step 1 is to the gaseous water stage (57.80 kcal/mol released), and step 2 is the change of gaseous water to liquid water, which releases another 10.52 kcal/mol.

5.6 HESS'S LAW

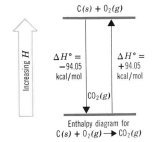

$C(s) + O_2(g)$

$\Delta H° = -94.05$ kcal/mol

$\Delta H° = +94.05$ kcal/mol

$CO_2(g)$

Enthalpy diagram for
$C(s) + O_2(g) \longrightarrow CO_2(g)$

We can use $\Delta H_f°$ data to calculate enthalpy changes that are hard to measure in other ways

If we have a value of $\Delta H°$ for a particular reaction we automatically have a value of $\Delta H°$ for the reverse reaction. Only the sign changes. Consider the combustion of carbon for which the thermochemical equation is

$$C(s) + O_2(g) \longrightarrow CO_2(g) \qquad \Delta H° = -94.05 \text{ kcal } (-393.5 \text{ kJ})$$

The law of conservation of energy tells us that $\Delta H°$ for the reverse reaction, although very hard to carry out, must be $+94.05$ kcal:

$$CO_2(g) \longrightarrow C(s) + O_2(g) \qquad \Delta H° = +94.05 \text{ kcal } (+393.5 \text{ kJ})$$

If this relationship between $\Delta H°$ values for forward and reverse reactions were not true, we could set up a perpetual motion machine. That's one that would create more energy than it used, and we would have the difference, free. (The U.S. Patent Office got so tired of receiving patent applications for these impossibilities that it took the step of insisting that all applications be accompanied by *working* models. None has come in since.) Here's how a perpetual motion machine would work in principle. Suppose it cost only 50 kcal/mol to break CO_2 back to C and O_2. If we get 94 kcal/mol by burning carbon and if it costs only 50 kcal/mol to recycle the system back to C + O_2, our profit is 44 kcal/mol. Over and over, we could let the elements burn, get 94 kcal/mol, save 50 kcal/mol for recycling and get 44 kcal/mol, free each time. We could use that to swat flies, bake cookies, or run some machine while we took our ease. Very nice—but impossible. We don't know *why* it's impossible except to say that nature is set up that way. All we have *discovered* is that it is impossible and, out of this discovery, almost endlessly repeated by would-be inventors of perpetual motion machines, came the law of conservation of energy.

We have discussed how we can use $\Delta H°$ data for given reactions to find $\Delta H°$ for the reverse reactions. Actually, we can calculate the heats of reactions for almost anything by simply adding or subtracting whole equations and their values of $\Delta H°$. All we need are enough basic data.

EXAMPLE 5.5 **Manipulating Thermochemical Equations**

Problem: Using the thermochemical equations provided below, show that the formation of carbon dioxide by the two-step process described gives the same result, chemically and thermally, as making carbon dioxide in one step directly from the elements by the following equation.

$$C(s) + O_2(g) \longrightarrow CO_2(g) \qquad \Delta H° = -94.05 \text{ kcal } (-393.5 \text{ kJ})$$

The two-step process involves the following thermochemical equations.
Step 1. Carbon reacts with oxygen to give carbon monoxide:

$$2C(s) + O_2(g) \longrightarrow 2CO(g) \qquad \Delta H° = -52.84 \text{ kcal } (-221.1 \text{ kJ})$$

Step 2. Carbon monoxide reacts with oxygen to give carbon dioxide:

$$2CO(g + O_2(g) \longrightarrow 2CO_2(g) \qquad \Delta H^\circ = -135.42 \text{ kcal } (-566.60 \text{ kJ})$$

Solution: We write the chemical equations for the two-step process as follows, one right beneath the other to make it easier to add them just as if they were two algebraic equations. The value of ΔH° for the sum of the two equations must likewise be the sum of the ΔH° values for the separate reactions.

$$2C(s) + O_2(g) \longrightarrow 2CO(g) \qquad \Delta H^\circ = -52.84 \text{ kcal } (-221.1 \text{ kJ})$$
$$\underline{2CO(g) + O_2(g) \longrightarrow 2CO_2(g) \qquad \Delta H^\circ = -135.42 \text{ kcal } (-566.60 \text{ kJ})}$$

$$2C(s) + \cancel{2CO(g)} + O_2(g) + O_2(g) \longrightarrow$$
$$\cancel{2CO(g)} + 2CO_2(g) \qquad \Delta H^\circ = -188.26 \text{ kcal } (-787.68 \text{ kJ})$$

Notice that $2CO(g)$ appears on both sides of the resulting equation, and therefore we can cancel these two symbols. We can always cancel identical formulas on opposite sides of the arrow in this manner *provided the chemicals are in identical states,* for example, both are gases (g). We also can always add formulas if they are of the same chemical in the same physical state; we can replace $O_2(g) + O_2(g)$ by $2O_2(g)$. These changes simplify the previous equation, and we have one for the direct, one-step production of $CO_2(g)$:

$$2C(s) + 2O_2(g) \longrightarrow 2CO_2(g) \qquad \Delta H^\circ = -188.26 \text{ kcal } (-787.68 \text{ kJ})$$

However, we have to divide all the coefficients by two to get them into their lowest whole numbers. When we do that we must divide ΔH° by 2 also.

$$C(s) + O_2(g) \longrightarrow CO_2(g) \qquad \Delta H^\circ = \tfrac{1}{2}(-188.26 \text{ kcal})$$
$$= -94.13 \text{ kcal } (-393.8 \text{ kJ})$$

The chemical results of the one-step and two-step processes are identical. The thermal results, shown also in the enthalpy diagram, are also essentially identical. The small discrepancy between -94.13 kcal (for two steps) and -94.05 (for one step) is partly caused by rounding-off and partly by the great difficulty in getting precise and accurate experimental data. Moreover, to three significant figures they are identical.

Enthalpy diagrams for
$C(s) + O_2 \rightarrow CO_2(g)$

Example 5.5 illustrates that ΔH° is a state function, one that is independent of the particular path or steps used to make a chemical change. Whether we take two steps or one, ΔH° comes out the same. The example also illustrates a general law of nature, **Hess's law of constant heat summation** (after Germain Henri Hess, 1802–1850, a professor in St. Petersburg, Russia).

Law of Constant Heat Summation (Hess's Law). For any reaction that can be written in steps, the standard heat of reaction is the sum of the standard heats of reaction for the steps.

In using Hess's law, we sometimes have to multiply or divide a thermochemical equation through by some whole number in order to make it possible to cancel a reactant by a product and get a net equation. Always remember that the same multiplication or division has to be done to the value of ΔH°, as we illustrated in Example 5.5.

In the next example we will see how we can use ΔH_f° from Table 5.2 and their associated chemical equations to figure out the standard heat of a given reaction. We will see that sometimes we can even reverse an available thermochemical equation in order to have an equation we can use in the process of adding and simplifying to obtain the specified reaction. When we reverse a thermochemical equation we must change the sign for its $\Delta H°$, as we learned at the start of this Section.

EXAMPLE 5.6   **Calculating With Hess's Law**

Problem: Carbon dioxide is a waste product of the combustion of methane (natural gas). Hydrogen might some day be made from water, the other product of this combustion, by the use of solar energy. One dream of energy planners is to recycle carbon dioxide back to methane, using solar energy. What is the standard heat of reaction for converting carbon dioxide and hydrogen into methane, CH_4? (Use $\Delta H_f°$ data from Table 5.2 only.)

$$CO_2(g) + 4H_2(g) \longrightarrow CH_4(g) + 2H_2O(\ell)$$

Solution: What we need is a set of thermochemical equations which, when added and simplified, give us the thermochemical equation for the desired reaction, the one specified in the statement of the problem. If we are limited to using only $\Delta H_f°$ data in Table 5.2, then the set of equations we want must all at least start out as thermochemical equations for the formation of one mole of each of the compounds in the desired equation. Thus the first task in a problem like this is to write these equations. From the beginning, though, we will write enthalpy values as $\Delta H°$ instead of as $\Delta H_f°$, because $\Delta H°$ is what we want to calculate for the desired equation. This is no problem, of course, because $\Delta H°$ and $\Delta H_f°$ are numerically equal when we show the formation of *one* mole of a compound from its elements. Thus we begin with the following thermochemical equations, one for each compound in the desired equation.

(1) For $CO_2(g)$: $C(s) + O_2(g) \longrightarrow CO_2(g)$ $\Delta H° = -94.05$ kcal
(From Table 5.2)

(2) For $H_2(g)$: No equation is necessary; $H_2(g)$ is an element in its standard state.

(3) For $CH_4(g)$: $C(s) + 4H_2(g) \longrightarrow CH_4(g)$ $\Delta H° = -17.889$ kcal
(From Table 5.2)

(4) For $H_2O(\ell)$: $H_2(g) + \frac{1}{2}O_2(g) \longrightarrow H_2O(\ell)$ $\Delta H° = -68.32$ kcal
(From Table 5.2)

Unfortunately, we can't simply add these as we did in Example 5.5 and expect to get the original equation as the net result. For example, because of the desired equation we have to have $CO_2(g)$ on the *left* side, not on the right side as in equation (1), before we add. But that's easy as we learned at the beginning of this Section. We just reverse equation (1) *and change the sign of $\Delta H°$.* In other words we'll write a thermochemical equation for the *decomposition* of each *reactant*, in this example, $CO_2(g)$, not one for its formation. [We hope we can eventually cancel $C(s)$ and $O_2(g)$ so they'll not appear among the products after we add and simplify.] Here is the reverse of equation (1) with the sign of $\Delta H°$ also reversed.

(1)-reversed $CO_2(g) \longrightarrow C(s) + O_2(g)$ $\Delta H° = +94.05$ kcal

Equations (3) and (4) have $CH_4(g)$ and $H_2O(\ell)$ on the right side where we want them, and we don't have to reverse these equations. Equation (3) also has a bonus; it has $H_2(g)$ on the left side where we want it before we add the equations. Equation (4), however, shows the formation of only 1 mol of $H_2O(\ell)$, and we need $2H_2O(\ell)$. But that is easily handled; we simply multiply everything in equation (4) by a factor of 2, including $\Delta H°$.

$$(4) \times 2 \qquad 2H_2(g) + O_2(g) \longrightarrow 2H_2O(\ell) \qquad \Delta H° = 2 \times (-68.32 \text{ kcal})$$
$$= -136.64 \text{ kcal}$$
$$\text{(to be rounded later)}$$

Now let's collect the equations, as modified, to see what we have.

From (1)-reversed $\quad CO_2(g) \qquad\qquad \longrightarrow C(s) + O_2(g) \quad \Delta H° = +94.05 \text{ kcal}$

From (3)-as is $\qquad C(s) + 2H_2(g) \longrightarrow CH_4(g) \qquad\qquad \Delta H° = -17.889 \text{ kcal}$

From (4) $\times$ 2 $\qquad 2H_2(g) + O_2(g) \longrightarrow 2H_2O(\ell) \qquad\quad \Delta H° = -136.64 \text{ kcal}$

Next, add:

$$CO_2(g) + \cancel{C(s)} + 2H_2(g) + 2H_2(g) + \cancel{O_2(g)} \qquad \Delta H° = [+94.05$$
$$\longrightarrow \cancel{C(s)} + \cancel{O_2(g)} + CH_4(g) + 2H_2O(\ell) \qquad\qquad -17.889$$
$$-136.64], \text{ or}$$
$$\Delta H° = -\ 60.48 \text{ kcal}$$

The net equation, after canceling and combining, is the original equation:

$$CO_2(g) + 4H_2(g) \longrightarrow CH_4(g) + 2H_2O(\ell) \qquad \Delta H° = -60.48 \text{ kcal} \ (-253.0 \text{ kJ})$$

EXERCISE 5.5 Calculate $\Delta H°$ for the synthesis of methyl alcohol, CH_3OH, shellac thinner and fondue fuel, from methane (natural gas) by this reaction:

$$2CH_4(g) + O_2(g) \longrightarrow 2CH_3OH(\ell)$$

Construct and use the thermochemical equations for the formation of $CH_4(g)$ and $CH_3OH(\ell)$ from their elements using the necessary $\Delta H°$ data from Table 5.2.

Another, quicker way of using $\Delta H_f°$ data in Hess's law calculations is to take advantage of the fact that the net $\Delta H°$ for any reaction must be the difference between the total enthalpies of formation of the products and those of the reactants. If the general equation is

$$aA + bB + \cdots \longrightarrow nN + mM + \cdots$$

we can find $\Delta H°$ by the following equation, called the **Hess law equation:**

$$\Delta H° = [n\ \Delta H_f°(N) + m\ \Delta H_f°(M) + \cdots] \ - \ [a\ \Delta H_f°(A) + b\ \Delta H_f°(B) + \cdots] \qquad (5.8)$$

What this Hess law equation amounts to is nothing more than that:

$\Delta H° =$ [sum of $\Delta H°$ for the formation of the products from the elements]
$\quad -$ [sum of $\Delta H°$ for the formation of the reactants from the elements] $\qquad (5.9)$

Example 5.7 shows how equation 5.8 may be used.

EXAMPLE 5.7 **Calculating With The Hess Law Equation**

Problem: What is $\Delta H°$ for the combustion in oxygen of ethyl alcohol, $C_2H_6O(\ell)$, to form $CO_2(g)$ and $H_2O(\ell)$? (Note: $C_2H_6O(\ell) = C_2H_5OH(\ell)$ in Table 5.2.)

Solution: Step 1. Write the balanced equation:

$$C_2H_6O(\ell) + 3O_2(g) \longrightarrow 2CO_2(g) + 3H_2O(g)$$

Step 2. Use the Hess law equation, 5.8, and ΔH_f° data.

$$\Delta H^\circ = [2\ \Delta H_f^\circ(CO_2)_g + 3\ \Delta H_f^\circ(H_2O)_g] - [\Delta H_f^\circ(C_2H_6O)_\ell + 3\ \Delta H_f^\circ(O_2)_g]$$

$$= \left[2\ \text{mol}\left(-94.05\ \frac{\text{kcal}}{\text{mol}}\right) + 3\ \text{mol}\left(-57.80\ \frac{\text{kcal}}{\text{mol}}\right)\right]$$

$$- \left[1\ \text{mol}\left(-66.356\ \frac{\text{kcal}}{\text{mol}}\right) + 3\ \text{mol}\left(0\ \frac{\text{kcal}}{\text{mol}}\right)\right]$$

$$= -295.1\ \text{kcal}\ (-1235\ \text{kJ}).$$

This is the heat of combustion of 1 mol of ethyl alcohol at standard conditions, because the equation shows 1 $C_2H_6O(\ell)$.

Remember, ΔH_f° is zero for all elements in their most stable forms.

EXERCISE 5.6 Calculate ΔH° for the following reactions.
(a) $2NO(g) + O_2(g) \longrightarrow 2NO_2(g)$
(b) $NaOH(s) + HCl(g) \longrightarrow NaCl(s) + H_2O(g)$

Burning something is often far easier than making it from its elements. Therefore, enthalpy data from combustions are sometimes more easily available than values of ΔH_f°. The **standard heat of combustion, $\Delta H_{combustion}^\circ$**, is the enthalpy of combustion for one mole of a compound under standard conditions. If we have all but one value of ΔH_f° for the Hess law equation (5.8), we can use $\Delta H_{combustion}^\circ$ to find ΔH° for applying this equation to calculate the last remaining value of ΔH_f°. Let's work an example to see how this is done.

EXAMPLE 5.8 **Using $\Delta H_{combustion}^\circ$ Data to Find ΔH_f°**

Problem: One of the building units for proteins such as those forming muscles and sinews is an amino acid called glycine, $C_2H_5NO_2$. The equation for its combustion is

$$4C_2H_5NO_2(s) + 9O_2(g) \longrightarrow 8CO_2(g) + 10H_2O(\ell) + 2N_2(g)$$

The value of its $\Delta H_{combustion}^\circ$ is -232.67 kcal/mol. Using these data and values of ΔH_f° for $CO_2(g)$ and $H_2O(\ell)$ from Table 5.2, calculate ΔH_f° for glycine.

Solution: The enthalpy change, ΔH°, for the combustion of 4 mol of glycine is

$$4\ \text{mol} \times \left(-232.67\ \frac{\text{kcal}}{\text{mol}}\right) = -930.68\ \text{kcal, the value that we}$$

must use for ΔH° in the Hess law equation. From Table 5.2 we find that $\Delta H_f^\circ = -94.05$ kcal/mol for $CO_2(g)$, and that $\Delta H_f^\circ = -68.32$ kcal/mol for $H_2O(\ell)$. We also know that the values of ΔH_f° for all elements in their most stable forms are all zero. We will use these data in the Hess law equation, which takes the following form for this problem.

$$\Delta H^\circ = [8\ \text{mol} \times \Delta H_f^\circ(CO_2)_g + 10\ \text{mol} \times \Delta H_f^\circ(H_2O)_l + 2\ \text{mol} \times \Delta H_f^\circ(N_2)_g]$$

$$- [4\ \text{mol} \times \Delta H_f^\circ(C_2H_5NO_2)_s + 9\ \text{mol} \times \Delta H_f^\circ(O_2)_g]$$

Or:

$$-930.68 \text{ kcal} = \left[\cancel{8 \text{ mol}} \times \left(-94.05 \frac{\text{kcal}}{\cancel{\text{mol}}} \right) + 10 \cancel{\text{ mol}} \times \left(-68.32 \frac{\text{kcal}}{\cancel{\text{mol}}} \right) \right.$$

$$\left. + 2 \text{ mol} \times (0) \right] - [4 \text{ mol} \times \Delta H_f^\circ (C_2H_5NO_2)_s + 9 \text{ mol} \times (0)]$$

$$= [-752.4 \text{ kcal} - 683.2 \text{ kcal}] - [4 \text{ mol} \times \Delta H_f^\circ (C_2H_5NO_2)_s]$$

Therefore,

$$4 \text{ mol} \times \Delta H_f^\circ (C_2H_5NO_2)_s = +930.68 \text{ kcal} - 752.4 \text{ kcal} - 683.2 \text{ kcal}$$

$$= -504.92 \text{ kcal}$$

And,

$$\Delta H_f^\circ (C_2H_5NO_2)_s = \frac{-504.92 \text{ kcal}}{4 \text{ mol}}$$

$$= -126.2 \text{ kcal/mol} \, (-528.0 \text{ kJ})$$

Thus, the standard heat of formation of glycine is −126.2 kcal/mol. Even though glycine cannot be made directly from its elements, we can still find out this value using other data that are more easily available, experimentally.

EXERCISE 5.7   The thermochemical equation for the combustion of sucrose (table sugar) is:

$$C_{12}H_{22}O_{11}(s) + 12O_2(g) \longrightarrow 12CO_2(g) + 11H_2O(\ell) \quad \Delta H^\circ = -1348.2 \text{ kcal} \, (-5640.9 \text{ kJ})$$

Using this equation together with the standard heats of formation from Table 5.2, calculate the standard heat of formation of sucrose.

Applications of Hess's law are not limited to use of heats of *combustion* or heats of *formation* from the elements. We can use any thermochemical data for any kind of reaction.

EXAMPLE 5.9   **Finding Enthalpy Changes from Thermochemical Data**

Problem: One of the reactions that occurs when an iron oxide found in iron ore is changed to pure iron is:

$$Fe_2O_3(s) + 3CO(g) \longrightarrow 2Fe(s) + 3CO_2(g)$$

Given the following thermochemical equations, find ΔH° for this reaction.

$$2Fe_2O_3(s) + 3C(s) \longrightarrow 4Fe(s) + 3CO_2(g)$$
$$\Delta H^\circ = +307.35 \text{ kcal} \, (1286.0 \text{ kJ})$$

$$CO_2(g) + C(s) \longrightarrow 2CO(g) \quad \Delta H^\circ = +41.21 \text{ kcal} \, (172.4 \text{ kJ})$$

Solution: The first of the thermochemical equations has the substances where we want them, Fe_2O_3 on the left and $Fe + CO_2$ on the right. The second equation has to be reversed and then scaled up to involve $3C(s)$ instead of $1C(s)$. In this way we will be able to cancel $C(s)$. When we reverse the second equation, therefore, we must multiply all its coefficients by 3 before adding it to the first.

$$2Fe_2O_3(s) + 3C(s) \longrightarrow 4Fe(s) + 3CO_2(g) \qquad \Delta H° = +307.35 \text{ kcal}$$

$$6CO(g) \longrightarrow 3C(s) + 3CO_2(g) \qquad \Delta H° = 3 \times (-41.21 \text{ kcal})$$

$$= -123.63 \text{ kcal}$$

$$2Fe_2O_3(s) + 6CO(g) \longrightarrow 4Fe(s) + 6CO_2(g) \qquad \Delta H° = 307.35 \text{ kcal} - 123.63 \text{ kcal}$$

$$= 183.72 \text{ kcal} (786.7 \text{ kJ})$$

The coefficients must now be divided by 2. When this is done, $\Delta H°$ must also be halved:

$$Fe_2O_3(s) + 3CO(g) \longrightarrow 2Fe(s) + 3CO_2(g) \qquad \Delta H° = 91.86 \text{ kcal} (384.3 \text{ kJ})$$

EXERCISE 5.8 Using data from the thermochemical equations provided, calculate $\Delta H°$ for the following reaction:

$$2NH_3(g) + 2O_2(g) \longrightarrow N_2O(g) + 3H_2O(\ell)$$

$$2NH_3(g) + 4SO_3(g) \longrightarrow N_2O(g) + 4SO_2(g) + 3H_2O(\ell) \qquad \Delta H° = -69.4 \text{ kcal} (-290 \text{ kJ})$$

$$2SO_3(g) \longrightarrow 2SO_2(g) + O_2(g) \qquad \Delta H° = +47.0 \text{ kcal} (197 \text{ kJ})$$

5.7 ELECTROMAGNETIC ENERGY

Electromagnetic energy is energy transmitted at the velocity of light in the form of oscillations in the strengths of electric and magnetic forces

Thus far in this chapter we have concentrated largely on one form of energy—heat—and its relationship to chemical changes. Many chemical reactions give off heat. Many others need a continuous flow of heat into the system for the reaction to occur. However, heat is by no means the only form of energy of interest in chemistry. Electrical energy, for example, is used to cause many useful chemical changes, as in using a silver compound to make elemental silver in silverplating. Conversely, many chemical reactions can be harnessed to generate an electrical current, as in all batteries. We will study electrical energy and chemical change more in Chapters Eighteen and Twenty.

Another form of energy of great value in chemistry is electromagnetic energy, popularly called light energy. Many chemical systems emit visible light as they react, for example, when anything burns. The firefly is able to accomplish a particularly interesting conversion of chemical energy into visible light. It has a chemical called luciferin, and one of the reactions of luciferin releases the light energy of the firefly's wink. Green plants have systems that absorb light from the sun, and plants use that energy to grow. Scientists who are specialists in making complex compounds that often have fragile molecules sometimes find that light energy is better than heat energy to bring about particular reactions. There is a large branch of chemistry with many practical applications called photochemistry in which light-induced reactions are studied.

When elements are heated to a high enough temperature, they emit visible light. An iron rod, for example, can be made red hot or white hot according to its temperature. The particular kinds of light energy and their intensities given off by glowing objects can be measured. Some of these data became very important clues to the atomic structure of matter, as we will study in Chapter Seven. Since electromagnetic energy, like heat energy, is important in many areas of chemistry, we need to lay a foundation here for some of these applications.

Electromagnetic energy is energy carried through space or matter by means of wavelike oscillations. These waves are like water waves in that something goes up and down but, unlike water waves, what oscillates is not matter. The oscillations are systematic fluctuations in the intensities of very tiny electrical and magnetic forces. The space in which these oscillations occur is called the electromagnetic field. Both electrical and magnetic forces are present in an electromagnetic wave, and each changes rhythmically with time. In other words, the value of each force goes through a maximum, then to zero and down to a minimum value, and back up again through zero to the maximum value. Figure 5.4 shows this for the electrical component. The magnetic component would be depicted in a plane at right angles to the plane carrying the electrical component.

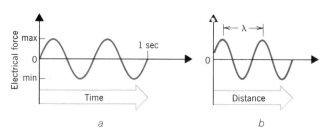

Figure 5.4
The electrical force associated with electromagnetic radiation fluctuates rhythmically. (*a*) Two cycles of fluctuation are shown and, therefore, the frequency is 2 Hz. (*b*) An electromagnetic radiation frozen in time. This curve shows how the electrical force varies along the direction of travel. The distance between two maximum values is the wavelength of the electromagnetic radiation.

Each oscillation is called one *cycle*. The successive series of these oscillations occurring through space from the origin of the light is called **electromagnetic radiation** (popularly, a light wave). The number of cycles per second is called the **frequency** of the electromagnetic radiation, and its symbol is ν (a Greek letter pronounced "new").

$$\text{frequency} = \nu = \text{cycles per second} \tag{5.10}$$

The SI unit of frequency is called the **hertz (Hz),** and 1 Hz = 1 cycle/sec.

As the radiation moves away from its source, the maximum values of the electrical and magnetic forces are regularly spaced in the field. The distance separating maximum values is called the radiation's **wavelength,** symbolized by λ (another Greek letter, "lambda"). See Figure 5.4*b*.

$$\text{wavelength} = \lambda = \text{meters per cycle} \tag{5.11}$$

If we multiply frequency by wavelength, the result has the units of velocity, meters per second (m/sec). The velocity of electromagnetic radiation in a vacuum is 3.00×10^8 m/sec, and this velocity is given the symbol c.

$$\underset{(\lambda)}{\frac{\text{meters}}{\text{cycle}}} \times \underset{(\nu)}{\frac{\text{cycles}}{\text{second}}} = \underset{(c)}{\frac{\text{meters}}{\text{second}}} = \text{velocity} \tag{5.12}$$

wavelength × frequency = velocity

$$\lambda \nu = c = 3.00 \times 10^8 \text{ m/sec} \tag{5.13}$$

Electromagnetic radiation comes in a large range of frequencies called the **electromagnetic spectrum** and shown in Figure 5.5. Each portion of the spectrum has a popular name. For example, radio waves are electromagnetic radiations having very low frequencies. Microwaves, which also have low frequencies, interact with molecules in food to give them a much higher kinetic en-

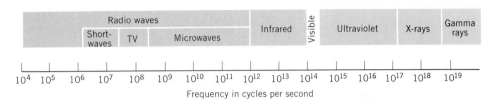

Figure 5.5
The electromagnetic spectrum.

ergy and, as a result, the food cooks. (Keep your hands and eyes out of the way of microwave radiation; it can cook you, too.) Infrared radiation consists of the range of frequencies that can make molecules of most substances vibrate internally. Each substance absorbs a uniquely different set of infrared frequencies. A plot of the frequencies absorbed versus the intensities of absorption is called an infrared spectrum. It can be used to identify a compound, because each infrared spectrum is as unique as a set of fingerprints. See Figure 5.6. Many substances will absorb visible and ultraviolet radiations in unique ways, too, and they have visible and ultraviolet spectra. Gamma rays are at the high-frequency end of the electromagnetic spectrum. They are produced by certain elements that are radioactive. X rays are very much like gamma rays, but they are usually made by special equipment. Both X rays and gamma rays penetrate living things easily.

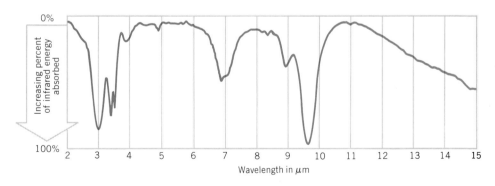

Figure 5.6
Infrared spectrum of methyl alcohol, wood alcohol (courtesy Sadtler Research Laboratories, Inc., Philadelphia, Pa.).

Only the extremely narrow band of frequencies between 4.3×10^{14} and 7×10^{14} Hz will be absorbed by the human eye to cause the sensation we call visible light. Each color we see is just a narrow band of frequencies in this range. See Figure 5.7. The visible spectrum is the series of colors from red through orange, yellow, green, blue, indigo, and violet. White light is a mixture of all these colors in roughly equal amounts. When white light is passed through a prism or a slit, its frequencies are spread apart. This phenomenon is called *refraction* when it is caused by a prism and *diffraction* when it is caused by a slit. (See Color Plate 2A.)

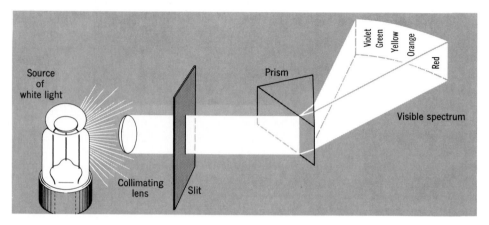

Figure 5.7
White light will be refracted by a glass prism to spread out the colors of the visible spectrum (from J.R. Holum. *Principles of Physical, Organic and Biological Chemistry,* 1968, John Wiley & Sons, New York, used by permission).

In 1900 Max Planck (1858–1947), a German physicist, launched one of the greatest upheavals in the history of science when he proposed that electromagnetic radiation is carried by tiny packets of energy called **photons.** Each photon pulses with a frequency, ν, and each travels with the speed of light. Planck proposed and Albert Einstein (1879–1955) confirmed that *the energy of a radiation is proportional to its frequency,* not to its intensity or brightness as had been believed up to that time.

$$\text{energy of a photon} = E = h\nu \tag{5.14}$$

where h is a proportionality constant now called Planck's constant. The energy of one photon is called one **quantum** of energy.

Planck and Einstein's discovery was really quite surprising. If a particular event requiring energy, such as photosynthesis in green plants, is initiated by the absorption of light, it is the frequency of the light that is important, not its intensity or brightness. An analogy is a group of pole-vaulters trying to get over a wall. If they have long enough poles, each can clear the wall. If the poles are too short, however, they can batter the wall with as much intensity they want, but they'll never clear the top. Not even doubling or tripling the number of vaulters with short poles will get anyone across.

5.8 SOLAR ENERGY

Solar energy is stored in fossil fuels and in complex molecules of all living things

Our richest source of electromagnetic energy is the sun. The sun sheds its energy onto our outer atmosphere at a rate of 1.3×10^{21} kcal/year (5.4×10^{24} J/year). Roughly 40 to 50% of this energy is promptly reflected back to space. The air, the oceans, and the landmasses absorb the rest. According to geologist M. King Hubbert of the U.S. Geological Survey, the solar energy absorbed *each year* is equivalent to 46 times the initial world reserves of minable coal or 425 times the initial world reserves of crude oil. Even if the world's population were to increase from 5 to 20 billion people, and the standard of living, expressed as the energy consumed per person per year, rose from 13 million to 2.5 billion kcal, the rate of energy consumption would still be less than 1% of that absorbed from the sun. (The U.S. energy consumption, in comparison, is presently about 75 to 80 million kcal per person per year.)

About 0.04% of the incoming solar energy is absorbed by chlorophyll in green plants and oceanic phytoplankton. The energy trapped by this green pigment drives a complex series of reactions called photosynthesis. **Photosynthesis** is the plant's use of solar energy to synthesize high-energy molecules from simple, low-energy molecules of carbon dioxide and water. The primary products are carbohydrates and oxygen. Carbohydrates are a family of compounds that includes sugar, starch, and cellulose (the fiber in cotton). If we let the formula CH_2O stand for the basic structural unit in carbohydrates, we may represent photosynthesis as follows.

Cotton, for example, is nearly pure cellulose. Rice is about 80% starch.

$$\text{solar energy} + CO_2 + H_2O \longrightarrow CH_2O + O_2 \qquad (5.15)$$

Animals cannot carry out photosynthesis. They have to get their high-energy molecules by eating either plants or animals that feed on plants.

When living things die, their complete decay returns carbon dioxide and water (plus simple nitrogen compounds and minerals) to the environment. However, complete decay is often prevented by lack of oxygen. Partially decayed remains that bear some resemblance to their original forms are called fossils. Over eons of time, as geological processes altered the continental landforms, the effects of heat and pressure on ancient plant and animal remains changed them into coal, petroleum, and natural gas. Because of their origins, these substances are called the **fossil fuels.** The fossil fuels are an extraordinarily rich legacy from the past. Their available chemical energy came originally from the sun by way of photosynthesis.

Coal is a black or brownish-black solid composed chiefly of carbon, compounds of carbon, and varying percentages of moisture and minerals. Nearly all deposits of coal contain sulfur and traces of hazardous elements such as arsenic, beryllium, fluorine, lead, mercury, selenium, and some radioactive ele-

Figure 5.8
Strip mine. The power plant in the background, at Fruitland, New Mexico is fueled by strip-mined coal taken from deposits in the foreground.

Molecules of carbon dioxide in the air are good absorbers of infrared energy. After absorbing energy, they reradiate it. Some energy stays in the air, some comes down to earth, and some escapes to outer space. The effect of carbon dioxide is to keep the air a trifle warmer, and that is roughly why the influence of carbon dioxide is called the greenhouse effect. Too high a level of carbon dioxide in the air would have too much of a warming effect. If the average temperature of the earth's atmosphere were just 2 to 3 degrees warmer, the world's glaciers would melt and the ocean level would rise 200 to 300 ft, which would destroy most of the world's largest cities. If the atmospheric temperature became just a few degrees cooler, a new ice age would be upon us. Since we began to burn fossil fuels at greatly increased rates, the level of carbon dioxide in the air has risen from 290 to 330 $\mu g/m^3$, and it could go to 400 $\mu g/m^3$ by the year 2000 to 2020. According to some specialists, we should not let it exceed 420 $\mu g/m^3$. Somewhat offsetting this warming effect is a cooling effect caused by increased levels of dusts and particles (from volcanoes, forest fires, dust storms, etc.). It is too early to tell how these opposing effects will balance, but right now it seems that, if anything, we are in a slight cooling trend.

ments. All of these are released into the atmosphere in various chemical forms when coal is burned. The extensive use of coal in place of nuclear fuels, then, is not completely risk-free, although it does not involve the risk of catastrophic accidents that suddenly threaten several thousands of lives. The ever-increasing rate of combustion of coal and other fossil fuels might in the long run, however, affect the climate of the world, as discussed in Special Topic 5.1.

"petro-," rock
"-oleum," oil

Petroleum, after whatever water that occurs with it has been removed, consists of a complex mixture of organic materials. These are mostly compounds composed of just carbon and hydrogen that are called, appropriately, hydrocarbons.

Natural gas is mostly methane, CH_4. It is present in petroleum, in some coal mines, and by itself in underground wells. Natural gas is usually the least polluting of all fuels because it has little if any sulfur or hazardous trace elements. (The gas from some wells does have pollutants and is called "sour gas.")

The world's reserves of fossil fuels cannot last forever. To replace them as sources of energy, we have to develop many new technologies, including several that use solar energy. The solar energy technologies presently being studied are: direct use of solar energy for water heating and space conditioning; wind energy conversion; conversion of solar heat energy to electricity; ocean thermal energy conversion; biomass conversion; and photovoltaic energy conversion.

The first technology listed—direct use of solar energy for heating water or for space conditioning—is being more and more widely used for residences.

Wind energy is classified as a form of solar energy because the uneven warming of the atmosphere by the sun is one cause of winds.

Mirrors can be used to collect and to focus to one small point the solar energy falling over a large land area and thus create temperatures high enough to operate the boiler of a power plant (Figure 5.9) or make hydrogen from water (Special Topic 5.2).

The oceans of the world absorb huge quantities of solar energy, which makes their water at the surface warmer than the water hundreds of feet down. This difference in temperature can be used to make a fluid flow through a turbine or generator and operate a power plant.

"Biomass" means any combustible materials made by plants. Biomass includes agricultural wastes such as straw, stalks, and manure as well as wastepaper, wood, and surplus grains. The conversion of biomass to fuel is now practiced on a small scale. A blend of gasoline and alcohol called gasohol can be used in today's automobiles. Since municipal trash includes great quantities of biomass and other burnable materials—tires, and plastics, for example—many cities in the world reduce their trash-disposal problems and generate municipal energy by burning their trash in special generators.

Special Topic 5.2 Energy From Hydrogen

Hydrogen might someday be a substitute for natural gas. When hydrogen burns, the product is water. Since the hydrogen would also be made from water, the fuel cycle would be perfect and the supply of hydrogen inexhaustible. But hydrogen is not a primary fuel or a primary source of energy. It has to be made, and that costs energy. The net gain would be highest if that bill were paid by solar energy because solar energy is also inexhaustible and relatively nonpolluting.

The scheme would be to use solar energy to make electricity, which would then be used to break water into hydrogen and oxygen by electrical means. These gases are relatively easily separated, and the hydrogen could be sent to consumers in existing natural gas pipelines and used in existing burners with some modifications.

Figure 5.9
High-temperature solar energy test facility at Sandia Laboratories near Albuquerque, New Mexico. In the foreground is a field of sun-tracking mirrors called heliostats, with each heliostat made of 25 16-ft² glass mirrors. The heliostats focus the sun's rays on a point in the tower where the maximum calculated temperature at the center of the beam is 2600 K. The maximum thermal energy is just under 20 GJ/hr for a power rating of 5 megawatts. (For comparison, large central power stations generally have power ratings of 500 to 1000 megawatts.)

Direct photovoltaic conversion is using sunlight to make electricity directly instead of using the heat from sunlight to replace a fossil fuel in a power plant. Sunlight falling on certain materials, such as pure silicon, generates electricity. The silicon is part of a solar cell. One idea now being studied is to put a large array of solar cells high enough into orbit around the earth to be constantly exposed to sunlight (Figure 5.10). The cells would convert solar energy into microwave radiation, which could be beamed even through clouds to a receiving antenna on earth where it would be converted to electricity.

At the moment, all of the solar technologies are more costly than traditional sources of fuel, but advances in engineering will bring those costs down. Finding alternatives to liquid fuels such as gasoline and diesel oil for transportation is the most difficult problem. The principal economically workable means of extending the supplies we have is conservation.

Figure 5.10
Satellite solar power station. In a synchronous orbit, this space satellite would be in a fixed position relative to a receiving antenna on earth. A solar collector array about 12 square miles in area could beam enough energy to earth to generate as much as 15,000 megawatts of electricity, enough to meet the needs of a city the size of New York in the year 2000.

5.9 CONSERVATION OF MASS-ENERGY

Einstein discovered a relationship between mass and energy that led to the law of conservation of mass-energy

Another important, although controversial, source of energy is nuclear power. To understand it, we need to make a small but important correction to the laws of conservation of mass and energy. These laws, as stated earlier, are not strictly true. The error is too small to be detected by weighing balances, but it reveals a very important fact about our world. Albert Einstein was the first to suspect that mass can actually be converted into energy and energy into mass. He correctly predicted that the relationship between the change in energy, ΔE, and the change in mass, Δm, is given by what we now call the **Einstein equation:**

$$\Delta E = \Delta mc^2 \tag{5.16}$$

where c is the velocity of light, 3.00×10^8 m/sec.

Because the velocity of light is very large (and its square, 9×10^{16}, is almost beyond grasping), an extremely small change in mass yields a huge quantity of energy. By the same token, modest changes in energy, such as those observed in chemical changes, involve changes in mass much too small to detect directly.

For example, the combustion of 1 mol of methane gives off 213 kcal (891 kJ):

$$CH_4 + 2O_2 \longrightarrow CO_2 + 2H_2O \qquad \Delta H° = -213 \text{ kcal/mol } (-891 \text{ kJ/mol})$$

The source of this energy is a *loss,* overall, of some mass. Let's calculate what that loss is by the Einstein equation.

a conversion factor from equation 5.1

$$\Delta m = \frac{\Delta E}{c^2} = \frac{(891 \text{ kJ})}{\left(3.00 \times 10^8 \ \frac{\cancel{m}}{\cancel{sec}}\right)^2} \times \left(\frac{1000 \ \cancel{J}}{1 \ \cancel{kJ}}\right) \times \left(\frac{1 \ \frac{\text{kg} \ \cancel{m^2}}{\cancel{sec^2}}}{1 \ \cancel{J}}\right)$$

$$= 9.90 \times 10^{-12} \text{ kg} = 9.90 \times 10^{-9} \text{g}$$

A loss of roughly 10 nanograms out of a total mass of 80 g can't be detected by weighing balances. We see why we can ignore the Einstein equation in working with the stoichiometry of any chemical change.

Where the Einstein equation is important is with nuclear changes. They are major sources of energy precisely because some matter changes into energy. Strict accuracy, therefore, requires that we restate the conservation laws by combining them into a **law of conservation of mass-energy.**

Law of Conservation of Mass-Energy. The sum of all the mass in the universe and of all the energy (expressed as an equivalent in mass) is a constant.

Like *all* laws of nature, this law is not known to be true without exception, simply because it is not possible to test it for all changes everywhere in the universe. Like all successful current laws, however, it has survived the tests that have thus far been made.

We want next to see how the Einstein equation can help us understand the energy available by nuclear changes. To reach that goal and to provide information essential to an understanding of the structure of atoms, we must first learn more about the composition of atomic nuclei, the topic of the next section.

5.10 ATOMIC NUCLEI

An atom has a dense core called a nucleus, which is surrounded by the atom's electrons

In the center of every atom is an exceedingly dense particle called the **nucleus.** Surrounding it are regions in which very light particles called the **electrons** of the atom reside. This is the basic picture we have of the structure of an atom: a nucleus surrounded by electrons. How the electrons are arranged determines the element's *chemistry,* but the *number* of electrons is set by one of the properties of the nucleus. We will concentrate our attention only on the nucleus in this section. (For the stories of the discoveries of the electron, the proton, and the nucleus, see Special Topics 5.3, 5.4, and 5.5.)

With one exception (hydrogen-1), the nuclei of all atoms consist of still smaller particles. The chief nuclear particles are the **proton** and the **neutron,** but there are many others. (These others are of no further interest in our study of chemistry.) Only atoms of one of the isotopes of hydrogen lack neutrons. Since they are found in nuclei, protons and neutrons are called **nucleons.**

The first hint that matter was electrical in nature came in 1834 when Michael Faraday, a British physicist, found that chemical changes would occur when an electrical current was sent through certain chemical solutions. Later in the century, some studies of the effects of electricity on matter were done with gas discharge tubes. These glass tubes contain a gas under low pressure, and they are fitted with metal wires called electrodes that can be plugged into a source of electricity. When the switch is thrown, the tube will glow. (Modern neon signs work this way.) The gas conducts the current in some way. The current moves from one wire, called the cathode, to the other, called the anode. What conducted the current from cathode to anode was called a cathode ray.

Cathode rays were found to consist of a stream of particles. They were not light waves. They could, for example, make a paddle wheel turn when it was placed in the way inside the tube. When a metal plate given a positive charge was put just outside the tube, the cathode rays bent toward the plate, which meant that the particles in the cathode ray were negatively charged. "Unlike charges attract." Regardless of the gas present in the tube or the metal in the electrodes, cathode rays of identical properties were observed. This meant that the particles in cathode rays were in *all* matter. They were fundamental subatomic particles, and were named electrons. Sir Joseph Thomson won the 1906 Nobel Prize in physics for this discovery.

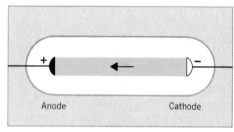

Gas discharge tube

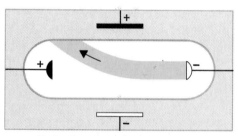

Deflection of a cathode ray
toward a positively charged plate.

One property of nucleons and electrons is that they have mass. Table 5.3 gives their relative masses, on the atomic weight scale. Notice how small the mass of the electron is compared to that of the nucleons. The electron has a mass of only about 1/1830 that of a proton or neutron. As far as atomic weights go, the electron's mass is negligible. No known element has atoms with more than 106 electrons. In one such atom the electrons would contribute only 0.0582 units to the relative mass, about 0.02% of the total atomic weight of that element. Except for the most precise calculations involving nuclear binding energies (Section 5.11), we ignore the electron's mass and we compute the mass of an atom as the sum of the masses of its nucleons. Each has a relative mass on the atomic weight scale very close to 1, and the simple numerical sum of the protons and neutrons in an atom of an isotope gives it the **mass number.** (Use this formal definition now to replace the interim definition we gave in Section 3.3.)

$106 \times 0.0005486 = 0.0582$

$$\text{mass number} = \text{number of protons} + \text{number of neutrons} \qquad (5.17)$$

Table 5.3

Properties of Subatomic Particles

Particle	Mass (amu)	Mass (g)	Electrical Charge	Symbol
Electron	0.0005486	9.109534×10^{-28}	$1-$	$_{-1}^{0}e$
Proton	1.007276	1.672649×10^{-24}	$1+$	$_{1}^{1}H$
Neutron	1.008665	1.674954×10^{-24}	0	$_{0}^{1}n$

Special Topic 5.4 The Discovery Of The Proton

After the discovery of the electron, gas discharge tubes (see Special Topic 5.3) were modified for additional experiments. In one series, the cathode was shaped like a disk with a hole in its center, and the tube's inner surface *behind* the cathode was coated with a chemical called a phosphor. Scientists had found that phosphors glow when struck by electron beams. (Today's TV tubes work on this principle. Different phosphors give off different colored lights.) In the new experiments, the phosphor was placed where the electron beam could not touch it, yet it still glowed. Something was moving through the hole in the cathode, in a direction opposite that of the cathode ray. It didn't take long to find out that a stream of particles was moving through the hole in the cathode and that those particles had a positive charge. (The beam, for example, could be deflected toward a negatively charged plate positioned outside the tube.) Unlike cathode rays, however, the masses of the particles varied depending on the gas present in the tube. When the tube had hydrogen gas, the mass of these positive particles was the lightest of all, yet it was still over 1800 times that of the mass of the electrons. When other gases were used, their masses always seemed to be some whole-number multiple of the mass observed with hydrogen. This suggested the possibility that clusters of the positively charged particles made from hydrogen atoms made up the posi-

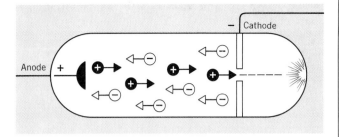

Positively charged particles are made when cathode rays (electrons) strike atoms of residual gas. They are attracted to the cathode, and some sail through the hole to strike the phosphor and generate a flash of light.

tively charged particles of other gases. The mechanism for making these particles, as the experimenters reasoned things out, was probably a collision between an electron and an atom of the residual gas. That collision knocked off an electron from the atom, leaving behind a particle with a positive charge. Since the lightest of all such particles came from hydrogen, it seemed likely that all gases—and all matter, in fact—were made of combinations of the particles in hydrogen. The hydrogen atom, minus an electron, thus seemed to be a fundamental particle in all matter, and that particle was therefore named the proton, after the Greek *proteios,* meaning "of first importance."

Like charges repel!

Two subatomic particles, the proton and the electron, carry electrical charge. You are familiar with electrical charge if you have ever felt the jolt or seen the spark when you touch a conductor (a metal object or a person) after walking across certain rugs when the air is very dry. The spark is a discharge of electrical charge. We know two kinds of charges, and you no doubt have experienced both. Have you ever tried to throw away a small piece of paper or plastic only to have it stick stubbornly to your fingers? Very annoying! The reason this happens is that you are carrying one kind of charge and the plastic or paper has picked up the opposite kind. *Opposite charges attract.* This is a very important rule in chemistry. Maybe you have seen your hair stand on end after drying it vigorously. Some see the same thing when they are outside on a hill or mountain during electrical disturbances. The individual hairs behave as if they repel each other, because each hair is carrying the same kind of electrical charge. *Like charges repel.* This is another very important rule. We are going to explain an astonishing amount of chemistry using these two rules that describe the behavior of electrical charges.

Attracting and repelling are opposites, and one way to signify opposites is to assign one a positive sign and the other a negative sign. The charge carried on a proton is, by definition, one unit of positive charge, $1+$. The electron has one unit of negative charge, $1-$. The neutron has no charge, so we say that its charge is zero (0). If free to move, protons repel each other and electrons repel each other, but protons and electrons attract each other.

All atomic nuclei are positively charged particles, because that is where the atom's protons are. *All isotopes (see Section 3.2) of the same element have iden-*

Special Topic 5.5 The Discovery Of The Atomic Nucleus

The effect of the cathode ray (an electron beam) on matter that led to the discovery of the proton suggested many similar experiments. The subatomic "bullets" did not all have to be made by gas discharge tubes, either. Elements had been discovered that spontaneously sent out showers of subatomic particles in a phenomenon called radioactivity. Some radioactive elements send out electrons, but others send out much larger particles having masses four times those of the proton and bearing two positive charges. These were called alpha particles.

Early in this century, Hans Geiger and Ernest Marsden, working under Ernest Rutherford at England's famous Cavendish laboratories, studied what happened when alpha rays hit thin metal foils. Most of the alpha particles sailed right on through as if the foils were virtually empty space. A significant number of alpha particles, however, were deflected at very

large angles. Some were even deflected backward as if they had hit stone walls. Rutherford was so astounded that he compared the effect to that of firing a 15-in. artillery shell at a piece of tissue paper and having it come back and hit the gunner. He reasoned that only something extraordinarily massive, compared to the alpha particle, could cause such an occurrence. From studying the angles of deflection of the particles, Rutherford determined that whatever it was in the foil that was so massive had to be positively charged. However, since most of the alpha particles went straight through, he further reasoned that the metal atoms in the foils must be mostly empty space. Rutherford's conclusion was that virtually all of the mass of an atom must be concentrated in a particle having a very small volume located in the center of the atom. He called this massive particle the atom's nucleus.

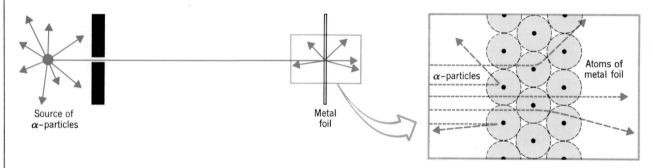

Alpha particles are scattered in all directions by a thin metal foil. Some hit something very massive head-on and are reflected backward. Many sail through. Some, making near misses with the massive "cores" (nuclei), are still deflected because alpha particles have the same kind of charge (+) as these cores. "Like charges repel."

tical numbers of protons. Every element can be characterized by a number unique to it, the number of protons in the nuclei of its atoms, which is called the **atomic number.**

$$\text{atomic number} = \text{number of protons} \qquad (5.18)$$

What makes isotopes of the same element different are the numbers of neutrons in their nuclei. This difference makes each isotope have its own mass number. Thus every isotope is fully defined by two numbers, the atomic number and the mass number. In writing equations involving particular isotopes, we show these numbers on the left of the atomic symbol. A left subscript shows the atomic number and a left superscript gives the mass number. Thus uranium-235 would be symbolized as follows:

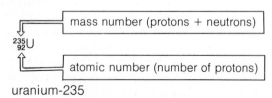

uranium-235

Special Topic 5.6   Solar Energy And Fusion

When electrons are stripped from nuclei at ultra-high temperatures, the resulting "gas" is called plasma. The temperature in the center of the sun is believed to be on the order of 10 million kelvins (10^7 K)—high enough to make plasma. The details are not well understood, but current speculation is that two types of fusion cycles occur in the sun to liberate energy—the proton-proton cycle and the carbon cycle.

Proton-Proton Cycle

$$^1_1H \; + \; ^1_1H \; \longrightarrow \; ^2_1H \; + \; ^0_1e \; + \; \nu$$

$$2\,^1_1H + 2\,^2_1H \longrightarrow 2\,^3_2He$$

$$^3_2He \; + \; ^3_2He \; \longrightarrow \; ^4_2He \; + \; ^1_1H \; + \; ^1_1H$$

Net: $\quad 4\,^1_1H \; \longrightarrow \; ^4_2He \; + \; 2\,^0_1e + 2\nu$

(The symbols ^0_1e and ν are, respectively, the positron and the neutrino—subatomic particles we will not discuss.)

Carbon Cycle

$$^1_1H \; + \; ^{12}_6C \; \longrightarrow \; ^{13}_7N$$

$$^{13}_7N \; \longrightarrow \; ^{13}_6C \; + \; ^0_1e \; + \; \nu$$

$$^{13}_6C \; + \; ^1_1H \; \longrightarrow \; ^{14}_7N$$

$$^{14}_7N \; + \; ^1_1H \; \longrightarrow \; ^{15}_8O$$

$$^{15}_8O \; \longrightarrow \; ^{15}_7N \; + \; ^0_1e \; + \; \nu$$

$$^{15}_7N \; + \; ^1_1H \; \longrightarrow \; ^{12}_6C \; + \; ^4_2He$$

Net: $\quad 4\,^1_1H \; \longrightarrow \; ^4_2He + 2\,^0_1e + 2\nu$

The net effect of the carbon cycle is the same as that of the proton-proton cycle. The fusion of four protons into one helium nucleus gives a calculated 4.28×10^{-12} J. The synthesis of 1 mol of helium nuclei would liberate 2.58×10^{12} J.

An atomic bomb explosion produces a temperature of about 5×10^7 K, high enough to initiate fusion. A hydrogen bomb consists of light nuclei with an atom bomb trigger to cause fusion.

EXERCISE 5.9   Write the symbol for the isotope of plutonium (Pu) with 146 neutrons.

EXERCISE 5.10   How many neutrons are in each atom of ^4_2He?

Now that we have some knowledge of an atom's composition we can write a better definition of the atom than the one handed down to us by Dalton. Atoms are electrically neutral particles having one nucleus and a number of electrons equal to the atomic number. The number of electrons in an atom must equal the number of protons (atomic number) in order for the electrical charges to "cancel," meaning that all the pluses added to all the minuses comes out to zero.

$$\text{atomic number} = \text{number of protons} = \text{number of electrons} \qquad (5.19)$$

The nuclei of some elements have combinations of protons and neutrons that are less stable than those of others. Some nuclei can be made more stable by forcing them to fuse together to form nuclei with larger mass numbers. This change is called nuclear **fusion.** It is the process that creates the energy sent out by the sun, and it is also the source of the energy of the hydrogen bomb. Other nuclei move toward greater stability by breaking apart. One way in which unstable nuclei break up is called radioactivity, which we will study in Chapter Twenty-one. The other way is called nuclear **fission,** which is the source of the energy of nuclear power plants and the atomic bomb. The huge yields of energy from either fusion or fission involve nuclear binding energies, which we will study next.

INDING ENERGIES

When the most precise values are used, the sum of the masses of the nucleons does *not* equal the mass of the nucleus

When the masses of all nucleons in one nucleus are added, the sum is generally a trifle *larger* than the actual mass of that nucleus. It is as if some mass was "lost" when the nucleus was formed. Actually, the extra mass was changed into energy that went into the surroundings. This energy is called the nuclear binding energy because its loss resulted in a more stable binding together of the nucleons. (Generally, systems become more stable when they lose energy.) The **binding energy** of a particular isotope is the energy equivalent of the difference in mass between the nucleus itself and the sum of the masses of its nucleons. We will calculate the binding energy for helium-4 to illustrate this definition.

For this calculation it will help if we use the name for one unit of relative mass introduced in Special Topic 3.1, the **atomic mass unit** (amu). Rounded to five significant figures,

$$1 \text{ amu} = 1.6606 \times 10^{-24} \text{ g}$$

To find the binding energy of a nucleus of a helium-4 atom, we have to find the difference between its observed mass and the mass calculated for two protons and two neutrons. The observed mass for one atom of helium-4 is 4.002603 amu; one *nucleus* of helium-4 has that mass less the mass of the atom's 2 electrons, each with a mass of 0.0005486 amu.

The observed mass was measured by using special instruments.

formula weight of helium-4 = 4.002603 amu
formula weight of 2 electrons,
or (2) × (0.0005486) = 0.001097 amu
formula weight of helium-4 nucleus = 4.001506 amu, the observed mass of
a helium-4 nucleus

However, if we calculate the formula weight of a helium-4 nucleus from the sum of the masses of its nucleons, we get a slightly higher value.

2 protons: 2 × 1.007276 amu = 2.014552 amu
2 neutrons: 2 × 1.008665 amu = 2.017330 amu
total relative mass of nucleons = 4.031882 amu, the "calculated"
mass of a helium-4 nucleus

The difference between the calculated and observed masses is 0.030376 amu (4.031882 amu − 4.001506 amu). Now we must calculate the energy equivalent of this difference in mass using Einstein's equation and the fact that 1 amu = 1.6606×10^{-24} g.

$$\Delta E = \Delta mc^2 = \underbrace{(0.030376 \text{ amu}) \left(\frac{1.6606 \times 10^{-24} \text{ g}}{1 \text{ amu}} \right) \left(\frac{1 \text{ kg}}{1000 \text{ g}} \right)}_{\Delta m \text{ in kilograms}} \underbrace{\left(3.00 \times 10^8 \frac{\text{m}}{\text{sec}} \right)^2}_{(\text{velocity of light})^2}$$

$$= 4.54 \times 10^{-12} \frac{\text{kg m}^2}{\text{sec}^2}$$

The units of this result are those of the joule (equation 5.1). Therefore, the binding energy of the nucleus of helium-4 is 4.54×10^{-12} J (1.09×10^{-12} cal). Since there are 4 nucleons in the helium-4 nucleus, the average is

$$\frac{4.54 \times 10^{-12} \text{ J}}{4 \text{ nucleons}} = 1.14 \times 10^{-12} \text{ J/nucleon}$$

If we could make Avogadro's number of these nuclei—only 4 g—we would see the net release of $6.02 \times 10^{23} \times 4.54 \times 10^{-12}$ J, or 2.73×10^{12} J of energy. That could keep a 100-watt light bulb lit for nearly 900 years. We're beginning to see the awesome energy available by fusion.

Similar calculations have been made for other isotopes. When the binding energies *per nucleon* are plotted versus atomic numbers (Figure 5.11), we see that the most stable isotopes are those having intermediate atomic numbers. Remember, the more binding energy per nucleon, the *more* stable the nucleus. Nuclear stability rises with atomic number until number 26, iron, is reached. Then it slowly drops. Below 26, we can expect to find nuclei that could become more stable by *fusion*. Therefore, we look for the possibility of fusion energy among the lightest elements. Above 26, among the heavy elements, we can expect to find nuclei that could become more stable by *fission*.

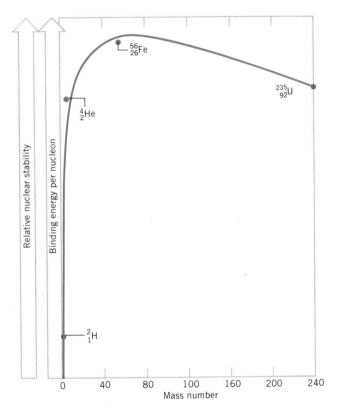

Figure 5.11
Nuclear binding energies per nucleon.

In addition to fusion and fission as possible ways for relatively less stable nuclei to change into more stable nuclei, there are several other less substantial changes that unstable nuclei can undergo. Many unstable nuclei adjust their numbers of neutrons and protons by emitting small particles, and the streams of these particles make up the dangerous atomic radiations you may have heard about. When nuclei send out various radiations, they are said to undergo **radioactive decay,** where ''radio-'' means that radiations are emitted and ''decay'' means that small changes in numbers of protons and neutrons occur. Decay proceeds until a stable array of protons and neutrons is reached and, therefore, the end product of radioactive decay is an atom of a different element. We will come back to the details about radioactive decay and atomic radiation in Chapter Twenty-one.

5.12 FUSION

Controlled fusion is likeliest among the nuclei of tritium and deuterium

No one yet knows if fusion will ever be successfully controlled. "Success" means generating more useful energy than is needed to initiate fusion. The basic scientific problems may be solved, but the engineering problems have barely been touched. One problem is that atoms having nuclei to be fused must first be stripped of their electrons. This can be done, but only by extraordinarily high temperatures, and that is costly. Another problem is that nuclei repel each other because they are like-charged. To hold them near each other long enough to get them to fuse demands properties possessed by no known materials. One means of containing hot nuclei is by using magnetic forces to create a magnetic "bottle," a cavity shaped by magnetic forces only and having no material walls (Figure 5.12). Shaping and holding magnetic forces exactly in the right configuration to prevent all leakage is a major difficulty. In another approach, lasers are used (Figure 5.13).

Isotopes of hydrogen are the likeliest candidate for successful fusion. Hydrogen is the only element whose isotopes have different names. Hydrogen-2 is

Figure 5.12
Fusion by magnetic confinement. (a) Cutaway view of a full-scale Ormak reactor nearing the end of its assembly. Two partially assembled sectors are in the foreground. (b) Cross-sectional view through the reactor (the lithium is used in the nuclear cycle that synthesizes tritium).

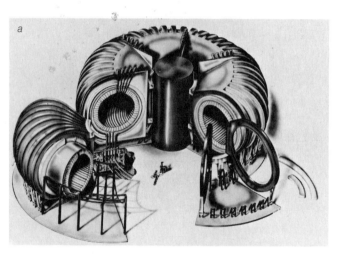

a

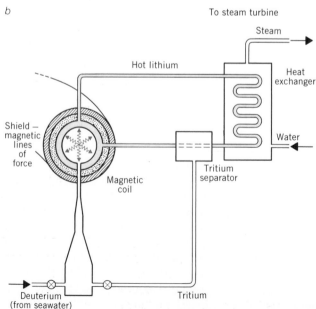

b

To steam turbine

Steam

Hot lithium

Heat exchanger

Water

Shield — magnetic lines of force

Magnetic coil

Tritium separator

Deuterium (from seawater)

Tritium

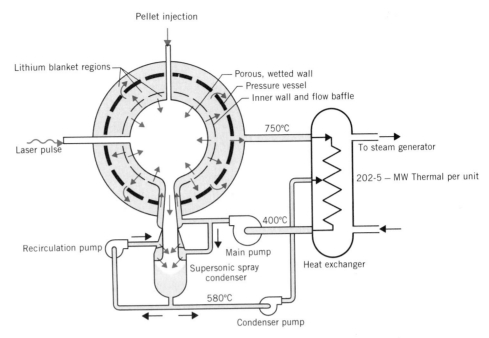

Figure 5.13
Fusion by laser implosion. The pellets bled in at the top are
hollow beads of thin glass filled with deuterium and tritium.
When a pellet is caught in intersecting pulses of a laser beam,
its contents are heated enough to lose their electrons. The
pulses cause the glass to flake off, much like a heat shield on
a reentering space capsule, and that action collapses the
capsules inward, creating the intense pressure needed to hold
the nuclei together long enough to fuse. At least, that's the
general idea of a project under investigation (U.S. Atomic
Energy Commission, WASH-1239, 1973).

called **deuterium** and hydrogen-3 is called **tritium.** Each isotope consists of
atoms with only one electron, and one electron is much easier to strip than are
two or more. Once stripped, the particles have only a 1+ charge, so they repel
each other much less than the bare nuclei of any higher elements would. We may
represent the fusion of tritium and deuterium to give helium-4 as follows:

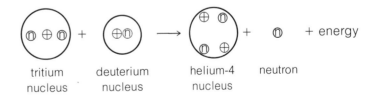

The tritium needed for fusion is itself made by one of two processes we will not
discuss. (The general topic of how isotopes can be made synthetically will be
discussed in Chapter Twenty-one.) Making tritium costs energy, and this cost
has to be figured into the calculations of the yields of energy obtainable by fusion. Depending on the way tritium is made, the calculated energy yield for each
helium-4 atom produced by the fusion of tritium and deuterium is 3.56×10^{-12}
J/He-4 atom (for one process) or 7.94×10^{-13} J/He-4 atom (by another
process). These yields translate into enough energy to keep a 100-watt light bulb
going for 150 to 700 years for each mole of He-4 formed.

Our chief source of deuterium is the ocean. All water contains some deuterium
as D_2O and DOH, the fully and partly deuterated forms of water, H_2O. According

For every 100,000 atoms of
natural hydrogen, 15 are deuterium, ^2_1H, and the rest are hydrogen, ^1_1H (sometimes called
protium).

to calculations by M. King Hubbert, if only 1% of the deuterium in the world's oceans were removed for fusion, it could supply 500,000 times the amount of energy in the combined total of all of the original fossil fuels in the world. The deuterium in only 0.005 km³ of ocean would have supplied all of the United States' energy needs for 1968. (The oceans' total volume is about 1.4×10^9 km³.) If scientists and engineers ever control fusion and its related pollution problems successfully, the world would have a supply of energy that would, for all practical purposes, be infinite. That is why energy planners believe that the research should be done, even though there is no guarantee it will work.

5.13 FISSION

Fission is a nuclear chain reaction that can be controlled

Fission is not spontaneous; it has to be initiated by streams of neutrons. An isotope that is capable of undergoing fission is called a **fissile isotope.** The important fissile isotopes are uranium-233, uranium-235, and plutonium-239, but only uranium-235 occurs naturally. Uranium metal is 0.71% U-235 and nearly all the rest is U-238.

The initiation of the fission of one atom of U-235 happens when the nucleus of that isotope captures one of the neutrons being produced by fission elsewhere in the sample. This neutron capture makes the nucleus very unstable, and it breaks apart. For example, neutron capture by an atom of uranium-235 might lead to the following nuclear change, where we use the symbol $_0^1n$ for the neutron.

The neutron is assigned an "atomic number" of zero, since it has no charge, and its symbol is $_0^1n$.

$$_0^1n \ + \ _{92}^{235}U \longrightarrow \ _{36}^{94}Kr \ + \ _{56}^{139}Ba \ + \ 3\,_0^1n$$

neutron   krypton-94   barium-139   additional neutrons

Actually, this equation represents only one of several possible ways in which atoms of uranium-235 can undergo fission. Once the U-235 nucleus captures a neutron, it momentarily becomes a nucleus of the U-236 atom. This nucleus can break apart in a number of ways, much as different crackers can be broken into different pieces. Generally, fission gives two isotopes of intermediate atomic mass numbers plus varying numbers of neutrons. These neutrons are free to collide with unchanged fissile atoms and initiate further fission events. The general equation for the fission of U-235 is

$$_{92}^{235}U + _0^1n \longrightarrow X + Y + 2.4\,_0^1n + 2.0 \times 10^{10} \text{ kJ/mol} \qquad (4.7 \times 10^9 \text{ kcal/mol})$$

where X represents elements with mass numbers between 85 and 110 and Y represents those between 125 and 150.

Because one fission can produce neutrons for initiating over two new fissions, and those fissions give neutrons for more than four more fissions and so on, fission is a chain reaction (see Figure 5.14). A **chain reaction** is a self-sustaining reaction in which products from one event cause one or more new events. The yield in energy from fission is so great that only 1 g of uranium-235 has the energy equivalent to 3 tons of soft coal or 13.2 barrels of oil. When enough mass, called the critical mass, of pure uranium-235 is brought together suddenly, a chain reaction affecting most of the atoms happens in a brief moment. The sudden release of the energy of fission causes an atomic explosion.

Pure U-235 is not used in nuclear power plants. Natural uranium is processed to enrich its concentration of U-235 from 0.71 to 2.4%. At that low concentration of U-235, no critical mass can form, and it is not possible for a nuclear power plant to explode like an atomic bomb. The bomb uses pure material—either U-235 or Pu-239. However, unless fission is carefully controlled in a nuclear

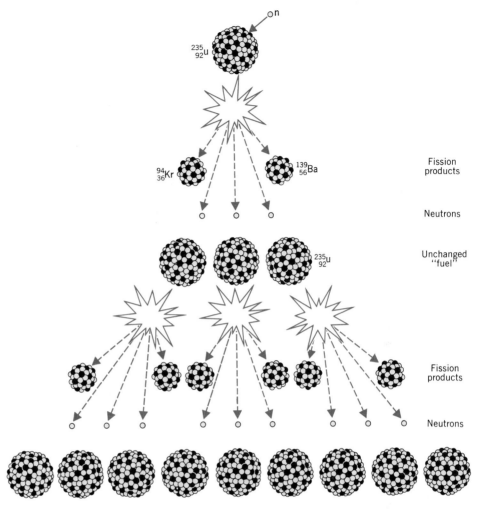

Figure 5.14
Nuclear chain reaction. Not all neutrons are captured. In an
atomic bomb, however, where the concentration of fissile isotope
is high and the volume occupied is large enough to have the
critical mass, the percent of neutrons captured is so high
that the nuclear reaction occurs everywhere in the mass and
it explodes. In nuclear reactors, control rods of nonfissile
material are inserted into or withdrawn from the fuel elements
to maintain a rate of fissioning that lets the heat be withdrawn
as rapidly as it is released.

power plant (Figure 5.15), it will generate enough heat to swamp the system's
ability to remove the heat. The danger, if this were to happen, is that a meltdown
of the reactor core would occur. If not brought under control, the intense heat
would melt the containment chamber housing the nuclear reactor, and a slug of
intensely radioactive matter would in all probability melt down through the containment chamber. This would release radioactive gases that would move out
and downwind. The possibility of such an accident is a cause of great controversy. Critics of nuclear power believe that the risks cannot be made small
enough. They cite the Three Mile Island incident in Pennsylvania in 1979 as evidence. People favoring nuclear power are satisfied that the risks are manage-
able, and they also cite the Three Mile Island incident to support their side.

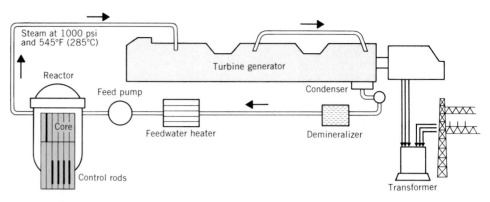

Figure 5.15
Nuclear reactor. Shown here is a boiling water reactor, the type used in most of the nuclear power plants in the United States. Water pumped around and through the fuel elements in the core carries away the heat of the nuclear chain reactions. Steam thus generated at 1000 psi (pounds per square inch) drives the turbines and is cooled before going back to the core (U.S. Atomic Energy Commission, WASH 1261, 1973).

Another vexing issue of nuclear power is the major problem of safely storing radioactive wastes. Some of the wastes must be held secure from human contact for hundreds of centuries, and critics argue that until this problem is solved, nuclear power should not be used. Proponents believe that technology will find a solution, but both sides agree that the problem is extremely difficult.

5.14 BREEDER REACTORS

In principle, the breeder reactor would extend the supplies of nuclear fuel for hundreds of years

The electron is assigned an "atomic number" of −1, since it bears that charge. Its symbol is $_{-1}^{0}e$.

Uranium-238, the common isotope of natural uranium, is not fissile, but it is fertile. A **fertile isotope** is one that can, by nuclear reactions, be converted into a fissile isotope. When the nucleus of an atom of uranium-238 captures a neutron, the following series of changes occurs:

$$_{0}^{1}n + {}_{92}^{238}U \longrightarrow {}_{92}^{239}U \longrightarrow {}_{93}^{239}Np + {}_{-1}^{0}e$$

electron

The initial capture produces an atom of U-239. Then an interesting event happens. The *nucleus* of U-239 ejects an electron. How could it, since there aren't any electrons in a nucleus? In effect, what happens is that a neutron changes into a proton and expels the electron that is left over. With one new proton, the atomic number becomes 93, that of neptunium-239. As Figure 5.16 shows, this loses another electron and becomes plutonium-239.

Uranium-238 is present in the fuel of power plants using uranium-235. In fact, it makes up 96 to 98% of the fuel. Some of the neutrons of the chain reactions of U-235 are captured by U-238. The end product is plutonium-239, which is a *fissile isotope*. The net effect is that of breeding an atom of a fissile isotope from a nonfissile, but fertile, isotope. This process would allow all of the atomic energy in uranium-238 to become available for use through plutonium-239. Various countries are developing the technology for making this a commercial operation, and reactors operating on these principles are called **breeder reactors.**

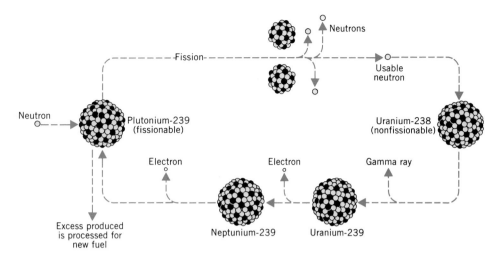

Figure 5.16
The breeding cycle whereby uranium-238 (nonfissile) is changed into plutonium-239 (fissile). (From J.R. Holum. *Topics and Terms in Environmental Problems*, 1978, Wiley-Interscience, New York, used by permission).

Even ordinary reactors breed some plutonium-239, and this material can be extracted from spent fuel. It can be used as fuel in ordinary reactors, or it can be diverted to make atomic bombs. This stark fact has been the reason behind the reluctance of the U.S. government to sell nuclear power plants. The customers, of course, then go elsewhere. India, for example, made its first atomic bomb from plutonium-239 extracted as a by-product in a nuclear power plant it bought from Canada.

Another problem with plutonium is that it is the most dangerous material ever discovered. The danger is caused by its radioactivity, and we will return to that aspect of nuclear energy in Chapter Twenty-one.

SUMMARY

Energy and Chemical Change. The law of conservation of mass-energy applies to all kinds of changes but, for chemical reactions, a simpler law of conservation of energy holds for any realistic precision in measurement. The net energy as heat that is produced or consumed by a chemical reaction can be measured with a calorimeter. The heat capacity of this instrument (or any object) is the heat energy released (or absorbed) per degree Celsius change in its temperature. The joule is the SI unit of energy but, for many heat measurements, the calorie (1 J = 4.184 cal, exactly) or kilocalorie is used. Every substance has its own value of specific heat, the heat needed to change the temperature of 1 g of it by 1 °C.

Energy of Reaction. The change in energy, ΔE, for a change in a system is a state function; it is completely independent of the path or the steps used to go from the initial to the final state. Part of the net ΔE may involve a flow of heat, and some may be in the form of work. How much of ΔE is in

the form of heat and how much is in the form of work does depend on the path or the steps used, but the total ΔE is independent of path. When no work is done, as in the operation of a bomb calorimeter, all the heat is ΔE, and ΔE = energy of the reaction.

Enthalpy of Reaction. When a reaction causes changes in the moles of gas and the work of expanding gases, it is better to work with ΔH, or change in enthalpy, instead of ΔE. $\Delta H = \Delta E$ + (expansion work). Where no expansion work is possible, then $\Delta H = \Delta E$. ΔH is a state function. Therefore, as Hess discovered, the overall value of ΔH for a change is the same if it happens in one or in a series of steps. When conditions of 25 °C and 1 atm are specified, ΔH is $\Delta H°$ and is called the standard heat of the reaction. A balanced equation becomes a thermochemical equation when the associated value of $\Delta H°$ is given. If such an equation shows the formation of 1 mol of a compound from its elements, all in their standard states, the corresponding $\Delta H°$ is $\Delta H_f°$, the standard heat of formation of the compound. (Some of the coefficients in such an equation may have to be fractions in order that the coefficient for compound being formed can be 1.) Any substance is in its standard state when, at 25 °C and 1 atm, it is in its most stable chemical form.

Calculating $\Delta H°$. The value of $\Delta H°$ for any reaction, whether or not it can actually be carried out, can be calculated from values of $\Delta H_f°$, if they are available, since $\Delta H° =$ [(sum of $\Delta H_f°$ of all products) − (sum of $\Delta H_f°$ of all reactants)]. Otherwise, $\Delta H°$ can be found using any combination of thermochemical equations that add up to the desired net reaction. The value of $\Delta H°$ for the reverse of a reaction is equal but opposite in sign to the value for the forward reaction, a fact guaranteed by the law of conservation of energy.

Electromagnetic Energy. This form of energy is transmitted at a speed of 3.00×10^8 m/sec (in a vacuum) by means of packets called photons. Associated with each photon is a frequency, ν, and a wavelength, λ. Their product, $\lambda \times \nu$, equals c, the speed of light. The energy of a photon is proportional to its associated frequency; $E = h\nu$, where h is Planck's constant. The success that the energy in any particular part of the electromagnetic spectrum has in initiating a chemical reaction depends solely on the frequency associated with its photons, not on how many or how often the photons strike the system, that is, how intense or bright the light is.

Solar Energy. The steady absorption of photons by green plants over eons of time triggered the photosynthetic growth of materials that later changed into the fossil fuels—coal, oil, and natural gas. Technologies that could be used—if developed—to take advantage of presently arriving solar energy include wind energy, biomass conversion, biomass conversion, ocean thermal energy, direct solar heat, direct conversion to electricity by solar cells, and indirect conversion using solar heat.

Atomic Nuclei and Nuclear Energy. Atomic nuclei, which are at the centers of atoms and surrounded by electrons, are made of nucleons—protons and neutrons. The number of protons is the atomic number; the sum of the protons and neutrons is the mass number. Each proton has a charge of 1 +; each neutron has zero charge. Therefore, the charge on a nucleus equals the atomic number. The atoms of an element that have different mass numbers make up the isotopes of the element. At the formation of a nucleus, a tiny quantity of mass from the fusing protons and neutrons is changed to energy—the binding energy. Isotopes differ in their binding energies per nucleon. Those of small mass number could, in principle, become more stable by fusion and release energy in the process. Those of high mass number could, again in principle, become more stable by fission, also releasing energy in the process. In practice, fusion that actually produces a net yield of energy will be realized only with deuterium or tritium, two isotopes of hydrogen. Fission, in practice, is possible with only one naturally occurring isotope, uranium-235. Fission occurs by means of a chain reaction involving neutrons. Neutrons can be captured by the nuclei in uranium-238, which is not fissile, and by a series of steps change U-238 into plutonium-239, a fissile isotope. This overall event is called breeding and occurs in breeder reactors. The safe containment of nuclear reactors during an accident, the adverse health effects of radioactive pollutants, and the indefinite safe storage of atomic wastes are major problems with atomic energy.

INDEX TO EXERCISES, QUESTIONS AND PROBLEMS
(Those numbered above 76 are review problems.)

REVIEW QUESTIONS

5.11. What is meant by "energy source"?

5.12. How are the calorie and the joule related?

5.13. How are the kilocalorie and the kilojoule related?

5.14. Which kind of substance will need more energy to undergo a rise of 5 degrees in temperature, something with a *high* or something with a *low* specific heat?

5.15. Which kind of substance will experience the greater rise in temperature upon receiving 10 kcal, something with a *high* or something with *low* specific heat?

5.16. Suppose that specific heat had been defined in terms of $\frac{kcal}{kg\ K}$. Would the numerical values in Table 5.1 change? If so, in what way?

5.17. Explain how water's high specific heat helps guard against swings in internal temperature as the outside temperature fluctuates.

5.18. How are the calorie and the Calorie related?

5.19. Why can't we know the *total* energy of, say, a 1-lb sack of sugar?

5.20. What do we mean by "system"?

5.21. What is meant by "surroundings"?

5.22. If a difference exists between the total energy of a system and its potential energy, what is that difference called?

5.23. How is ΔE formally defined?

5.24. What is the law of conservation of energy?

5.25. If ΔE for a system increases by, say, 100 kJ, what must be true about $\Delta E_{surroundings}$? Why?

5.26. When ΔE_{system} is positive for a chemical change, what descriptive name do we give that reaction?

5.27. What is the sign of ΔE_{system} for an exothermic change?

5.28. How can a system undergoing an exothermic change under conditions where no work is done become *warmer* if we define an exothermic change as one in which the system suffers a *loss* of energy?

5.29. Describe in your own words how ΔE_{system} is determined experimentally.

5.30. What is the difference between ΔH_{system} and ΔE_{system}?

5.31. How is ΔH_{system} defined formally?

5.32. How is ΔH_{system} for a chemical reaction measured?

5.33. Under what circumstances are ΔE_{system} and ΔH_{system} the same or nearly so?

5.34. In what way do the definitions of STP and "standard state" differ? (Review Section 3.9 for the meaning of STP.)

5.35. What are acceptable units for $\Delta H°$? For $\Delta H_f°$?

5.36. We stated that for *all* elements in their most stable forms, $\Delta H_f° = 0$. Is that a guess, a matter of defining it that way, or an experimentally measured fact?

5.37. Why do we specify the physical state, *g*, *ℓ*, *s*, or *solution*, when we give data for $\Delta H°$ or $\Delta H_f°$? Illustrate your answer with a specific example where such specifications make a difference.

5.38. What is Hess's law?

5.39. Would Hess's law be possible (and could we use enthalpy diagrams) if ΔH were not a state function? Explain.

5.40. In general terms, why do we call light "electromagnetic energy"?

5.41. What is meant by the "frequency of light"? What symbol is used for it, and what is the SI unit (and symbol) for frequency?

5.42. What do we mean by the "wavelength" of light? What symbol is used for it?

5.43. How is the frequency of a particular type of radiation related to the energy associated with it? (Give an equation, defining all symbols.)

5.44. Show that $E = \frac{hc}{\lambda}$.

5.45. What names do we give to various ranges of frequencies in the electromagnetic spectrum, beginning from the lowest frequency side?

5.46. Examine each of the following pairs and state which of the two has the higher *energy*.
(a) Microwaves and infrared.
(b) Visible light and infrared.
(c) Ultraviolet light and X rays.
(d) Visible light and ultraviolet light.

5.47. What is meant by "visible spectrum"?

5.48. What is a photon?

5.49. What is a quantum of energy?

5.50. What are the principal technologies classified under solar energy?

5.51. In approximate terms only, how do the world's reserves of coal and oil compare with the solar energy absorbed each year by our planet?

5.52. What is the function of chlorophyll in green plants?

5.53. In general terms, what is photosynthesis?

5.54. Name the three principal fossil fuels.

5.55. Which is the least polluting of the fossil fuels?

5.56. Why is each of the following classified as a form of solar energy: (a) wind, (b) ocean thermal gradients, and (c) biomass conversion?

5.57. What is the Einstein equation?

5.58. Why is the law of conservation of mass as opposed to the stricter law of conservation of mass-energy sufficient for the stoichiometry of chemical reactions?

5.59. What are the nucleons named in this chapter?

5.60. The sum of what two numbers gives the mass number for an isotope?

5.61. What do the nuclei of *all* isotopes of one element have in common?

5.62. In what way do the nuclei of isotopes of the same element differ?

5.63. What is the difference between a mass number and an atomic number?

5.64. Since atoms, by definition, are electrically uncharged particles, what must be the relation between the atomic number and the number of electrons in each atom of an element?

5.65. We do not call just any small, electrically uncharged particle an atom. What else must be true about atoms?

5.66. We do not call just any two atoms isotopes of each other. What must be true about them in particular?

5.67. Why isn't the sum of the masses of all nucleons in one nucleus equal to the mass of the actual nucleus?

5.68. Explain why fusion is easiest to cause between isotopes of hydrogen and not between isotopes having higher atomic numbers?

5.69. What is the difference between fission and fusion with

respect to (a) what happens, in general terms, and (b) which kinds of elements are likeliest candidates?

5.70. What is meant by a *fissile* isotope?

5.71. Which fissile isotope occurs in nature?

5.72. What fact about the fission of U-235 makes it a *chain* reaction?

5.73. One gram of U-235 is approximately equivalent in energy to how much coal? How much oil?

5.74. What does it mean when we say that U-238 is *fertile?*

5.75. How does a breeder reactor breed fuel?

5.76. What are the major issues raised by opponents of nuclear power and what do proponents say in response?

REVIEW PROBLEMS

5.77. In 458 kcal there are how many joules? Kilojoules?

5.78. How many kilocalories are in 5000 J?

5.79. The British thermal unit or BTU is a unit of energy much more commonly used in fuels engineering than either the kilocalorie or the joule. It is the energy needed to raise the temperature of 1 lb of water by one degree on the Fahrenheit scale. If $1 J = 9.48 \times 10^{-4}$ BTU, then 1 cal is equivalent to how many BTU?

5.80. For convenience in working with huge quantities of energy, the unit of the *quad* is used. In the late 1970s, for example, the United States consumed about 80 quads of energy per year. By definition, 1 quad = 10^{15} BTU (defined in problem 5.79). Using your results from problem 5.79, calculate how many kilocalories are in one quad.

5.81. A car weighing 4000 lb is traveling at 55 miles per hour. How much kinetic energy does it have in kilojoules? (Use conversion factors found in various tables in this book.)

5.82. A baseball weighing 5.0 oz traveling at 66 miles per hour has how much kinetic energy in kilojoules? (Use conversion factors found in various tables in this book.)

5.83. How much heat (in kilocalories) has to be extracted from 250 g of water to lower its temperature from 25 °C to 10 °C? (That would be like cooling a glass of lemonade.)

5.84. To bring 1.0 kg of water from 25 °C to 99 °C takes how much heat input (in kilocalories)? (That could be like making four cups of coffee.)

5.85. Fat tissue is 85% fat and 15% water. The complete breakdown of the fat itself converts it to CO_2 and H_2O, and releases about 9.0 kcal/g (of fat).
(a) How many kcal would be released by a loss of 1.0 lb of fat *tissue* in a weight-reduction program?
(b) A person running at 8.0 miles per hour expends about 500 kcal/hr of extra energy. How far would a person have to run to "burn off" 1.0 lb of fat *tissue* by this means alone?

5.86. A well-nourished person will add about 0.5 lb of fat tissue for each 3500 kcal of food energy taken in over and

above that needed. Suppose you decided to reduce your weight simply by omitting butter but keeping every other aspect of your diet and your activities the same. How many days would be needed to lose 1.0 lb of fat *tissue* by this strategy alone? The caloric content of butter is 9.0 kcal/g, and suppose you have been eating 0.25 lb of butter a day.

5.87. Write thermochemical equations for the formation of each of these compounds from its elements: (a) $Fe_2O_3(s)$, (b) $CH_3Cl(g)$, and (c) $CO(NH_2)_2(s)$ urea.

5.88. Write thermochemical equations for the formation of each of these compounds from its elements: (a) $HNO_3(\ell)$, (b) $SO_3(g)$, and (c) $NaOH(s)$.

5.89. Using the data in Table 5.2, calculate $\Delta H°$ for each of the following reactions.
(a) $2H_2O_2(\ell) \longrightarrow 2H_2O(\ell) + O_2(g)$
(b) $HCl(g) + NaOH(s) \longrightarrow NaCl(s) + H_2O(\ell)$
(c) $CH_4(g) + Cl_2(g) \longrightarrow CH_3Cl(g) + HCl(g)$
(d) $2NH_3(g) + CO_2(g) \longrightarrow CO(NH_2)_2(s) + H_2O(\ell)$
(urea)

5.90. From data in Table 5.2, calculate $\Delta H°$ for each of the following reactions.
(a) $CO_2(g) + 2H_2(g) \longrightarrow C(s) + 2H_2O(\ell)$
(b) $CH_4(g) + I_2(s) \longrightarrow CH_3I(g) + HI(g)$
(c) $2PbO(s) + C(s) \longrightarrow 2Pb(s) + CO_2(g)$
(d) $2NaHCO_3(s) \longrightarrow Na_2CO_3(s) + H_2O(\ell) + CO_2(g)$

5.91. From $\Delta H_f°$ data for the formation of $CO_2(g)$ and $H_2O(g)$, and the following data for the combustion of glucose— $C_6H_{12}O_6(s) + 6O_2(g) \rightarrow 6CO_2(g) + 6H_2O(\ell)$ $\Delta H°_{combustion} = -673$ kcal—estimate $\Delta H_f°$ of glucose, a primary product of photosynthesis.

5.92. Palmitic acid, $C_{16}H_{32}O_2$, is typical of the materials available from fats and oils insofar as chemical energy is concerned. The complete combustion of palmitic acid goes by the following equation:

$C_{16}H_{32}O_2(s) + 23O_2(g) \longrightarrow 16CO_2(g) + 16H_2O(\ell)$
$$\Delta H°_{combustion} = -2380 \text{ kcal}$$

Estimate the standard heat of formation of palmitic acid from these data and data in Table 5.2.

5.93. Sulfur trioxide, a trace air pollutant, reacts with water to produce corrosive sulfuric acid. What is $\Delta H°$ for this reaction: $SO_3(g) + H_2O(\ell) \rightarrow H_2SO_4(\ell)$? Use $\Delta H_f°$ data from Table 5.2.

5.94. Nitrogen dioxide, an air pollutant, combines with water to make nitric acid (HNO_3), a corrosive acid, and nitrogen monoxide (NO). What is $\Delta H°$ for this reaction: $3NO_2(g) + H_2O(\ell) \rightarrow 2HNO_3(\ell) + NO(g)$?

5.95. Iron oxides can be changed to iron metal by action of hot carbon. What is $\Delta H°$ for this reaction: $Fe_2O_3(s) + 3C(s) \rightarrow 2Fe(s) + 3CO(g)$?

5.96. Carbon dioxide can be removed from air by passing the air through solid granules of sodium hydroxide. What is $\Delta H°$ for this reaction: $CO_2(g) + NaOH(s) \rightarrow NaHCO_3(s)$?

5.97. Hydrogen chloride can be generated by heating a mixture of sulfuric acid and potassium chloride:

$$2KCl(s) + H_2SO_4(\ell) \longrightarrow 2HCl(g) + K_2SO_4(s)$$

Calculate $\Delta H°$ for this reaction from the following thermochemical equations. Give your answer in kilocalories and kilojoules.

$$HCl(g) + KOH(s) \longrightarrow KCl(s) + H_2O(\ell)$$
$$\Delta H° = -48.66 \text{ kcal}$$
$$H_2SO_4(\ell) + 2KOH(s) \rightarrow K_2SO_4(s) + 2H_2O(\ell)$$
$$\Delta H° = -81.83 \text{ kcal}$$

5.98. Calculate $\Delta H°$ for the following reaction for the preparation of an unstable acid, HNO_2, nitrous acid.

$$HCl(g) + NaNO_2(s) \longrightarrow HNO_2(\ell) + NaCl(s)$$

Use these thermochemical equations. Find the answer in kilocalories and kilojoules.

$$2NaCl(s) + H_2O(\ell) \longrightarrow 2HCl(g) + Na_2O(s)$$
$$\Delta H° = +121.25 \text{ kcal}$$
$$NO(g) + NO_2(g) + Na_2O(s) \longrightarrow 2NaNO_2(s)$$
$$\Delta H° = -102.09 \text{ kcal}$$
$$NO(g) + NO_2(g) \longrightarrow N_2O(g) + O_2(g)$$
$$\Delta H° = -10.20 \text{ kcal}$$
$$2HNO_2(\ell) \longrightarrow N_2O(g) + O_2(g) + H_2O(\ell)$$
$$\Delta H° = +8.21 \text{ kcal}$$

5.99. What frequency corresponds to a wavelength of 1.00×10^{-4} μm and in what range of the electromagnetic spectrum is it?

5.100. What are the limits of the visible spectrum in nanometers if they correspond to frequencies of 4.3×10^{14} Hz and 7×10^{14} Hz?

5.101. What is the mass number, the atomic number, and the number of neutrons per nucleus for (a) $^{235}_{92}U$, (b) carbon-13, and (c) an isotope of $^{54}_{26}Fe$ having one less neutron (and write its corresponding symbol)?

5.102. What is the atomic number, the mass number, and the number of neutrons per nucleus for (a) $^{54}_{24}Cr$, (b) tritium, and (c) an atom having the same mass number but one less proton than $^{66}_{30}Zn$ (write its symbol also)?

5.103. How many protons, electrons, and neutrons are in an atom of (a) $^{31}_{15}P$, (b) $^{119}_{50}Sn$, (c) $^{251}_{98}Cf$, and (d) $^{184}_{74}W$?

5.104. How many electrons, protons, and neutrons are in one atom of (a) $^{12}_{6}C$, (b) $^{64}_{29}Cu$, (c) $^{131}_{53}I$, and (d) $^{243}_{95}Am$?

5.105. Calculate the binding energy per nucleon of the deuterium nucleus whose mass is 2.0135 amu.

5.106. Calculate the binding energy per nucleon of the tritium nucleus whose mass is 3.01550 amu.

CHAPTER SIX

THE PROPERTIES OF ELEMENTS AND THE PERIODIC TABLE

6.1 THE NEED FOR ORGANIZATION

The large volume of chemical information gathered by scientists can only be understood if it is organized

In our lives we are faced constantly with a barrage of information of all sorts. New and seemingly unrelated facts bombard us daily. To make some sense of these facts, to be able to appreciate their significance, and to use them effectively, we must in some way organize the information we receive. Nowhere is this more true than in the sciences. The number of facts is enormous. Still, if we arrange them properly, important similarities, implications, and trends become clear and useful. In biology, for instance, the classification system for plants and animals has allowed biologists to study systematically the similarities and differences among the various species. A better understanding of evolution is but one of the benefits gained from these studies.

Organization is as important in chemistry as it is in biology. Indeed, the need for order led directly to the development of the modern periodic table (also often called the periodic chart), located on the inside front cover of this book. By the middle of the nineteenth century, scientists had gathered a wealth of empirical facts about the physical and chemical properties of the elements. They were not satisfied, though, with simply knowing that certain substances behaved in certain ways. They wanted to know *why* the facts they had discovered were true. Dalton's atomic theory, developed in the early 1800s, had helped provide some answers, but many questions still remained. Being able to answer them depended on finding some order in the jumble of chemical information.

A natural way to approach this problem was to categorize the elements according to their properties. If you think about it, we do much the same thing to expand our own knowledge of the world about us. We get to know the properties of substances by observing how things look, feel, smell, and act. As we gain experience, we learn to identify new materials and predict their properties by comparing them with others with which we are familiar and by noting similarities. For example, we have learned to associate luster and electrical conductivity with certain metals that we know. Then, when we come across a new substance that is shiny and conducts electricity, we feel fairly safe in assuming that it, too, is a metal.

To categorize elements according to their properties, scientists also searched

Smelling or tasting chemicals can be a very risky business. Don't do so unless instructed by your teacher.

The leg bone of a dog that had received an injection of a radioactive isotope of strontium about eight years earlier. This radioautograph is produced by placing the bone in contact with photographic film. Dark spots show high concentration of the radioactive strontium isotope.

for similarities. It was necessary to sort through many differences to find them. This is because the properties of the elements cover a broad range. Some are gases, some are liquids, and some are solids; some are very reactive and some seem totally unreactive. Nevertheless, similarities do exist.

Sometimes, the similarities among the elements can have frightening consequences. Strontium, for example, is very similar to calcium. Because of this, the dangerous radioactive isotope strontium-90 tends to accumulate in the bones of growing children along with the needed calcium. Another problem pair is zinc and cadmium. Zinc is the crystalline-looking coating used on galvanized steel (which is used to make things such as garbage cans). Because cadmium and zinc have similar properties, cadmium is often a contaminant in zinc. Zinc is (relatively) harmless. Cadmium, however, if absorbed by the body, can cause high blood pressure, heart disease, and even painful death.

In this chapter we will examine further some of the facts of chemistry and see how they led to the development of the periodic table. We will also see that by studying the way that chemical and physical properties vary according to an element's position in the periodic table we are better able to predict behavior and remember chemical facts. Equally important, we will see throughout the next several chapters that the chemical information contained in the periodic table serves as a foundation for theories that tell us about the internal structure of atoms. These theories provide explanations for the similarities and differences that the elements exhibit.

6.2 THE FIRST PERIODIC TABLE

Both Mendeleev and Meyer discovered that the properties of the elements repeat at regular intervals when the elements are arranged in order of increasing atomic weight

There were many early attempts to discover relationships among the chemical and physical properties of the elements. A number of different sequences of elements were tried in the search for some sort of order or pattern. A few of these arrangements came close, at least in some respects, to our current table, but either they were flawed in some way or they were presented to the scientific community in a manner that did not lead to their acceptance.

The periodic table we use today is based primarily on the efforts of a Russian chemist, Dmitri Ivanovich Mendeleev (1834–1907), and a German physicist, Julius Lothar Meyer (1830–1895). Working independently, these scientists developed similar periodic tables only a few months apart in 1869. Mendeleev is usually given the credit, however, because he had the good fortune to publish first.

Mendeleev was preparing a chemistry textbook for his students at the University of St. Petersburg. He found that when he arranged the elements in order of increasing atomic weight, similar chemical properties occurred over and over again at regular intervals. For instance, the elements lithium (Li), sodium (Na), potassium (K), rubidium (Rb), and cesium (Cs) have similar chemical properties. Each of them forms a water-soluble chlorine compound with the general formula MCl: LiCl, NaCl, KCl, RbCl, and CsCl. Moreover, the elements that immediately follow each of these elements also comprise a set with similar chemical properties. Beryllium (Be) follows lithium, magnesium (Mg) follows sodium, calcium (Ca) follows potassium, strontium (Sr) follows rubidium, and barium (Ba) follows cesium. All of these elements form a water-soluble chlorine compound with the general formula MCl$_2$: BeCl$_2$, MgCl$_2$, CaCl$_2$, SrCl$_2$, and BaCl$_2$. Based on extensive observations of this type, Mendeleev devised the original form of the periodic law. It stated that *the chemical and physical properties of the elements vary*

Periodic refers to the recurrence of properties at regular intervals.

in a periodic way with their atomic weights. Mendeleev used this law to construct his periodic table, which is shown in Figure 6.1.

The elements in Mendeleev's table are arranged in rows, called **periods,** in order of increasing atomic weight. When the rows are broken at the right places and stacked, the elements fall naturally into columns, called **groups** so that elements in any given column have similar chemical properties. Mendeleev's genius rested on his placing elements with similar properties in the same group even though this left occasional gaps in the table. For example, he placed arsenic (As) in Group V under phosphorus because its chemical properties were similar to those of phosphorus, even though this left gaps in Groups III and IV. Men-

Figure 6.1
Mendeleev's periodic table as it appeared in 1871. The numbers next to the symbols are atomic weights.

Period	Group I	Group II	Group III	Group IV	Group V	Group VI	Group VII	Group VIII
1	H 1							
2	Li 7	Be 9.4	B 11	C 12	N 14	O 16	F 19	
3	Na 23	Mg 24	Al 27.3	Si 28	P 31	S 32	Cl 35.5	
4	K 39	Ca 40	−44	Ti 48	V 51	Cr 52	Mn 55	Fe 56, Co 59 Ni 59, Cu 63
5	(Cu 63)	Zn 65	−68	−72	As 75	Se 78	Br 80	
6	Rb 85	Sr 87	?Yt 88	Zr 90	Nb 94	Mo 96	−100	Ru 104, Rh 104 Pd 105, Ag 108
7	(Ag 108)	Cd 112	In 113	Sn 118	Sb 122	Te 128	I 127	
8	Cs 133	Ba 137	?Di 138	?Ce 140	—	—	—	— —
9	—	—	—	—	—	—	—	
10	—	—	?Er 178	?La 180	Ta 182	W 184	—	Os 195, Ir 517 Pt 198, Au 199
11	(Au 199)	Hg 200	Tl 204	Pb 207	Bi 208	—		
12	—	—	—	Th 231	—	U 240	—	— — — —

deleev reasoned, correctly, that the elements that belonged in these gaps had simply not yet been discovered. Based on the location of these gaps, however, Mendeleev was able to predict with remarkable accuracy the properties of these yet-to-be-found substances. In fact, his predictions helped serve as a guide in the search for these missing elements. Table 6.1 compares Mendeleev's predicted properties of "eka-silicon" and some of its compounds with the actual properties measured for the element germanium.

Table 6.1
Comparison of Some Predicted Properties of Eka-Silicon and Observed Properties of Germanium

Property	Observed for Silicon (Si)	Predicted for Eka-Silicon (Es)	Observed for Tin (Sn)	Found for Germanium (Ge)
Atomic weight	28	72	118	72.59
Melting point (°C)	1410	high	232	947
Molar volume (cm³/mol)	12.1	13	16.2	13.6
Density (g/cm³)	2.33	5.5	7.28	5.35
Formula of oxide	SiO_2	EsO_2	SnO_2	GeO_2
Density of oxide (g/cm³)	2.66	4.7	6.95	4.23
Formula of chloride	$SiCl_4$	$EsCl_4$	$SnCl_2$	$GeCl_4$
Boiling point of chloride (°C)	57.6	100	114	84

Two elements, tellurium (Te) and iodine (I) caused Mendeleev some problems. The atomic weight of tellurium was known to be greater than that of iodine. Yet, if these elements were placed in the table according to their atomic weights, they would not fall into the proper groups required by their properties. Therefore, Mendeleev switched their order and in so doing violated his own periodic law. (Actually, he believed that his law was correct and the atomic weight of tellurium had been incorrectly measured, but this wasn't true.)

The table that Mendeleev developed is in many ways similar to the one we use today. One of the main differences, though, is that it lacks the elements in Group 0—helium (He) through radon (Rn). In Mendeleev's time, none of these elements had yet been found because they are very rare and because they have virtually no tendency to undergo chemical reactions. When they finally were discovered, beginning in 1894, another problem arose. Two more elements, argon (Ar) and potassium (K) did not fall into the groups required by their properties if they were placed in the table in the order required by their atomic weights. Another switch was necessary. Now there were two exceptions to Mendeleev's periodic law. Apparently, then, atomic weight was not the true basis for the periodic repetition of the properties of the elements. To determine what the true basis was, however, scientists had to await the discovery of the atomic nucleus, the proton, and atomic numbers.

6.3 THE MODERN PERIODIC TABLE

In our modern periodic table, the elements are arranged in the order of their increasing atomic number

In Section 5.10, we learned that the nucleus of an atom contains protons and neutrons and that it is surrounded by the atom's electrons. The number of protons in the nucleus, you recall, is the atom's atomic number. One of the gratifying dividends of the discovery of atomic numbers was that the elements in Mendeleev's table turned out to be arranged in precisely the order of increasing atomic number. In other words, if we take atomic numbers as the basis for arranging the elements in sequence, no annoying switches are required and the elements Te and I or Ar and K are no longer a problem. This leads to our present statement of the periodic law.

> **The Periodic Law.**   The chemical and physical properties of the elements vary in a periodic way with their atomic *numbers.*

The fact that it is the atomic number—the number of protons in the nucleus of an atom—that determines the order of elements in the table is very significant. We will see later that this has very important implications with regard to the relationship between the number of electrons in an atom and the atom's chemical properties.

Our modern periodic table, which appears on the inside front cover of the book is also reproduced in Figure 6.2. We will find many occasions to refer to the table, so it is important for you to become familiar with it and with some of the terminology applied to it.

As in Mendeleev's table, the elements are arranged in rows called **periods,** but here they are arranged in the order of increasing atomic number. For identification purposes the periods are numbered. We will find these numbers useful later on. Below the main body of the table there are two long rows of 14 elements each that actually belong in the main body of the table following La ($Z = 57$) and Ac ($Z = 89$) as shown in Figure 6.3. They are almost always placed below the table simply to conserve space. Fully spread out, the table requires too much room to be conveniently printed on one page. Notice that in the fully extended form of the table, with all the elements arranged in their proper locations, there is a great deal of empty space. An important requirement of a detailed atomic theory, which we will get to in the next chapter, is that it must explain not only the repetition of properties, but also why the empty space appears.

Again as in Mendeleev's table, the vertical columns are called **groups.** They are also frequently referred to as **families** of elements because of the similarities that exist among the members of each group. Notice that the columns are divided into A-groups and B-groups. Those labeled A, along with Group 0, are referred to as the **representative elements.** Elements that fall into the B-groups in the center of the table are called **transition elements.** The two long rows of elements below the table are called the **inner transition elements,** and each row is named after the element that it follows in the main body of the table. Thus ele-

We use the symbol Z to stand for atomic number.

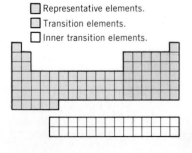
☐ Representative elements.
☐ Transition elements.
☐ Inner transition elements.

Atomic number → 1 H 1.0079 ← Atomic mass

Periods	IA	IIA	IIIB	IVB	VB	VIB	VIIB	VIII	VIII	VIII	IB	IIB	IIIA	IVA	VA	VIA	VIIA	Noble gases 0
1	1 H 1.0079																	2 He 4.00260
2	3 Li 6.941	4 Be 9.01218											5 B 10.81	6 C 12.011	7 N 14.0067	8 O 15.9994	9 F 18.998403	10 Ne 20.179
3	11 Na 22.98977	12 Mg 24.305											13 Al 26.98154	14 Si 28.0855	15 P 30.97376	16 S 32.06	17 Cl 35.453	18 Ar 39.948
4	19 K 39.0983	20 Ca 40.08	21 Sc 44.9559	22 Ti 47.90	23 V 50.9415	24 Cr 51.996	25 Mn 54.9380	26 Fe 55.847	27 Co 58.9332	28 Ni 58.70	29 Cu 63.546	30 Zn 65.38	31 Ga 69.72	32 Ge 72.59	33 As 74.9216	34 Se 78.96	35 Br 79.904	36 Kr 83.80
5	37 Rb 85.4678	38 Sr 87.62	39 Y 88.9059	40 Zr 91.22	41 Nb 92.9064	42 Mo 95.94	43 Tc (98)	44 Ru 101.07	45 Rh 102.9055	46 Pd 106.4	47 Ag 107.868	48 Cd 112.41	49 In 114.82	50 Sn 118.69	51 Sb 121.75	52 Te 127.60	53 I 126.9045	54 Xe 131.30
6	55 Cs 132.9054	56 Ba 137.33	57 *La 138.9055	72 Hf 178.49	73 Ta 180.9479	74 W 183.85	75 Re 186.207	76 Os 190.2	77 Ir 192.22	78 Pt 195.09	79 Au 196.9665	80 Hg 200.59	81 Tl 204.37	82 Pb 207.2	83 Bi 208.9804	84 Po (209)	85 At (210)	86 Rn (222)
7	87 Fr (223)	88 Ra 226.0254	89 †Ac 227.0278	104 Unq (261)	105 Unp (262)	106 Unh (263)												

*	58 Ce 140.12	59 Pr 140.9077	60 Nd 144.24	61 Pm (145)	62 Sm 150.4	63 Eu 151.96	64 Gd 157.25	65 Tb 158.9254	66 Dy 162.50	67 Ho 164.9304	68 Er 167.26	69 Tm 168.9342	70 Yb 173.04	71 Lu 174.967
†	90 Th 232.0381	91 Pa 231.0359	92 U 238.029	93 Np 237.0482	94 Pu (244)	95 Am (243)	96 Cm (247)	97 Bk (247)	98 Cf (251)	99 Es (254)	100 Fm (257)	101 Md (258)	102 No (259)	103 Lr (260)

Figure 6.2
The modern periodic table.

1 H																															2 He
3 Li	4 Be																									5 B	6 C	7 N	8 O	9 F	10 Ne
11 Na	12 Mg																									13 Al	14 Si	15 P	16 S	17 Cl	18 Ar
19 K	20 Ca	21 Sc															22 Ti	23 V	24 Cr	25 Mn	26 Fe	27 Co	28 Ni	29 Cu	30 Zn	31 Ga	32 Ge	33 As	34 Se	35 Br	36 Kr
37 Rb	38 Sr	39 Y															40 Zr	41 Nb	42 Mo	43 Tc	44 Ru	45 Rh	46 Pd	47 Ag	48 Cd	49 In	50 Sn	51 Sb	52 Te	53 I	54 Xe
55 Cs	56 Ba	57 *La	58 Ce	59 Pr	60 Nd	61 Pm	62 Sm	63 Eu	64 Gd	65 Tb	66 Dy	67 Ho	68 Er	69 Tm	70 Yb	71 Lu	72 Hf	73 Ta	74 W	75 Re	76 Os	77 Ir	78 Pt	79 Au	80 Hg	81 Tl	82 Pb	83 Bi	84 Po	85 At	86 Rn
87 Fr	88 Ra	89 †Ac	90 Th	91 Pa	92 U	93 Np	94 Pu	95 Am	96 Cm	97 Bk	98 Cf	99 Es	100 Fm	101 Md	102 No	103 Lr	104 Unq	105 Unp	106 Unh												

Figure 6.3
Extended form of the periodic table. The two long rows are placed in their proper places in the table.

ments 58 through 71 are called the **lanthanide elements,** and elements 90 through 103 are called the **actinide elements.**

Some of the groups have acquired common names. The Group IA elements, for example, are metals. They form compounds with oxygen that dissolve in water to give solutions that are strongly alkaline or caustic. As a result, these elements have become known as the **alkali metals** or simply the **alkalis.** The Group IIA elements are also metals. Their oxygen compounds are alkaline, too, but many compounds of the Group IIA elements are unable to dissolve in water and are found in deposits in the ground. Because of their properties and where they are found, the Group IIA elements became known as the **alkaline earth metals.**

On the right side of the table, in Group 0, are the **noble gases.** They used to be called **inert gases** until it was discovered that the heavier members of the group show a small degree of reactivity. The term *noble* is used when we wish to suggest a very limited degree of reactivity. Gold, for example, is sometimes referred to as a noble metal because so few chemicals are capable of reacting with it.

Finally, the elements in Group VIIA are called the **halogens,** derived from Greek words meaning sea or salt. Chlorine (Cl), for example, is found in familiar table salt, NaCl.

EXERCISE 6.1 Circle the correct choices in the following:
(a) Representative elements are: K, Cr, Pr, Ar, Al.
(b) A halogen is: Na, Fe, O, (Cl) Cu.
(c) An alkaline earth metal is: Rb, Ba, La, As, Kr.
(d) A noble gas is: H, Ne, F, S, N.
(e) An alkali metal is: Zn, Ag, Br, Ca, (Li).
(f) An inner transition element is: Ce, Pb, Ru, Xe, Mg.

6.4 METALS, NONMETALS, AND METALLOIDS

Elements are classified as metals, nonmetals, or metalloids according to such properties as electrical conductivity and luster

The periodic table is as important to chemists and chemistry students as good maps are to a traveler. The table organizes all sorts of chemical and physical information about the elements and their compounds. It allows us to study systematically the way properties vary with an element's position within the table and, in turn, makes the similarities and differences among the elements easier to understand and remember.

Even a casual inspection of samples of the elements reveals that some are familiar metals and that others equally familiar are not. Most of us are already familiar with metals such as lead, iron, or gold and nonmetals such as oxygen or nitrogen. A closer look at the nonmetallic elements, though, reveals that some of them, silicon and arsenic to name two, have properties that lie in between those of true metals and true nonmetals. These elements are called **metalloids.** Division of the elements into these three categories—metals, nonmetals, and metalloids—is not an even one, however (see Figure 6.4). Most of the elements are metals, slightly over a dozen are nonmetals, and only a handful are metalloids.

Metals

You probably know a metal when you see one. They tend to have a shine or luster that is easily recognized. For example, the color photograph of a freshly exposed surface of sodium in Color Plate 1C would most likely lead you to identify this ele-

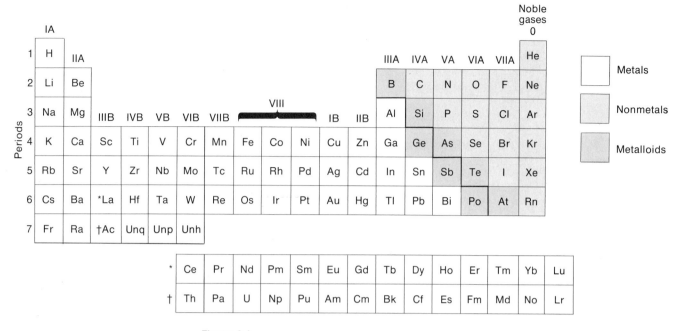

Figure 6.4
Distribution of metals, nonmetals, and metalloids in the periodic table.

ment as a metal even if you had never seen or heard of it before. We also know that metals conduct electricity. Few of us would poke an iron nail we were holding in our hand into an electrical outlet. In addition, we know that metals conduct heat very well. On a cool day, metals always feel colder to the touch than do neighboring nonmetallic objects because metals conduct heat away from your hand very rapidly. Nonmetals seem less cold because they can't conduct heat away as quickly and therefore their surfaces warm up faster.

Other properties that metals possess, to varying degrees, are malleability—their ability to be hammered or rolled into thin sheets—and ductility—their ability to be drawn into wire. For example, the production of sheet steel (Figure 6.5)

Figure 6.5
The malleability of iron is demonstrated in the production of sheet steel.

Figure 6.6
The ductility of copper allows it to be drawn into wire. Here copper wire passes through one die after another as it is drawn into thinner and thinner wire.

for automobiles and household appliances depends on the malleability of iron and steel, and the manufacture of electrical wire (Figure 6.6) is based on the ductility of copper.

Thin lead sheets are used for sound-deadening because the soft, easily deformed lead absorbs the sound vibrations.

Another important physical property that we usually think of when metals are mentioned is hardness. Some, such as chromium or iron, are indeed quite hard, but others, like copper and lead, are rather soft. The alkali metals are so soft that they can be cut with a knife, but they are also so chemically reactive that we rarely get to see them as free elements.

All of the metallic elements, except mercury, are solids at room temperature. Mercury's low freezing point ($-39\,°C$) and fairly high boiling point ($357\,°C$) make it useful as a fluid in thermometers. Most of the other metals have much higher melting points, and some are used primarily because of this. Tungsten, for example, has the highest melting point of any metal ($3400\,°C$ or $6150\,°F$), which explains its use as the filament in electric light bulbs. Cobalt and chromium are combined in alloys (mixtures of metals) such as "stellite" to make high-speed, high-temperature cutting tools for industry. This alloy retains its hardness even at high temperatures, and the cutting tools therefore remain sharp.

A bead of mercury on a porcelain surface.

The chemical properties of metals vary tremendously. Some, such as gold and platinum, are very unreactive toward almost all chemical agents. This property, plus their natural beauty and their rarity, makes them highly prized for use in jewelry. Other metals, however, are so reactive that few people except chemistry students ever have an opportunity to see them. For instance, the metal sodium reacts very quickly with oxygen or moisture in the air, and its bright metallic surface tarnishes almost immediately. On the other hand, compounds of sodium are quite stable and very common. Examples are table salt ($NaCl$), baking soda ($NaHCO_3$), lye ($NaOH$), and bleach ($NaOCl$). We will have more to say about the chemical properties of metals in Section 6.5.

Nonmetals

We see many objects each day that are clearly not metals. Some examples are plastics, wood, concrete, and glass. These aren't elements, though. Most often, we encounter the nonmetallic elements in the form of compounds or mixtures of compounds. There are, however, some nonmetals that are very important to us in their elemental forms. The air we breathe, for instance, contains mostly nitrogen, N_2, and oxygen, O_2. Both are gaseous, colorless, and odorless nonmetals. Since we can't see, taste, or smell them, however, it's difficult to experience their existence (although if you step into an atmosphere without oxygen, your body will very quickly tell you that something is missing!). Probably the most commonly *observed* nonmetallic element is carbon. We find it as the graphite in pencils, as coal, and as the charcoal used for barbecues. It also occurs in the more valuable form of diamonds. Although diamond and graphite differ in appearance, each is a form of elemental carbon.

Photographs of some of the nonmetallic elements appear in Color Plate 3. Their properties are almost completely opposite those of metals. Each of these elements lacks the characteristic appearance of a metal. They are poor conductors of heat and, with the exception of the graphite form of carbon, are also poor conductors of electricity. The electrical conductivity of graphite appears to be an accident of molecular structure, since the structures of metals and graphite are completely different.

Many of the nonmetals are solids at room temperature and atmospheric pressure, while many others are gases. All of the Group 0 elements are gases in which the particles consist of single atoms. The other gaseous elements—hydrogen, oxygen, nitrogen, fluorine, and chlorine—are composed of diatomic molecules. Their formulas are H_2, O_2, N_2, F_2, and Cl_2. As we learned in Section 3.9, a molecule is diatomic if it is composed of two atoms. Bromine and iodine are also diatomic, but bromine is a liquid and iodine is a solid at room temperature.

The nonmetallic elements lack the malleability and ductility of metals. A lump of sulfur will crumble if hammered or break apart if pulled on. Diamond cutters rely on the brittle nature of that nonmetallic solid when they split a gem quality stone by carefully striking it a quick blow with a sharp blade.

As with metals, nonmetals exhibit a broad range of chemical reactivity. Fluorine, for instance, is extremely reactive. It reacts readily with almost all the other elements. At the other extreme is helium, the gas used to inflate children's balloons and the Goodyear blimp. This element has never been made to react with anything. Chemists find helium useful when they want to provide a totally inert (unreactive) atmosphere inside some apparatus.

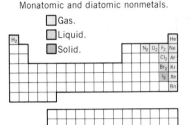

Monatomic and diatomic nonmetals.

☐ Gas.
☐ Liquid.
☐ Solid.

Mercury and bromine are the only two liquid elements at room temperature and pressure.

Metalloids

The properties of metalloids lie in between those of metals and those of nonmetals. This shouldn't surprise us since the metalloids are located between the metals and the nonmetals in the periodic table. In most respects, metalloids behave as nonmetals, both chemically and physically. However, in their most important physical property, their electrical conductivity, they somewhat resemble metals. Metalloids tend to be **semiconductors**—they conduct electricity, but not nearly as well as metals. This property, particularly as found in silicon and germanium, is responsible for the remarkable progress made during the last decade in the field of solid-state electronics. The newest hi-fi stereo systems, television receivers, and CB radios rely heavily on transistors made from semiconductors. Perhaps the most amazing advance of all has been the fantastic reduction semiconductors have allowed in the size of electronic components. To it, we owe the development of small and versatile hand-held calculators and microcomputers. The heart of these devices is a microcircuit printed on a tiny silicon chip (Figure 6.7).

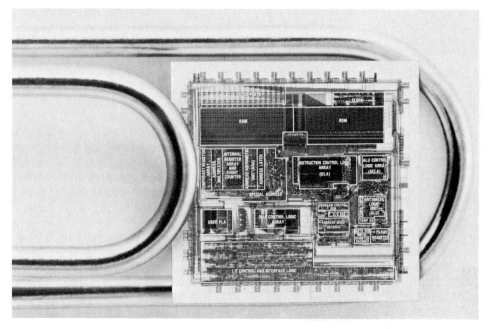

Figure 6.7
This microcircuit printed on a tiny silicon chip is compared to
a standard-sized paper clip. It is the Bell System's 30,000-
element MAC-4 "Computer-on-a-chip." Silicon's semiconductor
properties make possible microelectronic devices such as this.

6.5 SOME CHEMICAL PROPERTIES OF METALS

Metals react with nonmetals to produce compounds, called salts, that contain ions

A property possessed by nearly every element is the ability to combine with other elements to form compounds of varying degrees of complexity. However, not all combinations of elements appear possible. For example, sodium reacts just fine with chlorine to form sodium chloride (NaCl), but no compound is formed between sodium and iron. When we examine chemical properties of the elements, though, we find that certain generalizations are possible. It is worth looking at these now for a number of reasons. First, they provide a general framework within which we can study chemical behavior in greater detail later. Second, they illustrate how the periodic table can make it easier to remember chemical properties. And finally, they show us that to understand the chemical behavior of the elements, we need a theoretical model that explains what actually occurs between atoms when they react.

Metals react with nonmetals
An important property of metals is their ability to combine chemically with non-metals. In fact, except for some alloys, all compounds containing metals also contain nonmetals. A key aspect of most reactions between metals and non-metals is the transfer of one or more electrons from an atom of the metal to an atom of the nonmetal.

A huge amount of chemistry can be explained in terms of events no more complicated than the transfer, or the giving and receiving, of electrons. As we noted in Section 3.2, some kinds of atoms can lose electrons. Other kinds of atoms can accept electrons. However, these transfers occur at the expense of the atom's electrical neutrality. If an atom loses one electron, what is left behind is no longer an atom. It is a new particle with a net charge of 1 +. Similarly, if an atom gains an electron, a new particle with a net charge of 1− is formed. Since these particles are no longer atoms (by definition atoms are neutral), we give them a new name—ions. An **ion** is an electrically charged particle having at least one atomic nucleus but with too few or too many electrons to have a net charge of zero. A positive ion is called a **cation** (pronounced *CAT-ion*) and a negative ion is called an **anion** (pronounced *AN-ion*).[1]

The reaction of the two elements sodium and chlorine is a good illustration of the formation of ions. In this reaction, each sodium atom loses one electron and becomes a sodium ion, which has the symbol Na^+. Each chlorine atom accepts one electron to become a chloride[2] ion, which has the symbol Cl^-. We can express the net result of these two changes as follows:

$$Na \longrightarrow Na^+ + e^- \qquad \text{(electron, } e^-\text{, lost by Na)}$$
$$e^- + Cl \longrightarrow Cl^- \qquad \text{(electron, } e^-\text{, gained by Cl)}$$

Since opposite charges attract, the oppositely charged ions in solid NaCl attract each other, and these attractions hold the solid together. Because it is composed of positive and negative ions rather than neutral atoms, NaCl is classified as an **ionic compound.** Overall, however, the compound is electrically neutral because it is composed of an equal number of Na^+ and Cl^- ions. Compounds formed between metal ions and nonmetal ions are called salts, after the most common example, NaCl. We will see an even broader definition of the term *salt* in Section 6.8.

In ionic compounds, there are no molecules—we can't say that one particular Na^+ belongs to one particular Cl^-. The ions are simply stacked together in the most efficient way so that positive and negative ions can be as close to each other as possible. Because of this, we only specify the ratio in which the ions occur—*we always write empirical formulas for ionic compounds.* Thus the formula NaCl simply gives the 1:1 ratio of Na^+ to Cl^- ions. In terms of lab-sized quantities, a mole of NaCl contains a mole of formula units, where each formula unit consists of one Na^+ and one Cl^-.

You must be sure to learn the differences between the symbols for ions and those for atoms. What they indicate can be vital. For instance, Na^+ is the symbol of the sodium *ion*. It is a substance that we need in all parts of our bodies. However, Na (without the plus sign) is the symbol of the sodium *atom* as well as the element. If you were to swallow some of this substance, you would probably die or, at the very least, permanently scar your mouth and esophagous. Thus you should always be certain to include the electrical charge[3] when writing the symbol of the ion. The only exception will be when you are writing complete formulas, such as NaCl, in which the charges on the ions are omitted but "understood."

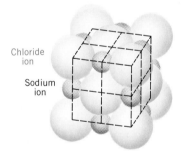

Chloride ion

Sodium ion

A crystal of sodium chloride (above). Below, packing of ions in NaCl. Notice that a cubic crystal is formed by the cubic packing of ions.

Metal-nonmetal compounds of the representative elements

By this time you have probably begun to wonder *why* atoms transfer electrons when they react. How are *you* expected to know that Na loses one electron and that Cl gains one when they combine with each other? And what happens in reactions between other elements? The answers to some of these questions will have to wait until the next chapter. For now, however, we can find some generalizations within the periodic table that can help us remember what kinds of ions

[1] The names, cation and anion, come from the way the ions behave when charged electrodes are dipped into a solution containing them. We will discuss this in detail in Chapter Eighteen.

[2] The names of negative ions are frequently formed by taking the name of the element and changing the ending to "ide." Positive ions simply take the name of the element followed by the word "ion." Don't worry too much about names now. We will say more about them in Chapter Eight.

[3] The symbol of ions with charges of 1+ or 1− are generally written without the numeral 1, for example, Na^+ and Cl^- instead of Na^{1+} and Cl^{1-}. The 1 is "understood."

Hydrogen is a special case. It is not a metal like the other Group IA elements and does not react by losing electrons.

are formed in certain reactions. For example, except for hydrogen, the neutral atoms of the Group IA elements always lose one electron when they react, thereby becoming ions with a charge of $1+$. Group IIA elements lose two electrons and form ions with a $2+$ charge. If we let M stand for a metal atom, then

$$M_{(Group\ IA)} \longrightarrow M^+_{(Group\ IA)} + e^-$$
$$M_{(Group\ IIA)} \longrightarrow M^{2+}_{(Group\ IIA)} + 2e^-$$

In Group IIIA the only important positive ion we need to consider now is that of aluminum, Al^{3+}. All of these ions are listed in Table 6.2. Notice that for the positive ions in this table the charge is equal to the group number. This doesn't work for all the elements, but it helps us remember what happens to the elements of Groups IA and IIA when they react.

Table 6.2
Some Ions Formed from Metals (M) and Nonmetals (X)

M^+	M^{2+}	M^{3+}	X^{2-}	X^-
Group IA	Group IIA	Group IIIA	Group VIA	Group VIIA
Li^+	Be^{2+}		O^{2-}	F^-
Na^+	Mg^{2+}	Al^{3+}	S^{2-}	Cl^-
K^+	Ca^{2+}		Se^{2-}	Br^-
Rb^+	Sr^{2+}		Te^{2-}	I^-
Cs^+	Ba^{2+}			

On the right-hand side of the table, we find that the halogens in Group VIIA form ions with a charge of $1-$ and the elements in Group VIA produce ions with a charge of $2-$. Notice that the number of negative charges on the ion is equal to the number of spaces to the right that we have to move to get to a noble gas.

Since all chemical compounds are electrically neutral, the ions in an ionic compound always occur in such a ratio that the total positive charge is equal to the total negative charge. This is why the formula for sodium chloride is NaCl—the $1:1$ ratio of Na^+ to Cl^- gives electrical neutrality. In calcium chloride, however, it takes two Cl^- to balance the charge of a single Ca^{2+}. Therefore, the formula for calcium chloride is $CaCl_2$.

The requirement of electrical neutrality allows us to write the formulas of ionic compounds very easily. There are only three rules to remember:

1 *The positive ion is always given first in the formula.* Nature doesn't demand this—it's just a custom that we always follow.
2 *The subscripts in the formula must produce an electrically neutral formula unit.* Nature does demand this.
3 *The subscripts should be the smallest set of whole numbers possible.* The reason for this rule is that we always write empirical formulas (simplest formulas) for ionic compounds.

Example 6.1 illustrates how we use these rules.

EXAMPLE 6.1 **Writing Formulas for Ionic Compounds**

Problem: Write the formulas for the ionic compounds formed from (a) Al and Cl, (b) Al and O, and (c) Ba and S.

Solution: In each case, the ions must be combined in a ratio that produces an electrically neutral formula unit with the smallest set of whole number subscripts.

(a) For these elements the ions are Al^{3+} and Cl^-. A neutral formula unit can be obtained by combining one Al^{3+} and three Cl^-. (The charge on Cl^- is $1-$; the 1 is understood.)

$$1(3+) + 3(1-) = 0$$

The formula is $AlCl_3$.

(b) For these elements the ions are Al^{3+} and O^{2-}. In the formula that we seek, there must be the same number of positive charges as negative charges. This number must be a whole number multiple of both 3 and 2. The smallest number that has both 3 and 2 as factors is 6, so there must be two Al^{3+} and three O^{2-} in the formula.

$$\begin{aligned} 2(3+) &= 6+ \\ 3(2-) &= \underline{6-} \\ \text{sum} \quad &\;\; 0 \end{aligned}$$

The formula is Al_2O_3.

(c) The ions here are Ba^{2+} and S^{2-}. Since the charges are equal but opposite, their ratio in the compound is 1 to 1. The formula is BaS.

EXERCISE 6.2   Write formulas for ionic compounds formed from (a) Na and F, (b) Na and O, (c) Mg and F, and (d) Al and S.

There is another rather simple way to obtain the formulas of the compounds in Example 6.1. The procedure is to make the subscript for one ion equal to the *number* of charges on the other. For example, for Al^{3+} and Cl^{1-} we can write

which gives

$$Al_1Cl_3 \quad \text{or simply} \quad AlCl_3$$

For the ions of Al^{3+} and O^{2-} we write

This gives

$$Al_2O_3$$

For the ions Ba^{2+} and S^{2-} this method would give the formula Ba_2S_2. Notice that both subscripts are divisible by 2. To obtain the empirical formula, we reduce the subscripts to the smallest set of whole numbers, which gives the correct formula, BaS. You might go back and try this method on Exercise 6.2.

Many of our most common chemicals are ionic compounds. We've mentioned NaCl, common table salt. Other ionic compounds that you may have encountered at one time or other are calcium chloride, $CaCl_2$, used to melt ice on walkways in the winter and to keep dust down on dirt roads in the summer (Figure 6.8), and calcium oxide, CaO, an important ingredient in cement.

Compounds containing ions composed of more than one element

The metal compounds that we have discussed thus far have been **binary compounds**—compounds formed between *two* different elements. There are many other ionic compounds that contain more than two elements. These substances contain **polyatomic ions**—ions that are themselves composed of two or

A compound is diatomic if it is composed of molecules that contain only two atoms. It is a binary compound if it contains two different elements, regardless of the number of each. $CaCl_2$ is a binary compound but it isn't diatomic.

Figure 6.8
Calcium chloride ($CaCl_2$) is spread on a dirt road to
keep down dust. This compounds absorbs moisture so readily
from the air that it actually forms an aqueous solution that
wets the road.

more elements. Table 6.3 lists some important polyatomic ions. The formulas of
ionic compounds formed from them are determined in just the same way as for
the binary ionic compounds. The ratio of the ions must be such that the formula is
electrically neutral.

Table 6.3
Some Polyatomic Ions

Ion	Name (Alternate Name in Parentheses)
NH_4^+	Ammonium ion
OH^-	Hydroxide ion
CN^-	Cyanide ion
NO_2^-	Nitrite ion
NO_3^-	Nitrate ion
OCl^-	Hypochlorite ion
ClO_2^-	Chlorite ion
ClO_3^-	Chlorate ion
ClO_4^-	Perchlorate ion
MnO_4^-	Permanganate ion
$C_2H_3O_2^-$	Acetate ion
CO_3^{2-}	Carbonate ion
HCO_3^-	Hydrogen carbonate ion (bicarbonate ion)[a]
SO_3^{2-}	Sulfite ion
SO_4^{2-}	Sulfate ion
HSO_4^-	Hydrogen sulfate ion (bisulfate ion)
CrO_4^{2-}	Chromate ion
$Cr_2O_7^{2-}$	Dichromate ion
PO_4^{3-}	Phosphate ion (orthophosphate ion)
HPO_4^{2-}	Monohydrogen phosphate ion
$H_2PO_4^-$	Dihydrogen phosphate ion

[a] Although "hydrogen carbonate ion" is formally correct, "bicarbonate ion" is what you will see
and hear the most. We'll use "bicarbonate" too.

EXAMPLE 6.2 **Writing Formulas Containing Polyatomic Ions**

Problem: Write the formula of the ionic compound formed from Ca^{2+} and PO_4^{3-}.

Solution: As before, we write the positive ion first and then exchange subscripts and numbers of charges.

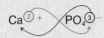

The formula is written with parentheses to show the number of PO_4^{3-} ions.

$$Ca_3(PO_4)_2$$

EXERCISE 6.3 Write formulas of ionic compounds formed from (a) Na^+ and CO_3^{2-}; (b) NH_4^+ and SO_4^{2-}; (c) a potassium ion and acetate ion; and (d) a strontium ion and nitrate ion.

Polyatomic ions are found in a great number of very important compounds. Some common ones are $CaSO_4$ (in plaster of paris), $NaHCO_3$ (baking soda), $NaOCl$ (liquid household bleach), $NaNO_2$ (sodium nitrite, a meat preservative), $MgSO_4$ (epsom salts), and $NH_4H_2PO_4$ (ammonium dihydrogen phosphate, a fertilizer).

Transition metals and post-transition metals

Thus far, we have looked briefly at the positive ions formed by those representative elements that, for a given period, appear before the transition elements. This is because these are the simplest ions to explain and to remember. The transition metals don't follow such a straightforward pattern. Many of them form more than one ion; iron, for example, can form two different ones, Fe^{2+} and Fe^{3+}. In fact, the ability to form more than one ion is often stated as a general property of the transition elements. Some of the most common ions of the transition elements are given in Table 6.4.

Table 6.4

Common Ions of Transition and Post-Transition Metals

Transition Metals	
Chromium	Cr^{3+}
Manganese	Mn^{2+}
Iron	Fe^{2+}, Fe^{3+}
Cobalt	Co^{2+}, Co^{3+}
Nickel	Ni^{2+}
Copper	Cu^+, Cu^{2+}
Zinc	Zn^{2+}
Silver	Ag^+
Cadmium	Cd^{2+}
Gold	Au^{3+}
Mercury	Hg_2^{2+} (actually two Hg^+ stuck together), Hg^{2+}
Post-Transition Metals	
Tin	Sn^{2+}, Sn^{4+}
Lead	Pb^{2+}, Pb^{4+}
Bismuth	Bi^{3+}

The **post-transition metals** are those metals that occur in the periodic table immediately following a row of transition elements. The two most common and important post-transition metals are tin (Sn) and lead (Pb). The ions that they form are also included in Table 6.4. Compounds formed by the transition and post-transition metals have formulas that we can obtain by our usual procedure.

6.6 PROPERTIES OF IONIC COMPOUNDS

Properties such as high melting point, brittleness, and the electrical behavior of salts are accounted for by the strong attractions and repulsions between ions

The properties of ionic compounds reflect the way ions interact with each other. The attractive forces between ions of opposite charge are very large, and in an ionic compound these oppositely-charged ions arrange themselves so that the attractions are a maximum and the repulsions between like-charged ions are a minimum. In a crystal at room temperature, the ions just rattle about within a cage of other ions. They are held too tightly to move away. Only at high temperatures do the ions have enough kinetic energy to break away from the pull of their neighbors, and therefore all ionic compounds are solids at room temperature and tend to have high melting points.

You have probably never seen an ionic substance melt. Most of them do so well above room temperature. For example, ordinary table salt, NaCl, melts at 801 °C. Some ionic substances melt only at extremely high temperatures. For instance, aluminum oxide, Al_2O_3, melts at about 2000 °C, and for this reason it is used in special bricks that line the inside walls of furnaces.

Another property of ionic compounds is that their solids are generally hard and brittle. Nonionic compounds like naphthalene (moth flakes) tend to be soft and easily crushed, but a crystal of rock salt is hard. When struck by a hammer, however, the salt crystal shatters. The slight movement of a layer of ions within an ionic crystal can suddenly place ions of the *same* charge next to one another, and the sudden large repulsive forces that result can split the solid as illustrated in Figure 6.9.

Figure 6.9
An ionic crystal shatters when struck because ions of like charge repel and force the crystal apart.

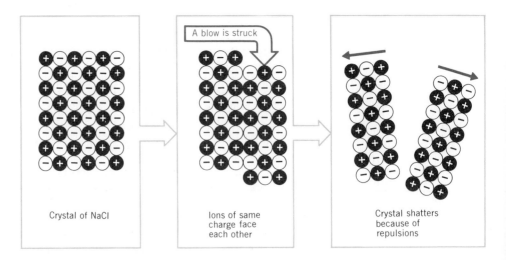

Crystal of NaCl

A blow is struck

Ions of same charge face each other

Crystal shatters because of repulsions

In fact, ion comes from the Greek word meaning "going particle."

When ionic compounds dissolve in water, their solutions conduct electricity. Electrical conductivity requires the movement of electrical charges. In water, the ions in compounds such as NaCl become separated and can move about more or less independently of each other. Many important chemical properties of ionic compounds depend on this.

In the solid state, ionic compounds do not conduct electricity. The attractive forces within the solid prevent the movement of ions through the crystal. Melting the compound has the same effect as dissolving it in water, however. Melting frees the ions so that they can move, and molten NaCl conducts electricity even though the solid does not.

Ionic compounds give solutions that conduct electricity and they conduct electricity when they are melted.

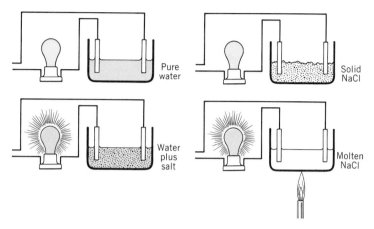

6.7 SOME CHEMICAL PROPERTIES OF NONMETALS

Nonmetals react with metals, other nonmetals, and metalloids

Although relatively few in number, the nonmetallic elements are found in more compounds than are the metals. They combine with metals to form ionic compounds, and they also react with each other and with metalloids to form nonionic, molecular compounds. A molecule, remember, consists of two or more atoms that are bound together very tightly so that they move about as a group just like a single particle.

Except for the noble gases (He through Rn), the nonmetals exist in molecular forms in the free state (i.e., when they are not in chemical combination with other, different elements). We have already seen that many are diatomic and are gases. Except for I_2, those that are solids have rather complex molecular structures, which we won't discuss now.

Remember: H_2, N_2, O_2, F_2, and Cl_2 are gases; Br_2 is a liquid; I_2 is a solid.

In the last section, we saw that atoms of the nonmetals gain electrons when they react with metals. The ions that they form have already been mentioned. When the nonmetals combine with each other, however, something entirely different takes place. Ions are not formed, yet the atoms are held together very strongly within the molecules that are produced. The molecules themselves are less tightly held together in molecular substances than are the ions in ionic compounds, and this explains why many molecular compounds tend to be soft and low melting.

The kinds of attractive forces, or "bonds," formed between atoms within molecules are of great importance. They are present in all common molecular substances: for example, H_2O, CO_2, SO_2 (sulfur dioxide—an air pollutant), CH_4 (methane—natural gas), C_2H_5OH (ethanol—grain alcohol), and C_8H_{18} (octane —found in gasoline). These bonds are also found in all of the molecules in living creatures, and they are what hold the atoms together within the polyatomic ions discussed in the last section.

The reason for the existence of bonds between uncharged atoms within molecules or polyatomic ions is not as simple to explain as the attraction between the ions in ionic compounds. To understand these bonds we must first have a more fundamental understanding of the internal structure of atoms. In the chapters ahead, the theoretical models that we shall build will have to account for both these nonionic forces within molecules and the formation of ions when metals and nonmetals react.

6.8 ACIDS AND BASES

Among the most important compounds containing metals and nonmetals are those that furnish hydrogen ions—acids—and those that give hydroxide ions—bases

Some common acids.

Some common bases.

$$H^+ + OH^- \longrightarrow H_2O$$

A method that we often use to establish some order among chemical facts is to categorize reactions and the chemicals that participate in them into certain classes. One important class of chemical reaction takes place between substances that we call acids and substances that we call bases—acids always react with bases in a predictable way.

Some of our most familiar chemicals as well as many of our most important laboratory reagents are acids or bases. The vinegar in a salad dressing, the sour juices of a lemon, and the liquid in the battery of an automobile are similar in at least one respect—they are all acids. The white crystals of lye in drain cleaners, creamy milk of magnesia, and a tablet of Tums® are all bases. Many acids and bases are also found within our bodies—who hasn't heard of stomach acid?

Acids are substances that release hydrogen ions, H^+, when placed in water. Actually, these H^+ ions become attached to water molecules and travel about as ions with the formula H_3O^+ (they are called **hydronium ions**). Nevertheless, the active ingredient is the H^+. **Bases** are substances that give hydroxide ions, OH^-, when placed in water. This is the way acids and bases were defined by the great Swedish chemist, Svante Arrhenius (1859–1927), winner of the third Nobel Prize in Chemistry in 1903.

A property of acids and bases that we discussed earlier is their reaction with each other, which destroys their acidic and basic characteristics. In Chapter One we mentioned that if hydrochloric acid and lye are mixed in just the right proportions, a solution containing nothing but ordinary table salt is formed. Hydrochloric acid, HCl, separates into ions in water. One of the ions is H^+ and the other is Cl^-. Lye is sodium hydroxide, NaOH, which separates into Na^+ and OH^- ions in water. When the two solutions are mixed, the H^+ and OH^- ions react to form a molecule of H_2O. Since the H^+ and OH^- disappear, the solution is no longer acidic or basic. All that is left is Na^+ and Cl^- ions, the ions found in table salt.

The reaction of an acid and a base is called neutralization because the H^+ and OH^- ions cancel or neutralize each other. We have already discussed the reaction between hydrochloric acid and sodium hydroxide,

$$HCl + NaOH \longrightarrow H_2O + NaCl$$

In Section 6.5 it was mentioned that the term *salt* is not reserved for just NaCl. We will use the term **salt** to mean any ionic compound that doesn't contain OH^-. This then permits us to make the general statement that whenever any acid reacts with any base, one product is water and the other is a salt. Another example is the reaction between nitric acid, HNO_3, and NaOH,

$$HNO_3 + NaOH \longrightarrow H_2O + NaNO_3$$

Sodium nitrate, $NaNO_3$, is ionic and is called a salt too. Perhaps you recognize the NO_3^- ion as one of the polyatomic ions listed in Table 6.3. Maybe you have also noticed that acid-base neutralization reactions provide another way of preparing salts beside the direct combination of the elements described earlier.

The tendency for elements to form acids or bases depends on their location in the periodic table. One of the controlling factors is the way an element's oxide reacts with water. In general, *metal oxides react with water to form bases; nonmetal oxides react with water to form acids.*

Metal oxides are called basic anhydrides. Anhydride means "without water."

Typical examples of the formation of a base are the reactions of sodium oxide, Na_2O, and calcium oxide, CaO, with water.

$$Na_2O + H_2O \longrightarrow 2NaOH \qquad \text{sodium hydroxide}$$
$$CaO + H_2O \longrightarrow Ca(OH)_2 \qquad \text{calcium hydroxide}$$

Calcium oxide, CaO, is an important ingredient in cement. When water is added to the cement the reaction above is one of many that occurs.

Nonmetal oxides are called acid anhydrides.

Typical examples of the formation of an acid are the reactions of CO_2 and SO_2 with water.

$$CO_2 + H_2O \longrightarrow H_2CO_3 \qquad \text{carbonic acid}$$
$$SO_2 + H_2O \longrightarrow H_2SO_3 \qquad \text{sulfurous acid}$$

Beautiful "Rainbow Lake" in Shenandoah Caverns, Virginia. Gradual erosion of limestone deposits by slightly acidic ground water creates caverns such as this.

Carbonic acid and sulfurous acid are too unstable to be isolated as pure compounds, but water solutions of them are quite common. Atmospheric CO_2 dissolved in groundwater exists partly as carbonic acid. It slowly dissolves limestone and is responsible for the large limestone caves found in various locations. Pollution of the air by SO_2 makes rain slightly acidic. This has caused a great deal of damage to marble statues in many parts of the world (marble is a form of limestone, $CaCO_3$).

For all acids, the hydrogen atoms that are available as H^+ are usually written first in the formula. Thus HNO_3 can furnish one H^+, H_2CO_3 and H_2SO_3 can furnish two H^+, and phosphoric acid, H_3PO_4, can give three H^+. Later we will encounter many acids that also contain one or more hydrogens that are not able to be released as H^+. An example is acetic acid, $HC_2H_3O_2$, the substance that gives vinegar its sour taste. This acid can only furnish one H^+—the one that appears at the beginning of its formula. In water, acetic acid separates into H^+ and $C_2H_3O_2^-$ ions. Since the hydrogen ion consists of a bare hydrogen nucleus—a proton—acids that can supply just one H^+ per molecule are often called **monoprotic acids.** Examples of monoprotic acids are HCl, HNO_3, and $HC_2H_3O_2$. Acids that can supply more than one are called **polyprotic acids.** We could be more specific and refer to H_2CO_3 and H_2SO_3 as **diprotic acids** and to H_3PO_4 as a **triprotic acid.** We will encounter these terms again in later chapters.

Strong acids and weak acids

In their pure form, acids such as HCl or HNO_3 are not ionic compounds. They exist as molecules. They release H^+ when placed in water because they react with water. An example is

$$HCl + H_2O \longrightarrow H_3O^+ + Cl^-$$

Recall that the H^+ are tied to water molecules in the solution and that H_3O^+ is therefore a more accurate way of representing H^+.

Not all acid molecules give up their hydrogen to water with equal ease. Some, like HCl, lose their H^+ very easily, but others, like HF, H_2CO_3, and H_2SO_3, hold onto their hydrogen more firmly. For example, when HF is dissolved in water, only a small fraction of the molecules placed in the solution actually exist in the form of H_3O^+ and F^- ions. This is because when F^- and H_3O^+ encounter each other in the solution, there is a strong tendency for them to recombine into HF molecules. In a solution of HF, therefore, there are actually two reactions taking place simultaneously.

$$HF + H_2O \longrightarrow H_3O^+ + F^-$$
$$H_3O^+ + F^- \longrightarrow HF + H_2O$$

When some HF is placed into water, it takes only an instant before both of these reactions are occurring at the same rate. Once that happens, the amounts of H^+ and F^- in the solution remain constant. It is a bit like a society with zero population growth (ZPG): the birthrate and death rate are equal and the size of

Dynamic equilibrium is one of the most important concepts in chemistry.

the population remains constant. Both the HF solution and the society with ZPG exist in what is known as a state of **dynamic equilibrium.** There is a constant change in the individuals that make up the population, but the size of the population—the total number of individuals—stays the same. In chemistry, we represent the two opposing reactions that produce the dynamic equilibrium by using double arrows, $\rightleftharpoons$. For HF we write

$$HF + H_2O \rightleftharpoons H_3O^+ + F^-$$

In any given solution of HF or HCl, the size of the H_3O^+ and F^- or H_3O^+ and Cl^- populations depends on how easily the molecules lose H^+ compared to the tendency for the ions to recombine. For HCl dissolved in water, there is virtually no tendency for H^+ and Cl^- to recombine. A mole of molecular HCl gives a mole of H^+ and a mole of Cl^- in the solution. For HF, however, the recombination of ions occurs quite easily. A solution of 1 mol of HF in 1 liter of water gives only about 0.02 mol of H^+ and 0.02 mol of F^- once equilibrium is reached. Thus, only about 2% of the HF exists as ions.

An acid, like hydrochloric acid, HCl, that completely separates into ions in water is called a **strong acid.** When only a small percentage of the acid in solution is in the form of ions, we call the substance a **weak acid.** Hydrofluoric acid, HF, is an example of a weak acid. There are not very many acids that are strong; most are weak.

EXERCISE 6.4   Write the equilibrium equation for the reaction of the weak acid, nitrous acid, HNO_2 with water.

Oxoacids

Strong Oxoacids

HNO_3
H_2SO_4
$HClO_4$
$HClO_3$

Acids that contain hydrogen, oxygen, plus another element are called **oxoacids.** Many common and important acids fall into this category, and Table 6.5 contains a list of some oxoacids of the nonmetals and metalloids. Those that are strong are marked with an asterisk. You should memorize the names and formulas of these strong acids because they are very useful acids in the laboratory. (Also, once you know the strong ones, you will know that any others are weak.)

Within the periodic table there are some trends in the strengths of acids that

Table 6.5

Some Oxoacids of the Nonmetals and Metalloids

Group IVA	Group VA	Group VIA	Group VIIA	
H_2CO_3 carbonic acid	*HNO_3 nitric acid	—	HFO	hypofluorous acid
	HNO_2 nitrous acid			
	H_3PO_4 phosphoric acid	*H_2SO_4 sulfuric acid	*$HClO_4$	perchloric acid
	$H_3PO_3^a$ phosphorous acid	H_2SO_3 sulfurous acid	*$HClO_3$	chloric acid
			$HClO_2$	chlorous acid
			$HClO$	hypochlorous acid
	H_3AsO_4 arsenic acid	H_2SeO_4 selenic acid	$HBrO_4$	perbromic acid
	H_3AsO_3 arsenious acid	H_2SeO_3 selenous acid	$HBrO_3$	bromic acid
			HIO_4	
			(H_5IO_6)	periodic acid[b]
			HIO_3	iodic acid

[a] This acid only gives two H^+ per molecule, not three as its formula might suggest.
[b] H_5IO_6 is formed from $HIO_4 + 2H_2O$.
* Strong acid.

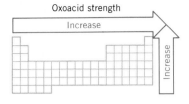

Oxoacid strength
Increase

are useful to remember. First, *for oxoacids with the same number of oxygens, the strength increases from bottom to top within a group.* Thus H_2SO_4 is stronger than H_2SeO_4 because sulfur is above selenium in Group VIA, and H_3PO_4 is stronger than H_3AsO_4 because phosphorus is above arsenic. *For oxoacids with the same number of oxygens, strength also increases from left to right in a period.* In period 3, phosphorus is to the left of sulfur, which is to the left of chlorine. Therefore, H_3PO_4 is weaker than H_2SO_4, which is weaker than $HClO_4$.

EXERCISE 6.5   Predict the stronger acid in each pair: (a) $HClO_4$ or $HBrO_4$; (b) H_3AsO_4 or H_2SeO_4.

The relative strengths of the oxoacids can also be correlated with the number of oxygen atoms in the molecule. *For a given nonmetal the acid strength increases with increasing number of oxygens.* Thus H_2SO_4 is stronger than H_2SO_3, and HNO_3 is stronger than HNO_2.

EXERCISE 6.6   Predict the order of acid strength for the four oxoacids of chlorine in Table 6.5.

Acids that do not contain oxygen

Strong Binary Acids

HCl
HBr
HI

The binary compounds between hydrogen and many of the nonmetals also are acidic. They are called **binary acids.** We have already mentioned HF and HCl. Table 6.6 lists the binary hydrogen compounds that are acids in water. Those that are strong are once again marked by an asterisk. You can add them to your list of strong acids.

Table 6.6
Binary Compounds of Hydrogen and Nonmetals Which Are Acidic[a]

Group VI	Group VII
(H_2O)	✗ HF hydrofluoric acid
✗ H_2S hydrosulfuric acid	✗* HCl hydrochloric acid
H_2Se hydroselenic acid	✗* HBr hydrobromic acid
H_2Te hydrotelluric acid	✗* HI hydriodic acid

* Strong acid.
[a] The names apply to the solutions of these compounds in water.

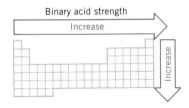

Binary acid strength
Increase

As with the oxoacids, the strengths of the binary acids increase from left to right in a period. Thus, HCl is stronger than H_2S. *Within a group, however, acid strength increases from top to bottom.* We have seen that HCl is stronger than HF.

Water itself is included in Table 6.6 because it actually reacts with itself to a very small degree.

$$H_2O + H_2O \rightleftharpoons H_3O^+ + OH^-$$

Pure water contains the same small concentration of both H_3O^+ and OH^- and is therefore neither acidic or basic. The reaction of water with itself, despite the very small extent to which it occurs, is *very* important. We will have much to say about it in Chapter Seventeen.

EXERCISE 6.7   Using only the periodic table, and without looking at Table 6.6, choose the stronger acid: (a) H_2Se or HBr; (b) H_2Se or H_2Te.

SUMMARY

The Periodic Table. The search for similarities and differences among the properties of the elements led Mendeleev to discover that when the elements are placed in (approximate) order of increasing atomic weight, similar properties recur at regular, repeating intervals. In the modern periodic table the elements are arranged in rows, called periods, in order of increasing atomic number. The rows are stacked so that elements in the columns, called groups or families, have similar chemical and physical properties. The A-group elements are called representative elements; the B-group elements are called transition elements. The two long rows of inner transition elements located below the main body of the table consist of the lanthanides, which follow La ($Z = 57$), and the actinides, which follow Ac ($Z = 89$). Certain groups are given family names: for instance, Group IA, the alkali metals (the alkalis); Group IIA, the alkaline earth metals; Group VIIIA, the halogens; Group 0, the noble gases.

Metals, Nonmetals, Metalloids. Most elements are metals; they occupy the lower left region of the periodic table (to the left of a line drawn approximately from boron, B, to astatine, At). Nonmetals are found in the upper right region of the table. Metalloids occupy a narrow band between the metals and nonmetals.

Metals have a characteristic luster, tend to be ductile and malleable, and conduct electricity. They react with nonmetals to form ionic compounds called salts. Nonmetals tend to be brittle, lack "metallic" luster, and are nonconductors of electricity. Many nonmetals are gases. Besides combining with metals, nonmetals combine with each other to form molecules without undergoing electron transfer. Bromine (a nonmetal) and mercury (a metal) are the two elements that are liquids at ordinary room temperature. Metalloids have properties intermediate between metals and nonmetals.

When ionic binary compounds are formed, electrons are transferred from a metal to a nonmetal. The metal atom becomes a positive ion; the nonmetal atom becomes a negative ion. The formulas of ionic compounds are controlled by the requirement that the compound must be electrically neutral. Many ionic compounds also contain polyatomic ions—ions that are composed of two or more atoms. Ionic compounds tend to be brittle, high melting, nonconducting solids. When melted or dissolved in water, however, they do conduct electricity.

Acids and Bases. Acids are substances that react with water to give hydronium ion, H_3O^+, which we usually abbreviate simply as H^+. Bases in water give hydroxide ion, OH^-. Neutralization of an acid by a base gives a salt plus water. Metal oxides react with water to form bases; nonmetal oxides react with water to give acids. Many acids are incompletely ionized in water and are called weak acids because a dynamic equilibrium exists between ionized and nonionized forms of the acid. Strong acids are 100% ionized in water.

Oxoacids, containing oxygen atoms in addition to hydrogen and another element, increase in strength as the number of oxygen atoms increases. Oxoacids having the same number of oxygens increase in strength from bottom to top within a group and from left to right across a period. Binary acids contain only hydrogen and another nonmetal. Their strength increases from top to bottom within a group and left to right across a period.

INDEX TO EXERCISES AND QUESTIONS

REVIEW QUESTIONS

6.8. Why is it important to look for similarities and differences among the chemical and physical properties of the elements?

6.9. What was it that Mendeleev observed that led him to develop his periodic table?

6.10. On the basis of their positions in the periodic table, why is it not surprising that strontium-90 replaces calcium in newly formed bones?

6.11. Why would you reasonably expect cadmium to be a contaminant in zinc but not in silver?

6.12. In the refining of copper, sizable amounts of silver and gold are recovered. Why is this not surprising?

6.13. Below are some data for the elements sulfur and tellurium.

	Sulfur	Tellurium
Atomic mass	32.06	127.60
Melting point (°C)	112.8	449.5
Boiling point (°C)	445	990
Formula of oxide	SO_2	TeO_2
Melting point of oxide (°C)	−72.7	733
Density (g/cm³)	2.07	6.25

Had the element selenium not been known in the time of Mendeleev, he would have called it eka-sulfur. Estimate its properties as an average of those of sulfur and tellurium.

6.14. What was the original form of the periodic law? Try to state it in your own words.

6.15. What is a "period"? What is a "group"?

6.16. Why were there gaps in Mendeleev's periodic table?

6.17. In the text, we identified two places in the periodic table where the atomic weight order was reversed. Using the table on the inside front cover, locate two other places where this occurs.

6.18. State the modern form of the periodic law using your own words.

6.19. Make a rough sketch of the periodic table and mark off those areas where you would find (a) the representative elements, (b) the transition elements, and (c) the inner transition elements.

6.20. With the knowledge gathered to date, is it likely that scientists will discover a new element, never before observed, having an atomic weight of approximately 73? Explain your answer.

6.21. Which of the following is an alkali metal: Ca, Cu, In, Li, S?

6.22. Which of the following is a halogen: Ce, Hg, Si, O, I?

6.23. Which of the following is a transition element: Pb, W, Ca, Cs, P?

6.24. Which of the following is a noble gas: Xe, Se, H, Sr, Zr?

6.25. Which of the following is a lanthanide element: Th, Sm, Ba, F, Sb?

6.26. Which of the following is an actinide element: Ho, Mn, Pu, At, Na?

6.27. Which of the following is an alkaline earth metal: Mg, Fe, K, Cl, Ni?

6.28. Give five physical properties that we usually observe for metals.

6.29. Why is mercury used in thermometers? Why is tungsten used in light bulbs?

6.30. What property of metals allows them to be drawn into wire?

6.31. Gold can be hammered into sheets so thin that some light can pass through them. What property of gold allows such thin sheets to be made?

6.32. What metals can you think of that are used for jewelry? Why isn't iron used to make jewelry?

6.33. Only two metals are colored (the rest are "white," like iron or lead). You have surely seen both of them. Which metals are they?

6.34. What nonmetals occur as monatomic gases (gases whose particles consist of single atoms)?

6.35. What nonmetals occur as diatomic molecules? Which are gases?

6.36. What are the two elements that exist as liquids at room temperature and pressure?

6.37. What physical property of metalloids distinguishes them from metals and nonmetals?

6.38. Sketch the shape of the periodic table and mark off those areas where we find (a) metals, (b) nonmetals, and (c) metalloids.

6.39. With what kind of elements do metals react?

6.40. What is an ionic compound? What holds an ionic compound together? Can we identify individual molecules in an ionic compound?

6.41. What is the difference between a binary compound and one that is diatomic? Give examples that illustrate this difference.

6.42. What rules do we apply when we write formulas for ionic compounds?

6.43. Write formulas for ionic compounds formed between (a) Na and Br, (b) K and I, (c) Ba and O, (d) Mg and Br, and (e) Ba and F.

6.44. Which of the following formulas are incorrect: (a) NaO_2, (b) RbCl, (c) K_2S, (d) Al_2Cl_3, (e) MgO_2?

6.45. From what you have learned in Sections 6.4 and 6.5, write correct balanced equations for the reaction between: (a) calcium and chlorine, (b) magnesium and oxygen, (c) aluminum and oxygen, and (d) sodium and sulfur.

6.46. Write formulas for the ionic compounds formed from: (a) K^+ and nitrate ion, (b) Ca^{2+} and acetate ion, (c) ammonium ion and Cl^-, (d) Fe^{3+} and carbonate ion, and (e) Mg^{2+} and phosphate ion.

6.47. Write formulas for the ionic compounds formed from: (a) Zn^{2+} and hydroxide ion; (b) Ag^+ and chromate ion; (c) Ba^{2+} and sulfite ion; (d) Rb^+ and sulfate ion; and (e) Li^+ and bicarbonate ion.

6.48. What is a post-transition metal?

6.49. With what kinds of elements do nonmetals combine?

6.50. In what major way do compounds formed between two nonmetals differ from those formed between a metal and a nonmetal?

6.51. Write the symbol for the ion formed when titanium (Ti) loses four electrons.

6.52. What are typical physical properties of ionic compounds?

6.53. Why does molten KCl conduct electricity even though solid KCl does not?

6.54. What happens when an ionic compound dissolves in water that allows its solution to conduct electricity?

6.55. Define *acid, base, neutralization,* and *salt.*

6.56. What is the name of the ion, H_3O^+?

6.57. A zero population growth society was used as an example of a dynamic equilibrium. Can you think of other situations from everyday life where a dynamic equilibrium could exist (at least temporarily)?

6.58. What is the difference between a weak acid and a strong acid?

6.59. Give the formulas of two strong oxoacids.

6.60. Give the formulas of two strong binary acids (binary compounds of hydrogen and a nonmetal).

6.61. Write the equation for the reaction of HNO_3 with water.

6.62. Write equations for the reaction of these weak acids with water: (a) HIO_3; and (b) $HCHO_2$ (formic acid, a monoprotic acid).

6.63. Complete and balance the following equations:
(a) $HNO_2 + KOH \longrightarrow$
(b) $HCl + Ca(OH)_2 \longrightarrow$
(c) $H_2SO_4 + NaOH \longrightarrow$
(d) $HClO_4 + Al(OH)_3 \longrightarrow$
(e) $H_3PO_4 + Ba(OH)_2 \longrightarrow$

6.64. Define *monoprotic acid, polyprotic acid,* and *diprotic acid.*

6.65. When diprotic acids, such as H_2CO_3, react with water to release their H^+, they do so in two steps, each of which is a dynamic equilibrium. Write chemical equations that would illustrate this for H_2CO_3.

6.66. Choose the stronger acid: (a) H_2S or H_2Se; (b) H_2Te or HI; (c) HIO_3 or HIO_4; (d) H_2SeO_4 or $HClO_4$.

6.67. Choose the stronger acid: (a) HBr or HCl; (b) H_2O or HF; (c) H_3AsO_3 or H_3AsO_4; (d) H_2S or HBr.

6.68. Write the equation for the reaction of water with itself.

6.69. Milk of magnesia contains $Mg(OH)_2$ and stomach acid contains HCl. Write the chemical reaction that occurs when you swallow milk of magnesia.

6.70. Many common antacids contain aluminum hydroxide, $Al(OH)_3$. Write an equation for the neutralization of the HCl of stomach acid by $Al(OH)_3$.

6.71. Suppose that a new element were discovered. Based on the discussions in this chapter, what properties (both physical and chemical) might be used to classify the element as a metal or a nonmetal?

6.72. Astatine, atomic number 85, is radioactive and does not occur in appreciable amounts in nature. On the basis of what you have learned in this chapter, answer the following:
(a) Would you expect At to be a metal, a nonmetal, or metalloid?
(b) What would be the formula for a compound formed from Na and At?
(c) What would you expect to be the formula of molecules of elemental At?
(d) Would elemental At be expected to be a solid, a liquid, or a gas?
(e) How would the acidity of HAt compare to HI?
(f) What might be a formula for an oxoacid of At?

CHAPTER SEVEN

ATOMIC STRUCTURE AND THE FORMATION OF IONIC COMPOUNDS

7.1 ATOMIC SPECTRA; ENERGY LEVELS IN ATOMS

The light given off by atoms that have absorbed energy shows that the electrons in an atom can possess only certain specific amounts of energy

The many chemical and physical properties that we observe for the elements and their compounds may seem, at first glance, to make understanding them hopelessly complex. Why, for instance, does metallic sodium react violently with water, whereas gold is resistant to attack by even concentrated acids? Why does ice float in liquid water? Why are heavy metals, such as mercury, poisonous? There's no end to such questions, of course, and in a beginning chemistry course we can only start to find some answers. We've already seen in the previous chapter that some of the complexity in chemistry can be simplified by organizing the elements into the periodic table. This shows us why, for example, strontium-90 collects in bones along with calcium—both elements are in the same periodic group and should be expected to have similar chemical properties. However, even more basic questions remain. For example, *why* do calcium and strontium have similar chemical properties that place them in the same group in the periodic table?

To answer even the most basic "why" questions of chemical behavior, we must once again look at the structure of the atom. As we learned earlier, the nucleus determines the mass of the atom and the number of electrons it must have to be electrically neutral. The nucleus does not, however, play a part in chemical reactions. When two or more atoms join together to form a compound, the nuclei of the atoms stay relatively far apart. Only the atoms' outer reaches—the areas inhabited by electrons—come in close contact. The chemical properties of the elements, then, must be determined by the electrons of the various atoms, and the similarities and differences in these properties must have something to do with the way these electrons are distributed around the particular nuclei.

How electrons are distributed about the nucleus is called the atom's **electronic structure.** The basic clue to the electronic structures of the various elements comes from the study of the light emitted when atoms of the elements are excited or energized. In Chapter Five, we saw that white light from the sun or even an ordinary incandescent light bulb is composed of electromagnetic radiation of all frequencies. When a beam of white light is passed through a prism, a continuous spectrum of colors much like a rainbow is produced. Light is also

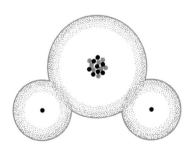

Atoms joined in water, H_2O. Nuclei stay far apart and only the outer parts of the atoms touch.

given off when a gas or some other substance is energized by passing an electric discharge through it or heating it in a flame. However, when a narrow beam of this light is passed through a prism, a continuous spectrum is *not* produced. Instead, only a few colors, displayed as a series of individual lines (see Figure 7.1), are observed. This series of lines is called the element's **atomic spectrum.** Because the frequencies of light given off by an excited atom of an element are unique to that element, each element has its own characteristic atomic spectrum.

Atomic spectra, like those in Color Plate 2A, are also called line spectra or emission spectra.

Figure 7.1
Production of a line spectrum.

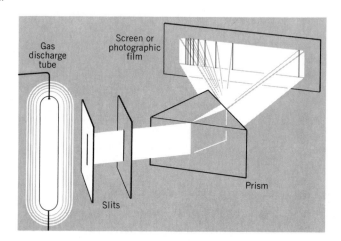

In Chapter Five we saw that there is a simple relationship between the frequency of light and its energy, $E = h\nu$. The fact that excited atoms emit light of only certain characteristic frequencies tells us that only certain characteristic energy changes take place within the atom. For instance, in the spectrum of hydrogen there is a red line that has a frequency of 4.57×10^{14} Hz. The energy of each photon of this light is 3.03×10^{-19} J. Therefore, when a hydrogen atom emits light of this frequency, the energy of the atom decreases by 3.03×10^{-19} J. What is very special here is that, whenever a hydrogen atom emits red light, its frequency is 4.57×10^{14} Hz and the energy change within the atom is *always exactly* 3.03×10^{-19} J, never more and never less. Atomic spectra, then, tell us that *when an excited atom loses energy, not just any arbitrary amount is lost.* Only certain specific energy changes can occur, which is why only certain specific frequencies of light are emitted. This is the only way atomic spectra can be explained.

How is it that atoms of a given element always undergo exactly the same specific energy changes? The answer seems to be that in an atom the electron can have only certain definite amounts of energy and no others. In the words of science, we say that the electron is restricted to certain **energy levels.** We also say that the energy of the electron is **quantized,** meaning once again that the electron's energy in a particular atom can have only certain values and no others.

The energy of an electron in an atom might be compared to the potential energy of a ball on a staircase (see Figure 7.2). The ball can only come to rest on a step, and on each step it will have some specific amount of potential energy. If the ball is raised to a higher step, its potential energy will be increased. When it drops to a lower step, its potential energy decreases. But each time the ball stops, it stops on one of the steps, never in between. Thus, the ball at rest can only have certain specific amounts of potential energy, which are determined by the energy levels of the various steps of the staircase. So it is with an electron in an atom. The electron can only have energies corresponding to the set of electron energy levels in the atom. When the atom is supplied with energy, as in a gas discharge tube, an electron is raised from a low energy level to a higher one. When the electron drops back, energy equal to the difference between the two

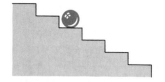

Figure 7.2
Ball on a staircase. The ball can have only certain amounts of potential energy when at rest.

The potential energy of the ball at rest is quantized.

levels is emitted as a photon. Because only certain energy jumps can occur, only certain frequencies can appear in the spectrum.

The existence of specific energy levels in atoms, as implied by atomic spectra, forms the foundation of all theories about electronic structure. Any model of the atom that attempts to describe the positions or motions of electrons must also account for atomic spectra.

7.2 THE BOHR MODEL OF THE ATOM

The first model of the atom to meet with some success imagined the electron to be revolving about the nucleus in orbits of fixed energy

The discovery that the energy of electrons was quantized led to attempts to develop theoretical models of the way electrons behave in atoms. The goals were to explain how electrons move, where they are located, and how they change energy to give off photons of light. Physicists were faced with a problem, however. None of the physical laws that seemed to govern the motion of large objects, like baseballs or people, were able to account for the strange behavior of electrons.

In 1913 Niels Bohr (1885–1962), a Danish physicist, proposed a theoretical model for the hydrogen atom. He choose hydrogen because its atoms are the simplest, having only one electron about the nucleus, and because it produces the simplest spectrum with the fewest lines. In his model, Bohr imagined the electron to move around the nucleus following fixed paths, or orbits, much as a planet moves around the sun. His model also restricted the sizes of the orbits and the energy that the electron could have in a given orbit. The equation Bohr derived for the energy of the electron included a number of physical constants such as the mass of the electron, its charge, and Planck's constant. It also contained an integer, n, that Bohr called a **quantum number.** Each of the orbits could be specified by its value of n. Combining all the constants together, Bohr's equation was

$$E = \frac{-k}{n^2} \qquad (7.1)$$

where E is the energy of the electron and k is the combined constant (its value is 2.18×10^{-18} J). The allowed values of n range from 1 to ∞, with all integers permitted (i.e., n could equal 1, 2, 3, 4, . . . , ∞). Therefore, the energy of the electron in any particular orbit could be calculated.

Because of the negative sign in equation 7.1, the lowest (most negative) energy value occurs when $n = 1$, which corresponds to the *first Bohr orbit*. This lowest energy state is called the **ground state.** According to Bohr's theory also, this orbit brings the electron closest to the nucleus.

When the hydrogen atom absorbs energy, as it does in a discharge tube, the electron is raised from the orbit having $n = 1$ to a higher orbit—to $n = 2$ or $n = 3$ or even higher. Then, when the electron drops back to a lower orbit, energy is emitted in the form of light (see Figure 7.3). Since the energy of the electron in a given orbit is fixed, a drop from one particular orbit to another, say from $n = 2$ to $n = 1$, always releases the same amount of energy, and the frequency of the light emitted because of this change in energy is always precisely the same.

Bohr's model of the atom was both a success and a failure. It successfully predicted the frequencies of the lines in the hydrogen spectrum; so there seemed to be some validity to the theory. Nevertheless, the model was a total failure for atoms with more than one electron. Still, though the theory met with only limited success, the introduction of the ideas of quantum numbers and fixed energy levels were important steps forward.

Figure 7.3
Emission of light by the hydrogen atom.

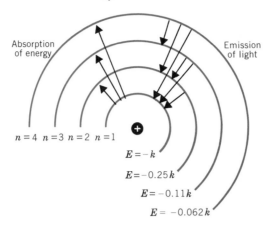

Absorption of energy

Emission of light

$n = 4$ $n = 3$ $n = 2$ $n = 1$

$E = -k$

$E = -0.25\,k$

$E = -0.11\,k$

$E = -0.062\,k$

7.3 THE WAVE NATURE OF MATTER

Very small particles show properties that we identify with waves

Bohr's efforts to develop a theory of electronic structure were doomed from the very beginning because the classical laws of physics—those known in his day—simply do not apply to particles as tiny as the electron. Since all of the objects that had been studied by scientists until that time were large and massive in comparison with the electron, no one had yet detected the limits of classical physics. Classical physics fails for atomic particles because matter is not really as our physical senses perceive it. Under appropriate circumstances, small bits of matter, such as an electron, behave not like solid particles, but instead like waves. This idea was first proposed in 1924 by a young French graduate student, Louis de Broglie.

De Broglie was awarded the Nobel prize in 1929.

In Chapter Five we saw that light waves are characterized by their wavelengths and their frequencies. The same is true of matter waves. De Broglie suggested that the wavelength of a matter wave, λ, is given by the equation:

$$\lambda = \frac{h}{mv} \tag{7.2}$$

where h is Planck's constant, m is the particle's mass, and v is its velocity.

The concept of a particle of matter behaving as a wave rather than as a solid object may at first seem difficult to comprehend. This book certainly seems solid enough, and if you drop it on your toe, it surely doesn't seem to be a wave, at least not as we generally think of waves in the ocean. The reason is that in de Broglie's equation (equation 7.2) the mass appears in the denominator. This means that heavy objects have extremely short wavelengths. The peaks of the matter waves for heavy objects are so close together that the wave properties go unnoticed and can't even be measured experimentally. But tiny particles with very low masses have much longer wavelengths; therefore, their wave properties become an important part of their overall behavior.

By now you may have begun to wonder if there is in fact any way to prove that matter has wave properties. Actually, these properties are easily shown by a common phenomenon that you have often observed. For example, if you drop two pebbles simultaneously into a quiet pond, ripples spread out from where the pebbles enter the water, as shown in Figure 7.4. When the two sets of ripples cross, there are places where the waves are *in phase*—the peak of one wave coincides with the peak of the other. At these points, the height of the water is equal to the sum of the heights of the two crossing waves. At other places the crossing waves cancel each other because the waves are *out of phase*. This reinforcement and cancellation of wave intensities, referred to, respectively, as con-

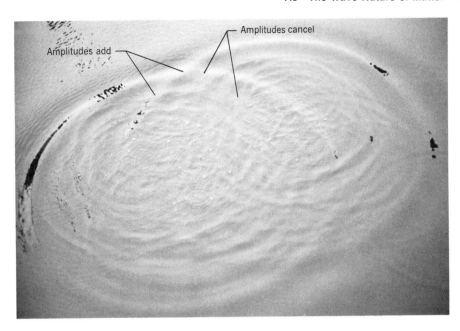

Amplitudes add

Amplitudes cancel

Figure 7.4
Diffraction of water waves produced by pebbles dropped
simultaneously into a pond.

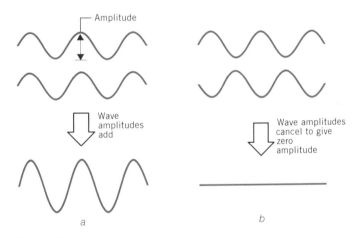

Amplitude

Wave
amplitudes
add

Wave amplitudes
cancel to give
zero
amplitude

a

b

Figure 7.5
(a) Waves in phase—constructive interference. (b) Waves out
of phase—destructive interference.

structive and destructive interference, is the phenomenon called **diffraction.** It is
examined more closely in Figure 7.5.

Diffraction occurs with water waves, light waves, or matter waves. For light
waves it can be shown in an experiment such as that illustrated in Figure 7.6. The
two beams of light formed by the closely spaced pair of slits interfere with each
other. On the screen, we see bands of light where the light waves reinforce each
other and bands of darkness where the waves cancel. The display of light and
dark bands is called a **diffraction pattern.**

Diffraction is a phenomenon that can only be explained as a property of waves.
We have seen how it can be demonstrated with water waves and light waves.
Experiments can also be done to show that electrons, protons, and neutrons
experience diffraction—in fact, that is the principle on which the electron microscope is based (Special Topic 7.1). These experiments prove that these small
particles of matter also have wave properties.

Figure 7.6
Production of a diffraction pattern.

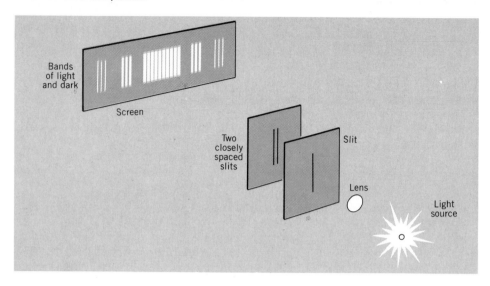

7.4 ELECTRON WAVES IN ATOMS

Electron waves in atoms can be identified by three quantum numbers

Current theories of electronic structure are based on the wave properties of the electron. In fact, the theory has been given the name **wave mechanics.** It is also called **quantum mechanics** because the theory predicts quantized energy levels. In 1926 Erwin Schrödinger (1887–1961), an Austrian physicist, became the first scientist to successfully apply the concept of the wave nature of matter to an explanation of electronic structure. His work, for which he won a Nobel prize in 1933, and the theory that has developed from it are highly mathematical. Fortunately, we need only a qualitative understanding of electronic structure, and the main points of the theory can be understood without all the math.

First, however, we must learn a little more about waves. There are basically two kinds of waves, traveling waves and standing waves. On a lake or ocean the wind produces waves whose crests and troughs move across the water's surface, as shown in Figure 7.7. The water moves up and down while the crests and troughs travel horizontally in the direction of the wind. These are examples of **traveling waves.**

A more important kind of wave for us is the standing wave. An example is the vibrating string of a guitar. When the string is plucked, its center vibrates up and down while the ends, of course, remain fixed. The crest, or point of maximum amplitude, of the wave occurs at one position. At the ends of the string are points of zero amplitude, called **nodes,** and their positions are also fixed. A **standing wave,** then, is one in which the crests and nodes do not change position.

As you know, many notes can be played on a guitar string by shortening its effective length with a finger placed at frets along the neck of the instrument. But even without shortening the string, we can play a variety of notes. For instance, if the string is touched momentarily at its midpoint at the same time it is plucked, the string vibrates as shown in Figure 7.8 and produces a tone that is an octave

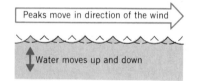

Figure 7.7
Traveling waves.

The wavelength of this wave is actually twice the length of the string.

Figure 7.8
Standing waves on a guitar string.

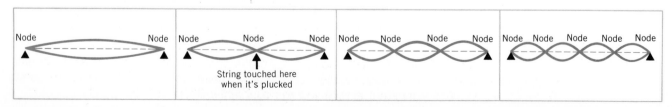

Special Topic 7.1   The Electron Microscope

The usefulness of a microscope in studying small specimens is limited by its ability to distinguish between closely spaced objects. We call this the resolving power of the microscope. Through optics, it is possible to increase the magnification and thereby increase the resolving power, but only within limits. These limits depend on the wavelength of the light that is used. Objects with diameters less than the wavelength of the light cannot be seen in detail. Since visible light has wavelengths of the order of about 200 to 800 nm, objects smaller than this can't be examined with a microscope that uses visible light.

The electron microscope uses electron waves to "see" very small objects. De Broglie's equation, $\lambda = h/mv$, suggests that if an electron, proton, or neutron has a very high velocity, its wavelength will be very small. In the electron microscope, electrons are accelerated to high speeds across high-voltage electrodes. This gives electron waves with typical wavelengths of about 0.006 to 0.001 nm that strike the sample and are then focused magnetically onto a fluorescent screen where they form a visible image. Because of other difficulties, the actual resolving power is about 1000 times this; therefore, in the end it is possible to see the detail of objects with sizes on the order of 6 to 1 nm.

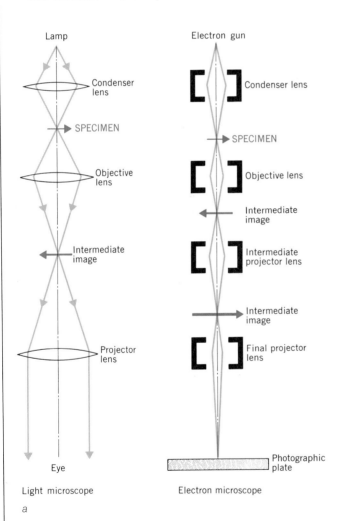

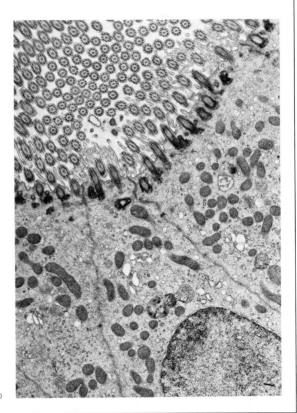

(a) Schematics of both the light microscope and the electron microscope. (From A. G. Marshall, *Biophysical Chemistry,* 1978, John Wiley and Sons, New York, used by permission.) (b) Electron micrograph of the retina of a rabbit, magnification 8,400×.

Notes played this way are called harmonics.   higher. The wave that produces this higher tone has a wavelength exactly half of that formed when the untouched string is plucked. In Figure 7.8 we see that other wavelengths are possible, too—each gives a different note.

If you examine Figure 7.8, you will see that there are some restrictions on the possible wavelengths. Not just any wavelength is possible because the nodes at either end of the string are in fixed positions. The *only* waves that can occur are those for which a half wavelength can be repeated *exactly* a whole number of times. Expressed another way, the length of the string is a whole-number multiple of half wavelengths. In a mathematical form we could write this as

$$L = n \left(\frac{\lambda}{2} \right)$$

where L is the length of the string, λ is a wavelength (therefore, $\lambda/2$ is half the wavelength), and n is an integer. We see that whole numbers (similar to quantum numbers) appear quite naturally in determining which vibrations can occur.

Knowing something about standing waves, we can now look at matter waves, concentrating on electron waves in particular. To study this topic, you may find it best to read through the entire discussion rather quickly to get an overview of the subject and then read it again more slowly. All of the information fits together in the end, somewhat like the pieces of a puzzle.

The theory of quantum mechanics tells us that, in the atom, electron waves are standing waves. Similar to the guitar string, electrons can have many different waveforms or wave patterns. Each of these waveforms, which are called **orbitals,** has a characteristic energy. Not all of the energies are different, but most are. Energy changes within an atom are simply the result of an electron changing from a wave pattern with one energy to a wave pattern with a different energy.

We will be interested in two properties of orbitals—their energies and their shapes. Their energies are important because we normally find atoms in their most stable states (their ground states), which occur when the electrons assume waveforms having the lowest possible energies. The shapes of the wave patterns (i.e., where their amplitudes are large and where they are small) are important because the theory tells us that the amplitude of a wave at any particular location is related to the likelihood of finding the electron there. This will be important when we study how and why atoms form chemical bonds to each other.

"Most stable" almost always means "lowest energy."

In much the same way that the characteristics of a wave on a guitar string can be related to an integer, wave mechanics tells us that the electron waves (orbitals) can be characterized by a set of *three* integer quantum numbers, n, ℓ and m. In discussing the energies of the orbitals, it is usually most convenient to sort the orbitals into groups according to these quantum numbers.

The quantum number n is called the **principal quantum number,** and all orbitals that have the same value of n are said to be in the same **shell.** The values of n can range from $n = 1$ to $n = \infty$. The shell with $n = 1$ is called the *first shell,* the shell with $n = 2$ is the *second shell,* and so forth. (The term "shell" comes from an early notion that atoms could be thought of as similar to onions, with the electrons being arranged in layers around the nucleus.) The various shells are also often identified by letters, beginning (for no significant reason) with K for the first shell ($n = 1$).

n	1	2	3	4	. . .
shell	K	L	M	N	. . .

The principal quantum number serves to determine the size of the electron wave—how far the wave effectively extends from the nucleus. In effect, it is related to how far from the nucleus we are likely to find the electron—the higher the value of n, the further is the electron's average distance from the nucleus. This quantum number is also related to the energy of the orbital. As n increases the energies of the orbitals also increase.

Bohr was lucky—he chose hydrogen to construct his theory.

Bohr's theory took into account only the principal quantum number n. His theory worked fine for hydrogen because it just happens to be the one element in which all orbitals having the same value of n also have the same energy. Bohr's theory failed for atoms other than hydrogen, however, because orbitals with the same value of n can have different energies when the atom has more than one electron.

ℓ is also called the azimuthal quantum number.

The **secondary quantum number,** ℓ, divides the primary electron shells into smaller groups of orbitals called **subshells** (or sublevels). The value of n determines the possible values of ℓ. For a given n, ℓ may range from $\ell = 0$ to $\ell = n - 1$. Thus when $n = 1$, ℓ can have only one value, 0. This means that when $n = 1$ there is only one subshell (the shell and subshell are really identical). When $n = 2$, ℓ can have values of 0 or 1 ($2 - 1$). This means that when $n = 2$ there are two orbital subshells. One subshell has $n = 2$ and $\ell = 0$, and the other has $n = 2$ and $\ell = 1$. Table 7.1 summarizes the relationship between n and the possible values of ℓ.

The number of subshells in a given shell is equal to n for that shell.

Table 7.1
Relationship Between n and ℓ

Value of n	Values of ℓ
1	0
2	0, 1
3	0, 1, 2
4	0, 1, 2, 3
5	0, 1, 2, 3, 4
. . .	. . .

Subshells could be identified by their value of ℓ. However, to avoid confusing the numerical values of n with those of ℓ, a letter code is normally used to specify the value of ℓ.

value of ℓ	0	1	2	3	4	5	. . .
letter designation	s	p	d	f	g	h	. . .

To designate a particular subshell, we write the value of its principal quantum number followed by the letter code for the subshell. For example, the subshell with $n = 2$ and $\ell = 1$ is the $2p$ subshell; the subshell with $n = 4$ and $\ell = 0$ is the $4s$ subshell. Notice that because of the relationship between n and ℓ, every shell has an s subshell ($1s$, $2s$, $3s$, etc.); all the shells except the first have a p subshell ($2p$, $3p$, $4p$, etc.); and all but the first and second shells have a d subshell ($3d$, $4d$, etc.); and so forth.

EXERCISE 7.1   What subshells would be found in the shells with $n = 3$ and $n = 4$?

Whereas the principle quantum number primarily describes the energy and size of an orbital, the secondary quantum number determines the shape of the orbital, which we will examine more closely later. Except for hydrogen, the subshells within a given shell differ slightly in energy, with the energy of the subshell increasing with increasing ℓ. This means that within a given shell the s subshell is lowest in energy, p is the next lowest, followed by d, then f, and so on.

m is used to explain additional lines that appear in the spectra of atoms when they emit light while in a magnetic field.

The third quantum number, m, is known as the **magnetic quantum number.** It splits the subshells into individual orbitals. This quantum number describes how an orbital is oriented in space relative to other orbitals. As with ℓ, there are restrictions as to the possible values of m; they can range from $-\ell$ to $+\ell$. When $\ell = 0$, m can only have the value 0 because $+0$ and -0 are the same. An s subshell, then, has but a single orbital. When $\ell = 1$, the possible values of m are -1, 0, and $+1$. A p subshell therefore has three orbitals. Following similar reasoning, we find that a *d* subshell has five orbitals and an *f* subshell has seven orbitals. The number of orbitals in a given subshell is easy to remember because they follow a simple arithmetic progression:

$$
\begin{array}{cccc}
s & p & d & f \\
1 & 3 & 5 & 7
\end{array}
$$

EXERCISE 7.2   How many orbitals are there in a *g* subshell?

We are finally ready, now, to look at the whole picture. The relationships among all three quantum numbers are summarized in Table 7.2. In addition, the relative energies of the subshells in an atom containing two or more electrons are depicted in Figure 7.9. There are several important features that should be noticed. First, note that each orbital on this energy diagram is indicated by a separate circle, one for an *s* subshell, three for a *p* subshell, and so forth. Second, notice that all the orbitals of a given subshell have the *same* energy. Third, note that, in going upward on the energy scale, the spacing between successive shells decreases as the number of subshells increases. This leads to the overlapping of shells having different values of *n*. For instance, the 4*s* subshell is lower in energy than the 3*d* subshell, 5*s* is lower than 4*d*, and 6*s* is lower than 5*d*. In addition, the 4*f* subshell is below the 5*d* subshell and 5*f* is below 6*d*.

We will see shortly that Figure 7.9 is very useful for predicting the electronic structure of atoms. Before this, however, we must examine another very important property of the electron called spin.

Table 7.2
Summary of Relationships Among *n*, ℓ, and *m*

Value of *n*	Value of ℓ	Values of *m*	Subshell	Number of Orbitals
1	0	0	1*s*	1
2	0	0	2*s*	1
	1	−1, 0, +1	2*p*	3
3	0	0	3*s*	1
	1	−1, 0, +1	3*p*	3
	2	−2, −1, 0, +1, +2	3*d*	5
4	0	0	4*s*	1
	1	−1, 0, +1	4*p*	3
	2	−2, −1, 0, +1, +2	4*d*	5
	3	−3, −2, −1, 0, +1, +2, +3	4*f*	7

7.5 ELECTRON SPIN

The magnetic properties of the electron, which can be explained by the rotation of electrical charge, limits the number of electrons per orbital to two

Earlier it was stated that an atom is in its most stable state (its ground state) when its electrons have wave patterns with the lowest possible energies. This occurs when the electrons "occupy" the lowest energy orbitals that are available. But what determines how the electrons "fill" these orbitals? Fortunately, there are some simple rules that can help. These govern both the maximum number of electrons that can be in a particular orbital and how orbitals having the same en-

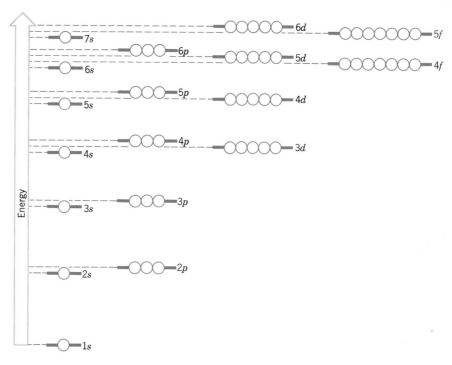

Figure 7.9
Approximate energy level diagram for atoms with two or more electrons.

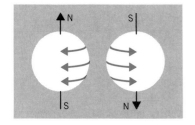

The electron can spin in either of two directions.

Pauli received the 1945 Nobel prize in physics for his discovery of the exclusion principle.

ergy can be filled. One important factor that influences them is the phenomenon known as electron spin.

The concept of **electron spin** is based on the fact that electrons behave as tiny magnets. This can be explained by imagining that an electron spins about its axis, much like a toy top. The revolving electrical charge of the electron creates its own magnetic field. The same effect is used to make electric motors work. The passage of electrical charge through the curved windings of an electric motor sets up magnetic forces that push and pull, thereby causing the rotor within the device to turn.

Electron spin gives us a fourth quantum number for the electron called the **spin quantum number,** m_s. Again like the toy top, the electron can spin in either of two directions. Therefore, the spin quantum number can take on either of two possible values: $m_s = +\frac{1}{2}$ or $m_s = -\frac{1}{2}$. The actual values of m_s and the reason that they are not integers isn't very important, but the fact that there are *only* two values is very significant.

In 1925 an Austrian physicist, Wolfgang Pauli (1900–1958), expressed the importance of electron spin in determining electronic structure. The **Pauli exclusion principle** states that no two electrons in the same atom may have identical values for all four quantum numbers. To understand what this means, suppose that two electrons occupy the 1s orbital of an atom. Each electron would have $n = 1$, $\ell = 0$, and $m = 0$. Since these three quantum numbers are the same for both electrons, the exclusion principle requires that their fourth quantum numbers (their spin quantum numbers) be different; one electron would have to have $m_s = +\frac{1}{2}$ and the other, $m_s = -\frac{1}{2}$. No more than two electrons can occupy the 1s orbital simultaneously because there are only two possible values of m_s. Thus, the Pauli exclusion principle limits the number of electrons in any orbital to two.

The limit of two electrons per orbital also limits the maximum electron population of the shells and subshells. For the subshells we have

Subshell	Number of Orbitals	Maximum Number of Electrons
s	1	2
p	3	6
d	5	10
f	7	14

The maximum electron population per shell is shown below.

Shell	Subshells	Maximum Shell Population[a]
1	1s	2
2	2s 2p	8
3	3s 3p 3d	18
4	4s 4p 4d 4f	32

[a] In general, the maximum electron population of a shell is $2n^2$.

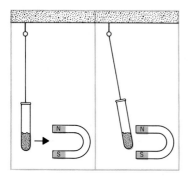

A paramagnetic substance is attracted to a magnetic field.

We have seen that two electrons occupying the same orbital must have different values of m_s. When this occurs, we say that the spins of the electrons are "*paired*" or simply that the electrons are *paired*. Such pairing leads to the cancellation of the electrons' magnetic effects because the north pole of one electron magnet is opposite the south pole of the other. Atoms having more electrons that spin in one direction than they do electrons that spin in the other are said to contain *unpaired* electrons. For these atoms, the magnetic effects do not cancel and the atoms themselves become tiny magnets which can be attracted to an external magnetic field. This weak attraction to a magnet of a substance containing unpaired electrons is called **paramagnetism.** The measurement of paramagnetism, then, provides experimental verification of the presence of unpaired electrons.

7.6 ELECTRON CONFIGURATIONS

The electron configuration of an element is obtained by placing electrons into the lowest available orbitals, while heeding the Pauli exclusion principle and spreading the electrons as much as possible among orbitals of identical energy

The Pauli exclusion principle and the energy level diagram in Figure 7.9 allow us to predict which orbitals in a particular atom will be populated by electrons and the number of electrons that will be found in each. This arrangement of electrons is called the atom's **electronic structure** or **electron configuration.** Knowing how to predict electron configurations is important because the arrangement of electrons controls an atom's chemical properties. Let's look at specific examples to see how all this works.

We will begin with the hydrogen atom whose atomic number, Z, is 1. A neutral hydrogen atom has one electron. In its ground state this electron will occupy the lowest energy orbital that's available, which is the 1s orbital. To indicate the population of a subshell we use a superscript with the subshell designation. Thus the electron configuration of hydrogen is written as

$$H \qquad 1s^1$$

Another way of expressing electron configurations that we will sometimes find useful is the **orbital diagram.** In it, each orbital is represented by a circle and arrows are used to indicate the individual electrons, heads up for spin in one direction and heads down for spin in the other. The orbital diagram for hydrogen is simply

$$H \quad ⬆$$
$$1s$$

Next, let's look at helium, for which $Z = 2$. This atom has two electrons, both of which are permitted to occupy the 1s orbital. The electron configuration of helium can therefore be written as

$$He \quad 1s^2 \qquad \text{or} \qquad He \quad ⬆⬇$$
$$1s$$

Notice that the orbital diagram clearly indicates that the electrons in the 1s orbital are paired.

We can proceed in the same fashion to predict successfully the electron configurations of most of the elements in the periodic table. For example, the next elements in the table are lithium, Li ($Z = 3$), and beryllium, Be ($Z = 4$), which have three and four electrons, respectively. After the 1s subshell is filled with two electrons, the next lowest energy orbital is the 2s. Therefore, we can represent the electronic structures of lithium and beryllium as

$$Li \quad 1s^2 2s^1 \qquad \text{or} \qquad Li \quad ⬆⬇ \quad ⬆$$
$$Be \quad 1s^2 2s^2 \qquad \text{or} \qquad Be \quad ⬆⬇ \quad ⬆⬇$$
$$\qquad\qquad\qquad\qquad\qquad\qquad\qquad 1s \quad 2s$$

Following beryllium we have boron, B ($Z = 5$). Referring to Figure 7.9, we see that the first four electrons of this atom complete the 1s and 2s subshells, so the fifth electron must be placed into the 2p subshell:

$$B \qquad 1s^2 2s^2 2p^1$$

In the orbital diagram for boron, the fifth electron can be put into any one of the 2p orbitals—which one doesn't matter because they are all of equal energy:

$$B \quad ⬆⬇ \quad ⬆⬇ \quad ⬆○○$$
$$\quad 1s \quad 2s \quad 2p$$

Notice, however, that when we give this orbital diagram we show *all* of the orbitals of the 2p subshell even though two of them are empty.

Next we come to carbon, which has six electrons. As before, the first four electrons complete the 1s and 2s orbitals. The remaining two electrons go in the 2p subshell to give

$$C \qquad 1s^2 2s^2 2p^2$$

Now, however, to give the orbital diagram, we have to make a decision as to where to put the two p electrons. (At this point you may have an unprintable suggestion! But try to bear up. It's really not all that bad.) To make this decision,

It doesn't matter which two orbitals are shown as occupied. Any of these are okay for carbon.

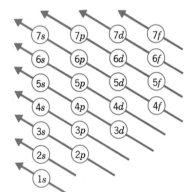

we use **Hund's rule.** This rule states that *when electrons are placed in a set of orbitals of equal energy they are spread out as much as possible to give as few paired electrons as possible.* Both theory and experiment have shown that following this rule leads to the electron arrangement with the lowest energy. For carbon, it means that the two *p* electrons are in separate orbitals and their spins are in the same direction:

$$C \quad \begin{array}{ccc} \underset{1s}{(\uparrow\downarrow)} & \underset{2s}{(\uparrow\downarrow)} & \underset{2p}{(\uparrow)(\uparrow)(\,)} \end{array}$$

Using the Pauli exclusion principle and Hund's rule we can now complete the orbital diagrams for the rest of the elements of the second period:

		1s	2s	2p
N	$1s^2 2s^2 2p^3$	$(\uparrow\downarrow)$	$(\uparrow\downarrow)$	$(\uparrow)(\uparrow)(\uparrow)$
O	$1s^2 2s^2 2p^4$	$(\uparrow\downarrow)$	$(\uparrow\downarrow)$	$(\uparrow\downarrow)(\uparrow)(\uparrow)$
F	$1s^2 2s^2 2p^5$	$(\uparrow\downarrow)$	$(\uparrow\downarrow)$	$(\uparrow\downarrow)(\uparrow\downarrow)(\uparrow)$
Ne	$1s^2 2s^2 2p^6$	$(\uparrow\downarrow)$	$(\uparrow\downarrow)$	$(\uparrow\downarrow)(\uparrow\downarrow)(\uparrow\downarrow)$

Figure 7.10
A way to remember the filling order of subshells. Write the subshell designations as shown and follow the diagonal arrows starting at the bottom.

We could, of course, continue to predict electron configurations using Figure 7.9 as a guide and following our filling rules. But how could *you* remember the sequence of energy levels if Figure 7.9 were not available for use? One device that can help is shown in Figure 7.10. The procedure is to follow the diagonal arrows beginning at the bottom. This tells us that the subshells should be filled in the following order: 1s, 2s, 2p, 3s, 3p, 4s, 3d, 4p, 5s, 4d, 5p, 6s, 4f, 5d, 6p, 7s, 5f, 6d, 7p, etc.

EXAMPLE 7.1   **Writing Electron Configurations**

Problem: Write the electron configuration for manganese, Mn ($Z = 25$).
Solution: First we determine how many electrons are in a manganese atom. Since Mn has $Z = 25$, there are 25 electrons. Following Figure 7.10 we begin with the 1s subshell and fill the rest of the subshells until we run out of electrons. Remember the maximum subshell populations: $s = 2$, $p = 6$, $d = 10$, and $f = 14$. This gives

$$Mn \quad 1s^2 2s^2 2p^6 3s^2 3p^6 4s^2 3d^5$$

Some people refer to write all subshells of a given shell together.

$$Mn \quad 1s^2 2s^2 2p^6 3s^2 3p^6 3d^5 4s^2$$

The orbital diagram for manganese would be

Mn $\underset{1s}{(\uparrow\downarrow)}$ $\underset{2s}{(\uparrow\downarrow)}$ $\underset{2p}{(\uparrow\downarrow)(\uparrow\downarrow)(\uparrow\downarrow)}$ $\underset{3s}{(\uparrow\downarrow)}$ $\underset{3p}{(\uparrow\downarrow)(\uparrow\downarrow)(\uparrow\downarrow)}$ $\underset{3d}{(\uparrow)(\uparrow)(\uparrow)(\uparrow)(\uparrow)}$ $\underset{4s}{(\uparrow\downarrow)}$

Notice that the manganese atom has five unpaired electrons.

EXAMPLE 7.2 **Writing Electron Configurations**

Problem: Predict the electron configuration of bismuth, Bi ($Z = 83$).

Solution: Once again we must follow the filling sequence of Figure 7.10. We get:

$$\text{Bi} \qquad 1s^2 2s^2 2p^6 3s^2 3p^6 4s^2 3d^{10} 4p^6 5s^2 4d^{10} 5p^6 6s^2 4f^{14} 5d^{10} 6p^3$$

If we group subshells of the same shell together,

$$\text{Bi} \qquad 1s^2 2s^2 2p^6 3s^2 3p^6 3d^{10} 4s^2 4p^6 4d^{10} 4f^{14} 5s^2 5p^6 5d^{10} 6s^2 6p^3$$

EXERCISE 7.3 Predict the electron configuration for (a) Mg, (b) Ge, (c) Cd, and (d) Gd. Group subshells of the same shell together.

EXERCISE 7.4 Write orbital diagrams for (a) Na, (b) S, and (c) V.

When considering the chemical properties of elements, we are rarely interested in electrons that are buried deep within the atom. Our attention is usually focused only on the electron configuration of the electrons in the outer shells of the atom. This is because the outer electrons are the ones that are involved in chemical reactions. The inner electrons, called the **core electrons,** of one atom are not exposed to the electrons of other atoms when chemical bonds are formed. To direct attention to these outer electrons, we often write electron configurations in an abbreviated, or shorthand, form. Consider, for instance, the elements in period 3 of the periodic table, taking sodium and magnesium as representative examples. The electron configurations of these elements are

$$\text{Na} \qquad 1s^2 2s^2 2p^6 3s^1$$
$$\text{Mg} \qquad 1s^2 2s^2 2p^6 3s^2$$

The outer electrons are in the third shell; the core ($1s^2 2s^2 2p^6$) is identical for both. This core configuration is the same as that of the noble gas, neon. To write the shorthand configuration for an element we indicate what the core is by placing in brackets the symbol of the noble gas whose electron configuration is the same as the core configuration. This is followed by the configuration of the outer electrons for the particular element. The noble gas used is almost always the one that occurs at the end of the period preceding the period containing the element whose configuration we wish to represent. Thus, for sodium and magnesium we would write

$$\text{Na} \qquad [\text{Ne}]3s^1$$
$$\text{Mg} \qquad [\text{Ne}]3s^2$$

EXAMPLE 7.3 **Writing Shorthand Electron Configurations**

Problem: What is the shorthand electron configuration of manganese?

Solution: Manganese is in period 4 and has 25 electrons (because $Z = 25$). The preceding noble gas is argon, Ar, which has an atomic number of 18. This means that an argon atom has 18 electrons. When we write the shorthand configuration, the first 18 electrons are represented by placing the symbol Ar in brackets. From Figure 7.10, this corresponds to completed $1s$, $2s$, $2p$,

3s, and 3p subshells. The remaining seven electrons are distributed as $4s^2 3d^5$. Therefore, the shorthand electron configuration for Mn is

Mn $[Ar]4s^2 3d^5$

Placing the electrons that are in the highest shell farthest to the right gives

Mn $[Ar]3d^5 4s^2$

EXERCISE 7.5   Write shorthand configurations for (a) P and (b) Sn.

7.7   Some Unexpected Electron Configurations

Filled and half-filled subshells have extra stability that sometimes affects electron configurations

The rules that you've learned for predicting electron configurations work most of the time—but not always. Appendix B gives the electron configurations of all of the elements as determined experimentally. Close examination reveals that there are quite a few exceptions to the rules. Fortunately, most of these exceptions are of little consequence to us because the elements involved are relatively rare and their chemistry will not be important to us in this course. Some of the exceptions are important, though, because they occur with common elements—notably, chromium and copper.

Following the rules, we would expect the following electron configurations:

Cr $[Ar]3d^4 4s^2$
Cu $[Ar]3d^9 4s^2$

However, the actual electron configurations, determined experimentally, are

Cr $[Ar]3d^5 4s^1$
Cu $[Ar]3d^{10} 4s^1$

The corresponding orbital diagrams are

Notice that for chromium, an electron is "borrowed" from the 4s subshell to give a 3d subshell that is exactly half-filled. For copper the 4s electron is borrowed to give a completely filled 3d subshell. A similar thing happens with silver and gold, which have filled 4d and 5d subshells, respectively.

Ag $[Kr]4d^{10} 5s^1$
Au $[Xe]4f^{14} 5d^{10} 6s^1$

Apparently, half-filled and filled subshells (particularly the latter) have some special stability that makes such borrowing energetically favorable. This subtle, but nevertheless important, phenomenon affects not only the ground state configuration of atoms but also the relative stabilities of some of the ions formed by the transition elements.

7.8 ELECTRON CONFIGURATIONS AND THE PERIODIC TABLE

The similarities among elements within groups and the structure of the periodic table can be explained by electron configurations

In the previous chapter, we saw that when the periodic table was constructed, atoms with similar chemical properties were arranged in vertical columns called groups. The basic structure and shape of the periodic table that results is one of the strongest empirical supports for the quantum theory that we have been using to predict electron configurations.

Consider, for example, the way the table is laid out (Figure 7.11). On the left there are *two* columns of elements, on the right there is a block of *six* columns, in the center there is a block of *ten* columns of elements, and below the table there are two rows consisting of *fourteen* elements each. These numbers—2, 6, 10, and 14—are *precisely* the number of electrons that the quantum theory tells us can occupy the s, p, d, and f subshells, respectively!

This would be an amazing coincidence if the theory were wrong.

Figure 7.11
Overall structure of the periodic table.

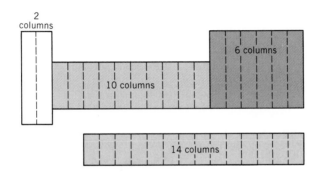

The similarities in group properties can also be explained. Because only the outer parts of atoms touch during chemical reactions, we might expect to find similarities in the outer-shell electron configurations of elements that have similar chemical properties—and we do. Let's look at the alkali metals, Group IA. Going by our rules, we get the following configurations:

Li	$1s^2 2s^1$
Na	$1s^2 2s^2 2p^6 3s^1$
K	$1s^2 2s^2 2p^6 3s^2 3p^6 4s^1$
Rb	$1s^2 2s^2 2p^6 3s^2 3p^6 3d^{10} 4s^2 4p^6 5s^1$
Cs	$1s^2 2s^2 2p^6 3s^2 3p^6 3d^{10} 4s^2 4p^6 4d^{10} 5s^2 5p^6 6s^1$

Each of these elements has one outer electron which is in an s subshell. We know that when they react, the alkali metals lose one electron to form ions with a charge of $1+$. For each, then, the electron that is lost is this outer s electron, and the electron configuration of the ion that is formed is the same as that of the preceding noble gas.

Li^+	$1s^2$		He	$1s^2$
Na^+	$1s^2 2s^2 2p^6$		Ne	$1s^2 2s^2 2p^6$
K^+	$1s^2 2s^2 2p^6 3s^2 3p^6$		Ar	$1s^2 2s^2 2p^6 3s^2 3p^6$

etc.

For the representative elements (those in A-groups), the only electrons that are normally important in controlling chemical properties are the ones in the outermost shell, that is, the occupied shell having the highest value of n. This outer shell is known as the **valence shell,** and the electrons in it are called **valence electrons.** (The term valence comes from the study of chemical bonding and relates to the combining capacity of an element, but that's not important here.) Even with the transition elements, we need only concern ourselves with the outermost shell and the d subshell just below. For example, iron has the configuration

$$\text{Fe} \qquad 1s^2 2s^2 2p^6 3s^2 3p^6 3d^6 4s^2$$

Only the $4s$ and $3d$ electrons play a role in the chemistry of iron. In general, the electrons below the outer s and d subshells of a transition element—the *core electrons*—are relatively unimportant. In every case these core electrons have the electron configuration of a noble gas.

For the representative elements there is a very simple way to determine the electron configuration of the valence shell using the periodic table. Consider the following two examples:

$$\text{C} \qquad 1s^2 2s^2 2p^2$$
$$\text{S} \qquad 1s^2 2s^2 2p^6 3s^2 3p^4$$

The valence shell is shown in color. Now, notice that carbon is in period 2 of the table. This number, 2, is the same as the value of n for carbon's valence shell. For sulfur, in period 3, the valence shell is the third shell, for which $n = 3$.

Next, notice that to get to carbon by moving from left to right in period 2 we first have to pass through the group of two columns. This corresponds to filling the $2s$ subshell. Then we have to go two spaces into the group of six columns to finally get to carbon. These two spaces are the number of electrons that go into the $2p$ subshell. For sulfur, in period 3, we pass through the group of two columns (hence $3s^2$) and then go four spaces into the group of six columns (hence $3p^4$).

We see that the periodic table serves as a useful guide in writing electron configurations. The value of n for the outer shell of elements in any given period is equal to the period number. Then, we simply move from left to right across the period, placing electrons into an s subshell as we cross through the block of two columns and into a p subshell as we cross the block of six columns. This is summarized in Figure 7.12.

Figure 7.12
Subshells that become filled as we cross periods.

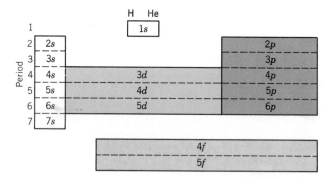

EXERCISE 7.6   Give the electron configuration of the valence shell of (a) N, (b) Si, and (c) Sr.

Earlier in this section we saw that the electrons that determine the chemical properties of iron, a transition element, are in the $4s$ and $3d$ subshells. To obtain the subshell populations of a transition element we can also use its location in the table as a guide. To get to iron in period 4, we first pass through the set of two columns ($4s^2$) and then enter the block of 10 columns. Here we fill a d subshell whose value of n is equal to the period number minus one (Figure 7.12). To reach iron, then, we move six spaces into this $(4 - 1) d$ or $3d$ subshell; thus the configuration of this subshell is $3d^6$.

EXAMPLE 7.4 **Writing Electron Configurations from the Periodic Table**

Problem: What is the electron configuration of zirconium, Zr ($Z = 40$)?

Solution: Zirconium is in period 5. The preceding noble gas is Kr. To reach Zr we pass across the first two columns ($5s^2$) and then through two spaces in the center. The last two electrons enter the $(5 - 1)d$ or $4d$ subshell. Thus, for Zr we get

$$Zr \quad [Kr]4d^2 5s^2$$

EXERCISE 7.7 Write the shorthand electron configuration for (a) Ni and (b) Ru.

EXAMPLE 7.5 **Writing Valance Shell Electron Configurations**

Problem: Predict the electron configuration of the valence shell of arsenic ($Z = 33$).

Solution: Arsenic is a period 4 element. The preceding noble gas is Ar. To reach As, we cross the s block to get $4s^2$, the d block to get $(4 - 1)d^{10}$ or $3d^{10}$, and then move three spaces into the p block to get $4p^3$. This gives

$$As \quad [Ar]3d^{10}4s^2 4p^3$$

Among the representative elements, completed subshells below the outer shell are unimportant in determining chemical properties. For elements such as arsenic we are only interested in the valence shell. The valence shell is the one with highest n, in this case, 4, so the configuration of the valence shell is $4s^2 4p^3$.

EXERCISE 7.8 What is the electron configuration for the valence shell of (a) Se, (b) Sn, and (c) I?

7.9 WHERE THE ELECTRON SPENDS ITS TIME

The s, p, d, and f orbitals have shapes and directional orientations that describe where and how electrons are most probably distributed

The same theory that tells us of the energies of atomic orbitals also describes the shapes of electron waves. How do we know these shapes are correct? We don't for sure, but many of the predictions that have been made using the theory seem to be born out by experiments. This gives the theoretical explanations strength and support. For example, the fact that the results of wave mechanics account for the shape of the periodic table so very well gives the theory a good deal of credibility.

The difficulty in describing where electrons are in an atom is due to the basic problem that we face when we attempt to picture a particle as a wave. There is nothing in our wordly experience that is comparable. The way we get around this perplexing conceptual problem, so that we can still think of the electron as a particle in the usual sense, is to speak in terms of the statistical probability of the electron being found at a particular place.

Heisenberg won the Nobel prize in physics for his work in 1932.

Describing the electron's position in terms of statistical probability is based on more than simple convenience. The German physicist, Werner Heisenberg, showed mathematically that there are limits to our ability to measure both a particle's velocity and its position at the same instant. This was Heisenberg's famous

uncertainty principle. The theoretical limitations on measuring speed and position are not significant for large objects. However, for small particles such as the electron, these limitations prevent us from ever knowing or predicting exactly where an electron will be at a particular instant. So we speak of probabilities instead.

Wave mechanics views the probability of finding an electron at a given point in space as equal to the square of the amplitude of the electron wave at that point. It seems quite reasonable to relate probability to amplitude, or intensity, because where a wave is intense its presence is strongly felt. The amplitude is squared because, mathematically, the amplitude can be either positive or negative, but probability only makes sense if it is positive. Squaring the amplitude assures us that the probabilities will be positive. We need not be very concerned about this point, however.

The concept of electron probability leads to two very important and frequently used ideas. One is that an electron behaves as if it were spread out around the nucleus in a sort of **electron cloud.** Figure 7.13 illustrates the way the probability of finding the electron varies for a 1s orbital. In those places where there are large numbers of dots, the amplitude of the wave is large and the probability of finding the electron is large.

Figure 7.13
Electron probability distribution for a 1s electron.

The other important idea is **electron density,** which relates to how much of the electron's charge is packed into a given volume. Because of its wave nature, the electron (and its charge) is spread out around the nucleus. In regions of high probability there is a high concentration of electrical charge and the electron density is large; in regions of low probability, the electron density is small. In looking at the way the electron density distributes itself in atomic orbitals we are interested in three things—the *shape* of the orbital, the *size* of the orbital, and the *orientation* of the orbital in space relative to other orbitals.

Electron density doesn't end abruptly at some particular distance from the nucleus. It gradually fades away. To define the size and shape of an orbital, it is useful to picture some imaginary surface enclosing, say 90% of the electron density of the orbital and on which the probability of finding the electron is everywhere the same. For the 1s orbital in Figure 7.13, we would find that if we go out a given distance from the nucleus in *any* direction, the probability of observing the electron would be the same. This means that all the points of equal probability would lie on the surface of a sphere, so we can say that the shape of the orbital is spherical. In fact, all s orbitals are spherical. As suggested earlier, their size increases with increasing *n*. This is illustrated in Figure 7.14. Notice that beginning

Figure 7.14
Size variation among s orbitals. Orbitals become larger with increasing *n*.

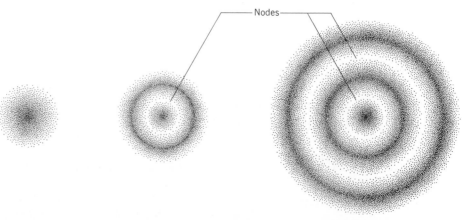

1s   2s   3s

with the 2s orbital, there are certain places where the electron density drops to zero. These are the nodes of the electron wave. It is interesting that electron waves have nodes just like the waves on a guitar string. For electron waves, however, the nodes consist of imaginary *surfaces* on which the electron density is zero.

The *p* orbitals are quite different from *s* orbitals, as shown in Figure 7.15. Notice that the electron density is equally distributed in two regions on opposite sides of the nucleus. Figure 7.15 shows the *two* lobes of *one* 2*p* orbital. The electron density of the lobes is concentrated about an imaginary line that passes through the nucleus. Between the two lobes, there is a nodal plane—an imaginary flat surface on which every point has an electron density of zero.

Figure 7.15
The probability distribution in a *p* orbital.

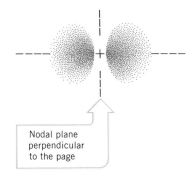

Nodal plane perpendicular to the page

A *p* subshell consists of three orbitals whose directions lie at 90° to each other along the axes of an imaginary *xyz* coordinate system (Figure 7.16). For convenience in referring to them, the orbitals are often labeled according to the axis along which they lie. The *p* orbital concentrated along the *x* axis is labeled p_x, and so forth. As with *s* orbitals, the sizes of the *p* orbitals increase with increasing *n*; as *n* increases they extend further from the nucleus.

The shapes of the *d* and *f* orbitals are even more complex than those of the *p* orbitals. We will discuss *d* orbitals later (Chapter Twenty) when we will need to know about their shapes, but we will have no need to consider the shapes of *f* orbitals. However, the shapes and directional properties of the *s* and *p* orbitals play an important role in determining molecular structure when atoms form chemical bonds with one another, as we will see in Chapter Nine.

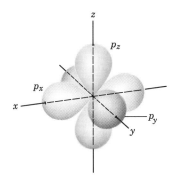

Figure 7.16
The orientation of the three *p* orbitals in a *p* subshell.

7.10 IONIZATION ENERGY

The energy required to strip an electron from an atom is called the ionization energy

Ionization energy is also called ionization potential.

In this and the following two sections, we will take a brief look at some of the properties of atoms that are related to their electron configurations and that are responsible for many of the chemical characteristics of the various elements. The first of these is **ionization energy (IE)** which is the energy needed to remove an electron from an isolated, gaseous atom. It is, in effect, a measure of how tightly the electrons are held by the atom and it is given in units of kJ/mol.

Atoms having more than one electron have more than one ionization energy. These energies correspond to the stepwise removal of electrons, one after the other. Lithium, for example, has three ionization energies because it has three electrons. To remove the outer electron from 1 mol of lithium atoms to give 1 mol of lithium ions, Li^+, requires 520 kJ; therefore, the first IE of lithium is 520 kJ/mol. The second IE for lithium is 7297 kJ/mol and the third IE is 11,810 kJ/mol. In general, successive ionization energies always increase because each subsequent electron is being pulled away from an increasingly more positive ion.

The magnitudes of IE are really quite enormous. The amount of energy required to remove the outer electron from 1 mol of lithium atoms (6.9 g Li) could raise the temperature of 20 liters of water by 6.2 °C. Such huge energy requirements indicate how very tightly atoms hold onto their electrons.

Within the periodic table there are trends in the way IE varies that are useful to know and to which we will refer in later discussions. In general, *ionization energy* **increases** *from bottom to top within a group and* **increases** *from left to right within a period* (see Figure 7.17). The fact that we can explain these variations in terms of electronic structure is important because it will help us to account for some of the chemical properties of the elements.

Compared to other atoms, lithium doesn't hold its outer electron very tightly!

Figure 7.17
Variation of ionization energy within the periodic table.

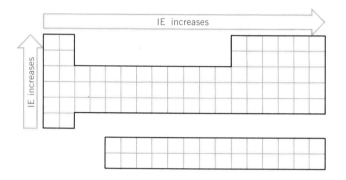

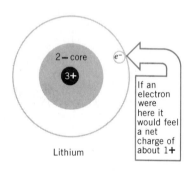
Lithium

Two key ideas relate IE to electronic structure. One is that electrons far from the nucleus are bound less tightly than those that are near the nucleus. This is because the attraction between oppositely charged particles decreases as the distance between them increases. The other idea is that the negative charges of the electrons in the inner shells help to offset the positive charge of the nucleus so that the electrons in the outer shell are affected by only a fraction of the full nuclear charge. The outer $2s$ electron of lithium, for example, is "shielded" from the $3+$ nuclear charge by the inner pair of $1s$ electrons. This outer $1s$ electron feels a net charge of only about $1+$ (the $3+$ charge of the nucleus being partially neutralized by the $2-$ charge of the two $1s$ electrons).

As we go from the top to the bottom of a group in the periodic table, the outer electron configurations remain the same except for the increase in the value of n for the valence shell. For lithium the outer shell configuration is $2s^1$; for sodium it is $3s^1$; for potassium it is $4s^1$; and so on. Earlier, we saw that the size of the orbital increases as n increases. As we go down a group, therefore, the outer electrons are further and further away from the nucleus. At the same time the effective charge felt by these outer electrons remains about the same because the increasing number of inner core electrons effectively shield the outer electrons from the increasing positive charge of the nucleus. The farther the outer electrons are from the nucleus, the less tightly they are held by the nearly constant effective charge, so the IE decreases.

Moving from left to right across a period, electrons are added to the same shell. The nuclear charge builds, but the inner core of electrons remains the same; only the outer shell becomes more populated. Since an electron is only slightly shielded by other electrons in the same shell, the outer shell electrons are affected by an increasing effective charge that binds them more and more tightly to the nucleus. Thus IE increases from left to right.

The results of these trends place those elements with highest IE in the upper right-hand corner of the periodic table. It is very difficult to cause these atoms to lose electrons. In the lower left-hand corner of the table are the elements that have loosely held outer electrons. These elements form positive ions rather easily, as we learned in the last chapter.

Stability of the noble gas configuration

Table 7.3 gives the successive ionization energies for the first 12 elements. Notice that for a given element, there is a gradual increase in IE until all of the electrons in the valence shell have been removed. Then there is a very sudden and dramatic increase in IE as the first inner shell is broken into. These huge increases in IE occur for inner electron configurations corresponding to the configuration of one of the nobel gases.

Table 7.3
Successive Ionization Energies in kJ/mol for Hydrogen through Magnesium

	1	2	3	4	5	6	7	8
H	1312							
He	2372	5250						
Li	520	7297	11,810					
Be	899	1757	14,845	21,000				
B	800	2426	3659	25,020	32,820			
C	1086	2352	4619	6221	37,820	47,260		
N	1402	2855	4576	7473	9442	53,250	64,340	
O	1314	3388	5296	7467	10,987	13,320	71,320	84,070
F	1680	3375	6045	8408	11,020	15,160	17,860	92,010
Ne	2080	3963	6130	9361	12,180	15,240	—	—
Na	496	4563	6913	9541	13,350	16,600	20,113	25,666
Mg	737	1450	7731	10,545	13,627	17,995	21,700	25,662

The data suggest that a great deal of stability is associated with an electron configuration corresponding to that of a noble gas. Indeed, this explains the extremely low degree of reactivity of the noble gas elements—their electron configurations are so stable that they do *not* tend to react in any way that would disturb their electronic structure. We will say more about this in Section 7.13.

7.11 ELECTRON AFFINITY

Energy is usually released when an electron is added to a neutral atom

In the last section we saw that energy is needed, and therefore work must be done, to remove an electron from an atom. Each electron in an atom is attracted to the nucleus, and this attraction must be overcome if an electron is to be removed. If an extra electron is added to an atom to create a negative ion, this electron will also come under the influence of the nucleus, and work will have to be done to remove it in order to make the atom neutral once again. The energy needed to remove an electron from a negative ion is equal in magnitude to the energy given off when the electron is added to the atom to create the ion. This energy is called the **electron affinity (EA).**

For nearly all the elements, the addition of one electron is exothermic, and therefore the EA is given as a negative value. This convention agrees with the one that we followed for determining the signs of ΔE and ΔH in Chapter Five. However, when a second electron is added, as in the formation of the oxide ion, O^{2-}, work must be done to force the electron into an already negative ion. Electron affinities for some common elements are given in Table 7.4.

Trends in EA are very similar to those for IE. EA increases from left to right in a period and increases from bottom to top in a group (see Figure 7.18). This

Table 7.4
Electron Affinities For Some Elements

Element	EA (kJ/mol)[a]	Process
Fluorine	−344	F + e^- ⟶ F^-
Chlorine	−349	Cl + e^- ⟶ Cl^-
Bromine	−325	Br + e^- ⟶ Br^-
Oxygen	−142	O + e^- ⟶ O^-
	+844	O^- + e^- ⟶ O^{2-}
Hydrogen	−72	H + e^- ⟶ H^-
Sodium	−50	Na + e^- ⟶ Na^-

[a] A negative sign means energy is evolved when an electron is added.

Figure 7.18
Variation of electron affinity within the periodic table.

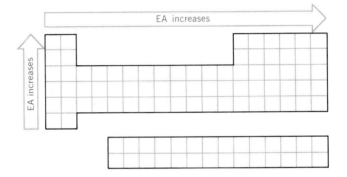

shouldn't be surprising because a valence shell that loses electrons easily will have little attraction for additional electrons. On the other hand, a valence shell that holds its electrons tightly will also tend to hold an additional electron tightly.

7.12 SIZES OF ATOMS AND IONS

Atomic and ionic size is determined by the balance between the attractions the electrons feel for the nucleus and the repulsions they feel for each other

The very nature of the wave concept of the electronic structure of the atom makes it difficult to define exactly what we mean by the "size" of an atom or ion. As we've seen, the electron cloud doesn't simply stop at some particular distance from the nucleus; instead it gradually fades away. Nevertheless, atoms and ions do behave in many ways as though they have characteristic sizes. For example, in a whole host of hydrocarbons—ranging from methane (CH_4, natural gas) to octane (C_8H_{18}, in gasoline) to many others—the distance between the nuclei of carbon and hydrogen atoms is virtually the same. This would suggest that carbon and hydrogen have the same relative sizes in each of these compounds.

The C—H distance in most hydrocarbons is about 1.10 Å.

It is difficult to appreciate how small atoms really are. They are incredibly tiny. We've already seen how little they weigh in Chapter Three. In size, atoms range from about 1.4×10^{-10} m to 5.7×10^{-10} m in diameter. Their radii, which is the usual way that size is specified, range from about 7.0×10^{-11} to 2.9×10^{-10} m. These are difficult numbers to comprehend. A million carbon atoms placed side by side in a line would extend a little less than 0.2 mm.

To express the sizes of atoms and ions, scientists have traditionally used a unit called the **angstrom** (symbol Å), defined as 1 Å = 1×10^{-8} cm (or 1 Å = 10^{-10} m). However, in many current scientific journals, atomic dimensions are given in SI units of picometers or nanometers (1 pm ≑ 10^{-12} m and 1 nm = 10^{-9} m). We will use angstroms simply because when atomic dimensions are expressed in angstroms the numbers are of a size that's easier to comprehend, but you may someday find it useful to remember the conversions:

$$1Å = 100 \text{ pm}$$

$$1Å = 0.1 \text{ nm}$$

The actual determination of atomic size is a difficult and complex task. We will touch on this subject briefly in Chapter Ten. For our purposes, though, it isn't really necessary to know how atomic radii are measured or even what the actual radii of atoms and ions are. However, it is useful to know something about the relative sizes of different atoms and how the sizes of ions compare with their neutral atoms.

As with IE and EA, atomic size varies in a more or less systematic way in the periodic table. As we go down a group, the atoms become larger because their outer electrons are in shells that have increasing values of n. We've seen that as n increases, the size of a given type of orbital increases also. Proceeding from left to right across a period, we find a gradual decrease in size. This occurs because the increasing effective charge felt by the outer electrons draws them closer to the nucleus.

The variation in size among the elements was one of the things that led Meyer to his discovery of the periodic law (Section 6.2). Actually, Meyer calculated the molar volumes for the elements from their densities. The molar volume is simply the volume occupied by 1 mol of the element. Figure 7.19 is a graph of molar volume versus atomic number. The relationship (an approximate one) between molar volume and size is easy to understand—a mole of large atoms will have a larger volume than will a mole of small atoms. The decrease in size across a period is evident; from Li to F, for example, or from Na to Cl. The increase in size down a period is even easier to see. For example, follow the molar volumes of the alkali metals or alkaline earth metals (Groups IA or IIA).

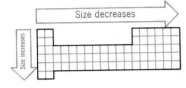

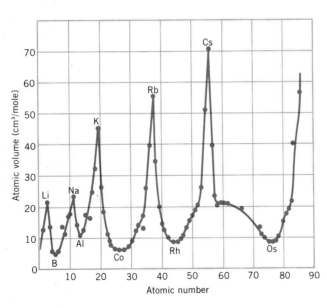

Figure 7.19
Graph of molar volume of the elements versus atomic number.

Across a row of the transition elements or inner transition elements, the size variations are less pronounced than they are among the representative elements. This is because the outer shell configuration remains the same while an inner shell is filled. Going from atomic numbers 21 to 30, for example, the outer elec-

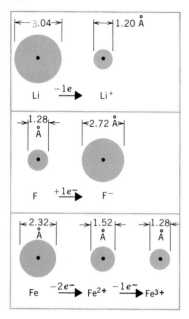

trons occupy the 4s subshell while the 3d subshell is gradually completed. Because the inner 3d electrons are quite effective at shielding the outer electrons, the effective charge felt by the 4s electrons changes very little as the nuclear charge increases, and only a small reduction in size occurs.

When atoms gain or lose electrons to form ions, rather significant size changes take place. The reasons for these changes are easy to understand and remember. *Adding electrons creates an ion that is larger than the neutral atom; removing electrons produces an ion smaller than the neutral atom.* When electrons are added, increased repulsions between the like-charged electrons cause them to spread out and occupy a larger volume. *Negative ions are always larger than the atoms from which they are formed.*

By similar reasoning we should expect that removing an electron from the valence shell would decrease electron-electron repulsions and thereby allow the remaining electrons to come closer together and be pulled closer to the nucleus. For example, the radius of Fe^{2+} is 0.76 Å whereas that of Fe^{3+} is 0.64 Å. When elements belonging to the A-groups form positive ions, the entire valence shell is generally emptied. (We will say more about this in a little while.) Removing the outer shell exposes the inner core of electrons which, naturally, has a smaller volume than the neutral atom. *Positive ions are always smaller than the atoms from which they are formed.* Figure 7.20 illustrates more accurately the trends in atomic size and shows how the sizes of ions are related to those of their neutral atoms.

Figure 7.20
Variations in atomic and ionic radii in the periodic table.

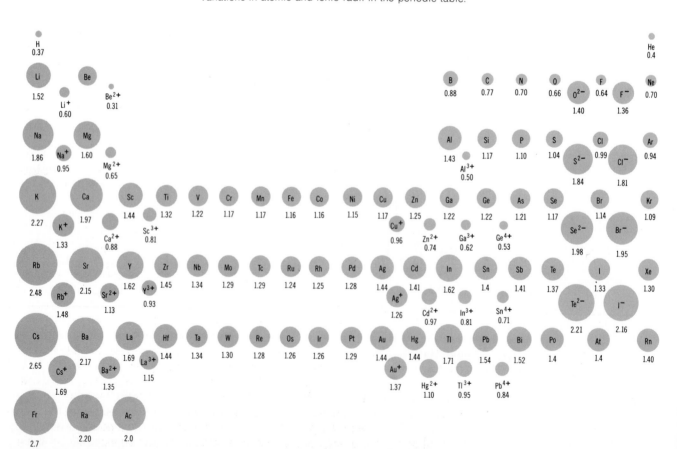

EXERCISE 7.9 Use the periodic table to choose the largest atom or ion in each set.
(a) Ge, Te, Se, Sn
(b) C, F, Br, Ga
(c) Fe, Fe^{2+}, Fe^{3+}
(d) O, O^{2-}, S, S^{2-}

7.13 FORMATION OF IONIC BONDS

When elements having low ionization energies react with elements having high electron affinities, ionic compounds are formed

When atoms combine to form compounds, they are held together in fixed proportions by forces of attraction that we call **chemical bonds.** Changes in these bonding forces are the underlying basis for all chemical reactions. Old bonds break and new ones form when chemicals react. For instance, when we digest meat, the bonds that hold the amino acids together in proteins are broken and the amino acid molecules are freed so that processes in our cells can recombine them to produce the specific proteins that we need for survival. Both the complex structures and the functions of starch, proteins, and DNA are determined not just by the atoms present, but also by the chemical bonds that hold the atoms together in their molecules.

There are two major classes of bonding forces—covalent bonding and ionic bonding. Ionic bonding is simpler to understand, so we will discuss it first. In Chapter Six, we saw that ionic compounds are formed when metals react with nonmetals. A common example is sodium chloride—table salt. We saw that each sodium atom loses one electron to form a sodium ion (Na^+) and each chlorine atom gains one electron to form a chloride ion (Cl^-). Once they are formed, the ions of sodium and chlorine cling together because their opposite charges attract one another. This type of attraction is called the **ionic bond.**

The reason for the attraction between Na^+ and Cl^- ions—the fact that opposite charges attract—is not difficult to understand. But why are electrons transferred between these and other atoms? Why does sodium form Na^+ and not Na^- or Na^{2+}; why does chlorine form Cl^- instead of Cl^+ or Cl^{2-}? A number of factors, taken together, determine whether an ionic compound will form and what its formula will be. These factors affect the numbers of electrons lost or gained, and they are all related to the total energy of the system of reactants and products.

For any stable compound to form from the elements there must be a net lowering of the energy. In other words, energy must be released. There are three major contributing factors that affect the energy in the formation of an ionic compound. One is the removal of electrons from one of the atoms (e.g., sodium). This requires an input of energy—the ionization energy. Another factor is the energy change that takes place when one or more electrons are added to atoms of the other element (e.g., chlorine). This is the electron affinity and, as we saw earlier, it is usually exothermic—it usually leads to a lowering of the energy.

If we are keeping track of the energy budget, both the IE and EA refer to isolated gaseous atoms becoming isolated gaseous ions. A crystal of salt, however, doesn't consist of isolated ions (see Figure 7.21). We've seen that the ions in salt are packed tightly together in a regular pattern, or *lattice.* This lattice has a lower energy than isolated ions because, as the ions come together, energy is released. (We know this is true because we would have to expend energy to separate them.) The imaginary process that forms the lattice from isolated ions leads to a lowering of the energy of the ions by an amount known as the **lattice energy.**

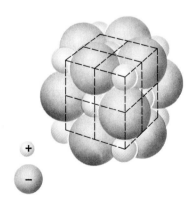

Figure 7.21
Packing of ions in NaCl.

The lattice energy has a major stabilizing influence on ionic compounds. Just the IE and EA together give a net increase in energy. However, the lattice energy produces a large energy-lowering effect that allows the formation of the ionic compound to yield a *net* lowering of the energy.

We can now explore what happens to atoms when they react. Right from the beginning, we can see why metals tend to form positive ions and nonmetals tend to form negative ions. At the left of the periodic table are the metals—elements with small IE and small EA. Relatively little energy input is needed to remove electrons from these elements to produce positive ions. At the upper right of the table are the nonmetals—elements with large IE and EA. It is very difficult to remove electrons from these elements, but relatively large amounts of energy are released if they gain electrons. On an energy basis, ionic bonding is favored over other kinds whenever atoms of small IE combine with atoms of large EA.

Now let's look at how the electronic structure of the elements affects the kinds of ions that they form. We will begin by examining what happens when sodium loses an electron. The electronic structure of Na is

$$\text{Na} \qquad 1s^2 2s^2 2p^6 3s^1$$

The electron that is lost is the one least tightly held. For sodium that is the single outer $3s$ electron. The electronic structure of the Na^+ ion, then, is

$$\text{Na}^+ \qquad 1s^2 2s^2 2p^6$$

From its position in the periodic table we know that Na has a low ionization energy. Therefore, the removal of the outer electron from sodium requires only a modest amount of energy (the first IE). Removal of any additional electrons is very difficult, however, because the electrons below the outer shell are very tightly held. Therefore, electron loss stops when the valence shell has emptied.

Similar situations exist for other metals. Calcium, for example, has the electron configuration

$$\text{Ca} \qquad 1s^2 2s^2 2p^6 3s^2 3p^6 4s^2$$

When it reacts, both $4s$ electrons are lost to give

$$\text{Ca}^{2+} \qquad 1s^2 2s^2 2p^6 3s^2 3p^6$$

Further loss of electrons doesn't occur because they would have to come from the core beneath the outer shell of the calcium atom. In both of these examples, notice that the ions produced have electron configurations that are the same as that of a noble gas (neon for Na^+ and argon for Ca^{2+}).

Now let's look at nonmetals such as chlorine or oxygen. These elements have high IEs; it's very difficult to remove their outer electrons. On the other hand, sizable amounts of energy are released when an electron is added to a nonmetal atom. Consequently, if electrons are available, these elements can form negative ions. For chlorine we have

$$\text{Cl} \qquad 1s^2 2s^2 2p^6 3s^2 3p^5$$

When an electron is gained, its configuration becomes

$$\text{Cl}^- \qquad 1s^2 2s^2 2p^6 3s^2 3p^6$$

Electron gain then ceases because to add another electron to Cl^- would involve placing the electron in a much higher energy level (the $4s$ subshell). When negative ions are formed, electron gain stops when the valence shell has acquired the electron configuration of a noble gas (argon when a chlorine atom gains an electron).

When oxygen reacts to form an ion, it becomes the oxide ion, O^{2-}.

$$\text{O } (1s^2 2s^2 2p^6) + 2e^- \longrightarrow O^{2-} (1s^2 2s^2 2p^6)$$

Again, electron gain stops when the ion has obtained a noble gas configuration.

The stability of the noble gas electron configuration, ns^2np^6, explains much about the formation of ions. *Atoms of most of the representative elements tend to gain or lose electrons until they have obtained a configuration that is the same as that of a noble gas.*

The situation among the transition elements is a bit more complicated than it is for the representative elements. In general, atoms of these elements have partially filled *d* subshells that are just slightly lower in energy than the outer *s* subshell. For example, the electron configuration of iron is

$$\text{Fe} \qquad [Ar]3d^64s^2$$

When iron reacts, it loses its 4*s* electrons fairly easily to give Fe^{2+}. Because the 3*d* subshell is close in energy to the 4*s*, it is not very difficult to remove still another electron to give Fe^{3+}.

$$Fe^{3+} \qquad [Ar]3d^5$$

A characteristic of the transition metals is that they can often form more than one ion. Frequently, they form an ion with a 2+ charge, which arises from the loss of the two outer *s* electrons. Ions with higher positive charges result when additional *d* electrons are lost. Unfortunately, it is not easy to predict exactly which ions can form for a given transition metal, nor is it simple to predict their relative stabilities. As your study of chemistry progresses, you will gradually learn which ions are formed by the common metals. In fact, you began this in Chapter Six.

EXAMPLE 7.6   **Electron Configurations of Ions**

Problem: How does the electron configuration of the valence shell of nitrogen change when it forms the N^{3-} ion?

Solution: The electron configuration for nitrogen is

$$\text{N} \qquad [He]2s^22p^3$$

To form N^{3-}, three electrons must be gained. These would enter the 2*p* subshell because it is the lowest available energy level. The configuration for the ion would be

$$N^{3-} \qquad [He]2s^22p^6$$

EXERCISE 7.10   How does the valence shell electron configuration of sulfur change when it forms the S^{2-} ion? How does the electron configuration of magnesium change when it forms Mg^{2+}?

7.14 ELECTRON BOOKKEEPING: LEWIS SYMBOLS

Dot symbols are used to show the valence electrons of an atom or ion

In the last section, we saw how the valence shells of atoms change when electrons are transferred during the formation of ions. In the next chapter we will see that some atoms share their outer electrons with each other when they form covalent bonds. In these discussions of bonding it is useful to be able to keep track of valence electrons. To help us do this, we use a simple bookkeeping device called Lewis symbols, named after the famous American chemist, G. N. Lewis (1875–1946).

To draw the **Lewis symbol** for an element, we write its chemical symbol surrounded by a number of dots (or other similar symbol), that represent the atom's valence electrons. For example, the element lithium, which has one valence electron in its 2s subshell, has the Lewis symbol

Li·

In fact, each element in Group IA would have a similar Lewis symbol because each has only one valence electron. The Lewis symbols for all of the Group IA elements are

Li·   Na·   K·   Rb·   Cs·

The Lewis symbols for the eight A-group elements of period 2 are[1]

Group	IA	IIA	IIIA	IVA	VA	VIA	VIIA	0
Symbol	Li·	·Be·	·Ḃ·	·Ċ·	·N̈·	·Ö:	·F̈:	:N̈e:

The elements in each group below those given will have Lewis symbols identical to the respective period 2 element except, of course, for the chemical symbol of the element. Notice that when an atom has more than four valence electrons, the additional electrons are shown to be paired with others. Also notice that *for the representative elements, the group number is equal to the number of valence electrons.*

EXAMPLE 7.7   **Writing Lewis Symbols**

Problem: What is the Lewis symbol for arsenic, As?

Solution: Arsenic is in Group VA and therefore has five valence electrons. The first four are placed about the symbol for the arsenic atom as

·A̤s·

The fifth electron is paired with one of the first four to give

·A̤s:

Note. Equally valid would be

·Äs·   or   :A̤s·   or   ·Äs·

EXERCISE 7.11   Write Lewis symbols for (a) Se, (b) I, and (c) Ca.

7.15   THE OCTET RULE

Representative elements tend to gain or lose electrons until they have eight in their outer shell

When atoms of elements in Groups IA and IIA lose electrons, the ions that form always have electron configurations that are identical to that of one or another of

---

[1] For beryllium, boron, and carbon, the number of unpaired electrons in the Lewis symbol doesn't agree with the number predicted from the atom's electron configuration. Boron, for example, has two electrons paired in its 2s orbital and a third electron in one of its 2p orbitals; therefore, there is actually only one unpaired electron in a boron atom. The Lewis symbols are drawn as shown, however, because when beryllium, boron, and carbon form bonds, they *behave* as if they have two, three, and four unpaired electrons, respectively.

the noble gases. For example, Na^+ has the electron configuration of Ne; Ba^{2+} has the same electron configuration as Xe.

Similarly, when nonmetals in Groups VA, VIA and VIIA gain electrons to form ions, they achieve the configurations of noble gases too. The chloride ion has the configuration of Ar; O^{2-} has the configuration of Ne, as does N^{3-}.

The electron configuration corresponding to the outer shell of a noble gas, ns^2np^6, is very stable. Its IE is high for two reasons: (1) the effective charge felt by the outer electrons in high and (2) the subshells are completely filled. When atoms react in such a way as to achieve this stable configuration, they have little remaining chemical reactivity. The tendency for atoms to react to achieve this outer shell of eight electrons is the basis for the **octet rule:** *an atom tends to gain or lose electrons until its outer shell consists of eight electrons.*

The octet rule helps us with our electron bookkeeping. If we use Lewis symbols to diagram the reaction between Na and Cl atoms, we have

$$\text{Na} \cdot \;\; + \;\; \cdot \ddot{\underset{\cdot\cdot}{\text{Cl}}} : \;\; \longrightarrow \;\; \text{Na}^+ \left[: \ddot{\underset{\cdot\cdot}{\text{Cl}}} : \right]^-$$

The valence shell of the Na is emptied, which leaves a new outer shell with eight electrons. The outer shell of Cl, which formerly had seven electrons, gains one electron so that it too has eight in its outer shell. The brackets are drawn about the chloride ion to show that the electrons (those represented by the dots) are the exclusive property of the chlorine.

We can diagram a similar reaction between calcium and chlorine.

$$: \ddot{\underset{\cdot\cdot}{\text{Cl}}} \cdot \;\;\longleftarrow \text{Ca} \longrightarrow \; \cdot \ddot{\underset{\cdot\cdot}{\text{Cl}}} : \;\; \longrightarrow \;\; \text{Ca}^{2+} + 2\left[: \ddot{\underset{\cdot\cdot}{\text{Cl}}} : \right]^-$$

EXAMPLE 7.8 **Using Lewis Symbols**

Problem: Using Lewis symbols, diagram the reaction that occurs between sodium and oxygen atoms to give Na^+ and O^{2-} ions.

Solution: First we draw the Lewis symbols for Na and O.

$$\text{Na} \cdot \qquad \cdot \ddot{\underset{\cdot\cdot}{\text{O}}} :$$

It takes two electrons to complete the octet around oxygen. Each Na can supply only one, so we need two Na atoms. Therefore,

$$\text{Na} \cdot \;\; + \;\; \cdot \ddot{\underset{\cdot\cdot}{\text{O}}} : \;\; + \;\; \cdot \text{Na} \;\; \longrightarrow \;\; 2\text{Na}^+ + \left[: \ddot{\underset{\cdot\cdot}{\text{O}}} : \right]^{2-}$$

Don't forget to put the brackets around the oxide ion.

EXERCISE 7.12 Diagram the reaction between magnesium and oxygen atoms to give Mg^{2+} and O^{2-} ions.

The octet rule is useful on many occasions, particularly with the representative elements. However, it fails for most of the transition elements and also for the ions of metals that follow a row of transition elements. Tin, for example, has an outer shell configuration $5s^25p^2$ and, when it forms the Sn^{2+} ion, it loses the two electrons from its $5p$ subshell. The Sn^{2+} ion, therefore, has the valence shell configuration, $5s^2$.

Despite these shortcomings, we will find the octet rule very useful for keeping track of electrons in molecules containing covalent bonds, which we will discuss in the next chapter.

SUMMARY

Atomic Spectra. The occurrence of line spectra tells us that atoms can emit energy only in discrete amounts and suggests that the energy of the electron is quantized—that is, the electron is restricted to definite energy levels in an atom. Neils Bohr recognized this and, although his theory was later shown to be incorrect, he was the first to introduce the idea of quantum numbers.

Matter Waves. The wave behavior of electrons and other tiny particles, which can be demonstrated by diffraction experiments, was introduced by de Broglie. Schrödinger applied this to the atom and launched the theory we call wave mechanics or quantum mechanics. This theory tells us that electron waves in atoms are standing waves whose crests and nodes are stationary. Each standing wave, or orbital, is characterized by three quantum numbers, n, ℓ, and m. Shells are designated by n, subshells by ℓ, and orbitals within subshells by m.

Electron Configurations. The Pauli exclusion principle, based on the concept of electron spin, limits orbitals to a maximum population of two electrons with paired spins. The electron configuration of an element in its ground state is obtained by filling orbitals beginning with the $1s$ subshell and following the Pauli exclusion principle and Hund's rule (electrons spread out as much as possible in orbitals of equal energy). Sometimes we represent electron configurations using orbital diagrams. Unexpected configurations occur for chromium and copper because of the extra stability of half-filled and filled subshells.

Periodic Table. We can divide the periodic table into sets of columns that can help us to write electron configurations. The valence shell configuration can be obtained from an element's period number and the sets of columns that must be crossed to reach the element in the table.

Orbital Shapes. All s orbitals are spherical; each p orbital consists of two lobes with a nodal plane between them. A p subshell has three p orbitals whose axes are mutually perpendicular and point along the x, y, and z axes of an imaginary coordinate system centered at the nucleus. In each orbital the electron is conveniently viewed as a sort of cloud with a varying electron density.

Atomic Properties. Ionization energy (IE) is the energy needed to remove an electron from a gaseous atom. Electron affinity (EA) is energy released when an electron is added to an atom. IE and EA increase from left to right and from bottom to top in the periodic table. Atomic radii decrease from left to right and bottom to top in the table. Negative ions are larger than the atoms from which they are formed; positive ions are smaller than the atoms from which they are formed.

Ionic Bonds. Lewis symbols, an electron bookkeeping device, are drawn by placing dots that represent valence electrons around the chemical symbol for an element. The formation of ionic bonds is favored when atoms with low ionization energies combine with elements with high EAs—that is, when metals combine with nonmetals. The chief stabilizing influence in the formation of ionic compounds is the lattice energy. When the representative elements form ions, they tend to achieve a noble gas configuration having eight electrons in the outer shell; this tendency is the basis for the octet rule. Lewis symbols can be used to analyze the electron transfer that occurs.

INDEX TO EXERCISES AND QUESTIONS

REVIEW QUESTIONS

7.13. Why are the nuclei of atoms relatively unimportant when the atoms react with each other?

7.14. What is an atomic spectrum? How does it differ from a continuous spectrum?

7.15. What do we mean by the term *electronic structure?*

7.16. What fundamental fact is implied by the existence of atomic spectra?

7.17. Describe Niels Bohr's model of the structure of the hydrogen atom.

7.18. In qualitative terms, how did Bohr's model account for the atomic spectrum of hydrogen?

7.19. What is the term used to describe the lowest energy state of an atom?

7.20. In what way was Bohr's theory both a success and a failure?

7.21. How does the behavior of very small particles differ from that of the larger, more massive objects that we meet in everyday life? Why don't we notice this same behavior for the larger, more massive objects?

7.22. Describe the phenomenon called diffraction. How can this be used to demonstrate that de Broglie's theory was correct?

7.23. What is the difference between a *traveling wave* and a *standing wave?*

7.24. What are the names used to refer to the theories that apply the matter-wave concept to electrons in atoms?

7.25. What is the term used to describe a particular waveform of a standing wave for an electron?

7.26. What are the two properties of orbitals in which we are most interested? Why?

7.27. What are the allowed values of the principle quantum number?

7.28. What is the value of n for (a) the K shell and (b) the M shell?

7.29. What is the letter code for a subshell with (a) $\ell = 1$, (b) $\ell = 3$, and (c) $\ell = 5$?

7.30. Give the values of n and ℓ for the following subshells. (a) 2s   (b) 3d   (c) 5f

7.31. For the shell with $n = 4$, what are the possible values of ℓ?

7.32. Why does every shell contain an s subshell?

7.33. What are the possible values of m for a subshell with (a) $\ell = 1$ and (b) $\ell = 3$?

7.34. How many orbitals are found in (a) an s subshell, (b) a p subshell, (c) a d subshell, and (d) an f subshell?

7.35. How many orbitals are there in an h subshell ($\ell = 5$)?

7.36. Within any given shell, how do the energies of the s, p, d, and f subshells compare?

7.37. How do the energies of the orbitals belonging to a given subshell compare?

7.38. What physical property of electrons leads us to propose that they spin like a toy top?

7.39. What is the name of the property exhibited by atoms that contain unpaired electrons?

7.40. What are the possible values of the spin quantum number?

7.41. What is the Pauli exclusion principle? What effect does it have on the populating of orbitals by electrons?

7.42. Give the complete set of quantum numbers for all of the electrons that could populate the 2p subshell of an atom.

7.43. Give the electron configurations of the elements in period 2 of the periodic table.

7.44. Predict the electron configuration of (a) S, (b) K, (c) Ti, and (d) Sn.

7.45. Predict the electron configuration of (a) As, (b) Cl, (c) Fe, and (d) Si.

7.46. Give the correct electron configuration of (a) Cr and (b) Cu.

7.47. Give orbital diagrams for (a) Mg and (b) Ti.

7.48. Give orbital diagrams for (a) As and (b) Ni.

7.49. How many unpaired electrons would be found in the ground state of (a) Mg, (b) P, and (c) V?

7.50. Write shorthand configurations for (a) Ni, (b) Cs, (c) Ge, and (d) Br.

7.51. Write shorthand configurations for (a) Al, (b) Se, (c) Ba, and (d) Sb.

7.52. How are the electron configurations of the elements in a given group similar?

7.53. Define the terms *valence shell* and *valence electrons.*

7.54. Give the configuration of the valence shell for (a) Na, (b) Al, (c) Ge, and (d) P.

7.55. Give the configuration of the valence shell for (a) Mg, (b) Br, (c) Ga, and (d) Pb.

7.56. In what general terms do we describe an electron's location in an atom?

7.57. Sketch the approximate shapes of s and p orbitals.

7.58. How does the size of a given type of orbital vary with n?

7.59. How are the p orbitals of a given p subshell oriented relative to each other?

7.60. What is a *nodal plane?*

7.61. Define *ionization energy.*

7.62. Choose the atom with the higher ionization energy in each pair.
(a) B or C   (b) O or S   (c) Cl or As

7.63. Why is the noble gas configuration so stable?

7.64. Explain *why* ionization energy increases from left to right in a period and decreases from top to bottom in a group.

7.65. Define *electron affinity.*

7.66. Choose the atom with the more exothermic EA in each pair.
(a) Cl or Br   (b) Se or Br   (c) Si or Ga

7.67. Why is the second electron affinity of an atom (M^- + $e^- \rightarrow M^{2-}$) always endothermic?

7.68. Choose the larger atom in each pair.
(a) Na or Si   (b) P or Sb

7.69. Why are the size changes among the transition elements more gradual than among the representative elements?

7.70. Choose the larger particle in each pair.
(a) Na or Na^+ (b) Co^{3+} or Co^{2+} (c) Cl or Cl^-

7.71. Write Lewis symbols for these atoms.
(a) Si   (b) Sb   (c) Ba   (d) Al

7.72. How is the tendency to form ionic bonds related to the IE and EA of the atoms involved?

7.73. What is the *lattice energy*? In what way does it contribute to the stability of ionic compounds?

7.74. Explain what happens to the electron configurations of Mg and Br when they react to form ionic $MgBr_2$.

7.75. Give the electron configurations of the ions expected to be formed by (a) Sr, (b) F, and (c) Al.

7.76. Why do many of the transition elements in period 4 form ions with a 2+ charge?

7.77. What is the *octet rule*? What is responsible for it?

7.78. Use Lewis symbols to diagram the reaction between (a) Ca and Br, (b) Al and O, and (c) K and S.

CHAPTER EIGHT

THE FORMATION OF MOLECULAR COMPOUNDS: COVALENT BONDING

8.1 ELECTRON SHARING

Sharing of electron pairs provides an alternative to ion formation when atoms form bonds

An ionic substance

A molecular substance

In Chapter Seven we learned that there are basically two types of chemical bonds—ionic bonds and covalent bonds. Ionic bonds were discussed first because the attractions that hold the ions together in compounds such as sodium chloride are easy to understand. Most of the substances that we encounter in our daily lives, however, are not ionic. Rather than existing as collections of electrically charged particles (ions), they occur as electrically neutral combinations of atoms called molecules. Water, for instance, consists of molecules composed of two hydrogen atoms and one oxygen atom. The formula of one particle of water is H_2O. Most substances consist of much larger molecules. For example, the formula for table sugar is $C_{12}H_{22}O_{11}$, and gasoline is composed of hydrocarbon molecules such as octane, C_8H_{18}. In this chapter we will discuss the forces—covalent bonds—that bind atoms together in these substances.

In the previous chapter we saw that for ionic bonding to occur the energy-lowering effect of the lattice energy must be greater than the combined energy-raising effect of the ionization energy (IE) and the electron affinity (EA). Many times this is not possible, particularly when the IE of all of the atoms involved is large, as happens when nonmetals combine with other nonmetals. In such instances, nature seeks a different way to lower the energy—electron sharing.

As a rough rule, ionic bonding occurs between metals and nonmetals, and covalent bonding occurs when nonmetals combine with nonmetals.

Let's look at what happens when two hydrogen atoms join together to form the H_2 molecule (Figure 8.1). As the two atoms approach each other, the electron of each atom begins to feel the attraction of both nuclei. This causes the electron density around each nucleus to shift toward the region between the two atoms. Therefore, as the distance between the nuclei decreases, there is an increase in the probability of finding either electron near either nucleus. In effect, then, each of the hydrogen atoms in this H_2 molecule now has a share of two electrons.

This sharing of electrons is called the **covalent bond.** The nuclei of the two hydrogen atoms are pulled toward each other by the electron cloud between them. Being of the same charge, however, the two nuclei also repel each other, as do the two electrons. In the bond that forms, therefore, the atoms are separated by that distance at which these attractions and repulsions are balanced. In the covalent bond of molecular hydrogen, for instance, all these forces are balanced

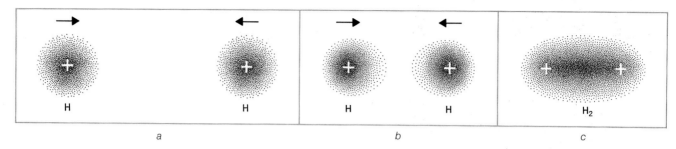

a   b   c

Figure 8.1
Formation of a bond between two hydrogen atoms. (*a*) Two
H atoms separated by a large distance. (*b*) As the atoms
approach each other, their electron densities begins to shift
to the region between the two nuclei. (*c*) The electron density
becomes concentrated between the nuclei.

when the nuclei of the two atoms are 0.75 Å apart. This separation is called the
bond length or **bond distance.**

Because the atoms in a covalent bond are held together, work must be done
(energy must be supplied) to separate them. When the bond is *formed,* an equivalent amount of energy is released, which means, of course, that the energy of
the atoms decreases. Figure 8.2 shows how the energy changes when two hydrogen atoms form H_2. We see that the minimum energy occurs at a bond distance of 0.75 Å and that a mole of hydrogen molecules is more stable than two
moles of hydrogen atoms by 435 kJ.

The formation of the covalent bond has lowered the energy.

Figure 8.2
Energy of two hydrogen atoms as
they approach each other.

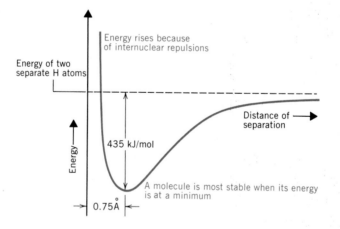

Before coming together, the separate hydrogen atoms have one electron in
their 1*s* orbitals. When the atoms join and these electrons are shared, the 1*s*
orbital of each atom has, in a sense, become filled. The electrons have also
become paired, as required by the Pauli exclusion principle—that is, the spin
quantum number, m_s, is $+\frac{1}{2}$ for one of the electrons and $-\frac{1}{2}$ for the other. In general, we almost always find that the electrons involved become paired when
atoms form covalent bonds. In fact, a covalent bond is sometimes referred to as
an **electron pair bond.**

To keep track of the electrons in covalent bonds, we often use Lewis symbols
in much the same way as we use them for ionic bonds. For the covalent bond, the
electrons that are shared are shown as a pair of dots between the symbols of the
bonded atoms. The formation of H_2 from hydrogen atoms, for example, can be
depicted as

Formulas constructed from the Lewis symbols for atoms are called Lewis formulas or Lewis structures.

$$H\cdot \ + \ H\cdot \ \longrightarrow \ H\!:\!H$$

Another device that is also very common is the use of a dash to represent the electron pair. For example, the hydrogen molecule could also be represented as

$$H\!-\!H$$

8.2 COVALENT BONDS AND THE OCTET RULE

The octet rule can often predict the number of covalent bonds that an atom will form

In Chapter Seven we saw that an electron configuration corresponding to that of a noble gas is very stable. When ions are formed, electrons tend to be gained or lost until a noble gas configuration is achieved. The stability of the noble gas configuration also influences the number of electrons an atom tends to acquire by sharing—it often controls the number of covalent bonds that an atom forms.

Hydrogen, with just one electron in its $1s$ orbital, can achieve a noble gas configuration (that of helium) by obtaining a share of one electron from another atom. The Lewis structure for H_2 implies that both atoms have access to both electrons in the bond.

(Two electrons can be counted around each of the H atoms.)

Since hydrogen obtains a stable valence shell configuration when it shares just one pair of electrons with another atom, hydrogen atoms only form one covalent bond.

The valence shells of the noble gases other than helium each contain eight electrons. The tendency to acquire such a configuration was the basis of the octet rule that we found useful in some of our discussions of ionic bonding. When atoms other than hydrogen form covalent bonds, they too tend to achieve an octet of electrons in their valence shells. They do this by acquiring electrons through sharing. The number of covalent bonds that such an atom forms is equal to the number of electrons needed to give eight (an octet) in the outer shell. For example, the halogens, Group VIIA of the periodic table, all have seven electrons in their valence shell. The Lewis symbol for a typical element of this group, chlorine, is

$$\cdot \ddot{\underset{\cdot\cdot}{C}}l\!:$$

We can see that only one electron is needed to complete an octet. Of course, chlorine can actually gain this electron and become a chloride ion. This is what it does when it forms ionic compounds such as sodium chloride (NaCl). But when chlorine combines with another nonmetal, the transfer of electrons is not energetically favorable. Therefore, in forming such compounds as HCl or Cl_2, chlorine gets the one electron it needs by forming a covalent bond.

$$H\cdot \ \overset{\frown}{+} \ \cdot \ddot{\underset{\cdot\cdot}{C}}l\!: \ \longrightarrow \ H\!:\!\ddot{\underset{\cdot\cdot}{C}}l\!:$$

$$:\!\ddot{\underset{\cdot\cdot}{C}}l\cdot \ \overset{\frown}{+} \ \cdot \ddot{\underset{\cdot\cdot}{C}}l\!: \ \longrightarrow \ :\!\ddot{\underset{\cdot\cdot}{C}}l\!:\!\ddot{\underset{\cdot\cdot}{C}}l\!:$$

The HCl and Cl_2 molecules could also be represented using the "dash" bond as

$$H\!-\!\ddot{\underset{\cdot\cdot}{C}}l\!: \quad \text{and} \quad :\!\ddot{\underset{\cdot\cdot}{C}}l\!-\!\ddot{\underset{\cdot\cdot}{C}}l\!:$$

There are many nonmetals that form more than one covalent bond. For example, the three most important elements in biochemical systems are carbon, nitrogen, and oxygen.

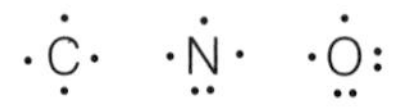

$$\cdot\overset{\cdot}{\underset{\cdot}{C}}\cdot \quad \cdot\overset{\cdot}{\underset{\cdot}{N}}\cdot \quad \cdot\overset{\cdot}{\underset{\cdot\cdot}{O}}:$$

These elements combine with hydrogen to form methane, ammonia, and water, respectively. Since hydrogen forms only one covalent bond, the formulas for the molecules of these compounds are easy to predict using the octet rule. Carbon has four electrons, so it needs to share four electrons to achieve an octet. This means that it forms covalent bonds with four hydrogen atoms. Using the same reasoning, we can see that nitrogen forms covalent bonds with three hydrogen atoms, and oxygen, with two.

These are examples of **structural formulas**—formulas that show how the atoms in a molecule are attached to each other. We will use structural formulas many times throughout this book.

In most of the compounds in which they occur, carbon forms four covalent bonds, nitrogen forms three, and oxygen forms two.

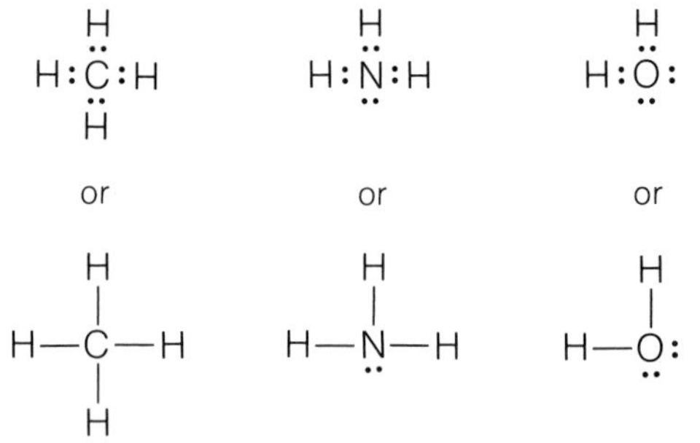

EXAMPLE 8.1 **Using the Octet Rule to Predict Formulas**

Problem: Use the octet rule to predict the formula of the compound formed from hydrogen and sulfur.

Solution: First we write Lewis symbols for each atom.

$$H\cdot \qquad \cdot\overset{\cdot}{\underset{\cdot\cdot}{S}}:$$

(Sulfur has six electrons because it is in Group VIA.)
Sulfur needs two electrons to complete its octet, so it bonds to two hydrogen atoms.

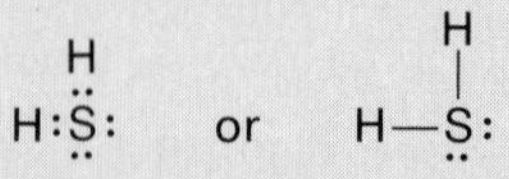

$$H:\overset{H}{\underset{\cdot\cdot}{S}}: \quad or \quad H-\overset{H}{\underset{\cdot\cdot}{S}}:$$

Thus, the molecule has the formula H_2S. This, by the way, is the substance that gives rotten eggs their foul odor.

EXERCISE 8.1 Use the octet rule to predict the formulas of compounds formed from (a) P and H and (b) S and F.

8.3 MULTIPLE BONDS

As many as two or three pairs of electrons may be shared between two atoms

The bond produced by the sharing of one pair of electrons between two atoms is called a **single bond.** So far, we have discussed only molecules that contain single bonds.

$$:\ddot{C}l—\ddot{C}l: \qquad H—H \qquad H—\underset{\displaystyle H}{\overset{\displaystyle H}{\underset{|}{\overset{|}{C}}}}—H$$

There are, however, many molecules in which more than one pair of electrons are shared between two atoms. For example, we know that nitrogen, the most abundant gas in the atmosphere, occurs in the form of diatomic molecules; that is, each of its molecules is composed of two nitrogen atoms, N_2. As we've just seen in the previous section, the Lewis symbol for nitrogen is

$$\cdot\ddot{N}\cdot$$

and each nitrogen atom needs three electrons to complete its octet. When the N_2 molecule is formed, each of the nitrogen atoms shares three electrons with the other.

$$:\!N\!\cdot \longleftrightarrow \cdot\!N\!:$$

The result is called a **triple bond.** Notice that in the Lewis formula for the molecule, the three shared pairs of electrons are placed between the two atoms. We count all of these electrons as though they belong to both of the atoms. Each nitrogen therefore has an octet.

8 electrons ⌐ ⌐ 8 electrons

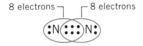

The triple bond also may be represented by three dashes. We would usually indicate the bonding in nitrogen as

$$:N\!\equiv\!N:$$

Double bonds also occur in molecules. A prime example is the CO_2 molecule, which contains a carbon atom that is bonded covalently to two separate oxygen atoms. We can diagram the formation of the bonds in CO_2 as follows:

$$:\ddot{O}\cdot \longleftrightarrow \cdot\ddot{C}\cdot\longleftrightarrow\cdot\ddot{O}: \longrightarrow :\ddot{O}::C::\ddot{O}:$$

The central carbon atom shares two of its electrons with each of the oxygen atoms, and each oxygen shares two electrons with carbon. The result is the formation of two **double bonds.** Once again, if we circle the valence shell electrons that "belong" to each atom, we can see that each has an octet.

⌐ 8 electrons ⌐

The structural formula for CO_2, using dashes, is

$$\overset{..}{\underset{..}{O}}=C=\overset{..}{\underset{..}{O}}$$

8.4 FAILURE OF THE OCTET RULE

The atoms in some molecules cannot obey the octet rule because there are either too few or too many electrons

Sometimes it is just impossible for all of the atoms in a molecule to obey the octet rule. This happens most often when an atom forms more than four bonds. Examples are PCl_5 and SF_6 in which there are five P—Cl bonds and six S—F bonds, respectively. Since each covalent bond requires the sharing of a pair of electrons, phosphorus and sulfur must exceed eight electrons in their outer shells. The Lewis formulas of these two molecules are shown below. In the next section we will discuss methods that you can learn to help you write structures such as these.

SF_6 is used as a gaseous insulator in high voltage electrical equipment.

8 electrons about each Cl
10 electrons about P

8 electrons about each F
12 electrons about S

If an atom forms more than four bonds, it must end up with more than four electron pairs in its valence shell.

Elements in period 2, such as carbon or nitrogen, never exceed an octet simply because their valence shells, having $n = 2$, can hold a maximum of only 8 electrons. Elements in periods below period 2, however, sometimes do exceed an octet because their valence shells can hold more than 8 electrons. For example, the valence shell for elements in period 3, for which $n = 3$, can hold a maximum of 18 electrons, and the valence shell for period 4 elements can hold as many as 32 electrons.

In some molecules (but not many), an atom has less than an octet. The most common examples are compounds of beryllium and boron.

$$\cdot Be\cdot + 2 \cdot \overset{..}{Cl}: \longrightarrow :\overset{..}{\underset{..}{Cl}}:Be:\overset{..}{\underset{..}{Cl}}:$$

(4e^- around Be)

$$\cdot B\cdot + 3 \cdot \overset{..}{Cl}: \longrightarrow :\overset{..}{\underset{..}{Cl}}:B:\overset{..}{\underset{..}{Cl}}:$$

(6e^- around B)

The method that we'll discuss in the next section doesn't work for these; they must simply be learned as exceptions.

8.5 DRAWING LEWIS STRUCTURES

In drawing Lewis structures we try, whenever possible, to have all atoms obey the octet rule

Being able to draw Lewis structures for molecules or polyatomic ions (polyatomic ions are held together by covalent bonds, too) is important because they often form the basis for much chemical reasoning. Also, we will soon see that quite accurate predictions can be made concerning the shapes of molecules if we know their Lewis structures.

The method of writing Lewis structures can be broken down into a number of steps. The first is to decide which atoms are bonded to each other so that we know where to put the dots. This is not always a simple matter. Many times the formula suggests the way the atoms are arranged because the central atom is usually written first. Examples are CO_2 and ClO_4^-, which have the *skeletal structures* (i.e., arrangements of atoms),

$$O \quad C \quad O \qquad \text{and} \qquad \begin{matrix} & O & \\ O & Cl & O \\ & O & \end{matrix}$$

Sometimes, obtaining the skeletal structure is not quite so simple. How, for example, would we predict that the skeletal structure of nitric acid, HNO_3, is

$$H \quad O \quad N \begin{matrix} O \\ \\ O \end{matrix} \quad \text{(correct)}$$

rather than some other structure such as one of the following?

$$\begin{matrix} & H & \\ O & N & O \\ & O & \end{matrix} \quad \text{or} \quad H \quad O \quad O \quad N \quad O \quad \text{(incorrect)}$$

The answer is not obvious. Nitric acid belongs to that group of substances called oxoacids (Section 6.8). It happens that the hydrogen atoms that are released from these molecules are always bonded to oxygen, which is in turn bonded to the other nonmetal atom. Therefore, recognizing HNO_3 as the formula of an oxoacid allows us to predict that the three oxygen atoms will be bonded to the nitrogen and the hydrogen will be bonded to one of the oxygens.

EXAMPLE 8.2   **Writing Skeletal Structures**

Problem: What is the probable skeletal structure of sulfuric acid, H_2SO_4?
Solution: From our discussion of HNO_3, we should expect that the four oxygens are bonded to the sulfur and that the two hydrogens are bonded to two of these oxygens. This would give

$$H \quad O \quad S \begin{matrix} O \\ \\ O \end{matrix} O \quad H$$

It is not important to which oxygens we attach the hydrogens.

EXERCISE 8.2   Predict reasonable skeletal structures for SO_2, NO_3^-, $HClO_3$, and H_3PO_4.

There are times when no reasonable basis can be found for choosing a particular skeletal structure. If you must make a guess, choose the most symmetrical arrangement of atoms.

Once you've decided on the skeletal structure, the next step in writing the Lewis structure is to count up all of the valence electrons to find out how many dots must appear in the final formula. Using the periodic table, locate the groups in which the elements in the formula occur to determine the number of valence electrons contributed by each atom. If the structure you wish to draw is that of an ion, add one additional valence electron for each negative charge or remove a valence electron for each positive charge.

EXAMPLE 8.3 **Counting Valence Electrons**

Problem: How many dots, representing electrons, must appear in the Lewis structures of SO_3, NO_3^-, and NH_4^+?

Solution:

SO_3	sulfur (Group VIA) contributes six electrons	$6 \times 1 =$	6
	oxygen (Group VIA) contributes six electrons each	$6 \times 3 =$	18
		Total	$24e^-$

NO_3^-	nitrogen (Group VA) contributes five electrons	$5 \times 1 =$	5
	oxygen (Group VIA) contributes six electrons each	$6 \times 3 =$	18
	add another electron for the $1-$ charge		1
		Total	$24e^-$

NH_4^+	nitrogen (Group VA)—five electrons	$5 \times 1 =$	5
	hydrogen (Group IA)—one electron each	$1 \times 4 =$	4
	subtract one electron for the $1+$ charge		-1
		Total	$8e^-$

EXERCISE 8.3 How many dots should appear in the Lewis structures of SO_2, PO_4^{3-}, and NO^+?

Once we know the total number of valence electrons, we place them in the skeletal formula, always in groups of two. Start by placing a pair in each bond. Then, if possible, complete the octets of the atoms attached to the central atom. Finally, check to see if the central atom has an octet. (Remember, however, that the maximum number of electrons about hydrogen is two.)

EXAMPLE 8.4 **Writing Lewis Structures**

Problem: Write the Lewis structure for the SO_4^{2-} ion.

Solution: We would expect the skeletal structure to be

$$\begin{array}{c} O \\ O \quad S \quad O \\ O \end{array}$$

There are $6 + 24 + 2 = 32$ electrons in the formula. First we place a pair in each bond, giving

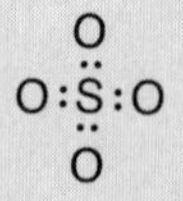

$$\begin{array}{c} O \\ O\!:\!\ddot{S}\!:\!O \\ O \end{array}$$

This leaves $32 - 8 = 24e^-$. We use these to complete the octets around the oxygens.

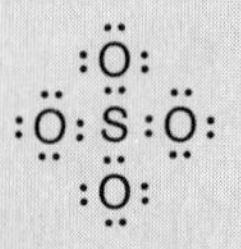

$$\begin{array}{c} :\ddot{O}: \\ :\ddot{O}\!:\!\ddot{S}\!:\!\ddot{O}: \\ :\ddot{O}: \end{array}$$

The final structure has 32 dots, and each atom obeys the octet rule. Since we are dealing with an ion, we should indicate its charge. We do this as follows:

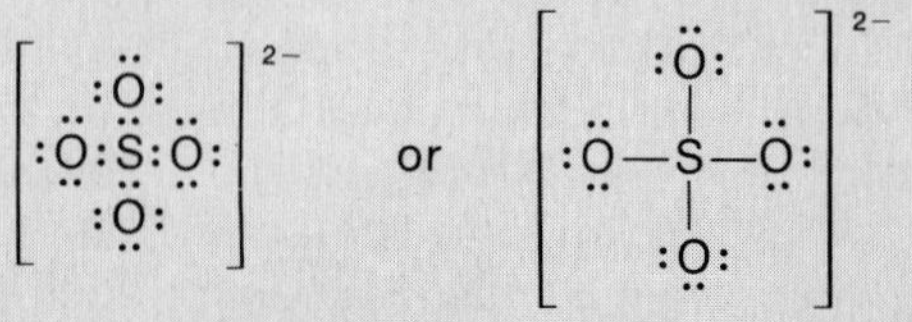

$$\left[\begin{array}{c} :\ddot{O}: \\ :\ddot{O}\!:\!\ddot{S}\!:\!\ddot{O}: \\ :\ddot{O}: \end{array} \right]^{2-} \quad \text{or} \quad \left[\begin{array}{c} :\ddot{O}: \\ | \\ :\ddot{O}\!-\!S\!-\!\ddot{O}: \\ | \\ :\ddot{O}: \end{array} \right]^{2-}$$

EXAMPLE 8.5   Writing Lewis Structures

Problem: What is the Lewis structure for the ClO_2^- ion?
Solution: The skeletal structure would be

$$O \quad Cl \quad O$$

In this ion there are $7 + 12 + 1 = 20$ electrons. First we put a pair in each bond.

$$O:Cl:O$$

Next, we complete the octets of the oxygens.

$$':\ddot{O}:Cl:\ddot{O}:$$

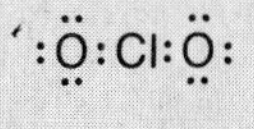

This leaves four electrons, which we place on the Cl.

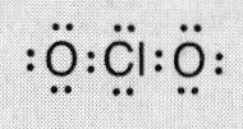

$$:\ddot{O}:\ddot{Cl}:\ddot{O}:$$

All atoms have an octet. To indicate the charge we write the structure of the ion as

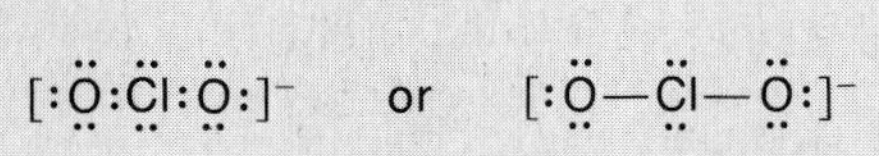

$$[:\ddot{O}:\ddot{Cl}:\ddot{O}:]^- \quad \text{or} \quad [:\ddot{O}-\ddot{Cl}-\ddot{O}:]^-$$

Sometimes, following the rules that we've been using, you will find that there are either too few electrons to complete the octets of all of the atoms, or there are electrons left over after all of the octets have been filled. When there are not enough electrons to give every atom an octet, multiple bonds must be created. When there are electrons left over, they are always placed on the central atom in the formula. The next three examples illustrate how this works.

EXAMPLE 8.6   Writing Lewis Structures

Problem: Write the Lewis structure for the air pollutant, SO_3 (sulfur trioxide).
Solution: The skeletal structure is

$$\begin{array}{cc} & O \\ O & S \\ & O \end{array}$$

Simply for convenience, we've arranged the oxygens symmetrically around the sulfur.

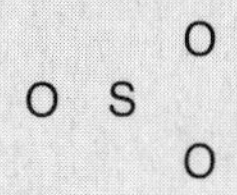

There are 24 valence electrons to be distributed in it. First, place two electrons in each bond.

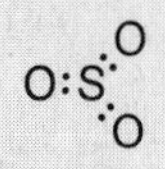

$$O:S \begin{array}{c} \cdot O \\ \cdot O \end{array}$$

There are $24 - 6 = 18e^-$ left. Next, we complete the octets of the oxygens.

$$:\ddot{O}:S \begin{array}{c} :\ddot{O}: \\ :\ddot{O}: \end{array}$$

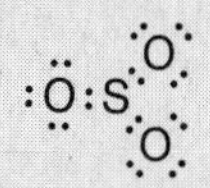

We're not finished yet, however, because there is not an octet about sulfur. We cannot simply add more dots because the total must be 24. The procedure that we follow, therefore, is to move a pair of electrons that we have shown to belong solely to an oxygen into a sulfur-oxygen bond so that they can also be counted as belonging to sulfur. In other words, we place a double bond between sulfur and one of the oxygens. It doesn't matter which oxygen we choose for this honor.

$$:\overset{..}{\underset{..}{O}}:S\overset{:\overset{..}{O}:}{\underset{:\overset{..}{O}:}{}}\quad\text{gives}\quad :\overset{..}{\underset{..}{O}}::S\overset{:\overset{..}{O}:}{\underset{:\overset{..}{O}:}{}}$$

We can also draw this structure as

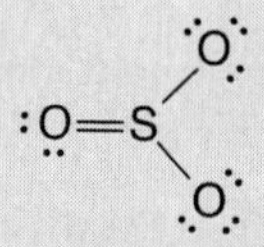

$$:\overset{..}{\underset{..}{O}}=S\overset{:\overset{..}{O}:}{\underset{:\overset{..}{O}:}{}}$$

Notice that each atom has an octet.

EXAMPLE 8.7 Writing Lewis Structures

Problem: What is the Lewis structure for the carbon monoxide molecule, CO?
Solution: The skeletal structure is simply

$$C\quad O$$

There is a total of $4 + 6 = 10$ valence electrons. We begin with a pair in the CO bond.

$$C:O$$

Now we try to complete the octets with the remaining eight electrons.

$$:\overset{..}{\underset{..}{C}}:O:$$

Obviously there aren't enough, which means that carbon monoxide must have a multiple bond of some sort. Therefore, we must shift enough electrons around so that oxygen also has an octet. Since the oxygen atom needs four more electrons, we must move two more pairs into the bond, which gives us a triple bond.

$$:\overset{..}{C}\,O:\quad\text{gives}\quad :C:::O:$$

Using dashes, the formula is

$$:C\equiv O:$$

EXAMPLE 8.8 Writing Lewis Structures

Problem: What is the Lewis structure for SF_4?
Solution: The skeletal structure is

$$\begin{array}{ccc} & F & \\ F & S & F \\ & F & \end{array}$$

and there must be $6 + 28 = 34$ electrons in the structure. First, we place two electrons in each bond, and then we complete the octets about the fluorine atoms. This uses 32 electrons.

There are still two electrons left, even though all the atoms have an octet. Remember that if there are electrons left over after completing all of the octets, *they are all placed on the central atom.* This gives us

We've redrawn the molecule to make room for the extra electrons on sulfur.

We see that, in the SF_4 molecule, sulfur violates the octet rule.

Figure 8.3 reviews the steps that we follow when we draw Lewis structures. Exercise 8.4 will give you some practice in applying them.

EXERCISE 8.4   Draw Lewis structures for OF_2, NH_4^+, SO_2, NO_3^-, ClF_3.

Figure 8.3
Summary of steps in writing Lewis structures.

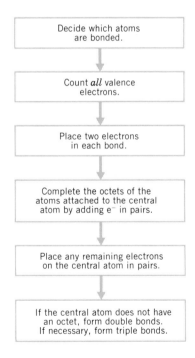

8.6   RESONANCE

The bonding in some molecules and ions cannot be adequately described by a single Lewis structure

Lewis structures are useful devices for keeping tabs on valence electrons, and most of the time the bonding pictures that they give seem reasonable with respect to properties that we can check experimentally. Two such properties are

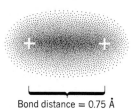

Bond distance = 0.75 Å

Figure 8.4
The hydrogen molecule.

bond length, the distance between the nuclei of the bonded atoms, and **bond energy,** the energy required to separate the bonded atoms to give neutral particles. For example, as we mentioned in Section 8.1, measurement has shown the H_2 molecule to have a bond length of 0.75 Å (Figure 8.4) and a bond energy of 435 kJ/mol, which means that it takes 435 kJ to break the bonds of one mole of H_2 molecules to give hydrogen atoms.

Bond length and bond energy depend on the amount of electron density in the bond between two atoms. The higher the electron density, the more tightly the nuclei are held and the closer they are drawn together. We therefore should expect to find a double bond to be shorter and stronger than a single bond because the electron density produced by two pairs of electrons is greater than that produced by one pair. Similarly, a triple bond should be shorter and stronger than a double bond. Such expectations are borne out by the data in Table 8.1, which gives typical bond lengths and bond energies for single, double, and triple carbon-carbon bonds.

Table 8.1
Average Bond Lengths and Bond Energies Measured for Carbon-Carbon Bonds

Bond	Bond Length (Å)	Bond Energy (kJ/mol)
C—C	1.54	368
C=C	1.34	698
C≡C	1.20	962

There are some molecules and ions for which we cannot write Lewis structures that agree with experimental measurements of bond length and bond energy. One example is the formate ion, HCO_2^-, which is produced by neutralizing formic acid, HCO_2H (the subtance that causes the stinging sensation in bites from fire ants). The skeletal structure for this ion is

Formic acid is an example of an organic acid. It has the structure

$$H—C—O—H$$

O
H   C
O

and, following the usual steps, we would write its Lewis structure as

$$\left[H—C \underset{\ddot{O}:}{\overset{\ddot{O}:}{<}} \right]^-$$

This structure suggests that one carbon-oxygen bond should be longer than the other, but experiment shows that they are identical. In fact, the C—O bond lengths are about halfway between that expected for a single bond and that expected for a double bond. The Lewis structure doesn't match reality, and there's no way to write one that does. It would require showing all of the electrons in pairs and, at the same time, showing 1.5 pairs of electrons per bond.

The way we get around problems like that of the formate ion is through the use of a concept called **resonance.** We view the actual structure of the molecule or ion that we cannot draw satisfactorily as a composite, or average, of a number of Lewis structures that we can draw. For example, for formate we write

No atoms have been moved, but the electrons have been redistributed.

$$\left[H—\overset{\ddot{O}:}{\underset{}{C}}—\ddot{O}: \right]^- \longleftrightarrow \left[H—C=\overset{:\ddot{O}:}{\underset{}{O}}: \right]^-$$

where we have simply shifted electrons around in going from one structure to another. The bond between the carbon and a particular oxygen is depicted as a single bond in one structure and as a double bond in the other. The average of these is 1.5 bonds—halfway between a single and a double bond—which is in agreement with experimental findings. In chemical terms, the actual structure of the ion is said to be a **resonance hybrid** *of the two contributing structures* that we've drawn. The double-ended arrow is drawn to show that both Lewis structures are representatives of the true hybrid structure.

The term resonance is often misleading to the beginning student. The word itself suggests that the actual structure flip-flops back and forth between the two structures shown. This is *not* the case! A mule, which is the hybrid offspring of a donkey and a horse, isn't a donkey one minute and a horse the next. Although it may have characteristics of both parents, a mule is a mule. A resonance hybrid has characteristics of its "parents," but it never has the exact structure of any one of them.

There is a simple way to determine when resonance should be applied to Lewis structures. If you find that you must move electrons to create a double bond while following the procedure developed in the previous section, the number of resonance structures is equal to the number of choices for the location of the double bond. For example, in drawing the Lewis structure for SO_3 following our procedure we reach this stage:

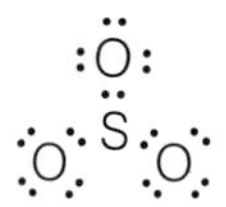

$$:\ddot{O}:$$
$$:\ddot{O}:\overset{..}{S}:\ddot{O}:$$

A double bond must be created to give sulfur an octet. Since it can be placed in any one of three positions, there are *three* resonance structures for this molecule.

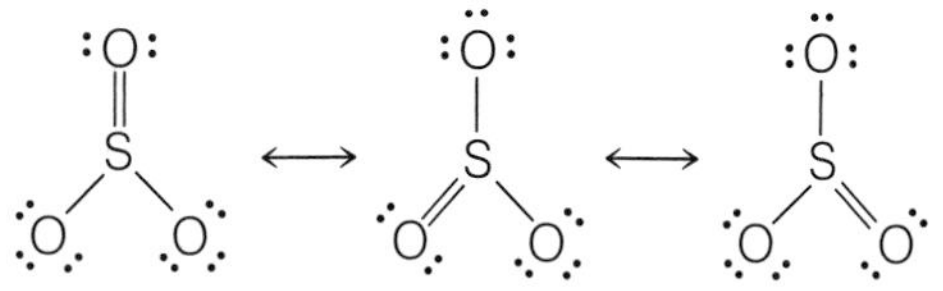

EXAMPLE 8.9   **Drawing Resonance Structures**

Problem: Draw resonance structures for SO_2.
Solution: The SO_2 molecule has $6 + 12 = 18$ electrons, and the expected skeletal structure is

$$O \quad S \quad O$$

Placing electrons following our procedure gives

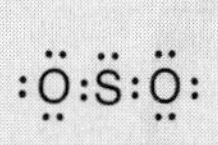

$$:\ddot{O}:\ddot{S}:\ddot{O}:$$

To give sulfur an octet we must move a pair of electrons to make a sulfur-oxygen double bond. There are two choices as to where this double bond can be placed, so there must be two resonance structures.

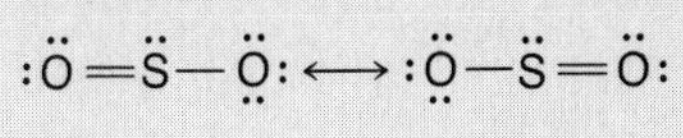

$$:\ddot{O}=\ddot{S}-\ddot{O}: \longleftrightarrow :\ddot{O}-\ddot{S}=\ddot{O}:$$

EXERCISE 8.5   Draw the resonance structures for the nitrate ion, NO_3^-.

8.7 COORDINATE COVALENT BONDS

Sometimes one atom supplies both of the electrons that are shared in a covalent bond

When a hydrogen atom forms a covalent bond with a chlorine atom, the pair of electrons that they share between them is made up of one electron from each of the two atoms.

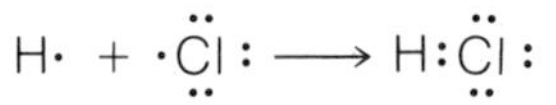

$$H\cdot \ + \ \cdot \ddot{\underset{..}{Cl}}: \ \longrightarrow \ H:\ddot{\underset{..}{Cl}}:$$

As we saw in Section 8.2, similar bonds are formed between a pair of chlorine atoms and when hydrogen forms bonds to carbon, nitrogen, and oxygen. Even in the N_2 molecule, the triple bond is formed from three electrons from each of the two nitrogen atoms. Sometimes, however, bonds are formed in which both electrons of the shared pair come from the same atom. Let's look at some examples.

When ammonia, NH_3, is placed into an acidic solution it picks up a hydrogen ion, H^+, and becomes NH_4^+. Let's keep tabs on all of the electrons when this happens.

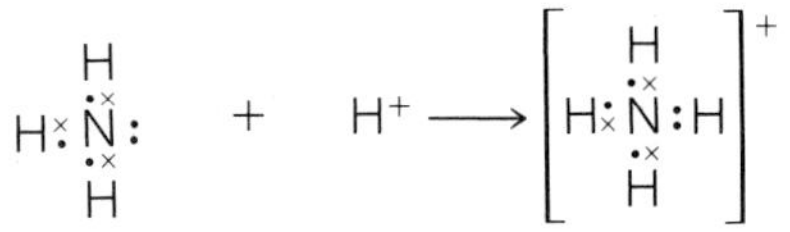

$$\underset{H}{\overset{H}{H \overset{\times}{:} \ddot{N} \underset{\times}{:}}} \quad + \quad H^+ \longrightarrow \left[\underset{H}{\overset{H}{H \overset{\times}{:} \ddot{N} \underset{\times}{:} H}} \right]^+$$

ammonia
(x's are H electrons)

The H^+ ion has a vacant valence shell that can accommodate two electrons. When the H^+ is bonded to the nitrogen of NH_3, the nitrogen donates both of the electrons to the bond. This type of bond, in which both electrons of the shared pair come from one of the two atoms, is called a **coordinate covalent bond.** Even though we make a distinction as to the origin of the electrons, once the bond is formed, it is really the same as any other covalent bond. We can't tell where the electrons in the bond came from. In NH_4^+, for instance, all four of the N—H bonds are equivalent once they have been formed, and no distinction is made between them. We usually write the structure of NH_4^+ as

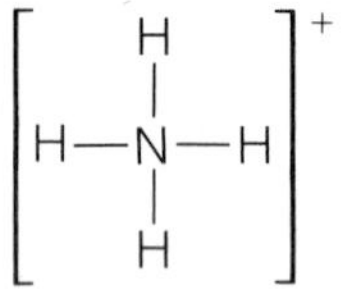

$$\left[\begin{array}{c} H \\ | \\ H-N-H \\ | \\ H \end{array} \right]^+$$

Another example of a coordinate covalent bond occurs when a molecule having an incomplete valence shell reacts with a molecule having electrons that aren't being used in bonding.

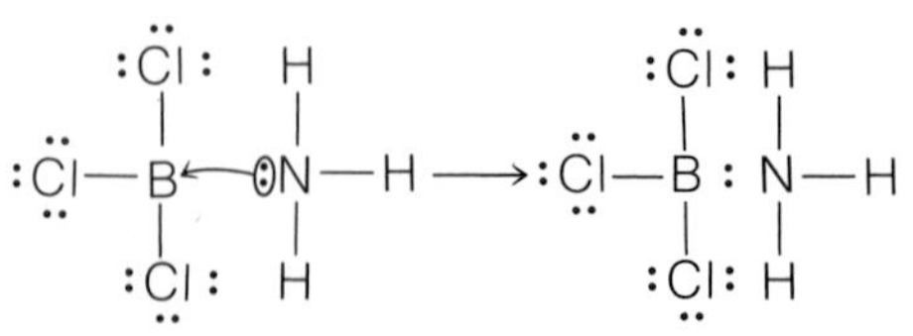

$$\underset{\underset{\ddot{Cl}:}{|}}{\overset{\overset{:\ddot{Cl}:}{|}}{:\ddot{Cl}-B}} \overset{H}{\underset{H}{\leftarrow} \overset{|}{\underset{|}{\overset{}{0}N-H}}} \longrightarrow \underset{\underset{\ddot{Cl}:}{|}}{\overset{\overset{:\ddot{Cl}:}{|}}{:\ddot{Cl}-B}} : \overset{H}{\underset{H}{\overset{|}{N-H}}}$$

An arrow sometimes is used to represent the donated pair of electrons in the coordinate covalent bond.

$$\underset{\underset{:\ddot{Cl}:}{|}}{\overset{\overset{:\ddot{Cl}:}{|}}{:\ddot{Cl}-B}} \longleftarrow \overset{H}{\underset{H}{\overset{|}{N-H}}}$$

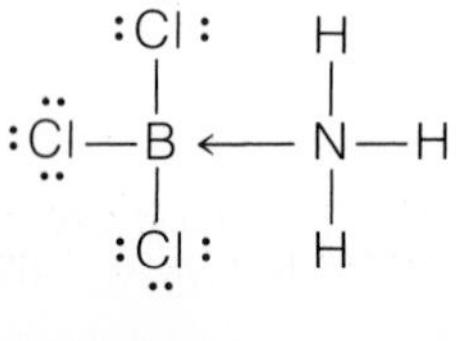

8.8 SHAPES OF MOLECULES: VSEPR THEORY

The three-dimensional shapes of molecules can be predicted if we assume that electron pairs in the valence shells of atoms stay as far apart as possible

The shapes of molecules are very important because many of their physical and chemical properties depend upon the three-dimensional arrangements of their atoms. For example, the functioning of enzymes, which are substances that control how fast biochemical reactions occur, requires that there be a very precise fit between one molecule and another. Even slight alterations in molecular geometry can destroy this fit and deactivate the enzyme, which in turn prevents the biochemical reaction involved from occurring. Nerve poisons work this way.

The best theoretical explanations of molecular shapes are based on quantum mechanics and the orbital pictures of electrons that we discussed in Chapter Seven. We will have more to say about these in Chapter Nine. There is a very simple theory, however, that is remarkably effective in predicting the shapes of molecules formed by the representative elements. It is called the **Valence Shell Electron Pair Repulsion Theory** (**VSEPR** theory, for short). Despite its long name, the principle behind the theory is very simple. The theory is based on the idea that valence shell electron pairs, being negatively charged, stay as far apart from each other as possible so that the repulsions between them are at a minimum. Let's look at an example.

Consider the $BeCl_2$ molecule. We've seen that its Lewis structure is expected to be

$$:\!\ddot{C}\!l\!:\!Be\!:\!\ddot{C}\!l\!:$$

But how are these atoms arranged? Is $BeCl_2$ linear or is it nonlinear—that is, do the atoms lie in a straight line, or do they form some angle less than 180°?

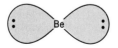

According to VSEPR theory, we can predict the shape of a molecule by looking at the electron pairs in the valence shell of the central atom. For $BeCl_2$, there are two pairs of electrons around the beryllium atom. The question is, how can they locate themselves to be as far apart as possible? The answer, of course, is that minimum repulsions will occur when the electron pairs are on opposite sides of the nucleus. We can represent this as

to suggest the approximate locations of the electron clouds of the valence shell electron pairs. In order for the electrons to be *in* the Be—Cl bonds, the Cl atoms must be placed where the electrons are; the result is that we predict that a $BeCl_2$ molecule should be linear,

$$Cl\!-\!Be\!-\!Cl$$

Let's look at another example, the BCl_3 molecule. Its Lewis structure is

$$:\!\ddot{C}\!l\!:$$
$$:\!\ddot{C}\!l\!:\!\ddot{B}\!:\!\ddot{C}\!l\!:$$

Here, the central atom has three electron pairs. What arrangement will lead to minimum repulsions? As you may have guessed, the electron pairs will be as far apart as possible when they are located at the corners of a triangle with the boron atom in the center.

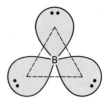

When we attach the Cl atoms we obtain a triangular molecule.

$$\begin{array}{c} Cl \\ | \\ B \\ Cl^{\diagdown}\quad^{\diagdown}Cl \end{array}$$

Actually, we say the shape is *planar triangular* because all four atoms lie in the same plane. Experimentally it's been proven that this is indeed the shape of BCl_3.

Figure 8.5 shows the shapes expected for different numbers of electron pairs around the central atom. Some of these geometric figures may be new to you, so you should try especially hard to visualize them.

EXAMPLE 8.10 **Predicting Molecular Shapes**

Problem: Carbon tetrachloride was once used as a cleaning fluid until it was discovered that it causes liver damage if absorbed by the body. What is the shape of the CCl_4 molecule?

Solution: To predict the molecular shape, we need the Lewis structure of the molecule.

$$:\ddot{C}l:$$
$$:\ddot{C}l:\overset{..}{C}:\ddot{C}l:$$
$$:\ddot{C}l:$$

There are four electron pairs in the valence shell of carbon. These would have a minimum repulsion when arranged tetrahedrally, so the molecule would be tetrahedral.

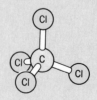

EXERCISE 8.6 What would be the shape of $SbCl_5$?

Molecular shapes when some electron pairs are not in bonds
In the molecules that we have considered so far, all the electron pairs of the central atom have been in bonds. This isn't always the case, though. Some mole-

Number of electron pairs	Shape		Example
Two pairs	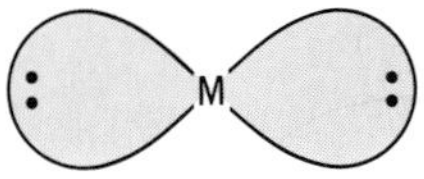	linear	BeCl$_2$ 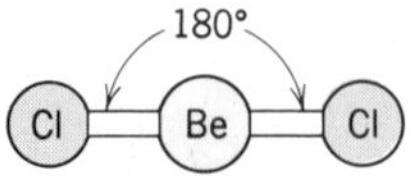
Three pairs	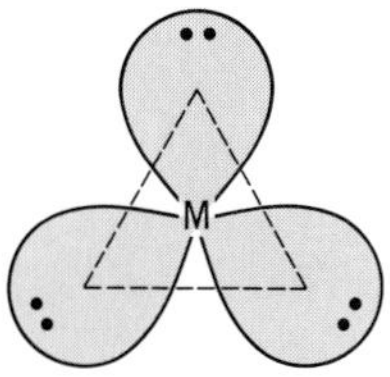	planar triangular	BCl$_3$ 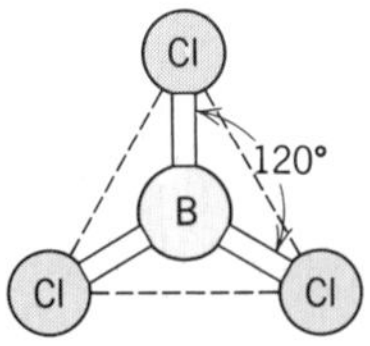
Four pairs	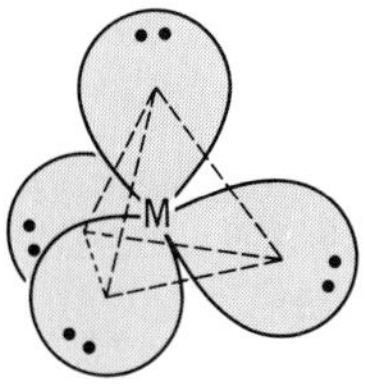	tetrahedral (A tetrahedron is pyramid-shaped. It has four triangular faces and four corners.)	CH$_4$ 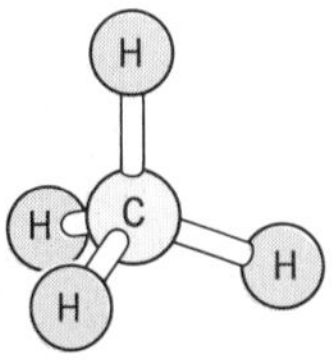
Five pairs	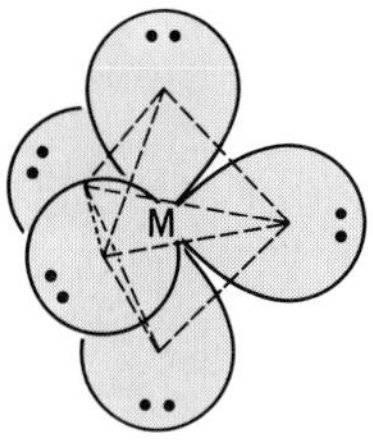	trigonal bipyramidal This figure consists of two three-sided pyramids joined by sharing a common face—the triangular plane through the center.)	PCl$_5$ 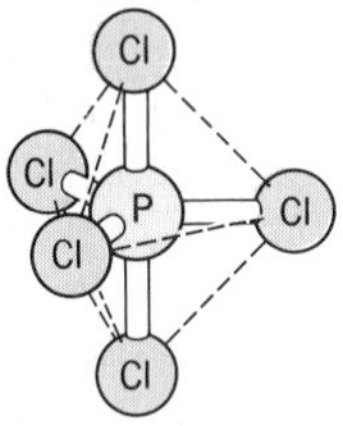
Six pairs	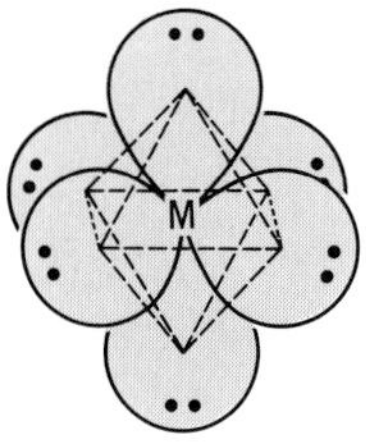	octahedral (An octahedron is an eight sided figure with *six* corners. It consists of two square pyramids that share a common square base.)	SF$_6$ 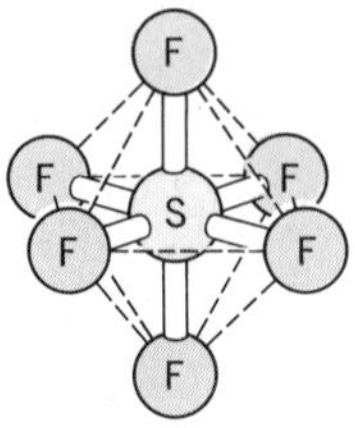

Figure 8.5
Shapes expected for different numbers of electron pairs.

cules have a central atom with one or more pairs of electrons that are not shared with another atom. Still, because they are in the valence shell, these unshared electron pairs—also called **lone pairs**—affect the geometry of the molecule. An example is SnCl$_2$.

$$:\overset{..}{\underset{..}{Cl}}:\overset{..}{Sn}:\overset{..}{\underset{..}{Cl}}:$$

There are three pairs of electrons around the tin atom—the two in the bonds plus the lone pair. As in BCl_3, the mutual repulsions of the three pairs will place them at the corners of a triangle.

Adding on the two chlorine atoms gives

We can't describe this molecule as triangular, even though that is how the electron pairs are arranged. *Molecular shape describes the arrangement of atoms, not the arrangement of electron pairs.* Therefore, we describe the shape of the $SnCl_2$ molecule as being **nonlinear** or **bent.**

Notice that when there are three electron pairs around the central atom, two different molecular shapes are possible. If all three electron pairs are in bonds, as in BCl_3, a molecule with a planar triangular shape is formed. If one of the electron pairs is a lone pair, as in $SnCl_2$, the arrangement of the atoms in the molecule is said to be nonlinear. The predicted shapes of both, however, are derived by first noting the triangular arrangement of electron pairs around the central atom and *then* adding the necessary number of atoms.

The most common type of molecule or ion has four electron pairs—an octet—in the valence shell of the central atom. When these electron pairs are used to form four bonds, as in methane (CH_4), the resulting molecule is tetrahedral, as we've seen. There are many examples, however, where some of the pairs are lone pairs. For example,

$$H-\overset{\cdot\cdot}{N}-H \qquad \text{1 lone pair}$$
$$\overset{\quad}{\underset{H}{|}}$$

$$:\overset{\cdot\cdot}{O}-H \qquad \text{2 lone pairs}$$
$$\overset{\quad}{\underset{H}{|}}$$

Figure 8.6 shows how the lone pairs affect the shapes of molecules of this type.

EXAMPLE 8.11   **Predicting Shapes of Molecules or Ions**

Problem: Do we expect the ClO_2^- ion to be linear or nonlinear?
Solution: To apply the VSEPR theory, we first need the Lewis structure for the ClO_2^- ion. Following our procedure we obtain

$$[:\overset{\cdot\cdot}{\underset{\cdot\cdot}{O}}-\overset{\cdot\cdot}{\underset{\cdot\cdot}{Cl}}-\overset{\cdot\cdot}{\underset{\cdot\cdot}{O}}:]^-$$

Counting electron pairs, we see that there are four around the chlorine.

Figure 8.6
Molecular shapes with four electron pairs about the central atom.

Number of Pairs in Bonds	Number of Lone Pairs	Structure	
4	0		Tetrahedral (Example, CH_4)
3	1		Pyramidal (Pyramid Shaped) (Example, NH_3)
2	2		Nonlinear, bent (Example, H_2O)

Four electron pairs (according to the theory) are always arranged tetrahedrally. For the electron pairs, this gives

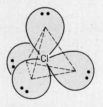

Now we add the two oxygens (it doesn't matter where in the tetrahedron).

We see that the O—Cl—O angle is less than 180° and the ion is therefore nonlinear.

When five electron pairs are present around the central atom, they are directed toward the corners of a trigonal bipyramid. Molecules such as PCl_5 have this geometry, as we saw in Figure 8.5. In the trigonal bipyramid, not all atoms have the same nearest-neighbor environments. Those at the top and bottom each have three neighbors at angles of 90°. Those in the triangular plane through the center of the molecule have two neighbors at 90° and two at 120°. When lone pairs occur in this structure, they always are located in the triangular plane. Figure 8.7 shows the kinds of geometries that we find for different numbers of lone pairs.

Figure 8.7
Molecular shapes with five electron pairs about the central atom.

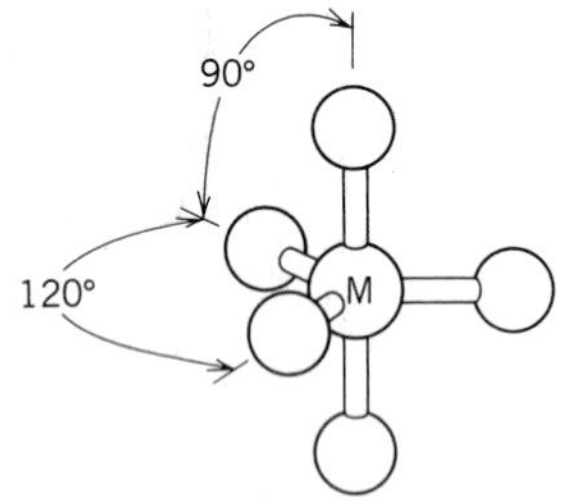

Number of Pairs in Bonds	Number of Lone Pairs	Structure	
5	0	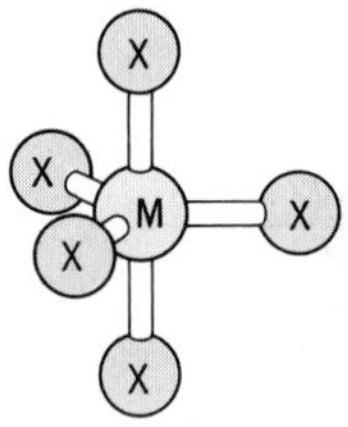	Trigonal bipyramidal (Example, PCl_5)
4	1	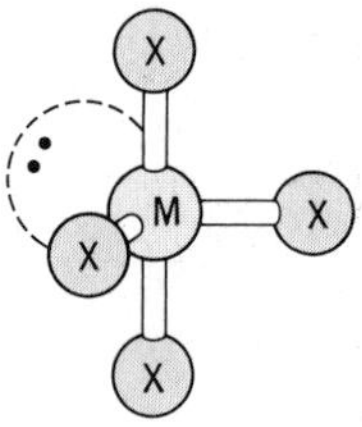	Unsymmetrical tetrahedron 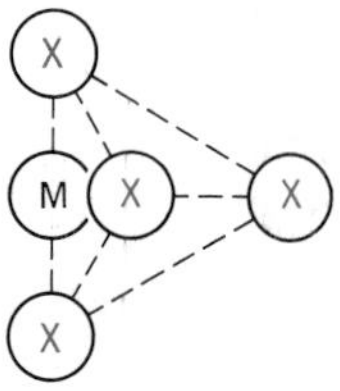 (Example, SF_4)
3	2	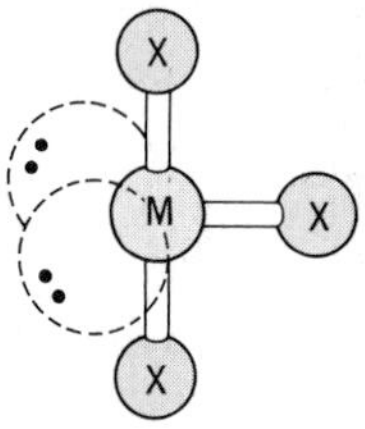	T-Shaped 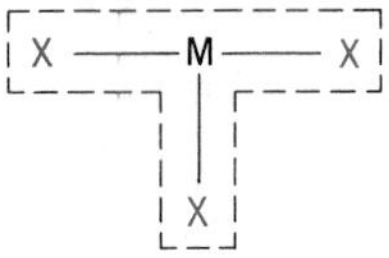 (Example, ClF_3)
2	3	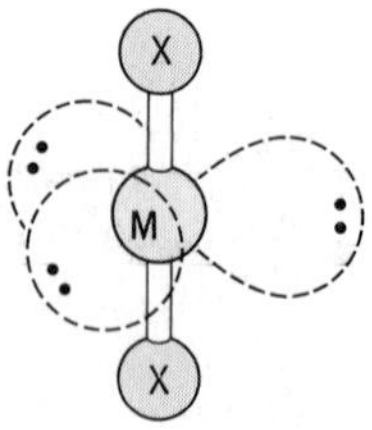	Linear (Example, I_3^-)

EXAMPLE 8.12   **Predicting Shapes of Molecules and Ions**

Problem: What is the geometry of XeF_2?
Solution: First we construct the Lewis structure.

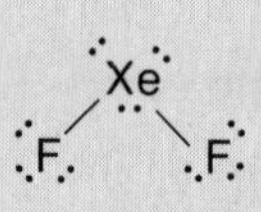

Next we count electron pairs around xenon—there are five of them. When there are five electron pairs, they are arranged in a trigonal bipyramid.

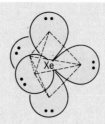

Now we must add the fluorine atoms. In a trigonal bipyramid, the lone pairs *always* occur in the triangular plane through the center, so the fluorines go on the top and bottom. This gives

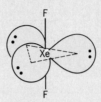

The three atoms, F—Xe—F, are arranged in a straight line — the molecule is linear.

No common molecule or ion with six electron pairs around the central atom has more than two of them as lone pairs.

Finally, we come to those molecules or ions that have six electron pairs around the central atom. When all are in bonds, as in SF_6, the molecule is octahedral. When one lone pair is present, the molecule or ion has the shape of a square pyramid, and when two lone pairs are present the molecule or ion has a square planar structure. These shapes are shown in Figure 8.8 on page 240.

EXAMPLE 8.13   **Predicting the Shapes of Molecules and Ions**

Problem: What is the probable geometry of the $XeOF_4$ molecule, which contains an oxygen and four fluorine atoms each bonded to xenon?

Solution: First we draw the Lewis structure.

There are six electron pairs around xenon, and they must be arranged octahedrally.

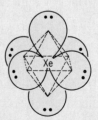

Next we attach the oxygen and four fluorines. The most symmetrical structure is

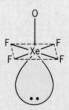

which we would describe as a square pyramid. Although the oxygen might, in principle, be placed in one of the positions at the corners of the square base (in place of one of the fluorines), the structure shown happens to be the actual structure of the molecule.

Figure 8.8
Molecular shapes found when there are six electron pairs about the central atom.

Number of Pairs in Bonds	Number of Lone Pairs	Structure	
6	0	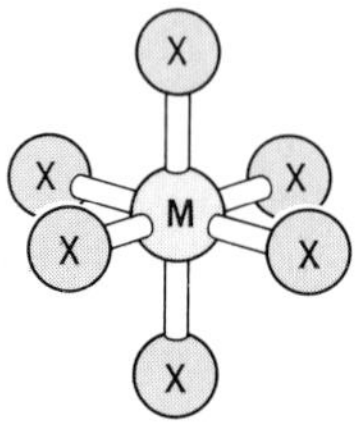	Octahedral (Example, SF_6)
5	1	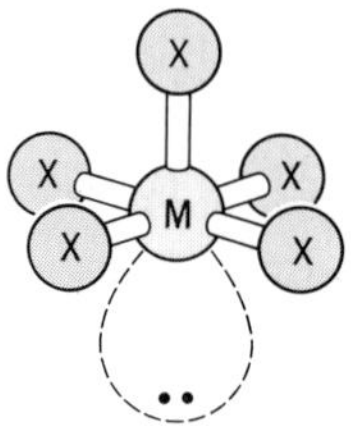	Square Pyramidal 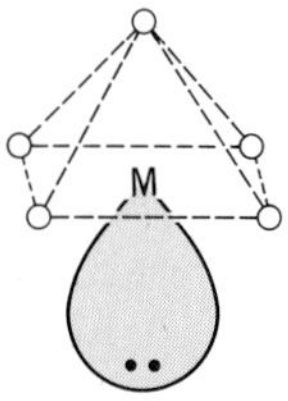 (Example, BrF_5)
4	2	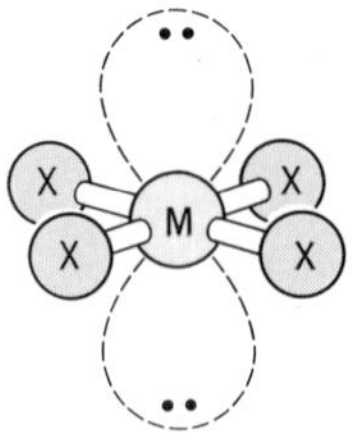	Square Planar (Example, XeF_4)

Shapes of molecules and ions with double or triple bonds

Our discussion to this point has dealt only with the shapes of molecules or ions having single bonds. Luckily, the presence of double or triple bonds does not complicate matters at all. In a double bond, both electron pairs must stay together between the two atoms; they can't wander off to different locations in the valence shell. This is also true for the three pairs of electrons in a triple bond. For the purposes of predicting molecular geometry, then, we can treat double and triple bonds just as we do single bonds. For example, the Lewis formula for CO_2 is

$$\ddot{O} = C = \ddot{O}$$

The carbon atom has no lone pairs. Therefore, the two groups of electron pairs that make up the double bonds are located on opposite sides of the nucleus and a linear molecule is formed. Similarly, we would predict the following shapes for SO_2 and SO_3.

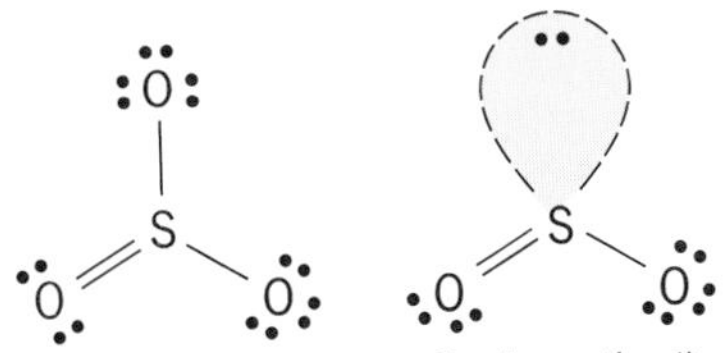

Planar triangular   Nonlinear (bent)

EXAMPLE 8.14   **Predicting the Shapes of Molecules and Ions**

Problem: The Lewis structure for the very poisonous gas, hydrogen cyanide, HCN, is

$$H—C≡N:$$

Is HCN a linear or nonlinear molecule?

Solution: The triple bond behaves like a single bond for the purposes of predicting overall geometry. Therefore, the triple and single bonds would locate themselves 180° apart and HCN would be linear.

EXERCISE 8.7   Predict the geometry of ClO_3^-, XeO_4, and OF_2.

EXERCISE 8.8   Predict the geometry of CO_3^{2-}.

8.9 POLAR BONDS AND POLAR MOLECULES

In many molecules, the electrons are not shared equally between the atoms

When two identical atoms form a covalent bond, as in H_2 or Cl_2, each has an equal share of the electron pair in the bond. The electron density at both ends of the bond is the same because the electrons are equally attracted to both nuclei. However, when different kinds of atoms combine, as in HCl, the attractions usually are not equal. Generally, one of the atoms attracts the electrons more strongly than the other.

The effect of unequal attractions for the bonding electrons is an unbalanced distribution of electron density within the bond. For example, it has been found that a chlorine atom attracts electrons more strongly than does a hydrogen atom. In the HCl molecule, therefore, the electron cloud is pulled more tightly around the Cl, and that end of the molecule experiences a slight buildup of negative charge. The electron density that shifts toward the chlorine is removed from the hydrogen, which causes the hydrogen end to acquire a slight positive charge.

In HCl, electron transfer is incomplete. The electrons are still shared, but unequally. The charges on either end of the molecule are less than full 1 + and 1 − charges—they are **partial charges,** normally indicated by the lower case Greek

Figure 8.9
Unsymmetrical distribution of
electrons in the HCl bond.

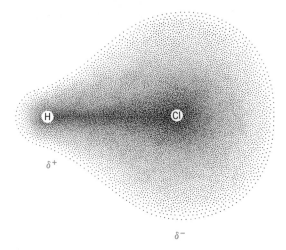

letter delta, δ (see Figure 8.9). (In HCl, for example, the hydrogen carries a partial charge of +0.17 and chlorine, a partial charge of −0.17.) Partial charges can also be indicated on Lewis structures. For example,

$$H-\overset{..}{\underset{..}{Cl}}:$$
$$\delta+ \quad \delta-$$

A bond that carries partial positive and negative charges at opposite ends is called a **polar bond.** The HCl molecule, for instance, is said to have a **polar covalent bond.** This polar bond in HCl results because the molecule has poles of opposite charge at either end; the molecule is also said to be a **dipole.** The extent of the polarity of a molecule is expressed quantitatively through its **dipole moment,** which is found by multiplying the charge on either end of the molecule by the distance between the charges.

Experimental measurement of dipole moments is one way of checking our theoretical models.

The H—Cl molecule has only one bond and, because the bond is polar, opposite ends of the molecule must have opposite electrical charges. In some molecules, however, the effects of the polar bonds cancel, and even though the individual bonds in the molecule may be polar, the molecule as a whole is nonpolar. Examples are symmetric molecules such as CO_2, BCl_3 and CCl_4. Figure 8.10 illustrates the way the individual bond dipoles cancel to give a net zero dipole moment for these molecules. The bond dipoles are shown here as arrows crossed at one end, ↦. The head of the arrow points in the direction of the negative end of the dipole.

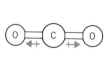
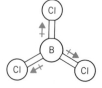

Figure 8.10
In symmetric molecules the bond dipoles cancel to give nonpolar molecules.

Molecular structure is important in determining whether or not a given molecule will be polar. For simplicity, suppose we represent a molecule by the formula MX_n, where M is the central atom and X stands for the atoms or groups of

MX₂ molecules with three lone pairs, and MX₄ molecules with two lone pairs will also be nonpolar if the X's are all the same in a given molecule.

atoms (*n* of them) bonded to *M*. If all of the *X*'s around *M* are the same (as in CO_2, BCl_3, and CCl_4) and *M* has no lone pairs in its valence shell, the structure of the molecule will be one of those found in Figure 8.5 (page 235). Such molecules are nonpolar. If there are lone pairs around the central atom, however, cancellation of the bond dipoles usually doesn't occur and the molecules will normally be polar. This is shown in Figure 8.11 for water and ammonia, both of which are very polar molecules.

Figure 8.11
When lone pairs occur on the central atom, the bond dipoles do not cancel and polar molecules result.

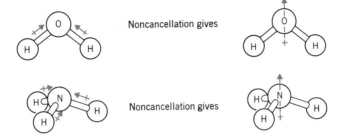

8.10 ELECTRONEGATIVITY

The relative abilities of bonded atoms to attract electrons determines how polar the bond is

The extent of molecular polarity is very useful to know because many physical properties, such as melting point and boiling point, are affected by it. This is because dipoles attract each other—the positive end of one dipole attracts the negative end of another (Figure 8.12). The strength of the attractions depends both on the amount of charge on either end and on the distance between these charges; that is, it depends on the dipole moment. Let's look at the way we might predict how polar a bond is.

Figure 8.12
Attractions between dipoles. Molecules tend to orient themselves so that the positive end of one dipole is near the negative end of another.

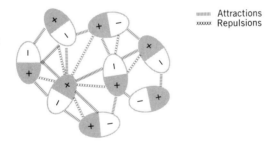

IIIIIII Attractions
xxxxx Repulsions

The relative attraction of an atom for the electrons in a bond is called the **electronegativity** of that atom. In HCl, for example, chlorine is more electronegative than hydrogen. The electron pair of the covalent bond spends more of its time around the more electronegative atom, which is why that end of the bond acquires a partial negative charge.

The concept of electronegativity has been put on a quantitative basis— numbers have been assigned, as shown in Figure 8.13. This information is useful because the *difference* in electronegativity provides an estimate of the degree of polarity of a bond. In addition, the relative magnitudes of the electronegativities indicate which end of the bond carries the negative charge. For instance, fluorine is more electronegative than chlorine. Therefore, we would expect an HF molecule to be more polar than an HCl molecule. In addition, hydrogen is less

Linus Pauling, winner of two Nobel prizes and recent advocate of vitamin C to prevent colds, was the first scientist to set up a table of electronegativity values.

H 2.1																	
Li 1.0	Be 1.5												B 2.0	C 2.5	N 3.1	O 3.5	F 4.1
Na 1.0	Mg 1.3												Al 1.5	Si 1.8	P 2.1	S 2.4	Cl 2.9
K 0.9	Ca 1.1	Sc 1.2	Ti 1.3	V 1.5	Cr 1.6	Mn 1.6	Fe 1.7	Co 1.7	Ni 1.8	Cu 1.8	Zn 1.7	Ga 1.8	Ge 2.0	As 2.2	Se 2.5	Br 2.8
Rb 0.9	Sr 1.0	Y 1.1	Zr 1.2	Nb 1.3	Mo 1.3	Tc 1.4	Ru 1.4	Rh 1.5	Pd 1.4	Ag 1.4	Cd 1.5	In 1.5	Sn 1.7	Sb 1.8	Te 2.0	I 2.2
Cs 0.9	Ba 0.9	La 1.1	Hf 1.2	Ta 1.4	W 1.4	Re 1.5	Os 1.5	Ir 1.6	Pt 1.5	Au 1.4	Hg 1.5	Tl 1.5	Pb 1.6	Bi 1.7	Po 1.8	At 2.0
Fr 0.9	Ra 0.9	Ac 1.0														

Lanthanides: 1.0–1.2
Actinides: 1.0–1.2

Figure 8.13
Table of electronegativities.

electronegative than either fluorine or chlorine, so in both of these molecules the hydrogen bears the positive charge.

$$H—\overset{\cdot\cdot}{\underset{\cdot\cdot}{F}}: \qquad\qquad H—\overset{\cdot\cdot}{\underset{\cdot\cdot}{Cl}}:$$
$$\delta+\ \ \delta- \qquad\qquad\quad \delta+\ \ \delta-$$

The concept of electronegativity also shows us that there is no sharp dividing line between ionic and covalent bonding. Ionic bonding and nonpolar covalent bonding simply represent the two extremes. Ionic bonding occurs when the difference in electronegativity between two atoms is very large; the more electronegative atom acquires essentially complete control of the bonding electrons. In a nonpolar covalent bond, there is no difference in electronegativity, so the pair of bonding electrons is shared equally.

$$Cs^+\ [\overset{\cdot\cdot}{\underset{\cdot\cdot}{:F:}}]^- \qquad\qquad\qquad :\overset{\cdot\cdot}{\underset{\cdot\cdot}{F}}\overset{\cdot\cdot}{\underset{\cdot\cdot}{:F}}:$$

"bonding pair" held   bonding pair
exclusively by fluorine   shared equally

The degree of polarity, which we might think of as the amount of ionic character of the bond, varies in a continuous way with changes in the electronegativity difference (Figure 8.14). The bond becomes more than 50% ionic when the electronegativity difference exceeds 1.7.

Figure 8.14
Variation of percent ionic character with electronegativity difference.

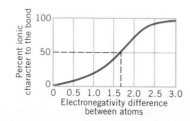

Metals have low electronegativities; nonmetals have high
electronegativities.

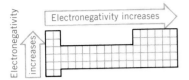

Within the periodic table, electronegativity increases from bottom to top within a group and from left to right across a period. Except for the noble gases, which are assigned electronegativities of zero, the trends follow those for the ionization energy. This should not be surprising, since atoms that lose electrons only with great difficulty clearly have a strong attraction for them. Elements in the same region of the table (e.g., the nonmetals) have similar electronegativities, and if they react with each other, they form bonds that are predominantly covalent. Elements from opposite sides of the table have large electronegativity differences and tend to form ionic bonds.

8.11 OXIDATION-REDUCTION REACTIONS

Many chemical reactions involve the transfer of electron density from one atom to another

On first thought, the reaction of sodium and chlorine to produce sodium chloride appears to be very different from the reaction of hydrogen and oxygen to form water. In NaCl, ions have been produced; in H_2O, covalent bonds have been formed. In at least one way, however, these reactions are similar—both involve a shift in electron density.

In the case of sodium chloride, an electron is transferred completely from Na to Cl when the Na^+ and Cl^- ions are formed. When H_2 and O_2 react, molecules with nonpolar bonds form H_2O, which has polar O—H bonds. In H_2, the hydrogen atoms are electrically neutral while in H_2O they carry a partial positive charge. Hydrogen therefore loses electron density when H_2O is formed, and oxygen gains electron density.

Many chemical reactions involve a shift of electron density from one atom to another. Collectively, such reactions are called **oxidation-reduction reactions,** or simply **redox reactions.** The term **oxidation** refers to the loss of electrons by one reactant, and **reduction** refers to the gain of electrons by another. For example, the reaction between sodium and chlorine involves a loss of electrons by sodium—oxidation of sodium—and a gain of electrons by chlorine—reduction of chlorine.

$$Na \longrightarrow Na^+ + e^- \quad \text{(oxidation)}$$
$$Cl_2 + 2e^- \longrightarrow 2Cl^- \quad \text{(reduction)}$$

We say that sodium is oxidized and chlorine is reduced.

Oxidation and reduction *always* occur together. No substance is ever oxidized unless something else is reduced. Otherwise, electrons would appear as a product of the reaction, and this is never observed. During a redox reaction, then, some substance *must* accept the electrons that another substance loses. This electron-accepting substance is called the oxidizing agent, and the substance that supplies the electrons is called the reducing agent. Sodium is a reducing agent, for example, when it supplies electrons to chlorine. In the process, sodium is oxidized. Chlorine is an oxidizing agent when it accepts electrons from the sodium, and when that happens, chlorine is reduced to chloride ion. One way to remember this is:

> The substance that is oxidized is the reducing agent.
> The substance that is reduced is the oxidizing agent.

Redox reactions are very common. Whenever you use a battery, a redox reaction occurs. The battery is constructed so that the electron transfer must take place through a wire. On its way, we use the energy provided by the electron transfer to light a flashlight or to power a pocket calculator. The metabolism of foods, which supplies our bodies with energy, also occurs by a series of redox reactions that use oxygen to convert carbohydrates and fats to carbon dioxide and water. Ordinary household bleach works by oxidizing substances that stain fabrics, making them easier to remove from the fabric or rendering them colorless.

EXAMPLE 8.15 **Identifying Oxidation and Reduction and Oxidizing and Reducing Agents**

Problem: Calcium and oxygen react to form calcium oxide, CaO, an ionic compound.

$$2Ca + O_2 \longrightarrow 2CaO$$

Which element is oxidized and which is reduced? What are the oxidizing and reducing agents?

Solution: When calcium reacts with oxygen, it forms Ca^{2+} ions, which means that the calcium atoms must *lose* electrons.

$$Ca \longrightarrow Ca^{2+} + 2e^-$$

This is oxidation; calcium is oxidized. Since calcium is oxidized, it must be the reducing agent.

When oxygen reacts with calcium, it forms O^{2-} ions, which means that the oxygen atoms must *gain* electrons.

$$O_2 + 4e^- \longrightarrow 2O^{2-}$$

This is reduction; oxygen is reduced. This means that oxygen is the oxidizing agent.

EXERCISE 8.9 Identify the substances oxidized and reduced and the oxidizing and reducing agents in the reaction of calcium and chlorine to form calcium chloride, $CaCl_2$, an ionic compound.

8.12 OXIDATION NUMBERS

An oxidation number is the charge that an atom in a compound would have if the electron pairs in the bonds belonged entirely to the more electronegative atoms

Because redox reactions are so important, chemists have developed a special bookkeeping system to count electrons when they are transferred. The system employs a device called oxidation numbers. Roughly speaking, an **oxidation number** is the charge that an atom would have if the electrons in each bond belonged entirely to the more electronegative element. For example, in assigning oxidation numbers to hydrogen and chlorine in the polar H—Cl molecule, we imagine that both electrons of the bond are in the sole possession of the more electronegative chlorine atom. If this were the case, then the charge on hydrogen would be 1+ and the charge on chlorine would be 1−. Therefore, in HCl

H   oxidation number = 1+
Cl   oxidation number = 1−

For covalently bonded molecules like HCl, of course, we know that the atoms never carry more than partial positive or negative charges. Nevertheless, oxidation numbers are assigned as if each compound were ionic. It is important to remember, therefore, that the oxidation numbers assigned to atoms in compounds *do not* have to correspond to the actual charges on the atoms—sometimes they do and sometimes they don't.

A term that is frequently used interchangeably with oxidation number is **oxidation state.** In the HCl molecule, hydrogen has an oxidation number of 1+ and is said to be in the 1+ oxidation state. Similarly, the chlorine in HCl is said to be in the 1− oxidation state. When a redox reaction occurs, the atoms involved undergo a change in oxidation number or oxidation state. We will see later that making note of changes in oxidation numbers is a convenient way to follow the electron transfer in these reactions. To do this, however, the oxidation numbers of the atoms in a compound must be able to be determined simply and quickly.

Fortunately, the assignment of oxidation numbers to atoms does not require continual reference to a table of electronegativity values. There are some simple rules and generalizations that make the task relatively easy.

First, we know that any free atom of an element carries no electrical charge. Therefore, we assign it an oxidation number of zero. When atoms of the same element combine to produce more complex molecular forms, such as O_2, N_2, P_4, or S_8, the bonds between the atoms are nonpolar. The atoms in any one of these molecules carry no partial charges because their electronegativities are identical—thus, their oxidation numbers are also zero. This gives us our first rule.

> Any element, when not combined with atoms of a different element, has an oxidation number of zero.

When an ionic compound is formed from two elements, complete electron transfer occurs. The atom with the lower electronegativity loses one or more electrons to the atom with the higher electronegativity. Consider sodium chloride, for example. The electrons in the Na—Cl "bond" are claimed exclusively by the chloride ion.

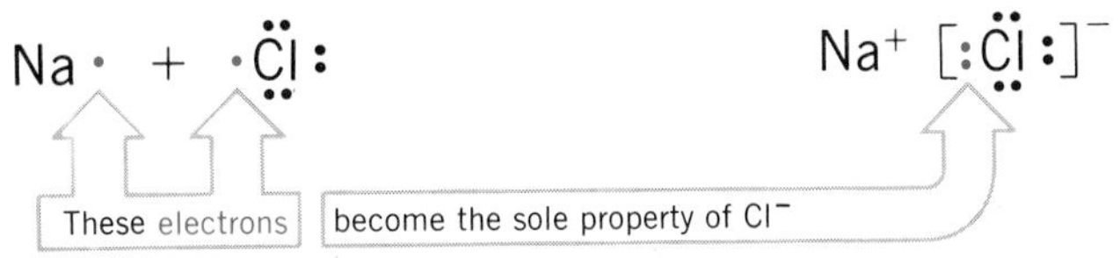

$$Na \cdot \ + \ \cdot \ddot{\underset{\cdot\cdot}{Cl}} : \qquad\qquad Na^+ \ [: \ddot{\underset{\cdot\cdot}{Cl}} :]^-$$

These electrons | become the sole property of Cl^-

The sodium and chloride ions are therefore assigned oxidation numbers equal to their charges.

$$Na^+ \quad \text{oxidation number} = 1+$$
$$Cl^- \quad \text{oxidation number} = 1-$$

In general,

> Any simple monatomic ion (one-atom ion) has an oxidation number equal to its charge.

In any compound (either ionic or covalent), the algebraic sum of all of the electrical charges is equal to zero. The formula of the compound always reflects this because it is written without a charge (e.g., NaCl, HCl, or $CaCl_2$). To conform with the electrical neutrality of a compound, the sum of the oxidation numbers of its atoms must always equal zero. Thus, for NaCl

$$
\begin{array}{lll}
\text{Na}^+ & \text{oxidation number} = 1+ \\
\text{Cl}^- & \text{oxidation number} = 1- \\
\hline
\text{NaCl} & \quad\quad\quad\quad\quad \text{sum} = 0
\end{array}
$$

and for $CaCl_2$

$$
\begin{array}{lll}
\text{Ca}^{2+} & \text{oxidation number} = 2+ \\
\text{Cl}^- & \text{oxidation number} = 1- \\
\text{Cl}^- & \text{oxidation number} = 1- \\
\hline
\text{CaCl}_2 & \quad\quad\quad\quad\quad \text{sum} = 0
\end{array}
$$

We will extend this concept to polyatomic ions, such as NO_3^- or SO_4^{2-}, and require that the sum of the oxidation numbers of their atoms add up to the *charge on the ion.* In general, then,

---

The sum of the oxidation numbers of all of the atoms in a formula must equal the charge written for the formula.

This summation rule is particularly useful when the oxidation numbers are known for all but one of the atoms in a formula. To make use of this rule, though, we must rely on some elements that always (or at least, *almost* always) have the same oxidation numbers in their compounds. The periodic table is helpful as an aid in remembering them.

In Table 6.2, on page 172, we saw that when the alkali metals (Group IA) and alkaline earth metals (Group IIA) react, they always form ions with a positive charge that is equal to their group number (1+ for the alkali metals; 2+ for the alkaline earth metals). Thus, the oxidation number of an alkali metal is always 1+ in compounds; that of an alkaline earth metal is always 2+. Aluminum, the only element that is important to us in Group IIIA, always produces Al^{3+} when it reacts.

Remember, a binary compound consists of two different elements.

On the right side of the periodic table are the nonmetals. When they are found in *binary compounds* with metals, they occur as ions with a negative charge that is equal to the number of electrons their atoms require to achieve an octet. For instance, the halogens, Group VIIA, each require one electron and form ions with a 1− charge. Therefore, whenever a halogen is encountered in a binary compound *with a metal,* its oxidation number is 1−. Examples are $NaCl$, MgI_2, $FeBr_2$, and $FeBr_3$. (Notice that, because the sum of the oxidation numbers must be zero for these compounds, the oxidation number of iron is 2+ in $FeBr_2$ and 3+ in $FeBr_3$.)

Referring to Table 6.2, we see that any Group VIA element will have an oxidation number of 2− in a binary compound with a metal (for example, O in Fe_2O_3 and S in PbS), and a Group VA element will have an oxidation number of 3− (for example, N in Mg_3N_2).

When nonmetals are combined with other nonmetals in molecules or in polyatomic ions such as NO_3^- or SO_4^{2-}, fewer generalizations about their oxidation numbers can be made. Only fluorine, oxygen, and hydrogen have relatively unchanging oxidation numbers from one compound to another.

Because fluorine is the most electronegative of all the elements, in its compounds it is *always* assigned an oxidation number of 1−. In other words, for the purposes of assigning oxidation numbers, we imagine that fluorine has total possession of the one electron that it acquires a share of when it forms a bond with another atom.

Oxygen, the second most electronegative element, is nearly always assigned an oxidation number of 2− in compounds. There are some exceptions,[1] but they

---

[1] When oxygen is combined with fluorine, as in OF_2 or O_2F_2, it is required to have a positive oxidation number. Other exceptions are peroxides, which have an oxygen-oxygen bond (e.g., hydrogen peroxide, H—O—O—H); the oxidation number of oxygen is 1−. In the superoxide ion, O_2^-, the oxidation number of each oxygen is $-\frac{1}{2}$.

do not occur often.

Finally, hydrogen is generally assigned an oxidation number of 1+ in its compounds. The only exceptions are when hydrogen occurs in a binary compound with a metal, such as in LiH, where it has an oxidation number of 1−. Exceptions in the case of hydrogen are also rare.

In compounds,
    F is always 1−
    O is almost always 2−
    H is almost always 1+

Now let's look at some examples that illustrate how we assign oxidation numbers to atoms in compounds and polyatomic ions.

EXAMPLE 8.16   **Assigning Oxidation Numbers**

Problem: Molybdenum disulfide, MoS_2, has a structure that allows it to behave as a dry lubricant, much like graphite. What are the oxidation numbers of the atoms in MoS_2?

Solution: This is a binary compound of a metal (molybdenum) with a Group VIA non-metal (sulfur). Therefore, we take the oxidation number of sulfur to be 2−. Then we can use the sum rule to find the oxidation number of molybdenum. Since MoS_2 is electrically neutral, the sum of the oxidation numbers must be zero.

$$
\begin{array}{lll}
\text{S} & 2 \text{ atoms} \times (2-) & = 4- \\
\text{Mo} & 1 \text{ atom} \times (x) & = \underline{x} \\
& & \text{sum} = 0
\end{array}
$$

The value of x must be 4+ for the sum to be 0. Therefore,

$$
\begin{array}{ll}
\text{S} & = 2- \\
\text{Mo} & = 4+
\end{array}
$$

EXAMPLE 8.17   **Assigning Oxidation Numbers**

Problem: Liquid laundry bleach contains hypochlorite ion, OCl^-, as its active ingredient. Assign oxidation numbers to the atoms in OCl^-.

Solution: Both oxygen and chlorine are nonmetals. In this case, it is oxygen whose oxidation number is assigned first—we take it to be 2−. We can then use the summation rule to find the oxidation number of Cl.

$$
\begin{array}{lll}
\text{O} & 1 \text{ atom} \times (2-) & = 2- \\
\text{Cl} & 1 \text{ atom} \times (x) & = \underline{x} \\
& & \text{sum} = 1-
\end{array}
$$

The value of x must be 1+ for the sum to be 1−. Therefore,

$$
\begin{array}{ll}
\text{O} & = 2- \\
\text{Cl} & = 1+
\end{array}
$$

EXAMPLE 8.18 Assigning Oxidation Numbers

Problem: A chemical used to remove unexposed silver compounds from photographic film is sodium thiosulfate, $Na_2S_2O_3$. What are the oxidation numbers of the atoms in this compound?

Solution: Once again we rely on the sum rule.

$$
\begin{aligned}
\text{Na} \quad &2 \text{ atoms} \times (1+) = 2+ \quad &\text{(Na is in Group IA)}\\
\text{S} \quad &2 \text{ atoms} \times (x) = 2x \\
\text{O} \quad &3 \text{ atoms} \times (2-) = \underline{6-} \quad &\text{(O is almost always } 2-)\\
&\qquad\qquad\qquad\quad \text{sum} = 0 \quad &\text{(summation rule)}
\end{aligned}
$$

We can solve for x algebraically.

$$(2+) + (2x) + (6-) = 0$$
$$2x = 4+$$
$$x = 2+$$

The oxidation number of sulfur is $2+$. Therefore,

$$
\begin{aligned}
\text{Na} &= 1+\\
\text{O} &= 2-\\
\text{S} &= 2+
\end{aligned}
$$

EXAMPLE 8.19 Assigning Oxidation Numbers

Problem: Acetone, C_3H_6O, is a common solvent in fingernail polish remover. What is the oxidation number of carbon in this compound?

Solution: Again we apply the summation rule.

$$
\begin{aligned}
\text{C} \quad &3 \text{ atoms} \times (x) = 3x \\
\text{H} \quad &6 \text{ atoms} \times (1+) = 6+ \quad &\text{(H is almost always } 1+)\\
\text{O} \quad &1 \text{ atom} \times (2-) = \underline{2-} \quad &\text{(O is almost always } 2-)\\
&\qquad\qquad\qquad\quad \text{sum} = 0 \quad &\text{(summation rule)}
\end{aligned}
$$

$$3x = 4-$$

$$x = \frac{4-}{3}$$

The oxidation number of carbon is $4/3-$. Note that oxidation numbers *do not* have to be whole numbers.[2]

Many compounds contain familiar polyatomic ions such as ammonium (NH_4^+), sulfate (SO_4^{2-}), nitrate (NO_3^-), and acetate ($C_2H_3O_2^-$). The charges on these ions can be considered to represent the net oxidation number of the atoms that make up the ion. We can use this sometimes to help assign oxidation numbers, as shown in Example 8.20.

---

[2] What we have calculated is the average oxidation number of carbon. There are ways of assigning oxidation numbers to the individual carbon atoms in the molecule. However, they require a knowledge of the structure of the molecule and therefore we won't explore them any further.

EXAMPLE 8.20   Assigning Oxidation Numbers

Problem: What is the oxidation number of chromium in the compound, $Cr(NO_3)_3$?
Solution: The nitrate ion has a charge of $1-$, which we take to be its net oxidation
number. Then we apply the summation rule.

$$\begin{array}{lll} Cr & 1 \text{ atom} \times (x) & = x \\ NO_3^- & 3 \text{ ions} \times (1-) & = \underline{3-} \\ & \text{sum} & = 0 \end{array}$$

The oxidation number of chromium must be $3+$.

EXERCISE 8.10   Assign oxidation numbers to each atom in $NiCl_2$, Mg_2TiO_4, $K_2Cr_2O_7$, SO_4^{2-}.

EXERCISE 8.11   What is the average oxidation number of carbon in isopropyl alcohol (rubbing alcohol)
C_3H_8O?

Oxidation numbers have several uses. One of them is in analyzing redox reactions to identify the substances that are oxidized and reduced. In general,

> Oxidation (electron loss) leads to an increase in oxidation number.
> Reduction (electron gain) leads to a decrease in oxidation number.

Let's see how this applies to the reaction of hydrogen with chlorine. To avoid ever confusing oxidation numbers with actual electrical charges, we write oxidation numbers directly above the symbols of the elements in the formula.

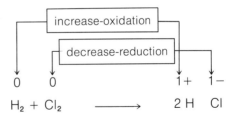

The changes in oxidation number tell us that hydrogen is oxidized and chlorine is reduced.

8.13   USING OXIDATION NUMBERS TO BALANCE EQUATIONS

**In redox equations, the total number of electrons gained
must equal the total number lost**

A useful feature of oxidation numbers is that they provide rather easy means for balancing redox equations. We have discussed balancing equations before, but these were relatively simple equations, with small coefficients, that could be balanced by inspection. Redox reactions, however, are sometimes complex, as illustrated by the equation for the oxidation of isopropyl alcohol to acetone:

$$3C_3H_8O + 2CrO_3 + 3H_2SO_4 \longrightarrow Cr_2(SO_4)_3 + 3C_3H_6O + 6H_2O$$

isopropyl   acetone
alcohol

Equations such as this are balanced by using the fact that the total number of electrons gained has to be equal to the total number lost. This must be true; otherwise, electrons would be a reactant or product. To count electrons gained or lost, oxidation numbers are used as if they were actual charges, regardless of whether the compounds involved are ionic or covalent.

To illustrate the principle, we can look at a very simple equation. (Actually, it's simpler to balance by inspection, but this time we'll do it the hard way.)

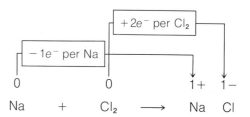

Notice that we've assigned oxidation numbers as before. Each Cl changes by one electron (Cl goes from 0 to 1−). Therefore, each Cl_2 gains a total of $2e^-$. Each Na loses only $1e^-$. We now *force* the number of electrons lost to be equal to the number gained. To do this we can double the number of Na that are reacting.

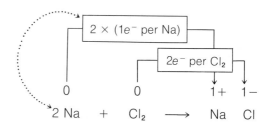

Now two electrons are gained and two are lost. Finally we place a 2 in front of NaCl to balance the Na.

$$2Na + Cl_2 \longrightarrow 2NaCl$$

Let's see how this works with a complex equation, like the oxidation of isopropyl alcohol to acetone. We can divide the procedure into a series of steps.

Step 1. *Write the correct formulas for all the reactants and products.*

$$C_3H_8O + CrO_3 + H_2SO_4 \longrightarrow Cr_2(SO_4)_3 + C_3H_6O + H_2O$$

Step 2. *Assign oxidation numbers to the atoms in the equation.*

$$\overset{2-\ 1+\ 2-}{C_3\ H_8\ O} + \overset{6+\ 2-}{Cr\ O_3} + \overset{1+\ 2-}{H_2\ \overset{\frown}{SO_4}} \longrightarrow \overset{3+\ 2-}{Cr_2\ \overset{\frown}{(SO_4)_3}} + \overset{4/3-\ 1+\ 2-}{C_3\ H_6\ O}$$

Because SO_4^{2-} appears the same on both sides of the equation, we have used its net oxidation number.

Step 3. *Identify which atoms change oxidation number.*

Inspection reveals that carbon and chromium change. Carbon goes from 2− to 4/3− and chromium goes from 6+ to 3+. We will now focus on the formulas containing these atoms.

Step 4. *Make the number of atoms that change oxidation number the same on both sides by inserting temporary coefficients.*

Both formulas containing carbon have three carbon atoms; the temporary coefficients of C_3H_8O and C_3H_6O are each 1. The $Cr_2(SO_4)_3$ contains two Cr while CrO_3 contains one. We place a temporary coefficient of 2 in front of CrO_3. Now there are the same numbers of C and Cr on both sides.

$$\overset{2-}{} \qquad \overset{6+}{} \qquad\qquad\qquad \overset{3+}{} \qquad \overset{4/3-}{}$$
$$1\ C_3H_8O\ +\ 2\ CrO_3\ +\ H_2SO_4\ \longrightarrow\ 1\ Cr_2(SO_4)_3\ +\ 1\ C_3H_6O\ +\ H_2O$$

Step 5. *Compute the **total** change in oxidation number for the oxidation and reduction that occur.*

For the elements that change, we calculate the *total* oxidation number that they contribute to each formula. In $1C_3H_8O$ there are 3 carbons, each contributing 2−; the total is 6−. In $2CrO_3$ there are 2 chromiums, each contributing 6+; the total is 12+, and so forth. Then we note how the total oxidation number for each element changes.

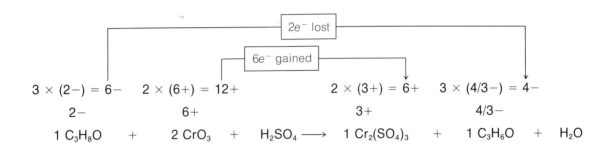

For carbon, there is a *total* loss of $2e^-$ (6− to 4− involves the loss of two *negative* electrons). For Cr there is a total gain of $6e^-$ (12+ to 6+).

Step 6. *Make the electron loss equal the electron gain by multiplication using appropriate factors.*

$$\text{For carbon}\quad (2e^-\ \text{loss}) \times 3 = 6e^-\ \text{loss}$$
$$\text{For chromium}\ (6e^-\ \text{gain}) \times 1 = 6e^-\ \text{gain}$$

The factors 3 and 1 are used to multiply the temporary coefficients.

$$\underset{3\times 1}{3\ \ C_3H_8O} +\ \underset{1\times 2}{2\ \ CrO_3} + H_2SO_4 \longrightarrow \underset{1\times 1}{1\ \ Cr_2(SO_4)_3} +\ \underset{3\times 1}{3\ \ C_3H_6O} + H_2O$$

Step 7. *Balance the remainder by inspection.*

We have now found the correct coefficients for all but two formulas. There are $3SO_4$ on the right, so we need a coefficient of 3 in front of H_2SO_4. Next, we look at the hydrogens. There are 30H on the left. On the right there are 18H in the acetone, so we must have 12 more in H_2O. Therefore, the coefficient of H_2O must be 6. This gives (finally!)

$$3C_3H_8O + 2CrO_3 + 3H_2SO_4 \longrightarrow Cr_2(SO_4)_3 + 3C_3H_6O + 6H_2O$$

The procedure may seem long, but after just a bit of practice you will find that it is not difficult.

EXERCISE 8.12 Balance the following redox equations.
(a) $KCl + MnO_2 + H_2SO_4 \longrightarrow K_2SO_4 + MnSO_4 + Cl_2 + H_2O$
(b) $KMnO_4 + FeSO_4 + H_2SO_4 \longrightarrow K_2SO_4 + MnSO_4 + Fe_2(SO_4)_3 + H_2O$
(c) $Zn + HNO_3 \longrightarrow Zn(NO_3)_2 + NH_4NO_3 + H_2O$
This last equation has some HNO_3 that changes oxidation number and some that does not. It may help to write the formula for HNO_3 twice so that you can deal with the two types separately.

$$Zn + HNO_3 + HNO_3 \longrightarrow Zn(NO_3)_2 + NH_4NO_3 + H_2O$$

8.14 NAMING CHEMICAL COMPOUNDS

The naming of chemical compounds follows a systematic procedure, internationally agreed upon by chemists, that allows only one formula to be associated with a name

When chemists talk among themselves, they often sound to nonchemists as though they are speaking a foreign tongue. Of course, chemists think the same thing when they overhear biologists. In fact, every profession or specialty has its own peculiar jargon—technical terms and phrases related to that particular field.

Included in the terms used by chemists are the names of chemical compounds. Many of these names have been around for a long time and are universally associated with a specific formula. Examples are water (H_2O) and ammonia (NH_3). Not all compounds are so well known, however. This, coupled with the fact that there are many, many compounds, led chemists around the world to agree upon a systematic method for naming them. As a result, we are able to write the correct formula for any compound given the correct name and vice versa.

In this section, we will discuss the *nomenclature* (naming) of simple **inorganic compounds**—those substances that, in general, would *not* be considered to be derived from methane (CH_4) or other hydrocarbon compounds. The hydrocarbons and compounds that come from them are called **organic compounds** (although they need not occur in living organisms). We will have more to say about them in the next chapter.

Binary compounds containing a metal and a nonmetal

Salts of a metal and nonmetal are named by giving the name of the metal first, followed by the name of the negative ion formed from the nonmetal. The latter is formed from the stem of the nonmetal name plus the suffix *-ide*. An example is our old friend, NaCl—sodium chlor*ide*. Table 8.2 lists some common monatomic negative ions and their names. Other examples of compounds formed from them are:

CaO	calcium oxide
ZnS	zinc sulfide
Mg_3N_2	magnesium nitride

The -ide suffix is usually used only for monatomic ions, although there are two common exceptions—*hydroxide* (OH^-) and *cyanide* (CN^-).

Table 8.2
Monatomic Negative Ions

H^-	hydride	N^{3-}	nitride	O^{2-}	oxide	F^-	fluoride
C^{4-}	carbide	P^{3-}	phosphide	S^{2-}	sulfide	Cl^-	chloride
Si^{4-}	silicide	As^{3-}	arsenide	Se^{2-}	selenide	Br^-	bromide
				Te^{2-}	telluride	I^-	iodide

Many metals, particularly transition and post-transition metals (metals in groups to the right of the transition metals), can be found in compounds in more than one oxidation state. Iron, a typical example, occurs as either Fe^{2+} or Fe^{3+}. Salts containing different iron ions have different formulas, so it is necessary to specify which iron ion is present when naming them. There are two ways of doing this. In the old system the suffix *-ous* is used to specify the lower oxidation state and *-ic* is used to specify the higher. With this method, we use the Latin stem for elements having symbols not derived from their English names.

Fe^{2+}	ferrous ion	$FeCl_2$	ferrous chloride
Fe^{3+}	ferric ion	$FeCl_3$	ferric chloride
Cu^+	cuprous ion	$CuCl$	cuprous chloride
Cu^{2+}	cupric ion	$CuCl_2$	cupric chloride

A list that contains additional examples is given in Table 8.3. Notice that mercury is an exception—we use the English stem when naming its ions.

Table 8.3

Metals that Form More than One Ion

Cr^{2+}	chromous	Mn^{2+}	manganous	Fe^{2+}	ferrous	Cu^+	cuprous
Cr^{3+}	chromic	Mn^{3+}	manganic	Fe^{3+}	ferric	Cu^{2+}	cupric
Hg_2^{2+}	mercurous[a]	Sn^{2+}	stannous	Pb^{2+}	plumbous		
Hg^{2+}	mercuric	Sn^{4+}	stannic	Pb^{4+}	plumbic		

[a] Notice that Hg_2^{2+} contains two joined Hg^+ ions.

The currently preferred method for naming ions of metals that can have more than one oxidation state in compounds is called the **Stock system.** Here we use the English name followed without a space by the numerical value of the oxidation number written as a Roman numeral in parentheses.

Fe^{2+}	iron(II)	$FeCl_2$	iron(II) chloride
Fe^{3+}	iron(III)	$FeCl_3$	iron(III) chloride
Cr^{2+}	chromium(II)	CrS	chromium(II) sulfide
Cr^{3+}	chromium(III)	Cr_2S_3	chromium(III) sulfide

Remember that the Roman numeral is the oxidation number of the metal; it is *not* necessarily a subscript in the formula. You must figure out the formula from the ion charges as discussed in Chapter Six. Even though the Stock system is now preferred, chemical companies still label bottles of chemicals using the old system. These old names also appear in the older scientific literature, which still holds much excellent data. Unfortunately, this means that you must know both systems.

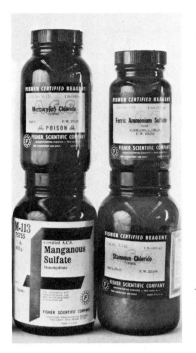

The older system of nomenclature is still used on the labels of many laboratory chemicals.

Alfred Stock (1876–1946), a German inorganic chemist, was one of the first scientists to warn the public of the dangers of mercury poisoning.

Copper(I) sulfate is Cu_2SO_4.
Copper(II) sulfate is $CuSO_4$.

EXERCISE 8.13 Name the compounds K_2S, Mg_3P_2, $NiCl_2$, and Fe_2O_3.

EXERCISE 8.14 Write formulas for these compounds:
(a) aluminum sulfide
(b) strontium fluoride
(c) titanium(IV) oxide
(d) chromous bromide

Binary compounds between two nonmetals

The compounds NO_2 and N_2O_4 would both be named nitrogen(IV) oxide if we followed the Stock system and our rules for assigning oxidation numbers. We can see, therefore, that we need a system that actually specifies the numbers of atoms in a molecule. Such a system has been developed which uses the following Greek prefixes:

mono-	= 1 (often omitted)	hexa-	= 6
di-	= 2	hepta-	= 7
tri-	= 3	octa-	= 8
tetra-	= 4	nona-	= 9
penta-	= 5	deca-	= 10

For example, NO_2 is nitrogen *di*oxide and N_2O_4 is *di*nitrogen *tetr*oxide (we drop the *a* for ease of pronunciation).
Some other examples are:

HCl	hydrogen chloride (*mono-* omitted)
CO	carbon monoxide
$AsCl_3$	arsenic trichloride
SF_6	sulfur hexafluoride

Notice that the element with the more positive oxidation number is named first.

EXERCISE 8.15 Name the following compounds using Greek prefixes when needed: PCl_3, SO_2, and Cl_2O_7.

Binary acids and their salts
In Chapter Six, we saw that binary acids are water solutions of hydrogen compounds of the nonmetals. These acids are named by adding the prefix *hydro-* and the suffix *-ic* to the stem of the nonmetal name followed by the word *acid*. For example, water solutions of hydrogen chloride and hydrogen sulfide are named as follows:

HCl	*hydro*chlor*ic acid*
H_2S	*hydro*sulfur*ic acid*

Salts of these acids contain the negative ion of the nonmetal and always end in *-ide*.

HCl, hydrochloric acid, gives chlor*ide* salts containing Cl^-.
H_2S, hydrosulfuric acid, gives sulf*ide* salts containing S^{2-}.

EXERCISE 8.16 Name the water solutions of these acids: HF, HBr.

Oxoacids and their salts
Oxoacids, such as H_2SO_4 or HNO_3, do not take the prefix *hydro-*. Since many nonmetals form oxoacids in which they have different oxidation numbers, these acids are named according to the particular oxidation state of the nonmetal they contain. If the nonmetal occurs in only two oxoacids, the one in which it has the higher oxidation number takes the suffix *-ic* and the one in which it has the lower oxidation number takes the suffix *-ous*.

H_2SO_4	sulfur*ic acid*	HNO_3	nitr*ic acid*
H_2SO_3	sulfur*ous acid*	HNO_2	nitr*ous acid*

The halogens can occur in as many as four different oxoacids. The oxoacid with the most oxygens has the prefix *per-,* and the one with the least has the prefix *hypo-*.

HClO   *hypo*chlorous acid (this formula is usually written HOCl)
HClO$_2$ chlor*ous* acid
HClO$_3$ chlor*ic* acid
HClO$_4$ *per*chloric acid

The neutralization of oxoacids produces negative polyatomic ions. There is a very simple relationship between the name of the polyatomic ion and its parent acid:

(1) *-ic* acids give *-ate* ions
   HNO$_3$ (nit*ric* acid)   NO$_3^-$ (nit*rate* ion)
(2) *-ous* acids give *-ite* ions
   H$_2$SO$_3$ (sulfur*ous* acid)   SO$_3^{2-}$ (sulf*ite* ion)

EXERCISE 8.17   The formula for arsenic acid is H$_3$AsO$_4$. What is the name of the salt Na$_3$AsO$_4$?

Acid salts

Polyprotic acids may be partially neutralized to give ions that still contain hydrogen. For example, H$_2$SO$_4$ can be partially neutralized to give the HSO$_4^-$ ion. These ions, which are themselves acidic, can be made to form compounds that we call **acid salts;** for example, NaHSO$_4$. These salts are named by specifying the number of hydrogens still present.

NaHSO$_4$ sodium hydrogen sulfate
NaH$_2$PO$_4$ sodium dihydrogen phosphate

For acid salts of diprotic acids, the prefix *bi-* is still often used.

NaHCO$_3$ sodium *bi*carbonate

Notice that the prefix *bi-* does *not* mean two; it means that there is a hydrogen in the compound.

EXERCISE 8.18   What is the formula of sodium bisulfite?

SUMMARY

Covalent Bonds. Electron sharing between atoms occurs when electron transfer is energetically too "expensive." Shared electrons attract the positive nuclei and thereby lead to a lowering of the energy of the atoms. Electrons generally become paired when they are shared. Lewis symbols for covalent bonds are drawn by using dots between chemical symbols to indicate the electrons. A dash can also represent a pair of electrons. An atom tends to share enough electrons to complete its valence shell. Except for hydrogen, the valence shell usually holds eight electrons, an octet. Single, double, and triple bonds involve sharing one, two, and three electron pairs, respectively. Some molecules don't obey the octet rule. Compounds containing Be and B often have less than an octet around these atoms. Other period 2 elements always obey the octet rule.

Elements in periods below period 2 can sometimes exceed an octet, particularly when they form more than four bonds.

Lewis Structures. Lewis structures give structural formulas for molecules and polyatomic ions. To draw a Lewis structure, first decide on the skeletal structure—the central atom is usually written first in the formula. Then count all valence electrons, taking into account the charge on the species (if any). Next, place two electrons in each bond and then add the rest by completing the octets of atoms attached to the central atom. If any electrons are left over, place them on the central atom in pairs. Then, if any atom has an incomplete octet, form double bonds.

Resonance. Two or more atoms of a given element in a molecule are chemically identical if they are attached to the same kinds of atoms or groups of atoms. If the Lewis structure suggests that they are chemically different, the actual structure of the molecule is a resonance hybrid of a number of contributing structures. Bond energy (the energy needed to separate the bonded atoms) and bond length are two experimentally measurable quantities that can be related to the number of pairs of electrons in the bond. For bonds between the same atoms, bond energy increases and bond length decreases going from single to double to triple bonds.

Coordinate Covalent Bonds. For "bookkeeping" purposes, we sometimes single out covalent bonds formed by the sharing of an electron pair that originates on one of the two atoms that are joined. An arrow is sometimes used to indicate the donated pair of electrons in the bond.

VSEPR Theory. Molecular geometry can be predicted by assuming that electron pairs in the valence shell of an atom stay as far apart as possible. Figures 8.5 through 8.8, along with the Lewis structure, generally give the correct structure of a molecule or polyatomic ion.

Polar Molecules and Electronegativity. When two atoms with different electronegativities share electrons, the bond between them is polar. In some polyatomic molecules, these polar bonds cancel to give nonpolar molecules. This occurs when the molecule is symmetrical. In nonsymmetrical molecules, polar bonds do not cancel, and polar molecules result. The extent of polarity is determined by the molecule's dipole moment—the magnitude of the partial charge on each end multiplied by the distance between the partial charges.

Oxidation-Reduction. Oxidation is the loss of electrons; reduction is the gain of electrons. Both always occur together in redox reactions. The substance oxidized is the reducing agent; the substance reduced is the oxidizing agent. Oxidation numbers can be assigned according to rules on pages 247–251. These help us to keep track of electron transfers. An increase in oxidation number is oxidation; a decrease is reduction. Balancing redox equations is accomplished by insuring that the electrons gained are equal to the electrons lost.

Naming Compounds. International agreement between chemists provides a system that allows us to write a single formula from a compound's name. For salts, the Stock system is preferred, but the older system must also be learned. Compounds between nonmetals use Greek prefixes to specify number. Binary acids are *hydro . . . ic acids* and give *-ide* anions. Oxoacids and their anions are related: *-ic acid* produces *-ate* ion; *-ous acid* produces *-ite* ion. Acid salts use the prefix *bi-*, or contain the name, *hydrogen.*

INDEX TO EXERCISES AND QUESTIONS

REVIEW QUESTIONS

8.19. In terms of energy, why doesn't ionic bonding occur when two nonmetals react with each other?

8.20. Describe what happens to the electron density around two hydrogen atoms as they come together to form an H_2 molecule.

8.21. What holds the two nuclei together in a covalent bond?

8.22. What happens to the energy of two hydrogen atoms as they approach each other?

8.23. What factors determine the bond distance in a covalent bond?

8.24. Why is the covalent bond also often called an electron pair bond?

8.25. Is bond formation endothermic or exothermic?

8.26. Use Lewis structures to diagram the formation of (a) Br_2, (b) H_2O, and (c) NH_3.

8.27. Use the octet rule to predict the formula of the simplest compound formed from hydrogen and (a) Se, (b) As, and (c) Si.

8.28. What would be the formula for the simplest compound formed from (a) P and Cl, (b) C and F, and (c) Br and Cl?

8.29. Why do period 2 elements never form more than four

covalent bonds? Why are period 3 elements able to exceed an octet when they form bonds?

8.30. What would be the skeletal structure for (a) $SiCl_4$, (b) PF_3, (c) PH_3, and (d) SCl_2?

8.31. How many dots must appear in the Lewis structures of the molecules in question 8.30?

8.32. Give Lewis structures for the molecules in Question 8.30.

8.33. Give Lewis structures for (a) CS_2, (b) CN^-, (c) SeO_3, and (d) SeO_2.

8.34. Give Lewis structures for (a) HNO_2, (b) $HClO_3$, and (c) H_2SeO_3.

8.35. Give Lewis structures for (a) NO^+, (b) NO_2^-, (c) $SbCl_6^-$, and (d) IO_3^-. Which contain multiple bonds?

8.36. Give Lewis structures for (a) TeF_4, (b) ClF_5, (c) XeF_2 and (d) XeF_4.

8.37. How many resonance structures could be written for the N_2O_4 molecule? Its skeletal structure is

```
O         O
    N  N
O         O
```

8.38. Define bond length and bond energy.

8.39. How should the N—O bond lengths compare in the ions NO_3^- and NO_2^-?

8.40. The energy required to break the H—Cl bond to give H^+ and Cl^- ions would not be called the H—Cl bond energy. Why?

8.41. How are bond energy and bond length related to the number of pairs of electrons shared between atoms in a bond?

8.42. Arrange the following in order of increasing C—O bond length: CO, CO_3^{2-}, CO_2.

8.43. Draw the resonance structures for CO_3^{2-}.

8.44. What is a coordinate covalent bond?

8.45. Once formed, how (if at all) does a coordinate covalent bond differ from an ordinary covalent bond?

8.46. BCl_3 has an incomplete valence shell. Show how it could form a coordinate covalent bond with a water molecule.

8.47. What arrangement of electron pairs are expected when the valence shell of the central atom contains (a) three pairs, (b) six pairs, (c) four pairs, and (d) five pairs of electrons?

8.48. Practice drawing the tetrahedron, trigonal bipyramid, and octahedron.

8.49. What molecular shape is expected when the central atom has in its valence shell (a) three bonding electron pairs and one lone pair, (b) four bonding electron pairs and two lone pairs, and (c) two bonding electron pairs and one lone pair?

8.50. Predict the shape of (a) FCl_2^+, (b) AsF_5, (c) AsF_3, and (d) SeO_2.

8.51. Predict the shape of (a) TeF_4, (b) $SbCl_6^-$, (c) NO_2^-, and (d) PO_4^{3-}.

8.52. Predict the shape of (a) IO_4^-, (b) ICl_4^-, (c) TeF_6, and (d) ICl_2^-.

8.53. What is a polar covalent bond? Define dipole moment.

8.54. Which of the following would be nonpolar molecules: (a) TeF_4, (b) AsF_5, (c) $SiCl_4$, (d) SeO_2, (e) SeO_3, (f) $HgCl_2$?

8.55. Which molecules in Question 8.54 would have polar bonds?

8.56. Define electronegativity.

8.57. Without referring to the table of electronegativity values, use the periodic table to choose the element in each list that has the highest electronegativity.
(a) Si, As, Ge, P
(b) P, Mg, Ba, Sb
(c) B, F, Te, P

8.58. Which of the following bonds is most polar? In each bond, choose the atom that carries the negative charge.
(a) Hg—I (b) P—Cl (c) Si—F (d) Mg—N

8.59. Which pair of elements from the following list should form the most ionic compound: Be, Ca, Sr, S, Se, Te?

8.60. Define oxidation and reduction.

8.61. In the reaction, $H_2 + Cl_2 \rightarrow 2HCl$, which substance is oxidized and which is reduced? Which is the oxidizing agent and which is the reducing agent?

8.62. Assign oxidation numbers to the atoms indicated by boldface type.
(a) **S**$^{2-}$ (b) **S**O_2 (c) **P**$_4$ (d) **P**H_3

8.63. Assign oxidation numbers to the atoms indicated by boldface type.
(a) **Cl**O_4^- (b) **Cr**Cl_3 (c) **Sn**S_2 (d) **Au**$(NO_3)_3$

8.64. Assign oxidation numbers to all of the atoms in the following compounds.
(a) Na_2HPO_4 (b) $BaMnO_4$ (c) $Na_2S_4O_6$ (d) ClF_2

8.65. Based on trends in electronegativity in the periodic table, what oxidation numbers would you assign to the atoms in
(a) BrCl and (b) SCl_2?

8.66. Balance the following redox equations using oxidation numbers.
(a) $HNO_3 + H_3AsO_3 \longrightarrow H_3AsO_4 + NO + H_2O$
(b) $NaI + HOCl \longrightarrow NaIO_3 + HCl + H_2O$
(c) $KMnO_4 + H_2C_2O_4 + H_2SO_4 \longrightarrow CO_2 + K_2SO_4 + MnSO_4 + H_2O$
(d) $H_2SO_4 + Al \longrightarrow Al_2(SO_4)_3 + SO_2 + H_2O$

8.67. Balance the following redox equations, using oxidation numbers.
(a) $K_2Cr_2O_7 + HCl \longrightarrow KCl + CrCl_3 + Cl_2 + H_2O$
(b) $NaIO_3 + NaI + HCl \longrightarrow NaCl + I_2 + H_2O$

(c) $Cu + HNO_3 \longrightarrow Cu(NO_3)_2 + NO + H_2O$
(d) $Cu + HNO_3 \longrightarrow Cu(NO_3)_2 + NO_2 + H_2O$

8.68. Balance the following redox equations using oxidation numbers.
(a) $Cu + H_2SO_4 \longrightarrow CuSO_4 + H_2O + SO_2$
(b) $SO_2 + HNO_3 + H_2O \longrightarrow H_2SO_4 + NO$
(c) $Zn + H_2SO_4 \longrightarrow ZnSO_4 + H_2S + H_2O$
(d) $I_2 + HNO_3 \longrightarrow HIO_3 + NO_2 + H_2O$

8.69. Balance the following equation. Note that the same substance is both oxidized and reduced.

$$I_2 + NaOH \longrightarrow NaI + NaIO_3 + H_2O$$

8.70. Name the following.
(a) CaS (d) Mg_2C
(b) NaF (e) Na_3P
(c) $AlBr_3$ (f) Li_3N

8.71. Name the following using both the old nomenclature system and the Stock system.
(a) $CrCl_3$ (d) Hg_2Cl_2
(b) Mn_2O_3 (e) SnO_2
(c) CuO (f) PbS

8.72. Name the following.
(a) SiO_2 (d) S_2Cl_2
(b) ClF_3 (e) P_4O_{10}
(c) XeF_4 (f) N_2O_5

8.73. The periodate ion has the formula, IO_4^-. What would be the name for the acid, HIO_4?

8.74. Name the following. If necessary, refer to Table 6.3 on page 174.
(a) $NaNO_2$ (d) $NH_4C_2H_3O_2$
(b) K_3PO_4 (e) $BaSO_4$
(c) $KMnO_4$ (f) $Fe_2(CO_3)_3$

8.75. What would be the formula for chromic acid?

8.76. Write formulas for the following.
(a) sodium monohydrogen phosphate
(b) lithium selenide
(c) sodium hydride
(d) chromic acetate
(e) nickel(II) cyanide
(f) manganese(VII) oxide
(g) stannic sulfide
(h) antimony pentafluoride
(i) dialuminum hexachloride
(j) tetraarsenic decaoxide
(k) magnesium hydroxide
(l) cupric bisulfate

8.77. Write formulas for the following.
(a) ammonium sulfide   (g) mercury(II) acetate
(b) chromium(III) sulfate   (h) barium bisulfite
(c) molybdenum(IV) sulfide   (i) silicon tetrafluoride
(d) tin(IV) chloride   (j) boron trichloride
(e) iron(III) oxide   (k) stannous sulfide
(f) calcium bromate   (l) calcium phosphide

8.78. Name the following oxoacids and give the names of the salts formed from them by neutralization with NaOH.
(a) $HOCl$ (b) HIO_2 (c) $HBrO_3$ (d) $HClO_4$

CHAPTER NINE

THE COVALENT BOND AND THE CHEMISTRY OF CARBON

9.1 MODERN THEORIES OF BONDING

Because Lewis structures cannot explain why or how covalent bonds are formed, we must look to other theories based on wave mechanics

In the last chapter we took a simplified look at covalent bonding based on the octet rule and the use of Lewis structures for electron bookkeeping. Lewis structures, however, tell us nothing about *why* covalent bonds are formed, *why* the octet rule is usually obeyed, and *why* it can sometimes be violated. The valence shell electron pair repulsion theory (VSEPR), as useful and accurate as it generally is at predicting molecular shapes, fails to tell us *how* the electron pairs in an atom's valence shell manage to avoid each other. Thus, as helpful and useful as these simple models are, we must look beyond them to understand more fully the covalent bond and the factors that determine molecular geometry.

In this chapter we will examine how atoms share electrons in covalent bonds and how this determines the shapes of molecules. We will also look at organic compounds, one of the most important classes of compounds whose atoms are linked by covalent bonds. These compounds, formed by the element carbon, owe their huge numbers, their chemical properties, and their molecular structures to covalent bonds of different kinds. Indeed, the chemistry of organic compounds could well be called the chemistry of covalent bonds.

Modern theories of bonding are based on the principles of wave mechanics, which gives us the electron configurations of atoms and the descriptions of the shapes of atomic orbitals in atoms. It should not be surprising to learn that these theories describe the bonds between atoms in terms of what happens to atomic orbitals when the atoms are joined together.

There are fundamentally two theories of covalent bonding that have evolved and, in many ways, they complement one another. They are called the **valence bond theory** (or VB theory, for short) and the **molecular orbital theory** (MO theory). The main difference between them is the way in which they construct a theoretical model of the set of bonds in a molecule. The valence bond theory imagines individual atoms, each with their orbitals and electrons, coming together and forming covalent bonds. The molecular orbital theory, on the other hand, doesn't look at how the molecule is formed. It just views a molecule as a collection of positive nuclei surrounded in some way by a set of molecular orbitals, in much the same way that an atom consists of a single nucleus surrounded by a set of atomic orbitals. In their simplest forms, the two theories appear to be rather different. However, it has been found that both theories can be extended and refined to give identical results.

9.2 VALENCE BOND THEORY

The overlap of atomic orbitals provides a way for a pair of electrons to be shared between two nuclei

In Section 8.1 we described the formation of the covalent bond in the hydrogen molecule, H_2. We saw that as two hydrogen atoms approach each other, the electron density shifts toward the region between the two nuclei. In the molecule, both electrons are able to move around both nuclei. Now, let's look at the formation of this covalent bond in terms of orbitals and electrons. In an isolated hydrogen atom, there is one electron in the atom's $1s$ atomic orbital. When a hydrogen molecule is formed, the atomic orbitals of two atoms merge so as to allow the electrons to move back and forth between the two nuclei. Valence bond theory gives us a way of describing how this merging of orbitals occurs.

According to VB theory, a bond between two atoms is formed by the overlap of two atomic orbitals, one provided by each atom. **Overlap** simply means that portions of the two orbitals share the same space. A *maximum* of two electrons, with their spins paired, can be shared between the overlapped orbitals. The strength of the bond thus formed—and therefore the extent to which the energy of the atoms is lowered—is determined, at least in part, by the extent to which orbitals overlap. Because of this, atoms tend to position themselves so that the maximum amount of overlap occurs. As we will see, this behavior is one of the major factors that control the shapes of molecules.

The formation of a hydrogen molecule according to VB theory is shown in Figure 9.1. As the two atoms approach each other, their $1s$ orbitals overlap, thereby giving the H—H bond. The description of the bond in this molecule that is provided by the VB theory is essentially the same as that discussed in Chapter Eight.

Separated H atoms   Overlapping of orbitals   Covalent bond in H_2

Figure 9.1
The formation of the hydrogen molecule according to valence bond theory.

Next, let's look at a molecule that is just a bit more complex than H_2—hydrogen fluoride, HF. Following the rules in Chapter Eight, we would write its Lewis structure as

$$H \colon \overset{\cdot\cdot}{\underset{\cdot\cdot}{F}} \colon$$

and we could diagram the formation of the bond as

$$H\cdot \ + \ \cdot \overset{\cdot\cdot}{\underset{\cdot\cdot}{F}} \colon \ \longrightarrow \ H \colon \overset{\cdot\cdot}{\underset{\cdot\cdot}{F}} \colon$$

The H—F bond is formed by the pairing of electrons—one from hydrogen and one from fluorine. To explain this according to VB theory, we must have two half-filled orbitals—one from each atom—that can be joined by overlap. (They must be half filled because we can't place more than two electrons into the bond.) To see clearly what must happen, it is best to look at the orbital diagrams of the valence shells of hydrogen and fluorine.

The requirements for bond formation are met by overlapping the half-filled 1s orbital of hydrogen with the half-filled 2p orbital of fluorine; there are then two orbitals plus two electrons whose spins can adjust so that they are paired. The formation of the bond is illustrated in Figure 9.2.

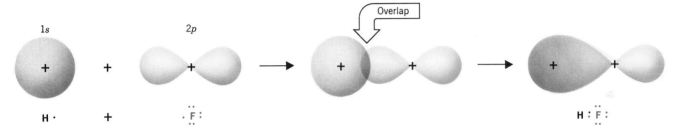

Figure 9.2
The formation of the hydrogen fluoride molecule according to valence bond theory. Only one of fluorine's 2p orbitals is shown.

The overlap of orbitals provides a means for sharing electrons, thereby allowing each atom to complete its valence shell. It is sometimes convenient to indicate this using orbital diagrams. For example, the diagram below shows how the fluorine atom completes its 2p subshell by acquiring a share of an electron from hydrogen.

F (in HF)   ⟨↑↓⟩   ⟨↑↓⟩⟨↑↓⟩⟨↑↓⟩   (colored arrow is the H electron)
               2s   2p

Since the Lewis and VB descriptions of the formation of the H—F bond both account for the completion of the atoms' valence shells, a Lewis structure can be viewed, in a very qualitative sense, as a shorthand notation for the valence bond description of the molecule.

Now, let's turn our attention to a more complicated molecule, hydrogen sulfide, H_2S. This is a nonlinear molecule, and experiment has shown that the H—S—H **bond angle**—the angle formed between the two H—S bonds—is about 92°.

The orbital diagram for sulfur's valence shell is

S   ⟨↑↓⟩   ⟨↑↓⟩⟨↑⟩⟨↑⟩
       3s   3p

Sulfur has two p orbitals that each contain only one electron. Each of these can overlap with the 1s orbital of a hydrogen atom, as shown in Figure 9.3. This overlap completes the valence shell of sulfur because each hydrogen provides one electron.

S (in H_2S)   ⟨↑↓⟩   ⟨↑↓⟩⟨↑⟩⟨↑↓⟩   (colored arrows are H electrons)

    3s   3p

In Figure 9.3, notice that when the 1s orbitals of the hydrogen atoms overlap with the p orbitals of sulfur, the best overlap occurs when the hydrogen atoms are located along the imaginary y and z axes drawn through the center of the sulfur atom. The angle formed between these two H—S bonds—the predicted bond angle—is 90°. This is very close to the actual bond angle of 92° found by experiment. Thus, the VB theory requirement for maximum overlap quite nicely explains the geometry of the hydrogen sulfide molecule.

Two orbitals from different atoms never overlap with opposite ends of the same p orbital simultaneously.

Figure 9.3
Bonding in H_2S. The hydrogen 1s orbitals must locate themselves so that they can overlap with the two partially filled 3p orbitals of sulfur.

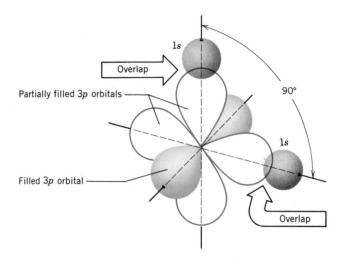

The overlap of p orbitals with each other is also possible. For example, we can consider the bonding in the fluorine molecule, F_2, to occur by the overlap of two 2p orbitals as shown in Figure 9.4. The formation of the other diatomic molecules of the halogens, all of which are held together by single bonds, could be similarly described.

Figure 9.4
The fluorine molecule in valence bond theory. The two completely filled p-orbitals on each fluorine atom are omitted for clarity.

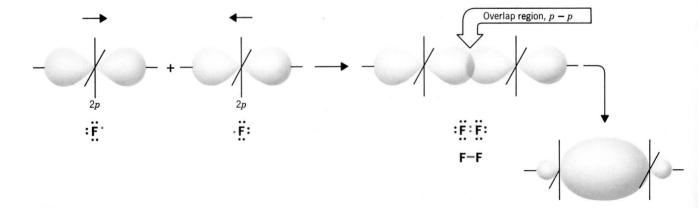

EXERCISE 9.1   Use the principles of VB theory to explain the bonding in HCl.

EXERCISE 9.2   The phosphine molecule, PH_3, has a pyramidal shape with H—P—H bond angles of 93.7°. Explain the bonding in PH_3 using VB theory.

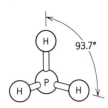

9.3 HYBRID ORBITALS

Mixing atomic orbitals gives new orbitals that are able to form stronger bonds

The approach that we have taken so far has worked well with some simple molecules. Their shapes are explained very nicely by the overlap of simple atomic orbitals. It does not take long, however, to find molecules with shapes and bond angles that fail to fit the model that we have developed. For example, methane, CH_4, has a shape that the VSEPR theory predicts (correctly) to be tetrahedral. The H—C—H bond angles in this molecule are 109.5°. No simple atomic orbitals are oriented at this angle with respect to each other. Therefore, before we can explain the bonds in more complicated molecules such as CH_4, we must take a closer look at how atomic orbitals interact with each other when bonds are formed.

Theoreticians who study wave mechanics have found that when atoms form bonds, their simple s, p, and d orbitals often *mix* to form new atomic orbitals, called **hybrid atomic orbitals.** These new orbitals have new shapes and directional properties. The reason for this mixing can be seen if we look at their shapes.

One kind of hybrid atomic orbital is formed by mixing an s orbital and a p orbital. This creates *two* new orbitals called **sp hybrid orbitals** (the *sp* is used to designate the kinds of orbitals from which the hybrid was formed). Their shapes and directional properties are illustrated in Figure 9.5. Notice that each of the hybrid orbitals has the same shape—each has one large lobe and another much smaller lobe. The large lobe extends further from the nucleus than either the s or p orbital from which the hybrid orbital was formed. This allows the hybrid orbital to overlap more effectively with an orbital on another atom when a bond is formed. In general, the greater the overlap of two orbitals, the stronger the bond. Therefore, hybrid orbitals allow atoms to form stronger, more stable bonds than would be possible if just simple atomic orbitals were used.

At a given internuclear distance, the greater "reach" of a hybrid orbital gives better overlap than either an s or p orbital.

Another point to notice in Figure 9.5 is that the large lobes of the two sp hybrid orbitals point in opposite directions—that is, they are 180° apart. If bonds are formed by overlap of these hybrids with orbitals of other atoms, the other atoms will occupy positions on opposite sides of this central atom. Let's look at a specific example, beryllium hydride, BeH_2.

The mathematics of wave mechanics predicts this 180° angle.

The orbital diagram for the valence shell of beryllium is

Be
      2s   2p

Figure 9.5
sp-Hybridization.

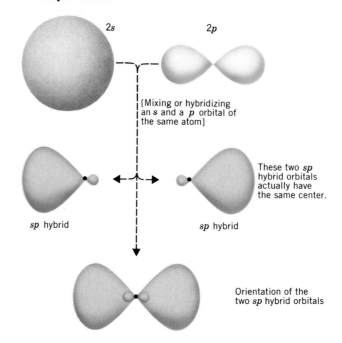

Note that the 2*s* orbital is filled and the three 2*p* orbitals are empty. For bonds to form between beryllium and two hydrogen atoms, the beryllium must have two orbitals that each contain only one electron. Since each hydrogen atom supplies one electron, a bond can't be formed from the *filled* 2*s* orbital of beryllium and the half-filled 1*s* orbital of hydrogen because there would then be three electrons in the bond and the theory only allows two. Therefore, when a beryllium atom forms bonds, its two valence electrons must become unpaired, and the resulting half-filled *s* and *p* orbitals become hybridized.

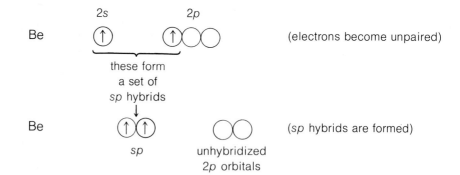

These *sp* hybrid orbitals are said to be *equivalent* because they have the same size, shape, and energy.

Now the 1*s* orbitals of hydrogen can overlap with the *sp* hybrids of beryllium as shown in Figure 9.6. Because the two *sp* hybrid orbitals of beryllium are identical in shape, the two Be—H bonds are identical except for the directions in which they point—we say that the bonds are *equivalent.* Since the bonds point in opposite directions, the molecule is linear. The orbital diagram for beryllium in this molecule is

Be (in BeH₂)   (colored arrows are H electrons)

          sp   unhybridized
                      2*p* orbitals

Figure 9.6
Bonding in BeH$_2$ according to
valence bond theory. Only the
larger lobe of each *sp* hybrid
orbital is shown.

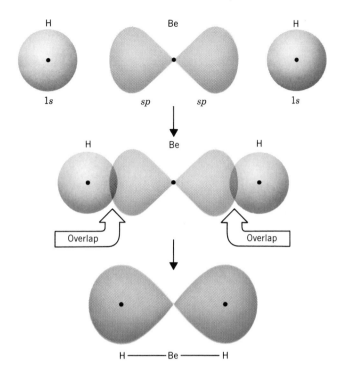

Figure 9.6
Bonding in BeH$_2$ according to
valence bond theory. Only the
larger lobe of each *sp* hybrid
orbital is shown.

Explaining the bonding in BeH$_2$ may seem to be a sort of theoretical "magic show"—a bit like pulling a rabbit out of a hat. It really isn't. A beryllium atom *does* indeed have a 1s^{2}2s^2 configuration and the Be *does* require two singly occupied orbitals with which the 1s orbitals of hydrogen can overlap. Unpairing electrons in the second shell would place one electron in the 2s orbital and one in a 2p orbital. It can be shown mathematically that the s and p orbitals combine by simple addition or subtraction of their electron waves to give the new hybrid orbitals. From a theoretical point of view, all the pieces of the puzzle fit together very well. If it does seem like magic, it is only because the theory must be treated very qualitatively in a course at this level. We use pictures because the actual math is too complex to deal with here.

We have used VB theory and the concept of hybrid orbitals to predict that BeH$_2$ should be a linear molecule. If we draw the Lewis structure for BeH$_2$, H:Be:H, and apply VSEPR theory, we would also conclude that BeH$_2$ is linear. Thus, the VSEPR theory and the VB theory using hybrid orbitals both predict the same shape for this molecule. As we will see, there is a very close correspondence between the structures predicted by VB theory and VSEPR theory, not only for linear molecules in which there are *sp* hybrids, but for molecules with other kinds of hybrids and molecular shapes as well.

Other hybrid orbitals

Hybrid atomic orbitals can be formed by mixing more than just two simple atomic orbitals. If an s orbital and two p orbitals combine, *three* hybrid orbitals, each similar in shape to the *sp* hybrids, are formed. They are called **sp^2 hybrid orbitals,** the superscript 2 specifying the *number* of p orbitals taking part in the formation of the hybrids. (A superscript of 1 is implied for the contribution of the s orbital.) Also, notice that the number of hybrids in the set is equal to the number of simple orbitals from which they are formed. The directional properties of *sp^2* hybrid orbitals are illustrated in Figure 9.7. All three large lobes, which have the same shape, are in the same plane and point toward the corners of a triangle.

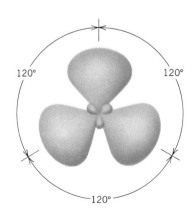

Figure 9.7
Directional properties of *sp^2* hybrid orbitals. The axes of the orbitals are in the same plane and they point at angles of 120°.

Boron trichloride, BCl_3, is a molecule in which the central boron atom uses sp^2 hybrids for bonding. A boron atom has the valence shell configuration

To form three bonds, boron must have three half-filled orbitals, so its 2s electrons must become unpaired. The resulting s and p orbitals then become hybridized.

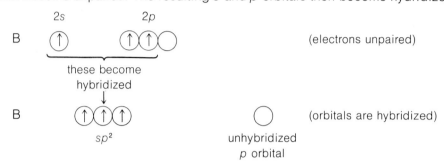

A chlorine atom has the valence shell configuration

$$Cl \quad (\uparrow\downarrow) \quad (\uparrow\downarrow)(\uparrow\downarrow)(\uparrow)$$
$$3s \qquad\qquad 3p$$

The half-filled 3p orbitals of each chlorine atom overlaps with one of the sp^2 hybrids of boron to give

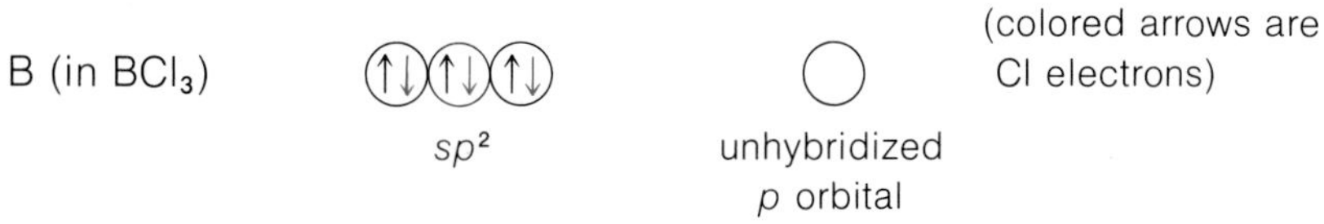

B (in BCl_3) $(\uparrow\downarrow)(\uparrow\downarrow)(\uparrow\downarrow)$ $(\bigcirc)$ (colored arrows are Cl electrons)

sp^2 unhybridized p orbital

These bonds are illustrated in Figure 9.8. Notice that all three bonds appear to be alike—they are equivalent. This agrees with experiment, which shows that all three B—Cl bonds have the same length and the same bond energy. The geometry of the molecule is also explained. Because the sp^2 hybrids point toward the corners of a triangle and because the chlorine atoms must be located so that their p orbitals can overlap with the hybrids, the BCl_3 molecule has a *planar triangular shape*. Once again, this is the same shape that we would predict using VSEPR theory and the Lewis structure of BCl_3.

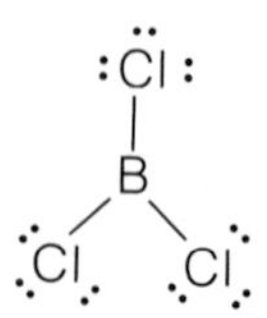

Table 9.1 lists some important types of hybrid orbitals. The directional properties of the various hybrids are shown in Figure 9.9. Let's look at an example showing how this information can be used to explain bonding and geometry.

Table 9.1

Hybrid Orbitals

Hybrid	Orbitals Mixed	Orientation in Space
sp	$s + p$	Linear
sp^2	$s + p + p$	Planar triangular
sp^3	$s + p + p + p$	Tetrahedral
sp^3d	$s + p + p + p + d$	Trigonal bipyramidal
sp^3d^2	$s + p + p + p + d + d$	Octahedral

Figure 9.8
Bonding in BCl₃. The four atoms as well as the bond axes all lie in the same plane.

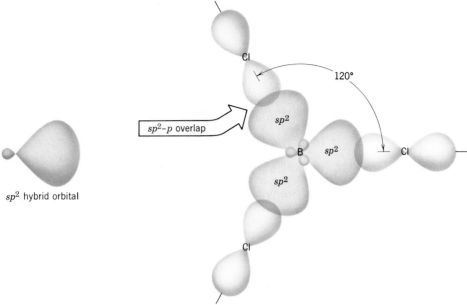

sp^2 hybrid orbital

$sp^2–p$ overlap

120°

Figure 9.9
The orientation of hybrid orbitals in space.

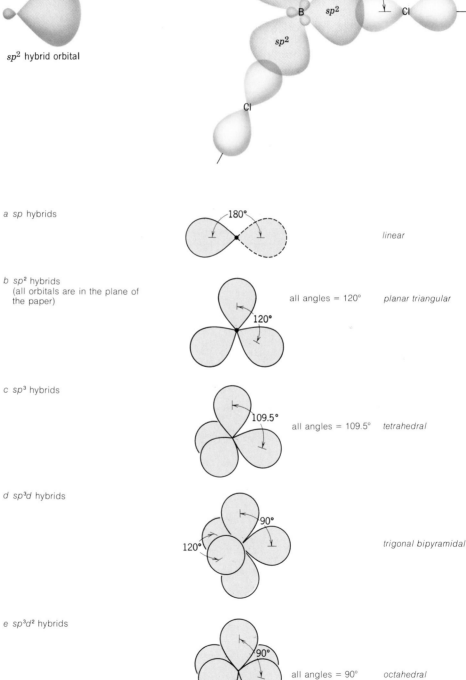

a sp hybrids

180°

linear

b sp² hybrids
   (all orbitals are in the plane of the paper)

all angles = 120°   planar triangular

120°

c sp³ hybrids

109.5°

all angles = 109.5°   tetrahedral

d sp³d hybrids

90°

120°

trigonal bipyramidal

e sp³d² hybrids

90°

all angles = 90°   octahedral

EXAMPLE 9.1 **Explaining Bonding with Hybrid Orbitals**

Problem: What kinds of hybrid orbitals would be found in methane, CH_4? What is the expected geometry of this molecule?

Solution: We begin by looking at the valence shell of carbon.

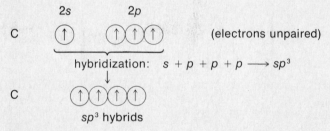

To form four C—H bonds, carbon needs four half-filled orbitals. The 2s electrons must therefore become unpaired, which places one of them in the unoccupied p orbital. Then the orbitals needed for bonding are hybridized.

$$C \quad \begin{array}{cc} 2s & 2p \\ \uparrow & \uparrow\ \uparrow\ \uparrow \end{array} \quad \text{(electrons unpaired)}$$

hybridization: $\quad s + p + p + p \longrightarrow sp^3$

$$C \quad \uparrow\ \uparrow\ \uparrow\ \uparrow$$

sp^3 hybrids

Finally we form the four bonds.

$$C \text{ (in } CH_4) \quad \uparrow\downarrow\ \uparrow\downarrow\ \uparrow\downarrow\ \uparrow\downarrow \quad \text{(colored arrows are H electrons)}$$

Since sp^3 hybrids point to the corners of a tetrahedron, the CH_4 molecule should be tetrahedral. Methane is, in fact, tetrahedral, and all four of its C—H bonds are equivalent.

In methane, carbon forms four single bonds with hydrogen atoms by using sp^3 hybrid orbitals. Carbon uses these same kinds of orbitals in *all* of its compounds in which it is bonded to four other atoms by single bonds. This makes the tetrahedral orientation of atoms around carbon one of the primary structural features of organic compounds, and organic chemists routinely think in terms of "tetrahedral carbon," shown in Figure 9.10.

Figure 9.10
The tetrahedral carbon. (a) The dashed lines are the axes of the bonds. (b) A ball-and-stick model of the CH_4 molecule. (c) A scale model of CH_4 that indicates the relative volumes occupied by the electron clouds.

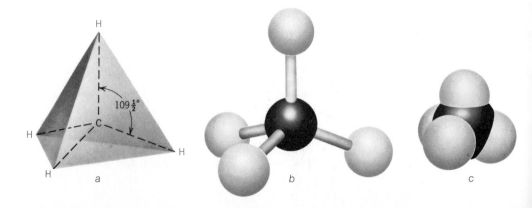

In many compounds, carbon atoms are bonded to other carbon atoms. The simplest such compound that contains only single bonds is called ethane, C_2H_6.

$$
\begin{array}{c}
\ \ \ \ \text{H} \ \ \ \text{H} \\
\ \ \ \ | \ \ \ \ \ | \\
\text{H}-\text{C}-\text{C}-\text{H} \\
\ \ \ \ | \ \ \ \ \ | \\
\ \ \ \ \text{H} \ \ \ \text{H}
\end{array}
$$

In this molecule, the carbons are bonded together by the overlap of sp^3 hybrid orbitals, as shown in Figure 9.11. One of the most important characteristics of this bond is that the overlap of the orbitals in the C—C bond is hardly affected at all if one portion of the molecule rotates relative to the other about the bond axis. Such rotation, therefore, is said to occur freely. This free rotation permits different possible relative orientations of the atoms in the molecule. These different relative orientations are called **conformations.** With complex molecules, the number of possible conformations is enormous. For example, Figure 9.12 on page 272 illustrates just three of the many possible conformations of pentane, C_5H_{12}, one of the low molecular weight organic compounds found in gasoline.

$$
\begin{array}{c}
\ \ \ \ \text{H} \ \ \ \text{H} \ \ \ \text{H} \ \ \ \text{H} \ \ \ \text{H} \\
\ \ \ \ | \ \ \ \ \ | \ \ \ \ \ | \ \ \ \ \ | \ \ \ \ \ | \\
\text{H}-\text{C}-\text{C}-\text{C}-\text{C}-\text{C}-\text{H} \ \ \ \text{(pentane)} \\
\ \ \ \ | \ \ \ \ \ | \ \ \ \ \ | \ \ \ \ \ | \ \ \ \ \ | \\
\ \ \ \ \text{H} \ \ \ \text{H} \ \ \ \text{H} \ \ \ \text{H} \ \ \ \text{H}
\end{array}
$$

Now let's look at an example of a molecule in which an atom must have more than eight electrons in its valence shell when it forms bonds.

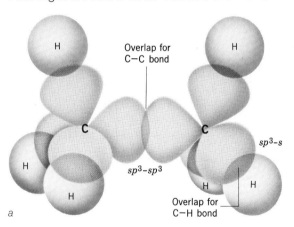

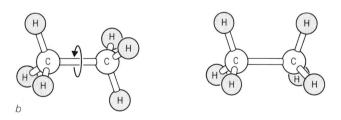

Figure 9.11
The bonds in the ethane molecule. (*a*) overlap of orbitals (*b*)
The degree of overlap of the sp^3 orbitals in the carbon-carbon
bond is not appreciably affected by the rotation of the two

$$
\begin{array}{c}
\ \ \ \ \text{H} \\
\ \ \ \ | \\
\text{H}-\text{C}- \text{ groups relative to each other about that bond.} \\
\ \ \ \ | \\
\ \ \ \ \text{H}
\end{array}
$$

Figure 9.12
Just three of the innumerable conformations of the carbon chain in pentane, C_5H_{12} (hydrogen atoms not shown). Free rotation about single bonds makes these different conformations possible.

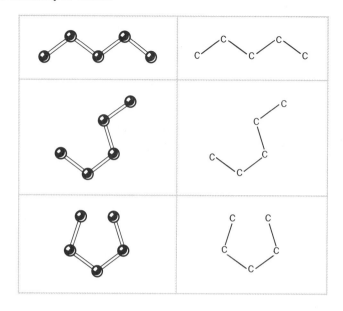

EXAMPLE 9.2   **Explaining Bonding with Hybrid Orbitals**

Problem: What kind of hybrid orbitals would be used by sulfur in sulfur hexa-fluoride? What would be the geometry of SF_6?

Solution: Once again we look at the configuration of the valence shell of the central atom.

To form six bonds, all of sulfur's valence electrons must be used. But how can six half-filled orbitals be formed? The answer is that we must dip into sulfur's unoccupied $3d$ subshell. Every atom has a $3d$ subshell, whether or not it is occupied. A free sulfur atom has an unoccupied $3d$ subshell that is fairly close in energy to its $3s$ and $3p$ subshells. When necessary (as in the formation of SF_6), sulfur can utilize its $3d$ orbitals to form hybrids. For sulfur its orbital diagram can be written

There is no $2d$ subshell—that is why period 2 elements never exceed an octet.

Unpairing all of the electrons to give six half-filled orbitals, followed by hybridization, gives six equivalent sp^3d^2 hybrid orbitals.

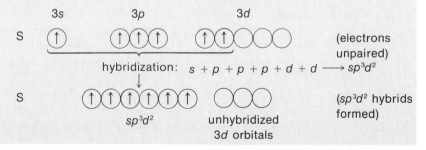

hybridization: $s + p + p + p + d + d \longrightarrow sp^3d^2$

Finally, the six S—F bonds are formed. The half-filled $2p$ orbital of each of the six fluorine atoms overlaps with one of the sp^3d^2 hybrids.

S (in SF$_6$) 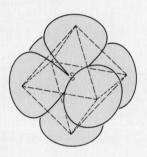 (colored arrows are F electrons)

The sp^3d^2 hybrids point to the corners of an octahedron, which explains the octahedral shape of the SF$_6$ molecule.

We have seen that if we know the kind of hybrid orbitals that the central atom uses in a molecule, we can predict the geometry of the molecule. But this can also be viewed from the other direction. *If we know the geometry of a molecule, we can predict the kind of hybrid orbitals that are used.* Since VSEPR theory works so well in predicting geometry, we can use it to help us obtain VB descriptions of bonding. For example, the Lewis structures of CH$_4$ and SF$_6$ are

In CH$_4$, there are four electron pairs around carbon. The VSEPR model tells us that they should be arranged tetrahedrally. The only hybrid orbitals that are tetrahedral are sp^3 hybrids, and we have seen that they explain the structure of this molecule well. Similarly, VSEPR theory tells us that the six electron pairs around sulfur should be arranged octahedrally. In Table 9.1, the only octahedrally oriented hybrids are sp^3d^2—the sulfur in the SF$_6$ molecule must use these hybrids.

EXERCISE 9.3   What kind of hybrid orbitals would be used by the central atom (underlined) in (a) $\underline{Si}H_4$ and (b) $\underline{P}Cl_5$.

Hybridization in molecules that have lone pairs of electrons

Methane, CH$_4$, is a tetrahedral molecule with sp^3 hybridization of the orbitals of carbon and H—C—H bond angles that are each equal to 109.5°. In ammonia, NH$_3$, the H—N—H bond angles are 107°, and in water the H—O—H bond angle is 104.5°. Both NH$_3$ and H$_2$O have H—X—H bond angles that are close to the bond angles expected for a molecule whose central atom has sp^3 hybrids. The use of sp^3 hybrids by oxygen and nitrogen, therefore, is often used to explain the geometry of H$_2$O and NH$_3$.

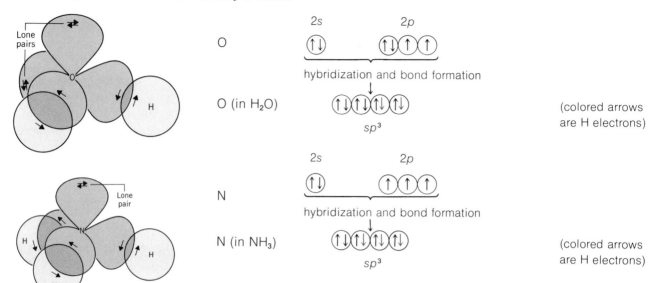

According to these descriptions, not all of the hybrid orbitals of the central atom must be used for bonding. Lone pairs (unshared pairs) of electrons can be accommodated in them too.

EXAMPLE 9.3 **Explaining Bonding with Hybrid Orbitals**

Problem: What kind of hybrid orbitals would sulfur have in sulfur tetrafluoride, SF_4?

Solution: As usual, we begin with the orbital diagram for the central sulfur atom.

In order to form four bonds, four half-filled orbitals must be available on sulfur. To create these, we must again use orbitals from sulfur's empty $3d$ subshell. First, let's rewrite the orbital diagram to show the vacant $3d$ subshell.

It is only necessary to unpair two electrons to give four half-filled orbitals. Then *all* of the orbitals that have electrons in them are hybridized.

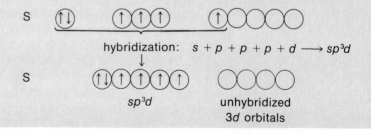

Now, the four S—F bonds can be formed by the overlap of a half-filled 2p orbital from each fluorine atom with one of the half-filled sp^3d hybrids.

S (in SF₄)   ⟨↑↓⟩⟨↑⟩⟨↑⟩⟨↑⟩⟨↑↓⟩   ◯◯◯◯   (colored arrows are F electrons)

        sp^3d 3d

Notice that one of the hybrid orbitals has a lone pair in it.

EXERCISE 9.4   What kinds of hybrid orbitals would the central atom (underlined) use in (a) P̲Cl₃ and (b) C̲lF₃.

9.4 COORDINATE COVALENT BONDS

A filled orbital and an empty orbital can overlap to form a coordinate covalent bond

In Section 8.7 we defined a coordinate covalent bond as a bond in which both the shared electrons are provided by just one of the joined atoms. For example, boron trifluoride, BF_3, can combine with an additional fluoride ion to form the tetrafluoroborate ion, BF_4^-, according to the equation

$$BF_3 + F^- \longrightarrow BF_4^-$$

tetrafluoroborate ion

Using x's for boron's electrons and dots for fluorine's electrons, this reaction can be shown as

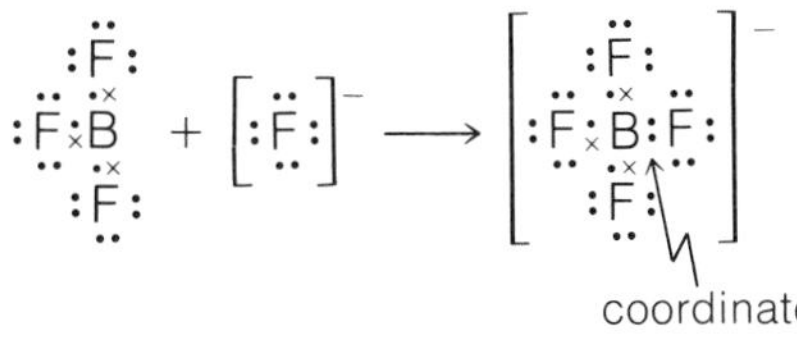

coordinate covalent bond

As we mentioned previously, the coordinate covalent bond is really no different from any other covalent bond once it has been formed. The distinction between them is made *only* for bookkeeping purposes. One place where such bookkeeping is useful is in keeping track of the orbitals and electrons used when atoms bond together.

The VB theory requirements for bond formation—two overlapping orbitals sharing two electrons—can be satisfied in two ways. One, as we have seen, is overlapping two half-filled orbitals. This gives us an "ordinary" covalent bond. The other is overlapping one filled orbital with one empty orbital. The shared pair of electrons is donated by the atom with the filled orbital and a coordinate covalent bond is formed.

EXAMPLE 9.4   **Explaining Coordinate Covalent Bonds**

Problem: Explain the bonding and predict the geometry for BF_4^-.
Solution: The orbital diagram for boron is

            B

                2s   2p

To form four bonds, we need four hybrid orbitals. These can be sp^3 hybrids formed from the 2s and 2p subshells. Boron has enough electrons to half fill only three of the hybrids. This gives

B   ⟨↑⟩⟨↑⟩⟨↑⟩⟨ ⟩
sp^3

Boron forms three ordinary covalent bonds with fluorine atoms plus one coordinate covalent bond with a fluoride ion.

B (in BF_4^-)   ⟨↑↓⟩⟨↑⟩⟨↑↓⟩⟨↑↓⟩
sp^3

— coordinate covalent bond —
both electrons from F^-
(colored arrows are F electrons)

Finally, we should expect the BF_4^- ion to have a tetrahedral shape because that is the geometry of sp^3 hybrids.

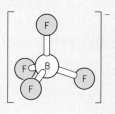

EXERCISE 9.5   What hybrid orbitals are used by phosphorus in PCl_6^-? Draw the orbital diagram for phosphorus in PCl_6^-. What is the geometry of PCl_6^-?

9.5   DOUBLE AND TRIPLE BONDS

The "sideways" overlap of p orbitals allows more than one pair of electrons to be shared between two atoms

The types of overlap of orbitals that we have described so far produces bonds in which the electron density is concentrated most heavily between the nuclei of the two atoms along an imaginary line that joins their centers. Any bond of this kind, whether formed from the overlap of s orbitals, p orbitals, or hybrid orbitals (Figure 9.13) is called a **sigma bond** or σ-**bond** (σ is the Greek letter *sigma*).

There is another way that p orbitals can overlap, which produces a bond in which the electron density is concentrated in *two* separate regions that lie on opposite sides of the imaginary line joining the two nuclei. This bond, called a **pi-bond** or π-**bond,** is shown in Figure 9.14. Notice that a π-bond, like a p orbital, consists of two parts. Each region of electron density makes up only half of a π-bond; it takes *both* of them to make up one π-bond.

The formation of a π-bond allows two atoms to share more than one pair of electrons between them. We can use ethene (also called ethylene), C_2H_4, as an example. The Lewis structure for this molecule contains a double bond.

Polyethylene, a common plastic, is made from C_2H_4.

$$\begin{array}{ccc} H & & H \\ \diagdown & & \diagup \\ & C{=}C & \\ \diagup & & \diagdown \\ H & & H \end{array}$$

We can explain the bonding in this molecule in the following way. First, the skeletal structure of the molecule—the molecular framework—is assembled by connecting the atoms together with σ-bonds that are formed by the overlap of hybrid orbitals. The number of hybrid orbitals needed by each atom depends on

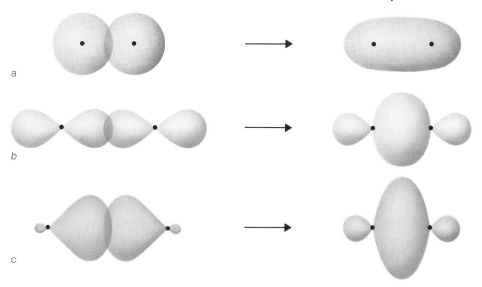

Figure 9.13
Sigma bonds. (*a*) From the overlap of *s*-orbitals. (*b*) From the end
to end overlap of *p*-orbitals. (*c*) From the overlap of hybrid
orbitals.

Figure 9.14
Formation of a π bond. Two
p-orbitals overlap sideways instead
of end to end. The electron density
is concentrated above and below
the bond axis.

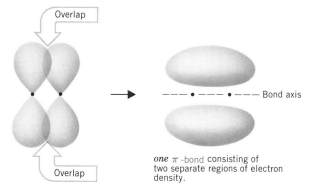

Overlap

Overlap

– – – • – – – • – – – Bond axis

one π -bond consisting of
two separate regions of electron
density.

the number of bonds it must form. In the case of ethene, each carbon atom is
bonded to three other atoms—two hydrogens and the other carbon. This means
that each carbon atom needs *three* hybrid orbitals for the σ-bonds. These can be
formed from the 2*s* orbital and two of the 2*p* orbitals.

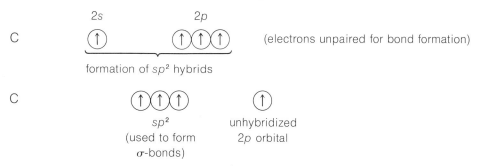

C — 2*s* — 2*p* — (electrons unpaired for bond formation)

formation of *sp*² hybrids

C — *sp*² (used to form σ-bonds) — unhybridized 2*p* orbital

Notice that the carbon atom has an unpaired electron in an unhybridized *p* orbital. This *p* orbital is oriented perpendicular to the triangular plane of the *sp*² hybrid orbitals, as shown in Figure 9.15 on page 278.

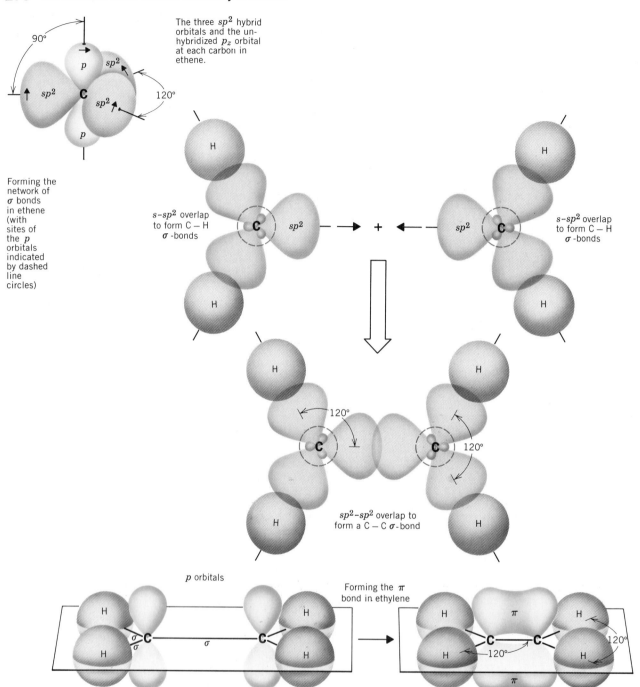

The three sp^2 hybrid orbitals and the unhybridized p_z orbital at each carbon in ethene.

Forming the network of σ bonds in ethene (with sites of the p orbitals indicated by dashed line circles)

s–sp^2 overlap to form C — H σ-bonds

s–sp^2 overlap to form C — H σ-bonds

sp^2–sp^2 overlap to form a C — C σ-bond

p orbitals

Forming the π bond in ethylene

Figure 9.15
The carbon-carbon double bond. (Adapted from J. R. Holum *Organic and Biological Chemistry,* 1978. John Wiley & Sons, New York.)

The C—H σ-bonds in C_2H_4 are formed by the overlap of the sp^2 orbitals of carbon with the $1s$ orbitals of hydrogen. The C—C σ-bond is formed by overlap of one sp^2 orbital from each of the two carbon atoms (also shown in Figure 9.15). Finally, the remaining unhybridized p orbitals, one from each carbon atom, also overlap to produce a π-bond.

In C_2H_4, the predicted bond angles are all 120°, which is the angle between the sp^2 hybrid orbitals. The actual bond angles are, in fact, very close to this. Our

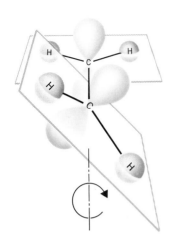

Figure 9.16
Restricted rotation about a double bond. Rotation of the CH_2 closest to us relative to that at the rear causes the unhybridized p-orbitals to become misaligned, thereby destroying the π bond.

description of the double bond, which requires the sideways overlap of the unhybridized p orbitals, also predicts that all six atoms should be in the same plane, which they are. In addition, notice that the electron pairs in the σ- and π-bonds successfully avoid each other, which minimizes their mutual repulsions. The σ-bond electrons are concentrated along the bond axis, and the π-bond electrons are concentrated in spaces above and below the bond axis.

One of the most important properties of double bonds is that rotation of one portion of a molecule relative to the rest about the axis of the double bond occurs only with great difficulty. In Figure 9.16 we see that as one $=C \big\langle {}^H_H$ group is rotated relative to the other about the carbon-carbon bond, the unhybridized p orbitals are no longer oriented so that they can overlap. This destroys the π-bond. In effect, rotation about a double bond involves bond breaking. In general, breaking a bond requires a large amount of energy, more than is available from the average kinetic energy of the molecules at room temperature. At room temperature, therefore, rotation around a double bond doesn't take place. As we will see, this property is particularly important in organic chemistry.

In almost every instance, a double bond consists of a σ-bond and a π-bond. Another example is the compound ethanal, also called formaldehyde (the substance used as a preservative for biological specimens and as an embalming fluid). The Lewis structure of this compound is

$$\begin{array}{c} H \\ \diagdown \\ C=\ddot{O} \\ \diagup \\ H \end{array}$$

As with ethene, the carbon forms sp^2 hybrids, leaving an unpaired electron in an unhybridized p orbital.

C (↑)(↑)(↑)   (↑)
          sp^2 p

The oxygen can also form sp^2 hybrids, placing electron pairs in two of them and an unpaired electron in the third. This means that the remaining unhybridized p orbital also has an unpaired electron.

        2s   2p
O   (↑↓)   (↑↓)(↑)(↑)   (ground state of oxygen)
        └─────────────────┘
       formation of sp^2 hybrids

O   (↑↓)(↑↓)(↑)   (↑)
                  sp^2 unhybridized
                                       p orbital

Figure 9.17 shows how the carbon, hydrogen, and oxygen atoms come together to form the molecule. As before, the basic framework of the molecule is formed by the σ-bonds. These determine the molecular shape. The carbon-oxygen double bond also contains a π-bond from the overlap of the unhybridized p orbitals.

Now let's look at a molecule containing a triple bond. An example is ethyne, C_2H_2, which is probably better known as acetylene (a gas used as a fuel for welding torches). This molecule has the Lewis structure

$$H—C≡C—H$$

In acetylene, each carbon needs two hybrid orbitals to form two σ-bonds—one to a hydrogen atom and one to the other carbon atom. These can be provided by mixing the 2s and one of the 2p orbitals to form sp hybrids. To help us visualize

Forming the network of sigma-bonds in methanal

$s-sp^2$ overlap to form C—H σ-bonds

Carbon

Oxygen

Outline of unhybridized p_z orbital

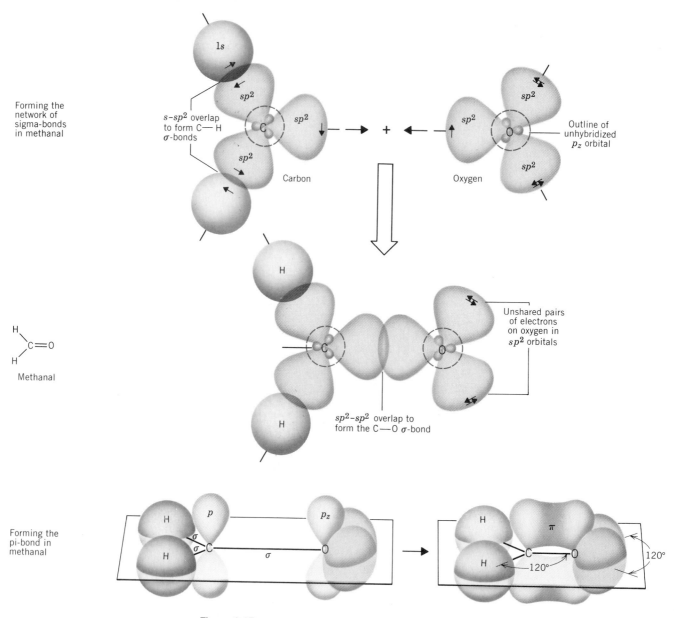

H

H

$$\begin{array}{c} H \\ {}^{}_{} \quad C = O \\ H \end{array}$$

Methanal

H

H

Unshared pairs of electrons on oxygen in sp^2 orbitals

sp^2-sp^2 overlap to form the C—O σ-bond

Forming the pi-bond in methanal

Figure 9.17
The carbon-oxygen double bond in methanal (formaldehyde).

Labeling the orbitals p_x, p_y, and p_z is just for convenience—they are really all equivalent.

the bonding, we will imagine that there is an *xyz* coordinate system centered at each carbon atom and that it is the $2p_x$ orbital that becomes mixed in the hybrid orbitals.

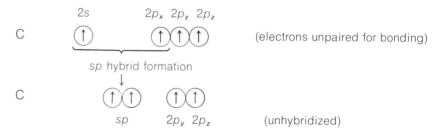

C (↑) 2s   (↑)(↑)(↑) $2p_x$ $2p_y$ $2p_z$ (electrons unpaired for bonding)

sp hybrid formation

C (↑)(↑) *sp*   (↑)(↑) $2p_y$ $2p_z$ (unhybridized)

Figure 9.18 shows how the molecule is formed. The *sp* orbitals point in opposite directions and are used to form the σ-bonds. The unhybridized $2p_y$ and $2p_z$ or-

Figure 9.18
The carbon-carbon triple bond in ethyne (acetylene).

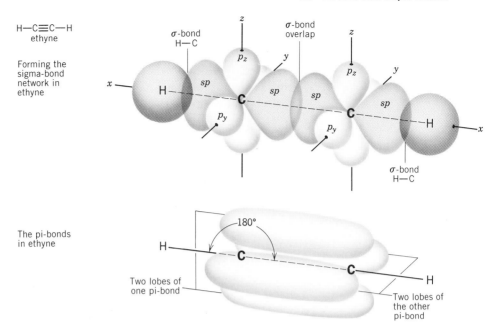

H—C≡C—H
ethyne

Forming the sigma-bond network in ethyne

The pi-bonds in ethyne

bitals are perpendicular to the C—C bond axis and overlap sideways to form *two* separate π-bonds, which surround the C—C σ-bond. Notice that we now have three pairs of electrons in three bonds—one σ-bond and two π-bonds—whose electron densities are concentrated in different places. The three electron pairs therefore manage to avoid each other as much as possible. Also notice that the use of sp hybrid orbitals for the σ-bonds allows us to predict that the molecule will be linear, so all four atoms should lay on the same straight line. This is, in fact, the structure that has been found for acetylene experimentally.

Similar descriptions can be used to explain the bonding in other molecules that have triple bonds. Figure 9.19, for example, shows how the nitrogen molecule, N_2, is formed. In it, too, the triple bond is composed of one σ-bond and two π-bonds.

Figure 9.19
The triple bond in nitrogen, N_2.

:N≡N:

Nitrogen

The sigma bond in N_2 is made by the overlap of two sp hybrid orbitals. Each nitrogen has an unshared pair of electrons in its other sp orbital.

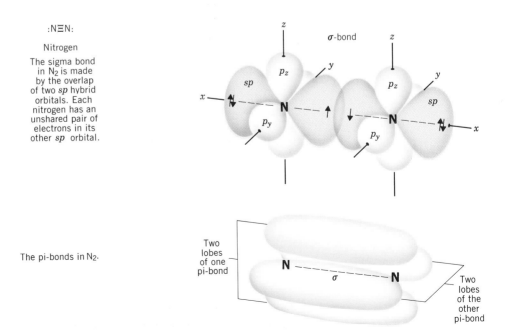

The pi-bonds in N_2.

9.6 MOLECULAR ORBITAL THEORY

The overlap of atomic orbitals produces two kinds of "molecular orbitals"—bonding orbitals, which help stabilize a molecule, and antibonding orbitals, which can destabilize a molecule

Molecular orbital theory takes the view that a molecule is really not too much different from an atom. They are different, of course, because a molecule has several positive nuclei instead of only one. They are similar, however, in that both an atom and a molecule have energy levels that correspond to various orbitals that can be populated with electrons. In atoms, these orbitals are called atomic orbitals; in molecules, they are called **molecular orbitals.** (We will frequently call them MO's.)

In general, the actual shapes and energies of the molecular orbitals of a molecule cannot be determined exactly. Nevertheless, theoreticians have found that reasonably good estimates of their shapes and energies can be obtained by combining the electron waves corresponding to the atomic orbitals of the atoms that make up the molecule. These waves interact by constructive or destructive interference just like other waves that we've seen. Their intensities are either added or subtracted when the atomic orbitals overlap. The way this occurs can be seen if we look at the overlap of a pair of 1s orbitals from two atoms in a molecule like H_2, shown in Figure 9.20. The *two* 1s orbitals combine when the molecule is formed to give *two* MO's. In one of them the electron density from the two orbitals is added together between the nuclei. This gives a buildup of negative charge and helps hold the nuclei near each other. Such an MO is said to be a **bonding molecular orbital.** In the other MO, cancellation of the electron waves reduces the electron density between the nuclei. The absence of much negative charge between the nuclei allows them to repel each other strongly, so this MO is called an **antibonding molecular orbital.** Antibonding MO's tend to destabilize a molecule when they are occupied.

Both the bonding and antibonding MO's formed by the overlap of s orbitals have their maximum electron density on an imaginary line that passes through the two nuclei. Earlier, we called bonds that have this property sigma bonds. Molecular orbitals like this are also designated as sigma (σ), an asterisk is used to indicate the MO's that are antibonding, and a subscript is written to report which atomic orbitals make up the MO. For example, the bonding MO formed by

> The number of MO's formed is always equal to the number of atomic orbitals that are combined.

Figure 9.20
Interaction of 1s atomic orbitals to produce bonding and antibonding molecular orbitals. These are σ-type orbitals because the electron density is symmetrical around the imaginary line that passes through both nuclei.

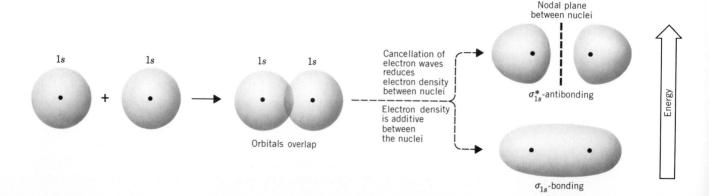

the overlap of 1s orbitals is symbolized as σ_{1s} and the antibonding MO is written as σ_{1s}^*.

Because they stabilize a molecule, bonding MO's are lower in energy than antibonding MO's. This is also depicted in Figure 9.20. When electrons populate molecular orbitals, they fill the lower energy, bonding MO's first. The rules that apply to filling MO's are the same as those that apply to filling atomic orbitals: *electrons spread out over orbitals of equal energy and two electrons can only occupy the same orbital if their spins are paired.*

Let's see how molecular orbital theory can be used to account for the existence of certain molecules, as well as the nonexistence of others. Figure 9.21 is an MO energy level diagram for H_2. The energies of the separate 1s orbitals are indicated at the left and right; those of the molecular orbitals are shown in the center. The H_2 molecule has two electrons, and both can be placed in the σ_{1s} orbital. The shape of this bonding orbital, shown in Figure 9.20, should be familiar. It's the same as the shape of the electron cloud that we described using the valence bond theory.

Figure 9.21
Molecular orbital energy level diagram for H_2.

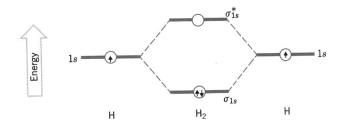

Next, let's consider what happens when two helium atoms come together. Why can't a stable molecule of He_2 be formed? Figure 9.22 is the energy diagram for He_2. Notice that both the bonding and antibonding orbitals are filled. In situations such as this there is a net destabilization because the antibonding MO is raised in energy, relative to the orbitals of the separated atoms, more than the bonding MO is lowered. Thus, the total energy of He_2 is greater than that of two separate He atoms, so the "molecule" immediately comes apart. In general, the effects of antibonding electrons (those in antibonding MO's) cancel the effects of an equal number of bonding electrons, and molecules with equal numbers of bonding and antibonding electrons are unstable. If we remove an antibonding electron from He_2 to give He_2^+, there would be a net excess of bonding electrons. The existence of the He_2^+ ion should therefore be possible. Actually, He_2^+ can be observed when an electric discharge is passed through a helium-filled tube.

He_2^+ has two bonding electrons and one antibonding electron.

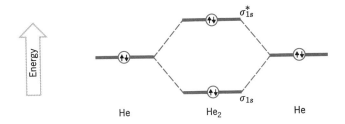

Figure 9.22
Molecular orbital energy level diagram for He_2.

According to the classical description, covalent bonds are formed by the sharing of *pairs* of electrons—one pair gives a single bond, two pairs give a double bond, and three pairs give a triple bond. To translate the MO description into these terms we can compute the **bond order** as

$$\text{bond order} = \frac{(\text{number of bonding electrons}) - (\text{number of antibonding electrons})}{2}$$

For the H_2 molecule, we have

$$\text{bond order} = \frac{2-0}{2} = 1$$

A bond order of 1 corresponds to a single bond. For He_2 we have

$$\text{bond order} = \frac{2-2}{2} = 0$$

A bond order of zero means no bond exists. The He_2 molecule is unable to exist. However, the He_2^+ ion is able to exist because its calculated bond order is $\frac{1}{2}$.

Diatomic molecules of period 2

The outer shell of a period 2 element consists of $2s$ and $2p$ subshells. When atoms of this period bond to each other, the atomic orbitals of these subshells interact strongly to produce molecular orbitals. The $2s$ orbitals, for example, overlap to form σ_{2s} and σ_{2s}^* molecular orbitals having essentially the same shapes as the σ_{1s} and σ_{1s}^* molecular orbitals, respectively. Figure 9.23 shows the shapes of the bonding and antibonding MO's produced when the $2p$ orbitals overlap. If we label those that point toward each other as $2p_x$, a set of bonding and antibonding MO's are formed that we can label as σ_{2p_x} and $\sigma_{2p_x}^*$. The $2p_y$ and

Figure 9.23

Production of molecular orbitals by the overlap of p-orbitals. (a) Two p_x orbitals that point at each other give bonding and antibonding σ-type MO's. (b) Perpendicular to the $2p_x$ orbitals are $2p_y$ and $2p_z$ orbitals that overlap to give two sets of bonding and antibonding π-type MO's.

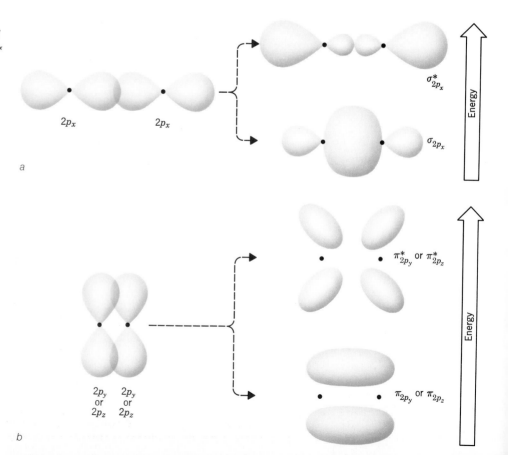

$2p_z$ orbitals, which are perpendicular to the $2p_x$ orbitals, then overlap sideways to give π-type molecular orbitals. They are labeled π_{2p_y} and $\pi^*_{2p_y}$ and π_{2p_z} and $\pi^*_{2p_z}$.

The approximate relative energies of the MO's formed from the second shell orbitals are shown in Figure 9.24. Using this energy diagram, we can predict the electronic structures of diatomic molecules of period 2. These *MO electron configurations* are obtained using the same rules that are applied to the filling of atomic orbitals in atoms.

1 No more than two electrons, with spins paired, can occupy any orbital.
2 Electrons fill the lowest energy orbitals that are available.
3 Electrons spread out as much as possible with spins unpaired over orbitals that have the same energy

Figure 9.24

Approximate relative energies of molecular orbitals in second period diatomic molecules.

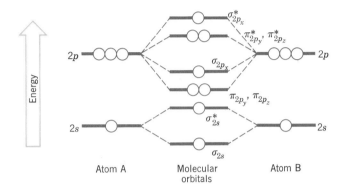

Applying these rules to the valence electrons of period 2 atoms gives the MO electron configurations shown in Table 9.2. Let's see how well the MO theory performs by examining data that is available for these molecules.

According to Table 9.2, MO theory predicts that molecules of Be_2 and Ne_2 should not exist at all because they have bond orders of zero. In beryllium vapor and in gaseous neon, no evidence of Be_2 or Ne_2 has ever been found. The MO

Table 9.2
Molecular Orbital Populations and Bond Orders for Period 2 Diatomic Molecules

	Li_2	Be_2	B_2	C_2	N_2	O_2	F_2	Ne_2
$\sigma^*_{2p_x}$	◯	◯	◯	◯	◯	◯	◯	⇅
$\pi^*_{2p_y}\ \pi^*_{2p_z}$	◯◯	◯◯	◯◯	◯◯	◯◯	↑ ↑	⇅⇅	⇅⇅
σ_{2p_x}	◯	◯	◯	◯	⇅	⇅	⇅	⇅
$\pi_{2p_y}\ \pi_{2p_z}$	◯◯	◯◯	↑ ↑	⇅⇅	⇅⇅	⇅⇅	⇅⇅	⇅⇅
σ^*_{2s}	◯	⇅	⇅	⇅	⇅	⇅	⇅	⇅
σ_{2s}	⇅	⇅	⇅	⇅	⇅	⇅	⇅	⇅
Number of bonding electrons	2	2	4	6	8	8	8	8
Number of antibonding electrons	0	2	2	2	2	4	6	8
Bond order	1	0	1	2	3	2	1	0
Bond energy (kJ/mol)	110	—	300	612	953	501	129	—
Bond length (Å)	2.67	—	1.58	1.24	1.09	1.21	1.44	—

(Energy axis label at left of table; arrow pointing up labeled "Energy")

theory also predicts that diatomic molecules of the other period 2 elements should exist because they all have bond orders greater than zero. These molecules have, in fact, been observed. Although lithium, boron, and carbon are complex solids under ordinary conditions, they can be vaporized, and, in the vapor, molecules of Li_2, B_2, and C_2 can be detected. Nitrogen, oxygen, and fluorine, as you already know, are gaseous elements that exist as N_2, O_2 and F_2.

In Table 9.2, we also see that the predicted bond order increases from boron to carbon to nitrogen and then decreases from nitrogen to oxygen to fluorine. As the bond order increases, the net number of bonding electrons increases, so more electron density is concentrated between the nuclei. This greater concentration of negative charge should bind the nuclei more tightly and therefore give a stronger bond. The attraction between the increased electron density and the positive nuclei should also draw the nuclei closer to the center of the bond, thereby decreasing the bond length. The *experimentally measured* bond energies and bond lengths given in Table 9.2 follow these predictions quite nicely.

Molecular orbital theory is particularly successful in explaining the electronic structure of the oxygen molecule. Experiments have shown that O_2 is paramagnetic; it contains two unpaired electrons. In addition, the bond length in O_2 is about what is expected for an oxygen-oxygen double bond. These data are not explained by valence bond theory. For example, if we write a Lewis structure for O_2 that shows a double bond and also obeys the octet rule, all the electrons appear in pairs.

Molecular oxygen is attracted weakly by a magnet.

$$:\ddot{O}::\ddot{O}:$$

(not acceptable based on experimental evidence because all electrons are paired)

On the other hand, if we indicate the unpaired electrons, the structure shows only a single bond and doesn't obey the octet rule.

$$:\ddot{O}:\ddot{O}:$$

(not acceptable based on experimental evidence because of the O—O single bond)

With molecular orbital theory, we don't have any of these difficulties. According to this theory, the two electrons that are placed in the π^* orbitals of O_2 will spread out over these orbitals with their spins unpaired because these orbitals have the same energy. The electrons in the two antibonding π orbitals cancel the effects of two electrons in the two bonding π orbitals, so the net bond order is 2 and the bond is a double bond.

EXERCISE 9.6   The MO energy level diagram for the NO, nitric oxide, molecule is essentially the same as that shown in Table 9.2 for the period 2 diatomic molecules. Indicate which MO's are populated in NO and calculate the bond order for the molecule.

9.7 DELOCALIZED MOLECULAR ORBITALS

When three or more orbitals are arranged so that they form a continuous overlapping sequence, the shared electrons can spread over more than two atoms

One of the least satisfying aspects of the way valence bond theory explains chemical bonding is the need to write resonance structures for certain molecules and ions. In Chapter Eight we discussed this in terms of Lewis structures, which we have since learned are really equivalent to shorthand notations for valence bond descriptions of molecules. For example, let's look at the formate ion, HCO_2^-, which we described in Chapter Eight as a resonance hybrid of the following two structures.

In terms of the overlap of orbitals, VB theory would picture the formate ion as shown in Figure 9.25. The σ-bond framework is formed by overlap of carbon's sp^2 hybrid orbitals with the $1s$ orbital of hydrogen and the p orbitals of the oxygen atoms. As you can see in Figure 9.25, the π-bond can be formed with either oxygen atom, which is how we obtain the two resonance structures.

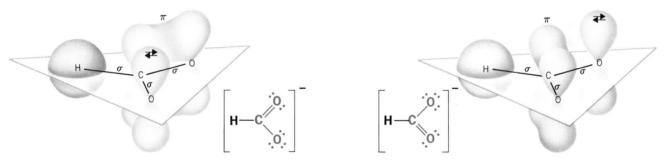

Figure 9.25
Valence bond descriptions of the resonance structures of the
formate ion, HCO_2^-. Only the p orbitals of oxygen that can form
π bonds to carbon are shown. The ion is planar because carbon
uses sp^2 orbitals for form the σ-bond framework.

Molecular orbital theory avoids the problem of resonance by recognizing that electron pairs can sometimes be shared among three or more overlapping orbitals. In the HCO_2^- ion, it allows all *three* p orbitals to overlap so as to form one large π-type molecular orbital that spreads over all three nuclei as shown in Figure 9.26. Since the π-electrons are not required to stay "localized" between just two nuclei, we say that the bond is **delocalized.** Delocalized π-type molecular orbitals permit a single description of the electronic structure of molecules and ions.

Figure 9.26
Molecular orbital theory allows the
electron pair in the π-type bond to
be spread out, or delocalized, over
all three atoms of the CO_2^- unit in
HCO_2^-.

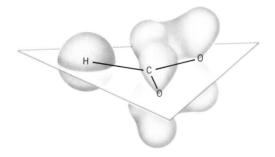

The delocalized nature of π-bonds that extend over three or more nuclei can be indicated using dotted lines rather than dashes when Lewis-type structural formulas are drawn. The formate ion, for example, can be drawn as

$$\left[H-C \underset{O}{\overset{O}{<}} \right]^{-}$$

Delocalized π-bonds are quite common in many kinds of molecules and ions. Some other examples are

$$\left[O \overset{O}{\underset{C}{\parallel}} O \right]^{2-} \qquad O \overset{O}{\underset{S}{\parallel}} O \qquad O \overset{S}{<} O$$

carbonate ion   sulfur trioxide   sulfur dioxide

An important example from the realm of organic chemistry is benzene, C_6H_6. As shown in Figure 9.27, this molecule consists of six carbon atoms arranged in a ring, with a hydrogen atom attached to each carbon. The σ-bond framework of benzene requires that the carbon atoms have sp^2 hybrid orbitals because each carbon forms three σ-bonds. This leaves each carbon atom with a half-filled unhybridized p orbital perpendicular to the plane of the ring. These p orbitals overlap to give a delocalized π-electron cloud that looks something like two donuts with the σ-bond framework sandwiched between them.

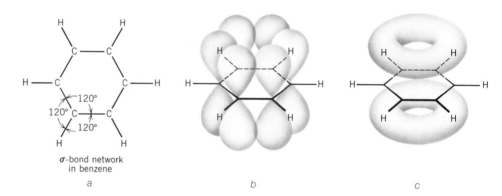

a   *b*   *c*

Figure 9.27
Benzene. (a) The σ-bond framework. All atoms lie in the same plane. (b) The unhybridized p orbitals at each carbon prior to side-to-side overlap. (c) The double donut-shaped space for the π electrons.

An important characteristic of delocalized bonds is that they make a molecule more stable than it would be if it had localized bonds. Benzene, for example, has a very stable ring structure. As we will discover later in this chapter, benzene does not behave as if it has ordinary double bonds, even though double bonds appear in its resonance structures.

(Two resonance structures for benzene)

For simplicity these structures are usually drawn as hexagons. It is assumed that at each corner there is a carbon that is also bonded to a hydrogen.

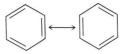

To indicate the delocalized π-electron cloud, the dashes representing the alternating double bonds are replaced with a circle.

9.8 COVALENT COMPOUNDS OF CARBON

Carbon forms strong covalent bonds to other carbon atoms while also forming strong covalent bonds to atoms of other nonmetals

A number of examples in the previous sections, several involving carbon, showed how electrons can be distributed around nuclei to give covalent bonds for holding molecules together. We learned that carbon's electrons can exist in three kinds of hybrid orbitals—sp, sp^2, and sp^3—as well as in unhybridized $2p$ orbitals. These can overlap with similar orbitals (or with s-orbitals) on the atoms of a small number of other elements, particularly atoms of the nonmetals hydrogen, oxygen, and nitrogen. Which nonmetal atoms are joined to which and by what kinds of bonds—single, double or triple—makes a great chemical and biological difference inside and outside the body. The consequences vary from the unusual chemical stability of polyethylene plastics to the devastating effects of ozone on plants, animals, and human lungs; from the gratifying decay resistance of nylon to the equally satisfying digestibility of proteins.

Proteins are complicated, however. If you are also taking a biology course (or have peeked into a classmate's biology text) you already have looked at the rather awesome molecular structures of proteins and other biological chemicals. It is wiser to start with what is simpler. What we will do here is to find out among simple molecules what difference it makes chemically to have π-bonds as well as σ-bonds; to have the closed circuit π-electron network of a benzene molecule; or to have a very polar covalent bond from carbon to oxygen or to nitrogen. (All these features occur in proteins, incidentally.)

Organic chemistry
With very few exceptions, the compounds of carbon are classified as **organic compounds.** All the rest are inorganic compounds, and the carbon compounds traditionally left in this class are a small number of salts such as the carbonates, bicarbonates, cyanides, cyanates, and carbides as well as the gaseous oxides of carbon, CO and CO_2. The word "organic," implying an organism, suggests life, and until about 150 years ago living things or their remains were the only sources of organic compounds. Since then, however, a few million organic compounds have been identified, and many have been made in the laboratory from inorganic carbon compounds. Organic compounds include most plastics and synthetic or natural fibers, nearly all dyes and drugs, the ingredients in most cosmetics and perfumes, and nearly all insecticides and herbicides. Among the naturally occurring organic compounds are the chemicals of heredity, the nucleic acids, as well as all foods, vitamins, and hormones. Thus, organic compounds are everywhere, and we must now ask, what makes so many possible?

Uniqueness of carbon

Carbon is like no other element. Its atoms, unlike all other kinds, not only bond to their own kind—other carbon atoms—but can do so theoretically in an unlimited extension of one carbon atom after another. A molecule found in gasoline (octane, for example) has seven consecutive, strong carbon-carbon σ-bonds.

octane, C_8H_{18}

Similar molecules, differing in the numbers of carbon atoms per molecule, are in diesel oil, kerosene, lubricating oils, and paraffin wax. In the molecules of the familiar plastic, polyethylene, hundreds of carbon atoms are joined one after another. Moreover, while carbon atoms bond strongly to each other, they can still form strong covalent bonds to atoms of other nonmetals such as hydrogen, oxygen, nitrogen, sulfur, and the halogens (Group VIIA). Not even silicon atoms have the versatile bonding abilities of carbon atoms, and silicon stands just below carbon in the periodic table.

The covalent bonds between carbon atoms or between carbon and hydrogen are not only strong, they are also nonpolar. The effect is to render compounds like octane and polyethylene exceptionally resistant to chemical attack by virtually all ionic compounds, at least at room temperature. Concentrated acids and bases do not react with octane or polyethylene at room temperature. (This inertness helps to explain the popularity of polyethylene as a container for food or hospital fluids.) Being nonpolar, the molecules in octane have no sites with either $\delta+$ or $\delta-$ charges that electrically charged ions would be attracted to and then react. The chemicals that do attack octane and similar compounds generally do so through neutral particles with unpaired electrons, such as oxygen molecules with two unpaired electrons (Table 9.2). Thus, octane reacts with oxygen, but only at a high temperature, and it burns to carbon dioxide and water.

$$2C_8H_{18} + 25O_2 \longrightarrow 16CO_2 + 18H_2O$$
octane

Functional groups

If all organic compounds were like octane, they would be quite useless as the basis for living things where at least some chemical reactions under mild conditions must occur. Most kinds of organic compounds, however, consist of *polar* molecules with sites of $\delta+$ and $\delta-$ where ions or other polar molecules can attack. Ethyl alcohol, acetic acid, acetone, ethylamine, and phenol all have a number of reactions involving ions or polar molecules. (Don't memorize these names and structures.)

etc.

polyethylene
(a typical chain will continue
for hundreds of carbons)

ethyl alcohol
(grain alcohol)

acetic acid
(in vinegar)

acetone
(fingernail
polish remover)

ethylamine
(in herring brine)

phenol (Lister's
antiseptic)

Each oxygen or nitrogen atom in these compounds bears a $\delta-$ because of their electronegativity. Each atom attached to the oxygen or nitrogen atom, C or H, bears a $\delta+$. Therefore, most of the chemical properties of these compounds involve the small cluster of atoms at oxygen or nitrogen. This cluster in acetic acid includes one C, two O atoms, and one H, $-\overset{\displaystyle O}{\overset{\|}{C}}-O-H$; the H-atom is responsible for the acidity of acetic acid. Acetic acid, for example, reacts as follows with the hydroxide ion, which might be supplied by sodium hydroxide, a base.

$$
H-\underset{\underset{\displaystyle H}{|}}{\overset{\overset{\displaystyle H}{|}}{C}}-\overset{\overset{\displaystyle O}{\|}}{C}-O-H + {}^-O-H \longrightarrow H-\underset{\underset{\displaystyle H}{|}}{\overset{\overset{\displaystyle H}{|}}{C}}-\overset{\overset{\displaystyle O}{\|}}{C}-O^- + H-O-H
$$

acetic acid   acetate ion

Octanoic acid, whose molecules have a long, octanelike portion, reacts in the same way with hydroxide ion because it also has the group that can donate a hydrogen ion, H^+.

$$
H-\overset{H}{\underset{H}{C}}-\overset{H}{\underset{H}{C}}-\overset{H}{\underset{H}{C}}-\overset{H}{\underset{H}{C}}-\overset{H}{\underset{H}{C}}-\overset{H}{\underset{H}{C}}-\overset{H}{\underset{H}{C}}-\overset{O}{C}-O-H + {}^-OH \longrightarrow H-\overset{H}{\underset{H}{C}}-\overset{H}{\underset{H}{C}}-\overset{H}{\underset{H}{C}}-\overset{H}{\underset{H}{C}}-\overset{H}{\underset{H}{C}}-\overset{H}{\underset{H}{C}}-\overset{H}{\underset{H}{C}}-\overset{O}{C}-O^- + H-O-H
$$

octanoic acid   octanoate ion

(Ions like the octanoate ion together with sodium ions make up ordinary soap.) Regardless of whatever else the structure has, if it has the same group as is present in acetic acid or octanoic acid, it will react in the same way with hydroxide ion. Notice that the octanelike section of the octanoic acid molecule didn't change at all. It didn't function in the reaction; only the end group functioned. A small molecular part of a molecule that does enter into chemical reactions is called a **functional group.**

Each kind of functional group defines a family of organic compounds. Acetic acid and octanoic acid are members of the carboxylic acid family. Table 9.3 gives the features of several other functional groups. Our ability to place the several million organic compounds into one or another of just a few families is one of the most important simplifying strategies in all of chemistry. We can classify organic compounds on the basis of structural similarities just as we classify plants and animals. The study of organic chemistry is not the study of individual compounds, taking them one at a time, but a study of the common properties of functional groups and how to change one into another. We can't carry out that study here, of course, so we will give attention to a very few illustrative examples. As goals for what follows, be sure to learn how to read the symbols used to represent structural formulas. (You may need this in a biology course before you get to it in a regular organic course.) Learn how to recognize a few functional groups by name and structure. Learn how to name the members of the alkane family, because the rules for naming them require only very simple modifications to apply to most other families. Finally, give some attention to those reactions that involve an oxidation, a reduction, or the presence of water either as a reactant or as a product. These kinds of reactions are among the most common in living things. As a necessary background to compounds with functional groups we have to learn about those that have none, the alkanes, one of the kinds of hydrocarbons.

Table 9.3
Some Important Classes of Organic Compounds

Class	Characteristic Structural Features of Molecules	Example
Hydrocarbons	Contain only carbon and hydrogen; may have carbon chains or carbon rings. Subclasses according to presence of multiple bonds	
	Alkanes: all single bonds	CH_3CH_3, ethane
	Alkenes: at least one double bond	$CH_3CH=CH_2$, ethene
	Alkynes: at least one triple bond	$HC\equiv CH$, ethyne
	Aromatic: at least one benzene-like ring system	benzene
Alcohols	At least one —OH joined to a tetrahedral carbon	CH_3CH_2OH, ethanol
Aldehydes	$\overset{O}{\overset{\|}{-C}}-H$ or aldehyde group	$CH_3\overset{O}{\overset{\|}{C}}H$, ethanal
Ketones	$-\overset{\|}{\underset{\|}{C}}-\overset{O}{\overset{\|}{C}}-\overset{\|}{\underset{\|}{C}}-$ or keto group	$CH_3\overset{O}{\overset{\|}{C}}CH_3$, propanone
Carboxylic acids	$\overset{O}{\overset{\|}{-C}}-OH$ or carboxyl group	$CH_3\overset{O}{\overset{\|}{C}}OH$, ethanoic acid
Amines	$-NH_2$, or $-NH-$ or $-\overset{\|}{\underset{\|}{N}}-$	CH_3NH_2, methylamine CH_3NHCH_3, dimethylamine

9.9 HYDROCARBONS

Alkanes and alkenes are two families of hydrocarbons

A small cluster of organic families consists only of carbon and hydrogen, and regardless of the bond types all are called **hydrocarbons.** Figure 9.28 outlines the principal families in this cluster, and the alkanes are the simplest of all.

Figure 9.28
Families of hydrocarbons.

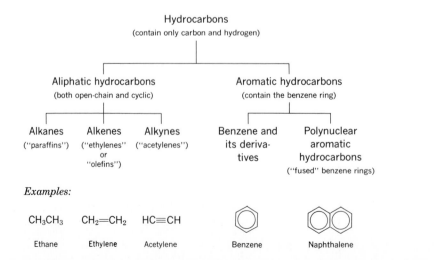

straight chain

branched chain

carbon ring

Alkanes

An **alkane** is a hydrocarbon having only single bonds; and any organic substance, regardless of family, having only single bonds is called a **saturated organic compound.** Our principal sources of hydrocarbons are natural gas, mostly methane, CH_4, and petroleum, a very complex mixture. One of the operations in refining petroleum is to separate it into mixtures of compounds that have boiling points near each other. These mixtures are called *fractions* of the petroleum, and each fraction works well as a solvent or as a fuel for a particular engine. Gasoline, for example, is a petroleum fraction consisting largely of alkanes having five to ten carbons per molecule. Kerosene, jet fuel, and diesel fuel are mixtures of alkanes having ten to eighteen carbons per molecule. The higher the carbon content, the higher the boiling point. Eicosane, $C_{20}H_{42}$, boils at 343 °C and actually is a solid at room temperature with a melting point of 37 °C. It is one compound in the mixture of low melting alkanes in paraffin wax, which is used to seal homemade jellies and to make candles.

The carbon atoms in alkane molecules may be joined one after another in a long string or chain to give **straight-chain** molecules as in octane. The chains may also branch to give **branched chain** structures. Sometimes the chain comes back on itself to close a **ring** of carbon atoms. Combinations of all these features often occur in the same molecule.

Condensed structural formulas

The ten simplest, straight-chain alkanes are listed in Table 9.4, where their structures are shown in *condensed* rather than expanded forms (those that show all bonds). These condensed forms are the usual way of representing organic structures, and we have to learn how to interpret them correctly. To change an expanded structure to its condensed form, merely group together all of the hydrogen atoms attached to a particular carbon atom next to the symbol of that atom, usually to the right. An example will show how this works.

Table 9.4
Straight-Chain Alkanes

Name	Number of Carbon Atoms	Molecular Formula[a]	Condensed Structural Formula	Bp (°C at Atmospheric Pressure)	Mp (°C)	Density (in g/ml at 20 °C)
Methane	1	CH_4	CH_4	−161.5		
Ethane	2	C_2H_6	CH_3CH_3	−88.6		
Propane	3	C_3H_8	$CH_3CH_2CH_3$	−42.1		
Butane	4	C_4H_{10}	$CH_3CH_2CH_2CH_3$	−0.5	−138.4	
Pentane	5	C_5H_{12}	$CH_3CH_2CH_2CH_2CH_3$	36.1	−129.7	0.626
Hexane	6	C_6H_{14}	$CH_3CH_2CH_2CH_2CH_2CH_3$	68.7	−95.3	0.659
Heptane	7	C_7H_{16}	$CH_3CH_2CH_2CH_2CH_2CH_2CH_3$	98.4	−90.6	0.684
Octane	8	C_8H_{18}	$CH_3CH_2CH_2CH_2CH_2CH_2CH_2CH_3$	125.7	−56.8	0.703
Nonane	9	C_9H_{20}	$CH_3CH_2CH_2CH_2CH_2CH_2CH_2CH_2CH_3$	150.8	−53.5	0.718
Decane	10	$C_{10}H_{22}$	$CH_3CH_2CH_2CH_2CH_2CH_2CH_2CH_2CH_2CH_3$	174.1	−29.7	0.730

[a] The molecular formulas of the open-chain alkanes fit the general formula C_nH_{2n+2}, where *n* is the number of carbons in the molecule.

EXAMPLE 9.5 **Condensing Structural Formulas**

Problem: Represent the expanded structure of isooctane by a condensed structure (Isooctane is the reference standard for a rating of 100 octane for gasoline.)

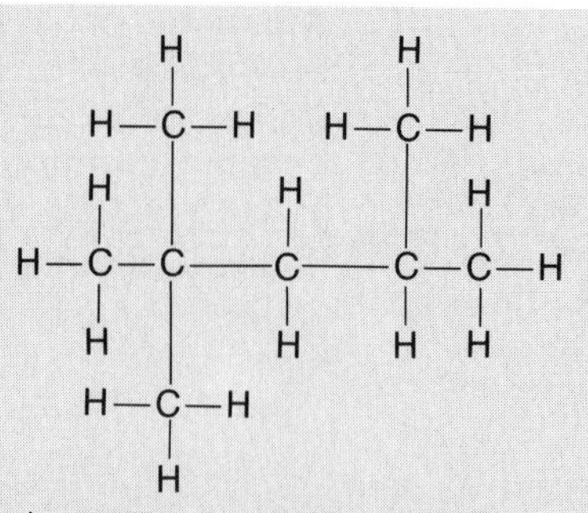

isooctane (expanded structure)

Solution: For any carbon holding three hydrogen atoms, write CH_3. For any carbon with two hydrogens, write CH_2. For any with just one hydrogen, write CH. In other words, we let all carbon-hydrogen bonds be "understood."

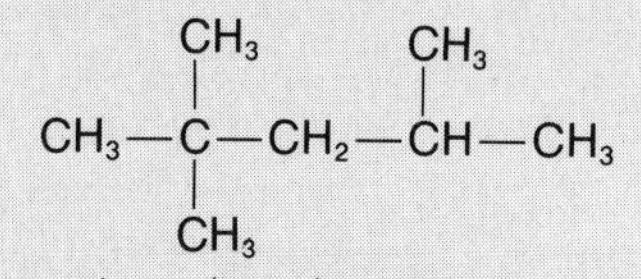

isooctane (condensed structure)

We don't even have to show carbon-carbon single bonds when they fall on the horizontal line through the molecule. For example, we could further condense the structure of isooctane as follows:

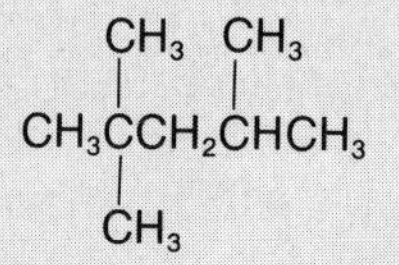

isooctane (condensed structure)

EXERCISE 9.7 Condense each of these structures.

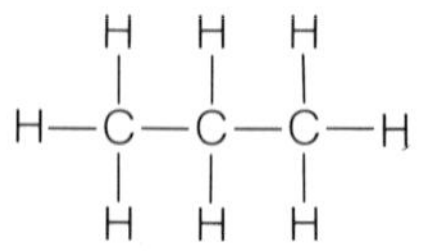

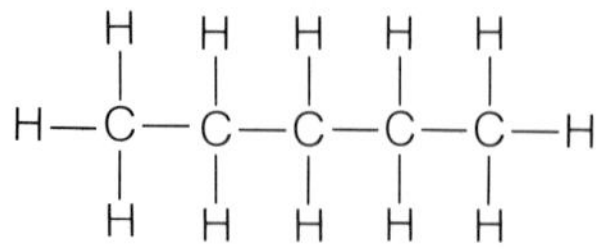

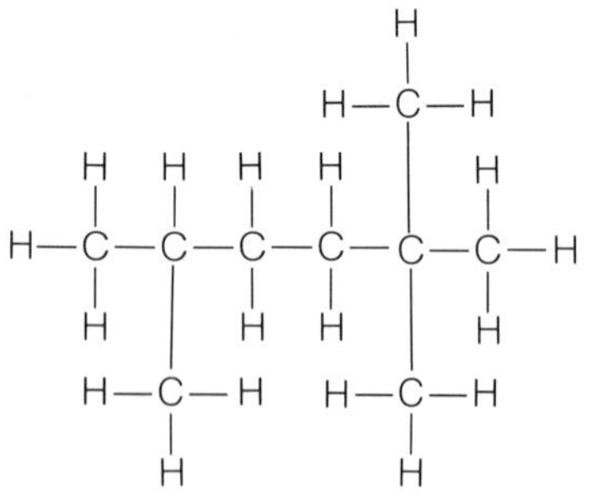

Cyclic compounds are so common that special condensed structural formulas for the rings of atoms in cyclic systems are widely used. When carbon atoms are the only atoms making up the ring, the ring is represented simply by a polygon. For example, cyclobutane is represented only by a square, cyclopentane by a regular pentagon, cyclohexane by a hexagon, and so forth. The progressively more condensed structures of cyclobutane are as follows (and chemists almost never use the first one shown).

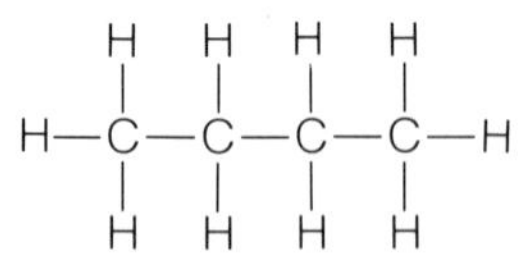

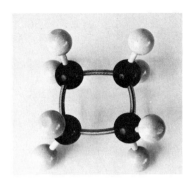

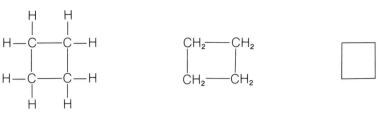

structures for cyclobutane

When a square is used to stand for cyclobutane, be careful to interpret each corner as having one carbon atom; only this carbon is "understood." Then each line forming one side of the polygon stands for a single bond from one carbon (one corner) to the next. Thus, the following segment of cyclobutane:

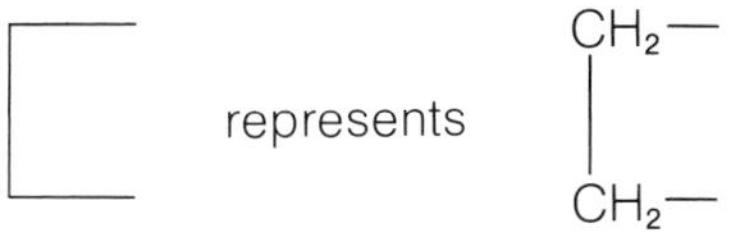

⬜ represents

$$\begin{array}{c} CH_2- \\ | \\ CH_2- \end{array}$$

Not only are the carbon atoms at the corners of rings understood *but so also are the hydrogen atoms attached to them.* Thus, the progressively more condensed structures of methylcyclopentane are as follows.

Cyclobutane.

Methylcyclopentane.

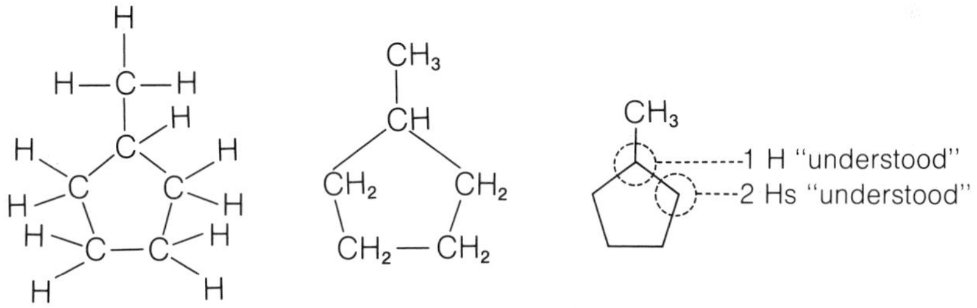

structures for methylcyclopentane

EXAMPLE 9.6   Understanding Condensed Structures

Problem: To make sure you can read condensed structures, expand this one:

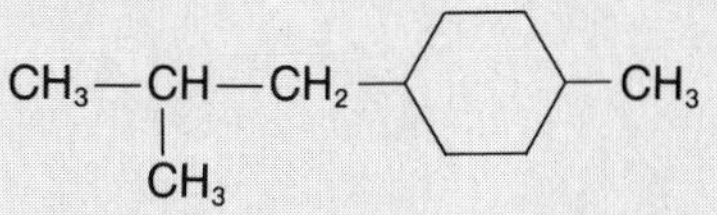

$$CH_3-CH-CH_2-\bigcirc-CH_3$$
$$\quad\quad\quad | $$
$$\quad\quad CH_3$$

Solution:

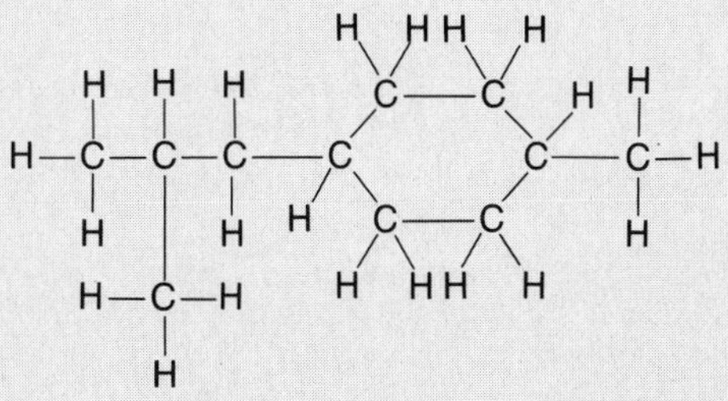

Give particular attention to the carbon atoms of the ring. They, like all carbon atoms in organic compounds, must have four covalent bonds. Those bonds that do not extend to neighboring atoms of the ring or to substituents on the ring are assumed to be holding as many hydrogens as needed to make a total of four bonds.

EXERCISE 9.8   Expand each of these structures.

a.

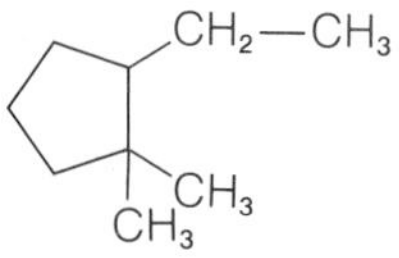

b.

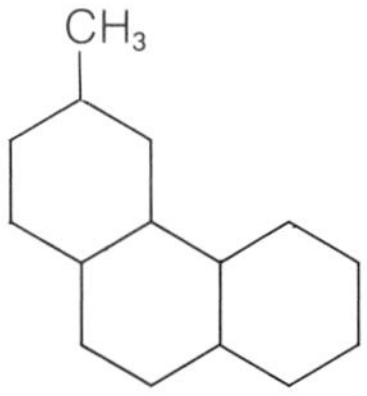

Showing the correct bond angles using either expanded or condensed structures is difficult, but leaving correct geometries to the informed imagination is perfectly acceptable. We know that at an sp^3 hybridized carbon the bond angles are 109.5°, for example, but we usually don't have to make a point of it.

When molecules have double or triple bonds, their parts of the structures must always show them. For example, the expanded structure of dipentene, used to give a lemon scent to waxes and polishes, condenses as follows.

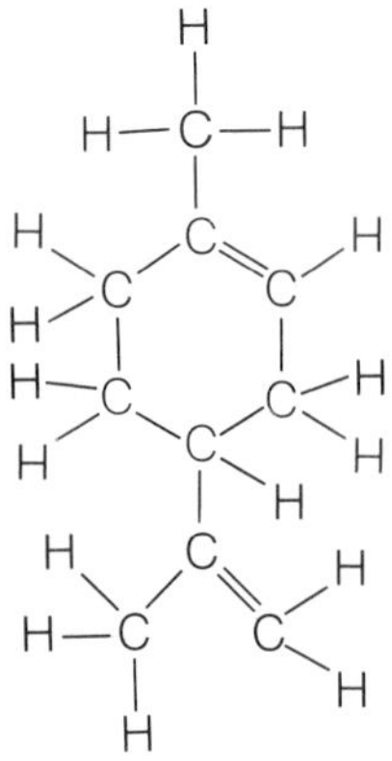

dipentene (expanded
structure)

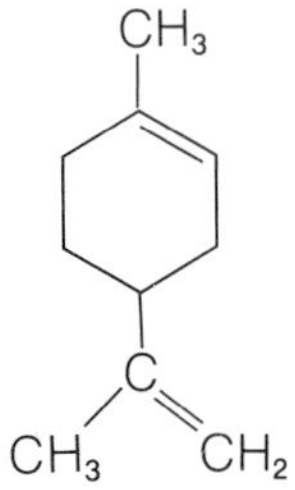

dipentene
(condensed structure)

Naming the alkanes

If the names of each of the several million organic compounds were chosen by whim, no one could remember them all. To handle this important problem, chemists have worked out a set of rules for systematically creating names from structural features. Presently a committee of the International Union of Pure and Applied Chemistry (IUPAC) supervises this work, and its recommendations are called the **IUPAC rules.** According to the IUPAC rules, the names of alkanes should be created as follows.

1 Use the ending, *-ane,* for all alkanes.
2 Attach a prefix to this ending to indicate the number of carbon atoms in the molecule. The prefixes are

meth-	1C	hex-	6C
eth-	2C	hept-	7C
prop-	3C	oct-	8C
but-	4C	non-	9C
pent-	5C	dec-	10C

(Prefixes for 11C and higher are also known.)

The names in Table 9.4 illustrate the use of these prefixes.
3 If the alkane is branched, consider it has having formed from the alkane corresponding to the longest continuous chain of carbons in the mole-

cule. For example, the branched alkane,

$$CH_3-CH_2-\overset{\overset{\displaystyle CH_3}{|}}{CH}-CH_2-CH_2-CH_2-CH_3$$

is viewed as coming from

$$CH_3-CH_2-CH_2-CH_2-CH_2-CH_2-CH_3 \quad \text{(heptane)}$$

by replacing a hydrogen on the third carbon from the left with a CH_3 group.

$$CH_3-CH_2-\overset{\overset{\displaystyle CH_3 \quad H}{\diagup}}{CH}-CH_2-CH_2-CH_2-CH_3 \longrightarrow CH_3-CH_2-\overset{\overset{\displaystyle CH_3}{|}}{CH}-CH_2-CH_2-CH_2-CH_3$$

The longest continuous chain of carbons in a branched chain is called the "parent" chain, and the parent's name in the above example is *heptane.*

4 To specify where the branch is, number the parent chain from whichever end reaches the carbon holding the branch soonest. For example, the correct numbering is

$$\underset{1}{CH_3}-\underset{2}{CH_2}-\underset{3}{\overset{\overset{\displaystyle CH_3}{|}}{CH}}-\underset{4}{CH_2}-\underset{5}{CH_2}-\underset{6}{CH_2}-\underset{7}{CH_3}$$

The location of the first branch must be given the lower of the two possible numbers.

Had we numbered from right to left, the carbon with the CH_3 branch would have been number 5, not 3.

5 Determine the correct name for each branch. Any branch consisting of only carbon and hydrogen and only single bonds is called an **alkyl group.** Each alkyl group is an alkane molecule minus one of its hydrogens, and the names of all alkyl groups end in *-yl.* The prefix is related to the name of the alkane itself. For example,

$$H-\overset{\overset{\displaystyle H}{|}}{\underset{\underset{\displaystyle H}{|}}{C}}-H \xrightarrow{\text{remove one H}} H-\overset{\overset{\displaystyle H}{|}}{\underset{\underset{\displaystyle H}{|}}{C}}- \quad \text{or} \quad CH_3-$$

methane   methyl

$$H-\overset{\overset{\displaystyle H}{|}}{\underset{\underset{\displaystyle H}{|}}{C}}-\overset{\overset{\displaystyle H}{|}}{\underset{\underset{\displaystyle H}{|}}{C}}-H \xrightarrow{\text{remove one H}} H-\overset{\overset{\displaystyle H}{|}}{\underset{\underset{\displaystyle H}{|}}{C}}-\overset{\overset{\displaystyle H}{|}}{\underset{\underset{\displaystyle H}{|}}{C}}- \quad \text{or} \quad CH_3CH_2-$$

ethane   ethyl

Two groups can be obtained from propane, $CH_3CH_2CH_3$:

$$CH_3CH_2CH_2- \quad \text{and} \quad CH_3\overset{}{\underset{|}{CH}}CH_3$$

propyl   isopropyl

Alkyl groups from the higher alkanes are also known, and each has its special name.

6 Use the name of the alkyl group as a prefix to the name of the parent alkane. Place the number locating the alkyl group in front of the resulting name, and separate the number from the name by a hyphen. Here is the correct name of the alkane we've been using as an illustration.

$$CH_3-CH_2-\overset{\overset{\displaystyle CH_3}{|}}{CH}-CH_2-CH_2-CH_2-CH_3$$

3-methylheptane

Propyl group.

Isopropyl group.

7 If two or more alkyl groups are on the parent chain, each must be named and located by number. Hyphens are used to separate numbers from parts of the name. For example.

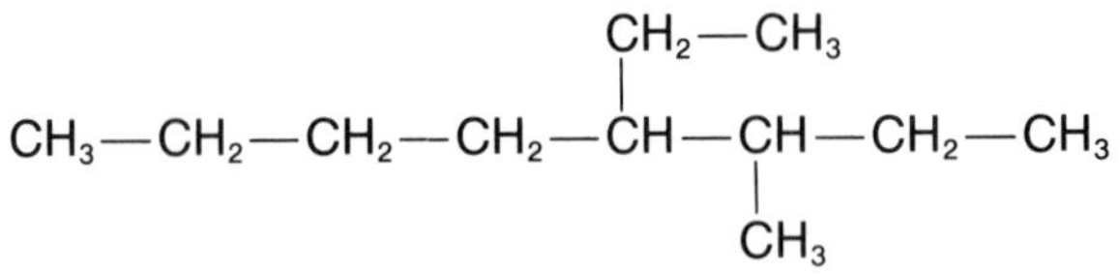

$$CH_3—CH_2—CH_2—CH_2—\overset{\overset{\displaystyle CH_2—CH_3}{|}}{CH}—\underset{\underset{\displaystyle CH_3}{|}}{CH}—CH_2—CH_3$$

Correct name:
3-methyl-4-ethyloctane
Also correct:
4-ethyl-3-methyloctane

IUPAC rules permit us to assemble the names of the alkyl groups either alphabetically or in order of increasing complexity.

8 If two or more alkyl groups on the parent chain are identical, then we use prefixes such as di- (for 2), tri- (for 3), tetra- (for 4), and so forth. The locations of the groups, however, must be still individually specified. Commas are used to separate one number from a number. For example,

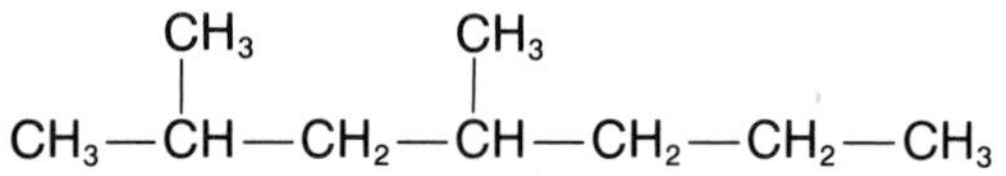

$$CH_3—\overset{\overset{\displaystyle CH_3}{|}}{CH}—CH_2—\overset{\overset{\displaystyle CH_3}{|}}{CH}—CH_2—CH_2—CH_3$$

Correct name:
2,4-dimethylheptane

9 Even when identical groups are on the same carbon, the number of that carbon must be repeated. For example,

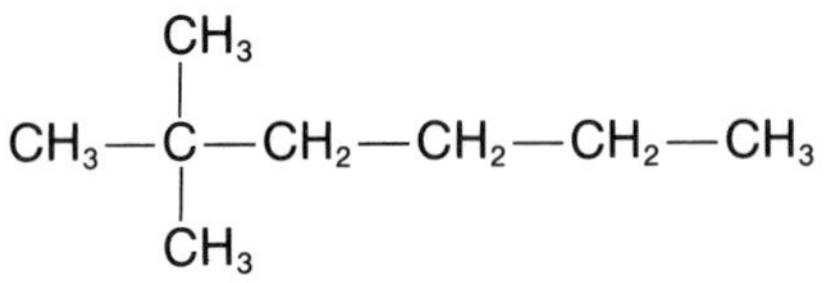

$$CH_3—\overset{\overset{\displaystyle CH_3}{|}}{\underset{\underset{\displaystyle CH_3}{|}}{C}}—CH_2—CH_2—CH_2—CH_3$$

Correct name: 2,2-dimethylhexane
Incorrect: 2-dimethylhexane
Incorrect: 2,2-methylhexane
Incorrect: 5,5-dimethylhexane

The rules we have given here won't cover all situations, but they will meet all our needs. (The definitive IUPAC rules for alkanes may be found in any recent edition of the *Handbook of Chemistry and Physics* [CRC Publishing Company].)

EXAMPLE 9.7 **Naming Alkanes by the IUPAC Rules**

Problem: What is the name of the following alkane?

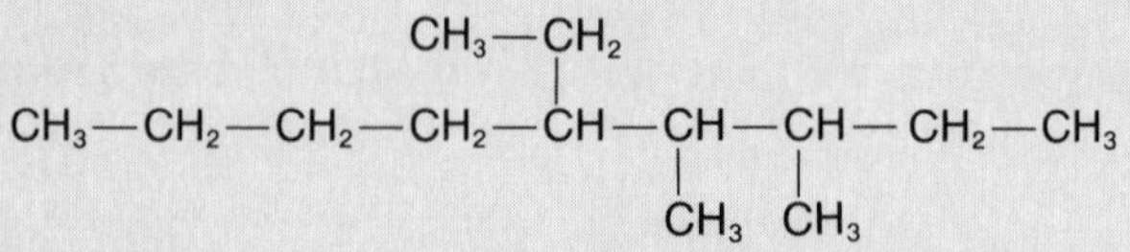

$$CH_3—CH_2—CH_2—CH_2—\overset{\overset{\displaystyle CH_3—CH_2}{|}}{CH}—\underset{\underset{\displaystyle CH_3}{|}}{CH}—\underset{\underset{\displaystyle CH_3}{|}}{CH}—CH_2—CH_3$$

Solution: First, find the parent alkane. It has 9 carbon atoms and the name *nonane.* Next, number the chain. If we number from right to left, we reach the first branch sooner than if we go from left to right. Now name the alkyl groups.

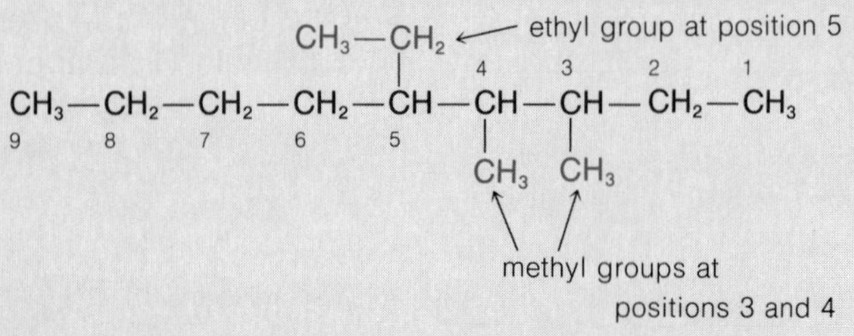

ethyl group at position 5

methyl groups at positions 3 and 4

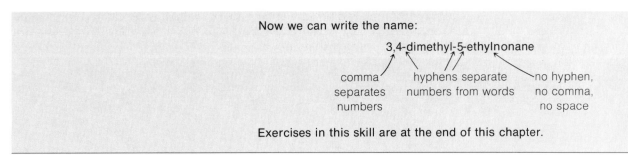

Now we can write the name:

3,4-dimethyl-5-ethylnonane

comma separates numbers

hyphens separate numbers from words

no hyphen, no comma, no space

Exercises in this skill are at the end of this chapter.

Isomerism

Butane and 2-methylpropane both have the same molecular formula, C_4H_{10}.

$$CH_3—CH_2—CH_2—CH_3$$

butane
bp −0.5 °C
mp −138.4 °C
density 0.622 g/ml (20 °C)

$$CH_3—\overset{\overset{\displaystyle CH_3}{|}}{CH}—CH_3$$

2-methylpropane
bp −11.7 °C
mp −159.6 °C
density 0.604 g/ml (20 °C)

Despite this, they have different physical properties. They are different compounds because their molecules have different structures. There are two ways to organize 4 carbon atoms and 10 hydrogen atoms into one molecule as seen in Figure 9.29. Compounds that have the same molecular formula but different structures are called **isomers.** Thus, butane and 2-methylpropane are isomers of each other. The general phenomenon of the existence of isomers is called **isomerism.**

Figure 9.29
The isomers of C_4H_{10}. A ball-and-stick model of butane is on the left and that of 2-methylpropane ("isobutane") is on the right.

Isomerism is one of the major reasons for the existence of so many organic compounds. As the number of carbons per molecule rises, the number of possible isomers becomes astronomical. For example, C_8H_{18} has 18 isomers; $C_{10}H_{22}$ has 75 isomers; $C_{20}H_{42}$ has 366,319 possible isomers (but not all have been made); and $C_{40}H_{82}$, someone has estimated, has 6.25×10^{13} possible isomers.

Because isomers that are in the alkane family have no functional groups and no polar sites, they have very similar chemical properties. When isomerism involves different functional groups, then the differences in both physical and chemical properties can be very large. Ethyl alcohol (bp 78.5 °C), and dimethyl ether (bp -24 °C), two isomers of C_2H_6O, not only have a large difference in boiling points but they display dramatically different reactions. Sodium reacts with ethyl alcohol, for example, but not with dimethyl ether.

$$CH_3-O-CH_3 \qquad \text{dimethyl ether, } C_2H_6O$$

$$2CH_3-CH_2-O-H + 2Na \longrightarrow 2CH_3-CH_2-O^- + 2Na^+ + H_2$$
$$\text{ethyl alcohol}$$
$$C_2H_6O$$

Alkenes

The **alkenes** are hydrocarbons with carbon-carbon double bonds. Each double bond consists of one stronger σ-bond and one weaker π-bond. The IUPAC names and structures of several alkenes are listed in Table 9.5. Some alkenes have two double bonds and are called dienes; some have three double bonds and are trienes, and so forth.

Table 9.5
Some Alkenes

Number of Carbons	Name	Structure	Boiling Point (in °C)
2	ethene	$CH_2{=}CH_2$	-104
3	propene	$CH_2{=}CH-CH_3$	-48
4	1-butene	$CH_2{=}CH-CH_2-CH_3$	-6
	cis-2-butene	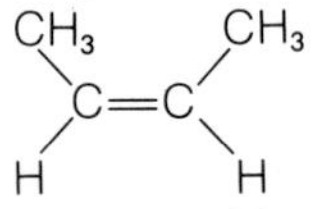	3.72
	trans-2-butene	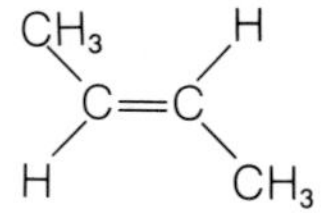	0.88
	1,3-butadiene	$CH_2{=}CH-CH{=}CH_2$	-4.4

IUPAC rules for naming alkenes are adaptations of the rules for alkanes. *The parent chain must include the double bond* even if it isn't the longest chain, and the numbering of the chain must start from whichever end lets us get to the double bond sooner. The name of the parent is that of the corresponding alkane but with an *-ene* instead of an *-ane* ending. Otherwise, the side chain alkyl groups are named and located just as with naming alkanes. Some examples of correctly named alkenes illustrate these rules.

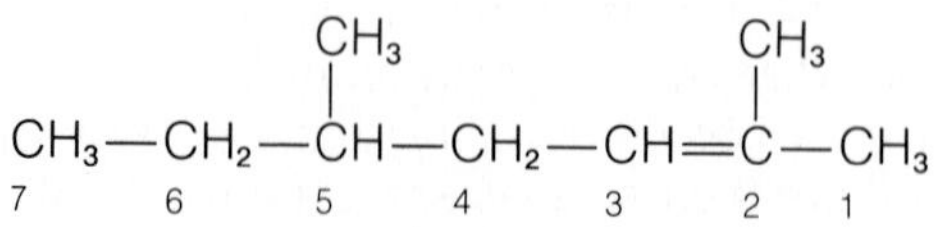

2,5-dimethyl-2-heptene

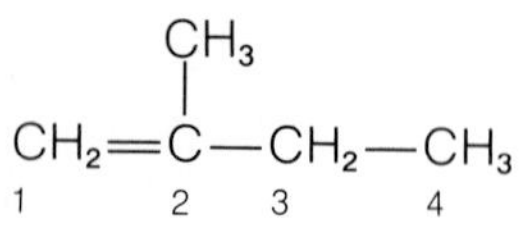

2-methyl-1-butene

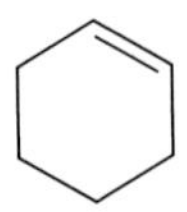

cyclohexene

Notice that the second "2" in 2,5-dimethyl-2-heptene is the number locating the first carbon of the double bond. We don't have to specify the number of the second carbon of the double bond because it cannot be any number other than the next one after 2. When two (or more) double bonds are present, each must be given a number in the name. For example,

$$CH_2{=}CH-CH{=}CH-CH_3$$
1,3-pentadiene

$$CH_2{=}CH-CH{=}CH-CH{=}CH_2$$
1,3,5-hexatriene

Geometric isomerism in alkenes

Two compounds in Table 9.5, *cis*-2-butene and *trans*-2-butene, illustrate a new way in which compounds can be isomers—geometric isomerism. These two alkenes are **geometric isomers** of each other, compounds not only with the same molecular formulas but also the same atom-to-atom sequences and yet different in the directions or geometries with which groups are pointed at a double bond (or at a carbon atom of a ring). As we learned earlier in this chapter, groups joined by a double bond cannot rotate with respect to each other. Rotation would destroy the side to side overlap of the *p*-orbitals that gave the π-bond. So the two methyl groups in 2-butene can be locked into place either on the same side of the double bond axes or on opposite sides. *Cis* means "on the same side" and *trans* means on opposite sides. This difference in geometry is enough to give the two 2-butenes different physical properties. Since both have the same functional group—the double bond—these isomers have the same kinds of chemical properties, however.

Cis-trans (geometric) isomerism is found among many ring compounds, too. For example,

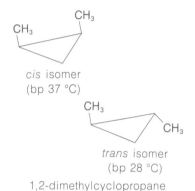

cis isomer
(bp 37 °C)

trans isomer
(bp 28 °C)

1,2-dimethylcyclopropane

Some chemical properties of alkenes

Most reactions of alkenes occur at the double bond, and alkenes have a rich and varied chemistry. Most of their reactions are **addition reactions.** For example, ethene will add hydrogen under special conditions (which we will not discuss).

$$
\begin{array}{c}
\text{H} \\
\end{array}
\underset{\text{ethene}}{
\begin{array}{c}
\text{H} \qquad\quad \text{H}\\
\diagdown\;\diagup\\
\text{C}=\text{C}\\
\diagup\;\diagdown\\
\text{H} \qquad\quad \text{H}
\end{array}}
+\ \text{H}-\text{H} \longrightarrow
\underset{\text{ethane}}{
\begin{array}{c}
\text{H}\ \ \text{H}\\
|\quad\ |\\
\text{H}-\text{C}-\text{C}-\text{H}\\
|\quad\ |\\
\text{H}\ \ \text{H}
\end{array}}
$$

One of the atoms in H_2 becomes attached to the carbon at one end of the double bond and the other hydrogen becomes joined to the other carbon. The π-bond makes this possible, because the electron pair in the π-bond can become the bonding pair in one of the two new sigma C—H bonds. While this is happening the two carbons are still held together by a σ-bond. The electron pair in the molecule of hydrogen becomes the second of the new C—H σ-bonds. By supplying this electron pair, the addition of hydrogen is a *reduction* of the double bond. Thus, the great difference made by a π-bond between two carbons is in opening the molecule to this kind of addition reaction.

Alkene double bonds will also add water molecules. (We will continue to ignore certain special conditions needed for many addition reactions because we want to focus attention on the overall possibilities, not the details.) Water will add to 2-butene as follows.

$$
\text{CH}_3-\text{CH}=\text{CH}-\text{CH}_3 + \text{H}-\text{O}-\text{H} \longrightarrow
\underset{\text{an alcohol}}{
\begin{array}{c}
\text{CH}_3-\text{CH}-\text{CH}-\text{CH}_3\\
|\quad\ \ |\\
\text{H}\ \ \ \text{OH}
\end{array}}
$$

(*cis* or *trans* isomers)

or, more condensed,

$$
\begin{array}{c}
\text{CH}_3\text{CH}_2\text{CHCH}_3\\
|\\
\text{OH}
\end{array}
$$

Don't memorize a specific reaction like this, but give your attention to the *changes* that occur. Another hydrogen appears at a carbon at one end of the double bond, and the —OH group from water goes to the other carbon. Only a single bond remains where the double bond once was. The addition of water to an alkene gives a member of the alcohol family, and this is a standard way of making alcohols.

We noted earlier that alkanes are inert toward concentrated sulfuric acid. Alkenes, in sharp contrast, react vigorously. For example,

$$
\underset{\text{ethene}}{\text{CH}_2{=}\text{CH}_2} + \underset{\text{sulfuric acid}}{
\begin{array}{c}
\text{O}\\
\|\\
\text{H}-\text{O}-\text{S}-\text{O}-\text{H}\\
\|\\
\text{O}
\end{array}}
\longrightarrow
\underset{\text{ethyl hydrogen sulfate}}{
\begin{array}{c}
\text{O}\\
\|\\
\text{CH}_3-\text{CH}_2-\text{O}-\text{S}-\text{O}-\text{H}\\
\|\\
\text{O}
\end{array}}
$$

Notice that one H from sulfuric acid ends up on one carbon of the original double bond (CH_2 changes to CH_3) and all the rest of the sulfuric acid molecule becomes attached to the other carbon. The product can form a sodium salt, and when the alkyl group is very long, these sodium salts are synthetic detergents.

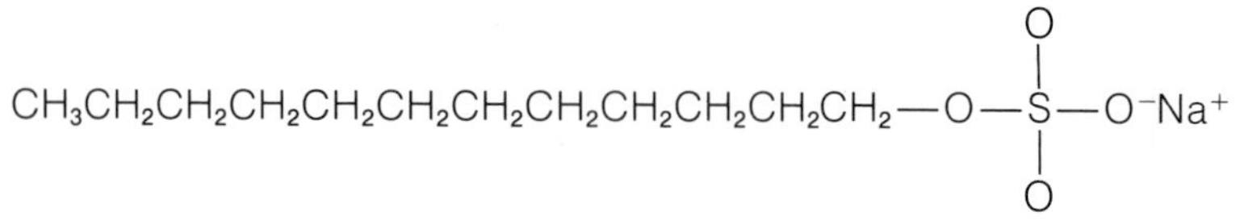

$$CH_3CH_2CH_2CH_2CH_2CH_2CH_2CH_2CH_2CH_2CH_2CH_2-O-\overset{\displaystyle O}{\underset{\displaystyle O}{\overset{\displaystyle \|}{\underset{\displaystyle \|}{S}}}}-O^-Na^+$$

a synthetic detergent

(We will learn more about detergents and how they work in Chapter 19.)

Because the double bond with its π-electrons is richer in electron density than sigma bonds alone can be, it is a functional group susceptible to attack by electron-taking reagents, namely, oxidizing agents. (Remember that an oxidizing agent *takes* electrons from something and becomes reduced.) Ozone, O_3, a constituent in smog, is a powerful oxidizing agent, and anything with alkene double bonds is torn apart by ozone. The reactions are complex and we won't go into the details. The principle is sufficient: electron-rich carbon-carbon double bonds are readily attacked by strong oxidizing agents.

Aromatic hydrocarbons

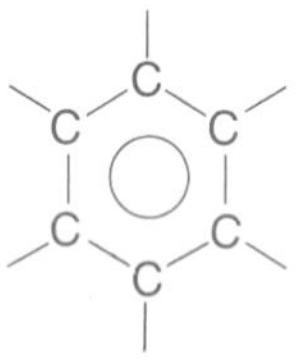

Thousands of compounds consist of molecules with the benzene ring. Benzene itself is the simplest example, and its molecules feature the delocalized molecular orbitals described in Section 9.7. Some of the earliest known compounds with benzene rings happened to have pleasant odors that people described as "aromatic." Oil of wintergreen (mint) and vanillin (vanilla extract) are two examples.

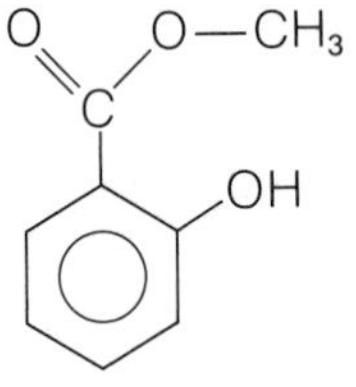

benzene ring

oil of wintergreen

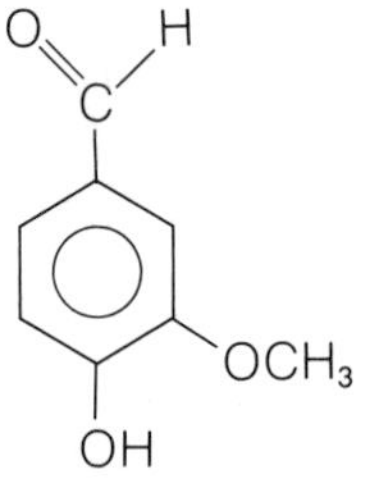

vanillin

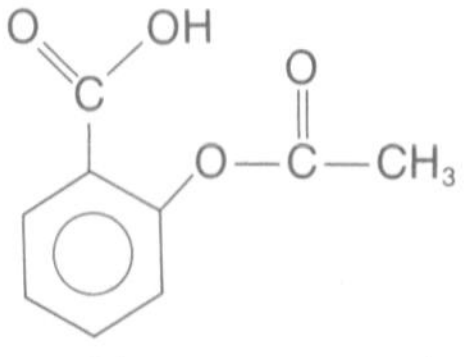

aspirin—an aromatic compound without an odor

Thus, chemists began to call any compound with a benzene ring an **aromatic compound,** although many are now known without any odors at all. Today "aromatic" is used to describe any compound with flat rings having delocalized π-electron networks and which give the same kinds of reactions as benzene itself.

The reactions of the benzene ring result in a hydrogen atom at a ring carbon being *substituted* by another atom or group of atoms provided by the other reactant. Benzene, in spite of a richness in π-electrons—more so than the alkenes—does not give addition reactions like the alkenes. Concentrated sulfuric acid reacts with benzene as follows, for example.

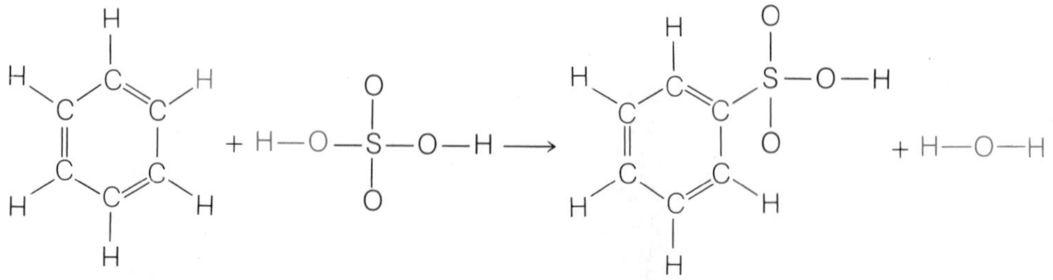

If benzene were to *add* sulfuric acid like an alkene, then the following would happen.

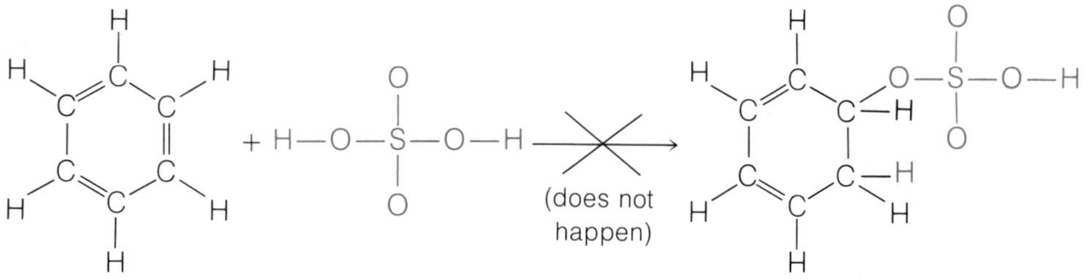

product of *adding* sulfuric
acid to one double bond

If this addition reaction happened, the ring would suffer a permanent loss of the cyclic, delocalized π-electron network characteristic of the benzene ring. The ring strongly resists any reaction that would cause that loss of stabilizing influence. In fact, in spite of the π-electron richness of the benzene ring benzene is exceptionally resistant to oxidizing agents, too. Thus, the delocalized π-electron network of a benzene ring makes an enormous difference in the kinds of reactions this ring will undergo. We learn from the kinds of reactions benzene will undergo what a strong stabilizing influence delocalization has.

9.10 ORGANIC COMPOUNDS OF OXYGEN

Important organic compounds of oxygen contain either the —OH group, a carbonyl group (C=O), or both

The functional groups we will survey briefly in this section are those in alcohols, ethers, aldehydes, ketones, and carboxylic acids.

Alcohols
An **alcohol** is any compound whose molecules have an —O—H group attached to a carbon that has three other *single* bonds.

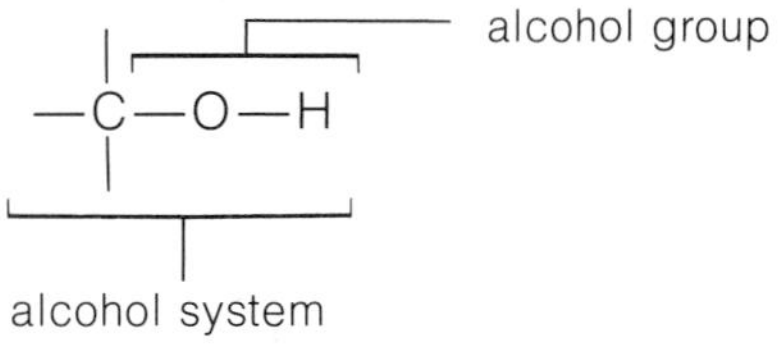

The simplest alcohol is methyl alcohol, also known as wood alcohol because it once was made by heating wood and driving off vapors that, when liquified, included methyl alcohol.

CH_3—OH
methyl alcohol
(methanol)

CH_3—CH_2—OH
ethyl alcohol
(ethanol)

CH_3—CH—CH_3 with OH above the CH
isopropyl alcohol
(2-propanol)

(We will use the names most commonly used for alcohols and show the IUPAC names in parentheses. Notice that the IUPAC ending for names of alcohols is *-anol* and that a number is used to locate the —OH group.) Ethyl alcohol is grain alcohol, the alcohol in alcoholic beverages. It is also the alcohol in gasohol. In the body, methyl and ethyl alcohols have radically different effects. Methyl alcohol causes blindness and often death. Ethyl alcohol causes a loss of coordination and of inhibitions (and too much ethyl alcohol over a long period ruins the liver). Isopropyl alcohol is the usual "alcohol" in rubbing alcohol.

Many compounds have two (or more) —OH groups per molecule. Both ethylene glycol and propylene glycol are used in antifreeze mixtures for car radiators. (Propylene glycol is less toxic.)

$$CH_2-CH_2 \atop || \atop OHOH$$ ethylene glycol (1,2-ethanediol)

$$CH_3-CH-CH_2 \atop || \atop OHOH$$ propylene glycol (1,2-propanediol)

$$CH_2-CH-CH_2 \atop ||| \atop OHOHOH$$ glycerol (1,2,3-propanetriol)

Glycerol is a thick, syrupy liquid used in moisturizing lotions. With three —OH groups per molecule, glycerol has a strong affinity for water. When we digest butterfat or olive oil, one of the products is glycerol. We will look at just one reaction of alcohols, but do so under carboxylic acids.

The —OH group has a great effect on solubility. Methane is insoluble in water but methyl alcohol dissolves in water in all proportions. The —OH group is polar, like water, and in Chapter Eleven we will see why being similar in polarities helps two compounds to dissolve in each other.

An alcohol group can hold a water molecule by a hydrogen bond:

$$CH_3-O \atop \diagdown_{\delta+}$$
$$H\cdots O \atop |$$
hydrogen bond H

Ethers

Ether molecules have two alkyl groups bonded to one oxygen.

$$CH_3-O-CH_3$$
dimethyl ether

$$CH_3-CH_2-O-CH_2-CH_3$$
diethyl ether

Diethyl ether was once the chief anesthetic used in surgery. As a family, ethers have scarcely any more reactions than alkanes. They burn easily, and a mixture of diethyl ether in air in the right proportions is an explosive combination. When diethyl ether is used as an anesthetic, great care must be used to prevent any electrical sparks from occurring.

Aldehydes and ketones

The **carbonyl group** is present in many families; the simplest are aldehydes and ketones.

$$\overset{O}{\underset{\diagup\diagdown}{\overset{\|}{C}}}$$
carbonyl group

$$-\overset{O}{\overset{\|}{C}}-H$$
aldehyde group

$$-\overset{}{\underset{|}{C}}-\overset{O}{\overset{\|}{C}}-\overset{}{\underset{|}{C}}-$$
ketone system

Formaldehyde is the simplest aldehyde. (We continue to use the common names, not the IUPAC names. You are not expected to memorize them.) It is a disinfectant and is used as a preservative for biological specimens. (The fragrance of the biology department when they are dissecting fetal pigs or other specimens may be formaldehyde odor.)

$$H-\overset{O}{\overset{\|}{C}}-H$$
formaldehyde (methanal)

$$CH_3-\overset{O}{\overset{\|}{C}}-H$$
acetaldehyde (ethanal)

$$CH_3-CH_2-CH_2-\overset{O}{\overset{\|}{C}}-H$$
butyraldehyde (butanal)

$$CH_3-\overset{O}{\overset{\|}{C}}-CH_3$$
acetone (propanone)

Butyraldehyde has an absolutely vile odor—the essence of locker rooms and stale socks that have abandoned all hope of ever meeting soap and water again. Acetone, the simplest member of the ketone family, is an important organic solvent that dissolves both water and less polar compounds. It can be bought at cosmetic counters as fingernail polish remover.

Aldehydes and ketones have many similar reactions because they have the carbonyl group. Members of both families will add hydrogen to the carbonyl group, for example. The reaction, a reduction, is very similar to the addition of hydrogen to the alkene double bond.

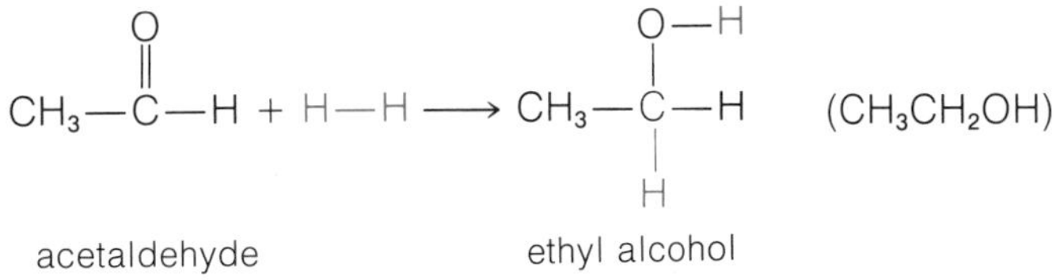

$$CH_3-\overset{\overset{\displaystyle O}{\|}}{C}-H + H-H \longrightarrow CH_3-\overset{\overset{\displaystyle O-H}{|}}{\underset{\underset{\displaystyle H}{|}}{C}}-H \quad (CH_3CH_2OH)$$

acetaldehyde   ethyl alcohol

What makes it useful to have separate families for aldehydes and ketones is their remarkably different behavior toward oxidizing agents. Aldehydes are easily oxidized, but ketones are not.

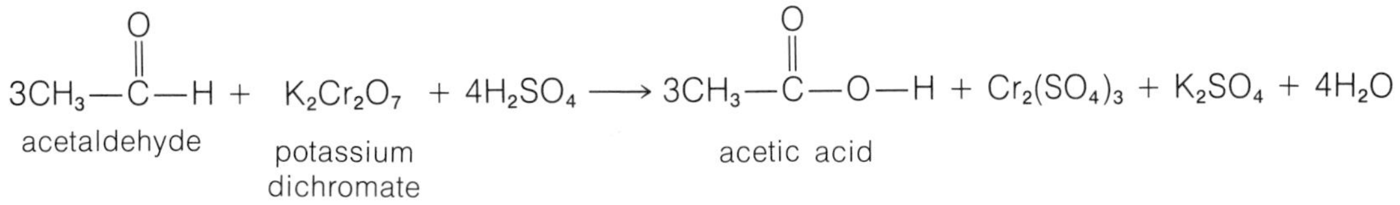

$$3CH_3-\overset{\overset{\displaystyle O}{\|}}{C}-H + K_2Cr_2O_7 + 4H_2SO_4 \longrightarrow 3CH_3-\overset{\overset{\displaystyle O}{\|}}{C}-O-H + Cr_2(SO_4)_3 + K_2SO_4 + 4H_2O$$

acetaldehyde   potassium   acetic acid
                    dichromate

The conditions that cause this oxidation leave acetone untouched. Even such a mild oxidizing agent as the silver ion, in the form of $Ag(NH_3)_2^+$, will oxidize the aldehyde group in glucose without touching anything else in the molecule.

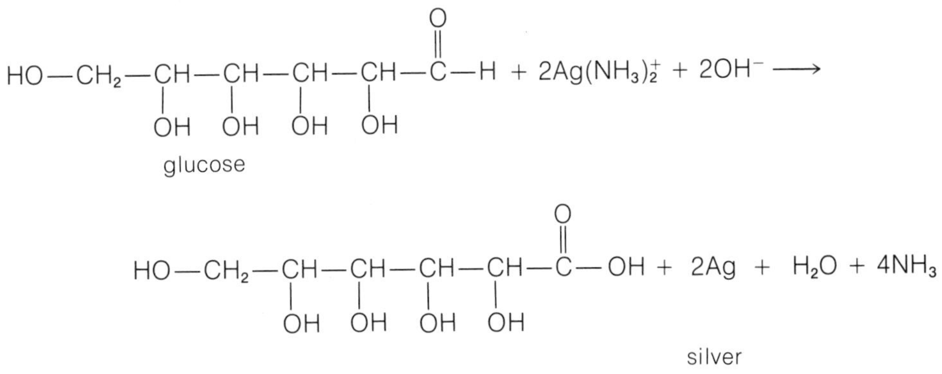

$$HO-CH_2-\underset{\underset{\displaystyle OH}{|}}{CH}-\underset{\underset{\displaystyle OH}{|}}{CH}-\underset{\underset{\displaystyle OH}{|}}{CH}-\underset{\underset{\displaystyle OH}{|}}{CH}-\overset{\overset{\displaystyle O}{\|}}{C}-H + 2Ag(NH_3)_2^+ + 2OH^- \longrightarrow$$

glucose

$$HO-CH_2-\underset{\underset{\displaystyle OH}{|}}{CH}-\underset{\underset{\displaystyle OH}{|}}{CH}-\underset{\underset{\displaystyle OH}{|}}{CH}-\underset{\underset{\displaystyle OH}{|}}{CH}-\overset{\overset{\displaystyle O}{\|}}{C}-OH + 2Ag + H_2O + 4NH_3$$

silver

Silver metal is the product of particular interest in this reaction, called the Tollens' reaction. It is the basis for silvering mirrors. If the inner glass surface of the container for the reaction is thoroughly clean, the silver comes out as a beautiful mirror.

Carboxylic acids

"Carboxyl" is coined from "*carbon*yl" + "hyd*roxyl*," and the carboxyl group characterizes the molecules of all **carboxylic acids.**

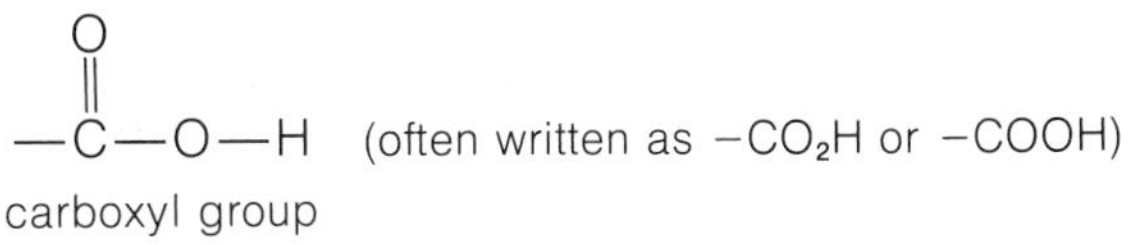

$$-\overset{\overset{\displaystyle O}{\|}}{C}-O-H \quad \text{(often written as } -CO_2H \text{ or } -COOH)$$

carboxyl group

We have noted before that formic acid gives the sting to an ant bite and that acetic acid is in vinegar. Butyric acid is responsible for the dreadful odor of rancid butter.

$$H-\overset{\overset{\displaystyle O}{\|}}{C}-O-H \qquad CH_3-\overset{\overset{\displaystyle O}{\|}}{C}-O-H \qquad CH_3-CH_2-CH_2-\overset{\overset{\displaystyle O}{\|}}{C}-O-H$$

formic acid   acetic acid   butyric acid
(methanoic acid)   (ethanoic acid)   (butanoic acid)

As we noted in Section 9.8, compounds with carboxyl groups react with metal hydroxides such as sodium hydroxide. One other kind of reaction of this group involves alcohols, and esters are the products.

Esters

When the H-atom on a carboxyl group is replaced by an alkyl group, the result is a member of the **ester** family.

$$-\overset{\overset{\displaystyle O}{\|}}{C}-O-\overset{|}{\underset{|}{C}}- \qquad CH_3-\overset{\overset{\displaystyle O}{\|}}{C}-O-CH_2-CH_2-\overset{\overset{\displaystyle CH_3}{|}}{CH}-CH_3 \qquad CH_3-\overset{\overset{\displaystyle O}{\|}}{C}-O-CH_2CH_2CH_2CH_2CH_2CH_2CH_2CH_3$$

ester   a typical ester—oil   another typical ester—oil
group   of banana fragrance   of orange fragrance

Esters of low formula weights have some of the most pleasant fragrances of all organic compounds. The delicate flavor of raspberries, for example, is caused by a mixture of nine esters together with other compounds.

Esters can be made in the laboratory by heating a carboxylic acid with an alcohol for a period of time. For example,

$$CH_3CH_2CH_2\overset{\overset{\displaystyle O}{\|}}{C}-O-H + H-O-CH_2CH_3 \xrightarrow{\text{heat}} CH_3CH_2CH_2\overset{\overset{\displaystyle O}{\|}}{C}-O-CH_2CH_3 + H-O-H$$

butyric acid   ethyl alcohol   ethyl butyrate   water
(pineapple fragrance)

When an ester is heated with a large molar excess of water it breaks back down to its original carboxylic acid and alcohol. This reaction, called the *hydrolysis* of an ester, is the same kind of reaction that occurs when we digest butterfat or olive oil or other fats and oils in the diet. Their molecules have ester groups that are hydrolyzed by digestion.

9.11 ORGANIC COMPOUNDS OF NITROGEN

Amines and amides are two important organic nitrogen families

If we think of taking a molecule of ammonia, NH_3, and replace one, two, or three of its hydrogens by alkyl groups, we will generate molecules of the **amines.**

The odor of low-formula-weight amines betrays spoilage in fish.

$$H-\overset{\overset{\displaystyle H}{|}}{N}-H \qquad CH_3-\overset{\overset{\displaystyle H}{|}}{N}-H \qquad CH_3-\overset{\overset{\displaystyle H}{|}}{N}-CH_3 \qquad CH_3-\overset{\overset{\displaystyle CH_3}{|}}{N}-CH_3$$

ammonia   methylamine   dimethylamine   trimethylamine

Like ammonia, the amines are basic compounds that react with acids to form salts. Methylamine, for example, reacts with sulfuric acid as follows.

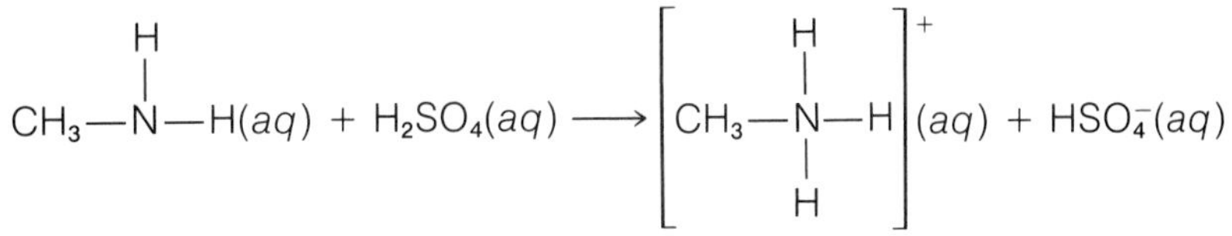

$$CH_3-\underset{\underset{\displaystyle H}{|}}{\overset{\overset{\displaystyle H}{|}}{N}}-H(aq) + H_2SO_4(aq) \longrightarrow \left[CH_3-\underset{\underset{\displaystyle H}{|}}{\overset{\overset{\displaystyle H}{|}}{N}}-H\right]^{+}(aq) + HSO_4^-(aq)$$

methylamine	sulfuric acid	methylammonium ion	hydrogen sulfate ion

Just how strongly basic the amines are will be discussed in Chapter 17.

If we put a carbonyl group directly on nitrogen, we generate the amide group, the functional group of the **amides.**

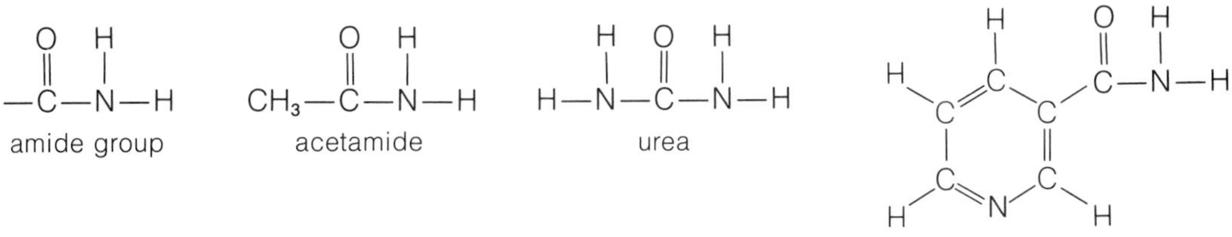

amide group   acetamide   urea

nicotinamide
(an essential B-vitamin)

The amides, like the esters, react with water to give carboxylic acids and either ammonia or an amine, and this hydrolysis of the amide group is the fundamental chemical reaction in the digestion of all proteins.

Urea is an organic waste excreted via the urine. It is also manufactured from ammonia and carbon dioxide to serve as a nitrogen fertilizer. In the soil it hydrolyzes to liberate ammonia.

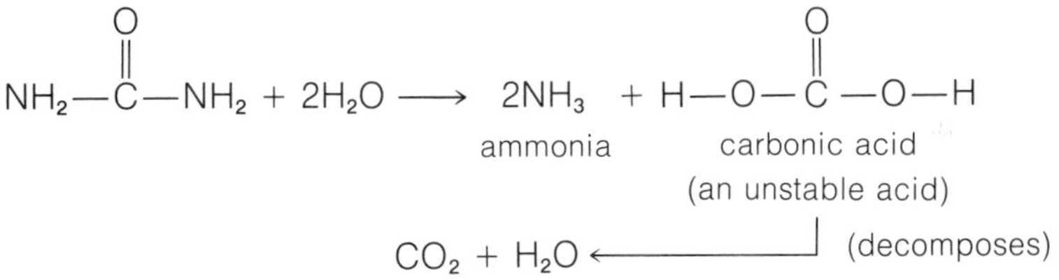

$$NH_2-\overset{\overset{\displaystyle O}{||}}{C}-NH_2 + 2H_2O \longrightarrow 2NH_3 + H-O-\overset{\overset{\displaystyle O}{||}}{C}-O-H$$

ammonia   carbonic acid
(an unstable acid)

$$CO_2 + H_2O \longleftarrow \rule{2cm}{0.4pt}\rfloor \text{ (decomposes)}$$

The ammonia then serves as a source of nitrogen in the soil for aiding plant growth.

Nicotinamide is an interesting compound not just because it is one of the essential B vitamins but also because its ring is not made entirely of carbon atoms. One nitrogen atom is present, and ring compounds whose rings have atoms other than carbon are called **heterocyclic compounds.** Many heterocyclic compounds occur in nature, and the hetero atom (which is N in nicotinamide) can be oxygen or sulfur, too.

With both the amides and the esters we see one of the chemical consequences of having a very polar group, the carbonyl group with its electronegative oxygen, attached to an electronegative atom, N or O. These combinations render the molecules of amides or esters open to chemical attack by the polar water molecule. Water does not attack alkanes. It *adds* to alkene double bonds because the π-bond is rather easily broken open. And water splits molecules apart at bonds between a carbonyl group and the O atom of an ester or the N atom of an amide. We have only opened a door here to the huge field of organic chemistry, opened it just enough to see that much depends on what nonmetals are bonded to which, and by what kinds of bonds.

SUMMARY

Valence Bond (VB) Theory. This is one of two modern theories of chemical bonding discussed in this chapter that are based on wave mechanics. According to VB theory, a bond is formed when a pair of electrons is shared between overlapping orbitals on two atoms. Mixed or hybrid orbitals provide better overlap than unhybridized orbitals. Overlap of s-s, s-p, end-to-end p-p, as well as overlap of hybrid orbitals produces σ-bonds. In complex molecules, the basic molecular framework is built with σ-bonds, which allow free rotation about the bond axis. Side-by-side overlap of two p-orbitals produces a π-bond. A double bond consists of one σ-bond and one π-bond and resists rotation about the bond axis. A triple bond consists of one σ-bond and two π-bonds.

Molecular Orbital (MO) Theory. This theory begins with the supposition that molecules, like atoms, have to be treated as collections of nuclei and electrons and that the electrons are in molecular orbitals of different energy that can spread over more than two nuclei. The molecular orbitals can be considered to form by constructive and destructive interference of the electron waves corresponding to the atomic orbitals of the atoms in the molecule. Bonding MO's concentrate electron density between nuclei; antibonding MO's remove electron density from between nuclei. The rules for the filling of MO's are the same as those for filling atomic orbitals. The ability of MO theory to describe delocalized orbitals avoids the need for resonance. Delocalization of bonds leads to a lowering of the energy and, therefore, more stable molecular structures.

Covalent Compounds of Carbon. Carbon's unique ability to form strong covalent bonds to nonmetal atoms, including other carbon atoms, is one reason for the great number of organic compounds. Which atoms are joined and by what kinds of bonds, σ- or π-bonds, gives rise to several functional groups. All molecules having the same functional group, regardless of chain length, are in the same family of organic compounds. The chemical reactions are characteristic of the functional group itself.

Hydrocarbons. Alkanes and alkenes consist of only carbon and hydrogen, but an alkene has a functional group, the double bond. Alkenes react at the double bond by the addition of such compounds as hydrogen, water, and sulfuric acid. Aromatic compounds like benzene that have flat rings with delocalized π-electron networks strongly resist addition reactions and oxidation. Benzene gives substitution reactions instead, because such reactions leave the delocalized network of electrons intact. Different arrangements of the same atoms occur often among organic compounds. These isomers may have similar properties if they have the same functional groups or if they differ only in cis or trans geometric relationships. When isomers involve different functional groups, then the chemical differences can be great.

Organic Compounds of Oxygen. Alcohols with their water-like —OH groups are much more soluble in water than alkanes. The carbonyl group, C=O, is found in several families. In aldehydes this group is easily oxidized to the carboxyl group. In ketones, oxidation is strongly resisted. The carboxylic acids are weak acids, but they can neutralize metal hydroxides. They also can be changed either to esters or to amides. Both of these can be hydrolyzed back to the parent acids. Ethers have few reactions.

Organic Compounds of Nitrogen. The amines are weakly basic compounds that neutralize acids. When a carbonyl group is attached to nitrogen, the result is an amide.

INDEX TO EXERCISES AND QUESTIONS

REVIEW QUESTIONS

9.9. What is the theoretical basis of both valence bond (VB) theory and molecular orbital (MO) theory?

9.10. What are the shortcomings of Lewis structures and the VSEPR theory that VB and MO theory attempt to overcome?

9.11. What is the main difference in the way VB and MO theories view the bonds in a molecule?

9.12. What is meant by orbital overlap?

9.13. How does VB theory explain electron sharing between atoms in a covalent bond?

9.14. What term is used to describe the mixing of atomic orbitals of the same atom?

9.15. Use drawings to describe how the covalent bond in H_2 is formed.

9.16. Use drawings to describe how valence bond theory would explain the formation of the H—Br bond in hydrogen bromide.

9.17. Hydrogen selenide is one of chemistry's most foul-smelling substances. Molecules of H_2Se have H—Se—H bond angles very close to 90°. How would valence bond theory explain the bonding in H_2Se?

9.18. Use drawings to show how valence bond theory explains the bonding in the F_2 molecule.

9.19. What are the characteristics of σ and π bonds?

9.20. Why do atoms usually prefer to use hybrid orbitals for bonding?

9.21. Sketch figures that illustrate the directional properties of the following hybrid orbitals: (a) sp (b) sp^2 (c) sp^3 (d) sp^3d (e) sp^3d^2.

9.22. Why do period 2 elements never use sp^3d or sp^3d^2 hybrid orbitals for bond formation?

9.23. What relationship is there, if any, between Lewis structures and valence bond descriptions of molecules?

9.24. Use orbital diagrams to explain how the $BeCl_2$ molecule is formed. What kind of hybrid orbitals does beryllium use in $BeCl_2$?

9.25. Why is free rotation permitted about a σ-bond axis but not about a π-bond axis?

9.26. Use orbital diagrams to describe the bonding in (a) $SnCl_4$ and (b) $SbCl_5$.

9.27. Draw Lewis structures for the following and use the geometry predicted by VSEPR theory for the electron pairs to determine what kind of hybrid orbitals the central atom uses in bond formation: (a) ClO_3^- (b) SO_3 (c) OF_2 (d) $SbCl_6^-$ (e) $BrCl_3$ (f) XeF_4.

9.28. What hybrid orbitals are used by tin in $SnCl_6^{2-}$? Draw the orbital diagram for Sn in $SnCl_6^{2-}$. What is the geometry of $SnCl_6^{2-}$?

9.29. A nitrogen atom, somewhat like carbon, can undergo sp hybridization and then become joined to carbon by a triple bond, $-C\equiv N:$. This triple bond consists of one σ- and two π-bonds.
(a) Write the orbital diagram for sp hybridized nitrogen as it would look before any bonds form.
(b) Using the carbon-carbon triple bond as the analogy, and drawing pictures to show which atomic orbitals overlap with which, show how the three bonds of the triple bond in $-C\equiv N:$ form.
(c) Again using sketches, describe all the bonds in hydrogen cyanide, $H-C\equiv N:$.

(d) What is the likeliest H—C—N bond angle in HCN?

9.30. How does VB theory treat the benzene molecule? (Draw sketches describing the orbital overlappings and the bond angles.)

9.31. Using orbital diagrams, describe how sp^3 hybridization occurs in each atom.
(a) carbon　(b) nitrogen　(c) oxygen

9.32. Sketch the way the orbitals overlap to form the bonds in each substance.
(a) CH_4　(b) NH_3　(c) H_2O

9.33. We explained the bond angles of 107° in NH_3 by using sp^3 hybridization of the central nitrogen. Had the original *unhybridized* p-orbitals of the nitrogen been used to overlap with 1s orbitals of each hydrogen, what would have been the H—N—H bond angles? Why?

9.34. The six-membered ring of carbons can hold a double bond but not a triple bond. Explain.

cyclohexene (exists)　cyclohexyne (unknown)

9.35. What facts about boron strongly suggested the need to consider sp^2 hybridization of its second shell atomic orbitals?

9.36. Using sketches describe sp^2 hybridization at
(a) boron and (b) carbon

9.37. A nitrogen atom can also undergo sp^2 hybridization when, for example, it becomes part of a carbon-nitrogen double bond as in $H_2C\!=\!NH$.
(a) Using a sketch, show the electronic configuration of sp^2 hybridized nitrogen just before the overlapping occurs to make this double bond.
(b) Using sketches (and the analogy to an alkene double bond) describe the two bonds of the carbon-nitrogen double bond.
(c) Describe the geometry of $H_2C\!=\!NH$ (using a sketch that shows all expected bond angles).

9.38. Phosphorus trifluoride, PF_3, has F—P—F bond angles of 97.8°.
(a) How would VB theory explain these data using hybrid orbitals?
(b) How would VB theory use unhybridized orbitals to account for these data?
(c) Do either of these models work very well?

9.39. The ammonia molecule, NH_3, can combine with a hydrogen ion, H^+, having an empty 1s orbital to form the ammonium ion, NH_4^+. (This is how ammonia can neutralize acid and therefore function as a base.) Sketch the geometry of the ammonium ion, indicating the bond angles.

9.40. If the central oxygen in the water molecule did not use sp^3 hybridized orbitals (or orbitals of any other kind of hybridization), what would be the bond angle in H_2O (assuming no angle-spreading force)?

9.41. Using sketches, describe the bonds and bond angles in ethene, C_2H_4.

9.42. Sketch the way the bonds form in ethyne, C_2H_2.

9.43. Cyclopropane is a triangular molecule with $C-C-C$ bond angles of 60°. Explain why the σ-bonds joining carbon atoms in cyclopropane are weaker than the carbon-carbon bonds in the noncyclic propane.

cyclopropane

propane

9.44. Why is the higher energy molecular orbital in H_2 called an antibonding orbital?

9.45. Using a sketch, describe the MO's of H_2 and their relation to their parent atomic orbitals.

9.46. Explain why He_2 does not exist but H_2 does.

9.47. How does MO theory account for the paramagnetism of O_2?

9.48. On the basis of molecular orbital theory, explain why Li_2 molecules can exist but Be_2 molecules cannot.

9.49. Use the molecular orbital energy diagram to predict which in each pair has the greater bond energy: (a) O_2 or O_2^+; (b) O_2 or O_2^-; (c) N_2 or N_2^+.

9.50. What is the bond order in (a) O_2^+, (b) O_2^-, and (c) C_2^+?

9.51. What relationship have we found between bond order and bond energy?

9.52. Sketch the shapes of π_{2p_x} and $\pi_{2p_x}^*$ molecular orbitals.

9.53. What is a delocalized molecular orbital?

9.54. Use a Lewis-type structure to indicate delocalized bonding in the nitrate ion.

9.55. What problem encountered by VB theory does MO theory avoid by delocalized bonding?

9.56. Draw the representation of the benzene that indicates its delocalized π system.

9.57. What effect does delocalization have on the stability of the electronic structure of a molecule?

9.58. Write the IUPAC names of the following hydrocarbons

(a) $CH_3CH_2CH_2CH_2CH_3$

(b) $CH_3CH_2CH_2CHCH_3$
          |
          CH_3

(c) $CH_3CHCH_2CHCH_2CH_3$
  |
  CH_3 |
          CH_3

(d) $CH_3CH_2CHCH_2CHCH_3$
            | |
            CH_3 CH_3

(e) $CH_3CH_2CH=CHCH_2CH_3$

(f) $CH_3CHCH=CHCH_3$

9.59. Write the condensed structures of the following compounds

(a) 2,2-dimethyloctane
(b) 1,3-dimethylcyclopentane
(c) 1,1-diethylcyclohexane
(d) 6-ethyl-5-isopropyl-7-methyl-1-octene

9.60. Each compound below has one functional group or is in the alkane family. What family is each in?

(a)

(b)

(c)   (d)

(e)   (f) $CH_3-O-CH_2-CH_3$

(g)   (h)

(i)   (j)

(k)   (l)

(m)   (n)

(o)

(p) $CH_3-CH_2-N-CH_2-CH_3$
              |
              H

(q)   (r)

(s) CH_3-CH_2-O-H (t) $CH_3-CH_2-CH_2-CH_2-O-H$

(u) $CH_3-CH_2-CH_2-CH_2-CH-CH_3$
                               OH
                                |

(v)

(w)

(x) $CH_3-CH_2-CH_2-CH_2-CO_2H$

9.61. Examine each of the following pairs of structures and decide if the two are of identical compounds, are isomers, or are unrelated.

(a) CH_3 and CH_3-CH_3
    |
    CH_3

(b)
$$CH_3 \\ \backslash \\ CH_2 \\ \diagdown CH_3$$ and $$CH_2 \\ \diagup \diagdown \\ H_3C \quad CH_3$$

(c) $CH_3CH_2—OH$ and $CH_3CH_2CH_2—OH$

(d) $CH_3CH=CH_2$ and $H_2C\!-\!\!-\!\!CH_2$ with CH_2 below forming a ring

(e) $H—\overset{\overset{\displaystyle O}{\|}}{C}—CH_3$ and $CH_3—\overset{\overset{\displaystyle O}{\|}}{C}—H$

(f) $CH_3\overset{\overset{\displaystyle CH_3}{|}}{C}HCH_3$ and $CH_3\overset{\overset{\displaystyle |}{C}H}{\underset{\underset{\displaystyle CH_3}{|}}{}}$

(g) $CH_3CH_2—NH_2$ and $CH_3—\underset{\underset{\displaystyle H}{|}}{N}—CH_3$

(h) $CH_3CH_2\overset{\overset{\displaystyle O}{\|}}{C}—O—H$ and $H—O—\overset{\overset{\displaystyle O}{\|}}{C}CH_2CH_3$

(i) $H—\overset{\overset{\displaystyle O}{\|}}{C}—O—CH_2CH_3$ and $CH_3CH_2—\overset{\overset{\displaystyle O}{\|}}{C}—O—H$

(j) $H—\overset{\overset{\displaystyle O}{\|}}{C}—O—CH_2CH_2OH$ and $HOCH_2CH_2—\overset{\overset{\displaystyle O}{\|}}{C}—OH$

(k) $CH_3\overset{\overset{\displaystyle O}{\|}}{C}CH_2CH_3$ and $CH_3CH_2\overset{\overset{\displaystyle O}{\|}}{C}CH_3$

(l) $CH_3—\underset{\underset{\displaystyle |}{C}H}{CH}—CH_3$ structures

(m) $CH_3—NH—\overset{\overset{\displaystyle O}{\|}}{C}—CH_3$ and $CH_3CH_2\overset{\overset{\displaystyle O}{\|}}{C}NH_2$

(n) $H—O—O—H$ and $H—O—H$

(o) $$\begin{array}{cc} CH_3 & CH_3 \\ \diagdown \ \ \diagup \\ C=C \\ \diagup \ \ \diagdown \\ CH_3 & H \end{array}$$ and $$\begin{array}{cc} CH_3 & H \\ \diagdown \ \ \diagup \\ C=C \\ \diagup \ \ \diagdown \\ CH_3 & CH_3 \end{array}$$

(p) $$\begin{array}{cc} CH_3—CH_2 & CH_3 \\ \diagdown \ \ \ \ \ \ \diagup \\ C=C \\ \diagup \ \ \diagdown \\ CH_3 & H \end{array}$$ and $$\begin{array}{cc} CH_3—CH_2 & H \\ \diagdown \ \ \ \ \ \ \diagup \\ C=C \\ \diagup \ \ \diagdown \\ CH_3 & CH_3 \end{array}$$

9.62. If at either *end* of a carbon-carbon double bond the two atoms or groups are identical, can the alkene exist as a pair of *cis* and *trans* isomers? (Examine again parts (o) and (p) of Question 9.61.)

9.63. What are the product(s) of the addition of hydrogen to *cis*-2-butene and *trans*-2-butene? Do isomers form? Explain.

9.64. Write the structure(s) of the product(s) in each of the following reactions of 2-butene.
(a) addition of H_2O (b) addition of H_2SO_4

9.65. Write the structure of the carboxylic acid that forms when the following aldehyde is oxidized.

$$CH_3—CH_2—\overset{\overset{\displaystyle O}{\|}}{C}—H$$

9.66. Write the structure of the negative ion of the salt that forms when sodium hydroxide neutralizes the following carboxylic acid.

$$CH_3—CH_2—\overset{\overset{\displaystyle O}{\|}}{C}—O—H$$

9.67. Which of the following will be split apart by a reaction with water, and what are the structures of the products that form?

$$CH_3—CH_2—O—CH_2—CH_3 \quad\quad CH_3—\overset{\overset{\displaystyle O}{\|}}{C}—O—CH_2—CH_3$$
$$\quad\quad\quad A \quad\quad\quad\quad\quad\quad\quad\quad\quad\quad\quad B$$

9.68. Which of these two compounds will react with concentrated sulfuric acid by *addition*? Write the structure of the product.

$$CH_3CH_2CH=CHCH_2CH_3$$
$$A$$
(B) benzene ring with $—CH_3$ substituent

9.69. Which of these two compounds will form a salt with hydrochloric acid? Write the structure of the salt's positive ion.

$$CH_3CH_2CH_3 \quad\quad CH_3CH_2\overset{\overset{\displaystyle H}{|}}{N}—H$$
$$\quad A \quad\quad\quad\quad\quad\quad\quad\quad B$$

9.70. Which of these two compounds will be split apart by a chemical reaction with water? Write the structures of the products.

$$CH_3—\overset{\overset{\displaystyle O}{\|}}{C}—\overset{\overset{\displaystyle H}{|}}{N}—H \quad\quad CH_3—CH_2—\overset{\overset{\displaystyle H}{|}}{N}—H$$
$$\quad\quad A \quad\quad\quad\quad\quad\quad\quad\quad\quad B$$

9.71. Which of these two compounds is easily oxidized? Write the structure of the organic product of the oxidation.

$$CH_3—\overset{\overset{\displaystyle O}{\|}}{C}—CH_3 \quad\quad CH_3—CH_2—\overset{\overset{\displaystyle O}{\|}}{C}—H$$
$$\quad\quad A \quad\quad\quad\quad\quad\quad\quad\quad B$$

CHAPTER TEN

LIQUIDS, SOLIDS, AND CHANGES OF STATE

10.1 PHYSICAL PROPERTIES VERSUS CHEMICAL PROPERTIES

Physical properties of substances are as important as chemical properties

Although much of the study of chemistry is devoted to chemical properties and reactions, the physical properties of substances often concern us more in our day-to-day activities. Farmers worry about whether expected precipitation will come in liquid or solid form. Rain (the liquid) is usually welcomed; hail (the solid) is almost universally feared. The fact that water expands when it solidifies may bring disaster (or at least a very expensive repair bill) to the automobile owner who has neglected to put antifreeze in the car's radiator if the outside air temperature dips very far below freezing. During the summer, the high rate of evaporation of paint solvent can make painting in the hot sun difficult. And in Denver, the traditional "three-minute egg" isn't nearly as well cooked as most people like, so it's boiled a bit longer.

These examples show how the physical properties of substances and transformations among the three states of matter influence our lives. One of our goals in chemistry is to understand what determines these properties. From this study, we stand to gain a more complete knowledge of the structure of matter and an understanding of how scientists have been able to design chemicals that possess the kinds of properties that we desire. We have already discussed the physical properties of gases in Chapter Two. Here, we will investigate the physical properties of liquids and solids and what happens when substances change from one state of matter to another.

Many common substances, such as synthetic fibers, were developed simply because they have desirable physical properties as well as chemical properties.

10.2 WHY GASES DIFFER FROM LIQUIDS AND SOLIDS

Physical properties depend on forces of attraction between molecules, which are generally strongest in solids and weakest in gases

When we studied the laws that govern the physical behavior of gases, no mention was made of a need to specify their chemical compositions. This is because the physical properties of gases are virtually independent of their chemical makeup. For example, one mole of any gas has nearly the same volume as one mole of any other gas, provided their temperatures and pressures are the same. We also know that the combined gas law works quite well for all gases.

The words "nearly" and "quite well" in the last two sentences of the preceding paragraph are very significant. The gas laws do not perfectly fit the physical behavior of *real gases*—that is, actual gases such as hydrogen, methane, carbon dioxide, or steam. Only a hypothetical ideal gas obeys the gas laws exactly. The gas laws are useful to us, though, because the properties of real gases are not far from "ideal." Only if very precise measurements are made do we normally notice a difference. Nevertheless, a difference does exist.

The model of a gas given by the kinetic theory is of a collection of rapidly moving particles, which themselves have essentially *no volume.* Another aspect of the particles envisioned by the theory is that there are *no attractive forces between them.* These postulates, among others, may actually be used to derive the ideal gas law, although the derivation is beyond the level of this book. The fact that real gases deviate from ideal gas behavior results because of slight over simplifications in the postulates of the theory.

As tiny as molecules are, they do possess some volume. In a gas, they can't be forever crowded closer and closer together. The space between them becomes filled with other gas molecules, and doubling the pressure simply cannot halve the volume. The pressure-volume law works least well, therefore, when the pressure on the gas is very high and the molecules are tightly packed.

Gas molecules also attract each other. Any gas will liquefy if it is cooled to a sufficiently low temperature. The molecules that fill the container when the substance is a gas, cling together when the liquid is formed. There must be attractions between the molecules in order to hold them to one another.

At high temperatures, the kinetic energies of colliding gas molecules are large enough that the molecules bounce off each other. At low temperatures, when the molecules are moving more slowly, they can stick together when they collide. Therefore, the gas laws work least well for real gases when the temperature is low.

In a gas, the molecules are far apart and the attractive forces between the widely spaced molecules are always very small, regardless of the composition of the gas. It is this similarity between the strengths of the attractive forces that gives rise to the similarities in the physical properties of gases.

In contrast to gases, the physical properties of different liquids and solids vary greatly. Some liquids evaporate easily and others seem never to evaporate. Some solids are extremely hard and have very high melting points, while others are easily broken and melt at low temperatures. In liquids and solids, the particles are much closer together than the particles in a gas. The attractive forces between the molecules, which we call **intermolecular attractions,** become much stronger as the distance between the molecules decreases. Therefore, the same kinds of intermolecular attractions that are uniformly weak in gases are much stronger in liquids and solids and play a dominant role in controlling all of the physical properties.

Later in this chapter, we will see that chemical composition and structure affect the magnitudes of the intermolecular attractions. Because these attractions so strongly influence physical properties, the way that liquids and solids behave is a function of their chemical makeup. Thus, even though all gases seem nearly the same, the properties of liquids and solids cover a broad range.

10.3 SOME GENERAL PROPERTIES OF LIQUIDS AND SOLIDS

**Physical properties such as volume, shape, surface
tension, and the ease of evaporation depend on the way
particles are arranged in liquids and solids**

We begin our study of liquids and solids by considering some general properties that are related to the way in which their particles are arranged. Some of these properties are recognized by everyone. For example, we all know that ice cubes retain their shapes when dropped into a glass but, when they melt, the liquid takes the shape of the vessel (Figure 10.1).

Figure 10.1
Solids, such as the ice cubes in the glass at the left, retain their shape when placed in a container. The same quantity of liquid takes the shape of the container.

The volume and shape of a gas, you recall, is the same as that of its container.

In both liquids and solids, the particles (either molecules or ions) are packed very closely together. In both states, the attractive forces maintain this tight packing when the substance is transferred from one container to another. Therefore, the volume of a liquid or a solid is independent of the size of the container. However, one of the major differences between a solid and a liquid is that the particles in a solid are held more or less rigidly in place, while the particles in a liquid may move about. Thus a solid retains its shape, but a liquid conforms to the shape of the container.

Incompressibility

The close packing of particles in liquids and solids has other important consequences, too. One of these is that liquids and solids are virtually **incompressible** —their volumes are hardly affected at all by the application of pressure. The reason is that there is very little empty space into which the particles can be squeezed. (This is in marked contrast to gases, which are very compressible because of the large amount of empty space between their molecules.) The engineering science of hydraulics is based on the incompressibility of liquids, and you count on it whenever you "step on the brakes" of a car. The pressure you apply to the pedal is transmitted by the brake fluid through a narrow tube to the brake shoes on the wheels.

Diffusion

Diffusion is another property that varies with the state of matter. It occurs rapidly in gases, slowly in liquids, and hardly at all in solids. For example, the scent of cologne on someone stepping into an elevator is quickly sensed by all present. A couple of spoonfuls of sugar in a cup of coffee, however, won't produce much immediate sweetening unless the mixture is stirred. But if left long enough, the coffee will slowly become sweetened throughout even without its being stirred.

In gases, molecules can diffuse rapidly because they travel relatively long distances between collisions—their movement from one place to another is not greatly hindered. In liquids, however, a given molecule suffers many collisions

Hydraulic machinery, such as this, uses the incompressibility of liquids to transmit forces that accomplish work.

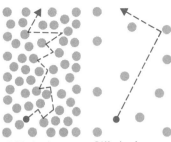

Diffusion in a liquid is slow because of many collisions between closely spaced particles

Diffusion in a gas is rapid because relatively few collisons occur between widely spaced molecules.

as it moves about and, therefore, it takes longer to move from place to place. Diffusion in solids is almost nonexistent at room temperature because the particles of a solid are held so tightly in place. At high temperatures, though, the particles in a solid sometimes have enough kinetic energy to jiggle their way past one another. Such high-temperature solid state diffusion is used in the production of transistors. It allows for small, carefully controlled amounts of some impurity—arsenic, for example—to be added to a semiconductor such as silicon or germanium. The process, called "doping," allows the conductivity of these materials to be precisely determined.

Surface tension

A characteristic phenomenon of the liquid state is surface tension. Among other things, it allows us to fill a water glass above the rim (see Figure 10.2), something most of us have done at one time or another. The phenomenon is caused by the difference between the attractions felt by molecules at a liquid's surface and those felt by molecules within the liquid. These attractions are illus-

Figure 10.2
A glass can be carefully filled with water above the rim.

trated in Figure 10.3. A molecule within the liquid is surrounded completely by other molecules. It therefore experiences essentially uniform attractions in all directions. A molecule at the surface, however, has neighbors beside and below it, but none above. Thus the attractions it feels are not uniform; instead, they occur mainly in the direction of the bulk of the liquid. As a result, the entire surface of the liquid experiences a pull toward the center, which makes it behave something like a "skin."

The only way to increase the surface area of the liquid—and by so doing make the "skin" larger—is to pull molecules to the surface from the interior. This requires energy because the molecules within the liquid experience greater total attractions than do those at the surface. **Surface tension,** then, is related to the amount of energy needed to increase the area of the surface.

Since energy is needed to expand the surface area of a liquid, energy is evolved if the liquid's surface becomes smaller. Any liquid, therefore, tends to minimize the area of its surface because, in so doing, it seeks its lowest energy. We are able to fill a glass above the rim because the surface tension of the water prevents it from spreading out, thereby increasing its surface area. If too much water is added, however, gravity finally overcomes the surface tension and the "skin breaks"—the water overflows.

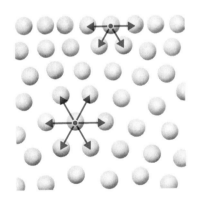

Figure 10.3
Molecules at the surface of a liquid are attracted toward the interior. Those within the liquid are uniformly attracted in all directions.

For liquids having strong intermolecular attractions, this "skin" is strong and the surface tension is high.

Children can build sand castles because the surface tension of water holds the moist grains of sand together.

Surface tension is responsible for the tendency of liquids to form spherical droplets. As we've all seen, if water is splashed onto a greasy or waxed surface, it beads. For a given volume of liquid, a sphere has the smallest surface area. Surface tension also causes moist grains of sand or soil to stick together. Farmers often squeeze a handful of soil to judge its moisture content. If it is dry, the soil will crumble; if there is much water in it, the soil will stick together and form a ball. Similarly, children quickly learn that sand under water will slip through their fingers. But if the sand is merely damp, the small grains cling to each other, and it is possible to build a sand castle. These things are possible because the water between the particles of sand or soil attempts to minimize its surface area and, by so doing, draws the grains of soil or sand together.

A property that we associate with liquids, especially water, is their ability to wet things. **Wetting** is the spreading of a liquid across a surface. Water, for example, will wet clean glass, such as the windshield of a car, by forming a thin film over the surface of the glass. Water won't wet an oily or greasy windshield, however. Instead, it forms tiny, and very annoying, beads of water.

In order for wetting to occur there must be attractive forces between the liquid and the surface that are strong enough to overcome the liquid's surface tension. A liquid that has a low surface tension wets objects more easily than does a liquid with a high surface tension. This is the reason detergents are used for such chores as doing laundry or washing floors. The detergents contain chemicals called **surfactants** that drastically lower the surface tension of water. They make the water "wetter," which allows the detergent solution to spread more easily across the surface to be cleaned.

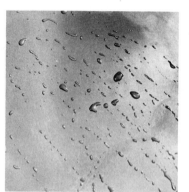

Water forms beads when it is sprinkled on a freshly waxed automobile.

Evaporation

Finally, we come to one of the most important properties of liquids—their tendency to evaporate. We have discussed how attractive forces act to hold the molecules within a liquid together. However, just as molecules in a gas have various kinetic energies, so do those in liquids. Some have low kinetic energies and move very slowly, while others, with higher kinetic energies, move faster. A small fraction of the molecules have very high kinetic energies and, therefore, very high velocities. If one of these high-speed molecules reaches the surface, and if it is moving fast enough, it may be able to escape the attractions of its neighbors. When this happens, we say that the molecule has entered the vapor or gaseous state—it has evaporated.

Through the process of evaporation, liquids gradually lose their high-speed molecules. Since these are also the molecules with the highest kinetic energies, the average kinetic energy of those molecules that remain in the liquid decreases. You might think of this as being similar to removing all the tall people from a classroom. When this is done, the average height of those that remain decreases.

Because average kinetic energy and temperature are directly related, lowering the average kinetic energy lowers the temperature. Therefore, liquids cool as

Evaporation of perspiration cools the body. That's why we perspire when we exercise.

they evaporate. A canvas bag filled with water, for instance, remains cool because some of the water seeps through the canvas and evaporates when it reaches the outside. This removes kinetic energy from the water and thus provides a cooling effect. Our bodies use evaporation to maintain a constant temperature. During warm weather or vigorous exercise, we perspire, and the evaporation of the perspiration cools the skin. You've probably also used the cooling effect caused by evaporation by blowing gently on the surface of a hot bowl of soup or a cup of coffee. The stream of air stirs the liquid, bringing more "hot" molecules to the surface, and removes these molecules from the vicinity of the liquid after they've evaporated.

10.4 VAPOR PRESSURE OF LIQUIDS

Molecules that evaporate from a liquid into a closed container exert a pressure that remains constant after the rates of evaporation and condensation become equal

If a liquid is allowed to evaporate from an open container, the volume of the liquid gradually decreases until all of it is gone. As they evaporate, the molecules of the liquid simply wander away into the atmosphere. If the container is sealed, however, the molecules cannot wander far (Figure 10.4a). They collect as a vapor over the liquid. As they bounce around in the vapor, the molecules collide with each other, with the walls of the container, and with the surface of the liquid itself. Those that strike the liquid's surface tend to stick because their energy becomes scattered among the surface molecules. It is somewhat like throwing a ping-pong ball into a large box of ping-pong balls. The incoming ball knocks the others about and thereby loses much of its own kinetic energy; so there is a high probability that it won't bounce out.

The rate at which vapor molecules return to the liquid—that is, the rate of their **condensation**—depends on how many of them there are in the vapor. As evaporation continues, more and more molecules enter the vapor, so more and more bounce back into the liquid. After a short time, the rate at which molecules return to the liquid becomes equal to the rate at which they leave. From that moment on the total number of vapor molecules will stay the same because, over any given period of time, the number of molecules leaving the liquid is the same as the number that return. The system has reached a state of dynamic equilibrium (Figure 10.4b). In many ways, it is similar to the dynamic equilibrium described for the reactions of weak acids with water (see Chapter Six).

The vapor that collects in a closed space above the liquid exerts a pressure, called the **vapor pressure,** just as any other gas would. From the very moment that the liquid begins to evaporate, there is a vapor pressure. This pressure

We will see that this is true as long as the temperature stays the same.

Figure 10.4
(a) A liquid begins to evaporate into a closed container. (b) Equilibrium is reached when the rate of evaporation equals the rate of condensation.

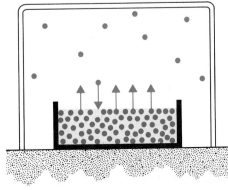

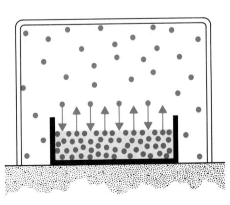

a

b

grows as the amount of vapor increases until finally equilibrium is reached. Then it remains constant. This final pressure of the vapor is known as the **equilibrium vapor pressure** of the liquid. In general, when we refer to the vapor pressure of a liquid, we really mean the equilibrium vapor pressure.

We can measure the vapor pressure of a liquid with an apparatus like that shown in Figure 10.5. Initially, both sides of the apparatus are open to the atmosphere, so that the pressure on both sides is the same. Then the left side is sealed, and a small amount of liquid is introduced into the flask. As some of the liquid evaporates, the pressure in the flask grows, forcing the fluid in the left side of the U-shaped tube downward. The increase in pressure, measured by the difference in levels in the U-tube, is equal to the pressure exerted by the vapor—the vapor pressure.

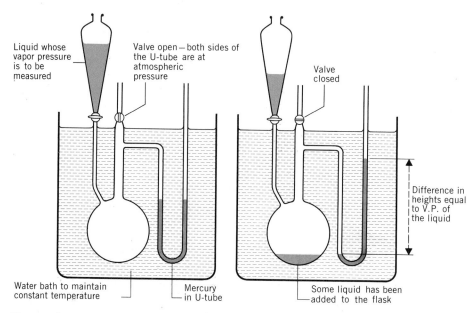

Figure 10.5
Measuring the vapor pressure of a liquid.

One interesting thing about vapor pressure is that its magnitude doesn't depend on the amount of liquid in the flask, just as long as some liquid remains when equilibrium is reached. Even though increasing the amount of liquid increases the surface area of the liquid, thereby raising the evaporation rate, the larger surface area also presents a larger target for returning vapor molecules; the rate of condensation increases too. The net result is that the rates of evaporation and condensation are increased *equally,* so no net change in vapor pressure is produced.

The vapor pressure is also independent of the size of the container. For example, if a liquid and its vapor are contained in a cylinder with a movable piston, withdrawing the piston increases the volume of the vapor and lowers the pressure. This means that fewer molecules strike a given surface area each instant, and the rate of condensation will have decreased. The rate of evaporation hasn't been changed, however, so more of the substance is evaporating than condensing. This condition will continue until there are enough molecules in the vapor to make the condensation rate again equal to the rate of evaporation, and at this point the vapor pressure will have returned to its original value. By similar reasoning, we can expect that decreasing a container's volume should increase the condensation rate without changing the rate of evaporation. Condensation, therefore, occurs faster than evaporation until equilibrium is finally restored and the vapor pressure drops to its original value again.

Substances with high vapor pressures, such as gasoline, evaporate quickly from open containers.

What we see here is that it is the evaporation rate that ultimately determines the vapor pressure. When the volume is changed, the number of molecules in the vapor will either increase or decrease until their rate of return balances their rate of evaporation.

There are two major factors that determine how rapidly molecules can escape from the liquid. One is the strength of the attractive forces between the molecules. If we compare two different liquids at the same temperature, both have the same distribution of kinetic energies, as illustrated in Figure 10.6. Suppose, however, that the attractive forces between the molecules in liquid *A* are weaker than those in liquid *B*. Less energy would then be needed for a molecule of *A* to escape from the surface than would be required for a molecule of *B*. Looking again at Figure 10.6, this means that a greater *total* fraction of *A* molecules would have enough energy to escape, so liquid *A* would evaporate more rapidly than liquid *B*. Therefore, liquid *A* will have a higher vapor pressure than liquid *B*, which is the same as saying that liquid *A* is more *volatile* than liquid *B*. We can conclude, then, that at a given temperature, liquids with high vapor pressures have weak attractive forces between their molecules. Similarly, liquids with low vapor pressures have strong attractive forces between their molecules.

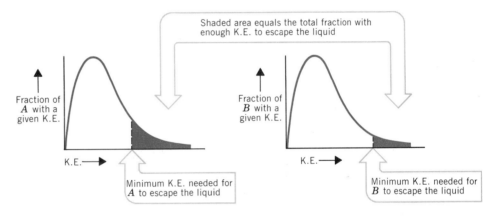

Figure 10.6
Kinetic energy distribution in two liquids, *A* and *B*, at the same temperature.

The other major factor that affects evaporation rate is temperature. As we increase the temperature of any given liquid, the average speed of the molecules becomes larger and their average kinetic energy increases. Figure 10.7 shows that, at a higher temperature, a greater total fraction of the molecules possess the kinetic energy required to overcome the intermolecular attractive forces. During any given moment more molecules can escape from the liquid, so the rate of evaporation is higher. This means that the vapor pressure is greater at the higher

Figure 10.7
Increasing the temperature increases the total fraction of molecules with enough kinetic energy to escape from the liquid.

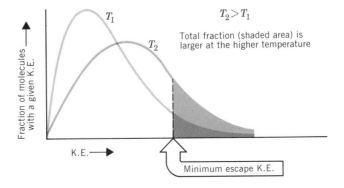

Figure 10.8
Variation of vapor pressure with temperature for some common liquids.

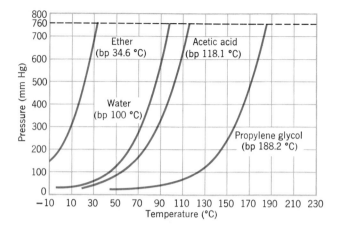

temperature. For any liquid, then, the vapor pressure increases with increasing temperatures. Figure 10.8 shows the vapor pressure of some common liquids and how they vary with temperature.

10.5 WHAT HAPPENS WHEN LIQUIDS BOIL

We know that this water is boiling because we see bubbles of steam headed for the surface.

The pressure of the vapor in the bubbles within a boiling liquid is equal to the pressure of the atmosphere

If you were asked to check whether a pot of water were boiling, what would you look for? Bubbles, naturally! When a liquid boils, large bubbles usually form at many places on the inner surface of the container and rise to the top. If you were to place a thermometer into the boiling water to measure its temperature, you would find it remains constant, regardless of how you adjust the flame under the pot. A hotter flame just makes the bubbles form faster. *Any liquid remains at a constant temperature while it is boiling.* This temperature is called the **boiling point** of the liquid.

If you measure the temperature of boiling water in Philadelphia, New York, or any place else that is nearly at sea level, your thermometer will read 100 °C or very close to it. However, if you try this experiment in Denver, Colorado, you will find that the water boils at about 95 °C. Denver is a mile above sea level, and the atmospheric pressure there is lower. We find, therefore, that the temperature of a boiling liquid depends on the atmospheric pressure.

Every liquid has its own characteristic boiling point at any given pressure. In addition, the boiling point is easy to measure because it remains constant as long as the pressure stays the same. These features make the boiling point a very useful physical property for identifying liquids. For example, if we find that a particular pure, clear, colorless liquid boils at 78 °C at a pressure of 1 atm, we know that it can *only* be one of those substances that boils at 78 °C when the pressure is 1 atm. We can be sure that the liquid is not water, even though it may look like water, because at this pressure water boils at 100 °C.

Besides being useful for identification purposes, the boiling point also tells us much about the nature of the molecules in a liquid. To understand this, however, we must first learn why liquids boil and why their boiling points vary with the atmospheric pressure.

In Section 10.4, we found that when we warm a liquid its vapor pressure increases. Eventually, the liquid reaches a temperature at which its vapor pressure is equal to the atmospheric pressure. At this point bubbles containing the liquid's vapor are able to form within the liquid. As a bubble grows, liquid evaporates from the inner surface of the bubble and the pressure of the vapor pushes the liquid aside and upward against the opposing pressure of the atmosphere, as shown in Figure 10.9. Bubbles of vapor are only stable if the pressure within them is equal to the atmospheric pressure. If the vapor pressure were less, the atmospheric pressure would collapse the bubbles. In scientific terms, then, the **boiling point** is the temperature at which the vapor pressure of the liquid is equal to the prevailing atmospheric pressure.

Now we can easily understand why water boils at a lower temperature in Denver than it does in New York City. Because the atmospheric pressure is lower in Denver, the water doesn't have to be made as hot in order for its vapor pressure to be equal to the atmospheric pressure. The lower temperature of boiling water at places with high altitudes, like Denver, makes it necessary to cook foods longer. In fact, on the top of a very high mountain, water boils at such a low temperature that foods won't cook in it at all. At the other extreme, a pressure cooker is a device that raises the pressure over the boiling water and thereby raises the boiling point. This makes foods cook more quickly.

On the top of Mt. Everest, water boils at only 69 °C.

EXERCISE 10.1   The atmospheric pressure at the top of Mt. McKinley in Alaska is 330 torr. Use Figure 10.8 on page 320 to estimate the boiling point of water at the top of this mountain.

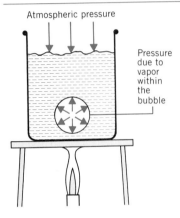

Atmospheric pressure

Pressure due to vapor within the bubble

Figure 10.9
The pressure of the vapor within a bubble in a boiling liquid pushes the liquid aside against the opposing pressure of the atmosphere

To make it possible to compare the boiling points of different liquids, chemists have chosen, somewhat arbitrarily, to record boiling points measured at a reference pressure of 1 atmoshere. The boiling point of a liquid at 1 atmosphere is called its **normal boiling point.** If a boiling point is reported without also mentioning the pressure at which it was measured, we assume it to be the normal boiling point.

In general, the boiling point is a property that reflects the strengths of the intermolecular attractive forces in a liquid. We have seen that when the attractive forces are high, the vapor pressure is low. Such a liquid must therefore be heated to a relatively high temperature to bring its vapor pressure up to the atmospheric pressure. If we examine the normal boiling points of water (100 °C) and acetone (56.2 °C), we can conclude that water molecules attract each other more strongly than do molecules of acetone. To understand why such differences exist, though, we have to learn about the kinds of attractive forces that occur between molecules. That is our next topic.

10.6 INTERMOLECULAR ATTRACTIONS

Dipole-dipole forces, London forces, and hydrogen bonds are the chief forces of intermolecular attraction

It should be apparent by now that the attractive forces that occur between molecules strongly influence the physical properties of liquids. They also play a key role in determining the physical properties of solids, and they are responsible for the nonideal behavior of real gases. Therefore, to understand the differences in behavior among various gases, liquids, or solids, we must learn about the kinds of intermolecular attractions and their relative strengths.

First, it is important to realize that the attractions *between* molecules are always much weaker than the attractions *within* molecules. For example, in a molecule of HCl, the hydrogen atom and chlorine atom are held together very tightly by a covalent bond. Neighboring HCl molecules are attracted to each other by much weaker forces. Therefore, when a particular chlorine atom moves, the hydrogen atom that is bonded to it is forced to follow along, and the HCl molecule remains intact as it moves about, as illustrated in Figure 10.10 on page 322.

Figure 10.10
Strong attractions exist between
H and Cl atoms within HCl
molecules. Weaker attractions exist
between neighboring HCl
molecules.

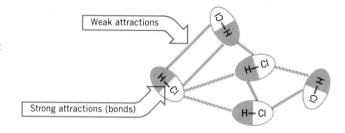

Weak attractions

Strong attractions (bonds)

Dipole-dipole attractions

In Chapter Eight we saw that HCl is an example of a polar molecule. Molecules like this have a partial positive charge at one end and a partial negative charge at the other. They tend to line up so that the positive end of one dipole is near the negative end of another. Thermal energy, however, causes the molecules to jiggle about, so the alignment isn't perfect. Nevertheless, there is still a net attraction between the polar molecules, as shown in Figure 10.11. We call these attractive forces **dipole-dipole attractions.** Generally, they are only about 1% as strong as covalent bonds.

Figure 10.11
Attractions between polar
molecules occur because the
molecules tend to align themselves
so that opposite charges are near
each other.

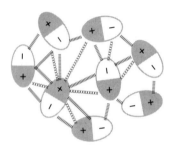

Attractions (⊷) are greater
than repulsions (⤬), so the
molecules feel a net attraction
to each other.

London forces

It is fairly easy to understand the mutual attractions between polar molecules, but it isn't at all apparent, at first glance, why nonpolar molecules should attract each other. We know that they do because they can be liquefied. Even atoms of the noble gases—such as argon, neon, and helium—can be condensed to a liquid; so there must be some attractions between the atoms to hold them together in the liquid state. These attractions, however, are much weaker than dipole-dipole attractions. This can be shown be comparing some boiling points. Table 10.1 gives data for two polar substances—HCl and H_2S—and two nonpolar substances—F_2 and Ar. All four substances are somewhat similar in that they have similar formula weights and the same number of electrons. Notice that the two nonpolar substances have very similar boiling points. The two polar substances also have similar boiling points. However, the nonpolar substances have boiling points that are at least 100 degrees lower than those of the polar sub-

Table 10.1

Boiling Points for Some Polar and Nonpolar Substances

Substance		Boiling Point (°C)	Formula Weight	Number of Electrons
HCl	} polar	−84.9	36	18
H_2S		−60.7	34	18
F_2	} nonpolar	−188.1	38	18
Ar		−185.7	40	18

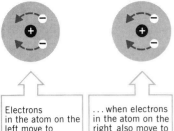

Electrons in the atom on the left move to the left...

...when electrons in the atom on the right also move to the left.

stances. We have seen that low boiling points indicate weak intermolecular attractions.

In 1930, Fritz London, a German physicist, offered a simple explanation for the weak attractions between nonpolar particles. In any atom or molecule, the electrons are not stationary—they are constantly moving. If we could examine the motion of the electrons in two neighboring particles, we would find that the movement of the electrons of one particle influences the movement of the electrons of the other. Because of their mutual repulsion, the electrons of the two particles try to stay as far apart as possible. As an electron of one particle gets near the other particle, electrons on the second particle move away as far as they can. To some extent, the electron density in both particles flickers back and forth in a synchronous fashion, as illustrated in Figure 10.12. This figure depicts instantaneous "frozen" views of the electron density. At any given instant, the electron density of a particle can be unsymmetrical—the motion of the electrons causes the particle to be a momentary or **instantaneous dipole.** Because the electron movement in one particle influences that in its neighbor, the formation of the instantaneous dipole causes another dipole to be formed in the particle alongside. This is called an **induced dipole.** The positive end of one dipole points at the negative end of the other, so there is an attraction between them. The attraction is momentary, however. The constant movement of the electrons causes the dipoles to vanish the next instant, perhaps to be formed a fraction of a second later in the opposite direction, when they experience another momentary attraction. Thus, the small, short-lived dipoles cause momentary tugs between the particles and thereby give rise to the attractive forces between them. These weak *instantaneous dipole-induced dipole forces* are called **London forces.**

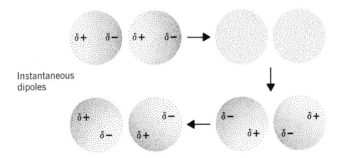

Instantaneous dipoles

Figure 10.12
Instantaneous or "frozen" views of the electron density in two neighboring particles.

London forces are the only kinds of attractions present between nonpolar molecules. They also are present between polar molecules (in addition to the regular dipole-dipole attractions), and they even occur between ions. However, because London forces are weak, their effects contribute relatively little to the overall attractions between ions.

The strengths of London forces depend on several factors. One of them is the size of the electron cloud of the particle. In general, when the electron cloud is large, the outer electrons are not held very tightly by the nucleus (or nuclei, if the particle is a molecule). This makes the electron cloud "mushy" and rather easily deformed (Figure 10.13), which is precisely the condition that most favors the formation of an instantaneous dipole or the creation of an induced dipole. Particles with large electron clouds therefore form short-lived dipoles more easily and experience stronger London forces than do similar particles with small electron clouds. The effects of size can be seen if we compare the boiling points of the halogens or the noble gases (Table 10.2). Those that are large have higher

Figure 10.13
A large electron cloud is easily deformed; a small electron cloud is not.

Table 10.2
Boiling Points of the Halogens and Noble Gases

Group VIIA	Boiling Point (°C)	Group O	Boiling Point (°C)
F_2	−188.1	He	−268.6
Cl_2	−34.6	Ne	−245.9
Br_2	58.8	Ar	−185.7
I_2	184.4	Kr	−152.3
		Xe	−107.1
		Rn	−61.8

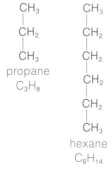

CH₃
|
CH₂
|
CH₃
propane
C_3H_8

CH₃
|
CH₂
|
CH₂
|
CH₂
|
CH₂
|
CH₃
hexane
C_6H_{14}

boiling points (and therefore stronger intermolecular attractions) than those that are small.

A second factor that affects the strength of London forces is the number of atoms in a molecule. For molecules containing the same elements, the London forces increase with the number of atoms. The hydrocarbons (see Table 10.3) are a good example. You should recall that they are composed of chains of carbon atoms with hydrogen atoms bonded to the carbon atoms all along the chain. If we compare two hydrocarbon molecules of different chain length—for example, C_3H_8 and C_6H_{14}—the one having the longer chain also has the higher boiling point. This means that the molecule with the longer chain length experiences the stronger attractive forces. This is so because it has more places along its length

Table 10.3
Boiling Points of Some Straight-Chain Hydrocarbons

Molecular Formula	Boiling Point (°C at 1 atm)
CH_4	−161.5
C_2H_6	−88.6
C_3H_8	−42.1
C_4H_{10}	−0.5
C_5H_{12}	36.1
C_6H_{14}	68.7
.	.
.	.
.	.
$C_{10}H_{22}$	174.1
.	.
.	.
.	.
$C_{22}H_{46}$	327

where it can be attracted to other molecules, as illustrated in Figure 10.14. Even if the strength of attraction at each location is about the same, the *total* attraction experienced by the longer C_6H_{14} molecule is greater than that felt by the shorter C_3H_8 molecule. In Table 10.3 you can see that some of the hydrocarbons have boiling points that are quite high, which shows that the cumulative effects of London forces can result in very strong attractions.

Hydrogen bonds
The last kind of attractive force that we will discuss is a special case of dipole-dipole attraction. When hydrogen is covalently bonded to a very small, highly electronegative atom such as fluorine, oxygen, or nitrogen, unusually strong dipole-dipole attractions are often observed. There are two reasons for this. First, because of the large electronegativity difference, the F—H, O—H, or

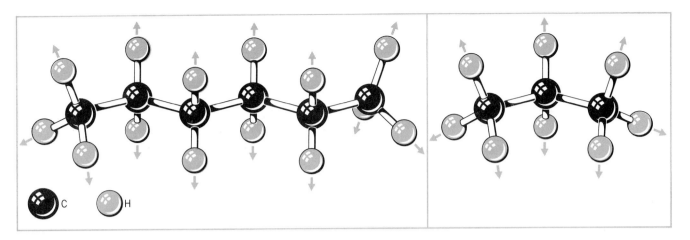

Figure 10.14
The C_6H_{14} molecule has more places along its length where it can be held than does the shorter C_3H_8 molecule.

Hydrogen bonding between water molecules in ice causes the molecules to be further apart in the solid than in the liquid. This makes ice less dense than liquid water, which is why icebergs like this float.

N—H bonds are very polar. The ends of these dipoles carry a substantial amount of charge. Second, because of the small size of the atoms involved, the charge on the end of a dipole is also highly concentrated. This makes it particularly effective at attracting the end of opposite charge on a neighboring dipole. These two factors combine to produce attractions, called **hydrogen bonds,** that are about five times stronger than normal dipole-dipole attractions. The strength of hydrogen bonds causes molecules that experience them to have some very unusual properties.

Liquids in which hydrogen bonding is important have vapor pressures that are unexpectedly low because their molecules must have large amounts of kinetic energy to escape the attractions of neighboring molecules at the surface. These low vapor pressures mean that such liquids must be heated to higher temperatures to bring their vapor pressures up to 1 atm, so their boiling points are unexpectedly high. This is apparent if we compare the boiling points of the compounds formed between hydrogen and the elements in Groups IVA, VA, VIA and VIIA (see Figure 10.15). Those that contain Group IVA elements are all symmetrical tetrahedral molecules and are therefore nonpolar. Their boiling points increase as we do *downward* in the group because the London forces become stronger as the molecules become larger. The same general trend is also found in Groups VA, VIA, and VIIA, except that NH_3, H_2O, and HF have unexpectedly high boiling points. These high boiling points are due to hydrogen bonding.

Figure 10.15
Boiling points of the hydrogen compounds of elements of Groups IVA, VA, VIA, and VIIA of the periodic table. The actual boiling points of HF, H_2O, and NH_3 are higher than expected, based on the trends shown by the other hydrogen compounds.

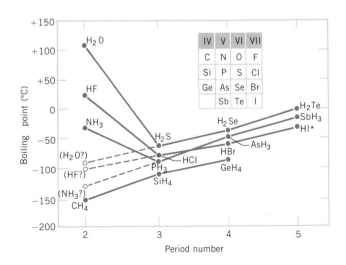

Many molecules in biological systems contain N—H and O—H bonds. Hydrogen bonding is often very important in determining the overall structures and shapes of these large molecules, as we will see in Chapter Nineteen.

10.7 CRYSTALLINE SOLIDS

In a crystal the atoms, molecules, or ions are arranged in a highly regular repeating pattern

Figure 10.16 is a photograph of crystals of one of our most familiar chemicals, sodium chloride—ordinary table salt. Notice that each particle is very nearly a perfect little cube. You might think that the manufacturers went to a lot of trouble and expense to make such uniformly shaped crystals. Actually, they could hardly avoid it. Whenever a solution of NaCl is evaporated, the crystals that form have edges that intersect at 90° angles. Cubes, then, are the norm, not the exception.

Figure 10.16
Crystals of table salt. The size of the tiny cubic sodium chloride crystals can be seen in comparison with a penny.

The regular, symmetrical shapes of snowflakes are caused by the highly organized packing of water molecules within crystals of ice.

When most substances freeze, or when they separate out as a solid from a solution that is being evaporated, they normally form crystals that have highly regular features. For example, the crystals have flat surfaces that meet at angles which are characteristic for a given substance. The regularity of these surface features reflects a high degree of order among the particles that lie within the crystal. This is true whether the particles are atoms, molecules, or ions.

The particles in crystals are arranged in patterns that repeat over and over again in all directions. The overall pattern that results is called a **crystal lattice.** Its high degree of regularity is the principle feature that makes solids different from liquids—a liquid lacks this long-range repetition of structure because the particles in a liquid are jumbled and disorganized as they move about.

Because there are millions of chemical compounds, it might seem that an enormous number of different kinds of lattices are possible. If this were true, studying solids would be hopelessly complex. Fortunately, however, the number of *kinds* of lattices that are mathematically possible is quite limited. This fact has allowed for a great deal of progress in understanding solid structures.

To describe the structure of a crystal it is convenient to view it as being composed of a huge number of simple, basic units called **unit cells.** By repeating this simple structural unit up and down, back and forth, in all directions, we can build the entire lattice. This is illustrated in Figure 10.17 for the simplest and most symmetrical of all unit cells called the **simple cubic.** This unit cell is a cube having atoms (or molecules or ions) at each of its eight corners. Stacking these unit cells gives a simple cubic lattice.

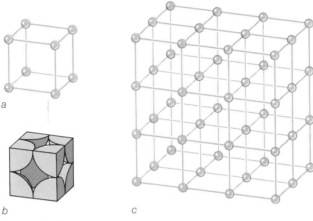

a

b   c

Figure 10.17

(*a*) A simple cubic unit cell showing the locations of the lattice positions. (*b*) A simple cubic unit cell with atoms having their nuclei at the corners. (*c*) A portion of a simple cubic lattice built by stacking simple cubic unit cells.

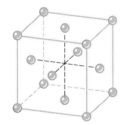

Figure 10.18
A face-centered cubic unit cell.

Two other cubic unit cells are also possible: face-centered cubic and body-centered cubic. The **face-centered cubic (fcc)** unit cell has identical particles at each of the corners plus another in the center of each face, as shown in Figure 10.18. Many common metals—copper, silver, gold, aluminum, and lead, for example—form crystals that have face-centered cubic lattices. Each of these metals has the same *kind* of lattice, but the sizes of their unit cells differ because the sizes of the atoms differ (Figure 10.19).

Figure 10.19
Two similar face-centered cubic unit cells. The atoms are arranged in the same way, but their unit cells have edges of different lengths because the atoms are of different sizes.

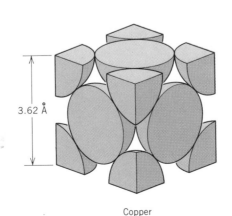

3.62 Å

Copper

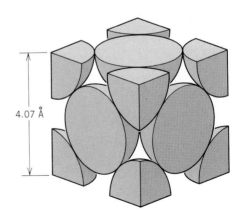

4.07 Å

Gold

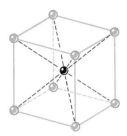

Figure 10.20
A body-centered cubic unit cell.

The **body-centered cubic (bcc)** unit cell has identical particles at each corner plus one in the center of the cell, as illustrated in Figure 10.20. The body-centered cubic lattice is also common among a number of metals—examples are chromium, iron, and platinum. Again, these are substances with the same *kind* of lattice, but the dimensions of the lattices reflect the sizes of the particular atoms.

Not all unit cells are cubic. Some have edges of different lengths or edges that intersect at angles other than 90°. Although you should realize that these other unit cells and the lattices they form exist, we will limit the remainder of our discussion to cubic lattices.

We have seen that a number of metals have fcc or bcc lattices. The same is true for many compounds. Figure 10.21, for example, is an illustration of a cutaway view of a portion of a sodium chloride crystal. The colored particles represent Na$^+$ ions. Notice that they are located at the lattice positions that correspond to a face-centered cubic unit cell. The Cl$^-$ ions fill the spaces between the Na$^+$ ions. Sodium chloride is said to have a face-centered cubic lattice, and the cubic shape of this lattice is the reason that NaCl crystals take on a cubic shape when they form.

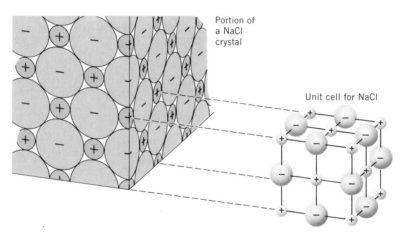

Portion of a NaCl crystal

Unit cell for NaCl

Figure 10.21
A face-centered cubic unit cell can be seen in the structure of NaCl.

Many of the alkali halides (Group IA-VIIA compounds), such as NaCl and KCl, crystallize with an fcc lattice. Since sodium chloride and potassium chloride both have the same kind of lattice, Figure 10.21 also could be used to describe the unit cell of KCl. The sizes of their unit cells are different, however, because K$^+$ is a larger ion than Na$^+$.

A major point to be learned from the preceding discussion is that a single lattice type can be used to describe the structures of many different crystals. For this reason, a handful of different lattice types is all we need to describe the crystal structure of every possible chemical element or compound.

X-RAY DIFFRACTION

Measuring the angles at which X rays of known wavelength are scattered by a crystal allows calculation of the distances between planes of atoms in a solid

When atoms are bathed in X rays, they absorb some of the radiation and then emit it again in all directions. In effect, each atom becomes a tiny X-ray source. If

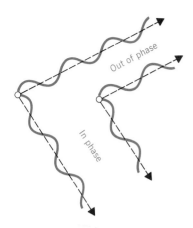

we look at radiation from two such atoms (Figure 10.22), we find that the X rays emitted are in phase in some directions and out of phase in others. In Chapter Seven we saw that such constructive and destructive interference produces a diffraction pattern.

In a crystal, there are enormous numbers of atoms evenly spaced throughout the lattice. When the crystal is bathed in X rays, intense beams of diffracted X rays caused by constructive interference appear in certain specific directions, while in other directions no X rays appear because of destructive interference. The diffraction pattern thus produced can be detected using photographic film, as is illustrated in Figure 10.23. (The film is darkened only where the X rays strike.)

Figure 10.22
X rays emitted from atoms are in phase in some directions and out of phase in other directions.

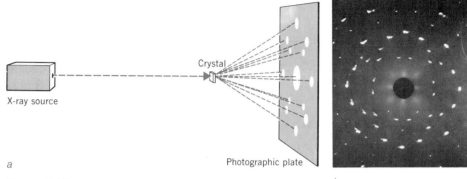

a

Photographic plate

b

Figure 10.23
X-ray diffraction. (a) The production of an X-ray diffraction pattern. (b) An X-ray diffraction pattern produced by sodium chloride.

In 1913, the British physicist William Henry Bragg and his son William Lawrence Bragg discovered that just a few variables control the appearance of such an X-ray diffraction pattern. These are shown in Figure 10.24, which illustrates the proper conditions necessary to obtain constructive interference of the X rays from successive layers of atoms (planes of atoms) in a crystal. A beam of X rays having a wavelength, λ, strikes the layers at an angle, θ. Constructive interference causes an intense diffracted beam to emerge at the same angle θ. The Braggs derived an equation relating λ, θ, and the distance between the planes of atoms, d,

$$n\lambda = 2d \sin \theta \qquad (10.1)$$

Bragg and his son shared the 1915 Nobel Prize for physics.

where n is an integer (i.e., n can equal 1 or 2 or 3, etc.). This equation, called the **Bragg equation,** is the basic tool used by scientists in their study of solid structures. Let's briefly examine how they use it.

Figure 10.24
Diffraction of X rays from successive layers of atoms in a crystal.

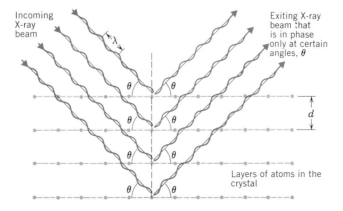

Special Topic 10.1   X Rays and Biochemistry

Biochemists have found that X-ray diffraction is extremely useful for studying the structures of large molecules. The most famous example of this use was the experimental determination of the molecular structure of DNA. DNA is found in the nuclei of cells and serves to carry an organism's genetic information. In 1953, using X-ray diffraction photographs of DNA fibers obtained by Rosalind Franklin and Maurice Wilkins, James Watson and Francis Crick came to the conclusion that the DNA structure consists of the now-famous double helix. We will examine this impor-

tant structure in more detail in Chapter Nineteen. Watson, Crick, and Wilkins shared the 1962 Nobel Prize for physiology and medicine for their discovery.

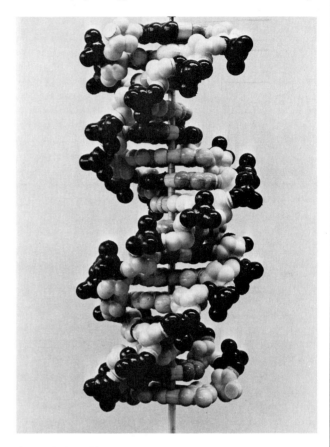

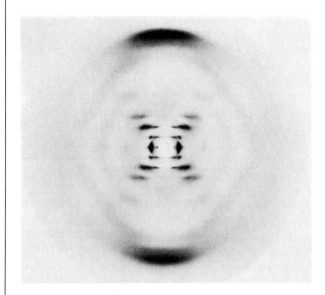

X-ray diffraction photograph of DNA.

A model of the DNA structure.

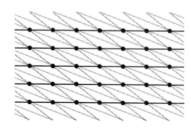

Figure 10.25
A two-dimensional pattern of points with many possible sets of parallel lines. In a crystal there are many sets of parallel planes.

In any crystal, many different sets of planes can be passed through the atoms. Figure 10.25 illustrates this idea in two dimensions for a simple pattern of points. When a crystal produces an X-ray diffraction pattern, many spots are observed because of diffraction from the many sets of planes. The physical geometry of the apparatus that is used to record the diffraction pattern allows the measurement of the angles at which the diffracted beams emerge from each distinct set of planes. Knowing the wavelength of the X rays, the values of n (which can be determined), and the measured angles, θ, the distances between planes of atoms, d, can be computed. The next step is to use the calculated interplanar distances to work backwards to deduce where the atoms in the crystal must be located so that layers of atoms are indeed separated by these distances. If this sounds like a difficult task, it is! Some sophisticated mathematics as well as computers are needed to accomplish it. The efforts, however, are well rewarded because the calculations give the locations of atoms within the unit cell and the distances between them. This information, plus a lot of chemical "common sense," is used by chemists to arrive at the shapes and sizes of the molecules in the crystal.

10.9 PHYSICAL PROPERTIES AND CRYSTAL TYPES

The physical properties of crystals are determined by the kinds of particles they are composed of and the attractive forces between them

You know from personal experience that solids exhibit a wide range of physical properties. Some, such as diamond, are very hard; others, such as naphthalene (moth flakes) or ice, are soft, by comparison, and are easily crushed. Some solids, such as salt crystals or iron, have high melting points, whereas others, like candle wax, melt at low temperatures. Some conduct electricity well, but others are nonconducting.

Physical properties such as these depend on the kinds and strengths of the attractive forces that hold the particles together in the solid. Even though we can't make exact predictions about such properties, some generalizations do exist. In making these generalizations we can divide crystals into several types according to the kinds of particles located at sites in the lattice and the kinds of attractions that exist between the particles.

We have already discussed some of the properties of **ionic crystals** in Chapter Six. We saw that they are hard, have high melting points, and are brittle. When they melt, the resulting liquid conducts electricity well. These properties reflect the strong attractive forces between ions of opposite charge as well as the repulsions that occur when ions of like charge are placed near each other.

Molecular crystals are solids in which the lattice sites are occupied either by atoms—as in solid argon or krypton—or by molecules—as in solid CO_2, SO_2, or H_2O. Such solids tend to be soft and have low melting points because the particles in the solid experience relatively weak intermolecular attractions. The crystals are soft because little effort is needed to separate the particles or to move them past each other. The solids melt at low temperatures because the particles need little kinetic energy to break away from the solid. If the crystals contain only individual atoms, as in solid argon or krypton, or if they are composed of nonpolar molecules such as carbon dioxide (dry ice), the only attractions between the particles are relatively weak London forces. In crystals containing polar molecules, such as sulfur dioxide, the major forces that hold the particles together are dipole-dipole attractions. In crystals such as water, the primary forces of attraction are due to hydrogen bonding.

Covalent crystals are solids in which lattice positions are occupied by atoms that are covalently bonded to other atoms at neighboring lattice sites. The result is a crystal that is essentially one gigantic molecule. These solids are sometimes called network solids because of the interlocking network of covalent bonds extending throughout the crystal in all directions. A typical example is the diamond, the structure of which is illustrated in Figure 10.26. Covalent crystals tend to be hard and to have very high melting points because of the strong attractions between covalently bonded atoms. Other examples of covalent crystals are quartz (SiO_2—typical grains of sand) and silicon carbide (SiC—a common abrasive used in sandpaper).

Metallic crystals have properties that are quite different from the other three types of crystals that we've discussed. They conduct heat and electricity well, and they have the luster that we characteristically associate with metals. A number of different models have been developed to describe metals. The simplest one views the crystal as having positive ions at the lattice positions which are surrounded by electrons in a cloud that spreads throughout the entire solid, as illustrated in Figure 10.27. The electrons in this cloud belong to no single positive ion, but rather to the crystal as a whole. Because the electrons aren't localized on any one atom, they are free to move easily, which accounts for the electrical conductivity of metals. By their movement, the electrons can also transmit

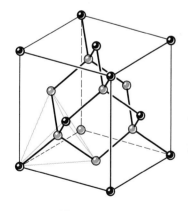

Figure 10.26
The structure of diamond. Notice that each carbon atom is covalently bonded to four others at the corners of a tetrahedron. This structure extends throughout an entire diamond crystal.

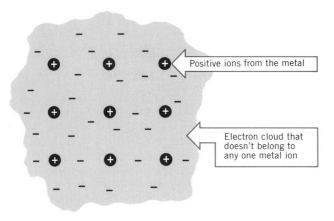

Figure 10.27
The "electron sea" model of a metallic crystal.

kinetic energy rapidly through the solid, so metals are also good conductors of heat. This model explains the luster of metals, too. When light shines on the metal, the loosely held electrons vibrate easily and readily re-emit the light with essentially the same frequency and intensity.

It is not possible to make many simple generalizations about the melting points of metals. We've seen before that some, such as tungsten, have very high melting points, while others, such as mercury, have quite low melting points. To some degree, the melting point depends on the charge of the positive ions in the metallic crystals. Group IA elements tend to exist as 1+ ions while Group IIA elements tend to lose two electrons to the lattice and form 2+ ions. The higher charge on the Group IIA ions would be expected to produce stronger attractions and give these metals higher melting points than their neighbors in Group IA. (For example, magnesium melts at 650 °C, while sodium melts at 98 °C.) Those metals with very high melting points, like tungsten, must have *very* strong attractions between their atoms. This suggests that there is probably some covalent bonding between them as well.

The different ways of classifying crystals and a summary of their general properties are given in Table 10.4.

EXAMPLE 10.1   Identifying Crystal Types

Problem: The metal osmium, Os, forms an oxide with the formula OsO_4. The soft crystals of OsO_4 melt at 40 °C, and the resulting liquid does not conduct electricity. In what form does OsO_4 probably exist in the solid?

Solution: The characteristics of the OsO_4 crystals—softness and low melting point—suggest that solid OsO_4 exists as molecular crystals that contain *molecules* of OsO_4. This is further supported by the fact that liquid OsO_4 does not conduct electricity, which is evidence for the lack of ions in the liquid.

EXERCISE 10.2   Boron nitride, which has the empirical formula BN, melts under pressure at 3000 °C and is as hard as a diamond. What is the probable solid-type for this compound?

EXERCISE 10.3   Crystals of elemental sulfur are easily crushed and melt at 113 °C to give a clear yellow liquid that doesn't conduct electricity. What is the probable crystal type for solid sulfur?

Table 10.4
Types of Crystals

Crystal Type	Particles Occupying Lattice Sites	Kinds of Attractive Forces	Typical Examples	Typical Properties
Ionic	Positive and negative ions	Attractions between ions of opposite charge	NaCl $CaCl_2$ $NaNO_3$	Hard; high melting points; nonconductors of electricity as solids, but conductors when melted
Molecular	Atoms or molecules	Dipole-dipole attractions London forces Hydrogen bonding	HCl N_2, CO_2 H_2O	Soft; low melting points; nonconductors of electricity in both solid and liquid
Covalent (network)	Atoms	Covalent bonds between atoms	Diamond SiC (silicon carbide) SiO_2 (sand, quartz)	Very hard; very high melting points, nonconductors of electricity
Metallic	Positive ions	Attractions between positive ions and an electron cloud that extends throughout the crystal	Cu Ag Fe Na Hg	Range from very hard to soft; melting points range from high to low; conduct electricity well in both solid and liquid; have characteristic luster

10.10 NONCRYSTALLINE SOLIDS

Some liquids cease to flow as they are cooled and never become crystalline when they solidify

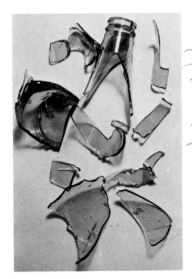

Pieces of broken glass have sharp edges, but their surfaces are not flat planes.

If a cubic salt crystal is broken, the pieces still have flat faces that intersect at 90° angles. On the other hand, if you shatter a piece of glass, the pieces often have surfaces that are not flat. Instead, they tend to be smooth and curved. This behavior illustrates a major difference between crystalline solids, like NaCl, and noncrystalline solids, or **amorphous solids,** such as glass.

The word amorphous is derived from the Greek word *amorphos,* which means "without form." Amorphous solids do not have long-range repetitive internal structures such as those found in crystals. In many ways they are more like liquids than solids. Examples are ordinary glass and many plastics. In fact, *glass* is sometimes used as a general term to refer to any amorphous solid.

Substances that form amorphous solids usually consist of long, chainlike molecules that are intertwined in the liquid state somewhat like long strands of cooked spaghetti. To form a crystal from the melted material, these long molecules would have to become untangled and line up in specific patterns. But as the liquid is cooled, the molecules move more slowly. Unless the liquid is cooled extremely slowly, the molecular motion decreases too rapidly for the untangling to take place, and the substance solidifies with the molecules still intertwined.

Compared to substances that produce crystalline solids, those that form amorphous solids behave quite oddly when they are cooled. Those that form crystals solidify at a constant temperature, as shown in Figure 10.28a. As the liquid is cooled it eventually reaches the substance's freezing point, and crystals begin to

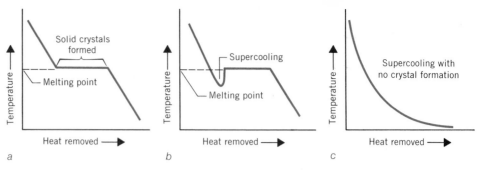

Figure 10.28
Cooling curves. (*a*) A substance that forms a crystalline solid.
(*b*) Supercooling followed by crystallization. (*c*) Supercooling of
an amorphous solid.

In an amorphous solid, long molecules are tangled and disorganized.

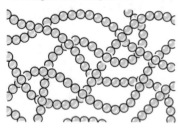

Some scientists prefer to reserve the term *solid* for crystalline substances, so they refer to glass as a "supercooled liquid."

form. Even though more heat continues to be removed, the temperature remains constant until all of the liquid has frozen. Only then does the temperature of the solid begin to drop.

Sometimes a liquid can actually be cooled below its freezing point. As the temperature approaches the freezing point, particles of the liquid may not be aligned in just the proper way for them to form a crystal. Thus, the temperature may continue to fall below the freezing point until, by chance, the particles in some small portion of the liquid suddenly find themselves properly arranged for a small crystal to form. This crystal grows rapidly, and the temperature rises again to the freezing point. The temperature then stays constant until all the liquid has frozen, as shown in Figure 10.28*b*. While the liquid has a temperature below its freezing point, it is said to be **supercooled.**

With substances that form amorphous solids, crystallization of the melted material never occurs because the molecules can't become untangled before they are frozen in place at low temperature. Figure 10.28*c* shows a typical cooling curve for this kind of substance. Cooling of the liquid continues until the substance no longer has the ability to flow, at which point we speak of it as a solid. A better term for an amorphous solid, however, is **supercooled liquid,** because it has never actually frozen.

Amorphous solids also soften gradually when they are heated. This is the reason you can heat glass tubing in a flame to soften it so that you can bend it. By contrast, if you warm an ice cube (crystalline water), it won't become soft gradually—at 0 °C it suddenly melts and drips all over you!

10.11 LIQUID CRYSTALS

Some liquids at certain temperatures possess qualities that we normally associate with the crystalline state

Certain liquids have structures that are not altogether disordered and disorganized. Although they are able to flow like liquids, they also possess crystallike ordered structures and display a few of the physical properties that are usually associated with crystalline solids. Because of this dual behavior, such substances are called **liquid crystals.**

The most widely studied types of liquid crystals are formed by substances that exhibit these special properties within a relatively narrow range of temperatures just above their melting points. They are called **thermotropic substances** because they behave as liquid crystals only within this temperature range. At higher temperatures they act like any other liquid.

Thermotropic substances are divided into three types according to the kind of semicrystalline structures they have. **Nematic** liquid crystals, Figure 10.29, have

Figure 10.29
Packing of rod-like molecules in a nematic liquid crystal.

Figure 10.30
Packing of rodlike molecules in a smectic liquid crystal.

Figure 10.31
Cholesteric liquid crystals are composed of layers of parallel rod-shaped molecules.

An electronic calculator with an LCD (liquid crystal display).

long rodlike molecules that are packed together like short pieces of uncooked spaghetti. The molecules can move backwards and forwards relatively easily, but they remain essentially parallel. **Smectic** liquid crystals also consist of rodlike molecules, but the parallel rods are arranged in layers, as illustrated in Figure 10.30. In this structure the layers can pass over one another, but the rods within the layers remain relatively fixed. The third class of thermotropic liquid crystals is formed by molecules similar in structure to cholesterol—a substance you have probably heard of in relation to hardening of the arteries. These **cholesteric** liquid crystals consist of layers of rodlike molecules in which the rods are arranged in a nematic fashion. Figure 10.31 shows how the parallel rods in one layer are oriented in a slightly different direction from those in the layers above and below.

One of the most interesting properties of liquid crystals, and the one that has made them the most useful, is their effect on light. For example, even very small changes in the temperature of a cholesteric substance give rather startling changes in color, as shown in Color Plate 4A. As the temperature rises, the distance between layers of molecules increases, which causes the color of the reflected light to change. A similar phenomenon can be observed in an oil slick on water—the color of the reflected light is determined by the thickness of the layer of oil on the water. The effect of temperature changes on cholesteric substances has some very interesting applications. One that is promising is thermal mapping of the body. In this procedure, the body is painted with the cholesteric substance. This allows the warm areas over arteries, veins, and certain organs to show up as regions with colors that are distinctly different from cooler areas alongside.

The optical properties of nematic liquid crystals can be affected by placing them in the electrical field between a pair of charged electrodes. This is the way the liquid crystal displays in electronic calculators and digital watches function. The display is made by sandwiching the liquid crystal between two transparent supporting layers (glass, for example) on which thin, transparent electrodes have been deposited. When a charge is applied to the electrodes, it causes the molecules of the liquid crystal to align themselves in such a way as to appear dark. The electrodes are shaped and arranged so that as certain ones are activated, particular numbers or letters appear.

There is some evidence that certain molecules in living systems are organized in patterns very similar to those of liquid crystals. The study of liquid crystals, then, may help in finding solutions to some important biochemical mysteries.

10.12 CHANGES OF STATE

The energy change that accompanies the change from one state to another gives a measure of the difference in the strengths of the intermolecular attractions in the two states

A **change of state** occurs when a substance is transformed from one state of matter to another. Examples are the melting of a solid or the evaporation of a liquid. Another kind of change of state is the conversion of a solid directly into a vapor. This type of change is called **sublimation.**

Solids have vapor pressures just as liquids do. In a crystal, the particles are not stationary. They vibrate back and forth about their equilibrium positions, which are located at the lattice sites. At a given temperature there is a distribution of kinetic energies, so some particles vibrate slowly while others vibrate with a great deal of energy. Some particles at the surface have high enough energies to break away from their neighbors and enter the vapor state. When particles in the vapor collide with the crystal, they can be recaptured, so condensation can

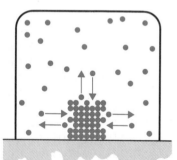

Molecules evaporate and condense on the crystal at equal rates when equilibrium is reached.

Delicate crystals of naphthalene condense directly from the vapor on the ice-cold surface of a watch glass.

occur too. Eventually, the concentration of particles in the vapor reaches a point where the rate of sublimation from the solid is the same as the rate of condensation, and a dynamic equilibrium is established. The pressure of the vapor that is in equilibrium with the solid is called the **equilibrium vapor pressure of the solid.**

The vapor pressure of a solid, like that of a liquid, is determined by the strengths of the attractive forces between its particles and by the temperature. Molecular crystals, with their relatively weak intermolecular attractions, tend to have high vapor pressures, and some of them sublime quite easily, particularly if the crystals are composed of nonpolar molecules. Common examples are dry ice (crystals of CO_2) and moth flakes (crystals of naphthalene). If exposed to the open air, both of these substances seem to disappear without a trace. At the other extreme are ionic crystals, in which the forces of attraction between the particles are very strong. Substances such as sodium chloride have extremely low vapor pressures and therefore have little tendency to sublime under ordinary conditions.

An important aspect of changes of state is that they can be made to occur under conditions in which an equilibrium exists between the states. Earlier in this chapter we saw that at a given pressure an equilibrium between two physical states, such as liquid and gas or liquid and solid, can only be established at one particular temperature. For example, we saw that at 100 °C the equilibrium vapor pressure of water is 1 atm. Another way of stating this is that when the pressure of the water vapor is 1 atm, equilibrium will *only* occur if the temperature of the liquid and vapor is 100 °C. Similarly, at a pressure of 1 atm, liquid and solid water (ice) can coexist in equilibrium only if the temperature of the mixture is 0 °C. Only at this temperature do melting and freezing occur at the same rate.

Another important feature of changes of state is that they are accompanied by energy changes. Events such as the melting of a solid, the evaporation of a liquid, or the sublimation of a solid are all endothermic. You have witnessed examples of this yourself. You have put ice into a glass of soda to keep it cool because ice absorbs heat from the soda as it melts. Similarly, the evaporation of perspiration cools your body because water absorbs heat from the body as it evaporates. And during the summer, ice cream trucks carry dry ice because, as the CO_2 sublimes, it absorbs heat and keeps the ice cream from melting.

In Chapter Five we found that energy changes that take place at constant pressure are called **enthalpy changes.** The energy absorbed or evolved in a change of state that occurs at constant pressure and temperature can therefore be described as a change in enthalpy. Usually these enthalpy changes are expressed on a "per mole" basis and are given special names that identify the kind of change involved. For example, the **molar heat of fusion,** ΔH_{fusion}, is the heat absorbed when one mole of a solid melts to give one mole of liquid at the same pressure and temperature. (Fusion is a word that means melting.) Similarly, the **molar heat of vaporization,** $\Delta H_{vaporization}$, is the heat absorbed when one mole of a liquid is converted to one mole of vapor, and the **molar heat of sublimation,** $\Delta H_{sublimation}$, is the heat absorbed by one mole of a solid when it sublimes to give a mole of vapor.

The values of ΔH for fusion, vaporization, and sublimation are all positive because the phase change involved in each case requires that molecules overcome, to some degree or other, the attractive forces of their neighbors. When a solid melts, the molecules must overcome the attractions that tend to hold them at their lattice sites in the crystal. They need energy to accomplish this. If the temperature of the substance is to remain constant, the molecules must acquire this energy from outside the crystal—energy must be absorbed. As a result, a cooling effect is observed in the surroundings. Earlier in this chapter, we saw that when a liquid evaporates, the intermolecular attractive forces within the liquid must be overcome, and it is not difficult to see that when a solid sublimes the molecules must overcome the attractive forces within the crystal.

The magnitude of ΔH for a change of state is determined by the strengths of the intermolecular attractive forces and by the amount that these forces change when the phase change occurs. For example, the melting of a given solid requires only relatively small energy inputs because the change involves only slight variations in the attractive forces between the molecules. The molecules are still quite close together after the solid melts, so, in effect, the forces are only weakened somewhat. Vaporization, however, requires rather large amounts of energy because, essentially, the molecules must be completely separated from one another. Thus, it is easy to see why ΔH_{fusion} is relatively small compared to $\Delta H_{vaporization}$. For the same substance, $\Delta H_{sublimation}$ is even larger than $\Delta H_{vaporization}$ because the attractive forces in the solid are stronger than they are in the liquid. Obviously, overcoming these stronger attractions requires more energy.

When we compare different substances, we find that liquids that have strong intermolecular attractions also have large values of $\Delta H_{vaporization}$ (Table 10.5). The heat of vaporization, which can be determined experimentally, is therefore a direct measure of the strengths of these attractions. Since the vapor pressure of a liquid is low if the attractive forces are large, there is also a relationship between $\Delta H_{vaporization}$ and the vapor pressure. At a given temperature, substances with large heats of vaporization tend to have low vapor pressures and strong intermolecular attractive forces.

A *fuse* used to protect an electrical circuit contains a wire that melts if too much current passes through it.

Table 10.5

Some Typical Heats of Vaporization

| Substance | $\Delta H_{vaporization}$ | | Kinds of Attractive Forces |
	kJ/mol	kcal/mol	
H_2O	+40.6	+9.70	Hydrogen bonding
NH_3	+21.7	+5.19	Hydrogen bonding
CCl_2F_2 (Freon-12)	+18.5	+4.42	Dipole-dipole
SO_2	+24.3	+5.81	Dipole-dipole
C_3H_8 (propane)	+16.9	+4.03	London forces

10.13 LE CHÂTELIER'S PRINCIPLE

Disturbing a system at equilibrium causes it to change in a way that counteracts the disturbing influence and brings the system to equilibrium again

In Section 10.4, we learned that when the temperature of a liquid is raised, its vapor pressure increases. We could also have reached this conclusion by analyzing what happens to the system in the following way. Initially, the liquid is in equilibrium with its vapor, which exerts a certain pressure. When the temperature is increased, equilibrium no longer exists because evaporation occurs more rapidly than condensation. Eventually, as the concentration of molecules in the vapor increases, the system reaches a new equilibrium in which there is more vapor. This greater amount of vapor exerts a larger pressure.

What happens to the vapor pressure of a liquid when the temperature is raised is an example of a general phenomenon. Whenever a system at equilibrium is subjected to a disturbance that upsets the equilibrium, the system changes in a way that will return it to equilibrium again. For a liquid-vapor equilibrium, such a disturbance is a change of temperature, as we saw in the preceding paragraph.

We will deal with many kinds of equilibria, both chemical and physical, from time to time. It would be very time consuming and sometimes very difficult to carry out a detailed analysis each time we wish to know the effects of some disturbance on the equilibrium system. Fortunately, there is a relatively simple and fast method of predicting the effect of a disturbance. It is based on a principle proposed in 1888 by a brilliant French chemist, Henry Le Châtelier (1850–1936).

> **Le Châtelier's Principle.** When a system in equilibrium is subjected to a disturbance that upsets the equilibrium, the system responds in a direction that tends to counteract the disturbance and restore equilibrium in the system.

Let's see how we can apply Le Châtelier's principle to a liquid-vapor equilibrium that is subjected to a temperature increase. To do this we have to ask ourselves: How do we go about increasing the temperature of a system? The answer, of course, is that we add heat to it. When the temperature is increased, it is the addition of heat that is really the disturbing influence.

If we write the liquid-vapor equilibrium in the form of a chemical equation, we can include the energy change as follows:

$$\text{heat} + \text{liquid} \rightleftharpoons \text{vapor}$$

This tells us that heat is absorbed by the liquid when it changes to the vapor and that heat is released when the vapor condenses to a liquid.

When heat is added to a liquid-vapor system that is at equilibrium, Le Châtelier's principle tells us that the system will try to adjust in a way that counteracts the disturbance. The system will attempt to change in a way that absorbs some of the heat that is added. This can happen if some liquid evaporates, because vaporization is an endothermic change. When liquid evaporates, the amount of vapor increases and the pressure rises. Thus, we have reached the correct conclusion in a very simple way.

A rise in temperature moves an equilibrium in the direction of an endothermic change.

Now, let's analyze what happens when the temperature of a liquid-vapor system at equilibrium is lowered. To lower the temperature, heat must be removed. The system responds in a direction that tends to replace the lost heat, so some vapor condenses and some heat is evolved. At the new equilibrium there will be more liquid and less vapor. Since there is less vapor, the pressure will be lower. Again, we come to the correct conclusion quite easily.

A decrease in temperature favors a net change that is exothermic.

EXERCISE 10.4   Use Le Châtelier's principle to predict how a temperature increase will affect the vapor pressure of a solid. (*Hint.* solid + heat $\rightleftharpoons$ vapor.)

10.14 PHASE DIAGRAMS

Graphical representations help us view the pressure-temperature relationships among the various phases of a substance

Sometimes it is useful to know whether a substance will be a liquid, a solid, or a gas at a particular temperature and pressure. A simple way of determining this is to use a **phase diagram**—a graphical representation of the pressure-temperature relationships that apply to the equilibria between the phases of the substance.

Figure 10.32 is the phase diagram for water. As we will see, it is not really as complicated as it appears at first glance. On it, there are three lines that intersect at a common point. These lines give temperatures and pressures at which equilibria between phases can exist. For example, line *AB* is the vapor pressure curve for the solid (ice). Every point on this line gives a temperature and a pressure at which ice and its vapor are in equilibrium. For instance, at $-10\,^{\circ}\mathrm{C}$ ice has a vapor pressure of 2.15 torr. This kind of information would be very useful to know if you wanted to design a system for making freeze-dried coffee. It tells you that ice will sublime at $-10\,^{\circ}\mathrm{C}$ if a vacuum pump can reduce the pressure above the ice to less than 2.15 torr.

Figure 10.32
The phase diagram for water, slightly distorted to emphasize certain features. Usually a phase diagram is drawn properly to scale so that its temperature and pressure scales may be read accurately.

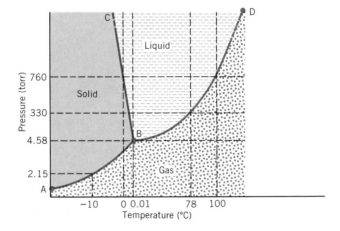

Line *BD* is the vapor pressure curve for liquid water. It gives the temperatures and pressures at which the liquid and vapor are in equilibrium. Notice that when the temperature is 100 °C, the vapor pressure is 760 torr. Therefore, this diagram also tells us that water will boil at 100 °C when the pressure is 1 atm (760 torr), because that is the temperature at which the vapor pressure equals 1 atm. At the top of Mt. McKinley, in Alaska, the atmospheric pressure is only about 330 torr. The phase diagram tells us that the boiling point of water there will be 78 °C.

The solid-vapor equilibrium line, *AB*, and the liquid-vapor line, *BD*, intersect at a common point, *B*. Because this point is on both lines, there is equilibrium between all three phases at the same time.

$$\text{liquid} \rightleftharpoons \overset{\text{vapor}}{\underset{}{\nearrow\ \ \nwarrow}} \text{solid}$$

The temperature and pressure at which this occurs is called the **triple point.** For water the triple point occurs at 0.01 °C and 4.58 torr. Every chemical substance has its own characteristic triple point, which is controlled by the balance of inter-molecular forces in the solid, liquid, and vapor.

In the SI, the triple point of water is used to define the Kelvin temperature of 273.16 K.

Line *BC*, which extends upward from the triple point, is the solid-liquid equilibrium line or *melting point line*. It gives temperatures and pressures at which solid-liquid equilibria occur. At the triple point, melting of ice occurs at +0.01 °C; at 760 torr, melting occurs at 0 °C. Thus, increasing the pressure on ice lowers its melting point.

The decrease in the melting point of ice that occurs when there is an increase in pressure can be predicted using Le Châtelier's principle and the knowledge that liquid water is more dense than ice. (More water molecules are packed into a given volume of liquid than in the same volume of solid water.) Let's consider the equilibrium

Ice floats because it is less dense than liquid water.

$$H_2O(s) \rightleftharpoons H_2O(\ell)$$

set up in an apparatus like that shown in Figure 10.33. If the piston is pushed in slightly, the pressure rises. According to Le Châtelier's principle, the system should respond in a way that will reduce the pressure. The only way that this can happen is for some of the ice to melt so that the ice-liquid mixture won't require as much space. Then the molecules won't push as hard against each other, and the pressure will drop. Thus, a pressure increase favors a volume decrease.

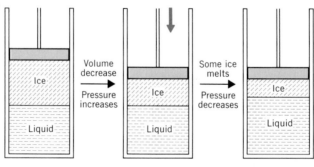

Figure 10.33
The effect of pressure on the H_2O (solid) $\rightleftharpoons$ H_2O (liquid) equilibrium.

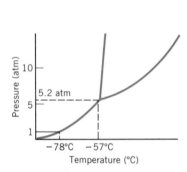

Figure 10.34
The phase diagram for CO_2.

Now, suppose we have ice at a pressure just below the solid-liquid line, *BC*. If we raise the pressure to a point just above the line, the ice will melt and become a liquid. This will happen even if we don't change the temperature. By changing the pressure *at constant temperature,* we change the system from a solid whose melting point is above that temperature to a liquid whose melting point is below that temperature. This can only be true if the melting point becomes lower as the pressure is raised.

Water is very unusual. Almost all other substances have melting points that increase with pressure. Consider, for example, the phase diagram for carbon dioxide shown in Figure 10.34. For CO_2 the solid-liquid line slants to the right. Also notice that solid carbon dioxide has a vapor pressure of 1 atm at −78 °C. This is the temperature of dry ice, which sublimes at atmospheric pressure.

Ice cream packed in dry ice is kept at −78 °C because that is the temperature at which CO_2 sublimes at 1 atm.

Besides specifying phase equilibria, the three intersecting lines on a phase diagram serve another important purpose—they define regions of temperature and pressure at which only a single phase can exist. For example, between lines *BC* and *BD* in Figure 10.32 are temperatures and pressures at which water exists as a liquid. At 760 torr, water will be a liquid anywhere between 0 °C and 100 °C. Below 0 °C at 760 torr, water will become a solid; above 100 °C at 760 torr, water will become a vapor. On the phase diagram for water, the phases that can exist in the different temperature-pressure regions are marked.

EXAMPLE 10.2 **Using a Phase Diagram**

Problem: What phase would we expect for water at 0 °C and 4.58 torr?
Solution: First we find 0 °C on the temperature axis. Then we move upward until we intersect a line corresponding to 4.58 torr. This intersection occurs in the "solid" region of the diagram. At 0 °C and 4.58 torr, then, water exists as a solid.

EXAMPLE 10.3 **Using a Phase Diagram**

Problem: What phase changes occur if the pressure on water at 0 °C is gradually increased from 2.15 torr to 800 torr?
Solution: At 0 °C and 2.15 torr, water exists as a gas. As we move upward along the 0 °C line, we first encounter the solid-vapor line. As we go above this line, the water will freeze and become ice. Continuing the climb at 0 °C, we next encounter the solid-liquid line at 760 torr. Above 760 torr the solid will melt. At 800 torr and 0 °C, the water will be liquid.

EXERCISE 10.5 What phase changes will occur if water at − 20 °C and 2.15 torr is heated to 50 °C under constant pressure?

EXERCISE 10.6 What phase will water be in if it is at a pressure of 330 torr and a temperature of 50 °C?

There is one final aspect of the phase diagram that has to be mentioned. The vapor pressure line for the liquid, which begins at point *B*, terminates at point *D*, which is known as the critical point. The temperature at *D* is called the **critical temperature,** T_c, and the pressure at *D* is called the **critical pressure,** P_c. Above the critical temperature, a distinct liquid phase can't exist, regardless of the pressure. Figure 10.35 illustrates what happens to a substance as it approaches its critical point.

Figure 10.35
What happens when a liquid is heated in a sealed container. (*a*) Below the critical temperature. (*b*) Above the critical temperature.

More dense liquid at the bottom can be detected by the interface between the phases

a

Densities of "liquid" and "vapor" have become the same — there is only one phase

b

In Figure 10.35*a*, we see a liquid in a container with some vapor above it. We can distinguish between the two phases because they have different densities, which causes them to bend light differently. This allows us to see the interface, or surface, between the more dense liquid and the less dense vapor.

This interface between liquid and gas is called a **meniscus.**

If this liquid is now heated, two things happen. First, more liquid evaporates. This causes an increase in the number of molecules per cubic centimeter of vapor which, in turn, causes the density of the vapor to increase. Second, the liquid expands (just like mercury does in a thermometer). This means that a given mass of liquid occupies more volume, so its density decreases. As the temperature of the liquid and vapor continue to rise, the vapor density and the liquid density approach each other. Eventually these densities become equal, and there no longer is any distinction between the liquid and the vapor—everything is the same (Figure 10.35b). The temperature at which this occurs is the critical temperature. At this temperature the pressure of the vapor is called the critical pressure. A substance with a temperature above its critical temperature is described as a **supercritical fluid.**

The values of the critical temperature and critical pressure are unique for every chemical substance. As with the other physical properties that we've discussed, the values T_c and P_c are controlled by the intermolecular attractions. Liquids with strong intermolecular attractions, such as water, tend to have high critical temperatures. Substances with weak intermolecular attractions tend to have low critical temperatures.

SUMMARY

General Properties. Although the chemical properties of substances are very important in many ways, their physical properties have a greater effect on our daily lives. Physical properties depend primarily on intermolecular attractions. All gases are very much alike because these attractions are weak. However, intermolecular attractions are much stronger in liquids and solids. Solids and liquids maintain a constant volume when transferred from one container to another, but only solids retain their shapes. Liquids and solids are virtually incompressible. Diffusion is slow in liquids and almost nonexistent in solids. Liquids tend to minimize their surface area because of their surface tension. The most important property of liquids is that they evaporate, which produces a cooling effect.

Vapor Pressure. When the rates of evaporation and condensation of a liquid are the same, the vapor exerts its equilibrium vapor pressure. Only changes in temperature can change the vapor pressure of a given liquid. The vapor pressure increases with increasing temperature because more molecules have enough kinetic energy to escape from the liquid at the higher temperature. In general, the vapor pressure will be high at a given temperature if the intermolecular attractions in the liquid are weak.

Boiling Point. A substance boils when its vapor pressure equals the prevailing atmospheric pressure. The normal boiling point of a liquid is the temperature at which its vapor pressure equals 1 atm. Substances with high boiling points have strong intermolecular attractions.

Intermolecular Attractions. Polar molecules attract each other primarily by dipole-dipole attractions, which arise because the positive end of one dipole attracts the negative end of another. Nonpolar molecules are attracted to each other by instantaneous dipole-induced dipole attractions called London forces. Hydrogen bonding, a special case of dipole-dipole attractions, occurs between molecules in which hydrogen is covalently bonded to a small, very electronegative element—principally, nitrogen, oxygen, and fluorine.

Crystalline Solids. Crystalline solids have very regular features that are determined by the highly ordered arrangements of particles within their lattices. The simplest portion of the lattice is the unit cell. Three cubic unit cells are possible—simple cubic, face-centered cubic, and body-centered cubic. Many different substances can have the same kind of lattice. Information about crystal structures is obtained from the crystal's X-ray diffraction pattern. Distances between planes of atoms can be calculated by the Bragg equation, $n\lambda = 2d \sin \theta$.

Crystal Types. Crystals can be divided into four general types: ionic, molecular, covalent, and metallic. Their properties depend on the kinds of particles within the lattice and on the attractions between the particles, as summarized in Table 10.4. A noncrystalline or amorphous solid is formed when a liquid is cooled without crystallization occurring. Such a solid has no sharply defined melting point and is called a supercooled liquid.

Liquid Crystals. Liquid crystals are substances that, within certain temperature ranges, display properties characteristic of both liquids and crystals. Nematic, smectic,

and cholesteric liquid crystals differ in the way their rodlike molecules are arranged.

Changes of State. Solids have vapor pressures and under appropriate conditions are able to sublime. Changes of state, such as melting, evaporation, and sublimation, can occur as equilibria. The enthalpy changes—ΔH_{fusion}, $\Delta H_{vaporization}$, and $\Delta H_{sublimation}$—are positive. Their magnitudes reflect the changes in the strengths of the intermolecular attractions when the corresponding changes of state occur. Le Châtelier's principle allows us to predict how a system will respond when its equilibrium is disturbed. Raising the temperature favors an endothermic change; raising the pressure favors a change that leads to a smaller volume.

Phase Diagrams. Temperatures and pressures at which equilibrium can exist between phases is given graphically in a phase diagram. The three equilibrium lines intersect at the triple point. The liquid-vapor line terminates at the critical point. At the critical temperature, a liquid has a vapor pressure equal to its critical pressure. Above the critical temperature a single phase exists. The equilibrium lines also divide the phase diagram into temperature-pressure regions in which a substance can exist in just a single phase.

INDEX TO EXERCISES, QUESTIONS, AND PROBLEMS
(Those numbered above 59 are problems.)

REVIEW QUESTIONS

10.7. List four items in your room whose physical properties account for their use in a particular application.

10.8. Under what conditions would we expect gases to obey the gas laws best?

10.9. Why is the behavior of gases affected very little by their chemical compositions?

10.10. Why are the intermolecular attractive forces stronger in liquids and solids than they are in gases?

10.11. Explain why liquids and gases differ in (a) compressibility and (b) rates of diffusion.

10.12. What is surface tension? Why do molecules at the

surface of a liquid behave differently than those within the interior?

10.13. What kinds of observable effects are produced by the surface tension of a liquid?

10.14. What relationship is there between surface tension and the intermolecular attractions in the liquid?

10.15. What is *wetting?* What is a *surfactant?* How does it function?

10.16. Why does evaporation lower the temperature of a liquid?

10.17. On the basis of the distribution of kinetic energies of the molecules of a liquid, explain why increasing the liquid's temperature increases the rate of evaporation.

10.18. Define *equilibrium vapor pressure*. Why do we call the equilibrium involved a *dynamic equilibrium?*

10.19. Explain why changing the volume of a container in which there is a liquid-vapor equilibrium has no effect on the vapor pressure.

10.20. What effect does increasing the temperature have on the vapor pressure of a liquid?

10.21. Below are the vapor pressures of some relatively common chemicals measured at 20 °C. Arrange these substances in order of increasing intermolecular attractive forces.

benzene, C_6H_6	80 torr
acetic acid, $HC_2H_3O_2$	11.7 torr
acetone, C_3H_6O	184.8 torr
ether, $C_4H_{10}O$	442.2 torr
water	17.5 torr

10.22. Define *boiling point* and *normal boiling point*.

10.23. Why does the boiling point vary with atmospheric pressure?

10.24. Mt. Kilimanjaro in Tanzania is the tallest peak in Africa (19,340 ft). The normal barometric pressure at the top of this mountain is about 345 torr. At what Celsius temperature would water be expected to boil there? (See Figure 10.8.)

10.25. Why is the boiling point a useful property with which to identify liquids?

10.26. Describe dipole-dipole attractions.

10.27. What are London forces? How are they affected by molecular size?

10.28. What are hydrogen bonds? How do they affect the boiling points of the compounds formed between hydrogen and the nonmetals?

10.29. Which is expected to have the higher boiling point, C_8H_{18} or C_4H_{10}? Explain your choice.

10.30. Ethanol and dimethyl ether have the same molecular formulas, C_2H_6O. Ethanol boils at 78.4 °C whereas dimethyl ether boils at −23.7 °C. Their condensed structural formulas are:

$$CH_3CH_2OH \qquad CH_3OCH_3$$
ethanol   dimethyl ether

Explain why the boiling point of the ether is so much lower than the boiling point of ethanol.

10.31. The boiling points of some common substances are given here. Arrange these substances in order of increasing strengths of intermolecular attractions.

ethanol C_2H_5OH	78.4 °C
ethylene glycol, $C_2H_4(OH)_2$	197.2 °C
water	100 °C
diethyl ether, $C_4H_{10}O$	34.5 °C

10.32. What surface features do crystals have that suggest a high degree of order among the particles within them?

10.33. What is a crystal lattice? What is a unit cell? What relationship is there between a unit cell and a crystal lattice?

10.34. Describe simple cubic, face-centered cubic, and body-centered cubic unit cells.

10.35. Make a sketch of a layer of sodium and chloride ions in a NaCl crystal. Indicate how the ions are arranged in a face-centered cubic pattern.

10.36. Only 11 different kinds of crystal lattices are possible. How can this be true, considering the fact that there are millions of different chemical compounds that are able to form crystals?

10.37. Write the Bragg equation and define the symbols.

10.38. Explain, qualitatively, how an X-ray diffraction pattern of a crystal and the Bragg equation provide information that allows chemists to figure out the structures of molecules.

10.39. Tin(IV) chloride, $SnCl_4$, has soft crystals with a melting point of −30.2 °C. The liquid is nonconducting. What type of crystal is formed by $SnCl_4$?

10.40. Magnesium chloride is ionic. What general properties are expected of its crystals?

10.41. Elemental boron is a semiconductor, is very hard, and has a melting point of about 2250 °C. What type of crystal is formed by boron?

10.42. Gallium crystals are shiny and conduct electricity. Gallium melts at 29.8 °C. What type of crystal is formed by gallium?

10.43. What kinds of particles are located at the lattice sites in a metallic crystal?

10.44. What kinds of attractive forces exist between particles in (a) molecular crystals, (b) ionic crystals, and (c) covalent crystals.

10.45. Why are covalent crystals sometimes called network solids?

10.46. What is an amorphous solid? What happens when a substance that forms an amorphous solid is cooled from the liquid state to the solid state?

10.47. What is supercooling?

10.48. What are the three types of thermotropic liquid crystals? How do they differ structurally?

10.49. Why do small temperature changes produce large changes in color in cholesteric liquid crystals?

10.51. What is sublimation?

10.50. Why is $\Delta H_{vaporization}$ larger than ΔH_{fusion}? How does $\Delta H_{sublimation}$ compare with $\Delta H_{vaporization}$?

10.52. Which would be expected to have a higher molar heat of vaporization, water (b.p., 100 °C) or alcohol (b.p., 78.4 °C)?

10.53. State Le Châtelier's principle in your own words.

10.54. For most substances, the solid is more dense than the liquid. Use Le Châtelier's principle to explain why the melting point of such substances should *increase* with increasing pressure.

10.55. Define *critical temperature* and *critical pressure*.

10.56. What phases are in equilibrium at the triple point?

10.57. Sketch the phase diagram for a substance that has a triple point at −10.0 °C and 0.25 atm, melts at −8.0 °C at 1 atm, and has a normal boiling point of 80 °C.

10.58. On the phase diagram of Question 10.57, below what pressure will the substance undergo sublimation?

10.59. Based on the phase diagram in Question 10.57, how does the density of the liquid compare to the density of the solid?

REVIEW PROBLEMS

10.60. Copper crystallizes with a face-centered cubic unit cell. The length of the edge of a unit cell is 3.62 Å. Sketch the face of a unit cell, showing the nuclei of the copper atoms at the lattice points. The atoms are in contact along the diagonal from one corner to another. The length of this diagonal is four times the radius of the copper atom. What is the atomic radius of copper?

10.61. Silver forms face-centered cubic crystals. The atomic radius of a silver atom is 1.44 Å. Draw the face of a unit cell with the nuclei of the silver atoms at the lattice points. The atoms are in contact along the diagonal. Calculate the length of an edge of this unit cell.

10.62. The molar heat of vaporization of water at 25 °C is 10.5 kcal/mol. How much energy would be required to vaporize 1.00 liter (1.00 kg) of water?

10.63. The molar heat of vaporization of acetone, C_3H_6O, is 7.23 kcal/mol at its boiling point. How much heat would be liberated by the condensation of 1.00 g of acetone?

CHAPTER ELEVEN

SOLUTIONS AND COLLOIDS

11.1 KINDS OF MIXTURES

Solutions, colloidal dispersions and suspensions are the three principal kinds of mixtures

Most of the things we see around us are mixtures, not elements or compounds. A **mixture** is anything made up of two or more elements or compounds that are *physically* combined in no particular proportion by mass. Buildings, trees, air, soil, rivers and lakes, cars, even friends and neighbors are all mixtures. While many of these are highly organized and have definite structural features, none has a strictly defined composition. For example, each of the cells in your body is a mixture—a mixture of mixtures, in fact, because several subcellular bodies, such as the cell nucleus, are present along with the membranes and cell fluids. These fluids change in composition minute by minute as nutrients move in and wastes move out. From our study of the properties of mixtures begun in this chapter, we will begin to see how these movements always occur in the right direction.

Because mixtures don't obey the law of definite proportions, their number in the universe is virtually infinite. Two substances can be mixed together in almost any ratio. For instance, sugar has a definite composition—$C_{12}H_{22}O_{11}$—and so does water—H_2O; but a mixture of these two substances may vary widely in composition. Some mixtures of sugar and water are too dilute to be tasted; others are as sweet and thick as syrup. Most of the *chemical* properties of a sugar solution can be figured out from the chemical properties of sugar and water individually. The *physical* properties, however, are something else. In this chapter we are going to concentrate mainly on how the components of mixtures interact to affect *physical* properties.

The best place to start a study of mixtures is with their simplest types: suspensions, colloidal dispersions and solutions. They are compared and contrasted in Table 11.1. Actual examples of any of these three types of mixtures will usually appear to be uniform, at least to the naked eye. Only solutions, however, are homogeneous in the sense that only in solutions are the intermixed particles at the atomic, ionic, or molecular levels of size. Thus, the relative sizes of the intermixed particles is the chief criterion for distinguishing these three types of mixtures.

Types of Matter	Examples
Elements	106
Compounds	a few million (and growing)
Mixtures	probably infinite

Table 11.1
Uniform Mixtures

Particle Sizes Become Larger →		
Solutions	**Colloidal Dispersions**	**Suspensions**
All particles are on the order of atoms, ions, or small molecules (0.1–1 nm)	Particles of at least one component are large clusters of atoms, ions, or small molecules or are very large ions or molecules (1–100 nm)	Particles of at least one component may be individually seen with a low-power microscope (100 nm)
Most stable to gravity	Less stable to gravity	Unstable to gravity
Most homogeneous	Also homogeneous but borderline	Heterogeneous but may appear to be uniform
Transparent (but often colored)	Often translucent or opaque; may be transparent	Often opaque but may appear translucent
No Tyndall effect	Tyndall effect	Not applicable (suspensions cannot be transparent)
No Brownian movement	Brownian movement	Not applicable (particles settle out)
Cannot be separated by filtration	Cannot be separated by filtration	Can be separated by filtration
Homogeneous	**to**	**Heterogeneous** →

Suspensions

A **suspension** is a mixture in which the particles of one (or more) of the substances are relatively large. These particles will have at least one dimension larger than 100 nanometers (100 nm or 100×10^{-9} m). Finely divided clay in water is an example. Unless a suspension is stirred or shaken, the large particles will sooner or later settle out.

Suspensions of solids in fluids can nearly always be separated by simple filtration through ordinary filter paper or by use of a centrifuge. In a centrifuge, a sample is whirled at high speed, which creates a gravitational-like force strong enough to pull the suspended matter down through the fluid (Figure 11.1). Blood is a suspension. Its red and white cells can be easily separated from the plasma by a centrifuge.

Colloidal dispersions

A **colloidal dispersion** is a mixture in which the particles of one (or more) substances are smaller than those in suspensions, but larger than those in solutions. They will have at least one dimension in the range of 1 to 10 nm (10^{-9} to 10^{-7} m). "Colloidal" comes from a Greek root meaning "gluelike." Old-fashioned glues form colloidal dispersions in water. Homogenized milk is a colloidal dispersion of butterfat in water that also contains the protein casein and many other substances. Several other examples of colloidal dispersions are given in Table 11.2.

A typical colloidal dispersion you might handle in the laboratory is starch and water. This mixture can be made so concentrated that it will look milky or even

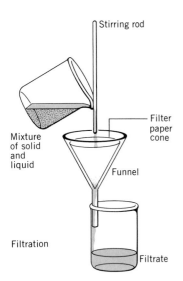

Stirring rod

Mixture of solid and liquid

Filter paper cone

Funnel

Filtration

Filtrate

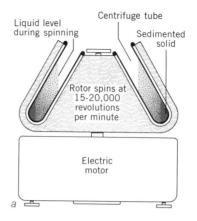

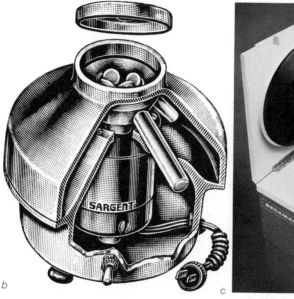

Figure 11.1

The centrifuge. (*a*) Cutaway showing how a typical, inclined-tube laboratory centrifuge works. (*b*) Photograph of a laboratory centrifuge. (*c*) An ultracentrifuge. This instrument is available with rotor speeds of 20,000 to 60,000 revolutions per minute. The chamber enclosing the centrifuge tubes can be evacuated to reduce air friction, and it may be cooled (or warmed) according to the experimental needs.

opaque (meaning no light will pass through). It can also be prepared so dilute that it will appear as clear and transparent as water—at least until you focus a strong beam of light on it. Then, where the beam is focused, the mixture will look milky, as seen in Figure 11.2. Colloidally dispersed particles are large enough to scatter light. Particles as small as atoms, ions or small molecules, however, do not scatter light, at least not as well. Therefore, you won't see this milkiness when a beam of light is focused on a solution such as sugar or salt in water because the molecules of sugar and water or ions of salt are so small. The scattering of light by colloidally dispersed particles is called the **Tyndall effect,** after British scientist John Tyndall (1820–1893). You have seen it if you have ever walked in the woods and seen sunlight "streaming" through openings in the forest canopy (Fig. 11.3). The beams of light result from the Tyndall effect as sunlight meets colloidally dispersed particles of smoke or microdroplets of liquids exuded by the trees.

Table 11.2
Colloidal Systems

Type	Dispersed Phase[a]	Dispersing Medium[b]	Common Examples
Foam	Gas	Liquid	Suds, whipped cream
Solid foam	Gas	Solid	Pumice, marshmallow
Liquid aerosol	Liquid	Gas	Mist, fog, clouds, certain pollutants in air
Emulsion	Liquid	Liquid	Cream, mayonnaise, milk
Solid emulsion	Liquid	Solid	Butter, cheese
Smoke	Solid	Gas	Dust and particulates in smog
Sol	Solid	Liquid	Starch in water, jellies,[c] paints
Solid sol	Solid	Solid	Black diamonds, pearls, opals, alloys

[a] The colloidal particles constitute the dispersed phase.
[b] The continuous matter into which the colloidal particles are scattered is called the *dispersing medium.*
[c] Sols that adopt a semisolid, semirigid form (e.g., gelatin desserts, fruit jellies) are called gels.

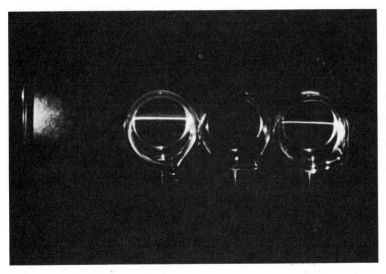

Figure 11.2
Tyndall effect. The view is of three beakers seen from the top with a source of focused light on the left. The middle beaker contains a solution. Those on either side hold dilute colloidal dispersions of starch in water. The beam is seen passing through the colloidal dispersions, but because the light is not scattered by anything in the solution, it is not seen passing through the middle beaker. (Photographed by David Crouch. From T. R. Dickson, *Introduction to Chemistry,* John Wiley & Sons, 1971. Used by permission.)

Figure 11.3
Tyndall effect as light breaks through a forest canopy.

The particles in a colloidal dispersion are generally too small to be trapped by ordinary filter paper. Milk, for example, will pass through filter paper unchanged, and so will a starch dispersion.

The stability of a colloidal dispersion—its resistance to the separation of the colloidal particles—is attributed to several factors. One is the small size of colloidal particles; the smaller they are the more stable is the dispersion and the less effective is gravity in causing a separation. Another factor is the constant erratic movements of colloidally dispersed particles in a fluid medium. The

Brown's discovery (1827) was early (and very dramatic) evidence for the existence of atoms and molecules and their kinetic motions.

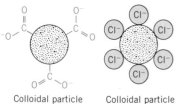

Colloidal particle with organic ionic groups

Colloidal particle with adsorbed chloride ions

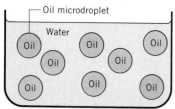

Oil in water emulsion

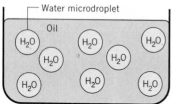

Water in oil emulsion

English botanist Robert Brown (1773–1858) was the first to notice these irregular motions when, using a microscope, he studied pollen grains suspended in water. Inside the grains he saw particles moving in erratic paths, and we now call the phenomenon of randomly moving colloidal particles the **Brownian movement**. This motion is caused by irregular buffeting received by colloidal particles from the kinetic motions of the molecules in the surrounding medium. It helps to keep colloidal particles from separating.

Another factor opposing gravity and helping to stabilize a dispersion is the electrically charged character of most colloidally dispersed particles. The particles in most **sols**—colloidal dispersions of solids in a fluid—carry just one kind of electrical charge, caused either by adsorbed ions or by ionic sites in molecules making up the particles. Like-charged sol particles naturally repel each other, and they do not grow to have masses too large to resist gravity or to be buffeted enough by the Brownian movement.

Most **emulsions**—colloidal dispersions of one liquid in another—are stabilized by a third component called an **emulsifying agent.** Mayonnaise, for example, is an oil in water emulsion of some edible oil in a dilute solution of an edible organic acid and stabilized by egg yolk. Molecules of protein in egg yolk form a "skin" around the microdroplets of oil. Margarine is a water in oil emulsion of microdroplets of water in an edible oil stabilized by a soybean product. Ordinary salad dressings (e.g., "Italian dressing") separate almost as soon as you stop shaking them because they have no emulsifying agents. Many sprays used in agriculture are oil emulsions that are diluted with water just before use.

Solutions

A **solution** is a homogeneous mixture in which all particles are the size of atoms, small molecules, or small ions. These particles have average diameters in the range of 0.1 to 1 nm. The various types of solutions and several examples are given in Table 11.3. Substances in solution cannot be separated by filtration, and they do not settle out because of gravity. Because the solute particles are so small, neither the Tyndall effect nor the Brownian movement can be seen with a solution. The emphasis in the remainder of this chapter will be on solutions.

Table 11.3
Solutions

Kinds	Common Examples
Gas in a liquid	Carbonated beverages (carbon dioxide in water)
Liquid in a liquid	Vinegar (acetic acid in water)
Solid in a liquid	Sugar in water
Gas in a gas	Air
Liquid in a gas	—
Solid in a gas[a]	—
Gas in a solid	Alloy of palladium and hydrogen[a]
Liquid in a solid	Benzene in rubber (for example, rubber "cement")
Solid in a solid	Carbon in iron (steel)

[a] Some doubt exists that this system can ever exist as a true solution.

11.2 SOLUBILITY

The nature of both solute and solvent sets limits to solubility—"likes dissolve likes"

It is easy to see how two gases can dissolve in each other. Their particles *are already separated* and moving randomly about. These particles simply diffuse among each other until they are completely mixed. Liquid or solid substances on the other hand dissolve in each other only if the forces that hold their own particles side by side are weak or, if strong, can somehow be replaced. In any case, the particles must give up the neighbors they had in their pure states and let new neighbors from the other components gather around. In some solutions, the solute particles occupy randomly distributed cavities called interstices (or interstitial spaces) between clusters of solvent molecules.[1] Atom or molecules of some gases are small enough to dissolve that way (see Figure 11.4). In most solutions, however, ions or molecules of both the solute and the solvent move about to make room for each other.

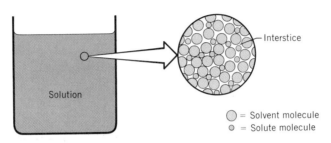

Figure 11.4
When some solutes dissolve, their molecules nestle into cavities called interstices between the molecules of the solvent.

How can we tell if two substances will form a solution? In general, if the forces of attraction between the particles in *both* the solute and the solvent are of comparable magnitude, the two can be mixed to make a solution. The particle-to-particle forces in the solution may be those either of a positive ion for a negative ion or of a dipole for a dipole, whether the dipoles be permanent or of the type involved with London forces.

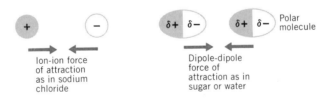

In the solvent, the particle-to-particle forces are dipole-dipole. Because water molecules are very polar, the dipole-dipole force in water is much stronger than it is in a relatively nonpolar solvent such as gasoline, benzene, or any hydrocarbon. A useful rule of thumb to predict the relative solubility of a solute in a solvent is *"likes dissolve likes."* "Likes" refers to relative polarities. Thus highly polar or ionic solutes generally dissolve far better in highly polar solvents than in nonpolar liquids. For example, salt (ionic) and sugar (polar molecular) are both soluble in water but not in benzene or gasoline. Nonpolar solutes dissolve well in nonpolar solvents. Thus greases and oils (both relatively nonpolar) are soluble in hydrocarbons such as benzene and gasoline. In contrast, if we mix something that is nonpolar, like gasoline, with something that is very polar, like water, no solution will form.

[1] Recall (Section 4.5) that the solvent is that part of a solution occurring as a *continuous* phase. The solute is the part that breaks up and becomes dispersed in the solvent. When both solute and solvent are liquids, however, which is called the solute and which the solvent is entirely a matter of one's personal preference.

EXAMPLE 11.1 **Working with the Rule "Likes Dissolve Likes"**

Problem: Methanol (wood alcohol) is a common solvent (and fuel for making fondue). It dissolves much better in water than in gasoline. Are the molecules of methanol polar or nonpolar?

Solution: Water is a polar solvent but gasoline is nonpolar. Since water is the better solvent for methanol, molecules of methanol must be "like" those of water—polar.

EXERCISE 11.1 Describe how you could use the solvents water and chloroform to determine whether a given white solid was soap (an ionic compound) or candle wax (a nonpolar molecular substance). Chloroform is much less polar than water.

EXERCISE 11.2 Ethyl alcohol, the "spirits" in alcoholic beverages, dissolves in *both* water and benzene (a nonpolar solvent). Which best describes molecules of ethyl alcohol—very polar, moderately polar, or nonpolar? Explain.

Hydration

Ionic and polar molecular compounds would be insoluble in water without the phenomenon of hydration. **Hydration** is the development of a cagelike layer of water molecules about a site on a particle that bears a full or partial charge. Figure 11.5 shows how, when a crystal of sodium chloride comes in contact with water, the polar water molecules help attract ions away from the crystal. A $\delta-$ on a water molecule attracts Na^+, a $\delta+$ on a different water molecule attracts Cl^-. Then in Figure 11.6, we see how the water molecules gather about the ions. The negative ends of the dipoles of water molecules point toward the positive sodium ions and are held in place by the attraction between unlike charges. Similarly, the positive ends of water molecules point toward the negative chloride ions. Each kind of ion becomes surrounded by a solvent cage. The ion's electrical charge is partly neutralized and becomes dispersed over the surface of the cage. Less strongly held and less organized layers of solvent gather around the inner cages. The ions are thus well insulated from each other. They no longer attract each other, at least not strongly enough to remake the crystal. The crucial

Figure 11.5
Forming a solution of salt in water. Polar water molecules bombard the surface of a crystal of sodium chloride whose ions become dislodged and hydrated.

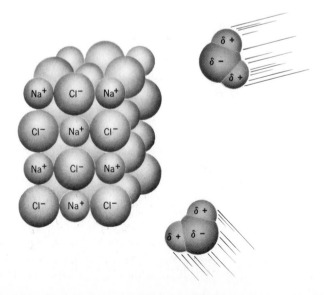

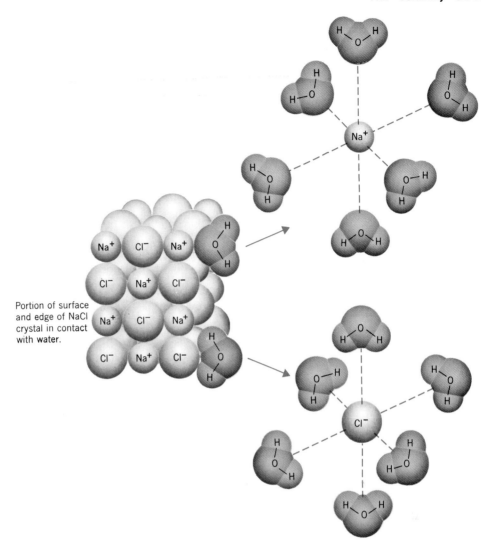

Figure 11.6
Hydration of ions. The ions of ionic substances that can dissolve
in water can trade forces of attraction to each other in the
crystal for forces of attraction to molecules of water. Likewise,
forces of attraction between water molecules can be replaced
by forces of attraction to ions.

service provided by the solvent molecules is that they give something for the
ions to be attracted to besides the neighboring ions in the crystal. The solvent, of
course, also becomes reorganized as some of its molecules abandon their other
solvent neighbors and become associated with the ions.

Nonpolar solvents such as gasoline won't dissolve ionic compounds, because
nonpolar molecules have no sites of partial charge ($\delta+$ or $\delta-$). They cannot set
up insulating cages about ions. They cannot *solvate* ions—to use the general
term that includes hydration as a special case when water is the solvent.

Even though water can hydrate ions, it doesn't dissolve all ionic compounds
equally well. Some are almost completely insoluble. Quite often substances in
which *both* the positive and negative ions have two or more charges are least
soluble in water. Thus $BaSO_4$ (barium sulfate consisting of Ba^{2+} and SO_4^{2-} ions)
and $AlPO_4$ (aluminum phosphate consisting of Al^{3+} and PO_4^{3-}) are only very
slightly soluble in water, but there are many exceptions. While NaCl (sodium
chloride) is soluble in water, AgCl (silver chloride) is not.

Solvation is to all solvents
what **hydration** is to water,
one specific solvent.

Figure 11.7
Solvation of polar molecular
compounds. Polar molecules can
trade forces of attraction to each
other for forces of attraction to the
molecules of a polar solvent, such
as water.

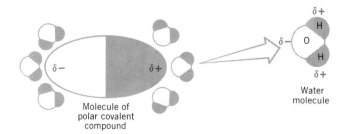

Figure 11.7
Solvation of polar molecular
compounds. Polar molecules can
trade forces of attraction to each
other for forces of attraction to the
molecules of a polar solvent, such
as water.

Molecular substances can also be hydrated, provided their molecules are very polar (see Figure 11.7). Sugar is an example. Hydration occurs around partially charged sites on neutral but polar sugar molecules. Methyl alcohol molecules do not solvate gasoline molecules, and these liquids do not dissolve in each other. Forces of attraction between polar molecules of methyl alcohol (Example 11.1) are too strong to be disrupted by the nonpolar molecules in gasoline.

The forces of hydration can be quite strong. When the solutions of many salts in water are carefully evaporated, not all of the water molecules leave. The names, formulas and uses of several examples are given in Table 11.4. As crystals of these compounds form out of an aqueous solution, some water molecules remain attached to the ions and become part of the crystal. This water is called **water of hydration** and, as indicated by the formulas in Table 11.4, the ratio of water molecules to the ions is a constant. Solids that have water of hydration are called **hydrates** and, because they obey the law of definite proportions, they are classified as compounds not as wet mixtures. Their formulas are always written to show the proportion of water present. Thus gypsum is represented by the formula $CaSO_4 \cdot 2H_2O$; a dot raised slightly above the line separates the formula of water with its coefficient from the rest of the formula. The parent substance in a hydrate is called the anhydrous form of the hydrate, where *anhydrous* means "without water." Thus $CaSO_4$ is the anhydrous form of $CaSO_4 \cdot 2H_2O$. (See Special Topic 11.1).

Table 11.4

Some Common Hydrates

Formulas	Names	Decomposition Modes and Temperatures[a]	Uses
$(CaSO_4)_2 \cdot H_2O$	Calcium sulfate sesquihydrate (plaster of paris)	$-H_2O$ (163 °C)	Casts, molds
$CaSO_4 \cdot 2H_2O$	Calcium sulfate dihydrate (gypsum)	$-2H_2O$ (163°)	Casts, molds, wallboard
$CuSO_4 \cdot 5H_2O$	Copper(II) sulfate pentahydrate (blue vitriol)	$-5H_2O$ (150°)	Insecticide
$MgSO_4 \cdot 7H_2O$	Magnesium sulfate heptahydrate (epsom salt)	$-6H_2O$ (150°)	Cathartic in medicine
		$-7H_2O$ (200°)	Used in dyeing and tanning
$Na_2B_4O_7 \cdot 10H_2O$	Sodium tetraborate decahydrate (borax)	$-8H_2O$ (60°) $-10H_2O$ (320°)	Laundry
$Na_2CO_3 \cdot 10H_2O$	Sodium carbonate decahydrate (washing soda)	$-H_2O$ (33.5°)	Water softener
$Na_2SO_4 \cdot 10H_2O$	Sodium sulfate decahydrate (Glauber's salt)	$-10H_2O$ (100°)	Cathartic
$Na_2S_2O_3 \cdot 5H_2O$	Sodium thiosulfate pentahydrate (photographer's hypo)	$-5H_2O$ (100°)	Photographic developing

[a] Loss of water is indicated by the minus sign before the symbol, and the loss occurs at the temperature given in parentheses.

Plaster of paris has been used for thousands of years in plastering. If we mix water with plaster of paris a chemical change occurs. Gypsum forms as a hard, crystalline mass.

$$(CaSO_4)_2 \cdot H_2O + 3H_2O \longrightarrow 2CaSO_4 \cdot 2H_2O$$

plaster of paris   gypsum

As the mixture sets, it expands slightly (about 0.5%) as small needlelike crystals grow outward. Because the expansion is slight and because gypsum is strong and hard, plaster of paris is widely used to make plaster casts. In order to work with it as it sets, we have to use more water than required by the reaction's stoichiometry. The excess eventually evaporates leaving small pores. The dry, finished product is about 45% air.

Gypsum is widely available. The 10 to 45 ft high dunes of the White Sands National Monument in New Mexico are almost pure gypsum. Deposits under the Montmartre in Paris have been mined for a long time—hence the name "plaster of paris" for the partly dehydrated form of gypsum.

EXAMPLE 11.2   Writing Formulas of Hydrates

Problem: Calcium hypochlorite, $Ca(OCl)_2$, present in commercial bleaching powders, can be bought as a trihydrate. Write the formula of the trihydrate.

Solution: First, write the formula of the parent compound, then a dot above the line, and then the formula of water preceded by the number of water molecules contained in the hydrate, 3 for a *tri*hydrate. Thus, the formula for calcium hypochlorite trihydrate is $Ca(OCl)_2 \cdot 3H_2O$.

EXERCISE 11.3   Iron(III) orthophosphate, $FePO_4$, sometimes used to furnish supplemental iron in the diet, comes as a dihydrate. Write its formula.

EXERCISE 11.4   A hydrated form of sodium citrate, $Na_3C_6H_5O_7 \cdot 5H_2O$, may be added to foods to control their acidity. What is the formula of its anhydrous form?

When heated, some hydrates *react* with their water of hydration and change to oxides, not anhydrous forms.

If a hydrate is heated, it will usually give up some or all of its water of hydration. For example:

$$2CaSO_4 \cdot 2H_2O \xrightarrow{125\ °C} (CaSO_4)_2 \cdot H_2O + 3H_2O$$

gypsum   plaster of paris

EXERCISE 11.5   As seen in Table 11.4, epsom salt (magnesium sulfate heptahydrate), changes to is monohydrate form at 150 °C. Write an equation for this reaction.

Granules of calcium chloride ($CaCl_2$) are often used to take moisture out of the air of damp rooms. They do this by forming the hydrate $CaCl_2 \cdot 6H_2O$. A chemical drying agent such as calcium chloride is called a **dessicant.**

EXERCISE 11.6   Calcium chloride can form three different hydrates depending on the conditions. Write the equation for the formation of its hexahydrate from the anhydrous salt.

The phenomenon of hydration occurs both in cells and in soil. Many proteins in both plant and animal cells are hydrated. Negative sites on colloidal-sized particles that are part of soil attract the positive dipoles of water molecules very strongly and help hold water in the soil. Many soil minerals occur as hydrates.

Factors affecting solubility

The usual way for describing the **solubility** of a compound in some solvent is to state the number of grams of compound that will dissolve in 100 g of the solvent at the specified temperature. Table 11.5 has several examples of solutions in which water is the solvent. Solutions that are prepared in the proportions given by the solubility data of the compound are saturated solutions at the specified temperature. One way to make sure a solution is saturated is to add some crystals of the solute. If the solution is saturated these crystals will rest undissolved on the bottom of the container. Their *appearance* may change over a period of time, however. Solute ions (or molecules) will come and go constantly between the crystals and the solution, but the system will be in equilibrium. A solution in which the dissolved and undissolved states of the solute are in equilibrium is a saturated solution.

$$\text{solute}_{undissolved} \rightleftarrows \text{solute}_{dissolved} \qquad (11.1)$$

Table 11.5
Solubilities of Some Substances In Water

Solute	Solubilities, in Grams per 100 Grams Water			
	0°C	20 °C	50 °C	100 °C
Solids				
Sodium chloride, NaCl	35.7	36.0	37.0	39.8
Sodium hydroxide, NaOH	42	109	145	347
Barium sulfate, $BaSO_4$	0.000 115	0.000 24	0.000 34	0.000 41
Calcium hydroxide, $Ca(OH)_2$	0.185	0.165	0.128	0.077
Gases				
Oxygen, O_2	0.006 9	0.004 3	0.002 7	0
Carbon dioxide, CO_2	0.335	0.169	0.076	0
Nitrogen, N_2	0.002 9	0.001 9	0.001 2	0
Sulfur dioxide, SO_2	22.8	10.6	4.3	1.8 (at 90 °C)
Ammonia, NH_3	89.9	51.8	28.4	7.4 (at 96 °C)

Temperature is one factor affecting solubilities. A change in the temperature of a saturated solution in contact with undissolved solute places a stress on the equilibrium of the system of equation 11.1. According to Le Châtelier's principle (Section 10.13), the equilibrium will shift in whichever direction will better absorb the stress. For example, a shift that can absorb heat—an endothermic shift—will be promoted by a rise in temperature. Most ionic substances do require additional heat and a higher temperature to dissolve in a solution that is already saturated. Therefore, the solubilities of most ionic substances increase with temperature. Figure 11.8 shows some examples.

The solubility of a gas in water decreases with increasing temperature, because the increased heat provides the energy for the endothermic change—the movement of the gas out of the dissolved state. See the data in Table 11.5. A glass of a carbonated drink goes "flat" as it warms up because the dissolved carbon dioxide leaves the solution. Next time you heat water in the laboratory, notice how dissolved air forms into tiny bubbles and leaves. One reason why game fish cannot survive in warm water is that too little oxygen is left in solution.

Figure 11.8
Solubility in water versus temperature for several substances.

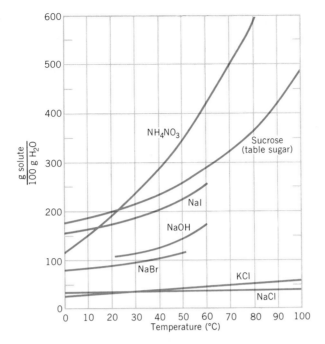

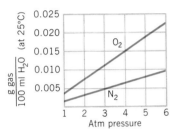

Figure 11.9
The effect of pressure on the solubility of nitrogen and oxygen in water.

"Suction" is simply creating a reduction in pressure over the fluid.

Lake trout cruise only the deepest and coolest holes during the hot weeks of July and August because shallower waters are too warm and oxygen-poor.

Pressure is another factor affecting solubility, but only when the solute is a gas. Figure 11.9 shows how the solubilities of two gases in water vary with pressure. Over a reasonable range of pressures, only gases will undergo significant changes in volume with changes in pressure. The relevant equilibrium equation is

$$\text{gas + solvent} \rightleftharpoons \text{solution} \tag{11.2}$$

This equilibrium will shift to the right with increasing pressure because such a change absorbs the volume-reducing stress of the additional pressure. This is another illustration of Le Châtelier's principle. Conversely, if a suction is created above a solution of a gas in a solvent, the equilibrium will shift to the left and the gas will leave the solution. Applying a suction (and a little heat) to a liquid is an effective way to degas the liquid.

The relationship between gas pressure and gas solubility is expressed by the **pressure-solubility law**, often called **Henry's law** after William Henry, 1775–1836, an English physician and chemist.

> **Pressure-Solubility Law (Henry's Law)** The concentration of a gas dissolved in a liquid at any given temperature is directly proportional to the partial pressure of the gas on the solution.

Stated mathematically, Henry's law is

$$C_g = k_g P_g \qquad \text{(at constant } T\text{)} \tag{11.3}$$

where C_g = the concentration of gas; k_g = the proportionality constant; P_g = the partial pressure of gas above the solution. Equation 11.3 is true only at relatively low concentrations and pressures, and it holds only for gases that do not react with the solvent.

EXAMPLE 11.3   **Calculating With The Pressure-Solubility Law (Henry's Law)**

Problem: At 20 °C the solubility of nitrogen in water is 0.015 g/liter when the partial pressure of the nitrogen is 580 torr. What will be the solubility of nitrogen in water at 20 °C when its partial pressure is 800 torr? (After studying this example, read Special Topic 11.2, "The Bends.")

Solution: Two approaches are possible. For example we can use the data to calculate the value of k_g.

$$k_g = \frac{C_g}{P_g} = \frac{0.015 \text{ g/liter}}{580 \text{ torr}} = 2.59 \times 10^{-5} \frac{\text{g/liter}}{\text{torr}}$$

Then we use this value of k_g to find C_g at 800 torr.

$$C_g = 2.59 \times 10^{-5} \frac{\text{g/liter}}{\text{torr}} \times 800 \text{ torr} = 0.021 \text{ g/liter}.$$

The other approach we leave as Exercise 11.8.

EXERCISE 11.7   How much nitrogen and oxygen, in grams, are dissolved in 100 g water at 20 °C when the water is saturated with air. In air saturated with water vapor at 20 °C the partial pressure of nitrogen is 586 torr and that of oxygen is 156 torr. The solubility of *pure* oxygen at 760 torr in water is 0.0043 g O_2/100 g H_2O and that of nitrogen is 0.0019 g N_2/100 g H_2O.

EXERCISE 11.8   Show how a method of ratios could be used to work the problem stated in Example 11.3, thus avoiding the calculation of the Henry's law constant, k_g.

Molal concentration and mole fractions

Many physical properties of solutions depend only on the physical presence of a solute—any solute—and not on its chemical nature. The chemical identities of the particles of solute and solvent do not matter; only the relative numbers of particles are important. Molal concentrations and the mole fraction are two ways of specifying these relative numbers.

Notice that it's per kilogram of *solvent*, not of solution. Don't confuse molal with molar.

Molal concentration, or **molality,** is the number of moles of solute per *kilogram of solvent.* Thus a 0.500 molal solution of sugar in water contains 0.500 mol sugar in 1000 g of water.

EXAMPLE 11.4   **Calculating Molal Concentration**

Problem: An experiment in plant biochemistry calls for an aqueous 0.150 molal sodium chloride solution. To prepare a solution of this concentration, how many grams of NaCl would be dissolved in 500 g water?

Solution: Step 1   Translate the label, "0.150 molal NaCl." A concentration of 0.150 molal NaCl means either of these two ratios:

$$\frac{0.150 \text{ mol NaCl}}{1000 \text{ g } H_2O} \quad \text{or} \quad \frac{1000 \text{ g } H_2O}{0.150 \text{ mol NaCl}}$$

Step 2   Calculate the moles of NaCl needed for 500 g of H_2O. Use the first ratio, because then the proper units will cancel to leave "mol NaCl" as the unit the answer must have.

$$500 \text{ g } H_2O \times \frac{0.150 \text{ mol NaCl}}{1000 \text{ g } H_2O} = 0.0750 \text{ mol NaCl}$$

Special Topic 11.2 The Bends (Decompression Sickness)

When people work in a space where the air pressure is quite high, they have to be very careful to go back to normal pressure slowly. Otherwise, they face the danger of the bends, severe pains in muscles and joints, fainting, and maybe even deafness, paralysis, or death. This is a risk in the lives of those who work in deep tunnels where air pressures are made high to help keep out water; of deep-sea divers; of astronauts whose space suits must not rupture during extravehicular activities; and of airline pilots (and passengers) whose planes must stay pressurized at high altitudes. Under high pressures the blood will dissolve more nitrogen and oxygen than under normal pressure; the gas pressure-solubility law (Henry's law) lets us calculate how much more. If the gas-enriched blood is then too quickly exposed to much lower pressures, the gases emerge in microbubbles that block the tiny blood capillaries and cause the symptoms. With slow decompression, the excess gases come out of solution slowly enough to be removed easily. For each atmosphere of pressure above the normal about 20 minutes of careful decompression is usually recommended. This allows time for the respiratory system to gather and expel the excess nitrogen, which can't leave any other way. Excess oxygen can be used up by normal metabolism.

Step 3 Convert mole of NaCl into grams. (The formula weight of NaCl is 58.5, meaning 58.5 g NaCl/mol NaCl.)

$$0.0750 \text{ \sout{mol NaCl}} \times \frac{58.5 \text{ g NaCl}}{1 \text{ \sout{mol NaCl}}} = 4.39 \text{ g NaCl}$$

Thus when 4.39 g NaCl is dissolved in 500 g H_2O the label on the bottle may read: "0.150 molal NaCl." Steps 2 and 3, of course, could be combined by setting up a series of conversion factors:

$$500 \text{ \sout{g H}}_2\text{\sout{O}} \times \frac{0.150 \text{ \sout{mol NaCl}}}{1000 \text{ \sout{g H}}_2\text{\sout{O}}} \times \frac{58.5 \text{ g NaCl}}{1 \text{ \sout{mol NaCl}}} = 4.39 \text{ g NaCl}$$

EXERCISE 11.9 As we shall soon study, water freezes at a lower temperature when it contains solutes. To study the effect of methyl alcohol on the freezing point of water, you might prepare a series of solutions of known molalities. How would you make 0.250 molal methyl alcohol (CH_3OH) using 2000 g water? (Calculate the number of grams of methyl alcohol needed.)

EXERCISE 11.10 The white, crystalline residue of dry sodium chloride weighed 5.26 g when 300 g of an aqueous *solution* was evaporated to dryness. Calculate the molality of the original solution.

The **mole fraction** of any one component in a solution (or any other mixture) is the ratio of the moles of that component to the total number of moles of all components present. Expressed mathematically:

$$X_A = \frac{n_A}{n_A + n_B + n_C + \cdots + n_X} \tag{11.4}$$

where X_A is the mole fraction of component A, and $n_A, n_B, n_C, \ldots, n_X$ are the numbers of moles of each component, $A, B, C, \ldots, X$, respectively. The sum of all mole fractions for a mixture must equal 1.

EXAMPLE 11.5 Calculating Mole Fractions

Problem: What are the mole fractions of each component in a solution consisting of 1.00 mol ethyl alcohol, 0.500 mol methyl alcohol, and 6.00 mol water?

Solution: Apply equation 11.4 to each component, in turn.

ethyl alcohol: $X_{ethyl\ alcohol} = \dfrac{1.00}{1.00 + 0.500 + 6.00} = 0.133$

methyl alcohol: $X_{methyl\ alcohol} = \dfrac{0.500}{1.00 + 0.500 + 6.00} = 0.067$

water: $X_{water} = \dfrac{6.00}{1.00 + 0.500 + 6.00} = 0.800$

total $= 1.000$

Notice that all the units in equation 11.4 cancel and, therefore, mole fractions have no units. Nevertheless, you should keep in mind the implied units for mole fractions—moles of component per total moles.

Sometimes we multiply a mole fraction by 100 and call the result the **mole percent**. In Example 11.5 there were 13.3 mole percent ethyl alcohol, 6.70 mole percent methyl alcohol, and 80.0 mole percent water.

EXERCISE 11.11 What are the mole fractions and mole percents of each component in an antifreeze solution made by mixing 500 g methyl alcohol (CH_3OH) with 500 g water?

EXERCISE 11.12 What are the mole fractions and mole percents of the solute and solvent in a solution labeled "0.750 molal NaCl"?

In working with mixtures of gases we may calculate mole fractions if we know the partial pressures of the components. Dalton's law of partial pressures (Section 2.5) and the ideal gas law equation ($PV = nRT$) make this calculation possible. The equation $PV = nRT$ gives us an expression for moles, n, of any one gas in the mixture when its partial pressure is P and the entire mixture has a volume V and a temperature T.

$$n = \frac{PV}{RT} \tag{11.5}$$

Using letters as subscripts to identify individual gases, equation 11.5 can be used to modify equation 11.4 when the mixture is of gases. All gases in the mixture, of course, have the same values of V, R, and T.

$$X_A = \frac{n_A}{n_A + n_B + n_C + \cdots + n_X}$$

$$= \frac{P_A V/RT}{P_A V/RT + P_B V/RT + P_C V/RT + \cdots + P_X V/RT}$$

We can factor out V/RT.

$$= \frac{[V/RT][P_A]}{[V/RT][P_A + P_B + P_C + \cdots + P_X]}$$

$$X_A = \frac{P_A}{P_A + P_B + P_C + \cdots + P_X} \tag{11.6}$$

However, according to Dalton's law of partial pressures, the sum of the partial pressures in the denominator of equation 11.6 equals the total pressure. Therefore, equation 11.6 reduces to a very simple expression for the mole fraction of a gas in a mixture of gases:

mole fraction of gas $A = X_A = \dfrac{P_A}{P_{total}}$ (11.7)

EXAMPLE 11.6   Calculating The Mole Fractions of Gases From Partial Pressures

Problem: What are the mole fractions of nitrogen and oxygen in air when their partial pressures are 160 torr for oxygen and 600 torr for nitrogen? (Assume no other gases are present.)

Solution: Step 1   Use equation 11.7 to find the mole fraction of the nitrogen.

$$X_{N_2} = \frac{P_{N_2}}{P_{total}}$$

$$X_{N_2} = \frac{600 \text{ torr}}{600 \text{ torr} + 160 \text{ torr}}$$

$$= 0.789 \text{ or } 78.9 \text{ mole percent}$$

Step 2   Repeat for oxygen:

$$X_{O_2} = \frac{160 \text{ torr}}{600 \text{ torr} + 160 \text{ torr}} = 0.211 \text{ or } 21.1 \text{ mol percent}$$

EXERCISE 11.13   What is the mole fraction of oxygen in exhaled air if its partial pressure is 116 torr and the total pressure is 760 torr? What is the mole percent?

11.3   VAPOR PRESSURE LOWERING

The vapor presure of any component of a mixture is usually lowered by the presence of the other components

The freedom of the particles of any given component in a solution to go into the vapor state is reduced by any collisions near the surface with particles of other components. Figure 11.10 illustrates this phenomenon for a two-component system in which one is a nonvolatile solute. Reduced freedom for the volatile substance to escape means that the vapor pressure for this component is less. When all parts of the mixture are molecular, not ionic, a simple relationship exists between the vapor pressure of one component and its mole fraction. This relationship is called the **vapor pressure-concentration law** or **Raoult's law** (after Francois Marie Raoult, 1830–1901, a French scientist):

Volatile means "can evaporate;" has a low boiling point. **Nonvolatile** means "cannot evaporate;" has a very high boiling point.

Figure 11.10
Lowering of vapor pressure by a nonvolatile solute. Because particles of the solute (·) are present, the opportunities for the escape of the solvent particles (●) are reduced and the solvent's vapor pressure is, therefore, reduced.

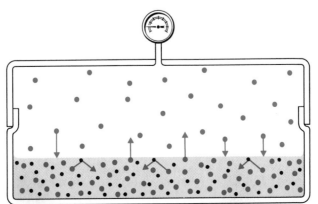

> **Vapor Pressure-Concentration Law (Raoult's Law)**   The vapor pressure of one component above a mixture of molecular compounds equals the product of its vapor pressure when pure (at the same temperature) and its mole fraction.

Mathematically, Raoult's law can be expressed as

$$P = P°X \qquad \text{(no components are ionic)} \qquad (11.8)$$

where P is the vapor pressure of the component above the solution and $P°$ is its vapor pressure when pure at the same temperature and X is the mole fraction of that component in the liquid phase. When all components of the mixture are volatile, then the value of P for any one component is its *partial pressure* above the solution. Any solution for which equation 11.8 would be exactly true for all components is defined as an **ideal solution.** Raoult's law should be regarded as an ideal that some solutions realize more closely than others.

EXAMPLE 11.7   Calculating With Raoult's Law

Problem: Acetone is a common solvent. Students in organic labs may use it to rinse out flasks because acetone easily dissolves water and many organic substances. At 20 °C acetone has a vapor pressure of 162 torr; water's vapor pressure at 20 °C is 17.5 torr. What is the vapor pressure of each component above a solution having 50.0 mol percent acetone and 50 mole percent water? Assume that the solution is ideal.

Solution: Since the data provide the mole percents, we change them to mole fractions and use the equation for Raoult's law.

For acetone: $P_{acetone} = 162 \text{ torr} \times 0.500 = 81.0 \text{ torr}$

For water: $P_{water} = 17.5 \text{ torr} \times 0.500 = 8.75 \text{ torr}$

EXERCISE 11.14   One of the early experiments in the organic lab is to distill a simple two-component solution such as a solution of cyclohexane and toluene. The two form a nearly ideal solution. At room temperature—20 °C—the vapor pressure of cyclohexane is 66.9 torr and that of toluene is 21.1 torr. If you begin such a distillation at 20 °C with a solution 50 mol percent in each component, what are the initial partial pressures of each component? What is the total vapor pressure of the solution? (Assume Dalton's law of partial pressures applies to the vapor above the solution.)

EXERCISE 11.15   Following Example 11.6 calculate the mole fractions of cyclohexane and toluene in the *vapor* above the solution described in Exercise 11.14 using the partial vapor pressure data you calculated. Which has the greater concentration of the more volatile component (cyclohexane), the vapor or the liquid?

As we suggested in Exercise 11.14, *the total vapor pressure above a two-component solution is the sum of the partial pressures of the components.* Figure 11.11 describes this graphically. The *total* vapor pressure of the two-component mixture is plotted versus composition on the upper line in Figure 11.11. Each of the lower lines is a plot of the *partial* vapor pressure of one component versus its mole fraction in the solution. If you take a rule marked in millimeters you can prove to yourself that, at any given value of composition, the distance from the base to one of the lower lines added to the distance from the base

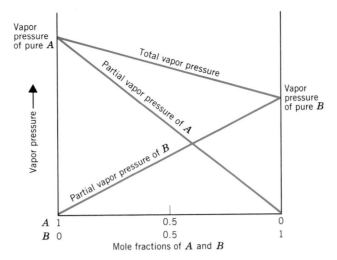

Figure 11.11
The vapor pressure of an ideal, binary solution of two volatile components.

to the other lower line gives the distance to the upper line. In doing this you are adding the individual partial pressures to find the total pressure.

Not very many real, two-component solutions come close to being ideal. Figure 11.12 shows two of the typical ways in which real two-component solutions behave. The partial vapor pressures do not show straight-line correlations with composition as in Figure 11.11. Therefore, the sums that give the upper line for the total pressure will, in some solutions, show an upward or *positive deviation,* and in others a downward or *negative deviation.*

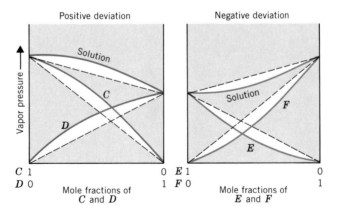

Figure 11.12
Deviations from ideal behavior of the total vapor pressure of real binary solutions of volatile components.

We will observe a negative deviation whenever the molecules of the two components have *stronger* attractions for each other than for their own kind. These stronger attractions restrain their escape into the vapor state from the solution and reduce the total vapor pressure.

Whenever molecules of two components have *weaker* attractions for each other than for their own kind, it is easier for each kind to escape into the vapor state from the solution. The total vapor pressure of such solutions will therefore be higher than that for an ideal solution, and we will see an upward or *positive deviation* in the total vapor pressure.

11.4 BOILING POINT ELEVATION

The boiling point of a solution with a nonvolatile solute is higher than that of the pure solvent

When the only volatile component of a solution is the solvent, then the vapor pressure of the *solution* will always be less than that of the pure solvent for any given temperature. This is shown in Figure 11.13. At the temperature where the pure solvent would normally boil, the vapor pressure above the *solution* is still not equal to the atmospheric pressure. To increase the vapor pressure and thereby make the *solution* boil requires a further rise in the temperature. Thus, we can say that a nonvolatile solute causes a boiling point elevation. Permanent-type antifreezes consist mostly of a nonvolatile, water-soluble liquid. When mixed with the water in a car's radiator, the solution has a higher boiling point than water. This keeps the radiator fluid from boiling away on those summer days when the engine runs hotter than usual. (However, these liquids are called anti-freezes, not antiboilers, for reasons given in Section 11.5.)

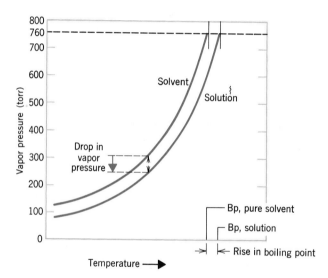

Figure 11.13
Vapor pressure versus temperature for a solvent and for a solution of a nonvolatile solute in this solvent.

bp = boiling point

The number of degrees of boiling point elevation caused by a nonvolatile so-lute is a simple function of its molal concentration when the solute is molecular, not ionic. Boiling point elevation is directly proportional to molality:

$$\Delta t \propto m$$

where $\Delta t = bp_{solution} - bp_{solvent}$ and m is the molal concentration of the solute. When we introduce a proportionality constant, here called the molal boiling point elevation constant with the symbol k_b, we have:

$$\boxed{\Delta t = k_b m} \tag{11.9}$$

Equation 11.9 works well for solutions at low concentration. Each solvent has its own molal boiling point elevation constant. The values of k_b for several sub-stances are given in Table 11.6. The units of k_b are those of $\Delta t/m$

$$\frac{\Delta t}{m} = \frac{°C}{\text{mol-solute/kg-solvent}} = \frac{°C \text{ kg-solvent}}{\text{mol-solute}}$$

Table 11.6
Boiling Point Elevation Constants (k_b) and Freezing Point Depression
Constants (k_f) for Various Solvents

Solvent	Bp (°C)	k_b	Mp (°C)	k_f
Water	100	0.51	0	1.86
Acetic acid	118.3	3.07	16.6	3.57
Benzene	80.2	2.53	5.45	5.07
Chloroform	61.2	3.63	—	—
Camphor	—	—	178.4	37.7

EXAMPLE 11.8   **Calculating With Molal Boiling Point Elevation Constants**

Problem: At what temperature will a 0.15 molal sugar solution boil? (Sugar is nonvolatile, of course.)

Solution: We calculate the *change* in boiling temperature, Δt, using equation 11.9.
From Table 11.6, we find that k_b is 0.51 $\dfrac{\text{°C kg-solvent}}{\text{mol-solute}}$

$$\Delta t = 0.51 \;\frac{\text{°C \cancel{kg-solvent}}}{\cancel{\text{mol-solute}}} \times 0.15 \;\frac{\cancel{\text{mol-solute}}}{\cancel{\text{kg-solvent}}}$$

$$= 0.077 \;\text{°C} \qquad \text{(molal units)}$$

Since pure water boils (at 760 torr) at 100.000 °C, the solution will boil at
100.000 °C + 0.077 °C = 100.077 °C. The elevation isn't much, but it can
be measured with good equipment.

EXERCISE 11.16   Estimate the boiling point of a permanent-type antifreeze that is 16 molal solution of ethylene glycol. (We can only estimate because we have to assume that equation 11.9 holds
even for a solution of this concentration. This solution corresponds to about equal weights
of ethylene glycol and water, a proportion commonly used in radiators.)

Equation 11.9 can also be used to find molal concentration if we measure Δt
and know the value of k_b for the solvent. Working backwards, we can then use a
measurement of Δt to determine a formula weight. Let's work an example.

EXAMPLE 11.9   **Calculating A Formula Weight From Boiling Point Elevation**

Problem: A solution made by dissolving 10.00 g of an unidentified compound in
100 g water boiled at 100.45 °C at 1 atmosphere. (The substance was
known to be essentially nonvolatile at this temperature, and it dissolved as
molecules, not ions.) What is the formula weight of the compound?

Solution: From Table 11.6, k_b for water is 0.51 $\dfrac{\text{°C kg-solvent}}{\text{mol-solute}}$. The value of Δt is:

$$\Delta t = 100.45 \;\text{°C} - 100.00 \;\text{°C}$$
$$= 0.45 \;\text{°C}$$

Using equation 11.9 we can find molal concentration, m:

$$0.45 \;\text{°C} = 0.51 \;\frac{\text{°C kg-solvent}}{\text{mol-solute}} \times m$$

Or,

$$m = \frac{0.45 \ \cancel{°C}}{0.51 \ \dfrac{\cancel{°C} \ \text{kg-solvent}}{\text{mol-solute}}} = 0.88 \ \frac{\text{mol-solute}}{\text{kg-solvent}}$$

From this molal concentration we can calculate the number of *moles* of solute in the solution.

$$0.88 \ \frac{\text{mol-solute}}{\cancel{\text{kg-solvent}}} \times 0.100 \ \cancel{\text{kg-solvent}} = 0.088 \ \text{mol-solute}$$

Since the solution contained 10.0 *grams* of solute we now know that 10.0 g of solute equals 0.088 mol of solute. Therefore, the ratio of grams to moles for the solute —the formula weight we seek —is

$$\frac{10.0 \ \text{g-solute}}{0.088 \ \text{mol-solute}} = 114 \ \text{g/mol}$$

Giving due consideration to significant figures, the formula weight of the unknown is between 110 and 120.

EXERCISE 11.17 A solution prepared by dissolving 11 g of a nonvolatile molecular solute in 100 g chloroform (bp 61.20 °C) was found to boil at 84.30 °C. What was the formula weight of the solute?

11.5 FREEZING POINT DEPRESSION

The freezing point of a solution is lower than that of the pure solvent

When a solvent is chilled, the average velocities of its molecules decrease until they take up sites in its growing crystals. When a *solution* is cooled until it starts to freeze, the forming crystals of solvent contain no molecules of the solute. Solute particles generally cannot fit into the crystal lattice of the solvent, so they are excluded. However, the solute particles are in the way, and they make it harder for the solvent molecules to assemble as a crystal. To counteract this, we have to slow the molecules of solvent down even more to get the crystals of the solvent to grow. We have to lower the temperature. This is how a solute depresses the freezing point of a solution below that of the pure solvent. For any given solvent, the number of degrees the freezing point is depressed depends only on the relative concentrations of solute and solvent particles, not on the chemical identity of the solute. Remember, we are restricting our discussion to the behavior of molecular, not ionic solutes. However, these solutes need not be nonvolatile in order to have the freezing point depression proportional to the molality (at relatively low concentrations):

$$\boxed{\Delta t = k_f m} \tag{11.10}$$

where

fp = freezing point

$\Delta t = fp_{\text{solvent}} - fp_{\text{solution}}$
m = moles solute/kg solvent (molal concentration)
k_f = molal freezing point depression constant in $\dfrac{°C \ \text{kg-solvent}}{\text{mol-solute}}$

Each solvent has its own freezing point depression constant, k_f. Several of these are given in Table 11.6.

EXAMPLE 11.10 **Calculating With Molal Freezing Point Depression Constants**

Problem: At what temperature will a 0.250 molal solution of sugar in water freeze?

Solution: We find in Table 11.6 that k_f for water is

$$1.86 \, \frac{°C \text{ kg-solvent}}{\text{mol-solute}}$$

Using equation 11.10:

$$\Delta t = 1.86 \, \frac{°C \text{ kg-solvent}}{\text{mol-solute}} \times 0.250 \, \frac{\text{mol-solute}}{\text{kg-solvent}}$$

$$\Delta t = 0.465 \, °C$$

The freezing point of water, 0.000 °C, is lowered by 0.465 °C. The *solution,* in other words, freezes at −0.465 °C.

EXERCISE 11.18 Estimate the freezing point of the antifreeze solution in Exercise 11.16. The actual freezing point is quite close to the calculated value.

The depression of a freezing point by a solute is the principle behind the use of salt in winter to melt the ice on city streets and the use of a very high-boiling liquid as a permanent-type antifreeze in a radiator cooling system. Ethylene glycol, the antifreeze in exercises 11.16 and 11.18, has a boiling point of 198 °C (at 760 torr). Therefore, it won't boil away when the engine runs hot. Methyl alcohol, which has a boiling point of 65 °C, could just as well protect a radiator's coolant against freezing—and at much less cost than ethylene glycol—but it boils away too readily. That is why in the days of its use in radiators, it was called temporary antifreeze.

Freezing point depression can also be used to determine formula weights. Let's use an example to learn how.

EXAMPLE 11.11 **Calculating A Formula Weight From Freezing Point Depression**

Problem: A solution made by dissolving 5.65 g of an unknown compound in 110 g of benzene froze at 4.39 °C. What was its formula weight?

Solution: From Table 11.6, k_f for benzene is $5.07 \, \frac{°C \text{ kg-benzene}}{\text{mol-solute}}$. The value of Δt, the freezing point depression, is

$$\Delta t = 5.45 \, °C - 4.39 \, °C = 1.06 \, °C$$

Using equation 11.10, we find the molal concentration, m:

$$1.06 \, °C = 5.07 \, °C \, \frac{\text{kg-benzene}}{\text{mol-solute}} \times m$$

$$m = \frac{1.06 \, °C}{5.07 \, °C \, \frac{\text{kg-benzene}}{\text{mol-solute}}} = 0.209 \, \frac{\text{mol-solute}}{\text{kg-benzene}}$$

From this molal concentration we can calculate the number of *moles* of solute in the solution where the mass of solvent was 110 g or 0.110 kg.

$$0.209 \, \frac{\text{mol-solute}}{\text{kg-benzene}} \times 0.110 \, \text{kg-benzene} = 0.0230 \, \text{mol-solute}$$

Since the solution also contained 5.65 g of solute, this mass of solute must equal 0.0230 mol of solute. Therefore, the ratio of grams of solute to moles of solute—the formula weight we seek—is

$$\frac{5.65 \, \text{g-solute}}{0.0230 \, \text{mol-solute}} = 246 \, \text{g/mol}$$

Therefore, the formula weight is 246.

EXERCISE 11.19 A solution made by dissolving 3.46 g of an unknown compound in 85.0 g of benzene froze at 4.13 °C. What is the formula weight of the unknown?

EXERCISE 11.20 If 3.46 g of the unknown of Exercise 11.19 were dissolved in 85 g of molten camphor, instead of benzene, at what temperature would the mixture freeze? Since large freezing point depressions can be measured with more precision than small freezing point depressions, which substance offers greater precision for the determination of formula weight by this method, camphor or benzene?

11.6 DIALYSIS AND OSMOSIS

If two solutions of different concentrations are separated by the right kind of membrane, the concentrations will change until they are equal

The phenomenon whereby a nonvolatile solute lowers the vapor pressure of a solvent, elevates its boiling point, and depresses its freezing point are called colligative properties. **Colligative properties** are those physical properties of a solution that depend only on the relative numbers of particles of solute and solvent, not on their chemical identities. We have one more important colligative property to discuss, osmosis.

The various membranes in living things that keep mixtures organized and separated from each other are semipermeable. They let water molecules pass through them, and they can selectively let small ions and molecules pass through, too. Large ions and molecules are stopped. This phenomenon, the selective passage of small molecules and ions through membranes, is called **dialysis,** and any membrane allowing such passage is called a dialyzing membrane. The limiting case of dialysis is osmosis. **Osmosis** occurs with membranes that allow *only* molecules of solvent through and do not let ions or molecules pass. Membranes allowing only osmosis are called osmotic membranes. Although such membranes are rare, they can be made.

One theory that explains how osmosis occurs is presented in Figure 11.14. The solvent on each side of the membrane (see part *b*), has its own "escaping tendency," which is just like its own vapor pressure. However, the less concentrated solution has the higher escaping tendency for the same reason it has the higher vapor pressure. That means that solvent molecules leave the dilute solution and go through the membrane more frequently than those from the concentrated solution pass in the opposite direction. Therefore, a net migration of solvent molecules is from the side with less *solute* per unit volume to the side having more.

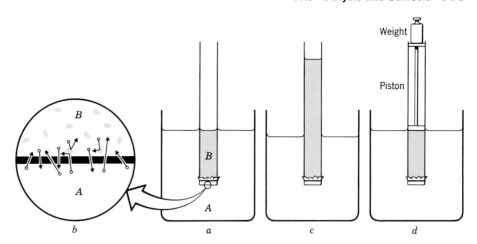

Figure 11.14

Osmosis and osmotic pressure. (a) Initially, the tube has a solution filled to a level even with that of the pure water in the beakers. An osmotic membrane closes the bottom of the tube. (b) Enlarged view of the membrane interfaces. Small dots represent molecules of water, the solvent. Larger dots are of molecules of solute. As drawn, for every 5 molecules of water that leave A only 3 return, leading to c. (c) The level in the tube has risen; that in the beaker has dropped. Osmosis has occurred. (d) A back pressure can prevent osmosis; the exact amount equals the osmotic pressure of the solution in the tube.

Eventually, the concentrations would become equal, unless a back pressure developed that stopped the net flow. The exact pressure needed to prevent this net flow in osmosis is called the **osmotic pressure.**

The symbol Π is capital pi, the capital of π.

Osmotic pressure is symbolized by Π. It is directly proportional to the molar concentration, M, of solute particles (not their identities):

$$\Pi \propto M \qquad (11.11)$$

Osmotic pressure is also proportional to the kelvin temperature (for the same reason that vapor pressure is proportional to T):

$$\Pi \propto T \qquad (11.12)$$

Since Π is proportional to both M and T, we may write

$$\Pi \propto MT$$

or

$$\Pi = MT \text{ (constant)} \qquad (11.13)$$

For very dilute solutions, the constant of proportionality is very close to the gas constant, R. Hence we may write

$$\Pi = MRT \qquad (11.14)$$

Of course $M = \text{moles/liter} = n/V$, where $n = \text{moles}$ and $V = \text{volume in liters}$. Substituting n/V for M and rearranging, we obtain an equation for osmotic pressure that is identical in form to the equation of state for an ideal gas

$$\boxed{\Pi V = nRT} \qquad (11.15)$$

The magnitude of osmotic pressure can be very high, even in dilute solutions. We'll work an example to show that.

EXAMPLE 11.12 **Calculating Osmotic Pressure**

Problem: A very dilute solution, 0.0010 *M* sugar in water is separated from pure water by an osmotic membrane. What osmotic pressure develops? (The temperature is 25 °C or 298 K.)

Solution: Use equation 11.15:

$$\Pi = \frac{nRT}{V} \left(\text{We will use } R = 0.0821 \frac{\text{liter atm}}{\text{mol K}}. \right)$$

Since n/V is in units of mol/liter, those of molar concentration:

$$\Pi = 0.0010 \frac{\text{mol}}{\text{liter}} \times 0.0821 \frac{\text{liter atm}}{\text{mol K}} \times 298 \text{ K}$$

$$= 0.024 \text{ atm}$$

Since 1 atm = 760 torr:

$$0.024 \text{ atm} \times \frac{760 \text{ torr}}{1 \text{ atm}} = 18 \text{ torr}$$

A pressure of 18 torr measured by a Torrcelli barometer corresponds to 18 mm Hg. Since the solution's density is essentially the same as the density of water, we could perform a calculation similar to the one in Example 2.3 in Chapter 2 to show that this osmotic pressure would support a column of this aqueous solution approximately 10 in. high. A 0.1 molar solution, still relatively dilute, would support a column 100 times as high, approximately 1000 in. or about 83 ft. Osmosis is part of the explanation for the rise of sap in trees.

When soil water has too high a concentration of dissolved salts, the plant can't get the water it needs. In fact it tends to *lose* water to the soil and wilts.

EXERCISE 11.21 An instrument called an osmometer measures osmotic pressures. We can use the osmotic pressure of a solution to calculate its molar concentration, n/V, by using equation 11.15. What is the molar concentration of a solution with an osmotic pressure of 26 torr at 20 °C. Hint: It might be easiest to convert torr to atm and use the value of R given in Example 11.12.

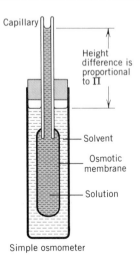

Capillary

Height difference is proportional to Π

Solvent

Osmotic membrane

Solution

Simple osmometer

Although membranes in living systems are not true osmotic membranes, they do hold back large colloidal-sized molecules or ions such as those of proteins and other natural polymers. When these are in a solution, an osmotic pressure can exist even across a dialyzing membrane. The contribution to a solution's total osmotic pressure made by the presence of colloidal-sized particles is called the colloidal osmotic pressure.

$$\text{total osmotic pressure} = \left(\begin{array}{c} \text{osmotic pressure} \\ \text{caused by} \\ \text{solutes} \end{array} \right) + \left(\begin{array}{c} \text{osmotic pressure caused} \\ \text{by colloidal particles—the} \\ \text{colloidal osmotic pressure} \end{array} \right)$$

When a dialyzing membrane is used instead of an osmotic membrane, then solute particles can move freely across the membrane. Only the colloidal osmotic pressure remains. Therefore the contributions to the total osmotic pressure caused by solutes vanishes when a dialyzing membrane is used.

Because colloidal particles of unequal concentration on opposite sides of a dialyzing membrane can contribute to the osmotic pressure, great care must be used in adding fluids to the blood by intravenous drip. The fluids both inside and outside red blood cells contain colloidal-size particles, but the total osmotic pressures of these fluids are the same. The osmotic pressure of the solution added to the blood must match that of the blood. Otherwise, the composition of the red blood cells will change as dialysis occurs. Which direction dialysis will

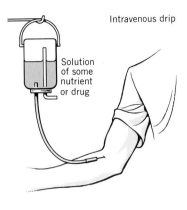
Intravenous drip

Solution
of some
nutrient
or drug

take depends on the relative concentrations inside and outside the red cells, as illustrated in Figure 11.15. Two solutions having identical osmotic pressures are called isotonic. If the two are not equal, the more concentrated solution is called hypertonic with respect to the other and the less concentrated solution is called hypotonic.

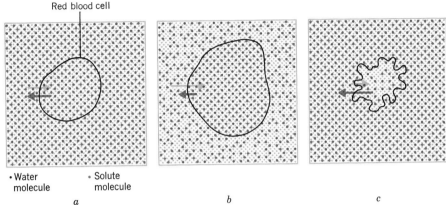

Red blood cell

• Water
molecule

• Solute
molecule

a   b   c

Figure 11.15
Dialysis and red blood cells. (a) The solutions both inside and outside the red cell are isotonic. (b) The solution outside of the cell is hypotonic with respect to the solution inside, perhaps because of an accidental injection of pure water. Dialysis occurs, bringing water inside the cell and making it swell. If this continues the cell will burst. (c) The solution outside the cell is now hypertonic with respect to the solution inside. Dialysis now occurs, and some water inside the cell moves outside making the cell shrivel.

One practical use of dialysis is the separation of substances in true solution from colloidal particles, a procedure that is often used in purifying colloidal substances. For this purpose, cellophane, collodion, or an animal bladder, may be used as the dialyzing membrane. Thus, if a mixture of colloidally dispersed starch and dissolved sodium chloride is placed in a dialyzing bag and water is circulated around it (see Figure 11.16), the ions will soon appear in the water outside the bag while the starch will remain in the bag. See Special Topic 11.3.

Figure 11.16
Dialysis as a means of purification. The molecules of starch (circles) are of colloidal size. The ions ($\oplus$ and $\ominus$) are small. The ions, together with molecules of water, can move through the dialyzing membrane. Eventually, all of the ions will be removed from the bag and the starch solution will be purified.

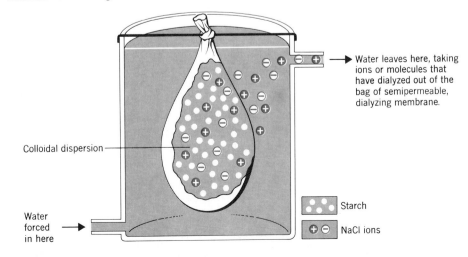

Water leaves here, taking
ions or molecules that
have dialyzed out of the
bag of semipermeable,
dialyzing membrane.

Colloidal dispersion

Water
forced
in here

Starch

$\oplus$ $\ominus$ NaCl ions

Our kidneys are organs that cleanse the blood of many of the wastes of metabolism. If the kidneys quit working, these wastes will cause death. Use of the artificial kidney is one strategy in this emergency. It works by dialysis, and the procedure is called hemodialysis. The blood is diverted from the body into a long, coiled cellophane tube—the dialyzing membrane. A solution called the dialysate is constantly circulated around the outside of this tube. The dialysate is made isotonic for all those substances dissolved in blood that we want to stay there. For example, if the concentration of sodium ion in the blood is exactly matched by its concentration in the dialysate, the rate at which these ions leave the blood by dialysis will exactly equal the rate at

which they return. Hence, the concentration of sodium ion in the blood will stay the same. However, the dialysate is kept very low in its concentrations of the wastes we want removed from the blood, for example, urea (the chief nitrogen waste). Urea's molecules are small (N_2H_4CO), and they easily move through a dialyzing membrane. As long as the dialysate is kept low in urea, the rate at which its molecules leave the blood is greater than the rate they can return. In this way the blood gives up this waste. The dialyzing membrane does not permit any loss of large molecules from blood (e.g., proteins), but these aren't wastes anyway. After the wastes are removed by dialysis, the blood is sent back into the body.

11.7   COLLIGATIVE PROPERTIES OF SOLUTIONS OF IONS

Mole for mole, ionic substances have two or three times the effect on colligative properties as molecular substances

Since the k_f for water is 1.86 °C kg-solvent/mol-solute, a 1 molal solution of sodium chloride might be expected to freeze at -1.86 °C. It doesn't. The freezing point depression of 1 molal NaCl is almost twice the expected amount. This solution freezes at -3.37 °C. We can understand why if we remember that colligative properties depend only on the concentrations of solute *particles,* not their chemical identities. When sodium chloride dissolves, it breaks up into ions. Thus, when one mole of sodium chloride dissolves, one mole of sodium ions and one mole of chloride ions are liberated into the solution:

s = solid state
ℓ = liquid state
g = gaseous state
aq = dissolved in water
      (*aq*ueous solution)

$$NaCl(s) \longrightarrow Na^+(aq) + Cl^-(aq)$$

Because of this ionization, a 1 molal solution of NaCl actually has a 2 molal concentration of solute particles. Theoretically, then, 1 molal sodium chloride ought to freeze at $2 \times (-1.86$ °C$) = -3.72$ °C. It doesn't freeze quite that low because its concentration is relatively high, and the ions are not really 100% free of each other. If the solution is made more and more dilute, however, the observed and calculated values of the freezing point come closer and closer together. As the solutions are made more dilute, the ions interact with each other less and less.

Just as ionic substances have larger effects, mole for mole, than molecular solutes on the depression of the freezing point and the vapor pressure, they also cause larger elevations of the boiling point and produce greater osmotic pressures. To describe the concentration of all particles that can cause colligative properties, including osmosis, it is common practice in physiology and biological chemistry to use the concepts of osmolarity or osmolality. A solution's **osmolarity** is its molar concentration of all solute particles, ions and molecules, that can affect osmosis, or the boiling point, or the melting point or the vapor pressure of the solution. **Osmolality** is the molal concentration of all particles that can cause these colligative properties. The osmolality of 0.001 molal sodium chloride is twice as much, 0.002 mol/1000 g solvent. The osmolarity of 0.01 molar Na_2SO_4 is 0.03 molar because each unit of Na_2SO_4 breaks up into three ions:

$$Na_2SO_4(s) \xrightarrow[\text{water}]{} 2Na^+(aq) + SO_4^{2-}(aq)$$

EXAMPLE 11.13 **Predicting A Colligative Property Of An Ionic Substance**

Problem: Estimate the boiling point of a 1.0 molal solution of magnesium chloride? When it dissolves magnesium chloride breaks up into ions as follows:

$$MgCl_2(s) \longrightarrow Mg^{2+}(aq) + 2Cl^-(aq)$$

Solution: Since 1 mol of $MgCl_2$ gives 3 mol of ions, its osmolality is not 1.0 molal but actually 3.0 molal when it comes to the concentration of solute particles. We will assume that at this concentration the ions are fully active, colligatively. The problem is now identical in kind to that of Example 11.8.

$$\Delta t = 0.51 \, \frac{°C \; \text{kg-solvent}}{\text{mol-solute}} \times 3.0 \, \frac{\text{mol-solute}}{\text{kg-solvent}}$$

$$= 1.5 \,°C$$

The boiling point of the solution is 100.0 °C + 1.5 °C = 101.5 °C.

EXERCISE 11.22 In Example 11.12, we calculated the osmotic pressure of a 0.0010 molar solution of glucose. What would be the osmotic pressure if the solution were 0.0010 molar in sodium chloride, assuming that the solute is 100% active colligatively?

11.8 NET IONIC EQUATIONS

The chemical properties of electrolytes in solution are best represented by ionic equations

We have just learned of some of the effects of ions on the physical properties of solutions. We now want to turn our attention back to certain chemical properties of ions.

Substances that can furnish ions are sometimes called **electrolytes.** Electrolytes can conduct electricity either in their molten state or in their dissolved state in water. Taking an electrocardiogram, a procedure in medical diagnosis, depends on the presence of electrolytes in body fluids to conduct weak electrical currents. Substances that cannot conduct electricity in solution or the molten state are **nonelectrolytes.**

The chemical effects of an electric current as it passes through an electrolyte will be studied in Chapter 18. Our study of most of the other chemical properties of electrolytes is spread throughout the rest of the book. However, one common property of electrolytes is that the ions of many pairs of them react completely and rapidly in solution. These reactions of ions are most easily studied with a new kind of equation, the net ionic equation. A **net ionic equation** is what remains when all the particles that do not change, the *spectator ions and molecules,* are removed from the molecular equation. For example, when aqueous solutions of lead nitrate and sodium sulfate are poured together, a precipitate, lead sulfate, appears leaving sodium nitrate in solution. We call the complete equation for this reaction a **molecular equation:**

Dissolved sulfate ions (as in soil water) can be detected by the appearance of a white precipitate of barium sulfate, $BaSO_4(s)$, when a few drops of a soluble barium compound such as $Ba(NO_3)_2(aq)$ are added.

$$\underset{\substack{\text{lead nitrate} \\ \text{(in solution)}}}{Pb(NO_3)_2(aq)} + \underset{\substack{\text{sodium sulfate} \\ \text{(in solution)}}}{Na_2SO_4(aq)} \longrightarrow \underset{\substack{\text{lead sulfate} \\ \text{(precipitate)}}}{PbSO_4(s)} + \underset{\substack{\text{sodium nitrate} \\ \text{(in solution)}}}{2NaNO_3(aq)}$$

(We say "molecular" equation even though none of these substances consists of molecules. All consist of ions. "Molecular" in this context simply means that the *complete formulas of all reactants and products are shown.*)

The net ionic equation for this reaction is:

$$Pb^{2+}(aq) \;+\; SO_4^{2-}(aq) \longrightarrow PbSO_4(s)$$

lead(II) ion   sulfate ion   lead sulfate
(in solution)   (in solution)   (precipitate)

With just a few facts, we can convert a molecular equation into its net ionic form. Let's see how that can be done for the example just given. The solution of lead nitrate actually contains no intact "molecules" of $Pb(NO_3)_2$. Instead, this substance behaves like all ionic substances when they dissolve in water (at least in moderately dilute solutions); it separates into ions as visualized in Figure 11.17a and by the following equation.

$$Pb(NO_3)_2(s) \xrightarrow[\text{as it dissolves in water}]{} Pb^{2+}(aq) \;+\; 2NO_3^-(aq)$$

lead ion   nitrate ion
(in solution)   (in solution)

Likewise, as sodium sulfate dissolves in water it breaks up as depicted also in Figure 11.17a and the following equation.

$$Na_2SO_4(s) \xrightarrow[\text{as it dissolves in water}]{} 2Na^+(aq) \;+\; SO_4^{2-}(aq)$$

sodium ion   sulfate ion
(in solution)   (in solution)

Figure 11.17
The reaction of $Pb(NO_3)_2$ (aq) with Na_2SO_4(aq). (a) These ionic compounds exist as separated ions in aqueous solutions. (b) When the two solutions of part (a) are poured together, the lead ions, Pb^{2+}, and the sulfate ions, SO_4^{2-}, collect together and form an insoluble precipitate, $PbSO_4(s)$. The sodium ions, Na^-, and the nitrate ions, NO_3^-, however, remain separated—as "spectators."

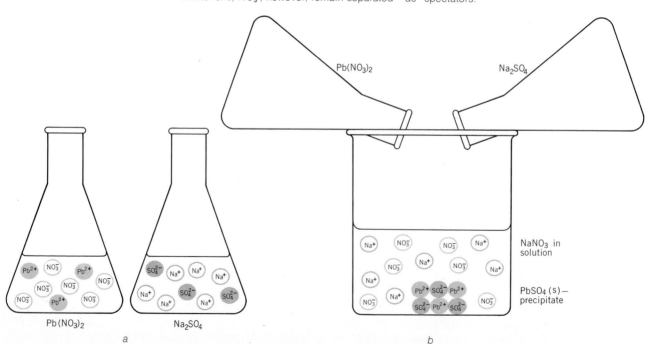

Therefore, when an aqueous solution of lead nitrate mixes with one of sodium sulfate, the already separated *ions* of these substances do the interacting. If we rewrite the molecular equation to show the separated ions, we have

$$Pb^{2+}(aq) + 2NO_3^-(aq) + 2Na^+(aq) + SO_4^{2-}(aq) \longrightarrow$$
[from Pb(NO₃)₂]   (from 2NaNO₃)

$$PbSO_4(s) + 2Na^+(aq) + 2NO_3^-(aq)$$
(dissolved NaNO₃)

One product, $PbSO_4(s)$, is not written as separated into its ions because it precipitates from the solution. See also Figure 11.17b. The other product, $NaNO_3(aq)$, must be written as its separated ions because this salt, an ionic compound, is soluble in water. When a salt dissolves in water, its ions become separated. Notice, however, that these same ions, $Na^+(aq)$ and $NO_3^-(aq)$, appear as *reactants*, too. *These ions do nothing chemically.* They are examples of "spectator" ions. Since they are in identical forms on both sides of the equation, and since they do not change chemically, *we may cancel them* from the equation. When this is done, all that remains is the net ionic equation. It is particularly useful for discussing the actual chemistry of a reaction because it focuses attention on the particular particles that interact and the new substances that form from them. We need the *molecular equation* for planning an experiment, because it identifies the actual reactants and let's us work out the amounts to use. We need the *ionic* equation, however, to discuss the fundamental chemistry involved. Two criteria have to be met to have a balanced ionic equation.

Another name for molecular equation is lab equation.

1 Material balance—all particles on one side of the equation must appear somewhere on the other.
2 Electrical balance—the net charge on the left equals the net charge on the right. Overall electrical neutrality for the solution requires an electrical balance.

We'll learn some rules for predicting solubilities in Section 16.9.

You're not expected, just from this discussion, to be able to convert any or all molecular equations into ionic equations. Such an ability requires considerable knowledge of chemistry. You have to know what substances will exist as ions in water, for example, or what substances are insoluble in water. Some of this knowledge will come in time, but for the moment the objective is simply to gain the ability to recognize an ionic equation, know its purpose, tell if it's balanced, and use it to describe some particular chemical property. Ionic equations are easy to write *when enough information is given*. Let's work an example.

EXAMPLE 11.14   **Writing Net Ionic Equations**

Problem: The light-sensitive coating on photographic film contains silver bromide (AgBr). It is made in a dark room by mixing aqueous solutions of sodium bromide (NaBr) and silver nitrate ($AgNO_3$). Both NaBr and $AgNO_3$ are ionic compounds that are soluble in water. When their solutions are mixed, AgBr precipitates. The other product, $NaNO_3$, is an ionic compound left in solution. Write a net ionic equation for this reaction.

Solution: The steps given here work for any problem like this.
Step 1 Write a balanced *molecular* equation.

$$NaBr(aq) + AgNO_3(aq) \longrightarrow AgBr(s) + NaNO_3(aq)$$

To expand a molecular formula means to break it up into the ions it consists of and write the formulas of the ions separately.

Step 2 Expand the formulas of the molecular equation (as the facts permit) to show all of the ions that are free to interact or to be spectators. This step requires facts given by the problem or (as you gain in knowledge and experience) remembered.

Whenever an *ionic* compound dissolves in water, we may show it as separated ions. The formulas for ionic precipitates however, cannot be expanded.

$$\underbrace{Na^+(aq) + Br^-(aq)}_{} + \underbrace{Ag^+(aq) + NO_3^-(aq)}_{} \longrightarrow$$

[expansion of NaBr(*aq*) permitted because it is water-soluble and ionic] [expansion of AgNO₃(*aq*) permitted for the same reason]

$$\underbrace{AgBr(s)}_{} + \underbrace{Na^+(aq) + NO_3^-(aq)}_{}$$

[expansion not allowed; AgBr is insoluble] [expansion of NaNO₃(*aq*) permitted; it is water-soluble and ionic]

Step 3 Cancel spectator particles and rewrite the equation.

$$Br^-(aq) + Ag^+(aq) \longrightarrow AgBr(s)$$

Step 4 Check to see that the equation is balanced. (1) The material balance is apparent. (2) The electrical balance is also apparent; a net charge of zero on the left balances the net charge of zero on the right.

EXERCISE 11.23 Aqueous solutions of sodium sulfide, Na_2S, and copper(II) nitrate, $Cu(NO_3)_2$, are mixed. (Each substance is an ionic compound.) A precipitate of copper(II) sulfide, CuS, forms at once. Left behind is a solution of sodium nitrate, $NaNO_3$. Write the net ionic equation for this reaction.

EXERCISE 11.24 The reaction of aqueous calcium nitrate, $Ca(NO_3)_2(aq)$, with aqueous sodium phosphate, $Na_3PO_4(aq)$ gives a precipitate of calcium phosphate, $Ca_3(PO_4)_2(s)$. Sodium nitrate (an ionic substance) stays dissolved in the solution. Write the molecular equation for the reaction. The net ionic equation is

$$3Ca^{2+}(aq) + 2PO_4^{3-}(aq) \longrightarrow Ca_3(PO_4)_2(s)$$

Not all of the formulas in an ionic equation need be those of ions or ionic compounds. For example, the reaction of hydrochloric acid (HCl) with sodium bicarbonate ($NaHCO_3$) gives sodium chloride, carbon dioxide, and water. Of these chemicals, $HCl(aq)$, $NaHCO_3$, and $NaCl$ are all water-soluble ionic compounds. Water and carbon dioxide, however, are nonionic, molecular substances *whose formulas we cannot expand.* The molecular equation for the reaction is

$$HCl(aq) + NaHCO_3(aq) \longrightarrow NaCl(aq) + H_2O(\ell) + CO_2(g)$$

After expanding all of the molecular formulas we can, we get the full ionic equation:

$$H^+(aq) + Cl^-(aq) + Na^+(aq) + HCO_3^-(aq)$$
$$\longrightarrow Na^+(aq) + Cl^-(aq) + H_2O(\ell) + CO_2(g)$$

This reduces to the net ionic equation

$$H^+(aq) + HCO_3^-(aq) \longrightarrow H_2O(\ell) + CO_2(g)$$

SUMMARY

Mixtures. Substances that are made of two or more elements or compounds physically intermixed in no definite proportions make up most of all matter. Of the uniform mixtures—solutions, colloidal dispersions, and suspensions—only the solution is homogeneous at the atomic, ionic, or (small) molecular level. Only the solution is indefinitely stable toward the force of gravity.

Suspensions. Particles having at least one dimension of 100 nm can be kept in suspension in a fluid, but filtration or the force of gravity will separate them from the fluid.

Colloidal dispersions. Particles having at least one dimension in the range of 1 to 100 nm settle slowly or not at all under gravity, and they cannot be trapped by ordinary filter papers. They are good light scattering agents as evidenced by the Tyndall effect. They do not pass through dialyzing membranes, but they contribute to the total osmotic pressure of a solution in proportion to their particle concentrations. They show the Brownian movement.

Solutions. Atoms and the smaller, ordinary-sized ions and molecules have diameters on the order of 0.1 to 1 nm. These cannot be trapped by filter paper; they do not scatter light; and only under very powerful gravitational-like forces will their kinetic movements be unable to keep them uniformly intermixed in a fluid medium. Solutes affect solvents by lowering their vapor pressures, raising their boiling points, lowering their freezing points, and making possible net solvent flows in dialysis or osmosis. If the solutes are electrolytes, they release ions in solution, and these solutions will conduct electricity.

Solubility. "Likes dissolve likes" is a way of predicting relative solubilities. ("Likes" refer to polarities.) Water is a good solvent for ionic substances because its molecules can surround and insulate ions—the phenomenon of hydration. When some ionic substances form in the presence of water, they incorporate definite quantities of water in their crystals. Such substances are called hydrates. The solubilities of most solids increase with increasing temperature. The solubilities of gases decrease with increasing temperatures. The increase of gas solubility with pressure is in accordance with Henry's law.

Colligative properties. The vapor pressure of one component in a liquid solution is lowered by the presence of other components in accordance with Raoult's law. This and other colligative properties depend only on the physical presence and concentration of the particles and not on their chemical identities. The osmolality and osmolarity of a dilute solution of an ionic substance is two or more times as great as that calculated on the basis of its molarity or molality, the exact relationship depending on the number of ions released per formula unit. The effects of ionic solutes on freezing point lowering, boiling point elevation, or vapor pressure lowering is similarly larger than those of molecular solutes.

Net ionic equations. We need a molecular equation in planning an actual experiment, but in discussing what happens *chemically* we need only the participants. The net ionic equation gives us this information uncluttered by uninvolved substances.

INDEX TO EXERCISES, QUESTIONS, AND PROBLEMS
(Those numbered above 66 are review problems.)

REVIEW QUESTIONS

11.25. In terms of chemical composition, what are the three kinds of matter?

11.26. What is the most important difference between elements and compounds, on the one hand, and mixtures, on the other, when compared in terms of composition?

11.27. What is the basis for distinguishing between solutions, colloidal dispersions, and suspensions?

11.28. When a salad dressing consisting of oil and vinegar is shaken very vigorously, it will appear to be *uniform*. Is it proper to describe it as homogeneous? Explain.

11.29. Would the mixture of question 11.28 be called a colloidal dispersion or a suspension? (Base your answer on what you know eventually happens to such a mixture.)

11.30. Why won't a solution give the Tyndall effect?

11.31. What causes the Brownian motion observable with colloidal dispersions?

11.32. Mayonnaise is an example of what kind of a colloidal dispersion?

11.33. Explain how electrical charges and relative sizes of the dispersed particles are factors in stabilizing sols.

11.34. What simple test could be used to tell if a clear, aqueous fluid contained colloidally dispersed material?

11.35. What does "likes" refer to in the rule "likes dissolve likes"?

11.36. Which would be a better solvent for iodine, I_2, water or carbon tetrachloride? Explain.

11.37. Iodine dissolves better in ethyl alcohol (as in "tincture of iodine") than in water. What does this tell us about molecules of alcohol?

11.38. Explain how liquid water can separate and dissolve the ions in sodium bromide but cannot dissolve molecules of cyclohexane.

11.39. Why are hydrates called compounds and not just wet mixtures?

11.40. What kinds of gases are the most soluble in water?

11.41. Why does the solubility of a solid or a liquid in water generally increase with temperature?

11.42. Using the language of equilibria, how do we define a saturated solution?

11.43. Is a saturated solution necessarily concentrated? Explain with examples.

11.44. If you were handed a liquid and told that it was (at that temperature) a saturated solution of sodium sulfate in water, what could you do to make sure that it is actually saturated and will stay saturated even if the temperature fluctuates a little?

11.45. If you cooled a clear, saturated solution of sodium thiosulfate (photographer's hypo) you would see one of two possible changes. What are they?

11.46. What causes a solution with a nonvolatile solute to have a lower vapor pressure than the solvent at the same temperature?

11.47. What kinds of data would you seek in order to find out if a binary solution of two miscible liquids were very nearly an ideal solution?

11.48. Aqueous solutions of ethyl alcohol show a positive deviation when their actual vapor pressures are compared to those calculated using Raoult's law. What does "positive deviation" mean (devise a figure as part of your answer); and how would it be explained in terms of molecule-molecule interactions)

11.49. Name the four colligative properties of solutions studied in this chapter.

11.50. When water is the solvent and the concentration of some solution is quite small, the molar concentration is essentially identical to the molal concentration. Show that this is true.

11.51. When an aqueous solution of sodium chloride starts to freeze, why don't the ice crystals contain ions of the salt?

11.52. Explain in your own words how the presence of, say, dissolved sugar in water causes (a) the boiling point of the solution to be higher than 100 °C; and (b) the freezing point to be lower than 0 °C.

11.53. What is the phenomenon of dialysis?

11.54. What is the difference between dialysis and osmosis?

11.55. In osmosis why *must* the net migration of solvent be from the side less concentrated in solute to the side more concentrated?

11.56. Why is it important that fluids administered to patients intravenously have osmotic pressures essentially the same as that of blood?

11.57. Besides cellular bodies, blood contains dissolved salts and some colloidally dispersed material (mostly proteins). When blood is separated from some other aqueous system by a *dialyzing* membrane, what gives it any osmotic pressure?

11.58. Two glucose solutions of unequal osmolarity are separated by an osmotic membrane. Which solution will *lose* water, the one with higher or lower osmolarity?

11.59. Osmosis and dialysis help get soil water into plant roots. For this to happen which should have the higher osmotic pressure, the soil water (containing dissolved minerals) or the aqueous fluid just inside the root surface? Explain.

11.60. A 1 molal solution of an unknown compound in water freezes at -1.84 °C. Is the compound likely to be ionic or molecular? Explain.

11.61. Even when the effect of breaking up into ions is considered, the freezing point of a concentrated salt solution isn't quite as low as predicted. Explain.

11.62. Why does the observed freezing point of a salt solution become more and more equal to the calculated freezing point as the solution is made more dilute?

11.63. Hydrochloric acid, $HCl(aq)$, reacts with potassium hydroxide, $KOH(aq)$ as follows:

$$HCl(aq) + KOH(aq) \longrightarrow KCl(aq) + H_2O(\ell)$$

Potassium hydroxide dissolves in water as K^+ and OH^-; $HCl(aq)$ as H^+ and Cl^-. Of the products, KCl is a water-soluble electrolyte; H_2O is un-ionized. Write the net ionic equation for this reaction.

11.64. Sodium carbonate (Na_2CO_3), a water-soluble electrolyte, reacts in water with dilute hydrobromic acid, HBr(aq), to give water, carbon dioxide, and sodium bromide. Sodium bromide, NaBr, is a water-soluble electrolyte. Write the molecular and the net ionic equations for this reaction.

11.65. Sodium chloride (NaCl), potassium bromide (KBr), sodium bromide (NaBr), and potassium chloride (KCl) are *all* water-soluble electrolytes. If you mixed 0.01 M solutions of NaCl and KBr, would a chemical reaction occur? If so,

write the net ionic equation. If not, explain why not.

11.66. Sulfur dioxide gas will form and bubble out of a solution made by pouring together aqueous solutions of sodium sulfite, $Na_2SO_3(aq)$, and hydrochloric acid, $HCl(aq)$. The net ionic equation before balancing is

$$H^+(aq) + SO_3^{2-}(aq) \longrightarrow H_2O(\ell) + SO_2(g)$$

Balance this equation and write a balanced molecular equation for the reaction.

REVIEW PROBLEMS

11.67. The solubility of methane, the chief component of bunsen burner gas, in water at 20 °C and 1.0 atm pressure is 0.025 g/liter. What will be its solubility at 1.4 atm and 20°C?

11.68. At 740 torr and 20 °C, nitrogen has a solubility in water of 0.018 g/liter. At 620 torr and 20 °C, its solubility is 0.015 g/liter. Does nitrogen obey the gas pressure—solubility law (Henry's law)? Do the necessary calculations.

11.69. A solution is prepared by mixing 60 g toluene (C_7H_8) and 60 g chlorobenzene (C_6H_5Cl). What is the mole fraction of each component?

11.70. A 50% (vol/vol) solution of methyl alcohol (CH_3OH) in water was used as an antifreeze in a car radiator. What is the mole fraction of each component? (Assume the density of methyl alcohol is 0.78 g/ml and that of water is 1.00 g/ml.)

11.71. A solution consisted of 33 mol percent methyl alcohol, 25 mol percent ethyl alcohol in water. What was the concentration of water in mole percent?

11.72. What is the mole percent of each component in the antifreeze solution of problem 11.70?

11.73. Air within the tiny air sacs (alveoli) in the lungs consists of nitrogen ($P_{N_2} = 570$ torr), oxygen ($P_{O_2} = 103$ torr), carbon dioxide ($P_{CO_2} = 40$ torr), and water vapor ($P_{H_2O} = 47$ torr). Assuming this mixture behaves as an ideal gas, what is the composition in mole percents. (Calculate to two significant figures.)

11.74. At the top of Mt. Everest (elevation 8.8 km) the atmospheric pressure is about 250 torr, a third of what it is at sea level. The partial pressure of oxygen is 53 torr and the partial pressure of nitrogen is 197 torr. (a) What is the composition of the air on Mt. Everest in mole percents? (b) What is the composition of the air in mole percents at sea level where P_{O_2} is 160 torr and P_{N_2} is 600 torr? (c) Comparing the data provided by your answers to parts a and b, what has to be the reason why it is hard to breathe without supplemental oxygen at high altitudes? (Do all calculations to two significant figures.)

11.75. At 20 °C the vapor pressure of pentane is 420 torr and the vapor pressure of heptane is 36 torr. Both are liquid

hydrocarbons found in gasoline. If a solution of the two is prepared having 30 mol percent pentane, what will be its total vapor pressure at 20 °C? (Calculate to two significant figures.)

11.76. Benzene and toluene help give lead-free gasoline good engine performance. At 40 °C the vapor pressure of benzene is 180 torr; that of toluene is 60 torr. Calculate the mole percents of each substance in a solution having a total vapor pressure of 96 torr. (Calculate to two significant figures.)

11.77. A solution is prepared by dissolving 18.0 g glucose (formula weight 180) in 1.00 kg water. (a) What is its molal concentration? (b) What is the mole fraction of glucose? (c) What is the mole fraction of water?

11.78. If you dissolved 10.0 g NaCl in 1.00 kg water, what would be its molal concentration? The volume of this solution is virtually identical with the original volume of the 1.00 kg of water. What, therefore, is the molar concentration of the solution? What would have to be true about any solvent for one of its dilute solutions to have essentially the same molar and molal concentrations?

11.79. The vapor pressure of water at 25 °C is 23.8 torr. Assuming that the solution of problem 11.77 is ideal, what would be its vapor pressure?

11.80. The vapor of pressure of water at 20 °C is 17.5 torr. Assuming the solution described in problem 11.78 to be ideal and that all the dissolved NaCl is 100% ionized, what would be the vapor pressure of that solution at 25 °C?

11.81. Ethylene glycol, $C_2H_6O_2$, is the basic component of some permanent antifreezes. In northern states and Canada, the protection of an automobile cooling system to −40 °F is generally sought each winter. (a) How many moles of ethylene glycol are needed per 1.00 kg water to insure this protection? (What molality of ethylene glycol in water would freeze at −40 °F?) (b) The density of ethylene glycol is 1.11 g/ml. What volume (in ml) of ethylene glycol per 1.00 kg water does your answer to part a amount to? (c) Look up and use conversion factors to find out how much ethylene glycol (in quarts) should be mixed with each quart of water to get protection to −40 °F.

11.82. You want to make homemade ice cream and you need a cooling temperature of − 10 °C (14 °F). What mixture of sodium chloride, NaCl, and water would freeze at − 10 °C, assuming that the salt will be 100% separated into its ions in this solution. Give your answer in grams of NaCl per 100 grams of water.

11.83. What would be the boiling point of 2.00 molal sugar in water? (Sugar is a molecular compound.) What would be the freezing point of this solution? (It is largely the sugar in ice cream that makes it harder to keep ice cream frozen than ice cubes.)

11.84. Glycerol ($C_3H_8O_3$, formula weight 92) is a liquid that for all practical purposes is nonvolatile. It is also very soluble in water. If 46.0 g of glycerol is dissolved in 250 g water.
(a) What is the boiling point of the solution at 760 torr?
(b) What is its freezing point?
(c) What is its vapor pressure at 25 °C? (The vapor pressure of pure water is 23.8 torr at 25 °C.)

11.85. A solution of 12.00 g of an unknown dissolved in 200.0 g benzene froze at exactly 3.45 °C. What was the formula weight of the solute?

11.86. A solution of 14 g of an unknown molecular compound in 1.0 kg benzene boiled at 81.7 °C. What was the formula weight of the unknown?

11.87. Which solution has the higher osmotic pressure, 10% NaCl or 10% NaI? (Both are wt/wt percents. Both compounds break up in water the same way—two ions per formula unit.)

11.88. What will be the osmotic pressure (in torr) of a 0.010 M solution in water of a molecular substance at 25 °C?

11.89. (a) Show that this equation is true:

$$\text{Molar mass} = \frac{[\text{mass in grams}] \times R \times T}{\Pi \times V}$$

(b) An aqueous solution of a protein was prepared in a concentration of 2.0 g/liter at 298 K. This solution had an osmotic pressure of 0.021 torr. What was the formula weight of the substance?

CHAPTER TWELVE

SIMPLE MOLECULES AND IONS OF NONMETALS

12.1 OCCURRENCE OF NONMETALLIC ELEMENTS

**Hydrogen, helium, and a few other nonmetals dominate
the composition of the universe**

The nonmetallic elements are in the upper right corner of the periodic table. They include all of the gaseous elements and one of the two liquid elements. The rest of the nonmetals are mostly brittle solids. Nearly half of the nonmetals have no chemical reactions whatsoever. Compared to highly prized and technologically important metals such as gold, silver, platinum, nickel, copper, chromium, iron, and many others, the nonmetals are drab, unexciting substances. Only one nonmetal, carbon in its diamond form, has any wide aesthetic appeal. Yet how significant these nonmetals are both on this planet and throughout the entire universe! We obtained strong hints of this importance in Chapter Nine, where we surveyed how carbon, nitrogen, hydrogen, and oxygen atoms are used to make organic molecules. In this chapter we will study the general occurrence and the relatively simple reactions and compounds of the nonmetals. Then, in Chapter Thirteen, we will see how a few nonmetals make up the huge molecules found in plastics and in crustal rocks. Finally, in Chapter Nineteen, we will learn more about the compounds of nonmetals that are essential to life itself.

The nonmetals in the universe, the sun, and planet earth
Scientists have estimated the relative abundances of the elements in the universe from the frequencies and intensities of the light emitted by stars, including our own sun, as well as from the compositions of innumerable meteorites (see Table 12.1). Hydrogen and helium, the elements with the lowest atomic weights, so dominate the other elements that all the rest almost seem like impurities in the universe!

The solar atmosphere is similarly dominated by hydrogen and helium, as you can see in Table 12.2. According to current thinking, hydrogen is the sun's fuel, and the nuclear fusion of hydrogen to helium generates the sun's energy, as discussed in Chapter Five. The fusion of hydrogen probably occurs in other stars, too, which may explain the relative abundance of hydrogen and helium everywhere in the universe.

According to the present theory, the rest of the elements formed by successive nuclear fusions and radioactive transformations that started with the hydrogen-helium conversion. The factor that seems to control the abundances of any of the elements is the relative stability of various combinations of nucleons—protons

Table 12.1

Elements in the Universe—The Ten Most Abundant

Element	Atomic Number	Abundance[a]
Hydrogen	1	4.0×10^8
Helium	2	3.1×10^7
Oxygen	8	2.2×10^5
Neon	10	8.6×10^4
Nitrogen	7	6.6×10^4
Carbon	6	3.5×10^4
Silicon	14	1.0×10^4
Magnesium	12	9.1×10^3
Iron	26	6.0×10^3
Sulfur	16	3.8×10^3

Data from B. Mason, *Principles of Geochemistry*, 3rd edition, 1966, John Wiley & Sons, Inc., New York.
[a] Number of atoms per 10,000 atoms of silicon.

Table 12.2

Elements in the Solar Atmosphere—The Ten Most Abundant

Element	Atomic Number	Abundance[a]
Hydrogen	1	3.2×10^8
Helium	2	5.0×10^7
Oxygen	8	2.9×10^5
Carbon	6	1.7×10^5
Nitrogen	7	3.0×10^4
Silicon	14	1.0×10^4
Magnesium	12	7.9×10^3
Sulfur	16	6.3×10^3
Iron	26	1.2×10^3
Sodium	11	6.3×10^2

Data from B. Mason, *Principles of Geochemistry*, 3rd edition, 1966, John Wiley & Sons, Inc., New York.
[a] Number of atoms per 10,000 atoms of silicon.

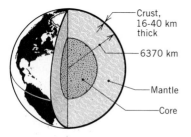

The earth's core, mantle, and crust.

About 90% of the volume of silicate rock is occupied by oxygen atoms.

and neutrons. The nuclei that have accumulated in the largest amounts are generally those of the nonradioactive isotopes of the elements of lower atomic weights. The heavier elements formed (and continue to form) during, and probably only during, super novas. A super nova is a catastrophic explosion and disintegration of a star that creates the extraordinary conditions needed for nuclear fusions. Super novas make stars shine briefly with a brilliance hundreds of millions times that of our sun. The higher elements that form during super novas then accumulate in meteorites, comets, planets, and moons.

When we examine the data for the elements that make up our own planet (Table 12.3), we find that there are more *atoms* of oxygen but iron leads on the basis of mass. Nearly all this iron is in the earth's core and mantle. In the crust oxygen is again the most prevalent element, whether measured by the number of atoms, the total mass, or the total volume, as you can see from the data in Table 12.4. In crustal rocks, oxygen accounts for over 60% of all the atoms and over 90% of the total volume. As we will study in Chapter Thirteen, oxygen atoms are held in place in rocks mostly by bonds to silicon atoms but, in this bonded form, the diameter of an oxygen atom is over three times that of a silicon atom.

Table 12.3

Elements in the Earth—The Ten Most Abundant

Element	Atomic Number	Atom Percent	Weight Percent
Oxygen	8	50	30
Iron	26	17	35
Silicon	14	14	15
Magnesium	12	14	13
Sulfur	16	1.6	1.9
Nickel	28	1.1	2.4
Aluminum	13	1.1	1.1
Calcium	20	0.7	1.1
Sodium	11	0.7	0.6
Chromium	24	0.01	0.3

Data on weight percents from B. Mason, *Principles of Geochemistry*, 3rd edition, 1966, John Wiley & Sons, Inc., New York.

Table 12.4
Elements in the Earth's Crust—The Ten Most Abundant

Element	Atomic Number	Atom Percent	Weight Percent	Volume Percent
Oxygen	8	61	47	93.8
Silicon	14	21	28	0.86
Aluminum	13	6.2	8.1	0.47
Hydrogen	1	3.0	0.14	—
Sodium	11	2.6	2.8	1.3
Iron	26	1.9	5.0	0.43
Magnesium	12	1.8	2.1	0.29
Calcium	20	1.6	3.6	1.0
Potassium	19	1.4	2.6	1.8
Titanium	22	0.2	0.44	—

Data from B. Mason, *Principles of Geochemistry*, 3rd edition, 1966, John Wiley & Sons, Inc., New York.

Most of the volume of a water molecule is that of the oxygen atom

Water covers about 71% of the earth's surface. Since oxygen takes up 16 out of the 18 units of atomic weight in water, the waters of our planet are 89% oxygen by mass. Oxygen atoms make up most of the volume of water molecules, too.

In the earth's atmosphere, which is just a mixture of gases (Table 12.5), nitrogen's net volume is about four times oxygen's. Therefore the mole ratio is also

Table 12.5
Composition of Clean, Dry Air at Sea Level

Component	Concentration
Major components	
Nitrogen, N_2	78 085% (vol/vol)
Oxygen, O_2	20.946% (vol/vol)
Minor components	Concentration (ppm)[a]
Oxides of carbon	
Carbon dioxide, CO_2	320
Carbon monoxide, CO	0.1
Oxides of nitrogen	
Dinitrogen oxide, N_2O	0.5
Nitrogen dioxide, NO_2	0.02
Oxide of sulfur	
Sulfur dioxide, SO_2	1
Noble gases	
Helium, He	5.24
Neon, Ne	18.18
Argon, Ar	9340 (0.934%, vol/vol)
Krypton, Kr	1.14
Xenon, Xe	0.087
Miscellaneous	
Ammonia, NH_3	0 to traces
Hydrogen, H_2	0.05
Methane, CH_4	2
Ozone, O_3	0.02–0.07 (seasonal variations)

Data from J. R. Holum, *Topics and Terms in Environmental Problems*, 1977, Wiley-Interscience, New York.
[a] This unit, ppm, means *parts per million*. For example, 320 ppm means 320 liters in 1 million liters of total volume.

about 4 to 1, since equal volumes of gas (at the same temperature and pressure) have equal numbers of moles. Nitrogen probably dominates in the air because it is chemically far less reactive than oxygen. Oxygen, not nitrogen, supports combustion, respiration, and decay, for example. Oxygen, not nitrogen, reacts to form solid oxides with metals from the earth's mantle and deep crustal regions as these metals work their way toward their eventual exposure to the atmosphere. The less reactive nitrogen is thus left as the principal component in air. Fortunately, for the sake of life on earth, the air's supply of oxygen is continuously replenished by the biosphere, the earth's realm of living things. Otherwise, the earth, like all other cosmic bodies that have been studied, would have no free, molecular oxygen (O_2) left. But, as we learned in Section 5.8, plants in the biosphere use sunlight, water, carbon dioxide, and minerals to make plant materials *and* molecular oxygen. This photosynthetic activity replaces the oxygen used up by all forms of oxidation. So far as can be determined, the rates of oxygen production and consumption on earth are now in exact balance.

The principal elements of the human body are given in Table 12.6. Only four nonmetallic elements—hydrogen, carbon, oxygen, and nitrogen—contribute over 96% of the mass of the body, with hydrogen providing the most atoms. Three other nonmetals—sulfur, phosphorus, and chlorine—are also high on the list of elements used to make the body.

Table 12.6
Elements in the Human Body[a]

More Than 1% By Atoms or by Weight			0.1–1% By Weight		Less than 0.1% By Weight	
Element	Atom Percent	Weight Percent	Element	Weight Percent	Element	Weight Percent
Hydrogen	61	9.3	Sulfur	0.64	Magnesium	0.04
Oxygen	25.7	62.8	Phosphorus	0.63	Iron	0.005
Carbon	10.6	19.4	Sodium	0.26	Zinc	0.0025
Nitrogen	2.4	5.1	Potassium	0.22	Copper	0.0004
Calcium	0.2	1.4	Chlorine	0.18	Tin	0.0002
					Manganese	0.0001
					Iodine	0.0001
					Molybdenum	0.00002
					Cobalt	0.000004
					Vanadium	0.000003

[a] Percents by weight are from B. Mason, *Principles of Geochemistry*, 3rd edition, 1966, John Wiley & Sons, Inc., New York. Mason's data include other very trace elements that are very likely impurities and not at all essential. His list omits fluorine, selenium, and chromium, which seem to be important to good health but in exceedingly trace quantities.

Nuclear stability, bonding ability, and relative abundances

Whether we journey to the outer reaches of space or plunge to the molecular level of life, a handful of nonmetallic elements of low atomic weights make up most of the universe. Why these and not others? Nuclear stability is one factor—perhaps a far more likely factor than chemical stability. Isotopes of the lightest elements are the most stable with respect to fission or radioactive breakdown. (Some reasons for this nuclear stability will be discussed in Chapter Twenty-one.) But what causes oxygen and silicon to make up so much of the crust of the earth, and what makes oxygen, carbon, nitrogen, and hydrogen so prevalent in living systems? Perhaps one answer is that atoms of these (and other) nonmetallic elements can form strong *covalent* bonds with each other.

Moreover, most of them can form double and triple bonds as well as single bonds. This strong bond-forming ability of the lighter nonmetallic elements no doubt contributed greatly to the directions taken by evolutionary processes during the most ancient of times. These processes tend to capitalize on any environmental factors, including basic chemical properties, that support survival.

Nonmetals and pi bonds

Carbon, oxygen, and nitrogen—three of our planet's most prevalent nonmetals—all have an unusual ability to form double bonds. In an atom of any of these elements, the p orbitals of the electrons available for bonding are in the second shell, not too far from the atom's nucleus. If we drop down to the third row of the periodic chart to silicon, sulfur, and phosphorus, the p orbitals of the bonding electrons are farther away from the nucleus in shell *three*. Therefore, when third-level p electrons go into pi molecular orbitals, they cannot be as effective in holding nuclei near each other as when the p electrons come from the second level. Thus, pi molecular orbitals formed by the overlap of third shell p orbitals do not make pi bonds as strong as those made from second shell p orbitals.

$$O = C = O$$

Carbon dioxide

The oxides of carbon and silicon illustrate that the types of bonds used to hold the atoms together can depend entirely on the level in which the bonding electrons reside. Carbon dioxide is a gas at room temperature and pressure. Silicon dioxide is a solid that in its purer forms makes up quartz and quartz sand. A molecule of the dioxide of carbon, a second-row element, has two double bonds, $O=C=O$, and therefore two pi bonds. The formula CO_2 is a true molecular formula, and carbon dioxide gas is made up of molecules. Silicon, just beneath carbon and in the third row of the periodic table, forms a dioxide whose formula is usually given as SiO_2, but that's only an empirical formula. No individual molecules of SiO_2 exist, and pi bonds are not present. The silicon-oxygen bonds are *single* bonds in a huge silicon-oxygen network of atoms. Evidently the p-orbitals of silicon atoms do not overlap with those of oxygen atoms to create pi bonds. (In the next chapter we will study various forms of silicon-oxygen compounds.)

Silicate system

12.2 HYDROGEN

Hydrides are compounds between hydrogen and metals or other nonmetals

Hydrogen doesn't fit into any chemical family. We would gain no particularly valuable insight by trying to force hydrogen into Group IA along with the alkali metals on the grounds that it forms an ion with a charge of 1+. Hydrogen can also exist as the hydride ion, H^-, but that doesn't make it resemble the elements of Group VIIA, the halogens, which also have ions with charges of 1−. Hydrogen really stands alone. We have put it in IA only to give it a place in the table.

An English chemist, Henry Cavendish (1731–1810), first recognized hydrogen as a separate substance. He made it much as we do today when we need only small amounts—by the action of a metal such as zinc on an acid (Figure 12.1).

(aq) = aqueous solution

$$Zn(s) + 2HCl(aq) \longrightarrow H_2(g) + ZnCl_2(aq)$$

hydrochloric
acid

The chemicals used in this method are far too costly and actually do not exist in enough quantities to make all the hydrogen annually needed by industry—about 100 billion mol in the United States. Water and methane are used instead as the chief sources of industrial hydrogen, most of which as we will soon see is used to make fertilizer, pipeline gas, and margarine. When water (as steam) is

Figure 12.1
Hydrogen gas bubbles vigorously from a beaker in which zinc metal reacts with hydrochloric acid.

Coke is made from coal.

passed over red-hot coke, a mixture of hydrogen and carbon monoxide called water gas forms.

$$H_2O(g) + C(s) \longrightarrow CO(g) + H_2(g)$$

carbon   carbon   hydrogen
(coke)   monoxide

water gas

Water gas can be separated by first cooling it to about $-200\ ^\circ$C with liquid air. Most of the carbon monoxide then condenses as the water gas changes to a slush from which the hydrogen, still a gas, can be pumped away.

When superheated steam is mixed with methane, the hydrogen in the molecules of both the water and methane is stripped.

$$CH_4(g) + H_2O(g) \longrightarrow CO(g) + 4H_2(g)$$

methane   water   carbon
monoxide

Isotopes of hydrogen

Hydrogen is the only element whose isotopes have their own names. The most abundant isotope is commonly called hydrogen, but *protium* may be used in special situations. Tritium, which is radioactive, forms in nuclear reactors. Only protium and deuterium occur naturally; and protium makes up 99.98% of elemental hydrogen. The chemistry of these two isotopes is very similar but, because deuterium atoms have twice the mass of protium atoms, some differences in chemistry occur. For example if a particular reaction causes a C—H bond to break, the same reaction at a C—D bond is slower. This difference in rates is called the **deuterium isotope effect.** Just replacing some of the hydrogen atoms in the molecules of the food you eat with deuterium atoms would slow down the rate of your body chemistry. This could be fatal.

Symbol	Name	Mass Number
H	(protium)	1
D	deuterium	2
T	tritium	3

	Bp (°C)	Mp (°C)
H_2	-252.77	-259.20
D_2	-249.49	-254.43

Chemical properties of hydrogen and hydrides

At room temperature and pressure, hydrogen doesn't react with anything except fluorine. However, mixtures of hydrogen and oxygen or of hydrogen and chlorine are very dangerous because a spark or ultraviolet light can set off violent reactions. Under the influence of heat, sometimes with the added requirement of increased pressure, hydrogen reacts with many other elements and oxides. Several hydrides are made by such reactions.

Compounds with high melting points despite low formula weights are usually made of ions.

Hydrides are binary compounds of hydrogen. There are three types: ionic, covalent, and metallic. Ionic hydrides are crystalline and saltlike, with high melting points. (Many decompose before they melt.) They form when metals from Groups IA and IIA are heated at high temperatures with hydrogen. For example:

$$Ca(s) + H_2(g) \longrightarrow CaH_2(s)$$

calcium   hydrogen   calcium hydride
(mp 816 °C, under H_2)

The ionic hydrides consist basically of the metal ion plus the hydride ion, H^-. All ionic hydrides react with water to form a metal hydroxide and hydrogen gas. For example:

$$NaH(s) + H_2O(\ell) \longrightarrow H_2(g) + NaOH(aq)$$

sodium
hydride

B_2H_6

The reaction is essentially between the hydride ion and water: $H^- + H_2O \rightarrow H_2 + OH^-$. Considerable heat is generated by this reaction, enough sometimes to ignite the evolving hydrogen. Therefore, to keep ionic hydrides safely, they must be stored in environments that are completely free of moisture and oxygen. The ionic hydrides are reducing agents because they readily transfer hydride ions, H^-. We learned in Section 8.11 that a reducing agent is something that can give up or donate electrons, and a hydride ion carries with it a pair of electrons when it transfers from one atom to another. Ionic hydrides, for example, will reduce oxygen to water.

$$2NaH + O_2 \longrightarrow Na_2O + H_2O$$

AlH_4^-

Lithium aluminum hydride, $LiAlH_4$, sodium borohydride, $NaBH_4$, and diborane (a boron hydride), B_2H_6, are all important reducing agents in organic chemistry.

In the body, a number of compounds transfer hydride ions directly to acceptor compounds involved in the reactions that let us use oxygen to generate energy for living. The B-vitamins, for example, are agents for hydride transfer.

Covalent hydrides are compounds of hydrogen and nonmetals or metalloids. Examples are methane (CH_4), ammonia (NH_3), water (H_2O), and the hydrogen halides such as HCl (hydrogen chloride). All except water are gases at room temperature, and they consist of molecules, not ions. If the covalent hydrides donate hydrogen at all, they donate it as H^+, not H^-. For example, H—Cl is an excellent donor of H^+, but CH_4 is not.

	Bp (°C)
CH_4	−164
NH_3	−33
HCl	−85
H_2O	100

Metal hydrides are compounds of hydrogen and transition metals. Many have definite formulas, such as those of nickel (NiH_2), iron (FeH_2), and uranium (UH_3), but many do not. They are little more than solutions of hydrogen in the metals, where H_2 molecules fit into the cavities between the atoms of the metal. They often behave as donors of gaseous H_2.

Uses of hydrogen

About two-thirds of all the production of hydrogen goes to make ammonia by the Haber process. Ammonia generally ranks in the top five of all chemicals in terms of total mass manufactured annually—about 34 to 35 billion lb (930 to 960 billion mol) in the United States in the late 1970s. Much ammonia is used directly as a fertilizer or in the manufacture of ammonium compounds such as ammonium nitrate (NH_4NO_3) or ammonium sulfate [$(NH_4)_2SO_4$], which also are fertilizers.

The **Haber process** is the direct combination of hydrogen and nitrogen under heat and pressure in the presence of a special mixture of metals.

$$N_2(g) + 3H_2(g) \xrightarrow[\text{heat}]{\text{pressure}} 2NH_3(g)$$

Notice that four volumes of reactants become two volumes of product in this reaction. High pressure forces this reduction in volume.

How the Haber catalyst works will be discussed in Section 14.9

The mixture of finely divided metals—mostly iron—used in the Haber process is an example of a catalyst. In general, a **catalyst** is any substance that increases the rate of a reaction without itself changing chemically. The phenomenon of reactions being speeded up by substances that themselves do not change is called **catalysis.** Very few important industrial chemical reactions happen without their own special catalysts. We will encounter many examples in our study and, in Chapter Fourteen, we will see in more detail how they work. The manufacture of ammonia by the Haber process would not be possible without the Haber catalyst. Since ammonia is such an important fertilizer, the implication of the Haber catalyst for worldwide food production and the related social and political issues simply cannot be calculated. Catalysts are even more important within our bodies. Virtually every reaction in the body requires a special catalyst called an **enzyme.** All enzymes belong to the protein family, and many involve B-vitamins and certain minerals, as we will study further in Chapter Nineteen. A number of serious inheritable diseases involve the absence of key enzymes.

Lead, mercury, and cyanide poisons work by inactivating enzymes—body catalysts.

Another use of hydrogen is to convert coal into natural gas—an application that may become more important in meeting future energy needs.

Manufacturers of margarine or of vegetable shortenings such as Crisco® and Spry® use hydrogen. The molecules of liquid vegetable oils—corn oil, peanut oil, cottonseed oil—have alkene groups that will combine with hydrogen over a nickel or platinum catalyst (see Sections 9.9 and 19.5).

alkene
group

The addition of hydrogen to some of the alkene groups in molecules of these oils changes the oils to solids.

12.3 OXYGEN

Stable oxides occur for nearly all of the elements, including two oxides of hydrogen, water and hydrogen peroxide

The members of the **oxygen family,** Group VIA, are given in Table 12.7. They all form hydrides and oxides with similar formulas, as you would expect of elements in the same family. Our study will be limited to the first two members of this group, oxygen and sulfur.

Elemental oxygen

You could probably go without food for 3 to 5 weeks and without water for a few days, but you could go without oxygen for only a few minutes. Oxygen's life-sustaining property naturally drew the attention of its earliest investigators—English philosopher and chemist, Joseph Priestly (1733–1804), Swedish chemist, Carl Wilhelm Scheele (1742–1786), and French chemist, Antoine Laurent Lavoisier (1743–1794). To Priestly goes the chief credit for the discovery of oxygen; but Lavoisier gave it its name.

Table 12.7
The Oxygen Family—Group VIA

Element	Symbol	Melting Point (°C)	Boiling Point (°C)	Appearance (at room temperature and atmospheric pressure)
Oxygen	O	−218	−183	Colorless gas
Sulfur	S	113[a]	445	Yellow, brittle solid
Selenium	Se	217	685	Bluish-gray metal
Tellurium	Te	452	1390	Silvery-white metal
Polonium	Po	254	962	An intensely radioactive metal

Some Compounds

Hydrides		Oxides	
Formula	Bp (°C)	Formula	Bp (°C)
H_2O	100	O_3 (ozone)	−112
		SO_2	−10
H_2S	−61	SO_3	−45
		SeO_2 (sublimes)	Mp > 300
H_2Se	−42	SeO_3	Mp 118
H_2Te	−2	TeO	Decomposes
—	—	TeO_3	Decomposes
		PoO_2	Decomposes

[a] For orthorhombic sulfur when heated rapidly.

Priestly made oxygen by heating mercuric oxide, HgO, and collecting the gas by the displacement of a liquid, water or mercury.

$$2HgO(s) \xrightarrow{\text{heat}} 2Hg(\ell) + O_2(g)$$

To make small quantities of oxygen in the laboratory today, potassium chlorate is heated.

$$2KClO_3(s) \xrightarrow{\text{heat}} 3O_2(g) + 2KCl(s)$$

potassium chlorate   potassium chloride

This reaction proceeds more smoothly and at a lower temperature if manganese dioxide (MnO_2) is added as a catalyst.

Air is the raw material for the industrial preparation of oxygen. First, the air is cooled and compressed to make liquid air, which is then carefully warmed. The nitrogen boils off first because the boiling point of nitrogen is −195.8 °C, just a few degrees lower than that of oxygen (−183.0 °C). The annual production of oxygen in the United States in the late 1970s was 32 to 36 billion lb (around 500 billion mol).

Oxygen is used to make steel from pig iron by the basic oxygen process, called BOP by the steel industry (Section 20.2). Liquid oxygen is used in rockets to burn the rocket fuel (Figure 12.2). In the health care field, oxygen and oxygen-enriched air are particularly important in respiratory care cases. Oxygen's greatest use, of course, is to support the natural processes of combustion, decay, and respiration.

Naturally occurring oxygen exists as a mixture of three isotopes with mass numbers of 16, 17, and 18. Oxygen-16 is the most abundant. Liquid and solid oxygen are both light blue in color, and both forms are paramagnetic. (We discussed the reason for this in Section 9.6.)

Industry uses the least expensive, most plentiful raw materials available.

Oxygen isotopes:
O-16, 99.759%
O-17 0.0374%
O-18 0.2039%

Figure 12.2
Liquid oxygen (LOX) was the oxidizing agent that combined with the fuel to give the energy for propelling American astronauts to the moon and back.

O_3

Ozone, O_3

Ozone is triatomic oxygen. When oxygen flows between metal surfaces carrying a high electrical charge, a low concentration of ozone forms.

$$3O_2 \longrightarrow 2O_3$$

The atmosphere contains traces of ozone ranging from 0.02 to 0.07 μg/m³, depending on the season and the latitude. Most ozone in the lower atmosphere where we live is made when lightning strikes through air.

Ozone is also present in smog, particularly in cities having ample sunshine. (See Special Topic 12.1.) In terms of its damage to plants, materials, and human health, ozone is one of the most serious pollutants in smog. At only 0.5 μg/m³, fortunately a level seldom attained, the physical activities of children and the elderly should be curtailed to reduce their inhaling the ozone deeply.

Ozone is one of the most powerful oxidants known. It attacks almost anything—fabrics, plants, lung tissue, rubber tires (see Figure 12.3). Since ozone also kills bacteria, it can be used to sterilize surgical instruments and disinfect drinking water. Some cities in Europe use ozone instead of chlorine for this purpose. The technology for using ozone, however, is much more expensive than that for chlorinating drinking water, and the tests needed to insure water quality are more difficult.

Some sterilizing lamps function by generating a low concentration of ozone by the action of ultraviolet light, which can break the bond in an oxygen molecule.

$$\underset{\text{oxygen}}{O_2} \xrightarrow[\text{UV light}]{} \underset{\substack{\text{oxygen} \\ \text{atoms}}}{2O}$$

An oxygen atom than collides with an oxygen molecule at the surface of some neutral particle, M, to form ozone.

$$O + O_2 + M \longrightarrow O_3 + M$$

Ozone forms in smog because of reactions that start with another pollutant, nitrogen monoxide (NO), which is formed by the direct combination of nitrogen and oxygen inside an automobile cylinder. As soon as the nitrogen monoxide in the hot exhaust hits the outside air, it reacts with more oxygen to make some nitrogen dioxide.

$$2NO(g) + O_2(g) \longrightarrow 2NO_2(g)$$

Some of this nitrogen dioxide is then split apart by the lower-energy ultraviolet rays of sunlight to give back nitrogen monoxide and, more important, *atoms* of oxygen.

$$NO_2(g) \xrightarrow[\substack{(\lambda = 366 \text{ nm for} \\ \text{best results})}]{\text{UV radiation}} NO(g) + \underset{\substack{\text{atomic} \\ \text{oxygen}}}{O(g)}$$

When atoms of oxygen collide with molecules of oxygen at the surface of some particle in air (M), ozone forms.

$$O + O_2 + M \longrightarrow O_3 + M$$

The daily buildup of ozone in urban areas follows a fairly regular pattern that parallels the buildup of NO and NO_2. Nitric oxide is the first pollutant to accumulate as cars spew out more and more exhaust during the morning rush hour. As its level rises, the NO starts to change to NO_2, so the NO_2 level starts to rise also. Just after the morning rush hour is over (and the sun is well up), the level of NO_2 peaks. But, by now, solar energy is beginning to split the NO_2 into NO and atoms of oxygen. Oxygen atoms react with oxygen molecules to form ozone. The ozone level reaches a peak about two hours after the peak level for NO_2. Eventually, the levels of NO, NO_2, and O_3 decline as reactions with other components in smog take place.

(The neutral particle accepts some of the energy of the collision. Otherwise, the ozone particle, having all the collision energy, would promptly break apart.) This interaction between oxygen and ultraviolet light also occurs in the stratosphere, where it screens out most of the sun's dangerous ultraviolet light. (See Special Topic 12.2.)

Figure 12.3
Ozone's effect on vegetation. Ozone attacks the chlorophyll, a green pigment, in pine needles. On the left, a healthy ponderosa pine in the San Bernardino National Forest, southern California. On the right, the same tree 10 years later, dying of smog.

Special Topic 12.2   Ozone in the Stratosphere

Unlike the ozone in smog, the ozone in the strato-sphere is essential to life on earth. The stratosphere is the part of the atmosphere between 16 and 40 km (10 to 25 miles) in altitude. Solar radiation is rich in high-energy ultraviolet rays of short wavelength (242 nm and below). When these enter the strato-sphere, they are absorbed by oxygen molecules. All ultraviolet radiation absorbed in this way is stopped from reaching us. The absorbed energy causes the ox-ygen molecules to split into atoms. These soon com-bine with oxygen molecules at some neutral surface (*M*) to generate ozone and heat. Ozone, however, can absorb ultraviolet energy at wavelengths between 240 and 320 nm, longer wavelengths (and therefore of lower energy) than that required to split oxygen mole-cules initially. Thus ozone is both made and destroyed in the stratosphere by reactions that use ultraviolet en-ergy and convert it into heat. At all times, a low level of ozone is constantly present. An ozone cycle exists that constantly removes ultraviolet radiation from incom-ing solar radiation. Without the ozone cycle, the human population would be exposed to the biologi-cally active ultraviolet radiation that causes skin cancer. Therefore, scientists were very alarmed in 1974 when evidence appeared that widely used pro-pellants in aerosol cans, the Freons, might interfere with the ozone cycle and deplete the stratospheric ozone concentration.

The Freons are a family of volatile, chemically stable and essentially odorless and tasteless chlorofluoro-carbons. For example, Freon-11, trichlorofluoro-methane (boiling point 24 °C, 75 °F), was once the propellant in 50 to 60% of all aerosol cans. When re-leased, the Freons mix in the atmosphere without chemically changing, become globally distributed, and migrate into the stratosphere. Up there, Freon molecules encounter ultraviolet light with wave-lengths of 185 to 227 nm that split carbon-chlorine

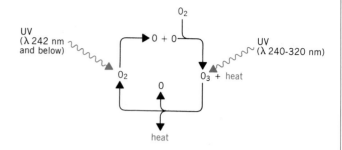

bonds. For example, Freon-11 breaks up as follows to give chlorine *atoms*.

$$\underset{\text{Freon-11}}{CCl_3F} + \underset{\text{energy}}{ultraviolet} \longrightarrow \cdot Cl_2F_2 + \underset{\substack{\text{chlorine} \\ \text{atom}}}{\cdot \ddot{\underset{..}{Cl}}:}$$

Chlorine atoms destroy ozone and give chlorine mon-oxide.

$$\cdot \ddot{\underset{..}{Cl}}: + \underset{\text{ozone}}{O_3} \longrightarrow \underset{\substack{\text{chlorine} \\ \text{monoxide}}}{ClO} + \underset{\text{oxygen}}{O_2}$$

To compound the troubles, chlorine monoxide reacts with the very oxygen atoms needed to make ozone.

$$ClO + O \longrightarrow Cl + O_2$$

Thus these last two reactions destroy ozone, remove an intermediate needed for ozone synthesis, and also regenerate one of the culprits (Cl).

The chemistry of the stratosphere is vastly more complicated than presented here, and no one knows with certainty what is happening to the ozone layer. A great deal of research is in progress. In the meantime, at the urging of the National Academy of Science, the use of Freons as aerosol propellants has been stopped in the United States.

Metal oxides

"Per," as in "peroxide," indi-cates a higher proportion of oxygen than in the normal oxide.

All of the metallic elements form oxides, usually directly, and some form perox-ides and superoxides. Lithium (Group IA) reacts directly with oxygen to give lith-ium oxide.

$$\underset{\text{lithium}}{4Li(s)} + O_2(g) \longrightarrow \underset{\text{lithium oxide}}{2Li_2O(s)}$$

Sodium rapidly reacts with atmospheric oxygen to give the peroxide.

$$2Na(s) + O_2(g) \longrightarrow \underset{\text{sodium peroxide}}{Na_2O_2(s)}$$

The remaining members of the alkali metals—potassium, rubidium, and cesium —form superoxides in air having the general formula MO_2 (where *M* is the metal). We will learn more about them in Chapter Twenty. To prevent these reactions, Group IA metals have to be stored submerged in an inert oil or in a dry nitrogen atmosphere.

The metals in Group IIA also change to their oxides by reacting with atmospheric oxygen. For example,

$$2Ca(s) + O_2(g) \longrightarrow 2CaO(s)$$
$$\text{calcium} \qquad\qquad\qquad \text{calcium}$$
$$\text{oxide}$$

When aluminum (in Group IIIA) reacts with atmospheric oxygen, the oxide coating on the surface of the aluminum is too thin to obscure the metal's bright luster. Yet the oxide so completely and tightly covers the surface that the aluminum is protected from further oxidation. This means that although aluminum, if kept free of its oxide coating, readily reacts with oxygen and water, it can be used in foil, cooking utensils, building materials, aircraft parts, and for hundreds of other purposes.

A few binary metal oxides consist of relatively nonpolar molecules and therefore have low melting points. Dimanganese heptoxide (Mn_2O_7), for example, melts at 6 °C; ruthenium tetroxide (RuO_4) melts at 25 °C; and osmium tetroxide (OsO_4) melts at 40 °C. (Interestingly, osmium metal itself melts at 3040 °C, fourth highest melting point of all the elements.) Most of the metal oxides, however, are very high-melting compounds and consist of aggregations of metal ions and oxide ions, O^{2-}. Thorium oxide, ThO_2, melts at 2950 °C. All the oxides of Group IIA metals melt above 1200 °C. Mixtures of some of these high-melting oxides are used to make the bricks that line ovens and furnaces in which pottery and china are baked. Bricks of these materials can take high temperatures even when exposed to air. Titanium oxide (TiO_2, mp 1870 °C) is the most common pigment in paints because it has an unusually high hiding power.

"Hiding power" is the ability of the paint pigment to cover up colored surfaces or designs.

As we learned in Section 6.8, most metal oxides are bases. They neutralize acids.

$$Na_2O(s) + 2HCl(aq) \longrightarrow 2NaCl(aq) + H_2O(\ell)$$

Some react with water to give metal hydroxides.

$$Na_2O(s) + 2H_2O(\ell) \longrightarrow 2NaOH(aq)$$

Amphoteros —Greek for "partly one and partly the other."

Some metal oxides will react with either acids or bases and are called **amphoteric oxides.** For example, aluminum oxide dissolves in acid with the liberation of aluminum ion.

$$Al_2O_3(s) + 6HCl(aq) \longrightarrow 2AlCl_3(aq) + 3H_2O(\ell)$$
$$\text{aluminum} \quad \text{hydrochloric} \qquad \text{aluminum}$$
$$\text{oxide} \qquad \text{acid} \qquad\qquad \text{chloride}$$

Aluminum oxide also reacts with and dissolves in concentrated sodium hydroxide.

$$Al_2O_3(s) + 2NaOH(aq) + 7H_2O(\ell) \longrightarrow 2Na[Al(H_2O)_2(OH)_4](aq)$$
$$\text{aluminum} \qquad \text{sodium} \qquad\qquad\qquad\qquad \text{sodium aluminate}$$
$$\text{oxide} \qquad \text{hydroxide}$$

A few metal oxides react with water to liberate acids. Chromium trioxide, for example, reacts as follows.

$$CrO_3(s) + H_2O(\ell) \longrightarrow H_2CrO_4(aq)$$
$$\text{chromium} \qquad\qquad\qquad \text{chromic}$$
$$\text{trioxide} \qquad\qquad\qquad\quad \text{acid}$$

These acidic metal oxides are uncommon, however, and occur only when the metal is in a very high oxidation state.

Nonmetal oxides

Some of the nonmetal oxides will be taken up individually elsewhere. We have already learned that many react with water to form acidic solutions (Section 6.8). Both oxides of sulfur, for example, change to acids in water.

$$SO_2(g) + H_2O(\ell) \longrightarrow H_2SO_3(aq)$$

sulfur   sulfurous
dioxide   acid

$$SO_3(g) + H_2O(\ell) \longrightarrow H_2SO_4(aq)$$

sulfur   sulfuric
trioxide   acid

Both sulfur oxides neutralize bases. For example,

$$SO_2(g) + NaOH(aq) \longrightarrow NaHSO_3(aq)$$

sodium
hydrogen sulfite

$$SO_3(g) + 2NaOH(aq) \longrightarrow Na_2SO_4(aq) + H_2O(\ell)$$

sodium sulfate

Carbon dioxide will also neutralize base.

$$CO_2(g) + NaOH(aq) \longrightarrow NaHCO_3(aq)$$

carbon   sodium   sodium
dioxide   hydroxide   bicarbonate

H_2O_2

Oxides of hydrogen

Two oxides of hydrogen are known—water and hydrogen peroxide (H_2O_2). Water is the most familiar chemical of reasonable purity that just about everybody sees every day. We have already studied some of its most distinctive properties, and others will be studied in later chapters.

Hydrogen peroxide, H_2O_2, is a colorless, unstable liquid that is particularly dangerous when pure. Almost anything, including dirt or heat, initiates its decomposition to water and oxygen. Even the products of the reaction catalyze it.

A reaction catalyzed by its own products is called an autocatalytic reaction.

$$2H_2O_2(\ell) \longrightarrow 2H_2O(\ell) + O_2(g)$$

The danger lies in the explosively expanding oxygen and in the ability of oxygen to support combustion. Thus 90% hydrogen peroxide was a part of the propellant system in Germany's V-1 missile in World War II. (The successor, the V-2 rocket, used liquid oxygen.) Even solutions of 20% hydrogen peroxide in water are dangerous, and they should be handled only by specialists. The "peroxide" sold in drugstores is 2 to 3% hydrogen peroxide, and it can be handled safely as a bleach and disinfectant. If you ever use it to clean a wound, you'll see it foam strongly because substances in blood catalyze its decomposition.

"Hydrolysis" means "a reaction with water."

Hydrogen peroxide is the "parent" compound of a family of substances with single O—O bonds called **peroxides.** Because their molecules have the O—O structural unit, the hydrolysis of peroxides such as the peroxydisulfate ion can be used to make hydrogen peroxide.

$$2H_2O(\ell) + (O_3S-O-O-SO_3)^{2-}(aq) \longrightarrow 2HSO_4^-(aq) + H-O-O-H(aq)$$

peroxydisulfate ion   hydrogen   hydrogen
sulfate ion   peroxide

The peroxydisulfate ion is obtained by passing direct current electricity through chilled solutions of Na_2SO_4 or $(NH_4)_2SO_4$.

Metal peroxides are salts containing the peroxide ion, O_2^{2-}. Some may be made by heating the oxide in air. For example,

$$2Na_2O(s) + O_2(g) \xrightarrow{\text{heat}} 2Na_2O_2(s)$$

sodium   sodium
oxide   peroxide

Sodium peroxide reacts readily with water to give hydrogen peroxide and sodium hydroxide.

$$Na_2O_2(s) + 2H_2O(\ell) \longrightarrow 2NaOH(aq) + H_2O_2(aq)$$

Freshly formed hydrogen peroxide decomposes particularly quickly in base to give oxygen and water. Because hydrogen peroxide and then oxygen can be generated by the action of water on sodium peroxide and other metal peroxides, the metal peroxides are strong bleaching and disinfecting agents.

12.4 SULFUR

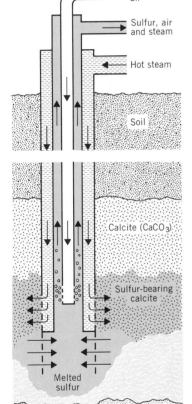

Figure 12.4
The Frasch process for extracting sulfur from deep deposits.

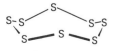

Crown configuration

In terms of total tonnage manufactured annually, sulfuric acid is the most important compound of sulfur

Sulfur is a brittle, bright yellow, nonmetallic element (Color Plate 3C) that occurs naturally in huge underground deposits in Louisiana, Texas, Sicily, and a few other places. The sulfur in these deposits is brought to the surface by the Frasch process, illustrated in Figure 12.4. Superheated steam is pumped into the deposit to melt the sulfur. Then the pressure of hot air and steam forces the molten sulfur to the surface in pipes that deliver it to cooling yards (Color Plate 4B).

Some important minerals are the sulfides of valuable metals, for example, lead sulfide or galena (PbS) and mercuric sulfide or cinnabar (HgS). Other minerals are sulfates, such as gypsum, $CaSO_4 \cdot 2H_2O$, and epsom salt, $MgSO_4 \cdot 7H_2O$. The "sand" dunes of White Sands National Monument (New Mexico) are nearly pure gypsum. In the living world, nearly all proteins contain sulfur in the form of —S—H and —S—S— groups.

Sulfur is also present in low but important concentrations in coal and petroleum. It occurs in coal as organic sulfur compounds and metal sulfides and in petroleum largely as hydrogen sulfide (H_2S). When sulfur-containing fuels are burned, sulfur dioxide is invariably a product, and it is a serious air pollutant. Some of the sulfur can be removed before the fuel is burned, however. For example, sulfur dioxide reacts with hydrogen sulfide in petroleum as follows.

$$2H_2S(g) + SO_2(g) \longrightarrow 2S(s) + 2H_2O(\ell)$$

hydrogen   sulfur
sulfide   dioxide

This process in petroleum refining is another source of industrial sulfur.

Allotropic forms of sulfur

Sulfur exists in several forms called allotropes. **Allotropes** are different crystalline or molecular forms of the same element. Oxygen and ozone, for example, are allotropes. Up to a temperature of 95.31 °C the most stable form of sulfur is orthorhombic sulfur, symbolized as S_α (Figure 12.5). It exists in beautiful crystals consisting of molecules of S_8 in which the atoms are joined in a crownlike ring. When S_α is heated above 95.4 °C, it changes to monoclinic sulfur, S_β, which exists as long crystalline needles (Figure 12.5). Monoclinic sulfur also consists of S_8 molecules, but they are not as well organized as in S_α crystals. Moreover, some of the rings have broken open to give S_8 chains. If the temperature of molten sulfur is brought to 159 °C, some of the S_8 chains join end to end to make molecules of S_{16}, S_{24}, S_{32}, and so forth. If the molten mixture of these molecules is poured into cold water, it solidifies into a third allotrope, plastic sulfur (or amorphous sulfur), which has no well-defined crystalline structure. Some S_8 rings reform and these, together with the long open chains, comprise plastic sulfur.

a   *b*

Figure 12.5
Two allotropic forms of sulfur. (*a*) Orthorhombic sulfur (*b*) Monoclinic sulfur.

When sulfur is produced by a chemical reaction, it forms in a very finely divided state called colloidal sulfur.

Chemical properties of sulfur
Sulfur will combine with most of the other elements if they are heated together but, in both environmental and economic terms, its reaction with oxygen is the most important.

$$S(s) + O_2(g) \longrightarrow SO_2(g)$$

A chemical *intermediate* is a chemical made not for direct sale so much as for its use to make another chemical.

Sulfur burns in air or oxygen with a bright blue flame, and the product, sulfur dioxide (SO_2), is both an environmental problem and a major chemical intermediate in industry.

Sulfur dioxide, sulfurous acid, and sulfites
Sulfur dioxide boils at $-10\,°C$ and freezes at $-72\,°C$. It has a sharp, acrid, disagreeable odor; if you ever inhale it, you will feel as if you were choking.

The major use for sulfur dioxide (and, therefore, for sulfur) is the manufacture of sulfuric acid, H_2SO_4. Sulfur dioxide is also a bleach, and it is used to whiten textiles, paper, wicker ware, and flour. If you look on packages of certain dried fruits, such as apricots and apples, you will see that sulfur dioxide is also used to destroy decay-causing organisms.

The solubility of sulfur dioxide in water is 12% at $15\,°C$. As noted in the discussion of nonmetal oxides, some dissolved sulfur dioxide reacts with water to give sulfurous acid, H_2SO_3. This diprotic acid is weak and unstable, and it cannot be made in pure form. It is known only as a dilute solution in water. When present as an air pollutant, sulfur dioxide is washed out of the atmosphere by rain. Rain made acidic by oxides of sulfur and nitrogen is called **acid rain,** and it is a widespread environmental problem. (See Special Topic 12.3 and Figure 12.6.)

Two kinds of salts of sulfurous acid are common: the hydrogen sulfites, which contain the HSO_3^- ion, and the sulfites, which contain the SO_3^{2-} ion. These salts are reducing agents, and all react with acids to liberate first sulfurous acid and then sulfur dioxide. The molecular and net ionic equations for these reactions, illustrated using the sodium salts, are as follows.

$$SO_2$$

119.5°

Rain as acidic as lemon juice has often fallen on lower Scandinavia and the northern parts of the United States and Great Britain. The record for acidity, rain as acidic as vinegar, was a downpour over Pitlochry, Scotland, in April 1974. Acid rain makes lakes too acidic to support desirable fish, and it is corrosive to exposed metals such as railroad rails, vehicles, and machinery and to stone building materials such as limestone and marble, both of which are largely calcium carbonate. The reaction of sulfur dioxide with wet limestone is one way to remove this pollutant from gases leaving furnaces that burn sulfur-containing coal or oil.

$$SO_2(g) + \underset{\substack{\text{calcium}\\\text{carbonate}\\(\textit{as in wet}\\\textit{limestone})}}{CaCO_3(s)} \xrightarrow{\text{moisture}} \underset{\substack{\text{calcium}\\\text{sulfite}}}{CaSO_3(s)} + CO_2(g)$$

(We looked at the solid waste disposal problem created by this technique in Problem 4.72.)

$$\underset{\substack{\text{sodium}\\\text{hydrogen}\\\text{sulfite}}}{NaHSO_3(aq)} + \underset{\substack{\text{hydrochloric}\\\text{acid}}}{HCl(aq)} \longrightarrow NaCl(aq) + \underset{\substack{\text{sulfurous}\\\text{acid}}}{H_2SO_3(aq)}$$
$$\longrightarrow SO_2(g) + H_2O(\ell)$$

Or,

$$HSO_3^-(aq) + H^+(aq) \longrightarrow H_2SO_3(aq) \longrightarrow SO_2(g) + H_2O(\ell)$$

From a sulfite salt:

$$\underset{\substack{\text{sodium}\\\text{sulfite}}}{Na_2SO_3(aq)} + 2HCl(aq) \longrightarrow 2NaCl(aq) + H_2SO_3$$
$$\downarrow$$
$$SO_2(g) + H_2O(\ell)$$

Or,

$$SO_3^{2-}(aq) + 2H^+(aq) \longrightarrow H_2SO_3(aq) \longrightarrow SO_2(g) + H_2O(\ell)$$

Figure 12.6
The polluted air of Germany's Rhein-Ruhr hastened the decay of this statue made of Baumberg sandstone at the Herten Castle in Westphalia, West Germany. Its appearance in 1908 after 206 years of exposure is on the left. On the right is its appearance in 1968 after only 60 more years. (Photos by Schmidt-Thomsen, Landesdenkmalamt, Westfallen-Lippe, Münster, West Germany.)

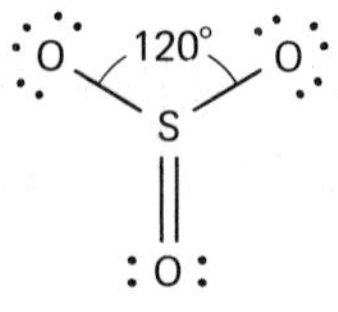

Sulfur trioxide, sulfuric acid, and the sulfates

Sulfur dioxide does not support combustion, but it can still be oxidized to sulfur trioxide by oxygen. In smoggy air a number of processes exist to aid that change. Some involve energy from sunlight and some involve the surfaces of microscopic solid particles that are also present in smog. The net result is the simple change

$$2SO_2(g) + O_2(g) \longrightarrow 2SO_3(g)$$

sulfur
trioxide

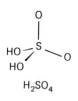

SO_3

Sulfur trioxide, which has an even more choking, stinging odor than sulfur dioxide, actually exists as a low-melting solid at room temperature. However, it easily sublimes and readily exists in the vapor state. It reacts with water to give sulfuric acid.

$$SO_3(s) + H_2O(\ell) \longrightarrow H_2SO_4(\ell)$$

sulfuric
acid

Sulfuric acid is a stable diprotic acid. Only when it is heated to 340 °C will it decompose to sulfur trioxide and water. It is a strong acid. The transfer of one of its two hydrogens to water—its first ionization—is essentially 100% complete in dilute aqueous solutions.

$$H_2SO_4(aq) + H_2O(\ell) \longrightarrow H_3O^+(aq) + HSO_4^-(aq)$$

sulfuric   hydronium   hydrogen
acid   ion   sulfate ion

H_2SO_4

Its second ionization, the transfer of its second hydrogen ion to water, is actually the ionization of the hydrogen sulfate ion. This reaction occurs to a much smaller extent than the first ionization.

$$HSO_4^-(aq) + H_2O(\ell) \longrightarrow H_3O^+(aq) + SO_4^{2-}(aq)$$

sulfate ion

(We will study the quantitative details of the weaker acids of Chapter Seventeen.)

For decades sulfuric acid has been the largest-volume industrial chemical in the United States. In the late 1970s the annual production of sulfuric acid was close to 70 billion lb (nearly 325 billion mol). Most sulfuric acid is made from sulfur by the **contact process.**

1 Sulfur is burned in air to give sulfur dioxide.

$$S(s) + O_2(g) \longrightarrow SO_2(g)$$

2. Sulfur dioxide is oxidized in contact with the catalyst, vanadium pentoxide (hence, the name *contact* process).

$$2SO_2(g) + O_2(g) \xrightarrow{V_2O_5} 2SO_3(g)$$

3 Gaseous sulfuric trioxide is trapped by bubbling it into concentrated sulfuric acid, which is a more effective trapping agent than water itself.

$$H_2SO_4(\ell) + SO_3(g) \longrightarrow H_2S_2O_7(s)$$

pyrosulfuric acid

4 Water is added to give sulfuric acid.

$$H_2O(\ell) + H_2S_2O_7(s) \longrightarrow 2H_2SO_4$$

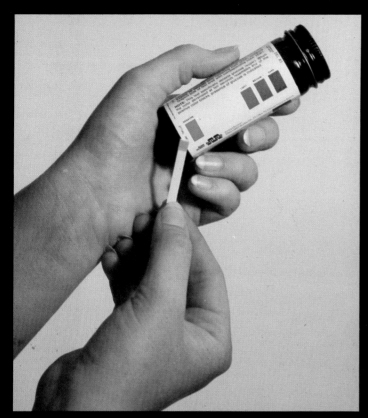

Plate 1B Iron pyrite. Its color accounts for its nickname, "fool's gold."

Plate 1A Color is a property that diabetics use when they test their urine for sugar using a kit such as this. This patient's urine is normal, as shown by the color match between the test strip and the color code on the test vial.

Plate 1C Freshly cut sodium (right) reveals a shiny metal surface that will quickly tarnish. Chlorine (below, left) is a gas at room temperature. Although table salt, NaCl, (below, right) is a compound of sodium and chlorine, two highly reactive and dangerous elements, it is itself essential to life. (Chlorine, courtesy Time/Life Books, Inc.)

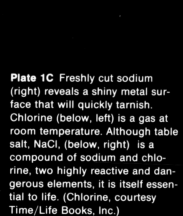

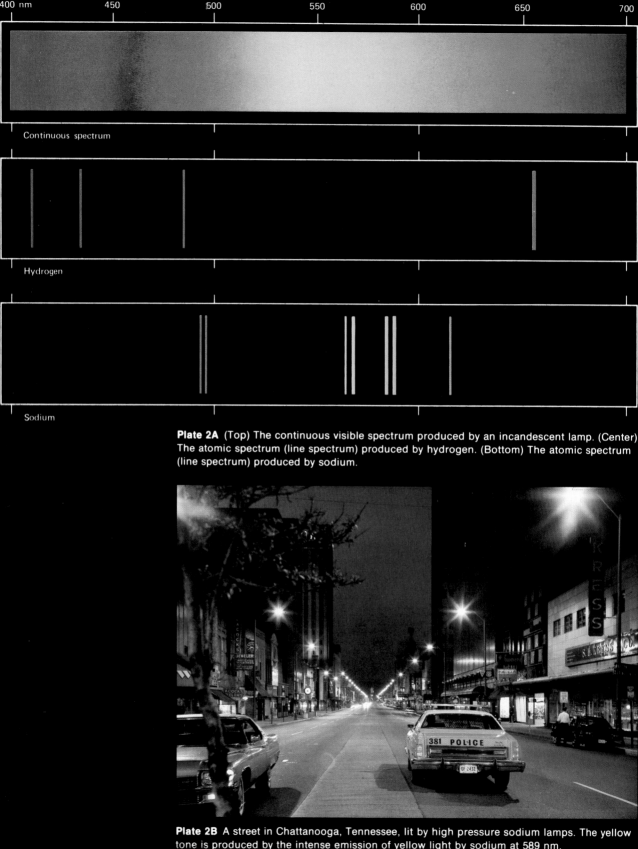

400 nm 450 500 550 600 650 700

Continuous spectrum

Hydrogen

Sodium

Plate 2A (Top) The continuous visible spectrum produced by an incandescent lamp. (Center) The atomic spectrum (line spectrum) produced by hydrogen. (Bottom) The atomic spectrum (line spectrum) produced by sodium.

Plate 2B A street in Chattanooga, Tennessee, lit by high pressure sodium lamps. The yellow tone is produced by the intense emission of yellow light by sodium at 589 nm.

Plate 3 Some nonmetallic elements. (A) Diamonds of gem quality. (B) Powdered graphite. Diamond and graphite are both forms of elemental carbon. (C) Crystals of sulfur. (D) Powdered red phosphorus, the form of phosphorus used on the striking surface of matchbooks. (E) Liquid bromine. Note the deep red color of the liquid. Bromine is very volatile, giving a red-orange vapor. (F) Iodine crystals. Solid iodine is volatile too, giving a purple vapor.

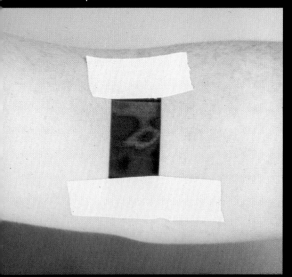

Plate 4A Thermal mapping with liquid crystals. The deep blue marks the warmest area of the skin directly over a blood vessel in a patient's arm.

Plate 4C Smog over a western city. The brown color of the haze on the horizon is caused by nitrogen dioxide, one of the major components of smog.

Plate 4B Sulfur. Enormous quantities of sulfur are processed to meet the industrial needs for this element, particularly for making sulfuric acid. This photo shows a mountain of sulfur brought to the surface from deep below the earth by the Frasch process.

Plate 5A Iodine, a halogen that is a solid at room temperature, sublimes when heated.

Plate 5B The noble gases, when placed in gas discharge tubes, give a variety of colors to displays that are commonly called "neon" signs.

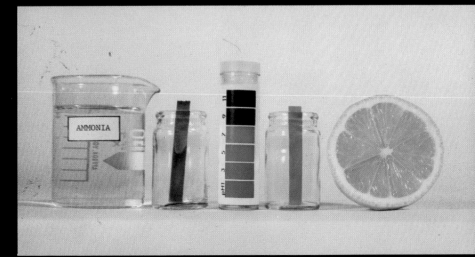

Plate 5C The same pH test paper can be used to indicate pH values over a wide range, from basic to acidic.

Plate 6A Electrolysis of a solution of potassium nitrate, KNO_3, in the presence of indicators. The initial yellow color indicates that the solution is neutral (neither acidic nor basic). As the electrolysis proceeds, H^+ produced at the anode along with O_2 causes the solution there to become pink. At the cathode, H_2 is evolved and OH^- ions are formed, which turns the solution around that electrode a bluish violet. After the electrolysis is stopped and the solution is stirred, the color becomes yellow again as the H^+ and OH^- ions formed by the electrolysis neutralize each other.

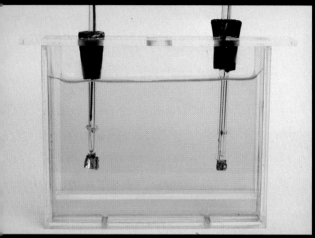

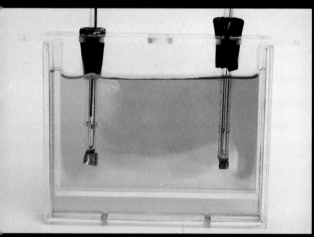

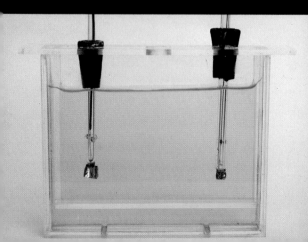

Plate 6B (Top) A coiled piece of copper wire next to a beaker containing a silver nitrate solution. (Center) When the copper wire is placed in the solution, copper dissolves, giving the solution its blue color, and metallic silver deposits as glittering crystals on the wire. (Bottom) After a while, much of the copper has dissolved and nearly all of the silver has deposited as the free metal.

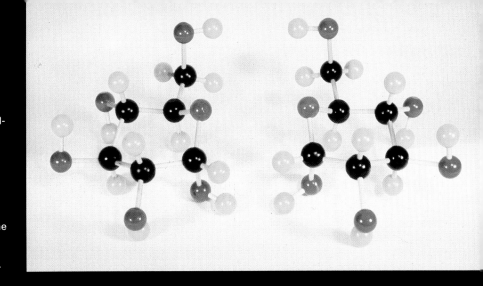

Plate 7C Naturally occurring aquamarine (top) and emerald
(bottom). When cut and polished they give beautiful gems. Both
are forms of the mineral beryl, $Be_3Al_2(SiO_3)_6$.

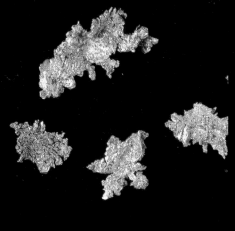

Plate 8A Samples of "native" gold—gold found naturally in the earth. If you look carefully, you can see pieces of quartz rock clinging to the gold. Can you imagine the thrill of finding these among a pile of worthless rock?

Plate 8B The gems sapphire (left) and ruby (right). Both are composed of nearly pure aluminum oxide, Al_2O_3. In the sapphire, the Al_2O_3 is contaminated by traces of iron and titanium, which give it its color. The ruby is Al_2O_3 that contains traces of chromium.

Plate 8C Each of these brightly colored compounds contains a complex ion of Co^{3+}. The variety of colors arises because of the different ligands (molecules or anions) that are bonded to the cobalt ion in the complexes.

Most major industrial operations involving chemicals and metals use sulfuric acid. Slightly over half is used in the manufacture of fertilizers, but the steel industry, the petroleum industry, and operations that make paints, dyes, drugs, plastics, textiles, batteries, explosives, and chemicals all use sulfuric acid. In fact, the bill for sulfuric acid in the United States is over $100 billion annually.

Sulfuric acid has several advantages over other acids for these purposes. It is the least expensive of the strong acids; it is stable and nonvolatile; at room temperature, it is a poor oxidizing agent, but this property improves when sulfuric acid is heated; it is a powerful dehydrating agent; and it is the only strong acid that can be easily prepared and shipped in an almost pure form—as 98% sulfuric acid. Some of its properties make concentrated sulfuric acid a dangerous chemical, unless it is handled by well-trained people. One of the dangers is that all of the reactions of concentrated sulfuric acid, including making a dilute solution in water, are highly exothermic, and the sharp rise in temperature as its reactions start further accelerates them. Concentrated sulfuric acid is also very dense and viscous, which causes it to cling to surfaces on which it is spilled. For these reasons, jugs of concentrated sulfuric acid should always be carried inside special rubber buckets large enough to contain the acid fully if the jug should break.

A dehydrating agent removes water from substances by a chemical reaction. Concentrated sulfuric acid will even remove water from sugar, $C_{12}H_{22}O_{11}$, leaving just carbon behind.

Two kinds of salts of sulfuric acid are common—the hydrogen sulfates containing the HSO_4^- ion and the sulfates having the SO_4^{2-} ion. Only the Group IA metals form hydrogen sulfates (sometimes called bisulfates). Sodium hydrogen sulfate (NaHSO$_4$, "sodium bisulfate") is an example. Its solution in water is acidic because the HSO_4^- ion ionizes as a hydrogen-ion donor. Many sulfates, several in the forms of their hydrates, are important commercial chemicals (see Table 12.8).

$$HSO_4^- + H_2O \rightleftarrows SO_4^{2-} + H_3O^+$$

Table 12.8
Some Important Sulfates

Substance	Common Name	Some Uses
Of Group IA Metals		
Na$_2$SO$_4$·10H$_2$O sodium sulfate decahydrate	Glauber's salt	Manufacture of glass; paper pulp; saline cathartic
K$_2$SO$_4$ potassium sulfate		Fertilizer
Of Group IIA Metals		
CaSO$_4$·2H$_2$O calcium sulfate dihydrate	Gypsum	Plasterboard; mortar

(CaSO$_4$)$_2$·H$_2$O
    calcium sulfate sesqui-
    hydrate
	Plaster of paris	Casts; impression moldings in dentistry

BaSO$_4$ barium sulfate	Barite	X-ray contrast medium (barium "cocktail") in radiology
Miscellaneous		
(NH$_4$)$_2$SO$_4$ ammonium sulfate	—	Fertilizer
CuSO$_4$·5H$_2$O copper sulfate pentahydrate	Blue vitriol	Fungicide; herbicide; textile dyeing; inks
ZnSO$_4$·7H$_2$O zinc sulfate heptahydrate	White vitriol	Dyeing; wood preservative

Hydrogen sulfide and sulfides

Hydrogen sulfide boils at -50 °C and freezes at -83 °C. It is a flammable, poisonous gas and the cause of the odor of rotten eggs. Fortunately, its bad odor allows it to be detected at levels far below those where it becomes acutely poisonous. Hydrogen sulfide can be made by heating a mixture of hydrogen and sulfur or by adding an acid to a metal sulfide, such as iron sulfide.

H_2S is more toxic than carbon monoxide.

$$FeS(s) + 2HCl(aq) \longrightarrow \underset{\substack{\text{hydrogen} \\ \text{sulfide}}}{H_2S(g)} + FeCl_2(aq)$$

The chief use of hydrogen sulfide in the laboratory is to detect and identify several metal ions that form sulfide precipitates having characteristic colors and solubilities. The most common way to make hydrogen sulfide for this purpose is by the reaction of water with thioacetamide.

$$\underset{\text{thioacetamide}}{CH_3 \overset{\overset{\textstyle S}{\|}}{-}C-NH_2} + H_2O \longrightarrow \underset{\text{acetamide}}{CH_3 \overset{\overset{\textstyle O}{\|}}{-}C-NH_2} + H_2S$$

Traces of hydrogen sulfide in the air or in certain foods cause silver to discolor. The dark tarnish is silver sulfide (Ag_2S). Hydrogen sulfide will also react with lead ions in lead-based paints and cause the paints to discolor as lead sulfide (PbS), a black compound, forms.

Sodium sulfide, Na_2S, is used to dehair animal hides and to make rubber. Like all metal sulfides, it liberates hydrogen sulfide in acid.

Hot solutions of Group IA sulfides will dissolve sulfur by forming ionic chains of sulfur atoms, S_x^{2-}, where x varies from 2 to about 10. These polysulfide solutions react with strong acids to give corresponding sulfanes, unstable compounds with the formula, H_2S_x. The sulfanes are yellow oils that eventually decompose to hydrogen sulfide and finely divide colloidal sulfur.

Sulfur in living systems

Sulfur occurs in most proteins in two forms—as the sulfhydryl group, H—S—, and as the disulfide group, —S—S—. Plants, but not animals, can make proteins from carbon dioxide, water, and inorganic molecules or ions, including sulfur in the form of sulfate ion. Plants cannot grow in sulfur-deficient soil, but such soil can be revitalized through the application of one of several fertilizers; for example, potassium sulfate (K_2SO_4), ammonium sulfate [$(NH_4)_2SO_4$], or simply elemental sulfur. Soil bacteria can change sulfur to sulfate ions, and this change also makes the soil slightly more acidic. Therefore, sulfur is often used on sulfate-poor soils that are initially too alkaline anyway.

Thiosulfates

If sulfur is added to a hot solution of sodium sulfite, it dissolves by reacting with the sulfite ion to make the thiosulfate ion, $S_2O_3^{2-}$.

$$S(s) + \underset{\substack{\text{sodium} \\ \text{sulfite}}}{Na_2SO_3(aq)} \xrightarrow{\text{heat}} \underset{\substack{\text{sodium} \\ \text{thiosulfate}}}{Na_2S_2O_3(aq)}$$

"Thio-" in the name means that a sulfur atom has replaced an oxygen atom; SO_4^{2-} is the sulfate ion and $S_2O_3^{2-}$ is the thiosulfate ion.

Photographers use "hypo," the pentahydrate of sodium thiosulfate ($Na_2S_2O_3 \cdot 5H_2O$), as a fixing agent in developing film. Photographic film contains silver salts such as silver bromide and iodide in the form of microscopically small crystals. When a picture is taken, these microcrystals are exposed to varying intensities of light. Exposure initiates chemical changes that make the

Special Topic 12.4 Poisoning by Ions of Heavy Metals

The special catalysts in living cells are all proteins and all contain at least some sulfhydryl groups, —SH. Ions such as Hg^{2+}, Cu^{2+}, and Pb^{2+} can combine more or less permanently with —SH groups and, if that happens to an enzyme, its ability to work as a catalyst is destroyed. Therefore, soluble salts of mercury, copper, lead, and silver, all metals with high atomic weights, are invariably poisons because they deactivate enzymes. Of course, these poisons will also do the same thing to enzymes in bacteria, or in bugs, fungi, and similar pests. Therefore, several salts of the heavy metals have long been used as pesticides. The mercury salts, however, have been banned in the United States because they became too widely distributed in the human food supply. Mercury in the body causes permanent injury to the nervous system. Mercury compounds were once used to change fur into felt for hats. The fabled mad hatter in Lewis Carroll's *Alice in Wonderland* presumably was a victim of mercury poisoning. Various lead compounds have been used as paint pigments, but this has now been curtailed for interior paints. Infants living in decaying tenements too often ate chips of flaking lead-based paints and suffered permanent brain damage.

exposed sites more susceptible than the unexposed sites to the *developer,* a chemical that reduces exposed silver ions to silver atoms. The greater the exposure, the greater the quantity of silver atoms and the darker the "negative" at that site. To prevent ("fix") the negative from further acting as a photographic film, its unreduced silver salts have to be washed out. Hypo solution dissolves silver salts. For example,

$$2S_2O_3^{2-}(aq) + AgBr(s) \longrightarrow Ag(S_2O_3)_2^{3-}(aq) + Br^-(aq)$$

This reaction extracts the silver ion from the insoluble silver halide and binds it as a water-soluble ion, $Ag(S_2O_3)_2^{3-}$. Now it can be washed out of the film.

12.5 NITROGEN

Ammonia, nitric acid, and salts of these compounds are commercially important compounds of nitrogen

The members of the **nitrogen family,** Group VA in the periodic table, are listed in Table 12.9. (See also Special Topic 12.5.) We will study only nitrogen and phosphorus in any detail.

Table 12.9
The Nitrogen Family—Group VA

Name	Symbol	Mp (°C)	Bp (°C)	Important Types of Compounds
Nitrogen	N	−210	−196	Nitrates (fertilizers, explosives) Oxides (air pollutants) Ammonia (fertilizer)
Phosphorus	P	44	281	Phosphates and polyphosphates (detergents, fertilizers)
Arsenic	As	815[a]	613[b]	Arsenates (pesticides)
Antimony	Sb	631	1750	Lead-antimony mixtures (alloys) for storage batteries and type metal
Bismuth	Bi	271	1560	In mixtures with other metals: low melting alloys for automatic fire alarms and sprinkler systems

[a] Under 28 atm
[b] Sublimation temperature

Special Topic 12.5 The Nitrogen Family, the Periodic System, and Physiological Properties

The properties of compounds of the nitrogen family warn us not to be too trusting in our use of the periodic table to compare elements of the same family and their compounds. Members of the same chemical family do have *some* chemical similarities. They tend to form compounds of the same general formula. The hydrides of Group VA, for example, have the formulas NH_3, PH_3, AsH_3, SbH_3, and BiH_3. The chief warning is that *physiological* properties—the chemical properties inside living systems—usually differ very dramatically among compounds of elements in the same family. Phosphorus and arsenic, for example, both form similar acids, phosphoric acid (H_3PO_4) and arsenic acid (H_3AsO_4), which give similar oxo ions: PO_4^{3-} and AsO_4^{3-}, HPO_4^{2-} and $HAsO_4^{2-}$, and $H_2PO_4^-$ and $H_2AsO_4^-$. However, these ions have radically different properties in the body. Who hasn't heard of arsenic poisoning? It is done with salts of these ions of arsenic acid or with various oxides of arsenic. But we cannot live without the ions of phosphoric acid, and phosphates are necessary fertilizers for the growth of plants. Thus systems that seem similar on the basis of formulas or family memberships can be dangerously different in the body.

Nitrogen

Nitrogen, a colorless, odorless, tasteless gas, makes up 78% of dry air. Virtually all of the 25 to 30 billion lb (400 to 490 billion mol) of nitrogen produced each year comes from liquified air. Nitrogen boils at −195.8 °C and freezes at −209.8 °C.

A rubber ball chilled to the temperature of liquid nitrogen will shatter when dropped.

A small amount of liquid nitrogen is commonly used in research laboratories as a low temperature coolant. Commercial uses of liquid nitrogen are increasing. For example, sperm from high-quality cattle, hogs, and sheep is kept indefinitely at the temperature of liquid nitrogen in commercial sperm banks. Farmers interested in breeding higher-quality herds can purchase the sperm to impregnate their best female animals. There are even firms that will quick-freeze your body in liquid nitrogen and store it, frozen, until someone finds the cure for what killed you (although no one has successfully completed the procedure, yet).

Most nitrogen is used in the Haber process (Section 12.2) to make ammonia, from which most other nitrogen compounds, including nitric acid, are made. Modern agriculture relies heavily on the application of nitrogen-containing fertilizers such as ammonia (NH_3), urea ($NH_2{-}\overset{\overset{\displaystyle O}{\|}}{C}{-}NH_2$) ammonium nitrate ($NH_4NO_3$) and ammonium sulfate [$(NH_4)_2SO_4$] (see Figure 12.7).

Nitrogen is an unreactive element under ordinary conditions. It takes 225 kcal/mol (941 kJ/mol) to break the triple bond in nitrogen. This is one of the highest of all known bond energies and is one reason why nitrogen enters into almost no chemical reactions under ordinary temperatures and pressures. However, several microorganisms in the soil have enzymes that catalyze **nitrogen fixation,** the conversion of molecular nitrogen from the air into ammonia that then is used by plants to make proteins.

Nitrogen isotopes:
N-14, 99.63%
N-15; 0.37%

Nitrogen was discovered in 1772 by British scientist Daniel Rutherford (1749–1819). Not long after, nitrogen was found in combined forms, particularly proteins, in all living things. Nitrogen occurs almost entirely as the N-14 isotope.

HNO_3

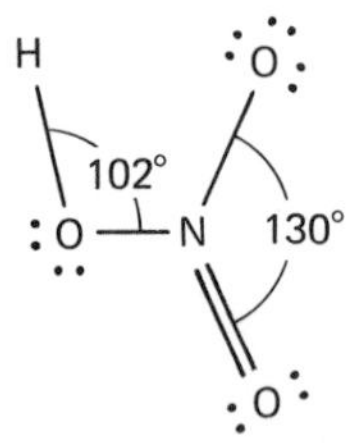

Nitric acid

Commercial "concentrated HNO_3" has a concentration of 16*M*, which means that it is 70% (wt/wt) HNO_3. The pure acid is unstable; when it is heated or exposed to sunlight, it decomposes to a mixture of water, oxygen, and oxides of nitrogen, chiefly nitrogen dioxide.

$$\underset{\text{nitric acid}}{4HNO_3(\ell)} \xrightarrow[\text{heat}]{} \underset{\text{nitrogen dioxide}}{4NO_2(g)} + O_2(g) + 2H_2O(\ell)$$

Figure 12.7
Liquid ammonia is injected into the soil before planting to
provide a nitrogen fertilizer.

Nitrogen dioxide is a red-brown gas that gives older solutions of nitric acid a red-dish color. Most commercial nitric acid is made from ammonia by the **Ostwald process,** developed in 1902 by Wilhelm Ostwald (Germany; 1853–1932; Nobel Prize, 1909). The first step in the Ostwald process is the heating of ammonia to 600 to 700 °C over a catalyst (a mixed rhodium-platinum gauze) to make nitric oxide, NO.

$$4NH_3(g) + 5O_2(g) \longrightarrow 4NO(g) + 6H_2O(g)$$
ammonia   nitric
oxide

This reaction generates much more heat than is needed to initiate it, and the temperature climbs to 1000 °C. The mixture is then cooled somewhat and air is let in to give it more oxygen. This oxidizes nitric oxide to nitrogen dioxide.

$$2NO(g) + O_2(g) \longrightarrow 2NO_2(g)$$
nitric   nitrogen
oxide   dioxide

The mixture of gases is sprayed with water, and nitrogen dioxide reacts with the water to give aqueous nitric acid and nitric oxide.

$$H_2O(\ell) + 3NO_2(g) \longrightarrow 2HNO_3(aq) + NO(g)$$
nitrogen   nitric   nitric
dioxide   acid   oxide

The nitric oxide is then recycled.

A *nitrating agent* introduces nitro groups, —NO_2, into molecules.

Nitric acid is a strong monoprotic acid, a strong oxidizing agent, and a nitrating agent. It neutralizes bases to form nitrate salts and water. For example,

$$HNO_3(aq) + KOH(aq) \longrightarrow KNO_3(aq) + H_2O(\ell)$$

nitric   potassium   potassium
acid   hydroxide   nitrate

A typical oxidation by nitric acid is its reaction with metallic copper. When concentrated nitric acid is used, the gaseous product is nitrogen dioxide.

$$Cu(s) + 4HNO_3(aq) \longrightarrow Cu(NO_3)_2(aq) + 2NO_2(g) + 2H_2O(\ell)$$

nitric acid   copper(II)   nitrogen
(concentrated)   nitrate   dioxide

When dilute nitric acid is used, the gaseous product is nitric oxide, NO.

$$3Cu(s) + 8HNO_3(aq) \longrightarrow 3Cu(NO_3)_2(aq) + 2NO(g) + 4H_2O(\ell)$$

nitric acid   nitric
(dilute)   oxide

The use of nitric acid as a nitrating agent may be illustrated by the synthesis of the explosive trinitrotoluene, or TNT.

H_2SO_4 (shown beneath the arrow) is the catalyst for this nitration.

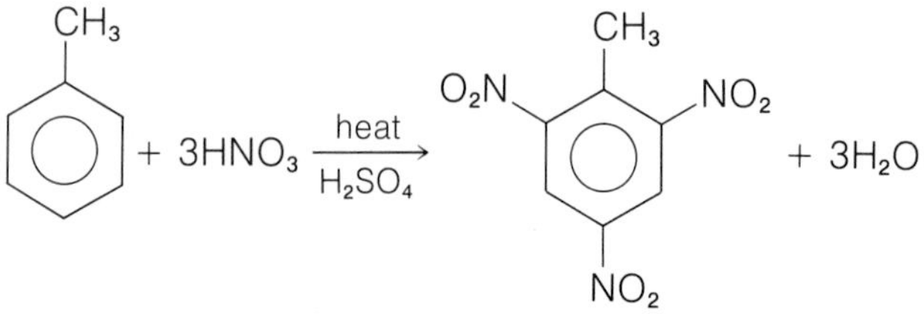

toluene   trinitrotoluene
TNT

Nitroglycerin, another important explosive, is made by nitrating glycerol.

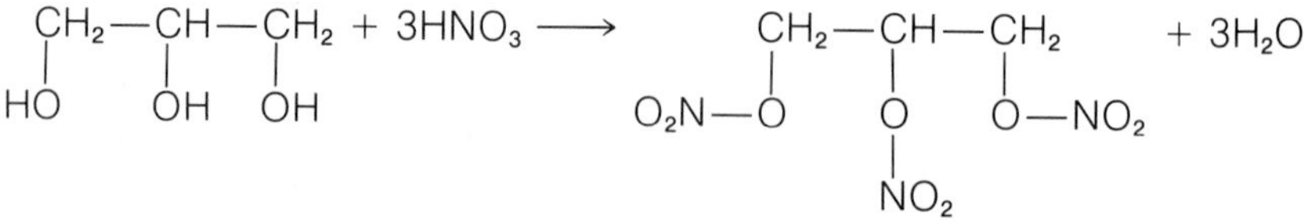

glycerol   nitroglycerin

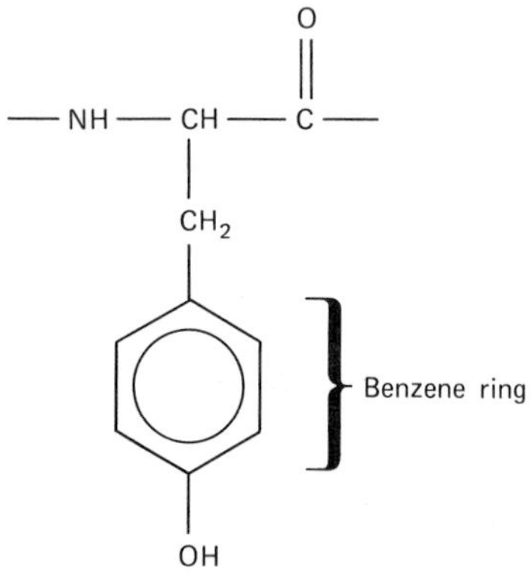

Section of a protein molecule

Benzene rings—which occur in the molecules of most proteins, including those of skin—are easily nitrated to give yellow pigments. If you splash concentrated nitric acid on your skin, flush the area at once with water, but still expect a bright yellow spot. The color can't be washed off, but it will wear off in about a week.

Oxides of nitrogen

In the five known oxides of nitrogen, listed in Table 12.10, the oxidation number of nitrogen varies from +1 to +5.

Dinitrogen monoxide (nitrous oxide), N_2O, may be made by carefully heating ammonium nitrate.

$$NH_4NO_3(s) \xrightarrow{\text{heat}} N_2O(g) + 2H_2O(\ell)$$

ammonium   dinitrogen
nitrate   monoxide

Table 12.10
Oxides of Nitrogen

Oxidation State of Nitrogen	Formula	Name	Color	Bp	Mp
+1	N_2O	Dinitrogen monoxide (nitrous oxide)	None	−89 °C	−91 °C
+2	NO	Nitrogen monoxide (nitric oxide)	None	−152	−164
+3	N_2O_3	Dinitrogen trioxide (nitrous anhydride)	Red-brown	decomposes	−102
+4	$2NO_2$ ⇅	Nitrogen dioxide[a]	Brown	−11[b]	
+4	N_2O_4	Dinitrogen tetroxide	None (as a solid)		−21
+5	N_2O_5	Dinitrogen pentoxide (nitric anhydride)	None	decomposes	32

[a] NO_2 and N_2O_4 exist in the presence of each other in both the liquid and the gaseous states. At 25 °C and 760 torr, N_2O_4 is favored in a mixture of these gases; its partial pressure is about 540 torr and that of NO_2 is about 220 torr. In the solid state only N_2O_4 is present.

[b] The temperature at which the *mixture's* vapor pressure equals the atmospheric pressure is −11 °C.

$$:N \equiv N - \ddot{\underset{\cdot\cdot}{O}}:$$
$$N_2O$$

If heated too strongly, dinitrogen monoxide breaks up into its elements.

$$2N_2O(g) \xrightarrow{heat} 2N_2(g) + O_2(g)$$

This oxide is an anesthetic used in dentistry. Hot rod and race car drivers sometimes use dinitrogen monoxide to get more power out of an engine.

$$Fuel + N_2O \longrightarrow CO_2 + H_2O + N_2$$

$$N_2(g) + \tfrac{1}{2}O_2(g) \rightarrow N_2O(g)$$
$$\Delta H_f^\circ = +18.65 \text{ kcal } (+78.03 \text{ kJ})$$

$$\cdot\ddot{N} = \ddot{\underset{\cdot\cdot}{O}}:$$
$$NO$$

Since the value of ΔH_f° for N_2O is positive, its decomposition in the engine gives extra energy and more zip to the car.

Nitrogen monoxide (nitric oxide), NO, forms when nitrogen and oxygen are heated to a high temperature. This condition exists where air is used to burn fuel in cylinders of engines and in the furnaces of industrial power plants. Therefore, nitrogen monoxide is a by-product of the combustion of fuel, and it is an important air pollutant. It also forms by the reaction of dilute nitric acid with relatively unreactive metals such as copper, which we studied earlier. Nitrogen monoxide is not stable in air or in pure oxygen. It changes to the dioxide.

Dinitrogen trioxide, N_2O_3, is a very unstable gas that readily splits apart.

$$N_2O_3(g) \longrightarrow NO(g) + NO_2(g)$$

In fact, for chemical studies of N_2O_3, a mixture of NO and NO_2 usually serves the purpose. When dinitrogen trioxide is bubbled into aqueous sodium hydroxide, sodium nitrite forms.

$$2NaOH(aq) + \underset{\substack{\text{dinitrogen} \\ \text{trioxide}}}{N_2O_3(g)} \longrightarrow \underset{\substack{\text{sodium} \\ \text{nitrite}}}{2NaNO_2(aq)} + H_2O(\ell)$$

Nitrogen dioxide, NO_2, and dinitrogen tetroxide, N_2O_4, occur together in an equilibrium.

$$\underset{\substack{\text{nitrogen} \\ \text{dioxide} \\ (reddish\text{-}brown)}}{2NO_2} \underset{heat}{\overset{cool}{\rightleftharpoons}} \underset{\substack{\text{dinitrogen} \\ \text{tetroxide} \\ (colorless)}}{N_2O_4}$$

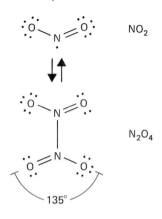

NO_2

N_2O_4

$135°$

Nitrogen dioxide is a poisonous gas. It is paramagnetic because each molecule has an unpaired electron. Dinitrogen tetroxide is diamagnetic. (Anything not paramagnetic is called diamagnetic, and all its electrons are paired.) Nitrogen dioxide, as we have already noted, forms when nitrogen monoxide in the exhaust gases of engines and furnaces mixes with the oxygen in air.

$$2NO(g) + O_2(g) \longrightarrow 2NO_2(g)$$

Thus one pollutant, NO, leads to another, NO_2, and the mixture of the two is symbolized by NO_x in references on air pollution. Nitrogen dioxide is largely responsible for the noticeable reddish-brown color of smog. (See Color Plate 4C.) Acids form when nitrogen dioxide reacts with water

$$2NO_2(g) + H_2O(\ell) \xrightarrow{\text{cold water}} \underset{\substack{\text{nitric} \\ \text{acid}}}{HNO_3(aq)} + \underset{\substack{\text{nitrous} \\ \text{acid}}}{HNO_2(aq)}$$

If the water is warm, the HNO_2 decomposes and the net reaction is

$$3NO_2(g) + H_2O(\ell) \xrightarrow{\text{warm water}} 2HNO_3(aq) + NO(g)$$

These acids also contribute to acid rain. When bubbled into a base, nitrogen dioxide changes to a mixture of nitrate and nitrite ions. For example,

$$2NO_2(g) + \underset{\substack{\text{sodium} \\ \text{hydroxide}}}{2NaOH(aq)} \longrightarrow \underset{\substack{\text{sodium} \\ \text{nitrate}}}{NaNO_3(aq)} + \underset{\substack{\text{sodium} \\ \text{nitrite}}}{NaNO_2(aq)} + H_2O(\ell)$$

Dinitrogen pentoxide (nitric anhydride), N_2O_5, is a white solid that easily sublimes at 32 °C. It can be made by the action of a powerful dehydrating agent such as tetraphosphorus decaoxide on nitric acid. The overall effect is the removal of one water molecule from every two molecules of nitric acid.

$$\underset{\substack{\text{nitric} \\ \text{acid}}}{4HNO_3(\ell)} + \underset{\substack{\text{tetraphosphorus} \\ \text{decaoxide}}}{P_4O_{10}(s)} \longrightarrow \underset{\substack{\text{metaphosphoric} \\ \text{acid}}}{4HPO_3(\ell)} + \underset{\substack{\text{dinitrogen} \\ \text{pentoxide}}}{2N_2O_5(s)}$$

Dinitrogen pentoxide reacts vigorously with water to regenerate nitric acid.

$$N_2O_5(s) + H_2O(\ell) \longrightarrow 2HNO_3(aq)$$

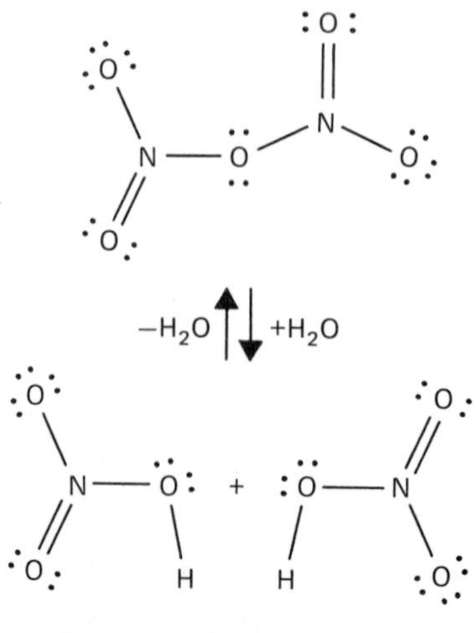

N_2O_5

$-H_2O$ $+H_2O$

$2HNO_3$

HNO₂

Nitrous acid and nitrites

Nitrous acid, HNO_2, is a weak, unstable acid, but it has stable salts. When nitrous acid is required for a reaction, a fresh solution is made just before it is needed by mixing cold, aqueous sodium nitrite with chilled hydrochloric acid or sulfuric acid. For example,

$$NaNO_2(aq) + HCl(aq) \xrightarrow{cool} HNO_2(aq) + NaCl(aq)$$

sodium nitrite   hydrochloric acid   nitrous acid

Nitrous acid has some very important reactions in organic chemistry that lead to dyes, pharmaceuticals, and chemical intermediates. In these applications, its solutions are kept at ice slush temperature because, at higher temperature, nitrous acid breaks down as follows:

$$3HNO_2(aq) \longrightarrow HNO_3(aq) + 2NO(g) + H_2O(\ell)$$

Sodium nitrite is a food additive. (See Special Topic 12.6.)

Reduced forms of nitrogen

In *reduced* forms of nitrogen, N has a negative oxidation number.

Ammonia, a weak base, is the most common reduced form of nitrogen. All the common acids are neutralized by ammonia. For example,

$$NH_3(aq) + HCl(aq) \longrightarrow NH_4Cl(aq)$$

hydrochloric acid   ammonium chloride

$$2NH_3(aq) + H_2SO_4(aq) \longrightarrow (NH_4)_2SO_4$$

sulfuric acid   ammonium sulfate

NH₃

Ammonia boils at $-33.4\ °C$ and freezes at $-77.7\ °C$. It is easily liquified, and liquid ammonia is used as a solvent for both inorganic and organic reactions. These reactions are generally carried out using widemouthed vacuum bottles called Dewar flasks to hold the solution. Liquid ammonia is also shipped in tanks for use as a fertilizer.

Liquid ammonia dissolves the Group IA and Group IIA metals, giving solutions with a beautiful blue color. Just as water can hydrate dissolved ions, molecules of liquid ammonia can ammoniate ions. In very dilute solutions of Group IA and IIA metals in liquid ammonia, both the metal ions *and the electrons* are ammoniated. Species, such as M_{am}^+ and e_{am}^- exist (where *am* means ammoniated—surrounded by molecules of the solvent, ammonia). Eventually, the ammoniated species interact permanently with ammonia to make solutions of metal amides. For example, a solution of sodium in liquid ammonia changes as follows.

$$2Na_{am}^+ + 2e_{am}^- + 2NH_3 \longrightarrow H_2(g) + 2NaNH_2$$

sodium amide

The reaction can be speeded up by adding traces of transition metal ions.

In liquid ammonia the amide ion, NH_2^-, is analogous to the hydroxide ion, OH^-, in liquid water. Each ion is a powerful base in its own solvent. Similarly, in liquid ammonia, the ammonium ion (NH_4^+) is analogous to the hydronium ion (H_3O^+) in liquid water. Each ion is an acidic species in its own solvent. Thus a whole range of acid-base chemistry exists for liquid ammonia, involving NH_4^+ and NH_2^- as the key acidic and basic species. Just as H_3O^+ will neutralize OH^-:

$$H_3O^+(aq) + OH^-(aq) \xrightarrow{H_2O(\ell)} 2H_2O(\ell)$$

so NH_4^+ will neutralize NH_2^-

$$NH_4^+(am) + NH_2^-(am) \xrightarrow{NH_3(\ell)} 2NH_3(\ell)$$

Special Topic 12.6 Nitrites and Nitrates in Food and Water

Nitrite and nitrate ions can be in the diet for a number of reasons. Drinking water taken from wells in farming areas where nitrates are heavily used as fertilizers often contains traces of both ions. (Some soil bacteria will also convert nitrate ions to nitrite ions.) Sodium nitrite is used as a preservative in several processed meats—bacon, ham, sausage, corned beef, beef jerky, frankfurters, and bologna—and some fish products. The nitrite ion in these meats inhibits the growth of a bacterium (*Clostridium botulinum*). This organism produces the botulinus toxin, which causes fatal food posioning. (Nitrite ion also fixes the reddish color of meat products, which otherwise would appear brownish.)

The nitrite ion is particularly dangerous to infants. It changes hemoglobin, the oxygen-carrying compound in blood, into forms that cannot hold oxygen, and infants manage that problem poorly. Nursing mothers whose water supply contains nitrite ion pass it to their infants.

Another danger associated with the nitrite ion is that under acidic conditions or at a high enough temperature it reacts with organic amines (compounds with NH_2 groups) to make nitrosamines having —NH—NO groups. Nitrosamines are among the most powerful cancer-causing agents known. They can form when nitrite-cured meats are fried or boiled, and many scientists have urged the U.S. Food and Drug Administration to ban the use of nitrites in meat. Others argue that food poisoning from the botulinus toxin would become a serious problem if nitrites in foods were banned. The result of the controversy has been a reduction in the allowed concentration of nitrites in meat products. Despite all of the possibilities of dangers from nitrites in food, the nitrite ion has yet to be proven to cause cancer in animals.

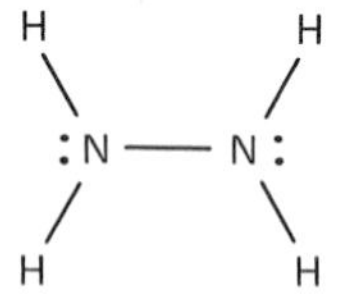

Hydrazine

Hydrazine, N_2H_4, is another reduced form of nitrogen. This colorless, toxic liquid, which has a boiling point of 113.5 °C and a melting point of 2 °C, is a weaker base than ammonia. It can be made by the reaction of ammonia with sodium hypochlorite.

$$2NH_3(aq) + NaOCl(aq) \longrightarrow N_2H_4(aq) + NaCl(aq) + H_2O(\ell)$$

The resulting solution may be distilled to give a 58.5% solution of hydrazine in water, which can be used to make pure hydrazine.

Because "chlorine" bleaches used in home laundries contain hypochlorite ion, they must never be mixed with household ammonia. The two react, as we just studied, to give hydrazine, which is toxic.

Hydrazine is a strong reducing agent. Liquid hydrazine and some of its organic derivatives are used as rocket fuels. When mixed with oxygen from liquid oxygen tanks, hydrazine reacts violently to produce rapidly expanding gases that provide the thrust to the rocket.

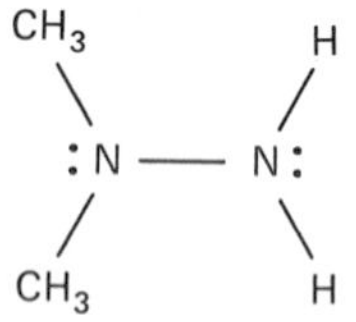

Dimethylhydrazine

$$N_2H_4(\ell) + O_2(g) \longrightarrow N_2(g) + 2H_2O(g)$$

The reaction between oxygen and dimethylhydrazine, an organic derivative of hydrazine, provided the thrust that carried U.S. astronauts successfully to the moon.

12.6 PHOSPHORUS

Phosphoric acid, related acids, their salts, and their organic derivatives are important compounds of phosphorus

Phosphorus has never been found free in nature, but several important minerals contain the phosphate ion. For example, phosphate rock contains calcium phosphate together with calcium hydroxide and, in some localities, calcium fluoride.

$$[Ca_3(PO_4)_2]\cdot Ca(OH)_2 \qquad [Ca_3(PO_4)_2]\cdot CaF_2$$

two minerals in phosphate rock

Phosphorus was one of the first elements not found free in nature to be discovered. In 1669, Hennig Brand, an alchemist in Hamburg, stumbled onto a way to obtain the white allotrope of phosphorus from the residue of evaporated urine. This allotrope glows in the dark and easily bursts into flame in air. At that time (even as now), finding something that would do that was a sensation. But it was not gold—the everlasting quest of the prechemists called alchemists. (Had Brand made gold from urine it certainly would have taken the romance out of gold rushes!)

Phosphorus, just below nitrogen in the periodic table, exists entirely as the P-31 isotope. In sharp contrast to nitrogen, phosphorus is a solid at room temperature. It has several allotropic forms. In the vapor state, phosphorus exists as molecules of P_4. These form when phosphorus is made by the reduction of phosphates by carbon in the presence of silica at a very high temperature.

$$2Ca_3(PO_4)_2(s) + 10C(s) + 6SiO_2(s) \xrightarrow{\text{heat}} P_4(g) + 10CO(g) + 6CaSiO_3(s)$$

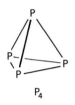

P_4

One of the solid allotropes, the very poisonous white phosphorus, forms when phosphorus vapor is condensed under water. The phosphorus atoms in the P_4 molecules of white phosphorus are located at the corners of a regular tetrahedron, and the bond angles are only 60°. If a normal P—P bond were made by the overlap of third-shell p-orbitals, the bond angle between one phosphorus atom and its two neighbors would be 90°. The system in P_4, then, is obviously strained—that is, there is poor overlap of orbitals—and white phosphorus is therefore very reactive. At 30 °C, just a little above room temperature, it bursts into flame in moist air. Therefore, it has to be stored under water where it is out of contact with air.

White phosphorus changes spontaneously to the most common allotrope, red phosphorus, when it is very carefully heated or exposed to light in the absence of air. The P_4 tetrahedra now open up, and red phosphorus is a mixture of branched chains of phosphorus atoms having various lengths. It is stable in air at room temperature. Many other allotropes are known, but they can be prepared only under very unusual conditions. Both red and white phosphorus are involved in the chemistry of igniting a match. (See Special Topic 12.7.)

Phosphorus will burn extremely brightly, if ignited in sufficient quantities of air, to give a cloud of white, powdery tetraphosphorus decaoxide, P_4O_{10} (see Figure 12.8).

$$4P(s) + 5O_2(g) \longrightarrow P_4O_{10}(s)$$

Red phosphorus is therefore used in smoke bombs, fireworks, and matches. If insufficient oxygen is available, tetraphosphorus hexaoxide forms.

$$4P(s) + 3O_2(g) \longrightarrow \underset{\substack{\text{tetraphosphorus} \\ \text{hexaoxide}}}{P_4O_6(s)}$$

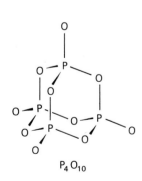

P_4O_{10}

P_4O_{10} is commonly called phosphorus pentoxide after its empirical formula, P_2O_5.

Phosphorus reacts with chlorine to give either phosphorus trichloride, a colorless, fuming liquid (bp 76 °C), or phosphorus pentachloride, a yellow, fuming, pungent solid (that sublimes at about 100 °C).

$$2P(s) + 3Cl_2(g) \longrightarrow \underset{\substack{\text{phosphorus} \\ \text{trichloride}}}{2PCl_3(\ell)}$$

$$2P(s) + 5Cl_2(g) \longrightarrow \underset{\substack{\text{phosphorus} \\ \text{pentachloride}}}{2PCl_5(s)}$$

These halides are used in organic chemistry as chlorinating agents. The two cor-

We take matches so much for granted that it is hard to imagine what life was like before safe and inexpensive matches were developed. The safety match was invented in Sweden in 1855 by J.E. Lundstrom. The head of a safety match is a mixture of an oxidizing agent, such as potassium chlorate ($KClO_3$), and antimony trisulfide (Sb_2S_3), bound together by glue. The striking surface on the matchbook cover or box is a mixture of red phosphorus and powdered glass held by glue. During the strike, some heat is produced that converts a tiny trace of red phosphorus to white phosphorus that instantly ignites. The heat generated by ignition activates the chemicals in the match head, and they react. Then the heat of that reaction ignites the paper or wood used to support the match head.

In the "strike-anywhere" kind of match, the match head end of the wooden stick is first dipped in hot wax. Then the head is formed from a mixture, held together by glue, of powdered abrasive (glass) and an oxidizing agent (e.g., $KClO_3$). At the very tip is a small amount of phosphorus sesquisulfide, P_4S_3, which is very easily ignited. Friction during the strike creates enough heat to cause the oxidizing agent to ignite the sulfide, and the brief flare of that reaction ignites the wax in the stick.

PCl₃

PCl₅

responding phosphorus bromides plus phosphorus triiodide are also known. They all react with water.

$$2PX_3(\ell) + 6H_2O(\ell) \longrightarrow 2H_3PO_3(aq) + 6HX(aq)$$

(X = Cl, Br, or I)   phosphorus acid

$$PX_5(s) + 4H_2O(\ell) \longrightarrow H_3PO_4(aq) + 5HX(aq)$$

(X = Cl or Br)   orthophosphoric acid

The hydride of phosphorus, phosphine (PH_3), is made by the reaction of calcium phosphide with water.

$$Ca_3P_2(s) + 6H_2O(\ell) \longrightarrow 2PH_3(g) + 3Ca(OH)_2(s)$$

calcium phosphide   phosphine

Phosphine is a poisonous gas with a terrible odor, and in air it burns spontaneously to give one of the forms of phosphoric acid.

$$PH_3(g) + 2O_2(g) \longrightarrow H_3PO_4(\ell)$$

phosphine   orthophosphoric acid

The phosphoric acids and their salts

When P_4O_{10} powder is dusted onto water, so much heat is generated that the system hisses and spits. Tetraphosphorus decaoxide combines so avidly with moisture that it is one of the best desiccating agents (water-removing agents) we have. We can't write a simple equation for this reaction, because it produces a family of three phosphoric acids that differ in the degree to which they have taken up water.

H_3PO_4

HPO_3	$H_4P_2O_7$	H_3PO_4
metaphosphoric acid	pyrophosphoric acid	orthophosphoric acid

The structures of the various forms of phosphoric acid are complicated by the presence of extended chains or rings, and we will return to this topic in Chapter Thirteen. Each form of phosphoric acid has a family of salts.

Orthophosphoric acid is the common type of phosphoric acid in commerce. Given a long enough exposure to water, the other forms of phosphoric acid change to it. When the name "phosphoric acid" is used, orthophosphoric acid is

Figure 12.8
The brilliant illumination from fireworks displays is provided in part by the combustion of phosphorus.

generally what is meant. The commercial "concentrated phosphoric acid" is a colorless, thick, syrupy liquid containing 85% H_3PO_4. Our study will continue with a survey of the common properties of orthophosphoric acid, which we will now call simply phosphoric acid as most chemists do.

Phosphoric acid is a moderately strong acid that can furnish three hydrogen ions. It is sometimes used to give a tart taste to manufactured soft drinks, but its largest use is in making phosphate fertilizers. (Problem 4.68 described this use.) Three kinds of salts of phosphoric acid are possible—dihydrogen phosphate salts having $H_2PO_4^-$ ions, monohydrogen phosphate salts having HPO_4^{2-} ions, and phosphate salts having PO_4^{3-} ions. Dihydrogen phosphate salts are *acid salts* because they have fairly readily available hydrogen ions. Thus calcium dihydrogen phosphate, $Ca(H_2PO_4)_2$, is used in baking powders as an *acidulant,* a substance that will begin to act as an acid when water is added. Sodium bicarbonate, $NaHCO_3$, is another component of baking powder. In water, the acidulant acts on the bicarbonate ion to release carbon dioxide. The generation of this gas creates the desired frothy texture in the baked item.

$$Ca(H_2PO_4)_2(aq) + 2NaHCO_3(aq) \longrightarrow 2CO_2(g) + 2H_2O(\ell) + CaHPO_4(aq) \\ + Na_2HPO_4(aq)$$

The net ionic equation is:

$$H_2PO_4^-(aq) + HCO_3^-(aq) \longrightarrow CO_2(aq) + H_2O(\ell) + HPO_4^{2-}(aq)$$

12.7 HALOGENS

Fluorine, chlorine, bromine, iodine, and astatine are the members of the halogen family

Some of the properties of the **halogen family** are given in Table 12.11. Except for the similarities in the formulas of some halogen salts, fluorine is not a very typical member of the family. Like the other elements that stand at the tops of their families in the periodic table, fluorine has unusual properties that are not seen farther down. At the bottom of the halogen family is astatine, which is intensely radioactive and is found only rarely in some uranium deposits. Our survey of the properties of the halogens will not include it.

Table 12.11
The Halogens—Group VIIA

Name	Symbol	Bp (°C)	Mp (°C)	Density (20 °C)
Fluorine	F	−188	−220	1.7 g/liter
Chlorine	Cl	−35	−101	3.2 g/liter
Bromine	Br	59	−7	3.1 g/ml
Iodine	I	184	114	4.9 g/ml
Astatine	At	Does not occur naturally; no stable isotope.		

Fluorine
Fluorine, a greenish gas, does not occur in the free state. Fluorite (CaF_2), cryolite (Na_3AlF_6), and apatite [$Ca_3(PO_4)_2 \cdot CaF_2$] are its chief ores, and only the F-19 isotope is present. Henri Moisson (France; 1852–1907) won the 1906 Nobel Prize in chemistry for discovering fluorine.

Fluorine reacts violently at room temperature with almost all other elements. Mixed with hydrocarbons, fluorine cracks carbon-carbon bonds and strips the hydrogens, leaving hydrogen fluoride, HF, and carbon tetrafluoride, CF_4. Water *burns* in fluorine to give HF and oxygen.

$$2F_2(g) + 2H_2O(\ell) \longrightarrow 4HF(g) + O_2(g)$$

Fluorine has to be stored in cylinders made of a special metal alloy.

Hydrogen fluoride
Hydrogen fluoride forms explosively when hydrogen and fluorine are mixed. To reduce this hazard, it is normally made by heating calcium fluoride with concentrated sulfuric acid.

$$CaF_2(s) + H_2SO_4(\ell) \longrightarrow 2HF(g) + CaSO_4(s)$$

calcium fluoride	sulfuric acid	hydrogen fluoride	calcium sulfate

Lead-lined containers have to be used for the reaction, because hydrogen fluoride destroys glass. This reaction can be controlled so as to etch glass objects. If we represent glass by the empirical formula of its chief constituent, silica (SiO_2), the reaction is

$$4HF(g) + SiO_2(s) \longrightarrow SiF_4(g) + 2H_2O(\ell)$$

Hydrogen fluoride boils at 19.9 °C and is therefore easily liquified, stored, and shipped as a liquid in sealed containers made of special steel. Its solution in water, hydrofluoric acid, HF(aq), is stored in bottles lined with wax or polyethylene or in cylinders made of a special steel.

Since fluorine is the most electronegative of all elements, the covalent bond in the H-F molecule is extremely polar, and HF molecules are strongly hydrogen bonded to each other.

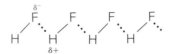

hydrogen bonding in HF

Hydrofluoric acid is a weak acid. Its solution contains the bifluoride ion, HF_2^-, which has no counterpart among the other halogens.

$$F^- + \overset{\delta+}{H} - \overset{\delta-}{F} \longrightarrow (F \cdot\cdot H \cdot\cdot F)^-$$

bifluoride
ion, HF_2^-

Hydrofluoric acid causes intensely painful tissue burns, and it and all its salts, the fluorides, are poisonous. They should be handled only by people well trained in working with them. Yet a trace of fluoride ion in drinking water during the years when the teeth are developing contributes to strong, acid-resistant teeth. Tooth enamel is mostly a mineral called hydroxyapatite. When fluoride ion is available, some of this mineral changes to fluoroapatite, which is harder and better able to

$$Ca_5(OH)(PO_4)_3 \qquad Ca_5F(PO_4)_3$$
hydroxyapatite   fluoroapatite

withstand tooth-decaying acids. Fluoride ion for strengthening tooth enamel can be supplied in drinking water but, if too much fluoride ion is in the water, the teeth will develop dark spots. Fluoride can also be provided in toothpaste or in special solutions that can be applied by a dentist. Tin(II) fluoride (Fluorostan®) is one widely used source of fluoride in toothpastes.

Organo fluorine compounds are an important part of the industrial organic chemicals industry. The most famous example is Teflon, the well-known nonstick coating used on frying pans (Section 13.3).

Chlorine

Chlorine isotopes:
Cl-35, 75.53%
Cl-37, 24.47%

This poisonous, yellowish-green gas was first isolated in 1774 by Carl Wilhelm Scheele (Sweden; 1742–1786) but it was not recognized as an element until 1810 by Humphrey Davy (England; 1778–1829). It consists of a mixture of two isotopes, chlorine-35 and chlorine-37. Chlorine occurs mostly as the chloride ion, Cl^-, in seawater and in deposits of sodium chloride (halite, NaCl) and potassium chloride (sylvite, KCl). Chlorine boils at -10.1 °C and therefore is easily liquified and shipped as a liquid.

Chlorine ranks in the top ten of all chemicals manufactured in the United States. In the late 1970s its annual production was about 22 billion lb (about 140 billion mol), most of it made by passing an electric current through aqueous sodium chloride.

$$2NaCl(aq) + 2H_2O(\ell) \xrightarrow[\text{direct electric current}]{\text{unstirred;}} Cl_2(g) + H_2(g) + 2NaOH(aq)$$

All these products are commercially valuable. The principle uses of chlorine are the chlorination of drinking water, the manufacture of many industrial chemicals, including pesticides, and the production of chlorine bleaches. If the solution of sodium chloride in water is stirred as electricity is passed through it, the Cl_2 and NaOH react to give hypochlorite, a bleach. In the *stirred* mixture, the net reaction is

$$NaCl(aq) + H_2O(\ell) \xrightarrow[\text{direct electric current}]{\text{stirred;}} \underset{\substack{\text{sodium}\\\text{hypochlorite}}}{NaOCl(aq)} + H_2(g)$$

How an electric current can cause chemical changes is discussed in Chapter Eighteen.

Hydrogen chloride and hydrochloric acid
Hydrogen chloride forms explosively when a mixture of hydrogen and chlorine is exposed to ultraviolet light or heat.

$$H_2(g) + Cl_2(g) \longrightarrow 2HCl(g)$$

Hydrogen chloride is manufactured by heating sodium chloride with sulfuric acid.

$$2NaCl(s) + H_2SO_4(\ell) \longrightarrow 2HCl(g) + Na_2SO_4(s)$$

Hydrogen chloride comes off as a gas (bp $-85\,°C$), and it is shipped under pressure in steel cylinders. Considerable hydrogen chloride is directly dissolved in water to make the aqueous solution called hydrochloric acid, $HCl(aq)$. The "concentrated hydrochloric acid" of commerce has a concentration of $12M$, the equivalent of 37% HCl. Hydrochloric acid is a strong acid; in dilute solutions it is completely ionized.

Hydrochloric acid is available in hardware stores as muriatic acid.

$$HCl(g) + H_2O(\ell) \longrightarrow H_3O^+(aq) + Cl^-(aq)$$

Its reactions are those of either the hydronium ion or the chloride ion and are studied elsewhere. The chloride ion is present in all the fluids of the body. Gastric juice in the stomach contains dilute hydrochloric acid.

Oxides and oxoacids of chlorine
Three oxides of chlorine are listed in Table 12.12. Two oxoacids of chlorine—hypochlorous acid and perchloric acid—are important.

Table 12.12

Oxides of Chlorine

Oxidation State of Chlorine	Formula	Name	Color	Bp (°C)	Mp (°C)
+1	Cl_2O	Dichlorine monoxide	Yellow-red	2.2	−116
+4	ClO_2	Chlorine dioxide	Yellow (gas) Red-brown (liquid)	11	−60
+7	Cl_2O_7	Dichlorine heptoxide (perchloric anhydride)	None	80[a]	−96

[a] A very explosive liquid.

When chlorine is bubbled into water, it reacts to give a mixture of hydrochloric acid and hypochlorous acid.

$$Cl_2(g) + H_2O(\ell) \longrightarrow \underset{\substack{\text{hydrochloric}\\\text{acid}}}{HCl(aq)} + \underset{\substack{\text{hypochlorous}\\\text{acid}}}{HOCl(aq)}$$

Mixed salts are those containing two or more different cations or anions; for example, Cl^- and OCl^- in $Ca(OCl)Cl$, bleaching powder.

If chlorine is bubbled into an aqueous slurry of calcium hydroxide, $Ca(OH)_2$, the product is a mixed salt—the chloride-hypochlorite of calcium called bleaching powder.

$$\underset{\substack{\text{calcium}\\\text{hydroxide}\\(slurry)}}{Ca(OH)_2(s)} + Cl_2(g) \longrightarrow \underset{\substack{\text{"bleaching}\\\text{powder"}}}{Ca(OCl)Cl(s)} + H_2O(\ell)$$

In the hypochlorite ion, ClO^-, chlorine is in its 1+ oxidation state. Hydrochloric acid reacts with hypochlorite salts to generate chlorine, and this reaction is a convenient source of small quantities of chlorine whenever this gas is to be used in an aqueous medium.

$$NaOCl(aq) + 2HCl(aq) \longrightarrow NaCl(aq) + Cl_2(g) + H_2O(\ell)$$

Perchloric acid, $HClO_4$, is the only reasonably stable oxoacid of chlorine, provided it is not made anhydrous. The normal form is a 72% solution. This acid is a strong acid and, when concentrated, it is also a *very* strong oxidizing agent. If made anhydrous, it reacts explosively on contact with cork, rubber, filter paper—anything organic. Interestingly, the perchlorate ion, ClO_4^-, in dilute solutions is one of the most inert ions, not even serving as an oxidizing agent.

Bromine and its major compounds

When Antone Balard (France; 1802–1876) first discovered the element bromine, he rummaged in the classical languages for a suitable name for this most vile-smelling element and found *bromos,* which is Greek for "stench." He had bubbled chlorine into waters from a marsh that happened to contain dissolved bromide ion. Chlorine can oxidize bromide ion to bromine.

$$Cl_2(g) + 2Br^-(aq) \longrightarrow 2Cl^-(aq) + Br_2(\ell)$$

Notice that this reaction means that chlorine is a stronger oxidizing agent than bromine. Each bromide ion loses an electron (oxidation), and a pair of bromide ions makes two electrons available to a chlorine molecule (reduction).

$$\text{oxidation} \quad 2\,\ddot{\underset{..}{Br}}:^- \longrightarrow\, :+ 2\,\ddot{\underset{..}{Br}}\cdot \longrightarrow \ddot{\underset{..}{Br}}\,\ddot{\underset{..}{Br}}:$$

$$\text{reduction} \quad :+ :\ddot{\underset{..}{Cl}}:\ddot{\underset{..}{Cl}}: \longrightarrow 2\,:\ddot{\underset{..}{Cl}}:\,^-$$

Following Balard's discovery, bromide ion was found in seawater, although in far lower concentrations than chloride ion. The chlorination of seawater or brines by a process like the one Balard used is one step in the commercial synthesis of bromine. (Brines are salt solutions, generally from wells, having salt concentrations greater than 3.5% (*w/w*). Some brines are richer than seawater in Br^-.)

Bromine isotopes:
Br-79; 50.54%
Br-81; 49.46%

Bromine is the only *liquid* nonmetallic element. It occurs in a nearly 50:50 mixture of two isotopes, Br-79 and Br-81. Bromine has a high density, 3.12 g/ml, a deep red-brown color, and a low enough boiling point (59 °C) to make it very volatile. Its vapors not only have an evil odor, they also attack the soft tissue of the nose and throat and even the tougher tissue of the hands. A bromine spill on the skin causes a severe burn. Therefore, this element must be handled with extreme caution at an efficient hood with the operator wearing good rubber gloves. (See Color Plate 3E.)

Bromine very much resembles chlorine in its chemical properties except that bromine has considerably lower reactivity. It will combine directly with most elements, but not quite as many as are attacked by chlorine. Bromine adds to carbon-carbon double and triple bonds (Section 9.9) like chlorine. One of the major uses of bromine is to make 1,2-dibromoethane, an additive in leaded gasoline that helps to keep lead from depositing in the cylinders of the engine.

$$CH_2\!\!=\!\!CH_2(g) + Br_2(\ell) \longrightarrow Br\!-\!CH_2\!-\!CH_2\!-\!Br(\ell)$$
ethene   1,2-dibromoethane

In a reaction very similar to that of chlorine, bromine reacts with water to give hypobromous and hydrobromic acids.

$$Br_2(\ell) + H_2O(\ell) \longrightarrow \quad HBr(aq) \quad + \quad HOBr(aq)$$
hydrobromic   hypobromous
acid   acid

Hypobromous acid, like hypochlorous acid, is unstable. It slowly decomposes to hydrobromic acid and oxygen.

$$2HOBr(aq) \longrightarrow 2HBr(aq) + O_2(g)$$

Hydrobromic acid, the aqueous solution of hydrogen bromide gas, is made by heating a mixture of a bromide salt with a nonvolatile, strong acid—a process similar to that used to prepare hydrochloric acid.

$$2NaBr(aq) + H_2SO_4(aq, 50\%) \xrightarrow{\text{heat}} 2HBr(g) + Na_2SO_4(aq)$$

The gaseous hydrogen bromide may be trapped in water or it may be dried and stored under pressure in a steel cylinder. Bromide ion is more easily oxidized than the chloride ion, and sulfuric acid is an oxidizing agent besides being an acid. Therefore, the sulfuric acid used must be more dilute than that used to make HCl. *Concentrated* sulfuric acid oxidizes HBr as follows.

$$2HBr(aq) + H_2SO_4(\text{concd}) \longrightarrow Br_2(\ell) + SO_2(g) + 2H_2O(\ell)$$

Hydrobromic acid, like hydrochloric acid, is a strong acid, giving all the reactions of the hydrogen ion and the bromide ion.

Oxides and oxoacids of bromine similar to some of those of chlorine are known, but most are unstable and hard to prepare.

Iodine and its major compounds

In the early nineteenth century, seaweed was collected, dried, and burned to give ashes that were a source of sodium and potassium salts. In 1811 Bernard Courtois (1777–1838) observed purple vapors rising from an extract of kelp ash that he had acidified and heated with sulfuric acid. The vapor condensed to black crystals on cold surfaces. (See Color Plate 5A.) This substance, later proven to be an element by Gay-Lussac and Davy, was named after the Greek *iodes*, meaning "purple." Iodine is still obtained from sodium iodide in the ashes of seaweed and kelp. For example, one oxidizing agent, manganese dioxide, acts as follows.

Only the I-127 isotope occurs naturally.

$$2NaI(aq) + 3H_2SO_4(aq) + \quad MnO_2(s) \quad \longrightarrow$$
sodium   sulfuric   manganese
iodide   acid   dioxide

$$I_2(s) \quad + \quad MnSO_4(aq) + 2NaHSO_4(aq) + 2H_2O(\ell)$$
iodine   manganese(II)
sulfate

Iodine can also be made from the iodide ion by the action of either bromine or chlorine.

$$2I^- + Br_2 \longrightarrow I_2 + 2Br^-$$

Or

$$2I^- + Cl_2 \longrightarrow I_2 + 2Cl^-$$

These two are redox reactions. Each iodide ion loses an electron, and a pair of iodide ions makes two electrons available. Each bromine or chlorine molecule accepts two electrons and changes to two bromide or chloride ions.

oxidation $\quad 2:\ddot{I}:^- \longrightarrow :I: + 2:\ddot{I}\cdot \longrightarrow :\ddot{I}:\ddot{I}:$

reduction $\quad :\ddot{X}:\ddot{X}: + : \longrightarrow 2:\ddot{X}:^-$

$(X = Cl \text{ or } Br)$

An excess of chlorine or bromine must be avoided. Excess chlorine, for example, will oxidize iodine to iodic acid, HIO_3, if water is present.

$$I_2(s) + 5Cl_2(g) + 6H_2O(\ell) \longrightarrow 2HIO_3(aq) + 10HCl(aq)$$

Bromine will also oxidize iodine to iodic acid. Fluorine cannot be used in the synthesis of iodine from aqueous sodium iodide because fluorine reacts violently with water. Fluorine is the most powerful oxidizing agent of all the halogens. Thus we have the halogens in the following order of strengths as oxidizing agents.

$$F_2 > Cl_2 > Br > I_2 \quad \text{(as oxidizing agents)}$$

When any halogen acts as an oxidizing agent, it is reduced to the corresponding halide ion. Halide ions are themselves reducing agents whose strengths as reducing agents are in the following order.

$$I^- > Br^- > Cl^- > F^- \quad \text{(as reducing agents)}$$

Iodine is not nearly as dangerous an element as the other halogens, although it is still a poison. Its vapors irritate the eyes, nose, and throat. A dilute solution of iodine in aqueous alcohol, called *tincture of iodine,* is a common antiseptic.

The body has to have the iodide ion in the diet in trace amounts to make the hormone thyroxin. Thyroxin is produced in the thyroid gland located in the neck. When iodide ion is missing from the diet for an extended period, a large, disfiguring swelling—a goiter—can develop in the neck.

Always use "iodized" salt instead of plain salt to be sure of getting enough iodide ion.

Iodine reacts with far fewer elements than bromine or chlorine do. A solution of hydrogen iodide in water is called hydriodic acid, $HI(aq)$, a strong acid like its counterparts, hydrobromic acid and hydrochloric acid. However, oxygen from the atmosphere easily oxidizes the iodide ion in hydriodic acid.

$$O_2(g) + 4HI(aq) \longrightarrow 2I_2(s) + 2H_2O(\ell)$$

Light energy initiates this reaction, and hydriodic acid has to be stored in brown bottles kept well capped.

Iodine forms some of the oxides and oxoacids similar to those of bromine and chlorine, but most are unstable. Iodic acid, HIO_3, it known both in crystalline form and in solution. Sodium iodate forms by the action of sodium hydroxide on iodine.

$$6NaOH(aq) + 3I_2 \longrightarrow NaIO_3(aq) + 5NaI(aq) + 3H_2O(\ell)$$
$$\ \text{sodium}\qquad \text{sodium}$$
$$\ \text{iodate}\qquad \text{iodide}$$

The iodates are oxidizing agents.

Periodic acid (pronounced per'-ī-ō-dic), another crystalline oxoacid of iodine, has enough stability to be made and stored. Depending on the degree of hydration, it exists as HIO_4, H_3IO_5, $H_4I_2O_9$, and H_5IO_6. The most common form is H_5IO_6, which is used as an oxidizing agent in organic chemistry.

Iodine has a +7 oxidation number in all these forms of periodic acid.

12.8 CARBON

The important inorganic compounds of carbon are carbonates, bicarbonates, and cyanides

The members of the **carbon family** are given in Table 12.13. They range from the brittle nonmetal, carbon, to the definitely metallic elements, tin and lead. Carbon, as we learned in Chapter Five, is the chief constituent of coal, the world's most abundant fossil fuel. It got into coal as well as petroleum and natural gas because carbon is the basis for all organic compounds, as we learned in Chapter Nine. In this section, however, we will consider only the compounds of carbon that most chemists regard as strictly mineral or inorganic—carbonates, bicarbonates, cyanides, and a few others.

Silicon and germanium are used to make electronic semiconductors for transistors. In the next chapter we will learn about compounds of silicon that give useful oils, gums, and synthetic rubbers.

Table 12.13
The Carbon Family—Group IVA

Element	Symbol	Mp (°C)	Bp (°C)
Carbon	C	3570	3470[a]
Silicon	Si	1414	2355
Germanium	Ge	959	2700
Tin	Sn	232	2275
Lead	Pb	327	1750

[a] Sublimation temperature.

Allotropic forms of carbon

The two crystalline forms of carbon are graphite and diamond. The carbon atoms in graphite occur in parallel sheets, as sketched in Figure 12.9. Each sheet is made of an extended network of carbon hexagons sandwiched between an enormous network of delocalized pi electrons. Because of this network, graphite conducts electricity, an unusual property for a nonmetal. The planes of atoms in graphite can slip by each other under ordinary pressures, which makes graphite useful as a dry lubricant.

The carbon atoms in diamond (Figure 12.9) are joined exclusively by sigma bonds projecting in a tetrahedral array from each atom. At ordinary pressures, diamond is the less stable allotrope of carbon. At especially high pressures, however, the opposite is true. Diamond has a density of 3.51 g/cm^3, which is higher than the density of graphite (2.22 g/cm^3). Therefore, high pressure favors the conversion of graphite to diamond because that change packs a given mass of carbon into a smaller volume. Diamonds suitable for making diamond saws and diamond-tipped drilling bits and for other commercial applications not requiring almost flawless beauty are now made by subjecting graphite to high pressure and temperature.

Carbon monoxide

The action of concentrated sulfuric acid on methanoic acid ("formic acid") is a convenient laboratory synthesis of carbon monoxide.

$$\underset{\substack{\text{formic} \\ \text{acid}}}{H-\overset{\overset{\textstyle O}{\|}}{C}-O-H} \xrightarrow[H_2SO_4 \text{ (concd)}]{} \underset{\substack{\text{carbon} \\ \text{monoxide}}}{CO(g)} + H_2O(\ell)$$

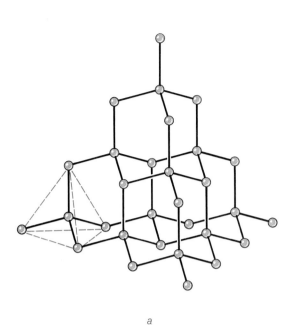

a

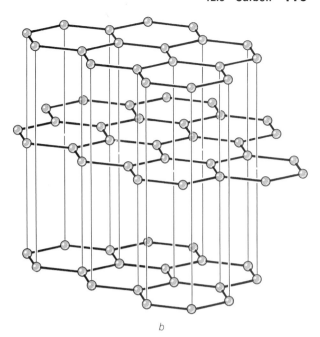

b

Figure 12.9
Allotropic forms of carbon. (a) Diamond structure (b) Graphite
structure.

This gas also is present in the gaseous mixture produced by the combustion of anything made of carbon, unless the supply of oxygen is plentiful enough to insure that all the carbon can be changed to carbon dioxide.

Carbon monoxide is a poison that is all the more dangerous because it is colorless, odorless, and tasteless. It acts by binding to hemoglobin in the blood about 200 times more strongly than oxygen can. Thus it prevents hemoglobin from taking up oxygen as the blood circulates through the lungs. Sublethal doses of carbon monoxide cause headaches and feelings of dullness or dizziness. They also put a strain on the heart because the heart works harder as it tries to deliver oxygen in the face of the reduced oxygen-carrying capacity of the hemoglobin.

Carbon monoxide is a reducing agent and, in the metallurgical industry, it is used to reduce certain metal oxides to their metals. For example,

$$FeO + CO \xrightarrow[\text{heat}]{} \underset{\text{iron}}{Fe} + CO_2$$

$$CuO + CO \xrightarrow[\text{heat}]{} \underset{\text{copper}}{Cu} + CO_2$$

Carbon monoxide reacts with some transition metals to give oily liquids called metal carbonyls. Iron pentacarbonyl, $Fe(CO)_5$, is one example, It is *pyrophoric,* meaning it spontaneously ignites and burns in air (giving Fe_2O_3 and CO_2). Nickel tetracarbonyl, $Ni(CO)_4$, which explodes at 60 °C, is a dangerous poison and is possibly an environmental pollutant. Coal contains traces of nickel; when coal is burned, some nickel tetracarbonyl forms and escapes.

Carbon dioxide and carbonic acid
Carbon dioxide is one end product of the complete combustion of organic substances. It can also be made by heating metal carbonates and bicarbonates. For

Some common names for calcium oxide are *lime, burnt lime, quicklime,* and *calx.*

example, when limestone—which is mostly calcium carbonate—is strongly heated, it decomposes to give carbon dioxide and a residue of calcium oxide.

$$CaCO_3(s) \longrightarrow CaO(s) + CO_2(g)$$

calcium   calcium   carbon
carbonate   oxide   dioxide

This is one of the chemical changes that take place when finely ground limestone and clay are strongly heated together to make portland cement. The cost of the energy is high, and cement making is called an *energy-intensive industry* because so much energy is consumed to make a unit of one of its products.

A saturated solution of carbon dioxide in water at 760 torr and 25 °C has a concentration of 0.033 M CO_2. Most of the gas exists in the solution as hydrated molecules of CO_2, but some is present as the weak, unstable acid, carbonic acid, H_2CO_3.

$$CO_2(aq) + H_2O(\ell) \rightleftharpoons H_2CO_3(aq)$$

carbonic
acid

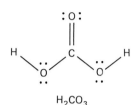

H_2CO_3

Carbon dioxide reacts quantitatively with metal hydroxides. For example, if the gas is bubbled into aqueous sodium hydroxide, it is trapped in the form of the bicarbonate ion, HCO_3^-, by the following reaction.

$$CO_2(g) + NaOH(aq) \longrightarrow NaHCO_3(aq)$$

Or

$$CO_2(g) + OH^-(aq) \longrightarrow HCO_3^-(aq)$$

HCO_3^-

Sodium bicarbonate is the monosodium salt of carbonic acid. Therefore, its formation from CO_2 by the preceding reaction is the same as if it had formed by the neutralization of one hydrogen ion from carbonic acid.

$$NaOH(aq) + H_2CO_3(aq) \longrightarrow NaHCO_3(aq) + H_2O$$

CO_3^{2-}

Because of this similarity, we can treat a solution of carbon dioxide in water as if it were entirely a solution of carbonic acid in water. *Carbon dioxide bubbled into water or released in water will neutralize base as if it were carbonic acid.* Carbonic acid will also react with carbonates to give bicarbonates.

Both the carbonate and bicarbonate ions will neutralize strong acids, and the products are carbon dioxide, water, and some salt. For example,

$$NaHCO_3(aq) + HCl(aq) \longrightarrow CO_2(g) + H_2O(\ell) + NaCl(aq)$$

The net ionic equation for this is:

$$HCO_3^-(aq) + H^+(aq) \longrightarrow CO_2(g) + H_2O(\ell)$$

Just a little warming of the solution expels most, if not all, of the carbon dioxide. All metal bicarbonates react this way with all strong acids, liberate carbon dioxide and water, and leave a salt either in solution or as an insoluble substance. All metal carbonates also react with strong acids to give the same products; the only difference is in the ratios of chemicals, not in what forms. For example,

$$Na_2CO_3(aq) + 2HCl(aq) \longrightarrow CO_2(g) + H_2O(\ell) + 2NaCl(aq)$$

The net ionic equation is:

$$CO_3^{2-}(aq) + 2H^+(aq) \longrightarrow CO_2(g) + H_2O(\ell)$$

Cyanides and hydrogen cyanide

Sodium cyanide, NaCN, releases the poisonous cyanide ion, CN^-, in water. This ion will combine with hydrogen ion to liberate hydrogen cyanide, a poisonous gas used in certain pest-control operations and in gas chambers that some governments use to carry out death sentences.

$(:C \equiv N:)^-$
cyanide ion

$$NaCN(aq \text{ or } s) + HCl(aq) \longrightarrow HCN(g) + NaCl(aq)$$

Or

$$CN^-(aq) + H^+(aq) \longrightarrow HCN(g)$$

Other inorganic carbon compounds

When sulfur vapor is passed over hot coke, carbon disulfide, CS_2, forms,

$$C(s) + 2S(g) \xrightarrow{\text{heat}} CS_2(\ell)$$

$\ddot{S}{=}C{=}\ddot{S}$
carbon disulfide

This liquid with a most disagreeable odor (bp 46 °C) is used principally as a solvent. It is extremely flammable; even boiling water can ignite it!

The carbides are compounds between carbon and metals. One type of carbide reacts with water to give ethyne ("acetylene"). Calcium carbide, CaC_2, is an example.

$$\underset{\substack{\text{calcium} \\ \text{carbide}}}{CaC_2} + 2H_2O \longrightarrow \underset{\substack{\text{ethyne} \\ \text{("acetylene")}}}{H{-}C{\equiv}C{-}H} + \underset{\substack{\text{calcium} \\ \text{hydroxide}}}{Ca(OH)_2}$$

C_2^{2-} is $(:C{\equiv}C:)^{2-}$
acetylide ion

Carbide lanterns, once used by underground miners and even as automobile headlamps, contained a supply of calcium carbide and a small reservoir of water. Illumination came from burning the acetylene that formed as water dripped onto the carbide.

Other carbides react with water to give methane. For example, magnesium carbide reacts as follows.

$$\underset{\substack{\text{magnesium} \\ \text{carbide}}}{Mg_2C} + 4H_2O \longrightarrow \underset{\substack{\text{magnesium} \\ \text{hydroxide}}}{2Mg(OH)_2} + \underset{\text{methane}}{CH_4}$$

The carbon in magnesium carbide behaves as if it is the C^{4-} ion.

Silicon carbide, SiC, is a covalent carbide made by heating silicon dioxide, SiO_2, with carbon at a very high temperature (2000 to 2600 °C). One crystalline form of silicon carbide is marketed as carborundum, an unusually effective abrasive used to make sandpapers and grinding wheels.

12.9 NOBLE GASES

The noble gas elements, Group 0, are helium, neon, argon, krypton, xenon, and radon

In some references the noble gas family is designated Group VIIIA instead of Group 0 in the periodic table.

The **noble gases,** listed in Table 12.14, are present in air in extremely small concentrations, although argon has a concentration of about 1%. They also occur in natural gas. Radon, the radioactive noble gas, is present in the minerals and underground formations that contain isotopes of uranium, thorium, and radium. The

Table 12.14
The Noble Gas Elements—Group 0

Element	Symbol	Atomic Number	Bp (°C)	Mp (°C)	Color Emitted from A Discharge Tube
Helium	He	2	−268.934	−272.2[a]	White to pink-violet
Neon	Ne	10	−246.048	−248.67	Red-orange
Argon	Ar	18	−185.7	−189.2	Violet-purple
Krypton	Kr	36	−152.30	−156.6	Pale violet
Xenon	Xe	54	−107.1	−111.9	Blue to blue-green
Radon	Rn	86	− 61.8	− 71	—

[a] Under a pressure of 26 atm.

The noble gases are also called the *rare gases*.

Some compounds of xenon are

XeF_2 XeO_3
XeF_4 XeO_4
XeF_6 $XeOF_4$
        XeO_2F_2

noble gases were once called the *inert gases* because, until 1962, they literally had no known chemistry. This aloofness from the turmoil of change, reminiscent of ancient nobility, also led to the name we are using, the noble gases.

In 1962 a complex yellow substance that was believed to be $XePtF_6$ was prepared. Soon thereafter, xenon tetrafluoride, XeF_4, was made by combining its elements at 400 °C. These two developments prompted scientists of many countries to study the possibilities of making other noble gas compounds, and now a number of fluorides, oxides, and mixed oxide-fluorides of xenon as well as krypton difluoride are known. Only the two most electronegative elements, fluorine and oxygen, give noble gas compounds, and only xenon and krypton are known to be reactive. (Radon is perhaps also reactive, but its nuclear instability has discouraged studies.)

The noble gases are prized largely for their physical properties and chemical inertness (except Xe and Kr). The space around arc-welding operations may be flooded with argon, for example, to keep air away. This keeps oxygen (or, in some cases, nitrogen) from reacting with the hot metal. If ordinary air, even at very low pressure, were inside light bulbs, the wire filaments would be rapidly destroyed by oxygen. Therefore, both incandescent and fluorescent bulbs contain argon instead of air. (Sometimes krypton or xenon is used.) Helium or argon is sometimes used to provide an inert environment for the chemicals of a reaction when they might react with either oxygen or nitrogen, The familiar "neon" signs are electric discharge tubes (fluorescent tubes) that contain one or another of the noble gases. The color of the sign depends on which noble gas is used (see Table 12.14 and Color Plate 5B).

Helium is used for dirigibles and balloons. Hydrogen, because of its lower density and greater lifting ability, would be better if it were not flammable. Helium removes that danger and still has good lifting power. Helium liquifies at −268.9 °C, just 4.3 K above absolute zero, but it never becomes a solid as it is further cooled. Below 2.17 K, liquid helium undergoes a startling change in properties. It loses its viscosity—its resistance to flow—and it even climbs the walls of its container! By using liquid helium to cool metal conductors of electricity, they become super conductors. All their resistance to the flow of electricity stops. Engineers are presently trying to figure out ways to take advantage of this. Up to a 20% voltage drop occurs during the transmission of electricity in wires over a long distance because of electrical resistance. If electricity could be carried by a superconductor, all of the loss in energy caused by this resistance would be saved.

SUMMARY

Nonmetals. The important nonmetallic elements are hydrogen, which does not fit well in any group, carbon (Group IVA), nitrogen and phosphorus (Group VA), oxygen and sulfur (Group VIA), and the elements in Groups VIIA (halogens) and 0 (noble gases). The nonmetals are located in the upper right corner and side of the periodic table. Throughout the universe just a few nonmetallic elements, notably hydrogen and helium, contribute the greatest numbers of atoms. Oxygen is especially prevalent on earth, and just a few nonmetals dominate the substances in living systems. These nonmetals have especially stable atomic nuclei; they can all form covalent bonds—many two and three—and some can form ions as well.

Hydrogen. Most hydrogen is prepared from water or methane, but small quantities can be made by letting a strong acid act on an active metal. The hydrides, binary compounds of hydrogen, occur in three types—ionic, covalent, and metallic. Ionic hydrides are saltlike in physical properties and react as donors of hydride ion, H^-. Covalent hydrides are molecular and donate hydrogen, if at all, as hydrogen, H^+. Many metallic hydrides have indefinite formulas, and some act as donors of hydrogen gas, H_2. The most important hydrides are those of nonmetals—the hydrides of carbon (CH_4, methane), nitrogen (NH_4, ammonia), oxygen (H_2O, water), and the halogens (HF, HCl, HBr, and HI). The largest use of hydrogen is to make ammonia from nitrogen by the Haber process.

$$2N_2(g) + 3H_2(g) \xrightarrow[\text{heat, pressure}]{\text{metal catalyst}} 2NH_3(g)$$

Oxygen. The oxygen family, Group VIA, consists of oxygen, sulfur, selenium, and tellurium. Oxygen's ability to support respiration, combustion, and decay makes it an essential element for life. For industrial needs, oxygen is obtained from liquid air. Its allotrope, ozone (O_3), occurs in smog (where it is dangerous) and in the stratosphere (where it is vital). Oxides of virtually all elements are known. Metal oxides are mostly ionic and high melting, and they can neutralize strong acids. A few metal oxides are amphoteric and some are actually acidic in water. Nonmetal oxides that can dissolve in and react with water are able to neutralize bases. Water and hydrogen peroxide are the two hydrides of oxygen. The peroxide, a bleach, decomposes to oxygen and water. Metal peroxides with the O_2^{2-} ion decompose to oxygen and a metal hydroxide.

Sulfur. Sulfur occurs in the free state, as sulfides, sulfates, some hydrogen sulfide, and in the —SH and —S—S— groups in proteins. It burns to give sulfur dioxide, a raw material for making sulfuric acid. Sulfur dissolves in solutions of metal sulfites to give the thiosulfate ion, $S_2O_3^{2-}$, and in solutions of metal sulfides or ammonium sulfide to give polysulfide ions, S_x^{2-}. When sulfur dioxide dissolves in water, some sulfurous acid (H_2SO_3) forms, a weak, unstable, diprotic acid having reasonably stable salts—the hydrogen sulfites and sulfites. These salts all react with strong acids to give sulfur dioxide plus sulfurous acid. Sulfur dioxide, released when sulfur-containing fossil fuels are burned, contributes to acid rain. In the contact process for making sulfuric acid, sulfur dioxide is oxidized in contact with a vanadium pentoxide catalyst to sulfur trioxide, which is changed to sulfuric acid by water. Sulfuric acid is the largest-selling industrial chemical and is used to make sulfate fertilizers, chemicals (including all of the hydrohalogen acids), and many other products. When sulfur occurs as a metal sulfide, the action of strong acids releases the hydride of sulfur, H_2S, hydrogen sulfide.

Nitrogen. Group VA, the nitrogen family, consists of nitrogen, phosphorus, arsenic, antimony, and bismuth. Only nitrogen occurs in the free state. Obtained from liquid air, nitrogen is used mostly to make ammonia (Haber process). Liquid ammonia is a fertilizer and a solvent. Ammonia is the chief source of nitric acid, made by the Ostwald process. Nitric acid, a strong, monoprotic acid, is used to make nitrates, explosives, and chemicals. It is not indefinitely stable and slowly decomposes to nitrogen dioxide and oxygen.

$$4HNO_3(aq) \longrightarrow 4NO_2(g) + O_2(g) + 2H_2O(\ell)$$

Nitrogen dioxide, NO_2, is red-brown gas that gives color to old supplies of concentrated nitric acid and to smog. Other oxides of nitrogen and their methods of synthesis are dinitrogen monoxide (N_2O; $NH_4NO_3 \xrightarrow{\text{heat}} N_2O + 2H_2O$); ni-

trogen monoxide (NO; $N_2 + O_2 \xrightarrow{\text{heat}} 2NO$); dinitrogen trioxide ($N_2O_3$; $NO + NO_2 \rightarrow N_2O_3$); a pair of oxides in equilibrium, NO_2 and N_2O_4 ($2NO + O_2 \rightarrow 2NO_2 \rightleftarrows N_2O_4$); and dinitrogen pentoxide (N_2O_5; $4HNO_3 + P_4O_{10} \rightarrow 2N_2O_5 + 2HPO_3$). Nitrous acid ($HNO_2$), an unstable monoprotic acid, exists in cold, aqueous solutions made by adding a strong acid to a metal nitrite. Hydrazine, N_2H_4, is a hydride of nitrogen used as a rocket fuel.

Phosphorus. Red phosphorus is the common and stable allotropic form. Phosphorus burns to tetraphosphorus decaoxide, which dissolves in water to make orthophosphoric acid (or other forms, depending on the extent of hydration).

$$4P + 5O_2 \longrightarrow P_4O_{10} \xrightarrow{6H_2O} 4H_3PO_4$$

Phosphoric acid is used chiefly to make various phosphate fertilizers or to make acidulants. Phosphorus also forms covalent compounds with the halogens having the general formulas PX_3 and PX_5, where X = a halogen.

Halogens. Fluorine, chlorine, bromine, and iodine (and one we ignored, the radioactive astatine) are the halogens, Group VIIA. None occurs in the free state. All can exist as negative ions (X^-), and in molecules of halogen oxides (seldom of much stability), hydrides (HX) and oxoacids (also seldom of much stability). All halogens occur in covalently bound forms, notably in bonds to hydrogen (HX) or to carbon in organohalogen compounds. The order of the electronegativity is the same as the order of their oxidizing ability.

$$F_2 > Cl_2 > Br_2 > I_2 \quad \text{(in electronegativity; in oxidizing ability)}$$

Their reduced forms, the halide ions, stand in the reverse order in ability to act as reducing agents.

$$I^- > Br^- > Cl^- > F^- \quad \text{(in reducing ability)}$$

Solutions of the hydrogen halides in water are all acidic, with HCl, HBr, and HI forming strong acids called hydrochloric, hydrobromic, and hydriodic acids, respectively. Hydrofluoric acid is a weak acid. Chlorine and its oxides and oxoacids are bleaches, and chlorine itself is used to disinfect drinking water. The chief oxoacids of chlorine are HOCl (hypochlorous acid) and $HClO_4$ (perchloric acid). Bromine and iodine both have chemical properties similar to those of chlorine, only they react (in the order given) much less readily than chlorine. The chloride ion is essential in all fluids of the body, and the body needs iodide ion to make a hormone and prevent goiter.

Carbon. Carbon heads Group IVA—carbon, silicon, germanium, tin, and lead. The chief inorganic compounds of carbon are two oxides (CO and CO_2), carbonic acid (H_2CO_3) and its salts, cyanides, and some carbides. Carbon monoxide (CO), a reducing agent, is used industrially to change certain metal ores to their metals. It is also a dangerous poison. Carbon dioxide (CO_2), one end product in the combustion of anything containing carbon, will dissolve in water, and some reacts to form unstable carbonic acid, H_2CO_3.

$$CO_2(g) + H_2O(\ell) \rightleftharpoons H_2CO_3(aq)$$

Therefore, aqueous carbon dioxide neutralizes metal hydroxides.

$$CO_2(g) + NaOH(aq) \longrightarrow NaHCO_3(aq)$$

All bicarbonates and carbonates react with strong acids to give carbon dioxide and water plus some salt. Metal cyanides react with strong acids to give hydrogen cyanide.

Noble Gases. The Group 0 elements—helium, neon, argon, krypton, xenon, and radon—give no chemical reactions, except xenon forms various fluorides, oxides, and mixed oxide-fluorides. The noble gases are used for their physical properties or to create an atmosphere free of oxygen for oxygen-sensitive substances.

INDEX TO QUESTIONS

REVIEW QUESTIONS

(Note. When the question asks for an equation, write a molecular equation unless directed otherwise.)

12.1. Which elements are the two most abundant in the universe?

12.2. According to current theory about the origin of the elements, what was the origin of (a) helium, and (b) the heavier elements?

12.3. In what three major forms does oxygen occur in or on the earth?

12.4. Which two elements dominate the composition of the earth's atmosphere and in what percentages do they occur?

12.5. How are the relatively high abundances of the nonmetallic elements of low atomic weights explained in current theory?

12.6. Why are second-row nonmetals more often involved in pi bonds than nonmetals in later rows?

12.7. Give the names, atomic symbols, and mass numbers of the isotopes of hydrogen.

12.8. What is the deuterium isotope effect?

12.9. What are the three kinds of hydrides? Give the formula and name of an example of each kind.

12.10. Give one physical and one chemical property in which ionic and covalent hydrides markedly differ.

12.11. What is the largest industrial use of hydrogen?

12.12. Write the equation for the Haber process.

12.13. Explain why a high pressure helps the Haber process.

12.14. What is meant by "catalyst"?

12.15. Give the names and the atomic symbols of the members of the oxygen family of elements.

12.16. Write the formula of each. (Remember that the formula of an ion must include the kind and amount of charge.)
(a) Hydride ion.   (b) Sulfite ion.
(c) Potassium amide.   (d) Hydrogen sulfate ion.
(e) Hydrochloric acid.   (f) Ammonium sulfate.
(g) Sulfuric acid.   (h) Ozone.
(i) Perchloric acid.   (j) Calcium carbide.

12.17. Write the formula of each. (Remember that the formula of an ion must include the kind and amount of charge.)
(a) Amide ion.   (b) Sodium oxide.
(c) Sulfate ion.   (d) Sodium hydride.
(e) Ammonium nitrate.   (f) Hydrobromic acid.
(g) Lithium amide.   (h) Nitric acid.
(i) Sodium sulfide.   (j) Hydrofluoric acid.

12.18. Write the formula of each. (Remember that the formula of an ion must include the kind and amount of charge.)
(a) Hydrogen sulfite ion.   (b) Sodium peroxide.
(c) Sodium amide.   (d) Thiosulfate ion.
(e) Calcium hydride.   (f) Hydriodic acid.
(g) Nitrous acid.   (h) Hypochlorous acid.
(i) Bifluoride ion.   (j) Hydrogen peroxide.

12.19. Write the name of each.
(a) H^- (b) HSO_3^- (c) $NaNH_2$
(d) H_2SO_4 (e) ClO^- (f) PCl_3
(g) H_3PO_4 (h) $HClO_4$ (i) H_2O_2
(j) N_2O (k) N_2O_3 (l) HNO_2

12.20. Write the name of each.
(a) NH_2^- (b) HSO_4^- (c) KH
(d) H_2SO_3 (e) ClO_3^- (f) PH_3
(g) $H_4P_2O_7$ (h) HNO_3 (i) $NaClO_4$
(j) NO (k) N_2O_5 (l) $NaNO_2$

12.21. Write the name of each.
(a) NH_3 (b) ClO_4^- (c) $(NH_4)_2CO_3$
(d) H_2S (e) PCl_5 (f) HPO_3
(g) $HBr(aq)$ (h) $HBr(g)$ (i) N_2H_4
(j) N_2O_4 (k) Cl_2O_7 (l) HCO_3^-

12.22. Name one important contribution each person made in chemistry.
(a) Cavendish.   (b) Haber.
(c) Priestly.   (d) Scheele.
(e) Lavoisier.   (f) Rutherford.
(g) Ostwald.   (h) Nobel.
(i) Davy.   (j) Balard.
(k) Courtois.

12.23. Complete and balance the following equations. Special experimental conditions are summarized over or under the arrow.
(a) $Zn(s) + HCl(aq) \longrightarrow$
(b) $H_2O(g) + C(s) \xrightarrow[\text{temperature}]{\text{high}}$
(c) $CH_4(g) + H_2O(g) \xrightarrow[\text{temperature}]{\text{high}}$
(d) $H_2(g) + F_2(g) \longrightarrow$
(e) $H_2(g) + Cl_2(g) \xrightarrow[\text{initiate}]{\text{heat to}}$
(f) $NaH(s) + H_2O(\ell) \longrightarrow$
(g) $Ca(s) + H_2(g) \longrightarrow$
(h) $Mg(s) + HCl(aq) \longrightarrow$
(i) $KH(s) + H_2O(\ell) \longrightarrow$

12.24. Complete and balance the following equations.
(a) $NaH(s) + O_2(g) \longrightarrow$
(b) $H^- + H_2O(\ell) \longrightarrow$
(c) $HgO(s) \xrightarrow[\text{temperature}]{\text{high}}$

(d) $KClO_3(s) \xrightarrow[\text{temperature}]{\text{high}}$
(e) $Na(s) + O_2(g) \longrightarrow$
(f) $Ca(s) + O_2(g) \longrightarrow$
(g) $Al(s) + O_2(g) \longrightarrow$
(h) $C(s) + O_2(g) \longrightarrow$
(i) $N_2(g) + O_2(g) \xrightarrow[\text{temperature}]{\text{high}}$

12.25. Complete and balance the following equations.
(a) $Na_2O(s) + H_2O(\ell) \longrightarrow$
(b) $Na_2O(s) + HCl(aq) \longrightarrow$
(c) $Na_2O(s) + O_2(g) \longrightarrow$
(d) $Al_2O_3(s) + HCl(aq) \longrightarrow$
(e) $Al_2O_3(s) + NaOH(aq) \longrightarrow$
(f) $K_2O(s) + HCl(aq) \longrightarrow$
(g) $K_2O(s) + H_2O(\ell) \longrightarrow$
(h) $Na_2O_2(s) + H_2O(\ell) \longrightarrow$
(i) $Al_2O_3(s) + HBr(aq) \longrightarrow$

12.26. Complete and balance the following equations.
(a) $SO_2(g) + H_2O(\ell) \longrightarrow$
(b) $SO_3(g) + H_2O(\ell) \longrightarrow$
(c) $CO_2(g) + H_2O(\ell) \longrightarrow$
(d) $NO_2(g) + H_2O(\ell) \longrightarrow$
(e) $N_2O_5(g) + H_2O(\ell) \longrightarrow$
(f) $SO_2(g) + NaOH(aq) \longrightarrow$
(g) $SO_3(g) + NaOH(aq) \longrightarrow$
(h) $CO_2(g) + NaOH(aq) \longrightarrow$
(i) $H_2O_2(\ell) \xrightarrow{\text{heat}}$

12.27. Write both the molecular and net ionic equations for the reaction of hydrochloric acid, $HCl(aq)$, with (a) $NaHSO_3(aq)$, and (b) $Na_2SO_3(aq)$.

12.28. Hydrobromic acid, $HBr(aq)$, reacts with hydrogen sulfites and sulfites in the same way as hydrochloric acid (previous question). Write the molecular and net ionic equations for the reaction of hydrobromic acid with (a) $NaHSO_3(aq)$, (b) $Na_2SO_3(aq)$, (c) $K_2SO_3(aq)$, and (d) $KHSO_3(aq)$.

12.29. Aqueous potassium hydroxide, $KOH(aq)$, reacts with nonmetal oxides in exactly the same way as aqueous sodium hydroxide, $NaOH(aq)$. Write molecular and net ionic equations for the reaction of aqueous potassium hydroxide with (a) $SO_2(g)$, (b) $SO_3(g)$, and (c) $CO_2(g)$.

12.30. Write the molecular and net ionic equations for the reaction of aqueous hydrochloric acid, $HCl(aq)$, with each substance.
(a) $NaOH(aq)$ (b) $NaHCO_3(aq)$
(c) $Na_2CO_3(aq)$ (d) $KOH(aq)$
(e) $K_2CO_3(aq)$ (f) $KHCO_3(aq)$
(g) $CaCO_3$ (h) $Ca(OH)_2$
(i) $Mg(OH)_2$ (j) $MgCO_3$

12.31. Hydrobromic acid, $HBr(aq)$, gives reactions very similar to those of hydrochloric acid (question 12.30). Write molecular and net ionic equations for the reaction of hydrobromic acid with each of the substances listed in question 12.30.

12.32. Nitric acid, $HNO_3(aq)$, reacts with the substances listed in question 12.30 in the same way as hydrochloric acid (except, of course, nitrate salts instead of chloride salts form). Write molecular and net ionic equations for the reaction of nitric acid with each of the substances in question 12.30.

12.33. Sulfuric acid, $H_2SO_4(aq)$, reacts with metal hydroxides, carbonates, and bicarbonates in the same way as hydrochloric, hydrobromic, and nitric acids. Assume that *both* hydrogen ions available from H_2SO_4 participate in the reaction and that salts such as Na_2SO_4 (not $NaHSO_4$) and $CaSO_4$ [not $Ca(HSO_4)_2$] are among the products. Write molecular and net ionic equations for the reaction of sulfuric acid with each substance listed in question 12.30.

12.34. Hydrochloric acid, hydrobromic acid, hydriodic acid, nitric acid, perchloric acid, and sulfuric acid all give very similar reactions with metal hydroxides, metal cyanides, metal bicarbonates, metal carbonates, metal hydrogen sulfites, and metal sulfites. Write the net ionic equation that *any* of the acids named will give with each of the following substances.
(a) $KOH(aq)$ (b) $KHCO_3(aq)$
(c) $K_2CO_3(aq)$ (d) $KHSO_3(aq)$
(e) $K_2SO_3(aq)$ (f) $KCN(aq)$

12.35. How is oxygen needed for industrial purposes prepared?

12.36. Give three major uses of oxygen.

12.37. What is the name and formula of the allotrope of oxygen?

12.38. How do ultraviolet light and neutral particles in air work together to convert oxygen to ozone?

12.39. Aluminum oxide is amphoteric. What does that mean? Give equations that illustrate this property.

12.40. List some uses of dilute hydrogen peroxide.

12.41. In what chemical forms does sulfur occur in nature?

12.42. What is the common allotrope of sulfur that is stable at room temperature?

12.43. What is the structure of the S_8 molecule in orthorhombic sulfur?

12.44. What happens structurally in the conversion $S_\alpha \rightarrow S_\beta$?

12.45. What happens to sulfur that has been heated to a temperature equal to or above 159 °C and then poured into cold water?

12.46. What is acid rain? What acids contribute to it and how do they get into rainfall?

12.47. The addition of dilute hydrochloric acid to a white powder that was known to be either $NaHSO_3$ or $NaHSO_4$ produced a gas that had a sharp, choking odor. What gas

formed and what was the solid? Write the molecular and net ionic equations for the reaction.

12.48. The addition of water to a white solid known to be either Na_2O or Na_2O_2 caused a basic solution to form. A gas evolved. When a burning match was thrust into the gas, the flame flared more brightly. What was the solid and what gas formed? Write the molecular equation.

12.49. Write the molecular equations for the individual steps in the *contact process*.

12.50. List the major uses of sulfuric acid in industry.

12.51. What advantages does concentrated sulfuric acid have over other strong acids for industrial applications?

12.52. A solid known to be either $NaHSO_4$ or Na_2SO_4 was dissolved in water. Then some solid Na_2CO_3 was added. The solution fizzed and a gas evolved. What was the gas? What was the original solid? Write a net ionic equation.

12.53. A solid was known to be either FeS or $FeSO_4$. When hydrochloric acid was added to it, a gas evolved with the odor of rotten eggs. What was the gas and what was the original solid. Write an equation.

12.54. What air pollutant causes silver to turn black? Give the name and formula.

12.55. What are the sulfanes and how are they made?

12.56. Give the names and formulas of two *compounds* that can be used to improve the sulfur content of soil.

12.57. Hot, aqueous solutions of what two kinds of sulfur compounds will dissolve elemental sulfur?

12.58. How does the thiosulfate ion dissolve silver bromide? Write a net ionic equation.

12.59. Elemental nitrogen needed for industrial purposes is obtained in what way?

12.60. Most industrial nitrogen is used for what purpose?

12.61. Give the names and formulas of four important nitrogen fertilizers.

12.62. What is nitrogen fixation?

12.63. What are the molecular equations for the Ostwald process?

12.64. After concentrated nitric acid has remained in a bottle exposed to sunlight for some time, its color turns from colorless to reddish brown. What happens (write the equation) to cause that change and what is responsible for the reddish-brown color?

12.65. Give the names of two important nitrogen-containing explosives.

12.66. Give the formal and common name and the formula of the oxide of nitrogen that:
(a) Gives nitric acid when dissolved in water.
(b) Forms N_2O_4 when cooled.
(c) Can be made by heating ammonium nitrate.
(d) Is unstable in air, changing to NO_2.
(e) Forms in an automobile cylinder by a reaction between nitrogen and oxygen.

(f) Is unstable and readily breaks up into two other oxides of nitrogen.
(g) Is a red-brown, poisonous gas.
(h) Gives the same reactions as a mixture of NO and NO_2.
(i) Is used by some auto racing enthusiasts to get more power out of the combustion of fuel.
(j) is paramagnetic. (There are two.)
(k) Is a solid at room temperature.
(l) Is recycled in the Ostwald process.
(m) Forms from the decomposition of nitrous acid in water.
(n) Reacts with aqueous sodium hydroxide to give a mixture of sodium nitrite and sodium nitrate.
(o) Is responsible for the characteristic color of heavy smog.

12.67. What are the names and formulas of two reduced forms of nitrogen?

12.68. When sodium dissolves in liquid ammonia, ammoniated forms of the sodium ion and the electron form. What does "ammoniated" mean in these uses?

12.69. Write molecular equations for the reactions of aqueous ammonia, $NH_3(aq)$, with each of these acids (assumed to be in *dilute* solutions.
(a) $HCl(aq)$ (b) $HBr(aq)$
(c) $HI(aq)$ (d) $H_2SO_4(aq)$, functioning as
(e) $HNO_3(aq)$ a diprotic acid

12.70. In the liquid ammonia system, what ion is the chief proton-donor and what is the chief proton-acceptor?

12.71. How is sodium amide prepared?

12.72. What reaction will the amide ion give with the ammonium ion in a liquid ammonia solution? Write the net ionic equation.

12.73. How is hydrazine prepared? (Write the molecular equation.)

12.74. Why is it particularly dangerous to let household bleach (the "chlorine" type) mix with household ammonia?

12.75. When hydrazine is used as a rocket fuel, what provides the thrust? As part of the answer, write the equation.

12.76. Which solid allotrope of phosphorus has to be stored beneath the surface of water and why?

12.77. Which solid allotrope of phosphorus is the most stable?

12.78. What is the structure of the molecules of phosphorus found in white phosphorus? How does this structure help us understand the high reactivity of white phosphorus in moist air?

12.79. How do red and white phosphorus differ structurally?

12.80. What reaction makes red phosphorus useful in smoke bombs? Write the equation, assuming a plentiful supply of oxygen is available.

12.81. What reactions occur between phosphorus and chlorine (depending on the supply of chlorine)? Write the equations and name the products.

12.82. Write the molecular equations for the reactions with water of (a) PCl_3, (b) PCl_5, (c) PBr_3, and (d) PBr_5.

12.83. Why is tetraphosphorus decaoxide called a powerful desiccating agent?

12.84. Give the names and formulas of the three forms of "phosphoric acid" that can form in the reaction of P_4O_{10} with water.

12.85. Which form (referring to question 12.84) is the most common (and the one whose formula is most frequently used in equations of the reactions of phosphoric acid)?

12.86. Write the formulas of (a) sodium dihydrogen phosphate, (b) potassium phosphate, (c) potassium monohydrogen phosphate, and (d) calcium phosphate.

12.87. Write the names of (a) H_3PO_4, (b) KH_2PO_4, (c) $MgHPO_4$, and (d) Na_2HPO_4.

12.88. What is an acidulant? Give the name and formula of one.

12.89. What are the names and molecular formulas of the halogens?

12.90. What is the order of the relative electronegativities of the halogens?

12.91. How are aqueous hydrohalogen acids, $HX(aq)$, made?

12.92. How does fluorine react with water? (Write the equation.)

12.93. How does hydrogen fluoride etch glass? (Write the equation.)

12.94. What is the principle behind the use of traces of fluoride ion in drinking water as a means of reducing the frequency of tooth decay among children?

12.95. Chlorine occurs in nature principally in what ways?

12.96. How is industrial chlorine made? (Write the equation.)

12.97. What is the reaction of chlorine with water? (Give the equation.)

12.98. How is bleaching powder made? (Write the equation.)

12.99. Give the names and formulas of the three oxides of chlorine.

12.100. What property makes perchloric acid particularly dangerous when it is pure?

12.101. If X and Y are both halogens and the following reaction takes place:

$$X_2 + 2Y^-(aq) \longrightarrow 2X^-(aq) + Y_2$$

(a) Which halogen is oxidized, X or Y?
(b) Which halogen is reduced, X or Y?
(c) If Y is Cl, can X be Br? Explain.
(d) If Y is Br, can X be Cl? Explain.
(e) If Y is I, can X be Br? Explain.

12.102. What is the equation for the reaction of bromine with water?

12.103. Sulfuric acid, if its concentration is controlled, can be used to make hydrogen bromide from sodium bromide. Write the equation for this reaction.

12.104. Referring to question 12.103, why can't concentrated sulfuric acid be used? (Write an equation.)

12.105. How is elemental iodine prepared? (Include an equation in the answer.)

12.106. What is the order of the strengths of the halogens as oxidizing agents?

12.107. The halide ions fall into what order in terms of their strengths as reducing agents?

12.108. What is tincture of iodine?

12.109. Why is the iodide ion important in the diet?

12.110. After prolonged standing on the reagent shelf, the contents of a bottle of hydriodic acid, $HI(aq)$, will become more and more darkly colored. Explain, writing an equation as part of the answer.

12.111 What are the names and formulas of two oxoacids of iodine?

12.112. Give the names and the atomic symbols of the elements in the carbon family.

12.113. What are the two allotropes of carbon and which is more stable at room pressure and temperature?

12.114. Why is graphite slippery?

12.115. How does carbon monoxide work as a poison?

12.116. Iron pentacarbonyl, $Fe(CO)_5$, is *pyrophoric*. What does that mean?

12.117. How is carbon monoxide used in the metallurgical industry? Give two equations as examples.

12.118. What is *lime* or *quick lime* and how is it made?

12.119. What does it mean when cement making is described as an energy-intensive industry?

12.120. Although relatively little of the carbon dioxide in an aqueous solution of this gas is present as carbonic acid, the solution can be treated as if all the carbon dioxide were in this form. Explain, using equations.

12.121. All metal carbonates and bicarbonates react alike with acids such as hydrochloric, sulfuric, and nitric acids. What are the net ionic equations for all these reactions? Write one for carbonates and another for bicarbonates.

12.122. What are the two types of carbides? Give a specific example of each and an equation for its reaction with water.

12.123. If we define the neutralization of an acid as any reaction that will change hydrogen ion, H^+, to anything else (e.g., H_2O, H_2), which of these substances can neutralize hydrochloric acid? For each one that does, write the molecular and net ionic equations.
(a) Zn   (b) ZnO   (c) $Zn(OH)_2$
(d) $Zn(NO_3)_2$ (e) $Zn(HCO_3)_2$ (f) $ZnCO_3$
(g) $ZnSO_3$ (h) $ZnSO_4$

12.124. What are names and atomic symbols of the members of the noble gas family?

12.125. Which noble gases can form compounds? Give the formulas of two examples.

12.126. Describe some uses of the noble gas elements.

12.127. Describe an unusual property of liquid helium.

12.128. What is the formula of each substance?
(a) Bleaching powder.   (b) Carborundum.
(c) Burnt lime.   (d) Water gas.
(e) Ozone.   (f) Nitrous oxide.
(g) Nitric oxide.   (h) Gypsum.
(i) Epsom salt.   (j) Glauber's salt.
(k) Monoclinic sulfur.   (l) Hypo.
(m) Hydrazine.   (n) Phosphate rock.
(o) Dry ice.   (p) Orthophosphoric acid.

CHAPTER THIRTEEN

LARGE MOLECULES

13.1 MACROMOLECULAR SUBSTANCES

Chemical and thermal stability, tensile strength, water tightness, fiber-forming ability, elasticity, and sheen are some of the properties associated with macromolecular substances

Nearly all of the substances we have studied thus far have had relatively low formula weights. Their ions or molecules have generally contained fewer than a hundred atoms. Both in nature and in the world of synthetics, however, there are many substances with formula weights ranging from the thousands to the hundred thousands whose molecules contain hundreds or even thousands of atoms. These extremely large molecules are called **macromolecules.**

You can see macromolecular substances almost everywhere you look. Trees derive their strength from lignins and cellulose, both of which consist of huge

Figure 13.1
No sharp spikes are needed on the shoes when football is played on artificial turf.

Figure 13.2
The availability of strong, super-lightweight synthetic fabrics for tents, packs, sleeping bags, and clothing has made backpacking a sport for thousands of people. This campsite is at the lake at the foot of Mendenhall glacier in Alaska.

molecules. The bones and muscles in your feet and whatever is under them—socks, leather, carpeting, concrete, wood, tile, or the ground itself—include macromolecular substances. A car's enamel paint, its windows and tires, its upholstery and dashboard—all are macromolecular. Your clothes, whether they are made of natural or synthetic fibers, consist of huge molecules. Records, tapes, skis, sails, boat hulls, fishing lines, tents, backpacks, hiking boots, running shoes—all sorts of recreational equipment—are made of molecules that are extra large. Perhaps you know someone with soft contact lenses, or a heart pacemaker, or a reconstructed hip joint, or even a new heart valve. These things are also made either wholly or mostly of macromolecular synthetics. In agriculture, flexible synthetic tubing is used in milking machines, irrigation systems and in the machinery that distributes pest control chemicals. Ultra-thin, flexible polyethylene sheeting with a slit for each plant can be used on rows of crops such as to-

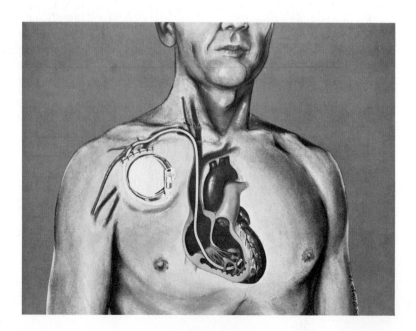

Figure 13.3
Endocardial implant or "pacemaker." The lead wire from the pacemaker is coated with silicone rubber.

Figure 13.4
Strawberry seedlings grow through the slits in this polyethylene film, which protects the plants from both heat and cold, helps the soil conserve moisture, and serves as a barrier to weeds and insect pests.

matoes or strawberries to help retain the soil's moisture and warmth and to retard the growth of weeds.

What is it about molecular size that fits macromolecular substances to so many uses? Is there some change in their chemistry that is the result of their size? To some extent molecular size does affect the *rates* of chemical reactions, but the *nature* of these reactions still depends on the functional groups present. Water, for example, or aqueous acids and bases generally react most rapidly with substances that they can also dissolve, but macromolecular substances are virtually insoluble in water or dilute aqueous solutions. Nevertheless, if the functional group is present, the chemistry of that group is present also, so size has nothing to do with the basic chemical properties of macromolecular substances. For example, the huge molecules in natural rubber have carbon-carbon double bonds. Ozone vigorously attacks these groups, and until tire manufacturers found out how to include antioxidant, ozone-resisting additives, tires cracked and deteriorated quickly in ozone-rich smog. The amide group is only very slowly attacked by dilute acids or bases whether it is in small molecules or in the macromolecules of nylon or proteins. Alkanes have no functional groups and are not attacked by dilute acids or bases or by either oxidizing or reducing agents. Not even concentrated sulfuric acid or sodium hydroxide react with alkanes, and among the alkanelike macromolecular substances are materials highly prized for this chemical inertness—polyethylene and polypropylene, for example. These substances are therefore ideal for use in food containers, disposable hospital items, wastebaskets, indoor-outdoor carpeting, and in parts of artificial heart valves.

Physical properties are the most sought-after features of macromolecular substances. Chemically, they must be as inert as possible to whatever substances make up the environment in which they will be used—the air at normal temperatures and humidity, for example, or food liquids, or body fluids. However, once the desired chemical stability is assured, then the physical properties become the sought-after features of a macromolecular substance. For instance, the Teflon coating on a frying pan must be chemically stable to the air and to food juices at relatively high temperatures. However, the specific quality that makes Teflon so desirable is a physical property—cooked foods don't stick to it. Similarly, it is nice that the digestive juices of moths can't chemically attack nylon,

Teflon coated frying pan

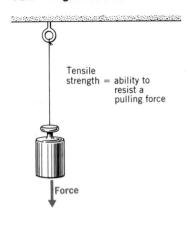

Tensile strength = ability to resist a pulling force

Force

but its superior tensile strength, whether wet or dry, and its attractive appearance—both physical properties—are what make nylon fiber a billion-dollar commodity. Dacron, unlike cotton, doesn't mildew, and Dacron now is the preferred fabric for making sails. The same overall strength as cotton is obtained at less weight, and Dacron stretches much less than cotton.

We have noted tensile strength as being one desirable physical property of macromolecular substances. General thermal stability is also important. The better the substance resists melting or becoming deformed at high temperatures or does not become brittle at low temperatures, the more valuable it becomes. For some uses we want a substance to form tight, water-impervious coatings. In other uses, we want a macromolecular compound to be capable of being drawn into fibers, and sometimes we want materials of great elasticity. Physical properties like these can be correlated with chemical structure, and studying some of these correlations is one goal of this chapter. To pursue this goal we need to learn some general terms.

Polymers

Poly-, many; *mono-*, one; *-meros*, parts.

Broadly classified, there are two kinds of macromolecular substances. In one kind the molecules have no simple feature that occurs over and over again throughout the molecule. Lignin, a material that imparts some of the strength to wood, is an example. We will not study this kind of macromolecular material any further. The other kind include those that consist of molecules in which a structural feature is repeated over and over again, and these are called polymers. **Polymers** are substances consisting of macromolecules that have repeating structural units. A polymer is made from a substance of low formula weight called a **monomer.** The reaction that converts a monomer into a polymer is called **polymerization.** For example, if we let A stand for a molecule of a monomer, then its polymerization can be represented by the following equation in which the coefficient of A, n, is a large number—several hundred to a few thousand.

$$nA \xrightarrow{\text{polymerization}} A—A—A—A—A—A—A—A—A—A—A—A—A—A—A—\text{etc.}$$
the polymer of A

As we discuss in the next sections, polyethylene and polypropylene are made from ethene ("ethylene") and propene ("propylene") in this way. Many useful polymers are actually **copolymers**—polymers made from two (or more) monomers. For example, if we let A and B be molecules with low formula weights that can react with each other, then one of the many ways in which they can copolymerize is as follows, where the coefficients n and m are very large numbers.

$$nA + mB \xrightarrow[\text{(one mode)}]{\text{copolymerization}} A—B—A—B—A—B—A—B—A—B—A—B—A—B—\text{etc.}$$
a copolymer of A and B

Figure 13.5 illustrates other general types of copolymers. As we will see in greater detail in Section 13.4, both nylon and Dacron are copolymers.

Linear copolymers

Alternate —A—B—A—B—A—B—A—B—A—B—A—B—A—B—

Block —A—A—A—B—B—A—A—A—B—B—A—A—A—B—B—

Random —A—B—B—A—A—A—B—A—B—B—A—A—A—B—A—B—

Graft copolymer

Cross-linked copolymer

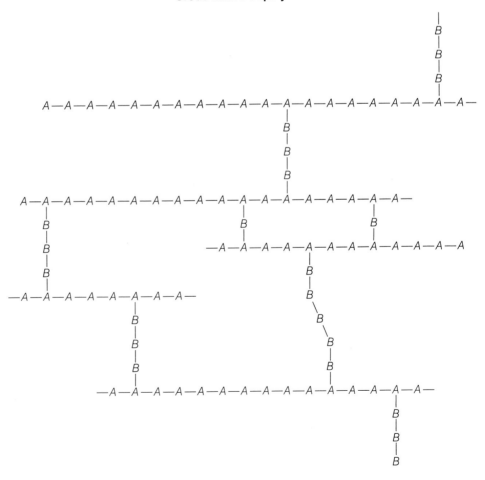

Figure 13.5
Some types of copolymers.

Many polymers occur in nature, for example, proteins, carbohydrates, and nucleic acids. We will study these polymeric substances in Chapter Nineteen, but to indicate how important and widespread they are in nature, consider that proteins make up all muscles, sinews, and the reinforcing networks in bones. The catalysts in the body, the enzymes, are proteins. Cellulose and starch, two carbohydrates, are both polymers of glucose ("dextrose"). Cotton is nearly pure cellulose; starch is used as food by both the plants that make it and animals that eat them. The nucleic acid family includes DNA, the polymeric substance that carries all the chemical information transmitted by heredity. Polymers are not restricted to organic substances. Among inorganic polymers are the silicate minerals found in the earth's crust in both rocks and soil.

Synthetic polymers have become almost as widely occurring as natural polymers. Some replace and help to conserve scarce natural resources. When World War II stopped shipments of silk and rubber to the United States and its allies, the newly developed nylon was able to replace silk and a major research effort was begun to find a way to make synthetic rubber. Today, a material identical with natural rubber can be made from petroleum. Whenever a polymer can be used in place of a metal, not only is the overall cost less but the metal is conserved for uses in which only it can serve.

The word *polymer* is a very broad term that only indicates something about its chemical structure. A **plastic,** on the other hand, is a finished or semifinished article made by molding, casting, extruding, drawing, or laminating a resin. A **resin** is a raw, unfabricated polymer used to make all or most of a plastic article, and it sometimes is made to undergo further chemical modification during the process that makes the plastic item. Some resins are used as protective coatings for surfaces. Alkyd resins, for example, are used as auto paints. Other resins are extruded into plastic rods or fibers, and some are molded into the desired final shape. Many plastic articles contain an inert **filler,** a substance added during polymerization or casting to prevent shrinkage. For example, silica—a form of silicon dioxide—is a filler in some plastics used for restorative work in dentistry, where no shrinkage can be allowed. Oily substances of relatively low formula weight called **plasticizers** are sometimes added to a resin to reduce hardness and brittleness in the final plastic. A plasticizer acts as an internal lubricant, but sometimes it migrates out of the article, especially when it becomes hot. You may have noticed that the steering wheel of a car or the plastic seat covers have a slightly stickier "feel" after the car has been parked for some time in the sun. The "new car smell" includes traces of plasticizers and solvents that are used in making interior components.

13.2 POLYMERIZATION

Linear, graft, and cross-linked polymers are the three major types

The simplest example of a polymerization is the conversion of ethylene to polyethylene.

$$n \, CH_2{=}CH_2 \xrightarrow{\text{polymerization}}$$
$$\underset{\text{ethylene}}{}$$

$$\text{etc.}{-}CH_2{-}CH_2{-}CH_2{-}CH_2{-}CH_2{-}CH_2{-}CH_2{-}CH_2{-}\text{etc.}$$
$$\underset{\text{polyethylene}}{}$$

The coefficient, n, is several hundred. The chain in the polymer continues until all of the n molecules of ethylene are strung out; and on the scale of the structural formula given, the chain would extend about another 20 ft to show one molecule typically found in this polymer. Within a given sample of polyethylene the value of n varies somewhat from molecule to molecule, so this polymer, like all polymers, is a mixture of slightly different molecules. However, all the molecules in any given polymer share the same basic structural features. To conserve space, a condensed structure is normally used to represent a polymer. For example, polyethylene is written as

$$-(CH_2-CH_2)_n-$$
polyethylene

It could have been written as $-(CH_2)_{2n}-$, but we want a condensed structure that indicates the particular parts of the polymer that were contributed by the monomer. The monomer for polyethylene has two carbons per molecule, not one.

Thus far we have not said what comes at the ends of the polyethylene chains. They are filled with pieces of the molecules of the initiator of polymerization. An **initiator** is like a catalyst, because it helps to promote polymerization, but unlike a catalyst it is chemically used up during the reaction. The fragments of the initiator molecules that are at the ends of the polymer chains are usually not shown because they make such a minor contribution to the overall structure. The initiator works by causing the reorganization of chemical bonds so that new bonds *between* the monomer molecules can form. We will not pursue the details of how all initiators work, but some general principles will help to show how they can get bonds to reorganize. Some initiators act by using a pair of electrons taken from one of the two bonds of the double bond to form a new bond and leave a positive charge on carbon. Let's use the symbol *In* to stand for the initiator; it might be in the form of In^+, or $In\cdot$, or $In:^-$. If it is a cation, In^+, its reaction with ethylene is as follows.

$$In^+ + CH_2=CH_2 \longrightarrow In-CH_2-CH_2^+$$

All this does is shift the positive charge from the initiator to carbon, but the new product is still very reactive, and it attacks more $CH_2=CH_2$.

$$In-CH_2-CH_2^+ + CH_2=CH_2 \longrightarrow In-CH_2-CH_2-CH_2-CH_2^+$$

The product is still highly reactive, and in the polymerizing mixture it has ample numbers of unpolymerized ethylene molecules around it for further reaction.

$$In-CH_2-CH_2-CH_2-CH_2^+ + CH_2=CH_2$$
$$\longrightarrow In-CH_2-CH_2-CH_2-CH_2-CH_2-CH_2^+$$

The chain continues to be extended by two-carbon units until by a random encounter with a negative ion the process stops.

$$In-CH_2-CH_2-(CH_2-CH_2)_{\overline{(n-2)}}CH_2-CH_2^+ + An^- \longrightarrow In-(CH_2-CH_2)_n-An$$
some
anion

If the initiator is of the type $In\cdot$ and has an unpaired electron, it will act on ethylene to take *one* electron from the double bond and leave one unpaired electron on carbon.

$$In\cdot + CH_2=CH_2 \longrightarrow In-CH_2-CH_2\cdot$$

The chain then grows by a similar reaction.

$$In-CH_2-CH_2\cdot + CH_2=CH_2 \longrightarrow In-CH_2-CH_2-CH_2-CH_2\cdot$$

In the next step a third ethylene unit will be added, and the chain will grow until a random encounter with $In\cdot$ or some similar species closes the chain.

$$In-CH_2-CH_2-(CH_2-CH_2)_{\overline{(n-2)}}CH_2-CH_2\cdot + In\cdot \longrightarrow In-(CH_2-CH_2)_n-In$$

Depending on how the conditions are controlled under which polymerization takes place, the growing chains can sometimes react with each other to transfer the site of chain-growth from the end of a molecule to some place down the chain. For example, when the growing chain has an unpaired electron at its end, it might react as follows.

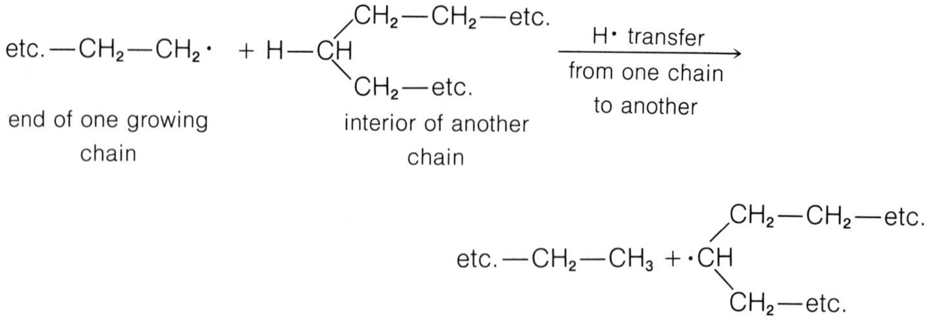

$$\underset{\substack{\text{end of one growing} \\ \text{chain}}}{\text{etc.}-CH_2-CH_2\cdot} \ + \ \underset{\substack{\text{interior of another} \\ \text{chain}}}{H-CH\!\!\begin{array}{c} {}^{\displaystyle CH_2-CH_2-\text{etc.}} \\ {}_{\displaystyle CH_2-\text{etc.}} \end{array}} \quad \xrightarrow[\substack{\text{from one chain} \\ \text{to another}}]{H\cdot \text{ transfer}}$$

$$\text{etc.}-CH_2-CH_3 + \cdot CH\!\!\begin{array}{c} {}^{\displaystyle CH_2-CH_2-\text{etc.}} \\ {}_{\displaystyle CH_2-\text{etc.}} \end{array}$$

Now there is a chain that can add ethylene units as a branch. For example,

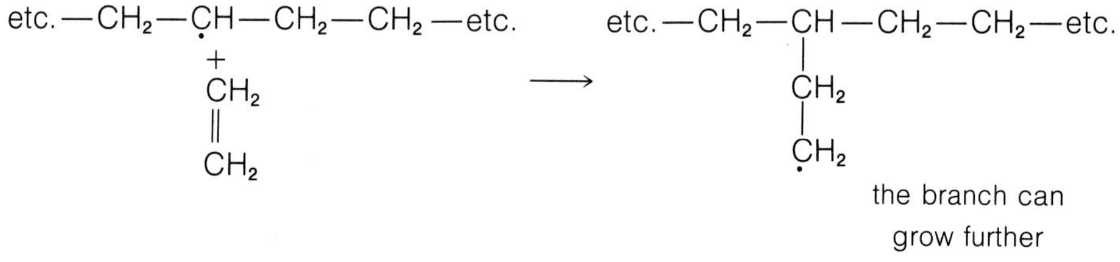

$$\underset{\substack{\displaystyle + \\ \displaystyle CH_2 \\ \displaystyle \| \\ \displaystyle CH_2}}{\text{etc.}-CH_2-\overset{\displaystyle \cdot}{CH}-CH_2-CH_2-\text{etc.}} \qquad \longrightarrow \qquad \underset{\substack{\displaystyle CH_2 \\ \displaystyle | \\ \displaystyle \overset{\displaystyle \cdot}{CH_2}}}{\text{etc.}-CH_2-CH-CH_2-CH_2-\text{etc.}}$$

the branch can
grow further

In terms of a general symbol we can represent a branched polymer as follows.

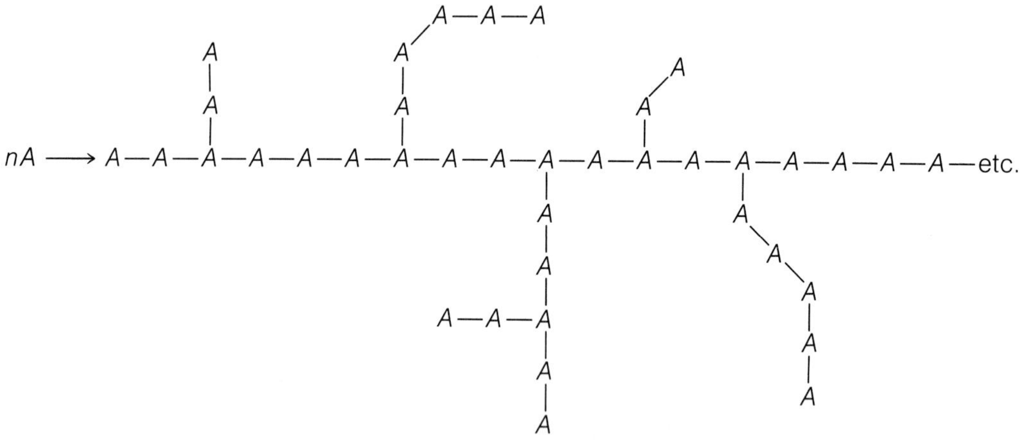

branched polymer

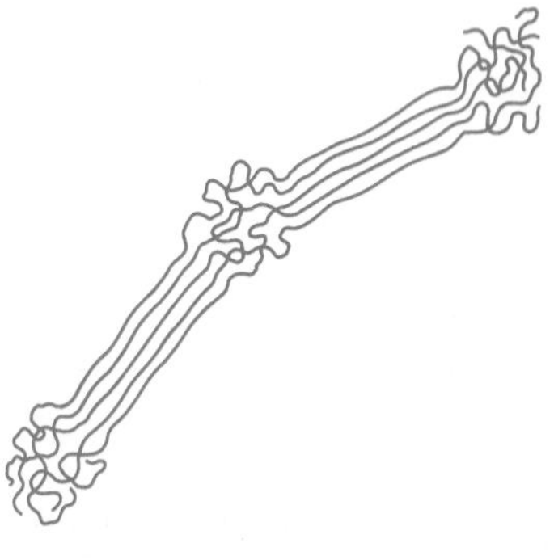

Figure 13.6
When a polymer forms and solidifies, the molecular chains will be lined up in a very orderly manner in some regions but in disordered tangles in other places.

Whether the molecules are linear or branched makes an important difference in how the polymer can be used in a finished article. The unbranched form of polyethylene, for example, has greater mechanical strength and a higher density than branched polyethylene. Its molecules can assemble much more closely together in the solid state than can highly branched molecules, much as trimmed logs can be stacked more neatly than untrimmed ones. However, polymer molecules do not become perfectly stacked together in the solid state in the way that sodium ions and chloride ions, for example, become arranged in a highly orderly manner in crystalline sodium chloride. Polymer chemists describe various polymers as having regions of crystallinity and regions of random, disordered arrangements, as illustrated in Figure 13.6. The regions of random arrays of molecules help to keep the finished plastic somewhat flexible and able to resist being shattered when struck or dropped.

Polypropylene is another important polymer. We will use it to illustrate further how polymer structures can be analyzed for recognizing their monomers.

EXAMPLE 13.1 Identifying Repeating Units in Polymers

Problem: Examine the structure of polypropylene, commonly used in indoor-outdoor carpeting, and identify the repeating units. Deduce the structure of the monomer.

$$\text{etc.} -CH_2-CH-CH_2-CH-CH_2-CH-CH_2-CH-CH_2-CH-\text{etc.}$$
$$\qquad\quad | \qquad\quad\, | \qquad\quad\, | \qquad\quad\, | \qquad\quad\, |$$
$$\qquad\quad CH_3 \qquad CH_3 \qquad CH_3 \qquad CH_3 \qquad CH_3$$

Solution: Notice the regularity; *every other carbon* has a CH_3 sidechain. To make sure that every monomer is alike, first draw a wavy line through every other carbon-carbon single bond. Between those lines are the repeating units:

$$-CH_2-CH-$$
$$\qquad\quad |$$
$$\qquad CH_3$$

$$-CH_2-CH \gamma CH_2-CH \gamma CH_2-CH \gamma CH_2-CH \gamma CH_2-CH-$$
$$\qquad\quad | \qquad\qquad | \qquad\qquad | \qquad\qquad | \qquad\qquad |$$
$$\qquad CH_3 \qquad\quad CH_3 \qquad\quad CH_3 \qquad\quad CH_3 \qquad\quad CH_3$$

Finally, use the *name* of the polymer as a guide to the name and the structure of the monomer. Each monomer unit has three carbons; it must come from propylene, $CH_2{=}CH$
$$\qquad\qquad\qquad\qquad\qquad\qquad\qquad\qquad\qquad\quad |$$
$$\qquad\qquad\qquad\qquad\qquad\qquad\qquad\qquad\qquad CH_3$$

EXERCISE 13.1 Write the structure of the monomer used to make Orlon®, a polymer used to make fibers for sweaters and other garments. The monomer has one carbon-carbon double bond. (Note, CN means $C{\equiv}N$, a cyano group.)

$$\qquad\quad CN \qquad\quad CN \qquad\quad CN \qquad\quad CN \qquad\quad CN$$
$$\qquad\quad | \qquad\qquad | \qquad\qquad | \qquad\qquad | \qquad\qquad |$$
$$\text{etc.} -CH_2-CH-CH_2-CH-CH_2-CH-CH_2-CH-CH_2-CH-\text{etc.}$$
$$\qquad\qquad\qquad\qquad\qquad\qquad\quad \text{Orlon}$$

13.3 ADDITION POLYMERS

Polyolefins, polyacrylates, and epoxy resins are all addition polymers

Addition polymers are those that form without their monomers losing any parts of their molecules during the polymerization reaction. Polyethylene, which we discussed in the preceding section, is the simplest example of an addition polymer. All that happens when ethylene polymerizes is that a pair of electrons in each molecule is used to make a new bond to another molecule. None of the atoms originally in the ethylene molecule is lost.

Polypropylene (Example 13.1) is another addition polymer. Unlike linear polyethylene, however, linear polypropylene has regularly spaced methyl groups along its entire chain. By the choice of the initiator, the polymerization of propylene can be directed to make all of these methyl groups project on the same side of the main chain (Figure 13.7), or make them project in alternating directions, or in random directions. The first form (Figure 13.7) makes the best fibers for indoor-outdoor carpeting.

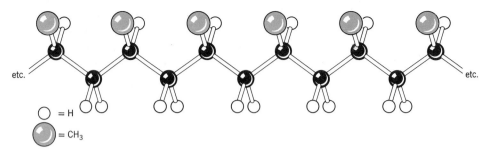

Figure 13.7
Propylene can be polymerized so that all of the side-chain groups, CH_3, project on the same side of the main chain.

Strong, flexible fibers can be made from resins whose molecules are not only very long but also have shapes that let the molecules align side by side, overlap end to end, twist into cables and intertwine to make fibers (see Figure 13.8). With polypropylene, this aligning and overlapping occurs best with the form illustrated in Figure 13.7 where the molecules are highly symmetrical. Polypropylene has an excellent flex life, which means that it is not easily weakened by bending. This property makes polypropylene very valuable in flexible parts of artificial heart valves. It is also used in making knitted surgical mesh, blood filters, and devices for hip joint repair for which both strength and chemical inertness are essential.

Figure 13.9
Foamed polystyrene is very commonly used to insulate buildings.

Polymer molecules overlap   Overlapping molecules twist together to form cables   Cables intertwine to form fibers

Figure 13.8
When molecules of polymers are very long and symmetrical about their axis, they can overlap and twist into cablelike systems that, in turn, can intertwine to give strong fibers.

Since members of the alkene family, like ethylene and propylene, have long been nicknamed *olefins,* polyethylene and polypropylene are also called **polyolefins.** A number of other important polyolefins are made from substituted ethylenes, alkenes whose molecules have some group other than a hydrogen atom at the double bond. (Propylene is a substituted ethylene, for example.) Polystyrene is made from styrene—phenylethene—and if an inert gas is blown into the polymerizing mixture the product is the familiar polystyrene foam (Figure 13.9).

$$n\,CH{=}CH_2 \longrightarrow \text{etc}-CH-CH_2-(CH-CH_2)_{\overline{n}}\,\text{etc.}$$

phenyl
group   styrene   polystyrene

Notice the regularity in its structure; the phenyl groups appear on alternate carbons. It is as if all of the styrene molecules face in the same direction at the moment that each is added to the growing chain.

Polyvinyl chloride (PVC) is the principal resin used to make clear, hard plastic bottles and plastic pipes. The monomer, vinyl chloride, unfortunately causes a rare form of liver cancer. Rigid standards of air quality for the working environment of people handling vinyl chloride are now in force.

$$CH_2{=}CH{-}Cl \qquad F_2C{=}CF_2$$

vinyl chloride tetrafluoroethene
(*chloroethene*)

The polymer of tetrafluoroethylene is called Teflon, a polymer of exceptional stability. Only molten sodium and potassium attack it. Its nonsticking and self-lubricating qualities not only make it valuable for coating cooking pots and frying pans but also for making surgical tubes and replacements for damaged arteries.

Figure 13.10
PVC or polyvinyl chloride is used to make hard, clear plastic bottles.

EXAMPLE 13.2 **Writing the Structure of an Addition Polymer**

Problem: Write the structure of polyvinyl chloride showing five repeating units. Then write its condensed structure. Assume that the same kind of regularity occurs as we described for polystyrene.

Solution: When any substituted alkene forms an addition polymer, one bond of the double bond in each molecule "swings out" to provide the bond to the adjacent monomer molecule. Since we expect the polymer structure to be regular like that of polystyrene, we start by writing down the required number of monomer structures all "facing" in the same way, but we omit the second bond of the double bond. We can write the side chain, which is only —Cl in this example, as projecting either up or down; it doesn't matter here.

$$\underset{}{CH_2{-}\overset{\displaystyle Cl}{|}\overset{}{CH}} \quad CH_2{-}\overset{\displaystyle Cl}{|}CH \quad CH_2{-}\overset{\displaystyle Cl}{|}CH \quad CH_2{-}\overset{\displaystyle Cl}{|}CH \quad CH_2{-}\overset{\displaystyle Cl}{|}CH$$

Next put in the bond that forms between the units:

$$-CH_2{-}\overset{\displaystyle Cl}{|}CH{-}CH_2{-}\overset{\displaystyle Cl}{|}CH{-}CH_2{-}\overset{\displaystyle Cl}{|}CH{-}CH_2{-}\overset{\displaystyle Cl}{|}CH{-}CH_2{-}\overset{\displaystyle Cl}{|}CH{-}\text{etc.}$$

polyvinyl chloride

The condensed structure is

$$-(CH_2{-}\overset{\displaystyle Cl}{|}CH)_n-$$

EXERCISE 13.2 Show five repeating units of Teflon, the polymer of tetrafluoroethylene. Also write its condensed structure.

EXERCISE 13.3 The monomer of Kel-F, fluorothene, is trifluorochloroethene:

$$\underset{F}{\overset{F}{\diagdown}}C{=}C\underset{Cl}{\overset{F}{\diagup}}$$

Show five repeating units of Kel-F and then write its condensed structure. Kel-F is used as a stain and water repellent on upholstery, drapes, and fabrics for clothing.

Figure 13.11
This lightweight Lucite® panel offers almost the abrasion resistance of glass at far less the mass.

Acrylates

Acrylic acid and several closely related compounds are the monomers for a large family of addition polymers called the acrylates. So susceptible are some of these monomers to polymerizing that they are difficult to handle and store. We learned about the polymer of acrylonitrile in Exercise 13.1. The polymer of meth-

$$CH_2=CH-CO_2H \qquad CH_2=CH-CO_2CH_3 \qquad CH_2=CH-C\equiv N$$

acrylic acid   methyl acrylate   acrylonitrile

$$\underset{\text{methyl } \alpha\text{-cyanoacrylate}}{CH_2=\overset{\overset{\displaystyle C\equiv N}{|}}{C}-CO_2CH_3} \qquad \underset{\substack{\text{methacrylic}\\\text{acid}}}{CH_2=\overset{\overset{\displaystyle CH_3}{|}}{C}-CO_2H} \qquad \underset{\substack{\text{methyl}\\\text{methacrylate}}}{CH_2=\overset{\overset{\displaystyle CH_3}{|}}{C}-CO_2CH_3}$$

acrylic acid is used as the hardening agent for epoxy glue. (These are the glues that come in *two* tubes.) The polymer of methyl methacrylate is used to make Plexiglas® and Lucite® plastic items. Methyl α-cyanoacrylate is a very unusual monomer because the initiator for its polymerization is water. All it takes is the moisture in the air or the moisture film on the surfaces of the articles being glued to set off a very rapid polymerization. This is why this monomer, called "super glue," can so swiftly glue your fingers together if you are careless in handling it. (You should always have some fingernail polish remover ready for ungluing your fingers if you have an accident in using super glue. And when you use a pin to puncture the tube, be sure that it is pointing away from you. In one reported accident, some super glue spurted into the user's eye gluing it shut!)

Figure 13.12
Three drops of cyanoacrylate adhesive ("superglue") form the only bond between two three-inch-in-diameter steel cylinders in the cable that supports this 1700-kg tractor.

EXERCISE 13.4   Show five repeating units of the structures of each of the following polymers. Assume that their structures have regularly spaced side chains.
(a) Poly(methacrylic acid).
(b) Poly(methyl methacrylate).
(c) Poly(methyl α-cyanoacrylate).

Elastomers

Polymers that are elastic like rubber are called **elastomers.** Natural rubber is a polymer of isoprene (2-methyl-1,3-butadiene). Isoprene can now be made from petroleum, and initiators have been found that induce it to polymerize with exactly the same geometry at the double bonds as in natural rubber.

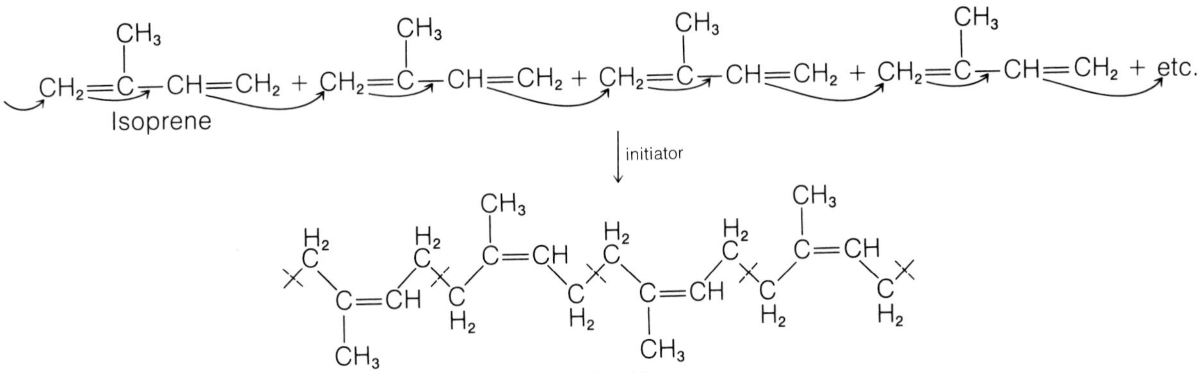

natural rubber
(Note the isoprene units between the dashed lines.)

Rubber and other elastomers can be made tougher and more resistant to abrasion through the introduction of cross-links into the polymer structure. The process, called *vulcanization* (after Vulcan, the Roman god of fire), involves the addition of certain chemicals, such as sulfur, to the heated polymer.

13.4 CONDENSATION COPOLYMERS

Polyamides and polyesters are typical condensation copolymers

A **condensation copolymer** is any polymer whose formation involves the splitting out of small molecules from the functional groups of the monomers. Suppose that *a* and *b* are two functional groups that can react with each other, but in reacting they split out as *a-b* from their original locations. For example, if *a* is —OH in a carboxyl group (—C(=O)—O—H) and *b* is —H in an amino group (NH_2—) or an alcohol group (—O—H), the groups with *a* and *b* can react as follows. (One product will be of the *a-b* type, H—OH.)

$$\underset{\text{carboxylic acid}}{R{-}\overset{O}{\overset{\|}{C}}{-}O{-}H} + \underset{\text{amide}}{H{-}NH{-}R'} \longrightarrow \underset{\text{amide}}{R{-}\overset{O}{\overset{\|}{C}}{-}NH{-}R'} + H{-}OH$$

$$R{-}\overset{O}{\overset{\|}{C}}{-}O{-}H + \underset{\text{alcohol}}{H{-}O{-}R'} \longrightarrow \underset{\text{ester}}{R{-}\overset{O}{\overset{\|}{C}}{-}O{-}R'} + H{-}OH$$

Reactions in which a small molecule such as H—OH split out are sometimes called condensation reactions. To use these reactions to make polymers, all we need are molecules with two or more functional groups. For example, a linear condensation copolymer forms if we use a molecule such as *a—X—a* and one such as *b—Y—b*:

$$a\text{-}X\text{-}a + b\text{-}Y\text{-}b + a\text{-}X\text{-}a + b\text{-}Y\text{-}b + a\text{-}X\text{-}a + b\text{-}Y\text{-}b + \text{etc.} \longrightarrow$$

etc.-X-Y-X-Y-X-Y-X-Y-etc. + *m a-b*

where *m* is a number whose size depends on how many monomers react. If one of the monomers has three functional groups instead of two, cross-links can form.

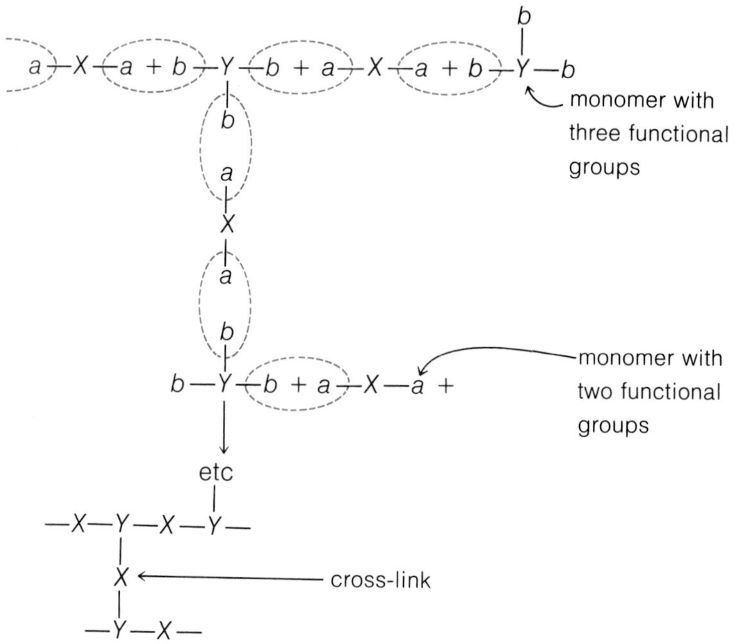

One of the largest families of condensation copolymers are the polyamides. Proteins, which we will study in Chapter Nineteen, are a naturally occurring family of polyamides. Among the synthetic polyamides the family of nylons includes the most famous. *Nylon* is a coined named. It applies to any member of the family of synthetic, long-chain, fiber-forming polyamides. The most common example is nylon-66, so numbered because each monomer molecule has six carbon atoms. We can visualize its formation from a dicarboxylic acid, adipic acid, and a diamine, hexamethylenediamine, as follows.

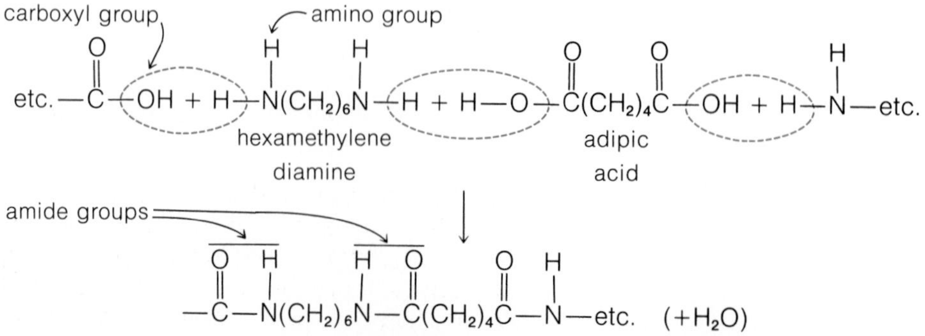

nylon-66

In order for nylon-66 fibers to be sufficiently strong, its molecules have to come from 50 to 90 molecules of *each* monomer. The long polyamide molecules in nylon-66 have no side chains to inhibit their lining side by side and overlapping in the lengthwise direction. Each molecule has several dozen regularly spaced amide groups, and between molecules there are therefore a large number of hydrogen bonds:

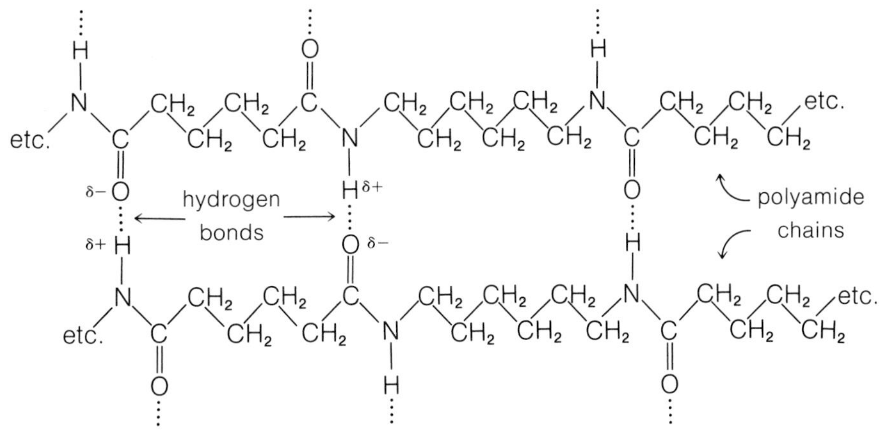

Although each hydrogen bond is a weak force of attraction, the total attractive force between molecules that are sufficiently long is great. An analogy is a zipper; each unit is very weak but all together they are very strong.

Another important family of condensation copolymers are the polyesters. Alkyd resins used for paints and car enamels are examples of very durable polyesters. Dacron, however, is perhaps the most well known of the polyesters. It is a linear, fiber-forming substance made from a dicarboxylic acid, terephthalic acid, and a dihydroxy alcohol, ethylene glycol:

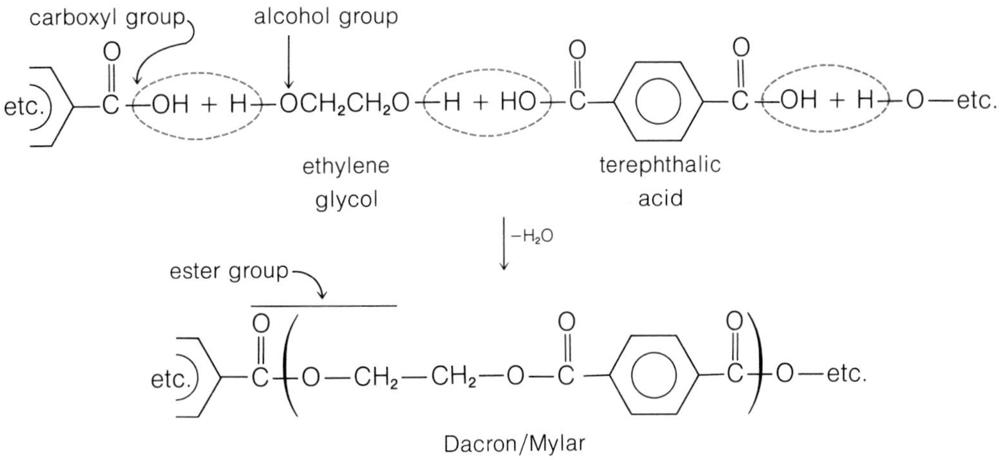

Dacron/Mylar

Dacron has most of the desirable properties of nylon. Garment makers prize it because fabrics made of Dacron can be set into permanent creases and pleats. Dacron has also been used to make parts for repairing or replacing segments of blood vessels. When Dacron is cast as a thin film, it is called Mylar®. Its exceptional resistance to tearing makes Mylar a choice material in the manufacture of both adhesive tapes and recording tapes.

Figure 13.13
Mylar® polyester film covers the body and wings of the Gossamer Albatross, the first human-powered, heavier-than-air, propeller-driven airplane to fly successfully for any appreciable distance. On June 12, 1979, Bryan Allen pedaled and piloted this airplane across the English Channel from England to France. The craft's designer, Paul MacCready, selected Mylar because it is extremely durable, it has a high resistance to tearing, and it does not deteriorate significantly with age, moisture, or temperature extremes.

13.5 INORGANIC POLYMERS

Polymers whose molecular chains are made of atoms other than carbon are usually much less flammable than organic polymers and they often can hold up better under temperature extremes

The organic polymers we have just studied have certain drawbacks. All are flammable, for example, and those containing nitrogen, such as polyacrylonitrile, give off poisonous hydrogen cyanide gas when they burn. In addition, organic polymers generally become inflexible, brittle, and inelastic at low temperatures. At higher temperatures, they deteriorate by slow oxidation. Organic solvents make organic polymers swell up. Only a few organic polymers are truly compatible with living tissue, and therefore only a few can be used to make devices to be implanted in the body. Many valuable inorganic polymers do not have these problems.

Inorganic polymers are those whose molecular "backbones" consist of atoms other than carbon atoms. In many, however, carbon-containing side chains are present.

Silicones

The synthetic inorganic polymers most widely used today are members of the family of silicones. In the silicones the molecular backbones are alternating atoms of silicon and oxygen. Each silicon atom also holds two organic groups, such as methyl groups. Silicone oils, gums, and rubber can be made by varying the starting materials.

Silicone oils are made by copolymerizing dichlorodimethylsilane, $(CH_3)_2SiCl_2$, and water in the presence of a trace amount of chlorotrimethylsilane as a chain-limiting agent.

$$CH_3-\underset{\underset{CH_3}{|}}{\overset{\overset{CH_3}{|}}{Si}}-Cl \;+\; Cl-\underset{\underset{CH_3}{|}}{\overset{\overset{CH_3}{|}}{Si}}-Cl \;+\; Cl-\underset{\underset{CH_3}{|}}{\overset{\overset{CH_3}{|}}{Si}}-Cl \;+\; \cdots\cdots \;+\; Cl-\underset{\underset{CH_3}{|}}{\overset{\overset{CH_3}{|}}{Si}}-CH_3$$

$$\overset{H\quad H}{\underset{O}{\diagdown\;\diagup}} \qquad \overset{H\quad H}{\underset{O}{\diagdown\;\diagup}} \qquad \overset{H\quad H}{\underset{O}{\diagdown\;\diagup}} \qquad \overset{H\quad H}{\underset{O}{\diagdown\;\diagup}}$$

chlorotri-
methylsilane   dichlorodimethylsilane

$$CH_3-\underset{\underset{CH_3}{|}}{\overset{\overset{CH_3}{|}}{Si}}-O-\underset{\underset{CH_3}{|}}{\overset{\overset{CH_3}{|}}{Si}}-O\left(\!\underset{\underset{CH_3}{|}}{\overset{\overset{CH_3}{|}}{Si}}-O\!\right)_{\!n}\underset{\underset{CH_3}{|}}{\overset{\overset{CH_3}{|}}{Si}}-CH_3 \quad (+HCl)$$

silicone oil

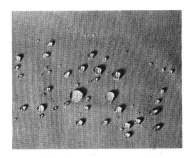

Figure 13.14
Water forms tight beads on a silicone-treated fabric, yet the fabric can still breathe.

Figure 13.15
A silicone board is all that separates the hand from the intense heat of the blowtorch.

If the chains are kept short enough, the product will be a liquid over a wide range of temperatures. It will be just as free-flowing at low temperatures as at room temperature, and it won't char or decompose at higher temperatures. These properties make silicon oils ideal for use as lubricants, motor oils, and hydraulic fluids in machines that must operate under extreme conditions.

The silicones do not react with water. In fact, they actually repel water. Water on a silicone-treated fabric or surface forms into tight beads and will not go through the pores between the threads of the material. Yet a silicone-treated fabric is still able to "breathe." Silicone oils are also used as water-repellent films on concrete surfaces. In medicine, silicone oils have been used as lubricants for treating burns and to augment soft tissue following breast surgery. Some silicone oils are even used in hand creams.

When the basic silicone chains are allowed to grow long enough, the product is a solid called silicone gum. Its chains can be cross-linked to make silicone rubber, a material that can withstand the heat of a blowtorch without melting or suffering any other damage.

Phosphazenes

The phosphazenes are a family of polymers whose molecules have alternating phosphorus and nitrogen atoms in the main chains.

$$-\underset{\underset{X}{|}}{\overset{\overset{X}{|}}{P}}=N-\underset{\underset{X}{|}}{\overset{\overset{X}{|}}{P}}=N-\underset{\underset{X}{|}}{\overset{\overset{X}{|}}{P}}=N-\underset{\underset{X}{|}}{\overset{\overset{X}{|}}{P}}=N-\underset{\underset{X}{|}}{\overset{\overset{X}{|}}{P}}=N-\underset{\underset{X}{|}}{\overset{\overset{X}{|}}{P}}=N- \quad\text{or}\quad \left(\!\underset{\underset{X}{|}}{\overset{\overset{X}{|}}{P}}=N\!\right)_{\!n}$$

phosphazines

The sidechains (X) may be any of a large variety of organic groups, and they need not be identical. When the side chain is the group, $-O-CH_2-CF_3$, the phosphazine is more strongly water-repellent than the silicones. Moreover, it will not react with living tissue, which makes it an excellent candidate for tubes used to replace blood vessels. The phosphazines are generally nonflammable. Some retain their flexibility at temperatures as low as $-90\,°C$. Some phosphazines are elastomers, and unlike rubber, they do not swell when they come in contact with organic solvents.

Polyphosphates

The polyphosphates are not really polymers because their chain lengths are very short. Their backbones are alternating phosphorus and oxygen atoms, and each phosphorus also holds two other oxygens. Sodium triphosphate (sometimes called sodium tripolyphosphate) is the chief "phosphate" in laundry detergents, but many areas have banned the use of detergent phosphates.

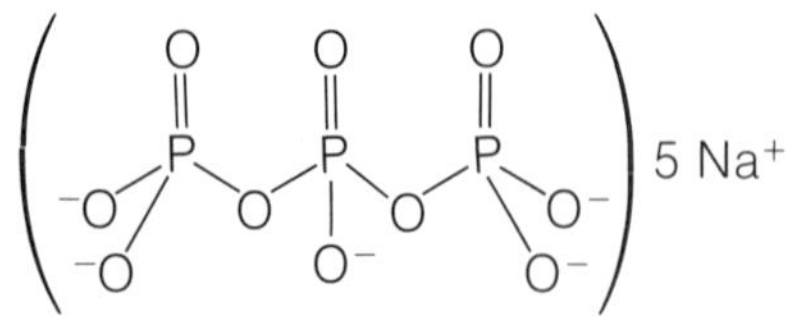

Na$_5$P$_3$O$_{10}$	Na$_3$P$_3$O$_9$	[Na$_3$P$_3$O$_9$]$_y$
sodium triphosphate	sodium metaphosphate	sodium metaphosphate
	(soluble form)	*(insoluble form)*

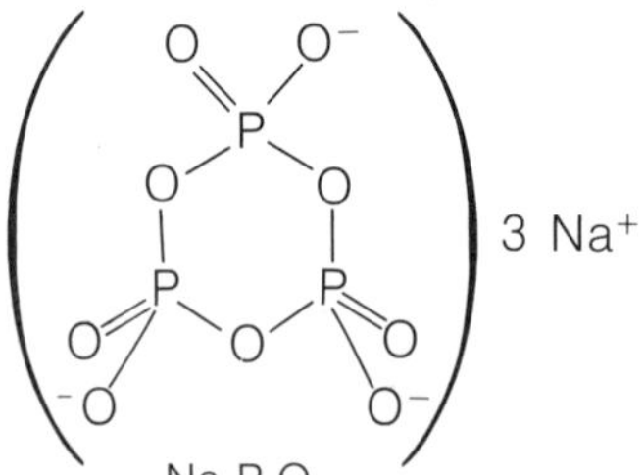

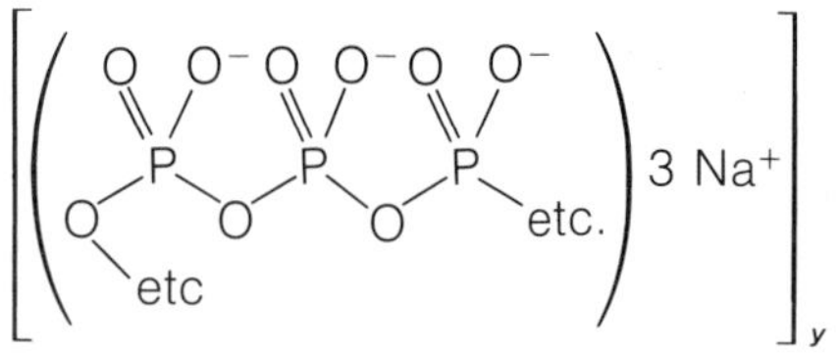

Phosphate-rich waste water entering nearby lakes promotes the growth of algae, and the lakes become unfit for recreational purposes. It is far easier (meaning far less costly) to ban detergent phosphates than to try to remove phosphates from waste water.

Sodium metaphosphate occurs chiefly in two forms, a water-soluble cyclic form and a water-insoluble linear structure. The former is used as a water softener. Sodium tripolyphosphate is also a water softener. Hard water is water in which ordinary soap does not readily form suds. Common soap consists of a mixture of the sodium salts of long-chain carboxylic acids. Sodium stearate is a typical example.

$$CH_3CH_2CH_2CH_2CH_2CH_2CH_2CH_2CH_2CH_2CH_2CH_2CH_2CH_2CH_2CH_2CH_2CO_2^-Na^+$$
$$\text{sodium stearate } [CH_3(CH_2)_{16}CO_2^-Na^+]$$

Although the sodium and potassium salts of these acids are soluble in water, their calcium, magnesium and iron salts are not. Hard water contains the ions of these metals which have been leached from soils and rocks by water as it moves through. The calcium ion will react with the stearate ion as follows:

$$2CH_3(CH_2)_{16}CO_2^-(aq) + Ca^{2+}(aq) \longrightarrow [CH_3(CH_2)_{16}CO_2]_2Ca(s)$$

stearate ion	calcium ion	calcium stearate
(in soap)	*(in hard water)*	*(insoluble soap "scum")*

Soft water has only 0 to 25 mg Ca^{2+}/liter, whereas very hard water has over 75 mg Ca^{2+}/liter.

The regularly spaced negative charges on the soluble metaphosphate or the triphosphate ions can bind Ca^{2+}, Mg^{2+}, and Fe^{2+} or Fe^{3+} ions more strongly than can the negative ions from soap. However, when held by phosphate systems, these metal ions do not form water-insoluble compounds. Instead the complexes are electrically charged, hydrated by water molecules, and kept in solution.

$$P_3O_{10}^{5-}(aq) + Ca^{2+}(aq) \longrightarrow CaP_3O_{10}^{3-}(aq)$$

triphosphate ion		complex of a calcium ion and a triphosphate ion

Thus the ions are kept from forming scums that interfere with the desired cleaning action of the soap.

13.6 SILICATES

Much of the rocks and soils of the earth's crust consists of silicon-oxygen tetrahedra arranged and joined in various polymeric ways

Macromolecular substances clothe us, house us, feed us, and enhance our recreational lives. They also make up most of the earth's crust. As we learned in the previous chapter, just 10 elements account for virtually 100% of the composition of the earth's crust. In Table 12.4 we saw that oxygen and silicon alone make up 82% of the atoms in the crust and 75% of the mass. And nearly 94% of the volume of the rocks and minerals in the crust is taken up by oxygen atoms alone. These atoms are held in place principally by bonds to silicon atoms in the form of various silicate minerals. Before discussing their structures, we need to distinguish among some terms. **Rocks** are complex mixtures of minerals. **Minerals** are specific solids in the earth's crust that have a more or less definite stoichiometry like individual compounds. An **ore** is a mineral that can be extracted and processed at a profit. The silicate minerals occur largely in igneous rocks, those that once were hot, thick fluids like lava. Rocks are changed to soil by weathering brought about by both chemical and physical agents and by the introduction of the remains of organic substances such as the remains of plants and animals. Although the bulk of any useful soil consists of silicate minerals, the quality of the soil depends heavily on the nature of the trace metal elements that are also present.

The silicate minerals are a large family that includes quartz, one of the purest of the silicates, as well as various feldspars, micas, and hornblendes. Table 13.1 gives the typical compositions of the major silicate minerals. In all of them, the dominant structural feature is the silicon-oxygen tetrahedron illustrated in Figure 13.16. Four relatively large oxygen atoms surround each silicon atom and are held to it by *single* hands. One single bond, of course, does not use up the bond-forming capacity of an oxygen atom. Therefore, an oxygen atom in one tetrahedron can extend another bond to a different silicon atom, one that is part of a neighboring tetrahedron. In fact, the tetrahedra *share* oxygen atoms. It is not possible to tell where one tetrahedron ends and the next starts. An analogy is a neatly stacked pile of oranges (oxygen atoms) at a supermarket. Each regularly spaced, unfilled cavity between the oranges could hold a small cherry (a silicon atom). Thus silicon-oxygen tetrahedra extend in three directions in a vast macromolecular network of great strength.

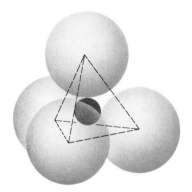

Figure 13.16
A silicon oxygen tetrahedron. One silicon atom nestles between four oxygen atoms.

Table 13.1
Chief Silicate Minerals

Name	Composition	Density (g/cm³)
Quartz	SiO_2	2.6
Feldspars		
Orthoclase	$(Na,K)AlSi_3O_8$	2.6
Plagioclase		
Sodium-rich	$NaAlSi_3O_8$	2.6
Calcium-rich	$CaAl_2Si_2O_8$	2.8
Micas	Complex silicates with Al, K, Mg, Fe and H_2O	2.9
(Biotites)		
Hornblendes	Complex silicates with Al, Ca, Mg and Fe	3.2
(Amphiboles)		
Augite	Complex silicates with Al, Ca, Mg and Fe	3.3
(Pyroxenes)		
Olivine	$(Mg,Fe)_2SiO_4$	3.3

Data from Arthur N. Strahler. *Planet Earth: Its Physical Systems Through Geologic Time,* 1972. Harper & Row, Publishers, Inc.

Every so often, there might be a metal ion other than silicon, a hydroxide ion, a hydrogen ion, or even a water molecule (as water of hydration) in one of the cavities. The degree to which other ions or water molecules are incorporated varies over a wide range from mineral to mineral. This explains, on the one hand, how

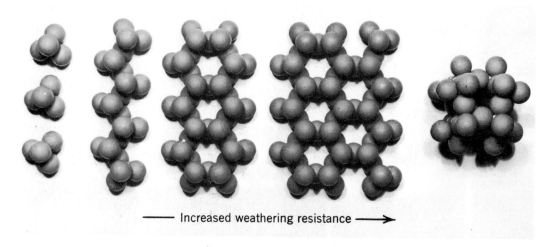

Increased weathering resistance ⟶

Arrangement of Si-O tetrahedra and representative minerals

Individual	Single chain	Double chain	Sheet	3-dimensional
Olivine	Pyroxene augite	Amphibole hornblende	Biotite (mica)	Quartz

		Oxygen–silicon ratio		
4	3	2.7	2.5	2

Figure 13.17
These models show the most common arrangements of silicon-oxygen tetrahedra in silicates and their relation to resistance to weathering. (From H. D. Foth, *Fundamentals of Soil Science,* 6th ed., 1978. John Wiley & Sons, Inc., New York. Used by permission.)

the rarer metal ions are held in silicates and, on the other hand, why it is seldom possible to write an exact formula for a particular silicate mineral.

Figure 13.17 shows five ways in which silicon-oxygen tetrahedra can be organized in various minerals. The more tightly packed the tetrahedra the more the mineral resists weathering. The tightest packing occurs in quartz, which has a ratio of oxygen to silicon of just 2 to 1, the lowest ratio of oxygen to silicon in the silicate family. Virtually no other ions or molecules occur in quartz to prevent its tetrahedra from being organized in very regular sequences that result in crystals of great beauty. In mica, the tetrahedra are organized in sheets, and this makes it possible to cleave mica very easily into flakes and leaflets.

Figure 13.18
Quartz crystals.

Figure 13.19
Mica, a form of silicate, that cleaves into thin sheets.

The formation of soil by the weathering of rocks involves the regrouping of molecule-sized fragments among the minerals. The silicate-based clays, so important in fertile soil, form by regroupings that bring aluminum ions, water molecules, and hydroxy groups into the systems. Figure 13.20 shows a model of the mineral kaolinite, just one of several varieties of clay. The layer-cake arrangement of atoms in kaolinite extensively uses hydrogen bonds involving entrapped water of hydration to hold the layers together. At the surfaces of kaolinite and other clay particles there are many electron-rich sites that help the soil hold polar

Figure 13.20
A model of the mineral kaolinite, one of the silicate clays. Shown here are three layers of silica tetrahedra held together by hydrogen bonds and electrostatic attractions of oppositely charged ionic centers. (From H. D. Foth. *Fundamentals of Soil Science,* 6th ed., 1978. John Wiley & Sons, Inc., New York. Used by permission.)

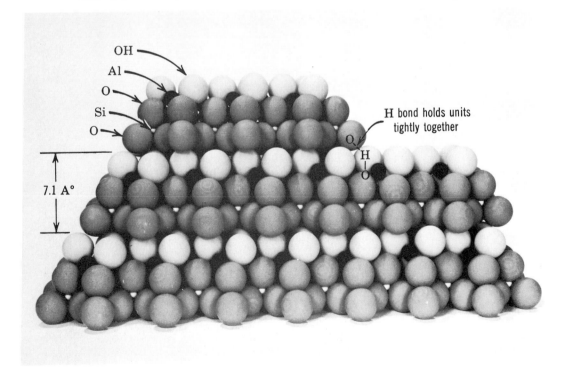

water molecules and hydrated cations that are needed for plant growth. The aluminum ions in clay help to hold anions such as nitrate, sulfate, phosphate and nitrite ions, which are also needed by plants.

SUMMARY

Macromolecular Substances. The chemical properties of macromolecular substances are determined by their functional groups and by the ability of a reagent to get at those groups. Generally, the larger its molecules, the less soluble is the substance and, therefore, the less able are ions or molecules of air, water, or dilute acids and bases to get into and react with the functional groups. Hence, macromolecular substances are usually much less reactive than are compounds of lower formula weights that have the same functional groups. The physical properties of macromolecular materials are determined by molecular chain lengths, degrees of chain branching, cross-links, the kinds of side chains and the regular spacing of these groups, and opportunities for hydrogen bonds to form.

Polymers. Macromolecular substances whose molecules have regularly repeating structural units are found among both organic and inorganic polymers, and among both natural and synthetic sources. The chief synthetic polymers are addition and condensation polymers. Making resins that can be successfully processed into finished or semifinished plastics or elastomers requires the correct choices of monomers, initiators, fillers, and plasticizers.

Addition Polymers. Some kinds of monomers can polymerize or copolymerize without losing any molecular pieces in the process. Alkenes and substituted alkenes (e.g., the acrylates) and dienes (such as isoprene) give addition polymers in this way. Polyolefins, polyacrylates, elastomers, and epoxies are in this family.

Condensation Polymers. Some pairs of monomers copolymerize in a process that also results in the loss of small pieces of their molecules as by-products (e.g., H—OH). The polyamides such as nylon and the polyesters such as Dacron form by condensation copolymerization reactions. In both families, the molecules have regularly spaced polar groups that provide forces of attraction between neighboring polymer chains as they line up side by side and overlap end to end. These forces combined with the symmetries of the chains in nylon and Dacron help to facilitate the formation of strong fibers and films. The silicones and phosphazines are inorganic polymers because their backbones are made of atoms other than carbon. However, their chains usually have organic groups. Despite these groups, inorganic polymers are remarkably resistant to combustion or to decomposition at high temperatures. They also retain their desirable properties at very low temperatures.

Silicates. Various combinations of silicon-oxygen tetrahedra plus metallic ions and water molecules make up the large family of silicate minerals, the chief components of the earth's crust. Residual charges on the surfaces of the tetrahedral arrays in soil minerals provide forces for holding water, cations, and anions that plants need for growth.

INDEX TO EXERCISES AND QUESTIONS

REVIEW QUESTIONS

13.5. In what ways is the term "macromolecular substance" broader than the term "polymer"?

13.6. What features seem to be essential if a polymer is to be useful for making strong fibers?

13.7. What is the difference between a resin and a plastic?

13.8 For what purpose would a filler be added to a polymerizing mixture?

13.9. Why would a plasticizer be incorporated into a plastic?

13.10. What is the function of an initiator and what happens to it?

13.11. In general terms, what is the difference between a polymer and a copolymer?

13.12. A synthetic polymer that competes with natural

rubber for making tires is called SBR, styrene-butyl-rubber. It is made by copolymerizing styrene (25%) and butadiene (75%). Although its molecules do not have the perfect regu-

$$C_6H_5CH=CH_2 \qquad CH_2=CH-CH=CH_2$$
styrene butadiene

larity that is found, for example, in polypropylene, the following feature is probably common and representative:

$$CH_2-CH=CH-CH_2-\underset{\underset{C_6H_5}{|}}{CH}-CH_2-CH_2-CH=CH-CH_2-CH_2-CH=CH-CH_2-$$

(a) Draw a circle about that part of the system contributed by styrene.
(b) Draw rectangles about the three butadiene units in the system.
(c) Would you expect SBR to be attacked by ozone? Why?

13.13. Rubber cements can be made by mixing some polymerized isobutylene with a solvent (e.g., benzene) to make a thick, sticky glue. When the solvent evaporates, the glue sets and holds. Write the structure of polyisobutylene in two ways.

$$CH_2=\underset{\underset{CH_3}{|}}{\overset{\overset{CH_3}{|}}{C}}$$
isobutylene

(a) Showing four successive repeating units.
(b) Condensed structure.

13.14. Polyvinyl acetate is used as a laminating agent in safety glass for windows. A thin, colorless film of this material is sandwiched between two sheets of glass. If the glass shatters, its pieces remain glued to the polyvinyl acetate and are not hurled about. Using four vinyl acetate units, write part of the structure of polyvinyl acetate. Write also its condensed structure.

$$CH_2=\underset{\underset{O_2CCH_3}{|}}{CH}$$
vinyl acetate

13.15. Tedlar is a polymer used to weatherproof building materials. It's a polymer of vinyl fluoride, $CH_2=CHF$. Show four repeating units of this polymer's structure and then write the condensed structure.

13.16. Saran Wrap® is essentially an addition polymer of dichloroethylene.

$$CH_2=\underset{\underset{Cl}{|}}{\overset{\overset{Cl}{|}}{C}}$$
dichloroethylene

Write four units of the structure of this polymer and then write its condensed structure.

13.17. During World War II, German scientists developed a blood plasma extender or substitute. It consisted of an aqueous solution of some salts together with an addition polymer, polyvinylpyrrolidinone. The monomer is N-vinylpyrrolidinone.

$$CH_2=CH$$

N-vinylpyrrolidinone

Write four units of the structure of this polymer.

13.18. Nylon-6 is another member of the nylon family.

$$-(NH-CH_2-CH_2-CH_2-CH_2-CH_2-\overset{\overset{O}{\parallel}}{C})_n-$$
Nylon-6
(*repeating unit*)

(a) Circle the polar groups in this structure that help neighboring molecules attract each other.
(b) Write three units of a molecule so as to show the repeating amide groups.

13.19. What effect on physical property does vulcanization have?

13.20. Which of the following combinations of monomers will make cross-linking possible, I or II?

$$x-M-x + y-Y-y \qquad or \qquad x-M-x + y-\underset{\underset{y}{|}}{Z}-y$$

(I) (II)

Explain. (Note, x and y are functional groups that can react and split out as molecules of x-y.)

13.21. What repeating sequence of atoms occurs in the main chains of silicones?

13.22. What undesireable properties are common among all polymers based on carbon backbones and that are not present when polymers based on silicon-oxygen or phosphorus-nitrogen units are used?

13.23. What structural features distinguish silicones from silicates?

13.24. Describe in general terms the ways in which variations in structural features lead to different kinds of silicate minerals.

13.25. In general terms, what is the relationship of rocks to minerals?

13.26. Silicate-based clays can retain both moisture and minerals. How do they do this; what forces from what sources make it possible?

13.27. Groundwater that filters through rocks or soils rich in limestone or dolomite picks up ions that make the water hard. What does "hard" mean in this context, and what ions are leached?

13.28. Sodium triphosphate can "remove" ions of Group IIA metals without forming a precipitate. How does this work?

13.29. If a little sodium carbonate is added to hard water, the water is softened, meaning that the ions responsible for hardness are removed. Reflecting on what you know about the water-insolubility of limestone, how does sodium carbonate work to soften water. (Write a molecular equation.)

CHAPTER FOURTEEN

RATES OF REACTION

14.1 SPEEDS AT WHICH REACTIONS OCCUR

**By studying the speed of a chemical reaction we can
learn details about how it occurs**

In the preceding two chapters we studied the chemical properties of some of the elements. We saw how they react and what kinds of compounds they form. In this and the next several chapters, we will take a closer look at chemical reactions to obtain a more thorough understanding of how and why they occur. We begin here by studying how fast reactions occur.

If we consider chemical reactions that are a part of everyday life, we will notice that the amount of time it takes for different reactions to occur varies considerably. For example, when a mixture of gasoline and air is fed into the cylinders of an automobile engine and ignited, a very rapid (almost instantaneous) reaction occurs that we use to propel the car. Most reactions that we observe are much slower than this, however. The cooking of foods, for instance, involves chemical reactions and, as any hungry person waiting for a hamburger knows, these reactions take a while. Fortunately for us, so do the biochemical reactions involved in digestion or metabolism. If biochemical reactions were as rapid as the combustion of gasoline vapor, our lives would be over in an instant.

Some reactions going on around us are quite slow. A fresh coat of paint that contains linseed oil "dries" slowly, first by evaporation of the solvent, and then by gradual reaction of the linseed oil resins in the paint with oxygen. This may take weeks or even months. The gradual curing of cement and concrete is another example. Although cement solidifies relatively quickly, chemical reactions within the mixture continue for years as the cement becomes harder and harder.

In scientific terms, the speed at which a reaction occurs is called the **rate of reaction.** It tells us how quickly (or slowly) the reactants disappear and the products form. Such information can be quite valuable. For instance, studying the rates of reactions and the factors that control these rates allows the manufacturers of chemical products to improve productivity and hold down operating costs by adjusting the conditions of a reaction so that it takes place as efficiently as possible. Of more fundamental importance to us, however, is the fact that, by studying the rates of reactions, we can learn a great deal about the chemical steps that lead ultimately from the reactants to the products.

When most chemical reactions take place, the change described by the balanced chemical equation—the net overall change—almost always occurs by a series of simpler chemical reactions. Consider, for example, the combustion of propane, C_3H_8:

$$C_3H_8(g) + 5O_2(g) \longrightarrow 3CO_2(g) + 4H_2O(g)$$

This reaction simply cannot occur in a single, simultaneous collision that involves one propane molecule and five oxygen molecules. Anyone who has ever played pool or billiards knows how seldom three balls come together with only one "click." It is easy to understand, then, how unlikely is the simultaneous collision of six molecules in three-dimensional space. In order for this reaction to proceed rapidly, which indeed it does, some very much more probable events must be involved. The series of individual steps leading to the overall observed reaction is called the **mechanism** of the reaction. Information about reaction mechanisms is one of the dividends that the study of reaction rates can give to us.

14.2   FACTORS THAT AFFECT REACTION RATE

The rate of a reaction depends on both the nature of the reactants and on the conditions under which the reaction occurs

The rates of nearly all chemical reactions are primarily controlled by five factors: (1) the chemical nature of the reactants; (2) the ability of the reactants to come in contact with each other; (3) the concentrations of the reactants; (4) the temperature of the reacting system; and (5) the availability of agents called catalysts that affect the rate of the reaction but are not themselves consumed.

The nature of the reactants
Fundamental differences in chemical reactivity, which are controlled by the tendencies toward bond formation, are a major factor in determining the rate of a reaction. Some reactions are just naturally fast and others are naturally slow. For example, a freshly exposed surface of sodium tarnishes almost instantly if exposed to air and moisture because sodium loses electrons so easily. Iron also reacts with air and moisture—it forms rust—but the reaction is much slower under identical conditions because iron simply doesn't lose electrons as easily as sodium.

The ability of the reactants to meet
Most reactions involve two or more reactants. Obviously, in order for the reaction to occur, the reactants must be able to come into contact with each other. This is one of the primary reasons that reactions are most often carried out in liquid solutions or in the gas phase. In these states, the reactants are able to intermingle on the molecular level, which allows their reacting particles—molecules or ions—to collide with each other easily.

It is easy to find common examples of how the ability of the reactants to meet affects the speed of a reaction. For instance, liquid gasoline burns fairly rapidly, but exactly how fast it burns depends on how quickly air can come into contact with the surface of the burning liquid. However, if the gasoline is vaporized and mixed with air before it is ignited, the combustion reaction is virtually instantaneous—the mixture explodes. An explosion is simply an extremely rapid reaction that generates hot gases that expand very quickly.

When the reactants are present in different phases—when one is a gas and the other is a liquid or a solid, for example—the reaction that occurs is called a **heterogeneous reaction.** In these reactions, the reactants are able to come into contact with each other only where they meet at the interface between the two phases. The size of this area of contact determines the rate of the reaction. This area can be increased by decreasing the sizes of the particles of the reactants, as anyone who has ever tried to start a campfire knows. Lighting a large log with

Special Topic 14.1   Plop, Plop . . . Fizz, Fizz

Acids react with bicarbonates and carbonates to generate carbon dioxide and water. The net reactions are

$$H^+(aq) + HCO_3^-(aq) \longrightarrow H_2O + CO_2(g)$$
$$2H^+(aq) + CO_3^{2-}(aq) \longrightarrow H_2O + CO_2(g)$$

But for these reactions to occur, the H^+ must come into direct contact with the HCO_3^- or CO_3^{2-} ions. The makers of the well-known analgesic tablets, Alka Seltzer®, take full advantage of this fact. Among the ingredients in these tablets are sodium bicarbonate

and citric acid. Crystals of these substances are able to exist in contact with each other without reacting because the hydrogen ions available from the citric acid cannot mingle with the bicarbonate ions in the sodium bicarbonate. Diffusion in solids is too slow for this to happen. But when the tablets are dropped into a glass of water (plop, plop), the citric acid and sodium bicarbonate dissolve, thus allowing the mixing of the ions and the production of carbon dioxide (fizz, fizz).

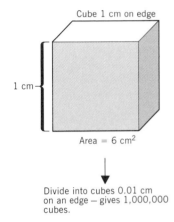

Cube 1 cm on edge

1 cm

Area = 6 cm²

Divide into cubes 0.01 cm on an edge — gives 1,000,000 cubes.

0.01 cm

Total area = 600 cm²

a match is very difficult. There is not enough contact between the wood and the oxygen in the air for combustion to occur easily. However, kindling—small twigs or slivers of wood—is relatively easy to light. Its greater surface area allows it to react more quickly. This is the reason that all good fire-builders always start with kindling.

If the particle sizes are extremely small, heterogeneous reactions can be explosive. Figure 14.1 shows the result of an explosion in a grain elevator that was used to store wheat. The explosion occurred when fine particles of grain dust mixed with air were accidentally ignited by a chance spark.

When all the reactants are in the same phase, the reaction that occurs is called a **homogeneous reaction.** Examples are the combustion of gasoline vapor and other gaseous reactions as well as those reactions that occur in liquid solutions. Most of the topics in the rest of this chapter focus on homogeneous systems.

The concentrations of the reactants

The rates of both homogeneous and heterogeneous reactions are affected by the concentrations of the reactants. For example, wood burns relatively quickly in air but extremely rapidly in pure oxygen. The hazards of a pure oxygen atmosphere were dramatically demonstrated in January 1967, when three astronauts died in

Figure 14.1
An explosion in a grain elevator in New Orleans, Louisiana, killed 35 people in December, 1977. It occurred when a spark ignited very fine dust in the grain silos. Extremely rapid combustion of the dust produced the explosive effect.

a flash fire that swept through their oxygen-filled Apollo spacecraft during a training exercise at Cape Kennedy (now Cape Canaveral).

The temperature of the system

Almost all chemical reactions occur faster at higher temperatures than they do at lower temperatures. If you have a Polaroid® camera, for instance, you may have noticed that the "instant picture" develops more quickly in warm weather than in cold weather. The reactions that develop the image occur faster at the higher temperature. You may have also noticed that insects begin to move more slowly in the autumn as the days become cooler. Insects are cold-blooded creatures whose body temperatures are determined by the temperatures of their surroundings. As they become cooler, their body chemistry slows down and they become sluggish.

The presence of catalysts

As we learned in Section 12.2, a catalyst is a substance that affects the rate of a reaction without being used up. The importance of catalysts of various kinds has been discussed in the preceding two chapters. Later in this chapter, we will study how catalysts function.

14.3   MEASURING THE RATE OF REACTION

The rate of a reaction is obtained by measuring the concentration of a reactant or a product at regular time intervals during the reaction

A **rate** is always expressed as a ratio in which units of time appear in the denominator. For example, if you have a job, you are probably paid at a certain rate, say five dollars per hour. Since the word *per* can be translated to mean *divided by*, this pay rate can be written as a fraction:

$$\text{rate of pay} = \frac{5 \text{ dollars}}{1 \text{ hour}}$$

In general, $x^{-1} = \frac{1}{x}$.

Since the fraction $\frac{1}{1 \text{ hour}}$ can also be written as hour^{-1}, your pay rate can be rewritten as

$$\text{rate of pay} = 5 \text{ dollars hour}^{-1}$$

You might think of your rate of pay as a change in your wealth. While you are working, your wealth changes (increases) at a rate of 5 dollars/hour.

As a chemical reaction occurs, the concentrations of the reactants and products change with time. The concentrations of the reactants decrease as the reactants are used up while the concentrations of the products increase as the products are formed. The rate of a chemical reaction, therefore, is expressed in terms of the rates at which the concentrations change. The unit of concentration that is used for this purpose is molar concentration, and the unit of time that is generally used is the second (abbreviated, s). This means that reaction rates are most frequently expressed with the units

$$\text{reaction rate} = \frac{\text{mol/liter}}{\text{s}}$$

Since 1/liter and 1/s can also be written as liter^{-1} and s^{-1}, the units for the rate can be expressed as

$$\text{reaction rate} = \text{mol liter}^{-1} \text{ s}^{-1}$$

Thus, if the concentration of a certain reactant were changing by 0.5 mol/liter each second, the rate of this reaction would be 0.5 mol liter^{-1} s^{-1}.

For nearly all chemical reactions, the rate does not remain constant as the reaction progresses. As mentioned in Section 14.2, the rate usually depends on the reactant concentrations and these change as the reaction proceeds. Table 14.1, for example, contains data on the decomposition of hydrogen iodide at a temperature of 508 °C:

$$2HI(g) \longrightarrow H_2(g) + I_2(g)$$

These data, which show the way the HI concentration changes with time, are plotted in Figure 14.2. The vertical axis gives the molar concentration of HI. Square brackets, [ ], indicate molar concentration in units of mol/liter.

Table 14.1
Data for the Reaction 2HI(g) $\longrightarrow$ H$_2$(g) + I$_2$(g) at 508 °C

Concentration of HI (mol/liter)	Time (s)
0.100	0
0.0716	50
0.0558	100
0.0457	150
0.0387	200
0.0336	250
0.0296	300
0.0265	350

In Figure 14.2, notice that the HI concentration drops fairly rapidly during the first 50 seconds of the reaction—the initial rate is relatively fast. Between 300 and 350 seconds, however, the concentration changes by only a small amount—the rate has slowed considerably. From this, we can see that the steepness of the concentration-time curve reflects the rate of the reaction—the steeper the curve, the higher the rate.

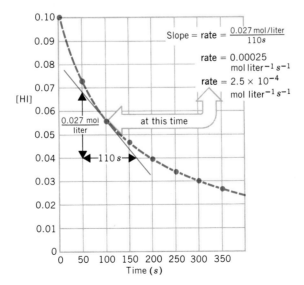

Figure 14.2
The change in the HI concentration with time for the reaction, $2HI(g) \rightarrow H_2(g) + I_2(g)$, at 508 °C. The data that are plotted are in Table 14.1.

If necessary, see Appendix A for a discussion of *slope*.

The exact rate at any particular time can be obtained by determining the slope of the line that is tangent to the concentration-time curve at that point. The slope, which is equal to the rate, is the ratio of the change in concentration to the change in time as read off the graph. In Figure 14.2, for example, the rate of the decomposition of hydrogen iodide is determined at a time 100 seconds from the start of the reaction. After the tangent to the curve is drawn, we measure a concentration change (0.027 mol/liter) and a time change (110 s) from the graph. The ratio of these quantities:

$$\frac{0.027 \text{ mol/liter}}{110 \text{ s}}$$

gives the rate. We see that, at this point in the reaction, the rate is 2.5×10^{-4} mol liter^{-1} s^{-1}. If you work Exercise 14.1, you will see that the rate of the reaction later on is indeed lower.

EXERCISE 14.1 Use the graph in Figure 14.2 to estimate the reaction rate 250 seconds from the start of the reaction.

Although we usually measure the rate at which the reactants disappear, it is sometimes more practical to measure the rate at which the products are formed. For example, in the decomposition of HI, it is easiest to monitor the I_2 concentration because I_2 vapor is purple. Therefore, as the reaction proceeds, more and more purple vapor appears and the color becomes more intense. Instruments are available that can be used to relate the amount of light absorbed by the I_2 vapor to the concentration of I_2.

The rate at which the products form can be related to the rate at which the reactants disappear if the equation for the reaction is known. For the reaction

$$2HI(g) \longrightarrow H_2(g) + I_2(g)$$

the HI concentration decreases twice as fast as the I_2 concentration increases because 2 mol of HI disappear for each mole of I_2 that is formed. It doesn't really matter, then, which substance we use to follow a reaction because the rates at which all of the concentrations change are interrelated.

EXERCISE 14.2 In the reaction, $N_2O_4(g) \rightarrow 2NO_2(g)$, if the rate of formation of NO_2 were 0.010 mol liter^{-1} s^{-1}, what would be the rate of decomposition for $N_2O_4(g)$?

14.4 CONCENTRATION AND RATE

An equation called the rate law expresses the way that the concentrations of the reactants affect the rate of a reaction

We learned in the last section that the rates of most reactions change when the concentrations of the reactants change. The way that the reaction rate and the concentrations of the reactants are related can be expressed in the form of an equation. In general, the rate is found to be proportional to the product of the concentrations of the reactants, each raised to a power. For example, if we have a chemical equation such as

$$A + B \rightarrow \text{products}$$

the rate of the reaction would be

$$\text{rate} \propto [A]^n[B]^m \tag{14.1}$$

The values of the exponents n and m must be determined by experiment—we will explore this further in a moment. The proportionality symbol, $\propto$, can be replaced by an equal sign if we introduce a proportionality constant, k, which is called the **rate constant** for the reaction. Equation 14.1 then takes the form

$$\text{rate} = k[A]^n[B]^m \tag{14.2}$$

This equation is called the **rate law** for the reaction. Once we have found values for k, n, and m, the rate law allows us to calculate the rate of the reaction if the concentrations are known. For example, at 0 °C, the initial rate of the reaction

$$H_2SeO_3 + 6I^- + 4H^+ \longrightarrow Se + 2I_3^- + 3H_2O \tag{14.3}$$

The units of the rate constant are always such that the calculated rate will have the units $mol\ liter^{-1}\ s^{-1}$.

has been found to have the rate law

$$\text{rate} = (5.0 \times 10^5\ liter^5\ mol^{-5}\ s^{-1})[H_2SeO_3]^1[I^-]^3[H^+]^2 \tag{14.4}$$

This equation allows us to calculate the rate of the reaction at 0 °C for any set of concentrations of H_2SeO_3, I^-, and H^+.

EXAMPLE 14.1   **Calculating Reaction Rate from the Rate Law**

Problem: What is the rate of reaction 14.3 when the reactant concentrations are as follows: $[H_2SeO_3] = 2.0 \times 10^{-2}\ M$, $[I^-] = 2.0 \times 10^{-3}\ M$, and $[H^+] = 1.0 \times 10^{-3}\ M$?

Solution: We simply substitute the concentrations into the rate law, equation 14.4. So that we can see how the units work out, we will write the rate constant's units in fraction form:

Remember, M means mol/liter.

$$\text{rate} = \frac{5.0 \times 10^5\ liter^5}{mol^5\ s} \times \left(\frac{2.0 \times 10^{-2}\ mol}{liter}\right) \times \left(\frac{2.0 \times 10^{-3}\ mol}{liter}\right)^3$$
$$\times \left(\frac{1.0 \times 10^{-3}\ mol}{liter}\right)^2$$

To perform the arithmetic, we first raise the concentrations *and* their units to the appropriate powers:

$$\text{rate} = \frac{5.0 \times 10^5\ liter^5}{mol^5\ s} \times \left(\frac{2.0 \times 10^{-2}\ mol}{liter}\right) \times \left(\frac{8.0 \times 10^{-9}\ mol^3}{liter^3}\right)$$
$$\times \left(\frac{1.0 \times 10^{-6}\ mol^2}{liter^2}\right) = \frac{8.0 \times 10^{-11}\ mol}{liter\ s}$$

The answer can also be written as

$$\text{rate} = 8.0 \times 10^{-11}\ mol\ liter^{-1}\ s^{-1}$$

Notice that the answer has the usual units for reaction rate.

EXERCISE 14.3   The rate law for the decomposition of HI is: rate = $k[HI]^2$. In Figure 14.2, the rate of the reaction was found to be $2.5 \times 10^{-4}\ mol\ liter^{-1}\ s^{-1}$ when the HI concentration was 0.0558 M. (a) What is the value of the rate constant for the reaction? (b) What are the units of k for this reaction?

Let's take a closer look, now, at reaction 14.3 and its rate law, equation 14.4. First, notice that the exponents to which the reactant concentrations must be raised in the rate law appear to be unrelated to any numbers associated with the reactants in the overall balanced equation. *There is, in fact, no way to know for sure what these exponents will be without doing experiments to determine them.*

The exponents in the rate law are not *always* different from the coefficients in the balanced chemical equation. Sometimes the coefficients and the exponents happen to be the same just by coincidence, as is the case with the rate law for the decomposition of HI:

$$2HI(g) \longrightarrow H_2(g) + I_2(g)$$
$$\text{rate} = k[HI]^2$$

There is no way of knowing this, however, without experimental data. You should never simply assume that they are the same—that's a trap into which many students fall.

The sum of exponents in the rate law is referred to as the **order** of the reaction.[1] For instance, the decomposition of gaseous N_2O_5 into NO_2 and O_2 has the rate law:

$$2N_2O_5 \longrightarrow 4NO_2 + O_2$$

$$\text{rate} = k[N_2O_5]^1$$

This reaction has only one reactant. The exponent of the N_2O_5 concentration is 1 and therefore the reaction is said to be *first order*. The rate law for the decomposition of HI has an exponent of 2 for the HI concentration; it, then, is a *second-order* reaction.

When the exponent is equal to 1 it is usually omitted. Therefore, when no exponent is written, assume that its value is 1.

When the rate law contains two or more reactants raised to powers, the order with respect to each reactant can be described. The rate law in equation 14.4 tells us that the reaction in 14.3 is *first order* with respect to H_2SeO_3, *third order* with respect to I^-, and *second order* with respect to H^+. The *overall order* of a reaction is the sum of the orders with respect to each reactant. Reaction 14.3 is a sixth-order reaction, overall.

$$1 + 3 + 2 = 6$$

The exponents in a rate law are usually whole numbers, but fractional and negative exponents are also found occasionally. A negative exponent means that the concentration term really belongs in the denominator and, as the concentration of the species increases, the rate of reaction decreases. Zero-order reactions also occur. These are reactions whose rate is independent of the concentration of the reactant.

EXERCISE 14.4   The reaction, $2HCrO_4^- + 5H^+ + 3HSO_3^- \rightarrow 2Cr^{3+} + 5H_2O + 3SO_4^{2-}$, has the rate law: rate = $k[HCrO_4^-][H^+][HSO_3^-]^2$. What is the order of the reaction with respect to each reactant? What is the overall order of the reaction?

Determining exponents of the rate law

We have stated several times in this chapter that the exponents of a rate law must be determined experimentally. *This is the only way for us to know for sure what these exponents are.*

To find the exponents of a rate law, we must study how changes in concentration affect the rate of the reaction. For example, suppose we have a reaction

$$A + B \longrightarrow \text{products}$$

for which the data in Table 14.2 had been obtained in a series of five experiments. The form of the rate law for the reaction would be

$$\text{rate} = k[A]^n[B]^m$$

For the first three sets of data, the concentration of B is constant. Changes in the rate are therefore due only to changes in the concentration of A. Now all we have

---

[1] The reason for describing the *order* of a reaction is that the mathematics involved in the treatment of the data are the same for all reactions having the same order. We will not go into this very deeply, but you should be familiar with the terminology used to describe the effects of concentration on the rates of reactions.

Table 14.2
Concentration/Rate Data for the Reaction:
$A + B \longrightarrow$ **Products**

Initial Concentration		
[A]	[B]	Initial Rate (mol liter^{-1} s^{-1})
0.10	0.10	0.20
0.20	0.10	0.40
0.30	0.10	0.60
0.30	0.20	2.40
0.30	0.30	5.40

to do is figure out what the order of the reaction must be with respect to A to give the observed changes.

Examining the data, we find that when [A] is doubled, the rate doubles; when [A] is tripled, the rate triples. The rate, then, is directly proportional to the concentration. This means that the exponent of [A] in the rate law must be 1. Thus, if we were to increase the concentration of A by a factor of 5, for example, $[A]^1$ would also become larger by a factor of 5 and the rate would increase by this same factor of 5.

In the final three sets of data, the concentration of A is held constant and the concentration of B is varied. Now it is the concentration of B that affects the rate. When [B] is doubled, the rate increases by a factor of 4; when [B] is tripled (from 0.1 to 0.3), the rate increases by a factor of 9 ($0.6 \times 9 = 5.4$). Notice that $4 = 2^2$ and $9 = 3^2$. In the rate law, therefore, [B] must be raised to the power, 2—that is, $[B]^2$. This means that if the concentration of B were increased by a factor of 5, the arithmetic used to compute the rate would involve squaring 5, and the rate would increase by a factor of 25.

Having determined the exponents of the concentration terms, we now know that the rate law for this reaction must be

$$\text{rate} = k[A]^1[B]^2$$

To calculate the value of k, we need only substitute rate and concentration data into this rate law for any one of the sets of data:

$$k = \frac{\text{rate}}{[A][B]^2}$$

Using the data from the first set,

$$k = \frac{0.20 \text{ mol liter}^{-1} \text{ s}^{-1}}{(0.10 \text{ mol liter}^{-1})(0.10 \text{ mol liter}^{-1})^2}$$
$$= \frac{0.20 \text{ mol liter}^{-1} \text{ s}^{-1}}{0.0010 \text{ mol}^3 \text{ liter}^{-3}}$$

Note that mol/mol^3 = mol^{-2}, and liter^{-1}/liter^{-3} = liter^2. Thus, the value of k, with its correct units, is

$$k = 2.0 \times 10^2 \text{ liter}^2 \text{ mol}^{-2} \text{ s}^{-1}$$

EXERCISE 14.5   Use the data from the other four experiments to calculate k for this reaction. What do you notice about the answers that you get?

The reasoning used to determine the order with respect to each reactant from experimental data can be summarized as shown in Table 14.3.

Table 14.3
Relation of the Order of a Reaction to Changes in Concentration and Rate

Factor by Which Concentration Is Changed	Factor by Which Rate Changes	Exponent on the Concentration Term Must Be:
2	$2 = 2^1$	1
3	$3 = 3^1$	1
4	$4 = 4^1$	1
2	$4 = 2^2$	2
3	$9 = 3^2$	2
4	$16 = 4^2$	2
2	$8 = 2^3$	3
3	$27 = 3^3$	3

EXAMPLE 14.2 **Determining the Exponents of a Rate Law**

Problem: The following data were collected on the decomposition of SO_2Cl_2 at a particular temperature.

$$SO_2Cl_2(g) \longrightarrow SO_2(g) + Cl_2(g)$$

Initial Concentration of SO_2Cl_2 (mol liter^{-1})	Initial rate (mol liter^{-1} s^{-1})
0.100	2.2×10^{-6}
0.200	4.4×10^{-6}
0.300	6.6×10^{-6}

What is the rate law and what is the value of the rate constant?

Solution: We expect the rate law to be

$$\text{rate} = k[SO_2Cl_2]^n$$

The initial rate doubles (from 2.2×10^{-6} to 4.4×10^{-6}) when the initial concentration doubles; the rate triples when the concentration triples. The reaction must therefore be first order. Thus

$$\text{rate} = k[SO_2Cl_2]^1$$

The value of k, calculated from the first set of data, is

$$k = \frac{\text{rate}}{[SO_2Cl_2]}$$

$$= \frac{2.2 \times 10^{-6} \text{ mol liter}^{-1} \text{ s}^{-1}}{0.10 \text{ mol liter}^{-1}}$$

$$= 2.2 \times 10^{-5} \text{ s}^{-1}$$

EXAMPLE 14.3 Determining the Exponents of a Rate Law

Problem: Isoprene, a monomer used to make natural rubber, forms a dimer called dipentene:

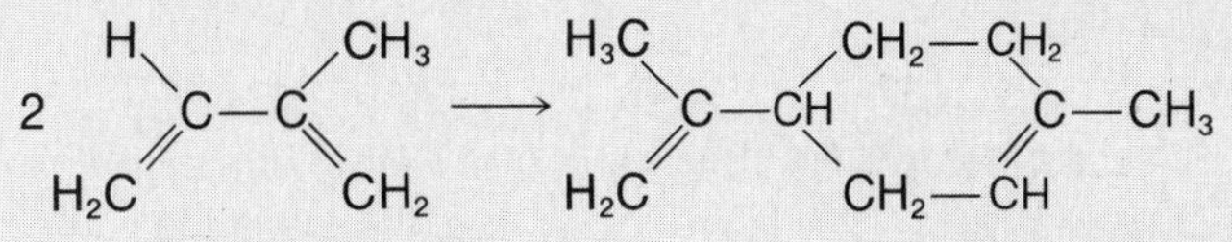

What is the rate law for this reaction, given the following data?

Initial Isoprene Concentration (mol liter^{-1})	Initial Reaction Rate (mol liter^{-1} s^{-1})
0.50	1.98
1.50	17.8

Solution: We expect the rate law to be of the form:

$$\text{rate} = k[\text{isoprene}]^n$$

Comparing the two experiments, the isoprene concentration in the second is three times larger than it is in the first. The rate in the second experiment is 17.8/1.98 or 8.99 times larger than it is in the first experiment. The value 8.99 is very nearly 9, which is 3^2. Therefore, the value of n is 2 and the rate law is

$$\text{rate} = k[\text{isoprene}]^2$$

EXAMPLE 14.4 Determining the Exponents of a Rate Law

Problem: The following data were measured for the reduction of nitric oxide with hydrogen:

$$2NO(g) + 2H_2(g) \longrightarrow N_2(g) + 2H_2O(g)$$

Initial Concentrations (mol liter^{-1})		Initial Rate (mol liter^{-1} s^{-1})
[NO]	[H$_2$]	
0.10	0.10	1.23×10^{-3}
0.10	0.20	2.46×10^{-3}
0.20	0.10	4.92×10^{-3}

What is the rate law for this reaction?

Solution: We expect the rate law to be

$$\text{rate} = k[\text{NO}]^n[\text{H}_2]^m$$

First, we compare experiments 1 and 2. When the H$_2$ concentration is doubled (while [NO] remains constant), the rate is also doubled. The reaction, then, is first order with respect to H_2. Next, we compare experiments 1 and 3. In these, the NO concentration is doubled while the H_2 concentration remains constant. Doubling [NO] increases the rate by a factor of 4.92/1.23 or 4.00. Thus, the reaction must be second order with respect to NO. Now we can write the rate law:

$$\text{rate} = k[NO]^2[H_2]$$

EXERCISE 14.6   Ordinary sucrose (table sugar) reacts with water to produce two simpler sugars, glucose and fructose, that have the same molecular formulas:

$$C_{12}H_{22}O_{11} + H_2O \longrightarrow C_6H_{12}O_6 + C_6H_{12}O_6$$
$$\text{sucrose} \qquad\qquad \text{glucose} \quad \text{fructose}$$

In a particular series of experiments the following data were obtained.

Initial Sucrose Concentration (mol liter^{-1})	Initial Rate (mol liter^{-1} s^{-1})
0.10	6.17×10^{-5}
0.20	1.23×10^{-4}
0.50	3.09×10^{-4}

What is the order of the reaction with respect to sucrose?

EXERCISE 14.7   A certain reaction that follows the equation, $A + B \rightarrow C + D$, gave the following data.

Initial Concentrations (mol liter^{-1})		Initial Rate (mol liter^{-1} s^{-1})
[A]	[B]	
0.40	0.30	1.0×10^{-4}
0.80	0.30	4.0×10^{-4}
0.80	0.60	1.6×10^{-3}

What is the rate law for the reaction? What is the value of the rate constant?

14.5 HALF-LIVES

The time required for the concentration of a reactant to be reduced to half of its initial value is called the half-life

A concept that provides a useful measure of the speed of a reaction, especially for first-order processes, is that of **half-life, $t_{1/2}$**—the amount of time required for half of a given reactant to disappear. When the half-life is short, a reaction is rapid because half of the reactant disappears quickly.

For any first-order reaction, the half-life is constant at any given temperature. It is not affected by the initial concentration of the reactant. A typical example is the change that radioactive isotopes undergo during radioactive "decay." (In fact, you have probably heard the term *half-life* used in reference to the life span of radioactive substances.)

Iodine-131, an unstable, radioactive isotope of iodine, undergoes a nuclear reaction that causes it to emit a type of radiation and transforms it into a stable isotope of xenon. The intensity of this radiation decreases, or *decays*, with time. Figure 14.3 is a graph of this decay. Notice that the time it takes for the first half of the ^{131}I to disappear is eight days. During the next eight days, half of the remaining ^{131}I disappears, and so on. Regardless of the amount begun with, it takes eight days for half of that amount of ^{131}I to disappear, which means that the half-life of ^{131}I is a constant.

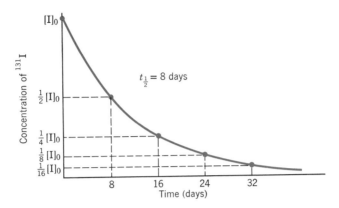

Figure 14.3
First-order radioactive decay of iodine-131. Initial concentration is $[I]_0$.

Iodine-131 is used in the diagnosis and treatment of thyroid disorders. The thyroid gland is the only part of the body that uses iodide ions. When a patient is given a small quantity of $^{131}I^-$ mixed with nonradioactive I^-, both are absorbed by the thyroid. This behavior allows for the testing of thyroid activity and is also used to treat certain types of thyroid cancer.

For any reaction, the half-life is related to the rate constant. If the reaction is first order, its $t_{1/2}$ can be calculated from its rate constant by using the equation:

$$t_{1/2} = \frac{0.693}{k} \tag{14.5}$$

The half-life of a second-order reaction *does* depend on the initial concentration of the reactants. We can see this by examining Figure 14.2 (page 457), which shows the decomposition of gaseous HI, a second-order reaction. The reaction begins with a hydrogen iodide concentration of 0.10 *M*. The concentration drops to half of this value, 0.050 *M*, in about 125 seconds. If we take 0.050 *M* as the next "initial" concentration, it drops to half its value, 0.025 *M*, 250 seconds later (at a total elapsed time of 375 seconds). Thus, halving the initial concentration, from 0.10 *M* to 0.05 *M*, causes a doubling of the half-life, from 125 to 250 seconds. For a second-order reaction like the decomposition of HI, the half-life is inversely proportional to the initial concentration of the reactant and is related to the rate constant by the equation

$$t_{1/2} = \frac{1}{k \times (\text{initial concentration of reactant})} \tag{14.6}$$

EXAMPLE 14.5   Using Half-lives

Problem: The half-life of iodine-131 is 8 days. What fraction of the initial iodine-131 would be present in a patient after 24 days if none of it were eliminated through natural bodily processes?

Solution: A period of 24 days is exactly three half-lives. Taking the fraction initially present as 1, we can set up a table:

Half-lives	0	1	2	3
Fraction	1	$\frac{1}{2}$	$\frac{1}{4}$	$\frac{1}{8}$

Half of the ^{131}I is lost in the first half-life, half of that disappears in the second half-life, and so on. Therefore, the fraction remaining after three half-lives is $\frac{1}{8}$.

EXAMPLE 14.6   **Calculating the Half-life**

Problem: The reaction, $2HI(g) \rightarrow H_2(g) + I_2(g)$, has the rate law, rate $= k[HI]^2$ with $k = 0.079$ liter $mol^{-1} s^{-1}$ at 508 °C. What is $t_{1/2}$ for this reaction when the initial concentration of HI is 0.050 M?

Solution: We have already found the answer to this problem by inspecting Figure 14.2, but let's calculate $t_{1/2}$ using equation 14.6, anyway.

$$t_{1/2} = \frac{1}{k \times \text{(initial concentration)}}$$

Substituting values given in the problem,

$$t_{1/2} = \frac{1}{(0.079 \text{ liter } mol^{-1} s^{-1})(0.050 \text{ mol liter}^{-1})}$$

$$t_{1/2} = 253 \text{ s}$$

Rounding to two significant figures gives $t_{1/2} = 250$ seconds, which is the same answer we obtained from Figure 14.2.

EXERCISE 14.8   In Exercise 14.6, the reaction of sucrose with water was found to be first order with respect to sucrose. The rate constant for the reaction under the conditions of the experiments was $6.17 \times 10^{-4} s^{-1}$. Calculate the value of $t_{1/2}$ for this reaction in minutes. How many minutes would it take for three-quarters of the sucrose to react? (*Hint.* What fraction of the sucrose remains?)

EXERCISE 14.9   Suppose that the value of $t_{1/2}$ for a certain reaction was found to be independent of the initial concentration of the reactants. What can you say about the order of the reaction?

14.6   THE EFFECT OF TEMPERATURE ON REACTION RATE

In any system, increasing the temperature increases the number of reactant molecules having the minimum kinetic energy that they need to react

In Section 14.2 we mentioned the fact that nearly all reactions proceed faster at higher temperatures. As a rule, the reaction rate increases by a factor of about 2 or 3 for each 10 °C rise in temperature, although the actual amount of increase differs from one reaction to another. To understand why temperature affects reaction rates in this way, we have to examine what actually happens to the molecules in a reaction system.

One of the simplest theories about the way various factors affect reaction rates is the **collision theory.** The basic postulate of this theory is that the rate of a reaction is proportional to the number of collisions per second among the reactant molecules. Anything that can increase the frequency of collisions should increase the rate. As reasonable as this sounds, it is impossible for every one of the collisions between the reactants to actually result in a chemical change. In a gas or a liquid, molecules of the reactants undergo an enormous number of collisions with each other each second. If each were effective, all reactions would be over in an instant. Of all of the collisions that occur, only a *very* small fraction

At the start of the reaction in Figure 14.2, only one out of every billion billion (10^{18}) collisions leads to a net reaction.

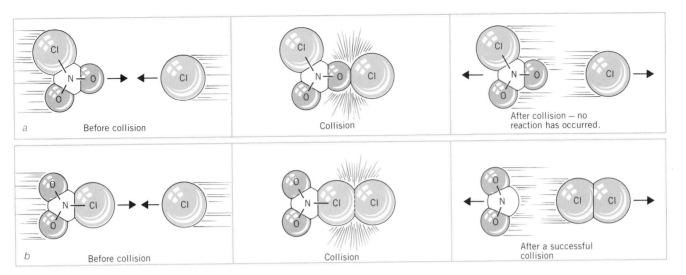

Figure 14.4
The importance of molecular orientation during a collision.
(a) A collision that cannot produce a Cl_2 molecule. (b) A collision that can produce NO_2 and Cl_2 molecules.

actually results in a chemical change. There are two principle reasons for this.

In many cases, when two reactants collide, their atoms must be oriented correctly in order for a reaction to occur. For example, later we will see that the reaction

$$2NO_2Cl \longrightarrow 2NO_2 + Cl_2$$

appears to proceed by a mechanism that involves two steps. One of these steps involves the collision of an NO_2Cl molecule with a chlorine atom:

$$NO_2Cl + Cl \longrightarrow NO_2 + Cl_2$$

The orientation of the NO_2Cl molecule is important in this collision, as is shown in Figure 14.4. In Figure 14.4a, no Cl_2 can be formed, but in Figure 14.4b the orientation is right for the collision to be effective.

The major reason that so few collisions actually lead to chemical change, however, is that the reacting molecules—even when correctly oriented—must possess between them a certain minimum kinetic energy (KE), called the **activation energy, E_a**. Very few molecules have this minimum energy.

Any chemical change involves a reorganization of chemical bonds. In general, old bonds are broken as new ones are formed. In order for this to happen during a collision, the nuclei of the reacting particles must somehow find themselves in the right locations. This requires that the colliding molecules experience a very energetic collision. For example, if two slow-moving molecules collide, the repulsions between their electron clouds causes the molecules to simply bounce apart. Only fast-moving molecules, which have large kinetic energies, can collide with enough force to enable their nuclei and electrons to overcome repulsions and thereby allow the necessary bond breaking and bond making.

Figure 14.5 shows why the rate of a reaction increases with temperature. Here we have the kinetic energy distributions for a collection of molecules at two different temperatures. Regardless of temperature, the minimum kinetic energy needed for an effective collision—the activation energy—is the same. At the higher temperature, the total fraction of molecules that have this necessary energy is greater than it is at the lower temperature. (This total fraction is indicated by the amount of shaded area under each curve.) In other words, at the higher

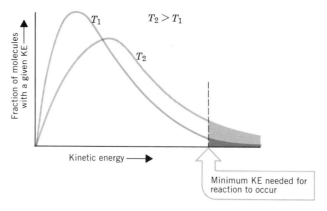

Figure 14.5
Kinetic energy distributions for a reaction mixture at two different temperatures. The size of the shaded areas under the curves is proportional to the total fraction of the molecules that possess the minimum activation energy.

temperature, a greater fraction of the total number of collisions that occur each second will result in a chemical change. The reactants will therefore disappear faster at the higher temperature.

Now let's take a close look at what happens when the reactant molecules come together in a collision. When they collide, the molecules slow down, stop, and then fly apart again. If a reaction occurs during the collision, the particles that separate are chemically different from those that collide. Regardless of what happens to them chemically, however, as the molecules slow down, the total kinetic energy that they possess decreases. Since energy can't disappear, this means that their total potential energy (PE) must increase.

The relationship between the activation energy and the total potential energy of the reactants and products can be expressed graphically by a potential-energy diagram. A typical diagram for a reaction is shown in Figure 14.6. The horizontal axis represents the extent to which the reactants have changed to the products. It follows the path taken by the reaction as reactant molecules come together and change to produce the product molecules that separate after collision. The activation energy appears as a potential energy "hill" between the reactants and products. Only molecules having energies at least as large as E_a are able to climb over the hill and produce the products.

Figure 14.6
Potential-energy diagram for an exothermic reaction.

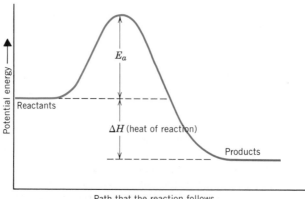

We can use the potential-energy diagram for a reaction to follow the progress of both an unsuccessful and a successful collision. As two reactant molecules collide, they slow down and their kinetic energy is changed to potential energy—they begin to climb the potential energy barrier toward the products. If their initial kinetic energies are less than E_a, they are unable to reach the top of the hill. Instead, they fall back toward the reactants. They come apart chemically unchanged and with their original kinetic energy; no net reaction has occurred (Figure 14.7a). On the other hand, if their combined kinetic energies are equal to or greater than E_a, they are able to pass over the activation energy barrier and form product molecules (Figure 14.7b).

The law of conservation of energy requires that PE + KE = constant during a collision.

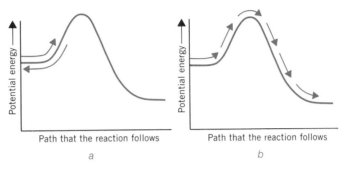

Path that the reaction follows

a

Path that the reaction follows

b

Figure 14.7
(*a*) An unsuccessful collision—molecules separate unchanged.
(*b*) A successful collision—activation energy barrier is crossed and the products are formed.

In Figure 14.6 we also find the heat of reaction—the difference between the potential energy of the products and the potential energy of the reactants. The reaction depicted in Figure 14.6 is exothermic because the products have a lower potential energy than the reactants. When a successful collision occurs, the decrease in potential energy appears as an increase in the kinetic energies of the emerging product molecules. The temperature of the system rises during an exothermic reaction because the average kinetic energy of the system increases.

Figure 14.8 is a potential-energy diagram for an endothermic reaction. In this case the products are at a higher potential energy than the reactants and, in terms of the heat of reaction, a net input of energy is needed to form the products. Endothermic reactions produce a cooling effect as they proceed because there is a net conversion of kinetic energy to potential energy. As the total kinetic energy decreases, the average kinetic energy decreases as well and the temperature drops.

In Chapter Five we saw that when the direction of a reaction is reversed, the sign of its ΔH is changed—a reaction that is exothermic in the forward direction is endothermic in the reverse direction, and vice versa. This suggests that, in general, reactions are reversible. If we look again at the energy diagram for a reaction that is exothermic in the forward direction, it is obvious that in the opposite direction it is endothermic. What differs most for the forward and reverse directions is the relative height of the activation energy barrier (Figure 14.9).

One of the main reasons for studying activation energies is that they provide information about what actually occurs during an effective collision between reactant molecules. For example, suppose that we were studying a reaction between the molecules A_2 and B_2 to form molecules of AB. One way for A_2 and B_2 to react during a collision is depicted in Figure 14.10. The A_2 and B_2 molecules come together and, at the moment of collision, the A—A and B—B bonds are

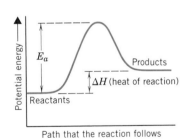

Figure 14.8
A potential-energy diagram for an endothermic reaction.

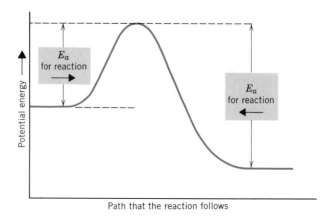

Figure 14.9
Activation energy barrier for the forward and reverse reaction.

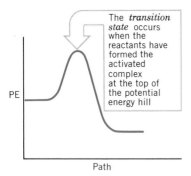

The *transition state* occurs when the reactants have formed the activated complex at the top of the potential energy hill

weakened as the new A—B bonds are formed. This brief moment during the reaction is called the **transition state.** It corresponds to the high point on the potential-energy diagram. The species that exists at that instant, with its partly formed and partly broken bonds, is called the **activated complex.**

The activation energy provides information about the relative importance of bond breaking and bond making during the formation of the activated complex. A very high activation energy, for instance, would suggest that bond breaking contributes very heavily in the formation of the activated complex because bond breaking is an energy-absorbing process.

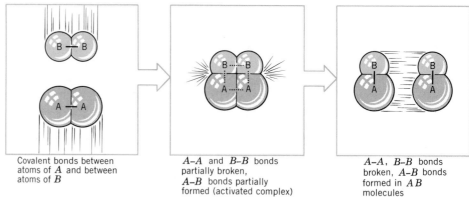

Covalent bonds between atoms of A and between atoms of B

A–A and B–B bonds partially broken, A–B bonds partially formed (activated complex)

A–A, B–B bonds broken, A–B bonds formed in AB molecules

Figure 14.10
Reaction between molecules of A_2 and B_2. This illustrates one way that molecules of AB could be formed by collisions between A_2 and B_2.

14.7 COLLISION THEORY AND REACTION MECHANISMS

The slow step in a reaction mechanism determines the overall rate of reaction

At the beginning of this chapter it was stated that one of the benefits to be gained from a study of reaction rates is information about the paths, or *mechanisms* followed by reactions. For most reactions, the individual chemical steps, called

elementary processes, that make up the mechanism cannot actually be observed. The mechanism that a chemist writes is really a theory about what occurs step-by-step as the reactants are converted to the products. Since the individual steps in the mechanism can't be observed, arriving at a chemically reasonable set of elementary processes is not at all a simple task. Making reasonable guesses requires a lot of "scientific intuition" and much more chemical knowledge than has presently been provided to you, so you need not worry about having to predict mechanisms for reactions. Nevertheless, to understsnd the science of chemistry better, it is worthwhile knowing how the study of reaction rates can provide clues to a reaction's mechanism.

In Section 14.6 we learned that the basic postulate of the collision theory is that the rate of a reaction is proportional to the number of collisions per second between the reactants. We also learned that, for a given set of conditions, the number of effective collisions is only a certain fraction of the total. If we could somehow double the total number of collisions, the number of effective collisions would be doubled also.

Let's suppose, now, that we know for a fact that a certain collision process takes place during a particular reaction. For example, suppose that the decomposition of NOCl into NO and Cl_2 actually involves collisions between NOCl molecules and Cl atoms as one step in the mechanism. In other words, the reaction

$$NOCl + Cl \longrightarrow NO + Cl_2 \qquad (14.7)$$

represents an elementary process. What would happen if we were to double the number of Cl atoms in the container? Since there would then be twice as many Cl atoms with which the NOCl molecules could collide, there should be twice as many NOCl-Cl collisions per second. This should, in turn, double the rate of the reaction given by equation 14.7. In other words, *doubling* the Cl concentration would *double* the rate of this elementary process. The rate law for equation 14.7 should therefore contain [Cl] raised to the first power.

Similarly, if we were to double the number of NOCl molecules in the container, there would be twice as many NOCl molecules with which Cl atoms could collide. As a result, the collision frequency should *double* and the rate of the reaction should *double*. This means that the NOCl concentration should also be raised to the first power in the rate law for this elementary process. Therefore, the rate law for equation 14.7 should be

$$\text{rate} = k[\text{Cl}][\text{NOCl}]$$

Notice that the exponents in the rate law for this *elementary process* are the same as the coefficients in the chemical equation for the elementary process.

Let's look at another elementary process, one involving collisions between like molecules:

$$2NO_2 \longrightarrow NO_3 + NO \qquad (14.8)$$

If the NO_2 concentration were doubled, there would be *twice* as many individual NO_2 molecules and each would have *twice* as many neighbors with which to collide. The number of NO_2-NO_2 collisions, then, would actually be increased by a factor of 4, which is 2^2. We've seen earlier that when a doubling of a reactant concentration increases the rate fourfold, the concentration of that reactant is raised to the second power in the rate law. Thus, the rate law for equation 14.8 should be

$$\text{rate} = k[\text{NO}_2]^2$$

Once again the exponent in the rate law for the *elementary process* is the same as the coefficient in the chemical equation.

The point that these two examples make is that the rate law for an elementary process can be predicted.

> The exponents in the rate law for an elementary process are equal to the coefficients of the reactants in the chemical equation for that elementary process.

It is very important to understand that this rule applies *only* to elementary processes. If all we know is an overall equation, we can only be sure of the exponents in the rate law if we determine them by doing experiments!

How does the ability to predict the rate law of an elementary process help chemists predict reaction mechanisms? To answer this question, let's look at two mechanisms. First, consider the gaseous reaction

$$2NO_2 \longrightarrow 2NO + O_2 \tag{14.9}$$

It has been experimentally determined that this is a second-order reaction having the rate law

$$rate = k[NO_2]^2$$

The mechanism that has been proposed for the reaction involves the elementary processes:

$$2NO_2 \longrightarrow NO_3 + NO$$
$$NO_3 \longrightarrow NO + O_2$$

(Notice that if these two reactions are added and NO_3 is cancelled from both sides, the net overall reaction, equation 14.9, is obtained.)

When a reaction such as this occurs in a series of steps, one step is very often much slower than the others. In this mechanism, for example, it is believed that the first step is slow and that once the NO_3 is formed, it decomposes quickly to give NO and O_2.

The slow step in a mechanism is called the **rate-determining step** or **rate-limiting step.** This is because the final products of the overall reaction cannot appear faster than the products of the slow step. In the mechanism for the decomposition of NO_2, then, the first reaction is the rate-determining step because the final products can't be formed any faster than the rate at which NO_3 is formed.

The rate-determining step is similar to a slow worker on an assembly line. Regardless of how fast the other workers are, the production rate depends on how quickly the slow worker does his or her job. The factors that control the speed of the rate-determining step therefore also control the overall rate of the reaction. This means that the rate law for the rate-determining step is directly related to the rate law for the overall reaction.

Because the rate-determining step is an elementary process, we can predict its rate law from the coefficients. The coefficient of NO_2 is 2, so the rate law for the first step would be the same as that found experimentally for the overall reaction. The mechanism, therefore, does not conflict with experimental evidence and could be correct.

Now let's look at the reaction

$$2NO_2Cl \longrightarrow 2NO_2 + Cl_2 \tag{14.10}$$

which is a first-order reaction that has the experimentally determined rate law

$$rate = k[NO_2Cl]$$

The rate-determining step for this reaction could not possibly involve collisions of two NO_2Cl molecules because if it did, the reaction would be second order. The actual mechanism here appears to be

$$NO_2Cl \longrightarrow NO_2 + Cl \quad \text{(slow)}$$
$$NO_2Cl + Cl \longrightarrow NO_2 + Cl_2 \quad \text{(fast)}$$

Note that the sum of the elementary processes (after cancelling Cl from both sides) gives the equation for the overall reaction (equation 14.10) and that the rate-determining step is first order because the coefficient of NO_2Cl is equal to one.

Although chemists may devise other experiments to help prove or disprove the accuracy of a mechanism, one of the strongest pieces of evidence is the experimentally measured rate law for the overall reaction. No matter how reasonable a particular mechanism may appear, if its elementary processes cannot yield a predicted rate law that matches the experimental one, the mechanism is wrong and must be discarded.

EXERCISE 14.10   The reaction of ozone, O_3, with nitric oxide, NO, to form nitrogen dioxide and oxygen:

$$NO + O_3 \longrightarrow NO_2 + O_2$$

is one of the reactions involved in the production of smog. It is believed to occur by a one-step mechanism (the reaction above). If this is so, what is the expected rate law for the reaction?

14.8 FREE RADICAL REACTIONS

Reactions that involve reactive species containing unpaired electrons tend to be very rapid and involve chain mechanisms

A **free radical** is a very reactive species that contains one or more unpaired electrons. Examples are chlorine atoms that are produced when a Cl_2 molecule absorbs a photon (light) of the appropriate energy:

Remember, $E = h\nu$.

$$Cl_2 + \text{light energy} \longrightarrow 2Cl\cdot$$

(A dot placed next to the symbol of an atom or molecule indicates that it is a free radical.) The high degree of reactivity of free radicals exists because of the tendency of electrons to pair through the formation of either ions or covalent bonds.

Free radicals are important in many gaseous reactions, including those responsible for the production of photochemical smog in urban areas. In biological systems, they appear to be responsible for the aging process as well as many of the harmful effects of radiation. Reactions involving free radicals have useful applications, too. Many polymerization reactions occur by mechanisms that involve free radicals as we learned in Section 13.2. In addition, free radicals play a part in one of the most important processes in the petroleum industry, thermal cracking. This reaction is used to break C—C and C—H bonds of long-chain hydrocarbons to produce the smaller molecules that give gasoline a higher octane rating. An example of a thermal-cracking reaction can be seen with butane. When butane is heated to 700 to 800 °C, one of the major reactions that occurs is

The central C—C bond is shown as $\dot{x}$ instead of a dash.

$$CH_3-CH_2\dot{x}CH_2-CH_3 \longrightarrow CH_3CH_2\cdot + CH_3CH_2{}^x$$

This reaction produces two ethyl radicals, $C_2H_5\cdot$.

Free radical reactions tend to have high initial activation energies because chemical bonds must be broken to form the radicals. This can be accomplished by light energy or heat. Once the free radicals are formed, however, the chemical reactions in which they are involved tend to be very rapid. In many cases, a free radical reacts with a reactant molecule to give a product molecule plus another free radical. Reactions that involve this step are called **chain reactions.**

Artist's concept of the Lunar Module maneuvering near the Apollo Command and Service Module. The explosive reaction of hydrogen and oxygen to form water provides the rocket thrust.

Not all chain reactions involve branching steps.

Many explosive reactions are chain reactions involving free radical mechanisms. One of the most studied reactions of this type is the formation of water from hydrogen and oxygen. The reactions involved in the mechanism can be described according to their roles in the mechanism. The chain reaction begins with an **initiation step** that produces free radicals:

$$H_2 + O_2 \xrightarrow{\text{hot surface}} 2OH\cdot \quad \text{(initiation)}$$

The chain continues with a **propagation step,** which produces the product plus another free radical:

$$OH\cdot + H_2 \longrightarrow H_2O + H\cdot \quad \text{(propagation)}$$

The reaction of H_2 and O_2 is explosive because the mechanism also contains **branching steps:**

$$\left.\begin{array}{l} H\cdot + O_2 \longrightarrow OH\cdot + O\cdot \\ O\cdot + H_2 \longrightarrow OH\cdot + H\cdot \end{array}\right\} \text{branching}$$

Thus, the reaction of one $H\cdot$ with oxygen leads to the production of two $OH\cdot$ plus another $H\cdot$. Every time an $H\cdot$ reacts with oxygen, then, there is an increase in the number of free radicals in the system. The free radical concentration grows rapidly, and the reaction rate becomes explosively fast.

Chain mechanisms also contain termination steps. In the reaction of H_2 and O_2, the wall of the reaction vessel serves to remove $H\cdot$, which tends to stop the chain process:

$$2H\cdot + \text{wall} \longrightarrow H_2 \text{ (termination)}$$

14.9 CATALYSTS

Catalysts accelerate chemical reactions by providing an alternative mechanism that has a lower activation energy

A catalyst, we have learned, is a substance that increases the rate of a chemical reaction without being used up itself. All of the catalyst added at the start of a reaction is present chemically unchanged after the reaction has gone to completion. The catalyst participates by changing the mechanism of the reaction. It provides a path to the products that has a lower activation energy than that of the uncatalyzed reaction, as illustrated in Figure 14.11. Since the activation energy following this new route is lower, a greater fraction of the reactant molecules have the minimum energy needed to react, so the rate of reaction increases.

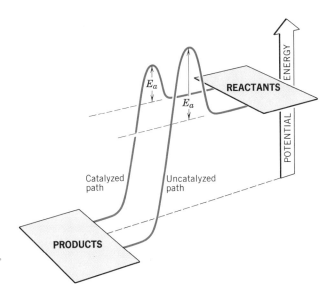

Figure 14.11
Effect of a catalyst on a reaction. The catalyst provides an alternative, low-energy path from the reactants to the products.

Catalysts can be divided into two groups—**homogeneous catalysts,** which are found in the same phase as the reactants, and **heterogeneous catalysts,** which exist as a separate phase. An example of homogeneous *catalysis* is found in the now outdated lead chamber process used for manufacturing sulfuric acid. To make sulfuric acid, sulfur is first burned to give SO_2, which is then oxidized to SO_3. Dissolving SO_3 in water gives H_2SO_4:

Catalysis is the action or effect produced by a catalyst.

$$S + O_2 \longrightarrow SO_2$$
$$2SO_2 + O_2 \longrightarrow 2SO_3$$
$$SO_3 + H_2O \longrightarrow H_2SO_4$$

The second reaction, oxidation of SO_2 to SO_3, occurs slowly. In the lead chamber process, the SO_2 is mixed with NO, NO_2, air, and steam in large lead-lined reaction chambers. The NO_2 readily oxidizes the SO_2 to give NO and SO_3. The NO is then reoxidized to NO_2:

$$NO_2 + SO_2 \longrightarrow SO_3 + NO$$
$$NO + \tfrac{1}{2}O_2 \longrightarrow NO_2$$

The NO_2 serves as a catalyst by being an oxygen carrier and by providing a low-energy path for the oxidation of SO_2 to SO_3. The steam in the reaction mixture reacts with the SO_3 as it is formed and produces the sulfuric acid.

A heterogeneous catalyst functions by promoting a reaction on its surface. One or more of the reactant molecules are adsorbed on the surface of the catalyst where interaction with the surface increases their reactivity. An example is the synthesis of ammonia from hydrogen and nitrogen by the Haber process, as we learned in Section 12.2.

$$3H_2(g) + N_2(g) \longrightarrow 2NH_3(g)$$

This is one of the most important industrial reactions in the world because ammonia and nitric acid (which is made from ammonia) are necessary for the production of fertilizers. The reaction takes place on an iron catalyst that contains traces of aluminum and potassium oxides. It is thought that hydrogen molecules and nitrogen molecules dissociate while being held on the surface of the catalyst. The hydrogen atoms then combine with the nitrogen atoms to form ammonia. Finally,

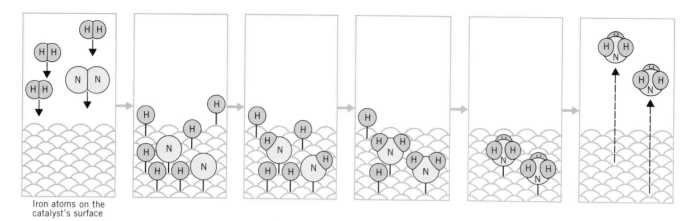

Iron atoms on the catalyst's surface

Figure 14.12
Catalytic formation of ammonia molecules on the surface of a catalyst. Nitrogen and hydrogen are adsorbed and dissociate into atoms that then combine to form ammonia molecules. In the final step, ammonia molecules leave the surface and the whole process can be repeated.

the completed ammonia molecule breaks away, freeing the surface of the catalyst for further reaction. This sequence of steps is illustrated in Figure 14.12.

Heterogeneous catalysts are used in many important commercial processes. The petroleum industry uses catalysts to crack hydrocarbons into smaller fragments and then reform them into the useful components of gasoline. The availability of catalysts allows refineries to produce gasoline, jet fuel, or heating oil in any ratio necessary to meet the demands of the marketplace.

The photograph on the left shows a catalytic cracking unit at an Exxon oil refinery in Argentina. On the right are beads of a reforming catalyst used in the production of high-octane gasoline.

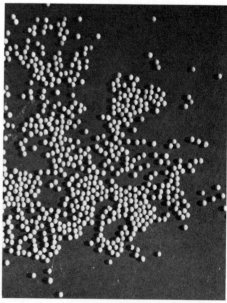

The use of catalytic converters has greatly reduced air pollution in cities, although severe pollution still exists in some places at certain times.

If you drive an automobile that requires unleaded gasoline, the car is equipped with a catalytic converter (Figure 14.13) designed to reduce the pollutants—primarily carbon monoxide and unburned hydrocarbons—in the exhaust. Air is mixed with the exhaust stream and passed over the catalyst on which the pollutants and oxygen are adsorbed. Carbon monoxide is oxidized to

One of more than 23 million catalytic converters built by General Motors Corporation since 1974. Cutaway view shows the beads on which the catalyst is supported and on which carbon monoxide and unburned hydrocarbons are oxidized to carbon dioxide and water. This unit also controls oxides of nitrogen.

Coleman catalytic heater that uses propane as its fuel.

The Mobil Oil Company has discovered a catalytic process for converting methanol to gasoline.

carbon dioxide, and hydrocarbons are oxidized to carbon dioxide and water. Newer catalysts that can remove nitric oxide, NO, from the exhaust by reducing it to N_2 and O_2 are also becoming available.

Unleaded gasoline must be used in cars equipped with catalytic converters because lead—present in gas as tetraethyl lead, $Pb(C_2H_5)_4$—interferes with the active sites on the catalytic surface. The lead "poisons" the surface by destroying its catalytic properties.

The poisoning of catalysts is a major problem in many industrial processes. Methanol (methyl alcohol, CH_3OH), for example, is a promising fuel. It can be made from coal and steam by the reaction

$$C \text{ (from coal)} + H_2O \longrightarrow CO + H_2$$

followed by

$$CO + 2H_2 \longrightarrow CH_3OH$$

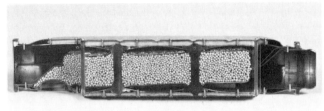

A catalyst for this process is copper(I) in solid solution with zinc oxide. However, traces of sulfur, a contaminant in coal, must be avoided because sulfur reacts with the catalyst and destroys its catalytic activity.

SUMMARY

Reaction Rates. The speed or rate of a reaction is controlled by five factors: (1) the nature of the reactants, (2) their ability to meet, (3) their concentrations, (4) the temperature, and (5) the presence of catalysts. The rates of heterogeneous reactions are determined largely by the area of contact between the phases; the rates of homogeneous

reactions are determined by the concentrations of the reactants. The rate of a reaction is measured by monitoring the change in reactant concentrations with time.

Rate Laws. The rate of a reaction is proportional to the product of the molar concentrations of the reactants, indicated by square brackets, [ ], each raised to an appropriate power. These exponents must be determined by experiments in which the concentrations are varied and the effects of the variations on the rate are measured. The proportionality constant, k, is called the rate constant. Its value depends on temperature but, of course, not on the concentrations of the reactants. The sum of the exponents in the rate law is the order (overall order) of the reaction.

Half-lives. The time required for half of a reactant to disappear is its half-life, $t_{1/2}$. For a first-order reaction, the half-life is constant; it is independent of the initial reactant concentration. The $t_{1/2}$ for a second-order reaction is inversely proportional to the initial concentration of the reactant.

Collision Theory. The rate of reaction depends on the number of effective collisions per second, which is only an extremely small fraction of the total number of collisions. One reason for the small number of successful collisions is that some collisions require special orientations of the reactant molecules. The major reason, however, is that the molecules must jointly possess a minimum activation energy, E_a. As the temperature increases, a larger fraction of reactant molecules have this necessary energy, so more collisions are effective and the reaction is faster. The E_a appears as an energy barrier on the potential-energy diagram for the reaction; the heat of reaction is the net potential energy difference between the reactants and the products. Since reactions are reversible, E_a for forward and reverse reactions can be identified. The species at the high point on the energy diagram is the activated complex and is in the transition state.

Reaction Mechanisms. The detailed sequence of elementary processes that lead to the net chemical change is the mechanism of the reaction. Since intermediates usually cannot be detected, the mechanism is a theory. Support for a mechanism comes from matching the predicted rate law for the mechanism with the rate law obtained from experimental data. For the rate-determining step (or any elementary process) the rate law has exponents equal to the coefficients.

Free Radical Reactions. Species with unpaired electrons tend to be very reactive because of nature's tendency to form electron-pair bonds. Free radicals can be formed from stable species by heat or light of the proper frequency. Reactions involving free radicals are often rapid chain reactions. Some chain reactions are branched—some steps produce more free radicals than they consume. These reactions tend to be explosive.

Catalysts. Catalysts are substances that affect the rate but are not consumed by the reaction. Homogeneous catalysts are in the same phase as the reactants and provide an alternative mechanism that has a lower activation energy than the uncatalyzed reaction. Heterogeneous catalysts provide a path of lower activation energy by having a surface on which the reactants are adsorbed and react.

REVIEW QUESTIONS

14.11. On the basis of what you learned in Chapter Ten, why do foods cook faster in a pressure cooker than in an open pot of boiling water?

14.12. Persons who have been submerged in very cold water and who are believed to have drowned sometimes can be revived. On the other hand, persons who have been submerged in warmer water for the same length of time have died. Explain this in terms of factors that affect the rates of chemical reactions.

14.13. What does rate of reaction mean in qualitative terms?

14.14. Give an example from everyday experience of (a) a very fast reaction, (b) a moderately fast reaction, and (c) a slow reaction.

14.15. From an economic point of view, why would industrial corporations want to know about the factors that affect the rate of a reaction?

14.16. What is an explosion?

14.17. Suppose we compared two reactions, one requiring the simultaneous collision of three molecules and the other requiring a collision between two molecules. From the standpoint of statistics, which reaction should be faster? Explain your answer.

14.18. What is the name that we give to the series of individual steps that lead to the net overall reaction?

14.19. State the five factors that affect the rates of chemical reactions.

14.20. What is a homogeneous reaction? Give an example.

14.21. What is a heterogeneous reaction? Give an example.

14.22. Why are chemical reactions usually carried out in solution?

14.23. What is the major factor that affects the rate of a heterogeneous reaction?

14.24. How does particle size affect the rate of a heterogeneous reaction? Why?

14.25. The rate of hardening of epoxy glue depends on the amount of hardener that is mixed into the glue. What factor affecting reaction rates does this illustrate?

14.26. What effect does temperature have on reaction rate?

14.27. What is a catalyst?

14.28. In cool weather, the number of chirps per minute from crickets diminishes. How can this be explained?

14.29. What units are used to express reaction rate?

14.30. In the reaction, $3H_2 + N_2 \rightarrow 2NH_3$, how does the rate of disappearance of hydrogen compare to the rate of disappearance of nitrogen? How does the rate of appearance of NH_3 compare to the rate of disappearance of nitrogen?

14.31. How do we represent molar concentration symbolically in a mathematical equation?

14.32. What is a rate law? What is the proportionality constant called?

14.33. What is meant by the order of a reaction?

14.34. What are the units of the rate constant for (a) a first-order reaction, (b) a second-order reaction, and (c) a third-order reaction?

14.35. How must the exponents in a rate law be determined?

14.36. Is there any way of using the coefficients in the balanced equation for a reaction to predict with certainty what the exponents are in the rate law?

14.37. If the concentration of a reactant is doubled and the reaction rate doubles, what must be the order of the reaction with respect to that reactant?

14.38. If the concentration of a reactant is doubled, by what factor will the rate increase if the reaction is second order with respect to that reactant?

14.39. What kind of experiments must be done to determine the exponents in the rate law for a reaction?

14.40. What is meant by the term *half-life?*

14.41. How is the half-life of a first-order reaction affected by the initial concentration of the reactant?

14.42. How is the half-life of a second-order reaction affected by the initial reactant concentration?

14.43. In general, to what extent are most reactions affected by a 10 °C rise in temperature?

14.44. What is the basic postulate of collision theory?

14.45. What two factors affect the effectiveness of molecular collisions in producing chemical change?

14.46. In terms of kinetic theory, why does an increase in temperature increase the reaction rate?

14.47. Draw the potential-energy diagram for an endothermic reaction. Indicate on the diagram the activation energy for both the forward and reverse reactions. Also indicate the heat of reaction.

14.48. Why does an endothermic reaction lead to a cooling of the reaction mixture if heat cannot enter from outside the system?

14.49. Define *transition state* and *activated complex.*

14.50. What is an elementary process and what is its relationship to a reaction mechanism?

14.51. What is a rate-determining step?

14.52. In what way is the rate law for a reaction related to the rate-determining step?

14.53. A reaction has the following mechanism:

$$2NO \longrightarrow N_2O_2$$
$$N_2O_2 + H_2 \longrightarrow N_2O + H_2O$$
$$N_2O + H_2 \longrightarrow N_2 + H_2O$$

What is the net overall change that occurs in this reaction?

14.54. If the reaction, $NO_2 + CO \rightarrow CO_2 + NO$, occurs by a one-step collision process, what would be the expected rate law for the reaction? The actual rate law is: rate = $k[NO_2]^2$. Could the reaction actually occur by a one-step collision between NO_2 and CO? Explain.

14.55. Oxidation of NO to NO_2—one of the reactions in the production of smog—appears to involve carbon monoxide. A possible mechanism is

$$CO + \cdot OH \longrightarrow CO_2 + H\cdot$$
$$H\cdot + O_2 \longrightarrow HOO\cdot$$
$$HOO\cdot + NO \longrightarrow \cdot OH + NO_2$$

Write the net chemical equation for the reaction.

14.56. Show that the following two mechanisms give the *same* net overall reaction.

Mechanism 1 $OCl^- + H_2O \longrightarrow HOCl + OH^-$
          $HOCl + I^- \longrightarrow HOI + Cl^-$
          $HOI + OH^- \longrightarrow H_2O + OI^-$

Mechanism 2 $OCl^- + H_2O \longrightarrow HOCl + OH^-$
          $I^- + HOCl \longrightarrow ICl + OH^-$
          $ICl + 2OH^- \longrightarrow OI^- + Cl^- + H_2O$

14.57. The experimental rate law for the reaction, $NO_2 + CO \rightarrow CO_2 + NO$, is rate = $k[NO_2]^2$. If the mechanism is

$$2NO_2 \longrightarrow NO_3 + NO \quad \text{(slow)}$$
$$NO_3 + CO \longrightarrow NO_2 + CO_2 \quad \text{(fast)}$$

show that the predicted rate law is the same as the experimental rate law.

14.58. What is a free radical? How can they be generated from stable molecules?

14.59. The reaction of hydrogen and bromine appears to follow the mechanism

$$Br_2 \xrightarrow{h\nu} 2Br\cdot$$
$$Br\cdot + H_2 \longrightarrow HBr + H\cdot$$
$$H\cdot + Br_2 \longrightarrow HBr + Br\cdot$$
$$2Br\cdot \longrightarrow Br_2$$

(a) Identify the initiation step in the mechanism.
(b) Identify any propagation steps.
(c) Identify the termination step.
The mechanism also contains the reaction:

$$H\cdot + HBr \longrightarrow H_2 + Br\cdot$$

How does this reaction affect the rate of production of HBr?

14.60. In the upper atmosphere, an ozone layer shields the earth from harmful ultraviolet radiation as discussed in Special Topic 12.2. The ozone is generated by the reactions:

$$O_2 + h\nu \longrightarrow 2O$$
$$O + O_2 \longrightarrow O_3$$

Release of chlorofluorocarbon aerosol propellants and refrigerant fluids such as Freon 12 (CCl_2F_2) into the atmosphere threatens to at least partially destroy the ozone shield by the reactions:

$$CCl_2F_2 \longrightarrow CClF_2 + Cl$$
$$Cl + O_3 \longrightarrow ClO + O_2 \quad (1)$$
$$ClO + O \longrightarrow Cl + O_2 \quad (2)$$

Explain how reactions 1 and 2 constitute a chain reaction that can destroy huge numbers of O_3 molecules and O atoms that might otherwise react to replenish the ozone.

14.61. How does a catalyst increase the rate of a chemical reaction?

14.62. What is a homogeneous catalyst? How does it function, in general terms?

14.63. What is a heterogeneous catalyst? How does it function?

14.64. What is the difference in meaning between the terms adsorption and absorption? (If necessary, use a dictionary.) Which one applies to heterogeneous catalysts?

14.65. What does the catalytic converter do in the exhaust system of an automobile? Why can't leaded gasoline be used in cars equipped with catalytic converters?

REVIEW PROBLEMS

14.66. The following data were collected for the reaction, $SO_2Cl_2 \rightarrow SO_2 + Cl_2$, at a certain temperature.

[SO_2Cl_2](mol/liter)	Time (s)
0.100	0
0.082	100
0.067	200
0.055	300
0.045	400
0.037	500
0.030	600
0.025	700
0.020	800

Make a graph of concentration versus time and determine the rate of the reaction at $t = 200$ seconds and $t = 600$ seconds.

14.67. The following data were collected on the following reaction at 530 °C.

$$CH_3CHO \longrightarrow CH_4 + CO$$

[CH_3CHO](mol/liter)	Time (s)
0.200	0
0.153	20
0.124	40
0.104	60
0.090	80
0.079	100
0.070	120
0.063	140
0.058	160
0.053	180
0.049	200

Make a graph of concentration versus time and determine the reaction rate after 60 seconds and after 120 seconds.

14.68. For the reaction, $2A + B \rightarrow 3C$, it was found that the rate of disappearance of B was 0.30 mol liter^{-1} s^{-1}. What was the rate of disappearance of A and the rate of appearance of C?

14.69. At a certain temperature, the rate of decomposition of N_2O_5:

$$2N_2O_5 \longrightarrow 4NO_2 + O_2$$

is 2.5×10^{-6} mol liter^{-1} s^{-1}. How fast are NO_2 and O_2 being formed?

14.70. The reaction, $2NO + O_2 \rightarrow 2NO_2$, has the rate law, rate $= k[NO_2]^2[O_2]$. At 25 °C, $k = 7.1 \times 10^9$ liter^2 mol^{-2} s^{-1}. What is the rate of reaction when $[NO_2] = 0.0010$ mol/liter and $[O_2] = 0.034$ mol/liter?

14.71. The decomposition of N_2O_5 has the rate law, rate $= k[N_2O_5]$. If $k = 1.0 \times 10^{-5}$ s^{-1}, what is the reaction rate when the N_2O_5 concentration is 0.0010 mol/liter?

14.72. The reaction, $NO + O_3 \rightarrow NO_2 + O_2$, has the rate law, rate $= k[NO][O_3]$. What is the order with respect to each reactant and what is the overall order of the reaction?

14.73. Biological reactions involve an interaction of an enzyme with a *substrate*, the substance that actually undergoes the chemical change. In many cases, the rate of reaction depends on the concentration of the enzyme but is independent of the substrate concentration. What is the order of the reaction with respect to the substrate in such instances?

14.74. The following data were collected on a reaction:

$$M + N \rightarrow P + Q$$

Initial Concentrations (mol/liter)		Initial Rate
[M]	[N]	(mol liter^{-1} s^{-1})
0.010	0.010	2.5×10^{-3}
0.020	0.010	5.0×10^{-3}
0.020	0.030	4.5×10^{-2}

What is the rate law for the reaction? What is the value of the rate constant?

14.75. At a certain temperature the following data were collected for the reaction, $2ICl + H_2 \rightarrow I_2 + 2HCl$.

Initial Concentrations (mol/liter)		Initial Rate
[ICl]	[H$_2$]	(mol liter^{-1} s^{-1})
0.10	0.10	0.0015
0.20	0.10	0.0030
0.10	0.050	0.00075

Determine the rate law and the rate constant for the reaction.

14.76. A certain first-order reaction has a rate constant, $k = 1.6 \times 10^{-3}$ s^{-1}. What is the half-life for this reaction?

14.77. The decomposition of NOCl, $2NOCl \rightarrow 2NO + Cl_2$, is a second-order reaction with $k = 6.7 \times 10^{-4}$ liter mol^{-1} s^{-1} at 400 K. What is the half-life of this reaction if the initial concentration of NOCl is 0.20 mol/liter?

14.78. Using the graph from Question 14.66, determine the time required for the SO_2Cl_2 concentration to drop from 0.100 mol/liter to 0.050 mol/liter. How long does it take for the concentration to drop from 0.050 mol/liter to 0.025 mol/liter? What is the order of this reaction? (*Hint.* How is the half-life related to concentration?)

14.79. Using the graph from Question 14.67, determine how long it takes for the CH_3CHO concentration to decrease from 0.200 mol/liter to 0.100 mol/liter. How long does it take the concentration to drop from 0.100 mol/liter to 0.050 mol/liter? What is the order of this reaction? (*Hint.* How is the half-life related to concentration?)

14.80. The half-life of a certain first-order reaction is 15 minutes. What fraction of the original reactant concentration will remain after 2.0 hours?

14.81. Strontium-90 has a half-life of 28 years. How long will it take for all of the strontium-90 presently on earth to be reduced to 1/32 of its present amount?

CHAPTER FIFTEEN

THERMODYNAMICS: A GUIDE TO WHAT'S POSSIBLE

15.1 INTRODUCTION

Studying the energy flow accompanying changes tells us what is possible and what is not

In the last chapter we studied the factors that control how fast chemical reactions proceed. Now we turn our attention to an even more basic question: What factors determine whether a reaction can take place at all? Obviously, any process, be it chemical or physical, must be possible before we can study its rate.

If we look around us, we see events of all sorts taking place. Plants grow, mature, and die. Leaves turn in autumn and fall to the ground. Rivers flow and stone walls crumble. Fuels burn to power machinery and, perhaps most important to us, biochemical reactions occur in our bodies that keep us alive and permit us to function—even to read chemistry books. But what is it about these processes that make them possible? Why do they occur at all? Why don't leaves fall up instead of down, and why doesn't the passage of time gradually turn heaps of sand into stone walls?

The answers to these questions can be found by understanding the principles of thermodynamics. As implied by its name, **thermodynamics** is the study of energy changes and the flow of energy from one substance to another—*thermo* suggesting heat and *dynamics* suggesting movement. You were already exposed to some of the principles of thermodynamics in Chapter Five when you learned to calculate enthalpy changes, or heats of reaction, although the label, "thermodynamics," wasn't used at that time. You should recall that enthalpy changes deal with the exchange of heat between a chemical system and its surroundings. Such transfers of heat, however, represent only one aspect of thermodynamics, as we will see in the pages ahead.

The answers to these questions aren't encouraging.

Although the principles of thermodynamics are important in science, they can also provide us with answers to some of the most perplexing questions that concern modern society. What, for instance, is the origin of the world's energy crisis, and what long-range solutions (if any) are possible? How can we best deal with the problem of pollution? Is it, in fact, possible to rid the world of pollution? We will explore these questions in the pages ahead as we develop the concepts of thermodynamics.

15.2 SPONTANEOUS CHANGE

Literally everything that happens begins with some spontaneous change somewhere

We take many spontaneous events for granted. What would you think if you dropped a book and it rose to the ceiling?

Many of the natural processes we witness each day occur without outside help. A book that slips from your hand will start to fall. The ice cubes in the cold drink you fix to quench your thirst on a warm day will gradually melt. Such events are examples of **spontaneous changes**—they occur by themselves without outside assistance. Once conditions are right for them to begin, they proceed on their own.

Some spontaneous changes occur very rapidly. An example is the set of biochemical reactions that take place when you accidentally touch something that is very hot—they cause you to jerk your hand away quickly. Other spontaneous events, such as the gradual erosion of a mountain, occur slowly and many years pass before a change is noticed. Still others occur at such an extremely slow rate under ordinary conditions that they appear not to be spontaneous at all. Gasoline-oxygen mixtures appear perfectly stable indefinitely at room temperature because under these conditions they react so very slowly. However, if heated, their rate of reaction increases and they can react explosively until all of one or the other is totally consumed.

Each day we also witness many events that are obviously not spontaneous. We may pass by a pile of bricks in the morning and later in the day find that they have become a brick wall. We know from experience that the wall didn't get there by itself. A pile of bricks becoming a brick wall is *not* spontaneous; it requires the intervention of a bricklayer. Similarly, the decomposition of water into hydrogen and oxygen is not spontaneous. We see water all the time and we know that it is stable. Nevertheless, we can cause water to decompose by passing an electric current through it in a process called *electrolysis*.

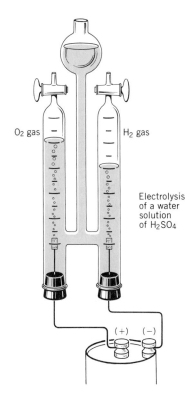

O₂ gas

H₂ gas

Electrolysis of a water solution of H_2SO_4

(+)   (−)

$$2H_2O(\ell) \longrightarrow 2H_2(g) + O_2(g)$$

This process will continue, however, only as long as the electric current is maintained. As soon as the supply of electricity is cut off, the decomposition ceases. This example demonstrates the difference between spontaneous and nonspontaneous changes. Once a spontaneous event begins, it has a tendency to continue until it is finished, A nonspontaneous event, on the other hand, can continue only as long as it receives some sort of outside assistance.

Nonspontaneous changes have another common characteristic. They can occur only when some spontaneous change has occurred first. The bricklayer consumes food, and a series of spontaneous biochemical reactions then occur that supply the necessary muscle power to build the wall. Similarly, the nonspontaneous electrolysis of water requires some sort of spontaneous mechanical or chemical change to generate the needed electricity. In short, all nonspontaneous events occur at the expense of spontaneous ones. Everything that happens can be traced, either directly or indirectly, to some spontaneous change.

The driving of nonspontaneous reactions to completion by linking them to spontaneous ones is an important principle in biochemistry.

15.3 ENTHALPY CHANGES AND SPONTANEITY

Exothermic changes tend to be spontaneous

Because spontaneous reactions are so important, it is necessary for us to understand the factors that favor spontaneity. We can begin by simply examining some everyday events such as those depicted in Figure 15.1. We know, for example, that a skateboard placed at the top of a hill will roll to the bottom all by itself. We also know that water flows downhill and that heat is generated when gasoline

Figure 15.1
Three common spontaneous events—skateboarders roll downhill, water cascades over waterfalls, and fuels burn in air.

burns. Each of these events is spontaneous. In fact, we expect them to happen and would be quite surprised if they didn't. Can we find some factor that all of them have in common?

If we study these events, we see that each of them leads to a lowering, or decrease, in the energy of the system. Both the skateboard and the water lose potential energy as they move from a higher to a lower altitude. Similarly, the chemical substances in the gasoline-oxygen mixture lose energy by evolving heat as they react to produce CO_2 and H_2O. Because these events are spontaneous, we conclude that *when a change lowers the energy of a system, it tends to occur spontaneously.* Since a change that lowers the energy of a system is said to be exothermic, we can state this factor another way—*exothermic changes have a tendency to proceed spontaneously.*

In the preceding statement the word *tendency* should be emphasized. We will see that not every exothermic change is spontaneous, nor is every endothermic one nonspontaneous. The important point is that an energy decrease works as *one factor* in favor of spontaneity.

We now have to ask: How does this relate to chemical reactions? In Chapter Five, we examined the energy changes that accompany chemical reactions. You should recall that the heat of reaction, or enthalpy change, ΔH, specifies the quantity of energy absorbed or released when a reaction occurs at constant temperature and pressure. If ΔH is measured at 25 °C and 1 atmosphere, it is called the standard heat of reaction, or standard enthalpy change, $\Delta H°$. You should also

> Most, but not all, chemical reactions that are exothermic occur spontaneously.

remember that a negative value for ΔH (or $\Delta H°$) signifies that the products possess less energy than the reactants and the reaction is exothermic. As a result, we can say that reactions for which ΔH is negative tend to proceed spontaneously. The enthalpy change thus serves as one of the thermodynamic factors that influence whether or not a given process will occur by itself. Because enthalpy changes are important in this way, you may find it worthwhile at this point to briefly review Sections 5.4 through 5.6.

15.4 ENTROPY AND SPONTANEOUS CHANGE

An increase in randomness favors a spontaneous change

For many years, scientists believed that only exothermic changes could occur spontaneously, despite the fact that they were all familiar with a common

Thermodynamics, like many other branches of science, is founded on a number of basic underlying facts of nature. They form the base on which thermodynamic principles are built and are called laws. The **first law of thermodynamics** can be stated in a number of ways, but they all reduce to the *law of conservation of energy*—energy can be neither created nor destroyed. The first law, as it is sometimes abbreviated, serves as the basis for the Hess's law calculations that you learned to perform in Chapter 5. The validity of Hess's law and the concept of enthalpy being a state function (i.e., ΔH being controlled only by the initial and final states of the system) depends on the law of conservation of energy. If energy could appear or disappear, there would be no reason to expect that the enthalpy change for some overall reaction could be obtained simply by adding the enthalpy changes of a series of thermochemical equations that combine to give the overall equation. You will meet the other laws of thermodynamics later in this chapter.

exception—the melting of ice. Because ice melts only if heat is supplied, the event is endothermic. Yet it is obviously spontaneous. On a warm day, ice melts quite well all by itself.

There are other examples of spontaneous endothermic changes. The evaporation of water is one. We have seen that the heat of vaporization is positive. Another example is the dissolving of certain salts in water. Since all of these events are both endothermic and spontaneous, they must have some factor in common that allows them to overcome the unfavorable change in energy.

Careful examination shows that in each of these events there is an increase in the randomness, or disorder, of the system. The water molecules in ice are arranged in a highly organized crystal pattern that allows little movement. As the ice melts, the water molecules become disorganized and are able to move more freely. The movements of these molecules become freer still when the water evaporates—instead of being confined to the liquid, they are able to roam throughout the entire atmosphere. The situation when certain salts dissolve in water is similar. The ions that form the salt are able to leave their highly ordered crystal arrangement and can then spread randomly throughout the solution.

Something that brings about randomness is more likely to occur than is something that brings about order. We can understand the reason for this by looking at the close relationship between randomness and statistical probability. Suppose, for instance, that we have a deck of playing cards separated according to suit and arranged numerically. We can see that the sequence of cards is certainly highly ordered. Now, if we throw the cards into the air, sweep them up, and restack them, we will almost surely find that they have become disordered. We expect this because, when the cards are tossed, there are many ways for them to become disordered, but there is only one way for them to come together again in their original sequence. On the basis of pure chance, therefore, a disordered sequence is far more probable than the ordered one with which we began.

This same principle applies to any physical or chemical event. However, the laws of chance are sometimes hidden by the changes in energy. (We will have more to say about this later.) If the energy changes could be ignored or made less important, statistical probability would become the primary driving force for change.

The cards are highly ordered

After they've been tossed they are disordered

The thermodynamic quantity that describes the degree of randomness of a system is called **entropy,** symbolized by the letter S. Like the enthalpy, H, the entropy of a system is a state function. It depends only on the initial and final states of the system. Neither the prior history of the system nor the path followed from the start of any process that affects the system to its completion need be considered. The change in entropy, ΔS, for any event is given by the equation

$$\Delta S = S_{final} - S_{initial}$$

As you can see, when S_{final} is greater than $S_{initial}$, ΔS is positive. Thus, when the randomness of a system increases, the entropy of the system increases.

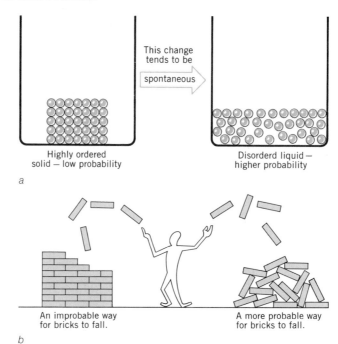

Figure 15.2
(a) If water molecules are dropped into the container, they would not be expected to land in just the right places to give the highly ordered solid. If the way they land is determined purely by chance, the disordered liquid is more likely to occur. (b) Tossing bricks into the air is more likely to give a pile of bricks than a brick wall.

Now let's look again at the melting of an ice cube (Figure 15.2). Of all the possible ways of placing water molecules into a container, the highly ordered arrangement within the crystalline ice cube has a very low probability, while the disordered arrangement of the molecules found in the liquid is more probable. An analogy would be throwing bricks in the air and observing how they land. The chance that they will fall one on the other to form a brick wall is very small—the brick wall is an improbable result. A more probable arrangement is a jumbled, disordered pile. Thus, when the ice melts, the system merely passes from a less probable distribution of the molecules to a more probable one. This is the same thing as saying that the system passes from a state of low entropy to one of higher entropy. We have seen that any event that moves in a direction that increases the randomness of the system (i.e., any change that results in a positive value for ΔS) also increases its probability of occurrence, so it tends to happen naturally. Therefore, we can say that *any event that is accompanied by an increase in entropy tends to be spontaneous.*

We're in a position now to attempt to answer some of the questions posed in the introduction to this chapter. One of the most far-reaching observations in science is incorporated in the **second law of thermodynamics,** which states, in effect, that whenever a spontaneous event takes place in our universe, it is accompanied by an overall increase in entropy. Because everything that happens relies on spontaneous changes of some sort, the entropy of the universe is constantly increasing. Now, much of our time and effort is spent in tidying up our own home and neighborhood. We seem constantly involved in trying to decrease disorder in our lives. We clean up our desks and sweep the floor. We put out the garbage, mow the lawn, and rake leaves in the autumn. Overall our activities are spontaneous because the biochemical reactions driven by the foods we eat and digest are spontaneous and allow us to do these things. Since the total entropy of the universe must be increased by our spontaneous activities, the increased order that we create for ourselves has to be balanced by an even larger increase in disorder somewhere in our surroundings—the environment.

Pollution involves the scattering of undesirable substances through our environment and is accompanied by an increase in entropy. It is a direct result of our efforts to create an orderly world. For example, if we burn our leaves to get rid of them, the smoke from our fire spreads through the neighborhood and disturbs our friends. When we expand our efforts beyond our home and attempt to provide order in our environment, our attempts are met by still more overall disorder. For instance, we use machinery to collect garbage so that we and our neighbors don't have to live in it. But the machinery, by burning fuel, generates air pollution that is even more difficult to clean up. It's a losing game. We are simply faced with the fact that we can never really eliminate pollution. All we can do is to try to keep it from places where it can do us great harm (Figure 15.3) and attempt to confine it to locations where it interferes with life as little as possible.

Figure 15.3
(a) Pollution and algae growth clog the shores of Lake Tahoe near the California–Nevada border in 1972. (b) Effluent from a chemical plant spills into the waters of Michigan.

a

b

The major oil companies are currently spending large sums to remove gasoline that has begun to leak from many old storage tanks buried underground at service stations 20 years or so old. This is a special problem in areas that depend on ground water for drinking and cooking.

The entropy effect on pollution is important to understand, particularly when we consider the consequences of releasing harmful substances into the environment. Such materials include the highly toxic elements released in trace amounts when coal or nuclear fuels are used, as well as many toxic insecticides such as DDT. When these chemicals are released, their spread is unavoidable because of the large entropy increase that occurs as they are scattered about. Eliminating them, once they have had an opportunity to disperse, is an almost impossible task because doing so requires an enormous expenditure of energy and *must* generate even more disorder around us. Only if the pollutant is extremely hazardous does it warrant the effort necessary to reduce its concentration in the environment, and even if this is done, we can never eliminate the pollutant entirely. We can only reduce our risk of injury. The surest way to overcome pollution, therefore, is not to create it in the first place. It is a trade-off between risk and benefit. For the benefits of activities that pollute (e.g., using fuels to make electricity), we try to reduce the risks as much as we judge possible and endure the rest.

Let's look briefly now at the energy crisis, another problem that comes to us because of the second law of thermodynamics. In one sense there really is no lack of energy. Energy cannot be destroyed (the law of conservation of energy). Our problems actually arise from the unavailability of energy. Every time we perform some spontaneous activity some energy becomes unavailable for future use because it is scattered about the universe (our environment) as heat. In other words, energy becomes disorganized or disordered when we use it. When we burn gasoline to take our car from one place to another some of the energy is lost as heat from the radiator, which cools the engine. As the car moves, more energy is lost through friction between the moving parts and between the tires and the pavement. When we step on the brake, friction between the brakes and the wheels generates still more heat which becomes lost to the environment.

In the design of electric cars, the kinetic energy of the auto is used to operate a generator that charges the batteries as the car slows down. This reduces the amount of energy lost as heat, since some of it is recovered to be used again.

In our day-to-day life, then, we don't consume energy; we merely convert it from a form where it is accessible, and where it can do work for us, to a form where we can no longer recover it. Our search for energy is really a search for usable forms of energy, whether it be stored in the bonds of chemical fuels, in nuclear fuels, or provided to us by the sun's rays.

15.5 THE THIRD LAW OF THERMODYNAMICS

The entropy of a pure crystal is zero at absolute zero

The entropy of a substance (i.e., the extent of its thermodynamic disorder) varies directly with the temperature of the substance. The lower the temperature, the lower the entropy. For example, at a pressure of 1 atmosphere and a temperature above 100 °C, water exists as a highly disordered gas with a very high entropy. If confined, the molecules of water vapor will be spread evenly throughout their container, and they will be in constant random motion (Figure 15.4). When the system is cooled, the water vapor eventually condenses to form a liquid. Although the molecules can still move somewhat freely, they are now confined to the bottom of the container. Their distribution in the container is not as random as it was in the gas and the entropy of the liquid is lower. Further cooling decreases the entropy even more, and below 0 °C, the water molecules join together to form ice, a crystalline solid. The molecules are now in a highly ordered state, particularly in comparison to that of the gas, and the entropy of the system is very low.

Yet even in the crystalline form, the order isn't perfect and the water molecules still have some entropy. There is enough thermal energy left to cause them to vibrate or rotate within the general area of their lattice sites. Thus at any particular instant, we would find the molecules near, but probably not exactly at, their equilibrium lattice positions (Figure 15.5). If we cool the solid further, we decrease the thermal energy and the molecules spend less time away from their equilib-

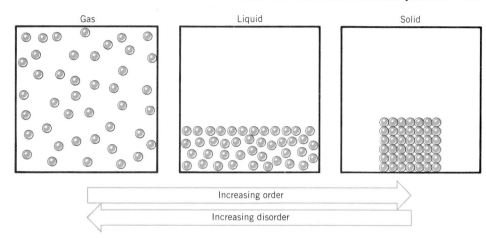

Figure 15.4
Degrees of disorder in the three states of matter.

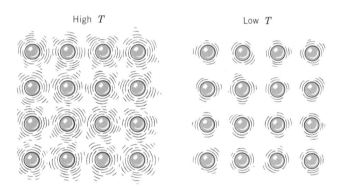

Figure 15.5
Greater disorder in a crystal at high temperature is caused by more violent molecular vibrations, which leads to a greater displacement of the atoms from their equilibrium lattice positions.

rium lattice positions. The order of the crystal increases and the entropy decreases. Finally, at absolute zero, the ice will be in a state of perfect order and its entropy will be zero. This leads us to the statement of the **third law of thermodynamics.** *At absolute zero, the entropy of a pure crystal is also zero.*

$$S = 0 \quad \text{at} \quad T = 0 \text{ K}$$

Because we know the point at which entropy has a value of zero, it is possible by measurement and calculation to determine the actual amount of entropy that a substance possesses at temperatures above 0 K. If the entropy of one mole of a substance is determined at a temperature of 298 K (25 °C) and a pressure of 1 atm, we call it the **standard entropy,** $S°$. Table 15.1 lists the standard entropies for a number of substances. Notice that entropy has the dimensions of energy/temperature (i.e., calories per kelvin or joules per kelvin). Once we know the entropies of a variety of substances we can calculate the **standard entropy change,** $\Delta S°$, for chemical reactions in much the same way as we calculated $\Delta H°$ in Chapter Five.

$$\Delta S° = (\text{sum of } S° \text{ of products}) - (\text{sum of } S° \text{ of reactants}) \quad (15.1)$$

If the reaction with which we are working corresponds to the formation of 1 mol of a compound from its elements, the $\Delta S°$ that we calculate can be referred to as the standard entropy of formation, $\Delta S_f°$. Values of $\Delta S_f°$ are not tabulated; if we need them, we must calculate them from values of $S°$.

Special Topic 15.3 Why the Units of Entropy are Energy/Temperature

Entropy is a state function, just like enthalpy. The value of ΔS doesn't depend on the "path" that is followed. In other words, we can proceed from one state to another in any way we like and ΔS will be the same. If we choose a path in which just a slight alteration in the system can change the direction of the process, we can measure ΔS directly. For example, at 25 °C an ice cube will melt and there is nothing we can do at that temperature to stop it. At 0 °C, however, it is simple to stop the melting process and reverse its direction. At 0 °C the melting of ice is said to be a reversible process. If we set up a change so that it is reversible, we can calculate the entropy change as $\Delta S = q/T$, where q is the heat added to the substance and T is the temperature at which the heat is added. We can understand this in the following way. In a system of molecules held together by attractive forces, disorder is introduced by adding heat, which makes the molecules move more violently. In a crystal, for instance, the molecules vibrate about their equilibrium lattice positions, so that at any instant we really have a "not-quite-ordered" collection of particles. Adding heat increases molecular motion and leads to greater instantaneous disorder. In other words, if we could freeze the motion of the molecules, like a photographer's flash freezes action, we would see that after we have added some heat there is greater disorder and therefore greater entropy. This entropy increase is *directly proportional to the amount of heat added.* However, this added heat makes a more noticeable and significant difference if it is added at low temperature, where little disorder exists, rather than at high temperature where there is already substantial disorder. For a given quantity of heat the entropy change is therefore *inversely proportional to the temperature at which the heat is added.* Thus, $\Delta S = q/T$, and entropy has units of energy divided by temperature (e.g., cal/K or J/K).

Table 15.1
Standard Entropies of Some Typical Substances (25 °C, 1 atm)

	Entropy, S°			Entropy, S°	
Substance	**cal/mol K**	**J/mol K**	**Substance**	**cal/mol K**	**J/mol K**
$Ag(s)$	41.32	(172.9)	$H_2O(g)$	45.11	(188.7)
$AgCl(s)$	58.5	(245)	$H_2O(\ell)$	16.72	(69.96)
$Al(s)$	6.77	(28.3)	$HCl(g)$	44.62	(186.7)
$Al_2O_3(s)$	12.19	(51.00)	$HNO_3(\ell)$	37.19	(155.6)
$C(s, \text{graphite})$	0.58	(2.4)	$H_2SO_4(\ell)$	37.5	(157)
$CO(g)$	47.30	(197.9)	$HC_2H_3O_2(\ell)$	38.2	(160)
$CO_2(g)$	51.06	(213.6)	$Hg(\ell)$	18.2	(76.1)
$CH_4(g)$	44.50	(186.2)	$Hg(g)$	41.8	(175)
$CH_3Cl(g)$	55.97	(234.2)	$K(s)$	38.30	(160.2)
$CH_3OH(\ell)$	30.3	(126.8)	$KCl(s)$	57.24	(239.5)
$CO(NH_2)_2(s)$	25.00	(104.6)	$K_2SO_4(s)$	42.0	(176)
$CO(NH_2)_2(aq)$	41.55	(173.8)	$N_2(g)$	45.77	(191.5)
$C_2H_2(g)$	48.00	(200.8)	$NH_3(g)$	46.01	(192.5)
$C_2H_4(g)$	52.54	(219.8)	$NH_4Cl(s)$	22.6	(94.6)
$C_2H_6(g)$	54.85	(229.5)	$NO(g)$	50.34	(210.6)
$C_2H_5OH(\ell)$	38.4	(161)	$NO_2(g)$	57.47	(240.5)
$Ca(s)$	36.99	(154.8)	$N_2O(g)$	52.58	(220.0)
$CaCO_3(s)$	22.2	(92.9)	$N_2O_4(g)$	72.7	(304)
$CaCl_2(s)$	27.2	(114)	$Na(s)$	36.72	(153.6)
$CaO(s)$	9.5	(40)	$NaCl(s)$	17.30	(72.38)
$Ca(OH)_2(s)$	18.2	(76.1)	$NaOH(s)$	15.34	(64.18)
$CaSO_4(s)$	25.5	(107)	$Na_2SO_4(s)$	35.7	(149.4)
$CaSO_4 \cdot \frac{1}{2}H_2O(s)$	31.2	(131)	$O_2(g)$	49.00	(205.0)
$CaSO_4 \cdot 2H_2O(s)$	46.36	(194.0)	$PbO(s)$	16.2	(67.8)
$Cl_2(g)$	53.29	(223.0)	$S(s)$	7.62	(31.9)
$Fe(s)$	6.5	(27)	$SO_2(g)$	59.40	(248.5)
$Fe_2O_3(s)$	21.5	(90.0)	$SO_3(g)$	61.24	(256.2)
$H_2(g)$	31.21	(130.6)			

EXAMPLE 15.1   Calculating $\Delta S°$ from Standard Entropies

Problem: Urea (from urine) hydrolyzes slowly in the presence of water to produce ammonia and carbon dioxide.

$$CO(NH_2)_2(aq) + H_2O(\ell) \longrightarrow CO_2(g) + 2NH_3(g)$$
urea

What is the standard entropy change, in cal/K, for this reaction when 1 mol of urea reacts with water?

Solution: The standard entropies for reactants and products are found in Table 15.1.

Substance	$S°$(cal/mol K)
$CO(NH_2)_2(aq)$	41.55
$H_2O(\ell)$	16.72
$CO_2(g)$	51.06
$NH_3(g)$	46.01

Following Equation 15.1 we have

$$\Delta S° = (S°_{CO_2} + 2S°_{NH_3}) - (S°_{CO(NH_2)_2} + S°_{H_2O})$$

$$= \left[1 \text{ mol} \times \left(\frac{51.06 \text{ cal}}{\text{mol K}}\right) + 2 \text{ mol} \times \left(\frac{46.01 \text{ cal}}{\text{mol K}}\right)\right]$$

$$- \left[1 \text{ mol} \times \left(\frac{41.55 \text{ cal}}{\text{mol K}}\right) + 1 \text{ mol} \times \left(\frac{16.72 \text{ cal}}{\text{mol K}}\right)\right]$$

$$= (143.08 \text{ cal/K}) - (58.27 \text{ cal/K})$$

$$= 84.81 \text{ cal/K}$$

EXERCISE 15.1   Calculate the standard entropy change for the following reactions:
(a) $CaO(s) + 2HCl(g) \longrightarrow CaCl_2(s) + H_2O(\ell)$
(b) $C_2H_4(g) + H_2(g) \longrightarrow C_2H_6(g)$

15.6   THE GIBBS FREE ENERGY

The Gibbs free energy gives us the net effect on spontaneity of the enthalpy and entropy changes

It's easy to see why stone walls need almost constant attention.

We have now seen that two factors—enthalpy and entropy—determine whether or not a given physical or chemical event will be spontaneous. Sometimes these two factors work together. For example, when a stone wall crumbles, its enthalpy decreases and its entropy increases. Since a decrease in enthalpy and an increase in entropy both favor a spontaneous change, the two factors complement one another. In other situations, the effects of enthalpy and entropy are in opposition. Such is the case, as we have seen, in the melting of ice or the evaporation of water. The endothermic nature of these changes tends to make them nonspontaneous, while the increase in the randomness of the molecules tends to make them spontaneous.

When enthalpy and entropy oppose one another, the results of the interactions between them are far from obvious. However, we can discover the net effect of the two factors through the use of another thermodynamic quantity called the

Gibbs free energy, G, named after Josiah Willard Gibbs (1839–1903), a famous U. S. scientist. The Gibbs free energy[1] is defined as

$$G = H - TS \tag{15.2}$$

For a change at constant temperature and pressure, the equation becomes

$$\Delta G = \Delta H - T\Delta S \tag{15.3}$$

Once again we have a state function, a quantity that is independent of the path of the event. This means that

$$\Delta G = G_{final} - G_{initial}$$

Reactions that occur with a free energy decrease to the system are sometimes said to be **exergonic.** Those that occur with a free energy increase are sometimes said to be **endergonic.**

What is of particular importance to us is the fact that *a change can only be spontaneous if it is accompanied by a decrease in free energy.* In other words, for a change to be spontaneous, G_{final} must be less than $G_{initial}$ and ΔG must be negative. What does this mean in terms of the signs of ΔH and ΔS?

When a change is exothermic and is also accompanied by an increase in entropy, both factors favor spontaneity.

$$\Delta H \text{ is negative } (-)$$
$$\Delta S \text{ is positive } (+)$$
$$\Delta G = \Delta H - T\Delta S$$
$$= (-) - T(+)$$

In such a change, ΔG will be negative regardless of the value of the absolute temperature, T (which can only have positive values). Therefore, the change will occur spontaneously at *all* temperatures.

On the other hand, if a change is endothermic and is accompanied by a decrease in entropy, both factors work against spontaneity.

$$\Delta H \text{ is positive } (+)$$
$$\Delta S \text{ is negative } (-)$$
$$\Delta G = \Delta H - T\Delta S$$
$$= (+) - T(-)$$

In this case, ΔG will be positive at all temperatures and the change will always be nonspontaneous.

When ΔH and ΔS have the same sign, the temperature becomes critical in determining whether or not an event is spontaneous. If ΔH and ΔS are both positive,

$$\Delta G = (+) - T(+)$$

Only at relatively high temperatures will the value of $T\Delta S$ be larger than the value of ΔH so that their difference, ΔG, is negative. A familiar example is the melting of ice.

$$H_2O(s) \longrightarrow H_2O(\ell)$$

Here is a change that we know is endothermic and occurs with an increase in entropy. At temperatures above 0 °C (when the pressure is 1 atm), ice melts because the $T\Delta S$ term is bigger than the ΔH term. At lower temperatures, ice doesn't melt because the smaller value of T gives a smaller value for $T\Delta S$ and the difference, $\Delta H - T\Delta S$, is positive.

For similar reasons, when ΔH and ΔS are both negative, ΔG will be negative only at relatively low temperatures. The freezing of water is an example.

$$H_2O(\ell) \longrightarrow H_2O(s)$$

Energy is released as the solid is formed and the entropy decreases. You know, of course, that water freezes spontaneously at low temperatures, that is, below 0 °C.

	ΔH	
	+	**−**
ΔS **+**	Spontaneous only at high T	Spontaneous at all T
ΔS **−**	Non-spontaneous at all T	Spontaneous only at low T

Figure 15.6
Summary of the effects of the signs of ΔH and ΔS on spontaneity.

[1] The origin of *free* in the term *free energy* is discussed in Section 15.8.

Figure 15.6 summarizes the effects on ΔG—and hence on the spontaneity of a physical or chemical event—of the signs of ΔH and ΔS.

15.7 STANDARD FREE ENERGIES

Standard free energies of formation can be used to obtain standard free energies of reaction.

When ΔG is determined at 25 °C (298 K) and 1 atm, we call it the **standard free energy change, $\Delta G°$.** There are a number of ways of obtaining $\Delta G°$ for a reaction. One of them is to compute $\Delta G°$ from $\Delta H°$ and $\Delta S°$.

$$\Delta G° = \Delta H° - (298 \text{ K}) \Delta S°$$

Experimental measurement of $\Delta G°$ is also possible, but we will discuss how this is done later in other chapters.

EXAMPLE 15.2   **Calculating $\Delta G°$ from $\Delta H°$ and $\Delta S°$**

Problem: Compute $\Delta G°$ for the hydrolysis of urea, $CO(NH_2)_2$,

$$CO(NH_2)_2(aq) + H_2O(\ell) \longrightarrow CO_2(g) + 2NH_3(g)$$

Solution: We can calculate $\Delta H°$ from standard heats of formation found in Table 5.2. From Hess's law we can write

$$\Delta H° = (\Delta H°_{CO_2(g)} + 2\Delta H°_{NH_3(g)}) - (\Delta H°_{CO(NH_2)_2(aq)} + \Delta H°_{H_2O(\ell)})$$

$$= \left[1 \text{ mol} \times \left(\frac{-94.05 \text{ kcal}}{\text{mol}}\right) + 2 \text{ mol} \times \left(\frac{-11.08 \text{ kcal}}{\text{mol}}\right)\right]$$

$$- \left[1 \text{ mol} \times \left(\frac{-76.30 \text{ kcal}}{\text{mol}}\right) + 1 \text{ mol} \times \left(\frac{-68.32 \text{ kcal}}{\text{mol}}\right)\right]$$

$$= (-116.21 \text{ kcal}) - (-144.62 \text{ kcal})$$

$$= +28.41 \text{ kcal}$$

In Example 15.1, we calculated $\Delta S°$ for this reaction to be $+84.81$ cal/K. Now, using the equation for the standard free energy change, we can compute $\Delta G°$. Note that we must be careful to express $\Delta H°$ and $\Delta S°$ in the same energy units.

$$\Delta G° = +28.41 \text{ kcal} - (298.2 \text{ K})(0.08481 \text{ kcal/K})$$
$$= +28.41 \text{ kcal} - 25.29 \text{ kcal} = +3.12 \text{ kcal}$$

Note that we have expressed the temperature to four significant figures to conform with the number of significant figures in the data (25.0 + 273.2 = 298.2).

EXERCISE 15.2   Use the data in Tables 5.2 and 15.1 to calculate $\Delta G°$ for the formation of iron oxide (hematite), present in rust.

$$4Fe(s) + 3O_2(g) \longrightarrow 2Fe_2O_3(s)$$

In Section 5.5 we found it useful to tabulate standard heats of formation, $\Delta H°_f$, because we could use them following Hess's law to calculate $\Delta H°$ for many different reactions. **Standard free energies of formation, $\Delta G°_f$, can be used in similar calculations to obtain $\Delta G°$.

$$\Delta G^\circ = (\text{sum of } \Delta G_f^\circ \text{ of products}) - (\text{sum of } \Delta G_f^\circ \text{ of reactants}) \quad (15.4)$$

The ΔG_f° values for some typical substances are found in Table 15.2. Example 15.3 shows how we can use them to calculate ΔG° for a reaction.

Table 15.2
Standard Free Energies of Formation of Typical Substances

Substance	ΔG_f° kcal/mol	ΔG_f° kJ/mol	Substance	ΔG_f° kcal/mol	ΔG_f° kJ/mol
Ag(s)	0.00	(0.00)	$H_2(g)$	0.00	(0.00)
AgCl(s)	−26.22	(−109.7)	$H_2O(g)$	−54.64	(−228.6)
Al(s)	0.00	(0.00)	$H_2O(\ell)$	−56.69	(−237.2)
$Al_2O_3(s)$	−376.77	(−1576.4)	HCl(g)	−22.77	(−95.27)
C(s, graphite)	0.00	(0.00)	$HNO_3(\ell)$	−19.10	(−79.91)
CO(g)	−32.81	(−137.3)	$H_2SO_4(\ell)$	−164.9	(−689.9)
$CO_2(g)$	−94.26	(−394.4)	$HC_2H_3O_2(\ell)$	−93.8	(−392.5)
$CH_4(g)$	−12.14	(−50.79)	Hg(ℓ)	0.00	(0.00)
$CH_3Cl(g)$	−14.0	(−58.6)	Hg(g)	+7.59	(+31.8)
$CH_3OH(\ell)$	−39.73	(−166.2)	K(s)	0.00	(0.00)
$CO(NH_2)_2(s)$	−47.12	(−197.2)	KCl(s)	−97.59	(−408.3)
$CO(NH_2)_2(aq)$	−48.72	(−203.8)	$K_2SO_4(s)$	−314.62	(−1316.4)
$C_2H_2(g)$	+50.0	(+209)	$N_2(g)$	0.00	(0.00)
$C_2H_4(g)$	+16.28	(+68.12)	$NH_3(g)$	−3.98	(−16.7)
$C_2H_6(g)$	−7.86	(−32.9)	$NH_4Cl(s)$	−48.73	(−203.9)
$C_2H_5OH(\ell)$	−41.77	(−174.8)	NO(g)	+20.72	(+86.69)
$C_8H_{18}(\ell)$	+4.14	(+17.3)	$NO_2(g)$	+12.39	(+51.84)
Ca(s)	0.00	(0.00)	$N_2O(g)$	+24.76	(+103.6)
$CaCO_3(s)$	−269.78	(−1128.8)	$N_2O_4(g)$	+23.49	(+98.28)
$CaCl_2(s)$	−179.3	(−750.2)	Na(s)	0.00	(0.00)
CaO(s)	−144.4	(−604.2)	NaCl(s)	−91.79	(−384.0)
$Ca(OH)_2(s)$	−214.33	(−896.76)	NaOH(s)	−91.4	(−382)
$CaSO_4(s)$	−315.56	(−1320.3)	$Na_2SO_4(s)$	−302.78	(−1266.8)
$CaSO_4 \cdot \frac{1}{2}H_2O(s)$	−343.02	(−1435.2)	$O_2(g)$	0.00	(0.00)
$CaSO_4 \cdot 2H_2O(s)$	−429.19	(−1795.7)	PbO(s)	−45.25	(−189.3)
$Cl_2(g)$	0.00	(0.00)	S(s)	0.00	(0.00)
Fe(s)	0.00	(0.00)	$SO_2(g)$	−71.79	(−300.4)
$Fe_2O_3(s)$	−177.1	(−741.0)	$SO_3(g)$	−88.52	(−370.4)

EXAMPLE 15.3 Calculating ΔG° from ΔG_f°

Problem: What is ΔG° for the combustion of ethyl alcohol (C_2H_5OH) to give $CO_2(g)$ and $H_2O(g)$?

Solution: First we need a balanced chemical equation.

$$C_2H_5OH(\ell) + 3O_2(g) \longrightarrow 2CO_2(g) + 3H_2O(g)$$

Using Equation 15.4,

$$\Delta G^\circ = [2\Delta G_f^\circ(CO_2) + 3\Delta G_f^\circ(H_2O)] - [\Delta G_f^\circ(C_2H_5OH) + 3\Delta G_f^\circ(O_2)]$$

As with ΔH_f°, $\Delta G_f^\circ = 0$ for an element in its normal state at 25 °C and 1 atm. Therefore,

$$\Delta G^\circ = [2 \text{ mol}(-94.26 \text{ kcal/mol}) + 3 \text{ mol}(-54.64 \text{ kcal/mol})]$$
$$- [1 \text{ mol}(-41.77 \text{ kcal/mol}) + 3 \text{ mol}(0 \text{ kcal/mol})]$$

$$\Delta G° = (-352.4 \text{ kcal}) - (-41.77 \text{ kcal})$$
$$= -310.6 \text{ kcal}$$

EXERCISE 15.3 Calculate $\Delta G°$ for the following reactions using the data in Table 15.2:
(a) $2NO(g) + O_2(g) \longrightarrow 2NO_2(g)$
(b) $Ca(OH)_2(s) + 2HCl(g) \longrightarrow CaCl_2(s) + 2H_2O(g)$

15.8 FREE ENERGY AND MAXIMUM WORK

ΔG is a measure of the maximum work that can be obtained from a reaction

One of the most important uses to which we put spontaneous chemical reactions is the production of useful work. Fuels are burned in gasoline or diesel engines to power automobiles and heavy machinery. Chemical reactions in batteries start our autos and run all sorts of modern electronic gadgets, including your hand calculator. And chemical reactions within our bodies pump blood, move muscles, transport nerve impulses, and do those other things that keep us alive and make everything else worthwhile.

When chemical reactions occur, however, their energy is not always harnessed to do work. For instance, if gasoline is burned in an open dish, the energy evolved is lost entirely as heat and no useful work is accomplished. Engineers, therefore, seek ways of capturing as much energy as possible in the form of work. One of their primary goals is to maximize the efficiency with which chemical energy is converted to work and to minimize the amount of energy transferred unproductively to the environment and lost as heat.

Scientists have discovered that the maximum conversion of chemical energy to work occurs if a reaction is carried out under conditions that are said to be reversible. A process is defined as **reversible** if its driving force is opposed by another force that is just a tiny bit weaker. An example of an almost-reversible process would be the use of a 12.00-V battery to charge an 11.99-V battery. The electricity from the 12.00-V battery is opposed by a "force" of 11.99 V.

A reversible process requires an infinite number of tiny steps. This takes forever to accomplish.

Unfortunately, along with the "good news" that we get maximum work if we carry out a reaction reversibly, there is the "bad news" that a truly reversible process proceeds at an extremely slow speed. Even though we get the maximum amount of work, we get it so slowly that it is of no use to us. The goal, then, is to approach reversibility for maximum efficiency, but to carry out the reaction at a speed that will deliver work at useful rates.

The Gibbs free energy change, ΔG, is the maximum amount of energy produced by a reaction that can be theoretically harnessed. It is that energy that need not be lost as heat and is therefore "free" to be used for work. Thus, by determining the value of ΔG, we can find out whether or not a given reaction will be an effective source of useful energy. Also, by comparing the actual amount of work derived from a given system with the ΔG values for the reactions involved, we can measure the efficiency of the system.

EXAMPLE 15.4 **Calculating Maximum Work**

Problem: Calculate the maximum work available, expressed in kilojoules, from the oxidation of 1 mol of octane, $C_8H_{18}(\ell)$, by oxygen to give $CO_2(g)$ and $H_2O(\ell)$ at 25 °C and 1 atm.

Solution: First we need a chemical equation,

$$C_8H_{18}(\ell) + 12\tfrac{1}{2}O_2(g) \longrightarrow 8CO_2(g) + 9H_2O(\ell)$$

Since the free energy change is equal to the maximum work available, we must then calculate $\Delta G°$. (The reaction described here occurs under those conditions at which standard free energies are determined—how fortunate!)

$$\Delta G° = [8\Delta G_f°(CO_2) + 9\Delta G_f°(H_2O)] - [\Delta G_f°(C_8H_{18}) + 12.5\Delta G_f°(O_2)]$$

Referring to Table 15.2, and using values in kJ/mol,

$$\Delta G° = [8(-394.4) + 9(-237.2)] - [1(+17.3) + 12.5(0.0)]$$

$$\Delta G° = (-5290) - (+17.3)$$
$$= -5307 \text{ kJ}$$

Thus, at 25 °C and 1 atm, we can expect no more than 5307 kJ of work from the oxidation of one mole of C_8H_{18}.

EXERCISE 15.4 Calculate the maximum work that could be obtained at 25 °C and 1 atm from the oxidation of 1.00 mol of Al by O_2 to give Al_2O_3.

15.9 FREE ENERGY AND EQUILIBRIUM

Equilibrium exists when $\Delta G = 0$

We have seen that when the value of ΔG for a given change is negative, the change will occur spontaneously. We have also seen that a change will be non-spontaneous when ΔG is positive. However, when ΔG is neither positive nor negative, the change will be neither spontaneous nor nonspontaneous—the system is in a state of equilibrium. This occurs when ΔG is equal to zero.

Let's again consider the freezing of water.

$$H_2O(\ell) \longrightarrow H_2O(s)$$

Because ΔS and ΔH for this "reaction" are both negative, ΔG is negative only at low temperatures (below 0 °C). Therefore, the reaction is spontaneous only at low temperatures. This agrees with our experience—we know that water freezes when it becomes very cold. At high temperatures, ΔG is positive, and the freezing is nonspontaneous. If ΔG is positive in the forward direction, it must be negative in the reverse direction. Therefore, at high temperatures (above 0 °C) it is melting that occurs spontaneously. Again this agrees with our experience.

Between high and low temperatures (at 0 °C), ΔG is equal to zero and an ice-water mixture exists in a condition of equilibrium. Neither freezing nor melting is spontaneous. As long as heat isn't added or removed from the system, the ice and liquid water can exist together indefinitely.

What does equilibrium mean as far as work is concerned? We have identified ΔG as a quantity that specifies the amount of work that is available from a system. Since at equilibrium, ΔG is zero, the amount of work available is zero also. Therefore, *when a system is at equilibrium, no work can be extracted from that system.* For an example, let's examine the common lead storage battery that we use to start our car.

When the battery is fully charged, there are virtually no products of the discharge reaction present. The chemical reactants, however, are present in large amounts. Therefore, the total free energy of the reactants far exceeds the total free energy of products and, since $\Delta G = G_{products} - G_{reactants}$, the ΔG of the system has a large negative value. This means that a lot of work is available. As the battery discharges, the reactants are converted to products and $G_{products}$ gets larger while $G_{reactants}$ gets smaller; thus ΔG becomes less negative, and less work is available. Finally, the battery reaches equilibrium. The total free energies of

It is *theoretically* impossible to obtain work from the melting or freezing of water at 0 °C and 1 atm. Why?

the reactants and the products have become equal, and $\Delta G = 0$. No further work can be extracted; the battery is dead.

Equilibrium in phase changes

When we have equilibrium in any system, we know that $\Delta G = 0$. For a phase change such as $H_2O(\ell) \rightarrow H_2O(s)$, equilibrium can only exist at one particular temperature at atmospheric pressure. In this instance, that temperature is 0 °C. Above 0 °C, only liquid water can exist, and below 0 °C all the liquid will freeze to give ice. This gives us an interesting relationship between ΔH and ΔS for a phase change. Since $\Delta G = 0$,

$$\Delta G = 0 = \Delta H - T\Delta S$$

Therefore,

$$\Delta H = T\Delta S$$

and

$$\Delta S = \frac{\Delta H}{T} \tag{15.5}$$

The magnitudes of ΔH and ΔS are relatively insensitive to temperature changes. However, ΔG is very temperature sensitive because $\Delta G = \Delta H - T\Delta S$.

Thus, if we know ΔH for the phase change and the temperature at which the two phases coexist, we can calculate ΔS.

Another interesting relationship that we can obtain is

$$T = \frac{\Delta H}{\Delta S} \tag{15.6}$$

Thus, if we know ΔH and ΔS, we can calculate the temperature at which equilibrium will occur.

EXAMPLE 15.5 **Calculating the Equilibrium Temperature for a Phase Change**

Problem: For the "reaction," $Br_2(\ell) \rightarrow Br_2(g)$, $\Delta H° = +31.0$ kJ/mol and $\Delta S° = 92.9$ J/mol K. Assuming that $\Delta H°$ and $\Delta S°$ are nearly temperature independent, calculate the approximate temperature at which $Br_2(\ell)$ will be in equilibrium with $Br_2(g)$ at 1 atm.

Solution: The temperature at which equilibrium exists is given by Equation 15.6,

$$T = \frac{\Delta H}{\Delta S}$$

If ΔH and ΔS do not depend on temperature,

$$T = \frac{\Delta H°}{\Delta S°}$$

Substituting the data given in the problem,

$$T = \frac{3.10 \times 10^4 \text{ J mol}^{-1}}{92.9 \text{ J mol}^{-1} \text{ K}^{-1}}$$

$$= 334 \text{ K}$$

The Celsius temperature is $334 - 273 = 61$ °C.

Notice that we were careful to express $\Delta H°$ in J, not kJ, so that the units would cancel properly.

EXERCISE 15.5 The heat of vaporization of mercury is 14.5 kcal/mol. For $Hg(\ell)$, $S° = 18.2$ cal/mol K, and for $Hg(g)$, $S° = 41.8$ cal/mol K. Calculate the temperature at which there is equilibrium between mercury vapor and liquid at a pressure of 1 atm.

15.10 PREDICTING THE OUTCOME OF CHEMICAL REACTIONS

$\Delta G°$ points the way in chemical reactions

In a phase change such as $H_2O(\ell) \rightarrow H_2O(s)$, equilibrium can exist for a given pressure, only at one particular temperature; for water at a pressure of 1 atm this temperature is 0 °C. At other temperatures, the "reaction" proceeds *entirely* to completion in one direction or another.

Figure 15.7 illustrates how the free energy changes when $H_2O(\ell)$ becomes $H_2O(s)$ at different temperatures. Below 0 °C, the free energy decreases continually until *all* the liquid has frozen. Above 0 °C, the free energy decreases in the direction of $H_2O(s) \rightarrow H_2O(\ell)$ and continues to drop until all the solid has melted. Above or below 0 °C the system is unable to establish an equilibrium mixture of liquid and solid. However, at 0 °C there would be no change in free energy if either melting or freezing occurred; therefore, there is no driving force for either change. As long as the system is insulated from warmer or colder surroundings, any particular mixture of ice and water is stable and a state of equilibrium exists.

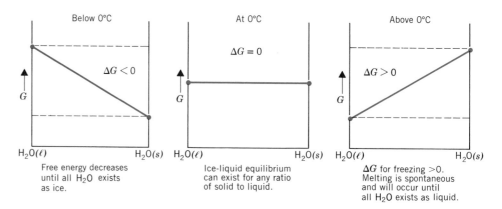

Figure 15.7
Free energy diagram for conversion of $H_2O(\ell)$ to $H_2O(s)$. At the left of each diagram, the system consists entirely of $H_2O(\ell)$. At the right is $H_2O(s)$. The horizontal axis represents the extent of conversion from $H_2O(\ell)$ to $H_2O(s)$.

The free energy changes that occur in most chemical reactions are more complex than those in phase changes. Figure 15.8 shows the variation in free energy for the decomposition of N_2O_4, a rocket fuel, into NO_2, an air pollutant.

$$N_2O_4(g) \longrightarrow 2NO_2(g)$$

Notice that in going from reactant to product, the free energy of the reaction mixture drops *below* that of either pure N_2O_4 or pure NO_2.

Any system will spontaneously seek the lowest point on its free energy curve. If we begin with pure $N_2O_4(g)$, some $NO_2(g)$ will be formed because proceeding in the direction of NO_2 leads to a lowering of the free energy. If we begin with pure $NO_2(g)$, a change also will occur. Going downhill on the free energy curve now takes place as the reverse reaction occurs [i.e., $2NO_2(g) \rightarrow N_2O_4(g)$]. Once the bottom of the "valley" is reached, the system has come to equilibrium. The composition of the mixture of N_2O_4 and NO_2 will remain constant because any change would require an uphill climb. Free energy increases are not spontaneous, so this doesn't happen.

Another thing to notice in Figure 15.8 is that some reaction takes place in the forward direction *even though $\Delta G°$ is positive*. For comparison, Figure 15.9 shows the shape of the free energy curve for a reaction having a negative $\Delta G°$.

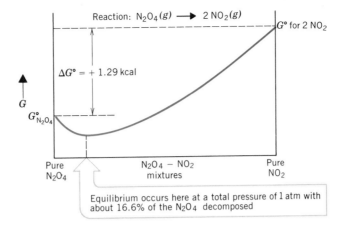

Figure 15.8
Free energy diagram for the decomposition of $N_2O_4(g)$. The minimum on the curve indicates the composition of the reaction mixture at equilibrium.

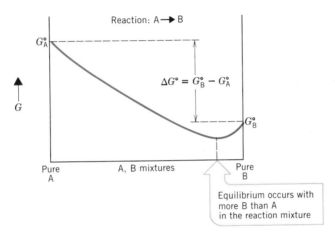

Figure 15.9
Free energy curve for a reaction having a negative $\Delta G°$. Since $G_B° < G_A°$, $\Delta G° < 0$.

We see here that at equilibrium there has been a much greater conversion of reactants to products.

In general, the value of $\Delta G°$ for most reactions is much larger numerically than the $\Delta G°$ for the $N_2O_4 \rightarrow 2NO_2$ reaction. The extent to which a reaction proceeds is *very* sensitive to the size of $\Delta G°$. If the $\Delta G°$ value for a reaction is reasonably large—about 5 kcal or more—almost no *observable* reaction will occur when $\Delta G°$ is positive. However, the reaction will go almost to completion if $\Delta G°$ is both large and negative.[2] From a practical standpoint, then, the size and sign of $\Delta G°$ serves as an indicator of whether an observable spontaneous reaction will occur. In the next chapter we will see how $\Delta G°$ can be used to estimate quantitatively the extent to which a reaction will have proceeded toward completion by the time it reaches equilibrium.

[2] To actually see a change take place, the rate of a spontaneous reaction must be reasonably fast. For example, the decomposition of the nitrogen oxides into N_2 and O_2 is thermodynamically spontaneous ($\Delta G°$ is negative) but their rates of decomposition are so slow that these substances appear to be stable and are obnoxious air pollutants.

Many biochemical reactions involved in the synthesis of large molecules, such as proteins, have positive values of $\Delta G°$. Thus, they require an input of free energy. Left to themselves, therefore, we wouldn't expect these reactions to occur. Yet they do take place. This is because they are "coupled" with one or more reactions that have a large enough negative $\Delta G°$ to give a net negative $\Delta G°$ for the overall reaction. Let's look at an example. In general terms, suppose we had the reactions,

(1) $A + B \longrightarrow C + D$ $\Delta G° = +10 \text{ kJ}$

(2) $D + E \longrightarrow F + G$ $\Delta G° = -15 \text{ kJ}$

Reaction (1) by itself wouldn't be expected to produce very much product. However, because one of the products, D, can be used in the second reaction, the two reactions are said to be "coupled." If we add reactions (1) and (2), we get,

$$A + B + D + E \longrightarrow C + D + F + G$$

Since substance D appears on both sides of the arrow, we cancel it from the equation. The net reaction is therefore

$$A + B + E \longrightarrow C + F + G$$

Recall that if we add two reactions to obtain a third, the ΔH for the third is the sum of the ΔH's for the two reactions that were added. The same principle applies for ΔG values. Our "coupled" reaction therefore has a $\Delta G°$ equal to $(+10 \text{ kJ}) + (-15 \text{ kJ}) = -5 \text{ kJ}$. Thus the net reaction has a negative $\Delta G°$ and should proceed toward completion to a large extent.

Biochemists diagram this coupling as

This diagram indicates that the reaction, $A + B \rightarrow C + D$, is being driven (i.e., caused to occur) by coupling it with the second reaction, $D + E \rightarrow F + G$. A specific biochemical example is found in problem 15.62 at the end of the chapter. The principle of coupling reactions is of great importance in nearly all biochemical systems.

EXAMPLE 15.6   Using $\Delta G°$ as a Predictor of the Outcome of a Reaction

Problem: Would we expect to be able to observe the following reaction?

$$NH_4Cl(s) \longrightarrow NH_3(g) + HCl(g)$$

Solution: First let's calculate $\Delta G°$ for the reaction using the data in Table 15.2. The procedure is the same as that discussed earlier.

$$\Delta G° = [\Delta G_f°(NH_3) + \Delta G_f°(HCl)] - \Delta G_f°(NH_4Cl)$$
$$= [(-3.98 \text{ kcal}) + (-22.77 \text{ kcal})] - (-48.73 \text{ kcal})$$
$$= +21.98 \text{ kcal}$$

Because $\Delta G°$ is large and positive we shouldn't expect to observe the spontaneous formation of any products.

EXERCISE 15.6   Use the data in Table 15.2 to determine whether the reaction, $SO_2(g) + \frac{1}{2}O_2(g) \rightarrow SO_3(g)$, should occur spontaneously.

EXERCISE 15.7   Use the data in Table 15.2 to determine whether we should expect to see the formation of $CaCO_3(s)$ in the reaction, $CaCl_2(s) + H_2O(g) + CO_2(g) \rightarrow CaCO_3(s) + 2HCl(g)$

SUMMARY

Spontaneity   Spontaneous changes occur without outside assistance. Nonspontaneous changes require continual help and can only occur if some other spontaneous event takes place first. The energy change is one factor that influences spontaneity. Exothermic changes, with their negative values of ΔH, tend to proceed spontaneously. Events that occur with an increase in randomness also tend to occur by themselves. Randomness is associated with statistical probability—the more random a collection of molecules, the greater the probability of its occurrence. The thermodynamic quantity associated with randomness is entropy. An increase in entropy favors a spontaneous change.

Second law of thermodynamics This states that the entropy of the universe increases whenever a spontaneous change occurs. All of our activities, being traced ultimately to spontaneous events, increase the total entropy or disorder in the world. We see the effects of this in environmental pollution. The energy crisis results because spontaneous events lead to a loss of energy to the environment where it becomes inaccessible for future use.

Third law of thermodynamics This states that the entropy of a pure crystal at 0 K is zero. This allows us to determine absolute amounts of entropy that substances have. Standard entropies, $S°$, are calculated for 25 °C and 1 atm (Table 15.1) and may be used to calculate $\Delta S°$ for reactions.

Gibbs free energy This allows us to determine the combined effects of enthalpy and entropy changes on the spontaneity of a reaction. A change will only be spontaneous when ΔG is negative. When ΔH and ΔS have the same sign the temperature becomes the critical factor in determining spontaneity.

When ΔG is determined at 25 °C and 1 atm it is the standard free energy change, $\Delta G°$. As with enthalpy changes, standard free energies of formation, $\Delta G_f°$, (Table 15.2) can be used to compute $\Delta G°$ for a change.

For any system the value of ΔG is equal to the maximum amount of useful work that is available. This maximum work can only be obtained if a change is performed in a reversible manner. All real changes are irreversible and we always obtain less work than is theoretically available—the remainder is lost as heat. When a system reaches equilibrium, $\Delta G = 0$ and no useful work can be obtained from the system. Phase changes can produce equilibria at a given pressure only at one temperature. The entropy change can be computed as $\Delta S = \Delta H/T$; the temperature at which equilibrium exists can be calculated from ΔS and ΔH, $T = \Delta H/\Delta S$.

In chemical reactions a minimum on the free energy curve occurs part way between pure reactants and products. This minimum can be approached from either reactants or products. When a reaction has a value of $\Delta G°$ that is both large and negative, it will appear to occur spontaneously because a lot of products will be formed by the time equilibrium is reached. If $\Delta G°$ is large and positive, no *observable* reaction will occur because only minute amounts of products will be present at equilibrium. The sign and magnitude of $\Delta G°$ can therefore be used to predict the outcome of a chemical reaction.

INDEX TO EXERCISES, QUESTIONS, AND PROBLEMS
(Those numbered above 37 are problems.)

REVIEW QUESTIONS

15.8. What is suggested by the term, thermodynamics?

15.9. What is a "spontaneous change"?

15.10. List five changes that you have encountered recently that occurred spontaneously. List five changes that are nonspontaneous but which *you* have caused to occur.

15.11. Which of the items that you listed in question 15.10 are exothermic and which are endothermic?

15.12. When solid potassium iodide is dissolved in water, a cooling of the mixture occurs because the solution process

is endothermic for these substances. Explain, in terms of what happens to the molecules and ions, why this mixing occurs spontaneously.

15.13. What is entropy?

15.14. Will the entropy change for each of the following be positive or negative?
(a) Moisture condenses on the outside of a cold glass.
(b) Raindrops form in a cloud.
(c) Gasoline vaporizes in the carburator of an automobile engine.

(d) Air is pumped into a tire.

(e) Frost forms on the windshield of your car.

(f) Sugar dissolves in coffee.

15.15. What is a state function?

15.16. On the basis of our definition of entropy, can you suggest why entropy is a state function?

15.17. State the second law of thermodynamics.

15.18. Since energy can't be destroyed, what are the prospects of capturing energy after we have used it so that we can use it again?

15.19. In animated cartoons, visual effects are often created (for amusement) which show events that ordinarily don't occur in real life because they are accompanied by huge entropy decreases. Can you think of an example of this? Explain why there is an entropy decrease in your example.

15.20. How is entropy related to pollution? What are our chances of eliminating pollution from the environment?

15.21. You have probably heard the term "thermal pollution," the introduction of unwanted heat into the environment. How is this related to the reasons for the energy crisis?

15.22. Without actually calculating $\Delta S°$, predict which of the following changes would be expected to occur with an increase in entropy.

(a) $I_2(s) \longrightarrow I_2(g)$

(b) $Cl_2(g) + Br_2(g) \longrightarrow 2BrCl(g)$

(c) $NH_3(g) + HCl(g) \longrightarrow NH_4Cl(s)$

(d) $CaO(s) + H_2O(g) \longrightarrow Ca(OH)_2(s)$

15.23. What is the third law of thermodynamics?

15.24. Would you expect the entropy of an alloy (a solution of two metals) to be zero at 0 K? Explain your answer.

15.25. Why does entropy increase with increasing temperature?

15.26. What is the equation expressing the change in the Gibbs free energy for a reaction occurring at constant temperature and pressure? Why do we concern ourselves with reactions at constant temperature and pressure?

15.27. Under what circumstances will a change be spontaneous:

(a) At all temperatures?

(b) At low temperatures but not at high temperatures?

(c) At high temperatures but not at low temperatures?

15.28. Under what circumstances will a change be nonspontaneous regardless of the temperature?

15.29. How is free energy related to useful work?

15.30. What is a reversible process?

15.31. When glucose is oxidized by the body to generate energy, this energy is stored in molecules of ATP (adenosine triphosphate). However, of the total energy released in the oxidation of glucose, only 38% actually becomes stored in the ATP. What happens to the rest of the energy?

15.32. Why are real, observable changes not considered to be reversible processes?

15.33. In what way is free energy related to equilibrium?

15.34. Considering the fact that the formation of a bond between two atoms is exothermic and is accompanied by an entropy decrease, explain why all chemical compounds decompose into individual atoms if heated to a high enough temperature.

15.35. When a warm object is placed in contact with a cold one, they both gradually come to the same temperature. On a molecular level, explain how this is related to entropy and spontaneity.

15.36. Sketch the shape of the free energy curve for a chemical reaction that has a positive $\Delta G°$. Indicate the composition of the reaction mixture corresponding to equilibrium.

15.37. Sketch a graph to show how the free energy changes during a phase change such as melting.

REVIEW PROBLEMS

15.38. Use the data from Table 5.2 to calculate $\Delta H°$ for the following reactions. On the basis of $\Delta H°$, which are favored to occur spontaneously?

(a) $CaO(s) + CO_2(g) \longrightarrow CaCO_3(s)$

(b) $C_2H_2(g) + 2H_2(g) \longrightarrow C_2H_6(g)$

(c) $3CaO(s) + 2Fe(s) \longrightarrow 3Ca(s) + Fe_2O_3(s)$

(d) $Ca(OH)_2(s) \longrightarrow CaO(s) + H_2O(\ell)$

(e) $2NaCl(s) + H_2SO_4(\ell) \longrightarrow Na_2SO_4(s) + 2HCl(g)$

15.39. Use the data from Table 5.2 to calculate $\Delta H°$ for the following reactions. On the basis of $\Delta H°$, which are favored to occur spontaneously?

(a) $2C_2H_2(g) + 5O_2(g) \longrightarrow 4CO_2(g) + 2H_2O(g)$

(b) $C_2H_2(g) + 5N_2O(g) \longrightarrow 2CO_2(g) + H_2O(g) + 5N_2(g)$

(c) $Fe_2O_3(s) + 2Al(s) \longrightarrow Al_2O_3(s) + 2Fe(s)$

(d) $NH_4Cl(s) \longrightarrow NH_3(g) + HCl(g)$

(e) $Ag(s) + KCl(s) \longrightarrow AgCl(s) + K(s)$

15.40. When a coin is tossed, there is a 50-50 chance of it landing either heads or tails. Suppose that you tossed four coins. On the basis of the different possible outcomes, what is the probability of all four coins coming up heads? What is the probability of an even heads-tails distribution?

15.41. Suppose that you had two equal-volume containers sharing a common wall with a hole in it. Suppose there were four molecules in this system. What is the probability that all four would be in one container at the same time? What is the probability of finding an even distribution of molecules between the two containers? (Hint: try question 15.40 first) What do the results suggest about why gases expand spontaneously?

15.42. Calculate $\Delta S°$ for the following reactions from the data in Table 15.1.
(a) $N_2(g) + 3H_2(g) \longrightarrow 2NH_3(g)$
(b) $CO(g) + 2H_2(g) \longrightarrow CH_3OH(\ell)$
(c) $2C_2H_6(g) + 7O_2(g) \longrightarrow 4CO_2(g) + 6H_2O(g)$
(d) $Ca(OH)_2(s) + H_2SO_4(\ell) \longrightarrow CaSO_4(s) + 2H_2O(\ell)$
(e) $S(s) + 2N_2O(g) \longrightarrow SO_2(g) + 2N_2(g)$

15.43. Calculate $\Delta S°$ for the following reactions using the data in Table 15.1.
(a) $Ag(s) + \frac{1}{2}Cl_2(g) \longrightarrow AgCl(s)$
(b) $H_2(g) + \frac{1}{2}O_2(g) \longrightarrow H_2O(g)$
(c) $H_2(g) + \frac{1}{2}O_2(g) \longrightarrow H_2O(\ell)$
(d) $CaCO_3(s) + H_2SO_4(\ell) \longrightarrow CaSO_4(s) + H_2O(g) + CO_2(g)$
(e) $NH_3(g) + HCl(g) \longrightarrow NH_4Cl(s)$

15.44. Calculate $\Delta S_f°$ for the following compounds.
(a) $C_2H_4(g)$
(b) $N_2O(g)$
(c) $NaCl(s)$
(d) $CaSO_4 \cdot 2H_2O(s)$
(e) $HC_2H_3O_2(\ell)$

15.45. Calculate $\Delta S_f°$ for the following compounds.
(a) $Al_2O_3(s)$
(b) $CaCO_3(s)$
(c) $N_2O_4(g)$
(d) $NH_4Cl(s)$
(e) $CaSO_4 \cdot \frac{1}{2}H_2O(s)$

15.46. Nitrogen dioxide, NO_2, an air pollutant, dissolves in rainwater to form a dilute solution of nitric acid. The equation for the reaction is:

$$3NO_2(g) + H_2O(\ell) \longrightarrow 2HNO_3(\ell) + NO(g)$$

Calculate $\Delta S°$ for this reaction.

15.47. Good wine will turn to vinegar if it is left exposed to air because the alcohol is oxidized to acetic acid. The equation for the reaction is:

$$C_2H_5OH(\ell) + O_2(g) \longrightarrow HC_2H_3O_2(\ell) + H_2O(\ell)$$

Calculate $\Delta S°$ for this reaction.

15.48. Phosgene, $COCl_2$, was used as a war gas during World War I. It reacts with the moisture in the lungs to produce HCl which causes the lungs to fill with fluid, leading to the death of the victim. $COCl_2$ has a standard entropy, $S° = 284$ J/mol K and $\Delta H_f° = -223$ kJ/mol. Use this information and the data in Table 15.1 to calculate $\Delta G_f°$ for $COCl_2(g)$.

15.49. Aluminum oxidizes rather easily, but forms a thin protective coating of Al_2O_3 that prevents further oxidation of aluminum beneath. Use the data for $\Delta H_f°$ (Table 5.2) and $S°$ to calculate $\Delta G_f°$ for $Al_2O_3(s)$.

15.50. Compute $\Delta G°$ for the following reactions using the data in Table 15.2.

(a) $SO_3(g) + H_2O(\ell) \longrightarrow H_2SO_4(\ell)$
(b) $2NH_4Cl(s) + CaO(s) \longrightarrow CaCl_2(s) + H_2O(\ell) + 2NH_3(g)$
(c) $CaSO_4(s) + 2HCl(g) \longrightarrow CaCl_2(s) + H_2SO_4(\ell)$
(d) $C_2H_4(g) + H_2O(g) \longrightarrow C_2H_5OH(\ell)$
(e) $Ca(s) + 2H_2SO_4(\ell) \longrightarrow CaSO_4(s) + SO_2(g) + 2H_2O(\ell)$

15.51. Compute $\Delta G°$ for the following reactions using the data in Table 15.2.
(a) $2HCl(g) + CaO(s) \longrightarrow CaCl_2(s) + H_2O(g)$
(b) $H_2SO_4(\ell) + 2NaCl(s) \longrightarrow 2HCl(g) + Na_2SO_4(s)$
(c) $3NO_2(g) + H_2O(\ell) \longrightarrow 2HNO_3(\ell) + NO(g)$
(d) $2AgCl(s) + Ca(s) \longrightarrow CaCl_2(s) + 2Ag(s)$
(e) $NH_3(g) + HCl(g) \longrightarrow NH_4Cl(s)$

15.52. Plaster of Paris, $CaSO_4 \cdot \frac{1}{2}H_2O(s)$, reacts with liquid water to form gypsum, $CaSO_4 \cdot 2H_2O(s)$. Write a chemical equation for the reaction and calculate $\Delta G°$ using the data in Table 15.2.

15.53. When phosgene, the war gas described in problem 15.48, reacts with water vapor, the products are $CO_2(g)$ and $HCl(g)$. Write an equation for the reaction and compute $\Delta G°$ using the value of $\Delta G_f°$ for $COCl_2$ that you calculated in Problem 15.48.

15.54. *Gasohol* is a mixture of gasoline and ethanol (grain alcohol), C_2H_5OH. Calculate the maximum work that could be obtained at 25 °C and 1 atm by burning 1 mol of C_2H_5OH.

$$C_2H_5OH(\ell) + 3O_2(g) \longrightarrow 2CO_2(g) + 3H_2O(g)$$

15.55. What is the maximum amount of useful work that could possibly be obtained at 25 °C and 1 atm from the combustion of 27.0 g of natural gas. $CH_4(g)$, to give $CO_2(g)$ and $H_2O(g)$?

15.56. Given the following,

$$4NO(g) \longrightarrow 2N_2O(g) + O_2(g) \qquad \Delta G° = -139.56 \text{ kJ}$$
$$2NO(g) + O_2(g) \longrightarrow 2NO_2(g) \qquad \Delta G° = -69.70 \text{ kJ}$$

calculate $\Delta G°$ for the reaction,

$$2N_2O(g) + 3O_2(g) \longrightarrow 4NO_2(g)$$

15.57. Given these reactions and their $\Delta G°$ values,

$$COCl_2(g) + 4NH_3(g) \longrightarrow CO(NH_2)_2(s) + 2NH_4Cl(s)$$
$$\Delta G° = -79.36 \text{ kcal}$$
$$COCl_2(g) + H_2O(\ell) \longrightarrow CO_2(g) + 2HCl(g)$$
$$\Delta G° = -33.89 \text{ kcal}$$
$$NH_3(g) + HCl(g) \longrightarrow NH_4Cl(s) \qquad \Delta G° = -21.98 \text{ kcal}$$

calculate $\Delta G°$ for the reaction,

$$CO(NH_2)_2(s) + H_2O(\ell) \longrightarrow CO_2(g) + 2NH_3(g)$$

15.58. Chloroform, formerly used as an anesthetic and now known to be a carcinogen (cancer-causing agent), has a heat of vaporization $\Delta H_{vap} = 31.4$ kJ/mol. The change, $CHCl_3(\ell) \rightarrow CHCl_3(g)$ has $\Delta S° = 94.2$ J/mol K. At what temperature do we expect $CHCl_3$ to boil (i.e., at what temperature will liquid and vapor be in equilibrium at 1 atm pressure)?

15.59. For the melting of aluminum, $Al(s) \rightarrow Al(\ell)$, $\Delta H° = 2.40$ kcal/mol and $\Delta S° = 2.27$ cal/mol K. Calculate the melting point of Al. (The actual melting point is 660 °C.)

15.60. Isooctane, an important constituent of gasoline, has a boiling point of 99.3 °C and a heat of vaporization of 9.01 kcal/mol. What is ΔS for the vaporization of 1 mol of isooctane?

15.61. Acetone (nail polish remover) has a boiling point of 56.2 °C. The change, $(CH_3)_2CO(\ell) \rightarrow (CH_3)_2CO(g)$ has $\Delta H^\circ = 31.9$ kJ/mol. What is ΔS° for this change?

15.62. Many biochemical reactions have positive values for ΔG° and seemingly should not be expected to be spontaneous. They occur, however, because they are chemically coupled with other reactions which have negative values of ΔG°. An example is the set of reactions which form the beginning part of the sequence of reactions involved in the metabolism of glucose, a sugar. Given these reactions and their corresponding ΔG° values,

glucose + phosphate $\longrightarrow$ glucose-6-phosphate + H_2O
$$\Delta G^\circ = +3.138 \text{ kcal}$$
ATP + $H_2O \longrightarrow$ ADP + phosphate $\Delta G^\circ = -7.700$ kcal

calculate ΔG° for the coupled reaction,

glucose + ATP $\longrightarrow$ glucose-6-phosphate + ADP

15.63. Determine whether the following reaction will be spontaneous:

$$C_2H_4(g) + 2HNO_3(\ell) \longrightarrow HC_2H_3O_2(\ell) + H_2O(\ell)$$
$$+ NO(g) + NO_2(g)$$

15.64. Ethyl alcohol, C_2H_5OH, has been suggested as an alternative to gasoline as a fuel. In Example 15.3 we calculated ΔG° for combustion of 1 mol of C_2H_5OH; in Example 15.4 we calculated ΔG° for combustion of 1 mol of octane. Let's assume that gasoline has the same properties as octane (one of its major constituents). The density of C_2H_5OH is 0.7893 g/ml; the density of octane, C_8H_{18}, is 0.7025 g/ml. Calculate the maximum work that could be obtained by burning 1 gallon (3.78 liters) of both C_2H_5OH and C_8H_{18}. On a volume basis, which is a better fuel?

CHAPTER SIXTEEN

EQUILIBRIUM

16.1 DYNAMIC EQUILIBRIA

When a chemical equilibrium is established, the concentrations of the reactants and products do not change with time because forward and reverse reactions occur at equal rates

One of our most important concepts in chemistry is that of dynamic equilibrium. We've already encountered this concept on several occasions earlier in the book. For example, we described a number of acids as "weak acids" because, when they are in solution, there is an equilibrium between the un-ionized molecules of acid and the ions produced by their reaction with water. For acetic acid solutions, for instance, we write the equation

$$HC_2H_3O_2(aq) + H_2O(\ell) \rightleftharpoons H_3O^+(aq) + C_2H_3O_2^-(aq)$$
$$\text{acetic acid}$$

This equation describes a condition in which two opposing reactions occur at equal rates. When acetic acid and water are first mixed, they begin to react and disappear while the products—H_3O^+ and $C_2H_3O_2^-$—begin to form. As the products accumulate, they begin to react to reform the reactants. Since the rate of a reaction normally depends on the concentrations of the reacting species, the rate of the forward reaction (read from left to right) decreases as the reactants disappear. At the same time, the rate of the reverse reaction (read from right to left) increases as the concentrations of the products rise. As a result, all of the concentrations approach steady values as the two rates approach each other. This is illustrated in Figure 16.1. Eventually, when the system reaches equilib-

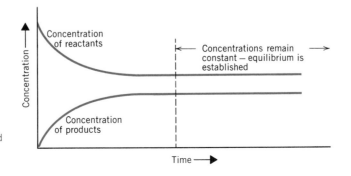

Figure 16.1

As a reaction proceeds, the concentrations of the reactants and products approach steady (constant) values.

It is an *equilibrium* because the concentrations don't change. It is *dynamic* because the opposing reactions never cease.

rium, the rate at which $HC_2H_3O_2$ molecules react with water molecules becomes the same as the rate at which H_3O^+ ions react with $C_2H_3O_2^-$ ions. Over a period of time, therefore, no change in concentrations of the various species is observed because they disappear and reform at the same rate. If the system is not disturbed, the concentrations will never change. This is what is implied by the set of double arrows, $\rightleftharpoons$, that we use to indicate an equilibrium.

Almost all chemical systems eventually reach a state of dynamic equilibrium, although sometimes this equilibrium is extremely difficult (or even impossible) to detect. This is because the amounts of either the reactants or products that are present at equilibrium are virtually zero. For instance, when a strong acid such as HCl is placed in water, it reacts so completely to produce H_3O^+ and Cl^- that there is no detectable amount of HCl molecules remaining—we say that the HCl is completely ionized. Similarly, in water vapor at room temperature, there are no detectable amounts of either H_2 or O_2 from the equilibrium

Sometimes an equal sign (=) is used in place of the double arrows.

$$2H_2O(g) \rightleftharpoons 2H_2(g) + O_2(g)$$

The water molecules are so stable that we can't detect whether any of them decompose. Even in cases such as these, however, it is often convenient to presume that an equilibrium does exist.

In this chapter we will study the equilibrium condition both qualitatively and quantitatively. Knowing the factors that affect the composition of an equilibrium system allows us to have some control over the outcome of a reaction. For example, the formation of the pollutant nitric oxide (NO) by reaction of N_2 and O_2 in a gasoline engine can be reduced, in a predictable way, by lowering the combustion temperature within the engine. In this chapter we will learn why. Because many biological substances are weak acids, knowledge of the principles of equilibrium helps biologists to understand how organisms control the acidity of fluids within them. We'll use the basic principles of equilibrium developed in this chapter to learn more about acids and bases in Chapter Seventeen.

16.2 REACTION REVERSIBILITY

An equilibrium mixture will have the same composition whether approached from either pure reactants or pure products

In the preceding chapter we saw that equilibrium occurs at the low point on the free energy diagram, below the free energy of either the pure reactants or the pure products. This is significant because it means that whether we begin with pure reactants, pure products, or any mixture of reactants and products, the system will always "slide downhill" on the free energy curve in the direction of this minimum.

For example, let's consider the reaction for the decomposition of N_2O_4 into NO_2

$$N_2O_4(g) \rightleftharpoons 2NO_2(g)$$

(The free energy curve for this reaction was shown in Figure 15.8 on page 499.) Suppose we set up the two experiments shown in Figure 16.2. In the first 1-liter flask we place 0.0350 mol of N_2O_4. Since no NO_2 is present, some of the N_2O_4 must *decompose* in order for the mixture to reach equilibrium. At equilibrium we would find that the concentration of N_2O_4 is 0.0292 mol/liter and that the concentration of NO_2 is 0.0116 mol/liter. In the second 1-liter flask we place 0.0700 mol of NO_2 (*precisely* the amount of NO_2 that would form if 0.0350 mol of N_2O_4 decomposed completely). In this second flask there is no N_2O_4 present initially, so that enough NO_2 molecules must *combine*, following the reverse reaction, until there is sufficient N_2O_4 to give equilibrium. When we measure the concentrations at equilibrium in the second flask, we find, once again, 0.0292 mol/liter of N_2O_4 and 0.0116 mol/liter of NO_2.

0.0700 mol of NO_2 could be formed from 0.0350 mol of N_2O_4.

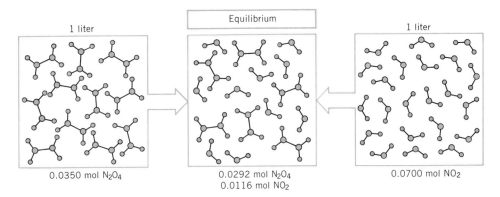

Figure 16.2
Reaction reversibility. The same equilibrium composition is reached from either the forward or reverse direction.

0.0350 mol N_2O_4

0.0292 mol N_2O_4
0.0116 mol NO_2

0.0700 mol NO_2

What we see here is that the same system composition results regardless of whether we begin with pure NO_2 or pure N_2O_4, just as long as the total amount of nitrogen and oxygen to be divided between these two substances is the same. For a given overall composition we always reach the same equilibrium concentrations whether equilibrium is approached from the forward or reverse direction. It is in this sense that chemical reactions are said to be reversible.

16.3 THE MASS ACTION EXPRESSION

At equilibrium the concentrations of reactants and products, expressed in mol/liter, are interrelated in a predictable way by the mass action expression

In our study of reaction rates, we learned that the rate of any given reaction depends on the reactant concentrations in a very specific way that is expressed by the rate law. In equilibrium systems there is also a specific relationship among molar concentrations—the molar concentrations of the reactants and the products.

Suppose we set up a number of experiments in which we carry out the same chemical reaction, each time at the same temperature, but with different initial amounts of reactants and products. For example, we could study the reaction of gaseous H_2 and I_2 to form gaseous HI. The reaction is

$$H_2(g) + I_2(g) \rightleftharpoons 2HI(g)$$

We allow each experiment shown in Figure 16.3 to come to equilibrium and then analyze the contents of each reaction vessel. As you can see in Figure 16.3 the equilibrium concentrations of H_2, I_2, and HI are different for each system, which is probably not surprising. What is amazing, however, is that there is a very simple relationship among the concentrations of the reactants and products in each case.

For each particular system, if we square the molar concentration of HI at equilibrium and then divide that by the product of the equilibrium molar concentrations of H_2 and I_2, we always obtain essentially the *same* numerical value, as shown in the last column of Table 16.1. The fraction used to calculate the values in the last column of Table 16.1 can be expressed as

$$\frac{[HI]^2}{[H_2][I_2]}$$

This fraction is known as the **mass action expression** or **reaction quotient** for the reaction.

The data in Table 16.1 would be obtained if we carried the reaction out at 440 °C. At this temperature the mass action expression for this reaction will always be equal to the same numerical value, 49.5, when H_2, I_2, and HI are in

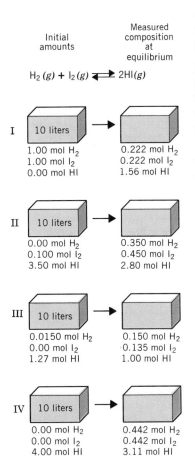

	Initial amounts	Measured composition at equilibrium
	$H_2 (g) + I_2 (g) \rightleftharpoons 2HI(g)$	

I 10 liters →

1.00 mol H_2
1.00 mol I_2
0.00 mol HI

0.222 mol H_2
0.222 mol I_2
1.56 mol HI

II 10 liters →

0.00 mol H_2
0.100 mol I_2
3.50 mol HI

0.350 mol H_2
0.450 mol I_2
2.80 mol HI

III 10 liters →

0.0150 mol H_2
0.00 mol I_2
1.27 mol HI

0.150 mol H_2
0.135 mol I_2
1.00 mol HI

IV 10 liters →

0.00 mol H_2
0.00 mol I_2
4.00 mol HI

0.442 mol H_2
0.442 mol I_2
3.11 mol HI

Figure 16.3
Four experiments to study the equilibrium between H_2, I_2, and HI at 440 °C.

Table 16.1
Equilibrium Concentrations and the Mass Action Expression

Experiment	Equilibrium Concentrations (mol/liter)			$\dfrac{[HI]^2}{[H_2][I_2]}$
	$[H_2]$	$[I_2]$	$[HI]$	
I	0.0222	0.0222	0.156	$(0.156)^2/(0.0222)(0.0222) = 49.4$
II	0.0350	0.0450	0.280	$(0.280)^2/(0.0350)(0.0450) = 49.8$
III	0.0150	0.0135	0.100	$(0.100)^2/(0.0150)(0.0135) = 49.4$
IV	0.0442	0.0442	0.311	$(0.311)^2/(0.0442)(0.0442) = \underline{49.5}$
				Average $= 49.5$

The letter Q is sometimes used to stand for the reaction quotient.

dynamic equilibrium. When they're not at equilibrium, the value of the mass action expression will be different from 49.5. The number 49.5 is a constant that characterizes the relation between the reactant and product concentrations at equilibrium. We call it the **equilibrium constant,** and we symbolize it as K_c (subscript c because we write the mass action expression using molar concentrations).

$$\frac{[HI]^2}{[H_2][I_2]} = K_c = 49.5 \text{ (at } 440 \text{ °C)} \qquad (16.1)$$

Equation 16.1 is the *equilibrium condition* for the reaction between H_2, I_2, and HI.

Because the mass action expression always equals K_c at equilibrium, Equation 16.1 becomes a criterion, or condition, that must be fulfilled for equilibrium to exist. Sometimes this equation is called the **equilibrium law** for the reaction.

We have seen that at 440 °C, $K_c = 49.5$ for the equilibrium between H_2, I_2, and HI. Had we performed our experiments at some other temperature, we would have found different sets of equilibrium concentrations. The calculated value of the mass action expression would have been different, but it would still be a constant at that temperature. The numerical value of K_c is different at different temperatures.

For the reaction, $CH_4(g) + H_2O(g) \rightleftharpoons CO(g) + 3H_2(g)$
$K_c = 1.78 \times 10^{-3}$ at 800 °C
$K_c = 4.68 \times 10^{-2}$ at 1000 °C
$K_c = 5.67$ at 1500 °C

An important fact about the mass action expression is that it can *always* be predicted from a knowledge of the balanced chemical equation for the reaction. Notice that this is quite a different situation from the rate law, which must always be determined experimentally. For the general reaction

$$dD + eE \rightleftharpoons fF + gG$$

where D, E, F, and G represent chemical formulas and d, e, f, and g are coefficients, the mass action expression is

$$\frac{[F]^f[G]^g}{[D]^d[E]^e}$$

The coefficients in the balanced equation are the same as the exponents in the mass action expression. The condition for equilibrium in this reaction would be

$$\frac{[F]^f[G]^g}{[D]^d[E]^e} = K_c$$

where the concentrations that satisfy the equation are equilibrium concentrations.

Notice that in writing the mass action expression the molar concentrations of the products are always placed in the numerator and those of the reactants appear in the denominator. Also note that after being raised to appropriate powers the concentrations are *multiplied*, not added. Using plus signs between concentration terms in the mass action expression is a common error that you should be sure to avoid.

EXAMPLE 16.1 **Writing the Equilibrium Law**

Problem: Write the equilibrium law for the reaction

$$N_2(g) + 3H_2(g) \rightleftharpoons 2NH_3(g)$$

This is the reaction used for the industrial preparation of ammonia, one of the world's most important chemicals (Section 12.2).

Solution: The coefficients become exponents. We have to remember to place the products in the numerator and the reactants in the denominator. Also, we must remember that when there is more than one reactant or product, the concentrations are multiplied, *not* added.

$$\frac{[NH_3]^2}{[N_2]^1[H_2]^3} = K_c$$

EXERCISE 16.1 Write the equilibrium law for the following:

$$2H_2(g) + O_2(g) \rightleftharpoons 2H_2O(g)$$
$$NH_3(aq) + H_2O(\ell) \rightleftharpoons NH_4^+(aq) + OH^-(aq)$$

The rule that we always write the concentrations of the products in the numerator of the mass action expression and the concentrations of the reactants in the denominator hasn't been established by nature. It is simply a custom that chemists have agreed on. Certainly, if the mass action expression is equal to a constant,

$$\frac{[HI]^2}{[H_2][I_2]} = K_c$$

its reciprocal is also equal to a constant.

$$\frac{[H_2][I_2]}{[HI]^2} = \frac{1}{K_c} = K_c'$$

The reason that chemists have chosen to stick to a set pattern—always placing the concentrations of the products in the numerator—is to avoid having to specify mass action expressions along with tabulated values of equilibrium constants. If we have the chemical equation for the equilibrium, we can always construct the correct mass action expression from it. For example, suppose we're told that at a particular temperature $K_c = 10.0$ for the reaction

$$2NO_2(g) \rightleftharpoons N_2O_4(g)$$

From the chemical equation we can write the correct mass action expression.

$$\frac{[N_2O_4]}{[NO_2]^2} = 10.0$$

The balanced chemical equation contains all the information we need to write the equilibrium law.

16.4 MASS ACTION EXPRESSIONS FOR GASEOUS REACTIONS

For gaseous reactions, the equilibrium law can be written using either concentrations or partial pressures

The pressure of a gas is proportional to its concentration. For example, 1 mol of O_2 in a 22.4-liter container at 0 °C exerts a pressure of 1 atm. We learned this in Chapter 3. The concentration of O_2 in the vessel is 1 mol/22.4 liter = 0.0446 mol/liter or 0.0446 M. If we were to double the number of moles of O_2 in the container, the pressure would also double. Two moles of O_2 in 22.4 liters at 0 °C would have a pressure of 2 atm and a molar concentration of 2 mol/22.4 liters = 0.0892 M (see Figure 16.4).

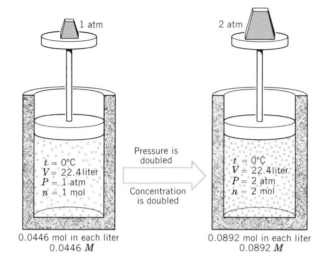

Figure 16.4
The molar concentration of a gas is directly proportional to the pressure of the gas. At a given temperature, packing twice as many molecules into the same volume doubles the pressure.

Because of this relationship between pressure and concentration, we can write the mass action expression for reactions between gases using either molar concentrations or partial pressures. (We must use *partial* pressures because we are dealing with mixtures of gases.) For example, the equilibrium law for the reaction to form ammonia

$$N_2(g) + 3H_2(g) \rightleftharpoons 2NH_3(g)$$

can be written in either of the following ways:

$$\frac{[NH_3(g)]^2}{[N_2(g)][H_2(g)]^3} = K_c \quad \left\{ \begin{array}{l} \text{because } \textit{concentrations} \text{ are} \\ \text{used in the mass action} \\ \text{expression} \end{array} \right.$$

or

$$\frac{p^2_{NH_3(g)}}{p_{N_2(g)}p^3_{H_2(g)}} = K_p \quad \left\{ \begin{array}{l} \text{because partial } \textit{pressures} \\ \text{are used in the mass action} \\ \text{expression} \end{array} \right.$$

At equilibrium, the molar concentrations can be used to calculate K_c, while the partial pressures of the gases at equilibrium can be used to calculate K_p.[1]

---

[1] In general, K_p and K_c do not have identical values. They are equal only if the sum of the coefficients of gaseous reactants equals the sum of the coefficients of gaseous products. While it is possible to calculate K_p from K_c, and vice versa, we won't do this here.

EXAMPLE 16.2   **Writing the Equilibrium Law for Gaseous Reactions**

Problem: What is the expression for K_p for the reaction,

$$N_2O_4(g) \rightleftharpoons 2NO_2(g)$$

Solution: For K_p we use partial pressures in the mass action expression. Therefore,

$$K_p = \frac{p^2_{NO_2}}{p_{N_2O_4}}$$

EXERCISE 16.2   Using partial pressures, write the equilibrium law for the reaction, $H_2(g) + I_2(g) \rightleftharpoons 2HI(g)$

16.5 EQUILIBRIUM CONSTANTS: WHAT THEY TELL US

The larger the value of K, the further the reaction will have proceeded toward completion when equilibrium is reached

A useful bonus of always writing the mass action expression with the product concentrations in the numerator is that the size of the equilibrium constant gives us an indication of how far the reaction has proceeded toward the formation of the products when equilibrium is reached. For example, the reaction

$$2H_2(g) + O_2(g) \rightleftharpoons 2H_2O(g)$$

has $K_c = 9.1 \times 10^{80}$ at 25 °C. This means that if there is an equilibrium between these reactants and product,

$$K_c = \frac{[H_2O]^2}{[H_2]^2[O_2]} = \frac{9.1 \times 10^{80}}{1}$$

By writing K_c as a fraction, $9.1 \times 10^{80}/1$, we can see that the only way for the numerator to be so much larger than the denominator is for the concentration of H_2O to be enormous in comparison to the concentrations of H_2 and O_2. This means that at equilibrium most of the hydrogen and oxygen atoms are found in H_2O; very few are present as H_2 and O_2. Thus, the large value of K_c indicates that the reaction between H_2 and O_2 goes essentially to completion.

Actually, you would need about 200,000 liters of water vapor at 25 °C just to find one molecule of O_2 and two molecules of H_2.

The reaction between N_2 and O_2 to give NO

$$N_2(g) + O_2(g) \rightleftharpoons 2NO(g)$$

has a very small equilibrium constant, $K_c = 4.8 \times 10^{-31}$ at 25 °C. The equilibrium law for this reaction is

$$\frac{[NO]^2}{[N_2][O_2]} = 4.8 \times 10^{-31}$$

Since $10^{-31} = 1/10^{31}$, we can write this as

$$\frac{[NO]^2}{[N_2][O_2]} = \frac{4.8}{10^{31}}$$

In air at 25 °C, the equilibrium NO concentration should be about 10^{-7} mol/liter

Here the denominator is huge compared to the numerator. The concentrations of N_2 and O_2 must therefore be very much larger than the concentration of NO. This means that in a mixture of N_2 and O_2 at this temperature, very little NO is formed. The reaction proceeds hardly at all toward completion before equilibrium is reached.

The relationship between the equilibrium constant and the extent to which the reaction proceeds toward the formation of products when equilibrium occurs can be summarized as follows:

K very large	Reaction proceeds far toward completion
$K \approx 1$	Concentrations of reactants and products are nearly the same at equilibrium
K very small	Hardly any products are formed

Note that we have omitted the subscript for K in this summary. The same qualitative predictions about the extent of reaction apply whether we use K_p or K_c.

One of the ways that we can use equilibrium constants is to compare the extents to which two or more reactions proceed to completion, as shown in Example 16.3. Care has to be exercised in making such comparisons, however, because unless the K's are greatly different, the comparison will only be valid if both reactions have the same number of reactant and product molecules appearing in their balanced chemical equations.

EXAMPLE 16.3   **Using K_c to Estimate Extent of Reaction**

Problem: Which of the following two reactions would tend to proceed furthest to completion when they reach equilibrium?

(1) $2NO(g) + O_2(g) \rightleftharpoons 2NO_2(g)$ $K_c = 3.4 \times 10^{13}$
(2) $4NH_3(g) + 5O_2(g) \rightleftharpoons 4NO(g) + 6H_2O(g)$ $K_c = 5 \times 10^{198}$

Solution: We know that the larger the value of K, the further the reaction proceeds toward completion. Since 5×10^{198} is much greater than 3.4×10^{13}, reaction (2) goes further to completion than reaction (1).

EXERCISE 16.3   Which of the following reactions will tend to proceed furthest toward completion?

(1) $2HBr(g) \rightleftharpoons H_2(g) + Br_2(g)$ $K_c = 7.0 \times 10^{-20}$
(2) $Si(s) + O_2(g) \rightleftharpoons SiO_2(s)$ $K_c = 2 \times 10^{142}$
(3) $C_2H_4(g) + H_2(g) \rightleftharpoons C_2H_6(g)$ $K_c = 1.2 \times 10^{19}$

16.6 CALCULATING EQUILIBRIUM CONSTANTS

Equilibrium constants can be calculated from measured equilibrium concentrations or from $\Delta G°$ for the reaction

Based on the preceding discussion, it's clear that equilibrium constants are useful quantities to know. For example, wouldn't it be helpful to know ahead of time that, for a particular reaction, only tiny amounts of products will be present when it gets to equilibrium? Then we wouldn't waste our time sitting around waiting for lots of products to be formed! The question is, how do we obtain equilibrium constants?

One obvious way to determine the value of the equilibrium constant is to carry out the reaction and actually measure the concentrations of reactants and products after equilibrium has been reached. As an example, let's look again at the decomposition of N_2O_4.

$$N_2O_4(g) \rightleftharpoons 2NO_2(g)$$

In Section 16.2, we saw that if 0.0350 mol of N_2O_4 is placed into a 1-liter flask at 25 °C, the concentrations of N_2O_4 and NO_2 at equilibrium will be

$$[N_2O_4] = 0.0292 \text{ mol/liter}$$

$$[NO_2] = 0.0116 \text{ mol/liter}$$

To calculate K_c for this reaction we substitute the equilibrium concentrations into the mass action expression.

$$\frac{[NO_2]^2}{[N_2O_4]} = K_c$$

$$\frac{(0.0116)^2}{(0.0292)} = K_c$$

Performing the arithmetic gives

$$K_c = 0.00461$$
$$= 4.61 \times 10^{-3}$$

Determining the value of an equilibrium constant this way seems very simple, but in practice there sometimes are difficulties. It isn't always easy to know when equilibrium has been reached, or to measure equilibrium concentrations directly. For example, if you mix H_2 and O_2 in a flask, nothing appears to happen. If you were young and innocent, you might be fooled into thinking that the reaction,

$$2H_2(g) + O_2(g) \rightleftharpoons 2H_2O(g)$$

has a very small K, because no H_2 or O_2 seems to disappear. Actually, $K_c = 9.1 \times 10^{80}$ for this reaction, and if you heat the mixture, you'll quickly discover that this reaction goes far toward completion. If the violent reaction doesn't scatter pieces of the flask about the lab, you'll now have a vessel filled with water vapor. Had we started with an exact 2:1 mol ratio of H_2 and O_2, their concentrations in the flask would be essentially zero and no known instrument could detect their presence, much less actually measure their concentrations. How, then, do we know the value of K_c for this reaction if we can't measure equilibrium concentrations?

The answer to this question comes from thermodynamics. In Chapter Fifteen we saw that when $\Delta G°$ for a reaction is negative, the minimum on the free energy curve, which corresponds to equilibrium, occurs near the products. Since substantial amounts of the products are present and very little of the reactants remain, such a reaction has a relatively large equilibrium constant. On the other hand, when $\Delta G°$ is positive, the free energy minimum occurs near the reactants. Very little products are present at equilibrium, and K for the reaction is small.

The relationship between $\Delta G°$ and K is more than merely a qualitative one. As you will see, it involves logarithms. If you feel a bit uncertain about logarithms (as many of your classmates do), it would be wise to review them in Appendix A.

From thermodynamics the following equation can be derived.

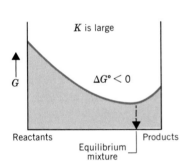

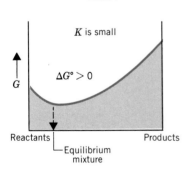

$$\boxed{\Delta G° = -RT \ln K} \qquad (16.2)$$

where R is the gas constant expressed in energy units (R = 8.314 J/mol K, or R = 1.987 cal/mol K), T is the absolute temperature in kelvins, and the final term, $\ln K$, is the *natural logarithm* of the equilibrium constant. Natural logarithms are logarithms having the base e = 2.71828. . . . In Equation 16.2 they ask the question, "To what power must we raise e to obtain the value of K?"

Equation 16.2 can also be written using common logarithms.

$$\boxed{\Delta G° = -2.303 \, RT \log K} \qquad (16.3)$$

The factor 2.303 converts common logs to natural logs.

$$2.303 \log x = \ln x$$

Equations 16.2 and 16.3 are valuable because they allow us to relate $\Delta G°$ to K quantitatively. For example, measurements of equilibrium constants allow us to calculate $\Delta G°$ and, therefore, equilibrium measurements serve as a source of thermodynamic data. We can also use these equations to calculate K from $\Delta G°$. Therefore, if we have a way of obtaining $\Delta G°$ for a reaction, as in a Hess's law-type calculation that you learned to do in Chapter Fifteen, we can calculate K and predict quantitatively how far toward completion a reaction will go.

There is one complicating factor involving Equations 16.2 and 16.3 of which you should be aware. For reactions involving gases, the K that is calculated is K_p, with partial pressures expressed in atmospheres. For reactions involving liquid solutions, the calculated K is K_c.

EXAMPLE 16.4 Calculating $\Delta G°$ from K

Problem: The brownish haze associated with air pollution is caused by nitrogen dioxide, NO_2, a red-brown gas. Nitric oxide, NO, is oxidized to NO_2 by oxygen.

$$2NO(g) + O_2(g) \rightleftharpoons 2NO_2(g)$$

The value of K_p for this reaction is 1.66×10^{12} at 25 °C. What is $\Delta G°$ for the reaction, expressed in joules? In kilojoules?

Solution: If a scientific calculator is available, it is easiest to use Equation 16.2,

$$\Delta G° = -RT \ln K_p$$

Substituting $R = 8.314$ J/mol K (because the answer must be in joules), $T = 298$ K, and $K_p = 1.66 \times 10^{12}$, we find that

$$\Delta G° = -(8.314 \times 298) \ln(1.66 \times 10^{12})$$

The natural log of $1.66 \times 10^{12} = 28.1$. Therefore,

$$\Delta G° = -(8.314 \times 298 \times 28.1)$$
$$= -69,600 \text{ J (to three significant figures)}$$

Expressed in kilojoules, $\Delta G° = -69.6$ kJ

EXAMPLE 16.5 Calculating K from $\Delta G°$

Problem: Sulfur dioxide reacts with oxygen when it passes over the catalyst in automobile catalytic converters. The product is SO_3.

$$2SO_2(g) + O_2(g) \rightleftharpoons 2SO_3(g)$$

For this reaction $\Delta G° = -33.4$ kcal at 25 °C. What is the value of K_p?

Solution: This time, let's use Equation 16.3.

$$\Delta G° = -2.303RT \log K_p$$

We will use $R = 1.987$ cal/mol K and convert $\Delta G°$ from kilocalories to calories so that calories will cancel.

$$\Delta G° = -33,400 \text{ cal}$$

Substituting the given values into our equation (taking $T = 298$ K)

$$-33,400 = -2.303(1.987)(298) \log K_p$$

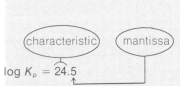

$\log K_p = \overset{\frown}{24.5}$

To take the antilog, find the mantissa in the log table. The characteristic becomes the exponent on the 10.

We next multiply both sides by −1 (to remove the negative sign) and then solve for the term, $\log K_p$.

$$\frac{33,400}{2.303(1.987)(298)} = \log K_p$$

$$24.5 = \log K_p$$

To obtain K_p we must take the antilog of 24.5. On a calculator we can do this by computing $10^{24.5}$, giving us

$$K_p = 3 \times 10^{24}$$

If you wish to use log tables to take the antilog of 24.5, find 0.5 within the table and notice that it's the log of 3. Again we obtain $K_p = 3 \times 10^{24}$.

EXERCISE 16.4   The reaction, $N_2(g) + 3H_2(g) \rightleftharpoons 2NH_3(g)$, has $K_p = 6.9 \times 10^5$ at 25 °C. Calculate $\Delta G°$ for this reaction in units of kcal.

EXERCISE 16.5   The reaction, $H_2(g) + I_2(g) \rightleftharpoons 2HI(g)$, has $\Delta G° = +3.3$ kJ at 25 °C. What is the value of K_p at 25 °C?

16.7 LE CHÂTELIER'S PRINCIPLE AND CHEMICAL EQUILIBRIA

If an equilibrium in a system is upset, the system will tend to react in a direction that will reestablish equilibrium

Although it is possible to perform calculations that tell us what the composition of an equilibrium system is, many times we really don't need to know exactly what the equilibrium concentrations are. Instead, we may simply want to know what actions we should take to maximize or minimize the amount of a given product or reactant. For instance, if we were designing gasoline engines, one of the things we would like to know is what could be done to reduce nitrogen oxide pollutants. Or, if we were preparing ammonia, NH_3, by the reaction of N_2 and H_2, we might want to know how to increase the yield of NH_3.

Le Châtelier's principle, introduced in Chapter Ten, provides us with the means for making qualitative predictions about changes in chemical equilibria. It does this in much the same way that it allows us to predict the effects of outside influences on equilibria that involve physical changes, such as liquid-vapor equilibria. Recall that Le Châtelier's principle states, in effect, that *if an outside influence upsets an equilibrium, the system responds in a direction that counteracts the disturbing influence such that equilibrium is reestablished.*

The following factors influence the relative amounts of products and reactants in a chemical system at equilibrium.

1 *Adding or removing a reactant or product.* If we add or remove a reactant or product, we will change one of the concentrations in the system. This means that the value of the mass action expression will change; it will no longer be equal to K_c and the system will not be at equilibrium. In order for the system to return to equilibrium, the concentrations will have to change in some way so that the mass action expression will be equal to K_c once again. This change is brought about by the chemical reaction proceeding either to the right or left. Le Châtelier's principle tells us which way the reaction goes. According to Le Châtelier's principle the reaction will proceed in a direction that will partially offset the change in concentration. For example, if we have the equilibrium,

$$3H_2(g) + N_2(g) \rightleftharpoons 2NH_3(g)$$

The reaction shifts in a direction that will remove a substance that's been added or replace a substance that's been removed.

and add some H_2, the system will respond by using up some of the H_2 by reaction with N_2. Therefore, some of the N_2 will also be consumed (because it reacts with the H_2) and some NH_3 will be formed. Thus, adding H_2 leads to the production of more NH_3 and we say that the equilibrium "shifts to the right." The equilibrium also shifts to the right if we remove some NH_3. In this case H_2 reacts with some N_2 to replace a portion of the NH_3 that has been removed.

Figure 16.5 illustrates how the concentrations change when we add H_2 or remove NH_3. At time t_1, we add some H_2 and its concentration suddenly increases. To restore equilibrium, some of the H_2 reacts with N_2 to form more NH_3. Therefore, the concentrations of H_2 and N_2 both drop and the concentration of NH_3 increases. The relative size of each change is determined by the stoichiometry of the reaction. As the balanced equation would predict, the H_2 concentration decreases three times as much as the N_2 concentration. Similarly, the NH_3 concentration increases by twice as much as the N_2 concentration decreases.

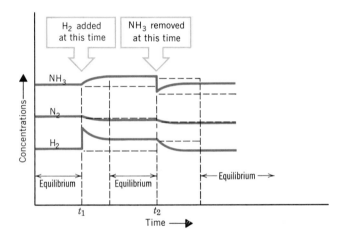

Figure 16.5

Effects of the addition of H_2 and removal of NH_3 on the equilibrium, $N_2(g) + 3H_2(g) \rightleftharpoons 2NH_3(g)$. When H_2 is added, some N_2 reacts and more NH_3 is formed as the system returns to equilibrium. When NH_3 is removed, more N_2 reacts with H_2 as the sytem again comes back to equilibrium.

A similar analysis of the changes in concentrations can be made when NH_3 is removed at time t_2. The NH_3 concentration drops suddenly and then increases as some of it is replaced and the system approaches equilibrium again. The concentrations of H_2 and N_2 both decrease because they are the substances that react to replace part of the NH_3 that was removed.

An important point to notice in Figure 16.5 is that not all of the H_2 that we've added is removed when the equilibrium is reestablished. Neither is all the NH_3 replaced when the reaction adjusts to the final equilibrium. The concentrations in the new equilibrium systems have all changed, but in such a way that the mass action expression is still equal to the same value of K_c.

2 *Changing the volume in gaseous reactions.* Changing the volume of a system composed of gaseous reactants and products changes the pressure. For example, if we reduced the volume by half, we would expect the pressure to double. Is there any way that the system could oppose this change?

Again, let's consider the equilibrium,

$$N_2(g) + 3H_2(g) \rightleftharpoons 2NH_3(g)$$

If the reaction proceeds to the right, two NH_3 molecules appear when four molecules (one N_2 and three H_2) disappear, and the number of molecules in the reaction vessel is reduced. You know by now that the fewer the number of molecules, the lower the pressure. Therefore, this equilibrium will respond to pressure increases caused by volume decreases by shifting to the right. The production of more NH_3 tends to lower the pressure and counter the effects of the volume decrease.

Now let's look at the equilibrium,

$$H_2(g) + I_2(g) \rightleftharpoons 2HI(g)$$

When this reaction occurs in either direction there is no change in the number of molecules of gas. This reaction, then, cannot respond to pressure changes, so changing the volume of the reaction vessel has virtually no effect on the equilibrium.

The simplest way to analyze the effects of a volume change on an equilibrium system is to count the number of molecules of *gaseous substances* on both sides of the equation. *An increase in pressure always drives the reaction in the direction of the fewest number of molecules of* **gas.**

As a final note here, pressure changes have essentially no effect on reactions involving only liquids or solids. Substances in these states are virtually incompressible, and reactions involving them have no way of counteracting pressure changes.

3 *Changes of temperature.* To understand how a chemical reaction at equilibrium responds to a change in temperature, we must know how the energy of the system changes when reaction occurs. For instance, the reaction to produce NH_3 from N_2 and H_2 is exothermic; heat is evolved when NH_3 molecules are formed ($\Delta H_f^\circ = -11.0$ kcal/mol or -46.0 kJ/mol from Table 5.2). If we include heat as a product in the equilibrium equation

$$N_2(g) + 3H_2(g) \rightleftharpoons 2NH_3(g) + \text{heat}$$

analyzing the effects of temperature changes becomes simple.

When we raise the temperature, we do so by adding heat from some external source, such as a bunsen burner. The equilibrium above can counter the temperature increase by absorbing some of the added heat. It does this by proceeding from right to left because the decomposition of some NH_3 to give more N_2 and H_2 is endothermic and consumes some of the heat that we've added. Thus, raising the temperature shifts this equilibrium to the left.

Increasing the temperature shifts a reaction in a direction that produces an endothermic change.

An important point to note here is that when we change the temperature, the concentrations change even though no chemical substances have been added or removed. The system comes to a new equilibrium in which the value of the mass action expression has changed, which means that K has changed. In other words, the value of K changes when we change the temperature.

The only factor that can change K_c or K_p is a change in temperature.

The equilibrium law for the formation of ammonia is

$$\frac{[NH_3]^2}{[N_2][H_2]^3} = K_c$$

K decreases with rising temperature for a reaction that is exothermic, when read from left to right.

When we raised the temperature, $[NH_3]$ decreased whereas $[N_2]$ and $[H_2]$ both increased. The new values for these equilibrium concentrations give a new K_c that is smaller at the higher temperature. As a general rule, K_c

As everyone knows, one of the most serious causes of air pollution is the automobile. All sorts of obnoxious chemicals are emitted in the exhaust, and there have been a variety of methods devised to control the amounts of these pollutants. For instance, most cars today are equipped with catalytic converters (see Chapter 14), which mix air with the exhaust gases and promote oxidation of carbon monoxide and unburned hydrocarbons to carbon dioxide. The use of catalysts to aid in the removal of nitrogen oxides has not been widespread, however, because of the extra cost and the further reduction of fuel economy that they involve.

When air is drawn into a car's engine, both N_2 and O_2 are present. During the combustion of the gasoline, oxygen reacts with the hydrocarbons in the fuel to produce CO_2, CO, and H_2O. However, N_2 and O_2 can also form NO.

$$N_2(g) + O_2(g) \rightleftharpoons 2NO(g)$$

At room temperature, K_c for this reaction is 4.8×10^{-31}. Its small value tells us that the equilibrium concentration of NO should be very small. Therefore, we don't find N_2 reacting with O_2 under ordinary conditions. The atmosphere, for instance, is quite stable.

The reaction of N_2 and O_2 to form NO is endothermic. Le Châtelier's principle tells us that at high temperatures, such as those found in the cylinders of a gasoline or diesel engine during combustion, this equilibrium should be shifted to the right. Therefore, at high temperatures some NO does form. Unfortunately, when the exhaust leaves the engine, it cools so rapidly that the NO can't decompose. The reaction rate at the lower temperature becomes too slow. The result is that some NO exists with the exhaust gases. Once in the atmosphere, the NO is oxidized to NO_2, which is responsible for the brownish haze often associated with severe air pollution. See Color Plate 4C.

One way to reduce the amount of NO pollution of the atmosphere is to reduce the amount of NO that's formed in automobile engines. Since the extent to which the reaction for the formation of NO proceeds to completion increases as the temperature is raised, it is clear that the amount of NO that is formed can be reduced by simply running the combustion reaction at a lower temperature. One method that has been used to accomplish this is to lower the compression ratio of the engine. This is the ratio of the volume of the cylinder when the piston is at the bottom of its stroke divided by the volume after the piston has compressed the air-fuel mixture. At high compression ratios, the air-fuel mixture is heated to a high temperature before it's ignited. After combustion, the gases are very hot, which favors the production of NO. Lowering the compression ratio lowers the maximum combustion temperature which decreases the tendency for NO to be formed. Unfortunately, lowering the compression ratio also lowers the efficiency of the engine, which makes for poorer fuel economy.

Another method for controling NO emissions that has been experimented with involves mixing water with the air-fuel mixture. Some of the heat from the combustion is absorbed by the water vapor, so the mixture of exhaust gases doesn't get as hot as it would otherwise. At these lower temperatures, the concentration of the NO in the exhaust is drastically reduced.

(or K_p) becomes smaller with increasing temperature for exothermic reactions. The opposite, of course, is true for endothermic reactions.

4 Effect of a catalyst. In Chapter Fourteen, we learned that a catalyst increases the rate of a reaction by lowering the activation energy. This change affects both the forward and reverse reactions in the same way; both rates are increased by the same factor. Therefore, adding a catalyst to a system at equilibrium simply makes both forward and reverse reactions occur faster by the same amount, and no change occurs in the composition of the system. The only effect a catalyst ever has on a chemical reaction is to bring it to equilibrium faster.

Catalysts cause reactions to come to equilibrium more rapidly.

EXAMPLE 16.6   Application of Le Châtelier's Principle

Problem: The reaction, $N_2O_4(g) \rightleftharpoons 2NO_2(g)$, is endothermic, with $\Delta H° = +56.9$ kJ. How will the amount of NO_2 at equilibrium be affected by (a) adding N_2O_4,

(b) lowering the pressure by increasing the volume of the container, (c) raising the temperature, and (d) adding a catalyst to the system. Which of these changes will alter the value of K_c?

Solution:

(a) Adding N_2O_4 will cause the equilibrium to shift to the right. *The amount of NO_2 will increase.*

(b) When the pressure in the system drops, the system responds by producing more molecules of gas, which will tend to raise the pressure and partially offset the change. Since more gas molecules are formed if some N_2O_4 decomposes, the *amount of NO_2 at equilibrium will increase.*

(c) Because the reaction is endothermic, we write the equation showing heat as a *reactant.*

$$\text{heat} + N_2O_4(g) \rightleftharpoons 2NO_2(g)$$

Since raising the temperature is accomplished by adding heat, the system will respond by absorbing heat. This means that the equilibrium will shift to the right. Therefore, when equilibrium is reestablished, *there will be more NO_2 present.*

(d) A catalyst has no effect on a chemical equilibrium. Catalysts only affect the speed of a reaction; they cause reactions to reach equilibrium more rapidly. Therefore, the amount of NO_2 at equilibrium will not be affected.

The only change that alters K_c is the temperature change. Raising the temperature (adding heat) will increase K_c for this endothermic reaction.

EXERCISE 16.6   Consider the equilibrium, $PCl_3(g) + Cl_2(g) \rightleftharpoons PCl_5(g)$, for which $\Delta H° = -21$ kcal. How will the amount of Cl_2 at equilibrium be affected by (a) adding PCl_3, (b) adding PCl_5, (c) raising the temperature, and (d) decreasing the volume of the container? How (if at all) will each of these changes affect K_p for the reaction?

16.8 EQUILIBRIUM CALCULATIONS

The equilibrium law serves as the focal point in performing all equilibrium calculations

Sometimes it is necessary to have more than merely a qualitative knowledge of equilibrium concentrations. For example, life forms are generally quite sensitive to the hydrogen ion concentration in their surroundings. If you wished to grow bacteria in a solution of a weak acid, it would be important to know (or to be able to calculate) the H^+ concentration in the solution.

Equilibrium calculations can be performed using either K_p or K_c for gaseous reactions, but for reactions in solution we must use K_c. Whether we deal with concentrations or partial pressures, however, the same basic principles apply. To keep things simple, we will restrict ourselves to calculations involving K_c and molar concentrations.

Overall, we can divide equilibrium calculations into two main categories: (1) calculating equilibrium constants from known equilibrium concentrations and (2) calculating one or more equilibrium concentrations using the known value of K_c.

Calculating K_c from equilibrium concentrations

In Section 16.6 we saw that if we know all of the equilibrium concentrations in a reaction mixture, it is a simple matter to substitute them into the mass action expression and calculate K_c. Sometimes, however, we must do a little work to figure out what all of the concentrations are, as shown in Example 16.7.

EXAMPLE 16.7 Calculating K_c from Equilibrium Concentrations

Problem: At a certain temperature, a mixture of H_2 and I_2 was prepared by placing 0.100 mol of H_2 and 0.100 mol of I_2 into a 1.00-liter flask. After a period of time the equilibrium,

$$H_2(g) + I_2(g) \rightleftharpoons 2HI(g)$$

was established. The purple color of the I_2 vapor was used to monitor the reaction, and from the decreased intensity of the purple color it was determined that, at equilibrium, the I_2 concentration had dropped to 0.020 mol/liter. What is the value of K_c for this reaction at this temperature?

Solution: To calculate the value of K_c, we must substitute the equilibrium concentrations of H_2, I_2, and HI into the mass action expression, because for this reaction at equilibrium

$$\frac{[HI]^2}{[H_2][I_2]} = K_c$$

But what are the equilibrium concentrations? It appears that we've been given only the value of $[I_2]$. To obtain the others, we must examine the initial concentrations (the concentrations of reactants and products that were initially placed into the reaction flask) and figure out how they changed as the reaction came to equilibrium. (We will have to perform similar analyses in many of our equilibrium calculations in this chapter and the next.) To help organize the information and to approach the problem systematically, we will arrange the data in a table, using the chemical equation as a guide. The completed table for this problem is shown below.

	$H_2(g)$ +	$I_2(g)$ $\rightleftharpoons$	$2HI(g)$
Initial concentrations	0.100 M	0.100 M	0.000 M
Changes in concentration	−0.080	−0.080	+2(0.080)
Equilibrium concentrations	0.020 M	0.020 M	0.160 M

Let's see how the entries in the table were obtained.

Initial Concentrations. Under the formulas we've written the initial *molar* concentrations—that is, the initial concentrations expressed in moles per liter. In making these entries, you must always remember to calculate the ratio of moles to liters (unless molarity is given). The volume is 1.00 liter in this problem; therefore, the initial concentrations of H_2 and I_2 were 0.100 mol/1.00 liter = 0.100 M. (Had the volume been 2.00 liters, the initial concentrations would have been 0.100 mol/2.00 liter = 0.0500 M.) A value of 0.000 is placed under HI because none of this substance was originally placed in the reaction flask.

Changes in Concentration. Because the original mixture was not at equilibrium, a chemical reaction had to occur until equilibrium was reached. Since no HI was present initially, the reaction had to proceed to the right; there has to be some HI in the flask at equilibrium. As the reaction proceeded to the right, all of the concentrations changed. To obtain the entries in the "change" row we have to do some figuring. We know, from the statement of the problem, that the equilibrium concentration of I_2 is

In any system, the initial concentrations are controlled by the person doing the experiment.

0.020 *M*. Since it began as 0.100 *M*, its concentration must have *decreased* by 0.080 mol per liter (0.100 − 0.080 = 0.020). In the table we've entered the change with a *minus* sign to indicate a *decrease* in concentration.

Once we know how one of the concentrations changed we can quickly figure how the others changed. The stoichiometry of the reaction controls this. Since H_2 and I_2 react in a one-to-one mole ratio, if 0.080 mol of I_2 reacted per liter, then 0.080 mol of H_2 must also have reacted per liter. The H_2 concentration must have decreased by the same amount as the I_2 concentration. Since two moles of HI are formed for each mole of I_2 that disappears, the HI concentration must have *increased* by *twice* the amount that the I_2 concentration has decreased—thus the plus sign and the factor of 2.

The changes in concentration are controlled by the stoichiometry of the reaction.

Equilibrium Concentrations. By using the minus and plus signs in the "change" row to show decreases and increases in concentration, the equilibrium concentrations always can be calculated as

$$\begin{pmatrix}\text{equilibrium}\\\text{concentration}\end{pmatrix} = \begin{pmatrix}\text{initial}\\\text{concentration}\end{pmatrix} + \begin{pmatrix}\text{change in}\\\text{concentration}\end{pmatrix}$$

For example,

$$[H_2] = (0.100) + (-0.080) = 0.020\ M$$

$$[HI] = (0.000) + (+0.160) = 0.160\ M$$

Once values have been obtained for all of the equilibrium concentrations, they are substituted into the mass action expression to calculate K_c.

$$K_c = \frac{[HI]^2}{[H_2][I_2]}$$

$$= \frac{(0.160)^2}{(0.020)(0.020)}$$

$$K_c = 64$$

In the preceding example, there are some key points that apply in constructing the concentration table under the chemical equation. These apply not only to this problem but to others that you will see that deal with equilibrium calculations. First, you must remember that the only values that we may substitute into the mass action expression in the equilibrium law are *equilibrium concentrations*—the values that appear in the last row of our table. Second, when we enter initial concentrations into the table, they should be in units of moles/liter. The initial concentrations are those present in the reaction mixture when it's prepared; we imagine that no reaction occurs until everything is mixed. Third, the changes in concentration always occur in the *same ratio* as the ratio of coefficients in the balanced equation. For example, if we were dealing with the equilibrium,

By using concentrations in mol/liter, we follow what happens to reactants and products in each liter of solution.

$$3H_2(g) + N_2(g) \rightleftharpoons 2NH_3(g)$$

and found that the $N_2(g)$ concentration decreased by 0.10 *M* during the approach to equilibrium, the entries in the "change" row would be as follows:

	$3H_2(g)$	+	$N_2(g)$	$\rightleftharpoons$	$2NH_3(g)$
change in concentration	$-3 \times (0.10)$		$-1 \times (0.10)$		$+2 \times (0.10)$
	$\downarrow$		$\downarrow$		$\downarrow$
	-0.30		-0.10		$+0.20$

If the initial concentration of some reactant is zero, its change must be positive (an increase) because the final concentration can't be negative.

Also note that all the reactant concentrations change in the same direction (both have minus signs) and that the product concentrations change in the opposite direction. Keep these ideas in mind as we construct the concentration tables for other equilibrium problems.

EXERCISE 16.7 In a particular experiment, it was found that when $O_2(g)$ and $CO(g)$ were mixed and allowed to react according to the equation

$$2CO(g) + O_2(g) \rightleftharpoons 2CO_2(g)$$

the O_2 concentration had decreased by 0.030 mol/liter when the reaction reached equilibrium. How did the concentrations of CO and CO_2 change?

EXERCISE 16.8 An equilibrium was established for the reaction

$$CO(g) + H_2O(g) \rightleftharpoons CO_2(g) + H_2(g)$$

at 500 °C. (This is an industrially important reaction for the preparation of hydrogen.) At equilibrium, the following concentrations were found in the reaction vessel: $[CO] = 0.180\ M$, $[H_2O] = 0.0411\ M$, $[CO_2] = 0.150\ M$, and $[H_2] = 0.200\ M$. What is the value of K_c for this reaction?

EXERCISE 16.9 A student placed 0.20 mol of $PCl_3(g)$ and 0.10 mol of $Cl_2(g)$ into a 1.00-liter flask at 250 °C. The reaction

$$PCl_3(g) + Cl_2(g) \rightleftharpoons PCl_5(g)$$

was allowed to come to equilibrium at which time it was found that the flask contained 0.12 mol of PCl_3.
(a) What were the initial concentrations of the reactants and product?
(b) What were the changes in concentration?
(c) What were the equilibrium concentrations?
(d) What is the value of K_c for this reaction?

Calculating equilibrium concentrations using K_c

The simplest calculation of this type occurs when all but one of the equilibrium concentrations are known, as illustrated in Example 16.8.

EXAMPLE 16.8 **Using K_c to Calculate Concentrations at Equilibrium**

Problem: The reversible reaction

$$CH_4(g) + H_2O(g) \rightleftharpoons CO(g) + 3H_2(g)$$

has been suggested as a possible source of hydrogen. The hydrogen would be shipped to its destination as CH_4 through our natural gas pipelines. At 1500 °C, an equilibrium mixture of these gases was found to have the following concentrations: $[CO] = 0.300\ M$, $[H_2] = 0.800\ M$, and $[CH_4] = 0.400\ M$. At 1500 °C, $K_c = 5.67$ for this reaction. What was the equilibrium concentration of $H_2O(g)$ in this mixture?

Solution: The first step, once we have the chemical equation for the equilibrium, is to write the equilibrium law for the reaction.

$$K_c = \frac{[CO(g)][H_2(g)]^3}{[CH_4(g)][H_2O(g)]}$$

The equilibrium constant and all of the equilibrium concentrations except that for H_2O are known. We simply substitute these values into the equilibrium law and solve for the unknown quantity.

$$5.67 = \frac{(0.300)(0.800)^3}{(0.400)[H_2O(g)]}$$

Multiplying both sides by $(0.400)[H_2O(g)]$ to clear fractions,

$$5.67(0.400)[H_2O(g)] = (0.300)(0.800)^3$$

$$2.27[H_2O(g)] = 0.154$$

$$[H_2O(g)] = \frac{0.154}{2.27}$$

$$[H_2O(g)] = 0.0678 \ M$$

EXERCISE 16.10 Ethyl acetate, $CH_3CO_2C_2H_5$, is an important solvent used in lacquers, adhesives, the manufacture of plastics, and even as a food flavoring. It is produced from acetic acid and ethanol by the reaction

$$CH_3CO_2H(\ell) + C_2H_5OH(\ell) \rightleftharpoons CH_3CO_2C_2H_5(\ell) + H_2O(\ell)$$

At 25 °C, $K_c = 4.10$ for this reaction. In a reaction mixture, the following equilibrium concentrations were observed: $[CH_3CO_2H] = 0.210 \ M$, $[CH_3CO_2C_2H_5] = 0.910 \ M$, and $[H_2O] = 0.00850 \ M$. What was the concentration of C_2H_5OH in the mixture?

A more complex type of calculation involves the use of initial concentrations and K_c to compute equilibrium concentrations. Although some of these problems can be so complicated that a computer is needed to solve them, we can learn the general principles involved by working on simple calculations. Even these, however, require a little applied algebra. This is where the concentration table, built up as in Example 16.7, can be very helpful.

EXAMPLE 16.9 **Using K_c to Calculate Equilibrium Concentrations**

Problem: The reaction

$$CO(g) + H_2O(g) \rightleftharpoons CO_2(g) + H_2(g)$$

has $K_c = 4.06$ at 500 °C. If 0.100 mol of CO and 0.100 mol of $H_2O(g)$ were placed in a 1.00-liter reaction vessel at this temperature, what were the concentrations of the reactants and products when the system reached equilibrium?

Solution: The key to solving this problem is that at equilibrium the mass action expression *must* equal K_c.

$$\frac{[CO_2][H_2]}{[CO][H_2O]} = 4.06$$

We must find values for the concentrations that satisfy this condition. Because we don't know what these concentrations are, we must represent them as unknowns, algebraically. This is where we use the concentration table. We need quantities to enter into "initial concentrations," "changes in concentration," and "equilibrium concentrations" rows. Let's figure them out and then build the table.

Initial Concentrations. The initial concentrations of CO and H_2O were each 0.100 mol/1.00 liter = 0.100 M. Since no CO_2 or H_2 were initially placed into the reaction vessel, their initial concentrations both were zero.

Changes in Concentration. Some CO_2 and H_2 had to form in order to reach equilibrium. This also means that some CO and H_2O must have reacted. How much? If we knew the answer, we could calculate the equilibrium concentrations. The changes in concentration are our unknown quantities.

We could just as easily have chosen x to be the number of moles/liter of H_2O that react or the moles/liter of CO_2 or H_2 that are formed. There's nothing special about having chosen CO to define x.

Let us allow x to be equal to the number of moles per liter of CO that reacted. The change in the concentration of CO would then be $-x$ (it is negative because the change *decreased* the CO concentration). Because CO and H_2O react one for one, the change in the H_2O concentration is also $-x$. Since one mole each of CO_2 and H_2 are formed from one mole of CO, the CO_2 and H_2O concentrations each increased by x (their changes are $+x$).

Equilibrium Concentrations. We obtain the equilibrium concentrations as:

$$\begin{pmatrix} \text{equilibrium} \\ \text{concentration} \end{pmatrix} = \begin{pmatrix} \text{initial} \\ \text{concentration} \end{pmatrix} + \begin{pmatrix} \text{change in} \\ \text{concentration} \end{pmatrix}$$

Now we can construct the concentration table.

	$CO(g)$	$+$ $H_2O(g)$	$\rightleftharpoons$ $CO_2(g)$	$+$ $H_2(g)$
Initial concentrations	0.100 M	0.100 M	0.0 M	0.0 M
Changes in concentration	$-x$	$-x$	$+x$	$+x$
Equilibrium concentrations	$0.100 - x$	$0.100 - x$	x	x

Note that the last line in this table merely tells us that the equilibrium CO and H_2O concentrations are equal to the number of moles per liter that were present initially minus the number of moles per liter that reacted. The equilibrium concentrations of CO_2 and H_2 are simply the number of moles per liter of each that formed, since no CO_2 or H_2 were present initially.

Next we substitute the quantities from the "equilibrium concentration" row into the mass action expression and solve for x.

$$\frac{[CO_2][H_2]}{[CO][H_2O]} = 4.06$$

Substituting,

$$\frac{(x)(x)}{(0.100 - x)(0.100 - x)} = 4.06$$

which we can write as

$$\frac{x^2}{(0.100 - x)^2} = 4.06$$

In this problem we can solve the equation for x most easily by taking the square root of both sides.

$$\frac{x}{(0.100 - x)} = \sqrt{4.06} = 2.01$$

Clearing fractions gives

$$x = 2.01(0.100 - x)$$

$$x = 0.201 - 2.01x$$

Collecting terms in x gives

$$x + 2.01x = 0.201$$

$$3.01x = 0.201$$

$$x = \frac{0.201}{3.01}$$

$$= 0.0668$$

Now that we know the value of x, we can calculate the equilibrium concentrations from the last row of our table.

$$[CO] = 0.100 - x = 0.100 - 0.0668 = 0.033 \, M$$

$$[H_2O] = 0.100 - x = 0.100 - 0.0668 = 0.033 \, M$$

$$[CO_2] = x = 0.0668 \, M$$

$$[H_2] = x = 0.0668 \, M$$

In the preceding example, we were able to simplify the solution by taking the square root of both sides of the algebraic equation that we obtained when we substituted equilibrium concentrations into the mass action expression. You can't expect to do this always, but keep an eye open for ways to simplify the algebra. None of the problems that you will encounter in this book will require very extensive mathematical skills.

EXERCISE 16.11 During an experiment, 0.200 mol of H_2 and 0.200 mol of I_2 were placed into a 1.00-liter vessel. The reaction

$$H_2(g) + I_2(g) \rightleftharpoons 2HI(g)$$

was allowed to come to equilibrium. For this reaction, $K_c = 49.5$. What were the equilibrium concentrations of H_2, I_2, and HI? (Hint: In constructing the change column, remember that *two* moles of HI are produced for each mole of H_2 or I_2 that reacts. If the H_2 concentration decreases by x, the HI concentration increases by $2x$.)

Many chemical reactions have equilibrium constants that are either very large or very small. For example, most of the weak acids that we find in biological systems have very small values for K_c. Only very tiny amounts of products form when these weak acids react with water.

When the K_c for a reaction is very small the equilibrium calculations can often be considerably simplified.

EXAMPLE 16.10 **Simplifying Equilibrium Calculations for Reactions with Small K_c**

Problem: Hydrogen, a potential fuel, is found in great abundance in water. Before the hydrogen can be used as a fuel, however, it must be separated from the oxygen; the water must be split into H_2 and O_2. One possibility is thermal decomposition, but this requires very high temperatures. Even at 1000 °C, $K_c = 7.3 \times 10^{-18}$ for the reaction

$$2H_2O(g) \rightleftharpoons 2H_2(g) + O_2(g)$$

If the H_2O concentration in a reaction vessel is set initially at 0.100 M, what will the H_2 concentration be when the reaction reaches equilibrium?

Solution: Once again, we know that at equilibrium

$$\frac{[H_2]^2[O_2]}{[H_2O]^2} = 7.3 \times 10^{-18}$$

We now set up the concentration table.

Initial Concentrations. The initial concentration of H_2O is 0.100 M; those of H_2 and O_2 are both 0.0 M.

Changes in Concentration. It is best to define x in terms of a reactant or product whose coefficient is 1. We will let x be the number of moles/liter of O_2 that are formed, so the O_2 concentration will increase by x. The H_2 concentration will increase by $2x$ because two H_2 molecules are formed for each O_2 that is formed. The H_2O concentration will decrease by $2x$ because two H_2O molecules react to give each O_2 molecule. This gives us

	$2H_2O(g)$ $\rightleftharpoons$	$2H_2(g)$ +	$O_2(g)$
Initial concentrations	0.100 M	0.0 M	0.0 M
Changes in concentration	$-2x$	$+2x$	$+x$
Equilibrium concentrations	$0.100 - 2x$	$2x$	x

> The coefficients of x can be the same as the coefficients in the balanced chemical equation.

When we substitute the equilibrium quantities into the mass action expression we get

$$\frac{(2x)^2 x}{(0.100 - 2x)^2} = 7.3 \times 10^{-18}$$

or

> $(2x)^2 x = (4x^2)x = 4x^3$

$$\frac{4x^3}{(0.100 - 2x)^2} = 7.3 \times 10^{-18}$$

This is a cubic equation (one term involves x^3) and is rather difficult to solve unless we can simplify it. In this case we are able to do so because the very small value of K_c tells us that hardly any of the H_2O will decompose. Whatever the actual value of x, we know that it is going to be very small. This means that $2x$ will also be small, so when this tiny value is subtracted from 0.100, the result will still be very, very close to 0.100. We will make an assumption, then, that the denominator will be essentially unchanged from 0.100 by subtracting $2x$; that is we will assume that $0.100 - 2x \approx 0.100$. This assumption greatly simplifies the math. We now have

$$\frac{4x^3}{(0.100)^2} = 7.3 \times 10^{-18}$$

$$4x^3 = 0.0100(7.3 \times 10^{-18}) = 7.3 \times 10^{-20}$$

$$x^3 = \frac{7.3 \times 10^{-20}}{4} = 1.8 \times 10^{-20}$$

$$x = \sqrt[3]{1.8 \times 10^{-20}}$$

$$= 2.6 \times 10^{-7}$$

Notice that the value of x that we've obtained is indeed very small. If we double it and subtract the answer from 0.100, we still get 0.100 when we round to the third decimal place.

$$0.100 - 2x = 0.100 - 2(2.6 \times 10^{-7}) = 0.09999948 \xrightarrow[\text{to}]{\text{rounds}} 0.100$$

This check verifies that our assumption was valid. Finally, we have to obtain the H_2 concentration. Our table gives

$$[H_2] = 2x$$

Therefore,

$$[H_2] = 2(2.6 \times 10^{-7}) = 5.2 \times 10^{-7}\ M$$

The simplifying assumption made in the preceding example is valid because a very small number is *subtracted* from a much larger one. We could also have neglected x (or $2x$) if it were a very small number that was being *added* to a much larger one. Remember that you can only neglect an x that's added or subtracted; you can never drop an x that occurs as a multiplying or dividing factor. Some examples are:

You would never be justified in neglecting this x. ⟶ $x(0.100 - x)$ / $(0.100 + x)$ ⟵ You are justified in neglecting this x, if x is small compared to 0.100.
⟶ x

As a rule of thumb, you can expect that these simplifying assumptions will be valid if the concentration from which x is subtracted, or to which x is added, is at least 1000 times greater than K. For instance, in the preceding example, $2x$ was subtracted from 0.100. Since 0.100 is much larger than $1000 \times (7.3 \times 10^{-19})$, we expect the assumption $0.100 - 2x \approx 0.100$ to be valid.

EXERCISE 16.12 In air at 25 °C and 1 atm, the N_2 concentration is 0.033 M and the O_2 concentration is 0.00810 M. The reaction

$$N_2(g) + O_2(g) \rightleftharpoons 2NO(g)$$

has $K_c = 4.8 \times 10^{-31}$ at 25 °C. Taking the N_2 and O_2 concentrations given above as initial values, calculate the equilibrium NO concentration that would exist in our atmosphere from this reaction at 25 °C.

16.9 PRECIPITATION REACTIONS

When solutions containing ions of an insoluble salt are combined, a precipitate of the salt forms until an equilibrium is reached

Nearly everyone has had the experience of flavoring a bowl of soup with a sprinkling of table salt—we all know that sodium chloride is soluble in water. Many other salts are water soluble too. Some familiar examples are magnesium sulfate, $MgSO_4$ (epsom salts), calcium chloride, $CaCl_2$ (a salt used to melt ice in winter), and sodium bicarbonate, $NaHCO_3$ (baking soda). However, there's also a large number of ionic compounds that have very low solubilities in water. In other words, very little of the solute is needed to give a saturated solution in a given volume of water. Calcium sulfate, $CaSO_4$ (plaster), calcium carbonate, $CaCO_3$ (limestone), and magnesium hydroxide, $Mg(OH)_2$ (milk of magnesia) are examples of compounds that *appear* to be virtually insoluble in water. We stress the word "appear" because even "insoluble" salts dissolve to a small extent to give a saturated solution. An equilibrium is established between the undissolved solid and the solute that is in solution. For instance, with $CaCO_3$ we would write this equilibrium:

Equilibrium between dissolved and undissolved solute exists in a saturated solution.

$$CaCO_3(s) \rightleftharpoons Ca^{2+}(aq) + CO_3^{2-}(aq)$$

On the right side of this equation we have written the ions of $CaCO_3$ separately.

 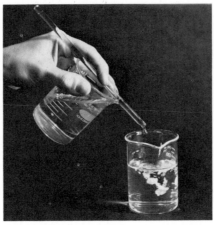 

Figure 16.6
Formation of a $CaCO_3$ precipitate as a 1.0-M $CaCl_2$ solution is
added to a 1.0-M Na_2CO_3 solution.

As we learned in Chapter Eleven, ionic compounds such as this are essentially completely dissociated in aqueous solution—that is, they exist in the form of separated ions.

In a saturated solution of $CaCO_3$, the concentrations of Ca^{2+} and CO_3^{2-} are very small because $CaCO_3$ has a low solubility. If two solutions are mixed, one containing a high concentration of Ca^{2+} and the other containing a high concentration of CO_3^{2-}, the ions will combine to form solid $CaCO_3$ until their concentrations drop to the low values found in a saturated $CaCO_3$ solution. This is pictured in Figure 16.6, which shows what happens when 1.0 M solutions of $CaCl_2$ and Na_2CO_3 are combined. If desired, the white precipitate of $CaCO_3$ that is formed can be separated from the solution by passing the mixture through filter paper. The paper retains the solid but allows the solution to pass through (see Figure 16.7).

In Chapter Eleven we saw that there are various ways to write equations for reactions like that between $CaCl_2$ and Na_2CO_3. The molecular equation for the reaction is

$$CaCl_2(aq) + Na_2CO_3(aq) \longrightarrow CaCO_3(s) + 2NaCl(aq)$$

This is an example of a **double replacement**, or **metathesis reaction.** It is called a double replacement reaction because carbonate replaces chloride in the calcium compound and chloride replaces carbonate in the sodium compound.

We can also write this equation in ionic form. Recall that the ionic equation is obtained by writing the ions of soluble salts separately. The formulas of insoluble compounds are written in molecular form.

$$Ca^{2+}(aq) + 2Cl^-(aq) + 2Na^+(aq) + CO_3^{2-}(aq) \longrightarrow CaCO_3(s) + 2Na^+(aq) + 2Cl^-(aq)$$

Canceling spectator ions gives the net ionic equation,

$$Ca^{2+}(aq) + CO_3^{2-}(aq) \longrightarrow CaCO_3(s)$$

This net ionic equation tells us that whenever we combine a solution of a soluble calcium salt with a solution of a soluble carbonate, we should expect to obtain a precipitate of $CaCO_3$.

The kind of information contained in the net ionic equation is very useful to know. If we needed $CaCO_3$ for some purpose and none was available, we could prepare it from solutions of any *soluble* calcium and carbonate compounds by a double replacement precipitation reaction.

Figure 16.7
Filtration is a procedure that separates a solid precipitate from a solution.

In Chapter Six, we saw that acid-base neutralization reactions provide a convenient method for preparing salts. Precipitation reactions offer a second way of obtaining these compounds. Use of this method, however, requires that we have some knowledge of solubilities. For example, if you wanted to prepare $CaCO_3$ from chemicals on a shelf, it would be very helpful to know which calcium and carbonate salts are soluble.

Below are some solubility rules for you to learn. Someday, long after finishing this course, you'll be surprised to discover how useful it is to know these rules.

Soluble Compounds

1 All compounds of the alkali metals (Group IA) are soluble.
2 All salts containing NH_4^+, NO_3^-, ClO_4^-, ClO_3^- and $C_2H_3O_2^-$ are soluble.
3 All chlorides, bromides, and iodides are soluble, *except* those of Ag^+, Pb^{2+}, and Hg_2^{2+} (mercury in the 1+ oxidation state).
4 All sulfates are soluble, *except* those of Pb^{2+}, Ca^{2+}, Sr^{2+}, and Ba^{2+}.

Insoluble Compounds

5 All hydroxides and metal oxides are insoluble, *except* those of Group IA and Ca^{2+}, Sr^{2+}, and Ba^{2+}. When oxides do dissolve, they give hydroxides (their solutions do not contain O^{2-} ions).

$$Na_2O + H_2O \longrightarrow 2NaOH$$

6 All compounds containing PO_4^{3-}, CO_3^{2-}, SO_3^{2-}, and S^{2-} are insoluble, *except* those of Group IA and NH_4^+.

These solubility rules allow us to predict the outcome of double replacement reactions, as shown in Example 16.11.

EXAMPLE 16.11   **Predicting Double Replacement Reactions**

Problem: Write molecular, ionic, and net ionic equations for the reaction between $Pb(NO_3)_2$ and $Fe_2(SO_4)_3$.

Solution: We begin with the molecular equation. Our reactants are

$$Pb(NO_3)_2 + Fe_2(SO_4)_3$$

To write the products, we interchange NO_3^- and SO_4^{2-}. However, we must be very careful to write the correct formulas of the products.[2] From the formulas of the reactants, we know that the positive ions here are Pb^{2+} and Fe^{3+}. The correct formulas of the products, then, are $PbSO_4$ and $Fe(NO_3)_3$.

Write equations in two steps.
1 Write correct formulas.
2 Balance the equation.

At this point, the unbalanced equation is

$$Pb(NO_3)_2 + Fe_2(SO_4)_3 \longrightarrow PbSO_4 + Fe(NO_3)_3$$

When it is balanced, we obtain the *molecular equation.*

$$3Pb(NO_3)_2 + Fe_2(SO_4)_3 \longrightarrow 3PbSO_4 + 2Fe(NO_3)_3$$

Next, we expand this to give the ionic equation in which soluble compounds are written in dissociated (separated) form as ions, and insoluble compounds are written in "molecular" form. To do this, we must examine

---

[2] Some students might be tempted (without thinking) to write $Pb(SO_4)_2$ and $Fe_2(NO_3)_3$, or even $Pb(SO_4)_3$ and $Fe_2(NO_3)_2$. This is a common error. Always be careful to figure out the charges of the ions that must be combined in the formula. Then take the ions in a ratio that gives a neutral formula unit.

the formulas using our solubility rules to determine which compounds are soluble.

$Pb(NO_3)_2$	soluble	(rule 2)
$Fe_2(SO_4)_3$	soluble	(rule 4)
$PbSO_4$	insoluble	(rule 4)
$Fe(NO_3)_2$	soluble	(rule 2)

Now we can write the ionic equation,

$$3Pb^{2+} + 6NO_3^- + 2Fe^{3+} + 3SO_4^{2-} \longrightarrow 3PbSO_4(s) + 2Fe^{3+} + 6NO_3^-$$

The net ionic equation is obtained by canceling spectator ions. This leaves

$$3Pb^{2+} + 3SO_4^{2-} \longrightarrow 3PbSO_4(s)$$

Finally, we can reduce the coefficients to

$$Pb^{2+} + SO_4^{2-} \longrightarrow PbSO_4(s)$$

EXAMPLE 16.12 **Predicting Double Replacement Reactions**

Problem: What reaction (if any) occurs between KNO_3 and NH_4Cl?
Solution: First we write the molecular equation:

$$KNO_3 + NH_4Cl \longrightarrow KCl + NH_4NO_3$$

Next we determine solubilities. The rules tell us that all four of these compounds are soluble. When we expand this, we obtain the ionic equation.

$$K^+ + NO_3^- + NH_4^+ + Cl^- \longrightarrow K^+ + Cl^- + NH_4^+ + NO_3^-$$

If we cancel spectator ions, nothing is left! This tells us that there is *no* net reaction. If we mix KNO_3 and NH_4Cl solutions, all we obtain is a mixture of the ions. No reaction occurs.

EXERCISE 16.13 Predict what reaction (if any) would occur upon mixing solutions of:
(a) $AgNO_3$ and NH_4Cl
(b) Na_2S and $Pb(C_2H_3O_2)_2$
(c) $BaCl_2$ and NH_4NO_3

EXERCISE 16.14 Acid-base neutralizations can produce precipitates, too. Write molecular, ionic, and net ionic equations for the reaction between $Ba(OH)_2$ and H_2SO_4.

16.10 HETEROGENEOUS EQUILIBRIA

The concentrations of reactants or products that are pure liquids or solids are constant and do not appear in the mass action expression

In a **homogeneous reaction**—or in a homogeneous equilibrium—all of the reactants and products are in the same phase. Equilibria among gases are homogeneous because all gases mix freely with each other and a single phase exists. There are also many equilibria in which reactants and products are dissolved in the same liquid phase. Examples are the ionization equilibria of weak acids, which we'll look at closely in Chapter Seventeen.

When more than one phase exists in a reaction mixture, we call it a **heterogeneous reaction.** Common examples are the burning of wood and the discharge

Special Topic 16.2 Boiler Scale and Hard Water

Precipitation reactions occur around us all the time and we hardly ever take notice—until they cause a problem. One common problem is caused by **hard water**, ground water that contains the "hardness ions," Ca^{2+}, Mg^{2+}, Fe^{2+}, or Fe^{3+} in concentrations high enough to form precipitates with ordinary soap. Soap normally consists of the sodium salts of long-chain organic acids derived from animal fats or oils. An example is sodium stearate, $NaC_{17}H_{35}CO_2$. The negative ion of the soap forms an insoluble "scum" with hardness ions, which reduces the effectiveness of the soap for removing dirt and grease. See also Section 13.5.

Hardness ions can be removed from water in a number of ways. One way is to add hydrated sodium carbonate, $Na_2CO_3 \cdot 10H_2O$, often called washing soda, to the water. The carbonate ion forms insoluble precipitates with the hardness ions; an example is $CaCO_3$.

$$Ca^{2+} + CO_3^{2-} \longrightarrow CaCO_3(s)$$

Once precipitated, the hardness ions are not available to interfere with the soap.

Another problem with hard water that contains bicarbonate ion is the precipitation of insoluble carbonates on the inner walls of hot water pipes. When heated, solutions containing bicarbonate ion lose CO_2.

$$2HCO_3^- \longrightarrow H_2O + CO_2(g) + CO_3^{2-}$$

Gases become less soluble as the temperature is

Boiler scale.

raised. As CO_2 is driven from the solution, the HCO_3^- is gradually converted to CO_3^{2-}, which is able to precipitate the hardness ions. This precipitate, which sticks to the inner walls of pipes and hot water boilers, is called boiler scale. In locations that have high concentrations of Ca^{2+} and HCO_3^- in the water supply, boiler scale is a very serious problem, as illustrated in the photograph above.

of an automobile battery (where there's a reaction between the plates in the battery and the sulfuric acid solution in contact with them). Another example is the thermal decomposition of sodium bicarbonate (baking soda) that occurs when it is sprinkled on a fire.

$$2NaHCO_3(s) \longrightarrow Na_2CO_3(s) + H_2O(g) + CO_2(g)$$

Many cooks keep a box of baking soda nearby because this reaction makes baking soda an excellent fire extenguisher for burning fats or oil. The fire is smothered by all the products.

Heterogeneous reactions are also able to reach equilibrium, just like homogeneous reactions can. If $NaHCO_3$ is placed in a sealed container so that no CO_2 or H_2O can escape, the gases and solids come to equilibrium.

$$2NaHCO_3(s) \rightleftharpoons Na_2CO_3(s) + H_2O(g) + CO_2(g)$$

Following our usual procedure, we can write the equilibrium law for this reaction as

$$\frac{[Na_2CO_3(s)][H_2O(g)][CO_2(g)]}{[NaHCO_3(s)]^2} = K$$

However, the equilibrium law for reactions involving pure liquids and solids can be written in an even simpler form. This is because the concentration of a pure liquid or solid is unchangeable—that is, for any pure liquid or solid, the *ratio* of moles of substance to volume of substance is a constant. For example, if we had a one-mole crystal of $NaHCO_3$, it would occupy a volume of 38.9 cm³. Two moles of $NaHCO_3$ would occupy twice this volume, 77.8 cm³ (Figure 16.8), but the ratio of moles to volume (i.e., the molar concentration) remains the same. For $NaHCO_3$, the concentration of the substance in the solid is

$$1 \text{ mol}/0.0389 \text{ liter} = 2 \text{ mol}/0.0778 \text{ liter} = 25.7 \text{ mol}/\text{liter}$$

This is the concentration of $NaHCO_3$ in the solid, regardless of the size of the solid sample. In other words, the concentration of $NaHCO_3$ is constant, provided that some of it is present in the reaction mixture.

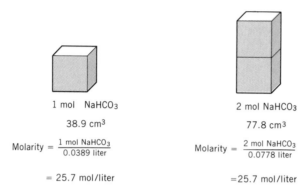

$$\text{Molarity} = \frac{1 \text{ mol } NaHCO_3}{0.0389 \text{ liter}}$$

$$= 25.7 \text{ mol}/\text{liter}$$

$$\text{Molarity} = \frac{2 \text{ mol } NaHCO_3}{0.0778 \text{ liter}}$$

$$= 25.7 \text{ mol}/\text{liter}$$

Figure 16.8
The concentration of a substance in the solid state is a constant. Doubling the number of moles also doubles the volume, but the *ratio* of moles to volume remains the same.

Similar reasoning shows that the concentration of Na_2CO_3 in pure solid Na_2CO_3 is a constant, too. This means that our equilibrium law now has three constants, K plus two of the concentration terms. It makes sense to combine all of the numerical constants together.

$$[H_2O(g)][CO_2(g)] = \frac{K[NaHCO_3(s)]^2}{[Na_2CO_3(s)]} = K_c$$

The equilibrium law for heterogeneous reactions is written without concentration terms for pure solids or liquids. Equilibrium constants that are given in tables, or calculated from $\Delta G°$, represent all of the constants combined.

EXAMPLE 16.13   **Writing the Equilibrium Law for Heterogeneous Reactions**

Problem: The air pollutant sulfur dioxide can be removed from a gas mixture by passing the mixture over calcium oxide. The equation is

$$CaO(s) + SO_2(g) \rightleftharpoons CaSO_3(s)$$

Write the equilibrium law for this reaction.

Solution: The concentrations of the two solids, CaO and $CaSO_3$, are incorporated into the equilibrium constant for the reaction. The only concentration term that should appear in the mass action expression is that of SO_2. The equilibrium law would be

$$\frac{1}{[SO_2(g)]} = K_c$$

EXERCISE 16.15   Write the equilibrium law for the following heterogeneous reactions.
(a) $2Hg(\ell) + Cl_2(g) \rightleftharpoons Hg_2Cl_2(s)$
(b) $NH_3(g) + HCl(g) \rightleftharpoons NH_4Cl(s)$

16.11 SOLUBILITY PRODUCT

A saturated solution of an "insoluble" salt is a heterogeneous equilibrium, for which the mass action expression is written as a product just of ion concentrations, each raised to an appropriate power

The solubility rules tell us that $PbCl_2$ and AgCl are both insoluble salts. However, if chloride ion is added slowly to a mixture containing both Pb^{2+} and Ag^+, nearly all of the silver is precipitated as AgCl before any $PbCl_2$ is formed. This is because silver chloride is much less soluble than lead chloride. To study such differences in solubility, we must examine solubility equilibria quantitatively. Let's begin by looking again at the compound $CaCO_3$. Recall that when this substance is placed in water, a small amount dissolves and we obtain the equilibrium

$$CaCO_3(s) \rightleftharpoons Ca^{2+}(aq) + CO_3^{2-}(aq)$$

As usual, we can write the equilibrium condition as

$$\frac{[Ca^{2+}][CO_3^{2-}]}{[CaCO_3(s)]} = K$$

This equilibrium, however, is a heterogeneous one. At the bottom of the solution rests pure solid $CaCO_3$. We saw earlier that the concentration of a substance within a pure solid is a constant. This means that the denominator of our mass action expression, $[CaCO_3(s)]$, is a constant that we can combine with K to obtain still another constant.

$$[Ca^{2+}][CO_3^{2-}] = \underset{\text{another constant}}{K[CaCO_3(s)]}$$

This new constant is called the **solubility product constant, K_{sp}**, because it is equal to a product of concentration terms. There's no denominator left in the mass action expression and the equilibrium condition is

$$[Ca^{2+}][CO_3^{2-}] = K_{sp}$$

The mass action expression for a solubility equilibrium is often called the **ion product** because it is a product of ion concentrations.

Many salts produce more than one of a given kind of ion when they dissociate. An example is silver sulfide, Ag_2S, the black film that forms on silver when it tarnishes. Each formula unit of Ag_2S gives two silver ions and one sulfide ion. Therefore, the equation representing the solubility equilibrium is

$$Ag_2S(s) \rightleftharpoons 2Ag^+(aq) + S^{2-}(aq)$$

As usual, the exponents on the concentration terms in the mass action expression are equal to the coefficients in the balanced equation. Thus, the solubility product expression for Ag_2S is

$$[Ag^+]^2[S^{2-}] = K_{sp}$$

EXERCISE 16.16   Write the K_{sp} expression for these equilibria:
(a) $BaCrO_4(s) \rightleftharpoons Ba^{2+} + CrO_4^{2-}$
(b) $Ag_3PO_4(s) \rightleftharpoons 3Ag^+ + PO_4^{3-}$

EXAMPLE 16.14   Calculating K_{sp} from Solubility Data

Problem: As we learned in Chapter Eleven, the light-sensitive ingredient in nearly all photographic film is AgBr. One liter of water is able to dissolve 7.1×10^{-7} mol of AgBr. What is K_{sp} for this salt?

Solution: In this problem we have been given the **molar solubility** of silver bromide—the number of moles of the solute that are needed to give one liter of a saturated solution. We can use this molar solubility, 7.1×10^{-7} mol/liter, to construct the concentration table for the equilibrium. Our equation is

$$AgBr(s) \rightleftharpoons Ag^+(aq) + Br^-(aq)$$

If AgBr is placed in pure water, the initial concentrations of Ag^+ and Br^- are zero. The molar solubility tells us that 7.1×10^{-7} mol of AgBr dissolves per liter. From the stoichiometry of the equation, 7.1×10^{-7} mol of AgBr will give 7.1×10^{-7} mol of Ag^+ and 7.1×10^{-7} mol of Br^-. The concentrations of these ions therefore increase by this amount. We can now calculate the equilibrium concentrations from the initial concentrations and the changes.

No entry appears under AgBr because its concentration term doesn't appear in the ion product.

	AgBr(s) $\rightleftharpoons$ Ag$^+$(aq)	+	Br$^-$(aq)
Initial concentrations		0.0 M	0.0 M
Changes in concentrations when AgBr dissolves	No entry in this column	$+7.1 \times 10^{-7}$ M	$+7.1 \times 10^{-7}$ M
Equilibrium concentrations		7.1×10^{-7} M	7.1×10^{-7} M

To calculate K_{sp}, we substitute the equilibrium concentrations into the K_{sp} expression.

$$K_{sp} = [Ag^+][Br^-]$$
$$= (7.1 \times 10^{-7})(7.1 \times 10^{-7})$$
$$K_{sp} = 5.0 \times 10^{-13}$$

EXAMPLE 16.15   Calculating K_{sp} from Solubility Data

Problem: The molar solubility of Ag_2S in pure water is 3.7×10^{-17} mol/liter. What is K_{sp} for Ag_2S?

Solution: The equilibrium equation is

$$Ag_2S(s) \rightleftharpoons 2Ag^+(aq) + S^{2-}(aq)$$

and the K_{sp} expression is

$$K_{sp} = [Ag^+]^2[S^{2-}]$$

Again, we have no Ag^+ or S^{2-} present in the water initially, so their initial concentrations are zero. When 3.7×10^{-17} mol of Ag_2S dissolves per liter, we will obtain 3.7×10^{-17} mol of S^{2-} and $2 \times (3.7 \times 10^{-17})$ mol of Ag^+. Now we can construct the concentration table.

$Ag_2S(s) \rightleftharpoons 2Ag^+(aq)$	$+$	$S^{2-}(aq)$
Initial concentrations	$0.0\ M$	$0.0\ M$
Changes in concentrations when Ag_2S dissolves	$2 \times (3.7 \times 10^{-17}\ M)$ $= +7.4 \times 10^{-17}\ M$	$+3.7 \times 10^{-17}\ M$
Equilibrium concentrations	$7.4 \times 10^{-17}\ M$	$3.7 \times 10^{-17}\ M$

Substituting the equilibrium concentrations into the K_{sp} expression gives

$$K_{sp} = (7.4 \times 10^{-17})^2(3.7 \times 10^{-17})$$
$$= 2.0 \times 10^{-49}$$

EXAMPLE 16.16   **Calculating K_{sp} from Solubility Data**

Problem: The molar solubility of $PbCl_2$ in a 0.10-M NaCl solution is 1.6×10^{-3} mol/liter. What is K_{sp} for $PbCl_2$?

Solution: Again, we begin by writing the equation for the equilibrium and the K_{sp} expression.

$$PbCl_2(s) \rightleftharpoons Pb^{2+}(aq) + 2Cl^-(aq)$$

$$K_{sp} = [Pb^{2+}][Cl^-]^2$$

As usual, we assume that any soluble salt is fully dissociated in water.

In this problem the $PbCl_2$ is being dissolved in a solution of NaCl, which already contains 0.10 M Cl^- (It also contains 0.10 M Na^+, but that's not important here because Na^+ doesn't appear in the K_{sp} expression.) The initial concentration of Pb^{2+} is zero because none was in the solution to begin with, but the initial concentration of Cl^- is 0.10 M. When the $PbCl_2$ dissolves in the NaCl solution, the Pb^{2+} concentration increases by $1.6 \times 10^{-3}\ M$ and the Cl^- concentration increases by $2 \times (1.6 \times 10^{-3}\ M)$. Now we can build our concentration table.

$PbCl_2(s) \rightleftharpoons Pb^{2+}(aq)$	$+$	$2Cl^-(aq)$
Initial concentrations	$0.0\ M$	$0.10\ M$
Changes in concentrations when $PbCl_2$ dissolves	$+1.6 \times 10^{-3}\ M$	$+2 \times (1.6 \times 10^{-3}\ M) =$ $+3.2 \times 10^{-3}\ M$
Equilibrium concentrations	$1.6 \times 10^{-3}\ M$	$0.10 + 0.0032 = 0.10\ M$ (to two significant figures)

Substituting the values for the equilibrium concentrations into the K_{sp} expression gives

$$K_{sp} = (1.6 \times 10^{-3})(0.10)^2$$
$$= 1.6 \times 10^{-5}$$

EXERCISE 16.17   One liter of water is able to dissolve 2.15×10^{-3} mol of PbF_2. What is K_{sp} for PbF_2?

EXERCISE 16.18 The molar solubility of $CoCO_3$ in a 0.10-M Na_2CO_3 solution is 8.0×10^{-12} mol/liter. What is K_{sp} for $CoCO_3$?

EXERCISE 16.19 The molar solubility of PbF_2 in a 0.10-M $Pb(NO_3)_2$ solution is 3.2×10^{-4} mol/liter. Calculate K_{sp} for PbF_2.

Besides calculating K_{sp} from solubility information, we can also compute solubility if we know the value of K_{sp}. Some typical K_{sp} values are in Table 16.2.

Table 16.2
Solubility-Product Constants

Salt	Ions Produced in Water	K_{sp}
Halides		
AgCl	$\rightleftharpoons Ag^+ + Cl^-$	1.6×10^{-10}
AgBr	$\rightleftharpoons Ag^+ + Br^-$	5.0×10^{-13}
AgI	$\rightleftharpoons Ag^+ + I^-$	1.5×10^{-16}
$PbCl_2$	$\rightleftharpoons Pb^{2+} + 2Cl^-$	1.6×10^{-5}
$PbBr_2$	$\rightleftharpoons Pb^{2+} + 2Br^-$	5×10^{-6}
PbI_2	$\rightleftharpoons Pb^{2+} + 2I^-$	8.3×10^{-9}
Hydroxides		
$Fe(OH)_2$	$\rightleftharpoons Fe^{2+} + 2OH^-$	1.6×10^{-14}
$Fe(OH)_3$	$\rightleftharpoons Fe^{3+} + 3OH^-$	1.1×10^{-36}
$Mg(OH)_2$	$\rightleftharpoons Mg^{2+} + 2OH^-$	1.2×10^{-11}
$Al(OH)_3$	$\rightleftharpoons Al^{3+} + 3OH^-$	3.7×10^{-15}
Sulfides		
CoS	$\rightleftharpoons Co^{2+} + S^{2-}$	3×10^{-26}
CuS	$\rightleftharpoons Cu^{2+} + S^{2-}$	8.5×10^{-45}
FeS	$\rightleftharpoons Fe^{2+} + S^{2-}$	3.7×10^{-19}
HgS	$\rightleftharpoons Hg^{2+} + S^{2-}$	4×10^{-53}
PbS	$\rightleftharpoons Pb^{2+} + S^{2-}$	3.4×10^{-28}
Ag_2S	$\rightleftharpoons 2Ag^+ + S^{2-}$	2.0×10^{-49}
Carbonates		
$BaCO_3$	$\rightleftharpoons Ba^{2+} + CO_3^{2-}$	8.1×10^{-9}
$CaCO_3$	$\rightleftharpoons Ca^{2+} + CO_3^{2-}$	8.7×10^{-9}
Ag_2CO_3	$\rightleftharpoons 2Ag^+ + CO_3^{2-}$	6.2×10^{-12}
Chromates		
$PbCrO_4$	$\rightleftharpoons Pb^{2+} + CrO_4^{2-}$	1.8×10^{-14}
Ag_2CrO_4	$\rightleftharpoons 2Ag^+ + CrO_4^{2-}$	9×10^{-12}
Sulfates		
$BaSO_4$	$\rightleftharpoons Ba^{2+} + SO_4^{2-}$	1.08×10^{-10}
$CaSO_4$	$\rightleftharpoons Ca^{2+} + SO_4^{2-}$	6×10^{-5}
$PbSO_4$	$\rightleftharpoons Pb^{2+} + SO_4^{2-}$	1.1×10^{-8}

EXAMPLE 16.17 **Calculating Molar Solubility from K_{sp}**

Problem: The value of K_{sp} for AgCl is 1.6×10^{-10}. What is the molar solubility of AgCl in pure water?

Solution: The equation is

$$AgCl(s) \rightleftharpoons Ag^+(aq) + Cl^-(aq)$$

for which

$$K_{sp} = [Ag^+][Cl^-]$$

All of the AgCl that dissolves breaks down into Ag^+ and Cl^-.

Since no Ag^+ or Cl^- is present initially, their initial concentrations are zero. If we let x be the molar solubility (i.e., the number of moles of AgCl that dissolve per liter), then the Ag^+ and Cl^- concentrations will each increase by x.

	AgCl(s) $\rightleftharpoons$ $Ag^+(aq)$ + $Cl^-(aq)$	
Initial concentrations	0.0 M	0.0 M
Changes in concentrations when AgCl dissolves	+x	+x
Equilibrium concentrations	x	x

Substituting,

$$K_{sp} = (x)(x) = 1.6 \times 10^{-10}$$

$$x^2 = 1.6 \times 10^{-10}$$

$$x = 1.3 \times 10^{-5}$$

The molar solubility of AgCl is 1.3×10^{-5} M in pure water.

EXAMPLE 16.18 **Calculating Solubility from K_{sp}**

Problem: Lead iodide, PbI_2, has $K_{sp} = 8.3 \times 10^{-9}$. What is the molar solubility of PbI_2 in water?

Solution: We have

$$PbI_2(s) \rightleftharpoons Pb^{2+}(aq) + 2I^-(aq)$$

$$K_{sp} = [Pb^{2+}][I^-]^2$$

The initial concentrations of Pb^{2+} and I^- are zero. If we let x be the molar solubility of PbI_2, then the Pb^{2+} concentration will increase by x and the I^- concentration will increase by $2x$.

The coefficients of x can be the same as the coefficients in the chemical equation.

	PbI$_2$(s) $\rightleftharpoons$ Pb^{2+}(aq) + 2I$^-$(aq)	
Initial concentrations	0.0 M	0.0 M
Changes in concentrations when PbI$_2$ dissolves	+x	+2x
Equilibrium concentrations	x	2x

Substituting,

$$K_{sp} = (x)(2x)^2 = 8.3 \times 10^{-9}$$

Squaring $2x$ gives $4x^2$. Therefore,

$$4x^3 = 8.3 \times 10^{-9}$$

$$x^3 = 2.1 \times 10^{-9}$$

$$x = 1.3 \times 10^{-3}$$

The molar solubility of PbI_2 is 1.3×10^{-3} mol/liter.

EXERCISE 16.20 What is the molar solubility of AgBr in water? (Obtain K_{sp} from Table 16.2.)

EXERCISE 16.21   What is the molar solubility of Ag_2CO_3 in water? (Obtain K_{sp} from Table 16.2.)

The common ion effect

Suppose that we established the equilibrium,

$$CaCO_3(s) \rightleftharpoons Ca^{2+}(aq) + CO_3^{2-}(aq)$$

and then added additional Ca^{2+} by dissolving some $CaCl_2$ in the solution. The resulting increase in the Ca^{2+} concentration will upset the equilibrium and, according to Le Châtelier's principle, the equilibrium will shift to the left to use up some of the Ca^{2+} that we added. This will cause some $CaCO_3$ to precipitate and less $CaCO_3$ will be left in solution when equilibrium is finally reestablished.

In this example, the Ca^{2+} in the final solution has two sources—the $CaCl_2$ that was added and the $CaCO_3$ that remains in the solution when equilibrium is reached again. Since calcium ion is common to both salts, it is called a **common ion.** The addition of the common ion lowers the solubility of the $CaCO_3$. The $CaCO_3$ is less soluble in the presence of $CaCl_2$ (or any other soluble calcium salt) than it is in pure water. This lowering of the solubility by the addition of a common ion is called the **common ion effect.**

Knowing about the common ion effect can be useful. If you wanted to remove Ca^{2+} from a solution (e.g., from hard water where the Ca^{2+} forms an insoluble precipitate with soap), you could add Na_2CO_3. By making sure that there is a relatively high concentration of CO_3^{2-} you could reduce the solubility of the $CaCO_3$ and thereby minimize the amount of free Ca^{2+} in the solution.

EXAMPLE 16.19   **Calculation Involving the Common Ion Effect**

Problem: What is the molar solubility of PbI_2 in a 0.10-M NaI solution? For PbI_2, $K_{sp} = 8.3 \times 10^{-9}$

Solution: The equilibrium is

$$PbI_2(s) \rightleftharpoons Pb^{2+}(aq) + 2I^-(aq)$$

and

$$K_{sp} = [Pb^{2+}][I^-]^2$$

The solution into which the PbI_2 is placed contains no Pb^{2+}, so the initial concentration of Pb^{2+} is zero. However, this solution does contain I^-. The salt, NaI, is fully dissociated in water, so the 0.10-M NaI solution contains 0.10 M Na^+ and 0.10 M I^-. Therefore, the initial concentration of I^- is 0.10 M. (As in Example 16.16, the solution also contains Na^+, but it isn't important because Na^+ is not involved in the equilibrium.) When x mol of PbI_2 now dissolves per liter, we will obtain x mol/liter of Pb^{2+} and $2x$ mol/liter of additional I^-. Setting up the concentration table,

	$PbI_2(s) \rightleftharpoons Pb^{2+}(aq) + 2I^-(aq)$	
Initial concentrations	0.0 M	0.10 M
Changes in concentrations when PbI_2 dissolves	$+x$	$+2x$
Equilibrium concentrations	x	$0.10 + 2x$

Because K_{sp} is so small, we can reasonably expect that x, or $2x$, will be very small. In problems of this type it is safe to assume that $0.10 + 2x \approx 0.10$.

Special Topic 16.3 Urinary Calculus Disease

In medicine a *calculus* is an abnormal nonliving aggregation of mineral salts in a framework or matrix of organic materials. Urinary calculi, commonly called kidney stones or bladder stones, depending on location, include calcium and magnesium salts. Why they develop in some people and not in others is not known. However, something involving the way the system manages its calcium and magnesium ions goes wrong. Present in body fluids are phosphate ions or near relatives (e.g., $H_2PO_4^-$ and HPO_4^{2-}). Calcium and magnesium phosphates are only slightly soluble in water. If increasing amounts of either calcium or magnesium ions gets into circulation, Le Châtelier's principle, as expressed in the common ion effect, works to make them precipitate, and the stone slowly grows. Bile stones include different salts, but their formation follows the same general principles. Mineral deposits that form in some joints in the condition known as gout are explained in a similar fashion, except that the negative ions come from uric acid, a breakdown product of nitrogen compounds called nucleic acids.

(You may recall that this is the same type of simplification that we made earlier in Section 16.8.) Substituting the final equilibrium quantities into the K_{sp} expression gives

$$K_{sp} = (x)(0.10)^2 = 8.3 \times 10^{-9}$$

$$x = \frac{8.3 \times 10^{-9}}{(0.10)^2}$$

$$x = 8.3 \times 10^{-7}$$

Note that $2x$, which equals 1.6×10^{-6}, is much smaller than 0.10, just as we anticipated when we made our simplification. Also, note that in a 0.10-*M* NaI solution, the molar solubility of PbI_2 is 8.3×10^{-7} *M*, compared to 1.3×10^{-3} *M* in pure water (Example 16.18).

In solving problems like Example 16.19 the greatest number of errors are made when students double the initial concentration because they see a coefficient of *2* before the I^- in the chemical equation. The I^- concentration in a 0.10-*M* solution of NaI is 0.10 *M*, *not* 0.20 *M*. The only time we use the coefficient of *2* from the chemical equation is in the "change" row; it is never in the "initial concentration" row.

EXERCISE 16.22 What is the molar solubility of AgI in 0.20 *M* NaI solution? (Obtain K_{sp} from Table 16.2.)

EXERCISE 16.23 What is the molar solubility of $Fe(OH)_3$ in a solution having a hydroxide ion concentration of 0.050 *M*?

Predicting when precipitation occurs

We can use K_{sp} to predict whether or not a precipitate will form in a given solution, provided that we know the concentrations of the ions that appear in the K_{sp} expression. For example, we know that in a saturated solution of $CaCO_3$ the ion product, $[Ca^{2+}][CO_3^{2-}]$, is exactly equal to K_{sp}. Examining a portion of this solution shows that it is stable; no precipitate is forming in it. An unsaturated solution of $CaCO_3$ contains concentrations of Ca^{2+} and CO_3^{2-} that are lower than those in a saturated solution, so in the unsaturated solution the ion product, $[Ca^{2+}][CO_3^{2-}]$, is less than K_{sp}. No precipitate will form in an unsaturated solution because it is actually capable of dissolving more of the salt.

Unsaturated, saturated, and supersaturated solutions were described in Section 4.5.

A supersaturated solution is one that contains more solute than is necessary for saturation. It is unstable and there is a tendency for the extra solute to precipitate. In a supersaturated solution of $CaCO_3$, the ion concentrations would be greater than in a saturated solution, and the ion product would be larger than K_{sp}. Comparison of K_{sp} to the ion product calculated for a given solution, therefore, serves to indicate whether or not a precipitate will be formed.

Precipitate will form:	ion product $> K_{sp}$ (supersaturated)
No precipitate will form:	ion product $= K_{sp}$ (saturated)
	ion product $< K_{sp}$ (unsaturated)

EXAMPLE 16.20 **Predicting Whether or Not a Precipitate Will Form**

Problem: A student wished to prepare 1.0 liter of a solution containing 0.015 mol of NaCl and 0.15 mol of $Pb(NO_3)_2$. She was concerned that a precipitate of $PbCl_2$ might form. The K_{sp} of $PbCl_2$ is 1.6×10^{-5}. Can she expect to observe a precipitate of $PbCl_2$ in this mixture?

Solution: First, the ion product for $PbCl_2$ should be written

$$[Pb^{2+}][Cl^-]^2$$

The solution that is being prepared will have the following concentrations:

$$[Pb^{2+}] = 0.15 \text{ mol}/1.0 \text{ liter} = 0.15 \, M$$

$$[Cl^-] = 0.015 \text{ mol}/1.0 \text{ liter} = 0.015 \, M$$

The value of the ion product is

$$[Pb^{2+}][Cl^-]^2 = (0.15)(0.015)^2$$
$$= 3.4 \times 10^{-5}$$

Since 3.4×10^{-5} is larger than K_{sp} (which equals 1.6×10^{-5}), a precipitate will form in the mixture.

EXERCISE 16.24 Will a precipitate of $CaSO_4$ form in a solution if the Ca^{2+} concentration is 0.0025 M and the SO_4^{2-} concentration is 0.030 M? For $CaSO_4$, $K_{sp} = 6 \times 10^{-5}$.

EXERCISE 16.25 Will a precipitate form in a solution containing 3.7×10^{-17} M S^{2-} and 4.0×10^{-17} M Ag^+? Use K_{sp} for Ag_2S from Table 16.2.

SUMMARY

Dynamic Equilibrium. When two opposing chemical reactions occur at equal rates, the system is in a state of dynamic equilibrium and the concentrations of reactants and products remain constant. For a given overall chemical composition, the amounts of reactants and products that are present at equilibrium are the same regardless of whether the equilibrium is approached from the direction of pure reactants, pure products, or any mixture of reactants and products.

Equilibrium Law. The mass action expression is a fraction. The concentrations of the products, raised to powers equal to their coefficients in the chemical equation, are multiplied together in the numerator. The denominator is constructed in the same way from the concentrations of the reactants raised to powers equal to their coefficients. At equilibrium, the mass action expression is equal to the equilibrium constant, K_c. If partial pressures of gases are used in the mass action expression, K_p is obtained. The magnitude of the equilibrium constant is roughly proportional to the extent to which the reaction proceeds to completion when equilibrium is reached.

Thermodynamics and Equilibrium. Equilibrium constants can be calculated from measured equilibrium concentrations or from $\Delta G°$.

$$\Delta G^\circ = -RT \ln K$$

or

$$\Delta G^\circ = -2.303 \, RT \log K$$

where $K = K_c$ for reactions in solutions and $K = K_p$ for gaseous reactions.

Le Châtelier's Principle. When an equilibrium is upset, a chemical change occurs in a direction that opposes the disturbing influence and brings the system to equilibrium again. Adding a reactant or product causes a reaction to occur that uses up part of what has been added. Removing a reactant or product causes a reaction that replaces part of what has been removed. Increasing the pressure (by reducing the volume) drives a reaction in the direction of the fewest number of moles of gas. Pressure has virtually no effect on equilibria involving only solids and liquids. Raising the temperature causes an equilibrium to shift in an endothermic direction. The value of K increases for reactions that are endothermic in the forward direction. A change in temperature is the *only* factor that changes K. Addition of a catalyst has no effect on an equilibrium.

Equilibrium Calculations. The initial concentrations in a chemical system are determined by the person who combines the chemicals at the start of the reaction. The changes in concentration are determined by the stoichiometry of the reaction. Only equilibrium concentrations can be substituted into the mass action expression so that it is equal to K_c. When a change in concentration is expected to be very small compared to the initial concentration, the change may be neglected and the equation corresponding to the equilibrium law can be simplified. In general, this simplification is valid if the initial concentration is at least 1000 times larger than K.

Precipitation Reactions. A double replacement (metathesis) reaction will occur if all of the reactant and product ions are not spectator ions. In writing the net ionic equation, insoluble salts are written in undissociated ("molecular") form. The solubility rules allow you to predict when a precipitation reaction will occur.

Heterogeneous Equilibria and Solubility Product. The mass action expression for a heterogeneous equilibrium never contains concentration terms for pure liquids or pure solids. When the heterogeneous equilibrium involves the solubility of a salt, the mass action expression is a product of ion concentrations raised to appropriate powers and the equilibrium constant, K_{sp}, is the solubility product constant. The molar solubility can be used to calculate K_{sp}, and K_{sp} can be used to calculate molar solubility. According to the common ion effect, a salt is less soluble in a solution containing one of its ions than it is in pure water. A solution containing a mixture of ions will produce a precipitate of a given salt only if the solution is supersaturated. Under these conditions the ion product is greater than the value of K_{sp} for the salt.

REVIEW QUESTIONS

16.26. Why are chemical equilibria called *dynamic* equilibria?

16.27. Sketch a graph showing how the concentrations of the reactants and products of a typical reaction vary with time.

16.28. What is meant when we say that chemical reactions are reversible?

16.29. Sketch the shape of a free energy curve for a reaction having a negative ΔG° (no numbers are necessary). Where does equilibrium occur? How does the system change if it contains only reactants? How does it change if it contains only products?

16.30. What relationship exists between the coefficients in an equation and the mass action expression?

16.31. How is the mass action expression related to K_c?

16.32. In general, what units must the quantities that are substituted into the mass action expression have?

16.33. What is an *equilibrium law*?

16.34. Write the mass action expression for the following reactions:
(a) $2PCl_3(g) + O_2(g) \rightleftharpoons 2POCl_3(g)$
(b) $2SO_3(g) \rightleftharpoons 2SO_2(g) + O_2(g)$
(c) $N_2H_4(g) + 2O_2(g) \rightleftharpoons 2NO(g) + 2H_2O(g)$
(d) $N_2H_4(g) + 6H_2O_2(g) \rightleftharpoons 2NO_2(g) + 8H_2O(g)$

16.35. Write the mass action expression for the following reactions:
(a) $H_2(g) + Cl_2(g) \rightleftharpoons 2HCl(g)$
(b) $\frac{1}{2}H_2(g) + \frac{1}{2}Cl_2(g) \rightleftharpoons HCl(g)$
How does K_c for reaction (a) compare with K_c for reaction (b)?

16.36. Write the equilibrium law for the reaction

$$2HCl(g) \rightleftharpoons H_2(g) + Cl_2(g)$$

How does K_c for this reaction compare with K_c for reaction (a) in question 16.35?

16.37. Write expressions for the equilibrium law using partial pressures for the reactions in question 16.34.

16.38. When a chemical equation and its equilibrium constant are given, why is it not necessary to also specify the form of the mass action expression?

16.39. At 225 °C, $K_p = 6.3 \times 10^{-3}$ for the reaction

$$CO(g) + 2H_2(g) \rightleftharpoons CH_3OH(g)$$

Would we expect this reaction to go nearly to completion?

16.40. Here are some reactions and their equilibrium constants.
(a) $2CH_4(g) \rightleftharpoons C_2H_6(g) + H_2(g)$ $K_c = 9.5 \times 10^{-13}$
(b) $CH_3OH(g) + H_2(g) \rightleftharpoons CH_4(g) + H_2O(g)$
     $K_c = 3.6 \times 10^{20}$
(c) $C_2H_4(g) + H_2O(g) \rightleftharpoons C_2H_5OH(g)$ $K_c = 8.2 \times 10^3$
Arrange these reactions in order of their increasing tendency to go toward completion.

16.41. How is the equilibrium constant related to the standard free energy change for a reaction?

16.42. Write an equation showing how you would convert from common logarithms to natural logarithms.

16.43. State Le Châtelier's principle *in your own words*.

16.44. How will the equilibrium

$$\text{heat} + CH_4(g) + 2H_2S(g) \rightleftharpoons CS_2(g) + 4H_2(g)$$

be affected by:
(a) The addition of $CH_4(g)$?
(b) The addition of $H_2(g)$?

(c) The removal of $CS_2(g)$?
(d) A decrease in the volume of the container?
(e) An increase in temperature?

16.45. The reaction, $CO(g) + 2H_2(g) \rightleftharpoons CH_3OH(g)$, has $\Delta H° = -18$ kJ. How will the amount of CH_3OH present at equilibrium be affected by:
(a) Adding $CO(g)$?
(b) Removing $H_2(g)$?
(c) Decreasing the volume of the container?
(d) Adding a catalyst?
(e) Increasing the temperature?

16.46. In questions 16.44 and 16.45, which change(s) will alter K_c?

16.47. Consider the equilibrium, $2NO(g) + Cl_2(g) \rightleftharpoons 2NOCl(g)$, for which $\Delta H° = -18.42$ kcal. How will the amount of Cl_2 at equilibrium be affected by:
(a) Removal of $NO(g)$?
(b) Addition of $NOCl(g)$?
(c) Raising the temperature?
(d) Decreasing the volume of the container?

16.48. Which of the equilibria in question 16.40 will *not* be affected by a change in the volume of the container?

16.49. What would have to be true about $\Delta H°$ in order for a reaction to have an equilibrium constant that is the same at all temperatures?

16.50. Study the solubility rules on page 529, then choose the compounds below that are soluble in water.
(a) $Ca(NO_3)_2$ (d) $AgNO_3$
(b) $FeCl_2$ (e) $BaSO_4$
(c) $Ni(OH)_2$ (f) $CuCO_3$

16.51. After studying the solubility rules on page 529, pick the compounds from the list below that are insoluble in water.
(a) $AgCl$ (d) $Ca_3(PO_4)_2$
(b) $Cr_2(SO_4)_3$ (e) $Al(C_2H_3O_2)_3$
(c) $(NH_4)_2CO_3$ (f) ZnO

16.52. What is another name for a double replacement reaction?

16.53. Write ionic and net ionic equations for these reactions.
(a) $(NH_4)_2CO_3(aq) + MgCl_2(aq)$
$$\longrightarrow 2NH_4Cl(aq) + MgCO_3(s)$$
(b) $CuCl_2(aq) + 2NaOH(aq) \longrightarrow Cu(OH)_2(s) + 2NaCl(aq)$
(c) $3FeSO_4(aq) + 2Na_3PO_4(aq)$
$$\longrightarrow Fe_3(PO_4)_2(s) + 3Na_2SO_4(aq)$$
(d) $2AgC_2H_3O_2(aq) + NiCl_2(aq)$
$$\longrightarrow 2AgCl(s) + Ni(C_2H_3O_2)_2(aq)$$

16.54. Complete the following and then write the net ionic equations. If all ions cancel, indicate that no reaction (N.R.) takes place
(a) $Na_2SO_3 + Ba(NO_3)_2 \longrightarrow$
(b) $K_2S + ZnCl_2 \longrightarrow$
(c) $NH_4Br + Pb(C_2H_3O_2)_2 \longrightarrow$
(d) $NH_4ClO_4 + Cu(NO_3)_2 \longrightarrow$

16.55. Write the equilibrium law for each of the following heterogeneous reactions.
(a) $2C(s) + O_2(g) \rightleftharpoons 2CO(g)$
(b) $2NaHSO_3(s) \rightleftharpoons Na_2SO_3(s) + H_2O(g) + SO_2(g)$
(c) $2C(s) + 2H_2O(g) \rightleftharpoons CH_4(g) + CO_2(g)$
(d) $CaCO_3(s) + 2HF(g) \rightleftharpoons CaF_2(s) + H_2O(g) + CO_2(g)$

16.56. Write the K_{sp} expression for these compounds.
(a) FeS   (c) $PbSO_4$
(b) Ag_2CO_3 (d) Bi_2S_3

(e) PbI_2 (f) $Cu(OH)_2$

16.57. Write the K_{sp} expression for these compounds.
(a) AgI   (d) $Al(OH)_3$
(b) CoS   (e) $ZnCO_3$
(c) $PbCrO_4$ (f) Sb_2S_3

16.58. What is the common ion effect?

16.59. With respect to K_{sp}, what conditions must be met if a precipitate is going to form in a solution?

REVIEW PROBLEMS

16.60. What is the value of $\Delta G°$ for a reaction having $K = 1$?

16.61. A potential reaction for conversion of coal to methane (natural gas) is

$$C(s) + 2H_2(g) \rightleftharpoons CH_4(g)$$

for which $\Delta G° = -12.14$ kcal. What is the value of K_p for this reaction at 25 °C? Does the value of K_p make studying this reaction as a method of natural gas production worthwhile?

16.62. One of the important reactions in living cells, from which the organism draws energy, is the reaction of adenosine triphosphate (ATP) with water to give adenosine diphosphate (ADP) plus free phosphate.

$$ATP + H_2O \rightleftharpoons ADP + phosphate$$

The value of $\Delta G'$ (the equivalent of $\Delta G°$, but at 37 °C) for this reaction is -8 kcal/mol. Calculate the equilibrium constant for the reaction.

16.63. What will be the value of the equilibrium constant for a reaction having $\Delta G° = 0$?

16.64. A certain reaction at 25 °C has $K_p = 10.0$. What is the value of $\Delta G°$ for this reaction?

16.65. Methanol, a potentially important fuel, can be made from CO and H_2. The reaction, $CO(g) + 2H_2(g) \rightleftharpoons CH_3OH(g)$, has $K_p = 6.25 \times 10^{-3}$ at 500 K. Calculate $\Delta G'$ for this reaction. (Note: We use the symbol $\Delta G'$ rather than $\Delta G°$ because the latter is the symbol for the free energy change at 25 °C.)

16.66. At 773 °C, a mixture of $CO(g)$, $H_2(g)$ and $CH_3OH(g)$ was allowed to come to equilibrium. The following equilibrium concentrations were then measured: $[CO] = 0.105\ M$, $[H_2] = 0.250\ M$, $[CH_3OH] = 0.00261\ M$. Calculate K_c for the reaction, $CO(g) + 2H_2(g) \rightleftharpoons CH_3OH(g)$.

16.67. Ethylene, C_2H_4, and water react under appropriate conditions to give ethanol. The reaction is

$$C_2H_4(g) + H_2O(g) \rightleftharpoons C_2H_5OH(g)$$

An equilibrium mixture of these gases at a certain temperature had the following concentrations: $[C_2H_4] = 0.0222\ M$, $[H_2O] = 0.0225\ M$, $[C_2H_5OH] = 0.150\ M$. What is the value of K_c?

16.68. At a certain temperature, the reaction, $CO(g) + 2H_2(g) \rightleftharpoons CH_3OH(g)$, has $K_c = 0.500$. If a reaction mixture at equilibrium contains 0.210 M CO and 0.100 M H_2, what is the concentration of CH_3OH?

16.69. We saw in Example 16.7 that $K_c = 64$ for the reaction, $N_2(g) + 3H_2(g) \rightleftharpoons 2NH_3(g)$, at a certain temperature. Suppose it was found that an equilibrium mixture of these gases contained 0.280 M NH_3 and 0.00840 M N_2. What was the concentration of H_2 in the mixture?

16.70. At high temperature, 0.500 mol of HBr was placed in a 1.00-liter container and allowed to decompose according to the reaction, $2HBr(g) \rightleftharpoons H_2(g) + Br_2(g)$. At equilibrium the concentration of Br_2 was measured to be 0.130 M. What is K_c for this reaction at this temperature?

16.71. A 0.100-mol sample of formaldehyde vapor, HCHO, was placed in a heated 1.00-liter vessel and some of it decomposed. The reaction is

$$HCHO(g) \rightleftharpoons H_2(g) + CO(g)$$

At equilibrium, the $HCHO(g)$ concentration was 0.080 mol/liter. Calculate the value of K_c for this reaction.

16.72. The equilibrium constant, K_c, for the reaction

$$SO_3(g) + NO(g) \rightleftharpoons NO_2(g) + SO_2(g)$$

was found to be 0.500 at a certain temperature. If 0.300 mol of SO_3 and 0.300 mol of NO were placed in a 2.00-liter container and allowed to react, what would be the equilibrium concentrations of each gas?

16.73. At a certain temperature the reaction, $CO(g) + H_2O(g) \rightleftharpoons CO_2(g) + H_2(g)$, has $K_c = 0.400$. Exactly 1.00 mol of each gas was placed in a 100-liter vessel and allowed to react. What were the equilibrium concentrations of each gas?

16.74. The reaction, $2HCl(g) \rightleftharpoons H_2(g) + Cl_2(g)$, has $K_c = 3.2 \times 10^{-34}$ at 25 °C. If a reaction vessel contains 2.00 mol/liter of HCl and is allowed to come to equilibrium, what will be the concentrations of H_2 and Cl_2?

16.75. The heterogeneous reaction, $2HCl(g) + I_2(s) \rightleftharpoons 2HI(g) + Cl_2(g)$, has $K_c = 1.6 \times 10^{-34}$ at 25 °C. Suppose 1.00 mol of HCl and solid I_2 was placed in a 1.00-liter container. What would be the equilibrium concentrations of HI and Cl_2 in the container?

16.76. Barium sulfate, $BaSO_4$, is so insoluble that it can be swallowed without significant danger, even though Ba^{2+} is toxic. In 1 liter of water, only 0.00243 g $BaSO_4$ will dissolve.
(a) How many moles of $BaSO_4$ will dissolve per liter?
(b) What are the Ba^{2+} and SO_4^{2-} concentrations in a saturated $BaSO_4$ solution?
(c) Calculate K_{sp} for $BaSO_4$.

16.77. Magnesium hydroxide, $Mg(OH)_2$, found in milk of magnesia, has a solubility of 8.35×10^{-3} g/liter.
(a) What is the solubility expressed in mol/liter?
(b) What are the Mg^{2+} and OH^- concentrations in mol/liter?
(c) Calculate K_{sp} for $Mg(OH)_2$.

16.78. The molar solubility of Ag_3PO_4 is 2.5×10^{-6} mol/liter. Calculate K_{sp} for this salt.

16.79. The molar solubility of $Ba_3(PO_4)_2$ is 8.9×10^{-9} mol/liter. Calculate K_{sp} for this salt.

16.80. Calcium sulfate is found in plaster. For $CaSO_4$, $K_{sp} = 6 \times 10^{-5}$. What is the solubility of $CaSO_4$ expressed in mol/liter?

16.81. Chalk is $CaCO_3$, whose $K_{sp} = 8.7 \times 10^{-9}$. What is the molar solubility of $CaCO_3$? How many grams of $CaCO_3$ will dissolve in 100 ml of water?

16.82. Calculate the molar solubility of lead iodide, PbI_2 ($K_{sp} = 8.3 \times 10^{-9}$).

16.83. What is the molar solubility of Ag_2SO_3 ($K_{sp} = 5 \times 10^{-14}$)?

16.84. What is the molar solubility of Ag_2CrO_4 in a 0.10 M $AgNO_3$ solution? For Ag_2CrO_4, $K_{sp} = 9 \times 10^{-12}$.

16.85. What is the molar solubility of $Mg(OH)_2$ in a 0.10-M NaOH solution? (NaOH is a strong electrolyte and is completely dissociated into Na^+ and OH^- in aqueous solution.) For $Mg(OH)_2$, $K_{sp} = 1.2 \times 10^{-11}$.

16.86. Will a precipitate of $PbCl_2$ form if 0.0100 mol of $Pb(NO_3)_2$ and 0.0100 mol of NaCl are dissolved together in 1.00 liter of solution?

16.87. Silver acetate, $AgC_2H_3O_2$, has $K_{sp} = 4 \times 10^{-3}$. Will a precipitate form if 0.010 mol of $AgNO_3$ and 0.30 mol of $Ca(C_2H_3O_2)_2$ are dissolved together in a total volume of 1.00 liter of solution?

CHAPTER SEVENTEEN

ACID-BASE EQUILIBRIA

17.1 BRØNSTED CONCEPT OF ACIDS AND BASES

Acids are proton donors and bases are proton acceptors

We have often discussed acids and bases in our study, beginning in Section 1.16 and continuing with their appearances in many later sections, worked examples and problems. In your lab work you have by now no doubt become familiar with many of the common acids and bases and their properties. They are extremely important substances, but they do arouse mixed emotions. The words "acid" and "lye" cause fear in the minds of many people. They associate acids with the corrosion of cars, garbage cans, and teeth. "Stomach acid" conjures nothing but nastiness and nausea, thanks to television ads that hawk this or that brand of "antacid." Arthur Conan Doyle heightened the drama of one of the escapes of Sherlock Holmes by having Kitty Winter hurl a vial of vitriol into the wicked Baron Gruner's face.

"Lye" is sodium hydroxide.

"Vitriol" is sulfuric acid.

Stomach acid is dilute hydrochloric acid.

Acids are neither good nor bad. Some of their *uses* are very dangerous, but most are valuable. We need some acid in the stomach or we can't digest meat properly. We need ascorbic acid (vitamin C) and nicotinic acid (a B vitamin) in our diets for good health. The many uses of sulfuric acid in manufacturing, as we have often noted, make it indispensable to any industrialized society. Lye also has many uses; one is to neutralize acids that are no longer needed. Common soap couldn't be made without lye.

In Section 6.8 we learned that Svante Arrhenius defined acids as substances that produce hydrogen ions in water and bases as compounds that give hydroxide ions in water. The neutralization of an acid and a base gives water and a salt. We also found that the Arrhenius view of acids, strictly speaking, is wrong because the hydrogen ion can't exist as a separate, distinct particle in water or any other solvent. Instead the acids known to Arrhenius give hydronium ions, H_3O^+, in water, not hydrogen ions, H^+. In the neutralization of an aqueous acid by aqueous sodium hydroxide, the hydroxide ion accepts a hydrogen ion *by direct transfer* from the hydronium ion; it does not remove separately existing hydrogen ions. The distinction between combining with individual H^+ particles and accepting them by direct transfer from a donor is important. In 1923 Johannes Brønsted (1874–1947), a Danish scientist, established the importance of this distinction in proposing new and broader definitions of acids and bases.[1] Much

---

[1] Thomas Lowry (1874–1947), a British scientist working independently of Brønsted, developed much the same ideas but did not carry them as far. Some references call this view of acids and bases the Brønsted-Lowry theory.

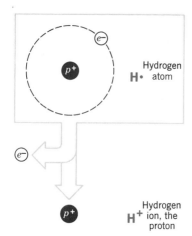

The hydrogen atom and the hydrogen ion.

proton-transfer
occurs here

H_3O^+

of this chapter is devoted to the Brønsted view of acids, bases, acid-base equilibria, and neutralization. Because of our background from Chapter Sixteen of equilibria and equilibrium constants we can now study quantitative aspects of acid-base equilibria. A good, problem-solving knowledge of these is essential for any further studies in chemistry, biology, nutrition, pharmacy, clinical chemistry, most of the medical disciplines, and many other fields. In some of the worked examples of this chapter we will learn how equilibrium concentrations of the hydronium ion are life-and-death matters in human health.

In developing his definitions, Brønsted chose to use the name **proton** for H^+ or hydrogen ion. This tends to emphasize the nature of the particle and deemphasize the earlier association with Arrhenius's definitions. Brønsted realized that in any acid-base neutralization, regardless of the solvent, all that happens is the direct *transfer* of a proton from a donor to an acceptor. He therefore defined acids and bases as follows.

> **Brønsted Definitions of Acids and Bases.**
> An **acid** is a proton donor.
> A **base** is a proton acceptor.

The hydronium ion clearly qualifies as an acid and the hydroxide ion as a base when they react in water.

$$H_3O^+(aq) + OH^-(aq) \longrightarrow 2H_2O(\ell)$$

However, Brønsted's definitions apply to a great many other ions and molecules. For example, when hydrogen chloride, $HCl(g)$, combines with ammonia, $NH_3(g)$, a proton transfers from the donor or acid, hydrogen chloride, to the acceptor or base, ammonia.

$$HCl(g) + NH_3(g) \longrightarrow NH_4Cl(s)$$
ammonium chloride

Here there is no solvent, yet there is acid-base neutralization.

These two reactions illustrate that the electrical charge on the reactants plays no part in their role as Brønsted acids or bases. Regardless of the charge, the acid is whichever particle gives up a proton and the base is the particle that accepts it. We will work an example that shows how to identify the acid and base species in a reaction because this skill is needed throughout the chapter.

EXAMPLE 17.1   **Identifying Brønsted Acids and Bases**

Problem: Sodium hydrogen sulfate is used in the manufacture of certain kinds of cement and to clean oxide coatings on metals. Its anion, HSO_4^-, reacts as follows with the anion of trisodium phosphate, PO_4^{3-}.

$$HSO_4^-(aq) + PO_4^{3-}(aq) \longrightarrow HPO_4^{2-}(aq) + SO_4^{2-}(aq)$$

Which reactants are the Brønsted acid and base in this reaction?

Solution: If you study the formulas in this equation, you will see that the only thing that happens is the transfer of a proton. (Remember that "proton" means H^+ when we write formulas and compute net electrical charges.) The proton-donor is HSO_4^-, and this ion is the Brønsted acid. The proton-acceptor is PO_4^{3-}, which is the Brønsted base.

EXERCISE 17.1 When aqueous solutions of sodium bicarbonate and sodium dihydrogen phosphate are mixed in the proper mole ratios of solutes, the following reaction occurs.

$$HCO_3^-(aq) + H_2PO_4^-(aq) \longrightarrow H_2CO_3(aq) + HPO_4^{2-}(aq)$$

Identify the Brønsted acid and base among the reactants. (One product, H_2CO_3 or carbonic acid, is unstable and spontaneously breaks down to CO_2 and H_2O.)

Strong and weak acids

In Section 6.8 we introduced the concepts of strong and weak acids. Here we will review this concept but extend it to include the Brønsted acids and bases. The proton in an acid is held by a chemical bond whose strength varies from acid to acid. For example, it is quite weak in hydrogen chloride, $HCl(g)$, and much stronger in hydrogen cyanide, $HCN(g)$. Therefore, when different acids interact with the same base, chemical equilibria form having widely varying equilibrium constants. When hydrogen chloride is bubbled into water, water molecules serve as the proton acceptors or base.

HCN is H—C≡N.

$$HCl(g) + H_2O(\ell) \rightleftharpoons Cl^-(aq) + H_3O^+(aq) \qquad K_c = \frac{[H_3O^+][Cl^-]}{[HCl][H_2O]}$$
hydrogen chloride   hydrochloric acid

Such a high percentage of the initial $HCl(g)$ molecules donate their protons to water molecules that the value of the molar concentration of the remaining HCl—[HCl] in the denominator of K_c—drops nearly to zero. Therefore, the value of K_c is too high for any meaningful application; it is estimated as being about 10^7. Whenever the concentrations of the products at equilibrium greatly exceed the concentrations of the reactants we can say that "the equilibrium favors the products," or that "the position of the equilibrium is on the right" (as written).

When hydrogen cyanide is bubbled into water, fewer than one HCN molecule in a thousand transfers a proton to water at equilibrium.

$$HCN(g) + H_2O(\ell) \rightleftharpoons CN^-(aq) + H_3O^+(aq) \qquad K_c = \frac{[H_3O^+][CN^-]}{[HCN][H_2O]}$$

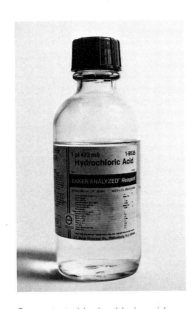

Concentrated hydrochloric acid.

The value of this K_c is on the order of 10^{-10}, a very low value, and we can describe this kind of equilibrium by saying that "the equilibrium favors the reactants," or that "the position of the equilibrium is well on the left" (as we have written it).

Because acids differ widely in their abilities to donate protons to the same base, we need equilibrium constants to compare their strengths—a topic we will discuss in later sections. Since equilibrium constants are seldom memorized, however, it is highly useful in discussions of acids (and bases) to have some qualitative descriptions of their strengths. A **strong acid,** for example, is any that is a good proton donor. Hydrogen chloride is a strong acid. Hydrogen cyanide is a **weak acid** because it is a poor proton-donor.

Once a strong acid has dissolved in water and has given up its protons to water molecules, then the proton-donating species in solution is the hydronium ion, not the original solute molecule. For example, in hydrochloric acid—$HCl(aq)$—the principal proton donor is $H_3O^+(aq)$, not $HCl(g)$. The hydronium ion itself is a strong Brønsted acid, and hydrochloric acid is called a strong acid because it is such an excellent producer of H_3O^+. "Strong acid" used in such a context means that the solute exists in water in essentially a 100-percent ionized state. Thus, we have two different but related meanings for "strong" when speaking of strong acids. A strong acid is a good proton-donor, and a strong (aqueous) acid is any whose ionization is virtually 100%. Just remember that in a strong (aqueous) acid solution we have H_3O^+ ions as the true Brønsted acid species. The hydronium ion, in fact, is the strongest acid we can have in water.

Any stronger acid, for example, $HCl(g)$, gives up its proton to water to make H_3O^+.

Only a few acids are strong acids in water. Their names and formulas are given in Table 17.1, and they should be memorized. Learn also that phosphoric acid, $H_3PO_4(aq)$, is a moderate acid, meaning a moderately strong acid, in water; in a 0.1 M solution its percentage ionization is about 30%. With very few exceptions anything else going by the name of "acid" is a weak acid in water, and there are hundreds of them. Boric acid, $H_3BO_3(aq)$, carbonic acid, $H_2CO_3(aq)$, sulfurous acid, $H_2SO_3(aq)$, nitrous acid, $HNO_2(aq)$, hydrofluoric acid, $HF(aq)$, and virtually all of the organic carboxylic acids (e.g., acetic acid, $HC_2H_3O_2$) are weak acids. Their percentage ionizations in water are generally less than 1% in 0.1 M solutions. (The advantage of learning the short list of strong and moderate acids is that by implication you then automatically know that hundreds of other acids are relatively weak in water.)

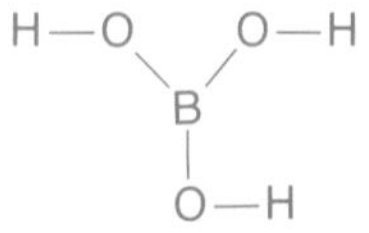

H—O   O—H
     \ /
      B
      |
      O—H
   boric acid

Table 17.1
The Common Strong Aqueous Acids and Bases

Strong Acids	
Perchloric acid	$HClO_4(aq)$
Chloric acid	$HClO_3(aq)$
Hydriodic acid	$HI(aq)$
Hydrobromic acid	$HBr(aq)$
Sulfuric acid	$H_2SO_4(aq)$
Hydrochloric acid	$HCl(aq)$
Nitric acid	$HNO_3(aq)$
Strong Bases	
Very soluble in water	
Sodium hydroxide	$NaOH(aq)$
Potassium hydroxide	$KOH(aq)$
Slightly soluble in water	
Magnesium hydroxide	$Mg(OH)_2(aq)$
Calcium hydroxide	$Ca(OH)_2(aq)$
Barium hydroxide	$Ba(OH)_2(aq)$

memorize

Strong and weak bases

A **strong base** is a strong proton acceptor, a chemical that will accept and *strongly bind* a proton. A **weak base,** while somewhat a proton acceptor, cannot hold the proton well. The water molecule is a weak base in this sense. Although it readily accepts a proton from hydrogen chloride as hydrochloric acid forms, it only weakly holds the proton in the form of the hydronium ion.

The hydroxide ion is the only common strong base. It will even take a proton from an acid as weak as hydrogen cyanide.

$$HCN(aq) + OH^-(aq) \rightleftharpoons CN^-(aq) + H_2O(\ell)$$

The position of this equilibrium is almost entirely on the right. Just as the hydronium ion is actually the strongest Brønsted acid we can have in water, so the hydroxide ion is the strongest Brønsted base we can have in water. Anything stronger reacts with water to generate hydroxide ion. For example, the amide ion is a stronger base than the hydroxide ion. When this ion is supplied to water in the form of some salt, for example, $NaNH_2(s)$, it reacts with water as follows:

$$\underset{\text{amide ion}}{NH_2^-(s)} + H_2O(\ell) \rightleftharpoons \underset{\text{ammonia}}{NH_3(aq)} + OH^-(aq)$$

The position of this equilibrium is so far on the right that the reaction goes essentially 100% from left to right.

Table 17.1 also gives the common strong (aqueous) bases, those that ionize virtually to 100% in water. All furnish the hydroxide ion, but three of them are so slightly soluble in water that they cannot provide a high molar concentration of $OH^-(aq)$. However, we must emphasize again that the terms "strong" and "weak" as applied to aqueous acids or bases say nothing about molar concentrations of $H_3O^+(aq)$ or $OH^-(aq)$. They refer only to *percentage* ionization of whatever quantity of solute actually gets into solution.

The most common weak, aqueous base besides water itself is the ammonia molecule in aqueous ammonia, $NH_3(aq)$. It is a much stronger Brønsted base than water, but it is still a weak base compared to the hydroxide ion. In aqueous ammonia there is an equilibrium in which some formation of hydroxide ion occurs:

$$NH_3(aq) + H_2O(\ell) \rightleftharpoons NH_4^+(aq) + OH^-(aq)$$

We will go into quantitative details about this equilibrium in a later section, but the position of the equilibrium strongly favors the *reactants.*

The organic derivatives of ammonia, the amines, are weak bases, too. Those that dissolve in water react only slightly with water to establish the following kinds of equilibria.

$$\underset{\text{methylamine}}{CH_3NH_2(aq)} + H_2O(\ell) \rightleftharpoons \underset{\substack{\text{methylammonium} \\ \text{ion}}}{CH_3NH_3^+(aq)} + OH^-(aq)$$

$$\underset{\text{dimethylamine}}{CH_3NHCH_3(aq)} + H_2O(\ell) \rightleftharpoons \underset{\substack{\text{dimethylammonium} \\ \text{ion}}}{CH_3\overset{+}{N}H_2CH_3(aq)} + OH^-(aq)$$

These amines with low formula weights have strong ammonia-like odors.

$$\underset{\substack{\text{trimethylamine}}}{CH_3-\underset{\underset{CH_3}{|}}{N}-CH_3(aq)} + H_2O(\ell) \rightleftharpoons \underset{\substack{\text{trimethylammonium} \\ \text{ion}}}{CH_3-\underset{\underset{CH_3}{|}}{\overset{+}{N}H}-CH_3(aq)} + OH^-(aq)$$

The positions of all these equilibria are far on the left in favor of the reactants. Therefore the percentage of amine molecules that have generated hydroxide ions is low—generally less than 1%—in all cases.

Amphoteric compounds

We have seen water function as a Brønsted base:

$$HCl(g) + H_2O \rightleftharpoons Cl^- + H_3O^+$$

And we have seen it react as a Brønsted acid:

$$NH_2^- + H_2O \rightleftharpoons NH_3 + OH^-$$

Compounds that can be either an acid or a base depending on the other substance present are called **amphoteric compounds.** Some ions are also amphoteric, for example, the bicarbonate ion. Toward the hydronium ion, it is a base:

"Amphoteros"—Greek for partly one and partly the other.

$$HCO_3^-(aq) + H_3O^+(aq) \longrightarrow \underset{\substack{\text{carbonic acid} \\ \text{(unstable)}}}{H_2CO_3(aq)} + H_2O(\ell)$$

Toward the hydroxide ion, the bicarbonate ion is an acid:

$$HCO_3^-(aq) + OH^-(aq) \longrightarrow CO_3^{2-}(aq) + H_2O(\ell)$$

Conjugate acid-base relationships

Let us think again about the equilibrium that is present when a weak acid such as HCN is dissolved in water.

$$HCN(aq) + H_2O(\ell) \rightleftharpoons H_3O^+(aq) + CN^-(aq)$$

Since HCN is a weak acid we know that only a very low percentage of CN^- ions are present, and that the position of the equilibrium is on the left. We also know that H_3O^+ is a strong Brønsted acid. Thus, there really are *two* acids present in this equilibrium, one weaker (HCN) and the other stronger (H_3O^+). There are also two Brønsted bases, one weaker (H_2O) and the other stronger (CN^-). In fact the cyanide ion, when furnished by a soluble, fully dissociated salt such as sodium cyanide, reacts almost quantitatively with any strong aqueous acid as follows.

$$\underset{\substack{\text{stronger} \\ \text{acid}}}{H_3O^+(aq)} + \underset{\substack{\text{stronger} \\ \text{base}}}{CN^-(aq)} \longrightarrow \underset{\substack{\text{weaker} \\ \text{base}}}{H_2O(\ell)} + \underset{\substack{\text{weaker} \\ \text{acid}}}{HCN(aq)}$$

The cyanide ion is called the conjugate base of hydrogen cyanide. The two, CN^- and HCN, form a conjugate acid-base pair. Members of any **conjugate acid-base pair** have formulas that differ only by one H^+. Thus, H_3O^+ and H_2O are a conjugate acid-base pair; H_3O^+ is the conjugate acid of H_2O and H_2O is the conjugate base of H_3O^+. It will be helpful in later discussions of acids and bases to be able to write the formulas of the second member of a conjugate acid-base pair when the first is given.

EXAMPLE 17.2   **Writing the Formula of a Conjugate Base**

Problem: What is the conjugate base of nitric acid, HNO_3?
Solution: A conjugate base can always be found by removing one H^+ from a given acid. Removing one H^+, both the atom and the charge, from HNO_3 leaves NO_3^-. The nitrate ion, NO_3^- is the conjugate base of HNO_3.

EXAMPLE 17.3   **Writing the Formula of a Conjugate Base**

Problem: The hydrogen sulfate ion, HSO_4^- has a conjugate base. Write its formula.
Solution: Deleting an H^+ from HSO_4^- leaves SO_4^{2-}. [Notice that the charge goes from $1-$ to $2-$ because $(1-) - (1+) = (2-)$.]

EXAMPLE 17.4   **Writing the Formula of a Conjugate Acid**

Problem: What is the formula of the conjugate acid of OH^-?
Solution: To find a conjugate acid we add one H^+ to the formula of the given base. Adding H^+ to OH^- gives H_2O. Water is the conjugate acid of the hydroxide ion.

EXAMPLE 17.5   **Writing the Formula of a Conjugate Acid**

Problem: The phosphate ion, PO_4^{3-}, is a rather strong Brønsted base. What is the formula of its conjugate acid?
Solution: Adding H^+ to the formula PO_4^{3-} gives HPO_4^{2-}. The conjugate acid of PO_4^{3-} is HPO_4^{2-}. Always be sure that the electrical charge is correctly shown.

EXERCISE 17.2 Write the formula of the conjugate base for each of the following Brønsted acids.

(a) H_2O (e) $H_2PO_4^-$
(b) HI (f) HPO_4^{2-}
(c) HNO_2 (g) H_2
(d) H_3PO_4 (h) NH_4^+

EXERCISE 17.3 Write the formula of the conjugate acid of each of the following Brønsted bases.

(a) HO^- (e) NH_2^-
(b) SO_4^{2-} (f) NH_3
(c) PO_4^{3-} (g) $H_2PO_4^-$
(h) HPO_4^{2-}

$$ (d)\ CH_3 \overset{\overset{\displaystyle O}{\|}}{-C} - O^- $$

EXAMPLE 17.6 **Identifying Conjugate Acid-Base Pairs in Equilibria**

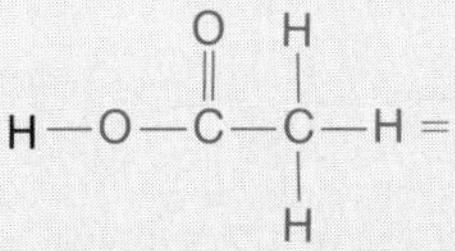

$$ H - O - \overset{\overset{\displaystyle O}{\|}}{C} - \overset{\overset{\displaystyle H}{|}}{\underset{\underset{\displaystyle H}{|}}{C}} - H = HC_2H_3O_2 $$

acetic acid

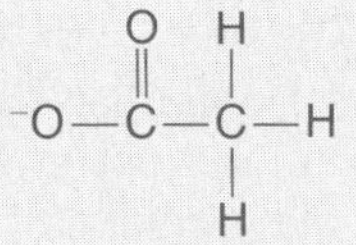

$$ {}^-O - \overset{\overset{\displaystyle O}{\|}}{C} - \overset{\overset{\displaystyle H}{|}}{\underset{\underset{\displaystyle H}{|}}{C}} - H = C_2H_3O_2^- $$

acetate ion

Problem: Identify the conjugate pairs of acids and bases in the following equilibrium.

$$ C_2H_3O_2^-(aq) + H_3O^+(aq) \rightleftharpoons HC_2H_3O_2(aq) + H_2O(\ell) $$

acetate ion   hydronium   acetic acid
ion

Solution: To identify conjugate acid-base pairs, always look for formulas that differ by only one H^+. Remember that in such a pair the base always has one less H^+ than its conjugate acid. One of the common ways of designating the members of a conjugate acid-base pair in an equilibrium is to connect them by a line, as follows.

conjugate pair

$$ C_2H_3O_2^- + H_3O^+ \rightleftharpoons HC_2H_3O_2 + H_2O $$

conjugate pair

EXERCISE 17.4 If some of the strong cleaning agent, trisodium phosphate, were mixed with household vinegar, which contains acetic acid, the following equilibrium would be established. (The products are favored.) Identify the pairs of conjugate acids and bases.

$$ PO_4^{3-}(aq) + HC_2H_3O_2(aq) \rightleftharpoons HPO_4^{2-}(aq) + C_2H_3O_2^-(aq) $$

phosphate   acetic acid   monohydrogen   acetate ion
ion   phosphate ion

17.2 SELF-IONIZATION OF WATER

Water self-ionizes to give low concentrations of hydronium and hydroxide ions

Water reacts with itself to a very slight but important extent to establish the following equilibrium

$$ H_2O(\ell) + H_2O(\ell) \rightleftharpoons H_3O^+(aq) + OH^-(aq) $$

The position of this equilibrium lies strongly on the left.

Ion-product constant of water, K_w

The equilibrium constant for the self-ionization of water can be written as follows.

$$K_c = \frac{[H_3O^+][OH^-]}{[H_2O]^2} = 3.25 \times 10^{-18} \qquad (17.1)$$

In *pure* water the values of $[H_3O^+]$ and $[OH^-]$ must equal each other because the ions form in identical numbers. At 25 °C the molar concentration of each is 1.00×10^{-7} mol/liter and the molar concentration of water itself is 55.5 mol/liter. If we compare 10^{-7} with 55.5, we can see that the self-ionization of water changes its molar concentration by essentially nothing even if we rounded to seven significant figures. Therefore, the value of $[H_2O]$ is a constant. (The error this causes in calculating K_c is about 4×10^{-7} percent.) For this reason equation 17.1 is simplified by defining a new constant, $K_c \times [H_2O]^2$, called the **ion-product constant of water, K_w**.

Table 17.2

K_w at Various Temperatures

Temperature (°C)	K_w
0	1.5×10^{-15}
10	3.0×10^{-15}
20	6.8×10^{-15}
25	1.0×10^{-14}
30	1.5×10^{-14}
40	3.0×10^{-14}
50	5.5×10^{-14}
60	9.5×10^{-14}

$$\boxed{K_w = [H_3O^+][OH^-]} \qquad (17.2)$$

The value of K_w varies with temperature, as the data in Table 17.2 show. Since at 25 °C $[H_3O^+] = [OH^-] = 1.00 \times 10^{-7}$ mol/liter,

$$K_w = (1.00 \times 10^{-7})(1.00 \times 10^{-7})$$

$$K_w = 1.00 \times 10^{-14} \ (25\ °C) \qquad (17.3)$$

We will use this value in all our calculations unless some other temperature is specified that requires us to use another value.

Criteria for acidic, basic, and neutral solutions

In the context of any discussion about aqueous acids and bases a neutral solution is any in which the molar concentrations of hydronium and hydroxide ions are equal. We then have the following criteria for applying the terms acidic, basic, or neutral to an aqueous system.

Neutral solution	$[H_3O^+] = [OH^-]$
Acidic solution	$[H_3O^+] > [OH^-]$
Basic solution	$[H_3O^+] < [OH^-]$

EXAMPLE 17.7 Calculating $[H_3O^+]$ from $[OH^-]$ or $[OH^-]$ from $[H_3O^+]$

Problem: A sample of blood was analyzed to have $[H_3O^+] = 4.60 \times 10^{-8}\ M$. Find the molar concentration of OH^-, and decide whether the sample was acidic, basic, or neutral.

Solution: From equation 17.2 we know that

$$K_w = 1.00 \times 10^{-14} = [H_3O^+][OH^-]$$

Therefore, for this sample

$$1.00 \times 10^{-14} = (4.60 \times 10^{-8})[OH^-]$$

$$[OH^-] = \frac{1.00 \times 10^{-14}}{4.60 \times 10^{-8}} = 2.17 \times 10^{-7}\ \text{mol/liter}$$

Comparing $[H_3O^+]$ with $[OH^-]$ shows that $[OH^-] > [H_3O^+]$. Therefore, the blood is very slightly basic.

EXERCISE 17.5 If you found that an aqueous solution of baking soda, $NaHCO_3$, had a molar concentration of hydroxide ion of 7.80×10^{-6} mol/liter, what was its molar concentration of hydronium ion? Was the solution acidic, basic, or neutral?

Using H⁺ as a symbol for H₃O⁺

The next topics we will discuss will be simpler if we shift to the shorthand abbreviation of H_3O^+ introduced in Section 6.8, namely H^+. We will use H^+ to represent H_3O^+ in almost all future discussions of acids and bases.

17.3 THE pH CONCEPT

The value of $-\log [H^+]$ is called the pH of an aqueous solution

In Example 17.7 we learned that the molar concentration of H^+ in blood is very low and that blood is slightly basic. The number 4.60×10^{-8}, however, is awkward to use, particularly when we want to compare the relative acidities of two weakly acidic solutions. For example, to compare this value of H^+ concentration with a H^+ concentration of 8.30×10^{-8} mol/liter, our eyes have to search two places. We have to compare the exponents on the 10, then after noting that they are the same, we have to compare 4.60 to 8.30. Yet very low values of H^+ concentration often appear in discussions of weakly acidic or weakly basic fluids. The pH concept was invented to handle this problem.

The **pH** of a solution is the negative power to which 10 must be raised to describe the molar concentration of H^+.

$$[H^+] = 10^{-pH} \qquad (17.4)$$

Since $[H^+] = 1.00 \times 10^{-7}$ mol/liter in pure water (25 °C), the pH of pure water is 7.00. If we take the negative logarithms of both sides of equation 17.4, we get an equivalent expression for pH that is easier to use in pH calculations.

$$\boxed{pH = -\log [H^+]} \qquad (17.5)$$

Since the self-ionization of water gives hydroxide ions as well as hydrogen ions, there is an analogy to the pH concept for hydroxide ions, the **pOH** of a solution. The equation analogous to 17.4 is

$$[OH^-] = 10^{-pOH} \qquad (17.6)$$

The form more useful in calculations is

$$\boxed{pOH = -\log [OH^-]} \qquad (17.7)$$

An important relationship exists between pH, pOH, and K_w. We know that

$$[H^+][OH^-] = K_w = 1.00 \times 10^{-14}$$

Therefore,

$$(10^{-pH})(10^{-pOH}) = 1.00 \times 10^{-14}$$

Because we *add* exponents when we multiply exponential expressions,

$$(-pH) + (-pOH) = (-14.00)$$

Or, changing all the signs

$$pH + pOH = 14.00 \qquad \text{(at 25 °C)} \qquad (17.8)$$

In any aqueous solution of *any* solute at 25 °C, the sum of the pH and the pOH will always be 14.00. Thus, if we find the value of one, we can easily calculate the value of the other. However, pH values are normally used more often than pOH values.

Because pH occurs as a negative exponent (or a negative log), *the lower the pH, the higher the molar concentration of* H^+. Since a pH of 7.00 represents a

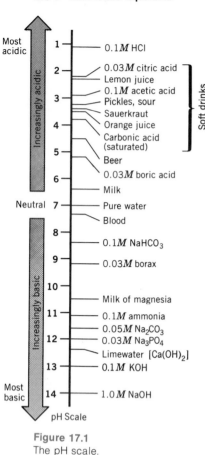

Most
acidic

Increasing acidly

1 — 0.1M HCl

2 — 0.03M citric acid
Lemon juice
0.1M acetic acid
3 — Pickles, sour
Sauerkraut
4 — Orange juice
Carbonic acid
(saturated)
5 — Beer
0.03M boric acid
6 — Milk

Neutral 7 — Pure water
Blood
8 —
0.1M NaHCO$_3$
9 — 0.03M borax
10 —
Milk of magnesia
11 — 0.1M ammonia
0.05M Na$_2$CO$_3$
12 — 0.03M Na$_3$PO$_4$
Limewater [Ca(OH)$_2$]
13 — 0.1M KOH

Most
basic
14 — 1.0M NaOH

pH Scale

Soft drinks

Increasingly basic

Figure 17.1
The pH scale.

neutral solution, any pH less than 7.00 means that a solution is acidic and a pH greater than 7.00 means that a solution is basic.

If pH < 7.00	Acidic solution
If pH > 7.00	Basic solution
If pH = 7.00	Neutral solution

Table 17.3 gives the relationships between pH, [H$^+$], [OH$^-$], and pOH. The pH for a number of common substances is given on a pH scale in Figure 17.1. Two kinds of calculations involving pH are particularly important: finding the pH from a value of [H$^+$] and calculating [H$^+$] from pH.

Table 17.3
pH, [H$^+$], [OH$^-$], and pOH

pH	[H$^+$][a]	[OH$^-$][a]	pOH	
0	1	1×10^{-14}	14	
1	1×10^{-1}	1×10^{-13}	13	
2	1×10^{-2}	1×10^{-12}	12	
3	1×10^{-3}	1×10^{-11}	11	Acidic solutions
4	1×10^{-4}	1×10^{-10}	10	
5	1×10^{-5}	1×10^{-9}	9	
6	1×10^{-6}	1×10^{-8}	8	
7	1×10^{-7}	1×10^{-7}	7	Neutral solution
8	1×10^{-8}	1×10^{-6}	6	
9	1×10^{-9}	1×10^{-5}	5	
10	1×10^{-10}	1×10^{-4}	4	Basic solutions
11	1×10^{-11}	1×10^{-3}	3	
12	1×10^{-12}	1×10^{-2}	2	
13	1×10^{-13}	1×10^{-1}	1	
14	1×10^{-14}	1	0	

Increasing acidity
Increasing basicity

[a] Concentrations are in mole/liter. Note, at pH of zero or pOH of zero, $1 \times 10^{-0} = 1$. The data are for a temperature of 25 °C.

EXAMPLE 17.8 Calculating pH from [H$^+$]

Problem: Because rain washes the air of pollutants, the lakes in many parts of the world are experiencing changes in pH. In a New England state the water in one lake was found to have [H$^+$] = 3.2×10^{-5} mol/liter. Calculate the pH of the lake's water and decide whether it is acidic or basic.

Solution:[2] We know that pH = $-$log [H$^+$] and that the value of [H$^+$] is 3.2×10^{-5} mol/liter. Therefore,

$$pH = -\log(3.2 \times 10^{-5})$$

Since the log of any product is the *sum* of the logs of the numbers being multiplied,

$$pH = -[\log(3.2) + \log(10^{-5})]$$

[2] The solution will use logarithms. The rule for significant figures in logs is that the decimal places in the log of a number equals the significant figures in the number. For example,

$$\log 3.2 \times 10^{-5} = -4.49$$

In the log, -4.49, the first "4" is only used to locate the decimal point in the number 3.2×10^{-5} (which is the same as 0.000032). The "49" in -4.49 tells us that the log has two significant figures to match the two significant figures in the "3.2" of 3.2×10^{-5}.

The log of 3.2 is 0.51; the log of 10^{-5} is -5. Therefore,

$$pH = -[(0.51) + (-5)]$$
$$= -[-4.49]$$
$$= 4.49$$

The pH is less than 7, so the lake's water is acidic.

EXERCISE 17.6 The concentration of H^+ in the blood of a patient with untreated diabetes was 7.24×10^{-8} mol/liter. What was the pH of the blood? Would the blood be described as slightly acidic or slightly basic?

The pH of normal blood is about 7.35. If it were to drop to 7.0 or rise to 7.9 the individual would die. Thus, small changes in the pH of blood produce serious medical emergencies. In Exercise 17.6 the pH of the untreated diabetic person had drifted downward from 7.35 to 7.14 warranting emergency treatment to stop and reverse the trend.

EXAMPLE 17.9 **Calculating [H⁺] from pH**

Problem: Calcareous soil is rich in calcium carbonate. The pH of such soil generally ranges from just over 7 to as high as 8.3. What value of $[H^+]$ corresponds to a pH of 8.3? Is the soil slightly acidic or slightly basic?

Solution: We know that $pH = -\log [H^+]$ and the given value of pH is 8.3. Therefore,

$$8.3 = -\log [H^+]$$

Or, from simply switching signs,

$$-8.3 = \log [H^+]$$

Therefore,

$$[H^+] = \text{antilog} (-8.3)$$

In other words we must find the number whose log is (-8.3). Log tables, however, give positive decimal numbers. So to use a log table we have to write -8.3 as $(-9) + (0.7)$, that is, as its next more negative whole number plus a positive decimal number that adds up to -8.3. Therefore,

$$[H^+] = [\text{antilog} (-9)] \times [\text{antilog} (0.7)]$$
$$= (10^{-9}) \times (5)$$

Written conventionally,

$$[H^+] = 5 \times 10^{-9} \text{ mol/liter}$$

Since this is less than 1.00×10^{-7} mol/liter, the calcareous soil is slightly basic. (If finding antilogs or logs is a problem, turn to Appendix A for a review that also has some drill exercises. If you have a good hand-held calculator with "+/−", "log" and "10^x" buttons, use the instruction manual that came with the calculator to learn how to find logs and antilogs.)

EXAMPLE 17.10 **Calculating [H⁺] from pH**

Problem: A backpacker suffering from severe high-altitude sickness was removed by helicopter to a nearby hospital. His blood was found to have a pH of 7.54. What was the $[H^+]$ of the patient's blood?

Solution: We will use the same basic steps as in Example 17.9. We first use the general equation for pH:

$$pH = -\log[H^+]$$

$$7.54 = -\log[H^+]$$

Or

$$-7.54 = \log[H^+]$$

$$\log[H^+] = (-8) + (0.46)$$

Therefore,

$$[H^+] = [\text{antilog}(-8)] \times [\text{antilog}(0.46)]$$
$$= (10^{-8}) \times 2.9$$

Writing this result in the usual way, we have

$$[H^+] = 2.9 \times 10^{-8} \text{ mol/liter}$$

Thus, the patient's blood has become slightly more basic than normal. It happened because the backpacker overbreathed and expelled excessive amounts of carbon dioxide. We do have to remove the carbon dioxide produced by the body, but only at a suitable rate. Carbon dioxide participates in keeping the pH of the blood from drifting upward by neutralizing OH^-,

$$CO_2(aq) + OH^-(aq) \longrightarrow HCO_3^-(aq)$$

EXERCISE 17.7 Find the values of $[H^+]$ that correspond to each of the following values of pH. (Remember that the answers should be rounded to two significant figures.)
(a) 2.25 (approximate pH of lemon juice)
(b) 3.40 (approximate pH of sauerkraut)
(c) 10.50 (pH of milk of magnesia, a laxative)
(d) 3.50 (pH of orange juice on the average)
(e) 11.10 (pH of dilute household ammonia)

EXERCISE 17.8 Go back over the results of Exercise 17.7 and tell if each solution is acidic or basic.

One of the deceptive aspects of the pH concept comes from its exponential definition. A change of just 1 pH unit means a *tenfold* change in $[H^+]$. When the pH is 0, $[H^+] = 1$ mol/liter, and you need only a liter of solution (about a quart) to have a mole of hydrogen ions. At pH 1, $[H^+] = 10^{-1}$ mol/liter, and you need 10 liters or about a medium sized wastebasket of solution to have 1 mol of hydrogen ions. At pH 5 you'd need the largest railroad tankcar you've ever seen to hold the volume of solution containing 1 mol H^+, because they're now at such a low concentration. If the pH of the water flowing over Niagara Falls were 10, you'd have to watch the falls about an hour to see 1 mol of H^+ go by. And at pH 14, very alkaline, you'd need a quarter of the volume of Lake Erie to find 1 mol of H^+.

Measuring pH
pH is measured either with a pH meter—the most accurate method—or by dipping an indicator paper into the solution and comparing its new color with a color code. See Color Plate 5C. As we learned in Section 4.7 and Special Topic 4.1, dyes whose colors in aqueous solutions depend on the pH are called **acid-base indicators.** Table 17.4 gives several examples. Litmus is available as "litmus paper," either red or blue. Below pH 4.7 litmus is red and above 8.3 it is blue. The transition for the color change occurs over a pH range of 4.7 to 8.3 with the center of that change at 6.5, very nearly a neutral pH. Some commercial test pa-

Table 17.4

Common Acid-Base Indicators

Name	Approximate pH Range	Color Change (lower to higher pH)
Methyl green	0.2–1.8	yellow to blue
Thymol blue	1.2–2.8	yellow to blue
Methyl orange	3.2–4.4	red to yellow
Ethyl red	4.0–5.8	colorless to red
Bromocresol purple	5.2–6.8	yellow to purple
Bromothymol blue	6.0–7.6	yellow to blue
Cresol red	7.0–8.8	yellow to red
Thymol blue	8.0–9.6	yellow to blue
Phenolphthalein	8.2–10.0	colorless to pink
Thymolphthalein	9.4–10.6	colorless to blue
Alizarin yellow R	10.1–12.0	yellow to red
Clayton yellow	12.2–13.2	yellow to amber

pers (e.g., Hydrion®) are impregnated with several dyes. The containers of these test papers carry the color code. When we study how to measure the concentration of some acid or base by chemical analysis, we will see why it is useful to have a variety of indicators that change colors at widely different pH values.

Values of pH can be determined to within ±0.01 pH unit with the better models of pH meters. See Figure 17.2. These instruments also have to be used when solutions are themselves highly colored. Another advantage of a pH meter is that it can be used to follow the change in pH of a solution while some reaction is occurring.

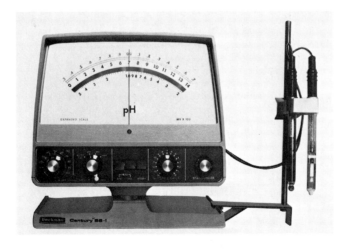

Figure 17.2
pH meter.

17.4 ACID AND BASE IONIZATION CONSTANTS

The relative strengths of acids and bases are described quantitatively by acid and base ionization constants, K_a and K_b

We have learned that the strengths of weak acids and bases vary and that some are more able than others to affect the pH of a solution. To compare the strengths of acids and bases we use modified equilibrium constants, and we will work with weak acids first.

To make the discussion applicable to all weak acids, let's represent any weak Brønsted acid by the symbol HA (where the A denotes "acid," or it may be regarded as a potential anion). It doesn't matter if HA is electrically neutral (e.g., acetic acid, $HC_2H_3O_2$), positively charged (e.g., the ammonium ion, NH_4^+) or negatively charged (e.g., the hydrogen sulfate ion, HSO_4^-). All are Brønsted acids, and we want to consider the ionization of only one proton at a time.

$$HA \quad + \quad H_2O \rightleftharpoons H_3O^+ \quad + \quad A^-$$

weak Brønsted   conjugate
acid   base

For example,

$$HC_2H_3O_2 + H_2O \rightleftharpoons H_3O^+ + C_2H_3O_2^-$$
$$NH_4^+ + H_2O \rightleftharpoons H_3O^+ + NH_3$$
$$HSO_4^- + H_2O \rightleftharpoons H_3O^+ + SO_4^{2-}$$

At equilibrium there will be a particular *equilibrium* concentration of each species, and only these values are used in the equation for the true equilibrium constant:

$$K_c = \frac{[H_3O^+][A^-]}{[HA][H_2O]} \tag{17.9}$$

For weak acids the percentage of ionization is low, and therefore there is a completely negligible change in the molar concentration of water, $[H_2O]$. (We commented on this fact in Section 17.2 in connection with K_w.) We can take the value of $[H_2O]$ to be a constant regardless of the initial concentration of the acid. Therefore, we can modify equation 17.9 as follows.

$$K_c \times [H_2O] = \frac{[H_3O^+][A^-]\cancel{[H_2O]}}{[HA]\cancel{[H_2O]}} = \text{a new constant}$$

The new constant is called the **acid ionization constant, K_a**. Switching from H_3O^+ to H^+, the equation for K_a is

$$\boxed{K_a = \frac{[H^+][A^-]}{[HA]} \quad \text{(for } HA \rightleftharpoons H^+ + A^-\text{)}} \tag{17.10}$$

K_a values for several acids are given in Table 17.5.

EXAMPLE 17.11   **Writing Equations for K_a for Brønsted Acids**

Problem: What is the equation expressing K_a, the acid ionization constant, for acetic acid ($HC_2H_3O_2$), an organic acid present in wine that has gone sour.

Solution: We first have to write the equation for the equilibrium, and since we use H^+ for H_3O^+, the equation will omit H_2O. We just assume that this proton-acceptor is present. The equation for the ionization of acetic acid is

$$HC_2H_3O_2(aq) \rightleftharpoons H^+(aq) + C_2H_3O_2^-(aq)$$

We next write the equation for K_a using the formulas in the equilibrium equation. Remember that the products always appear in the numerator and the reactant(s) in the denominator.

$$K_a = \frac{[H^+][C_2H_3O_2^-]}{[HC_2H_3O_2]}$$

Table 17.5

K_a Values for Acids

Name	Formula	K_a (25 °C)[a]
Perchloric acid	$HClO_4$	large
Hydriodic acid	HI	large
Hydrobromic acid	HBr	large
Sulfuric acid	H_2SO_4	large
Hydrochloric acid	HCl	large
Nitric acid	HNO_3	large
HYDRONIUM ION	H_3O^+	—
Sulfurous acid	H_2SO_3	1.54×10^{-2} (18 °C)
Hydrogen sulfate ion	HSO_4^-	1.20×10^{-2}
Phosphoric acid	H_3PO_4	7.52×10^{-3}
Formic acid	HCO_2H	1.77×10^{-4}
Citric acid[b]	$C_6H_8O_7$	7.10×10^{-4}
Nitrous acid	HNO_2	4.6×10^{-4} (12.5 °C)
Hydrofluoric acid	HF	3.53×10^{-4}
Barbituric acid[c]	$C_4H_4N_2O_3$	9.8×10^{-5}
Ascorbic acid[d]	$C_6H_8O_6$	7.94×10^{-5}
Acetic acid	$HC_2H_3O_2$	1.76×10^{-5}
Carbonic acid	H_2CO_3	4.30×10^{-7}
Hydrogen sulfite ion	HSO_3^-	1.02×10^{-7} (18 °C)
Hydrogen sulfide (aq)	H_2S	9.1×10^{-8} (18 °C)
Dihydrogen phosphate ion	$H_2PO_4^-$	6.23×10^{-8}
Hypochlorous acid	$HOCl$	2.95×10^{-8} (18 °C)
Ammonium ion	NH_4^+	5.65×10^{-10}
Hydrocyanic acid	HCN	4.93×10^{-10}
Bicarbonate ion	HCO_3^-	5.61×10^{-11}
Hydrogen sulfide ion	HS^-	1.1×10^{-12} (18 °C)
Monohydrogen phosphate ion	HPO_4^{2-}	2.2×10^{-13}

(left margin, vertical: INCREASING ACID STRENGTH, with upward arrow)

[a] Data are from *Handbook of Chemistry and Physics*, 57th edition. CRC Press, Cleveland, Ohio. K_a values are for ionization of one proton from the named acid in 0.1 to 0.01 normal solutions at 25 °C except where noted. The acids are arranged in their order of decreasing strengths.

[b] Citric acid is triprotic; K_a is for the first proton.

[c] Barbituric acid is monoprotic.

[d] Ascorbic acid (vitamin C) is diprotic; K_a is for the first proton.

EXERCISE 17.9   Write the K_a equation for nitrous acid, HNO_2. (Nitrous acid may form in trace concentrations in the stomach by the interaction of stomach acid [HCl(*aq*)] with nitrite preservatives in processed meats. This has caused concern among many scientists because nitrous acid is suspected of being a carcinogen.)

EXERCISE 17.10   Write the K_a equation for the monohydrogen phosphate ion, HPO_4^{2-}. (Veterinarians sometimes use the sodium salt of this compound as a laxative for calves and foals.)

There are basically two kinds of calculations involving K_a that we have to be concerned about. One is calculating the K_a value of an acid from experimentally measured data for equilibrium concentrations. A pH, for example, after a simple calculation, gives an *equilibrium* concentration of $[H^+]$. The other kind of calculation is using a K_a value to find the pH in a solution of a weak acid.

EXAMPLE 17.12 Calculating K_a Values From pH Data

$$\begin{matrix} O \\ \| \\ H-C-O-H \end{matrix} = HCO_2H$$

formic acid

$$\begin{matrix} O \\ \| \\ H-C-O^- \end{matrix} = HCO_2^-$$

formate ion

$[H^+]$ means $[H^+]_{at\ equilibrium}$
$[HCO_2^-]$ means $[HCO_2^-]_{at\ equilibrium}$
$[HCO_2H]$ means
$[HCO_2H]_{at\ equilibrium}$

"Formica" is Latin for *ant*. Formic acid is partly responsible for the pain of an ant sting.

Problem: Formic acid, HCO_2H, is a monoprotic acid. In a 0.10 M solution of formic acid, the pH was found to be 2.38 at 25 °C. Calculate K_a for formic acid at this temperature.

Solution: In any problem like this begin by writing the equation for the equilibrium and then set up the expression for K_a.

$$HCO_2H \rightleftharpoons H^+ + HCO_2^- \qquad K_a = \frac{[H^+][HCO_2^-]}{[HCO_2H]}$$

formic acid   formate ion

To calculate K_a we need to know $[H^+]$, but we are given pH instead of $[H^+]$. So we have to follow the steps of Example 17.10 to calculate $[H^+]$ from pH. Remember that $[H^+]$ in the expression for K_a is the molar concentration of H^+ at *equilibrium*, but that is exactly what the pH value indirectly gives us.

$$2.38 = -\log [H^+]$$

Or

$$-2.38 = \log [H^+] = -3 + 0.62$$

Therefore,

$$[H^+] = [\text{antilog} (-3)] \times [\text{antilog} (0.62)]$$
$$= 10^{-3} \times 4.2$$

Written conventionally,

$$[H^+] = 4.2 \times 10^{-3} \text{ mol/liter}$$
$$= 0.0042 \text{ mol/liter (the form most useful in what follows)}.$$

This value of $[H^+]$ must also be the value of $[HCO_2^-]$, since the ionization produces them in a 1 to 1 ratio. Now we can set up a concentration table as we did in Chapter 16 in which we write the data beneath the formulas.

	$HCO_2H(aq)$	$\rightleftharpoons$	$H^+(aq)$	+	$HCO_2^-(aq)$
Initial concentrations	0.100		0		0 (*Note 1*)
Changes in concentration caused by the ionization	−0.0042		+0.0042		+0.0042 (*Note 2*)
Final concentrations at equilibrium	(0.100 − 0.0042) = 0.096)		0.0042		0.0042

Note 1. The *initial* concentration of the acid is what the label tells us, 0.100 M; it ignores whatever ionization takes place. However, by ignoring this we have to set the *initial* values of $[H^+]$ and $[HCO_2]$ equal to zero. (Actually, there is always a little bit of H^+ concentration from the ionization of water, but it is so small in relation to the H^+ concentration given by the acid that we ignore it. We will follow that policy for all the acids we will study.)

Note 2. The values in the table for the *changes* in concentration come from the calculated value of H^+ concentration. For every molecule of H^+ that forms by ionization there is one less molecule of the initial acid. The minus sign in −0.0042 in the column for HCO_2H is used because the initial concentration of the acid is *reduced* by the ionization. The last row of data give us the equilibrium concentrations that we now use to calculate K_a.

$$K_a = \frac{(4.2 \times 10^{-3})(4.2 \times 10^{-3})}{0.096}$$

$$= 1.8 \times 10^{-4} \text{ (our answer)}$$

We cannot emphasize too strongly that the equation for K_a works only for the values of the concentrations *at equilibrium,* not the initial values before the equilibrium becomes established. We get the *initial* value of the acid by reading the label on the bottle. We get the *equilibrium* values through the use of the pH.

EXERCISE 17.11 When butter becomes rancid, its foul odor is caused, in part, by butyric acid, an organic acid. A 0.0100 M solution of butyric acid has a pH of 3.40 at 20 °C. Calculate the K_a of butyric acid at this temperature. (Use any symbol you want for butyric acid and its conjugate base, for example, HBu and Bu⁻.)

$$CH_3CH_2CH_2\overset{\overset{\displaystyle O}{\|}}{C}-O-H = HC_4H_7O_2$$
butyric acid

$$CH_3CH_2CH_2\overset{\overset{\displaystyle O}{\|}}{C}-O^- = C_4H_7O_2^-$$
butyrate ion

EXERCISE 17.12 At 60 °C the pH of 0.0100 M butyric acid is 2.98. Calculate the K_a of butyric acid at this temperature.

EXAMPLE 17.13 **Calculating the Values of [H⁺] and pH for a Solution of a Weak Acid from the K_a Value**

Problem: The concentration of a sample of vinegar was analyzed to be 0.750 M acetic acid, $HC_2H_3O_2$. Calculate the values of [H⁺] and pH of this sample.

Solution: We can use the same general approach as given in Example 17.12, beginning with writing the chemical equilibrium and then constructing a table of concentrations.

	$HC_2H_3O_2(aq) \rightleftharpoons$	$H^+(aq) +$	$C_2H_3O_2^-(aq)$
Initial concentrations	0.750	0	0
Changes in concentrations caused by the ionization	$-x$	$+x$	$+x$ (See Note 1 below.)
Final concentrations at equilibrium	$(0.750 - x)$ $= 0.750$	x	x (See Note 2 below.)

Note 1. We let $(+x)$ stand for the increase in the concentration of both H⁺ and $C_2H_3O_2^-$ and therefore the initial concentration of $HC_2H_3O_2$ is reduced by the same amount.

Note 2. A very important simplification occurs here. The acid is weak as shown by the magnitude of its K_a. Because it is weak, we know that the change in its initial concentration caused by the ionization, $(-x)$, is going to be very small in relation to the much larger initial value (0.750). This means that substracting (x) from 0.750 to find the true equilibrium concentration of the acid will change 0.750 by so little that after we have rounded to the correct number of significant figures we will obtain 0.750 or be within a few percent of 0.750. Therefore, we ignore (x) in $(0.750 - x)$ and

Special Topic 17.1 Calculating [H⁺] from K_a Without Making Assumptions

In Example 17.13 the equilibrium concentrations before applying the simplification of *Note 2* was:

$$[HC_2H_3O_2] = (0.750 - x)$$

Instead of simplifying this to set $(0.750 - x) = 0.750$, let us use $(0.750 - x)$ in the expression for K_a and solve for x by the quadratic equation. This equation tells us that if $ax^2 + bx + c = 0$, then:

$$x = \frac{-b \pm [b^2 - 4ac]^{1/2}}{2a}$$

$$K_a = \frac{[H^+][C_2H_3O_2^-]}{[HC_2H_3O_2]} = \frac{(x)(x)}{(0.750 - x)} = 1.76 \times 10^{-5}$$

$$x^2 = (0.750 - x)(1.76 \times 10^{-5})$$
$$= 1.32 \times 10^{-5} - 1.76 \times 10^{-5}(x)$$

$$x^2 + (1.76 \times 10^{-5})x - (1.32 \times 10^{-5}) = 0$$

$$x = \frac{\{-(1.76 \times 10^{-5}) \pm [(1.76 \times 10^{-5})^2 - 4(1)(-1.32 \times 10^{-5})]^{1/2}\}}{2(1)}$$

$$= \frac{-1.76 \times 10^{-5} \pm [3.10 \times 10^{-10} + 5.28 \times 10^{-5}]^{1/2}}{2}$$

$$= \frac{-1.76 \times 10^{-5} \pm [5.28 \times 10^{-5}]^{1/2}}{2}$$ (rounded from 5.280031×10^{-5})

$$= \frac{-1.76 \times 10^{-5} \pm 7.27 \times 10^{-3}}{2}$$ (We ignore the negative root because the value of [H⁺] cannot be less than zero.)

$$= \frac{7.25 \times 10^{-3}}{2}$$ (rounded from 7.2524×10^{-3})

$$[H^+] = 3.63 \times 10^{-3} \text{ mol/liter}$$ (rounded from 3.625×10^{-3})

The simplification used in Example 17.13 gave the same answer. Thus the simplifying assumption is not only valid, but it saves an awful lot of work!

---

assume that $(0.750 - x) \approx 0.750$. Special Topic 17.1 gives the calculation that proves that we can make this simplification.

Now that we have values for the equilibrium concentrations, we can use the K_a expression to find [H⁺].

$$K_a = \frac{[H^+][C_2H_3O_2^-]}{[HC_2H_3O_2]} = 1.76 \times 10^{-5}$$

Or

$$\frac{(x)(x)}{0.750} = 1.76 \times 10^{-5}$$

$$x^2 = (0.750)(1.76 \times 10^{-5})$$
$$= 1.32 \times 10^{-5}$$

$$x = \sqrt{1.32 \times 10^{-5}}$$

Thus,

$$[H^+] = x = 3.63 \times 10^{-3} = [C_2H_3O_2^-]$$

Having found that at equilibrium $[H^+] = 3.63 \times 10^{-3}$ mol/liter, we next have to calculate the pH.

$$pH = -\log [H^+]$$
$$= -\log (3.63 \times 10^{-3})$$
$$= -[\log 3.63 + \log 10^{-3}]$$
$$= -[0.560 + (-3)] = -[-2.44]$$

pH = 2.44 (our second answer)[3]

---

[3] Although the rule in footnote 2 says that we should have three places after the decimal point in this answer (because we known [H⁺] to three significant figures), chemists almost always round values of pH to two places after the decimal point anyway. There are just enough uncertainties in making actual measurements of pH that three places following the decimal point are seldom justified. Rounding to two places after the decimal point will also be carried out when we later find calculated values of quantities analogous to pH or pOH—quantities such as pK_a, pK_b, and pK_w.

When a computer program was used to do the calculations for the problem in Example 17.13 without making any simplifying assumptions and including the very small contribution to the value of $[H^+]$ from the self-ionization of water, the results were $[H^+] = 3.62437 \times 10^{-3}$ mol/liter and pH = 2.44033 (rounding neither result). These values would round to 3.62×10^{-3} mol/liter and 2.44, respectively, which are very close to or exactly the same as the answers we obtained in our solution. In other words, the simplification described in *Note 2* of Example 17.13 introduced essentially no error. These simplifications are so important in K_a calculations that we should review them. In K_a calculations involving weak acids in dilute solutions

1 We can ignore the contribution to $[H^+]$ from the self-ionization of water.

2 $[HA]_{equilibrium} = [HA]_{initial} - [H^+]_{equilibrium}$
$= [HA]_{initial}$

too small and dropped

The problems in this book have been selected in such a way that these assumptions work, but do not think that they are universally applicable. In extremely dilute solutions of weak acids or with acids with very low values of K_a we cannot ignore the contribution to $[H^+]$ from the self-ionization of water. And the second assumption that $[HA]_{equilibrium} = [HA]_{initial}$ works less and less well as the value of K_a moves up to 10^{-3} or higher. Most chemists have a general rule for evaluating how well the assumptions work; if you find that the calculated value of (x) is smaller than $\frac{1}{10}$ of the value of $[HA]_{initial}$ then the assumptions worked well. Otherwise, if (x) is greater than $\frac{1}{10}$ the value of $[HA]_{initial}$ then the quadratic solution described in Special Topic 17.1 must be used.

EXERCISE 17.13 Nicotinic acid, $HC_6H_4NO_2$, is a B vitamin. Its K_a is 1.4×10^{-5}. What is $[H^+]$ and the pH of a 0.010 M solution?

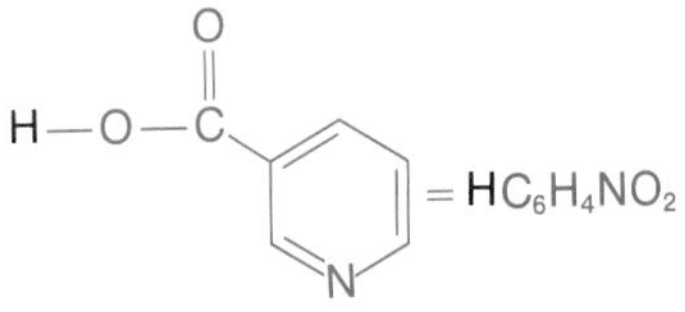

nicotinic acid

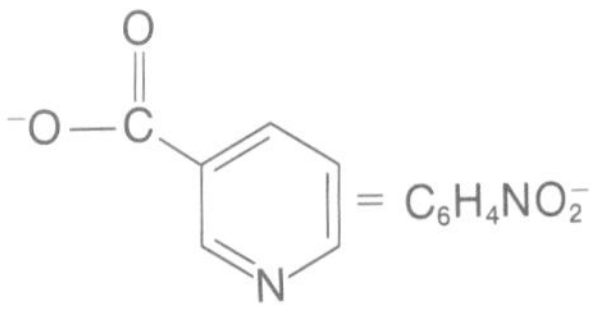

nicotinate ion

Now that we have studied acid ionization constants, we have a better basis than percentage ionization for classifying acids as weak, moderate, or strong.

$K_a < 10^{-3}$	Weak acid
$K_a = 1$ to 10^{-3}	Moderate acid
$K_a > 1$	Strong acid

Base ionization constants, K_b

Strong bases, such as sodium hydroxide, are ionic. Like other ionic compounds, their dissociation in water is essentially complete. Weak Brønsted bases, on the other hand, are usually molecules or ions that react with water to remove a proton from water and generate a hydroxide ion. For example,

$$NH_3(aq) + H_2O(\ell) \rightleftharpoons NH_4^+(aq) + OH^-(aq)$$
$$CO_3^{2-}(aq) + H_2O(\ell) \rightleftharpoons HCO_3^-(aq) + OH^-(aq)$$

Weak bases vary considerably in their abilities to accept protons from water molecules, and we use a modified form of the equilibrium constant to compare these base strengths. We can represent any base by the symbol "B", and write the general equilibrium involving this base and water as follows:

The base, B, may be uncharged as in NH_3, or it may be charged as in OH^- or HCO_3^-.

$$\underset{\substack{\text{Brønsted} \\ \text{base}}}{B(aq)} + H_2O(\ell) \rightleftharpoons \underset{\substack{\text{conjugate} \\ \text{acid}}}{BH^+(aq)} + OH^-(aq)$$

The **base ionization constant** for this reaction, K_b, is defined by the following equation.

$$K_b = \frac{[BH^+][OH^-]}{[B]} \tag{17.11}$$

EXAMPLE 17.14 **Writing Expressions for K_b for Brønsted Bases**

Problem: Hydrazine, N_2H_4, is a weak base like ammonia. (It is also a rocket fuel.) Write the equation for K_b for this base.

Solution: We first have to write the equation for the acid-base equilibrium. Since in all these problems we assume that water is the solvent, in K_b problems it is therefore also the Brønsted acid. So we set down the given base and water as the reactants, and we write the conjugate acid of the base plus the hydroxide ion as the products.

$$N_2H_4(aq) + H_2O(\ell) \rightleftharpoons N_2H_5^+(aq) + OH^-(aq)$$

Now we can write the expression for K_b, omitting $[H_2O]$ and remembering that products always are in the numerator.

$$K_b = \frac{[N_2H_5^+][OH^-]}{[N_2H_4]}$$

EXERCISE 17.14

Aniline is a raw material for aniline dyes as well as for several pharmaceuticals. It is an aromatic amine and a very weak base.

Write the expressions for K_b for each of the following Brønsted bases.
(a) CN^- cyanide ion
(b) $C_2H_3O_2^-$ acetate ion
(c) $C_6H_5NH_2$ aniline

K_b values for several bases are given in Table 17.6. *The smaller the K_b the weaker is the base,* and a weak base is any base with a K_b value less than 10^{-3}.

There are two kinds of calculations that involve K_b. One kind is the calculation of a K_b value from pOH (or from pH, since pOH = 14.00 − pH). The other kind of calculation is to find the equilibrium concentrations of $[H^+]$ or $[OH^-]$ from values of K_b and $[B]_{initial}$.

Table 17.6
K_b Values for Bases

	Name	Formula	K_b (at 25 °C)[a]
	HYDROXIDE ION	OH^-	—
	Phosphate ion	PO_4^{3-}	4.5×10^{-2}
	Sulfide ion	S^{2-}	3.2×10^{-3} (18 °C)
	Carbonate ion	CO_3^{2-}	1.78×10^{-4}
	Cyanide ion	CN^-	2.03×10^{-5}
	Ammonia	NH_3	1.77×10^{-5}
	Hypochlorite ion	ClO^-	1.90×10^{-7} (18 °C)
	Monohydrogen phosphate ion	HPO_4^{2-}	1.61×10^{-7}
	Hydrogen sulfide ion	HS^-	6.2×10^{-8} (18 °C)
	Sulfite ion	SO_3^{2-}	5.49×10^{-8} (18 °C)
	Bicarbonate ion	HCO_3^-	2.33×10^{-8}
	Acetate ion	$C_2H_3O_2^-$	5.68×10^{-10}
	Ascorbate ion	$C_6H_7O_6^-$	1.26×10^{-10}
	Barbiturate ion	$C_4H_3N_2O_3^-$	1.0×10^{-10}
	Fluoride ion	F^-	2.83×10^{-11}
	Nitrite ion	NO_2^-	7.7×10^{-12} (12.5 °C)
	Citrate ion	$C_6H_7O_7^-$	1.41×10^{-11}
	Formate ion	HCO_2^-	5.65×10^{-11}
	Dihydrogen phosphate ion	$H_2PO_4^-$	1.33×10^{-12}
	Sulfate ion	SO_4^{2-}	8.33×10^{-13}
	Hydrogen sulfite ion	HSO_3^-	3.64×10^{-13} (18 °C)
	WATER	H_2O	—
	Nitrate ion	NO_3^-	very small
	Chloride ion	Cl^-	very small
	Hydrogen sulfate ion	HSO_4^-	very small
	Bromide ion	Br^-	very small
	Iodide ion	I^-	very small
	Perchlorate ion	ClO_4^-	very small

INCREASING BASE STRENGTH ↑

[a] K_b values were calculated from K_a values in Table 17.5 (except for ammonia). For these cal-culations, the values of K_w that were used were 3.54×10^{-15} at 12.5 °C, 5.60×10^{-15} at 18 °C, and 1.00×10^{-14} at 25 °C.

EXAMPLE 17.15   **Calculating K_b from pH**

Problem: Methylamine, CH_3NH_2, is one of the many substances that contributes to the pungent odor of herring brine. In 0.100 M CH_3NH_2 the pH is 11.80. What is the K_b of methylamine?

Solution: Doing this calculation is very similar to finding K_a from pH (Example 17.12). The first step is to write the equilibrium equation and the expression for K_b

$$CH_3NH_2(aq) + H_2O(\ell) \rightleftharpoons CH_3NH_3^+(aq) + OH^-(aq)$$

$$K_b = \frac{[CH_3NH_3^+][OH^-]}{[CH_3NH_2]}$$

If we can find the equilibrium value of $[OH^-]$, we will also get the equilib-rium value of $[CH_3NH_3^+]$ since both their coefficients are 1 in the equation. The easiest way to find the equilibrium value of $[OH^-]$ is to use the pH to find the pOH and then to calculate $[OH^-]$ from pOH by equation 17.7,

$$pOH = -\log [OH^-].$$

$$pOH = 14.00 - 11.80 = 2.20$$

$$2.20 = -\log [OH^-]$$

$$[OH^-] = \text{antilog} (-2.20)$$
$$= \text{antilog} (-3 + 0.80)$$
$$= (\text{antilog } 0.80) \times 10^{-3}$$

$$[OH^-] = 6.3 \times 10^{-3} \text{ mol/liter or } 0.0063 \text{ mol/liter}$$

Now we assemble the data beneath the formulas in the equilibrium equation

	$CH_3NH_2(aq)$ +	$H_2O(\ell) \rightleftharpoons$	$CH_3NH_3^+(aq)$ +	$OH^-(aq)$
Initial concentrations	0.100		0	0
Changes in concentrations caused by the ionization	−0.0063		+0.0063	+0.0063
Final concentration at equilibrium	(0.100 − 0.0063) = 0.094		0.0063	0.0063

Now we can calculate K_b:

$$K_b = \frac{(0.0063)(0.0063)}{(0.094)}$$
$$= 4.2 \times 10^{-4} \text{ (our answer)}$$

EXERCISE 17.15 Few substances are more effective for relieving intense pain than morphine. Morphine is an alkaloid (meaning an "alkali-like compound"), and the alkaloids in general are weakly basic amines. In 0.010 M morphine the pH is 10.10. Calculate the K_b of morphine.

EXAMPLE 17.16 **Calculating [OH⁻], pOH, and pH from K_b**

Problem: Calculate the values of [OH⁻], pOH, and pH for a 0.20 M solution of ammonia.

Solution: The best approach for this kind of problem is just like that used for finding [H⁺] from K_a in Example 17.13. And we can make simplifying assumptions analogous to the two we used in that example.

Assumption 1. We can ignore the contribution to [OH⁻] from the self-ionization of water since it contributes negligibly to the overall concentration of OH⁻.

Assumption 2. $[B]_{equilibrium} = [B]_{initial}$ for weak bases in moderately concentrated solutions.

	$NH_3(aq)$ +	$H_2O(\ell) \rightleftharpoons$	$NH_4^+(aq)$ +	$OH^-(aq)$
Initial concentrations	0.20		0	0
Changes in concentrations caused by the ionization	−x		+x	+x
Final concentrations at equilibrium	(0.20 − x) = 0.20		x	x

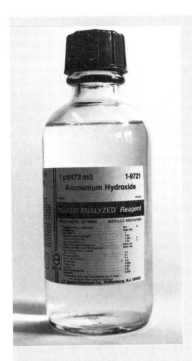

Concentrated aqueous ammonia.

Since (x) is small in relation to (0.20), the value of $(0.20 - x)$ will equal 0.20 when correctly rounded (*Assumption 2*). Now we can calculate the value of (x) from K_b.

$$K_b = \frac{[NH_4^+][OH^-]}{[NH_3]}$$

$$= \frac{(x)(x)}{0.20} = 1.8 \times 10^{-5} \text{ (rounded from Table 17.6)}$$

$$x^2 = (0.20)(1.8 \times 10^{-5}) = 3.6 \times 10^{-6}$$

$$x = \sqrt{3.6 \times 10^{-6}}$$
$$= 1.9 \times 10^{-3} \text{ or } 0.0019$$

Therefore,

$$[OH^-] = 1.9 \times 10^{-3} \text{ mol/liter}$$

Notice that the value of (x) turned out to be much smaller than 0.20, which makes assumption 2 valid. Now we have to find pOH. It is easier to calculate pH *after* we calculate pOH because pH + pOH = 14.00.

$$pOH = -\log [OH^-]$$
$$= -\log [1.9 \times 10^{-3}]$$
$$= -[\log (1.9) + \log (10^{-3})]$$
$$= -[0.28 + (-3)]$$
$$= -[-2.72]$$

$$pOH = 2.72$$

Therefore,

$$pH + 2.72 = 14.00$$

$$pH = 14.00 - 2.72$$

$$pH = 11.28$$

EXERCISE 17.16 In Example 17.16 the pH of a 0.20 *M* solution of ammonia was calculated. Suppose we dilute that solution to 0.020 *M*. Now what are the values of pH, pOH, and [OH$^-$]?

The low value of K_b for ammonia means that the principal basic species in aqueous ammonia is molecular ammonia itself, not the hydroxide ion. You will sometimes see aqueous ammonia called "ammonium hydroxide" both in references and on some bottle labels, but no such compound exists in a pure state. "Ammonium hydroxide" should be understood to mean "aqueous ammonia."

17.5 pK_a AND pK_b

The value of $-\log K_a$ is called pK_a and the value of $-\log K_b$ is called pK_b

The values of K_a and K_b for weak acids and bases, like the values of [H$^+$] for weakly acidic solution, are in negative exponential forms. In the discussion of the pH concept we commented on the difficulty of comparing numbers in such forms. In analogy to the pH concept we have similar expressions for K_a and K_b, their negative logarithms.

$$pK_a = -\log K_a \qquad (17.12)$$

$$pK_b = -\log K_b \qquad (17.13)$$

Because pK_a and pK_b are defined as *negative* logarithms, *the higher the* pK_a, *the weaker the acid,* and *the higher the* pK_b *the weaker the base.*

EXAMPLE 17.17   Finding pK_a from K_a

Problem: The K_a for acetic acid is 1.76×10^{-5} at 25 °C. What is its pK_a at this temperature?

Solution:

$$pK_a = -[\log K_a]$$
$$= -[\log (1.76 \times 10^{-5})]$$
$$= -[\log (1.76) + \log (10^{-5})]$$
$$= -[(0.246) + (-5)] = -[-4.754]$$

$pK_a = 4.75$; (rounded see footnote 3, page 562).

EXERCISE 17.17   What is the pK_a for nitrous acid, HNO_2, if $K_a = 4.6 \times 10^{-4}$.

EXAMPLE 17.18   Using pK_a Values to Compare Strengths of Acids

Problem: Hypoiodous acid, HIO, has a pK_a of 10.6; the pK_a of hypobromous acid, HBrO, is 8.69. Which of these is the weaker acid?

Solution: Remember the reciprocal relation: the higher pK_a, the weaker the acid. Therefore, HIO is the weaker acid.

EXERCISE 17.18   The pK_a of hydrocyanic acid (HCN) is 9.31 and that of acetic acid is 4.75. Which is the stronger acid?

EXAMPLE 17.19   Finding pK_b from K_b

Problem: The K_b of ammonia is 1.77×10^{-5} (at 25 °C). What is its pK_b?

Solution:

$$pK_b = -\log [K_b]$$
$$= -[\log (1.77 \times 10^{-5})]$$
$$= -[\log (1.77) + \log (10^{-5})]$$
$$= -[(0.248) + (-5)] = -[-4.75]$$

$pK_b = 4.75$ (rounded as per footnote 3)

EXERCISE 17.19   The K_b for ethylamine, $CH_3CH_2NH_2$, is 6.41×10^{-4}. Calculate its pK_b.

An important and simple relationship exists between K_a and K_b for conjugate acid-base pairs:

$$\boxed{K_a K_b = K_w \quad \text{(for a conjugate acid-base pair)}} \tag{17.14}$$

To prove that equation 17.14 is true, all we have to do is to substitute the expressions for K_a and K_b into equation 17.14 and show that, after identical terms are canceled, what is left is simply $[H^+][OH^-] = K_w$ (equation 17.2 except that we have substituted H^+ for H_3O^+).

Using HA to represent the acid and A^- its conjugate base, the chemical equilibria involved and the corresponding expressions for K_a and K_b are as follows.

The equilibrium for the acid is

$$HA \rightleftharpoons H^+ + A^- \qquad K_a = \frac{[H^+][A^-]}{[HA]}$$

The equilibrium for the conjugate base is

$$A^- + H_2O \rightleftharpoons HA + OH^- \qquad K_b = \frac{[HA][OH^-]}{[A^-]}$$

Now let's multiply the expressions for K_a and K_b and cancel whatever terms we can:

$$K_a \times K_b = \frac{[H^+][\cancel{A^-}]}{[\cancel{HA}]} \times \frac{[\cancel{HA}][OH^-]}{[\cancel{A^-}]}$$
$$= [H^+][OH^-]$$
$$= K_w$$

What makes equation 17.14 particularly useful is that if we know K_a for an acid, we can calculate K_b for its conjugate base. Or, if we have K_b for a base, we can find K_a for its conjugate acid.

$$K_a = \frac{K_w}{K_b} \qquad K_b = \frac{K_w}{K_a}$$

For a further simplification of the calculations involving equation 17.14, let's take the negative log of K_w and call it **pK_w**.

$$\boxed{pK_w = -\log K_w} \tag{17.15}$$

Now we can reformulate the relationship in equation 17.14 as follows:

$$\log (K_a K_b) = \log K_w$$

Or

$$\log K_a + \log K_b = \log K_w$$

If we change all the signs,

$$(-\log K_a) + (-\log K_b) = (-\log K_w)$$

But this is the same as

$$\boxed{pK_a + pK_b = pK_w} \tag{17.16}$$

At 25 °C, $K_w = 1.00 \times 10^{-14}$. Therefore, at 25 °C $pK_w = 14.00$ (rounding to the second decimal place as discussed at the end of footnote 3, page 562). For this temperature,

At 20 °C $pK_w = 14.17$.
At 30 °C $pK_w = 13.83$.

$$\boxed{pK_a + pK_b = 14.00 \text{ (at 25 °C)}} \tag{17.17}$$

EXAMPLE 17.20   **Finding pK_a from pK_b or pK_b from pK_a for Conjugate Acid/Base Pairs**

Problem: The pK_a of acetic acid at 25 °C is 4.76. What is the pK_b of its conjugate base, the acetate ion, $C_2H_3O_2^-$? Write the equilibrium equation in which the conjugate base acts as a base in water, and write the expression for its K_b.

Solution:

$$pK_a + pK_b = 14.00 \text{ (using } pK_w = 14.00 \text{ unless we specify otherwise)}$$

$$4.76 + pK_b = 14.00$$

$$pK_b = 14.00 - 4.76$$

$$pK_b = 9.24 \text{ for the acetate ion}$$

The equilibrium for which this value applies is one in which the acetate ion, the base, interacts with water as the acid:

$$C_2H_3O_2^- + H_2O \rightleftharpoons HC_2H_3O_2 + OH^-$$

$$K_b = \frac{[HC_2H_3O_2][OH^-]}{[C_2H_3O_2^-]}$$

EXERCISE 17.20   Hydrocyanic acid (HCN) is a very weak acid with a pK_a of 9.31 at 25 °C. Calculate the pK_b of its conjugate base, CN^-. Write the equilibrium equation for the interaction of this base with water and then write the expression for K_b.

If you worked Exercise 17.20 you found that the conjugate base (CN^-) of a weak Brønsted acid (HCN) is a fairly strong base. The pK_b of CN^- is 4.69, which makes the cyanide ion a stronger base than ammonia (pK_b 4.75). In other words, if a Brønsted acid, HA, is relatively weak, its conjugate base, A^-, will be relatively strong. Of course, this reciprocal relationship in the comparative strengths of an acid and its conjugate base must hold true for all conjugate acid-base pairs since, as we have seen,

$$K_a = \frac{K_w}{K_b} \quad \text{and} \quad K_b = \frac{K_w}{K_a}$$

where K_w is a constant at any specified temperature. *The weaker the acid, the stronger is its conjugate base.* Exercise 17.20 illustrated this. We can also say *the stronger the acid, the weaker is its conjugate base.* For example, hydrochloric acid, HCl(aq), is a very strong acid. Therefore, the chloride ion, Cl^-, must be a very weak base. In fact, the conjugate bases of all the strong acids (Table 17.1) are very weak, too weak even to think of them as Brønsted bases.

If our comparison begins by referring to a particular base, then we can say that *the stronger the base, the weaker is its conjugate acid.* For example, we know that the hydroxide ion is a very strong base; and we know that its conjugate acid, water, is a very weak acid. Finally, *the weaker the base, the stronger is its conjugate acid.* For example, ammonia is a weak base; therefore, we know that its conjugate acid, the ammonium ion, must be a good Brønsted acid. Its pK_a is 9.25, making the ammonium ion a stronger acid than hydrogen cyanide (pK_a 9.31). We will see how useful these reciprocal relationships can be in the next section.

17.6   HYDROLYSIS OF SALTS

Many salts react with water to give aqueous solutions that are acidic or basic, depending on the salt

In many laboratory situations an aqueous solution of some salt is prepared only to have it to be either acidic or basic. Unless we are aware of this possibility, we might unknowingly prepare a salt solution that, being acidic or basic, is more corrosive to metals than expected, or could harm living things, or might make food or drink or aqueous medications unfit for human use, or could hold a catalyst (e.g., H^+) that would affect a reaction rate. Many salts react with water—the reaction is called the **hydrolysis of salts**—to upset the 1 to 1 mol ratio of $[H^+]$ to $[OH^-]$ initially present.

We will first develop a simple strategy for making a rough prediction about the tendency (if any) of a salt to hydrolyze. If the application of that strategy tells us that the solution will indeed be either acidic or basic, then we have to be able to calculate the pH (or pOH) of the salt solution. We will learn how to make such a calculation in this section, too.

To find out if a given salt will hydrolyze and what kind of a solution, acidic or basic, it produces if it does, we have to study the abilities of the cation and the anion from the salt to act as a Brønsted acid or a Brønsted base *toward water*. Suppose that the cation can donate a proton to water and make some hydronium ion, but the anion cannot to the same degree accept a proton from water to make some hydroxide ion. There will then be more H^+ than OH^- in the solution, and the salt has caused the acidic solution. Ammonium chloride, NH_4Cl, is a common example of such a salt. Its aqueous solutions are weakly acidic—by how much we will calculate later. At the end of the last section we learned that the cation in NH_4Cl, namely NH_4^+, is a Brønsted acid toward water:

$$NH_4^+ + H_2O \rightleftharpoons NH_3 + H_3O^+$$

The anion in NH_4Cl, namely Cl^-, is the conjugate base of a very strong acid, $HCl(aq)$, and therefore Cl^- is not a Brønsted base. Thus, of the two ions in NH_4Cl, one (NH_4^+) generates some hydronium ion but the other (Cl^-) does nothing to counteract this by generating hydroxide ion. Therefore, an aqueous ammonium chloride solution tests acidic.

Most salts involve metal cations, not ammonium ions. For such salts we have to take a slightly different approach in evaluating the salt's ability to hydrolyze. A cation such as Na^+ can be neither a donor of a proton (it has no H^+) nor an acceptor (two like-charged ions repel each other). So we have to ask if the metal ion will remove *and bind* OH^- from an H_2O molecule, leave H^+ free in solution, and thereby make the solution acidic. For example, can the sodium ion do the following to a water molecule and change to un-ionized "molecules" of NaOH?

$$Na^+ + H_2O \xrightarrow{?} NaOH + H^+$$

We know, however, that NaOH is a *strong* base and is fully ionized in water. If any NaOH did form by this reaction, the OH^- ion in it would be free and would be neutralized by H^+. In other words, Na^+ cannot upset the 1 to 1 mol ratio of OH^- to H^+ in pure water. We can extend these arguments to include the metal ions in any salt of any of the strong bases in Table 17.1. These are metal ions from Group IA and IIA elements, and they do not become involved in hydrolysis. (Beryllium in Group IIA is an exception.) Such metal ions in salts serve as "spectator" ions. They are just counter-ions for the anions of the salt. If the salt of any strong aqueous base hydrolyzes, the change in pH of the solution will be caused by the anion, not the metal cation. In Example 17.21 we will learn how to identify an anion as a Brønsted base even without Table 17.6. All we need to remember is the list of strong acids, Table 17.1, as we will see.

Common metal ions that do not participate in the hydrolysis of a salt are Li^+, Na^+, and K^+ Mg^{2+}, Ca^{2+}, Sr^{2+}, and Ba^{2+}

EXAMPLE 17.21   **Predicting If a Salt Hydrolyzes and If the Solution is Acidic, Basic, or Neutral**

Problem: Trisodium phosphate, Na_3PO_4, is a common cleaning agent for walls and floors. Does it hydrolyze in water and, if so, is the solution acidic or basic?

Solution: The cation, Na^+, is from Group IA and does not participate. The anion, PO_4^{3-}, is a Brønsted base. Without even using Table 17.6 we can make that statement because the conjugate acid, HPO_4^{2-}, is not one of the strong acids (Table 17.1). Therefore, HPO_4^{2-} must be a weak acid. (The *moderate* acid is H_3PO_4, not HPO_4^{2-}.) If HPO_4^{2-} is a weak acid, then its conjugate base, PO_4^{3-} and the anion in trisodium phosphate, must be a relatively strong base. Therefore, we can predict that the phosphate ion will be the active participant in the hydrolysis of this salt:

$$PO_4^{3-}(aq) + H_2O(\ell) \rightleftharpoons HPO_4^{2-}(aq) + OH^-(aq)$$

Because OH^- is generated without compensating H^+, the solution of the salt is basic.

A solution of trisodium phosphate in water can be so alkaline that you ought to wear rubber gloves if you use it as a cleaning agent.

Notice how in Example 17.21 we didn't need a table of Brønsted bases to make our prediction. We used our knowledge of just a few memorized facts—which acids and bases are the strong aqueous acids and bases—to figure out what we needed to know. Let's work an example on just this part of the strategy: how to use the facts about the strong aqueous acids and bases to decide if a given anion is a Brønsted base.

EXAMPLE 17.22   **Judging If an Anion Is a Brønsted Base in Water**

Problem: In predicting if sodium nitrite, $NaNO_2$, hydrolyzes, we have to decide if the nitrite ion, NO_2^-, is a Brønsted base—assuming we can't look this up in a table. How do we make that decision?

Solution: The conjugate acid of NO_2^- is HNO_2. Even if we don't remember the name of this acid, we know from its formula that it is not on the list of strong (or moderate) acids. Therefore, we conclude it must be a weak acid. If HNO_2 is a weak acid, then its conjugate base, NO_2^-, must be something of a moderate or strong Brønsted base. (To learn exactly how strong we do need a table of ionization constants, of course.)

EXERCISE 17.21   Determine without the use of tables (other than Table 17.1) if each ion is a Brønsted base toward water.
(a) CO_3^{2-} (d) HSO_4^-
(b) HPO_4^{2-} (e) NO_3^-
(c) S^{2-} (f) Br^-

EXERCISE 17.22   Will sodium acetate, $NaC_2H_3O_2$, hydrolyze and, if so, will its solution be acidic or basic?

EXERCISE 17.23   Will potassium chloride hydrolyze and, if so, will its solution be acidic or basic?

EXERCISE 17.24   Ammonium nitrate, NH_4NO_3 is a nitrogen fertilizer. What effect will the application of an aqueous solution of this fertilizer have on the pH of soil? Will it tend to raise the pH or lower it or leave it unchanged?

Now that we have a strategy for predicting if a salt will hydrolyze, we next have to learn how to calculate the pH of a solution in which such hydrolysis occurs. For any salt in which the cation is a metal ion in Group IA or IIA (except beryllium) and the anion is recognized as a Brønsted base, the calculation is exactly like the determination of the pH of a solution of a Brønsted base given its value of K_b and its molar concentration.

EXAMPLE 17.23 Calculating the pH of a Salt Solution

Problem: Sodium acetate hydrolyzes to give a basic solution (Exercise 17.22). What will be the pH of a 0.150 M solution of sodium acetate?

Solution: Now that we know that the sodium ion cannot affect the values either of $[H^+]$ or of $[OH^-]$, the problem becomes simply one of finding the pH of a solution of the Brønsted base present, the acetate ion, $C_2H_3O_2^-$. We learned the steps for doing that in Example 17.16. Applying the same steps to $C_2H_3O_2^-$, we first write the acid-base equilibrium and then assemble the data.

	$C_2H_3O_2^-(aq) + H_2O(\ell) \rightleftharpoons HC_2H_3O_2(aq) + OH^-(aq)$		
Initial concentrations	0.150	0	0
Changes in concentrations caused by the hydrolysis	$-x$	$+x$	$+x$
Final concentrations at equilibrium	$(0.150 - x)$ $= 0.150$	x	x

In setting $(0.150 - x)$ equal to (0.150) we are making the same assumption made in Example 17.16. We are assuming that the value of x will be too small to change the value of 0.150 when $(0.150 - x)$ is rounded to the correct number of significant figures. This may introduce an error of a few percent, but an error of this magnitude is negligible. Now we write the expression for K_b and calculate its value from the data in the last row of the table.

$$K_b = \frac{[HC_2H_3O_2][OH^-]}{[C_2H_3O_2^-]} = 5.68 \times 10^{-10} \text{ (Table 17.6)}$$

Or

$$\frac{(x)(x)}{0.150} = 5.68 \times 10^{-10}$$

$$x^2 = (0.150)(5.68 \times 10^{-10})$$

$$x = 9.23 \times 10^{-6}$$

Thus, the value of $[OH^-]$ is 9.23×10^{-6} mol/liter. To find the value of the pH, the easiest way is to use $[OH^-]$ to find pOH first and then calculate pH from the equation pH + pOH = 14.00.

$$pOH = -\log (9.23 \times 10^{-6})$$
$$= 5.03$$

Therefore,

$$pH = 14.00 - 5.03$$
$$= 8.97$$

A 0.150 M solution of sodium acetate has a pH of 8.97 and therefore this solution is slightly basic.

EXAMPLE 17.24 Calculating the pH of a Salt Solution

Problem: We have learned that ammonium chloride hydrolyzes to give an acidic solution. What is the pH of 0.100 M NH_4Cl?

Solution: We know that the chloride ion does not participate in the hydrolysis. Its conjugate acid, $HCl(aq)$, is a very strong acid; consequently, Cl^- is a very weak base—actually not a base at all in water. So we work only with the ammonium ion and write the equilibrium equation for its acid-base reaction with water. Then we will assemble the data.

	$NH_4^+(aq)$ + $H_2O(\ell)$ $\rightleftharpoons$ $NH_3(aq)$ + $H_3O^+(aq)$		
Initial concentrations	0.100	0	0
Changes in concentrations caused by the hydrolysis	$-x$	$+x$	$+x$
Final concentrations at equilibrium	$(0.100 - x)$ $= 0.100$	x	x

The expression for K_a, using H^+ for H_3O^+ as usual, is

$$K_a = \frac{[NH_3][H^+]}{[NH_4^+]} = 5.65 \times 10^{-10} \text{ (Table 17.5)}$$

$$\frac{(x)(x)}{0.100} = 5.65 \times 10^{-10}$$

$$x^2 = 5.65 \times 10^{-11}$$

$$x = 7.52 \times 10^{-6}$$

Notice that $(x) \lll 0.100$.

Therefore,

$$[H^+] = 7.52 \times 10^{-6} \text{ mol/liter}$$

$$pH = -\log [H^+]$$
$$= -\log 7.52 \times 10^{-6}$$
$$= 5.12$$

Thus, the pH of 0.100 M NH_4Cl is 5.12 and this solution is slightly acidic.

EXERCISE 17.25 What is the pH of a 1.00 M solution of ammonium chloride?

17.7 POLYPROTIC ACIDS

Each ionizable hydrogen of a polyprotic acid has its own acid ionization constant

Carbonic acid is a diprotic acid, and there is a separate K_a value for each of the ionizable hydrogens. The first acid ionization constant, K_{a_1}, is for the following equilibrium.

$$H_2CO_3(aq) \rightleftharpoons H^+(aq) + HCO_3^-(aq) \qquad K_{a_1} = \frac{[H^+][HCO_3^-]}{[H_2CO_3]} = 4.30 \times 10^{-7}$$

carbonic acid   bicarbonate ion

The second acid ionization constant, K_{a_2} is for this equilibrium:

$$HCO_3^-(aq) \rightleftharpoons H^+(aq) + CO_3^{2-}(aq) \qquad K_{a_2} = \frac{[H^+][CO_3^{2-}]}{[HCO_3^-]} = 5.61 \times 10^{-11}$$

bicarbonate   carbonate
ion   ion

In general, for any diprotic acid K_{a_2} is always much less than K_{a_1} because it is much easier to take a positive ion (H^+) from a neutral molecule (e.g., H_2CO_3) than from a negatively charged ion (e.g., HCO_3^-). Typically, K_{a_2} is less than K_{a_1} by a factor of between 10^{-4} and 10^{-5} as the data in Table 17.7 show. The third acid ionization constant for a triprotic acid is similarly much less than the second, as the data for phosphoric acid, H_3PO_4, in Table 17.7 show.

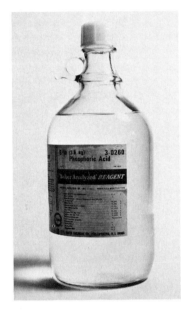

85% phosphoric acid.

Table 17.7
K_a and pK_a For Common Polyprotic Acids

Acid		K_a
Carbonic acid	$H_2CO_3 \rightleftharpoons H^+ + HCO_3^-$	4.30×10^{-7}
(25 °C)	$HCO_3^- \rightleftharpoons H^+ + CO_3^{2-}$	5.61×10^{-11}
Hydrogen sulfide	$H_2S \rightleftharpoons H^+ + HS^-$	$9.1\ \times 10^{-8}$
(18 °C)	$HS^- \rightleftharpoons H^+ + S^{2-}$	$1.1\ \times 10^{-12}$
Phosphoric acid	$H_3PO_4 \rightleftharpoons H^+ + H_2PO_4^-$	7.52×10^{-3}
(25 °C)	$H_2PO_4^- \rightleftharpoons H^+ + HPO_4^{2-}$	6.23×10^{-8}
	$HPO_4^{2-} \rightleftharpoons H^+ + PO_4^{3-}$	$2.2\ \times 10^{-13}$
Sulfuric acid	$H_2SO_4 \rightleftharpoons H^+ + HSO_4^-$	large
(25 °C)	$HSO_4^- \rightleftharpoons H^+ + SO_4^{2-}$	1.20×10^{-2}
Sulfurous acid	$H_2SO_3 \rightleftharpoons H^+ + HSO_3^-$	1.54×10^{-2}
(18 °C)	$HSO_3^- \rightleftharpoons H^+ + SO_3^{2-}$	1.02×10^{-7}
ORGANIC ACIDS		
Ascorbic acid	K_{a_1} (24 °C)	7.94×10^{-5}
(Vitamin C)	K_{a_2} (16 °C)	1.62×10^{-12}

HO   CH$_2$—OH
  \   /
   CH
HO
   \
    ═
   /
HO
   ‖
   O

Citric acid[a]	K_{a_1}	7.10×10^{-4}
(18 °C)	K_{a_2}	1.68×10^{-5}
	K_{a_3}	$6.4\ \times 10^{-6}$

[a] See page 584 for the structure of citric acid.

Because the second acid ionization constant, K_{a_2}, is so much smaller than the first (and any succeeding constants, such as K_{a_3}, are even smaller), the pH of a dilute solution of a weak diprotic or triprotic acid can be calculated just from the value of K_{a_1}. The contributions to the value of $[H^+]_{equilibrium}$ and to the pH from the second or higher ionizations are so small that after rounding the answer to the correct number of significant figures their contributions are negligible. We will work an example for a hypothetical diprotic acid, H_2A, that shows how to use just K_{a_1} to find $[H^+]_{equilibrium}$. We will also show in this worked example that the value of $[A^{2-}]_{equilibrium}$ is numerically equal to the second acid ionization constant, K_{a_2}.

EXAMPLE 17.25   **Calculating $[H^+]$ and $[A^{2-}]$ at Equilibrium in a Solution of a Weak Diprotic Acid, H_2A**

Problem: If the value of K_{a_1} is 1.00×10^{-5} and of K_{a_2} is 1.00×10^{-9} for the diprotic acid H_2A, what are the values of $[H^+]$ and $[A^{2-}]$ at equilibrium in 0.100 *M* H_2A?

Solution: We will set up the equilibrium equation for the first ionization and prepare the table of data in the usual way. As in earlier examples, we will use (x) to represent the *change* in the initial concentrations caused by the ionization, but we will for the moment completely ignore the second ionization. Then we will add a new kind of row to the table to enter the correction (y) to the values of (x). This is the point at which we do consider the second ionization, that of HA^-. For a moment it will look as if (y) really complicates the problem, but now we will see how the large difference between K_{a_1} and K_{a_2} lets us make some very helpful simplifications.

	$H_2A(aq)$	$\rightleftharpoons$ $H^+(aq)$	$+$ $HA^-(aq)$
Initial concentrations	0.100	0	0
Changes in concentrations ignoring the second ionization	$-x$	$+x$	$+x$
Corrections to changes in concentration caused by the second ionization	0 (*Note 1*)	$+y$	$-y$ (*Note 2*)
Final concentrations at equilibrium	$(0.100 - x)$	$(x + y)$	$(x - y)$ (*Note 3*)
Values after simplifications (*Note 3*)	0.100	x	x

Note 1. There is no correction here because H_2A does not participate in the *second* ionization.

Note 2. The second ionization *adds* something $(+y)$ to the concentration of $[H^+]$ at the expense $(-y)$ of $[HA^-]$.

Note 3. Since $K_{a_2} \lll K_{a_1}$, the value of (y) is very small compared to the value of (x). Therefore, after rounding correctly, $(x + y)$ and $(x - y)$ will equal (x), or within a few percent of (x). Similarly, the value of (x) is very low compared to 0.100, and $(0.100 - x)$ will equal (0.100), after rounding. Now we use the values in the last row of the table and the value of K_{a_1} to find $[H^+]$ (which, of course, is x).

The ionization of HA^- adds (y) to $[H^+]$ but subtracts (y) from $[HA^-]$.

$$K_{a_1} = \frac{[H^+][HA^-]}{[H_2A]} = \frac{(x)(x)}{0.100} = 1.00 \times 10^{-5}$$

$$x^2 = (0.100)(1.00 \times 10^{-5}) = 1.00 \times 10^{-6}$$

$$[H^+] = x = 1.00 \times 10^{-3} \text{ mol/liter (one answer we seek)}$$

Next we will calculate the molar concentration of A^{2-} and thereby show that numerically it equals the value of K_{a_2}. The expression for K_{a_2} is

$$K_{a_2} = \frac{[H^+][A^{2-}]}{[HA^-]} \quad \text{(from } HA^- \rightleftharpoons H^+ + A^{2-})$$

But both $[H^+]$ and $[HA^-]$ equal (x) in the table we constructed for the first ionization. Therefore

$$K_{a_2} = \frac{(x)[A^{2-}]}{(x)} = [A^{2-}]$$

Since $K_{a_2} = 1.00 \times 10^{-9}$,

$$[A^{2-}] = 1.00 \times 10^{-9} \text{ mol/liter}$$

EXERCISE 17.26   Hydrogen sulfide, the gas responsible for the odor of rotten eggs, is a weak diprotic acid. For H_2S, $K_{a_1} = 9.1 \times 10^{-8}$ and $K_{a_2} = 1.1 \times 10^{-12}$. Calculate $[H^+]$ and $[S^{2-}]$ in 0.050 M H_2S. Then calculate the pH of the solution.

17.8 BUFFERS

A buffered solution is one that contains solutes that protect it against changes in pH if strong acids or bases are added to it

A downward drift in the pH of blood from 7.35 toward 7.00 is called *acidosis*. An upward drift toward 8.00 is called *alkalosis*. Either condition is a serious medical emergency. Either condition, for example, interferes with respiration. What normally prevents acidosis and alkalosis are solutes in the blood that neutralize hydrogen ions or hydroxide ions as quickly as they are produced. For example, because there is always some carbon dioxide dissolved in blood the blood always has some carbonic acid produced in the following equilibrium reaction.

$$CO_2(aq) + H_2O(\ell) \rightleftharpoons H_2CO_3(aq)$$
$$\text{carbonic acid}$$

And the blood always has some dissolved bicarbonate ion also. The carbonic acid can neutralize hydroxide ions and thereby prevent a rise in pH. The bicarbonate ion, a Brønsted base, can neutralize hydrogen ion and prevent a drop in pH. Because the blood has both a base-neutralizer, carbonic acid, and an acid-neutralizer, the bicarbonate ion, we say that it is buffered against changes in pH. The solutes that perform this service constitute the **buffer.** Usually a buffer is made up of two principal solutes.[4] One is a weak acid, HA, and the other is a soluble salt of that acid such as NaA, which is the source of the Brønsted base, A^-. The weak acid can also be the ammonium ion furnished by a salt such as ammonium chloride. Then the Brønsted base is aqueous ammonia. We will confine our discussion to the HA/NaA type of buffer systems.

To neutralize extra H^+: $A^- + H^+ \longrightarrow HA$
To neutralize extra OH^-: $HA + OH^- \longrightarrow A^- + H_2O$

There is almost no area of experimental work in chemistry or its applications in microbiology, cell biology, or the sciences of foods, soils, nutrition, clinical analysis, or molecular biology in which buffers and their practical applications are not important. Special Topic 17.2 describes how acid-base equilibria are involved in breathing. Unless the body manages well the concentrations of CO_2 and H^+ in blood, these equilibria are upset and breathing is impaired.

---

[4] Sometimes an amphoteric solute provides the effect of *both* components needed for a buffer. One part of the amphoteric molecule can neutralize H^+ and another part can neutralize OH^-.

When one of your cells has done some "work," it has used oxygen and made carbon dioxide. To continue to function it needs to get more oxygen and to get rid of the carbon dioxide. Remarkably, meeting one need helps to satisfy the other. The carbon dioxide goes from the cell where it was made to the blood stream where it combines in a red blood cell with water:

$$\underset{\substack{\text{leaving} \\ \text{the cell}}}{CO_2} + H_2O \rightleftharpoons \underset{\substack{\text{carbonic} \\ \text{acid}}}{H_2CO_3}$$

The ionization of H_2CO_3 helps to trap the CO_2:

$$H_2CO_3 \rightleftharpoons \underset{\substack{\text{in red} \\ \text{cell}}}{HCO_3^-} + \underset{\substack{\text{newly} \\ \text{made}}}{H^+}$$

The newly made H^+ is exactly what is needed to help drive oxygen out of the red cell and into the tissue cell where it is needed. In oxygen-rich hemoglobin, there is the equilibrium

$$\underset{\substack{\text{newly} \\ \text{made}}}{H^+} + \underset{\substack{\text{arriving} \\ \text{from lungs}}}{HbO_2^-} \rightleftharpoons \underset{\substack{\text{will go} \\ \text{back to} \\ \text{lungs}}}{HHb} + \underset{\substack{\text{available} \\ \text{to a cell} \\ \text{needing it}}}{O_2}$$

(In this equilibrium HHb represents hemoglobin, the oxygen-carrier in blood, as a weak acid, and HbO_2^-, oxyhemoglobin, is the form in which oxygen is transported from the lungs. $HHbO_2$ is a stronger acid than HHb, and is shown in its ionized form on the left.) Following Le Châtelier's principle, as the H^+ concentration increases from the ionization of H_2CO_3, this equilibrium is shifted to the right to release oxygen. Thus,

by sending out CO_2, a working cell not only gets rid of a waste, it helps to draw O_2 into itself where it is needed. The net effect of the three previous equilibria, all moving to the right, is:

$$\underset{\substack{\text{leaving} \\ \text{cell}}}{CO_2} + H_2O + \underset{\substack{\text{arriving} \\ \text{at cell}}}{HbO_2} \longrightarrow$$

$$\underset{\substack{\text{leaving} \\ \text{cell region} \\ \text{for lungs}}}{HCO_3^-} + \underset{\substack{\text{returning} \\ \text{to lungs}}}{HHb} + \underset{\substack{\text{released} \\ \text{to the cell} \\ \text{needing it}}}{O_2}$$

After a moment of blood flowage, HCO_3^- and HHb arrive back at the lungs where more O_2 awaits them. The partial pressure of O_2 is greater in the lungs than anywhere else in the body, and it helps to drive the previous reaction backwards—at the lungs.

$$\underset{\substack{\text{arriving} \\ \text{from air}}}{O_2} + \underset{\substack{\text{returning} \\ \text{from cell} \\ \text{region}}}{HHb} + \underset{\substack{\text{arriving} \\ \text{from cell} \\ \text{region}}}{HCO_3^-} \longrightarrow$$

$$\underset{\substack{\text{going out} \\ \text{to cell} \\ \text{region}}}{HbO_2^-} + H_2O + \underset{\substack{\text{leaving} \\ \text{the body} \\ \text{in exhaled} \\ \text{air}}}{CO_2}$$

The net effect at the lungs is to load hemoglobin with oxygen in the form of HbO_2^- and to expel CO_2. These equilibria are all sensitive to the value of $[H^+]$, and hence to the pH of the blood. Thus, the buffers in blood are vitally important to the system's ability to breathe normally.

There are three kinds of problems in connection with buffers. First, find the pH that can be held constant by a given buffer system; second, find the effect on the pH of adding some strong acid or base to a buffered solution; and, third, make the calculations necessary for preparing a buffered solution having a desired pH.

Henderson-Hasselbalch equation

To find the pH maintained by a buffer system, we can use the equation for K_a for the weak acid in the buffer. We can also use the kinds of simplifications often made in this chapter as well as often used in practical laboratory work. Let's assume that the buffer system is a solution containing the weak acid, HA, and some salt of its anion, A^-. The expression for K_a for the weak acid is

$$K_a = \frac{[H^+][A^-]}{[HA]}$$

To find $[H^+]$, this equation can be rearranged to give

$$[H^+] = K_a \times \frac{[HA]}{[A^-]} \tag{17.18}$$

Because HA is weak, very little of it will have dissociated at equilibrium, even if HA were the only solute present. However, the solution also contains A^- from the

The suppression of the ionization of HA by A^- is an example of the common ion effect.

dissolved salt. The presence of A^- will suppress the already slight ionization of HA by shifting the following equilibrium to the left.

$$HA \rightleftharpoons H^+ + A^-$$

The net result is that the value of $[HA]_{equilibrium}$ will be very nearly equal to the value of $[HA]_{initial}$. Furthermore, the value of $[A^-]_{equilibrium}$ will also be very nearly identical with $[A^-]_{initial}$ as provided by the salt, NaA, because there is almost no contribution to $[A^-]_{equilibrium}$ from the slight ionization of the acid. Two very good assumptions that we can make, therefore, are

$$[HA]_{equilibrium} = [HA]_{\substack{from\ initial\ acid \\ concentration}} = [acid]$$

$$[A^-]_{equilibrium} = [A^-]_{\substack{from\ initial\ salt \\ concentration}} = [anion]$$

We will therefore substitute the terms [acid] and [anion] into equation 17.18 for the terms [HA] and [A^-] with the understanding that [acid] means the *initial* concentration of the weak acid in the buffer and that [anion] means the *initial* concentration of anion directly provided by the salt in the buffer system.

$$[H^+] = K_a \times \frac{[acid]}{[anion]} \tag{17.19}$$

Before working an example, let's convert equation 17.19 into a form for finding the pH instead of [H^+] for a buffered solution. We can make this change by finding the negative logarithms of both sides of equation 17.19:

$$-\log[H^+] = -\log K_a - \log \frac{[acid]}{[anion]}$$

Or,

$$pH = pK_a - \log \frac{[acid]}{[anion]}$$

If we note that

$-\log a/b = +\log b/a$

$$-\log \frac{[acid]}{[anion]} = +\log \frac{[anion]}{[acid]}$$

then we have the final form of what is popularly known, particularly in the life sciences, as the **Henderson-Hasselbalch equation**[5]:

$$\boxed{pH = pK_a + \log \frac{[anion]}{[acid]}} \tag{17.20}$$

Two factors govern the pH of a buffered solution, first, the pK_a, and, second, the *ratio* of the initial molar concentrations of the anion and the acid. If we prepare the buffered solution so that this ratio equals 1, then the pH of the solution will equal the pK_a of the acid. Thus, the chief factor in determining the general area on the pH scale where a buffer will hold a solution is the pK_a of the weak

When [anion] = [acid], $\log \frac{[anion]}{[acid]} = \log 1 = 0.$

---

[5] You may sometimes see the Henderson-Hasselbalch equation written as

$$pH = pK_a + \log \frac{[salt]}{[acid]}$$

In other words, [salt] is used instead of [anion], as if the two were always identical. This works, however, only when the cation in the salt is of the form M^+ (as, for example, Na^+, K^+, or NH_4^+) so that each formula unit of the salt gives only *one* anion. If the cation of the salt is of the form M^{2+} (for example, Ca^{2+}), then *two* anions are released by the ionization of only one formula unit of the salt. With such salts the value of [anion] in moles per liter is twice the value of [salt] in moles per liter. It is safest to stick with the form of equation 17.20.

Acid-Base Equilibria

acid. Then by selecting the ratio of [anion] to [acid] we can move the pH of the buffered solution to one side or the other of the value of pK_a.

EXAMPLE 17.26 **Calculating the pH of a Buffered Solution by the Henderson-Hasselbalch Equation**

Problem: To study the effect of a weakly acidic culture medium on the growth of a certain strain of bacteria, a microbiologist prepared a buffer solution by making it 0.11 M $NaC_2H_3O_2$ (sodium acetate) and also 0.090 M $HC_2H_3O_2$ (acetic acid). What was the pH of this solution?

Solution: From Example 17.17, $pK_a = 4.75$ for acetic acid. From the given concentrations, [acid] = 0.090 M and [anion] = 0.11 M. (Each formula unit of $NaC_2H_3O_2$ gives one anion, $C_2H_3O_2^-$ since sodium salts are fully dissociated in water.) Using these data in the Henderson-Hasselbalch equation gives

$$pH = 4.75 + \log \frac{(0.11)}{(0.090)}$$

$$= 4.75 + \log 1.2$$
$$= 4.75 + 0.079$$
$$= 4.83$$

This solution is buffered to hold the pH at 4.83 even if small quantities of acids or bases enter it, such as by the biochemical activities of the bacteria being studied.

EXERCISE 17.27 To compare the pH of the buffered solution of Example 17.26 with individual unbuffered solutions of the same salts, calculate the pH values for each of the following solutions.
(a) 0.11 M $NaC_2H_3O_2$ (as in Example 17.23)
(b) 0.090 M $HC_2H_3O_2$ (as in Example 17.13)

EXERCISE 17.28 How is the pH related to the pK_a of the weak acid if
(a) the ratio of [anion] to [acid] is 10 to 1;
(b) the ratio of [anion] to [acid] is 1 to 10?

EXERCISE 17.29 Calculate the pH of a buffered solution made up as 0.015 M sodium acetate and 0.10 M acetic acid. The pK_a for acetic acid is 4.75.

The effectiveness of a buffer

How well does a buffer system work? Suppose that 0.010 mol of a strong base such as sodium hydroxide were added to 1 liter of the buffer solution described in Example 17.26. This will change the concentration of the acid by neutralizing some of it. The new value of [acid] will be (0.090 M − 0.010 M) = 0.080 M. Of course, this neutralization makes the same amount of change in the value of [anion]. The new value of [anion] will be (0.11 M + 0.010 M) = 0.12 M. Using the Henderson-Hasselbalch equation, we find that the new pH will be

$$pH = pK_a + \log \frac{[anion]}{[acid]}$$

$$= 4.75 + \log \frac{(0.12)}{(0.080)}$$

$$= 4.75 + \log 1.5 = 4.75 + 0.18$$
$$= 4.93$$

Notice that the pH changed by only 0.10 pH unit even though enough base was added to reduce the value of [acid] by 11 percent and to increase the value of [anion] by the same amount. Had we added 0.010 mol NaOH to 1 liter of pure water, the pH would have changed from 7.00 to 12, a difference of 5 pH units. We can see, on the one hand, that a buffer does not hold a pH *exactly* constant, but on the other hand it does a very good job of limiting changes in pH to very low amounts.

EXERCISE 17.30   Suppose a trace of acid were added to the buffer of Example 17.26 to change the value of [acid] by just 1 percent. This added acid would be neutralized by acetate ion, so the value of [salt] would drop by 1 percent of its initial value. What is the new pH?

No buffer has unlimited capacity. The capacity of a buffer is the amount of acid or base it can handle before the pH of the solution changes by more than a pH unit. If you add enough strong acid to neutralize all of the buffer's basic component, then additional strong acid will make the pH drop very rapidly. The addition of enough strong base will similarly overwhelm a buffer. The buffer capacity depends on the actual concentrations of its Brønsted acid and Brønsted base. The more of each, the more strong acid or base can be added without changing the solution's pH by more than a few tenths of a pH unit.

Preparation of a buffer

To prepare a solution that will be buffered at a particular pH, we begin with the Henderson-Hasselbalch equation. In practice, we generally pick the acid member of the buffer pair by its pK_a because the equation tells us that the desired pH can't help being near the pK_a. The problem then is to find the ratio of [anion] to [acid] whose log, when added to pK_a, gives the pH we want. The usual guideline is that the pK_a should have such a value that the desired pH = $pK_a \pm$ 1. If the value of [anion]/[acid] is 10 (because the ratio is 10 to 1), then pH = pK_a + 1. If the value of [anion]/[acid] is 1/10, the pH = pK_a − 1. Generally, the *range* in the ratio of [anion] to [acid] should be maintained within the limits of 10 to 1 down to 1 to 10. Outside this range the buffer's effectiveness declines.

$$\log \frac{(10)}{(1)} = \log 10^1 = 1$$

$$\log \frac{(1)}{(10)} = \log 10^{-1} = -1$$

EXAMPLE 17.27   **Preparing a Buffer Solution**

Problem: A solution buffered at pH 5.00 is wanted. Will the pair, acetic acid and sodium acetate, work for this purpose and, if so, what ratio of acetate ion to acetic acid is needed?

Solution: We first have to check the pK_a of acetic acid to see if the desired pH is within $\pm$1 pH unit of pK_a. Since pK_a = 4.75 (Example 17.17), these solutes will work. Now we apply the Henderson-Hasselbalch equation to find the proper ratio of [anion] to [acid].

$$pH = pK_a + \log \frac{[anion]}{[acid]}$$

$$5.00 = 4.75 + \log \frac{[anion]}{[acid]}$$

$$\log \frac{[anion]}{[acid]} = 5.00 - 4.75 = 0.25$$

$$\frac{[anion]}{[acid]} = \text{antilog } 0.25$$

$$= 1.8$$

The answer, 1.8, is the ratio of the molar *concentrations* of the anion and the acid. However, that ratio is in this example identical to the ratio of the moles of anion to moles of acid as we can see by this analysis of the units:

$$\frac{[\text{anion}]}{[\text{acid}]} = \frac{\dfrac{\text{mole anion}}{\text{liter solution}}}{\dfrac{\text{mole acid}}{\text{liter solution}}} = \frac{\text{mole anion}}{\text{mole acid}}$$

Thus, the answer tells us that a solution containing a ratio of 1.8 mol of the acetate ion to 1.0 mol of acetic acid will be buffered at pH 5.00. To obtain that ratio we could take 0.18 mol of $NaC_2H_3O_2$ to 0.10 mol of $HC_2H_3O_2$, or if we wanted a larger capacity buffer, we could take 1.8 mol of $NaC_2H_3O_2$ to 1.0 mol of $HC_2H_3O_2$ in the same volume. (If the buffer is to be used with some living organism, then we have to be careful about other factors, such as relative toxicities of the buffer system's compounds.)

EXERCISE 17.31 A nutrition scientist needed an aqueous buffer for a pH of 3.90. Would formic acid and its salt, sodium formate, make a good pair for this purpose? If so, what mole ratio of the anion of this salt, HCO_2^-, to the acid, HCO_2H, is needed? For formic acid, $pK_a = 3.75$.

17.9 ACID-BASE TITRATION

One equivalent weight of an acid supplies one mole of H^+ and one equivalent weight of a base neutralizes one mole of H^+

In Chapter Four we studied the overall procedure for acid-base titration and saw how titration data can be used to calculate the molar concentration of some "unknown." In principle the experimental goal of a well-conducted titration is to stop the addition of the standard solution, the one whose concentration is accurately known, at exactly that point where the quantity of its solute is stoichiometrically equivalent to the solute in the "unknown." The point in the titration when this relationship is exactly true is called the **equivalence point.** The equivalence point is not necessarily identical with the end point, however. The **end point,** where the titration actually is ended, occurs when the indicator changes color. One of the important decisions the analyst must make for each titration is to pick an indicator whose color change occurs at the pH that the titrated solution will have at the equivalence point. Remember that an acid-base neutralization produces a salt, and many salts hydrolyze in water to give acidic or basic solutions, depending on the particular salt. For example, the titration of sodium hydroxide by acetic acid gives a solution of sodium acetate at the equivalence point.

$$NaOH(aq) + HC_2H_3O_2(aq) \longrightarrow NaC_2H_3O_2(aq) + H_2O(\ell)$$

Because the acetate ion is a Brønsted base, but the sodium ion does nothing to the pH of a solution, a solution of sodium acetate is basic. We therefore do not want an indicator whose color changes either on the acidic side or at pH 7.00. You can see by this example that to carry out a well-run titration requires careful planning and a knowledge of the properties not just of acids and bases but of their salts as well.

In this section we will give further study to the stoichiometric relationships in acid-base titrations, to the concept of an equivalent weight, to the theory of acid-base indicators and their proper selection, and to the patterns of pH-changes in the titrations of weak and strong acids and bases.

Equivalent weights

If you look back to Example 4.15 on page 104 you will find the procedure for calculating the molarity of an "unknown" sodium hydroxide solution from titration data and the molarity of the standard acid. That procedure is very straightforward, and it can always be used in acid-base titrations. It is the best procedure to use, at least until you have become really comfortable about the principles involved. Most chemical analysts, however, use a variation in the calculations that takes into account the ratio of coefficients in the neutralization equation in a different way. While it is an extra topic to learn, in the long run it simplifies the calculations. Moreover, the concept involved, the equivalent weight, is widely used, particularly in chemical applications in biology and medicine.

One **equivalent weight,** usually called simply one **equivalent (Eq.)** of an acid is the mass in grams that *in a specified reaction* delivers one mole of hydrogen ion. This means that for all monoprotic acids, HA, the equivalent weight is identical with the formula weight. One mole of $HCl(aq)$, for example, has a mass of 36.46 g, and hence one Eq of $HCl(aq)$ has the same mass.

$$1 \text{ mol } HCl = 1 \text{ Eq } HCl = 36.46 \text{ g } HCl$$

In general,

$$1 \text{ mol } HA = 1 \text{ Eq } HA \text{ for all monoprotic acids.}$$

For di- and triprotic acids, however, the mass in grams that delivers one mol of hydrogen ion varies with the specific reaction. Thus, equivalent weights are *not* fixed like formula weights. For example, sulfuric acid might be used as a monoprotic acid if only the first proton is neutralized:

$$H_2SO_4(aq) + NaOH(aq) \longrightarrow NaHSO_4(aq) + H_2O(\ell)$$

If this acid is used in this manner, then its equivalent weight equals its formula weight. However, both hydrogens can be and usually are used in a neutralization:

$$H_2SO_4 \ (aq) + 2NaOH(aq) \longrightarrow Na_2SO_4(aq) + 2H_2O(\ell)$$

Now one mole of sulfuric acid delivers two moles of hydrogen ion, so it takes only half a mole of this acid to give one mole of hydrogen ion. For any diprotic acid, H_2A, for which *both* hydrogens are neutralized we can write:

$$1 \text{ mol } H_2A = 2 \text{ Eq } H_2A$$

Therefore, for sulfuric acid when both hydrogens are neutralized we can write:

$$1 \text{ mol } H_2SO_4 = 2 \text{ Eq } H_2SO_4 = 98.08 \text{ g } H_2SO_4$$

Or

$$1 \text{ Eq } H_2SO_4 = 49.04 \text{ g } H_2SO_4$$

Thus, for any diprotic acid used so that both protons are neutralized, the equivalent weight is one-half the formula weight.

Phosphoric acid, H_3PO_4, is a triprotic acid, and if all three hydrogens are neutralized then its equivalent weight is one-third the formula weight, or

$$1 \text{ mol } H_3PO_4 = 3 \text{ Eq } H_3PO_4$$

Therefore, the equivalent weight is

$$\frac{98.00 \text{ g } H_3PO_4}{1 \text{ mol } H_3PO_4} \times \frac{1 \text{ mol } H_3PO_4}{3 \text{ Eq } H_3PO_4} = 32.67 \frac{\text{g } H_3PO_4}{\text{Eq } H_3PO_4}$$

Figure 17.3 illustrates the relationships between the formula weights and equivalent weights of three common acids.

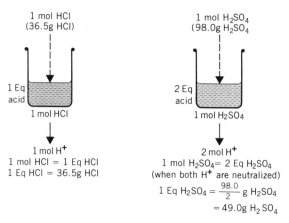

Figure 17.3
Equivalents of representative acids. Each calculation assumes that the full proton-donating capacity of the acid will be used.

EXAMPLE 17.28 **Calculating the Equivalent Weight of an Acid**

Problem: Citric acid, the compound responsible for the tart taste of lemons and other citrus fruits, has a molecular formula of $H_3C_6H_5O_7$ (written to indicate that it is a triprotic acid). What is its equivalent weight when used in the following reaction?

$$H_3C_6H_5O_7(aq) + 3NaOH(aq) \longrightarrow Na_3C_6H_5O_7(aq) + 3H_2O(\ell)$$

Solution: Since it is used as a triprotic acid, it has three equivalents per mole. After finding that its formula weight is 192.1, we can then convert grams/mole to grams/Eq as follows:

$$\frac{192.1 \text{ g } H_3C_6H_5O_7}{1 \text{ mol } H_3C_6H_5O_7} \times \frac{1 \text{ mol } H_3C_6H_5O_7}{3 \text{ Eq } H_3C_6H_5O_7} = 64.03 \frac{\text{g } H_3C_6H_5O_7}{\text{Eq } H_3C_6H_5O_7}$$

One equivalent of citric acid has a mass of 64.03 g when it is used as triprotic acid.

When we consider equivalent weights for bases we have to regard their H^+-neutralizing work *in a specified reaction*. Bases with one H^+-neutralizing group per formula unit have equivalent weights identical with their formula weights. Examples are sodium hydroxide and potassium hydroxide. A base such as calcium hydroxide, $Ca(OH)_2$, might be involved in a reaction in which each formula unit neutralizes just one H^+, and then its equivalent weight also equals its formula weight. Generally, however, it is used to take advantage of both OH^- ions:

$$Ca(OH)_2(aq) + 2HCl(aq) \longrightarrow CaCl_2(aq) + 2H_2O(\ell)$$

In this reaction one mole of $Ca(OH)_2$ neutralizes two moles of H^+. Hence one-half mole of $Ca(OH)_2$ neutralizes one mole H^+. For $Ca(OH)_2$ and similar Group IIA hydroxides, when both OH^- ions are used,

$$1 \text{ mol } M(OH)_2 = 2 \text{ Eq } M(OH)_2$$

(where "M" stands for a dipositive metal ion such as Ca^{2+} or Mg^{2+}). Since the formula weight of calcium hydroxide is 74.09,

$$\frac{74.09 \text{ g } Ca(OH)_2}{1 \text{ mol } Ca(OH)_2} \times \frac{1 \text{ mol } Ca(OH)_2}{2 \text{ Eq } Ca(OH)_2} = 37.05 \frac{\text{g } Ca(OH)_2}{\text{Eq } Ca(OH)_2}$$

Bicarbonates and carbonates are commonly used to neutralize acids. If sodium bicarbonate is used to neutralize hydrochloric acid:

$$NaHCO_3(aq) + HCl(aq) \longrightarrow NaCl(aq) + H_2O(\ell) + CO_2(g)$$

then it is obvious that one mole of $NaHCO_3$ neutralizes one mole of hydrogen ion. Therefore, the equivalent weight of sodium bicarbonate equals its formula weight.

$$1 \text{ mol } NaHCO_3 = 1 \text{ Eq } NaHCO_3$$

When sodium carbonate is used, full advantage normally is taken of its ability to neutralize acid:

$$Na_2CO_3(aq) + 2HCl(aq) \longrightarrow 2NaCl(aq) + CO_2(g) + H_2O(\ell)$$

According to this equation, one mole of sodium carbonate neutralizes the two moles of H^+ given by $2HCl(aq)$. Therefore one-half mole of sodium carbonate neutralizes one mole of H^+, and the equivalent weight of this base is one-half its formula weight.

$$1 \text{ mol } Na_2CO_3 = 2 \text{ Eq } Na_2CO_3$$

Figure 17.4 visualizes the calculation of equivalent weights of bases.

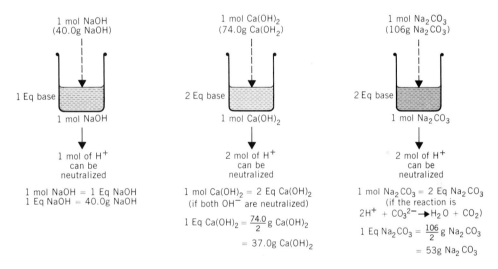

Figure 17.4
Equivalents of representative bases.

EXAMPLE 17.29 **Calculating the Equivalent Weight of a Base**

Problem: When small amounts of carbon dioxide gas are needed in the lab, hydrochloric acid may be added to chips of marble (calcium carbonate).

$$2HCl(aq) + CaCO_3(s) \longrightarrow CaCl_2(aq) + H_2O(\ell) + CO_2(g)$$

What is the equivalent weight of the base, $CaCO_3$, in this reaction?
Solution: Notice that one mole of the base neutralizes two moles of H^+. Therefore,

$$1 \text{ mol } CaCO_3 = 2 \text{ Eq } CaCO_3$$

We apply this relationship to the formula weight of calcium carbonate, 100.1, as follows:

$$\frac{100.1 \text{ g CaCO}_3}{1 \text{ mol CaCO}_3} \times \frac{1 \text{ mol CaCO}_3}{2 \text{ Eq CaCO}_3} = 50.05 \frac{\text{g CaCO}_3}{\text{Eq CaCO}_3}$$

One equivalent of $CaCO_3$ has a mass of 50.05 g.

EXERCISE 17.32 Calculate the equivalent weights of each of the following acids and bases as used in the reactions given.
(a) $HNO_3(aq) + KOH(aq) \longrightarrow KNO_3(aq) + H_2O(\ell)$
(b) $H_2SO_3(aq) + 2NaOH(aq) \longrightarrow Na_2SO_3(aq) + 2H_2O(\ell)$
(c) $Mg(OH)_2(s) + 2NaHSO_4(aq) \longrightarrow MgSO_4(aq) + Na_2SO_4(aq) + 2H_2O(\ell)$

The ability to calculate equivalent weights is the first step toward a new kind of expression for the concentration of a solution of an acid or a base. We have used "moles/liter" in many situations; we are heading toward an analogous expression, "equivalents/liter." Suppose, for example, we want to prepare one liter of a solution with a concentration of 0.350 Eq H_2SO_4/liter. We have to obtain 0.350 Eq H_2SO_4 and dissolve it in water to make the final volume one liter. To get 0.350 Eq H_2SO_4 we need to be able to calculate grams from equivalents; the question here is "how many grams of H_2SO_4 make up 0.350 Eq H_2SO_4?" We now assume (unless carefully told otherwise) that the equivalent weight of a diprotic acid is half its formula weight—that *both* hydrogens in H_2SO_4 will be used, in other words. The problem is a straightforward use of conversion factors, as we will see in the following worked example.

EXAMPLE 17.30 **Calculating Grams from Equivalents**

Problem: How many grams of sulfuric acid make 0.350 Eq H_2SO_4?
Solution: We found the equivalent weight of sulfuric acid to be 49.04 g. In other words,

$$1 \text{ Eq H}_2SO_4 = 49.04 \text{ g H}_2SO_4$$

Our choices of conversion factors are, therefore,

$$\frac{1 \text{ Eq H}_2SO_4}{49.04 \text{ g H}_2SO_4} \quad \text{or} \quad \frac{49.04 \text{ g H}_2SO_4}{1 \text{ Eq H}_2SO_4}$$

Therefore, to calculate the grams of H_2SO_4 in 0.350 Eq H_2SO_4, we do the following calculation:

$$0.350 \text{ Eq H}_2SO_4 \times \frac{49.04 \text{ g H}_2SO_4}{1 \text{ Eq H}_2SO_4} = 17.2 \text{ g H}_2SO_4$$

If we dissolved 17.2 g H_2SO_4 (or 0.350 Eq H_2SO_4) in enough water to make the final volume 1.00 liter, then the concentration is 0.350 Eq H_2SO_4/liter.

Another important calculation is finding the number of equivalents of an acid or a base in a given mass of substance. This is also a straightforward use of conversion factors as seen in the next worked example.

EXAMPLE 17.31 Calculating Equivalents from Grams

Problem: A sample of 1.38 g Ca(OH)$_2$ exactly neutralized a certain quantity of stomach acid (dilute HCl) in a test of various "antacids." How many equivalents of Ca(OH)$_2$ are in 1.38 g Ca(OH)$_2$, assuming that the following reaction occurs:

$$Ca(OH)_2(s) + 2HCl(aq) \longrightarrow CaCl_2(aq) + 2H_2O(\ell)$$

Solution: Since both OH$^-$ ions in Ca(OH)$_2$ are used,

$$1 \text{ mol Ca(OH)}_2 = 2 \text{ Eq Ca(OH)}_2$$

The formula weight of Ca(OH)$_2$ is 74.1. Therefore, we can do the required calculation as follows:

$$1.38 \text{ g Ca(OH)}_2 \times \frac{1 \text{ mol Ca(OH)}_2}{74.1 \text{ g Ca(OH)}_2} \times \frac{2 \text{ Eq Ca(OH)}_2}{1 \text{ mol Ca(OH)}_2} = 0.0372 \text{ Eq Ca(OH)}_2$$

Thus 1.38 g Ca(OH)$_2$ contains 0.0372 Eq Ca(OH)$_2$.

EXERCISE 17.33 (a) How many grams of H$_3$PO$_4$ are in 0.368 Eq H$_3$PO$_4$ when all three H$^+$ ions in H$_3$PO$_4$ are neutralized? (b) How many equivalents of H$_2$SO$_4$ are in 3.26 g H$_2$SO$_4$ in a reaction where both H$^+$ ions in this acid are neutralized?

The normality of a solution

The ratio of the number of equivalents per liter (or 1000 ml) of a solution is called the **normality** of the solution: its symbol is N.

mol/liter = M (molarity)
Eq/liter = N (normality)
"Normal" is to Eq/liter as "molar" is to mol/liter.

$$\text{Normality } (N) = \frac{\text{Eq solute}}{\text{liter solution}} = \frac{\text{Eq solute}}{1000 \text{ ml solution}} \qquad (17.21)$$

A solution containing 0.350 Eq H$_2$SO$_4$/1000 ml would have the label "0.350 N H$_2$SO$_4$," pronounced "0.350 *normal* H$_2$SO$_4$."

EXAMPLE 17.32 Calculating the Normality of a Solution From the Mass of Solute and the Volume of Solution

Problem: If 10.00 g of KOH is dissolved in water to make a final volume of 250.0 ml of solution, what is the normality of this solution?

Solution: Since we have to find the ratio of *equivalents* per liter, we first have to calculate the number of equivalents in 10.00 g of KOH. This base will neutralize 1 mol of H$^+$ per one mole of its formula unit, and therefore its equivalent weight equals its formula weight.

$$1 \text{ mol KOH} = 1 \text{ Eq KOH} = 56.00 \text{ g KOH}$$

Therefore,

$$10.00 \text{ g KOH} \times \frac{1 \text{ Eq KOH}}{56.00 \text{ g KOH}} = 0.1786 \text{ Eq KOH}$$

Now the ratio of Eq/liter:

$$\frac{0.1786 \text{ Eq KOH}}{250.0 \text{ ml KOH}(aq)} \times \frac{1000 \text{ ml KOH}(aq)}{1.000 \text{ liter KOH}(aq)} = 0.7144 \text{ Eq KOH/liter KOH}(aq)$$

$$= 0.7144 \; N \text{ KOH}$$

EXERCISE 17.34 Calculate the normality of each of the following solutions.
(a) 10.00 g of NaOH in 500.0 ml of solution.
(b) 0.3578 g of H_2SO_4 in 100.0 ml of solution.
(c) 3.578 g of Na_2CO_3 in 250 ml of solution.
(In all parts, assume that the neutralizing capacities of the substances are utilized to their fullest.)

Using data from acid-base titrations

The concept of normality is generally used only when the full neutralizing capacities of acids and bases are used. This means that the two H^+ ions in diprotic acids are used to neutralize a base; that three H^+ ions in a triprotic acid are used; that both OH^- ions in a base such as $Ca(OH)_2$ are neutralized, and so forth. If these assumptions are followed, then in *all* acid-base neutralizations the equivalence point in a titration occurs when the following equation is true.

Number of equivalents of acid = number of equivalents of base

Or

$$Eq_a = Eq_b \qquad \text{(definition of equivalence point)} \qquad (17.22)$$

The data from a titration, of course, are not directly in equivalents; instead, the data are milliliters of solutions and either molarities or normalities. Analysts who like the simplification of the normality concept (after learning it!) employ a simple equation for doing calculations with titration data, which we will now derive.

From the definition of normality we have the following relationships for the acid and the base.

V_a (volume of acid) and V_b (volume of base) are in liters here.

$$N_a = \frac{Eq_a}{V_a}$$

$$N_b = \frac{Eq_b}{V_b}$$

Each can be rearranged as follows.

$$Eq_a = N_a \times V_a$$

$$Eq_b = N_b \times V_b$$

Therefore, at the equivalence point (since $Eq_a = Eq_b$):

Since the units in equation 17.23 will cancel, V_a and V_b may be in milliliters now.

$$\boxed{N_a \times V_a = N_b \times V_b} \qquad (17.23)$$

Equation 17.23 holds regardless of the unit of volume provided that the same unit is used on both sides. In routine acid-base titrations the simplicity of equation 17.23 justifies the use of normalities instead of molarities for handling the calculations.

EXAMPLE 17.33 **Calculating Concentrations of "Unknowns" from Titration Data**

Problem: Suppose you have prepared a solution that is approximately 0.1 *N* NaOH in order to do a series of analyses of vinegar samples (dilute acetic acid). Before these analyses, however, you have to standardize the base. The supply room has a supply of standard HCl(*aq*) at 0.1067 *N*. Suppose you transfer 25.00 ml of the base to a flask, add a suitable indicator, and titrate the base until the indicator changes color. (The procedure was described in Special Topic 4.1.) If 22.48 ml of 0.1067 N HCl(aq) is required, what is the normality of the base?

Solution: Assemble the data and use equation 17.23.

$$N_a = 0.1067; \quad V_a = 22.48 \text{ ml}; \quad N_b = ?; \quad V_b = 25.00 \text{ ml}$$

$$(0.1067) \times (22.48) = N_b \times (25.00)$$

$$N_b = 0.09594 \text{ (the answer)}$$

EXERCISE 17.35 A state inspector selected a bottle of vinegar suspected of being below strength and sent it to you to analyze. Using 0.09594 N NaOH (from Example 17.33), you found that a 5.00 ml sample of vinegar required 29.20 ml of the standard base to be fully neutralized. What was the normality of the vinegar? (The acceptable concentration of household vinegar is 0.7 to 0.8 N acetic acid.)

Selecting the best acid-base indicator

In the titration of dilute solutions of sodium hydroxide and acetic acid, the solution at the equivalence point contains sodium acetate. The pH of that solution will be between 8 and 9, depending on the concentration, as we calculated in Example 17.23. The acid-base indicator we want for this titration therefore should undergo its characteristic color change in this range, not at pH 7. Similarly, if we titrated dilute solutions of ammonia and hydrochloric acid, the solution at the equivalence point contains ammonium chloride. The pH of that solution will be between 5 and 6, depending on the exact concentration, as we found in Example 17.24. Hence, this titration should use an indicator whose color change occurs in the range of 5 to 6, not at 7.00. Thus to do a good titration we have to know enough about the available indicators to make a suitable selection.

Most acid-base indicators are weak, organic acids. Let's represent one by the general symbol H*In*. In the acid-form, H*In*, it has one color; its conjugate base, *In*$^-$, has a different color, the more strikingly different the better. The usual equilibrium equation for weak acids applies to indicators, too.

$$\text{H}In \rightleftharpoons \text{H}^+ + In^-$$

acid-form (one color) base-form (another color)

Being a weak acid, H*In* has a K_a and a pK_a value.

$$K_{In} = \frac{[\text{H}^+][In^-]}{[\text{H}In]}$$

$$pK_{In} = -\log K_{In}$$

Since the indicator is an acid, it will also be neutralized in the titration. Therefore, we want to use as little of the indicator as possible to reduce its contribution to the volume of acid (or base) consumed in the titration. Of course, a trace quantity of indicator will work only if its two colors are very intense and the color change is still very obvious even in an extremely dilute solution.

Midway in the transition from the indicator's color in its acid-form to its color in its base-form, the concentrations of H*In* and *In*$^-$ will be equal, because the indicator will be exactly one-half neutralized. When [H*In*] = [*In*$^-$], then the acid ionization constant for the indicator reduces as follows.

$$K_{In} = \frac{[\text{H}^+][In^-]}{[\text{H}In]} = [\text{H}^+]$$

Taking negative logarithms of both sides gives us

$$-\log K_{In} = -\log [\text{H}^+]$$

Or

$$pK_{In} = pH_{\text{at equivalence point}}$$

This important result tells us that once we have calculated the pH that the solution should have at its equivalence point we automatically know the value of pK_{In} that the indicator ought to have. Therefore, the way to pick an indicator is to use one whose pK_{In} equals or is as close as possible to the pH calculated for the solution that is present at the equivalence point. This will insure that the actual end point—the change in color—will happen at the equivalence point.

One factor that has been ignored thus far is the ability of the human eye to distinguish between slight changes in shades of color. As discussed in Special Topic 17.3, it takes as much as a change in two pH units for some indicators (more for litmus) before the color is actually observed to switch from the color of the indicator's acid-form to the color of its base-form. Tables of acid-base indicators (for example, Table 17.4) therefore report the approximate pH range over which the color changes rather than the value of pK_{In}.

Titration Curves

When the pH of a solution at different stages during a titration is plotted against the volume of standard solution added, we obtain a *titration curve*. Figure 17.5, for example, gives a titration curve for the titration of a dilute solution of a strong acid such as hydrochloric acid with a dilute solution of a strong base such as sodium hydroxide. Up until just before the equivalence point the pH changes slowly. Then the last 0.01 ml of the standard base needed to reach the equivalence point makes the pH leap from 4.4 to 7. If another 0.01 ml of the base is added to go slightly beyond the equivalence point, the pH jumps from 7 to 9.6. In other words, from 24.99 ml of base to 25.01 ml of base the pH changes by 5.2 units. Phenolphthalein is therefore an acceptable indicator because its pH range is 8.2 (colorless) to 10.0 (pink). While the pH range for bromothymol blue (6.0 to 7.6) or cresol red (7.0 to 8.8) seem better, neither gives quite as definite a visible change in color as phenolphthalein. A change from colorless to pink (phenolphthalein) is far more noticeable than a change from yellow to blue (for bromothymol blue) or yellow to red (cresol red).

Just one drop delivers about 0.05 ml.

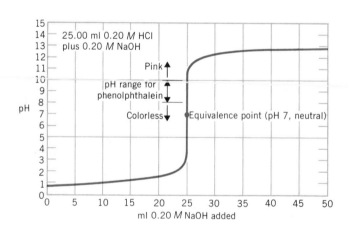

Figure 17.5
Titration curve for the titration of a strong acid (0.20 M HCl) with a strong base (0.20 M NaOH).

Figure 17.6 gives the titration curve for the titration of a typical weak acid with a typical strong base. The pH at the equivalence point is 8.87, which makes phenolphthalein a nearly perfect indicator. Notice again how rapidly the pH changes at points very close to the equivalence point.

Suppose that the color of HIn is red and the color of its anion (In^-) is yellow. In terms of the eye's abilities, the ratio of [HIn] to [In^-] has to be roughly at least as large as 10 to 1 for it to see the color as definitely red. And this ratio has to be at least as small as 1 to 10 in order for the eye to see the color as definitely yellow. Therefore, the color transition at the end point of the titration happens as the ratio of [HIn] to [In^-] changes from 10/1 to 1/10. We can calculate the relation of pH to pK_a for the indicator at the start of this range as follows. We know that

$$K_{In} = \frac{[H^+][In^-]}{[HIn]}$$

Therefore,

$$K_{In} = [H^+]\frac{(10)}{(1)}$$

$$pK_{in} = pH - \log 10$$

Or

$$pH = pK_{in} + 1$$

At the end of the range when [In^-]/[HIn] is 1/10 we have

$$K_{In} = [H^+]\frac{(1)}{(10)}$$

$$pK_{in} = pH - \log(10^{-1})$$

$$pH = pK_{In} - 1$$

Therefore, the *range* of pH over which a definite change in color occurs is

$$pH = pK_{in} \pm 1$$

A range of ± 1 pH units about the value of pK_{in} corresponds to a span of 2 pH units over which the eye will definitely notice the color change.

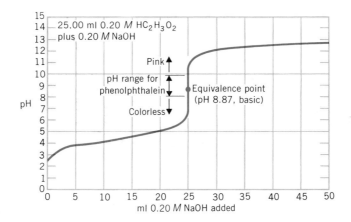

Figure 17.6
Titration curve for the titration of a weak acid (0.20 M acetic acid) with a strong base (0.20 M NaOH).

Figure 17.7 shows the titration curve for the titration of a typical weak base with a strong acid. Now the equivalence point is less than 7, at 6.62, and phenolphthalein no longer is the best indicator. Now bromothymol blue is a good indicator; its pH range is 6.0 to 7.6. However, as Figure 17.7 shows, even methyl orange (range, 3.2–4.4) would work.

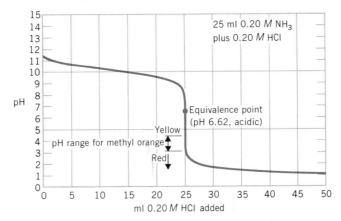

Figure 17.7
Titration curve for the titration of a weak base (0.20 M NH$_3$) with a strong acid (0.20 M HCl)

17.10 LEWIS THEORY OF ACIDS AND BASES

G. N. Lewis defined an acid as an electron pair acceptor and a base as an electron pair donor

In the Brønsted view of acids and bases, the key reaction of neutralization is simply a proton-transfer from the acid to the base. However, there are a number of other reactions with the "look and feel" of neutralization that don't involve protons. In considering these reactions, G. N. Lewis saw a way to broaden the concept of acids and bases beyond the Brønsted view. **Lewis acids and bases** are defined as follows.

> **Lewis Definitions of Acids and Bases**
> 1 An acid is any ionic or molecular species that can accept a pair of electrons in the formation of a bond.
> 2 A base is any ionic or molecular species that can donate a pair of electrons to form a bond.
> 3 Neutralization is the formation of an electron-pair bond between the donor (base) and the acceptor (acid).

The reaction between ammonia and boron trifluoride, both of which are gases, is the classic example of a Lewis acid-base reaction.

ammonia — Lewis base; boron trifluoride — Lewis acid; ammonia-boron trifluoride complex

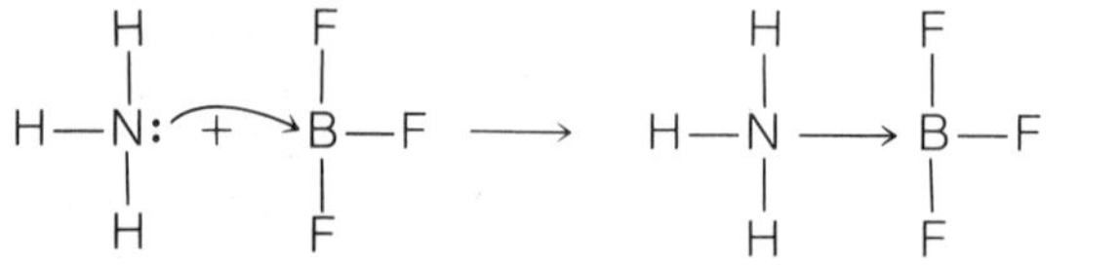

The ammonia molecule has an unshared pair of electrons and is a Lewis base. The boron trifluoride molecule has only a sextet of outer-shell electrons, not an octet, and it is a Lewis acid. When ammonia and boron trifluoride combine, a coordinate covalent bond forms between nitrogen and boron, which we can represent by an arrow as we learned in Section 8.7.

The Lewis concept includes both the Arrhenius and Brønsted acids and bases as special cases. The proton, for example, is a Lewis acid; the hydroxide ion is a Lewis base:

$$H^+ + :\ddot{\underset{\cdot\cdot}{O}}{-}H^- \longrightarrow H_2O$$

Arrhenius acid	Arrhenius base
Brønsted acid	Brønsted base
Lewis acid	Lewis base

The reaction of sulfur trioxide with water to give sulfuric acid can be seen as a Lewis acid-base reaction:

Lewis base; Lewis acid; sulfuric acid H_2SO_4

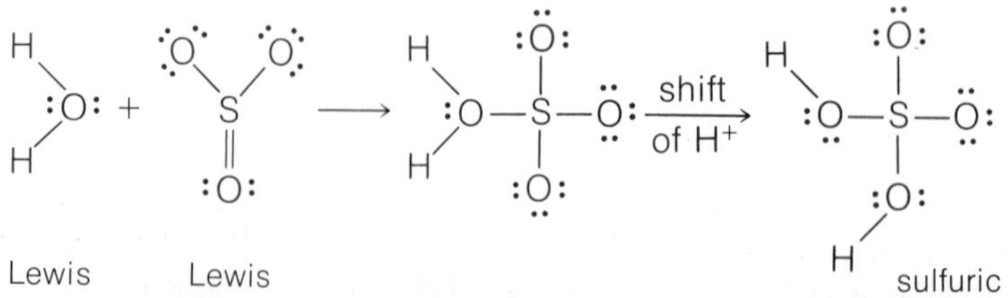

EXERCISE 17.36

This behavior of OH⁻ to CO₂ means that a standard solution of NaOH or KOH exposed to the air, which contains CO₂, will not long remain at the normality given on the bottle label.

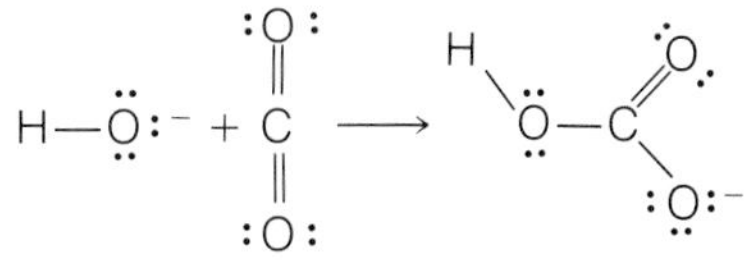

When carbon dioxide is bubbled into aqueous sodium hydroxide, the gas is instantly trapped as the bicarbonate ion. (In fact, this reaction makes it hard to store standard NaOH solutions in contact with air in which traces of CO_2 are always present.) The reaction can be written as follows.

bicarbonate ion

Which is the Lewis acid and which is the Lewis base?

The hydration of an ion is also a Lewis acid-base reaction. For example, Al^{3+} draws six molecules of water to itself:

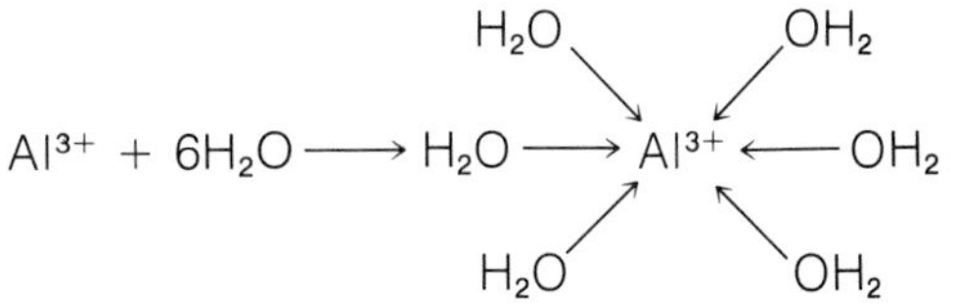

When the metal ion has a charge as high as 3+, and often when the charge is 2+, the ion attracts electron density from the surrounding solvent molecules so much that bonds holding protons are weakened. These *hydrated* ions are now actually Brønsted acids. A proton can be donated to a water molecule to make H_3O^+, and the solution becomes more acidic.

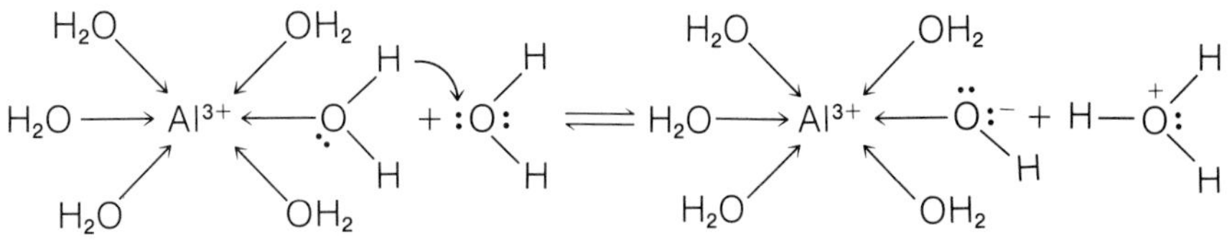

In this way solutions of salts between the anions of strong acids and *any* metallic ions with charges of 2+ and 3+ will test acidic in water. The chief exceptions are the Group IIA metals below beryllium. The larger the positive charge on the central ion, the more acidic can a salt solution be. A 0.1 M solution of aluminum chloride has a pH of 3, which makes aluminum chloride almost as good a proton-generator as 0.1 M acetic acid.

EXERCISE 17.37 We have learned that BF_3 is a Lewis acid. How does the position of aluminum in the periodic table help explain why $AlCl_3$ is also a good Lewis acid?

SUMMARY

Self-Ionization of Water. In pure water an equilibrium exists: $H_2O \rightleftharpoons H^+ + OH^-$; and $[H^+] = [OH^-]$, a relationship that defines a neutral solution. A modified K_c called K_w, the ion-product constant of water, is equal to the product

$[H^+][OH^-]$. $K_w = 1.00 \times 10^{-14}$ at 25 °C. When $[H^+] > [OH^-]$, the solution is acidic; when $[H^+] < [OH^-]$, the solution is basic.

pH Concept. To make it easier to compare values of $[H^+]$ when they are very small, they are converted to values of pH by the equation, $pH = -\log [H^+]$. Low values of $[OH^-]$ may be compared by analogous pOH values, where $pOH = -\log [OH^-]$. At 25 °C $pH + pOH = 14.00$. When $pH < 7$, the solution is acidic, and when $pH > 7$, it is basic.

Brønsted Concept of Acids and Bases. At the heart of all acid-base reactions in water is the transfer of H^+, the hydrogen ion or the proton. A donor of H^+, a Brønsted acid, gives one to an acceptor, a Brønsted base. The members of a conjugate acid-base pair differ only by one H^+. To compare the relative strengths of these acids, acid ionization constants. K_a, are used or their modified forms, pK_a. $K_a = \dfrac{[H^+][\text{conjugate base of acid}]}{[\text{acid}]}$ and $pK_a = -\log K_a$. The nature of negative logarithms is such that the *lower* the pK_a, the stronger the acid. Polyprotic acids have individual values of K_a for each successive loss of a proton. Similar expressions exist for Brønsted bases.

$$K_b = \frac{[OH^-][\text{conjugate acid of the base}]}{[\text{base}]} \quad \text{and} \quad pk_b =$$

$-\log K_b$. The lower the pK_b, the stronger the base. The K_a of an acid and the K_b of its own conjugate base are related: $K_a K_b = K_w$ and $pK_a + pK_b = pK_w$. This means that the stronger an acid is, the weaker is its conjugate base. And the stronger a base, the weaker its conjugate acid. Some species are amphoteric and will be proton-donors to good acceptors and acceptors to good donors. The direction that a proton-transfer reaction takes follows the rule that a stronger acid-base pair will always change to be replaced by a weaker acid-base pair.

Hydrolysis of Salts. When water itself, as the solvent, is directly involved either as a proton-donor or a proton-acceptor to an ion of a dissolved salt, the balance between $[H^+]$ and $[OH^-]$ will change. In such reactions, called hydrolyses of salts, the medium will become acidic if the cation of the salt is a Brønsted acid (e.g., NH_4^+) or a Lewis acid (e.g., Al^{3+}) but the anion is not comparably basic. A salt will cause its aqueous solution to be basic if its anion is a

Brønsted or Lewis base but its cation is not comparably acidic.

Buffers. The addition of small quantities of strong acids or bases to a solution holding a conjugate acid/base pair, the buffer system, will change the pH only slightly. Generally, the acid of the pair is relatively weak but can still neutralize OH^-. The conjugate base, while also weak, can still neutralize H^+. The Henderson-Hasselbalch equation is used to find the pH of a buffer solution:

$$pH = pK_a + \log \frac{[\text{anion}]}{[\text{acid}]}.$$

Acid-Base Titrations. The equivalence point in a titration occurs when the number of equivalents of acid equals the number of equivalents of base. How many grams of an acid or a base constitute one equivalent of the compound depends on the specific reaction. When the full hydrogen-ion donating ability of a di- or triprotic acid is used, then the equivalent weight is one-half the formula weight for the diprotic acid and one-third the formula weight for the triprotic acid. Likewise, when the full hydrogen-ion neutralizing ability of a base is used, then the equivalent weight will be equal to or a simple fraction of the formula weight. The ratio of the number of equivalents of an acid (or a base) to the volume of the solution in liters is called the normality, N, of the solution. The value of N gives the number of equivalents per liter. The actual pH of a solution at its equivalence point in a titration may not equal 7.00, because the salt produced by the titration may hydrolyze. The indicator is picked so that its own pK_{In} is within ± 1 pH unit of the pH of the solution of the salt made in the titration. In repeated routine analytic work with acid-base titrations, the analyst usually calculates normalities from the equation $N_a \times V_a = N_b \times V_b$.

Lewis Concept of Acids and Bases. Species with central atoms lacking octets (e.g., BF_3, $AlCl_3$)—Lewis acids—can form covalent bonds with the central atoms having unshared electron pairs of other species—Lewis bases. Metal ions with charges of $3+$, and many with charges of $2+$, are typical Lewis acids, too. Water is a common Lewis base. The hydration of such metal atoms gives hydrated ions that can function as Brønsted acids. Their aqueous solutions usually test acidic.

INDEX TO EXERCISES, QUESTIONS, AND PROBLEMS
(Those numbered above 64 are review problems.)

REVIEW QUESTIONS

17.38. Label each species in these equations as stronger or weaker acids and bases (in the Brønsted sense). In every equilibrium the substances on the right side of the arrows are favored.
(a) $H_2SO_3 + NH_3 \rightleftharpoons HSO_3^- + NH_4^+$
(b) $HCO_3^- + H_3O^+ \rightleftharpoons H_2CO_3 + H_2O$
(c) $PO_4^{3-} + H_2O \rightleftharpoons HPO_4^{2-} + OH^-$
(d) $HCO_3^- + HO^- \rightleftharpoons CO_3^{2-} + H_2O$
(e) $OH^- + NH_4^+ \rightleftharpoons H_2O + NH_3$
(f) $H_2O + HCl(g) \rightleftharpoons H_3O^+ + Cl^-$

17.39. Give the names and formulas of seven strong acids.

17.40. Give the names and formulas of five strong aqueous bases.

17.41. Complete the following equations to represent proton-transfer equilibria. (Assume only *single* proton-transfer events *from* the first species.)
(a) $HSO_4^- + H_2O \rightleftharpoons$
(b) $H_2CO_3 + NH_3 \rightleftharpoons$
(c) $HSO_4^- + HSO_3^- \rightleftharpoons$
(d) $H_2PO_4^- + CO_3^{2-} \rightleftharpoons$
(e) $H_2PO_4^- + PO_4^{3-} \rightleftharpoons$
(f) $H_3O^+ + NH_3 \rightleftharpoons$
(g) $HSO_4^- + NH_3 \rightleftharpoons$

17.42. Connect conjugate acid-base pairs in the equilibria of 17.41 by lines.

17.43. How does the equation for the ion-product constant of water differ from that of the true equilibrium constant for water's self-ionization.

17.44. How are acidic, basic, and neutral solutions in water defined
(a) in terms of $[H^+]$ and $[OH^-]$?
(b) in terms of pH?

17.45. The value of K_w increases as the temperature of water is raised.
(a) What effect will increasing temperature have on the pH of pure water? (Will the pH increase, decrease or stay the same?)
(b) As the temperature of pure water rises will the water become more acidic, more basic, or remain neutral?
(c) Carefully explain your answers to both (a) and (b)

17.46. Benzoic acid, $C_6H_5CO_2H$, is an organic acid whose sodium salt, $C_6H_5CO_2Na$, has long been used as a safe food additive to protect beverages and many foods against harmful yeasts and bacteria. The acid is monoprotic. Write the equation for its K_a.

17.47. Write the equation for the equilibrium that the benzoate ion (Question 17.46) would produce in water as it functions as a Brønsted base. Then write the expression for the K_b of the conjugate base of benzoic acid.

17.48. When sulfur dioxide, an air pollutant from the burning of sulfur-containing coal or oil, dissolves in water some reacts to form sulfurous acid, H_2SO_3:

$$SO_2 + H_2O \rightleftharpoons H_2SO_3$$

(a) Use the data in Table 17.5 to decide whether this acid is a weak, moderate, or strong acid?
(b) Write the expression for K_{a_1} and K_{a_2}.
(c) Carbon dioxide, also produced by the burning of coal or oil, will similarly react with water to produce carbonic acid, H_2CO_3. Write the equilibria and K_a expressions of its K_{a_1} and K_{a_2}.
(d) Which is the stronger acid: sulfurous or carbonic?

17.49. If water were treated as any other weak diprotic Brønsted acid, what would be the equation expressing its K_{a_1}? How does that expression differ from K_w? Write the expression for K_{a_2} for water.

17.50. What is the conjugate *base* of each of these species?
(a) CH_4 (methane, natural gas)
(b) OH^- (hydroxide ion)
(c) NH_3 (ammonia)
(d) CH_3-O-H (methyl alcohol; treat as a monoprotic acid with the H on the oxygen as the transferable proton)
(e) $H-C\equiv C-H$ (acetylene; treat as a monoprotic acid)

17.51. What is the conjugate *acid* of each of these species?
(a) H^- (hydride ion, a species that *transfers* but is never free during biological oxidations)
(b) $CH_3CH_2-O^-$ (the ethoxide ion, an important base used in organic syntheses because it's more widely soluble in organic solvents than OH^-, and it's stronger than OH^-)
(c) CH_3-O-H (methyl alcohol; use the oxygen as the H^+ accepting site)
(d) S^{2-} (the sulfide ion, an important analytical reagent)
(e) $CH_3-\ddot{N}H_2$ (methylamine; the nitrogen is the H^+ accepting site)

17.52. The percent dissociation of a weak acid (HA) is defined by this equation:

$$\% \text{ dissociation} = \frac{[H^+]_{equilib.}}{[HA]_{initial.}} \times 100.$$

Show that when you know K_a of the weak acid and its molar concentration, the percent ionization can be found by this expression:

$$\text{percent ionization} = \sqrt{\frac{K_a}{[HA]_{initial}}} \times 100$$

What assumption is made in this proof?

17.53. The pK_a of hydrocyanic acid (HCN) is 9.31 and of hydrofluoric acid (HF) is 3.45. Which is the stronger Brønsted base: CN^- or F^-?

17.54. Aspirin is the sodium salt of acetylsalicylic acid whose K_a value is 3.27×10^{-4}. Will a solution of aspirin in water test acidic, basic, or neutral?

17.55. Write ionic equations that illustrate how each pair of compounds can serve as a buffer pair.
(a) H_2CO_3 and $NaHCO_3$ (The "carbonate" buffer in blood.)
(b) NaH_2PO_4 and Na_2HPO_4 (The "phosphate" buffer in blood.)
(c) NH_4Cl and NH_3

(d) Show that for the buffer pair of part (c) that the Henderson-Hasselbalch equation takes the following form, where K_a refers to the ammonium ion, NH_4^+.

$$pH = pK_a + \log \frac{[NH_3]_{initial}}{[NH_4^+]_{initial}}$$

17.56. Which buffer would be better able to hold a steady pH at the addition of strong acid: buffer 1 or buffer 2? Explain.
Buffer 1: a solution containing 0.10 M NH_4Cl and 1 M NH_3
Buffer 2: a solution containing 1 M NH_4Cl and 0.10 M NH_3

17.57. What would make the titrated solution at the equivalence point in an acid-base titration have a pH not equal to 7.00? How does this affect the choice of an indicator?

17.58. Explain why ethyl red would be a better indicator than phenolphthalein in the titration of dilute ammonia by dilute hydrochloric acid.

17.59. What would be a good indicator for titrating potassium hydroxide with hydrobromic acid? Explain.

17.60. In the titration of an acid with a base what condition concerning the quantities of reactants ought to be true at the equivalence point?

17.61. A base in the Brønsted concept is defined by what it can *accept*. In the Lewis definition, a base is defined by what it can *donate*. In both views, NH_3 is a base. Explain.

17.62. Which of these are Lewis bases and which are Lewis acids? (a) H_2O (b) Cu^{2+} (c) F^- (d) H^+ (e) OH^-

17.63. A solution of copper sulfate, $CuSO_4$, tests slightly acidic. How are the extra hydrogen ions generated?

17.64. The pH of a 20% solution of lead nitrate, $Pb(NO_3)_2$, is about 4. Explain why it isn't 7.

REVIEW PROBLEMS

17.65. At the temperature of the human body, 37 °C, the value of K_w is 2.42×10^{-14}. Calculate $[H^+]$ and $[OH^-]$, pH and pOH, and pK_w. What is the relation between pH, pOH, and pK_w at this temperature? Is water neutral at this temperature?

17.66. Deuterium oxide, D_2O, ionizes like water. At 20 °C its K_w or ion-product constant analogous to that of water is 8.93×10^{-16}. Calculate $[D^+]$ and $[OD^-]$ in deuterium oxide at 20 °C. Calculate also the pD, the pOD, and the pK_w values.

17.67. Assuming 100% ionization of HCl in dilute solutions, what is the pH of 0.01 M HCl?

17.68. If nitric acid is 100% ionized in a 0.00500 M solution, what is the pH of this solution?

17.69. A sodium hydroxide solution is prepared by dissolving 6.00 g NaOH in 1.00 liter of solution. Assuming that 100% ionization occurs, what is the pOH and the pH of this solution?

17.70. A solution was made by dissolving 0.056 g KOH in 100 ml final volume. If KOH is fully broken up into its ions, what is the pOH and the pH of this solution?

17.71. A certain brand of beer had a hydrogen ion concentration equal to 1.9×10^{-5} mol/liter. What is the pH of this beer?

17.72. A soft drink was put on the market with $[H^+] = 1.4 \times 10^{-5}$ mol/liter. What is its pH?

17.73. What will be the percent ionization in a 0.15 M solution of HF? (See Question 17.52.) What is the pH of this solution?

17.74. What is the percent ionization in 1.0 M acetic acid. (See Question 17.52.) What is the pH of this solution?

17.75. Periodic acid, HIO_4, is an important oxidizing agent and a moderately strong acid. In a 0.10 M solution $[H^+] = 3.8 \times 10^{-2}$ mol/liter. Calculate the K_a and pK_a for periodic acid.

17.76. Chloroacetic acid, $Cl-CH_2-CO_2H$, is a stronger monoprotic acid than acetic acid. In a 0.10 M solution of this acid the pH was 1.96. Calculate the K_a and pK_a for chloroacetic acid.

17.77. *para*-Aminobenzoic acid, PABA, is a powerful sunscreening agent whose salts are used widely in sun tanning and screening lotions. The parent acid, which we may symbolize by H-Paba, is a weak acid with a pK_a of 4.92 (25 °C). What will be $[H^+]$ and pH of a 0.030 M solution of this acid?

17.78. Barbituric acid, HBar, was discovered by Adolph von Baeyer (of Baeyer aspirin fame) and named after a friend, Barbara. It is the parent compound of widely used sleeping drugs, the barbiturates. Its pK_a is 4.01. What will be the $[H^+]$ and pH of a 0.050 M solution.

17.79. Ethylamine, $CH_3CH_2NH_2$ has a strong, pungent odor similar to that of ammonia. Like ammonia, it is a Brønsted base. A 0.10 M solution has a pH of 11.86. Calculate the K_b and pK_b for the ethylamine, and find the pK_a for its conjugate acid, $CH_3CH_2NH_3^+$.

17.80. Hydrazine, N_2H_4, has been used as a rocket fuel. Like ammonia, it is a Brønsted base. A 0.15 M solution has a pH of 10.70. What is K_b and pK_b for hydrazine and the pK_a of its conjugate acid?

17.81. Codeine, a cough-suppressant extracted from crude opium, is a weak base with a pK_b of 5.79. What will be the pH of a 0.020 M solution? (Use (Cod) as a symbol for codeine.)

17.82. A hydroxy derivative of ammonia, hydroxylamine (NH_2OH), is a weak base with a pK_b of 7.97. What will be the pH of a 0.25 M solution? It reacts with water as follows: $NH_2OH + H_2O \rightleftharpoons NH_3OH^+ + OH^-$.

17.83. Many drugs that are natural Brønsted bases are put into aqueous solution as their much more soluble salts with strong acids. The powerful pain-killer, morphine, for example, is very slightly soluble in water, but morphine nitrate is quite soluble. We may represent morphine by the symbol (Mor) and its conjugate acid as H-(Mor)$^+$. The pK_b of morphine is 5.79. What is the pK_a of its conjugate acid, and what will be the calculated pH of a 0.20 M solution of morphine nitrate.

17.84. Quinine, an important drug in treating malaria, is a weak Brønsted base that we may represent as (Qu). At 25 °C its pK_b is 5.48. To make it more soluble in water, it is put into a solution as its conjugate acid, which we may represent as H-(Qu)$^+$; e.g., as quinine chloride, H-(Qu)Cl. What is the calculated pH of a 0.15 M solution of quinine chloride?

17.85. Phosphorus acid, H_3PO_3, is actually a diprotic acid; $K_{a_1} = 1.0 \times 10^{-2}$ and $K_{a_2} = 2.6 \times 10^{-7}$. What will be [H$^+$], pH, and [HPO$_3^{2-}$] in a 1.0 M solution? (The simplifying assumption of Example 17.25 may be used because the concentration is relatively high.)

17.86. Tellurium, in the same family as sulfur, forms an acid analogous to sulfuric acid and called telluric acid. It exists, however, as H_6TeO_6 (which looks like $H_2TeO_4 + 2H_2O$). It is a diprotic acid. $K_{a_1} = 2.1 \times 10^{-8}$ and $K_{a_2} = 6.5 \times 10^{-12}$. Calculate [H$^+$], pH, and [H$_4TeO_6^{2-}$] in a 0.25 M solution.

17.87. How many moles of sodium acetate, $NaC_2H_3O_2$, would have to be added to 1 liter of 0.15 M acetic acid ($pK_a = 4.76$) to make the solution a buffer for pH 5.00.

17.88. How many moles of sodium formate, $NaCHO_2$, would have to be dissolved in 1.0 liter of 0.12 M formic acid ($pK_a = 3.75$) to make the solution a buffer for pH 3.80?

17.89. What ratio of molar concentrations of NH_4Cl and NH_3 would buffer a solution at pH 9.25? For NH_4^+, $pK_a = 9.25$. (See Question 17.55d.)

17.90. How many moles of ammonium chloride would have to be dissolved in 500 ml of 0.20 M NH_3 to prepare a solution buffered at pH 10.00? (See Problem 17.89.)

17.91. If 50 ml of 0.10 M formic acid is titrated with 0.10 M sodium hydroxide, what is the pH at the equivalence point? (Be sure to take into account the change in volume during the titration.) What would be a good indicator for this titration?

17.92. If 25 ml of 0.10 M aqueous ammonia is titrated with 0.10 M hydrobromic acid, what will be the pH at the equivalence point? What would be a good indicator?

17.93. The stock supply of 0.958 N HCl was considered too concentrated for a particular analysis, and 100.0 ml was diluted to 500.0 ml. What was the new normality? (Just think in terms of the equivalents taken into the new volume.)

17.94. Concentrated hydrochloric acid is marketed as a 12.0 M solution. To prepare 1.00 liter of 0.500 M solution what volume of the 12 M HCl would have to be diluted to 1.00 liter? (Find the equivalents of HCl in 1 liter of 0.500 M HCl and take enough of 12 M HCl to obtain that amount.)

17.95. Before accepting delivery of a large consignment of acetic acid that was claimed to be 17.5 M (the concentrated acetic acid of commerce), the firm's testing laboratory took a 1.00 ml sample, diluted it to 25.00 ml and titrated it with 0.502 N NaOH. The titration to an ethyl red end point used 34.84 ml base. Was the shipment acceptable? Find the normality and molarity of the acetic acid.

17.96. Concentrated sulfuric acid, the largest volume bulk chemical sold in the United States, is sold at a concentration of 18.0 molar. For routine analysis of samples, 1.00 ml portions are withdrawn and diluted to 25.00 ml for titration against aqueous sodium hydroxide. The base, freshly standardized each day, was found to be 0.122 N one morning. If the acid being analyzed is 18.0 M as advertised, what volume of the base would be needed in this titration? The reaction is

$$H_2SO_4 + 2NaOH \longrightarrow Na_2SO_4 + 2H_2O$$

17.97. To determine the normality of a sodium hydroxide solution suppose you used what happened to be the only standard acid in the laboratory, 0.2000 M H_2SO_4.
(a) What is the *normality* of this acid if you used it for a titration according to the following reaction:

$$H_2SO_4 + NaOH \longrightarrow NaHSO_4 + H_2O$$

(b) However, if you used this standard acid for a titration according to the following equation, what would you have to use as the value of the acid's normality?

$$H_2SO_4 + 2NaOH \longrightarrow Na_2SO_4 + H_2O$$

17.98. When it came time to prepare a fresh supply of dialysate for an artificial kidney machine, the only source of one of the needed chemicals, Na_2CO_3, was a 0.025 M solution. If the directions called for 100 mEq Na_2CO_3, how many ml of 0.025 M Na_2CO_3 would be needed; 1 Eq = 1000 mEq. (Assume 2 Eq Na_2CO_3/mol Na_2CO_3.)

17.99. A clinic received a new shipment of morphine (see Problem 17.83). After two or three uses, the clinical chemist suspected that the morphine was impure. This drug can be titrated with hydrochloric acid according to the equation:

$$Mor + HCl \longrightarrow HMorCl$$

The molecular formula of morphine is $C_{17}H_{19}NO_3$. A sample of this "morphine" with a mass of 1.230 g was titrated with 0.00100 N HCl, and 3.75 ml were neutralized by the morphine in the sample.
(a) How many moles of HCl were used up by the reaction?

(b) How many moles of morphine reacted?

(c) How many grams of morphine were actually in the sample?

(d) What percent of the sample was morphine?

17.100. Aspirin is acetylsalicylic acid with a molecular formula of $C_9H_8O_4$. It can be titrated as a monoprotic acid by sodium hydroxide:

$$C_9H_8O_4 + NaOH \longrightarrow NaC_9H_7O_4 + H_2O$$

A pharmacy received a quantity of powdered aspirin that the pharmacist suspected had been adulterated by starch (which is neutral toward acids and bases). A sample of the questionable aspirin with a mass of 2.346 g was titrated with 0.00350 N NaOH, and the aspirin in the sample neutralized 3.11 ml of the base.

(a) How many moles of NaOH were used up by the aspirin?

(b) How many moles of aspirin reacted?

(c) How many grams of aspirin were actually in the sample?

(d) What percent of the sample was aspirin?

CHAPTER EIGHTEEN
ELECTROCHEMISTRY

18.1 ELECTRICITY AND CHEMICAL REACTIONS

Oxidation-reduction reactions can be brought about by electricity or used to produce electricity

Oxidation and reduction—the loss and gain of electrons—occur in many chemical systems. The rusting of iron, the photosynthesis that takes place in the leaves of trees and other green plants, and the conversion of foods to energy in the body are all examples of chemical changes that involve the transfer of electrons from one chemical species to another. When such reactions can be made to cause electrons to flow through a wire or when a flow of electrons makes a redox reaction happen, the processes are referred to as **electrochemical changes.** The study of these changes is called **electrochemistry.**

The applications of electrochemistry are widespread. Batteries, which produce electrical energy by means of chemical reactions, are used to power toys, flashlights, electronic calculators, pacemakers that maintain the rhythm of the heart, radios, tape recorders, and even some automobiles. In the laboratory, electrical measurements enable us to monitor chemical reactions of all sorts, even those in systems as tiny as a living cell. In industry, many important chemicals—including liquid bleach (sodium hypochlorite) and lye (sodium hydroxide, which is used to manufacture soap)—are manufactured by electrochemical reactions. In fact, if it were not for electrochemical reactions, the important structural metals aluminum and magnesium would be only laboratory curiosities. Most people would only see them in small amounts in museums.

In this chapter we will study the factors that affect the outcome of electrochemical changes. Besides the practical applications, fundamental information about chemical reactions is available from electrical measurements—for example, free energy changes and equilibrium constants. We will see, therefore, that electrochemistry is a very versatile tool for investigating chemical and biological systems.

General Motors' experimental electric car, the Electrovette. Note the battery pack installed in the rear of the vehicle.

18.2 ELECTROLYSIS

Electricity can provide the necessary energy to cause otherwise nonspontaneous reactions to occur

When electricity is passed through a molten ionic compound or through a solution containing ions—an electrolyte—a chemical reaction called **electrolysis** occurs. A typical electrolysis apparatus, referred to as an **electrolysis cell** or **electrolytic cell,** is shown in Figure 18.1. This particular cell contains molten sodium chloride. (A substance undergoing electrolysis must be molten or in solution so that its ions can move freely and conduction can occur.) Inert electrodes—electrodes that won't react with the molten NaCl—are dipped into the cell and then connected to a source of direct current (D.C.) electricity.

Sodium chloride melts at 801 °C.

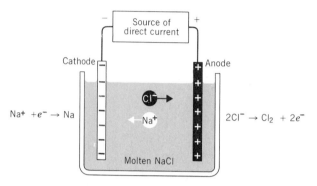

Figure 18.1
Electrolysis cell in which the passage of an electric current decomposes molten sodium chloride into metallic sodium and gaseous chlorine. Unless the products are kept apart, they react on contact to reform NaCl.

When the electricity starts to flow, chemical changes begin to take place. At the positive electrode, the **anode,** *oxidation* occurs as electrons are pulled from negatively charged chloride ions. The D.C. source pumps these electrons through the external electrical circuit to the negative electrode, the **cathode,** where *reduction* takes place as the electrons are picked up by positively charged sodium ions. These reactions can be written

At the melting point of NaCl, metallic sodium is a liquid.

$$Na^+(\ell) + e^- \longrightarrow Na(\ell) \qquad \text{(cathode)}$$
$$2Cl^-(\ell) \longrightarrow Cl_2(g) + 2e^- \qquad \text{(anode)}$$

In any electrochemical cell,

> *oxidation* always occurs at the *anode.*
> *reduction* always occurs at the *cathode.*

When sodium chloride undergoes electrolysis (when sodium chloride is *electrolyzed*), no electrons actually pass through the molten NaCl between the electrodes. Instead, there is a movement of ions through the melt. The positive sodium ions gradually move toward the negative electrode, and the negative chloride ions move toward the positive electrode. In the immediate neighborhood of each electrode, a layer of ions of opposite charge to that of the electrode is formed. Positive ions surround the cathode and negative ions surround the anode. Electrical conduction in the external circuit is only possible because of the oxidation-reduction (redox) reactions that take place at the electrodes.

To obtain the **cell reaction**—the overall reaction for the electrolysis—we simply add the individual electrode reactions together. Before we can do this, however, we must make sure that the number of electrons gained is equal to the number lost, as must be the case in any redox change. We can see that the cathode reaction must occur twice to take up the two electrons that are lost when the anode reaction occurs once. This means that the coefficients of the cathode reaction must be multipled by 2.

$$2Na^+(\ell) + 2e^- \longrightarrow 2Na(\ell) \qquad \text{(cathode)}$$
$$2Cl^-(\ell) \longrightarrow Cl_2(g) + 2e^- \qquad \text{(anode)}$$

Adding these gives

$$2Na^+(\ell) + 2Cl^-(\ell) + 2e^- \longrightarrow 2Na(\ell) + Cl_2(g) + 2e^-$$

Since two electrons appear on both sides of the equation, they can be canceled (just like spectator ions). The final cell reaction, then, is

$$2Na^+(\ell) + 2Cl^-(\ell) \xrightarrow{\text{electrolysis}} 2Na(\ell) + Cl_2(g)$$

As you know, table salt is quite stable. It doesn't normally decompose because the reaction of sodium and chlorine to form sodium chloride is highly spontaneous. We therefore write the word *electrolysis* above the arrow in the equation to show that electricity is the driving force for this nonspontaneous reaction.

Now let's examine some electrolysis reactions in aqueous solution. These are more difficult to predict because the oxidation and reduction of water can also occur. This happens, for example, when electrolysis is carried out in a solution of potassium nitrate. The products of the electrolysis are hydrogen and oxygen, as shown in Figure 18.2. At the cathode, water is reduced.

> When electrons appear as a reactant, the process is reduction; when they appear as a product, it is oxidation.

$$2H_2O(\ell) + 2e^- \longrightarrow H_2(g) + 2OH^-(aq) \qquad \text{(cathode)}$$

At the anode, water is oxidized.

$$2H_2O(\ell) \longrightarrow O_2(g) + 4H^+(aq) + 4e^- \qquad \text{(anode)}$$

If acid-base indicators are placed in the solution, color changes confirm that the solution becomes basic around the cathode and acidic around the anode as the electrolysis proceeds (see Color Plate 6A). In addition, the gases H_2 and O_2 can be collected.

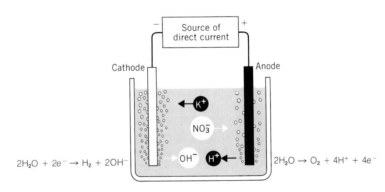

Figure 18.2
Electrolysis of a solution of potassium nitrate gives hydrogen gas and oxygen gas as products.

The overall cell reaction can be obtained as before. Since the number of electrons lost has to equal the number gained, the cathode reaction must occur twice each time the anode reaction occurs once.

$$4H_2O(\ell) + 4e^- \longrightarrow 2H_2(g) + 4OH^-(aq)$$
$$2H_2O(\ell) \longrightarrow O_2(g) + 4H^+(aq) + 4e^-$$

After adding, we combine the coefficients for water and cancel the electrons from both sides. This gives

$$6H_2O(\ell) \longrightarrow 2H_2(g) + O_2(g) + 4H^+(aq) + 4OH^-(aq)$$

Notice that hydrogen ions and hydroxide ions are produced in equal numbers. If the solution is stirred, they combine to form water. The net change that takes place, then, is

$$2H_2O(\ell) \xrightarrow{\text{electrolysis}} 2H_2(g) + O_2(g)$$

At this point you may have begun to wonder whether the potassium nitrate serves any function. If electrolysis is attempted without the KNO₃ (i.e., using pure distilled water), nothing happens. There is no current flow, and no H₂ or O₂ are produced. Obviously, then, the potassium nitrate must have some function.

The presence of KNO₃ (or some other electrolyte) is necessary to maintain electrical neutrality in the vicinity of the electrodes. If the KNO₃ were not present and the electrolysis were to occur, the solution around the anode would become filled with H⁺ ions, with no negative ions to counter the charge. Similarly, the solution surrounding the cathode would become filled with hydroxide ions with no nearby positive ions. Nature simply doesn't allow this to happen.

When KNO₃ is in the solution, K⁺ ions move toward the cathode where they mingle with the OH⁻ ions as they are formed. The NO₃⁻ move toward the anode and mingle with the H⁺ ions as they are produced there. In this way, at any moment, each small region of the solution contains the same number of positive and negative charges.

Water molecules are oxidized and reduced when the electrolysis of a potassium nitrate solution is carried out because K⁺ is more difficult to reduce than water and because NO₃⁻ is more difficult to oxidize than water. This is not always the case with solutes other than KNO₃. For instance, when a solution of copper(II) bromide is electrolyzed, the cathode becomes coated with a reddish-brown deposit as the blue color of copper ion in the surrounding solution fades. At the same time, the solution around the anode acquires a reddish-brown color. These observations tell us that copper ions are reduced to metallic copper at the cathode and bromide ions are oxidized to molecular bromine at the anode. The electrode reactions are

$$Cu^{2+}(aq) + 2e^- \longrightarrow Cu(s) \quad \text{(cathode)}$$
$$2Br^-(aq) \longrightarrow Br_2(aq) + 2e^- \quad \text{(anode)}$$

and the net cell reaction is

$$Cu^{2+}(aq) + 2Br^-(aq) \xrightarrow{\text{electrolysis}} Cu(s) + Br_2(aq)$$

The fact that this reaction occurs tells us that Cu²⁺ is more easily reduced than water and that Br⁻ is more easily oxidized than water.

The behavior of an ion toward reduction or oxidation by electrolysis is the same regardless of the source of the ion. Therefore, once we know what happens to a solution of copper(II) bromide we can predict, at least partially, what will happen whenever a solution of a salt containing copper(II) ion or a salt containing bromide ion is electrolyzed. Solutions of CuCl₂, Cu(NO₃)₂, CuSO₄, and Cu(C₂H₃O₂)₂ all contain Cu²⁺ and, upon electrolysis, will give metallic copper at the cathode. Similarly, solutions of KBr, NaBr, FeBr₂ and CaBr₂ all contain Br⁻ and will give Br₂ at the anode when electrolyzed.

The reaction that can occur most easily is the reaction that we actually observe.

EXAMPLE 18.1 Predicting the Outcome of an Electrolysis

Problem: From the results of the electrolysis of solutions of KNO₃ and of CuBr₂, predict the products that will appear if a solution of Cu(NO₃)₂ is electrolyzed. Give the net cell reaction.

Solution: A solution of Cu(NO₃)₂ contains Cu²⁺ and NO₃⁻ ions. The Cu²⁺ will move toward the cathode and NO₃⁻ will move toward the anode. The electrolyses described previously in this section tell us that the Cu²⁺ is more easily reduced than water at the cathode and that water is more easily oxidized than the NO₃⁻ at the anode. The electrode reactions, therefore, will be

$$Cu^{2+}(aq) + 2e^- \longrightarrow Cu(s) \qquad \text{(cathode—reduction of } Cu^{2+})$$

$$2H_2O(\ell) \longrightarrow O_2(g) + 4H^+(aq) + 4e^- \qquad \text{(anode—oxidation of } H_2O)$$

To obtain the cell reaction, we must multiply the cathode reaction by 2 so that the numbers of electrons gained and lost are equal. This gives

$$2Cu^{2+}(aq) + 2H_2O(\ell) \longrightarrow 2Cu(s) + O_2(g) + 4H^+(aq)$$

EXERCISE 18.1   From the electrolysis reactions described in this section, predict the products that will form when a solution of KBr undergoes electrolysis. Write the equation for the net cell reaction.

18.3   HALF-REACTIONS: BALANCING REDOX EQUATIONS

Dividing a reaction into two parts that are balanced separately provides a simple method for obtaining a net ionic equation for a redox reaction

In the last section we saw that the oxidation and reduction reactions that occur at the electrodes can be added to obtain the overall, or net, cell reaction for an electrolysis. The critical requirement is that the number of electrons gained in the reduction has to be the same as the number lost in the oxidation. Sometimes the coefficients have to be multiplied by some factor to make the number of electrons equal before the two equations can be added.

The individual oxidation and reduction reactions written in the last section are called **half-reactions**—two half-reactions add to give the entire net reaction. The procedure used for combining half-reactions to obtain the net reaction is so useful that it forms the basis for a simple, systematic approach to balancing redox equations. The procedure is called the **ion-electron method.** Applying this method gives the net ionic equation (i.e., the ionic equation, ignoring any spectator ions that might be present) for a redox reaction. However, before using the method (which is the simplest way to see how it works), we must first discuss some facts about redox reactions in aqueous solutions.

In many oxidation-reduction reactions, H^+ or OH^- ions play an important role, as do water molecules. For example, if $K_2Cr_2O_7$ and $FeSO_4$ solutions are mixed, it is observed that the pH of the solution rises as dichromate ion, $Cr_2O_7^{2-}$, oxidizes iron(II). This is because the reaction uses up H^+ as a reactant and it produces H_2O as a product. In other reactions, OH^- is consumed while in still others, H_2O is a reactant. Another fact is that in many cases the products (or even the reactants) of a redox reaction will differ depending on the pH of the solution. For example, in an acidic solution $Cr_2O_7^{2-}$ oxidizes Fe^{2+} to give Cr^{3+} and Fe^{3+} as products. On the other hand, if the reaction is carried out in basic solution, the reactants are written as CrO_4^{2-} and $Fe(OH)_2$. This is because $Cr_2O_7^{2-}$ changes to CrO_4^{2-} in basic solution and because $Fe(OH)_2$ is insoluble. The products of the reaction are CrO_2^- and $Fe(OH)_3$.

Because H^+ or OH^- can be consumed or produced by a redox reaction, and because the products can change during the reaction if the solution changes from acidic to basic (or vice versa), redox reactions are generally carried out in solutions containing a substantial excess of either acid or base. Therefore, before we can apply the ion-electron method, we have to know whether the reaction occurs in an acidic or a basic solution. (This information will always be given to you in this book.)

Balancing redox equations for acidic solutions

As we just learned, $Cr_2O_7^{2-}$ reacts with Fe^{2+} to give Cr^{3+} and Fe^{3+} as products. This information gives the skeleton equation,

$$Cr_2O_7^{2-} + Fe^{2+} \longrightarrow Cr^{3+} + Fe^{3+}$$

We can then use the following steps to find the balanced net ionic equation.

Step 1. Divide the skeleton equation into half-reactions.
    We create the beginnings of two half-reactions. Except for hydrogen and oxygen, the same elements have to appear on both sides of a given half-reaction.

$$Cr_2O_7^{2-} \longrightarrow Cr^{3+}$$
$$Fe^{2+} \longrightarrow Fe^{3+}$$

Step 2. Balance atoms other than H and O.
    There are two Cr atoms on the left and only one on the right in the first half-reaction, so a 2 is placed in front of Cr^{3+}. The second half-reaction is already balanced in terms of atoms, so nothing need be done to it.

$$Cr_2O_7^{2-} \longrightarrow 2Cr^{3+}$$
$$Fe^{2+} \longrightarrow Fe^{3+}$$

Step 3. Balance oxygen atoms by adding H_2O to the side that needs O.
    There are seven oxygen atoms on the left of the first reaction. These are balanced by adding $7H_2O$ (which contains 7 oxygen atoms) to the right. (We have just discovered that water is a product in this reaction.)

$$Cr_2O_7^{2-} \longrightarrow 2Cr^{3+} + 7H_2O$$
$$Fe^{2+} \longrightarrow Fe^{3+}$$

Step 4. Balance hydrogen by adding H^+ to the side that needs H.
    The first half-reaction has 14H on the right; we add $14H^+$ to the left. When you do this step (or others) *be careful to write the charges on the ions.* If they are omitted, you will not obtain a balanced equation in the end.

$$14H^+ + Cr_2O_7^{2-} \longrightarrow 2Cr^{3+} + 7H_2O$$
$$Fe^{2+} \longrightarrow Fe^{3+}$$

Now each half-reaction is balanced in terms of atoms. Next we will balance the charge.

Step 5. Balance the charge by adding electrons.
    First compute the net electrical charge on each side. For the first half-reaction we have

$$\underbrace{14H^+ + Cr_2O_7^{2-}}_{\text{net charge} = (14+) + (2-) = 12+} \longrightarrow \underbrace{2Cr^{3+} + 7H_2O}_{\text{net charge} = 2(3+) = 6+}$$

The difference between the net charges on each side is equal to the number of electrons that must be added to the most positive (or least negative) side. In this half-reaction, $(12+) - (6+) = 6+$. This means that we have to add $6e^-$ to the left side.

$$6e^- + 14H^+ + Cr_2O_7^{2-} \longrightarrow 2Cr^{3+} + 7H_2O$$

This half-reaction is now complete—it is balanced in terms of both atoms and charge.

Check the net charge after adding e^-—it must be the same on both sides.

    The second half-reaction ($Fe^{2+} \to Fe^{3+}$) is easy to balance. We simply add one electron to the right.

$$Fe^{2+} \longrightarrow Fe^{3+} + e^-$$

Now this half-reaction is completed too.

Step 6. Make the electrons gained equal to the electrons lost and then add the two half-reactions.

At this point we have the two half-reactions,

$$6e^- + 14H^+ + Cr_2O_7^{2-} \longrightarrow 2Cr^{3+} + 7H_2O$$
$$Fe^{2+} \longrightarrow Fe^{3+} + e^-$$

Six electrons are gained in the first half-reaction but only one electron is lost in the second. Therefore, the second half-reaction must be multiplied by 6 before it can be added to the first.

$$6e^- + 14H^+ + Cr_2O_7^{2-} \longrightarrow 2Cr^{3+} + 7H_2O$$
$$6(Fe^{2+} \longrightarrow Fe^{3+} + e^-)$$

(Sum) $6e^- + 14H^+ + Cr_2O_7^{2-} + 6Fe^{2+} \longrightarrow 2Cr^{3+} + 7H_2O + 6Fe^{3+} + 6e^-$

Step 7. Cancel anything that is the same on both sides.
This is the final step (at last!). Six electrons are canceled from each side to give the final equation.

$$14H^+ + Cr_2O_7^{2-} + 6Fe^{2+} \longrightarrow 2Cr^{3+} + 7H_2O + 6Fe^{3+}$$

Notice that *both* charge and atoms balance.
In some reactions you may have H_2O or H^+ on both sides—for example, $6H_2O$ on the left and $2H_2O$ on the right. Cancel as many as you can. For example,

$$\ldots + 6H_2O \longrightarrow \ldots + 2H_2O$$

gives

$$\ldots + 4H_2O \longrightarrow \ldots$$

The following is a summary of the steps used for balancing redox reactions that occur in an acidic solution.

Ion-Electron Method—Acid Solution

Step 1. Divide equation into half-reactions.
Step 2. Balance atoms other than H and O.
Step 3. Balance O by adding H_2O.
Step 4. Balance H by adding H^+.
Step 5. Balance net charge by adding e^-.
Step 6. Make e^- gain and loss equal; then add half-reactions.
Step 7. Cancel anything that's the same on both sides.

At this point you have probably noticed that the ion-election method gives a balanced redox equation without the use of oxidation numbers. This is an important advantage of the method. In fact, when you use the ion-electron method, you should be careful to avoid oxidation numbers altogether.

EXAMPLE 18.2   **Using the Ion-Electron Method**

Problem: Balance the following equation. The reaction occurs in an acidic solution.

$$MnO_4^- + H_2SO_3 \longrightarrow SO_4^{2-} + Mn^{2+}$$

Solution:
Step 1:
$$MnO_4^- \longrightarrow Mn^{2+}$$
$$H_2SO_3 \longrightarrow SO_4^{2-}$$

Step 2: This step is not required for this particular reaction because all atoms other than H and O are in balance.

Step 3:
$$MnO_4^- \longrightarrow Mn^{2+} + 4H_2O$$
$$H_2O + H_2SO_3 \longrightarrow SO_4^{2-}$$

Step 4:
$$8H^+ + MnO_4^- \longrightarrow Mn^{2+} + 4H_2O$$
$$H_2O + H_2SO_3 \longrightarrow SO_4^{2-} + 4H^+$$

Step 5:
$$5e^- + 8H^+ + MnO_4^- \longrightarrow Mn^{2+} + 4H_2O$$
$$H_2O + H_2SO_3 \longrightarrow SO_4^{2-} + 4H^+ + 2e^-$$

Step 6:
$$2\,(5e^- + 8H^+ + MnO_4^- \longrightarrow Mn^{2+} + 4H_2O)$$
$$5\,(\qquad H_2O + H_2SO_3 \longrightarrow SO_4^{2-} + 4H^+ + 2e^-)$$
$$\overline{10e^- + 16H^+ + 2MnO_4^- + 5H_2O + 5H_2SO_3 \longrightarrow 2Mn^{2+} + 8H_2O + 5SO_4^{2-} + 20H^+ + 10e^-}$$

Step 7: Cancel $10e^-$, $16H^+$, and $5H_2O$ from each side. The final equation is

$$2MnO_4^- + 5H_2SO_3 \longrightarrow 2Mn^{2+} + 3H_2O + 5SO_4^{2-} + 4H^+$$

Notice that the equation has the same number of atoms on each side and the same net charge on each side. It is therefore a balanced equation.

EXERCISE 18.2 What is the balanced equation for the following reaction in acidic solution?

$$Cu + NO_3^- \longrightarrow Cu^{2+} + NO$$

Balancing redox equations for basic solutions

In basic solutions the concentration of H^+ is very small; the dominant species are H_2O and OH^-. Strictly speaking, these should be used to balance the half-reactions. However, the simplest way to obtain a balanced equation for a redox reaction that occurs in a basic solution is to pretend that the solution is acidic. We can begin to balance the equation using the seven steps we just learned. Then we can use a simple three-step procedure to convert the equation to the correct form for a basic solution.

Suppose, for example, that we wanted to balance the following equation for a basic solution:

$$SO_3^{2-} + MnO_4^- \longrightarrow SO_4^{2-} + MnO_2$$

Following Steps 1 through 7 for acidic solutions gives us

$$2H^+ + 3SO_3^{2-} + 2MnO_4^- \longrightarrow 3SO_4^{2-} + 2MnO_2 + H_2O$$

Conversion of this equation to one appropriate for a basic solution takes advantage of the fact that H^+ and OH^- react in a 1 to 1 ratio to give H_2O. The procedure is as follows:

Step 1. Add the same number of OH^- as there are H^+ to both sides of the equation.
The equation for acidic solution has $2H^+$ on the left, so we add $2OH^-$ to each side. This gives

$$2OH^- + 2H^+ + 3SO_3^{2-} + 2MnO_4^- \longrightarrow 3SO_4^{2-} + 2MnO_2 + H_2O + 2OH^-$$

Step 2. Change OH^- and H^+ to H_2O.
The left side has $2OH^-$ and $2H^+$. We combine these to give $2H_2O$.

$$2H_2O + 3SO_3^{2-} + 2MnO_4^- \longrightarrow 3SO_4^{2-} + 2MnO_2 + H_2O + 2OH^-$$

Step 3. Cancel any H_2O that you can.
In this equation, one H_2O can be canceled from both sides. The final equation, balanced for a basic solution, is

$$H_2O + 3SO_3^{2-} + 2MnO_4^- \longrightarrow 3SO_4^{2-} + 2MnO_2 + 2OH^-$$

EXERCISE 18.3 Balance the following equation for basic solution:

$$MnO_4^- + C_2O_4^{2-} \longrightarrow MnO_2 + CO_3^{2-}$$

Writing balanced half-reactions, and combining them to obtain a final net reaction, is important in electrochemistry because the redox half-reactions that we write are the same as the electrode reactions that occur at the anode or cathode. For example, electrolysis of a basic solution containing chloride ion gives hypochlorite ion at the anode. The balanced half-reaction obtained by the ion-electron method is

$$2OH^- + Cl^- \longrightarrow ClO^- + H_2O + 2e^-$$

This is actually the chemical change that takes place at the anode. Hydroxide ions and chloride ions are both consumed as hypochlorite ions are produced.

18.4 QUANTITATIVE CHANGES DURING ELECTROLYSIS

Experimental measurements of electric current and time can be used to calculate the amount of chemical change during electrolysis

Michael Faraday.

Much of the early research in electrochemistry was performed by a British scientist named Michael Faraday (1791–1867). It was he who coined the terms *anode, cathode, electrode, electrolyte,* and *electrolysis.* In about 1833, Faraday discovered that the amount of chemical change that occurs during electrolysis is directly proportional to the amount of electricity that is passed through an electrolysis cell. For example, the reduction of copper ion at a cathode is given by the equation

$$Cu^{2+}(aq) + 2e^- \longrightarrow Cu(s)$$

To deposit one mole of metallic copper requires two moles of electrons. To deposit two moles of copper requires four moles of electrons, and that takes twice as much electricity.

A unit normally used in electrochemistry to mean one mole of electrons is the **faraday** $(\mathscr{F})$, named in honor of Michael Faraday.

$$1\mathscr{F} = 1 \text{ mol } e^-$$

The half-reaction for an oxidation or reduction, therefore, relates the amount of chemical substance consumed or produced to the number of faradays that the electric current must supply.

To use the faraday we must relate it to electrical measurements that can be made in the laboratory. The SI unit of electric current is the **ampere (A)** and the SI unit of charge is the **coulomb (C).** A coulomb is the amount of charge that passes by a given point in a wire when an electric current of one ampere flows for one second. This means that coulombs are the product of amperes of current multiplied by seconds. Thus

There are electronic devices that measure amperes directly.

$$1 \text{ coulomb} = 1 \text{ ampere} \times 1 \text{ second}$$

$$\boxed{1 \text{ C} = 1 \text{ A} \cdot \text{s}}$$

A · s means amperes × seconds. The dot means "multiplied by."

For example, if a current of 4 A flows for 10 s, 40 C pass by a given point in the wire.

$$(4 \text{ A}) \times (10 \text{ s}) = 40 \text{ A} \cdot \text{s}$$
$$= 40 \text{ C}$$

Experimentally, it has been determined that 1 faraday (1 mol of electrons) carries a charge of 96,494 C. We will use this number rounded to three significant figures.

$$1\mathscr{F} = 96,500 \text{ C} \qquad \text{(to three significant figures)}$$

Now we have a way to relate laboratory measurements to the amount of chemical change that occurs during an electrolysis. Measuring current and time allows us to calculate the number of coulombs. From this we can get faradays, which we can then use to calculate the amount of chemical change produced.

EXAMPLE 18.3 **Quantitative Problems on Electrolysis**

Problem: How many grams of copper are deposited on the cathode of an electrolytic cell if an electric current of 2.00 A is run through a solution of $CuSO_4$ for a period of 20.0 min?

Solution: First we convert minutes to seconds.

$$20.0 \; \cancel{min} \times \left(\frac{60 \; s}{1 \; \cancel{min}} \right) = 1200 \; s$$

Then we multiply the current by the time to obtain the number of coulombs (1 A · s = 1 C).

$$(1200 \; s) \times (2.00 \; A) = 2400 \; A \cdot s$$
$$= 2400 \; C$$

Since $1\mathscr{F} = 96{,}500$ C,

$$2400 \; \cancel{C} \times \left(\frac{1\mathscr{F}}{96{,}500 \; \cancel{C}} \right) = 0.0249\mathscr{F}$$

Next we need the equation for the reduction of copper ion so that we can relate faradays to moles of copper. This equation is

$$Cu^{2+} + 2e^- \longrightarrow Cu$$

It provides us with the conversion factors

$$\frac{1 \; mol \; Cu}{2\mathscr{F}} \quad \text{and} \quad \frac{2\mathscr{F}}{1 \; mol \; Cu}$$

Using the first of these along with the atomic weight of copper,

$$0.0249\mathscr{F} \times \left(\frac{1 \; \cancel{mol \; Cu}}{2\mathscr{F}} \right) \times \left(\frac{63.5 \; g \; Cu}{1 \; \cancel{mol \; Cu}} \right) = 0.791 \; g \; Cu$$

The electrolysis will deposit 0.791 g of copper on the cathode.

EXAMPLE 18.4 **Quantitative Problems on Electrolysis**

Problem: Electroplating is an important application of electrolysis. How much time would it take in minutes to deposit 0.500 g of metallic nickel on a metal object using a current of 3.00 amperes? The nickel is reduced from the 2+ oxidation state.

Solution: First of all, we need an equation for the reduction. Since the nickel is reduced to the free metal from the 2+ state, we can write

$$Ni^{2+} + 2e^- \longrightarrow Ni$$

This gives the conversion factors

$$\frac{1 \; mol \; Ni}{2\mathscr{F}} \quad \text{and} \quad \frac{2\mathscr{F}}{1 \; mol \; Ni}$$

We wish to deposit 5.00 g of Ni. This must be converted to moles.

$$0.500 \ \text{g Ni} \times \left(\frac{1 \ \text{mol Ni}}{58.7 \ \text{g Ni}}\right) = 0.00852 \ \text{mol Ni}$$

Next we determine the number of faradays required.

$$0.00852 \ \text{mol Ni} \times \left(\frac{2\mathcal{F}}{1 \ \text{mol Ni}}\right) = 0.0170\mathcal{F}$$

Then we calculate the number of coulombs.

$$0.0170\mathcal{F} \times \left(\frac{96{,}500 \ \text{C}}{1\mathcal{F}}\right) = 1640 \ \text{C} \quad \text{(to three significant figures)}$$

$$= 1640 \ \text{A} \cdot \text{s}$$

This tells us that the product of current multiplied by time equals 1640 A · s. The current is 3.00 A. Dividing 1640 A · s by 3.00 A gives the time required in seconds, which can then be converted to minutes.

$$\left(\frac{1640 \ \text{A} \cdot \text{s}}{3.00 \ \text{A}}\right) \times \left(\frac{1 \ \text{min}}{60 \ \text{s}}\right) = 9.11 \ \text{min}$$

The time required is 9.11 min.

EXAMPLE 18.5   **Quantitative Problems on Electrolysis**

Problem: What current is needed to deposit 0.500 g of chromium metal from a solution of Cr^{3+} in a period of 1.00 hr?

Solution: This problem is quite similar to Example 18.4. First we write the equation for the reaction.

$$Cr^{3+} + 3e^- \longrightarrow Cr$$

This tells us that $3\mathcal{F}$ are needed for each mole of Cr produced. Now we can calculate the number of faradays required.

$$0.500 \ \text{g Cr} \times \left(\frac{1 \ \text{mol Cr}}{52.0 \ \text{g Cr}}\right) \times \left(\frac{3\mathcal{F}}{1 \ \text{mol Cr}}\right) = 0.0288\mathcal{F}$$

Then we convert faradays to coulombs.

$$0.0288\mathcal{F} \times \frac{96{,}500 \ \text{C}}{1\mathcal{F}} = 2780 \ \text{C} \quad \text{(to three significant figures)}$$

$$= 2780 \ \text{A} \cdot \text{s}$$

Since we want the metal to be deposited in 1 hr (3600 s), the current is

$$\frac{2780 \ \text{A} \cdot \text{s}}{3600 \ \text{s}} = 0.772 \ \text{A}$$

The current required is 0.772 A.

EXERCISE 18.4   How many *moles* of hydroxide ion will be produced at the cathode during the electrolysis of water with a current of 4.00 A for a period of 200 s? The cathode reaction is

$$2e^- + 2H_2O \longrightarrow H_2 + 2OH^-$$

EXERCISE 18.5   How long (in minutes) will it take a current of 10.0 A to deposit 3.00 g of gold from a solution of $AuCl_3$?

EXERCISE 18.6   What current must be supplied to deposit 3.00 g of gold from a solution of $AuCl_3$ in 20.0 min?

18.5 APPLICATIONS OF ELECTROLYSIS

Commercial applications of electrolysis include electroplating; the production of aluminum, magnesium and sodium; the refining of copper; and the synthesis of sodium hydroxide and sodium hypochlorite

Besides being a useful tool in the chemistry laboratory, electrolysis has many important industrial applications. In this section we will briefly examine the chemistry of electroplating and the production of some of our most common chemicals.

Electroplating

Electroplating—the application of a thin ornamental or protective coating of a metal—is a common technique for improving the appearance and durability of metal objects. For instance, a thin, shiny coating of metallic chromium is applied over steel automobile bumpers to make them attractive and to prevent rusting of the steel. Silver and gold plating is applied to jewelry made from less expensive metals, and silver plating is common on eating utensils (knives, forks, spoons, etc.). These thin metallic layers, generally 0.03 to 0.05 mm (0.001 to 0.002 inches) thick, are usually applied by electrolysis.

Figure 18.3 illustrates a typical apparatus used for plating a metal such as silver. Silver ion in the solution is reduced at the cathode where it is deposited as metallic silver on the object to be plated. At the anode, silver from the metal bar is oxidized, replenishing the supply of the silver ion in the solution. As time passes, silver is gradually transferred from the bar at the anode onto the object at the cathode.

The exact composition of the electroplating bath varies, depending on the metal to be deposited, and can affect the appearance and durability of the finished surface. For example, silver deposited from a solution of silver nitrate ($AgNO_3$) does not stick to other metal surfaces very well. However, if it is deposited from a solution of silver cyanide ($AgCN$), the coating adheres well and is bright and shiny. Other metals that are electroplated from a cyanide bath are gold and cadmium. Nickel, which can also be applied as a protective coating, is plated from a nickel sulfate solution, and chromium is plated from a chromic acid (H_2CrO_4) solution.

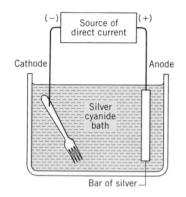

Figure 18.3
Apparatus for electroplating silver.

Production of aluminum

Until the latter part of the nineteenth century, aluminum was an uncommon metal—only the rich could afford aluminum products. A student at Oberlin College, 21-year-old Charles M. Hall, learned of this and began a series of experiments in an attempt to invent a cheap method of extracting the metal from its compounds. The difficulty that he faced was that aluminum is a very reactive element. It is difficult to produce as a free element by usual chemical reactions. Efforts to produce aluminum by electrolysis were unproductive because its anhydrous salts were difficult to prepare and its oxide, Al_2O_3, has such a high melting point (over 2000 °C) that no practical method of melting it could be found. In 1886 Hall discovered that Al_2O_3 dissolves in a mineral called cryolite, Na_3AlF_6, to give a conducting mixture with a relatively low melting point from which aluminum could be produced electrolytically.

A diagram of the apparatus used to produce aluminum is shown in Figure 18.4. Aluminum ore, called bauxite, contains Al_2O_3. The ore is purified and the Al_2O_3 is then added to the molten cryolite electrolyte in which it dissolves and dissociates. At the cathode, the aluminum ions are reduced to produce the free metal, which forms as a layer of molten aluminum below the less dense electrolyte. At the carbon anodes, oxide ion is oxidized to give free O_2.

Aluminum is used as a structural metal and in such products as aluminum foil, electrical wire, alloys, kitchen utensils.

A large cell can produce as much as 900 lb of aluminum per day.

$$Al^{3+} + 3e^- \longrightarrow Al(\ell) \qquad \text{(cathode)}$$
$$2O^{2-} \longrightarrow O_2(g) + 4e^- \qquad \text{(anode)}$$

An overhead crane moves a crucible of molten aluminum which has just been tapped from the electrolytic cells, or pots, in this potline.

The net cell reaction is

$$4Al^{3+} + 6O^{2-} \longrightarrow 4Al(\ell) + 3O_2(g)$$

The oxygen produced at the anode attacks the carbon electrodes (producing CO_2), so the electrodes must be replaced frequently.

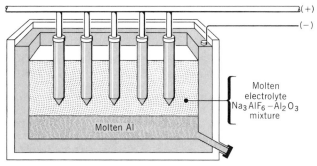

Figure 18.4
Diagram of the apparatus used to produce aluminum electrolytically by the Hall process.

The production of aluminum consumes enormous amounts of electricity and is therefore very costly, not only in terms of dollars but also in terms of energy resources. For this reason, recycling of aluminum should receive a high priority as we seek to minimize our use of energy.

Production of magnesium

Magnesium is a metal that has found a number of structural uses because of its light weight. Around the home, for example, you may have a magnesium alloy ladder. Magnesium is also the wire inside flashbulbs. It produces a brilliant flash of light when it reacts with oxygen in the bulb to give MgO. The hydroxide of magnesium, $Mg(OH)_2$, is the creamy substance in milk of magnesia.

The major source of magnesium is sea water. On a mole basis, Mg^{2+} is the third most abundant ion in the ocean, exceeded only by sodium ion and chloride ion. To obtain magnesium, sea water is made basic, which precipitates $Mg(OH)_2$. This is separated by filtration and then dissolved in hydrochloric acid.

$$Mg(OH)_2 + 2HCl \longrightarrow MgCl_2 + 2H_2O$$

The solution of $MgCl_2$ is evaporated, and the resulting solid is melted and electrolyzed. Free magnesium is deposited at the cathode and chlorine gas is produced at the anode.

Production of sodium

Sodium is prepared by the electrolysis of molten sodium chloride. The principles of this were discussed in Section 18.1. From a practical standpoint, the metallic sodium and the chlorine gas that is formed must be kept apart, otherwise they react violently to reform NaCl. The Downs cell, illustrated in Figure 18.5 on page 612, accomplished this separation.

The products of the electrolysis of molten NaCl are both commercially important. Chlorine is used to purify water and to manufacture many solvents as well as plastics such as polyvinyl chloride (PVC). Sodium is used in the manufacture of tetraethyllead (an additive for gasoline), as a coolant in nuclear reactors, and in the production of sodium vapor lamps (see Color Plate 2). Sodium lamps, which produce a bright yellow color, have the advantage of giving off most of their energy in a portion of the spectrum that humans can see. In terms of useful light output versus energy input, they are about 15 times more efficient than ordinary incandescent lightbulbs and about three to four times more efficient than the bluish-white mercury vapor lamps used for street lighting.

The town of Oyster Bay, N.Y., expects to cut electricity costs by 40 percent by replacing incandescent street lights with sodium lamps.

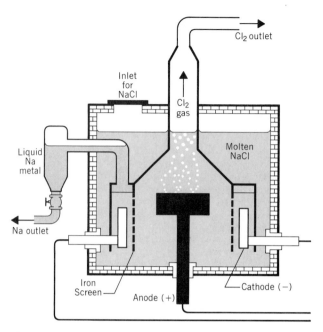

Figure 18.5
Cross section of the Downs cell used for the electrolysis of molten sodium chloride. The cathode is a circular ring that surrounds the anode. The electrodes are separated from each other by an iron screen. During the operation of the cell, molten sodium collects at the top of the cathode compartment from which it is periodically drained. The chlorine gas bubbles out of the anode compartment and is collected.

Refining of copper

One of the most interesting and economically attractive applications of electrolysis is the purification or refining of metallic copper. When copper is first removed from its ore, it is about 99 percent pure. The impurities—mostly silver, gold, platinum, iron, and zinc—decrease the electrical conductivity of the copper significantly enough that it must be further refined before it can be used in electrical wire. Figure 18.6 illustrates how this is done.

The impure copper is used as the anode in an electrolysis cell that contains a solution of copper sulfate and sulfuric acid as the electrolyte. The cathode is a thin sheet of very pure copper. When the cell is operated at the correct voltage, only copper and impurities more easily oxidized than copper (iron and zinc) dissolve at the anode. The less active metals simply settle to the bottom of the container. At the cathode, copper ion is reduced, but the zinc ions and iron ions remain in solution because they are more difficult to reduce than copper. Gradu-

Figure 18.6
Purification of copper by the use of electrolysis.

A worker lowers a set of slablike copper anodes into an electrolytic tank where they will be suspended between thin copper cathode starter sheets. When an electric current is passed between the electrodes the anodes dissolve and pure copper is deposited on the cathodes. It takes about 28 days for one of the anode bars to be completely dissolved.

ally, the impure copper anode dissolves and the copper cathode grows. This copper is now about 99.95 percent pure. The sludge—called *anode mud*—that accumulates in the vessel is removed periodically and the value of the silver, gold, and platinum recovered from it virtually pays for the entire refining operation.

Electrolysis of brine

One of the most important commercial electrolysis reactions, introduced in Section 12.7, is the electrolysis of aqueous sodium chloride solutions called *brine*. An apparatus that could be used for this in the laboratory is shown in Figure 18.7. At the cathode, water is more easily reduced than sodium ions, so hydrogen is evolved.

$$2e^- + 2H_2O \longrightarrow H_2(g) + 2OH^-(aq) \qquad \text{(cathode)}$$

At the anode, chloride ions are oxidized to chlorine because they are more easily oxidized than water.

$$2Cl^-(aq) \longrightarrow Cl_2(g) + 2e^- \qquad \text{(anode)}$$

The net reaction is

$$2H_2O + 2Cl^-(aq) \xrightarrow{\text{electrolysis}} H_2(g) + Cl_2(g) + 2OH^-(aq)$$

The chloride ions in the NaCl solution are replaced by hydroxide ions and in this way NaCl is gradually converted to NaOH. This net reaction is a commercial source of three important chemicals: hydrogen (which is used to make ammonia and hydrogenated vegetable oils), chlorine (which is used to purify drinking water and to make plastics and solvents), and sodium hydroxide (which is used to make soap and paper and to purify aluminum ores).

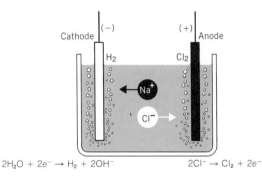

Figure 18.7
Electrolysis of brine.

$$2H_2O + 2e^- \rightarrow H_2 + 2OH^- \qquad\qquad 2Cl^- \rightarrow Cl_2 + 2e^-$$

If the solution is stirred while the electrolysis of the brine is taking place, the chlorine produced at the anode reacts with the hydroxide ion formed at the cathode.

$$Cl_2 + 2OH^- \longrightarrow Cl^- + OCl^- + H_2O$$

Continued electrolysis gradually converts all of the chloride ion to hypochlorite ion (OCl^-), and the sodium chloride solution is changed to a solution of sodium hypochlorite. When this is diluted, it is sold as liquid laundry bleach (e.g., Clorox®).

18.6 GALVANIC CELLS

Two half-reactions, one an oxidation and the other a reduction, can be set up so that the electron transfer must occur through an external electrical circuit

If you have silver fillings in your teeth, you may have experienced a strange and perhaps even unpleasant sensation while accidentally biting on a piece of aluminum foil. That sensation was caused by a very mild electric shock produced by a voltage difference between your metal fillings and the aluminum. In effect, you created a battery. Batteries not too much different from this serve to power all sorts of gadgets, as mentioned at the beginning of this chapter. The energy for this comes from spontaneous redox reactions in which the electron transfer is forced to take place through a wire. Cells that provide electricity in this way are called **galvanic cells,**[1] after Luigi Galvani (1737–1798), an Italian anatomist who discovered that electricity can cause the contraction of muscles.

If a shiny piece of metallic copper is placed into a solution of silver nitrate, a spontaneous reaction occurs. A grayish white deposit is formed on the copper, and the solution itself becomes pale blue. This is shown in Color Plate 6B. The reaction that takes place is

$$2Ag^+(aq) + Cu(s) \longrightarrow Cu^{2+}(aq) + 2Ag(s)$$

No usable energy can be harnessed from this process, however, because the energy change that accompanies the reaction is lost as heat.

However, the same chemical reaction can occur and produce usable energy if the two half-reactions involved in this net reaction are made to occur in separate containers. An apparatus to accomplish this—a galvanic cell—is shown in Figure 18.8. On the left, a silver electrode is dipping into a solution of $AgNO_3$; on the right, a copper electrode is dipping into a $Cu(NO_3)_2$ solution. The two solutions are connected by a salt bridge, the function of which will be described shortly, and by an external electrical circuit. When the circuit is completed, reduction of Ag^+ to Ag occurs in the beaker on the left and oxidation of copper occurs in the beaker on the right. Because of the nature of the reactions taking place, we can identify the silver electrode as the cathode and the copper electrode as the anode.

$$Aq^+(aq) + e^- \longrightarrow Ag(s) \qquad \text{(reduction—cathode)}$$
$$Cu(s) \longrightarrow Cu^{2+}(aq) + 2e^- \quad \text{(oxidation—anode)}$$

When these reactions take place, electrons left behind by oxidation of the copper travel through the external circuit as an electric current to the cathode where they are picked up by the silver ions, which are thereby reduced.

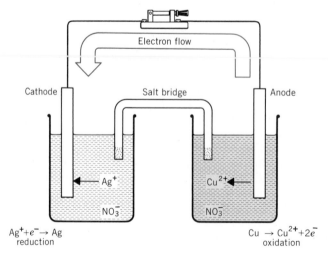

Figure 18.8
A galvanic cell.

---

[1] They are also often called voltaic cells, after Alessandro Volta (1745–1827), the inventor of the battery.

In order for a galvanic cell to work, the solutions in both compartments, or **half-cells,** must remain electrically neutral. However, when copper is oxidized, the solution surrounding the electrode becomes filled with Cu^{2+} ions. Similarly, when Ag^+ ions are reduced, NO_3^- ions are left behind in the solution. As we discussed in Section 18.2, these redox reactions can only take place if there are enough ions of opposite charge present to compensate for the ions that the reactions produce. The salt bridge shown in Figure 18.8 serves to maintain this necessary electrical neutrality. A **salt bridge** is a tube filled with an electrolyte, commonly KNO_3 or KCl, and fitted with porous plugs at either end. During the cell reaction, either negative ions diffuse from the salt bridge into the copper half-cell or Cu^{2+} ions diffuse into the salt bridge to keep that half-cell electrically neutral. At the same time, the silver half-cell is kept electrically neutral by the diffusion of positive ions from the salt bridge or NO_3^- ions into the salt bridge. Without the salt bridge, then, no electrical current would be produced by the galvanic cell. Electrolytic contact—contact by means of a solution containing ions—must be maintained in order for the cell to function.

In an electrolysis cell, the cathode carries a negative charge and the anode carries a positive charge. In a galvanic cell, these charges are reversed. At the anode of the cell in Figure 18.8, copper atoms leave the electrode and enter the solution as Cu^{2+} ions. The electrons that are left behind give the anode a negative charge. At the cathode, electrons are joining Ag^+ ions to produce neutral atoms, but the effect is the same as if Ag^+ ions became part of the electrode, so the cathode acquires a positive charge. The difference in charge between cathode and anode is what causes the electric current to flow when the circuit is complete. The important thing to remember is that it is the nature of the chemical reaction, not the charge, that determines whether we label an electrode as a cathode or an anode.

Electrolytic cell
    cathode is negative. (reduction)
    anode   is positive. (oxidation)
Galvanic cell
    cathode is positive. (reduction)
    anode   is negative. (oxidation)

Even though the charges on the cathode and anode differ between electrolytic cells and galvanic cells, the ions in solution always move in the same direction. A cation is a positive ion that always moves away from the anode toward the cathode. In both types of cells, positive ions move toward the cathode. They are attracted there by the negative charge on the cathode in an electrolysis cell; they diffuse toward the cathode in our galvanic cell to balance the charge of negative ions left behind when the Ag^+ ions are reduced. Similarly, anions are negative ions that move away from the cathode and toward the anode. They are attracted to the positive anode in an electrolysis cell, and they diffuse toward the anode in our galvanic cell to balance the charge of the Cu^{2+} ions entering the solution.

EXAMPLE 18.6   **Describing Galvanic Cells**

Problem: The following spontaneous reaction occurs when metallic zinc is dipped into a solution of copper sulfate.

$$Zn(s) + Cu^{2+}(aq) \longrightarrow Zn^{2+}(aq) + Cu(s)$$

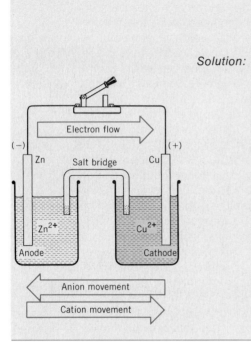

Describe a galvanic cell that could take advantage of this reaction. What are the half-cell reactions? Make a sketch of the cell and label the cathode and anode, the charges on each electrode, the direction of ion flow, and the direction of electron flow.

Solution: Two half-cells are needed. One must contain a zinc electrode that dips into a solution containing Zn^{2+} [e.g., $Zn(NO_3)_2$ or $ZnSO_4$]. The other must contain a copper electrode that dips into a solution containing Cu^{2+} [e.g., $Cu(NO_3)_2$ or $CuSO_4$].

Dividing the overall reaction into half-reactions, we get

$$Zn \longrightarrow Zn^{2+} + 2e^- \quad \text{(oxidation)}$$
$$Cu^{2+} + 2e^- \longrightarrow Cu \quad \text{(reduction)}$$

Since zinc is oxidized, it is the anode and carries a negative charge; copper is the cathode and is positive. Anions move toward the anode and cations move toward the cathode. Electrons travel from the negative anode to the positive cathode via the external circuit.

EXERCISE 18.7 Sketch and label a galvanic cell that makes use of the following spontaneous redox reaction.

$$Mg(s) + Fe^{2+}(aq) \longrightarrow Mg^{2+}(aq) + Fe(s)$$

18.7 CELL POTENTIALS AND REDUCTION POTENTIALS

The voltage across the electrodes of a galvanic cell can be attributed to the difference in the tendencies of the two half-cells to undergo reduction

Voltage or **electromotive force (emf)** can conveniently be thought of as the force with which an electric current is pushed through a wire. It is measured in terms of an electrical unit called the **volt (V)**. Strictly speaking, voltage is a measure of the amount of energy that can be delivered by a coulomb of electrical charge as it passes through a circuit. A current flowing under an emf of 1 volt can deliver 1 joule of energy per coulomb.

$$1 \text{ V} = 1 \text{ J/C} \qquad (18.1)$$

We will find this definition important in Section 18.9.

If current is drawn from a cell, some of the cell's emf is lost overcoming its own internal resistance.

The maximum emf of a galvanic cell is called the **cell potential, E_{cell}**. It is measured with a device that draws negligible current during the measurement. E_{cell} depends on the composition of the electrodes and the concentration of the ions in each of the half-cells. For reference purposes, the **standard cell potential,** symbolized $E°_{cell}$, is the potential of the cell when all of the ion concentrations are 1.00 M, the temperature is 25 °C, and any gases that are involved in the cell reaction are at a pressure of 1 atm.

Cell potentials are rarely larger than a few volts. For example, the standard cell potential for the galvanic cell constructed from silver and copper electrodes shown in Figure 18.9 is only 0.46 V, and one cell in an automobile battery produces only about 2 V. Batteries that generate higher voltages contain a number of cells arranged in series so that their emf's are additive.

A problem that has plagued humanity ever since the discovery of methods for obtaining iron and other metals from their ores has been corrosion—the reaction of a metal with substances in the environment. The rusting of iron, in particular, is a serious problem because iron and steel have so many uses. The process appears to require two important ingredients, oxygen and water, and involves electrochemical reactions. One of these reactions is the oxidation of iron.

$$Fe(s) \longrightarrow Fe^{2+}(aq) + 2e^-$$

The Fe^{2+} ions enter the water and move to the surface where they react with oxygen and water to form a hydrated form of ferric oxide, Fe_2O_3, which is rust. At the same time, hydrogen ions in the water are reduced.

$$H^+ + e^- \longrightarrow H$$

The hydrogen atoms either combine to form H_2, or they react with oxygen to form H_2O. These reactions occur because hydrogen has a higher reduction potential than iron.

One way to prevent the rusting of iron is to coat it with another metal. This is done with "tin" cans, which are actually steel cans that have been coated with a thin layer of tin. However, if the layer of tin is scratched and the iron beneath is exposed, the corrosion is accelerated because iron has a lower reduction potential than tin—the iron becomes the anode in an electrochemical cell and is easily oxidized.

Another way to prevent corrosion is called *cathodic protection*. It involves placing the iron in contact with a metal that is more easily oxidized. This other metal is then the anode and tends to be oxidized. This keeps

Zinc wire being placed beside the Alaskan pipeline. When electrically connected to the pipeline, it prevents corrosion of the pipe.

the iron itself from corroding. The photograph here shows how zinc wire is laid alongside the Alaskan pipeline, to which it is electrically connected. This wire protects the pipeline from corroding by serving as a *sacrificial anode*. Similar zinc anodes are placed on the exposed underwater metal surfaces of ships and boats to protect them from the corrosive effects of sea water. Galvanizing, the coating of iron with zinc, protects such common objects as chain-link fences and metal garbage pails.

Reduction potentials

As we will see, it is useful to imagine that the measured overall cell potential arises from a competition between the two half-cells for electrons. We think of each half-cell reaction as having a certain natural tendency to proceed as a reduction, the magnitude of which is expressed by its **reduction potential** (or **standard reduction potential** when the temperature is 25 °C, concentrations are 1 M, and the pressure is 1 atm). When two half-cells are connected, the one with the larger reduction potential—the one with the greater tendency to undergo reduction—acquires electrons from the half-cell with the lower reduction potential, which is therefore forced to undergo oxidation. The measured cell potential actually represents the magnitude of the *difference* between the reduction potential of one half-cell and the reduction potential of the other. In general,

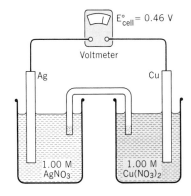

Figure 18.9
Cell designed to generate the standard cell potential.

$$E^\circ_{cell} = \begin{pmatrix} \text{standard reduction} \\ \text{potential of} \\ \text{substance reduced} \end{pmatrix} - \begin{pmatrix} \text{standard reduction} \\ \text{potential of} \\ \text{substance oxidized} \end{pmatrix} \quad (18.2)$$

As an example, let's look at the silver-copper cell. From the cell reaction,

$$2Ag^+(aq) + Cu(s) \longrightarrow Cu^{2+}(aq) + 2Ag(s)$$

we can see that silver is reduced and copper is oxidized. If we compare the two possible reduction half-reactions,

$$Ag^+(aq) + e^- \longrightarrow Ag(s)$$
$$Cu^{2+}(aq) + 2e^- \longrightarrow Cu(s)$$

the one for Ag^+ must have a greater tendency to proceed than the one for Cu^{2+} because it is the silver ion that is actually reduced. This means that the standard reduction potential of Ag^+, $E°_{Ag^+}$, must be larger than the standard reduction potential of Cu^{2+}, $E°_{Cu^{2+}}$. If we knew the values of $E°_{Ag^+}$ and $E°_{Cu^{2+}}$, we could calculate $E°_{cell}$ with Equation 18.2 by subtracting the smaller reduction potential from the larger one.

$$E°_{cell} = E°_{Ag^+} - E°_{Cu^{2+}}$$

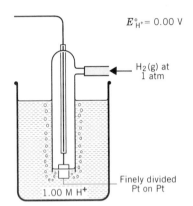

$E°_{H^+} = 0.00$ V

$H_2(g)$ at 1 atm

Finely divided Pt on Pt

1.00 M H^+

Figure 18.10
The hydrogen electrode. The half-reaction is $2H^+(aq) + 2e^- \rightleftharpoons H_2(g)$.

Unfortunately there is no way to measure the standard reduction potential of an isolated half-cell. All we can measure is the difference that is produced when two half-cells are connected. Therefore, in order to assign values to the various standard reduction potentials, a reference electrode has been arbitrarily chosen and its standard reduction potential has been assigned a value of 0.00 V. This reference electrode is called the **hydrogen electrode,** illustrated in Figure 18.10. Gaseous hydrogen at a pressure of 1 atm is bubbled over a platinum electrode that is coated with very finely divided platinum, which serves as a catalyst for the electrode reaction. This electrode is surrounded by a solution whose temperature is 25 °C and in which the hydrogen ion concentration is 1.00 M. The half-cell reaction that occurs at the platinum surface is

$$2H^+(aq, 1.00\ M) + 2e^- \rightleftharpoons H_2(g, 1\ \text{atm}) \qquad E°_{H^+} = 0.00\ \text{V at 25 °C}$$

The double arrows indicate that the reaction is reversible—whether it occurs as oxidation or reduction depends on the reduction potential of the half-cell with which it is paired.

Figure 18.11 illustrates the hydrogen electrode connected to the copper half-cell to form a galvanic cell. To obtain the cell reaction, we have to know what is oxidized and what is reduced—we need to know which is the cathode and which is the anode. We can determine this by measuring the charges on the electrode, because we know that the cathode in a galvanic cell is the positive electrode and the anode is the negative electrode. When we use a voltmeter to measure the potential of the cell, we find that proper measurements are obtained

Remember, in a galvanic cell, the cathode is (+) and the anode is (−).

Figure 18.11
A galvanic cell composed of copper and hydrogen half-cells.
The cell reaction is: $Cu^{2+}(aq) + H_2(g) \rightarrow Cu(s) + 2H^+(aq)$.

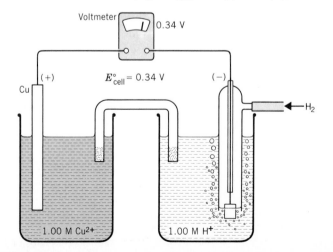

Voltmeter 0.34 V

(+) $E°_{cell} = 0.34$ V (−)

Cu

H_2

1.00 M Cu^{2+} 1.00 M H^+

only if the terminal labeled (+) is connected to the copper electrode and the terminal labeled (−) is connected to the hydrogen electrode. Thus, copper must be the cathode, and Cu^{2+} is reduced to Cu when the cell operates. Similarly, hydrogen must be the anode, and H_2 is oxidized to H^+. The half-reactions and cell reaction, therefore, are

$$Cu^{2+}(aq) + 2e^- \longrightarrow Cu(s) \qquad \text{(cathode)}$$
$$\underline{H_2(g) \longrightarrow 2H^+(aq) + 2e^- \qquad \text{(anode)}}$$
$$Cu^{2+}(aq) + H_2(g) \longrightarrow Cu(s) + 2H^+(aq) \qquad \text{(cell reaction)}$$

Using Equation 18.2, we can express E°_{cell} in terms of $E^\circ_{Cu^{2+}}$ and $E^\circ_{H^+}$.

$$E^\circ_{cell} = \underbrace{E^\circ_{Cu^{2+}}}_{\substack{\text{reduction potential} \\ \text{of substance reduced}}} - \underbrace{E^\circ_{H^+}}_{\substack{\text{reduction potential} \\ \text{of substance oxidized}}}$$

The measured standard cell potential is 0.34 V and $E^\circ_{H^+}$ equals 0.00 V. Therefore,

$$0.34 \text{ V} = E^\circ_{Cu^{2+}} - 0.00 \text{ V}$$

Relative to the hydrogen electrode, then, the standard reduction potential of Cu^{2+} is 0.34 V.

Now let's look at a galvanic cell set up between a zinc electrode and the hydrogen electrode, as shown in Figure 18.12. This time the voltmeter must have its positive terminal connected to the hydrogen electrode and its negative terminal connected to the zinc electrode—hydrogen is the cathode and zinc is the anode. This means that hydrogen ion is being reduced and zinc is being oxidized. The half-reactions and cell reaction are

$$2H^+(aq) + 2e^- \longrightarrow H_2(g) \qquad \text{(cathode)}$$
$$\underline{Zn(s) \longrightarrow Zn^{2+}(aq) + 2e^- \qquad \text{(anode)}}$$
$$2H^+(aq) + Zn(s) \longrightarrow H_2(g) + Zn^{2+}(aq) \qquad \text{(cell reaction)}$$

From Equation 18.2, the standard cell potential is

$$E^\circ_{cell} = E^\circ_{H^+} - E^\circ_{Zn^{2+}}$$

Substituting into this the measured standard cell potential of 0.76 V and $E^\circ_{H^+} = 0.00$ V,

$$0.76 \text{ V} = 0.00 \text{ V} - E^\circ_{Zn^{2+}}$$
$$E^\circ_{Zn^{2+}} = -0.76 \text{ V}$$

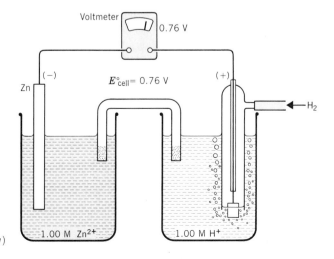

Figure 18.12
A galvanic cell composed of zinc and hydrogen half-cells. The cell reaction is:
$Zn(s) + 2H^+(aq) \rightarrow Zn^{2+}(aq) + H_2(g)$

The reduction potential of zinc is negative because the tendency of zinc to be reduced is less than that of H^+—that's why Zn is oxidized when it is paired with the hydrogen electrode.

The standard reduction potentials of many half-reactions can be compared to that for the hydrogen electrode in the manner described above. Table 18.1 contains a list of values obtained for some typical half-reactions. They are arranged in decreasing order—the half-reactions at the top have the greatest tendency to occur as reduction, while those at the bottom have the least tendency to occur as reduction.

Table 18.1
Standard Reduction Potentials Measured at 25 °C

Half-Reaction	$E°$ (volts)
$F_2(g) + 2e^- \rightleftharpoons 2F^-(aq)$	+2.87
$PbO_2(s) + SO_4^{2-}(aq) + 4H^+(aq) + 2e^- \rightleftharpoons PbSO_4(s) + 2H_2O$	+1.69
$2HOCl(aq) + 2H^+(aq) + 2e^- \rightleftharpoons Cl_2(g) + 2H_2O$	+1.63
$MnO_4^-(aq) + 8H^+(aq) + 5e^- \rightleftharpoons Mn^{2+}(aq) + 4H_2O$	+1.49
$PbO_2(s) + 4H^+(aq) + 2e^- \rightleftharpoons Pb^{2+}(aq) + 2H_2O$	+1.46
$BrO_3^-(aq) + 6H^+(aq) + 6e^- \rightleftharpoons Br^-(aq) + 3H_2O$	+1.44
$Au^{3+}(aq) + 3e^- \rightleftharpoons Au(s)$	+1.42
$Cl_2(g) + 2e^- \rightleftharpoons 2Cl^-(aq)$	+1.36
$O_2(g) + 4H^+(aq) + 4e^- \rightleftharpoons 2H_2O$	+1.23
$Br_2(aq) + 2e^- \rightleftharpoons 2Br^-(aq)$	+1.07
$NO_3^-(aq) + 4H^+(aq) + 3e^- \rightleftharpoons NO(g) + 2H_2O$	+0.96
$Ag^+(aq) + e^- \rightleftharpoons Ag(s)$	+0.80
$Fe^{3+}(aq) + e^- \rightleftharpoons Fe^{2+}(aq)$	+0.77
$I_2(s) + 2e^- \rightleftharpoons 2I^-(aq)$	+0.54
$NiO_2(s) + 2H_2O + 2e^- \rightleftharpoons Ni(OH)_2(s) + 2OH^-(aq)$	+0.49
$Cu^{2+}(aq) + 2e^- \rightleftharpoons Cu(s)$	+0.34
$SO_4^{2-}(aq) + 4H^+(aq) + 2e^- \rightleftharpoons H_2SO_3(aq) + H_2O$	+0.17
$2H^+(aq) + 2e^- \rightleftharpoons H_2(g)$	0.00
$Sn^{2+}(aq) + 2e^- \rightleftharpoons Sn(s)$	−0.14
$Ni^{2+}(aq) + 2e^- \rightleftharpoons Ni(s)$	−0.25
$Co^{2+}(aq) + 2e^- \rightleftharpoons Co(s)$	−0.28
$PbSO_4(s) + 2e^- \rightleftharpoons Pb(s) + SO_4^{2-}(aq)$	−0.36
$Cd^{2+}(aq) + 2e^- \rightleftharpoons Cd(s)$	−0.40
$Fe^{2+}(aq) + 2e^- \rightleftharpoons Fe(s)$	−0.41
$Cr^{3+}(aq) + 3e^- \rightleftharpoons Cr(s)$	−0.74
$Zn^{2+}(aq) + 2e^- \rightleftharpoons Zn(s)$	−0.76
$2H_2O + 2e^- \rightleftharpoons H_2(g) + 2OH^-(aq)$	−0.83
$Al^{3+}(aq) + 3e^- \rightleftharpoons Al(s)$	−1.66
$Mg^{2+}(aq) + 2e^- \rightleftharpoons Mg(s)$	−2.37
$Na^+(aq) + e^- \rightleftharpoons Na(s)$	−2.71
$Ca^{2+}(aq) + 2e^- \rightleftharpoons Ca(s)$	−2.76
$K^+(aq) + e^- \rightleftharpoons K(s)$	−2.92
$Li^+(aq) + e^- \rightleftharpoons Li(s)$	−3.05

Since a substance that is reduced is an *oxidizing agent*, the substances located to the left of the double arrows are all oxidizing agents. The best oxidizing agents are those at the top of the table (e.g., fluorine). When one half-reaction occurs as a reduction, another must occur as an oxidation. Substances that are oxidized are *reducing agents*. Reducing agents are located to the right of the double arrows and the best reducing agents are those found at the bottom of the table (e.g., lithium).

EXAMPLE 18.7   **Calculating Half-Cell Potentials**

Problem: We mentioned earlier that the standard cell potential of the silver-copper galvanic cell has a value of 0.46 V. The cell reaction is

$$2Ag^+(aq) + Cu(s) \longrightarrow 2Ag(s) + Cu^{2+}(aq)$$

and we have seen that the reduction potential of Cu^{2+}, $E^\circ_{Cu^{2+}}$, is 0.34 V. What is the value of $E^\circ_{Ag^+}$, the reduction potential of Ag^+?

Solution: The substance that is reduced is Ag^+, which means that Cu is oxidized. According to Equation 18.2, then,

$$E^\circ_{cell} = E^\circ_{Ag^+} - E^\circ_{Cu^{2+}}$$

Substituting in the values for E°_{cell} and $E^\circ_{Cu^{2+}}$ gives

$$0.46 \text{ V} = E^\circ_{Ag^+} - 0.34 \text{ V}$$

$$E^\circ_{Ag^+} = 0.46 \text{ V} + 0.34 \text{ V}$$
$$= 0.80 \text{ V}$$

EXERCISE 18.8   The galvanic cell described in Exercise 18.7 has a standard cell potential of 1.96 V. The standard reduction potential of Fe^{2+} corresponding to the half-reaction $Fe^{2+} + 2e^- \longrightarrow$ Fe is -0.41 V. Calculate the standard reduction potential of magnesium. Check your answer by referring to Table 18.1.

18.8   USING STANDARD REDUCTION POTENTIALS

Remembering that the half-reaction having the higher reduction potential occurs as a reduction allows us to predict the outcomes of redox reactions and calculate cell potentials

The half-reactions and their standard reduction potentials found in Table 18.1 can be used in a number of ways to give information about galvanic cells. Equally useful is the fact that they can provide information about the outcome of redox reactions, whether they occur in a galvanic cell or not.

Example 18.8 below shows how reduction potentials can be used to predict the standard cell potential and the overall cell reaction in a galvanic cell.

EXAMPLE 18.8   **Predicting the Cell Reaction and Cell Potential**

Problem: A galvanic cell was constructed using electrodes made of lead and lead dioxide (PbO_2) with sulfuric acid as the electrolyte. The half-reactions and their reduction potentials in this system are

$$PbO_2(s) + 4H^+(aq) + SO_4^{2-}(aq) + 2e^- \rightleftharpoons PbSO_4(s) + 2H_2O$$
$$E^\circ_{PbO_2} = 1.69 \text{ V}$$

$$PbSO_4(s) + 2e^- \rightleftharpoons Pb(s) + SO_4^{2-}(aq) \qquad E^\circ_{PbSO_4} = -0.36 \text{ V}$$

What is the cell reaction and what is the standard potential of the cell?

Solution: In a competition for electrons, the half-reaction having the higher (more positive) reduction potential undergoes reduction. The half-reaction with the lower reduction potential is therefore forced to proceed as an oxidation. This means that here the first half-reaction given will occur as reduction and the second will be forced to reverse its direction. In the cell, the reactions are

$$PbO_2(s) + 4H^+(aq) + SO_4^{2-}(aq) + 2e^- \longrightarrow PbSO_4(s) + 2H_2O \quad \text{(reduction)}$$
$$Pb(s) + SO_4^{2-}(aq) \longrightarrow PbSO_4(s) + 2e^- \quad \text{(oxidation)}$$

Adding the two half-reactions and canceling electrons gives the cell reaction,

$$PbO_2(s) + 4H^+(aq) + Pb(s) + 2SO_4^{2-}(aq) \longrightarrow 2PbSO_4(s) + 2H_2O$$

(This is the reaction that takes place in lead storage batteries used to start cars.) The cell potential can be obtained by using Equation 18.2.

$$E^\circ_{cell} = (E^\circ \text{ of substance reduced}) - (E^\circ \text{ of substance oxidized})$$

Since the first half-reaction occurs as a reduction and the second as an oxidation,

$$E^\circ_{cell} = E^\circ_{PbO_2} - E^\circ_{PbSO_4}$$
$$= (1.69 \text{ V}) - (-0.36 \text{ V})$$
$$= 2.05 \text{ V}$$

EXAMPLE 18.9 **Predicting the Cell Reaction and Cell Potential**

Problem: What would be the cell reaction and the standard cell potential of a galvanic cell employing the following half-reactions?

$$Al^{3+}(aq) + 3e^- \rightleftharpoons Al(s) \qquad E^\circ_{Al^{3+}} = -1.66 \text{ V}$$
$$Cu^{2+}(aq) + 2e^- \rightleftharpoons Cu(s) \qquad E^\circ_{Cu^{2+}} = 0.34 \text{ V}$$

Which half-cell would be the anode?

Solution: Once again, the half-reaction with the more positive reduction potential occurs as a reduction; the other occurs as an oxidation. In this cell, then, Cu^{2+} is reduced and Al is oxidized. To obtain the cell reaction, we add the two half-reactions, remembering that the electrons must cancel.

$$
\begin{array}{lll}
3(Cu^{2+}(aq) + 2e^- \longrightarrow Cu(s) \qquad) & \text{(reduction)} \\
\underline{2(\qquad\quad Al(s) \longrightarrow Al^{3+}(aq) + 3e^-)} & \text{(oxidation)} \\
3Cu^{2+}(aq) + 2Al(s) \longrightarrow 3Cu(s) + 2Al^{3+}(aq) & \text{(cell reaction)}
\end{array}
$$

The anode in the cell is aluminum because that is where oxidation takes place.

To obtain the cell potential, we substitute into Equation 18.2.

$$E^\circ_{cell} = E^\circ_{Cu^{2+}} - E^\circ_{Al^{3+}}$$
$$= (0.34 \text{ V}) - (-1.66 \text{ V})$$
$$= 2.00 \text{ V}$$

Notice that we multiply the half-reactions by factors to make the electrons cancel, but *we do not multiply the reduction potentials by these factors.*[2] To obtain the cell potential, we simply subtract one reduction potential from the other.

---

[2] Reduction potentials have the units, volts, which are joules *per* coulomb. The same number of joules are available for each coulomb of charge regardless of the total number of electrons shown in the equation. Therefore, reduction potentials are never multiplied by factors before they are subtracted to give the cell potential.

EXERCISE 18.9 What is the overall cell reaction and the standard cell potential of a galvanic cell employing the following half-reactions?

$$NiO_2(s) + 2H_2O + 2e^- \rightleftharpoons Ni(OH)_2(s) + 2OH^-(aq) \qquad E°_{NiO_2} = 0.49 \text{ V}$$
$$Fe(OH)_2(s) + 2e^- \rightleftharpoons Fe(s) + 2OH^-(aq) \qquad E°_{Fe(OH)_2} = -0.88 \text{ V}$$

(These are the reactions in an Edison cell, a type of rechargeable storage battery.)

EXERCISE 18.10 What is the overall cell reaction and the standard cell potential of a galvanic cell employing the following half-reactions?

$$Cr^{3+}(aq) + 3e^- \rightleftharpoons Cr(s) \qquad E°_{Cr^{3+}} = -0.74 \text{ V}$$
$$MnO_4^-(aq) + 8H^+(aq) + 5e^- \rightleftharpoons Mn^{2+}(aq) + 4H_2O \qquad E°_{MnO_4^-} = +1.49 \text{ V}$$

Reduction potentials can also be used to predict the spontaneous reaction between the substances given in two half-reactions, even when these substances are not in a galvanic cell. The procedure is very simple because we know that the half-reaction having the higher reduction potential always undergoes reduction while the other is forced to undergo oxidation.

EXAMPLE 18.10 **Predicting a Spontaneous Reaction**

Problem: What spontaneous reaction will occur if Cl_2 and Br_2 are added to a solution containing Cl^- and Br^-?

Solution: There are two possible reduction reactions.

$$Cl_2 + 2e^- \longrightarrow 2Cl^-$$
$$Br_2 + 2e^- \longrightarrow 2Br^-$$

Referring to Table 18.1, we find that Cl_2 has a more positive reduction potential (1.36 V) than does Br_2 (1.07 V). This means that Cl_2 will be reduced. When reaction occurs the half-reactions will be

$$Cl_2 + 2e^- \longrightarrow 2Cl^-$$
$$2Br^- \longrightarrow Br_2 + 2e^-$$

The net reaction will be

$$Cl_2 + 2Br^- \longrightarrow Br_2 + 2Cl^-$$

Experimentally, we find that chlorine does indeed oxidize bromide ion to bromine, and this fact is used in the synthesis of bromine (Section 12.7).

When reduction potentials are listed in order of decreasing magnitude as they are in Table 18.1, the reactants and products of spontaneous redox reactions are easy to spot. If we choose any pair of half-reactions, the one higher up in the table has the higher reduction potential and occurs as a reduction. The other half-reaction occurs as an oxidation. The reactants are found on the left side of the higher half-reaction and on the right side of the lower half-reaction.

EXAMPLE 18.11 **Predicting a Spontaneous Reaction**

Problem: Referring to Table 18.1, predict the reaction that will occur when Ni and Fe are added to a solution that contains both Ni^{2+} and Fe^{2+}.

Solution: Ni^{2+} has a more positive (less negative) reduction potential than Fe^{2+}. Therefore, Ni^{2+} will be reduced and Fe (on the right of the lower reaction) will be oxidized.

$$Ni^{2+}(aq) + 2e^- \longrightarrow Ni(s) \qquad \text{(reduction)}$$
$$\underline{Fe(s) \longrightarrow Fe^{2+}(aq) + 2e^- \qquad \text{(oxidation)}}$$
$$Ni^{2+}(aq) + Fe(s) \longrightarrow Ni(s) + Fe^{2+}(aq) \qquad \text{(net reaction)}$$

EXERCISE 18.11   Use the positions of the half-reactions in Table 18.1 to predict the spontaneous reaction when Br^-, SO_4^{2-}, H_2SO_3, and Br_2 are mixed in an acidic solution.

Since we can predict the spontaneous reaction that will take place among a mixture of reactants, it should be possible to predict whether or not a particular reaction, as written, can occur spontaneously. This can be done by calculating the cell potential that corresponds to the reaction in question.

For any functioning galvanic cell, the measured cell potential is a positive number. To obtain the cell potential for a spontaneous reaction in our previous examples, we subtracted the reduction potentials in a way that gave a positive answer. Therefore, if we compute the cell potential for a particular reaction based on the way it is written and it comes out positive, the reaction is spontaneous. If the cell potential comes out negative, however, the reaction is nonspontaneous—in fact, it's spontaneous in the reverse direction.

EXAMPLE 18.12   **Determining Whether a Reaction Is Spontaneous**

Problem: Determine whether the following reactions are spontaneous as written. If they are not, give the reaction that is spontaneous.

(a) $Cu(s) + 2H^+(aq) \longrightarrow Cu^{2+}(aq) + H_2(g)$
(b) $3Cu(s) + 2NO_3^-(aq) + 8H^+(aq) \longrightarrow 3Cu^{2+}(aq) + 2NO(g) + 4H_2O$

Solution: (a) The half-reactions involved in this reaction are

$$Cu(s) \longrightarrow Cu^{2+}(aq) + 2e^-$$
$$2H^+(aq) + 2e^- \longrightarrow H_2(g)$$

The H^+ is reduced and Cu is oxidized. Therefore, from Equation 18.2,

$$E^\circ_{cell} = E^\circ_{H^+} - E^\circ_{Cu^{2+}}$$

Substituting values from Table 18.1,

$$E^\circ_{cell} = (0.00 \text{ V}) - (0.34 \text{ V})$$
$$= -0.34 \text{ V}$$

Copper will not dissolve in HCl because the only oxidizing agent is H^+.

The calculated cell potential is negative and therefore reaction (a) is *not* spontaneous in the forward direction. The spontaneous reaction is

$$H_2(g) + Cu^{2+}(aq) \longrightarrow Cu(s) + 2H^+(aq)$$

(b) The half-reactions involved are

$$Cu(s) \longrightarrow Cu^{2+}(aq) + 2e^-$$
$$NO_3^-(aq) + 4H^+(aq) + 3e^- \longrightarrow NO(g) + 2H_2O$$

The Cu is oxidized while the NO_3^- is reduced. According to Equation 18.2,

$$E^\circ_{cell} = E^\circ_{NO_3^-} - E^\circ_{Cu^{2+}}$$

Copper will dissolve in HNO_3 because it contains NO_3^- as an oxidizing agent.

Substituting values from Table 18.1 gives

$$E^\circ_{cell} = (0.96 \text{ V}) - (0.34 \text{ V})$$
$$= 0.62 \text{ V}$$

Since the calculated cell potential is positive, this reaction is spontaneous in the forward direction.

EXERCISE 18.12 Which of the following reactions occur spontaneously in the forward direction?
(a) $Br_2(aq) + Cl_2(g) + 2H_2O \longrightarrow 2Br^-(aq) + 2HOCl(aq) + 2H^+(aq)$
(b) $3Zn(s) + 2Cr^{3+}(aq) \longrightarrow 3Zn^{2+}(aq) + 2Cr(s)$

18.9 CELL POTENTIALS AND THERMODYNAMICS

Standard cell potentials can be used to calculate free energy changes and equilibrium constants

The fact that cell potentials allow us to predict the spontaneity of redox reactions is no coincidence. There is a relationship between the cell potential and the free energy change for a reaction. In Chapter Fifteen we saw that ΔG for a reaction is a measure of the maximum useful work that can be obtained from a chemical reaction.

$$-\Delta G = (\text{maximum work}) \tag{18.3}$$

In an electrical system, work is supplied by the electric current that is pushed along by the potential of the cell. It can be calculated from the equation

$$\text{maximum work} = n\mathscr{F}E_{cell} \tag{18.4}$$

where n is the number of moles of electrons transferred, $\mathscr{F}$ is the faraday constant (96,500 coulombs per mole of electrons), and E_{cell} is the potential of the cell in volts. To see that Equation 18.4 gives *work* (which has the units of energy) we can analyze the units. In Equation 18.1 we saw that 1 V = 1 joule/coulomb. Therefore,

$$\text{maximum work} = (\cancel{\text{mol }} e^-) \times \left(\frac{\cancel{\text{coulombs}}}{\cancel{\text{mol }} e^-}\right) \times \left(\frac{\text{joule}}{\cancel{\text{coulomb}}}\right)$$
$$= \text{joules}$$

Note that when E_{cell} is positive, ΔG will be negative and the cell reaction will be spontaneous. We've used this idea earlier in predicting spontaneous cell reactions.

Equating 18.3 and 18.4 gives us

$$\boxed{\Delta G = -n\mathscr{F}E_{cell}} \tag{18.5}$$

If we are dealing with the standard cell potential, we can calculate the standard free energy change.

$$\boxed{\Delta G^\circ = -n\mathscr{F}E^\circ_{cell}} \tag{18.6}$$

EXAMPLE 18.13 **Calculating the Standard Free Energy Change**

Problem: Calculate ΔG° for the following reaction, given that its standard cell potential is 0.320 V at 25 °C.

$$NiO_2 + 2Cl^- + 4H^+ \longrightarrow Cl_2 + Ni^{2+} + 2H_2O$$

Solution: Since two Cl^- are oxidized to Cl_2, two electrons are transferred. Interpreting the coefficients as moles, 2 mol of electrons are transferred, so $n = 2$. Using Equation 18.6, we have

$$\Delta G° = -(2 \text{ mol } e^-) \times \left(\frac{96,500 \text{ } \cancel{C}}{\text{mol } e^-}\right) \times \left(\frac{0.320 \text{ J}}{\cancel{C}}\right)$$

$$= -61,800 \text{ J} \quad \text{(rounded to three significant figures)}$$

$$= -61.8 \text{ kJ}$$

EXERCISE 18.13 Calculate $\Delta G°$ for the reaction that takes place in the galvanic cell described in Exercise 18.9 on page 623.

One of the most useful applications of electrochemistry is the measurement of equilibrium constants. In Chapter Sixteen, we saw that $\Delta G°$ was related to the equilibrium constant (K_p for gaseous reactions and K_c for reactions in solution). Now we've seen that $\Delta G°$ is related to $E°_{cell}$. Therefore, $E°_{cell}$ and the equilibrium constant are also related. The equation is

$$\boxed{E°_{cell} = \frac{0.0592}{n} \log K_c} \tag{18.7}$$

The constant 0.0592 has units of volts and applies to reactions that occur at 25 °C. As before, n is the number of moles of electrons transferred in the reaction.

EXAMPLE 18.14 **Calculating Equilibrium Constants from $E°_{cell}$**

Problem: Calculate K_c for the reaction in Example 18.13.
Solution: The reaction had $E°_{cell} = 0.320$ V and $n = 2$. Substituting these values into Equation 18.7 gives

$$0.320 \text{ V} = \frac{0.0592 \text{ V}}{2} \log K_c$$

$$\frac{2(0.320)}{(0.0592)} = \log K_c$$

$$10.8 = \log K_c$$

$10^{10.8} = 6 \times 10^{10}$ Taking the antilog gives

$$K_c = 6 \times 10^{10}$$

EXERCISE 18.14 The calculated standard cell potential for the reaction

$$Cu^{2+}(aq) + 2Ag(s) \rightleftharpoons Cu(s) + 2Ag^+(aq)$$

is $E°_{cell} = -0.46$ V. Calculate K_c for this reaction. Would you expect a relatively large amount of products to form?

18.10 THE EFFECT OF CONCENTRATION ON CELL POTENTIALS

Altering the concentrations of the ions in the half-cells affects the cell potential in a way that depends on the logarithm of the mass action expression for the cell reaction

When all of the ion concentrations in a cell are one molar, the cell potential is equal to the standard potential. When the concentrations change, however, so does the potential. For example, in an operating cell or battery, the potential gradually drops as the reactants are consumed. The cell approaches equilibrium, and when it gets there the potential has dropped to zero—the battery is dead.

The effect of concentration on the cell potential can be calculated from the **Nernst equation,**

Walter Nernst (1864–1941) was a noted electrochemist and also a discoverer of the third law of thermodynamics.

$$E_{cell} = E^\circ_{cell} - \frac{0.0592}{n} \log Q \qquad (18.8)$$

where E_{cell} is the actual cell potential, E°_{cell} is the standard potential, Q is the reaction quotient or mass action expression for the reaction, and n is the number of moles of electrons transferred in the reaction.

EXAMPLE 18.15   **Calculating the Effect of Concentration on E_{cell}**

Problem: What is the potential of a copper-silver cell similar to that in Figure 18.8 if the Ag^+ concentration is 1.0×10^{-3} M and the Cu^{2+} concentration is 1.0×10^{-4} M? The standard potential of the cell is 0.46 V and the cell reaction is

$$2Ag^+(aq) + Cu(s) \longrightarrow 2Ag(s) + Cu^{2+}(aq)$$

Solution: First, let's write the correct form of the Nernst equation for this cell. Since two electrons are lost when one Cu atom is oxidized to give one Cu^{2+} ion, $n = 2$. The correct mass action expression for this reaction, as usual, has the product concentration in the numerator.

$$Q = \frac{[Cu^{2+}]}{[Ag^+]^2}$$

Notice that we have omitted the concentrations of the solids just as we did when we wrote similar expressions in Chapter Sixteen. The Nernst equation for this system, then, is

$$E_{cell} = E^\circ_{cell} - \frac{0.0592 \text{ V}}{2} \log \frac{[Cu^{2+}]}{[Ag^+]^2}$$

Substituting in the values given in the problem, we have

$$E_{cell} = 0.46 \text{ V} - \frac{0.0592 \text{ V}}{2} \log \frac{1.0 \times 10^{-4}}{(1.0 \times 10^{-3})^2}$$

$$= 0.46 \text{ V} - 0.0296 \text{ V} \log (1.0 \times 10^2)$$
$$= 0.46 \text{ V} - 0.0296 \text{ V} (2.00)$$
$$= 0.46 \text{ V} - 0.0592 \text{ V}$$
$$= 0.40 \text{ V}$$

The cell potential is 0.40 V.

EXAMPLE 18.16   **Calculating the Effect of Concentration on E_{cell}**

Problem: A cell employs the following half-reactions:

$$Ni^{2+}(aq) + 2e^- \rightleftharpoons Ni(s) \qquad E^\circ_{Ni^{2+}} = -0.25 \text{ V}$$
$$Cr^{3+}(aq) + 3e^- \rightleftharpoons Cr(s) \qquad E^\circ_{Cr^{3+}} = -0.74 \text{ V}$$

Calculate the cell potential if $[Ni^{2+}] = 1.0 \times 10^{-4}$ and $[Cr^{3+}] = 2.0 \times 10^{-3}$ M.

Solution: First we need the cell reaction. Nickel ion is reduced because it has the higher reduction potential. This means, of course, that chromium is oxidized. Making electron gain equal electron loss gives

$$\begin{array}{ll} 3(Ni^+(aq) + 2e^- \longrightarrow Ni(s)) & \text{(reduction)} \\ 2(\quad\quad Cr(s) \longrightarrow Cr^{3+}(aq) + 3e^-) & \text{(oxidation)} \\ \hline 3Ni^{2+}(aq) + 2Cr(s) \longrightarrow 3Ni(s) + 2Cr^{3+}(aq) & \text{(cell reaction)} \end{array}$$

Notice that six electrons are transferred overall, so $n = 6$. Now we can write the Nernst equation.

$$E_{cell} = E^\circ_{cell} - \frac{0.0592}{6} \log \frac{[Cr^{3+}]^2}{[Ni^{2+}]^3}$$

Next we need E°_{cell}. Since Ni^{2+} is reduced,

$$\begin{aligned} E^\circ_{cell} &= E^\circ_{Ni^{2+}} - E^\circ_{Cr^{3+}} \\ &= (-0.25\ V) - (-0.74\ V) \\ &= 0.49\ V \end{aligned}$$

Substituting values into the equation, we get

$$\begin{aligned} E_{cell} &= 0.49\ V - \frac{0.0592\ V}{6} \log \frac{(2.0 \times 10^{-3})^2}{(1.0 \times 10^{-4})^3} \\ &= 0.49\ V - 0.010\ V \log (4.0 \times 10^6) \\ &= 0.49\ V - 0.010\ V\ (6.60) \\ &= 0.49\ V - 0.066\ V \\ E_{cell} &= 0.42\ V \end{aligned}$$

EXERCISE 18.15   In a certain zinc-copper cell

$$Zn(s) + Cu^{2+}(aq) \longrightarrow Zn^{2+}(aq) + Cu(s)$$

the ion concentrations are $[Cu^{2+}] = 0.0100\ M$ and $[Zn^{2+}] = 1.00\ M$. What is the cell potential? The standard cell potential is 1.10 V.

18.11 PRACTICAL APPLICATION OF GALVANIC CELLS

Galvanic cells provide portable sources of electric power

One of the principal uses of galvanic cells is the generation of useful electrical energy. In this section we will briefly examine some of the more common present-day uses of galvanic cells and some promising future applications.

The lead storage battery

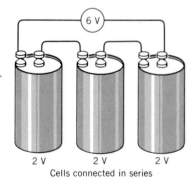

6 V

2 V　　　2 V　　　2 V
Cells connected in series

The common lead storage battery used to start an automobile is composed of a number of galvanic cells, each having an emf of about 2 V, connected in series so that their voltages will add. Most automobile batteries used today contain six such cells and give about 12 V, but 6-, 24-, and 32-V batteries are also available.

The construction of a typical lead storage battery is shown in Figure 18.13. The anode is composed of several plates of lead, and the cathode is composed of several plates of lead dioxide. The electrolyte is sulfuric acid. When the battery is discharging—for example, when it is delivering current to start a car—the electrode reactions are

$$\begin{array}{ll} PbO_2(s) + 4H^+(aq) + SO_4^{2-}(aq) + 2e^- \longrightarrow PbSO_4(s) + 2H_2O & \text{(cathode)} \\ Pb(s) + SO_4^{2-}(aq) \longrightarrow PbSO_4(s) + 2e^- & \text{(anode)} \end{array}$$

The net reaction taking place in each cell is

$$PbO_2(s) + Pb(s) + \underbrace{4H^+(aq) + 2SO_4^{2-}(aq)}_{2H_2SO_4} \longrightarrow 2PbSO_4(s) + 2H_2O$$

Special Topic 18.2 Measurement of Concentration

The Nernst equation tells us that the emf of a cell changes with the concentrations of the ions involved in the cell reaction. One of the most useful consequences of this is that emf measurements provide a way of determining concentrations. Scientists have developed a number of electrodes that can be dipped into one solution after another and whose emf is affected in a reproducible way by the concentration of only one ionic species out of the many that may be present in a solution. An example is the glass electrode used with pH meters. It consists of a glass tube with a very thin glass membrane at one end. The tube is filled with a solution of hydrochloric acid. A silver wire coated with silver chloride is immersed in this solution. The emf of the glass electrode is sensitive to the hydrogen ion concentration. It can also be made sensitive to the concentration of ammonium ion.

Other electrodes can be constructed that are sensitive to nonionic substances, too. An interesting biological application is the coating of the electrode with a gel in which is embedded an enzyme—a biochemical catalyst that itself is sensitive to one or perhaps only a few kinds of molecule. For example, if the enzyme is urease, it catalyzes the decomposition of urea.

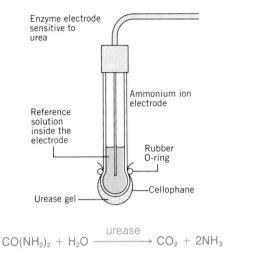

$$CO(NH_2)_2 + H_2O \xrightarrow{\text{urease}} CO_2 + 2NH_3$$

The ammonia produced reacts with water and forms NH_4^+, which can be detected with an electrode that is selectively sensitive to ammonium ion. In this way the electrode is actually able to monitor, in a secondhand way, the urea concentration in a solution. Other electrodes have been developed that are capable of measuring the concentrations of ethanol, cholesterol and several other substances in body fluids. In fact, a glucose electrode is commercially available.

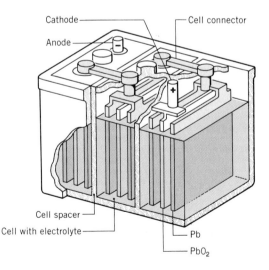

Figure 18.13
A six-volt lead storage battery is composed of three cells connected in series. Each cell produces about 2V.

As the cell discharges, the sulfuric acid concentration decreases because it is used up in the cell reaction. This provides a convenient means of checking the state of a battery. Since the density of a sulfuric acid solution decreases as its concentration drops, the concentration can be determined very simply by measuring the density with a hydrometer (see Figure 8.14). A hydrometer consists of a rubber bulb that is used to draw the battery fluid into a glass tube containing a float. The depth to which the float sinks is inversely proportional to the density of the liquid—the deeper the float sinks, the lower is the density and the weaker is the charge on the battery. The narrow neck of the float is usually marked to indicate the state of charge of the battery.

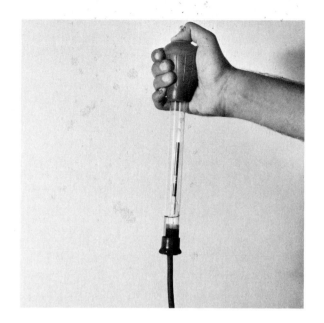

Figure 18.14
A battery hydrometer. Battery acid is drawn into the glass tube. The depth to which the float sinks in the acid is inversely proportional to the strength of the battery.

The principal advantage of the lead storage battery is that the cell reactions that occur spontaneously during discharge can be reversed by the application of voltage from an external source. In other words, the battery can be recharged by electrolysis. The reaction is

$$2PbSO_4(s) + 2H_2O \xrightarrow{\text{electrolysis}} PbO_2(s) + Pb(s) + 4H^+(aq) + 2SO_4^{2-}(aq)$$

The major disadvantages are that these batteries are very heavy and they usually must be vented, which means that the corrosive sulfuric acid can spill out if one of these batteries is upset.

The zinc-carbon dry cell

The ordinary, relatively inexpensive, 1.5-V zinc-carbon dry cell used to power flashlights, tape recorders, and the like is not really dry. A cutaway view of the internal construction of this type of cell is shown in Figure 18.15. The outer shell is made of zinc and serves as the anode—its exposed outer surface at the bottom is the negative end of the battery. The cathode—the positive terminal of the battery—consists of a carbon (graphite) rod surrounded by a moist paste of graphite powder, manganese dioxide, and ammonium chloride.

The anode reaction is simply the oxidation of the zinc.

$$Zn(s) \longrightarrow Zn^{2+}(aq) + 2e^- \quad \text{(anode)}$$

The cathode reaction is complex, and a mixture of products is formed. One of the major reactions is

$$2MnO_2(s) + 2NH_4^+(aq) + 2e^- \longrightarrow Mn_2O_3(s) + 2NH_3(aq) + H_2O \quad \text{(cathode)}$$

The ammonia formed at the cathode reacts with some of the Zn^{2+} produced from the anode to form an ion having the formula $Zn(NH_3)_4^{2+}$. Because of the complexity of the cathode half-cell reaction, no simple overall cell reaction can be written.

The major advantages of the dry cell are its relatively low price and the fact that it normally works without leaking. One disadvantage is that the cell loses its ability to function rather rapidly under heavy current drain. This happens because the products of reaction can't diffuse away from the electrodes very easily. If unused for a while, however, these batteries can rejuvenate themselves some-

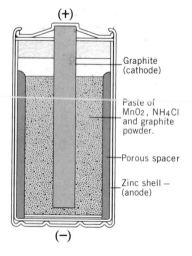

(+)

Graphite (cathode)

Paste of MnO_2, NH_4Cl and graphite powder.

Porous spacer

Zinc shell (anode)

(−)

Figure 18.15
A cross section of a zinc-carbon dry cell.

what. Another disadvantage is that they can't be recharged. Devices that claim to recharge zinc-carbon dry cell batteries simply drive the reaction products away from the electrodes, which allows them to function again. This doesn't work very many times, however. Before too long the zinc casing will develop holes and the battery must be discarded.

The common alkaline flashlight battery also uses Zn and MnO_2 as reactants, but under basic conditions. The half-cell reactions are

$$Zn(s) + 2OH^-(aq) \longrightarrow ZnO(s) + H_2O + 2e^- \qquad \text{(anode)}$$
$$2MnO_2(s) + H_2O + 2e^- \longrightarrow Mn_2O_3(s) + 2OH^-(aq) \qquad \text{(cathode)}$$

and the voltage is about 1.54 V. It has a longer shelf-life and is able to deliver higher currents for longer periods than the less expensive zinc-carbon cell.

The nickel-cadmium storage cell

The nickel-cadmium storage cell is the *nicad* battery that powers rechargeable electronic calculators, electric shavers, and power tools. It produces a potential of about 1.4 V, which is slightly lower than the zinc-carbon cell. The electrode reactions in the cell are

$$Cd(s) + 2OH^-(aq) \longrightarrow Cd(OH)_2(s) + 2e^- \qquad \text{(anode)}$$
$$2e^- + NiO(s) + 2H_2O \longrightarrow Ni(OH)_2(s) + 2OH^-(aq) \qquad \text{(cathode)}$$

The nickel-cadmium battery is rechargeable and can be sealed to prevent leakage, which is particularly important in electronic devices. For the storage of a given amount of electrical energy, however, the nicad battery is considerably more expensive than the lead storage battery.

Fuel cells

The production of usable energy by the combustion of fuels is an extremely inefficient process. Modern electric power plants are able to harness only about 35 to 40 percent of the energy theoretically available from oil, coal, or natural gas. The gasoline or diesel engine has an efficiency of only about 25 to 30 percent. The rest of the energy is lost to the surroundings as heat. This is the reason that a car must have an effective cooling system.

Fuel cells are electrochemical cells that "burn" fuel under conditions that are much more nearly thermodynamically reversible than simple combustion. They therefore achieve much greater efficiencies—75 percent is quite feasible. Figure 18.16 illustrates a hydrogen-oxygen fuel cell. The electrolyte, a hot concentrated solution of potassium hydroxide in the center compartment is in contact with two porous electrodes. Porous carbon and thin porous nickel electrodes

If we could "burn" our petroleum in fuel cells to produce energy, consumption could be halved.

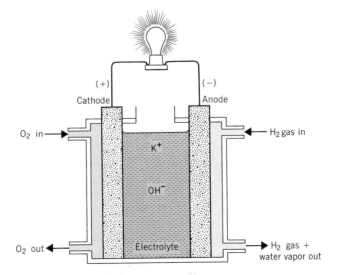

Figure 18.16
Hydrogen-oxygen fuel cell.

have been used. Gaseous H_2 and O_2 under pressure are circulated so as to come in contact with the electrodes. At the cathode, oxygen is reduced.

$$O_2(g) + 2H_2O + 4e^- \longrightarrow 4OH^-(aq) \qquad \text{(cathode)}$$

At the anode, hydrogen is oxidized to water.

$$H_2(g) + 2OH^-(aq) \longrightarrow 2H_2O + 2e^- \qquad \text{(anode)}$$

Part of the water formed at the anode leaves as steam mixed with the circulating hydrogen gas. The net cell reaction, after making electron loss equal to electron gain, is

$$2H_2(g) + O_2(g) \longrightarrow 2H_2O$$

The advantages of the fuel cell are obvious. There is no electrode material to be replaced, as there is in an ordinary battery. The fuel can be fed in continuously to produce power. In fact, hydrogen-oxygen fuel cells have been used on spacecraft for just this reason. They also provide another benefit—drinking water for the astronauts. The disadvantages at the present time include their high cost and the rather large size that is necessary to produce useful quantities of electrical power. Nevertheless, the potential efficiency of these cells and the prospects of generating the hydrogen and oxygen fuels by solar energy make fuel cells an exciting possibility as practical energy-generating devices for the future.

SUMMARY

Electrolysis. In an electrolytic cell, a flow of electricity causes reduction at a negatively charged cathode and oxidation at a positively charged anode. Ion movement rather than electron transport occurs in the electrolyte. The electrode reactions are determined by which species is most easily reduced and which is most easily oxidized. In the electrolysis of water, an electrolyte must be present to maintain electrical neutrality at the electrodes.

Ion-Electron Method for Balancing Equations. To obtain a net ionic equation for a redox reaction in aqueous solution, the equation is divided into half-reactions that are balanced individually before being recombined to give the final equation. Oxygen is balanced by adding H_2O to the side needing O; hydrogen is balanced by adding H^+ to the side needing H; and the net charge on both sides of a half-reaction is balanced by adding electrons. The electron gain and loss are made equal and the half-reactions are added. For reactions in basic solution, OH^- is added to both sides to remove H^+ by the "reaction," $H^+ + OH^- \rightarrow H_2O$.

Quantitative Problems on Electrolysis. A faraday is 1 mol of electrons and 96,500 C equals 1 faraday. The product of current (amperes) and time (seconds) gives coulombs. These relationships and the half-reactions occurring at anode or cathode permit us to relate the amount of chemical change to measurements of current and time.

Applications of Electrolysis. Electroplating, the production of aluminum and magnesium, the refining of copper, and the electrolysis of molten and aqueous sodium chloride are examples of practical applications of electrolysis.

Galvanic Cells. A spontaneous redox reaction, in which the individual half-reactions occur in separated half-cells, can force the electron transfer to occur through an external electrical circuit. Reduction occurs at the positively charged cathode and oxidation takes place at the negatively charged anode. The half-cells must be connected electrolytically to complete an electrical circuit. A salt bridge accomplishes this—it permits electrical neutrality to be maintained. The emf or voltage produced by a cell is equal to the standard cell potential when all ion concentrations are 1 M. The cell potential can be considered to be the difference between the reduction potentials of the half-cells. The half-cell with the higher reduction potential undergoes reduction and forces the other to undergo oxidation. Although the reduction potential of an isolated half-cell can't be measured, values are assigned by choosing the hydrogen electrode as a reference electrode having a reduction potential of 0.00 V. Species more easily reduced than H^+ have positive reduction potentials; those less easily reduced have negative reduction potentials. Reduction potentials can be used to predict the cell reaction and to calculate $E°_{cell}$. They can also be used to predict spontaneous redox reactions not occurring in galvanic cells and to predict whether or not a given reaction is spontaneous.

Thermodynamics and Cell Potentials. The values of $\Delta G°$ and K_c for a reaction can be calculated from $E°_{cell}$. The

Nernst equation relates the cell potential to the standard cell potential and the mass action expression. It allows the cell potential to be calculated for ion concentrations other than 1 M.

Practical Galvanic Cells. The lead storage battery and the nickel-cadmium battery are rechargeable. The zinc-carbon cell—the common dry cell—is not. The common al-

kaline battery uses essentially the same reactions as the less expensive dry cell. Fuel cells consume fuel that can be fed continuously. For the production of usable energy, they offer much higher thermodynamic efficiencies than can be obtained by burning the fuel in conventional power plants or internal combustion engines.

REVIEW QUESTIONS

18.16. What is an electrochemical change?

18.17. How do we define anode and cathode?

18.18. What electrical charges do the anode and the cathode carry in an electrolytic cell?

18.19. How does electrical conduction occur in an electrolyte?

18.20. Why must NaCl be melted before it is electrolyzed to give Na and Cl_2?

18.21. What does the term *inert electrode* mean?

18.22. Write the anode, cathode, and overall reactions for the electrolysis of molten NaCl.

18.23. Write half-reactions for the oxidation and the reduction of water.

18.24. What happens to the pH of the solution near the cathode and anode during the electrolysis of KNO_3?

18.25. What function does KNO_3 serve in the electrolysis of a KNO_3 solution?

18.26. Compare the ease of oxidation between (a) NO_3^- and H_2O, (b) Br^- and H_2O, and (c) NO_3^- and Br^-.

18.27. Compare the ease of reduction between (a) K^+ and H_2O, (b) Cu^{2+} and H_2O, and (c) K^+ and Cu^{2+}.

18.28. What products would we expect at the electrodes if a solution containing both KBr and $Cu(NO_3)_2$ were electrolyzed?

18.29. Balance the following half-reactions occurring in an acidic solution. Indicate whether each is an oxidation or a reduction.
(a) $BiO_3^- \longrightarrow Bi^{3+}$
(b) $NO_3^- \longrightarrow NH_4^+$
(c) $Pb^{2+} \longrightarrow PbO_2$
(d) $Cl_2 \longrightarrow ClO_3^-$

18.30. Balance the following half-reactions occurring in a basic solution. Indicate whether each is an oxidation or a reduction.
(a) $Fe \longrightarrow Fe(OH)_2$
(b) $SO_2Cl_2 \longrightarrow SO_3^{2-} + Cl^-$
(c) $Mn(OH)_2 \longrightarrow MnO_4^{2-}$
(d) $H_4IO_6^- \longrightarrow I_2$

18.31. Balance these reactions occurring in an acidic solution.
(a) $S_2O_3^{2-} + OCl^- \longrightarrow Cl^- + S_4O_6^{2-}$
(b) $NO_3^- + Cu \longrightarrow NO_2 + Cu^{2+}$
(c) $IO_3^- + AsO_3^{3-} \longrightarrow I^- + AsO_4^{3-}$
(d) $SO_4^{2-} + Zn \longrightarrow Zn^{2+} + SO_2$

18.32. Balance these reactions occurring in a basic solution.
(a) $CrO_4^{2-} + S^{2-} \longrightarrow S + CrO_2^-$
(b) $MnO_4^- + C_2O_4^{2-} \longrightarrow CO_2 + MnO_2$
(c) $ClO_3^- + N_2H_4 \longrightarrow NO + Cl^-$
(d) $NiO_2 + Mn(OH)_2 \longrightarrow Mn_2O_3 + Ni(OH)_2$

18.33. What is a faraday? What relationships relate faradays to current and time measurements?

18.34. What is electroplating? Sketch an apparatus to electroplate silver.

18.35. Describe the Hall process for producing metallic aluminum. What electrochemical reaction is involved?

18.36. Describe how magnesium is recovered from sea water. What electrochemical reaction is involved?

18.37. How is metallic sodium produced? What are some uses of metallic sodium?

18.38. Describe the electrolytic refining of copper. What economic advantages offset the cost of electricity for this process?

18.39. Describe the electrolysis of aqueous sodium chloride. How do the products of the electrolysis compare for stirred and unstirred reactions?

18.40. What is a galvanic cell? What is a half-cell?

18.41. What is the function of a salt bridge?

18.42. In the copper-silver cell, why must the Cu^{2+} and Ag^+ solutions be kept in separate containers?

18.43. What processes take place at the anode and cathode in a galvanic cell? What electrical charge do the anode and cathode carry in a galvanic cell?

18.44. Explain how the movement of ions is the same in both galvanic and electrolytic cells.

18.45. When magnesium metal is placed into a solution of copper sulfate, the magnesium dissolves to give Mg^{2+} and copper metal is formed. Write an ionic equation for this reaction. Describe how you could use the reaction in a galvanic cell. Which metal, copper or magnesium, is the cathode?

18.46. Aluminum will displace tin from solution according to the equation: $2Al(s) + 3Sn^{2+}(aq) \rightarrow 2Al^{3+}(aq) + 3Sn(s)$. What would be the individual half-cell reactions if this were the cell reaction in a galvanic cell? Which metal would be the anode and which the cathode?

18.47. At first glance, the following equation may appear to be balanced: $MnO_4^- + Sn^{2+} \rightarrow SnO_2 + MnO_2$. What is wrong with it?

18.48. What is the difference between a cell potential and a standard cell potential?

18.49. How are standard reduction potentials combined to give the standard cell potential for a spontaneous reaction?

18.50. What is emf? What are its units?

18.51. What ratio of units gives volts?

18.52. What are the units of amperes $\times$ volts $\times$ seconds?

18.53. Is it possible to measure the emf of an isolated half-cell? Explain your answer.

18.54. Describe the hydrogen electrode. What is the value of its standard reduction potential?

18.55. What do the positive and negative signs of reduction potentials tell us?

18.56. If $E^{\circ}_{Cu^{2+}}$ had been chosen as a reference electrode with a potential of 0.00 V, what would the reduction potential of the hydrogen electrode be relative to it?

18.57. From the positions of the half-reactions in Table 18.1, determine whether the following reactions are spontaneous.
(a) $2Au^{3+} + 6I^- \longrightarrow 3I_2 + 2Au$
(b) $3Fe^{2+} + 2NO + 4H_2O \longrightarrow 3Fe + 2NO_3^- + 8H^+$
(c) $3Ca + 2Cr^{3+} \longrightarrow 2Cr + 3Ca^{2+}$

18.58. Using a voltmeter, how can you tell which is the anode and which is the cathode in a galvanic cell?

18.59. The cell reaction during the discharge of a lead storage battery is

$$Pb(s) + PbO_2(s) + 4H^+(aq) + 2SO_4^{2-}(aq)$$
$$\longrightarrow 2PbSO_4(s) + 2H_2O$$

The standard cell potential is 2.05 V. What is the correct form of the Nernst equation for this reaction?

18.60. What are the anode and cathode reactions during the discharge of a lead storage battery? How can a battery produce an emf of 12 V if the cell reaction has a standard potential of only 2 V?

18.61. What reactions occur in the dry cell?

18.62. Give the half-cell reactions and the cell reaction in a nicad battery.

18.63. What advantages do fuel cells offer over conventional means of obtaining electrical power by the oxidation of fuels?

REVIEW PROBLEMS

18.64. How many faradays are required (a) to produce 1 mol of Cu from Cu^{2+}, and (b) to produce 3 mol of Cr from Cr^{3+}?

18.65. How many faradays are needed to oxidize Cr^{3+} to produce 1 mol of $Cr_2O_7^{2-}$? (*Hint:* Write a balanced half-reaction for an acidic solution.)

18.66. How many grams of Cl_2 would be produced in the electrolysis of NaCl by a current of 2.50 A for 40.0 min?

18.67. How many grams of $Fe(OH)_2$ are produced at an iron anode when a basic solution undergoes electrolysis at a current of 12.0 A for 10.0 min?

18.68. How many milliliters of gaseous H_2, measured at STP, would be produced at the cathode in the electrolysis of water with a current of 1.50 A for 5.00 min?

18.69. A solution of NaCl in water was electrolyzed with a current of 3.00 A for 10.0 min. How many milliliters of

0.100 M HCl would be required to neutralize the resulting solution?

18.70. How many hours would it take to produce 25.0 g of metallic chromium by the electrolytic reduction of Cr^{3+} with a current of 1.25 A?

18.71. How many hours would it take to generate 25.0 g lead from $PbSO_4$ during the charging of a storage battery using a current of 0.50 A? The half-reaction is: $PbSO_4 + 2e^- \rightarrow Pb + SO_4^{2-}$.

18.72. How many amperes would be needed to produce 48.0 g of magnesium during the electrolysis of molten $MgCl_2$ in 1.00 hr?

18.73. A large electrolysis cell that produces metallic aluminum from Al_2O_3 by the Hall process is capable of yielding 900 lb (409 kg) of aluminum in 24 hr. What current is required for this?

18.74. Use the data in Table 18.1 to calculate the standard cell potential for the reaction

$$NO_3^- + 4H^+ + 3Fe^{2+} \longrightarrow 3Fe^{3+} + NO + 2H_2O$$

18.75. Use the data in Table 18.1 to calculate the cell potential for the reaction

$$Cd^{2+} + Fe \longrightarrow Cd + Fe^{2+}$$

18.76. From the half-reactions below, determine the cell reaction and standard cell potential.

$BrO_3^- + 6H^+ + 6e^- \rightleftharpoons Br^- + 3H_2O$ $E^\circ_{BrO_3^-} = 1.44$ V
$I_2 + 2e^- \rightleftharpoons 2I^-$ $E^\circ_{I_2} = 0.54$ V

18.77. What is the standard cell potential and the net reaction in a galvanic cell that has the following half-reactions?

$MnO_2 + 4H^+ + 2e^- \rightleftharpoons Mn^{2+} + 2H_2O$ $E^\circ_{MnO_2} = 1.23$ V
$PbCl_2 + 2e^- \rightleftharpoons Pb + 2Cl^-$ $E^\circ_{PbCl_2} = -0.27$ V

18.78. What will be the spontaneous reaction among H_2SO_3, $S_2O_3^{2-}$, HOCl, and Cl_2? The half-reactions involved are

$2H_2SO_3 + 2H^+ + 4e^- \rightleftharpoons S_2O_3^{2-} + 3H_2O$
$E^\circ_{H_2SO_3} = +0.40$ V

$2HOCl + 2H^+ + 2e^- \rightleftharpoons Cl_2 + 2H_2O$
$E^\circ_{HOCl} = 1.63$ V

18.79. What will be the spontaneous reaction among Br_2, I_2, Br^- and I^-? Use the data in Table 18.1.

18.80. Will the following reaction occur spontaneously?

$$SO_4^{2-} + 4H^+ + 2Br^- \longrightarrow Br_2 + H_2SO_3 + H_2O$$

Use the data in Table 18.1 to answer this question.

18.81. Use the data below to determine whether the reaction

$$S_2O_8^{2-} + Ni(OH)_2 + 2OH^- \longrightarrow 2SO_4^{2-} + NiO_2 + 2H_2O$$

will occur spontaneously.

$NiO_2 + 2H_2O + 2e^- \rightleftharpoons Ni(OH)_2 + 2OH^-$
$E^\circ_{NiO_2} = 0.49$ V

$S_2O_8^{2-} + 2e^- \rightleftharpoons 2SO_4^{2-}$
$E^\circ_{S_2O_8^{2-}} = 2.01$ V

18.82. Calculate ΔG° for the reaction

$$2MnO_4^- + 6H^+ + 5HCO_2H \longrightarrow 2Mn^{2+} + 8H_2O + 5CO_2$$

for which $E^\circ_{cell} = 1.69$ V.

18.83. Calculate ΔG° for the following reaction *as written*.

$$2Br^- + I_2 \longrightarrow 2I^- + Br_2$$

18.84. Calculate K_c for the reaction, $Ni^{2+} + Co \longrightarrow Ni + Co^{2+}$. Use the data in Table 18.1. Assume $T = 298$ K.

18.85. The reaction, $2AgI + Sn \rightarrow Sn^{2+} + 2Ag + 2I^-$, has a calculated $E^\circ_{cell} = -0.015$ V. What is the value of K_c for this reaction?

18.86. The cell reaction, $NiO_2(s) + 4H^+(aq) + 2Ag(s) \rightarrow Ni^{2+}(aq) + 2H_2O + 2Ag^+(aq)$ has $E^\circ_{cell} = 2.48$ V. What will be the cell potential at a pH of 6.00 when the concentrations of Ni^{2+} and Ag^+ are each 0.10 M?

18.87. $E^\circ_{cell} = 0.135$ V for the reaction

$$3I_2(s) + 5Cr_2O_7^{2-}(aq) + 34H^+(aq) \longrightarrow 6IO_3^-(aq) + 10Cr^{3+}(aq) + 17H_2O$$

What is E_{cell} if $[Cr_2O_7^{2-}] = 0.10$ M, $[H^+] = 0.010$ M, $[IO_3^-] = 0.0010$ M, and $[Cr^{3+}] = 0.00010$ M?

CHAPTER NINETEEN

CHEMISTRY OF LIVING SYSTEMS

19.1 MOLECULAR BASIS OF LIFE

**Materials, energy, and information are the three
irreducible requirements for life at any level of existence**

Biochemistry is the systematic study of the chemicals of living systems, their organization into cells, and the principles of their chemical interactions as they participate in the processes of life. The molecules of living systems are lifeless, yet life has a molecular basis. When they are isolated from their cells or when they are studied in their living environments, the chemicals at the foundation of life obey all of the known laws of chemistry and physics. However, in isolated states, not one single compound of a cell has life. The cell, as an intricate organization of interacting chemicals, is as important to life as the chemicals themselves. The cell is the smallest unit of matter that lives and that can make a new cell exactly. Yet even a small change in a cell's environment—a change in temperature of a few degrees, or of pH, pressure, humidity, or the appearance of hostile chemicals—any of these changes can quickly cause a cell to be nothing more than a lifeless bag of chemicals that soon will spill its contents.

An ancient writer said that we once were dust and to dust we will return. The marvelous interlude between dust and dust, called life, endures only as long as the organism can maintain its dynamic organization of parts and can defend it against the natural tendencies toward disintegration and decay. Both a cell and an organism are finely ordered, high-energy systems compared to dust and ashes scattered to the four winds. We will not speculate here on how life began, but as far back as we know, as well as in the present time, life begets life. A new creature is shaped and fashioned by an older creature. One life, if a plant, takes ordinary substances such as water, carbon dioxide, and soil minerals and uses them to grow larger, and to make seeds or other fruiting systems. As the seeds await their destinies, the mature plant lives out its brief existence. After it dies, decay again returns the substance of the plant to the much lower energy, much more disorganized forms of carbon dioxide, water, and soil minerals.

Animals must begin with substances more organized and more energy-rich than the nutrients plants use. Therefore, animals feed on plants or on other animals. Whether it occurs in a plant or an animal, however, all of life has three absolutely vital needs—materials, energy, and information. Our very limited purpose in this chapter is to study the molecular basis of fulfilling these needs. Our background from Chapters Nine and Thirteen permits only a cursory view, but we will exploit that background where it best can be used: the structural features of proteins, lipids, carbohydrates, and nucleic acids. The first three substances provide both materials and energy; the fourth, the nucleic acids, carry, transmit, and translate all of the genetic information needed by cells and whole organisms. The chemical information encoded in the structures of nucleic acids, for example, insures that each member of each species of living things has the unique set of enzymes for that individual. Enzymes, as we have noted earlier, are the catalysts without which no reactions of a cell could occur, and enzymes are proteins.

We will also look briefly at the structural features and uses of lipids (e.g., animal fats and vegetable oils) and carbohydrates (e.g., sugar and starch). Their molecules are also complicated, and you will not be expected to memorize all of their structures. We will study those structures, however, on the grounds that we cannot say that we know what these substances are until we have become somewhat familiar with their structures. With this as an overview of where we are going, let's start back at the beginning with a topic vital to the study of all biologically important organic substances: molecular shape or chirality.

19.2 MOLECULAR CHIRALITY

$$O$$
$$\parallel$$
$$-C-NH_2 + HO-H$$
amide
group

$$O$$
$$\parallel$$
$$\longrightarrow -C-OH \quad + \quad H-NH_2$$
carboxyl   ammonia
group

$$C=C + H-H$$
alkene
group

$$\longrightarrow -C-C-$$
$$\quad\quad | \quad |$$
$$\quad\quad H \quad H$$

Two isomers of C_2H_6O:
CH_3-O-CH_3 dimethyl ether
CH_3-CH_2-OH ethanol

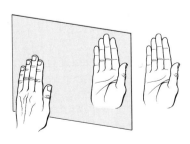

The image of the left hand reflected in the mirror appears the same as the right hand.

Most of the molecules in living systems have a peculiar "handedness" or chirality

In Chapters Nine and Thirteen we commented on the fact that the chemical properties of organic substances are determined by their functional groups. Molecules with amide groups, for example, can be hydrolyzed to smaller molecules. Alcohols can be oxidized or converted into esters. Because alkanes have no functional groups, they have the fewest kinds of chemical reactions. Where a carbon-carbon double bond occurs, the molecule can add hydrogen. All of these facts and types of reactions belong to biological chemicals, too, but there is another feature that is quite remarkable and altogether a matter of life and death for the organism. With very few exceptions, molecules of biological chemicals must not only have the proper functional groups but also the right overall geometrical shapes in order to be useful to the organism. In this section we will study how molecules can have unique shapes and what difference that makes.

We learned in Chapter Nine that many compounds are related to each other as isomers; they have identical molecular formulas but different structural formulas. We will now study the most subtle way in which two molecules can be isomers—by having nothing more than an opposite "handedness." By handedness we mean the property of having a unique molecular geometry that permits the existence of two isomers whose molecular shapes are related as a left hand is related to a right hand. We know how two hands are alike, but now it is important to think about how they are different. Hold your two hands in front of you with their palms facing each other. Now imagine that a mirror is halfway between them. (Try this with a real mirror, also.) Compare the reflection in the mirror with the actual hand behind the mirror. They are the same, aren't they? (We have to ignore wrinkles and fingerprints.) One hand is the mirror image of the other. Yet the right and left hands are different, a fact you quickly discover whenever you try to put the wrong glove on one hand.

Two different molecules can also be related as the right hand is related to the left hand. If a molecular model is made of one molecule and placed before a mirror, the reflection you see is the molecular model of the other molecule. To look ahead a bit, if we crudely imagine an enzyme to be like a glove, then the ability of the enzyme molecule to fit to and catalyze a reaction of either of these two molecules, each with its own handedness, will depend entirely on that peculiar handedness. In this way the handedness of molecules is a crucial factor in almost all reactions in living cells.

A left-hand glove won't fit the right hand.

"Chirality" comes from *cheir,* Greek for "hand."

Your left hand and your right hand are related as object-to-mirror image but they do not superimpose.

A thumb-tack and its mirror image are identical; they are superimposable.

Not all substances have molecules with the property of handedness, and what we now have to do is develop a way to recognize when this structural feature is possible. For this purpose we need a test for deciding whether two things—two hands or two molecules—are identical or are different even when they are as similar as a right and a left hand.

The fundamental test for handedness is **superimposability.** To see how it works, bring your left and right hands facing each other again. Now have them touch—palm to palm, thumb to thumb. What you are doing is "imposing them." Superimposing goes beyond that, but it can be done only by the imagination. (Maybe that's why it is called *super*imposing.) Pretend to blend the right hand into the left to put it completely *inside* but not through the left hand. Will all of the parts perfectly match at the same time? The answer is "No." The fingers may seem to be blending all right, but notice that the fingernail of one finger superimposes the fingerprint region of the opposite finger. The palms do not superimpose, either. One palm comes out on the backside of the other hand. What if you try to turn one hand over to get the palms to superimpose? Then the thumbs point in opposite directions. The left hand and the right hand simply do not superimpose. In the final analysis, *only if two objects pass the test of superimposability may we call them identical.*

Try the same test with two plain thumbtacks. It is not hard to line them up so that one tack looks exactly like the other's reflection in the mirror. However, unlike the hands, the two tacks superimpose. If they were models of molecules, they could not be called isomers. They are identical.

The technical term for handedness is **chirality.** Two plain thumbtacks do not have chirality, but your left hand and your right hand do. Two gloves in a pair also have chirality, and we know that each glove fits best only to one hand. The molecules of every enzyme in the body have chirality. If an enzyme's task is to catalyze a reaction of a substance whose molecules are also chiral, then the enzyme will work only on those molecules having a matching chirality—like a glove fitting readily only to one of the two hands.

Let's move from the world of gloves, hands, and tacks back to the realm of molecules to discover what particular features of molecular structure make a molecule chiral. We will limit our study to organic compounds and the chirality of their molecules that is caused by a special structural condition at a *carbon* atom. In Chapter Twenty we will see that chirality is not limited to carbon compounds. *If a molecule has one carbon atom that is bonded to four different atoms or groups of atoms, then the molecule is chiral.* Any carbon holding four different atoms or groups of atoms is called a **chiral carbon,** and we say that this carbon is the *center of chirality* for the molecule. To see what these statements mean, study Figure 19.1 very carefully. The changes taking place from parts (a) to (d) are the replacements of hydrogen atoms in CH_4 by more and more different atoms. (We could have used groups such as methyl, ethyl, and so forth, but using halogen atoms keeps the figures simple.) Two models of each compound are shown, and the two are placed so that one serves as an *object* and the other as the image of that object called the *mirror image.* Thus, if you made a model of methane, CH_4, out of a kit, placed it in front of a mirror, and used the reflection in the mirror as the guide for making the second model, then you would be making a model of the mirror image. Notice carefully that in parts (a) through (c) the object model and its mirror image superimpose. In parts (b) and (c) the curved arrows show how one model has to be lifted and turned in space before imposing it on the other model, but this is allowed. In part (d), however, the object model of CHClBrI has a mirror image model with which it cannot superimpose. The carbon in this compound is attached to four different atoms. In the earlier compounds, the carbon atoms held at least two identical atoms—two, three, or four hydrogens in the examples chosen. *Only the molecules having four different atoms (or groups) bonded to the same carbon are not superimposable on their mirror images and therefore are chiral molecules.*

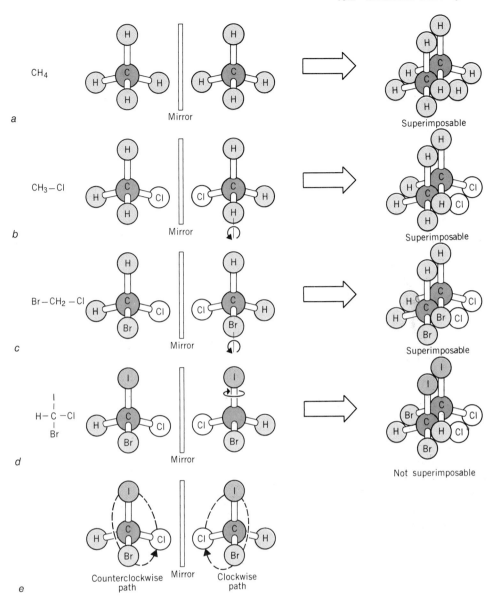

Figure 19.1
Four different atoms or groups attached to the same atom make
that atom a chiral center in a molecule.

Notice in part (e) the difference between the object and mirror image models
of part (d). With each model in part (e), imagine that you are looking down the
bond that goes from carbon to hydrogen, much the same as if you were looking
down a steering column of a car. Imagine that the other atoms, I, Br, and Cl, are
arranged around the steering wheel itself. The only difference between the two
models is that your eyes must move counterclockwise to make the "trip" from I to
Br to Cl in one model and clockwise to make the identical trip in the other. The
two models of part (d) are not identical because they are not superimposable.
Yet they are different in a most subtle way. They do not just have the same molecular formula, they also have the same basic organization of parts, the same
atoms attached to the same carbon. The only difference is in the relative orientations in space of the atoms (or groups) attached to the carbon.

Special Topic 19.1 Optical Activity

Since the molecules of enantiomers have identical internal geometries—bond angles and lengths—and the same atoms or groups are joined to the same carbon atoms, enantiomers must also have identical polarities. Therefore, enantiomers must and do have identical boiling points, melting points, densities, and other physical properties—except for one difference. *Solutions of enantiomers behave oppositely toward polarized light.*

If you and a friend each have a pair of Polaroid® sunglasses, get them out and try these experiments. The light that gets through the lens of one of these glasses is plane polarized light, which means that its vibrations are only in one plane, not in every plane. Put a second Polaroid® lens over one lens in the first pair of glasses, and then slowly turn one with respect to the other, holding them up to a light. The accompanying figure shows this being done to two pieces of Polaroid film. Find a position of the two lenses where the light that can get through *both* is the brightest. Now rotate one lens 90° with respect to the other. Notice how the light getting through both first gradually grows dimmer and then goes out. Turning one lens is like pinching the vibrations out of a guitar string until there is no sound. Turn one lens another 90°; the brightness comes back, doesn't it? Turn it still another 90°, and it goes out again; go still another 90° and back the brightness comes.

Now imagine the two films positioned to let no light through. Here's a part you will have to imagine because it takes special equipment to show it. If you placed a solution of an enantiomer between the two films it would restore at least some of the light's brightness. You would have to rotate one film a certain measurable number of degrees to shut out the

No light gets through where two Polaroid® lenses overlap and are oriented properly with respect to each other.

light again, say 10° *clockwise.* If you repeated this experiment with the other enantiomer, the same thing would happen—except you would have to rotate one film 10° *counterclockwise.* This is the only physical difference between enantiomers. They affect polarized light the same number of degrees but in opposite directions. Brightness is restored by the enantiomer because the electron clouds of its molecules force the plane polarized light to experience a twist or rotation in the plane of its vibration. One enantiomer rotates the plane in one direction; the other, an identical number of degrees in the opposite direction. This ability of an enantiomer to rotate the plane of plane polarized light is called *optical activity.* Each enantiomer is said to be optically active, and one enantiomer is sometimes called an *optical isomer* of the other.

"Stereo-" comes from *stereos,* Greek for "solid," meaning three-dimensional.

The general name for isomers that differ only in spatial orientation is **stereoisomers.** There are two broad kinds of stereoisomers. We learned about one in Chapter Nine, geometric or *cis-trans* isomers, in which the difference between two isomers was only in the directions that two groups point at a double bond or on a cyclic molecule. The stereoisomers that we have introduced in this chapter are called **optical isomers** because of a peculiar effect an optical isomer has on polarized light (see Special Topic 19.1). We can say that the two isomers shown in part (*d*) of Figure 19.1 are optical isomers of each other. Another term is also used, because the term "optical isomers" includes more than one kind. (We must leave the details to more advanced courses.) Optical isomers, whose molecules are related as an object is related to its mirror image but that cannot be superimposed, are called **enantiomers.**

"Enantio-" comes from *enantios,* Greek for "opposite," and "meros" is from the Greek for "parts."

EXAMPLE 19.1 Finding Chiral Carbon Atoms

Problem: Lactic acid is the cause of sourness in sour milk. Does its molecule have a chiral carbon? More than one? Pinpoint it (them).

$$H-\overset{\overset{\displaystyle H}{|}}{\underset{\underset{\displaystyle H}{|}}{C}}-\overset{\overset{\displaystyle H}{|}}{\underset{\underset{\displaystyle O}{|}}{C}}-\overset{\overset{\displaystyle O}{\|}}{C}-O-H \quad \text{lactic acid}$$

$$\underset{\displaystyle H}{}$$

Solution: Examine each carbon at a time. To be chiral a carbon must (a) hold *four* groups, not three or fewer; and (b) hold four *different* groups. The carbon on the left has four groups, but three (actually atoms) are the same, —H. The carbon on the other end has only three atoms or groups directly joined to it. The middle carbon has four atoms or groups, and they are different:

$$H-\overset{\overset{\displaystyle H}{|}}{\underset{\underset{\displaystyle H}{|}}{C}}-, \quad , \quad -\overset{\overset{\displaystyle O}{\|}}{C}-O-H, \quad \text{and} \quad \overset{|}{\underset{OH}{}}$$

Thus this carbon is chiral.

EXERCISE 19.1 Examine each structure to find its chiral carbons (if any). Put an asterisk beside each one, and make a list of the structures of the four different atoms or groups attached to it.

—NH$_2$ amino group

$$\overset{O}{\underset{\displaystyle}{\|}}$$
—C—OH carboxylic acid group

$$\overset{O}{\underset{\displaystyle}{\|}}$$
— C —H aldehyde group

(a)
$$H-N-\overset{\overset{\displaystyle H}{|}}{\underset{\underset{\displaystyle H}{|}}{C}}-\overset{\overset{\displaystyle O}{\|}}{C}-O-H$$
Glycine, an amino acid and a building block of proteins

(b)
$$H-N-\overset{\overset{\displaystyle H}{|}}{\underset{\underset{\displaystyle}{|}}{C}}-\overset{\overset{\displaystyle O}{\|}}{C}-O-H$$
$$H-C-H$$
$$\underset{\displaystyle H}{}$$
Alanine, another amino acid from protein

(c)
$$H-O-\overset{\overset{\displaystyle H}{|}}{\underset{\underset{\displaystyle H}{|}}{C}}-\overset{\overset{\displaystyle H}{|}}{\underset{\underset{\displaystyle OH}{|}}{C}}-\overset{\overset{\displaystyle O}{\|}}{C}-H$$
Glyceraldehyde, an intermediate in the breakdown of sugar by the body

Having discovered that two substances can be different compounds despite having molecules as closely similar as enantiomers are, the natural question becomes, "So what?" Since enantiomers must have identical functional groups, we might easily conclude that they must have identical chemical properties. And we would be right—but only up to a vitally important point, so vital in fact that it is literally a matter of life and death to us and all living species!

Enantiomers have identical chemical properties toward one broad class of reactants, those whose ions or molecules are not chiral. Such reactants might be water, sodium, hydrochloric acid, hydrogen gas, bromine, or potassium permanganate. These reactants do not have chiral molecules or ions, and one enantiomer experiences the same reactions with them as the other enantiomer. However, if the molecules or ions of a reactant or even a catalyst (an enzyme, for example) are chiral, then one enantiomer reacts very differently than the other.

An interesting example of different chemical properties of enantiomers involves our sense of taste. Taste is a chemical sense because the taste signal to the brain depends on a chemical interaction between the taste buds and the substance being tasted. The substances in the taste buds responsible for the primary interaction have chiral molecules. Asparagine is a compound for which there are two enantiomers. One asparagine enantiomer is made by a herb called vetch, and it has a sweet taste. The other enantiomer is made by asparagus and has a bitter taste. The two enantiomers react very differently with the taste buds, yet the only difference in their molecules is their handedness or chirality.

asparagine

The difference that chirality makes to the physical properties of enantiomers was discussed in Special Topic 19.1.

In the rest of this chapter we will see more examples illustrating that molecular *shape* is just as critical a factor in the body's ability to use a compound as anything else. Our diets include three large families of food materials—proteins, lipids, and carbohydrates. Almost all of their molecules are chiral and complement the chirality of our enzymes. We now turn our attention to these families.

19.3 AMINO ACIDS, POLYPEPTIDES, AND PROTEINS

A protein is a macromolecular substance made of polymers of α-amino acids

Proteins are found in all cells and make up about half of the dry weight of the human body. There are several major types, as described in Special Topic 19.2. Proteins hold you together (skin); give you levers (muscles and tendons); reinforce your bones (like steel in concrete); patrol your circulatory system as disease-fighting antibodies; carry essential nutrients such as oxygen, long-chain acids, amino acids, and certain minerals; catalyze all of the chemical reactions in your body (enzymes); and give you a communications system, both neural (nerves) and humoral (certain of the hormones). Yet, despite all these varieties of function, all proteins have many common features. All proteins, for example, consist mostly and sometimes entirely of polypeptide molecules. Therefore, to understand the structures of proteins, we first have to study polypeptides.

Actually, polypeptides are copolymers because different monomers are involved.

All **polypeptides** are polymers, and their monomers are a set of about 20 different α-**amino acids.**

α-amino acids (general structure)[1]

[1] This structure is called an amino acid although technically it has neither an amino group (NH_2) nor a carboxyl group (CO_2H) as such. However, these two groups are the parents of the electrically charged groups in amino acids because a proton, H^+, has transferred from the carboxyl group (an acid) to the amino group (a base).

Table 19.1
Common Amino Acids

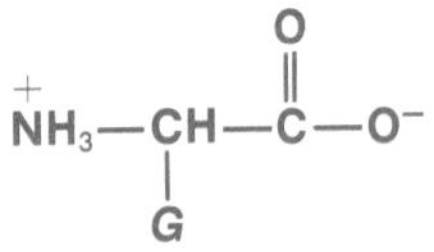

$$\overset{+}{N}H_3 - CH - \overset{\overset{\displaystyle O}{\|}}{C} - O^-$$

(with G on the side chain of the CH)

	G (Side Chain)	Name
Side chain is nonpolar	—H	Glycine
	—CH$_3$	Alanine
	—CH(CH$_3$)$_2$	Valine
	—CH$_2$CH(CH$_3$)$_2$	Leucine
	—CHCH$_2$CH$_3$ (with CH$_2$ below)	Isoleucine
	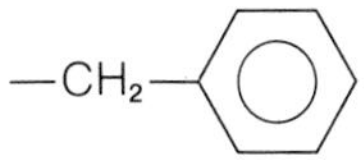 —CH$_2$—⟨phenyl⟩	Phenylalanine
	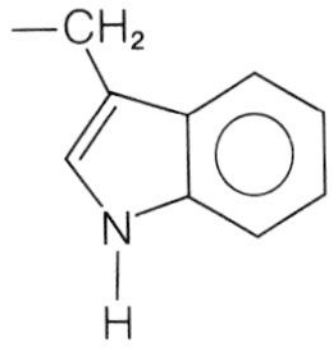 —CH$_2$—⟨indole⟩	Tryptophan
	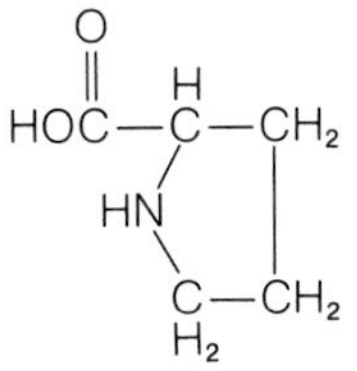 (complete structure) Proline ring	Proline
Side chain has a hydroxyl group	—CH$_2$OH	Serine
	—CH$_2$OH (with CH$_2$ below)	Threonine
	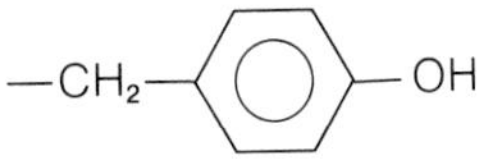 —CH$_2$—⟨phenyl⟩—OH	Tyrosine
Side chain has a carboxyl group (or an amide)	—CH$_2$CO$_2$H	Asparatic acid
	—CH$_2$CH$_2$CO$_2$H	Glutamic acid
	—CH$_2$CONH$_2$	Asparagine
	—CH$_2$CH$_2$CONH$_2$	Glutamine
Side chain has a basic amino group	—CH$_2$CH$_2$CH$_2$CH$_2$NH$_2$	Lysine
	—CH$_2$CH$_2$CH$_2$NH—C(=NH)—NH$_2$	Arginine
	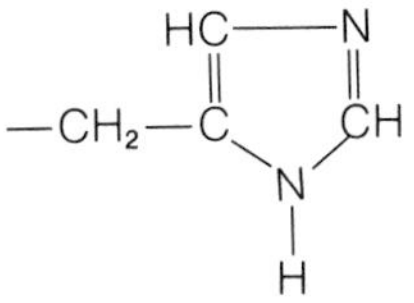 —CH$_2$—⟨imidazole⟩	Histidine
Side chain contains sulfur	—CH$_2$S—H	Cysteine
	—CH$_2$CH$_2$SCH$_3$	Methionine

EXAMPLE 19.2   Writing the Structure of an α-Amino Acid

Problem: Use the information in Table 19.1 to write the structure of glutamic acid.
Solution: First, write everything except the side chain; that part is the same in all
α-amino acids.

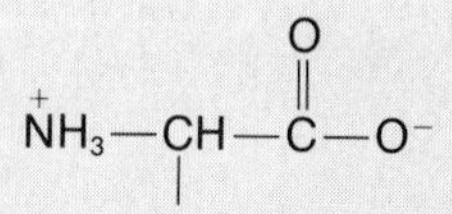

$$\overset{+}{N}H_3-CH-\overset{\displaystyle \overset{O}{\|}}{C}-O^-$$

Next, attach the side chain, given as $-CH_2CH_2CO_2H$ in Table 19.1, to the
α-position

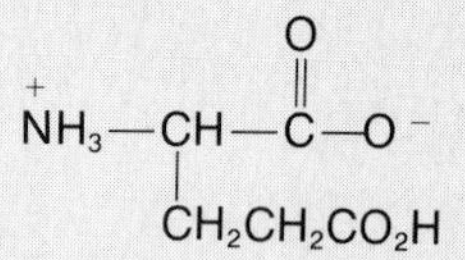

$$\overset{+}{N}H_3-CH-\overset{\displaystyle \overset{O}{\|}}{C}-O^-$$
$$CH_2CH_2CO_2H$$
glutamic acid

EXERCISE 19.2   Use the information in Table 19.1 to write the structure of lysine, an amino acid essential to
growth and health but that we cannot make ourselves from other amino acids. It has to be
in the diet. (Malnutrition in little children is widespread wherever the diet depends largely
on corn or maize, as in Latin America. The protein in corn is very deficient in lysine.)

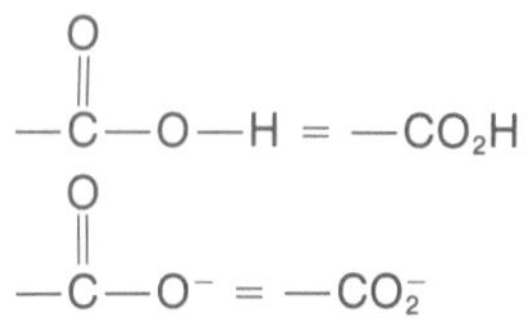

$$-\overset{\displaystyle \overset{O}{\|}}{C}-O-H = -CO_2H$$
$$-\overset{\displaystyle \overset{O}{\|}}{C}-O^- = -CO_2^-$$

The G— in this structure stands for an organic group (or for H— in glycine).
There are 20 common amino acids, listed in Table 19.1, that make up the set of
monomers for polypeptides. The alpha-carbon in every amino acid except glycine is a chiral carbon, and all chiral amino acids are chiral in the same way (as
illustrated in Figure 19.2). Therefore, all proteins are also chiral.
    Amino acids have the essential feature for making a linear condensation co-
polymer (Section 13.4): two functional groups that can react with each other. The
linking of one amino acid unit to another occurs by the splitting out of a water
molecule between the $\overset{+}{N}H_3$— group of one amino acid and the $-CO_2^-$ group of

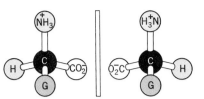

Figure 19.2
Amino acid entantiomers. Only the α-amino acids whose molecules all have the form shown on the right can be used by the body to make proteins. If somehow your diet suddenly included the amino acids with the configuration on the left, you would die of starvation.

another. If that happens to just two amino acids, the product is called a **dipeptide,** and the new bond that joins one amino acid unit to the other is called a **peptide bond.**

an α-amino acid   another α-amino acid

a dipeptide

The order in which the amino acids are taken makes a difference to the structure of the product. For example, if we take glycine (G = H) and alanine (G = CH₃) in the following order, we obtain a dipeptide in which the methyl group is next to the carboxyl group.

glycine   alanine

However, if we take the other possible order, first alanine and then glycine, we get a dipeptide with the methyl group next to the $\overset{+}{N}H_3$— group.

alanine   glycine

Taking amino acids in different orders gives different products, and this is one of the principal ways in which proteins are different, too. The amino acids used to make them can be taken in different orders. (Looking ahead, the chemistry of heredity controls the order in which amino acids are linked together to make the proteins that are unique to an individual.)

EXERCISE 19.3 Write the structures of the two dipeptides that can be made from alanine and valine.

At the ends of dipeptide chains are functional groups where additional amino acids can be added to lengthen the chains. For example, we could add a cysteine unit at the $\overset{+}{N}H_3$— end of the dipeptide made from glycine and alanine:

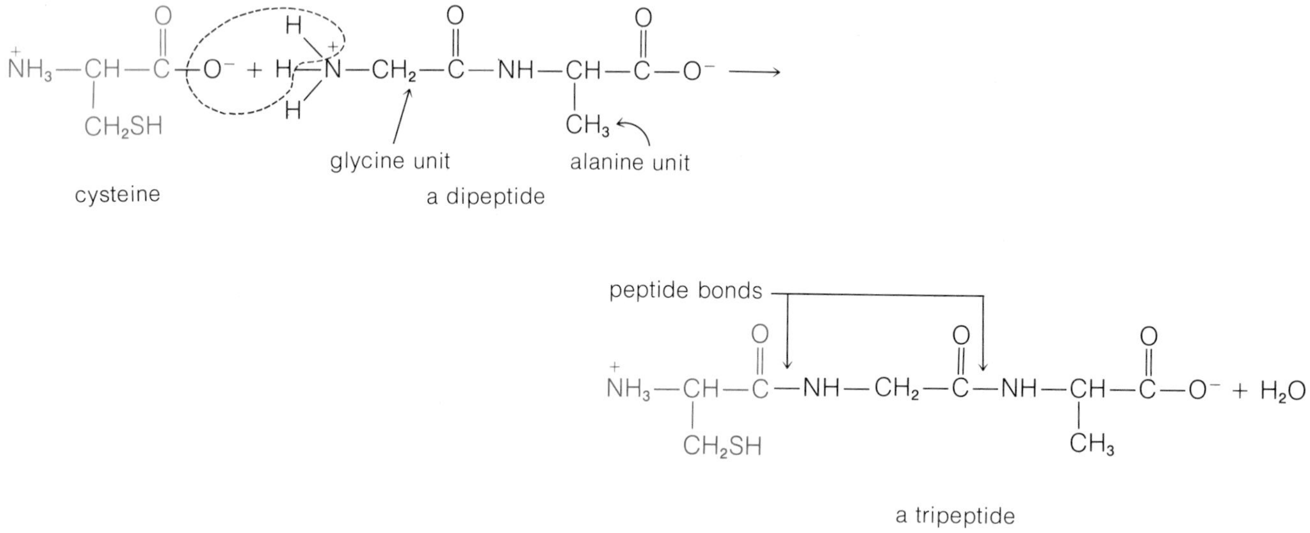

cysteine   a dipeptide

glycine unit   alanine unit

a tripeptide

Or we could add the cysteine unit to the other end:

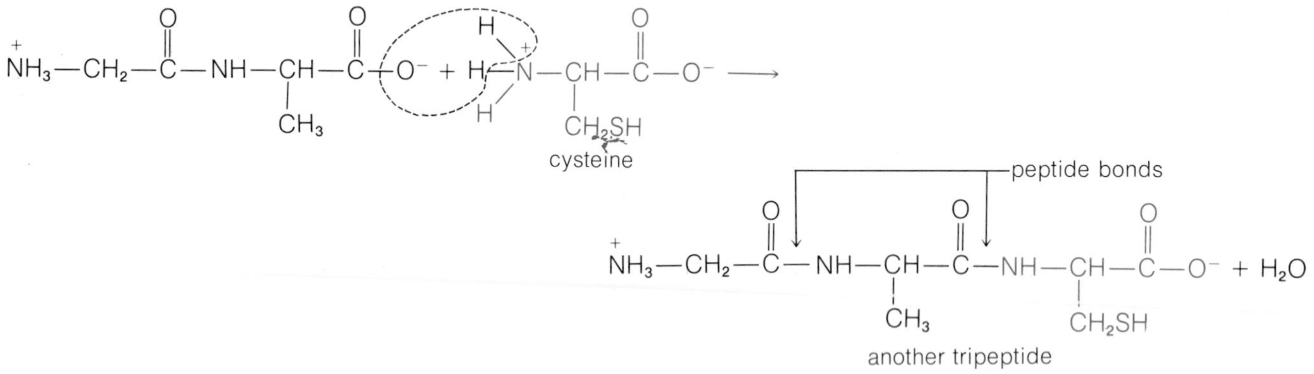

cysteine

peptide bonds

another tripeptide

EXERCISE 19.4 Write the structures of the two tripeptides that could be made from cysteine and the following dipeptide.

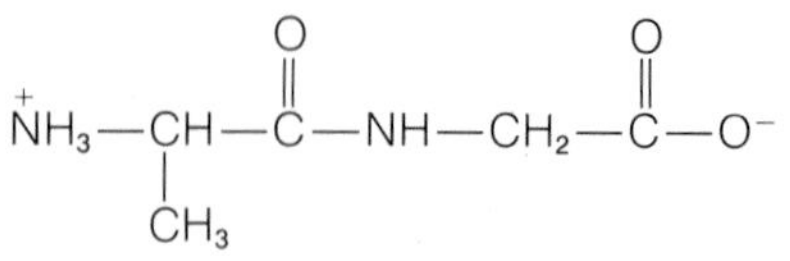

EXERCISE 19.5 Write the structure of a tripeptide made of three alanine units. Are there more than one that can be made?

At the ends of the tripeptides we just made, there are still the functional groups needed to lengthen the chains further. Each tripeptide can be made into a tetrapeptide; each of these can be converted to a pentapeptide, and so forth. Given 20 different amino acids, the number of possible combinations becomes enormous. For example, someone has calculated that if you used each of the 20 amino acids just once, there are over 2.4×10^{18} different ways of organizing them. This is more than 2.4 billion billion isomers in which the only differences are the sequences of the side chains. All have the identical sequence of atoms along the "backbones" of the molecules in which the unit, $-NH-CH-\overset{\displaystyle O}{\overset{\displaystyle \|}{C}}-$, repeats itself over and over:

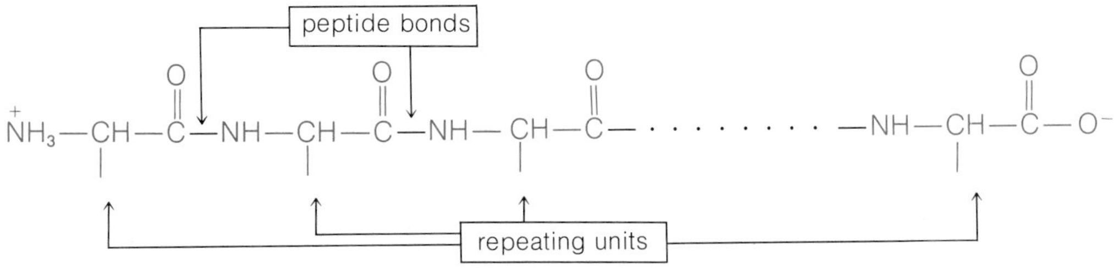

All polypeptides are identical to this extent. They differ in how long the chains are, in the kinds of side chains present, and in the order in which the side chains appear along the main chain. Attaching G groups to the main chain then gives us the following general picture of the primary structure of a polypeptide.

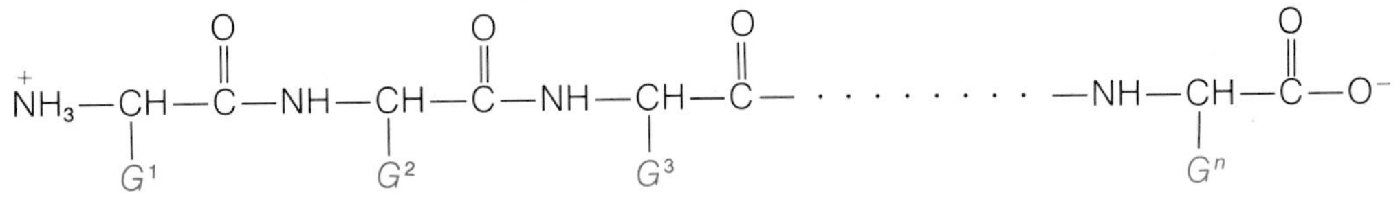

primary structure of a polypeptide

EXERCISE 19.6 Write the structure of a tetrapeptide in which the following amino acids occur in the order given: alanine, valine, glutamic acid, and glycine. (*Hint.* Make a chain, four repeating units long, without the side chains. Then find the structures of the side chains in Table 19.1 and attach them in the right order.)

Insulin is a hormone needed for the use of glucose in the body. Without insulin you would have diabetes.

Polypeptides vary greatly in size. Insulin is made of 51 amino acids, and it has a formula weight of 5733. A virus that infects tobacco plants, the tobacco mosaic virus, is made from about 336,500 amino acids and has a formula weight of about 40 million. Hemoglobin, the oxygen-carrying protein in blood, has 574 residues and a formula weight of 64,500.

Many proteins consist of the molecules of just one polypeptide. Many others are clusters of two or more different polypeptides. Still other proteins not only have more than one polypeptide but they also have incorporated within them a nonprotein organic molecule or a metal ion. Thus the term **protein** is a broad term that we apply to any macromolecular substance having a specific biological function that consists mostly or entirely of polypeptide molecules. Hemoglobin (Figure 19.3) illustrates all of these complications. Each hemoglobin molecule consists of a cluster of four polypeptide molecules each holding a nonprotein molecule called heme. Each heme molecule includes one ion of iron, Fe^{2+} (Figure 19.4). Heme gives the characteristic color to blood, and the heme units are the actual carriers of oxygen molecules.

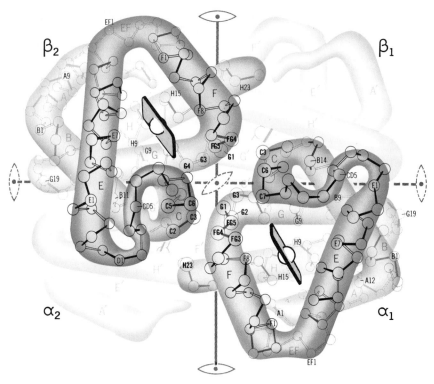

Figure 19.3
Hemoglobin. Hemoglobin consists of four polypeptide chains.
Each of the tubelike segments shown here encloses a chain that
is twisted in a spiral helix (see also Figure 19.5). These spirals
are further bent into the forms shown. Then four such forms
have aggregated into one large complex. The flat disks are
heme units shown in Figure 19.4. (From R. E. Dickerson and
I. Geis. *The Structure and Action of Proteins,* © 1969.
W. A. Benjamin, Inc., Menlow Park, CA. All rights reserved.
Used by permission. The letters inside circles identify specific
amino acids referred to in this publication.)

As seen in Figure 19.3, each polypeptide in hemoglobin has a particular
shape. The overall shape that a polypeptide has in a living system is as important to its biological function as the fundamental sequence of amino acids. When
a cell manufactures a polypeptide for some purpose, each newly finished mole-

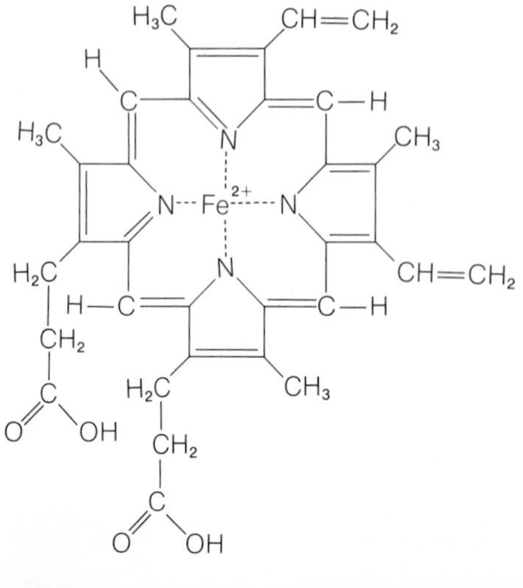

Figure 19.4
Heme.

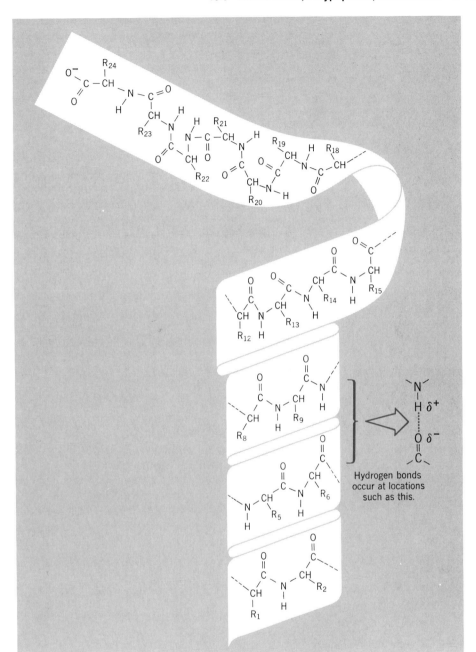

Figure 19.5
The α-helix, a feature of protein structure that contributes to the overall shape of polypeptides. Hydrogen bonds stabilize this right-handed coil. (From G. H. Haggis, D. Michie, A. R. Muir, K. B. Roberts and P. M. B. Walker. *Introduction to Molecular Biology*, © 1964. John Wiley & Sons).

cule automatically coils and folds into one shape or another depending on the forces of attraction and repulsion that arise from the polar groups present. The carbonyl and NH— units at the peptide bond are two such groups and, because they are regularly spaced and can have hydrogen bonds between them, these units are particularly important to the final shape of the polypeptide. For example, many polypeptides automatically coil into a right-handed helix as soon as they form—a structural feature called the **α-helix** (illustrated in Figure 19.5). A hydrogen bond forms between the following groups:

Each hydrogen bond is a weak force of attraction, but in a properly coiled polypeptide there are a large number of them and they give the helix considerable stability.

The helices are not ramrod straight like the spring of a screen door. They have side chains that get in the way of such perfection. The side chains not only have bulk but also are more or less polar, depending on the individual groups. For example, the first eight amino acids in Table 19.1 have nonpolar side chains. Those with sulfur at the bottom of the table are also relatively nonpolar. The rest have —OH groups, —NH₂ groups, or —NH— groups, and all can participate in hydrogen bonding. In the aqueous medium of a cell, these groups will attract water molecules. We call them **hydrophilic groups** ["water-loving" from *hydro-* (water) and *philos* (loving)]. The nonpolar groups in an aqueous medium will tend to avoid water as much as possible, and they are called **hydrophobic groups** ["water-hating" from *hydro-* (water) and *phobia* (hatred of)]. The same principles apply here that we introduced with the "likes dissolve likes" rule for predicting solubilities (Section 11.2). Although helices do form when polypeptides are made, in the aqueous environment of the cell they further bend and twist in a effort to expose the hydrophilic groups as much as possible to water molecules in the surroundings and to bury the hydrophobic groups inside the twisted system, as much as possible away from water molecules. Thus, in the structure of hemoglobin in Figure 19.3, each polypeptide unit is not only coiled into helices; these helices are further folded and bent.

If anything happens to alter the shapes of the polypeptide units in a protein, the protein normally loses all of its ability to perform its biological function. This loss of function through loss of molecular shape is called **denaturation.** Heat alone can denature proteins, and this happens to egg albumin when an egg is cooked. Even if egg albumin ("egg white") is vigorously whipped to make meringue for pies, it is denatured. A number of poisons work by denaturing enzyme proteins. The alcohol solution used to sterilize clinical thermometers denatures proteins in disease-causing bacteria. Soap kills bacteria in the same way. Table 19.2 gives a collection of conditions that denature various proteins.

Our diet gets proteins from dairy products, eggs, all forms of meat, and most vegetables and cereal products, including bread. When we digest proteins, all that happens is that they react with water and break up into their individual amino acids (see Figure 19.6). Then these migrate from the digestive system into the

All alkyl groups, such as methyl (CH₃—), ethyl (CH₃CH₂—), and so forth, are hydrophobic.

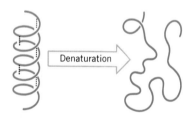

A polypeptide loses its shape in denaturation.

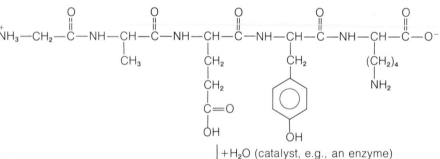

Figure 19.6
Protein digestion. The hydrolysis of this pentapeptide illustrates what happens when a protein is digested. Of all the covalent bonds present, only the peptide bonds are affected.

Glycine　　Alanine　　Glutamic acid　　Tyrosine　　Lysine

Table 19.2
Chemicals and Conditions that Cause Protein Denaturation

Denaturing Agent	How the Agent May Operate
Heat	Disrupts hydrogen bonds by making molecules vibrate too violently. Produces coagulation as in the frying of an egg.
Solutions of urea $(NH_2-\overset{\overset{O}{\|}}{C}-NH_2)$	Disrupt hydrogen bonds. Since it is amidelike, urea can form hydrogen bonds of its own.
Ultraviolet radiation	Appears to operate the same way that heat operates (e.g., sunburning).
Organic solvents (e.g., ethyl alcohol, acetone, isopropyl alcohol)	May interfere with hydrogen bonds in protein, since alcohol molecules are themselves capable of hydrogen bonding. Quickly denatures the proteins of bacteria, thus killing them (e.g., disinfectant action of ethyl alcohol, 70% solution).
Strong acids or bases	Can disrupt hydrogen bonds. Prolonged action of aqueous acids or bases leads to actual hydrolysis of proteins.
Detergent	May affect hydrogen bonds.
Salts of heavy metals (e.g., salts of the ions Hg^{2+}, Ag^+, Pb^{2+})	Ions combine with SH groups. These ions usually precipitate proteins.
Violent whipping or shaking	May form surface films of denatured proteins from protein solutions (e.g., beating egg white into meringue).

From J. R. Holum. *Fundamentals of General, Organic and Biological Chemistry,* © 1978. John Wiley & Sons.

blood stream that carries them to cells needing amino acids. Some are used to make polypeptides to repair or replace proteins. Others are used to make other substances, and those not needed are broken down for chemical energy.

19.4   ENZYMES, VITAMINS, AND TRACE ELEMENTS

Without their individual enzymes as catalysts, almost none of the thousands of reactions in the body would go fast enough for life

Enzymes are proteins that serve as the catalysts in living things. Many of these proteins include a nonprotein molecule called a **coenzyme.** Then neither the polypeptide portion nor the coenzyme alone is catalytically active; both are needed. Some of the B-vitamins are essential parts of coenzymes. So without the B-vitamins several enzymes in every cell of the body could not exist, and the reactions they catalyze could not occur. Thiamine, for example, is a B-vitamin needed to make a coenzyme called thiamine pyrophosphate.

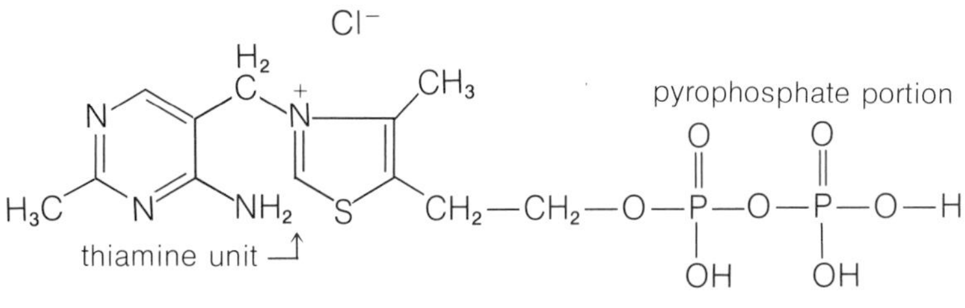

thiamine pyrophosphate

The enzymes using this coenzyme handle certain reactions of carbohydrates. Without thiamine you would suffer from beriberi, a disease that affects the nervous system and leads to the wasting away of muscles. In general terms, a **vitamin** is an organic compound, found in traces in various foods, that we cannot make ourselves and whose absence from the diet leads to a vitamin-deficiency disease. We do not know whether all of the vitamins are specifically needed to make coenzymes, but the B-vitamins form a group that are.

Many enzymes require a metal ion to work, and the list of these ions constitutes the **trace elements** that we must have in our diets. Various enzymes need, for example, chromium as Cr^{3+}, manganese as Mn^{2+}, iron as Fe^{2+}, cobalt as Co^{3+}, copper as Cu^{2+}, and zinc as Zn^{2+}. These form coordination compounds with organic molecules of the kinds that we will study in the next chapter. Heme (Figure 19.4) is a coordination compound of Fe^{2+}. They are called *trace* elements because only traces are needed—a few milligrams a day. Selenium, molybdenum, nickel, silicon, tin, and vanadium may also be required by humans.

The molecule that an enzyme acts upon is called the **substrate** for the enzyme. Each enzyme works best for only one kind of substrate. For example, trypsin is an enzyme used to help digest polypeptides. It catalyzes the hydrolysis of peptide bonds but only those that are adjacent to particular side chains in the polypeptide. Other protein-digesting enzymes work best on peptide bonds near other side chains. The catalytic activities of enzymes are so vital that no reaction in the body could take place rapidly enough to sustain life without them.

Enzymes are specific for their own substrates because all or some part of the substrate has a unique molecular shape that is complementary to, or matches,

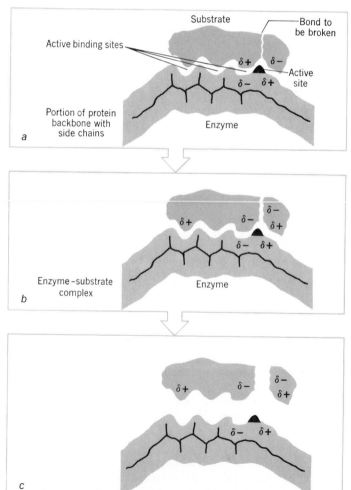

Figure 19.7

Lock-and-key model of enzyme action. (*a*) Enzyme and substrate have complementary shapes. Polar sites help to attract the two together. (*b*) The active catalytic site on the enzyme works on the substrate to help change it into the products of the reaction. As the substrate changes chemically, its polar sites also change. (*c*) Molecules of the product leave the surface of the enzyme. The active catalytic site is often supplied by a B-vitamin. (From J. R. Holum, *Fundamentals of General, Organic and Biological Chemistry,* © 1978. John Wiley & Sons, New York.)

the shape of a section of the enzyme molecule. As enzyme and substrate molecules come together, the two nestle into an **enzyme-substrate complex.** This has to form before parts of the enzyme molecule can work catalytically on the substrate (see Figure 19.7). An analogy is the tumbler lock and its key, and the theory of how enzymes work is sometimes called the lock-and-key theory. Only one key can fit into and permit a tumbler lock to turn, and molecules of only one substrate (or one family of similar substrates) will fit to the surface of their enzyme. Even if the only thing "wrong" with a potential substrate is its chirality, the enzyme cannot accept it, form the enzyme-substrate complex, and act on the substrate (see Figure 19.8).

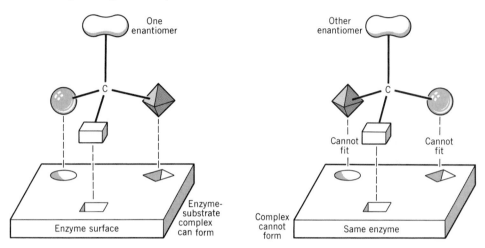

Figure 19.8
Since an enzyme is chiral, it cannot accept the wrong enantiomer of a chiral substrate. Simple geometric forms are used here to illustrate the difference that chirality can make in getting two chiral objects related as enantiomers to fit into a chiral surface.

Because enzymes are mostly polypeptide and therefore easily denatured, they are vulnerable to denaturing agents. The nerve poisons, including several pesticides, denature enzymes of the nervous system. Copper, mercury, and lead compounds are poisons because heavy metal ions irreversibly bind to enzymes and denature them.

19.5 LIPIDS—FATS AND OILS

Lipids are substances in plants and animals that are relatively soluble in nonpolar solvents such as ether and carbon tetrachloride

Lipids are defined by a physical property rather than a chemical structure. Any substance in plants or animals that is somewhat soluble in nonpolar solvents but insoluble in water is classified as a **lipid.** Knowing that "likes dissolve likes," this property tells us that lipid molecules must be mostly, if not entirely, hydrocarbonlike. We can see this in the molecules of the simplest family of lipids: the waxes. Beeswax and various fruit waxes are in this family and, except for an ester group, the molecules in waxes are entirely alkanelike, as the structures in Figure 19.9 show.

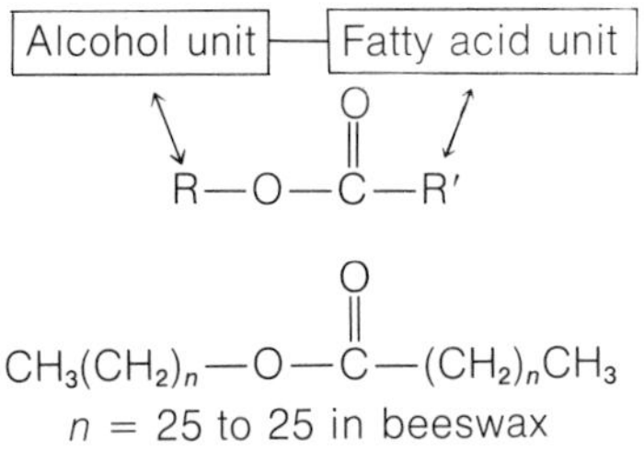

The animal fats and the vegetable oils make up the most important kinds of lipids found in our diets. Figure 19.10 gives the basic plan and general structure of their molecules together with a typical example. Fats and oils are structurally very similar. Both are triacylglycerols—esters of glycerol and three (usually dif-

Alcohol unit ─── Fatty acid unit

$$R-O-\overset{\overset{\displaystyle O}{\|}}{C}-R'$$

$$CH_3(CH_2)_n-O-\overset{\overset{\displaystyle O}{\|}}{C}-(CH_2)_nCH_3$$

$n = 25$ to 25 in beeswax

Figure 19.9
Structural features of waxes.

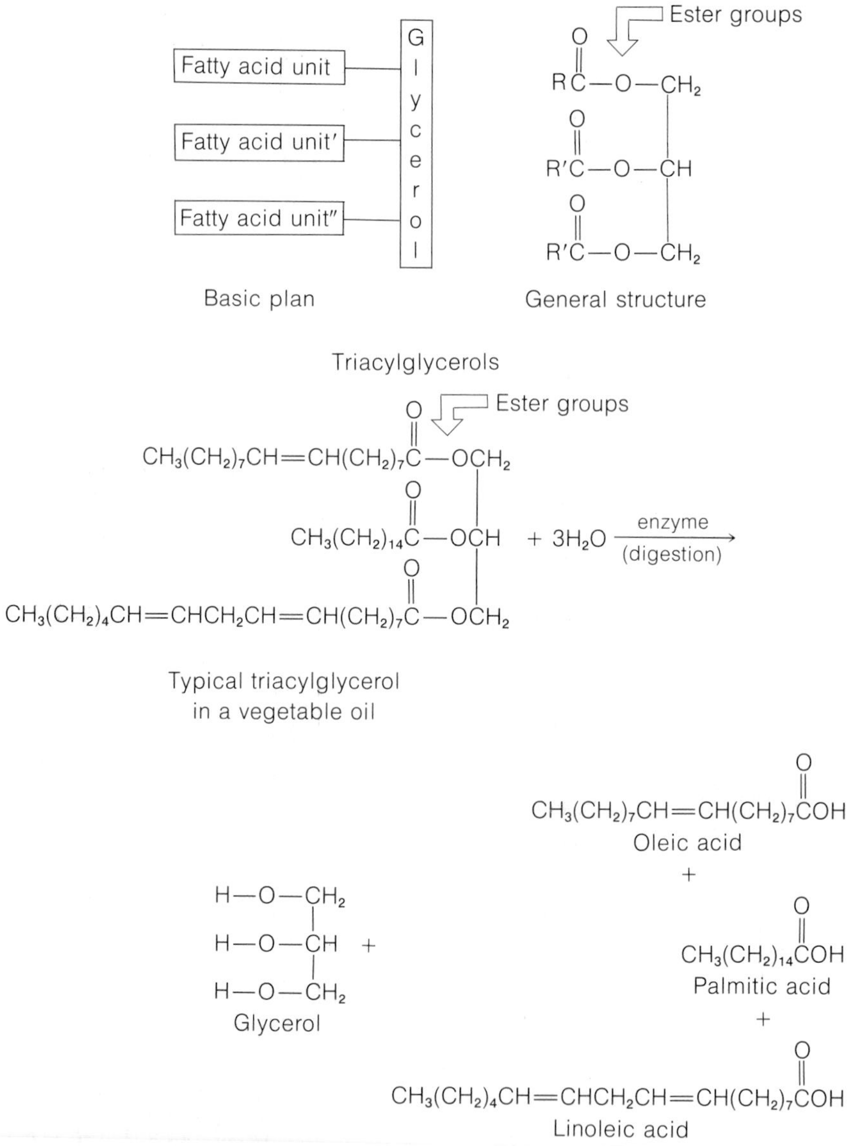

Basic plan   General structure

Triacylglycerols

$$CH_3(CH_2)_7CH{=}CH(CH_2)_7\overset{O}{\overset{\|}{C}}{-}OCH_2$$

$$CH_3(CH_2)_{14}\overset{O}{\overset{\|}{C}}{-}OCH + 3H_2O \xrightarrow[\text{(digestion)}]{\text{enzyme}}$$

$$CH_3(CH_2)_4CH{=}CHCH_2CH{=}CH(CH_2)_7\overset{O}{\overset{\|}{C}}{-}OCH_2$$

Typical triacylglycerol
in a vegetable oil

$$CH_3(CH_2)_7CH{=}CH(CH_2)_7\overset{O}{\overset{\|}{C}}OH$$

Oleic acid

+

$$CH_3(CH_2)_{14}\overset{O}{\overset{\|}{C}}OH$$

Palmitic acid

+

$$H{-}O{-}CH_2$$
$$H{-}O{-}CH \;+$$
$$H{-}O{-}CH_2$$

Glycerol

$$CH_3(CH_2)_4CH{=}CHCH_2CH{=}CH(CH_2)_7\overset{O}{\overset{\|}{C}}OH$$

Linoleic acid

Figure 19.10
Triacylglycerols—structural
features (top) and the reaction of
digestion.

Butterfat, lard, and tallow are
typical animal fats.

Olive oil, corn oil, peanut oil,
cottonseed oil, and linseed oil
are typical vegetable oils.

$$\overset{O}{\overset{\|}{-C}}{-}O{-}\overset{|}{C}{-} = \text{ester group}$$

ferent) long-chain carboxylic acids. In lipid chemistry these acids are called **fatty acids,** and seven of the most common ones are given in Table 19.3. Notice that besides being long-chain, they have no chain branches, and each one has an even number of carbon atoms. These features arise from the use in nature of two-carbon units, taken from molecules of acetic acid, to synthesize the fatty acids. Some of the fatty acids have one or more alkene double bonds, and the chief structural difference between animal fats and vegetable oils is that the oils have more double bonds per molecule than the fats. When television advertising praises the virtues of this or that brand of "polyunsaturated" oil, it refers to this structural difference, because an alkene group is called an unsaturated group. ("Unsaturated" means that you can add something, like hydrogen.)

Because of their ester groups, all triacylglycerols react with water, as shown in Figure 19.10. Since these substances do not dissolve in water, this reaction is extremely slow. However, with the help of enzymes and naturally produced detergents called bile salts, we can digest triacylglycerols very easily. The digestion of fats and oils is nothing more than the hydrolysis of the ester groups to give glycerol and a mixture of fatty acids. These products enter into circulation, and

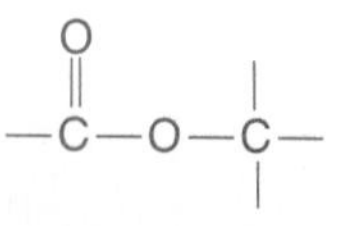

Table 19.3
Common Fatty Acids

Saturated acids

$$CH_3CH_2CH_2CH_2CH_2CH_2\ CH_2CH_2CH_2CH_2CH_2\overset{\overset{\displaystyle O}{\|}}{C}OH = CH_3(CH_2)_{10}\overset{\overset{\displaystyle O}{\|}}{C}OH$$
  lauric acid

$$CH_3CH_2CH_2CH_2CH_2CH_2CH_2CH_2CH_2CH_2CH_2CH_2\overset{\overset{\displaystyle O}{\|}}{C}OH = CH_3(CH_2)_{12}\overset{\overset{\displaystyle O}{\|}}{C}OH$$
  myristic acid

$$CH_3CH_2CH_2CH_2CH_2CH_2CH_2CH_2CH_2\ CH_2CH_2CH_2CH_2CH_2\overset{\overset{\displaystyle O}{\|}}{C}OH = CH_3(CH_2)_{14}\overset{\overset{\displaystyle O}{\|}}{C}OH$$
  palmitic acid

$$CH_3CH_2CH_2CH_2CH_2CH_2CH_2CH_2CH_2CH_2CH_2CH_2CH_2CH_2CH_2CH_2\overset{\overset{\displaystyle O}{\|}}{C}OH = CH_3(CH_2)_{16}\overset{\overset{\displaystyle O}{\|}}{C}OH$$
  stearic acid

Unsaturated acids

$$CH_3CH_2CH_2CH_2CH_2CH_2CH_2CH_2CH=CHCH_2CH_2CH_2CH_2CH_2CH_2CH_2\overset{\overset{\displaystyle O}{\|}}{C}OH$$
  oleic acid

$$CH_3CH_2CH_2CH_2CH_2CH=CHCH_2CH=CHCH_2CH_2CH_2CH_2CH_2CH_2CH_2\overset{\overset{\displaystyle O}{\|}}{C}OH$$
  linoleic acid

$$CH_3CH_2CH=CHCH_2CH=CHCH_2CH=CHCH_2CH_2CH_2CH_2CH_2CH_2CH_2\overset{\overset{\displaystyle O}{\|}}{C}OH$$
  linolenic acid

special proteins in blood serve as carrier molecules to transport the otherwise blood-insoluble fatty acids. Depending of the body's needs at the moment, the fatty acids might be processed in the liver to generate chemical energy. They have more chemical energy per molecule than any other product of digestion. Otherwise, the fatty acids might be placed into storage in the cells of fatty tissue to be recalled later when the body's needs for energy aren't being met by the diet. We also use fatty acids to make the membranes of cells and to make certain hormones.

One of the commercial uses of fatty acids is in the form of their sodium salts. These salts are the substances that make up ordinary soap. To manufacture soap, fats or oils are heated together with concentrated aqueous sodium hydroxide. The reaction is similar to the hydrolysis of an ester group, except that the base gives the fatty acids in the form of their salts. This is a general reaction of esters and is called *saponification* (Latin *sapo,* soap). Ethyl acetate, a simple ester, can be saponified:

$$CH_3-\overset{\overset{\displaystyle O}{\|}}{C}-O-CH_2-CH_3 + NaOH(aq) \longrightarrow CH_3-\overset{\overset{\displaystyle O}{\|}}{C}-O^-Na^+ + HO-CH_2-CH_3$$

  ethyl acetate   sodium   sodium   ethyl alcohol
                                    hydroxide   acetate

Soap is an example of a surfactant, a compound that lowers the surface tension of water, as we studied in Section 10.3. How soap and certain synthetic surfactants work is described in Special Topic 19.3.

Not all lipids are esters. Remember: all that is needed to be a lipid is to be sufficiently hydrocarbonlike to be insoluble in water and relatively soluble in nonpolar solvents. Thus cholesterol and several hormones are also in the lipid group.

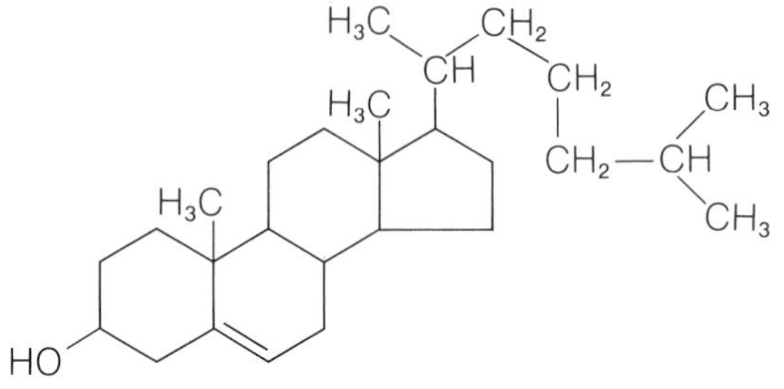

cholesterol

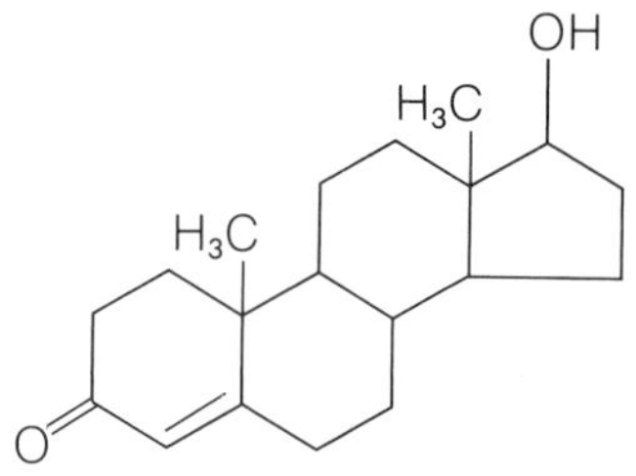

testosterone

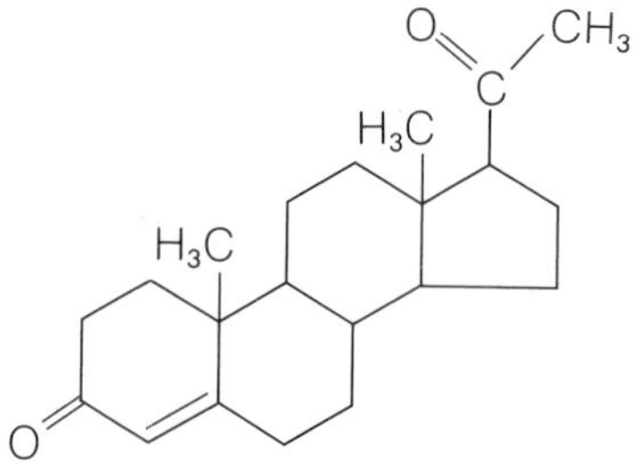

progesterone

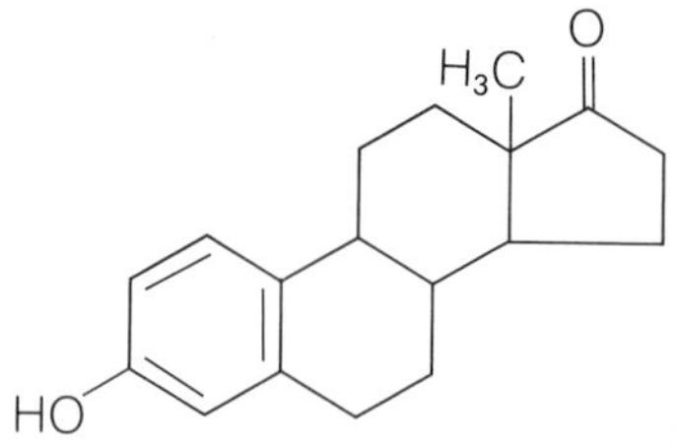

estrone

Cholesterol is an essential constituent in cell membranes, and it is the body's raw material for making sex hormones. If too much cholesterol is made, it tends to plug up blood capillaries and contribute to heart disease. The relationship between cholesterol and heart disease, however, is still under extensive study. Testosterone, a male sex hormone, regulates the development of reproductive organs and other sex characteristics including (some believe) a tendency toward aggressive behavior. Progesterone and estrone are female sex hormones. Some synthetic relatives of these hormones are used in the "pill" for human fertility control.

19.6 CARBOHYDRATES

The glucose structure is the most widely used structural feature among all carbohydrates

Where lipids are among the least polar substances in our diets, **carbohydrates** are among the most polar. Their molecules are loaded with alcohol groups, and all of the carbohydrates of relatively low formula weights are soluble in water. Only when we reach macromolecular polymers of glucose such as cellulose and starch do we find carbohydrates with little or no solubility in water. The carbohydrate family includes the sweets in our diets as well as the starches.

All but one family of carbohydrates, the **monosaccharides,** react with water. The monosaccharides include glucose, and they cannot be further broken down by the action of water. The relationships among the other carbohydrates can best be explained by "word" equations, equations in which we use the names instead of the formulas of complicated structures. The disaccharides, for example, include table sugar and milk sugar, and all disaccharides react with water to give two monosaccharide molecules. Polysaccharides are hydrolyzed to several hundred monosaccharide molecules.

Special Topic 19.3 How Detergents Work

Detergents are generally formulated as mixtures of compounds with specialized functions. The principal cleansing agent is a surfactant, and surfactants are usually salts of organic acids having large, hydrocarbonlike groups. The structures below represent typically used surfactants.

Usually, soil is held to skin or fabrics by a thin, invisible layer of greasy material. Since this is nonpolar, the hydrocarbon tails of detergent ions "dissolve" in them, as seen in (a). The ionic "heads" of the anions stay in and are hydrated by the water. With a little agitation and warmth, the grease layer breaks up and becomes pincushioned by detergent anions, as seen in (b). These pincushioned globules all bear sizable negative charges, and all are *like-charged*, as seen in (c). Since like charges repel, the globules can't merge back together, and since they are hydrated they easily wash down the drain with the wash water.

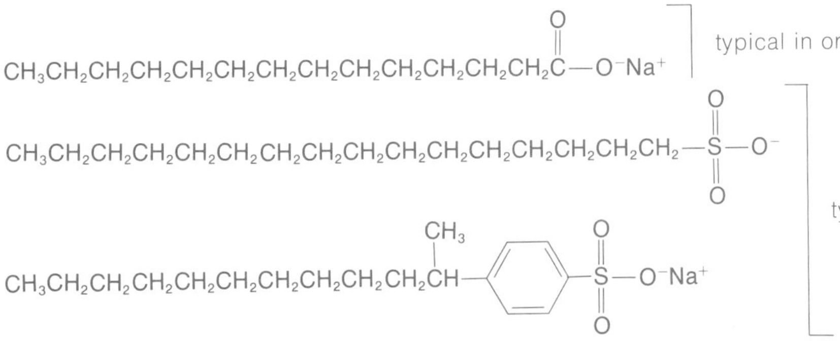

typical in ordinary soap

typical in synthetic detergents

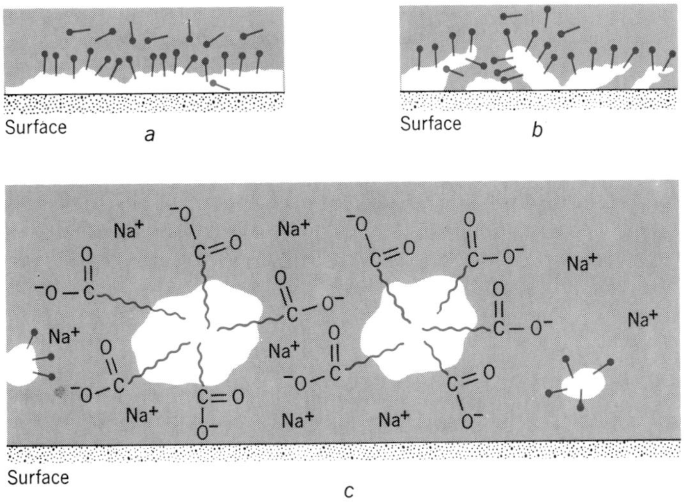

Surface *a*

Surface *b*

Surface *c*

Disaccharides

$$\text{sucrose} + H_2O \longrightarrow \text{glucose} + \text{fructose}$$
(table sugar)   (dextrose)   (levulose)

$$\text{lactose} + H_2O \longrightarrow \text{glucose} + \text{galactose}$$
(milk sugar)

$$\text{maltose} + H_2O \longrightarrow \text{glucose} + \text{glucose}$$
(malt sugar)

Polysaccharides

The coefficient *n* is large and variable; these equations are not balanced.

$$\text{starch} + nH_2O \longrightarrow \text{glucose}$$
$$\text{glycogen} + nH_2O \longrightarrow \text{glucose}$$
$$\text{cellulose} + nH_2O \longrightarrow \text{glucose}$$

These equations dramatize the importance of glucose units in all carbohydrates. All of the substances whose reactions were just described, except cellulose, are digestible in our diets. And cellulose is in the diet as part of the fibrous materials in vegetables. We do not have an enzyme for catalyzing the digestion of cellulose, as many microorganisms do. Glucose is absolutely vital to human life. In situations such as starvation when it is not in the diet, the body actually makes glucose from other substances in cells. The brain relies on glucose, minute by minute, as its chief source of chemical energy and, to supply it, the bloodstream constantly carries glucose in circulation. If the diet becomes too rich in relation to the body's needs, then we can convert glucose into fatty acids. Eating too many sweets, as we all know, makes us gain weight.

In solution, glucose exists in three forms in equilibrium with each other. Two are cyclic compounds, and one is open chain. These are shown in Figure 19.11. In the open-chain form, the glucose molecule has one aldehyde group, five alcohol groups, and four chiral carbon atoms. The curved arrows, in the illustration, show how bonds reorganize and a hydrogen atom shifts over to give the cyclic forms. This ring closure produces another alcohol group—at carbon number 1—and ring closure can occur in one of two ways. The new —OH group comes out either on the top side of the molecule or the bottom side. ("Top side" means the side having the —CH₂OH group of carbon number 6.) In the form called

Unlike its use in amino acid chemistry, α has a geometric meaning in carbohydrate chemistry (as does β).

α-glucose, this new —OH group at C—1 projects downward; in the form called β-glucose, it projects on the top side. If you turn to Color Plate 7A you will see on the left a molecular model of naturally occurring glucose in its alpha form. On the right is a model of the exact mirror image of the first. With a little imagination you can see that neither model could be turned or rotated in space to make the two superimposable. These models therefore represent enantiomers. Both the beta and the open-chain forms have enantiomers also. Our bodies can use the enantiomer pictured on the left in the color plate, but not the one on the right. The

Figure 19.11

Glucose. At the top are the three forms of glucose present in an equilibrium in water. The curved arrows, by the open form, show how bonds reorganize for ring closure. The new —OH that closure creates at carbon-1 can come out on top (β-form) or project to the bottom (α-form). The six-membered rings of the cyclic forms are not actually flat. They are puckered in the forms shown beneath the flat forms.

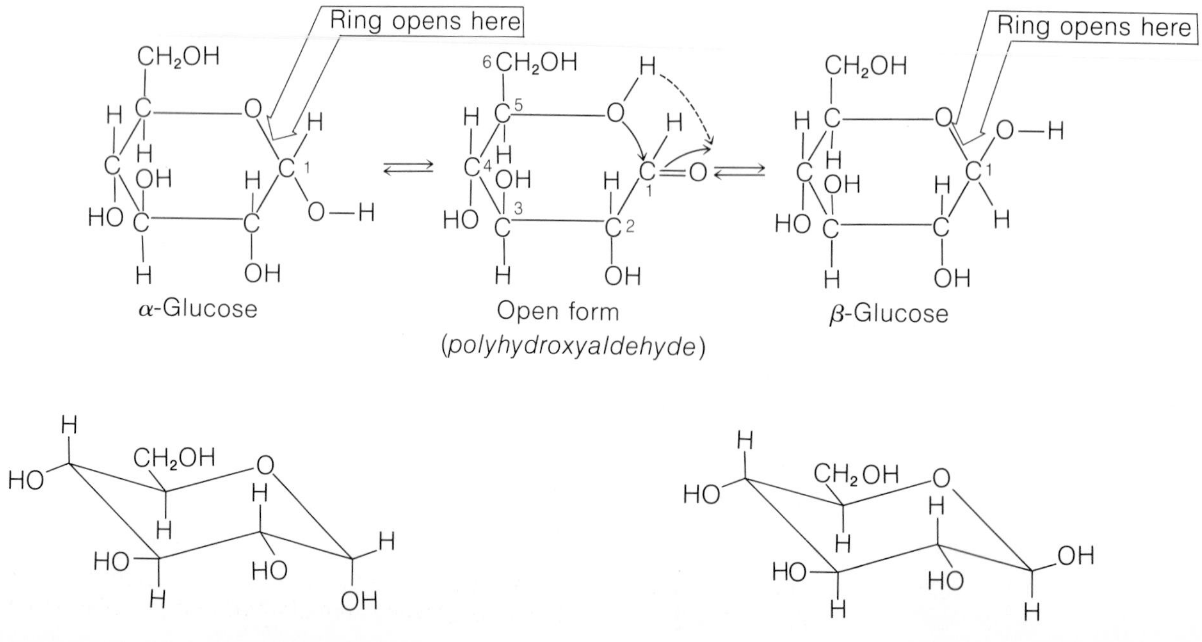

enzymes we have simply cannot fit together with the wrong enantiomer to catalyze any useful reaction.

Maltose is the simplest disaccharide because it is made of two glucose units bound to the same oxygen atom. When we digest maltose, water reacts to split its molecules apart at that point.

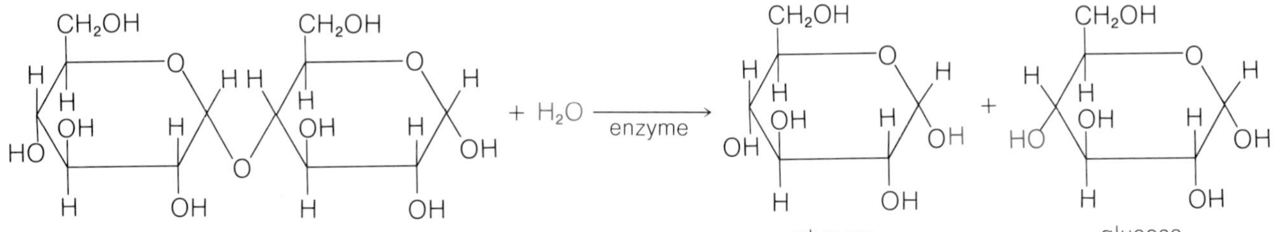

Maltose occurs in corn syrup and is made by the partial hydrolysis of starch.

maltose + H₂O → (enzyme) → glucose + glucose

Notice that the oxygen that links the two glucose units together in maltose projects downward on the sides of the rings opposite their —CH₂OH groups. This direction is called the alpha direction, and maltose is described as having an *alpha link* between its two units. The disaccharide called cellobiose has a beta-link between two glucose units, and if we had an enzyme for catalyzing the hydrolysis (digestion) of cellobiose we could then use the resulting glucose. But this seemingly small difference in geometry—having the oxygen link project upward instead of downward—is important. We don't have the requisite enzyme. However, we do have enzymes for hydrolyzing the oxygen linkages in both lactose (even though the oxygen linkage is beta here) and sucrose. Lactose is the sugar present in milk—4 to 5 percent in cow's milk and 5 to 8 percent in human milk. Sucrose is the sugar extracted from sugar beets or sugar cane for use on the table and in baking.

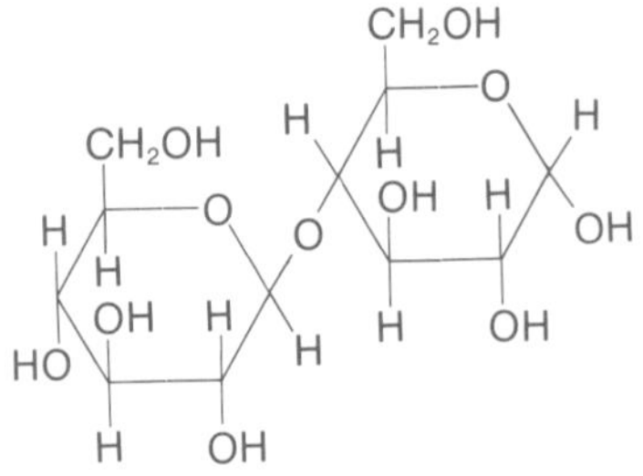

cellobiose

Lactose has a β-linkage from its galactose to its glucose unit. Sucrose has an α-linkage from its glucose to its fructose unit.

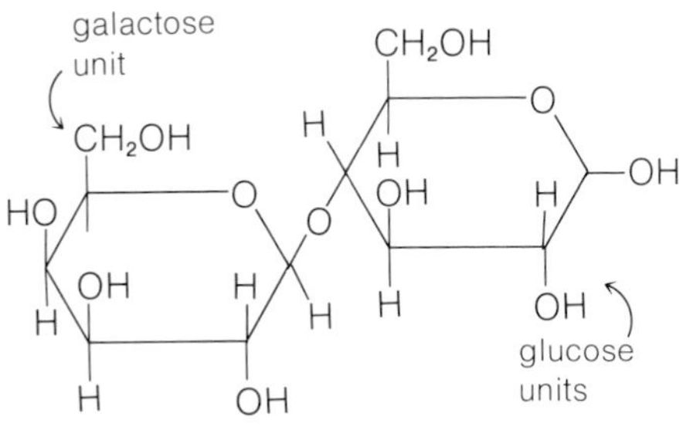

lactose

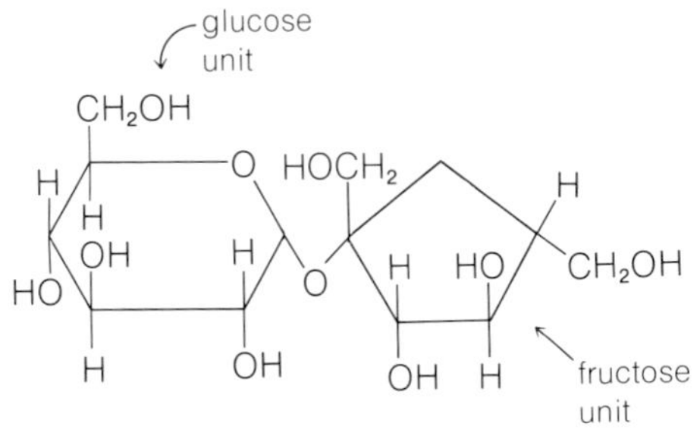

sucrose

Starch is a mixture of two polymers of α-glucose, amylose and amylopectin, shown in Figure 19.12. We have enzymes for catalyzing the hydrolysis (digestion) of all of the bonds that link the glucose units to each other. We lack such enzymes for cellulose, a polymer of β-glucose also shown in Figure 19.12. The only difference between amylose and cellulose is in the geometry of the linkages between glucose units.

Starch is a plant material, and seeds or tubers of plants are rich in this polysaccharide because the plant uses the glucose units for chemical energy particularly during germination and growth. Animals and humans also store chemical

Cotton is nearly 100 percent pure cellulose.

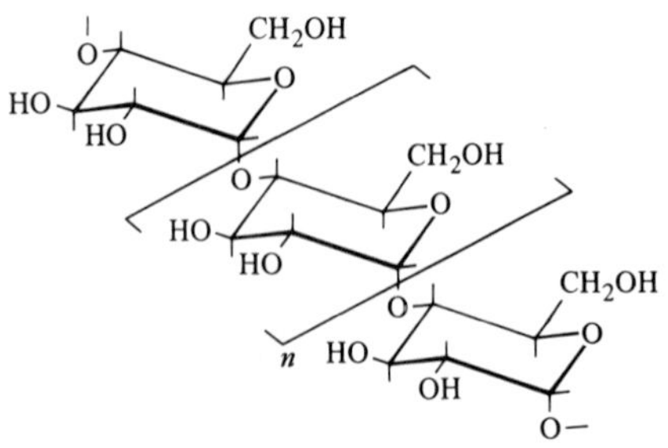

Amylose

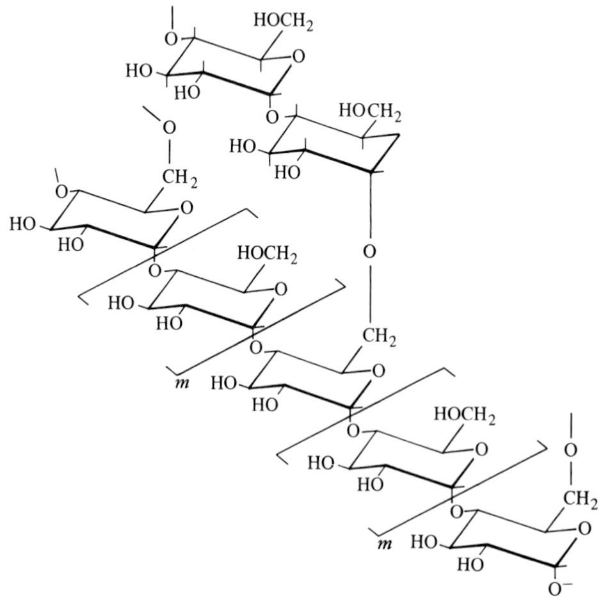

Amylopectin

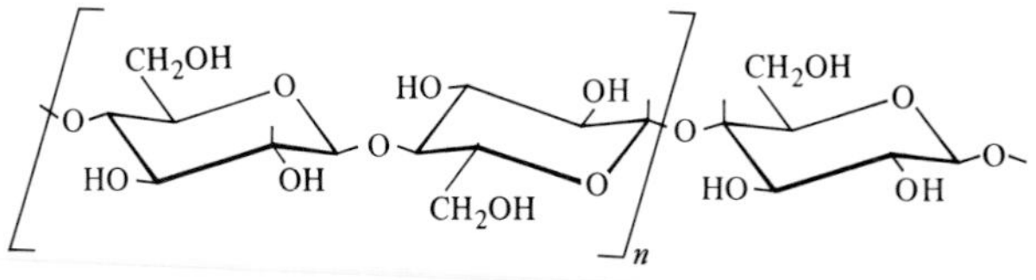

Cellulose

Figure 19.12
Common polysaccharides. Natural starches are about 10 to 20%
amylose and 80 to 90% amylopectin. Their formula weights
range from 50,000 to several million. (A formula weight of
1 million corresponds to about 6000 glucose units per molecule.)
Cotton is nearly pure cellulose and, in cotton, the cellulose
molecules have from 2000 to 9000 glucose units. A cotton
fiber gets its strength from hydrogen bonds that hold overlapping molecules side by side.

energy in the form of a polysaccharide called glycogen. Glycogen is similar to
amylopectin—it is a polymer of α-glucose, but it has a number of chain
branches. Between meals or during more extended periods of fasting, the body
draws on its glycogen reserves to maintain the supply of circulating glucose.

19.7 NUCLEIC ACIDS AND HEREDITY

The sequence of side chains on molecules of DNA controls the sequence of side chains on molecules of enzymes

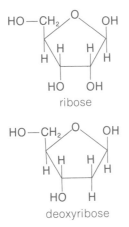

ribose

deoxyribose

Important as enzymes are to all of the reactions in a living system, the nucleic acids comprise an even more important family because nucleic acids direct the synthesis of enzymes. If the cell is like a chemical factory and enzymes are the managers, then nuclei acids are the master blueprints for all operations. Proteins and carbohydrates provide raw materials and energy; nucleic acids supply all of the internal information.

Nucleic acids are polymers and they occur in two broad types: **RNA** or ribonucleic acids and **DNA** or deoxyribonucleic acids. DNA is the actual chemical of a gene, the individual unit of heredity and the chemical basis through which we inherit our physical and biological characteristics. Our purpose in this section is to study the general structural features of the nucleic acids, how DNA molecules (genes) direct their own duplications, and how they guide the assembling of amino acids in a unique sequence.

Figure 19.13 gives a short segment of a DNA molecule together with simpler representations. As with any polymer, we focus our attention first on the main chain, the "backbone," and then on any side chains. The main chains of all nu-

Figure 19.13

Nucleic acids. Shown is a segment of a DNA chain featuring each of the four bases common to DNA. If the sites marked by the asterisks each held an —OH group, the segment would be from RNA (provided also that uracil replaced thymine). One distinct gene generally has about 1500 pairs of bases (of a double helix), and several genes can be strung out on one DNA molecule. The inset shows a simplified version of the strand. (From J. R. Holum, *Fundamentals of General, Organic and Biological Chemistry strand.* © 1978, John Wiley & Sons.)

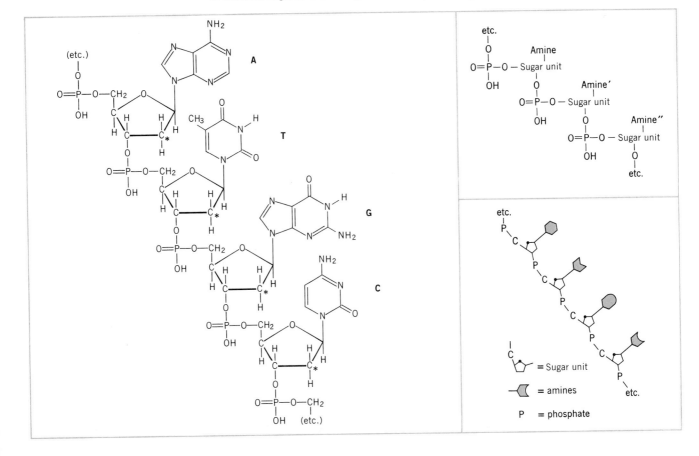

cleic acids have alternating units derived from phosphoric acid and a simple monosaccharide. DNA and RNA differ in which monosaccharide is used. In RNA it is ribose (hence the R in RNA), and in DNA it is deoxyribose (and the D in DNA). The only difference between these two sugar units is one —OH group. ("Deoxy" means lacking in one —OH group.) Thus, in both DNA and RNA, we have a system of

phosphate—sugar—phosphate—sugar—phosphate—sugar—etc.

The sugar units differ in the two kinds of nucleic acids. Now let's look at the side chains.

The side chains in the nucleic acids are far fewer in number than the side chains in amino acids. Moreover, both kinds of nucleic acids have three of the same side chains and, besides these, RNA and DNA each have one more. The side chains are all cyclic amines. They can be represented in simpler pictures of nucleic acid structures by geometric shapes because their molecular shapes have much to do with the function of these polymers, as we will soon see. It is also convenient to represent these side chain amines simply by single letters. One is adenine, which we may represent by *A*; another is guanine, *G*; the third that is common to both RNA and DNA is cytosine, *C*. The fourth amine is thymine, *T*, in DNA and uracil, *U*, in RNA. (A few other amines are also known to be present in certain nucleic acids, but the five we have named are by far the most common.) The uniqueness of any given molecule of DNA or any segment of a DNA molecule that is a gene is the sequence in which the side chain amines are lined up along the backbone. All of the hereditary information of an individual resides in the sequence of side chain amines in its DNA molecules.

Crick-Watson theory

In 1953, F. H. C. Crick in England and J. D. Watson, an American, used X ray diffraction studies of crystalline DNA, published by Rosalind Franklin and Maurice Wilkins, to propose a structure of DNA. They also proposed a mechanism to explain how DNA could direct its own reproduction. They deduced that DNA occurs as a pair of intertwined strands called a double helix, illustrated in Figure 19.14. Color Plate 7B shows a molecular model of part of a DNA double helix. The chirality of the sugar units sets the twist of the helices, and the two strands are held together by hydrogen bonds. Here is where the shapes of the side chain amines have such an important function. These amines occur in pairs having complementary shapes with functional groups in exactly the right locations for setting up hydrogen bonds between members of the pairs. Figure 19.15 shows how adenine, *A*, forms a pair with thymine, *T*. (Adenine can also form a pair with uracil, *U*, which is not shown.) Guanine, *G*, can also form a hydrogen-bonded pair with cytosine, *C*. Opposite every adenine side chain on one of the two strands in a DNA double helix is a thymine side chain on the other strand. Opposite every guanine side chain is a cytosine side chain on the other strand. The hydrogen bonds between these pairs of amines are the forces that hold the strands together in the double helix.

Adenine
A

Thymine
T
(from DNA)

Uracil
U
(from RNA)

Guanine
G

Cytosine
C

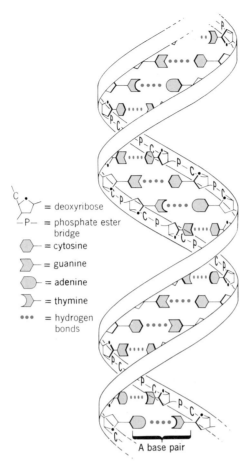

= deoxyribose

—P— = phosphate ester bridge

= cytosine

= guanine

= adenine

= thymine

••• = hydrogen bonds

A base pair

Figure 19.14

The DNA double helix. Hydrogen bonds between pairs of amines hold the strands together. (To better emphasize how the base pairs "fit," each is turned 90° from its true orientation in which the plane of each base molecule is perpendicular to the page. Only one dotted line or hydrogen bond is shown between each pair of amines to keep the picture simple. However, there actually are two or three hydrogen bonds for each pair seen in Figure 19.15.) (From J. R. Holum, *Fundamentals of General, and Organic and Biological Chemistry,* © 1978, John Wiley & Sons).

Figure 19.15

Hydrogen bonding between pairs of amines in DNA. The schematic in the inset shows how the pairs fit into a double helix.

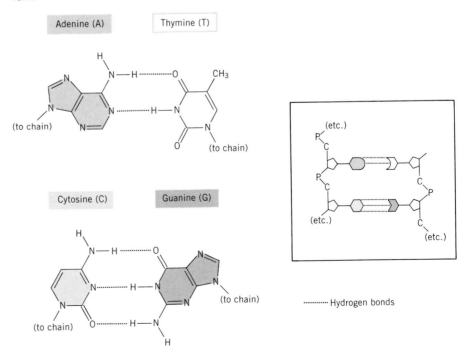

Adenine (A)   Thymine (T)

Cytosine (C)   Guanine (G)

(to chain)

·········· Hydrogen bonds

structural features of
monomers for DNA

Just before a cell divides, all of its DNA double helices must undergo exact **replication**—exact reproduction in duplicate. Then the two new cells produced by cell division will each have the identical set of genes that the original had. The accuracy of replication is the result of the pairing of amines. In Figure 19.16, the monomers for new DNA strands are presented simply by the letters A, T, G, and C. (Actually, each monomer is a rather large molecule consisting of one phosphate unit joined to one sugar unit which, in turn, holds one amine.) As seen in Figure 19.16, replication involves a separation of the strands and a lining up of the monomer units along exposed side chains. If T was paired to A at one point in the original double helix, then in one of the new double helices, at the identical location, T will be paired to a new A. Over on the other newly forming double helix, an original A will be paired to a new T. Only A and T can pair; only G and C can pair as new double helices emerge.

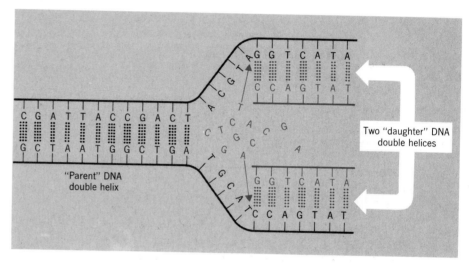

Figure 19.16
Replication of DNA. Only the basic plan is described by this highly schematic diagram. The monomer units are shown only by their one-letter symbols, A, T, G, and C. They join and make new DNA strands in an order determined by opportunities of fitting, by hydrogen bonds, to complementary amines on the original strands. (From K. R. Atkins, J. R. Holum and A. N. Strahler. *Essentials of Physical Science,* © 1978. John Wiley & Sons).

From gene to enzyme

Each enzyme is made under the direction of its own gene, a segment of a DNA molecule. In broad outline the sequence from DNA to enzyme is

DNA *(gene)* $\xrightarrow{\text{transcription}}$ RNA $\xrightarrow{\text{translation}}$ Enzyme $\xrightarrow{\text{catalysis}}$ Other compounds and reactions in cells

transcription (genetic message is read off in the nucleus and transferred to RNA)

translation (genetic message, now on RNA outside the nucleus, is used to direct a particular assembly of amino acids into a polypeptide)

Transcription, the first step, brings monomer units for RNA out of the cellular fluid and lines them up along a single DNA (gene) section in the order in which they can form hydrogen-bonded pairs. Each adenine unit, *A*, on a DNA segment will pair to a uracil unit, *U*; each thymine unit on the DNA strand will pair with an adenine unit in the RNA being made. Similarly, *G* and *C* will form pairs. In this way the sequence of amines in the DNA segment controls the sequence of amines in the RNA being made. Figure 19.17 shows this aspect of the work of genes. The RNA being made from a gene is called *messenger RNA,* or *m*RNA, because it now will carry the genetic message to a place in the cell where amino acids are assembled into polypeptides.

U is used instead of *T* in making RNA.

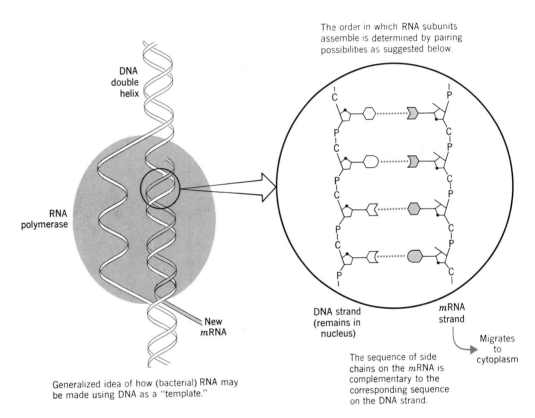

The order in which RNA subunits assemble is determined by pairing possibilities as suggested below.

DNA double helix

RNA polymerase

New *m*RNA

Generalized idea of how (bacterial) RNA may be made using DNA as a "template."

DNA strand (remains in nucleus)

*m*RNA strand

Migrates to cytoplasm

The sequence of side chains on the *m*RNA is complementary to the corresponding sequence on the DNA strand.

Figure 19.17
Transcription—showing how the amine sequence on DNA controls the amine sequence on RNA. The specific kind of RNA shown here is called messenger RNA, *m*RNA, because it carries the genetic message from DNA in the cell's nucleus to the sites where the cell makes protein. RNA polymerase is the enzyme for transcription. (From J. R. Holum, *Elements of General and Biological Chemistry,* Fifth Edition, © 1979. John Wiley & Sons).

Translation, the final step from gene to enzyme, is the use of the amine sequence on messenger RNA to force amino acids to line up in the right order to make a specific polypeptide—an enzyme, for example. For this purpose the cell uses another kind of RNA called *transfer RNA* or *t*RNA. For each amino acid, the cell makes one unique *t*RNA, and the function of the *t*RNA is to pick up its amino acid and transport it to the place where the amino acid will be joined to a growing polypeptide strand. The *t*RNA molecule holds the amino acid in one place, and in another part of the *t*RNA molecule there are side chain amines that can fit to side chain amines on the *m*RNA strand (see Figure 19.18). It is as if the *t*RNA molecule has two hands. One carries an amino acid, and the other reaches for a particular place on the messenger RNA. This second hand, to continue the analogy, has just three fingers and each finger is a particular side chain amine. Each *t*RNA has its own unique set of three such amines. We could even imagine a little groping as the three amine "fingers" seek three complementary amine fingers on *m*RNA. Only when the exactly correct pairing of side chains by hydrogen bonds becomes possible can this groping stop. Each sequence of *three* consecutive side chains on *m*RNA is called a genetic **codon,** because it codes for one particular amino acid. Each *t*RNA has its own complementary sequence of three consecutive side chain amines. Thus, the *t*RNA molecules can themselves line up along an *m*RNA strand in only one sequence, and that lining up insures that the amino acids line up in just one sequence, too. Figure 19.19 illustrates how translation works.

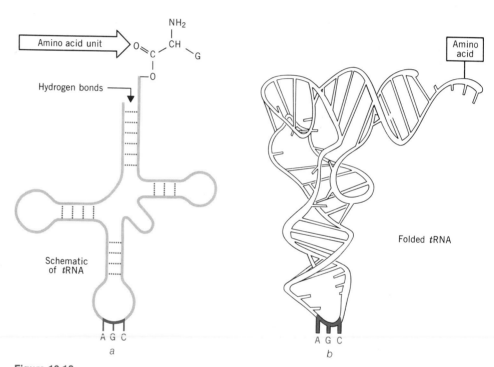

Figure 19.18
Transfer RNA or *t*RNA. Each *t*RNA consists of a relatively short strand of RNA that is folded and twisted into regions of parallel strands and regions of loops. The parallel regions are held together by hydrogen bonds between paired amines. (These bonds are shown as dotted lines in (*a*) and as straight bars in (*b*). Three amines, *A*, *G*, and *C*, have been chosen to illustrate the feature in this *t*RNA that is able to "recognize" by base pairing a complementary series of three bases—a codon—on messenger RNA.

Genetic diseases

If a gene is defective or missing, then an enzyme will be defective or missing. About 2000 diseases in humans are the result of faulty or absent genes. Cystic fibrosis is one; albinism is another. Albinos lack an enzyme essential for making pigments for eyes and skin.

(a) A growing polypeptide is shown here just two amino acid units long. A third amino acid, glycine, carried by its own tRNA is moving into place at the codon GGU. (Behind the assemblage shown here is a large particle called a ribosome, not shown, that stabilizes the assemblage and catalyzes the reactions).

(b) The tRNA-glycine unit is in place, and now the dipeptide will transfer to the glycine, form another peptide bond, and make a tripeptide unit.

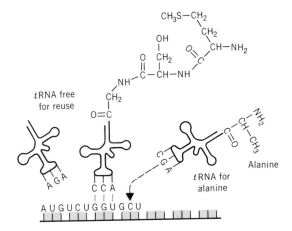

(c) The tripeptide unit has formed, and the next series of reactions that will add another amino acid unit can now start.

(d) One tRNA leaves to be reused. Another tRNA carrying a fourth amino acid unit, alanine, is moving into place along the mRNA strand to give an assemblage like that shown in part (b). Then the tripeptide unit will transfer to alanine as another peptide bond is made and a tetrapeptide unit forms. A cycle of events thus occurs as the polypeptide forms.

Figure 19.19
Polypeptide synthesis at an mRNA strand.

If errors occur during the replication of DNA, then the newly made cells will have defective genes and they may die. On the other hand, they may survive as mutant cells which could lead to a tumor, a cancer, or a birth defect if it happens in a fetus. Many chemicals have the potential for reacting with side chain amines on DNA to cause errors in replication or in transcription. These include several pesticides. X rays and other radiations, which we will study in Chapter Twenty-One, also damage nucleic acids and can cause cancer. A pregnant woman must be particularly careful about becoming exposed to chemicals or radiations that might harm the fetus. A huge amount of well-being depends on events at the molecular and genetic levels of life.

SUMMARY

Chiral Molecules. Most of the substances at the molecular level of life consist of chiral molecules. Generally, nature produces only one of the two enantiomers of each of these compounds. The difference in "handedness" or chirality makes enantiomers have quite different chemical reactions with reagents that also are chiral (but no difference when reagents are not). The superimposability of molecular models of two structures, one built to be an object in front of a mirror and the other to be its mirror image, is the ultimate test of chirality.

Amino Acids and Proteins. Twenty amino acids are the monomers for proteins and all but one, glycine, consist of chiral molecules. All are alpha-amino acids, but each has a different side chain. Peptide bonds link amino acid residues together in polypeptides. Once a polypeptide is made in a cell, it automatically assumes a particular overall shape. Much of its length might twist into a helix, which then might undergo further folding and bending to tuck hydrophobic side chains into the inside away from water and leave hydrophilic side chains exposed to the solution. Sometimes all there is to a protein is a single polypeptide but, more often, two or more different polypeptides and maybe some nonprotein molecules clump together before a complete protein is made. If the polypeptides in a protein lose their shapes, the protein is denatured. The digestion of a protein is the hydrolysis of its peptide bonds.

Enzymes, Vitamins, and Minerals. One family of proteins, the enzymes, catalyzes virtually all of the reactions in the body. Many enzymes require a coenzyme, an organic molecule made out of a B-vitamin, and some enzymes need a metal ion ("mineral"). Vitamins cannot be made internally; they must be provided in the diet. Without them, a number of vitamin-deficiency diseases can develop. The first event in an enzyme-catalyzed reaction is the formation of an enzyme-substrate complex in which molecular shapes must be complementary for enzyme and substrate to come together.

Lipids. Natural substances in living systems that are the most soluble in nonpolar solvents are also the most hydrocarbonlike. Common families in this large and diverse group of lipids are the waxes, as well as the animal fats and vegetable oils—the triacylglycerols. In triacylglycerol molecules, there are three ester groups involving three different fatty acids and glycerol. These are the substances released when fats and oils are digested. The vegetable ("polyunsaturated") oils have more alkene double bonds per molecule than animal fats. Some lipids have no ester groups; examples are cholesterol and certain hormones.

Carbohydrates. The glucose unit is the most frequently occurring unit in carbohydrates that have nutritional importance. Glucose itself, a polyhydroxyaldehyde, exists in solution in three forms in equilibrium—two cyclic forms that differ only in the direction of an —OH group and an open-chain form. Monosaccharides, such as glucose, cannot be hydrolyzed to smaller molecules. Three disaccharides are important in the diet: maltose, lactose, and sucrose. By digestion, maltose breaks down to glucose units, lactose to glucose and galactose, and sucrose to glucose and fructose. Starch, a mixture of amylose and amylopectin, is a polymeric form of α-glucose, and it is made by plants. Glycogen, another polymer of α-glucose, is structurally similar to amylopectin, and it is the means of glucose-storing in animals. Cellulose, a polymer of β-glucose, is found in cotton fibers as well as in cell's walls of plants.

Nucleic Acids. Nucleic acids direct the formation of the correct amino acid sequence in an enzyme. All of the genetic messages implied by the term "heredity" are encoded on molecules of DNA by a sequence of amines. These amines are side chains on a chain of alternating phosphate-pentose units. The four bases used in DNA are adenine (A), thymine (T), guanine (G), and cytosine (C). DNA exists most of the time as a double helix, two intertwined strands held by the weak forces of hydrogen bonds between units of A on one strand and T on the other, as well as G on one strand and C on the other. A and T form a base pair; G and C form another in DNA. When DNA replicates just before cell division, identical DNA copies are made,

one for each of the two new cells. When DNA directs the formation of RNA during transcription, the sequence of amines on DNA turns up as a matching sequence in *m*RNA (except that uracil, U, takes the place of thymine, T). Then this form of the message on RNA directs the assembling of amino acids in their right order to make an enzyme.

INDEX TO EXERCISES AND QUESTIONS

REVIEW QUESTIONS

19.7. What are the three minimum essentials that must be available to an organism for its life?

19.8. What family of compounds is the principal supplier of construction materials for an organism?

19.9. What kinds of substances are the major suppliers of chemical energy for life?

19.10. The information needed to organize and run a living system is encoded on what kinds of substances?

19.11. Two isomers of opposite chirality are called what?

19.12. What fundamental test will always tell whether molecular models represent different compounds?

19.13. What major kind of *chemical* difference exists between enantiomers?

19.14. What *structural* difference occurs between enantiomers?

19.15. Write the structural formula of the simplest alkane that has chiral molecules.

19.16. Write the structure of the simplest chloroalkane that has chiral molecules.

19.17. Which of these compounds have chiral molecules? (Place an asterisk by any chiral carbons.)
(a) CH_3CH_2OH (ethyl alcohol)
(b) $CH_3-CH=CH-CH-CH_3$
                        |
                       CH_3
(c) $HO-CH-CH_2-OH$ ascorbic acid (vitamin C)

(d) $H-O-CH_2-CH-CH-CH_2-\overset{\overset{\displaystyle O}{\|}}{C}-H$ deoxyribose (open-chain form), with OH, OH below the two middle carbons.

19.18. List seven ways in which proteins work in the body.

19.19. What family of compounds are the monomers for proteins?

19.20. In what structural way are the monomers of proteins all alike?

19.21. In what structural way do the monomers of proteins differ?

19.22. Using information in Table 19.1, write the structures of (a) phenylalanine, (b) lysine, (c) threonine, and (d) a tetrapeptide in which these amino acids appear in the order given: glycine, phenylalanine, lysine, and threonine.

19.23. What are all of the structural ways in which two polypeptides can be different?

19.24. A weak bond, the hydrogen bond, is very important in polypeptide structure. In what way?

19.25. What structural feature(s) are common to hydrophilic groups?

19.26. What kinds of side chains in amino acids are hydrophobic?

19.27. What are the forces or factors that cause a newly made polypeptide to adopt a specific overall shape?

19.28. What is the function of hemoglobin in the blood?

19.29. Describe how hemoglobin is more than simply a polypeptide.

19.30. A protein can suffer total loss of its ability to function without any peptide bonds breaking. Explain how that can happen.

19.31. Explain, in general terms, how mercury poisoning works.

19.32. What general function do the B-vitamins have?

19.33. What criteria must be met before something in the diet is called a vitamin?

19.34. Molecular shape is particularly important in the work of enzymes. Explain *how* shape is a factor in enzyme activity.

19.35. A lock-and-key analogy works with enzymes. Describe how this analogy is used.

19.36. How are lipids defined?

19.37. What structural feature is the chief difference between animal fats and vegetable oils?

19.38. Cholesterol is not an ester, yet it is classified as a lipid. Why?

19.39. Gasoline is very soluble in nonpolar solvents, but it isn't classified as a lipid. Why not?

19.40. Which of these two fatty acids is unlikely to be a product of the digestion of an animal fat and why?

$$CH_3(CH_2)_7CH\!=\!CH(CH_2)_7\overset{\displaystyle O}{\overset{\displaystyle \|}{C}}\!-\!OH$$
A

$$\underset{\displaystyle B}{CH_3\overset{\displaystyle CH_3}{\overset{\displaystyle |}{C}}H(CH_2)_6CH\!=\!CH(CH_2)_7\overset{\displaystyle O}{\overset{\displaystyle \|}{C}}\!-\!OH}$$

19.41. What does "polyunsaturated" mean when used to describe a lipid?

19.42. What uses does the human body make of fatty acids? Name two.

19.43. Why is glucose classified as a monosaccharide?

19.44. Glucose is a high-melting solid suggesting that its molecules are polar. What groups are the source of this polarity?

19.45. Are α-glucose and β-glucose enantiomers? Explain your answer.

19.46. Explain what the designations *alpha* and *beta* mean in each situation.
(a) In glucose molecules.
(b) In the oxygen linkage between glucose units in maltose.

19.47. Using the *names* of the substances involved, not their structures, write equations for the digestion of (a) amylose, (b) maltose, (c) amylopectin, (d) sucrose, and (e) lactose.

19.48. Why can't humans digest cellulose?

19.49. How do plants store glucose units for their energy needs?

19.50. In what form do animals store glucose units?

19.51. For what purpose do we maintain reserves of glycogen?

19.52. In general terms, how does the body solve the problem of getting particular amino-acid sequences rather than a mixture of randomly organized sequences when it has to make a particular polypeptide?

19.53. In what specific structural way does DNA carry genetic messages?

19.54. DNA occurs as a double helix. In terms of their side-chain bases, how are the two strands related?

19.55. How are the two strands in a DNA double helix held side by side?

19.56. How do DNA and RNA differ structurally?

19.57. Why are genetic messages "written" in three-letter triplets instead of sequences of single bases?

19.58. What forms from DNA replication?

19.59. If errors occur in replication, what are some possible consequences?

19.60. What are some agents that can cause errors in replication?

19.61. In general terms, what happens when a genetic message is transcribed?

19.62. If RNA does *not* exist as a double helix, in what way is the ability of its side chains to pair with other bases important?

CHAPTER TWENTY

METALS AND METALLURGY

20.1 METALLIC CHARACTER AND THE PERIODIC TABLE

The closer an element is to the lower left corner of the periodic table, the more metallic are its properties

In Chapter Six we discussed some of the general physical and chemical properties of metals. Now that you have acquired a background in topics such as electronic structure and bonding, we can look again at metals to gain a better understanding of their similarities and differences.

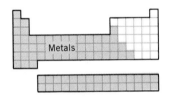

The metals, you will recall, make up about 75% of the elements. We encounter some of them every day. Metals are commonly used as structural materials—iron and aluminum, for example—and in coins and jewelry—copper, silver, gold, and platinum. Other common metals, such as sodium and calcium, are generally seen only in compounds. Many metals are rare and have only limited applications, so we are hardly ever aware of their impact on our lives, even though some of them are necessary components of certain enzymes that keep us alive.

The chemical and physical properties of metals, like those of the nonmetals, are determined by their electronic structures. We have seen that nonmetals react with each other to form compounds by sharing electrons in covalent bonds. These bonds allow the nonmetals to complete their valence shells. Covalent bonding does not occur, to any appreciable extent, among metal atoms, however, because they have too few electrons to form enough bonds to give a completed octet. This fact, plus the low ionization energies that metals generally have, are responsible for the metallic lattice in which electrons are lost to the lattice as a whole. The properties of the metallic lattice were described in Chapter Ten.

We have also seen that metals tend to react with nonmetals to form ionic compounds. Usually, not too much energy is needed to remove an electron from a metal. In addition, metals have few valence electrons, so only a small number must be removed to give the metal a stable electron configuration. Therefore, the total expenditure of energy needed to form the positive ion is relatively small and can be recovered by the lattice energy. Thus, stable ionic compounds can be formed.

Although metals tend to form positive ions when they react with nonmetals, the completeness of the electron transfer is not the same in every case. In Chapter Eight we saw that the ionic character of a bond—that is, the extent to which the bond is ionic or the extent to which it is polar—depends on the difference in the

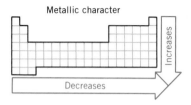

Metallic character

Increases

Decreases

electronegativity of the two bonded atoms. Because of the way electronegativity varies in the periodic table—increasing from left to right across a period and from bottom to top in a group—the further away from the lower left corner of the table a metal is located, the more chemically "nonmetallic" it becomes.

Within the periodic table, metallic character decreases from left to right across a period and increases from top to bottom in a group. In period 3, for example, sodium and magnesium are metals with typically basic oxides—their oxides react with acids, but not with bases. Aluminum is also a metal, but its oxide is amphoteric—it reacts with both acids and bases.

$$Al_2O_3 + 6H^+ \longrightarrow 2Al^{3+} + 3H_2O$$
$$Al_2O_3 + 2OH^- \longrightarrow 2AlO_2^- + H_2O$$

The fact that Al_2O_3 reacts with bases means that it is at least somewhat acidic. Since acidic oxides are normally associated with nonmetals, we can say that aluminum is "less metallic" than sodium or magnesium. Moving further to the right in period 3 we come to silicon, which is a metalloid. Then we come to the clearly nonmetallic elements—phosphorus, sulfur, chlorine, and argon.

The increase in metallic character going down a group can be seen most clearly in Group IVA. At the top of the group is carbon, a nonmetal whose oxide, CO_2, dissolves in water to give carbonic acid, H_2CO_3. Next are silicon and germanium, both of which are metalloids having the semiconductor properties typical of metalloids. Then come tin and lead, which are metals. Both show some amphoteric properties, but tin more so than lead. Thus tin is less metallic than lead.

20.2 METALLURGY

The science of extracting metals from their naturally occurring compounds and then preparing them for practical use employs many important chemical reactions

Even before the beginning of recorded history, metals played an important role in the growth and development of society. Archaeological evidence indicates that gold was used in making eating utensils and ornaments as early as 3500 B.C. Silver was discovered at least as early as 2400 B.C., and iron and steel have been used as construction materials since about 1000 B.C. From these earliest times, methods for extracting metals from their ores—minerals containing some metal-bearing component in useful amounts—have gradually evolved as more and more knowledge of the behavior of metals was acquired. Modern **metal-** the science and technology of metals, is primarily concerned with the procedures and chemical reactions used to separate metals from their ores and to prepare them for use.

Gathering metals and their compounds
Metals are obtained from a variety of sources. Most of them come from ore deposits in the earth, although some are taken from the sea. In almost all cases, the metals occur chemically combined with other elements. However, some unreactive metals do occur naturally in the free state. A well-known example is gold (see Color Plate 8A).

The two most abundant cations in sea water are those of sodium and magnesium. Their concentrations are about 0.6 M for Na^+ and about 0.06 M for Mg^{2+}. Sodium chloride, from which sodium is extracted by electrolysis (Section 18.5), can be obtained in a fairly pure form by the evaporation of sea water in large ponds (see Figure 20.1). About 13% of the salt produced in the United States comes from this source. The rest is mined from vast underground deposits, some of which lie quite deep below the earth's surface.

Separation of magnesium from sea water takes advantage of the low solubility

In this photo, taken deep within the famous Avery Island salt dome, the ceilings, sidewalls and floor are pure rock salt.

Figure 20.1
Salt crystals are harvested from a sea water evaporation pond
in California.

of magnesium hydroxide. A batch of sea water is made basic by dissolving lime
in it. Lime is calcium oxide, which is prepared by decomposing calcium carbonate.

$$CaCO_3(s) \xrightarrow{\text{heat}} CaO(s) + CO_2(g)$$
$$\text{lime}$$

Near the ocean, sea shells, which are composed chiefly of $CaCO_3$, provide a
cheap source of this raw material.

When the lime is dissolved in sea water, it reacts to form calcium hydroxide.

$$CaO(s) + H_2O \longrightarrow Ca^{2+}(aq) + 2OH^-(aq)$$

The hydroxide ion precipitates magnesium hydroxide, $Mg(OH)_2$, which is insoluble.

$$Mg^{2+}(aq) + 2OH^-(aq) \longrightarrow Mg(OH)_2(s)$$

This precipitate is then filtered from the sea water and dissolved in hydrochloric
acid.

$$Mg(OH)_2(s) + 2H^+(aq) + 2Cl^-(aq) \longrightarrow Mg^{2+}(aq) + 2Cl^-(aq) + 2H_2O$$

Evaporation of the resulting solution yields magnesium chloride, which is dried,
melted, and finally electrolyzed to give free magnesium metal.

Another potential source of metals from the sea is the mining of manganese
nodules from the ocean floor. **Manganese nodules** are lumps about the size of
an orange that contain significant amounts of manganese (almost 25%) and iron
(about 15%). In some places, they occur in large numbers, as the photo in Figure
20.2 shows. Although the procedures and equipment used to recover these nodules from the ocean floor are very expensive, the huge number of nodules that is
believed to exist makes the concept of deep-ocean mining attractive, more so as
land-based sources of these metals become depleted.

When the source of a metal is an ore that is dug from the ground, considerable
amounts of sand and dirt are usually unavoidably mixed in with the ore. Therefore, the ore is normally concentrated before the metal is separated from it to reduce the volume of material that must be processed. How this is done depends

An open pit iron mine in the
Mesabi Range in Minnesota.

Figure 20.2
A photograph of manganese nodules lying on the ocean floor.
The water depth here is about 2900 m (9500 ft).

Gold panning in Little Meadow
Creek, site of the discovery of
an 18 pound gold nugget in 1799.

on the physical and chemical properties of the ore itself, as well as those of the impurities.

In some cases, the unwanted rock and sand, called **gangue,** can be removed simply by washing the material with a stream of water. This washes away the waste and leaves the enriched ore behind. Some iron ores, for example, are treated in this way. This procedure also forms the basis for the well-known technique called "panning for gold" that you've probably seen in movies. A sample of sand that might also contain gold is placed into a shallow pan filled with water. As the water is swirled around, it washes the less dense sand over the rim of the pan, but leaves any of the more dense bits of gold behind.

Flotation is a method commonly used to enrich the sulfide ores of copper and lead. First, the ore is crushed, mixed with water, and ground into a souplike slurry. The slurry is then transferred to flotation tanks (Figure 20.3) where it is mixed with detergents and oil. The oil adheres to the particles of sulfide ore, but not to the particles of sand and dirt. Next, air is blown through the mixture. The

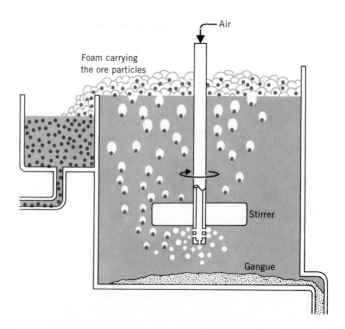

Figure 20.3
A flotation apparatus.

rising air bubbles become attached to the oil-coated ore particles and bring them to the surface where they are held in a froth. The detergents in the mixture stabilize the bubbles long enough for the froth and its load of ore particles to be skimmed off. Meanwhile, the sand and dirt settle to the bottom of the tanks, from which they are removed and disposed of.

Many ores must undergo a second round of pretreatment before the metal can be recovered from them. For example, after enrichment, sulfide ores are usually heated in air. This procedure, called **roasting,** converts the sulfides to oxides. This is done because oxides are more conveniently reduced than sulfides. Typical reactions that occur during roasting are

$$2Cu_2S + 3O_2 \longrightarrow 2Cu_2O + 2SO_2$$
$$2PbS + 3O_2 \longrightarrow 2PbO + 2SO_2$$

As you can see, a major by-product of roasting is sulfur dioxide. This cannot simply be released into the atmosphere because it would create a major pollution problem. One way of removing it from the exhaust gases is to allow it to react with calcium oxide, which is made by decomposing limestone ($CaCO_3$).

$$CaCO_3 \xrightarrow{\text{heat}} CaO + CO_2$$
$$CaO + SO_2 \longrightarrow CaSO_3$$

Another way of disposing of the SO_2 is to oxidize it to SO_3. The SO_3 can be converted to sulfuric acid and then sold.

Aluminum ore, called *bauxite,* must also be pretreated before it can be processed. Bauxite contains aluminum oxide, Al_2O_3, but a number of impurities are also present. To remove these impurities, use is made of aluminum oxide's amphoteric behavior. The ore is mixed with a concentrated sodium hydroxide solution, which dissolves the Al_2O_3.

$$Al_2O_3(s) + 2OH^-(aq) \longrightarrow 2AlO_2^-(aq) + H_2O$$

The major impurities, however, are insoluble in base, so when the mixture is filtered, the impurities remain on the filter while the aluminum-containing solution passes through. The solution is then acidified, which precipitates aluminum hydroxide.

$$AlO_2^-(aq) + H^+(aq) + H_2O \longrightarrow Al(OH)_3(s)$$

When the precipitate is heated, water is driven off and the oxide is formed.

$$2Al(OH)_3 \longrightarrow Al_2O_3 + 3H_2O$$

This purified Al_2O_3 then becomes the raw material for the Hall process discussed in Section 18.5.

Separating metals from their compounds.
In general, producing a free metal from one of its compounds involves reduction. This is because, with very few exceptions, metals in compounds have positive oxidation states. They must therefore gain electrons to become free elements. The nature of the reducing agent that is needed to provide these electrons depends on how difficult the reduction process is. If the free metal itself is very reactive, only electrolysis can provide enough energy to cause the decomposition of its compounds. This is the reason active metals such as sodium, magnesium, and aluminum are produced electrolytically. Less active metals such as lead or copper require less active-reducing agents to displace them from their compounds, and chemical reducing agents can do the job.

A plentiful, and therefore inexpensive, reducing agent used to produce a number of metals is carbon. Coal is usually the source of this carbon. When coal is heated, volatile components are driven off and **coke** is formed. Coke is composed almost completely of carbon. Carbon is an effective reducing agent for

Between each pair of vertical dividers in a coke oven battery is an individual oven in which coal is heated and converted to coke. Here we see a coke plant operator preparing to clean the oven before charging it up again.

metal oxides because it combines with the oxygen to form carbon dioxide. For example, after it is roasted, lead oxide is mixed with coke and heated.

Heating an ore with a reducing agent is called smelting.

$$2PbO(s) + C(s) \xrightarrow{\text{heat}} 2Pb(\ell) + CO_2(g)$$

The high thermodynamic stability of CO_2 serves as one driving force for this reaction. The loss of CO_2 is another.

Copper oxide ores can also be reduced with carbon.

$$2CuO + C \xrightarrow{\text{heat}} 2Cu + CO_2$$

This step is unnecessary for some copper sulfide ores if the conditions under which the ore is roasted are properly controlled. For example, heating an ore that contains Cu_2S in air can convert some of the Cu_2S to Cu_2O.

$$2Cu_2S + 3O_2 \xrightarrow{\text{heat}} 2Cu_2O + 2SO_2$$

At the appropriate point, the supply of oxygen is cut off, and the mixture of Cu_2S and Cu_2O reacts further to give metallic copper.

$$Cu_2S + 2Cu_2O \xrightarrow{\text{heat}} 6Cu + SO_2$$

Without question, the most important use of carbon as a reducing agent is in the production of iron and steel. The chemical reactions take place in a huge tower called a **blast furnace** (see Figure 20.4). Some blast furnaces are as tall as a 15-story building and produce up to 2400 tons of iron a day. They are designed for continuous operation, so that the raw materials can be added at the

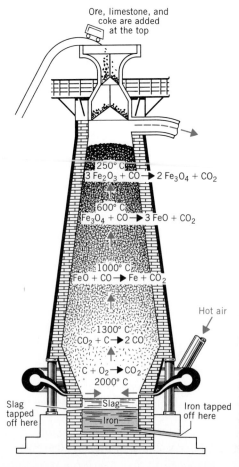

Figure 20.4
A typical blast furnace for the reduction of iron ore. The photograph shows Bethelem Steel's 105-foot blast furnace "J" at its Lackawana, New York plant. This furnace has a 29-foot hearth diameter and an annual capacity of 600,000 tons of pig iron.

Ore, limestone, and coke are added at the top

250° C
$3 Fe_2O_3 + CO \rightarrow 2 Fe_3O_4 + CO_2$

600° C
$Fe_3O_4 + CO \rightarrow 3 FeO + CO_2$

1000° C
$FeO + CO \rightarrow Fe + CO_2$

1300° C
$CO_2 + C \rightarrow 2 CO$

$C + O_2 \rightarrow CO_2$
2000° C

Hot air

Slag tapped off here

Slag

Iron

Iron tapped off here

top and molten iron can be tapped off at the bottom. Once started, a typical blast furnace may run continuously for two years or longer before it is worn out and must be rebuilt.

The material put into the top of the blast furnace is called the *charge.* It consists of a mixture of iron ore, limestone, and coke. A typical iron ore consists of an iron oxide—Fe_2O_3, for example—plus impurities of sand and rock. The coke is added to reduce the iron oxide to the free metal. The limestone is added to react with the high-melting impurities to form a **slag,** which has a lower melting point. The slag can then be drained off as a liquid at the base of the furnace.

To understand what happens in the furnace, it is best to begin with the reactions that take place near the bottom. Here, heated air is blown into the furnace where carbon (from the coke) reacts with oxygen to form carbon dioxide.

$$C + O_2 \longrightarrow CO_2$$

The reaction is very exothermic, and the temperature in this part of the furnace rises to nearly 2000 °C. It is the hottest region of the furnace. The hot CO_2 rises and reacts with additional carbon to form carbon monoxide.

$$CO_2 + C \longrightarrow 2CO$$

This reaction is endothermic, which causes the temperature higher up in the furnace to drop to about 1300 °C. As the carbon monoxide rises through the charge, it reacts with the iron oxides and reduces them to the free metal. The reactions are

$$3Fe_2O_3 + CO \longrightarrow 2Fe_3O_4 + CO_2$$
$$Fe_3O_4 + CO \longrightarrow 3FeO + CO_2$$
$$FeO + CO \longrightarrow Fe + CO_2$$

As the charge settles toward the bottom, molten iron trickles down and collects in a well at the base of the furnace.

The high temperature in the furnace causes the limestone in the reaction mixture to decompose to give calcium oxide.

$$CaCO_3 \longrightarrow CaO + CO_2$$
limestone

The calcium oxide reacts with impurities such as silica (SiO_2) in the sand to form the slag.

$$CaO + SiO_2 \longrightarrow CaSiO_3$$
calcium
silicate
(slag)

The molten slag also trickles down through the charge. It collects as a liquid layer on top of the more dense molten iron. Periodically, the furnace is tapped, and the iron and slag are drawn off. The iron, which still contains some impurities, is called pig iron.[1] It is usually treated further to produce steel. The slag itself is a valuable by-product. It is used to make insulating materials and is one of the chief ingredients in the manufacture of portland cement.

Preparing metals for use

Before metals can be used, most of them must be purified, or refined, after they are reduced to the metallic state. For example, the metallic copper that comes from the smelting process is about 99% pure, but before it can be used in electrical wiring, it must be purified further. The electrolytic refining of copper, you recall, was described in Chapter Eighteen.

To make one ton of iron requires 1.75 tons or ore, 0.75 tons of coke, and 0.25 tons of limestone.

This blast of hot air is what gives the blast furnace its name.

[1] The name *pig iron* comes from an early method of casting the molten iron into bars for shipment. The molten metal was run through a central channel that fed into sand molds. The arrangement looked a little like a litter of pigs feeding from their mother.

Molten steel is poured from a basic oxygen furnace.

The conversion of pig iron to steel is the most important commercial refining process. When it comes from the blast furnace, pig iron consists of about 95% iron, 3 to 4% carbon, and smaller amounts of phosphorus, sulfur, manganese, and other elements. Steel contains much less carbon as well as certain other ingredients in very definite proportions. Converting pig iron to steel, therefore, involves removing the impurities and much of the carbon and adding other metals in precisely controlled amounts.

Today, most steel is made either by the open hearth method or the basic oxygen process. In an **open hearth furnace** (Figure 20.5) the hearth, or floor, of the furnace is exposed to the burning gases that are used to heat the steel. The furnace is charged with weighed portions of scrap iron, iron ore (Fe_2O_3), and limestone. These are heated by a blast of hot air and burning fuel. Molten pig iron is then added and the hot gases are circulated over the top of the molten mass while the chemical reactions that convert it to steel are taking place. The Fe_2O_3 reacts with carbon in the mixture, forming bubbles of CO_2. The limestone decomposes and the resulting calcium oxide reacts with impurities to form a slag. Other metals, such as manganese or chromium, are added to give the steel special properties, such as hardness or resistance to corrosion. Finally, after about five to eight hours, the steel is ready and the furnace is tapped. The steel is then fabricated into the various forms required by steel users.

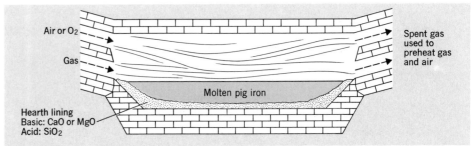

Figure 20.5
Cross section of an open hearth furnace.

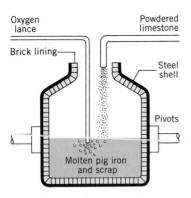

Figure 20.6
Basic-oxygen furnace used for the production of steel.

The **basic oxygen process** for making steel uses a pear-shaped vessel that is mounted on pivots. It is lined with an insulating layer of special bricks, as shown in Figure 20.6. The charge consists of about 30% scrap iron and scrap steel and about 70% molten pig iron, which melts the scrap. An *oxygen lance*—a tube through which pure oxygen is blown—is dipped into the molten metal. The oxygen burns off the excess carbon and converts the iron to steel. Powdered limestone is also added to form a slag with the impurities. After the steel is ready, the reaction vessel can be tipped to pour out its contents. This method of making steel is very fast. A batch of steel weighing 300 tons can be made in less than an hour. Because of this speed, the basic oxygen process has replaced the open hearth method as the principal steel-making process.

20.3 THE REPRESENTATIVE METALS; GROUPS IA AND IIA

The metals of Groups IA and IIA are very reactive, with close chemical and physical similarities among the members of each group

Now we examine some of the specific properties and characteristics of the metallic elements. We begin by studying the representative metals—those found in the A-groups of the periodic table. Because of the placement of elements with similar properties in the same column of the periodic table, a discussion of chemical and physical properties is most meaningful if the elements are considered one group at a time. In many ways, the chemistry of one member of a group is reflected in the chemistry of the others as well. This is particularly true for the elements in Groups IA and IIA.

The metals of Group IA

The elements of Group IA are hydrogen, lithium, sodium, potassium, rubidium, cesium, and francium. Hydrogen, at the top of the group, is a nonmetal, and its chemistry was discussed earlier in Chapter Twelve. As we saw, hydrogen really doesn't fit well into any group. It owes its location in Group IA to its electron configuration, $1s^1$, rather than to its chemical properties. All the rest of the Group IA elements are metals—the **alkali metals.**

Because almost all of their salts are fairly soluble in water, the alkali metals are often found in nature as ions in aqueous solution, primarily in sea water and in deep brine wells. There are also large underground deposits of salt and some surface deposits in dry salt lakes. In addition, some clays contain alkali metal ions along with aluminum, silicon, and oxygen (Section 13.6).

The most abundant of the alkali metals are sodium and potassium. By mass, they rank sixth and seventh among the elements in the earth's crust. They are also the most biologically important metals of Group IA. Both sodium ions and potassium ions are important in animals, but potassium ions are much more important than sodium ions in plants.

Lithium, which is relatively rare, has shown promise in the treatment of certain mental disorders. Lithium carbonate, for example, is used as a drug for treating manic depression.

Both rubidium and cesium are rare and have little commercial importance. Francium has only a fleeting existence because all of its isotopes are radioactive and have very short half-lives.

Some of the properties of the alkali metals are summarized in Table 20.1. The elements are all soft, and they all have low melting points. This is because their atoms each have only one valence electron, so the metallic lattices that they form contain only singly charged cations. These ions are rather weakly attracted to their surrounding "sea" of electrons, which means that the lattice is easily deformed and that a small amount of thermal energy is all that is necessary to allow the atoms to overcome the attractions and enter the liquid state.

Francium's longest-lived isotope is ^{223}Fr, which has a half-life of only 22 minutes.

Table 20.1

Some Properties of the Alkali Metals

Element	Electron Configuration	Ionization Energy (kJ/mol)	Melting Point (°C)	Reduction Potential (V)
Lithium	[He] $2s^1$	520.1	180.5	−3.05
Sodium	[Ne] $3s^1$	495.7	97.8	−2.71
Potassium	[Ar] $4s^1$	418.7	63.7	−2.92
Rubidium	[Kr] $5s^1$	402.9	39.0	−2.93
Cesium	[Xe] $6s^1$	375.6	28.6	−2.92
Francium	[Rn] $7s^1$	—	—	—

We have already learned that the alkali metals have low ionization energies because of their electron configurations. This makes it easy to remove the single outer electron from one of their atoms, and the free metals are easily oxidized. Their very negative reduction potentials reflect this. (The metals are easily oxidized, so their ions are difficult to reduce.)

Sodium reacts violently with water. The heat of reaction ignites the sodium metal and the hydrogen gas produced by the reaction.

Because the alkali metals are so difficult to reduce, preparation of the free elements by reduction of their 1+ ions is generally carried out by the electrolysis of a molten salt, such as $NaCl$ or $LiCl$. Interestingly, potassium can be made by passing sodium vapor over molten KCl. Even though potassium is more difficult to reduce than sodium, the reaction

$$KCl(\ell) + Na(g) \longrightarrow NaCl(\ell) + K(g)$$

is able to proceed because potassium is more volatile than sodium. The potassium vapor is swept away so that equilibrium is never attained. Metallic rubidium and cesium can be prepared in the same way.

Compounds and Reactions. The Group IA metals are the most reactive of all the metals. They are such powerful reducing agents that they readily reduce water. When placed into water, they react violently, releasing hydrogen and forming the hydroxide of the metal.

$$2M(s) + 2H_2O \longrightarrow 2M^+(aq) + 2OH^-(aq) + H_2(g)$$

Some of the most interesting reactions of the alkali metals are those that occur with oxygen. All of these metals react rapidly with oxygen when they are exposed to air, but as we learned in Section 12.3 the products of the reactions differ for the different metals. Lithium reacts to give a "normal" oxide containing the O^{2-} ion.

$$4Li + O_2 \longrightarrow 2Li_2O$$

Sodium, however, reacts with O_2 to form a peroxide that contains the O_2^{2-} ion.

$$2Na + O_2 \longrightarrow \underset{\substack{\text{sodium} \\ \text{peroxide}}}{Na_2O_2}$$

The rest of the alkali metals form *superoxides* containing the paramagnetic O_2^- ion.

$$K + O_2 \longrightarrow KO_2 \qquad \text{potassium superoxide}$$
$$Rb + O_2 \longrightarrow RbO_2 \qquad \text{rubidium superoxide}$$
$$Cs + O_2 \longrightarrow CsO_2 \qquad \text{cesium superoxide}$$

Although lithium is the only alkali metal to form a normal oxide by direct combination of the elements, normal oxides of the other alkali metals can be prepared by indirect means.

When lithium oxide or any of the other normal oxides react with water, a hydroxide is formed.

$$Li_2O + H_2O \longrightarrow 2LiOH$$

Sodium peroxide reacts with water to give hydrogen peroxide as well.

$$Na_2O_2 + 2H_2O \longrightarrow 2NaOH + H_2O_2$$

The superoxides react with water to generate oxygen. For example,

$$2KO_2 + 2H_2O \longrightarrow 2KOH + O_2 + H_2O_2$$

Potassium superoxide has been used in recirculating breathing equipment, in which air that is exhaled by the user is purified, recirculated, and breathed again. Reaction of the KO_2 with moisture in the exhaled air generates oxygen and the KOH removes carbon dioxide.

$$KOH + CO_2 \longrightarrow KHCO_3$$

Besides the oxides, some other important compounds of the alkali metals are the halides, the hydroxides, and the carbonates. The halides serve as raw materials for producing the free metals as well as most of the remaining alkali metal compounds.

Because sodium chloride is so plentiful—and is therefore such an inexpensive raw material—sodium compounds tend to be far more commercially important than compounds of the other alkali metals. For example, sodium hydroxide, also called *lye* or *caustic soda,* is made from sodium chloride by electrolysis (Section 18.5). Annual production of sodium hydroxide is about 12 million tons, which makes it industry's most important strong base.

Sodium carbonate, also called *soda ash,* is another very useful chemical. Approximately 9 million tons of it are produced each year. About half of this is used to make glass, and the rest is used in making other chemicals, paper, and detergents and in water softening. About 80% of the sodium carbonate produced each year is mined from deposits of *trona* ore, $Na_2CO_3 \cdot NaHCO_3 \cdot 2H_2O$. The rest is manufactured from salt by the **Solvay process.**

The raw materials for the Solvay process are sodium chloride, ammonia, and calcium carbonate—limestone. The limestone is heated to produce carbon dioxide.

Potassium carbonate, K_2CO_3, is called potash and is present in large amounts in wood ashes.

$$CaCO_3 \xrightarrow{\text{heat}} CaO + CO_2$$

This CO_2 is bubbled into a cold solution of sodium chloride and ammonia. The dissolved CO_2 reacts with water to form carbonic acid, which is partially neutralized by the ammonia.

$$CO_2 + H_2O \longrightarrow H_2CO_3$$
$$H_2CO_3 + NH_3 \longrightarrow NH_4^+ + HCO_3^-$$

At low temperatures, sodium bicarbonate is less soluble than sodium chloride, so $NaHCO_3$ precipitates out, leaving chloride ion and ammonium ion in the solution. The net overall reaction is

$$NaCl(aq) + NH_3(aq) + CO_2(aq) + H_2O \longrightarrow NaHCO_3(s) + NH_4Cl(aq)$$

Ammonium chloride, which is recovered from the reaction mixture after the $NaHCO_3$ is removed by filtration, is then allowed to react with the lime (CaO) produced earlier by the decomposition of the limestone.

$$2NH_4Cl + CaO \longrightarrow CaCl_2 + 2NH_3 + H_2O$$

This reaction regenerates ammonia, which is recycled. Meanwhile, the sodium bicarbonate is heated to give sodium carbonate and carbon dioxide, which is also recycled.

$$2NaHCO_3 \longrightarrow Na_2CO_3 + CO_2 + H_2O$$

The final products are Na_2CO_3 and $CaCl_2$.

Solubilities of Salts. In the laboratory, sodium and potassium salts are common reagents because nearly all of them are soluble in water. Therefore, if a particular anion is needed for a reaction in solution, a chemist almost always reaches for a bottle of its sodium or potassium salt. However, salts of these metals with a given anion are not equally soluble. An interesting and useful generalization is that the sodium salt of a *strong acid* is often more soluble than the potassium salt, whereas the potassium salt of a *weak acid* is often more soluble than the sodium salt. Table 20.2 contains some data that illustrates this.

If you need NO_3^- in a reaction, you can use either $NaNO_3$ or KNO_3.

Spectra. When ions of an alkali metal are introduced into a flame, brilliant colors are produced that are characteristic of the element's atomic spectrum. Sodium salts, for example, produce a bright yellow flame. If you have ever softened glass rod or glass tubing by heating them in a bunsen burner flame, you

Table 20.2
Molar Solubilities of Some Sodium and Potassium Salts[a]

Strong Acids		Sodium Salt	Solubility (mol/100 g H$_2$O)	Potassium Salt	Solubility (mol/100 g H$_2$O)
Hydrochloric	(HCl)	NaCl	0.61	KCl	0.46
Perchloric	(HClO$_4$)	NaClO$_4$	1.49	KClO$_4$	0.01
Nitric	(HNO$_3$)	NaNO$_3$	0.86	KNO$_3$	0.12
Weak Acids					
Acetic	(HC$_2$H$_3$O$_2$)	NaC$_2$H$_3$O$_2$	1.45	KC$_2$H$_3$O$_2$	2.58
Tartaric	(H$_2$C$_4$H$_4$O$_6$)	Na$_2$C$_4$H$_4$O$_6$	0.035	K$_2$C$_4$H$_4$O$_6$	0.64
Citric	(H$_3$C$_6$H$_5$O$_7$)	Na$_3$C$_6$H$_5$O$_7$	0.25	K$_3$C$_6$H$_5$O$_7$	0.56

[a] Compared at the same temperature.

Remember, Na$_2$CO$_3$ is used in making glass.

have seen this yellow color. It is produced by sodium ions that are vaporized from the hot glass.

The yellow flame produced by sodium is so easily recognized that it forms the basis of a **flame test** for that element. For example, if a sample is suspected of containing sodium, a clean wire is dipped into a solution of the substance and then held in the flame of a bunsen burner. A bright yellow color in the flame confirms the presence of sodium; the absence of the bright yellow flame means that the sample contains no sodium.

Each alkali metal imparts a different color to a flame. Sodium compounds, as we have noted, produce a bright yellow color. The other alkali metals, for which a flame test is normally used, are potassium and lithium. Potassium salts give a pale violet color to a flame, and lithium salts give a beautiful, deep red color. The pale violet of potassium is sometimes difficult to see, particularly if sodium is also present. The yellow from the sodium masks the violet produced by potassium. Viewing the flame through *cobalt glass,* a glass with a deep blue color, filters out the yellow and allows the pale violet color of the potassium flame to be seen.

Some fireplace logs are soaked in solutions of metal salts to produce brightly colored flames when they are burned.

The metals of Group IIA

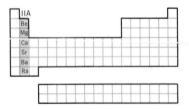

The elements of Group IIA are beryllium, magnesium, calcium, strontium, barium, and radium. They are called the **alkaline earth metals.** Calcium and magnesium are the most abundant metals of this group, and they are also the most biologically important ones. For example, a magnesium ion lies at the center of the chlorophyll molecule. Because of chlorophyll's role in photosynthesis, it is ultimately responsible for the maintenance of virtually all forms of life on this planet. Similarly, calcium is an important component of the bones of animals and the hard shells of shellfish.

As we learned earlier, a relatively large concentration of magnesium is present in sea water. Calcium is also present in sea water, and marine organisms take calcium ions from the sea to make their calcium carbonate shells. On land, the alkaline earth metals are found in various mineral deposits. Largest among them are limestone (calcium carbonate, CaCO$_3$) and *dolomite* (a mixed calcium-magnesium carbonate, generally written as CaCO$_3$·MgCO$_3$). Many of these limestone deposits occur below the earth's surface, and ground water trickling through them often creates spectacular caverns. Another important calcium mineral is gypsum, CaSO$_4$·2H$_2$O, from which plaster is made. Beryllium is found in the mineral beryl, Be$_3$Al$_2$(SiO$_3$)$_6$. You have probably seen beryl crystals such as *emerald* and *aquamarine* (see Color Plate 7C). Strontium and barium occur chiefly as their sulfates and carbonates. Radium is isolated from uranium ores because it is a product of the radioactive decay of uranium.

The *lime* that is used on lawns or gardens is really pulverized limestone or dolomite.

Some properties of the alkaline earth metals are given in Table 20.3. Each atom of these elements has a pair of rather loosely held outer electrons that is lost to the metallic lattice. The resulting doubly charged cations are bound more tightly to the surrounding sea of electrons than are the singly charged cations in crystals of the alkali metals. Therefore, the Group IIA metals have higher melting points and are harder than their neighbors in Group IA.

Table 20.3
Some Properties of the Alkaline Earth Metals

Element	Electron Configuration	Ionization Energy (kJ/mol)		Melting Point (°C)	Reduction Potential (V)[a]
		First	Second		
Beryllium	[He] $2s^2$	899	1757	1278	−1.85
Magnesium	[Ne] $3s^2$	737	1450	651	−2.37
Calcium	[Ar] $4s^2$	590	1145	843	−2.76
Strontium	[Kr] $5s^2$	549	1059	769	−2.89
Barium	[Xe] $6s^2$	503	960	725	−2.90
Radium	[Rn] $7s^2$	509	975	700	−2.92

[a] For $M^{2+}(aq) + 2e^- \rightarrow M(s)$.

The ionization energies of the alkaline earth metals are relatively low, but they are higher than those of the alkali metals. As a result, even though the Group IIA metals are easily oxidized, they are not quite as effective as reducing agents as are the alkali metals. This is reflected in the reduction potentials of the alkaline earth metals, which are not as negative as those of the alkali metals.

The alkaline earth metals are generally prepared by electrolysis because their compounds are difficult to reduce chemically. Magnesium and beryllium, which are less reactive than the elements below them, are the only metals of this group that are prepared in large quantities. The production of magnesium has been described earlier. Beryllium is prepared by electrolysis of a molten mixture of beryllium chloride, $BeCl_2$, and sodium chloride. The sodium chloride is added to serve as an electrolyte because beryllium chloride itself is a poor conductor of electricity.

Beryllium is used in alloys with copper and bronze to give them hardness.

Compounds and Reactions. The Group IIA metals are quite reactive; they are very easily oxidized. Their reduction potentials (Table 20.3) indicate that the ease of oxidation increases going down the group, and the elements below magnesium—calcium, strontium, barium, and radium—react with water to liberate hydrogen. For example,

$$Ca + 2H_2O \longrightarrow Ca(OH)_2 + H_2$$

However, the reactions of these metals with water are less violent than are the reactions of the alkali metals with water.

In their compounds, the alkaline earth metals always exist in the 2+ oxidation state. In general, compounds of calcium, strontium, and barium are distinctly ionic—that is, their compounds exhibit properties that are typical of ionic substances. Most magnesium compounds are also ionic, although magnesium forms some compounds, called organo magnesium compounds, in which there is covalent bonding between magnesium and another atom. An example is $Mg(C_2H_5)_2$ in which two ethyl groups are covalently bonded to a magnesium. (Since magnesium is less electronegative than carbon, it is still assigned an oxidation number of 2+.)

In the case of beryllium, there is no evidence that Be^{2+} ions actually exist. Beryllium compounds, such as $BeCl_2$, are covalent. This is attributed to the small size and highly concentrated charge that a Be^{2+} ion would have. As illustrated in Figure 20.7, if a Be^{2+} ion were placed next to an anion, its high concentration of positive charge would distort the anion's electron cloud and draw the electron density into the region between the Be^{2+} and the anion. Because this would con-

A beryllium-copper alloy leaf spring from a signal switch that is used on a truck trailer. Its size is compared to that of an ordinary paper clip.

centrate the electron density between the two nuclei, it would produce, in effect, a covalent bond. Larger positive ions—Mg^{2+}, for example—are considerably less effective in distorting the electron cloud of an anion because the positive charge of the cation is spread over a larger volume. As a result, beryllium chloride is covalent, but magnesium chloride is ionic.

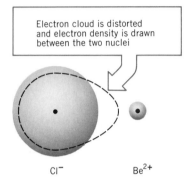

Electron cloud is distorted and electron density is drawn between the two nuclei

Cl^- Be^{2+}

Figure 20.7

A small Be^{2+} ion would have such a high concentration of positive charge that it would distort the electron cloud of an anion such as Cl^-. The electron density drawn between the nuclei gives a great deal of covalent character to the bond.

Some important compounds of the alkaline earth metals include the chlorides, oxides, hydroxides, carbonates, and sulfates. All of the chlorides are water soluble and, except for $BeCl_2$, they are all ionic. Calcium chloride is unusual because of its high affinity for moisture. If calcium chloride is left exposed to humid air, it absorbs so much water that the crystals actually dissolve to form a concentrated solution of $CaCl_2$. This property is called **deliquescence**—calcium chloride is said to be *deliquescent*. Calcium chloride can be purchased in hardware stores (although not always under its chemical name) for use in removing moisture from areas of high humidity such as damp basements.

In Chapter Six we learned that $CaCl_2$ is spread on dirt roads because the moisture it absorbs keeps the dust down.

The oxides of the alkaline earth metals can be formed by direct combination of the elements.

$$2M + O_2 \longrightarrow 2MO$$

Finely divided BeO is very toxic if inhaled.

When magnesium burns, the reaction produces heat plus a great deal of light. Flashbulbs contain fine magnesium wire in an atmosphere of oxygen. When the wire is heated electrically, it ignites and gives a brilliant burst of light as it burns.

The usual way of preparing the oxide of a Group IIA element is by the thermal decomposition of its carbonate. We have already seen this reaction several times in the decomposition of calcium carbonate. Heating the hydroxide of the metal also produces the oxide.

$$Ca(OH)_2 \xrightarrow{\text{heat}} CaO + H_2O$$

Except for beryllium and magnesium, the Group IIA oxides react with water to give the hydroxide. This is the reverse of the reaction above.

$$CaO + H_2O \longrightarrow Ca(OH)_2$$

Calcium oxide is called *lime* and produces *slaked lime* (calcium hydroxide) when treated with water. This is one of the important chemical reactions that take place when cement hardens.

Magnesium oxide resists attack by water. In fact, magnesium metal is protected from attack by water by a thin film of magnesium oxide that forms on its surface. To make magnesium hydroxide, the ingredient in milk of magnesia, a solution containing Mg^{2+} must be made basic. This precipitates the white $Mg(OH)_2$.

$$Mg^{2+}(aq) + 2OH^-(aq) \longrightarrow Mg(OH)_2(s)$$

Have you ever swallowed this form of magnesium hydroxide?

The solubilities of the alkaline earth hydroxides increase going down the group. Magnesium hydroxide is very insoluble, calcium hydroxide is moderately

$BaSO_4$ $K_{sp} = 1.08 \times 10^{-10}$
$CaSO_4$ $K_{sp} = 6 \times 10^{-5}$

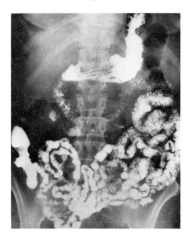

Barium sulfate, which is opaque to X rays, defines the path of the small intestine in a patient who swalled a suspension of this solid in water.

insoluble, and barium hydroxide is moderately soluble. It is interesting that the solubilities of the sulfates vary in the opposite direction. Barium sulfate, for example, has an extremely low solubility. It is used medically to obtain X-ray photographs of the digestive tract. A patient drinks a suspension of $BaSO_4$ in water and then an X-ray photograph is taken. The path of the patient's digestive tract is clearly visible on the film because the $BaSO_4$ is opaque to X rays. Even though barium is very toxic to the human body, barium sulfate is safe to drink because its solubility is so low. Hardly any Ba^{2+} is absorbed by the body as the $BaSO_4$ passes through the digestive system.

Although calcium sulfate is also considered to be insoluble, its solubility is not as low as that of barium sulfate. The dihydrate, $CaSO_4 \cdot 2H_2O$, is called gypsum, which is one of the ores of calcium. Gypsum is one of the most important compounds of the Group IIA elements. When it is heated, it loses water.

$$2CaSO_4 \cdot 2H_2O \longrightarrow (CaSO_4)_2 \cdot H_2O + 3H_2O$$

The product, called *plaster of paris,* is often written as $CaSO_4 \cdot \frac{1}{2}H_2O$ because there is only one-half mole of water per mole of calcium sulfate. When mixed with water, the $CaSO_4 \cdot \frac{1}{2}H_2O$ crystals absorb water and reform the dihydrate. This is a chemical reaction and it is exothermic. If you have ever had a broken bone set in a plaster cast, you may have noticed how warm the cast became as it began to harden. The hardening occurs because tiny crystals fuse together to give a solid mass as the plaster and water react. This reaction is also used to make *plaster board* or *sheet rock,* one of the most common materials used in constructing the interior walls of buildings. See also Section 11.2 and Special Topic 11.1

Magnesium sulfate is very soluble in water. It can be purchased under the name epsom salts, $MgSO_4 \cdot 7H_2O$. This substance is used as a fertilizer, in fireproofing fabrics, and in the tanning of leather.

Spectra. Like the alkali metals, certain of the alkaline earth metals give characteristic colors to flames. Calcium salts give an orange-red color, strontium salts produce a bright red (crimson) flame, and barium salts give a yellow-green color. These colors are intense enough to serve as flame tests.

20.4 THE REPRESENTATIVE METALS; GROUPS IIIA, IVA, AND VA

The metals in groups to the right of the transition elements are less reactive than the metals of Groups IA and IIA, and the heavier of them have two oxidation states

Except for aluminum, the metals in Groups IIIA, IVA, and VA are called posttransition metals because they follow the row of transition elements in their respective periods. The post-transition metals are, in general, considerably less reactive than the metals in Groups IA and IIA. They each have a completed *d* subshell just below the valence shell. Because the *d* electrons are less than 100% effective in screening the nuclear charge, their outer electrons are held more tightly than the outer electrons of the alkali or alkaline earth metals.

The metals of Group IIIA

The elements of Group IIIA are boron, aluminum, gallium, indium, and thallium. They have no general group name, as do the elements of Groups IA and IIA. Except for boron, which is a metalloid, all the rest of the elements of this group are metals. Some of their properties are given in Table 20.4. Notice that gallium has a melting point of only 29.8 °C—it will melt in the palm of your hand. It also has a very high boiling point. This wide liquid range has made gallium useful for certain types of thermometers.

The only really important metal in Group IIIA, as far as we are concerned, is aluminum. It is the most common metallic element in the earth's crust. Aluminum

Table 20.4
Some Properties of the Group IIIA Elements

Element	Electron Configuration	Ionization Energy (kJ/mol)			Melting Point (°C)	Reduction Potential (V)[a]
		First	Second	Third		
Boron	[He] $2s^2 2p^1$	801	2427	3660	2200	—
Aluminum	[Ne] $3s^2 3p^1$	577	1816	2744	660	−1.66
Gallium	[Ar] $3d^{10} 4s^2 4p^1$	578	1971	2950	29.8	−0.53
Indium	[Kr] $4d^{10} 5s^2 5p^1$	559	1813	2690	157	−0.33
Thallium	[Xe] $4f^{14} 5d^{10} 6s^2 6p^1$	589	1961	2860	303	+0.72

[a] For $M^{3+}(aq) + 3e^- \rightarrow M(s)$.

occurs in bauxite, its major ore, as a hydrated oxide, $Al_2O_3 \cdot nH_2O$; in the mineral *corundum* as an anhydrous oxide, Al_2O_3; in various silicate minerals; and in cryolite, Na_3AlF_6. The free element, produced by electrolysis in the Hall process (Section 18.5), is valued as a structural metal because of its high strength and low density ("light weight").

Aluminum metal is easily oxidized; its reduction potential is −1.66 V. Fortunately, the reaction between aluminum and oxygen produces a tough oxide coating that protects the aluminum from further attack. In fact, *anodized aluminum* has an oxide coating that is made deliberately thick by electrolysis. Because of the way it is formed, this coating is porous enough to accept and hold printing inks.

Aluminum is a powerful reducing agent because of its ease of oxidation. One of the most spectacular reactions of aluminum is the **thermite reaction** in which iron oxide is reduced to the free metal.

$$2Al + Fe_2O_3 \longrightarrow Al_2O_3 + 2Fe$$

This is an extremely exothermic reaction ($\Delta H° = -847.6$ kJ), and the heat that is evolved produces molten iron that is hot enough to weld iron or steel parts together.

Aluminum is amphoteric. It dissolves in both acids and bases to liberate hydrogen. The reactions are often written as

$$2Al + 6H^+ \longrightarrow 2Al^{3+} + 3H_2 \quad \text{(acidic solution)}$$
$$2Al + 2OH^- + 2H_2O \longrightarrow 2AlO_2^- + 3H_2 \quad \text{(basic solution)}$$

The reaction of aluminum with base explains why oven cleaners containing lye (NaOH) can't be used on aluminum pots and pans!

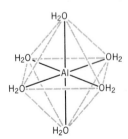

The nature of the aluminum species in these solutions is more complex than indicated by the equations, however. In acidic or neutral solutions, the aluminum ion exists in a hydrated form, $Al(H_2O)_6^{3+}$, in which the six water molecules are bound rather tightly to the Al^{3+} ion in an octahedral arrangement. When aluminum salts are crystallized from water, they usually contain the hydrated aluminum ion within their crystals.

Solutions containing aluminum salts, such as $AlCl_3$ or $Al_2(SO_4)_3$, are not neutral, as we saw in Section 17.10. Instead, they are acidic. The attraction of the Al^{3+} ion for the electrons of the oxygen atoms in the surrounding water molecules draws electron density from the O-H bonds. This tends to further polarize the already polar O-H bonds of the water molecules, which increases the amount of positive charge on the hydrogen atoms. This means that these hydrogens can be more easily removed as H^+ ions, which become attached to water to form hydronium ions, as illustrated in Figure 20.8. The equilibrium is

$$Al(H_2O)_6^{3+} + H_2O \rightleftharpoons Al(H_2O)_5OH^{2+} + H_3O^+$$

The addition of a base to a solution of an aluminum salt precipitates a gelatin-like hydroxide. This can be formulated as a neutralization reaction.

$$(H_2O)_5Al^{3+} \longleftarrow \ddot{O} = H \quad \ddot{O} \underset{H}{\overset{H}{\diagup}} \longrightarrow \left[(H_2O)_5Al \longleftarrow \ddot{O} \underset{H}{\overset{}{|}} \right]^{2+} + \left[\underset{H}{\overset{H}{|}} O-H \right]^{+}$$

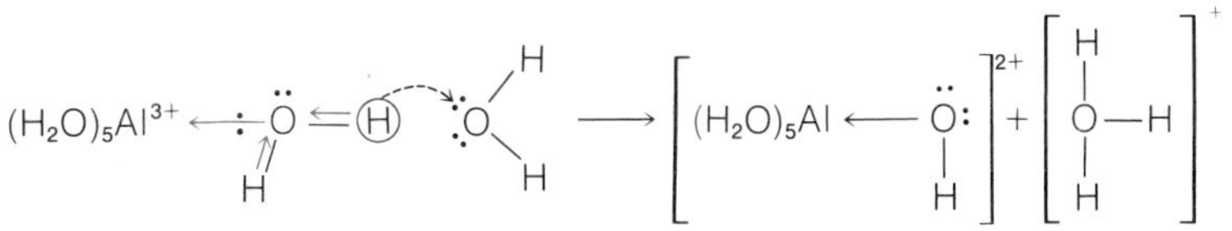

Figure 20.8
The aluminum ion draws electrons to itself from the oxygen atoms of the neighboring water molecules. This further polarizes the O—H bonds. This polarization, shown here for one of the six water molecules, makes it easier for an H^+ to be transferred to a molecule of water in the surrounding solvent, thereby producing H_3O^+ in the solution.

$$Al(H_2O)_6^{3+} + 3OH^- \longrightarrow Al(H_2O)_3(OH)_3 + \underset{water}{3H\,OH}$$

Aluminum hydroxide is amphoteric because it dissolves in either acid or base.

$$Al(H_2O)_3(OH)_3 + OH^- \longrightarrow Al(H_2O)_2(OH)_4^- + H_2O$$
$$Al(H_2O)_3(OH)_3 + H_3O^+ \longrightarrow Al(H_2O)_4(OH)_2^+ + H_2O$$

Notice that $Al(H_2O)_2(OH)_4^-$ is equivalent (in terms of total atoms) to $AlO_2^- + 4H_2O$. Writing the formula AlO_2^- for the species in basic solution is really shorthand for the more complicated species that are actually in the solution.

Among the important compounds of aluminum are the oxide, the halides, and the sulfates. Aluminum oxide has the formula, Al_2O_3, but it occurs in two different crystalline forms that differ greatly in their chemical reactivity. If the gelatinous hydroxide is dehydrated, γ-Al_2O_3 is formed.

$$2Al(H_2O)_3(OH)_3 \xrightarrow{\text{heat}} \gamma\text{-}Al_2O_3 + 9H_2O$$

This form of the oxide readily dissolves in both acidic and basic solutions. If γ-Al_2O_3 is heated to temperatures above 1000 °C, its crystal structure changes to that of α-Al_2O_3. This form of the oxide is very resistant to chemical attack. Naturally occurring α-Al_2O_3 is called *corundum*. Its crystals are very hard, and it is commonly used as an abrasive in sandpaper. Aluminum oxide has an extremely high melting point (2045 °C), and it is used to make special bricks for the interiors of furnaces.

Large crystals of corundum that contain traces of certain other metals are valued as gems. For example, sapphire is Al_2O_3 with very small amounts of iron and titanium; ruby is Al_2O_3 with trace amounts of chromium (see Color Plate 8B).

Aluminum sulfate, another commercially important chemical, can be made by dissolving aluminum oxide (from bauxite) in sulfuric acid.

$$Al_2O_3 + 3H_2SO_4 \longrightarrow Al_2(SO_4)_3 + 3H_2O$$

A well-formed crystal of potassium alum, $KAl(SO_4)_2 \cdot 12H_2O$.

Aluminum sulfate is important in making paper and dying fabrics, and appreciable amounts of $Al_2(SO_4)_3$ are also used to process sewage.

When a solution that contains equal numbers of moles of aluminum sulfate and potassium sulfate is evaporated, crystals having the composition $KAl(SO_4)_2 \cdot 12H_2O$ are formed. Similar crystals are obtained from solutions that contain both aluminum sulfate and ammonium sulfate. Their formula is $NH_4Al(SO_4)_2 \cdot 12H_2O$. These are examples of **double salts**—crystals that contain the components of two different salts in a definite ratio. Double salts that have the general formula $M^+M^{3+}(SO_4)_2 \cdot 12H_2O$ are called **alums.** Potassium alum, $KAl(SO_4)_2 \cdot 12H_2O$, is used to treat cotton fibers to make them absorb dyes more easily. Sodium alum, $NaAl(SO_4)_2 \cdot 12H_2O$, is used in certain baking powders.

The anhydrous halides of aluminum are interesting because of the existence of covalent bonding, particularly in the vapor. Aluminum bromide exists as a dimeric species with the formula, Al_2Br_6. Aluminum chloride in the vapor exists in a similar molecular form, Al_2Cl_6. The structures of these molecules are illustrated in Figure 20.9. Their formation from AlX_3 ($X = Cl$ or Br) can be viewed as an attempt by aluminum to complete its octet.

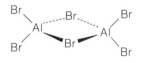

Figure 20.9
The structure of Al_2Br_6. Each aluminum atom is surrounded by four bromine atoms in a tetrahedral fashion. This is also the structure of Al_2Cl_6.

In the dimer, aluminum is surrounded by four electron pairs and the halogen atoms are arranged approximately tetrahedrally, as we would expect from VSEPR theory.

When anhydrous aluminum chloride or bromide is dissolved in water, their molecules dissociate and hydrated aluminum ions, $Al(H_2O)_6^{3+}$, and halide ions are produced. These reactions are extremely exothermic.

Gallium, Indium, and Thallium. The most noteworthy aspect of the chemistry of these metals is the increasing tendency, going down the group, to form an ion with a 1+ charge. For thallium, both the 1+ and 3+ oxidation states are well known. This increasing stability of the lower oxidation state going down a group is a characteristic of the chemistry of all of the post-transition metals.

Thallium salts are extremely poisonous. They were once used as a poison in ant traps.

The metals of Groups IVA and VA
The metals of Group IVA are tin and lead. Bismuth is the only metal in Group VA. Some of the properties of these metals are given in Table 20.5. Tin is generally found in nature as its oxide, SnO_2, and can be reduced to the free metal by reaction with carbon (coke).

$$SnO_2 + C \longrightarrow Sn + CO_2$$

Lead occurs as a sulfide, PbS, in ores called *galena*. Roasting produces the oxide, which is then reduced by carbon to give the free metal.

$$2PbS + 3O_2 \longrightarrow 2PbO + 2SO_2$$
$$2PbO + C \longrightarrow 2Pb + CO_2$$

Table 20.5
Some Properties of Tin, Lead, and Bismuth

Element	Electron Configuration	Melting Point (°C)	Oxidation States (Most stable in bold type)
Tin	$[Kr]\ 4d^{10}5s^25p^2$	232	2+, **4+**
Lead	$[Xe]\ 4f^{14}5d^{10}6s^26p^2$	328	**2+**, 4+
Bismuth	$[Xe]\ 4f^{14}5d^{10}6s^26p^3$	271	**3+**, 5+

Bismuth sometimes is found in nature as the free metal, but it usually occurs as either the oxide or sulfide. Roasting, followed by reduction with carbon, gives free bismuth.

$$2Bi_2S_3 + 9O_2 \longrightarrow 2Bi_2O_3 + 6SO_2$$
$$2Bi_2O_3 + 3C \longrightarrow 4Bi + 3CO_2$$

Neither tin, lead, nor bismuth is very reactive, and some of their uses as free metals are based on this fact. Tin is used in various alloys such as bronze (tin and copper) and as a coating on steel—the ordinary tin can. Elemental tin can exist in either of two allotropic crystalline forms, depending on the temperature. At temperatures above 18 °C, tin occurs as a normal, metallic crystal called white tin. At lower temperatures, however, tin gradually changes to a nonmetallic form called gray tin. Tin objects kept in cold climates for long periods sometimes develop lumps as the crystal structure of the metal begins to change. At one time it was believed that some organism was attacking the tin, and the phenomenon was called "tin disease."

Allotropes of sulfur were described in Section 12.4.

The Latin name for lead is *plumbum,* and the terms *plumbing* and *plumber* come from the early use of lead for pipes and pipe joints. The metal is also used to make batteries and solder, and to manufacture tetraethyllead, $Pb(C_2H_5)_4$, a gasoline octane booster.

$Pb(C_2H_5)_4$ is in "leaded" gasoline.

Bismuth is one of only a few substances that expand slightly when they freeze. This property makes bismuth ideal for making accurate castings because it expands to fill all the details of the mold. The other principal use of bismuth is in making alloys with low melting points. An example is *Wood's metal,* an alloy containing 50% bismuth, 25% lead, 12.5% tin, and 12.5% cadmium. This alloy has a melting point of only 70 °C and is used in overhead sprinkler systems. A fire will trigger the system automatically by melting the alloy before the temperature has risen very high.

One of the major features of the chemistry of tin, lead, and bismuth is the occurrence of two oxidation states. The lower oxidation state becomes increasingly more stable compared to the higher one going from top to bottom in a group. For instance, both tin and lead form compounds in the 2+ and 4+ oxidation states. These correspond to the loss of the outer pair of *p* electrons, followed by the further loss of the pair of *s* electrons, respectively. Compounds of tin(II) and tin(IV) are both common, but tin(II) compounds are easily oxidized to the 4+ state. On the other hand, lead(IV) compounds are powerful oxidizing agents, which means that they tend to acquire electrons and become reduced to the more stable 2+ state.

The relative stabilities of the oxidation states can be seen in the reactions of tin and lead with oxygen. Tin reacts with oxygen to form SnO_2 rather than SnO. On the other hand, lead normally reacts with oxygen to form PbO (called *litharge*), although lead will react further with air at high temperatures to give Pb_3O_4, which is called *red lead*. Lead(IV) oxide, PbO_2, is a powerful oxidizing agent and the lead is easily reduced to the 2+ state.

Pb_3O_4 is often used as a protective undercoat when painting iron and steel structures.

Bismuth, which has the valence shell configuration $6s^26p^3$, forms compounds in the 3+ and 5+ oxidation states. These states correspond, at least in principle, to the loss of first the three *p* and then the two *s* electrons, respectively. Of the two, the 3+ state is the more stable. Compounds such as sodium bismuthate, $NaBiO_3$, which contain bismuth in the 5+ state, are extremely powerful oxidizing agents. For example, in an acidic solution, Mn^{2+} is oxidized to MnO_4^- by the bismuthate ion.

$$14H^+ + 5BiO_3^- + 2Mn^{2+} \longrightarrow 2MnO_4^- + 5Bi^{3+} + 7H_2O$$

Among the halides, tin forms both SnX_2 and SnX_4 compounds (where X stands for a halogen). One of the most important of these is stannous fluoride, SnF_2, the fluoride ingredient in toothpaste. The tin(IV) halides are all covalent. An example is $SnCl_4$, a colorless liquid that has a melting point of $-33\,°C$ and a boiling point of $114\,°C$. The covalent nature of the Sn-Cl bonds in $SnCl_4$ can be explained in the same way as the covalent Be-Cl bonds in $BeCl_2$. The small size and high charge that would be expected for a Sn^{4+} ion would distort the electron cloud of the chloride ion to such an extent that the electron density drawn between the two nuclei would constitute a covalent bond.[2]

Lead forms halides in the 2+ state, but in the 4+ state, the compounds $PbBr_4$ and PbI_4 do not exist. Lead(IV) is such a powerful oxidizing agent that the Br^- and I^- would be oxidized by it. In fact, $PbCl_4$ decomposes easily by an internal redox reaction.

$$\overset{4+}{P}\overset{1-}{b}Cl_4 \longrightarrow \overset{0}{Cl_2} + \overset{2+}{Pb}Cl_2$$

(with arrows labeled "(reduction)" and "(oxidation)")

Bismuth(III) fluoride is ionic, but the other halides of bismuth, such as $BiCl_3$, are covalent. When placed in water, $BiCl_3$ hydrolyzes—that is, it reacts with water—to give a precipitate having the formula $BiOCl$ (bismuthyl chloride).

$$BiCl_3 + H_2O \longrightarrow BiOCl(s) + 2H^+ + 2Cl^-$$

Other bismuth salts such as the nitrate, $Bi(NO_3)_3$, and sulfate, $Bi_2(SO_4)_3$, also react with water to give the bismuthyl ion, BiO^+.

20.5 THE TRANSITION METALS; GENERAL CHARACTERISTICS AND PERIODIC TRENDS

Metals with partially filled *d*-subshells often exhibit more than one oxidation state, have compounds that are colored, and form complex ions easily

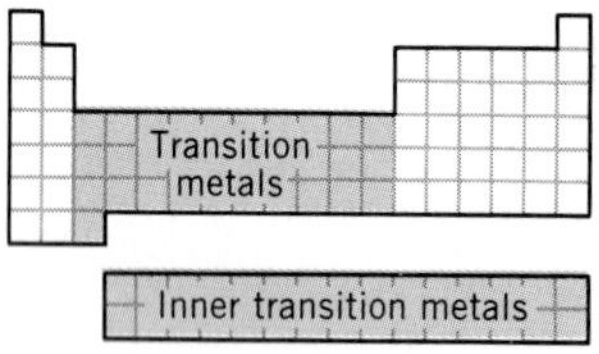

All of the elements between Group IIA and Group IIIA of the periodic table are metals. They are usually divided into two main categories. The **transition elements** or **transition metals** are the block of elements generally placed in the body of the table; they consist of the elements in the B-groups plus Group VIII. The **inner transition elements** are those in the two long rows normally placed just below the main body of the table. The elements in the first of the long rows are called the **lanthanides** because they follow lanthanum, atomic number 57. Those in the second long row are called the **actinides** because they follow actinium, atomic number 89. The lanthanides and actinides are rare elements and their chemistry is not particularly important to us in our present studies, so we will have little more to say about them. However, since many of the transition elements are common structural metals, and some of their ions play major roles in certain biochemical reactions, it is worthwhile to learn about them.

The lanthanides are also called the rare earth metals.

---

[2] Of course, we can also view the bonds in $SnCl_4$ as ordinary covalent bonds formed in the usual way.

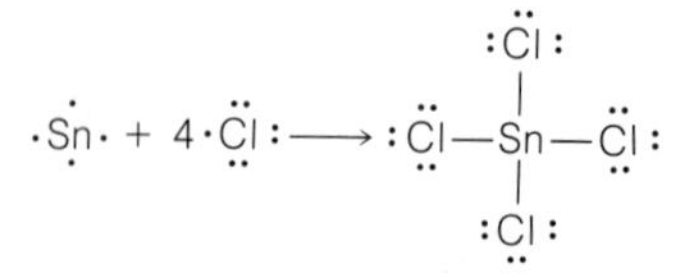

In either case, the result is the same.

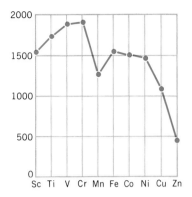

Figure 20.10
Melting points in °C of the period 4 transition elements.

As is true with all of the other elements that we have studied, the chemical and physical properties of the transition metals depend on their electron configurations. In Chapter Seven, we saw that as we cross a period from left to right, a d subshell is gradually filled as we pass through a row of these elements. Except for the metals in Group IIB, atoms of the transition elements all have partially filled d subshells. This is the main feature that distinguishes them from the representative elements.[3] In fact, the transition elements are often called the *d-block* elements.

There are several properties that many of the transition elements have in common. For example, they generally tend to be hard and to have high melting points. This is particularly true of the elements near the center of each row, as shown in Figure 20.10. The elements having the highest melting points also have the maximum number of unpaired d electrons, which suggests that the d electrons are probably involved to some extent in covalent bonding within the metallic lattice.

Another general characteristic of the transition metals is the occurrence of multiple oxidation states. The 2+ state is common to many of these elements because many of their atoms have a pair of electrons in their outermost s subshell. Some examples are manganese, iron, and cobalt.

$$Mn \quad [Ar]\ 3d^5 4s^2$$
$$Fe \quad [Ar]\ 3d^6 4s^2$$
$$Co \quad [Ar]\ 3d^7 4s^2$$

Each forms an ion by the loss of both $4s$ electrons (Mn^{2+}, Fe^{2+}, Co^{2+}). The underlying $3d$ subshell is fairly close in energy to the $4s$ subshell, however, and not very much energy is needed to remove still another electron to give an ion with a 3+ charge. We will take a closer look at oxidation states later in this section.

Complex ions are often simply called complexes.

Still another property of the transition metals is the tendency of their ions to combine with neutral molecules or anions to form **complex ions.** One example is the reaction of the copper(II) ion with ammonia to form a complex ion with the formula $Cu(NH_3)_4^{2+}$.

$$Cu^{2+}(aq) + 4NH_3(aq) \rightleftharpoons Cu(NH_3)_4^{2+}(aq)$$
$$\text{pale blue} \qquad\qquad\qquad\qquad \text{deep blue}$$

The charge on a complex ion is just the sum of the charges of the molecules and ions from which the complex is formed.

As suggested by the double arrows, complex ions such as this are often somewhat unstable and are able to decompose. The number of complex ions formed by the transition elements is enormous, and their study is a major specialty of chemistry. We will say more about them later in this chapter.

One of the most interesting properties of transition metal compounds is that they are often colored. For instance, all of the compounds of chromium are colored. In fact, chromium gets its name from the Greek word *chroma,* which means color. Many complex ions also have beautiful colors, as shown in Color Plate 8C.

Variations in atomic size

One of the properties of the transition elements that varies in a more or less systematic way within the periodic table is atomic size (atomic radius). Across a period there is only a relatively small change in size because the outer s electrons are shielded quite well from the gradually increasing nuclear charge by the electrons that are being added to the underlying d subshell. In Figure 20.11, we see that there is a gradual decrease in size from left to right, with a minimum near the center of each row of transition elements.

Going down a column, we find an unusual variation in size. From the first row to the second the atomic radius increases. This is the same phenomenon that we

Variations in atomic size within groups were discussed in Section 7.12.

[3] The Group IIB elements have electron configurations corresponding to $(n-1)d^{10}ns^2$ outside of a noble gas core, and some chemists prefer not to consider them to be true transition elements because their d subshells are complete.

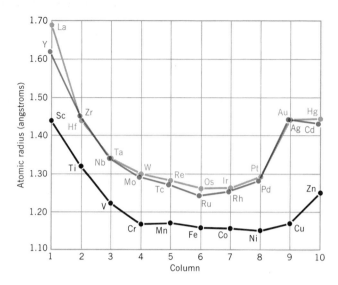

Figure 20.11
Size variations among the transition elements.

observe going down any group of representative elements. Notice, however, that going from the second to the third row of transition elements there is virtually no size change at all, except in Group IIIB. The reason for this is a phenomenon called the **lanthanide contraction.**

Lanthanum, in Group IIIB, is larger than yttrium, as we would expect based on the way size varies down a group of the representative elements. But now let's consider what happens moving from left to right across periods 5 and 6. Going from yttrium to zirconium there is an increase of 1 in the nuclear charge and, as usual, there is a slight decrease in size. Going from lanthanum to hafnium in period 6, however, we must pass through the entire series of lanthanide elements. This increases the nuclear charge by a total of 15. Because the 4f electrons that are added as the lanthanides are completed are not totally effective at shielding the outer electrons from the nucleus, the outer electrons of hafnium experience an extra large effective nuclear charge. This draws the electrons closer to the nucleus and makes hafnium unexpectedly small. Thus, the increase in size that we might have otherwise anticipated on going from zirconium to hafnium is canceled because the outer electrons of hafnium are pulled closer to the nucleus than expected. This unexpected shrinking is the lanthanide contraction. The result of it is that both zirconium and hafnium are nearly identical in size. In Figure 20.11 we see that other similarities in size exist between the second and third members of the groups that occur to the right of hafnium.

Two major consequences of the lanthanide contraction are that the transition elements in period 6 are very dense, and they are very resistant to oxidation. They have unusually high densities because their atoms have nearly twice as much mass as do the atoms of the elements above them, but their sizes are virtually the same. Twice the mass occupying about the same volume leads to about twice the density.

The resistance of the period 6 transition metals to oxidation (loss of electrons) occurs because their outer electrons feel a very large effective nuclear charge. This means that the electrons are tightly held and are difficult to remove. The resistance to oxidation of metals like platinum and gold makes them commercially useful and adds to their overall value. Gold, for example, is used to plate electrical contacts in low voltage circuits where even small amounts of corrosion would block the flow of current.

Compounds and oxidation states

If we compare the *formulas* of compounds formed by the B-group elements—the transition elements—with those formed by the elements in the corresponding A groups, we find certain similarities. For example, the maximum oxidation state for elements in Groups IIIB through VIIB is equal to the group number. This is also true for the elements in Groups IIIA through VIIA. Similar relationships hold true for compounds of the elements in Groups I and II, although both copper and gold (Group IB) are also commonly found in oxidation states higher than 1+. As a result, elements having the same group number form compounds having the same general formula when the elements are in their highest oxidation states. Some examples are shown in Table 20.6.

Table 20.6
Similarities in Formulas of Compounds of the A- and B-Group Elements

Group Number	Oxidation State	A-Group Compound	B-Group Compound
I	1+	$NaCl$	$CuCl$
II	2+	$CaCl_2$	$ZnCl_2$
III	3+	Al_2O_3	Sc_2O_3
IV	4+	CO_2	TiO_2
V	5+	Na_3PO_4	Na_3VO_4
VI	6+	Na_2SO_4	Na_2CrO_4
VII	7+	$KClO_4$	$KMnO_4$

Although similarities exist among the *formulas* of some of the compounds in the A and B groups, most of the chemical and physical properties of the A- and B-group elements are quite different. Therefore, it is not wise to attempt to extend the analogy between the groups any further.

The nine elements that are collectively labeled Group VIII bear no similarities to the other elements in the periodic table in terms of the general formulas of their compounds. In addition, greater horizontal similarities than vertical similarities are found within this group. For instance, the similarities among iron, cobalt, and nickel are greater than those among iron, ruthenium, and osmium. These horizontal sets of elements are generally spoken of as *triads*. Iron, cobalt, and nickel constitute the *iron triad*, for example.

With this as background, let's look now at some of the specific chemical and physical properties of the more important transition metals.

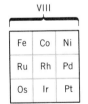

VIII		
Fe	Co	Ni
Ru	Rh	Pd
Os	Ir	Pt

20.6 **PROPERTIES OF SOME IMPORTANT TRANSITION METALS**

The functioning of our bodies and of our technological society depends on the variety of chemical and physical properties exhibited by the transition elements

If you pause for a moment and look around, you will see a number of important transition metals. The paint on the wall probably contains titanium, the coins in your pocket are copper or a copper-nickel alloy, and iron and steel alloys are surely not far from you. Iron and a host of other transition metals are present in various quantities—some of them in only trace amounts—in your body. Certain transition elements are clearly crucial for our comfort and even our existence. In this section we will focus our attention on the specific properties of some of the more important ones.

Titanium and Vanadium. Titanium is difficult to extract from its compounds because it reacts with carbon, oxygen, and nitrogen at the high temperatures normally required by metallurgical processes. Titanium is a very useful metal, however, because it is strong, lightweight (its density is only about 60% of that of iron), and resistant to corrosion. The most important compound of titanium is its brilliant white oxide, TiO_2. Titanium dioxide is a common pigment in paint and is used as a brightener in paper.

Another titanium compound is titanium tetrachloride, $TiCl_4$. It is a clear, colorless liquid with a boiling point of 136 °C and is composed of covalently bonded molecules similar to $SnCl_4$. The U.S. Navy has used $TiCl_4$ to make smoke screens because it reacts almost instantly with moist air to create a dense fog of TiO_2 and HCl.

$$TiCl_4(g) + 2H_2O(g) \longrightarrow TiO_2(s) + HCl(g)$$

Vanadium, like titanium, is also difficult to extract from its compounds. Its principal use is in alloys. When added to steel, it makes the metal more ductile and resistant to shock. Vanadium is also used in nuclear reactors because it is highly "transparent" to neutrons. The most important compound of vanadium is the oxide, V_2O_5, which is used as a catalyst in the contact process for the manufacture of sulfuric acid (Section 12.4).

Chromium. This familiar metal is white, lustrous, hard, brittle, and very resistant to corrosion. These properties make it excellent as both a decorative and a protective coating over other metals such as brass, bronze, and steel. Almost everyone is familiar with the *chrome plate* that is deposited electrolytically on automobile parts such as bumpers. Chromium is also used in large amounts to produce *stainless steel*—a type of steel alloy that is resistant to corrosion. A typical stainless steel contains about 18% chromium, 10% nickel, plus small amounts of manganese, carbon, phosphorus, sulfur, and silicon, all combined with iron.

In compounds, chromium can exist in a number of different oxidation states. The most common are 2+, 3+, and 6+. The chromium(II) ion (chromous ion) is the least stable. The pale blue Cr^{2+} ion is very easily oxidized to the 3+ state, which is the most stable oxidation state of chromium. In water, Cr^{3+} ion actually exists as the violet $Cr(H_2O)_6^{3+}$ ion, and many chromium(III) salts owe their color to the presence of this hydrated ion in their crystals. An example is violet chrome alum, $KCr(SO_4)_2 \cdot 12H_2O$, which is formed when solutions containing K_2SO_4 and $Cr_2(SO_4)_3$ are gradually evaporated.

In the 3+ state, chromium is amphoteric. Addition of a base to a violet solution containing $Cr(H_2O)_6^{3+}$ precipitates the pale blue-violet gelatinous hydroxide. Making the mixture more basic causes the precipitate to dissolve, yielding a deep green solution. The reactions are similar to those of aluminum, which were discussed in the last section.

$$\underset{\text{violet}}{Cr(H_2O)_6^{3+}} + 3OH^- \longrightarrow \underset{\text{blue-violet}}{Cr(H_2O)_3(OH)_3(s)} + 3H_2O$$

$$\underset{\text{blue-violet}}{Cr(H_2O)_3(OH)_3} + OH^- \longrightarrow \underset{\text{green}}{Cr(H_2O)_2(OH)_4^-} + H_2O$$

The third important oxidation state of chromium is 6+. The red-orange oxide, CrO_3, is a powerful oxidizing agent and is the acid anhydride of chromic acid, H_2CrO_4. Chromic acid is the primary species in aqueous solutions at very low pH

The U.S. Navy used $TiCl_4$ to make smokescreens during World War II, which is when this photograph was taken of a destroyer laying smoke during action.

(highly acidic solutions). At higher pH, two other species predominate, the yellow *chromate ion,* CrO_4^{2-}, and the red-orange *dichromate ion,* $Cr_2O_7^{2-}$. There is an equilibrium between CrO_4^{2-} and $Cr_2O_7^{2-}$ that can be written[4]

$$2CrO_4^{2-} + 2H^+ \rightleftharpoons Cr_2O_7^{2-} + H_2O$$

In solutions that are acidic, this equilibrium is shifted to the right and dichromate ion predominates. The principal ion present in basic solutions is CrO_4^{2-} because the reaction is shifted to the left as the H^+ concentration decreases. All of the chromium(VI) species are good oxidizing agents, although their strongest tendency to serve as an oxidizing agent occurs in acidic solutions.

Manganese. Manganese has many properties that are similar to those of iron. It corrodes in moist air, for example, and it dissolves in dilute acids with the evolution of hydrogen.

$$Mn(s) + 2H^+(aq) \longrightarrow Mn^{2+}(aq) + H_2(g)$$

Its chief uses are as an additive to steel and in the preparation of other alloys such as manganese bronze, a copper-manganese alloy.

> Manganese has a 3+ state that is stable in complex ions.

The most important oxidation states of manganese are 2+, 4+, 6+, and 7+. The most stable is the 2+ state, which is formed by the removal of the outer $4s$ electrons from the manganese atom.

$$Mn([Ar]3d^54s^2) \longrightarrow Mn^{2+}([Ar]3d^5) + 2e^-$$

The Mn^{2+} ion thus has a half-filled $3d$ subshell—a configuration that is particularly stable.

The least stable oxidation state of manganese is 7+, which has a strong tendency to be reduced and therefore is a powerful oxidizing agent. The most common compound of manganese(VII) is *potassium permanganate,* $KMnO_4$, which dissolves in water to give deep purple solutions containing the MnO_4^- ion. In an acidic solution, MnO_4^- is usually reduced to Mn^{2+}, which has a very pale pink color that is practically invisible if the solution is dilute.

$$\underset{\substack{\text{deep}\\\text{purple}}}{MnO_4^-} + 8H^+ + 5e^- \longrightarrow \underset{\substack{\text{almost}\\\text{colorless}}}{Mn^{2+}} + 4H_2O$$

Many analytical procedures use $KMnO_4$ as an oxidizing agent in titrations. As the MnO_4^- is added to a reaction mixture from a buret, the Mn^{2+} ion produced by the reduction of MnO_4^- has hardly any effect on the color of the solution. When the

---

[4] The formation of $Cr_2O_7^{2-}$ from CrO_4^{2-} is easier to understand if the equilibrium is written

$$2HCrO_4^- \rightleftharpoons H_2O + Cr_2O_7^{2-}$$

Using Lewis structures, we see that removal of the components of water joins the chromium atoms by an oxygen bridge.

reaction is complete, however, the next drop of $KMnO_4$ solution that is added produces a noticeable color. This signals the endpoint. In effect, then, $KMnO_4$ serves as its own indicator in a titration.

When permanganate ion is reduced in neutral or basic solutions, reduction ceases at the 4+ state with the formation of *manganese dioxide,* MnO_2.

$$MnO_4^- + 2H_2O + 3e^- \longrightarrow MnO_2(s) + 4OH^-$$

Manganese dioxide is a common compound from which many other manganese compounds are made. It tends to be **nonstoichiometric** —the ratio of oxygen to manganese is somewhat variable and is not exactly 2 to 1 as indicated by its formula. Manganese dioxide is found in the manganese nodules scattered on the ocean floor that were described earlier in this chapter. According to one theory, microorganisms may have extracted manganese from the sea water and deposited it as MnO_2 in the nodules.

When MnO_2 is added to molten potassium hydroxide and oxidized with air or potassium nitrate, the green *manganate ion,* MnO_4^{2-}, is formed. This ion is stable only in very basic solutions. When acidified, a portion of the MnO_4^{2-} is oxidized while the rest is reduced. The products are MnO_4^- and MnO_2.

$$3MnO_4^{2-} + 4H^+ \longrightarrow 2MnO_4^- + MnO_2 + 2H_2O$$

A reaction such as this, in which one portion of a substance is oxidized by the rest is called **disproportionation.** In a disproportionation reaction, the same species undergoes both oxidation and reduction.

Iron. You are probably more familiar with this metal than with any other. It is relatively inexpensive, and iron and its alloys have such useful properties that they have been put to more uses than any other metal.

Iron is the second most abundant metal, next to aluminum, and it is the fourth most abundant element in the earth's crust. The molten core of the earth is thought to be composed mostly of iron and nickel. In the pure state, iron is white and lustrous, but it is not especially hard. It is also quite reactive. It reacts with nonoxidizing acids such as HCl or H_2SO_4 to generate hydrogen.

$$Fe(s) + 2H^+(aq) \longrightarrow Fe^{2+}(aq) + H_2(g)$$

Its reaction with moisture and air is particularly bothersome because the corrosion product—rust, $Fe_2O_3 \cdot xH_2O$—doesn't adhere to the metal. Instead, it falls away, exposing fresh metal to attack. The process of corrosion was discussed in some detail in Chapter Eighteen. Besides oxygen, iron also reacts readily with other nonmetals such as chlorine, sulfur, phosphorus, and carbon when it is heated.

Iron forms three oxides, FeO, Fe_3O_4, and Fe_2O_3, which all tend to be nonstoichiometric. Crystalline FeO is difficult to prepare. When heated, it undergoes disproportionation to give Fe and Fe_2O_3. The oxide Fe_3O_4 contains iron in both the 2+ and 3+ oxidation states. It occurs in nature as a magnetic ore called *magnetite.* The magnetic properties of this ore make its separation from the gangue rather easy—it is simply pulled out with a magnet. The high iron content of this oxide makes it a very desirable iron ore. The most common iron ore is the red-orange oxide, Fe_2O_3, which is called *hematite.*

The two principal oxidation states of iron, 2+ and 3+, are both relatively stable. Unlike the transition metals earlier in period 4, iron forms no compounds in an oxidation state that would, in principle, involve all eight of its $3d$ and $4s$ electrons. The highest oxidation state observed for iron is 6+, and that state is rare (and, therefore, unimportant to us). What we see here is an example of how the lower oxidation states of the transition metals become increasingly more stable than the higher ones as we move from left to right across a period.

Cobalt. Cobalt is a hard, bluish white metal that is used mostly in catalysts and

Recall that MnO_2 is one of the reactants in the dry cell.

$$3FeO \xrightarrow{\text{heat}} Fe + Fe_2O_3$$

alloys. For example, it is combined with chromium and tungsten in an alloy called *stellite,* which retains its hardness even when hot. This property makes stellite useful for high-speed cutting tools (e.g., drill bits) used to machine steel.

Chemically, cobalt is somewhat less active than iron, although it dissolves slowly in acids such as HCl. There are two important oxidation states of cobalt, 2+ and 3+. In water, the most stable species is Co^{2+}, which actually exists as the pink $Co(H_2O)_6^{2+}$ ion. The 3+ oxidation state is unstable except in the presence of molecules or anions with which it can form complex ions.

Nickel. Nickel is one of modern technology's most useful metals. In the pure state, it resists corrosion, and metals such as iron or steel are frequently given a thin protective coating of nickel by electrolysis. When added to iron, nickel makes the metal more ductile and resistant to corrosion. We saw earlier that nickel and chromium are the chief additives to iron in making stainless steel. Nickel also makes steel resistant to impact, a property that is particularly desirable in armor plate. Combined with copper, nickel produces an alloy called *monel* that is very resistant to corrosion and is very hard and strong. Propeller shafts made of monel are very desirable in boats that operate in the corrosive environment of sea water. Nickel is also used as a catalyst for hydrogenation of organic compounds that contain double bonds.

Nickel can be purified by an interesting method called the Mond process. When nickel is warmed in the presence of carbon monoxide, it forms a compound called *nickel tetracarbonyl,* which has the formula, $Ni(CO)_4$. One of the properties of this compound is that it is a liquid with a high vapor pressure so that, as it is formed, its vapor is carried away by the stream of carbon monoxide. When the nickel tetracarbonyl vapor is passed over a very hot surface, it decomposes into metallic nickel and carbon monoxide, which is recycled.

Nickel tetracarbonyl is one of a number of compounds formed by carbon monoxide with the transition metals. Another example is $Fe(CO)_5$. These are unusual compounds because the oxidation state of the metal is zero.

Nickel is a moderately active metal, and it dissolves in nonoxidizing acids such as HCl to give Ni^{2+} and H_2. Many nickel salts are green because of the presence of the $Ni(H_2O)_6^{2+}$ ion. Like many other transition metal ions, Ni^{2+} forms a large number of complex ions.

The most stable oxidation state of nickel is the 2+ state. Higher oxidation states are powerful oxidizing agents because of their strong tendency to be reduced to the 2+ state. You may recall that the nickel cadmium cell described in Section 18.11 uses NiO_2 as the cathode. The cell reaction is

$$Cd(s) + NiO_2(s) + 2H_2O \longrightarrow Cd(OH)_2(s) + Ni(OH)_2(s)$$

Copper, Silver, and Gold. Copper, silver, and gold are often called the **coinage metals** because they nave been used for that purpose since ancient times. Each has an outer electron configuration of $(n\text{-}1)d^{10}ns^1$. The loss of the single s-electron gives the 1+ oxidation state, which is the reason for the IB designation for the group. In the case of copper, a second electron can be lost, and many compounds contain Cu^{2+}. Gold loses another two electrons fairly easily, so it tends to form Au^{3+} rather than Au^+.

All of the Group IB metals are commercially valuable. Copper is a metal that every U.S. citizen would recognize because it is used to make the penny. It has a very high electrical and thermal conductivity and, for this reason, is used in electrical wiring. In fact, the largest concentration of copper in the world is said to be the copper wire that lies under the streets of New York City. Copper is also fairly resistant to corrosion and is used widely as pipe to carry hot and cold water in buildings.

Silver has the highest thermal and electrical conductivity of any metal. Its value as a coinage metal, however, makes it too expensive to be used often as an

The "nickel" coin is a copper-nickel alloy.

$Ni(CO)_4$ is extremely toxic.

Nickel compounds added to glass give the glass a green color.

Copper and gold are the only two colored metals.

Cu $3d^{10}4s^1$
Ag $4d^{10}5s^1$
Au $5d^{10}6s^1$

Special Topic 20.2 Photography

One of the very useful properties of silver compounds —particularly the halides—is their sensitivity toward light; they tend to decompose. All practical photographic processes, both black-and-white and color, make use of this fact. Most films and photographic papers are coated with a gelatin emulsion that contains silver bromide or silver bromide mixed with silver iodide. These insoluble halides are prepared by precipitation reactions in the dark. The silver halides are then mixed with the gelatin and applied to one side of the film or paper. In color films and papers, several different layers of emulsions are used that contain special dyes as well as silver bromide.

When photographic film or paper is exposed to light, tiny grains of silver bromide crystals absorb the light energy and become "activated." These crystals are more easily reduced to free metallic silver than are the crystals in unexposed parts of the image. The photographic development process involves treating the film with a mild reducing agent. This causes metallic silver to be deposited at the sites of the activated crystals. The particles of metallic silver are very tiny and because of their size they appear black.

Another step in photographic processing involves removing the unexposed silver bromide from the film. If this were not done, daylight would gradually darken the entire photograph. To remove the silver bromide, the film is soaked in a solution containing thiosulfate ion, $S_2O_3^{2-}$. (Photographers call this solution, *hypo.*) This ion forms a stable, soluble complex ion with silver. The reaction is

$$AgBr(s) + 2S_2O_3^{2-}(aq) \longrightarrow Ag(S_2O_3)_2^{3-}(aq) + Br^-(aq)$$

After it is formed, the complex ion is washed from the emulsion.

Although other complex reactions are involved in processing color film, reduction of the activated crystals of silver halide and the removal of the unexposed silver halide by thiosulfate are key steps in producing the photographic image.

How mirrors are silvered is discussed on page 305.

electrical conductor. Silver has a very high luster and, when polished, reflects light very well. This has made it valuable for jewelry and for coating the backs of mirrors. One of silver's most important applications is in photography. Silver compounds tend to be unstable and sensitive to light. (See Special Topic 20.2.)

Everyone knows of gold's value as bullion and as a decorative metal for jewelry. As mentioned earlier, gold is also used occasionally to plate electrical contacts because of its low chemical reactivity. Pure gold is very soft and is particularly ductile and malleable. Gold leaf, used for decorative lettering in signs, is made by pounding gold into very thin sheets. It is so thin that some light is able to pass through it. A stack of 11,000 gold leaflets is only 1 mm thick.

Free elemental copper, called *native copper,* is found in the upper part of Michigan.

In general, the elements of Group IB are less reactive than the other metals that we've discussed so far, and their reactivity decreases going down the group. All three are found as free metals in nature, although copper and silver are usually found in compounds. Although the free metals have little tendency to react with oxygen, metallic silver has a strong affinity for sulfur. Silverware tarnishes by reaction of the metal with small amounts of H_2S in the air to produce black Ag_2S.

The reduction potentials of copper, silver, and gold are positive, which means that they will not dissolve in nonoxidizing acids such as hydrochloric acid or dilute sulfuric acid with the evolution of hydrogen. Copper and silver do dissolve in oxidizing acids such as nitric acid. In this case, the metal is oxidized by the nitrate ion, which is reduced to one or more of the oxides of nitrogen.

$$Cu + 4H^+ + 2NO_3^- \longrightarrow Cu^{2+} + 2NO_2 + 2H_2O$$
$$Ag + 2H^+ + NO_3^- \longrightarrow Ag^+ + NO_2 + H_2O$$

Aqua regia consists of one part of concentrated HNO_3 and three parts concentrated HCl.

Gold is so unreactive that even concentrated HNO_3 fails to attack it. A mixture of HNO_3 and HCl, called **aqua regia** by the alchemists, will dissolve gold slowly. The chloride ion helps the nitrate ion to oxidize the gold by stabilizing the Au^{3+} in the form of a complex ion with the formula, $AuCl_4^-$.

$$Au + 6H^+ + 3NO_3^- + 4Cl^- \longrightarrow AuCl_4^- + 3NO_2 + 3H_2O$$

Each of the elements of Group IB has a strong tendency to form complex ions. Many copper compounds are blue because they contain the $Cu(H_2O)_4^{2+}$ ion. An example is copper sulfate, which forms blue crystals with the composition $CuSO_4 \cdot 5H_2O$. When heated, these crystals lose water and crumble into a white powder—anhydrous $CuSO_4$. This powder regains moisture if left exposed to the air and turns blue again. Addition of ammonia to a pale blue solution containing $Cu(H_2O)_4^{2+}$ gives the deep blue $Cu(NH_3)_4^{2+}$ ion.

One of the water molecules in $CuSO_4 \cdot 5H_2O$ is hydrogen bonded to the sulfate ion.

$$Cu(H_2O)_4^{2+} + 4NH_3 \rightleftharpoons Cu(NH_3)_4^{2+} + 4H_2O$$
$$\text{pale blue} \qquad\qquad\qquad \text{deep blue}$$

This reaction is often used to test for the presence of copper because the intense blue of the ammonia complex is much more easily seen than the pale blue of the $Cu(H_2O)_4^{2+}$.

Complex ions are also used in the test for silver. The silver halides (except AgF) are insoluble in water, so addition of chloride ion to a solution containing Ag^+ precipitates AgCl. If ammonia is then added, the silver chloride dissolves, and the complex ion, $Ag(NH_3)_2^+$ is formed. The equilibria involved are

Silver iodide is used to "seed" clouds to bring on rain.

$$AgCl(s) \rightleftharpoons Ag^+ + Cl^-$$
$$Ag^+ + 2NH_3 \rightleftharpoons Ag(NH_3)_2^+$$

As NH_3 is added, the second equilibrium is shifted to the right, which decreases the Ag^+ concentration. This causes the first equilibrium to shift to the right in an attempt to replenish the Ag^+. As a result, silver chloride dissolves.

The final step in the test for Ag^+ is to acidify the solution containing $Ag(NH_3)_2^+$. Addition of the acid removes NH_3 by forming ammonium ion.

$$NH_3 + H^+ \longrightarrow NH_4^+$$

The $Ag(NH_3)_2^+$ decomposes to replace the NH_3, which causes the concentration of free Ag^+ in the solution to rise. As the Ag^+ is produced, it recombines with the chloride ion in the solution and AgCl is reprecipitated.

Zinc, Cadmium, and Mercury. The elements of Group IIB all have completed d subshells beneath their outer pair of s electrons. When they react, only their outer s electrons are involved. Zinc and cadmium are both quite reactive and form ions with a charge of 2+. This is their only oxidation state. Mercury is less reactive than either zinc or cadmium and is found in two oxidation states, 1+ and 2+. The 1+ state is characterized by the dimeric ion, Hg_2^{2+}. It consists of two mercury atoms joined by a covalent bond and carries a net charge of 2+.

Zn $3d^{10}4s^2$
Cd $4d^{10}5s^2$
Hg $5d^{10}6s^2$

Zinc is one of industry's most important metals. It occurs in the ore, *zinc blende,* which is a zinc sulfide (ZnS) ore. The zinc is recovered by first roasting the ore to give zinc oxide, which is then reduced with carbon. The metal is frequently used as a protective coating over other metals, particularly steel. Zinc is a very reactive metal, and it combines with moisture and carbon dioxide to form a combined hydroxide-carbonate coating called a basic carbonate, $Zn_2(OH)_2CO_3$. This coating protects the metallic zinc beneath from further corrosion. Since zinc is more easily oxidized than iron, if the zinc coating is scratched or partially worn away, it is the zinc that is oxidized in preference to the iron—the two metals form a galvanic cell in which iron is the cathode and zinc is the anode. Zinc is also used in various alloys. Examples are brass—an alloy of copper and zinc—and bronze—an alloy containing copper, tin, and zinc. Another common use of zinc is in the manufacture of the zinc-carbon dry cell (the common flashlight battery).

Coating iron with zinc is called galvanizing.

This is another example of cathodic protection.

Cadmium is less abundant than zinc and is usually found as an impurity in zinc ores. The free metal is soft and moderately active. It is used as a protective coating over some metals and in making nickel-cadmium batteries.

Beads of mercury.

Cadmium and mercury are both toxic, but zinc is needed by the body.

Mercury is important because it is a liquid at temperatures both above and below room temperature (it melts at $-38.9\ °C$ and boils at $357\ °C$). You've seen it as the silvery liquid in thermometers. It occurs naturally in an ore called *cinnabar*, HgS. Roasting of the ore gives the oxide which is easily decomposed to give the free metal.

Metallic mercury is able to dissolve many other metals. The solutions that are formed are called *amalgams*. Dentists use a silver amalgam (containing an excess of silver) to fill teeth. Gold also forms amalgams easily, and mercury is used to separate gold from its ores. The ore is mixed with mercury which dissolves the gold. Because mercury is so dense, the rocks and other debris float on its surface and can be easily removed. Then the mercury is distilled away, leaving the gold behind. The mercury is then condensed and used again.

Chemically, zinc and cadmium are quite similar in many respects, but they differ considerably from mercury. For example, both zinc and cadmium readily dissolve in dilute nonoxidizing acids such as HCl or H_2SO_4, but mercury is unaffected by these acids. One of the main differences between zinc and cadmium is that zinc is amphoteric. Solutions containing Zn^{2+} give $Zn(OH)_2$ when a base is added. The hydroxide dissolves, however, if the mixture is made even more basic.

$$Zn(OH)_2(s) + 2OH^-(aq) \longrightarrow Zn(OH)_4^{2-}(aq)$$

The metal itself dissolves in base with the evolution of hydrogen.

$$Zn(s) + 2OH^-(aq) + 2H_2O \longrightarrow Zn(OH)_4^{2-}(aq) + H_2(g)$$

Cadmium, however, is unaffected by base. For this reason, cadmium can be used as a protective coating over metals that must be exposed to basic conditions.

Zinc and cadmium form many complex ions. For example, both $Zn(OH)_2$ and $Cd(OH)_2$ dissolve in concentrated aqueous ammonia to form complex ions having the general formula, $M(NH_3)_4^{2+}$.

$$Zn(OH)_2 + 4NH_3 \rightleftharpoons Zn(NH_3)_4^{2+} + 2OH^-$$
$$Cd(OH)_2 + 4NH_3 \rightleftharpoons Cd(NH_3)_4^{2+} + 2OH^-$$

Zinc oxide ointment is used as a sunscreen.

Hg$_2$Cl$_2$ has been used in agriculture to control root maggots on onions and cabbage.

Many zinc compounds are very useful. Zinc chloride, which is exceptionally soluble in water (38.1 mol/liter at $25\ °C$), has a range of uses that extend from embalming, to fireproofing lumber, to the refining of petroleum. Zinc oxide, a white powder, is used in various creams (sun screens) and to make quick-setting dental cements. Zinc sulfide is interesting because it can be used to prepare *phosphors*—substances that glow when bathed in ultraviolet light or cathode rays (high energy electrons). Such phosphors are used, for example, on the inner surface of television picture tubes and in devices for detecting atomic radiations (Section 21.3).

Mercury compounds are much less ionic than those of zinc and cadmium. For example, mercury(II) chloride (mercuric chloride), $HgCl_2$, is molecular and does not dissociate fully in water. It is actually a weak electrolyte and constitutes an exception to the rule that metal-nonmetal compounds dissociate completely when they are dissolved in water.

Mercury(I) chloride (mercurous chloride), Hg_2Cl_2, is very insoluble in water. Its low solubility has permitted therapeutic uses as an antiseptic and as a treatment for syphilis before the discovery of penicillin. Very little mercury is retained by the body because so little of the Hg_2Cl_2 is able to dissolve. Mercury(II) chloride, however, is quite soluble in water and is very toxic—one or two grams is generally fatal.

20.7 COMPLEX IONS

Living organisms and many common consumer products make use of complex ions formed from metal ions and a host of different molecules and anions

In the last section, we saw a number of examples of complex ions of the transition metals.[5] These substances, you recall, are formed when neutral molecules or anions become bonded to a metal ion. Examples are $Cu(NH_3)_4^{2+}$, $Ag(NH_3)_2^+$, and $Zn(OH)_4^{2-}$. As stated earlier, the number of such complexes is huge, and the study of their properties, reactions, structures, and bonding is a major specialty of chemistry.

One of the reasons why there is so much interest in complex ions is that they are important in many chemical systems that we encounter every day. For instance, many common household products and food additives are effective because they form complex ions with metals. Furthermore, most metal ions that play an active role in biochemical reactions are found to be at the center of complex ions. Two examples are the iron that is found in the hemoglobin in our blood and the magnesium that is present in the chlorophyll in green plants.

In this section we will study some of the biological and commercial applications of complex ions and we will learn about the kinds of molecules and anions that form complex ions with metals. Before we can do this, however, we must learn some of the terminology that is used in discussing these substances.

In general, molecules and anions that can form bonds to metal ions are called **ligands.** For example, ammonia molecules are the ligands in the complex, $Cu(NH_3)_4^{2+}$. Similarly, hydroxide ions are the ligands in $Zn(OH)_4^{2-}$. Notice that in the formulas of these complex ions, the metal ion is written first, followed by the ligands. This sequence is always followed.

Sometimes the formula of a complex ion is written with the metal ion and its ligands enclosed within brackets and the charge written outside—for example, $[Cu(NH_3)_4]^{2+}$. The purpose of this is to emphasize that the ligands are attached to the metal ion and are not free to roam about. Brackets are frequently used when the complex can be isolated as a salt. For instance, chromium forms a complex ion with the formula $Cr(H_2O)_5Cl^{2+}$ that can be isolated as a chloride salt. The formula of the salt is written as $[Cr(H_2O)_5Cl]Cl_2$, which clearly shows that five water molecules and a chloride ion are attached to the chromium ion and that the other two chloride ions in the salt provide electrical neutrality for the compound.

As a rule, ligands are Lewis bases—that is, they have one or more unshared pairs of electrons that they can use to form coordinate covalent bonds. In fact, complex ions and their salts are often called **coordination compounds** because one way to view the bonds between the metal ion and the ligands is as coordinate covalent bonds. Consider, for instance, the formation of the complex $Cu(NH_3)_4^{2+}$.

$$Cu^{2+} + 4NH_3 \longrightarrow Cu(NH_3)_4^{2+}$$

Each ammonia molecule has a lone pair of electrons that it can donate to the copper ion.

[5] Although most of the complex ions that we will discuss are formed by transition metals, many of the representative metals form complexes too. An example is the ion $Al(H_2O)_6^{3+}$, which was discussed in Section 20.3.

This concept of the bonding in complex ions has been largely replaced by new and better theories, as we will see in Section 20.10. However, the words that chemists use in discussing complex ions still retain much of the "flavor" of the coordinate covalent bond. We speak of ligands as being "coordinated" to a metal ion, and we will later define a term called *coordination number*. We also refer to the atom that a ligand uses to attach itself to the metal ion as the *donor atom*. Nitrogen, for example, is the donor atom in the ammonia molecule.

Ligands

Ligands are generally classified according to the number of atoms that they contain that are able to bind to a metal. Substances with only one such atom are called **monodentate ligands** because they have only one "tooth" with which to "bite" a metal ion. Common monatomic anions that serve as ligands include the halide ions—fluoride, chloride, bromide and iodide. Many polyatomic anions are also monodentate ligands. Examples are nitrite (NO_2^-), cyanide (CN^-), hydroxide (OH^-), thiocyanate (SCN^-), and thiosulfate ($S_2O_3^{2-}$).

Monatomic anions, of course, must be monodentate.

The most common molecule that is a monodentate ligand is water. Whenever the salt of a metal dissolves in water, it dissociates into its ions, which are immediately surrounded by molecules of water. Often, these water molecules become rather firmly attached to the metal ion, and a complex ion is formed. We've seen this earlier with the aluminum ion, which forms $Al(H_2O)_6^{3+}$, and with copper(II) ion, which forms $Cu(H_2O)_4^{2+}$. When salts of these ions are crystallized from water, the complex ions remain intact and become part of the crystal.

The chemical properties of most metal ions are really the chemical properties of their complex ions.

Ammonia is another common monodentate ligand. As we've seen, the lone pair of electrons on the NH_3 molecule enable it to coordinate to metal ions. Most nitrogen compounds have a lone pair of electrons on the nitrogen atom and are thus able to serve as ligands. This is particularly true for certain important biological molecules, as we will see shortly.

One of the things that provides so much variation and complexity in the study of coordination compounds is the existence of many ligands that have two or more atoms that can simultaneously become attached to a metal ion. These are called **polydentate ligands.** When they bind to a metal ion, a ring structure is formed. This is shown below for two common **bidentate ligands**—ligands with two donor atoms—ethylenediamine and oxalate ion.

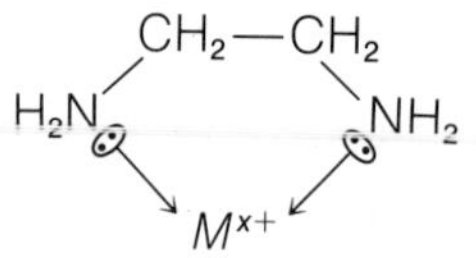

ethylenediamine
(usually abbreviated, en)

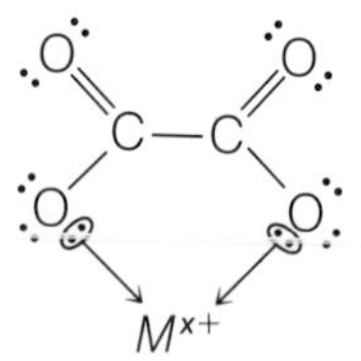

oxalate

Complex ions containing rings of atoms formed by attaching polydentate ligands to a metal ion are called **chelates,** from the Greek, *chele,* meaning crab's claw. The ligand "grasps" the metal ion somewhat like a crab grasps its prey.

No one can say that chemists have no imagination!

The formation of chelate rings by polydentate ligands produces complexes that are nearly always considerably more stable than those formed with similar monodentate ligands. For example, $Ni(en)_3^{2+}$ is much more stable than $Ni(NH_3)_6^{2+}$, even though both complexes have six nitrogen atoms bound to a Ni^{2+} ion. This phenomenon is called the

The reason for the chelate effect appears to be related to the probability of the ligand becoming removed from the vicinity of the metal ion. If one end of a bidentate ligand comes loose from the metal ion, the donor atom cannot wander very far because the other end of the ligand is still attached to the metal. There is a high probability that the loose end will become reattached to the metal ion before the other end can let go, so the ligand appears to be bound tightly. With a monodentate ligand, however, there is nothing to hold the ligand near the metal ion if it becomes detached. The ligand can easily wander off into the surrounding solution and be lost. As a result, a monodentate ligand doesn't behave as if it is as firmly attached to the metal ion as a polydentate ligand.

There are many ligand molecules that contain more than two donor atoms. Their ring structures are even more complex than those formed by bidentate ligands. One such substance that has many applications is ethylenediaminetetraacetic acid, generally abbreviated *EDTA*.

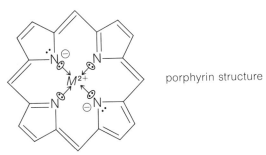

(Colored atoms are the donor atoms.)

EDTA

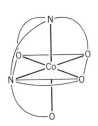

Figure 20.12
The structure of the Co(EDTA)⁻ ion. To simplify the presentation of the structure, the atoms that join the donor atoms are indicated by curved lines. Thus the line connecting the two nitrogen atoms stands for the atoms —CH₂—CH₂—, and the line that connects an oxygen and a nitrogen stands for the atoms

$$\begin{array}{c} O \\ \parallel \\ -CH_2-C-. \end{array}$$

Both nitrogens of this ligand serve as donor atoms, and when the hydrogen atoms of the carboxyl groups ($-\overset{\overset{\displaystyle O}{\parallel}}{C}-OH$) are removed as H^+, the oxygens that become negatively charged can also serve as donor atoms. Thus the $EDTA^{4-}$ ion has six donor atoms and very effectively wraps itself around metal ions. The structure of an EDTA complex of cobalt, formed by the reaction

$$Co^{3+} + EDTA^{4-} \longrightarrow Co(EDTA)^-$$

is illustrated in Figure 20.12.

EDTA is an important ligand because it is not particularly toxic and can therefore be used in foods to tie up metal ions that tend to catalyze oxidation and spoilage. If you look at the labels of many salad dressings, you will find that one of the ingredients is $CaNa_2EDTA$. The calcium salt is used to prevent the EDTA from extracting Ca^{2+} from the body. Many shampoos contain Na_4EDTA to aid in softening water. The EDTA binds ions such as Ca^{2+} that are present in hard water and prevents them from interfering with the action of the soaps in the shampoo. EDTA is also added in small amounts to whole blood to prevent clotting, which would be hastened by the presence of free Ca^{2+} ions. EDTA can also be used to remove toxic metal ions such as Pb^{2+} from the body.

A polydentate ligand structure that is found in many biologically important complexes is the *porphyrin structure*. It provides a square planar arrangement of four nitrogen atoms that can hold a metal ion in the center.

porphyrin structure

This same basic ligand structure, with only minor modifications and different organic side chains attached to the outside of the ring network, is found in hemoglobin, myoglobin, chlorophyll, and vitamin B_{12} (also called cobalamin). In hemoglobin (Figures 19.3 and 19.4) and in myoglobin there is an Fe^{2+} ion in the center of the ring. The porphyrin structure that results is called a *heme* group. In chlorophyll, the metal ion in the ring is Mg^{2+}; and in vitamin B_{12}, the metal ion is Co^{2+}. Figure 20.13 illustrates the structure of cyanocobalamin, the form of vitamin B_{12} found in vitamin pills.

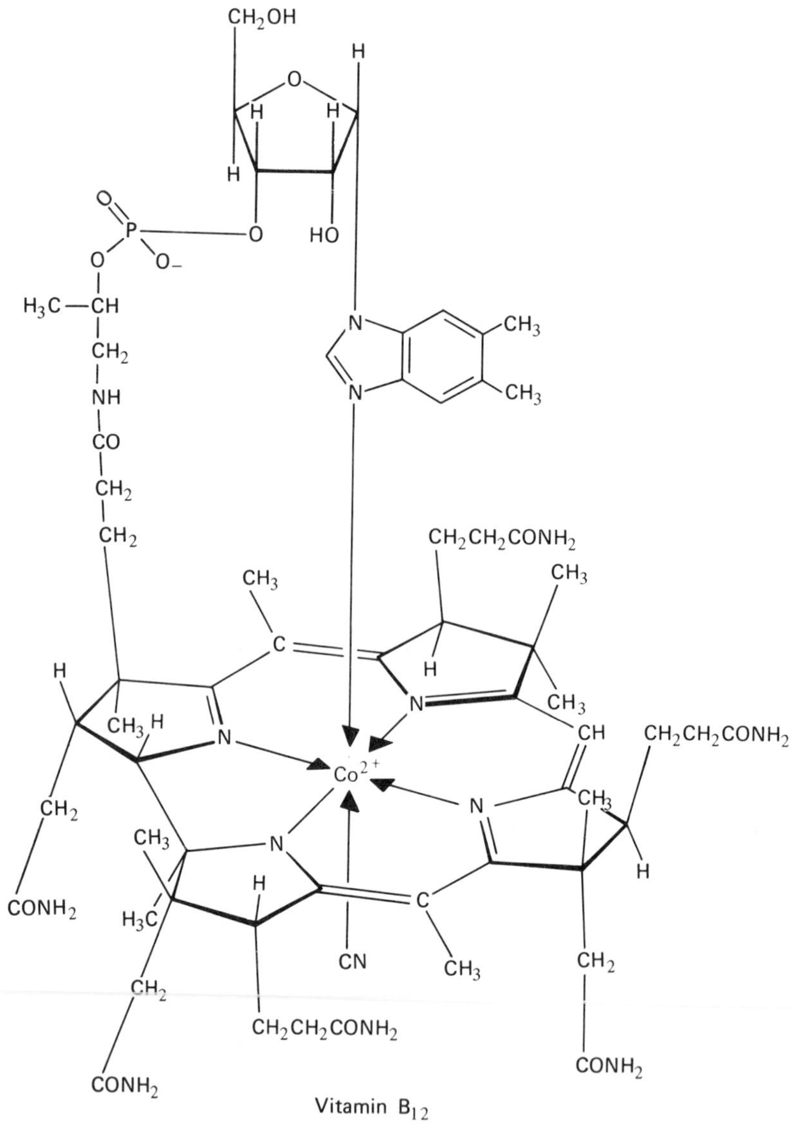

Figure 20.13
The structure of cyanocobalamin. Notice the cobalt atom in the center of the square planar arrangement of nitrogen atoms.

Vitamin B_{12}

20.8 COORDINATION NUMBER AND STRUCTURE

The structure of a complex ion depends on the number of ligand donor atoms that are bound to the metal ion

The number of donor atoms that can surround a metal ion in a complex is referred to as the ion's **coordination number.** It is determined by a number of factors, including the electronic structure and charge of the metal ion and the nature of the ligands. The coordination number can vary from one metal to another and can even differ for the same metal. One of the properties of a complex ion that depends on the coordination number is its structure. Although a variety of coordination numbers are observed for different complexes, we will only examine coordination numbers of four and six, which are the most important.

Two common geometries occur when four ligand atoms are bonded to a metal ion—tetrahedral and square planar. These are illustrated in Figure 20.14. The tetrahedral geometry is usually found with metal ions that have completed d subshells such as Zn^{2+}. The complexes $Zn(NH_3)_4^{2+}$ and $Zn(OH)_4^{2-}$ are examples.

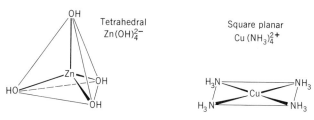

Figure 20.14
Tetrahedral and square planar geometries that occur for complexes in which the metal has a coordination number of four.

Square planar geometries are observed for complexes of Cu^{2+}, Ni^{2+}, and especially Pt^{2+}. Examples are $Cu(NH_3)_4^{2+}$, $Ni(CN)_4^{2-}$ and $PtCl_4^{2-}$. Because they are considerably more stable than the others, the most well-studied square planar complexes are those of Pt^{2+}.

The most common coordination number for complex ions is six. Examples are $Al(H_2O)_6^{3+}$, $Ni(en)_3^{2+}$, and $Co(EDTA)^-$. With few exceptions, all complexes with a coordination number of six are octahedral. This holds true for those formed from both monodentate and polydentate ligands, as illustrated in Figure 20.15. In describing the shapes of complexes, most chemists use the simplified drawing of the octahedron shown in Figure 20.16.

Studying complex ions is easier if you are able to draw their structures.

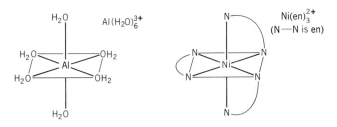

Figure 20.15
Octahedral complexes can be formed with either monodentate ligands or polydentate ligands.

Figure 20.16
Drawing the structure of an octahedral complex.

Step 1. Draw a parallelogram to represent a square viewed in perspective.	*Step 2.* Draw a vertical line upward from the center.	*Step 3.* Draw a vertical line of the same length downward. Erase the portion that would be hidden by the plane.	*Step 4.* Draw four lines to the center of the "square"	*Step 5.* Fill in the symbols for the atoms. For $CrCl_6^{3-}$ we would have the following structure.

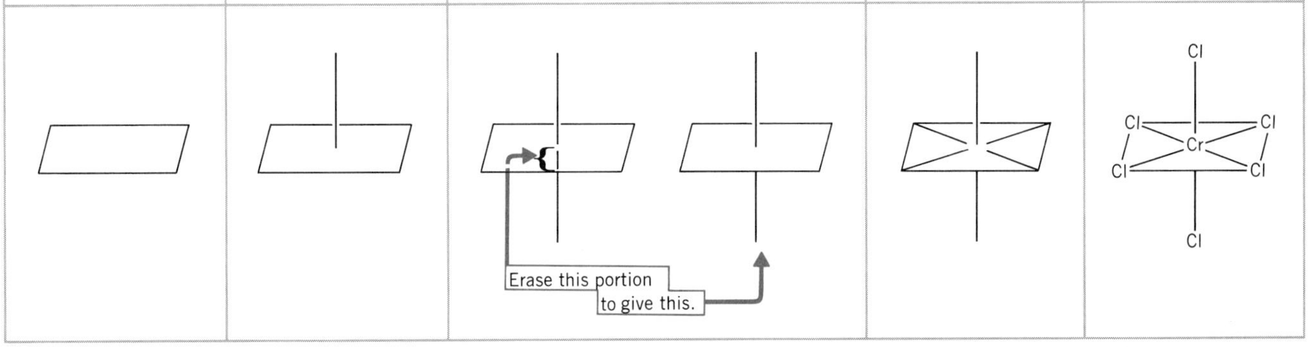

20.9 ISOMERS OF COORDINATION COMPOUNDS

Several kinds of isomerism are possible for coordination compounds, including geometric and optical isomerism

In Chapter Nine, during our discussion of organic chemistry, we learned that it is possible for two substances to have the same chemical formula but distinctly different properties. Such substances, you recall, are said to be *isomers*. Isomerism is not a phenomenon that is restricted to organic compounds. For example, three different solids, each with its own characteristic color and other properties, can be isolated from a solution of chromium(III) chloride. All three have the same formula, $CrCl_3 \cdot 6H_2O$. Experiments have shown that these solids are actually the salts of three different complex ions. Their formulas are

$[Cr(H_2O)_6]Cl_3$	violet
$[Cr(H_2O)_5Cl]Cl_2 \cdot H_2O$	blue-green
$[Cr(H_2O)_4Cl_2]Cl \cdot 2H_2O$	green

There are more ways for isomerism to occur for complex ions than for organic compounds. Several reasons for this exist. In organic compounds, carbon atoms hold either two, three, or four atoms but, in complex ions, coordination numbers ranging from two to more than eight are observed. Also, among the higher coordination numbers—for example, the common coordination number of six—the number of possible arrangements of ligands is greater than when only four groups are attached to carbon. Another reason is that complex ions are isolated as salts. They have both a cation and an anion that can contribute to the number of isomers, as in $CrCl_3 \cdot 6H_2O$ above. Carbon compounds, on the other hand, are almost always neutral molecules, which eliminates this possibility.

The most interesting kind of isomerism among coordination compounds is stereoisomerism—differences among compounds that arise as a result of the various possible orientations of their atoms in space. One form of stereoisomerism is geometric isomerism. As an example, consider the square planar complexes having the formula $Pt(NH_3)_2Cl_2$. There are two ways to arrange the ligands around the platinum. In one isomer, the *cis* isomer, the chloride ions are next to each other and the ammonia molecules are next to each other. In the other isomer, the *trans* isomer, identical ligands are opposite each other.

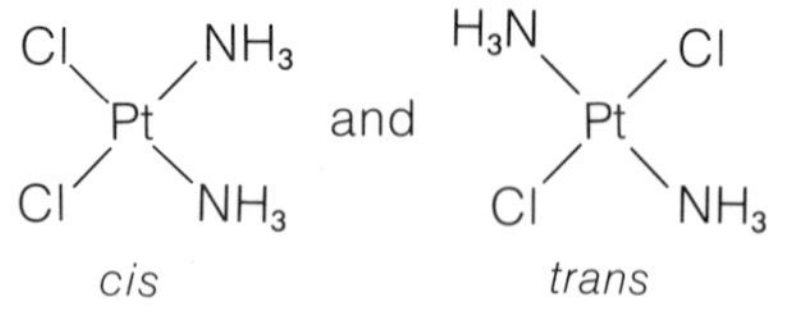

cis   and   trans

Notice that the terms *cis* and *trans* express the same kinds of structural information for complex ions as they do for organic compounds. *Cis* means "on the same side" and *trans* means "on opposite sides."

Geometric isomerism also occurs for octahedral complexes. For example, consider the ions $Cr(H_2O)_4Cl_2^+$ and $Cr(en)_2Cl_2^+$. Both can be isolated as *cis* and *trans* isomers.

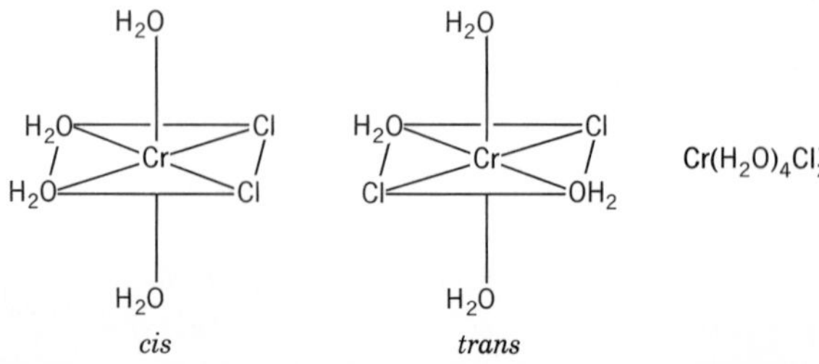

cis   trans $Cr(H_2O)_4Cl_2^+$

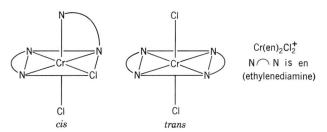

In the *cis* isomer, the chloride ligands are both on the same side of the metal ion; in the *trans* isomer, the chloride ligands are on opposite ends of a line that passes through the center of the metal ion.

The second type of stereoisomerism observed for complex ions is optical isomerism; certain complex ions are chiral just as certain organic compounds are chiral. In Section 19.2, we saw that a chiral structure is not superimposable with its mirror image. Two isomers exist that bear the same relationship to each other as your left hand bears to your right hand. They are almost—but not quite—the same; their mirror images do not fit over each other exactly.

For organic compounds, we saw that there is a simple test for the existence of chirality. If four different groups are attached to carbon, the structure is chiral. There is no such simple test for complex ions, however, and the structures must be analyzed by comparing them to their mirror images. If a complex and its mirror image are superimposable, both are exactly the same and the complex isn't chiral. However, if they are nonsuperimposable, they are optical isomers.

The most common examples of chirality among coordination compounds occur with octahedral complexes that contain two or three bidentate ligands—for instance, $Co(en)_2Cl_2^+$ and $Co(en)_3^{3+}$. For the complex, $Co(en)_3^{3+}$, the two enantiomers are shown in Figure 20.17. For the complex, $Co(en)_2Cl_2^+$, only the *cis* isomer is chiral, as described in Figure 20.18.

Enantiomers are nonsuperimposable mirror-image isomers (Section 19.2).

Figure 20.17
The two isomers of $Co(en)_3^{3+}$. One is the mirror image of the other, but they cannot be superimposed no matter how they are rotated.

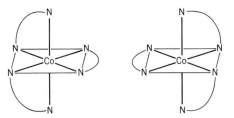

Figure 20.18
Isomers of $Co(en)_2Cl_2^+$. The mirror image of the *trans* isomer can be superimposed exactly on the original. The *cis* isomer is chiral, however, because its mirror image cannot be superimposed on the original.

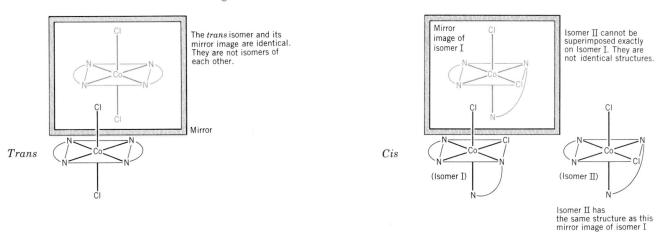

20.10 BONDING IN COMPLEX IONS

The colors and magnetic properties of complex ions are explained by the way that the ligands affect the energies of the *d* orbitals

Among the most characteristic properties of the complex ions of the transition metals are their colors and magnetic properties. A given metal ion, for example, can form complexes with different ligands to give a rainbow of colors. This is shown in Color Plate 8C for some of the complexes of cobalt. Because the transition metal ions generally have incompletely filled *d* subshells, we expect to find many of them with unpaired electrons. Their ions should therefore be paramagnetic. But for a given metal ion, the number of unpaired electrons is not always the same from one complex to another. For instance, Fe^{2+} has four of its six $3d$ electrons unpaired in $Fe(H_2O)_6^{3+}$, but all six of its $3d$ electrons are paired in $Fe(CN)_6^{4-}$.

To be successful, any theory that attempts to explain the bonding in complex ions must also explain their colors and magnetic properties. One of the simplest theories that does this is the **crystal field theory.** (The name of the theory comes from its original use in explaining the behavior of metal ions in crystals. It was discovered later that the theory works well for complex ions, too.)

Crystal field theory considers the effects of the ligands on the energies of the *d* orbitals of the central metal ion. Before we can proceed further, therefore, we must examine the shapes and directional properties of the *d* orbitals. These are shown in Figure 20.19. We can see that *d* orbitals are more complex than *p* orbitals. Four of the *d* orbitals have four lobes. The fifth one, the d_{z^2} orbital, has two lobes that point along the *z* axis plus a doughnut-shaped electron cloud around the center.

Of primary importance to us are the directions in which the lobes of the *d* orbitals point. Notice that three of the *d* orbitals—those labeled d_{xy}, d_{xz}, and d_{yz}—have their lobes pointing between the *x*, *y*, and *z* axes. The other two—the $d_{x^2-y^2}$ and the d_{z^2} orbitals—concentrate their electron density directly along the *x*, *y*, and *z* axes.

The labels for the d-orbitals come from the mathematics of quantum mechanics.

In an atom or an isolated ion, all of the *d* orbitals of a given subshell have the same energy. This means that an electron has the same energy regardless of which *d* orbital it occupies. What we wish to study now is the way that the energies of the orbitals change when a complex ion is formed. Let's suppose that we form an octahedral complex by bringing six ligands toward the metal ion along the *xyz* axes. In general, ligands are either negative ions or they are dipoles with their negative ends pointing toward the metal. Therefore, any electrons that are in the *d* orbitals that point at the ligands—electrons in the $d_{x^2-y^2}$

Figure 20.19
The shapes and directional properties of the five *d* orbitals of a *d* subshell.

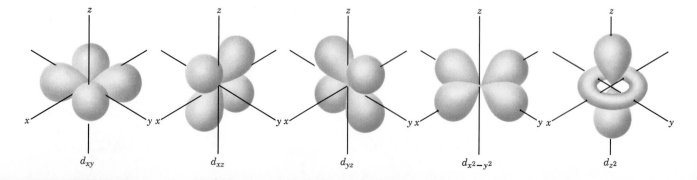

d_{xy} d_{xz} d_{yz} $d_{x^2-y^2}$ d_{z^2}

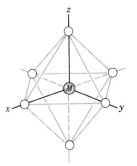

An octahedral complex ion with ligands along the x, y, and z axes

and d_{z^2} orbitals—will be strongly repelled by the negative charge of the ligands and their energy will increase. Electrons that are in d orbitals that point between the ligands—electrons in the d_{xy}, d_{xz}, and d_{yz} orbitals—will be repelled less strongly. Their energy will increase, too, but not as much. As a result of these repulsions, the energies of the electrons in the $d_{x^2-y^2}$ and d_{z^2} orbitals will be raised *more* than the energies of the electrons in the d_{xy}, d_{xz}, and d_{yz} orbitals, as shown in Figure 20.20.

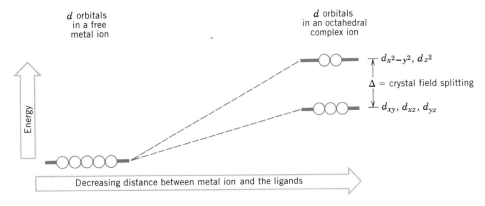

Figure 20.20
The change in the energies of the d orbitals as an octahedral complex is formed.

In Figure 20.20, we see that there are two sets of d orbitals in an octehedral complex that have different energies. The difference in energy between them, which is called the **crystal field splitting,** is usually specified by the Greek letter, Δ (delta). The magnitude of Δ in a given complex depends on several factors, including the charge on the metal ion, the distance between the metal ion and the ligands, and the nature of the ligands themselves.

Now let's look at a specific example that illustrates how crystal field theory is applied. Consider the complex $Cr(H_2O)_6^{3+}$, in which chromium is in the 3 + oxidation state. The electron configuration of a chromium atom is

$$Cr \quad [Ar]3d^5 4s^1$$

Removing three electrons to give Cr^{3+} produces

$$Cr^{3+} \quad [Ar]3d^3$$

In the $Cr(H_2O)_6^{3+}$ ion, these three 3d electrons will occupy orbitals that give the complex the lowest possible energy, which means that they will occupy the lower set of three orbitals (d_{xy}, d_{xz}, d_{yz}). As usual, the electrons must follow Hund's rule and the Pauli exclusion principle, as shown in Figure 20.21a.

Figure 20.21
(a) Electron distribution in the $Cr(H_2O)_6^{3+}$ ion. (b) Light energy raises an electron from the lower energy set of orbitals to the higher energy set.

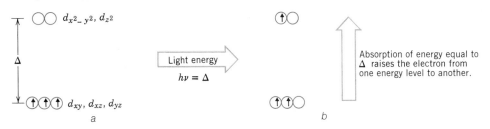

$E = h\nu$

Cr(NH₃)₆³⁺ absorbs blue light and its solutions appear yellow.

When light is absorbed by this complex ion, the energy of the light wave raises an electron from the lower-energy set of d orbitals to the higher-energy set, as shown in Figure 20.21b. The amount of energy that is required to do this depends on the magnitude of the crystal field splitting, Δ. Since the energy of a light wave is related to its frequency, the frequency (and color) of the light absorbed also depends on the magnitude of Δ.

The ion $Cr(H_2O)_6^{3+}$ absorbs light that has a frequency of 5.22×10^{14} Hz, which is the frequency of yellow light. When this color is removed from white light, the color of the light that remains is violet. For this reason, solutions containing $Cr(H_2O)_6^{3+}$ have a violet color.

Changing the ligands that are attached to a metal ion changes the magnitude of the crystal field splitting. For example, replacing H_2O by NH_3 gives the ion, $Cr(NH_3)_6^{3+}$. This ion absorbs light of a higher frequency, and therefore of a higher energy. This means that Δ in $Cr(NH_3)_6^{3+}$ is larger than Δ in $Cr(H_2O)_6^{3+}$. Because the crystal field splitting produced by various ligands is different, the same metal ion is able to form a variety of complexes having a large range of colors.

A ligand that produces a large Δ in complexes with one metal ion also produces a large Δ in complexes with another metal. For example, the cyanide ion is a very effective ligand and always gives a very large Δ, regardless of the metal ion to which it is bound. Ammonia is less effective than cyanide, but more effective than water. Thus, ligands can be arranged in order of their effectiveness at producing a large crystal field splitting. This sequence is called the **spectrochemical series.** Such a series containing some common ligands arranged in order of their decreasing strength is

$$CN^- > en > NH_3 > H_2O > C_2O_4^{2-} > OH^- > F^- > Cl^- > Br^- > I^-$$

For a given metal ion, cyanide ion produces the largest Δ and iodide ion produces the smallest Δ.

Finally, let's see how crystal field theory explains the magnetic properties of complex ions. Suppose that we have an octahedral complex that has four d electrons. The first three electrons can be placed into the lower set of three orbitals, as shown for Cr^{3+} in Figure 20.19. When we come to the fourth electron, however, we must decide whether to place it with the others in the lower energy set of orbitals or to keep it unpaired and place it into the higher set.[6] Placing the fourth electron in the lower level tends to give it a lower energy (greater stability), but some of this stability is lost because it takes energy to force two electrons to become paired and occupy the same orbital. On the other hand, placing the fourth electron into a higher-energy orbital relieves us of the burden of pairing the electrons, but it also tends to give the electron a higher energy. Thus "pairing" and "placement" work in opposite directions in the way they affect the energy of the electron arrangement in the complex.

The critical factor in determining whether the fourth electron enters the lower level and becomes paired with an electron that already occupies one of the orbitals, or whether it enters the upper level with the same spin as the other d electrons is the magnitude of Δ. If Δ is large, the greater stability is achieved if the electrons are paired in the lower level. If Δ is small, however, the greater stability results when the electrons are unpaired and occupy separate orbitals. This is illustrated in Figure 20.22.

At the beginning of this section we stated that the $Fe(H_2O)_6^{2+}$ ion has four unpaired electrons but the $Fe(CN)_6^{4-}$ ion has no unpaired electrons. Now we can explain why this is so. The electron configuration of Fe^{2+} is

$$Fe^{2+} \quad [Ar]3d^6$$

---

[6] We've never had to make this kind of decision before because the energies of the levels in atoms were always widely spaced. In complex ions, however, the two d-orbital energy levels are fairly close in energy.

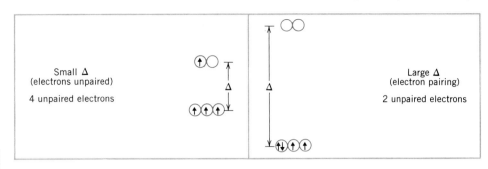

Figure 20.22
The effect of Δ on the electron distribution in a complex with four d electrons. When Δ is small, the electrons remain unpaired. When Δ is large, the lower energy level becomes filled as electrons are paired.

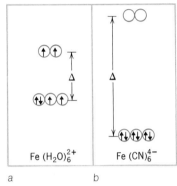

Figure 20.23
Distribution of d electrons in $Fe(H_2O)_6^{2+}$ and $Fe(CN)_6^{4-}$. The magnitude of Δ for the cyanide complex is greater than for the water complex. This allows maximum pairing of electrons in $Fe(CN)_6^{4-}$.

Water is a ligand that produces only a moderately strong crystal field splitting, so greater stability results if the electrons are spread out as much as possible over all five d orbitals (Figure 20.23a). This gives four unpaired electrons. On the other hand, cyanide ion is a ligand that gives a very large Δ. In the case of $Fe(CN)_6^{4-}$, the greatest stability results if the electrons are paired as much as possible in the lower energy level (Figure 20.23b). When this occurs, all of the electrons become paired, and there are no unpaired electrons left over.

The principles of the crystal field theory that we have developed for octahedral complexes can be extended to other geometries as well. Although the relative energies of the d orbitals are different, the applications of the theory are basically the same.

SUMMARY

Metals. Metals form a metallic lattice; they have too few electrons to form enough covalent bonds to give an octet. Metals form ionic compounds with nonmetals. Within the periodic table, elements become more metallic moving from right to left in a period and from top to bottom in a group.

Metallurgy. Some unreactive metals, such as gold, are found free in nature, but most are combined with other elements. Sodium and magnesium are extracted from sea water. Manganese nodules on the ocean floor are a potential source of manganese and iron. Usually, when an ore is dug from the ground, the metal-bearing component must be enriched by a pretreatment step that removes much of the gangue. Flotation is often used with lead and copper ores. Aluminum's amphoteric character is exploited in purifying bauxite.

Metals are obtained from their compounds by reduction. Active metals must generally be prepared by electrolysis. Carbon is a common chemical reducing agent because it is plentiful and inexpensive. The production of iron takes place in a blast furnace using a charge of iron ore, limestone, and coke (which is made by heating coal). Molten

iron and slag flow to the bottom of the furnace. Steel is made from iron by removing impurities and adding special metals. The open hearth furnace has been replaced by the basic oxygen process as the most-used method for making steel.

Representative Metals. Sodium and potassium are the most common of the alkali metals, all of which are very reactive. They react with water to give H_2. With O_2, lithium gives Li_2O, sodium gives Na_2O_2, and the other alkali metals give MO_2 (superoxides). Sodium carbonate, used in glass, can be made from $CaCO_3$ and NaCl using NH_3 in the Solvay process. The potassium salts of weak acids are often more soluble than the sodium salts; the sodium salts of strong acids are often more soluble than the potassium salts. A flame is colored yellow by sodium, violet by potassium, and red by lithium.

Calcium and magnesium are the most abundant alkaline earth metals; they occur in sea water and in limestone and dolomite deposits. Gypsum is an important mineral of calcium. Beryl, an ore of beryllium, forms gem crystals of emerald and aquamarine. The alkaline earth metals are less reactive than Group IA metals, but Ca, Sr, and Ba displace hydrogen from water. $BeCl_2$ is covalent and NaCl is added as an electrolyte to molten $BeCl_2$ when Be is recovered by electrolysis. $CaCl_2$ is deliquescent. Oxides of the Group IIA metals can be made by the decomposition of their hydroxides or their carbonates. Solubilities of the Group IIA hydroxides increase going down the group, but the solubilities of the sulfates vary in the opposite direction. Partially dehydrated gypsum, $CaSO_4 \cdot \frac{1}{2}H_2O$, is plaster of Paris; $MgSO_4 \cdot 7H_2O$ is epsom salts. A flame is colored orange by calcium, crimson by strontium, and green by barium.

Aluminum, the most abundant metal in the earth's crust, is the only important metal of Group IIIA. It is a good reducing agent and is amphoteric, which means that it dissolves in both acids and bases. In water it forms the hydrated ion $Al(H_2O)_6^{3+}$, which is acidic. $Al_2(SO_4)_3$ forms double salts with K_2SO_4 and $(NH_4)_2SO_4$ called alums, for example, $KAl(SO_4)_2 \cdot 12H_2O$. Aluminum oxide can exist in two forms. Ruby and sapphire are corundum (α-Al_2O_3) crystals that contain traces of other metal ions. The anhydrous halides have the formula Al_2X_6. From gallium to indium to thallium there is an increasing tendency to form the 1+ oxidation state.

Tin and lead (Group IVA) and bismuth (Group VA) have a low degree of reactivity. They are prepared by reducing their oxides with carbon. Tin forms two allotropes, a metallic one at high temperature and a nonmetallic one at low temperature. Tin and lead have oxidation states of 2+ and 4+. The lower state becomes more stable than the higher state going down the group. Bismuth has two oxidation states, 3+ and 5+. The 5+ state is a powerful oxidizing agent. Tin(II) halides are ionic, but tin(IV) halides are molecular. $PbBr_4$ and PbI_4 are not known because Pb^{4+} is such a powerful oxidizing agent.

Transition Metals. These are the elements in the center of the periodic table. They have partially filled d subshells (except for Zn, Cd, and Hg); they are generally hard and have high melting points; they exhibit multiple oxidation states; their ions form many complex ions; and many of their compounds are colored. The size of transition metal atoms decreases gradually from left to right. The third row elements are nearly the same size as those in the row above because of the lanthanide contraction. This causes the third row elements to be very dense and unreactive. The columns are labeled B groups because parallels exist between formulas of A-group and B-group compounds. Group VIII consists of three horizontal triads.

Titanium is an important structural metal; vanadium is used mostly in alloys and catalysts. $TiCl_4$ is molecular and reacts with water to form TiO_2. Titanium dioxide is the most important compound of titanium. It is the white pigment used in paint.

Chromium has properties that make it an excellent decorative and protective coating over other metals. It is an ingredient in stainless steel. All of its compounds are colored. $Cr(H_2O)_6^{3+}$ is amphoteric. CrO_3 is a good oxidizing agent. Salts containing CrO_4^{2-} and $Cr_2O_7^{2-}$ can be prepared.

Manganese, a fairly active metal, has common oxidation states of 2+, 4+, 6+ and 7+. The most stable is the 2+ state. MnO_4^- is a very good oxidizing agent and serves as its own indicator in titrations. MnO_2, a common compound of manganese, is nonstoichiometric, and is found in manganese nodules. It can be oxidized under basic conditions to MnO_4^{2-}.

Iron is the most commercially important transition element. It is quite reactive, and both its 2+ and 3+ state are stable. Iron forms three oxides—FeO, Fe_2O_3, and Fe_3O_4 (magnetite).

Cobalt is used mostly in alloys and catalysts. It has two oxidation states, 2+ and 3+. The 3+ state is stable in complex ions.

Nickel is used in stainless steel, armor plate, and monel. It can be purified by the Mond process because $Ni(CO)_4$ is volatile. Nickel's most stable oxidation state is 2+. NiO_2 is used in the nickel-cadmium cell.

Copper, silver, and gold are the coinage metals. Silver has the highest electrical and thermal conductivity of any metal. One of its major uses is in photography. All three metals are quite unreactive. They all can have a 1+ oxidation state. The most stable states of Cu and Au are 2+ and 3+, respectively. Gold is found free in nature but copper and silver are usually found in compounds. All three form complex ions easily. The test for Ag^+ uses NH_3 to dissolve AgCl, forming $Ag(NH_2)_2^+$. Acidifying the solution destroys the complex and reprecipitates AgCl.

Zinc is found in the ore zinc blende. Cadmium is a contaminant in zinc ores. Mercury occurs as cinnabar, HgS. Zinc is an active metal whose corrosion products form a protective coating on the metal. Zinc is amphoteric, but cadmium is not. In many other ways, zinc and cadmium are chemically similar, but they differ from mercury. Many mercury compounds are covalent. Mercury metal is a liquid that is able to dissolve some other metals to form amalgams. Zinc and cadmium only form a 2+ oxidation state; mercury forms Hg_2^{2+} and Hg^{2+}. These elements also form many complex ions.

Complex Ions. Coordination compounds contain complex ions formed from a metal ion and a number of ligands. Ligands are Lewis bases and may be monodentate, bidentate, or in general, polydentate, depending on the number of donor atoms that they contain. Water is the most common monodentate ligand. A common bidentate ligand is ethylenediamine (en) and a common polydentate ligand is ethylenediaminetetraacetic acid (EDTA), which has six donor atoms. Complexes formed by polydentate ligands, called chelates, are usually more stable than similar complexes formed by monodentate ligands.

Complexes with a coordination number of four are either tetrahedral or square planar; those with a coordination number of six are almost always octahedral. Geometric isomers—*cis* and *trans* isomers—can occur in square planar and octahedral complexes. Octahedral complexes with three bidentate ligands are chiral, as are the *cis* isomers of octahedral complexes with two bidentate ligands.

Crystal Field Theory. The negative charge of the ligands causes the $d_{x^2-y^2}$ and d_{z^2} orbitals to be raised in energy more than the d_{xy}, d_{xz}, and d_{yz} when an octahedral complex is formed. This splits the d subshell into two energy levels that are separated by an energy gap, Δ, which is called the crystal field splitting. Electrons that populate the lower of the two levels can be raised in energy by absorbing light from the visible spectrum. We see the remaining light. This leads to the colors of complex ions.

In certain cases, there is a choice of whether to spread all of the d electrons out over all five orbitals with the maximum number of unpaired electrons or to pair them as much as possible in the lower set of three orbitals. When Δ is large, electrons fill the lower set of three orbitals before beginning to fill the upper set. When Δ is small, the electrons are unpaired as much as possible.

INDEX TO REVIEW QUESTIONS

REVIEW QUESTIONS

20.1. Why do metals form a metallic lattice?

20.2. Choose the more metallic element in each set.
(a) Ga or Tl   (b) Ba or Tl   (c) Mg or K

20.3. Which compound would be most ionic, $MgCl_2$, $AlCl_3$, or RbCl?

20.4. Aluminum oxide is amphoteric. What does this mean and why does it imply that Al is less metallic than Mg, whose oxide is not amphoteric?

20.5. What is an *ore*? What is *metallurgy*?

20.6. Why are sodium and magnesium extracted from sea water?

20.7. Give the chemical reactions for the separation of Mg^{2+} from sea water and the preparation of $MgCl_2$.

20.8. What is lime? How is it made from $CaCO_3$?

20.9. What are manganese nodules? Why are they a potentially attractive source of metals?

20.10. What is *gangue*?

20.11. Why can gold be separated from sand by *panning*?

20.12. Describe the *flotation process*.

20.13. Write chemical equations for the roasting of Cu_2S and PbS in air. How can SO_2 from the roasting be recovered?

20.14. Write chemical equations showing how bauxite, impure Al_2O_3, is purified.

20.15. Why is reduction rather than oxidation necessary to convert the metal in a compound into the free metal?

20.16. Sodium, magnesium, and aluminum are produced by electrolysis rather than by reduction with a chemical reducing agent. Why?

20.17. Why is carbon such a useful industrial reducing agent? How is *coke* made?

20.18. Write chemical equations for the reduction of PbO and CuO with carbon.

20.19. Cu_2S can be converted to Cu without adding a reducing agent. Explain this using appropriate chemical equations.

20.20. Why is a blast furnace called a *blast* furnace?

20.21. Describe the charge that is added to a blast furnace.

20.22. Describe the chemical reactions in the blast furnace that involve the reduction of Fe_2O_3 by carbon.

20.23. What is *slag*? How is it formed in the blast furnace? What are some of its uses?

20.24. What does *refining* mean in metallurgy?

20.25. What is the difference between pig iron and steel?

20.26. Describe the *open hearth furnace*. What reactions take place during the conversion of pig iron to steel?

20.27. Describe the *basic oxygen process*. Why has it replaced the open hearth process as the chief method for making steel?

20.28. What are the group names for the elements in Groups IA and IIA?

20.29. What are the sources of the Group IA metals?

20.30. Why are the alkali metals soft and low-melting?

20.31. How is metallic sodium prepared? Why can molten KCl be reduced by Na vapor?

20.32. Complete and balance the following equations.
(a) $Li + O_2 \longrightarrow$
(b) $Na + O_2 \longrightarrow$
(c) $Rb + O_2 \longrightarrow$
(d) $Na + H_2O \longrightarrow$

20.33. Why is KO_2 used in a recirculating breathing apparatus?

20.34. Complete and balance the following equations.
(a) $Na_2O_2 + H_2O \longrightarrow$
(b) $Na_2O + H_2O \longrightarrow$
(c) $KO_2 + H_2O \longrightarrow$
(d) $NaOH + CO_2 \longrightarrow$

20.35. What is the formula of (a) caustic soda and (b) lye?

20.36. What is the principal use of Na_2CO_3? From what ore is it obtained? How is Na_2CO_3 made in the Solvay process (write chemical equations)?

20.37. Which salt would you expect to be more soluble in water in each of the following pairs: (a) $NaClO_4$ or $KClO_4$, (b) $NaC_2H_3O_2$ or $KC_2H_3O_2$, (c) $KC_2H_3O_2$ or $KClO_4$?

20.38. What color flame is caused by (a) Na^+ (b) K^+ (c) Li^+?

20.39. Which alkaline earth metals are most important biologically?

20.40. What is (a) dolomite, (b) gypsum, and (c) beryl?

20.41. What is the source of radium?

20.42. Why are the alkaline earth metals harder and higher-melting than the alkali metals? How does the reactivity of the Group IIA metals compare to that of the Group IA metals?

20.43. How are the alkaline earth metals prepared?

Which alkaline earth metals react with cold water? Give a typical equation. How does this explain why Mg and Be are the only commercially important metals from Group IIA?

20.45. How is beryllium metal prepared?

20.46. Why is $BeCl_2$ covalent, but $MgCl_2$ is ionic? How does the tendency to form covalent bonds vary among the Group IIA elements?

20.47. How do the solubilities of the Group IIA sulfates and hydroxides change going down the group?

20.48. Define *deliquescent*?

20.49. Give three reactions for the formation of MgO.

20.50. How can $Ca(OH)_2$ and $Mg(OH)_2$ be prepared?

20.51. Give uses for the following.
(a) $CaSO_4 \cdot 2H_2O$ (e) $CaCl_2$
(b) $CaCO_3$ (f) $Mg(OH)_2$
(c) $BaSO_4$ (g) Be
(d) $MgSO_4 \cdot 7H_2O$ (h) Mg

20.52. How is plaster of paris made? Describe its reaction with water.

20.53. What color flame is produced by Ca^{2+}, Sr^{2+}, and Ba^{2+}?

20.54. Which are the post-transition metals?

20.55. Write equations for the reactions of aluminum with acids and with bases.

20.56. What is *anodized* aluminum? Aluminum is a very reactive metal. Why doesn't it experience rapid and extensive corrosion in air?

20.57. Write an equation for the *thermite reaction*.

20.58. Why is the $Al(H_2O)_6^{3+}$ ion acidic? Write equations that show what happens to $Al(H_2O)_6^{3+}$ when base is added.

20.59. What differences are there in the properties of α-Al_2O_3 and γ-Al_2O_3? What is ruby?

20.60. What is an *alum*? What are uses of potassium alum and sodium alum?

20.61. Sketch the structure of the dimer of $AlCl_3$. How are the chlorine atoms arranged around the aluminum?

20.62. How do the stabilities of the oxidation states of Ga, In, and Tl vary going down Group IIIA?

20.63. Write equations for the preparation of tin, lead, and bismuth metals.

20.64. What are the allotropes of tin? What are their properties?

20.65. What are some uses of tin, lead, and bismuth? What is Wood's metal? Why is bismuth used to make castings?

20.66. How do the stabilities of the oxidation states of tin and lead compare?

20.67. Why is $SnCl_4$ covalent, but $SnCl_2$ is ionic?

20.68. $BiCl_5$ cannot be made, but $BiCl_3$ does exist. Explain this on the basis of what you've learned about the stabilities of the oxidation states of bismuth.

20.69. What happens when $BiCl_3$ is dissolved in water?

20.70. What oxidation states are observed for (a) Na, (b) Al, (c) Bi, (d) Sn, (e) Ba, and (f) Tl?

20.71. Make a sketch of the periodic table. Mark off those regions where you would find (a) the transition elements, (b) the inner transition metals, (c) the lanthanides, (d) the actinides, and (e) the representative metals.

20.72. Give three properties of the transition metals that distinguish them from the representative metals.

20.73. Why is 2+ a common oxidation state among the transition elements?

20.74. What is a complex ion? What would be the formula and the charge on a complex ion formed from Cu^{2+} and four CN^- ions?

20.75. Why are many of the transition metals able to exist in more than one oxidation state?

20.76. How do the sizes of the transition metal atoms vary from left to right across a period?

20.77. What is the *lanthanide contraction*? How does it affect the properties of the transition elements in period 6?

20.78. What similarities exist between elements in the A and B groups in the periodic table? Give three examples.

20.79. What is a *triad*?

20.80. Why is titanium a commercially useful metal? What is the most important compound of titanium?

20.81. What are two principal uses of vanadium?

20.82. Give two uses of chromium?

20.83. In what form does Cr^{3+} exist in aqueous solution?

20.84. Write equations showing the amphoteric behavior of Cr^{3+} ion in water.

20.85. What is chrom alum? How can it be made?

20.86. What are the common oxidation states of chromium? Which is most stable?

20.87. What are the metals usually found in stainless steel?

20.88. What relationships exist among CrO_3, H_2CrO_4, CrO_4^{2-} and $Cr_2O_7^{2-}$?

20.89. What is the most stable oxidation state of manganese? How does the reactivity of Mn compare to that of Cr?

20.90. What properties of the MnO_4^- ion make it useful in redox titrations.

20.91. Under what conditions is the manganate ion, MnO_4^{2-}, stable?

20.92. What is a *nonstoichiometric* compound?

20.93. Write half-reactions for the reduction of MnO_4^- in (a) acidic and (b) basic solutions.

20.94. What is a *disproportionation* reaction? Write an equation for the disproportionation of MnO_4^{2-} in acidic solution.

20.95. What are the two oxidation states of iron?

20.96. What are the formulas of the oxides of iron?

20.97. What is magnetite? What makes it a valuable iron ore?

20.98. What are the properties and uses of cobalt? What oxidation states are found for cobalt?

20.99. Why is nickel an important metal? What metals are in the "nickel" used to make coins? What is monel?

20.100. Write chemical equations for the reactions described in the Mond process for purifying nickel.

20.101. What is the oxidation number of nickel in $Ni(CO)_4$? What is the most important oxidation state of nickel?

20.102. Give examples of practical applications of metallic copper, silver, and gold.

20.103. What oxidation states are observed for Cu, Ag, and Au?

20.104. Write equations for the reactions of manganese, iron, and nickel with hydrochloric acid.

20.105. Write equations showing the reactions of copper and silver with nitric acid.

20.106. What is *aqua regia*? Why is it able to dissolve gold?

20.107. Why are copper sulfate crystals blue?

20.108. Give chemical equations for the reactions involved in the analytical test for Ag^+.

20.109. What test can be used to detect the presence of copper ion in a solution?

20.110. What oxidation states are observed for zinc, cadmium, and mercury?

20.111. What are the ores for zinc, cadmium, and mercury?

20.112. Even though zinc is an active metal, it corrodes very slowly. Why?

20.113. How does a coating of zinc protect iron from corrosion, even if the coating is partially worn away?

20.114. Why is cadmium sometimes used as a protective coating over steel instead of zinc?

20.115. What is an *amalgam*?

20.116. How do the chemical properties of zinc, cadmium, and mercury compare?

20.117. Hg_2Cl_2 is much less toxic than $HgCl_2$ if taken internally. Why?

20.118. What happens to metallic zinc, cadmium, and mercury when they are treated with (a) HCl or (b) NaOH?

20.119. In general, what types of substances serve as ligands in complex ions?

20.120. Why are substances that contain complex ions called *coordination compounds*?

20.121. How do ethylenediamine and oxalate ion form chelate rings?

20.122. What is a monodentate ligand? What is a bidentate ligand? Sketch the structure of EDTA.

20.123. In what way is the porphyrin structure important in biological systems?

20.124. Sketch the structure of an octahedral complex containing monodentate ligands.

20.125. NTA is the abbreviation for *nitrilotriacetic acid,* a substance that was used at one time in detergents. Its structure is

The four donor atoms of this ligand are shown in color. Sketch the structure of an octahedral complex containing this ligand. Assume that two water molecules are also coordinated to the metal and that each oxygen donor atom is *cis* to the nitrogen atom in the complex.

20.126. What is *coordination number*?

20.127. What structures are generally observed for complexes in which the metal ion has a coordination number of four?

20.128. What is the chelate effect?

20.129. Sketch the isomers of the square planar complex $Pt(NH_3)_2ClBr$.

20.130. The complex $Pt(NH_3)_2Cl_2$ can be obtained as two distinct isomers. Make a model of a tetrahedron and show that if this complex were tetrahedral, two isomers would be impossible.

20.131. The complex $Co(NH_3)_3Cl_3$ can exist in *two* isomeric forms. Sketch them.

20.132. Sketch the chiral isomers of $Cr(en)_2Cl_2^+$. Is there a nonchiral isomer?

20.133. Sketch the chiral isomers of $Co(C_2O_4)_3^{3-}$. ($C_2O_4^{2-}$ is oxalate ion—a bidentate ligand.)

20.134. Sketch and label the five *d* orbitals.

20.135. Explain why an electron in the $d_{x^2-y^2}$ or d_{z^2} orbital in an octahedral complex will experience greater repulsions due to the ligands than an electron in a d_{xy}, d_{xz}, or d_{yz} orbital.

20.136. Explain how the same metal ion in the same oxidation state can given complex ions of different colors.

20.137. Which complex would be expected to absorb light of highest frequency: $Cr(H_2O)_6^{3+}$, $Cr(en)_3^{3+}$ or $Cr(CN)_6^{3-}$?

20.138. The magnitude of Δ affects the number of unpaired electrons in a complex in which the metal ion has four *d* electrons. For what other numbers of *d* electrons is this also true?

20.139. Sketch *d*-orbital energy diagrams for $Fe(H_2O)_6^{3+}$ and $Fe(CN)_6^{3-}$ and indicate the number of unpaired *d* electrons that would be expected in each of them.

20.140. What does the term *spectrochemical series* mean? How is it determined?

CHAPTER TWENTY-ONE

NUCLEAR REACTIONS AND THEIR ROLE IN CHEMISTRY

21.1 RADIOACTIVITY

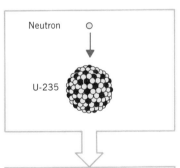

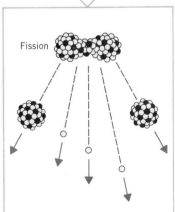

The nuclei of several naturally occurring isotopes decay spontaneously by emitting alpha particles, beta particles, or gamma rays

In Chapter Five we learned about the enormous amounts of energy available from two types of nuclear reactions—fission and fusion. However, only one naturally occurring isotope—uranium-235—will undergo fission, and only two isotopes—hydrogen-2 (deuterium) and hydrogen-3 (tritium)—are even considered as candidates for practical fusion operations in future power plants. Yet there are other kinds of nuclear reactions besides fission and fusion. One of the most important of these is radioactivity.

Radioactivity, or **radioactive decay,** is the emission of one or more kinds of radiation from an isotope whose atoms have unstable nuclei. A radioactive isotope is called a **radionuclide.** Radioactivity is caused by the inherent instability of certain combinations of protons and neutrons in nuclei. About 50 of the approximately 350 naturally occurring isotopes as well as all of the synthetic isotopes are radioactive. Among the naturally occurring radionuclides, only three kinds of radiations have been observed—alpha, beta, and gamma radiation.

Alpha radiation

Alpha radiation consists of streams of the bare nuclei of helium atoms, called **alpha particles.** The symbol for an alpha particle is ^4_2He, where 4 is the mass number (2 protons + 2 neutrons) and 2 is the atomic number (the number of protons). Each alpha particle has a charge of 2+, but the charge is omitted from the symbol.

Alpha particles are the most massive of the particles involved in natural radiations. When they are ejected (Figure 21.1), alpha particles break out of their unstable "home" nuclei and shoot through the atoms' electron orbitals into the surrounding space. They emerge with velocities of up to one-tenth the velocity of light. Because alpha particles are relatively massive compared to atoms in matter, they seldom go very far. In air, for instance, they will travel only a few centimeters before they collide with air molecules. They lose their energy in these collisions, and they pick up electrons and change to neutral helium atoms. Similarly, alpha particles cannot pass through the body's outer layer of dead skin cells, although intense exposure to them will cause a skin burn. However, if isotopes that emit alpha radiation are carried inside the body on dust particles or food, they can cause far more damage to the body's softer tissues. This damage might lead to cancer, tumors, chromosome breaks or, if they enter a fetus or cause damage to germ cells (ova or sperm), they might cause birth defects.

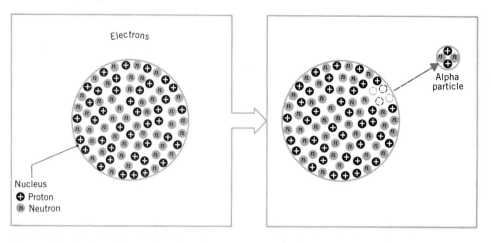

Figure 21.1
Emission of an alpha particle.

One source of alpha radiation is the most abundant isotope of uranium, uranium-238 or $^{238}_{92}U$. When a uranium-238 nucleus emits an alpha particle, it sends out two protons and two neutrons in one package. The nucleus thereby loses two units in atomic number and four units in mass number. The new nucleus left behind is that of a thorium isotope, $^{234}_{90}Th$.

To symbolize a nuclear reaction, we use a special equation called a **nuclear equation.** (These were first introduced in Chapter Five, but little explanation was given because the emphasis there was on energy yields rather than on the specific isotopes involved.) The nuclear equation for the decay of uranium-238 is

$$^{238}_{92}U \longrightarrow {}^{234}_{90}Th + {}^{4}_{2}He$$

Nuclear equations and chemical equations are not balanced in the same way. Unlike the changes that happen in chemical reactions, in nuclear reactions atoms change from one element into those of another. A nuclear equation is balanced when two conditions are met.

1 The sums of the mass numbers on each side are equal.
2 The sums of the atomic numbers on each side are equal.

In the equation for the decay of uranium-238, notice that the sum of the atomic numbers on the right side, 90 + 2, is the same as the atomic number on the left, 92. Likewise, the mass numbers on each side balance: 238 = 234 + 4.

Electrical charges on ions are not indicated in nuclear equations because the particles eventually lose their charges by picking up or giving off orbital electrons. For example, the thorium particle has two too many electrons for electrical neutrality at the instant of its formation because it formed by the loss of two protons. Similarly, the alpha particle lacks two electrons. These particles eventually lose or gain electrons as needed to become neutral atoms. As mentioned earlier, the alpha particles gain electrons and become neutral helium atoms, and the thorium particles lose their extra electrons and become neutral thorium atoms.

Beta radiation
Beta radiation consists of streams of electrons, which often are called **beta particles** when the subject being discussed is nuclear chemistry or physics. The symbol of a beta particle in a nuclear equation is $^{0}_{-1}e$ because the electron's mass number is 0 and its charge is 1−. Tritium is a beta emitter.

$$\underset{\text{tritium}}{^{3}_{1}H} \longrightarrow \underset{\text{helium-3}}{^{3}_{2}He} + \underset{\substack{\text{beta particle}\\(\textit{electron})}}{^{0}_{-1}e}$$

A beta particle comes from an atom's *nucleus,* not from its normal supply of orbital electrons (Figure 21.2). Electrons do not exist as such in the nucleus, of course, but the nuclear reaction creates them out of neutrons. This process leaves behind an extra proton. Using the symbol $_0^1 n$ for the neutron because its mass number is 1 and its charge is 0, we can write the following nuclear equation for the formation of a beta particle.

Some references symbolize the proton in a nuclear equation as $_1^1 H$.

$$_0^1 n \longrightarrow \; _{-1}^0 e \; + \; _1^1 p$$

neutron	beta particle	proton
(*in nucleus*)	(*emitted*)	(*left in the nucleus*)

Since electrons are more than 7000 times less massive than alpha particles, they are less likely to collide with nuclei in the matter through which they pass. Beta particles can go as far as 300 cm in dry air, much farther than alpha particles. However, only extremely high-energy beta radiations will penetrate the skin. The penetrating ability of a radiation is a function of its energy, and different beta-emitting radionuclides eject beta particles with different energies.

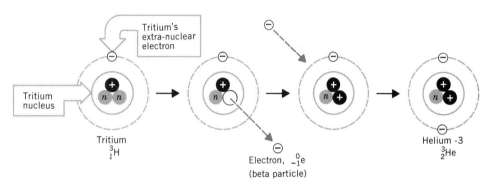

Figure 21.2
Emission of a beta particle.

Gamma radiation

Gamma radiation consists of high-energy photons (Section 5.7) that originate within certain nuclei. Atomic nuclei have various discrete energy states of their own much as the atom as a whole has energy states associated with the orbital electrons. Therefore, a nucleus absorbs energy by changing to a higher nuclear energy state. One way an excited nucleus can return to a lower nuclear energy state is to release a photon of gamma radiation having energy associated with the high-energy end of the electromagnetic spectrum. **X rays,** like gamma rays, consist of high-energy photons, but their energies are usually less than those of gamma rays. Moreover, X rays are nearly always artificially produced, whereas gamma radiation is a naturally occurring phenomenon. X rays are made by directing a high-energy electron beam at a metal target. Orbital electrons are thereby knocked out of the target's atoms leaving behind empty or half-filled orbitals called *holes.* If a hole is created at a low orbital energy level, then an orbital electron at a higher energy level drops down to fill the hole, creating a new hole higher up. An electron at a still higher energy level drops into it, making still another hole. Thus a cascade of electronic transitions occurs, and those of highest energy result in the emission of photons in the X-ray region of the spectrum. Thus the chief difference between X rays and gamma rays is that X rays come from transitions involving electronic energy levels and gamma rays are from transitions involving nuclear energy levels.

The electron-volt is the unit of energy most widely used for describing the energies of gamma rays, X rays or other atomic radiations. One **electron-volt,** symbolized as 1 **eV,** is the energy an electron receives when it is accelerated under the influence of 1 volt. It is related to the joule as follows:

$$1 \text{ eV} = 1.6 \times 10^{-19} \text{ J}$$

Because the electron-volt is so small, the kiloelectron-volt (1 keV) and the megaelectron-volt (1 MeV) are the units most commonly used.

$$1 \text{ keV} = 10^3 \text{ eV}$$
$$1 \text{ MeV} = 10^6 \text{ eV}$$

The gamma radiation from cobalt-60, a radionuclide widely used in cancer therapy, consists of photons with energies of 1173 keV and 1332 keV (1.173 MeV and 1.332 MeV). X rays used in diagnosis typically have energies of 100 keV or less.

In nuclear equations, a photon of gamma radiation has the symbol $^0_0\gamma$ (or often, just γ). Because they have very high energies but neither mass nor charge, gamma rays and X rays are among the most penetrating of all radiations. The thickness of the absorbing material needed to reduce by 10% the intensity of gamma radiation with an energy of 1 MeV is 3 cm for lead, about 50 cm for human body tissue, and 400 m for dry air. Table 21.1 has a summary of the penetrating abilities of alpha, beta and gamma radiations.

Table 21.1

Penetrating Abilities of Some Common Radiations[a]

Type of Radiation	Common Sources	Approximate Energy When from These Sources	Approximate Depth of Penetration of Radiation into:		
			Dry Air	Tissue	Lead
Alpha rays	Radium-226 Radon-222 Polonium-210	5 MeV	4 cm	0.05 mm[c]	0
Beta rays	Tritium Strontium-90 Iodine-131 Carbon-14	0.05 to 1 MeV	6 to 300 cm[b]	0.06 to 4 mm[c]	0.005 to 0.3 mm
			Thickness to Reduce Initial Intensity by 10%		
Gamma rays	Cobalt-60 Cesium-137 Decay products of radium-226	1 MeV	400 m	50 cm	30 mm
X Rays Diagnostic Therapeutic	— —	Up to 90 keV Up to 250 keV	120 m 240 m	15 cm 30 cm	0.3 mm 1.5 mm

[a] Data from J. B. Little, *The New England Journal of Medicine*, Vol. 275, pages 929–938, 1966.
[b] The range of beta particles in air is about 30 cm per MeV. Thus a 2-MeV beta particle has a range of about 60 cm in air.
[c] The protective layer of skin is about 0.07 mm thick. To penetrate it, alpha particles need about 7.5 MeV of energy and beta particles about 0.07 MeV.

EXAMPLE 21.1 Balancing Nuclear Equations

Problem: Cesium-137, $^{137}_{55}$Cs, is one of the radioactive wastes that form during fission in a nuclear power plant or an atomic bomb explosion. This isotope decays by emitting both beta and gamma radiation. Write the complete nuclear equation for the decay of cesium-137.

Solution: Set up the nuclear equation using the information given and leaving blanks for any information we have to figure out.

$$^{137}_{55}\text{Cs} \longrightarrow {}^{0}_{-1}\text{e} + {}^{0}_{0}\gamma + \underset{\substack{\text{atomic symbol} \\ \text{atomic number}}}{\overset{\text{mass number}}{\underline{\quad\quad}}}$$

beta   gamma
particle radiation

To find the atomic symbol for the isotope that forms we first have to find its atomic number. Then we can go to a table of the elements or to the periodic table and look up its symbol. If we let Z stand for the atomic number we seek, we can calculate its value from the knowledge that the sum of the atomic numbers on each side of the equation must equal. Therefore,

$$55 = -1 + 0 + Z$$
$$Z = 56$$

The element with the atomic number of 56 is barium. We find which isotope of barium by calculating the mass number from the knowledge that the sum of the mass numbers on the two sides of the nuclear equation must equal. Letting A be the mass number of the barium isotope,

$$137 = 0 + 0 + A$$
$$A = 137$$

Thus the other product is specifically the barium-137 isotope, and the balanced nuclear equation is

$$^{137}_{55}\text{Cs} \longrightarrow {}^{0}_{-1}\text{e} + {}^{0}_{0}\gamma + {}^{137}_{56}\text{Ba}$$

EXERCISE 21.1 Marie Curie (1867–1934; Poland) earned one of her two Nobel prizes for isolating the element radium, which soon became widely used for treating cancer. Radium-226, $^{226}_{88}\text{Ra}$, is an alpha and gamma emitter. Write a balanced nuclear equation for the decay of radium-226 showing the correct symbol for the isotope that forms.

EXERCISE 21.2 Strontium-90 is one of the many radionuclides present in the radioactive wastes of operating nuclear power plants. Strontium-90 is a beta emitter. Write the balanced nuclear equation for its decay.

Positrons are positive electrons.   A number of synthetic isotopes emit **positrons,** particles having the mass of an electron but a positive instead of a negative charge. The symbol for the positron is $^{0}_{1}\text{e}$. It forms by the conversion of a proton into a neutron (Figure 21.3). Positrons do not last long. When a positron hits an electron—which, in view of the abun-

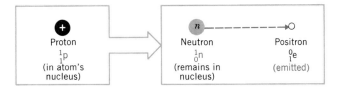

Figure 21.3
The emission of a positron changes a proton into a neutron.

dance of electrons in all matter, is likely to occur very soon after the positron leaves the nucleus—the positron and the electron destroy each other. The annihilation collision produces two photons of gamma radiation (Figure 21.4) called *annihilation radiation photons.* Each has an energy of 511 keV.

$$\underset{\substack{\text{electron} \\ (\textit{orbital})}}{{}^{0}_{-1}\text{e}} + \underset{\text{positron}}{{}^{0}_{1}\text{e}} \longrightarrow \underset{\substack{\text{gamma} \\ \text{radiation}}}{2\,{}^{0}_{0}\gamma}$$

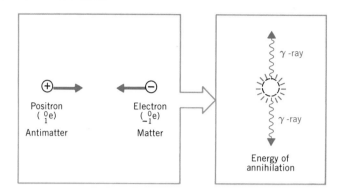

Figure 21.4
The collision of a positron with an electron annihilates both and generates gamma radiation.

Cobalt-54 decays by positron emission to give a stable isotope of iron.

$$^{54}_{27}\text{Co} \longrightarrow {}^{54}_{26}\text{Fe} + {}^{0}_{1}\text{e}$$

Because a positron destroys a particle of ordinary matter (an electron), it is called a particle of **antimatter.** To be called antimatter, a particle must have a counterpart among the particles of ordinary matter, and the two must annihilate each other when they collide. For example, a neutron destroys an antineutron.

Neutron emission is another kind of nuclear reaction. Krypton-87, for example, decays as follows:

$$^{87}_{36}\text{Kr} \longrightarrow {}^{86}_{36}\text{Kr} + {}^{1}_{0}\text{n}$$

Thus this kind of decay does not lead to an isotope of a different element.

Electron capture is still another kind of nuclear reaction. It is very rare among natural isotopes but common among synthetic radionuclides. Vanadium-50 nuclei, for example, can capture orbital K-electrons and change to nuclei of stable atoms of titanium:

$$^{50}_{23}\text{V} + \underset{\substack{\text{orbital}\\K\text{-electron}}}{{}^{0}_{-1}\text{e}} \longrightarrow {}^{50}_{22}\text{Ti} + \text{X rays}$$

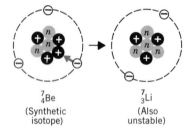

$^{7}_{4}\text{Be}$
(Synthetic isotope)

$^{7}_{3}\text{Li}$
(Also unstable)

Figure 21.5
When electron capture occurs, an orbital electron collapses into the nucleus changing a proton into a neutron.

The net effect of electron capture is the conversion of a proton into a neutron (Figure 21.5), which we can symbolize by a nuclear equation:

$$\underset{\substack{\text{proton}\\(in\ the\\nucleus)}}{{}^{1}_{1}\text{p}} + \underset{\substack{\text{electron}\\\text{captured}\\\text{from the}\\K\text{-shell}}}{{}^{0}_{-1}\text{e}} \longrightarrow \underset{\substack{\text{neutron}\\(in\ the\\nucleus)}}{{}^{1}_{0}\text{n}}$$

Electron capture doesn't change an atom's mass number, only its atomic number. It also leaves a hole in the K-shell, and the atom emits photons of X rays as other orbital electrons drop down to fill the hole. Moreover, the nucleus that has just captured an orbital electron may emit a gamma ray photon.

Radioactive disintegration series
Very frequently, the decay of one radioactive isotope leads to another radioactive isotope, not one that is stable. This isotope might, in turn, decay to still another radionuclide, and the process can continue through a series of nuclear reactions until a stable isotope finally forms. Such sequences of nuclear reactions are called **radioactive disintegration series.** Four are known. Uranium-238 is at the head of one, shown in Figure 21.6. Notice that some of the half-lives are extremely short but others are very long. (We studied half-lives in Section 14.5.) Table 21.2 gives the half-lives of several natural and synthetic radionuclides.

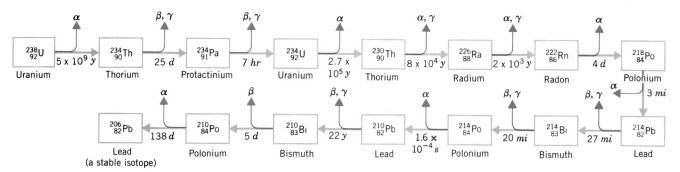

Figure 21.6
The uranium-238 radioactive disintegration series. The half-life of each isotope is beneath the arrow leading away from the isotope; y = years; m = months; d = days; mi = minutes; and s = seconds.

Table 21.2

Typical Half-Lives

Element	Isotope	Half-Life	Radiations or Mode of Decay
Naturally Occurring Radionuclides			
Potassium	$^{40}_{19}K$	1.3×10^9 years	beta, gamma
Tellurium	$^{123}_{52}Te$	1.2×10^{13} years	electron capture
Neodymium	$^{144}_{60}Nd$	5×10^{15} years	alpha
Samarium	$^{149}_{62}Sm$	4×10^{14} years	alpha
Rhenium	$^{187}_{75}Re$	7×10^{10} years	beta
Radon	$^{222}_{86}Rn$	3.82 days	alpha
Radium	$^{226}_{88}Ra$	1590 years	alpha, gamma
Thorium	$^{230}_{90}Th$	80,000 years	alpha, gamma
Uranium	$^{238}_{92}U$	4.51×10^9 years	alpha
Synthetic Radionuclides			
Hydrogen	$^{3}_{1}H$ (tritium)	12.26 years	beta
Oxygen	$^{15}_{8}O$	124 seconds	positron
Phosphorus	$^{32}_{15}P$	14.3 days	beta
Technetium	$^{99m}_{43}Tc$	6.02 hours	gamma
Iodine	$^{131}_{53}I$	8.07 days	beta
Cesium	$^{137}_{55}Cs$	30 years	beta
Strontium	$^{90}_{38}Sr$	28.1 years	beta
Americium	$^{243}_{95}Am$	7.37×10^3 years	alpha

Nuclear stability

Both the proton and the neutron are nucleons.

There is no mathematical equation involving the numbers of protons and neutrons in a nucleus that can be used to predict whether an isotope will be stable or, if unstable, what its half-life is. In Section 5.11, we learned that the greater the binding energy per nucleon the more stable is the nucleus. When we applied this criterion we found that isotopes having atomic numbers near 26 (iron) should have the most stable nuclei. Yet several isotopes of iron, all synthetic, are unstable. One—iron-53—has a half-life of only 8.5 minutes, but another—iron-60—has a half-life of 3×10^5 years. In the absence of a mathematical equation for predicting nuclear stability, scientists have developed several "rules of thumb" to aid them in picking those candidates for synthesis that stand some chance of having half-lives long enough to allow the isotope to be detected and studied.

One such rule uses an array of all known stable and unstable isotopes. Each isotope is given a location in the array according to its number of protons and neutrons, as shown in Figure 21.7. Element 83 (bismuth) is the last one to have at least one stable nucleus, and the array is not plotted beyond 83. The two curved lines are drawn so as to enclose all stable nuclei, and between these lines lies what is called the **band of stability.** Not all of the isotopes in this band are stable, but those that are not have half-lives long enough to enable detection.

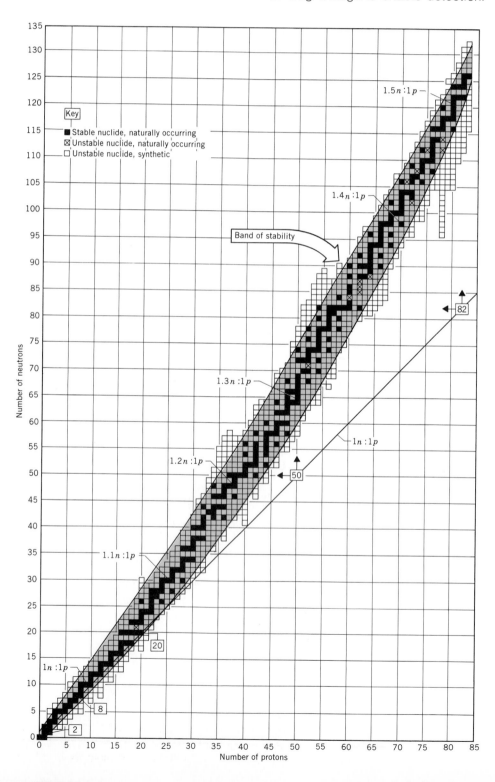

Figure 21.7
The band of stability.

Any isotope not represented anywhere on the array of Figure 21.7 quite likely has a half-life at least as short as 10^{-8} second, too short for detection and observation even if some nuclear reaction produces one. Therefore, as a rule of thumb, the search for isotopes lying at some distance from the band of stability will likely be unsuccessful. For example, an isotope with 50 neutrons and 60 protons or one with 60 protons and 50 neutrons would be too unstable to justify the time and money for an attempt to make them.

Another rule of thumb about stable isotopes is the **odd-even rule:** When the numbers of neutrons and protons in a nucleus are both even, the isotope is far more likely to be stable than when both numbers are odd. Out of the 264 stable isotopes, only 5 have odd numbers of *both* protons and neutrons, whereas 157 have even numbers of both. The rest have an odd number of one and an even number of the other. In Figure 21.7, notice that the horizontal rows on which the black squares most commonly occur correspond to *even* numbers of neutrons. Similarly, the vertical columns on which the black squares most often fall correspond to *even* numbers of protons.

The odd-even rule can be explained by the fact that, like orbital electrons, both protons and neutrons are spinning particles. When two like particles are spinning in opposite directions, their spins are said to be paired and their combined energy is less than when they are spinning in the same direction. Of course, the lower the combined energy, the more stable is the arrangement. When the numbers of protons and neutrons are both even, all proton spins and all neutron spins are paired. Such nuclei are therefore the most stable. The least stable nuclei are those that have odd numbers of both protons and neutrons because, in these nuclei, one proton and one neutron must have unpaired spins.

Another rule of thumb concerning nuclear stability involves "magic numbers." These numbers are a part of a tentative theory about nuclear stability that we will have to leave to more advanced references, but we can at least mention some of the predictions of that theory. According to one prediction, isotopes having specific numbers of protons or neutrons—the **magic numbers**—ought to be more stable than the rest. The magic numbers are 2, 8, 20, 50, 82, and 126. Figure 21.7 shows where the magic numbers fall (except for 126 protons). When both the number of protons and the number of neutrons are the *same* magic number, the isotope is particularly stable. These isotopes are ^4_2He, $^{16}_8\text{O}$, and $^{40}_{20}\text{Ca}$. One stable isotope of lead, $^{208}_{82}\text{Pb}$, has 82 protons and 126 neutrons, both of which are magic numbers. The existence of magic numbers supports the theory that there are nuclear energy levels just as there are electron energy levels. Electron orbitals have a similar set of "magic numbers" of their own—the maximum number of electrons that can be in the principal energy levels. These numbers are 2, 8, 18, 32, 50, 72, and 98 for principal levels 1, 2, 3, 4, 5, 6, and 7, respectively.

One last observation concerning the array in Figure 21.7 involves the ratio of neutrons to protons. With increasing numbers of protons, the ratio of neutrons to protons gradually moves farther and farther away from a simple 1:1 ratio (represented by the straight line). Protons repel each other, and as their number increases, more and more neutrons are apparently needed to maintain nuclear stability.

21.2 TRANSMUTATION

When a high-speed particle is captured by a nucleus, the nucleus may be permanently changed to that of another element

The change of one isotope into another is called **transmutation.** Radioactive decay is just one way that transmutations happen. They can also be caused by bombarding the atoms of an isotope with high-energy particles such as alpha particles from natural alpha-emitters, neutrons from atomic reactors, and protons made by the removal of electrons from hydrogen. The electrically charged nature of alpha particles and protons makes it possible for them to be accelerated to ultrahigh energies in special accelerators. With sufficient energy, bombarding particles will shoot through the electron orbitals and bury themselves in nuclei. Beta particles (electrons) are too strongly repelled by the orbital electrons to be frequently successful in getting through to the nucleus. Therefore, beta particles are generally not used to cause transmutations.

"Compound" here refers only to the idea of "combination," not to a chemical.

When a nucleus captures a bombarding particle it becomes a **compound nucleus.** At the moment of impact, it has all of the energy of the bombarding particle and is therefore unstable. This additional energy is soon distributed among all of the nuclear particles, but eventually the nucleus has to get rid of the excess energy. It can do this by ejecting a high-energy particle such as a neutron, a proton, or an electron or by emitting gamma radiation. If a particle is ejected, a transmutation occurs.

Ernest Rutherford (Special Topic 5.5) observed the first transmutation to occur under laboratory conditions. He allowed alpha particles from a natural radionuclide to pass through a chamber containing nitrogen atoms. He discovered that another radiation, one that could be detected at far greater distances than alpha particles could reach, was produced. This new radiation consisted of high-energy protons, $_1^1p$, and Rutherford concluded that these protons must have resulted from the conversion of nitrogen nuclei into oxygen nuclei. He reasoned that the capture of an alpha particle by a nitrogen nucleus produced a compound nucleus, one of fluorine-18, that soon emitted a proton to form an atom of oxygen-17 (Figure 21.8).

The asterisk (*) symbolizes that a high-energy form of the nucleus—a compound nucleus—is present.

$$_2^4He \; + \; _7^{14}N \; \longrightarrow \; _9^{18}F^* \; \longrightarrow \; _8^{17}O \; + \; _1^1p$$

| alpha particle | nitrogen nucleus | fluorine (*compound nucleus*) | oxygen (*a rare but stable isotope*) | proton (*high-energy*) |

Alpha particles and protons will pick up speed and energy if they are attracted toward a negatively charged object. The awareness of this fact led to the development of a number of particle accelerators, gigantic machines such as the cyclotrons, bevatrons, and synchocyclotrons. The first transmutation caused by experimentally accelerated particles was the conversion of lithium-7 to helium-4. The bombarding particles were protons made by using energy to strip electrons from ordinary hydrogen. The protons were accelerated and then focused onto a

Figure 21.8
Artificial transmutation of nitrogen into oxygen. When the nucleus of nitrogen-14 captures an alpha particle it becomes a compound nucleus of fluorine-18. This expels a proton and becomes the nucleus of oxygen-17.

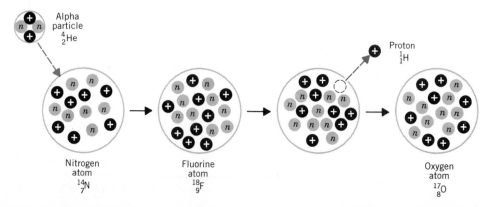

lithium target. The capture of a proton by a lithium nucleus creates a compound nucleus of beryllium, which promptly breaks in half to give two alpha particles. (These eventually pick up electrons and become helium, as we have noted.)

$$\underset{\text{proton}}{^1_1\text{p}} + \underset{\text{lithium}}{^7_3\text{Li}} \longrightarrow \underset{\text{beryllium}}{^8_4\text{Be}^*} \longrightarrow \underset{\text{alpha particles}}{2\,^4_2\text{He}}$$

How a compound nucleus decays is independent of the bombarding particle but dependent on how much energy it has. For example, a high-energy, compound nucleus of an atom of aluminum-27 could form from any of the following reactions.

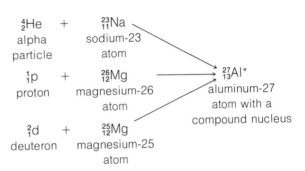

$$\begin{array}{ll}
\underset{\substack{\text{alpha}\\\text{particle}}}{^4_2\text{He}} + \underset{\substack{\text{sodium-23}\\\text{atom}}}{^{23}_{11}\text{Na}} \\[2em]
\underset{\text{proton}}{^1_1\text{p}} + \underset{\substack{\text{magnesium-26}\\\text{atom}}}{^{26}_{12}\text{Mg}} \longrightarrow \underset{\substack{\text{aluminum-27}\\\text{atom with a}\\\text{compound nucleus}}}{^{27}_{13}\text{Al}^*} \\[2em]
\underset{\text{deuteron}}{^2_1\text{d}} + \underset{\substack{\text{magnesium-25}\\\text{atom}}}{^{25}_{12}\text{Mg}}
\end{array}$$

A deuteron is a deuterium atom without its orbital electron. It is like a proton, but it has a mass number of 2, not 1.

Once formed, the high-energy aluminum-27 nucleus "forgets" how it was made. Therefore, its mode of decay is a function only of how much *energy* it holds as a result of the collision that made it. Any of the following decay modes are open to it, depending on its energy. They illustrate how the synthesis of such a large number of synthetic radionuclides has been possible.

$$\underset{\text{aluminum-27}}{^{27}_{13}\text{Al}^*} \xrightarrow[\substack{\text{a function of the}\\\text{nuclear energy}}]{\text{mode of decay is}} \left\{ \begin{array}{l}
\underset{\substack{\text{aluminum-27}\\(\textit{now stable})}}{^{27}_{13}\text{Al}} + \underset{\substack{\text{gamma}\\\text{radiation}}}{^0_0\gamma} \\[2em]
\underset{\substack{\text{magnesium-26}\\(\textit{stable})}}{^{26}_{12}\text{Mg}} + \underset{\substack{\text{proton}\\\text{radiation}}}{^1_1\text{p}} \\[2em]
\underset{\substack{\text{aluminum-26}\\(\textit{unstable})\\t_{1/2} = 7.4 \times 10^5 \text{ yr}}}{^{26}_{13}\text{Al}} + \underset{\substack{\text{neutron}\\\text{radiation}}}{^1_0\text{n}} \\[2em]
\underset{\substack{\text{magnesium-25}\\(\textit{stable})}}{^{25}_{12}\text{Mg}} + \underset{\substack{\text{neutron}\\\text{radiation}}}{^1_0\text{n}} + \underset{\substack{\text{proton}\\\text{radiation}}}{^1_1\text{p}} \\[2em]
\underset{\substack{\text{sodium-23}\\(\textit{stable})}}{^{23}_{11}\text{Na}} + \underset{\substack{\text{alpha}\\\text{radiation}}}{^4_2\text{He}}
\end{array} \right.$$

From the standpoint of practical applications, some of the most useful transmutations have been those that produce radioactive elements with sufficiently long half-lives to be isolated and used in research, medicine, and technology. Over 1000 radionuclides have been made by transmutations, and many are isotopes that do not occur in nature. (There are nearly 900 open squares in Figure 21.7 designating known synthetic radionuclides for elements up through atomic number 83.) Many naturally occurring isotopes exist above 83, but all have long half-lives. Many of the other possible isotopes above 83 may well have existed at one time in nature, but had half-lives too short in relation to the age of the planet to have survived into the twentieth century. For example, neptunium-237, which has a half-life of over 2 million years, undoubtedly existed naturally at one time, but the estimated age of the earth is over 4 billion years, long enough for essentially all of the neptunium to have decayed. We know about neptunium-237 today only because it can be made in accelerators. None of the known elements from neptunium (number 93) and up occurs naturally; all are synthetic. Most of these are alpha emitters because the ejection of alpha particles is the most direct way to move a nuclide closer to the band of stability.

The elements from 93 and higher are called the **transuranium elements.** They make up most of the **actinide series** in the periodic chart (from atomic number 90 to 103). The synthesis of elements 104 through 106 has been accomplished, but they have not yet been officially named. There is strong evidence from the analysis of meteorites that element 114 once existed in the solar system.

21.3 DETECTING AND MEASURING RADIATIONS

The ability of radiations to generate ions in matter makes their detection and measurement possible

Physicists have developed many ways to detect, record, and measure radiations from radionuclides. Several are based on one of the most important properties of radiations: their ability to generate ions in the matter that they penetrate. All of the radiations that we have thus far described are **ionizing radiations.** When ions are produced in a gas, even momentarily, the gas becomes a conductor of electricity. A Geiger-Muller tube is one radiation detector based on this fact (Figure 21.9). It is particularly useful in detecting beta and gamma radiation. When a pulse of radiation enters the thin-walled window of the tube, it creates ions in the gas inside. This allows a pulse of electricity to flow from the cathode, the metallicized surface of the tube, to the anode, a wire electrode. This flow of current activates an instrument that amplifies the current and records the pulse. The tube, and the associated equipment, is called a *Geiger counter.*

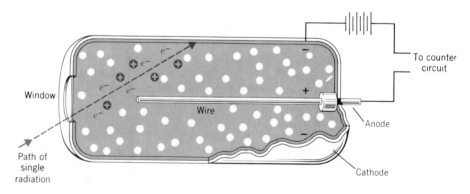

Figure 21.9
The Geiger-Muller tube. The white dots in the tube represent atoms of argon at very low pressure.

Actual clouds form from supersaturated water vapor in which condensation is started by dust particles.

For visualizing the actual tracks of particles of radiation, several kinds of *cloud chambers* have been invented. When an enclosed space is briefly made supersaturated in the vapor of some fluid—water or alcohol, for example—while it is exposed to radiation, microdroplets of condensed vapor will form along any path taken by a particle, as seen in Figure 21.10. The existence of some of the fundamental atomic particles was inferred by analyses of photographed tracks produced by cloud chambers.

Scintillation counters (Figure 21.11) are devices that permit an investigator to see when a collision occurs between a particle and a special surface on the

Figure 21.10
Cloud chamber photo. Where the heavy tracks intersect there
were collisions between neutrons and an oxygen atom and a
carbon atom.

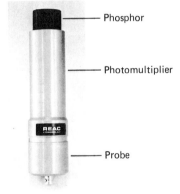

Phosphor

Photomultiplier

Probe

Figure 21.11
Scintillation probe. Energy received
from radiations striking the
phosphor at the top of this probe is
processed by the photomultiplier
unit and sent to an instrument
(not shown) for recording.

counter. This surface is coated with a substance that gives off a tiny light flash
when it is hit by a particle of radiation. For example, if the coating contains a zinc
sulfide phosphor, then alpha particles cause visible scintillations. The scintillations can be magnified with electronic equipment and automatically counted.

A *dosimeter* is used to measure the total quantity of radiation received by a
surface during a specified interval of time. Some dosimeters (Figure 21.12) use
photographic plates kept completely shielded from ordinary light but that are
sensitive to atomic radiations, including gamma and X rays. The amount of dark-
ening in the developed film is proportional to the dose of radiation received and
can be measured. Dense tissues, such as bones, reduce the intensity of X rays or
gamma rays. Therefore, if either radiation passes through the body toward a photographic plate, shadows of bones will be cast on the plate and will show up as a
familiar "X ray" picture. For a revolutionary advance in X ray technology, see
Special Topic 21.1.

Figure 21.12
Dosimeter. (*a*) Badge dosimeters.
The millirem doses of various kinds
of radiations for predetermined
periods of exposure are carefully
logged to help in preventing over-
exposures to people whose jobs require them to use radioactive materials or X rays. (*b*) The heart of
the dosimeter (as seen from the
back) is this piece that can be
slipped out for analysis. By using
different filters, different parts of the
badge register exposures to radiations of various energies and penetrating abilities—X rays, beta rays,
and gamma rays.

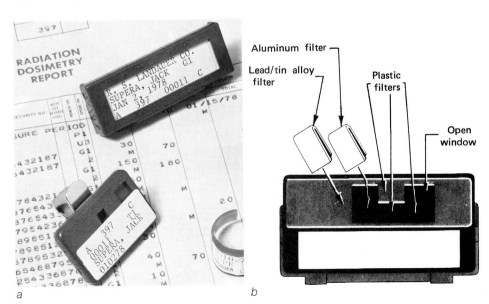

Aluminum filter

Lead/tin alloy
filter

Plastic
filters

Open
window

a

b

In the late 1970s a revolutionary X-ray technique was developed that allows the production of X-ray pictures of cross-sections of the body. The method, called computer-assisted tomography (CAT), uses X-ray tubes that rotate around the body and send in extremely brief pulses of X rays from all angles across one cross-sectional area of the patient (Figure 21.13). The pulses received on the opposite side are processed by a computer that generates an X-ray image such as shown in Figure 21.14. Someone has likened the technique to getting a full view of the cross section of a cherry pit without breaking into the cherry. A. M. Cormack (United States) and G. N. Hounsfield (England) shared the 1979 Nobel prize in medicine for the development of the CAT scanner, which has been called the greatest development in X-ray techniques since the discovery of X rays themselves. Scientists at the Mayo Clinic are presently working on a successor to the CAT scanner, dubbed the DSR (for dynamic spatial reconstructor).It will provide much greater detail, sometimes in three dimensions, of the interiors of organs.

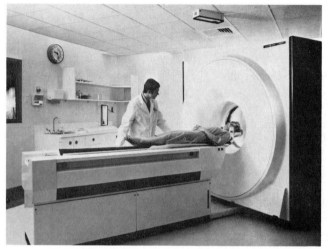

Figure 21.13
Instrument for computerized axial scanning (CAT). Shown here in General Electric's CT/T Total Body Scanner that can complete an entire scan of the head or the body in as little as 4.8 seconds.

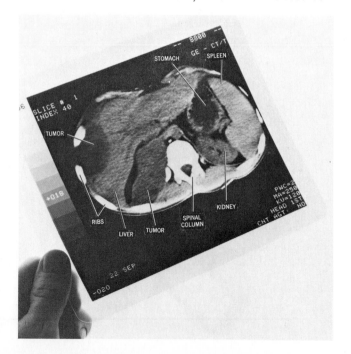

Figure 21.14
This cross-sectional view through a patient showing two tumors was obtained with a CAT scanner.

To describe the quantity or dose of radiation absorbed by some material, units of *absorbed dose* (or simply dose) have been devised. The most common unit of dose in the United States is the **rad** (from radiation absorbed dose), abbreviated **rd.** One rad corresponds to the absorption of 10^{-5} J per gram of tissue. The SI unit of absorbed dose is called the **gray, Gy,** and 1 gray corresponds to 1 joule of energy absorbed per kilogram of absorbing material.

The gray is named after Harold Gray, a British radiologist.

$$1 \text{ rd} = 10^{-5} \text{ J/g}$$
$$1 \text{ Gy} = 1 \text{ J/kg}$$
$$1 \text{ Gy} = 100 \text{ rad}$$

One of the drawbacks of the rad (or the gray) is that the damage caused by absorbed radiation is a function not just of its energy but also of the kind of radiation carrying the energy. Neutrons, for example, are ten times as dangerous as beta radiation of the same energy and intensity. To take these differences into account, scientists devised a unit called the **rem** (from radiation equivalent for man). A dose in rems is calculated by multiplying the dose in rads by a factor that reflects the effectiveness of the *kind* of radiation in causing damage. When doses in rems are calculated, then the overall danger to which anyone has been subjected by being exposed to different kinds of radiation hitting different kinds of tissue can be estimated by simply adding up the rems for each radiation. The lethal, whole-body, radiation dose for humans is about 300 rem. In contrast, a typical chest X ray involves a dose of 7 millirem (7×10^{-3} rem).

The rem is a very small quantity of energy if we think of the energy only in terms of heat. Yet, workers who might be exposed to radiations are warned not to receive more than 0.3 rem in a week. Obviously, the danger of overstepping this guideline cannot lie in the associated heat energy. Instead, the danger comes from the ability of absorbed radiation to create unstable ions and free radicals. (A free radical, as we learned in Section 14.8, is any particle with an unpaired electron.) For example, water can interact with radiations as follows.

$$(H-\ddot{O}-H) + \text{radioactive particle or gamma ray} \longrightarrow (H-\dot{O}-H)^+ + {}_{-1}^{0}e \text{ electron}$$

The new cation, $(H-\dot{O}-H)^+$, is unstable, and one way it can break down is

$$(H-\dot{O}-H)^+ \longrightarrow H^+ \text{ proton} + :\dot{O}-H \text{ hydroxyl radical}$$

The H$^+$ might pick up a stray electron and become H·, a hydrogen atom. Both H· and :O—H are examples of free radicals, and both are decidedly unstable. What they might do depends on what other molecules are in the vicinity. The formation of free radicals and unstable ions can set off a series of unwanted chemical changes inside a body cell that can lead to cancer, tumors, or other equally grave results. For this chemical reason, absorbed radiations can inflict injury all out of proportion to their actual energies. The actual energy in a dose of 600 rem causes the ionization of only one water molecule for every 36 million present, yet 600 rem is well over the lethal dose.

When the radiation department of a hospital or clinic buys radioactive material, it is particularly interested in the activity of the sample. *Activity* refers to the number of nuclear disintegrations per second occurring in the sample. The SI unit of activity is called the **becquerel (Bq),** which is equal to one disintegration or other kind of nuclear transformation per second. It was named after Antoine Becquerel (1852–1908), a French physicist who won the Nobel prize in 1903 for discovering radioactivity.

An older, pre-SI unit of activity that is still widely used was named after Marie Curie, the discoverer of radium (see Exercise 21.1). One **curie, Ci,** equals the rate of disintegrations that occur in a sample of 1.0 g of radium, which is 37 billion disintegrations per second.

$$1 \text{ Ci} = 3.7 \times 10^{10} \text{ disintegrations/s}$$

Thus,

$$1 \text{ Ci} = 3.7 \times 10^{10} \text{ Bq}$$

If a hospital owns a radioactive source rated as 1.5 Ci, the source delivers $1.5 \times 3.7 \times 10^{10} = 5.6 \times 10^{10}$ Bq.

21.4 APPLICATIONS OF RADIOACTIVITY

Isotope dilution analysis, neutron activation analysis, tracer studies, and both geological and archeological dating are applications of radioactivity in science, technology, and medicine

Radionuclides have the chemical properties that are fundamentally the same as those of the stable isotopes of the same element. For instance, strontium-90, a beta and gamma emitter, behaves *chemically* like strontium-88, the most abundant stable isotope of strontium. Similarly, iodine-131, a beta emitter, has chemical properties identical to those of iodine-127, a stable isotope and the only isotope of iodine that occurs naturally. In fact, the only chemical differences ever noticed between isotopes have been in the relative rates of their reactions, not in the kinds of reactions. We mentioned the deuterium isotope effect in Section 12.2, for example. Radioactive isotopes are therefore useful not for any unique chemical properties but for their radiations. One application is tracer analysis.

Tracer analysis

In **tracer analysis,** a radionuclide's radiation is used to follow the movements or to determine the locations of the radionuclide. It is like putting a cowbell on a cow. The bell doesn't change the cow in any way, but it signals the cow's location. The radiation from iodine-131, for example, lets a physician follow the rate of uptake of this radionuclide (in the form of the iodide ion) by the thyroid gland, as we studied in Section 14.5.

The ability of a radionuclide to decay is totally unaffected by its state of chemical combination.

Tracer analysis can be used to pinpoint the location of a brain tumor. Blood capillaries in a brain tumor lose their ability to retard the migration of small ions from the blood stream. A brain tumor will therefore show up as a "hot spot" when radiation detectors are focused on the head of a patient who has been given a small amount of technetium 99m[1] in the form of the pertechnitate ion, TcO_4^-. This strong gamma emitter is one of the most widely used radionuclides in medicine today (Special Topic 21.2).

Tracer analysis is also used in industry. For example, large oil pipelines are cleaned by sending remote controlled carts, called "go-devils," rigged with cleaning equipment through them. If a go-devil gets jammed somewhere, finding it could be very costly if lots of pipe had to be dug up. However, if the go-devil also carries a small amount of a gamma emitter, its precise location can be found using radiation detectors.

Biochemists have often used tracer analysis to find out how a living organism makes particular molecules. For example, when biochemists worked on the question of how the body makes cholesterol, they knew that one raw material of

[1] The *m* in 99m stands for metastable, which is a condition of being poised at the brink of a major energy-releasing change. It is similar to a ball at the edge of a high, narrow shelf or a pack of mountain snow that needs only a sharp sound to start an avalanche.

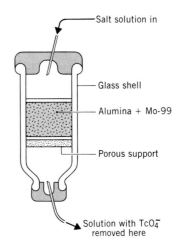

- Salt solution in
- Glass shell
- Alumina + Mo-99
- Porous support
- Solution with TcO_4^- removed here

A physician wanting a dilute solution of technetium-99*m* uses an isotonic sodium chloride solution (0.9 g/dl) to leach (or "milk") pertechnitate ion, TcO_4^-, made of Tc-99*m*, from a special apparatus called the "cow." (See the accompanying figure.) This generator contains molybdenum-99 in the form of the molybdate ion, MoO_4^{2-}, adsorbed onto alumina granules. Molybdenum-99 decays by beta emission and has a half-life of 67 hr.

$$^{99}_{42}Mo \longrightarrow \, ^{99m}_{43}Tc + \, ^{0}_{-1}e$$

The molybdenum-99 is made at a facility with an atomic reactor. The neutrons from controlled fission are allowed to bombard molybdenum-98 (also in the form of MoO_4^{2-}). This generates molybdenum-99.

$$^{98}_{42}Mo + \, ^{1}_{0}n \longrightarrow \, ^{99}_{42}Mo + \, ^{0}_{0}\gamma$$

Once charged with molybdenum-99, the apparatus is sheathed in lead and sent to the clinic or hospital. There, each morning, the radiology staff "milks the cow" of its accumulated supply of technetium-99*m*. After about a week's use, the apparatus is sent back to the reactor.

Technetium-99*m* is an almost ideal radionuclide for use in diagnostic work. Its half-life (6.02 hr) is not so short that it will disappear before it can be gotten into the patient and measurements can be taken. Yet the half-life is short enough so that most of its decay occurs *during* the diagnostic work when its dangerous gamma radiation can provide the needed benefit—detection—instead of after the work when no benefit from the radiation is possible. The product of the decay of technetium-99*m* is technetium-99, which has a half-life so long (212,000 yr) that its activity is too low to be of concern. Moreover, technetium, in any of its forms used in diagnosis, is routinely elimi-

nated by the body. (Ideally, of course, the decay product of a radionuclide used in diagnosis would be completely stable and promptly eliminated.) Technetium-99*m* has a further advantage in that all of its radiation is gamma radiation. Because this is very penetrating, *all* of the radiation serves to provide the benefit of detection. This, combined with the high activity of even very small samples, means that the patient need be given minimal exposure to the risks of radiations while enjoying the benefits.

Although the CAT scanner (Special Topic 21.1) has taken over about 95% of the diagnostic work for locating brain tumors, technetium-99*m* and other radionuclides continue to be valuable for studying the degree to which various organs—liver, spleen, kidneys, and lungs, for example—are functioning normally. Whereas the CAT scanner is useful in locating anatomical features, radionuclides are useful in assessing bodily functions.

cholesterol was the acetate ion. When this ion was made with carbon-14 as the atom holding the two oxygen atoms, the cholesterol actually produced had carbon-14 atoms (beta emitters) at each of the positions indicated by the black dots in the following structure.

$$CH_3-\overset{\overset{\displaystyle O}{\|}}{C}-O^-$$

acetate ion with
carbon-14 (·)

cholesterol
(· = C-14)

(The exact locations of the carbon-14 atoms were deduced by systematically breaking down the radionuclide-labeled cholesterol into small molecules each of which was further decomposed, one carbon at a time, until all of the carbon-14 had been found.) When carbon-14 was used to make the CH_3 group in the acetate ion, and this labeled ion was used to make cholesterol, then all of the other carbon atoms were carbon-14. This research helped scientists learn how to manipulate and suppress cholesterol synthesis in people whose bodies produce too much.

Neutron activation analysis

Neutron activation analysis is another use of radioactivity. This is a method of analysis of trace impurities in substances by activating the atoms of the impurities to make them gamma emitters and then measuring the frequencies and intensities of the gamma radiations. The actual activation is caused by neutron bombardment. Neutron capture by atoms of the impurity gives compound nuclei that emit gamma radiation but do not at the same time change to nuclei of different elements. Thus the sample is not destroyed by the analysis. We can represent neutron activation by the following general equation.

A = mass number
X = isotope symbol
Z = atomic number

$$\underset{\substack{\text{isotope}\\\text{of trace}\\\text{impurity}}}{^{A}_{Z}X} + \underset{\text{neutron}}{^{1}_{0}n} \longrightarrow \underset{\substack{\text{compound}\\\text{nucleus}}}{^{(A+1)}_{Z}X^{*}} \longrightarrow \underset{\substack{\text{stable form}\\\text{of a new}\\\text{isotope}\\\text{of } X}}{^{(A+1)}_{Z}X} + \underset{\substack{\text{gamma}\\\text{radiation}}}{^{0}_{0}\gamma}$$

Of course, other nuclei besides those of the impurities are activated, but each isotope has a unique gamma emission spectrum. The *frequencies* of the emitted gamma radiations reveal what trace element is present as an impurity. The *intensities* of these gamma rays can be used to calculate the concentration of the impurity. So sensitive is neutron activation analysis that it can be used to determine concentrations as low as 10^{-9}%. Even if a museum has only a lock of hair from an historic figure believed to have died from arsenic poisoning hundreds of years ago, neutron activation analysis can be used to detect and measure the arsenic in the hair and thus confirm or disprove the belief. And it will still have the lock of hair!

Isotope dilution analysis

Isotope dilution analysis is still another application of radioactivity. For example, the volume of a fluid that might be very difficult to measure directly can be determined by this method. A very small quantity of the fluid is mixed with a small but known quantity of radionuclide. After thorough mixing, another small quantity of the fluid is analyzed for radioactivity and, from the dilution of the intensity of the radioactivity, the original volume can be calculated. By this method the blood volume of an animal can be measured without sacrificing the animal. Suppose a weighed sample of the blood—say, 1.0 g—is mixed with a trace quantity of a radionuclide in a chemical form that is fully compatible with the animal's biochemistry. This gives the 1.0-g sample a certain *specific activity,* the becquerels per gram. Next the sample is returned to the animal's circulatory system where it mixes completely with the blood. If the specific activity were, for example, 1000 Bq/g before mixing, and became only 10 Bq/g after being mixed with the animal's blood, it means that the 1-g sample was diluted by a factor of 100. Since the original 1.0 g sample became part of 100 g of blood, the original quantity of blood was 99 g.

All it takes for isotope dilution analysis is an isotope that is chemically compatible with the substance being analyzed and a means of getting the substance thoroughly mixed. The method has been used with some success in estimating the volume of water in underground water supplies.

Radiological dating

Radiological dating of geological deposits or archeological finds is one of the best known uses of radioactivity. In this method the concentration of a particular radionuclide is measured and then its half-life is used to calculate the age of the specimen. The technique involves some important assumptions. One is that half-lives have not changed during the existence of the planet. In geological dating, another assumption is that *all* of the radionuclide found in the specimen formed by the decay of another. For example, at the end of the uranium-238 disintegration series (Figure 21.6) is a stable isotope of lead, lead-206. If we can assume that a particular rock deposit contained no lead-206 when the deposit formed, then if today it has both uranium-238 and lead-206 it is reasonable to assume that all of the lead-206 came from the decay of uranium-238. The half-life of uranium-238 is 4.51×10^9 years, which is by far the longest half-life in the uranium-238 series. Nothing else in the series significantly affects the production of lead-206. Therefore, if a sample of rock were found to contain uranium-238 and lead-206 in exactly a 1:1 ratio, it would mean that the rock had been in existence for one half-life period of uranium-238, 4.51×10^9 years. (Actually, the technique is too uncertain to allow that many significant figures; one significant figure is all that is appropriate.) If ratios of lead-206 to uranium-238 other than 1:1 are found, then calculations (which we will not study) can be used to determine the age of the specimen. There are several other pairs of isotopes that can be used for geological dating.

For dating organic remains, such as wooden objects from ancient tombs, scientists can use carbon-14 analysis. Carbon-14 is a beta emitter. Today, as well as throughout recorded history, it has been produced in the atmosphere at a steady rate by cosmic rays (Special Topic 21.3). High-energy neutrons in cosmic rays can transmute nitrogen atoms into carbon-14 atoms as indicated by the following equation (and Figure 21.15).

$$\underset{\substack{\text{neutron} \\ (\textit{in cosmic} \\ \textit{ray})}}{^{1}_{0}\text{n}} + \underset{\substack{\text{nitrogen atom} \\ \text{in the} \\ \text{atmosphere}}}{^{14}_{7}\text{N}} \longrightarrow \underset{\substack{\text{compound} \\ \text{nucleus}}}{^{15}_{7}\text{N*}} \longrightarrow \underset{\substack{\text{carbon-14} \\ \text{atom} (\textit{a} \\ \textit{beta emitter})}}{^{14}_{6}\text{C}} + \underset{\substack{\text{high-energy} \\ \text{proton}}}{^{1}_{1}\text{p}}$$

The newly formed carbon-14 migrates downward into the lower atmosphere where it is eventually oxidized to carbon dioxide. In this form it enters the earth's biosphere through photosynthesis in plants. The plants are eaten by animals, and eventually both plants and animals die and decompose. Undecayed carbon-14 atoms are recycled in the biosphere, and the half-life of carbon-14 is long enough (5730 years) to allowed for many such recyclings. Moreover, more is being made by cosmic rays. The net effect is a grand equilibrium in which all living plants and animals have a known and reasonably constant ratio of carbon-14 to carbon-12. As long as the plant or animal is alive, that ratio is main-

For every 100 atoms of uranium-238 initially in a sample, after radioactive decay for one half-life period, 50 atoms would remain and 50 atoms of lead would have formed—a ratio of 1:1.

The *biosphere* is the region of the earth where life occurs.

Figure 21.15
The formation of carbon-14 following neutron capture by a nitrogen-14 nucleus

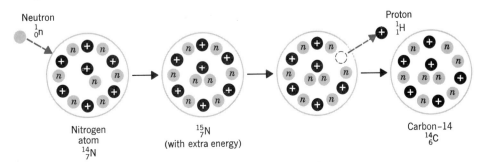

tained. As soon as it dies, however, it no longer takes in any carbon-14. Its remains have as much of this isotope as they will ever have. Because carbon-14 decays at a steady rate, the ratio of carbon-14 to carbon-12 in the remains changes at a known, steady rate, and this ratio can be measured. For example, if the ratio has been cut in half, then the specimen has been dead for one half-life period or 5730 years. Because of a number of uncertainties in the method, artifacts older than about 7000 years cannot yet be dated with acceptable precision by carbon-14 analysis alone.

Background radiation and radiation protection

Scientists, medical personnel, and technicians who work with radionuclides or X ray machines must provide as much protection as practical from repeated exposures to radiations. However, no one can completely escape low-level exposures because sources of radioactivity are everywhere. Together, they make up what is called the natural **background radiation,** and this gives an average of about 100 millirems of exposure per year per person in the United States. Background radiation includes cosmic rays from above and radiations from radionuclides in the earth. The top 40 cm of soil holds an average of 1 g of radium per square kilometer. Carbon-14 is present in all the food we eat. Inside the body of an adult human, about 5×10^5 nuclear disintegrations occur every minute. Additional exposures come from X rays used for diagnosis or therapy, from pollutants released by nuclear power plants, or from pollutants that are present in the smoke from coal-burning power plants. Table 21.3 gives a summary of the amount of radiation in various parts of the background.

Coal usually contains very minute concentrations of radium and thorium, both radioactive.

Although we cannot escape background radiation, we ought to minimize exposures from any other sources. People who routinely work with radioactive materials should use radiation shielding materials and, insofar as it is practical, they should stay away from the source of the radiation. As the data in Table 21.1 show, gamma radiation and X rays are effectively shielded only by very dense materials such as lead. Low-density materials such as cardboard, plastic, or aluminum are poor shielders, although concrete (if thick enough) is a good and relatively inexpensive material for shielding.

Keeping one's distance from a radioactive source is effective in providing radiation protection because the intensity of radiation diminishes with the *square* of the distance from the source. Thus, if a worker doubles the distance from the source, the intensity of the exposure will be reduced by one-fourth. The relationship between distance and intensity is given by the **inverse square law:**

$$\text{radiation intensity} \propto \frac{1}{d^2}$$

where d is the distance from the source. If the intensity, I_1, is known at distance d_1, then the intensity, I_2, at distance d_2 can be calculated by the following equation.

$$\frac{I_1}{I_2} = \frac{d_2^2}{d_1^2}$$

Table 21.3
Average Radiation Doses Received Annually by the United States Population

Source	Dose (mrem)
Natural background (whole body dose equivalent)	100
Medical sources	
Diagnostic X rays	20–50
Therapeutic X rays	3–5
Radioisotopes	0.2
Other sources	
Fallout (from atmospheric testing, 1954–1962)	1.5 (5 mrem maximum in early 1960s)
Radioactive pollutants introduced into the environment from nuclear power plants (1970), United States	0.85
Miscellaneous	2.0[a]
Maximum annual limit proposed for the average, general population (by the International Commission on Radiological Protection, 1970), from all sources *exclusive of medical and background*[b]	170

[a] The one-pack-a-day smoker's lungs get at least 40 mrem per year more dose equivalent. Naturally occurring radon, a radioactive, gaseous decay product of naturally occurring radium, decays to radioactive lead that deposits on earth and foliage (including broad-leaf tobacco plants).

[b] The Advisory Committee on the Biological Effects of Ionizing Radiation of the National Academy of Sciences—National Research Council sharply criticized this recommendation in its report in late 1972. Americans, on the average, receive 100 mrem from natural background according to the committee. To allow the average exposure to climb to 170 mrem would cause a 2% increase in the spontaneous cancer rate.

EXAMPLE 21.2 **Using the Inverse Square Law**

Problem: At 1.5 m from a radioactive source the radiation intensity was 40 units. If the operator moves to a distance of 4.5 m from the source, what will be the radiation intensity?

Solution: We substitute into the equation just given.

$$\frac{40 \text{ units}}{I_2} = \frac{(4.5 \text{ m})^2}{(1.5 \text{ m})^2}$$

$$I_2 = \frac{(40 \text{ units})(1.5)^2 m^2}{(4.5)^2 m^2}$$

$$= 4.4 \text{ units}$$

EXERCISE 21.3 If an operator, at 10 m from a source, is exposed to 1.4 units of radiation, what will be the intensity of the radiation if he moves to 1.2 m from the source?

SUMMARY

Radioactivity. Alpha, beta, and gamma radiations are the three kinds of radiations produced by naturally occurring radionuclides. When a nuclide ejects an alpha particle, the nucleus of a helium atom ($^4_2$He), the nuclide loses four units of mass and drops two units in atomic number. When a nuclide loses a beta particle, an electron ($^0_{-1}$e), its atomic number increases by one unit but its mass number stays the same. (The beta particle forms when a neutron changes to a proton.) In gamma radiation, only high-energy electromagnetic radiation leaves the nuclide. Alpha, beta, and gamma radiations are also emitted by various synthetic nuclides. Some radionuclides emit positrons ($_{1}^{0}e$, "positive electron"), and some decay by internal electron capture. This is accompanied by the production of X rays, and it converts a nuclear proton into a neutron (lowering the atomic number by one unit). Nuclear equations describing nuclear reactions are balanced when the mass numbers and atomic numbers respectively balance.

Nuclear Stability. Radioactive decay often leads to a new nucleus that is still unstable. In the four known radioactive disintegration series, several successive radionuclides form and then decay until stable isotopes form. All stable nuclides fall within a band of stability that reflects stable ratios of neutrons to protons. Nuclei with magic numbers of protons or neutrons (2, 8, 20, 50, 82, 126) tend to be more stable than others. Nuclei with even numbers of both protons and neutrons are more stable than others. The half-life period is a convenient measure of a radionuclide's stability.

Transmutation. Various kinds of radiations can be used to bombard atomic nuclei. When a nucleus captures a bombarding particle, it becomes a high-energy compound nucleus that then decays, and the decay product is often a radionuclide with a half-life long enough to permit its use in some application. By transmutations, nuclides of elements from 93 through 106 have been made, and some higher nuclides have been detected.

Detecting and Measuring Radiations. Radiation detecting and measuring instruments—Geiger counters, scintillation counters, and dosimeters—take advantage of the formation of ions when radiations interact with matter. Several units have been devised to be used in answer to various questions about a particular radiation. For example,

"How *active* is a source?" The curie (Ci) and the becquerel (Bq) can be used in the answer.

$$1 \text{ Ci} = 3.7 \times 10^{10} \text{ disintegrations/sec}$$
$$1 \text{ Bq} = 1 \text{ disintegration/sec}$$

"How much energy is carried by radiation?" The electron-volt (eV), and its multiples (keV and MeV), is the energy unit for this question. One electron-volt is equivalent to 1.6×10^{-9} J, an extremely small amount of energy. "How much energy is absorbed by a unit mass of tissue?" The SI unit of absorbed dose, the gray (Gy), delivers 1 J/kg, and it takes 100 rads, an older unit, to equal 1 Gy. "How can we compare doses absorbed by different tissues that respond differently to the same dose in grays?" The answer is given in units of rems. The number of rems in a dose is calculated from the number of rads multiplied by factors that reflect the effectiveness of the kind of radiation in causing damage.

Applications. Because radiations are easily detected and because radionuclides have the same chemistry as their stable isotopes, radionuclides can be used in such applications in tracer analysis, activation analysis, isotope dilution analysis, and the dating of archeological finds.

Radiation Protection. The existence of background radiation means that everybody is exposed to some radiation. Protection against additional exposure is provided by using dense shielding materials (e.g., lead, thick concrete), avoiding overuse of radiations, and by taking advantage of the inverse square law of radiation intensity. The intensity of the radiation varies inversely with the *square* of the distance from the source; doubling the distance means a fourfold drop in intensity.

INDEX TO EXERCISES, QUESTIONS, AND PROBLEMS
(Those numbered above 48 are review problems.)

REVIEW QUESTIONS

21.4. What is a radionuclide?

21.5. What three kinds of radiation have been observed from naturally occurring radionuclides?

21.6. Give the composition of (a) alpha particle, (b) beta particle, (c) positron, and (d) deuteron.

21.7. Why is the penetrating ability of alpha radiation less than that of beta or gamma radiation?

21.8. If electrons do not exist as such in an atomic nucleus, how can an electron in beta radiation come from a nucleus?

21.9. Complete these nuclear equations by writing the symbols of the missing particles.

(a) $_{82}^{211}Pb \longrightarrow _{-1}^{0}e + \underline{\hspace{1cm}}$

(b) $_{32}^{68}Ge \xrightarrow{\text{electron capture}} + \underline{\hspace{2cm}}$

(c) $_{86}^{220}Rn \longrightarrow _{2}^{4}He + \underline{\hspace{1cm}}$

(d) $_{10}^{19}Ne \longrightarrow _{1}^{0}e + \underline{\hspace{1cm}}$

21.10. Write the symbols of the missing particles to complete these nuclear equations.
(a) $^{245}_{96}Cm \longrightarrow {}^{4}_{2}He + \underline{\hspace{1cm}}$
(b) $^{140}_{56}Ba \longrightarrow {}^{0}_{-1}e + \underline{\hspace{1cm}}$
(c) $^{58}_{29}Cu \longrightarrow {}^{0}_{1}e + \underline{\hspace{1cm}}$
(b) $^{177}_{73}Ta \xrightarrow[\text{electron capture}]{} \underline{\hspace{1cm}}$

21.11. Write a balanced nuclear equation for each of these changes.
(a) Alpha emission from plutonium-242.
(b) Beta emission from magnesium-28.
(c) Positron emission from silicon-26.
(d) Electron capture by argon-37.

21.12. Write the balanced nuclear equation for each of these nuclear reactions.
(a) Electron capture by iron-55.
(b) Beta emission by potassium-42.
(c) Positron emission by ruthenium-93.
(d) Alpha emission by californium-251.

21.13. What is the nuclear equation for each of these nuclear reactions?
(a) Beta emission from aluminum-30.
(b) Alpha emission from einsteinium-252.
(c) Electron capture by molybdenum-93.
(d) Positron emission by phosphorus-28.

21.14. Give the nuclear equation for each of these changes.
(a) Positron emission by carbon-10.
(b) alpha emission by curium-243.
(c) Electron capture by vanadium-49.
(d) Beta emission by oxygen-20.

21.15. Write the symbols, including the atomic number and mass number, for the radionuclides that would give each of these products.
(a) Fermium-100 by alpha emission.
(b) Bismuth-211 by beta emission.
(c) Neodymium-141 by positron emission.
(d) Tantalum-179 by electron capture.

21.16. Each of these nuclides forms by the decay mode specified. Write the symbol of the radionuclide parents, giving both the atomic number and the mass number.
(a) Rubidium-80 formed by electron capture.
(b) Antimony-121 formed by beta emission.
(c) Chromium-50 formed by positron emission.
(d) Californium-253 formed by alpha emission.

21.17. Iodine-123 is better than iodine-131 for diagnostic work in assessing thyroid function. It has a half-life of 13.3 hr (compared to 8 days for I-131), and it decays by electron capture, not beta emission. (a) Why do these properties make I-123 better than I-131 for diagnostic work? (Consult special topic 21.2, also.) (b) Write the equation for the decay of I-123.

21.18. Write the symbol of the nuclide that forms from cobalt-58 when it decays by electron capture.

21.19. With respect to their modes of formation, how do gamma rays and X rays differ?

21.20. What happens in electron capture to generate X rays?

21.21. What is the band of stability?

21.22. Since the majority of unstable nuclides fall inside the band of stability, what is this band used for?

21.23. Both barium-123 and barium-140 are radioactive, but which is likelier to have the longer half-life? Explain your answer.

21.24. Tin-112 is a stable nuclide but indium-112 is radioactive and has a very short half-life (14 min). What does tin-112 have that indium-112 does not have to account for this difference in stability?

21.25. Lanthenum-139 is a stable nuclide, but lanthenum-140 is unstable ($t_{1/2}$ 40 hr). What "rule of thumb" concerning nuclear stability is involved?

21.26. As the atomic number increases, the ratio of neutrons to protons increases. What does this suggest as a factor in nuclear stability?

21.27. Radionuclides of high atomic number are far more likely to be alpha emitters than those of low atomic number. Offer an explanation for this.

21.28. Although lead-164 has two magic numbers, 82 protons, and 82 neutrons, it is unknown, whereas lead-208 is known and stable. What problem exists for lead-164 to account for the fact it cannot exist?

21.29. When vanadium-51 captures a deuteron (^2_1d), what compound nucleus forms? Write the symbol. This compound nucleus expels a proton (^1_1p). Write the nuclear equation for the overall change caused by bombarding vanadium-51 with deuterons.

21.30. The bombardment of fluorine-19 by alpha particles generates sodium-22 and neutrons. Write the nuclear equation for this transmutation including the intermediate compound nucleus.

21.31. Gamma ray bombardment of bromine-81 causes a transmutation in which neutrons are one product. Write the symbol of the other product.

21.32. Neutron bombardment of cadmium-115 results in neutron capture and the release of gamma radiation. Write the nuclear equation.

21.33. When manganese-55 is bombarded by protons, neutrons are released. What else forms? Write the nuclear equation.

21.34. What specific property of radiations is used by the Geiger counter?

21.35. Dangerous doses of radiations can actually involve very small quantities of energy. Explain.

21.36. What radiation units (common and SI) are used to describe the activity of a sample of radioactive material?

21.37. What energy unit is commonly used to describe the energy of a particle or a photon involved in radiation?

21.38. What is the name of the radiation unit that equals 1 disintegration/sec?

21.39. What radiation units (common and SI) are used to describe the quantity of energy absorbed by a mass of tissue?

21.40. A sample giving 3.7×10^{10} disintegrations/sec has what activity in Ci and in Bq?

21.41. In general terms only, how is the rem related to the rad? (What consideration made the development of the rem useful?)

21.42. Why should a radionuclide used in diagnostic work have a short half-life? If the half-life is too short, what problem arises?

21.43. Why isn't an alpha emitter used in diagnostic work?

21.44. In general terms, explain how neutron activation analysis is used and how it works.

21.45. How do cosmic rays produce carbon-14 in the atmosphere?

21.46. The ratio of carbon-14 to carbon-12 is (or is assumed to be) constant in a living organism but changes after the organism dies. Why is it constant during life?

21.47. What is one assumption in the use of the uranium-to-lead ratio for dating ancient geologic formations?

21.48. List some of the sources of exposure to radiations experienced by everybody.

REVIEW PROBLEMS

21.49. If we begin with 3.00 mg of iodine-131 ($t_{1/2}$ 8.07 day), how many mg of I-131 remains after 6 half-life periods?

21.50. A 9 nanogram sample of technetium-99m will still have now much of this radionuclide left after 4 half-life periods (about 1 day)?

21.51. Potassium-40, which has a half-life of 1.3×10^9 yr, decays by beta and gamma emission to argon-40, a stable isotope. If this decay occurs in dense rocks, the argon cannot escape. A 500-mg sample of such rock was analyzed and found to have 2.45×10^{-6} mol of argon-40 and 2.45×10^{-6} mol of potassium-40. How old was the rock? (And what assumption has to be made about the origin of the argon-40?)

21.52. If a sample of dense rock was found to contain 1.16×10^{-7} mol of argon-40, then how much potassium-40 would also have to be present in the same sample for the rock to be 1.3×10^9 years old? (The argon-40 comes from potassium-40 by a decay having a half-life of 1.3×10^9 yr.)

21.53. If exposure from a distance of 1.60 m gave a worker a dose of 8.4 rem, then how far should the worker move away from the source to reduce the dose to 0.50 rem for the same period?

21.54. During work with radioactive source, a worker was told that he would receive 50 mrem of dose at a distance of 4.0 m during 30 min of work. What would happen to the dose received if the worker moved closer, to 0.50 m for the same period?

APPENDIX A
REVIEW OF MATHEMATICS

A.1 EXPONENTIALS

Very large and very small numbers are often expressed as powers of ten. This is called exponential notation or scientific notation. Several examples are given in Table A.1, and in the last column the small number up and to the right of each 10 is called the exponent. In the number 10^4, 4 is the exponent, and we say that the number is 10 raised to the 4th power.

What a positive exponent means

Suppose you come across a number such as 6.4×10^4. This is just an alternative way of writing 64,000. In other words, the number 6.4 is multiplied by 10 *four* times.

$$6.4 \times 10^4 = 6.4 \times 10 \times 10 \times 10 \times 10 = 64,000$$

Instead of actually writing out all of these 10s, notice that the decimal point is simply moved four places to the right in going from 6.4×10^4 to 64,000. Here are a few other examples. Study them to see how the decimal point changes.

$$53.476 \times 10^2 = 5,347.6$$

$$0.0016 \times 10^5 = 160$$

$$0.000056 \times 10^3 = 0.056$$

Here is another example that more forcefully shows the value of using an exponential form of a number.

602,000,000,000,000,000,000,000
$$= 6.02 \times 10 \times 10 \times 10 \times 10 \times 10 \times 10 \times 10$$
$$\times 10 \times 10 \times 10 \times 10 \times 10 \times 10 \times 10 \times 10$$
$$\times 10 \times 10 \times 10 \times 10 \times 10 \times 10 \times 10 \times 10$$

$$= 6.02 \times 10^{23}$$

(This number is called Avogadro's number, which is the number of atoms in 12 g of carbon.)

Table A.1

Number	Exponential Form
1	1×10^0
10	1×10^1
100	1×10^2
1000	1×10^3
10,000	1×10^4
100,000	1×10^5
1,000,000	1×10^6
0.1	1×10^{-1}
0.01	1×10^{-2}
0.001	1×10^{-3}
0.0001	1×10^{-4}
0.00001	1×10^{-5}
0.000001	1×10^{-6}
0.0000001	1×10^{-7}

In some calculations it is helpful to reexpress large numbers in exponential forms. To do this, count the number of places you would have to move the decimal point to the left to put it just after the first digit in the number. For example, you would have to move the decimal point four places left in the following number.

$$\underset{4\ 3\ 2\ 1}{60530}$$

Therefore, we can rewrite 60530 as 6.0530×10^4. Thus the number of moves equals the exponent. Now try these problems.

EXERCISE 1

Write each number in expanded form.
(a) 1.2378×10^{10}
(b) 0.000588×10^6
(c) 45.6×10^5

EXERCISE 2

Write each number in exponential form.
(a) 13,000,000,000,000
(b) 15.68
(c) 6,000,456.568

Answers to Exercises 1 and 2
1. (a) 12,378,000,000   (b) 588
   (c) 4,560,000
2. (a) 1.3×10^{13} (b) 1.568×10^1
   (c) 6.000456568×10^6

What a negative exponent means

Imagine you have just encountered a number such as 1.4×10^{-4}. This is an alternative way of writing $\dfrac{1.4}{10 \times 10 \times 10 \times 10}$. Thus a negative exponent tells us how many times to *divide* by ten. Dividing by ten, of course, can be done by simply moving the decimal point. Thus, 2.4×10^{-4} can be reexpressed as 0.00024. Notice that the decimal point is four places to the *left* of the 2. Here are some examples of numbers expressed as negative exponentials and their equivalents.

$$3567.9 \times 10^{-3} = 3.5679$$

$$0.01456 \times 10^{-2} = 0.0001456$$

$$45691 \times 10^{-3} = 45.691$$

To change a number smaller than 1 into a negative exponential, we count the number of places the decimal has to be moved to put it to the right of the first digit. The number that we count is the negative exponent. For example, you would have to move the decimal point five places to the right to put it to the right of the first digit in

$$0.000068901$$

This number can be reexpressed as 6.8901×10^{-5}. Try these problems and check your answers.

EXERCISE 3

Write each number in expanded form.
(a) 1.45×10^{-4}
(b) 0.00568×10^{-3}
(c) 45623.34×10^{-4}

EXERCISE 4

Write each number in exponential form.
(a) 0.1004
(b) 0.0000000000000428
(c) 0.00400056

Answers to Exercises 3 and 4
1. (a) 0.000145   (b) 0.00000568
   (c) 4.562334
2. (a) 1.004×10^{-1} (b) 4.28×10^{-14}
   (c) 4.00056×10^{-3}

Multiplying numbers written in exponential form

The real value of exponential forms of numbers comes in carrying out multiplications and divisions of large and small numbers. Suppose we want to multiply 1000 by 10000. Doing this by a long process fills the page with zeros (and some 1s). But notice the following relationships.

numbers	1000	×	10000		
or exponential forms	$[10 \times 10 \times 10]$	×	$[10 \times 10 \times 10 \times 10]$	=	10,000,000
	10^3	×	10^4	=	10^7
exponents	3	+	4	=	7

If we *add* 3 and 4, the exponents of the exponential forms of 1000 and 10000, we get the exponent for the answer. Suppose we want to multiply 4160 by 20000. If we reexpress each number in exponential form we have

$$[4.16 \times 10^3] \times [2 \times 10^4]$$

This now simplifies to

$$(4.16) \times (2) \times [10^3 \times 10^4]$$

The answer can be easily worked while, at the same time, the decimal point isn't lost. The *exponents are added* to give the exponent we want in the answer, and the numbers 4.16 and 2 are *multiplied*. The product is 8.32×10^7. Thus there are three steps in multiplying numbers that are more conveniently written in exponential form.

1. First write the numbers in exponential forms (if they aren't already).
2. Next *multiply* the numbers before the 10s.
3. Finally *add* the exponents of the 10s *algebraically*.

Here are some examples.

EXAMPLE 1
Using Exponentials

Problem: What is 3650×0.00000568?
Solution: Step 1. $3650 = 3.650 \times 10^3$
            $0.00000568 = 5.68 \times 10^{-6}$

Step 2. 3650×0.00000568
$$= [3.650 \times 10^3] \times [5.68 \times 10^{-6}]$$
$$= (3.650 \times 5.68) \times (10^3 \times 10^{-6})$$
$$= 20.732 \times 10^{[3+(-6)]}$$
$$= 20.7 \times 10^{-3} \quad \text{(rounded)}$$

EXAMPLE 2
Using Exponentials

Problem: What is 10045×0.025058?
Solution: This works out as follows:

$$[1.0045 \times 10^4] \times [2.5058 \times 10^{-2}]$$
$$= (1.0045) \times (2.5058) \times 10^{(4-2)}$$
$$= 2.5179 \times 10^2$$

For practice, try these problems.

EXERCISE 5

Calculate the products of the following.
(a) $6.20 \times 10^{23} \times 1.50 \times 10^{-4}$
(b) $0.003 \times 0.002 \times 0.004$
(c) $0.00045 \times 45,000,000$
Answers
(a) 9.30×10^{19} (b) 24×10^{-9} or 2.4×10^{-8} (c) 20.25×10^3 or 2.025×10^4

Dividing numbers written in exponential form

Suppose we want to divide 10,000,000 by 1000. Doing this by long division would be primitive to say the least. Notice the following relationships.

$$\frac{10,000,000}{1000} = \frac{10^7}{10^3} = 10^4 = 10000$$

The exponent in the result is $(7 - 3)$. So, to divide numbers in exponential form, we *subtract* exponents. Before continuing, work these problems.

EXERCISE 6

Carry out the following divisions and express the answers in exponential forms.
(a) $1000 \div 10000$
(b) $0.001 \div 0.0001$
(c) $100 \div 0.001$
Answers
(a) $10^3 \div 10^4 = 10^{(3-4)} = 10^{-1}$
(b) $10^{-3} \div 10^{-4} = 10^{[(-3)-(-4)]} = 10^{(-3+4)} = 10^1$
(c) $10^2 \div 10^{-3} = 10^{[(2)-(-3)]} = 10^{(2+3)} = 10^5$

To extend this operation one step, we consider how

to divide 0.00468 by 0.0000400. This may be expressed by using exponentials as follows.

$$\frac{4.68 \times 10^{-3}}{4.00 \times 10^{-5}} = \frac{4.68}{4.00} \times 10^{[(-3)-(-5)]}$$
$$= 1.17 \times 10^{[-3+5]}$$
$$= 1.17 \times 10^2$$

This calculation illustrates the three steps for dividing numbers.

1. Write the numbers in exponential form.
2. Carry out the division of the numbers standing before the 10s.
3. Algebraically subtract the exponents, the exponent in the denominator from the exponent in the numerator.

Now try these exercises.

EXERCISE 7

Carry out the following divisions.
(a) $9593 \div 362,000$
(b) $0.035 \div 0.70$
(c) $450 \div 0.090$
Answers
(a) 2.65×10^{-2} ($9.593 \times 10^3 \div 3.62 \times 10^5$)
(b) 5.0×10^{-2}
(c) 5.0×10^3

Suppose you carry out the steps on the following:

$$\frac{1908 \times 0.00456 \times 45,200,000}{0.1664 \times 2,340,000,000 \times 456}$$

In exponential forms, it is easiest to collect all of the 10s with their exponents in one place and the other numbers together.

$$\frac{(1.908) \times (4.56) \times (4.52) \times (10^3) \times (10^{-3}) \times (10^7)}{(1.664) \times (2.34) \times (4.56) \times (10^{-1}) \times (10^9) \times (10^2)}$$
$$= 2.21 \times 10^{[(3-3+7)-(-1+9+2)]}$$
$$= 2.21 \times 10^{[7-(10)]}$$
$$= 2.21 \times 10^{-3} \quad \text{(rounded)}.$$

EXERCISE 8

In applying the ideal gas equation studied in Chapter 3, you might have to carry out the following calculation.

$$\text{Pressure} = \frac{0.150 \times 0.0821 \times 373}{0.750} \text{ atm}$$

What is the pressure?
Answer. 6.12 atm, rounded

The pocket calculator and exponentials

In scientific work the concept of an exponential is very important because it is much easier to handle either very large or very small numbers in exponential forms. The review we've provided in this appendix is intended to help you refresh your memory from secondary school about these kinds of numbers. No doubt you know that almost any pocket calculator gives you the ability to multiply and divide large or small numbers whether they are reexpressed as exponentials or not. We encourage you to study the *Instruction Manual* for your pocket calculator and read Section A.4 of this appendix so that you can do these kinds of calculations easily. However, we caution you that doing this without a basic understanding of exponentials will leave many traps in your path to skill in performing chemistry calculations.

A.2 LOGARITHMS

Common logarithms

The **common logarithm** or **log** of a number N to the base 10 is the exponent to which 10 must be raised to equal the number N. If

$$N = 10^x$$

then the common logarithm of N, or log N, is simply x.

$$\log N = x$$

There is another kind of logarithm called a natural logarithm that we'll take up later. However, when we use the terms "logarithm" or "log N" we always mean *common* logarithms related to the base 10.

Some examples of logs are as follows. When

$N = 10$	$\log N = 1$	because	$10 = 10^1$
$= 100$	$= 2$		$100 = 10^2$
$= 1000$	$= 3$		$1000 = 10^3$
$= 0.1$	$= -1$		$0.1 = 10^{-1}$
$= 0.00001$	$= -5$		$0.00001 = 10^{-5}$

The logs of these numbers were easy to figure out by just using the definition of logarithms, but usually a log isn't this simple. For example, what is the value of x in the following equations?

$$4.23 = 10^x$$

$$\log 4.23 = x$$

To determine the value of x we have to use a table of logarithms—such as Table A.2—or a hand-held calculator with the capability of using logs. Unless you understand how logs work, however, the calculator will get you into considerable trouble. This is why learning how to use logs and log tables is a necessary development.

To see how to use a log table, we will work an example. However, all log tables give only one part of a log, and a log consists of two parts. The value of log 423,000, for example, is 5.6263—meaning that

$$423,000 = 10^{5.6263}$$

$$\log 423,000 = 5.\overset{}{6263}$$

mantissa (0.6263)

characteristic (5)

The characteristic of a log consists of the digits in front of the decimal point; the mantissa of a log is made up of all the digits after the decimal. *Log tables give mantissas only* (and you have to put in the decimal point yourself). The characteristic is something you figure out, but this is easy because it's the whole-number exponent of 10 when the number N is written in exponential form. Here is one place where being able to write numbers in exponential notation is essential. We have to write (or at least be able to visualize) 423,000 as 4.23×10^5, and the characteristic of the log of this number is simply 5, the exponent. We use the log table to find the mantissa, and *this is always the log of a number between 1 and 10* (after inserting the decimal point). The exponent (characteristic) and the mantissa are then added *algebraically* to give the log.

The reason for this procedure flows out of the properties of exponents when tied to the definition of a logarithm. We know, for example, that the exponent for $10^x \times 10^y$ is $(x + y)$, meaning that

$$10^x \times 10^y = 10^{(x+y)}$$

We also know that we can express 423,000 as

$$423,000 = 10^x \times 10^5 = 10^{(x+5)}$$

Now x has to be some number between 0 and 1, because 423,000 is a number between 100,000 ($10^0 \times 10^5 = 10^5$) and 1,000,000 ($10^1 \times 10^5 = 10^6$); in fact, $423,000 = 4.23 \times 100,000$. Now we can use our definition of a log. If

$$423,000 = 4.23 \times 100,000 = 10^{(x+5)}$$

then

$$\log (4.23 \times 100,000) = \log 10^{(x+5)}$$

$$= x + 5$$

The "5" in $(x + 5)$ goes with the 100,000 (10^5) and the value of x goes with 4.23. We find this value in the log table;

$$x = 0.6263$$

and

$$\log 423{,}000 = 0.6263 + 5$$
$$= 5.6263$$

In other words,

$$423{,}000 = 10^{0.6263} \times 10^5$$
$$= 10^{5.6263}$$

To summarize, when

$$N = M \times 10^x \text{ (where } M \text{ is between 1 and 10)}$$

then

$$\log N = x + \log M$$

where x is the characteristic of $\log N$, and $\log M$ (preceded by a decimal point) is the mantissa.

Let's work some examples to see how to use a log table.

EXAMPLE 3
Using a Log Table to Find a Common Logarithm

Problem: What is the value of log 40?

Solution: The first step with any number not between 1 and 10 is to express it in exponential form:

$$40 = 4 \times 10^1$$

Therefore, $\log 40 = 1 + \log 4$. (The characteristic of the log is 1.) To find the value of log 4, let your finger move down the left-hand column of the log table to "40." Interpret this to mean "4.0". Now move your finger over so that it is in the column headed by "0" at the top of the table. This brings you to 4.00, and your finger is now over 6021. Interpret this to be 0.6021. (Remember: the table omits all of the decimal points in front of mantissa.) Finally, add the "1"—the characteristic—to 0.6021, the mantissa.

$$\log 40 = 1.6021$$

EXERCISE 9

What are the values of log 50? Of log 800?
Answers. log 50 = 1.6990   log 800 = 2.9031

EXAMPLE 4
Using a Log Table to Find a Common Logarithm

Problem: What is the value of log 423,000?

Solution: In exponential form, $423{,}000 = 4.23 \times 10^5$. Therefore

$$\log 423{,}000 = 5 + \log 4.23$$

To find the log of 4.23, go to the log table and move down the left-hand column to 42 (meaning 4.2), then over to the column headed by a "3". Now we are at 4.23. This gives the mantissa of 4.23, 0.6263. Therefore, the answer is

$$\log 423{,}000 = 5.6263$$

EXERCISE 9

Find the logs of each of the following numbers.
(a) 625   (b) 12.900
Answers. (a) 2.7959   (b) 1.1106

EXAMPLE 5
Using a Log Table to Find a Common Logarithm of a Number Less than 1

Problem: What is the value of log 0.000524?

Solution: As usual, we express the number in exponential form: 5.24×10^{-4}. Also as usual, we find the log of 5.24; this is 0.7193. What we now must remember is that we have to add the characteristic, which is -4 (and is treated as an exact number), to the mantissa algebraically:

$$-4 + 0.7193 = -3.2807$$

Remembering this *algebraic* addition is the only extra thing in working with the logs of numbers less than 1.

EXERCISE 10

Find the logs of each of the following numbers.
(a) 0.00000294   (b) 0.872
Answers. (a) -5.5317 (b) -0.0595

In the examples used thus far, the number standing before the 10 in the exponential forms always had no more than three digits. What do we do if there are more? We might want to look up, for example, the value of log 4.238 because we're manipulating some numbers such as 4238 (4.238×10^3). The mantissa we want will be between 0.6263 (for 4.23) and 0.6274 (for 4.24). It is not hard to estimate where the mantissa for 4.238 will be, but instead we'll adopt a simpler rule that will take care of all of the needs in this book. We will round a number like 4.238 or 4.244 to 4.24 *before* looking up the mantissa. (Our needs, in other words, will never exceed three significant figures.)

Logs are powerful tools for handling complex multiplications and divisions, and for these uses we have the following rules:

For multiplications:

$$\text{If } N = A \times B$$
$$\log N = \log A + \log B$$

Appendix A

Table A.2
Logarithms

	0	1	2	3	4	5	6	7	8	9
10	0000	0043	0086	0128	0170	0212	0253	0294	0334	0374
11	0414	0453	0492	0531	0569	0607	0645	0682	0719	0755
12	0792	0828	0864	0899	0934	0969	1004	1038	1072	1106
13	1139	1173	1206	1239	1271	1303	1335	1367	1399	1430
14	1461	1492	1523	1553	1584	1614	1644	1673	1703	1732
15	1761	1790	1818	1847	1875	1903	1931	1959	1987	2014
16	2041	2068	2095	2122	2148	2175	2201	2227	2253	2279
17	2304	2330	2355	2380	2405	2430	2455	2480	2504	2529
18	2553	2577	2601	2625	2648	2672	2695	2718	2742	2765
19	2788	2810	2833	2856	2878	2900	2923	2945	2967	2989
20	3010	3032	3054	3075	3096	3118	3139	3160	3181	3201
21	3222	3243	3263	3284	3304	3324	3345	3365	3385	3404
22	3424	3444	3464	3483	3502	3522	3541	3560	3579	3598
23	3617	3636	3655	3674	3692	3711	3729	3747	3766	3784
24	3802	3820	3838	3856	3874	3892	3909	3927	3945	3962
25	3979	3997	4014	4031	4048	4065	4082	4099	4116	4133
26	4150	4166	4183	4200	4216	4232	4249	4265	4281	4298
27	4314	4330	4346	4362	4378	4393	4409	4425	4440	4456
28	4472	4487	4502	4518	4533	4548	4564	4579	4594	4609
29	4624	4639	4654	4669	4683	4698	4713	4728	4742	4757
30	4771	4786	4800	4814	4829	4843	4857	4871	4886	4900
31	4914	4928	4942	4955	4969	4983	4997	5011	5024	5038
32	5051	5065	5079	5092	5105	5119	5132	5145	5159	5172
33	5185	5198	5211	5224	5237	5250	5263	5276	5289	5302
34	5315	5328	5340	5353	5366	5378	5391	5403	5416	5428
35	5441	5453	5465	5478	5490	5502	5514	5527	5539	5551
36	5563	5575	5587	5599	5611	5623	5635	5647	5658	5670
37	5682	5694	5705	5717	5729	5740	5752	5763	5775	5786
38	5798	5809	5821	5832	5843	5855	5866	5877	5888	5899
39	5911	5922	5933	5944	5955	5966	5977	5988	5999	6010
40	6021	6031	6042	6053	6064	6075	6085	6096	6107	6117
41	6128	6138	6149	6160	6170	6180	6191	6201	6212	6222
42	6232	6243	6253	6263	6274	6284	6294	6304	6314	6325
43	6335	6345	6355	6365	6375	6385	6395	6405	6415	6425
44	6435	6444	6454	6464	6474	6484	6493	6503	6513	6522
45	6532	6542	6551	6561	6571	6580	6590	6599	6609	6618
46	6628	6637	6646	6656	6665	6675	6684	6693	6702	6712
47	6721	6730	6739	6749	6758	6767	6776	6785	6794	6803
48	6812	6821	6830	6839	6848	6857	6866	6875	6884	6893
49	6902	6911	6920	6928	6937	6946	6955	6964	6972	6981
50	6990	6998	7007	7016	7024	7033	7042	7050	7059	7067
51	7076	7084	7093	7101	7110	7118	7126	7135	7143	7152
52	7160	7168	7177	7185	7193	7202	7210	7218	7226	7235
53	7243	7251	7259	7267	7275	7284	7292	7300	7308	7316
54	7324	7332	7340	7348	7356	7364	7372	7380	7388	7396

Table A.2
Logarithms (*continued*)

	0	1	2	3	4	5	6	7	8	9
55	7404	7412	7419	7427	7435	7443	7451	7459	7466	7474
56	7482	7490	7497	7505	7513	7520	7528	7536	7543	7551
57	7559	7566	7574	7582	7589	7597	7604	7612	7619	7627
58	7634	7642	7649	7657	7664	7672	7679	7686	7694	7701
59	7709	7716	7723	7731	7738	7745	7752	7760	7767	7774
60	7782	7789	7796	7803	7810	7818	7825	7832	7839	7846
61	7853	7860	7868	7875	7882	7889	7896	7903	7910	7917
62	7924	7931	7938	7945	7952	7959	7966	7973	7980	7987
63	7993	8000	8007	8014	8021	8028	8035	8041	8048	8055
64	8062	8069	8075	8082	8089	8096	8102	8109	8116	8122
65	8129	8136	8142	8149	8156	8162	8169	8176	8182	8189
66	8195	8202	8209	8215	8222	8228	8235	8241	8248	8254
67	8261	8267	8274	8280	8287	8293	8299	8306	8312	8319
68	8325	8331	8338	8344	8351	8357	8363	8370	8376	8382
69	8388	8395	8401	8407	8414	8420	8426	8432	8439	8445
70	8451	8457	8463	8470	8476	8482	8488	8494	8500	8506
71	8513	8519	8525	8531	8537	8543	8549	8555	8561	8567
72	8573	8579	8585	8591	8597	8603	8609	8615	8621	8627
73	8633	8639	8645	8651	8657	8663	8669	8675	8681	8686
74	8692	8698	8704	8710	8716	8722	8727	8733	8739	8745
75	8751	8756	8762	8768	8774	8779	8785	8791	8797	8802
76	8808	8814	8820	8825	8831	8837	8842	8848	8854	8859
77	8865	8871	8876	8882	8887	8893	8899	8904	8910	8915
78	8921	8927	8932	8938	8943	8949	8954	8960	8965	8971
79	8976	8982	8987	8993	8998	9004	9009	9015	9020	9025
80	9031	9036	9042	9047	9053	9058	9063	9069	9074	9079
81	9085	9090	9096	9101	9106	9112	9117	9122	9128	9133
82	9138	9143	9149	9154	9159	9165	9170	9175	9180	9186
83	9191	9196	9201	9206	9212	9217	9222	9227	9232	9238
84	9243	9248	9253	9258	9263	9269	9274	9279	9284	9289
85	9294	9299	9304	9309	9315	9320	9325	9330	9335	9340
86	9345	9350	9355	9360	9365	9370	9375	9380	9385	9390
87	9395	9400	9405	9410	9415	9420	9425	9430	9435	9440
88	9445	9450	9455	9460	9465	9469	9474	9479	9484	9489
89	9494	9499	9504	9509	9513	9518	9523	9528	9533	9538
90	9542	9547	9552	9557	9562	9566	9571	9576	9581	9586
91	9590	9595	9600	9605	9609	9614	9619	9624	9628	9633
92	9638	9643	9647	9652	9657	9661	9666	9671	9675	9680
93	9685	9689	9694	9699	9703	9708	9713	9717	9722	9727
94	9731	9736	9741	9745	9750	9754	9759	9763	9768	9773
95	9777	9782	9786	9791	9795	9800	9805	9809	9814	9818
96	9823	9827	9832	9836	9841	9845	9850	9854	9859	9863
97	9868	9872	9877	9881	9886	9890	9894	9899	9903	9908
98	9912	9917	9921	9926	9930	9934	9939	9943	9948	9952
99	9956	9961	9965	9969	9974	9978	9983	9987	9991	9996

For divisions:

$$\text{If } N = \frac{A}{B}$$

$$\log N = \log A - \log B$$

For exponentials:

$$\text{If } N = A^b$$

$$\log N = b \times \log A$$

EXERCISE 11

Find the values of the logarithms of each of the following numbers.
(a) $N = 525 \times 0.00346$
(b) $N = 422 \div 0.0451$
(c) $N = 2.68^3$
(d) $N = 81^{1/2}$
Answers. (a) $2.7202 + (-2.4609) = 0.2593$ (b) $2.6253 - (-1.3458) = 3.9711$ (c) $3 \times 0.4281 = 1.2843$ (d) $\frac{1}{2} \times 1.9085 = 0.9543$)

Doing multiplications and divisions by means of logarithms works only if we can use a log found by any of the above operations to figure out the actual number that it is a log of. This number is called the **antilogarithm** or the **antilog.** Thus if

$$N = 10^x$$

and

$$\log N = x$$

then

$$N = \text{antilog of } x$$

Finding an antilog is easy. We just use the log table "in reverse." Suppose we found that the log of some number, say, $81^{1/2}$ is 0.9543—as we did in part (d) of the previous exercise. What number has 0.9543 as its log? This is basically what we ask when we look for an antilog. Go to the log table and find 9543 among the columns of mantissas. The nearest mantissa to it is 9542, the mantissa for the number 9.0. Thus 9.0 is the antilog of 0.9543 (and we know that 9 is the square root of 81, and should be written as 9.0 to have as many significant figures are there are in 81.)

EXAMPLE 6
Finding Antilogs

Problem: What is the antilog of 3.9711?
Solution: The characteristic is 3; therefore, 10^3 must be part of the antilog. A mantissa of 9711 comes closest to 9713 in the log table, and this corre-

sponds to 9.36. (Don't forget the decimal point; mantissas are always for numbers between 1 and 10.) Therefore,

$$\text{antilog } 3.9711 = 9.36 \times 10^3$$

EXAMPLE 7
Finding Antilogs for Negative Logarithms

Problem: What is the antilog of -4.7899?
Solution: It is now vitally important to remember how this log, -4.7899, was obtained. We would have looked up a mantissa that when *algebraically added* to -5 gives -4.7899. In other words,

$$[-5 + (\text{some number})] = -4.7899$$

This "some number" is the mantissa we seek, and its value is $(5 - 4.7899) = 0.2101$. The antilog of -4.7899 is really the antilog of $[-5 + 0.2101]$ so the antilog must have 10^{-5} as part of it. In the log tables the number whose log is 0.2101 is (or is closest to) 1.62. Therefore,

$$\text{antilog } (-4.7899) = 1.62 \times 10^{-5}$$

EXERCISE 11

Find the antilog of each of the following numbers.
(a) -8.4422
(b) -0.2468
(c) -1.2233
Answers. (a) 3.61×10^{-9} [Since $-8.4422 = -9 + 0.5578$; and the antilog of $0.5578 = 3.61$.] (b) 5.67×10^{-1} [Since $-0.2468 = -1 + 0.7532$; and the antilog of 0.7532 is 5.67.] (c) 5.98×10^{-2} [Since $-1.2233 = -2 + 0.7767$; and the antilog of 0.7767 is 5.98.]

EXAMPLE 8
Solving a Problem in Division By Using Logs

Problem: What is 0.0456 divided by 0.00382?

Solution: Remember that the $\log \dfrac{A}{B} = \log A - \log B$

$$\log 0.0456 = \log (4.56 \times 10^{-2})$$
$$= -2 + 0.6590$$
$$= -1.3410$$

$$\log 0.00382 = \log (3.82 \times 10^{-3})$$
$$= -3 + 0.5821$$
$$= -2.4179$$

Therefore,

$$\log \frac{0.0456}{0.00382} = -1.3410 - (-2.4179)$$
$$= 1.0769$$

Next, we find the antilog of 1.0769 to get the answer. We can express it this way.

$$\text{antilog } 1.0769 = 10^1 \times 10^{.0769}$$
$$= (10^1) \times (1.19)$$
$$= 1.19 \times 10^1, \text{ our answer}$$

EXAMPLE 9
Solving a Problem in Multiplication and Roots Using Logs

Problem: What is the value of $423 \times 0.456^{1/4}$?

Solution: We solve this by first finding the log of this expression, remembering that we add logs when we want to multiply and we find logs of numbers involving exponentials by multiplying the log of the number by its exponent.

$$\log 423 = \log(4.23 \times 10^2) = 2.6263$$

$$\log 0.456^{1/4} = \frac{1}{4} \times \log 0.456$$
$$= \frac{1}{4} \times \log(4.56 \times 10^{-1})$$
$$= \frac{1}{4} \times (-1 + 0.6590)$$
$$= \frac{1}{4} \times (-0.3410)$$
$$= -0.0853$$

Therefore,

$$\log(423 \times 0.456^{1/4}) = 2.6263 + (-0.0853) = 2.5410$$

Next we take the antilog:

$$\text{antilog } 2.5410 = 10^2 \times 10^{0.5410}$$
$$= 10^2 \times 3.48$$

Our answer is 3.48×10^2. Since taking roots of numbers, particularly roots higher than square roots, is very difficult, you can see one reason why mathematicians and scientists welcomed the invention of logarithms.

Long strings of multiplications and divisions can be handled through logarithms simply by adding or subtracting logs and then taking the antilog of the result.

EXERCISE 12

Carrying out the necessary operations, use logarithms to find the value of the following.

(a) $\dfrac{4520 \times 0.000362}{89.2}$

(b) $\dfrac{92^{1/8} \times 0.759 \times 298,000}{56.8 \times 0.0467}$

Answers. (a) 1.83×10^{-2} (From: $[3.6551 + (-3.4413)] - 1.9504 = -1.7366 = -2 + 0.2634$ (b) 1.50×10^5 (From $\left[\frac{1}{8} \times 1.9638 + (-0.1198) + 5.4742\right] - [1.7543 + (-1.3307)] = 0.2455 - 0.1198 + 5.4742 - 1.7543 + 1.3307 = 5.1763$

$$\text{antilog } 5.1763 = 10^5 \times 10^{0.1763}$$
$$= 1.50 \times 10^5$$

Natural logarithms

We have just seen how common logarithms are found and used. They are a special case—made special by our need to use the decimal system—for a general idea in mathematics:

$$\text{when} \quad N = B^x$$

then $\quad \log_B N = x \quad$ (read: "log N to the base B equals x")

where B is called the *base*. Therefore, common logs are base 10 logs.

In many applications in science another base is needed, the base e, where

$$e = 2.7182818\ldots \quad \text{(to show just 8 significant figures)}$$

To appreciate the need for this base, you must have a good background in calculus. We can't assume that you have yet acquired such a background, but we can still use the results. Logarithms to the base e are called **natural logarithms,** and their symbol is **ln.** Thus if N is our number, then ln N is the natural logarithm of the number.*

Just as the logarithm of 10 to the base 10 is 1, so the logarithm of e to the base e is also 1. It can also be shown that

$$\log_e 10 = \ln 10 = 2.303$$

or

$$e^{2.303} = 10$$

Tables of natural logarithms are available, and their values can easily be "called up" with good hand-held calculators. They are used in the same way as common logarithms. For most needs, tables of natural logs aren't necessary because of a simple relationship between natural and common logs:

$$\ln N = 2.303 \log N$$

---

* If you have had calculus, then if ln $x = y$, and we take the derivative of x with respect to y, the result is

$$\frac{d(\ln x)}{dy} = \frac{1}{x}$$

Without the base e we would have no function in mathematics whose derivative is $1/x$, and such a function is often needed both in mathematics and in its uses in all branches of science.

In other words, to find ln N, simply find the common log of N and multiply it by 2.303.

A.3 GRAPHING

One of the most useful ways of describing the relationship between two quantities is by means of a graph. It allows us to obtain an overall view of how one quantity changes when the other changes. Constructing graphs from experimental data, or from data calculated from an equation, is therefore a common operation in the sciences, so it is important that you understand how to present data graphically. Let's look at an example.

Suppose that we wanted to know how the volume of a given quantity of a gas varies as we change its pressure. In Chapter Two we find an equation that gives the pressure-volume relationship for a fixed amount of gas at a constant temperature,

$$PV = \text{constant} \qquad \text{(A-1)}$$

where P = pressure and V = volume. For calculations, this equation can be used in two ways. If we solve for the volume, we get

$$V = \frac{\text{constant}}{P} \qquad \text{(A-2)}$$

Knowing the value of the constant for a given gas sample at a given temperature allows us to calculate the volume that the gas occupies at any particular pressure. Simply substitute in the values for the constant and the pressure and compute the volume.

Equation A-1 can also be solved for the pressure.

$$P = \frac{\text{constant}}{V} \qquad \text{(A-3)}$$

From this equation the pressure corresponding to any particular volume can be calculated.

Suppose, now, that we want to see graphically how the pressure and volume of a gas are related as the pressure is raised from 1.0 atm to 10.0 atm in steps of 1.0 atm. Let's also suppose that for this sample of gas the value of the constant in our equation is 0.25 liter atm. To construct the graph we first must decide how to label and number the axes.

Usually, the horizontal axis, called the *abscissa*, is chosen to correspond to the *independent variable*— the variable whose values are chosen first and from which the values of the *dependent variable* are determined. In our example, we are choosing values of pressure (1.0 atm, 2.0 atm, etc.), so we will label the abscissa "pressure." We will also mark off the axis evenly from 1.0 atm to 10.0 atm, as shown in Figure A.1. Next we put the label "volume" on the vertical axis (the *ordinate*). Before we can number this axis, however, we need to know over what range the vol-

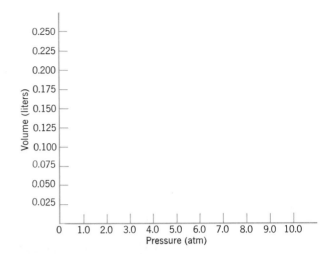

Figure A.1
Choosing and labeling the axes of a graph.

ume will vary. Using Equation A-2 we can calculate the data in Table A.3, and we see that the volume ranges from a low of 0.025 liters to a high of 0.25 liters. We can therefore mark off the ordinate evenly in increments of 0.025 liter, starting at the bottom with 0.025 liter. Notice that in labeling the axes we have indicated the units of P and V in parentheses.

Next we plot the data as shown in Figure A.2. To clearly show each plotted point, we use a small circle that has its center located at the coordinates of the point. Then we draw a *smooth* curve through the points.

Sometimes, when plotting experimentally measured data, the points do not fall exactly on a smooth line (Figure A.3). Nature generally is not irregular even though that's what the data appear to suggest. The fluctuations are usually due to experimental error of some sort. Therefore, rather than draw an irregular line connecting all the data points (the dashed line), we draw a smooth curve that passes as close as possible to all of the points, even though it may not actually pass through any of them.

Table A.3
Pressure-Volume Data

Pressure (atm)	Volume (liters)
1.0	0.25
2.0	0.125
3.0	0.083
4.0	0.062
5.0	0.050
6.0	0.042
7.0	0.036
8.0	0.031
9.0	0.028
10.0	0.025

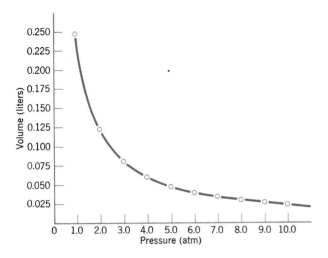

Figure A.2
Plotting points and drawing the curve.

Slope

One of the properties of curves and lines on a graph is their *slope* or steepness. Consider Figure A.4, which is a straight line drawn on a set of *xy* coordinate axes. The slope of the line is defined as the change in *y*, Δy, divided by the change in *x*, Δx.

$$\text{slope} = \frac{\Delta y}{\Delta x} = \frac{y_2 - y_1}{x_2 - x_1}$$

Curved lines have slopes too, but the slope changes from point to point on the curve. To obtain the slope graphically, we can draw a line that is tangent to the curve at the point where we want to know the slope. The slope of this tangent is the same as the slope of the curve at this point. In Figure A.5 we see a curve for which the slope is determined at two different points. At point *M* the curve is rising steeply and $\Delta y/\Delta x$ for the tangent is large. At point *N* the curve is rising less steeply, and $\Delta y/\Delta x$ for the tangent at *N* is small. The slope at *M* is larger than the slope at *N*.

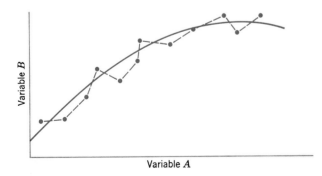

Figure A.3
Experimental data do not all fall on a smooth curve when they are plotted. A smooth curve is drawn as close to all of the data points as possible, rather than the jerky dashed line.

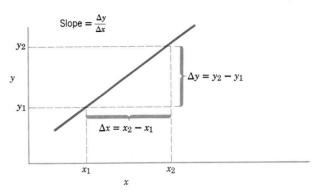

Figure A.4
Determining the slope of a straight line.

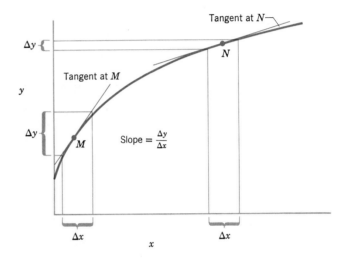

Figure A.5
For a given Δx, Δy is much larger at *M* than it is at *N*. Therefore, the slope at *M* is larger than at *N*.

A.4 ELECTRONIC CALCULATORS

Performing arithmetic calculations is commonplace in all of the sciences, and electronic calculators have made these operations simple and routine. So-called scientific calculators are inexpensive (some cost less than this book) and are able to perform complex functions at the touch of a button. But these electronic wizards are only as versatile as the person who presses the keys.

This section is not intended to teach you how to use your particular calculator—each calculator comes with its own instruction booklet. Instead, our goal is to point out common errors that students make with some kinds of calculators and to explain how to speed up certain types of calculations.

Most scientific calculators use an algebraic entry system in which keys are pressed in the order that the quantities and mathematical operations appear in the computation. Thus

$$2 \times 3 = 6$$

is entered by pressing keys in this order: 2, ×, 3, =. We will indicate this by using boxes to represent the keys.

$$\boxed{2}\ \boxed{\times}\ \boxed{3}\ \boxed{=}$$

When the keys are pressed in this sequence, the answer, 6, appears on the display. In explaining how to use the calculator for certain calculations, we will assume an algebraic entry system.

Scientific notation

Earlier in this appendix and in Section 1.9 we saw that numbers expressed as

$$2.3 \times 10^5$$

$$6.7 \times 10^{-12}$$

are said to be in *exponential notation* or *scientific notation*. Scientific calculators are able to do arithmetic with numbers such as these, provided they are entered correctly.

Many calculators have a key labeled *EE* or *EXP* that is used to enter the exponential portion of numbers in scientific notation. If we read the number 2.3×10^5 as "two point three times ten to the fifth," the *EE* (or *EXP*) key should be read as ". . . times ten to the. . . ." Thus the sequence of keys that enter this number is

$$\boxed{2},\boxed{.},\boxed{3},\boxed{EE},\boxed{5}$$

Try this on your own calculator and check to be sure that the number is correctly displayed after you've entered it. If not, check the instruction booklet that came with the calculator or ask your instructor for help.

When entering a number such as 6.7×10^{-12}, be sure to press the "change sign" or "$+/-$" key after pressing the *EE* key. Most calculators will not give the correct entry if the "minus" key is pressed instead of the "$+/-$" key. For example, try the following sequences of keystrokes on your calculator.

$$\boxed{6}\ \boxed{.}\ \boxed{7}\ \boxed{EE}\ \boxed{+/-}\ \boxed{1}\ \boxed{2}$$

Clear the display and then try this sequence.

$$\boxed{6}\ \boxed{.}\ \boxed{7}\ \boxed{EE}\ \boxed{-}\ \boxed{1}\ \boxed{2}$$

Press the "equals" key and see what happens.

Chain calculations

It is not uncommon to have to calculate the value of a fraction in which there are a string of numbers that must be multiplied together in the numerator and another string of numbers to be multiplied together in the denominator. For example,

$$\frac{5.0 \times 7.3 \times 8.5}{6.2 \times 2.5 \times 3.9} = ?$$

Many students will compute the value of the numerator and write down the answer, then compute the value of the denominator and write it down, and finally divide the value of the numerator by the value of the denominator (whew!). Although this gives the correct answer, there is a simpler way to do the arithmetic that doesn't require writing down any intermediate values. The procedure is as follows:

1. **Enter the first value from the numerator (5.0).**
2. **Any of the other values from the numerator are entered by multiplication and any of the values from the denominator are entered by division.**

The following sequence gives the answer.

$$5.0 \times 7.3 \times 8.5 \div 6.2 \div 2.5 \div 3.9 = 5.13234 \ldots$$

Notice that *each* value from the denominator is entered by division. It also doesn't matter in what sequence the numbers are entered. The following gives the same answer.

$$5.0 \div 6.2 \times 7.3 \div 2.5 \div 3.9 \times 8.5 = 5.13234 \ldots$$

EXERCISE 13

Compute the value of the following fractions:

(a) $\dfrac{3 \times 4 \times 8}{6 \times 2}$

(b) $\dfrac{2.54 \times 1.47 \times 8.11}{9.23 \times 41.3}$

(c) $\dfrac{1}{3.0 \times 2.5 \times 7.3}$

(d) $\dfrac{(2.3 \times 10^{-5}) \times (3.1 \times 10^8)}{(4.3 \times 10^4) \times (8.4 \times 10^{-2}) \times (5.9 \times 10^6)}$

Answers. (a) 8 (b) 0.0794 (c) 0.018 (d) 3.3×10^{-7}
Note that we have followed the rules for significant figures given in Section 1.8.

APPENDIX B

THE ELECTRON CONFIGURATIONS OF THE ELEMENTS

Atomic Number				Atomic Number				Atomic Number			
1	H		$1s^1$	36	Kr	[Ar]	$4s^2 3d^{10} 4p^6$	71	Lu	[Xe]	$6s^2 4f^{14} 5d^1$
2	He		$1s^2$	37	Rb	[Kr]	$5s^1$	72	Hf	[Xe]	$6s^2 4f^{14} 5d^2$
3	Li	[He]	$2s^1$	38	Sr	[Kr]	$5s^2$	73	Ta	[Xe]	$6s^2 4f^{14} 5d^3$
4	Be	[He]	$2s^2$	39	Y	[Kr]	$5s^2 4d^1$	74	W	[Xe]	$6s^2 4f^{14} 5d^4$
5	B	[He]	$2s^2 2p^1$	40	Zr	[Kr]	$5s^2 4d^2$	75	Re	[Xe]	$6s^2 4f^{14} 5d^5$
6	C	[He]	$2s^2 2p^2$	41	Nb	[Kr]	$5s^1 4d^4$	76	Os	[Xe]	$6s^2 4f^{14} 5d^6$
7	N	[He]	$2s^2 2p^3$	42	Mo	[Kr]	$5s^1 4d^5$	77	Ir	[Xe]	$6s^2 4f^{14} 5d^7$
8	O	[He]	$2s^2 2p^4$	43	Tc	[Kr]	$5s^2 4d^5$	78	Pt	[Xe]	$6s^1 4f^{14} 5d^9$
9	F	[He]	$2s^2 2p^5$	44	Ru	[Kr]	$5s^1 4d^7$	79	Au	[Xe]	$6s^1 4f^{14} 5d^{10}$
10	Ne	[He]	$2s^2 2p^6$	45	Rh	[Kr]	$5s^1 4d^8$	80	Hg	[Xe]	$6s^2 4f^{14} 5d^{10}$
11	Na	[Ne]	$3s^1$	46	Pd	[Kr]	$4d^{10}$	81	Tl	[Xe]	$6s^2 4f^{14} 5d^{10} 6p^1$
12	Mg	[Ne]	$3s^2$	47	Ag	[Kr]	$5s^1 4d^{10}$	82	Pb	[Xe]	$6s^2 4f^{14} 5d^{10} 6p^2$
13	Al	[Ne]	$3s^2 3p^1$	48	Cd	[Kr]	$5s^2 4d^{10}$	83	Bi	[Xe]	$6s^2 4f^{14} 5d^{10} 6p^3$
14	Si	[Ne]	$3s^2 3p^2$	49	In	[Kr]	$5s^2 4d^{10} 5p^1$	84	Po	[Xe]	$6s^2 4f^{14} 5d^{10} 6p^4$
15	P	[Ne]	$3s^2 3p^3$	50	Sn	[Kr]	$5s^2 4d^{10} 5p^2$	85	At	[Xe]	$6s^2 4f^{14} 5d^{10} 6p^5$
16	S	[Ne]	$3s^2 3p^4$	51	Sb	[Kr]	$5s^2 4d^{10} 5p^3$	86	Rn	[Xe]	$6s^2 4f^{14} 5d^{10} 6p^6$
17	Cl	[Ne]	$3s^2 3p^5$	52	Te	[Kr]	$5s^2 4d^{10} 5p^4$	87	Fr	[Rn]	$7s^1$
18	Ar	[Ne]	$3s^2 3p^6$	53	I	[Kr]	$5s^2 4d^{10} 5p^5$	88	Ra	[Rn]	$7s^2$
19	K	[Ar]	$4s^1$	54	Xe	[Kr]	$5s^2 4d^{10} 5p^6$	89	Ac	[Rn]	$7s^2 6d^1$
20	Ca	[Ar]	$4s^2$	55	Cs	[Xe]	$6s^1$	90	Th	[Rn]	$7s^2 6d^2$
21	Sc	[Ar]	$4s^2 3d^1$	56	Ba	[Xe]	$6s^2$	91	Pa	[Rn]	$7s^2 5f^2 6d^1$
22	Ti	[Ar]	$4s^2 3d^2$	57	La	[Xe]	$6s^2 5d^1$	92	U	[Rn]	$7s^2 5f^3 6d^1$
23	V	[Ar]	$4s^2 3d^3$	58	Ce	[Xe]	$6s^2 4f^1 5d^1$	93	Np	[Rn]	$7s^2 5f^4 6d^1$
24	Cr	[Ar]	$4s^1 3d^5$	59	Pr	[Xe]	$6s^2 4f^3$	94	Pu	[Rn]	$7s^2 5f^6$
25	Mn	[Ar]	$4s^2 3d^5$	60	Nd	[Xe]	$6s^2 4f^4$	95	Am	[Rn]	$7s^2 5f^7$
26	Fe	[Ar]	$4s^2 3d^6$	61	Pm	[Xe]	$6s^2 4f^5$	96	Cm	[Rn]	$7s^2 5f^7 6d^1$
27	Co	[Ar]	$4s^2 3d^7$	62	Sm	[Xe]	$6s^2 4f^6$	97	Bk	[Rn]	$7s^2 5f^9$
28	Ni	[Ar]	$4s^2 3d^8$	63	Eu	[Xe]	$6s^2 4f^7$	98	Cf	[Rn]	$7s^2 5f^{10}$
29	Cu	[Ar]	$4s^1 3d^{10}$	64	Gd	[Xe]	$6s^2 4f^7 5d^1$	99	Es	[Rn]	$7s^2 5f^{11}$
30	Zn	[Ar]	$4s^2 3d^{10}$	65	Tb	[Xe]	$6s^2 4f^9$	100	Fm	[Rn]	$7s^2 5f^{12}$
31	Ga	[Ar]	$4s^2 3d^{10} 4p^1$	66	Dy	[Xe]	$6s^2 4f^{10}$	101	Md	[Rn]	$7s^2 5f^{13}$
32	Ge	[Ar]	$4s^2 3d^{10} 4p^2$	67	Ho	[Xe]	$6s^2 4f^{11}$	102	No	[Rn]	$7s^2 5f^{14}$
33	As	[Ar]	$4s^2 3d^{10} 4p^3$	68	Er	[Xe]	$6s^2 4f^{12}$	103	Lw	[Rn]	$7s^2 5f^{14} 6d^1$
34	Se	[Ar]	$4s^2 3d^{10} 4p^4$	69	Tm	[Xe]	$6s^2 4f^{13}$				
35	Br	[Ar]	$4s^2 3d^{10} 4p^5$	70	Yb	[Xe]	$6s^2 4f^{14}$				

APPENDIX C

ANSWERS TO EXERCISES AND ODD-NUMBERED REVIEW PROBLEMS

If your answer in a calculation is close to but not exactly the answer given here, the difference might be caused by rounding off at a different time in the calculation than was done in the preparation of these answers.

Chapter 1

1.1   (a) m × m × m = m³ (b) m/s (meter per second)
1.2   (a) nanometer (b) centimeter (c) kilometer (d) picometer (e) millimeter
1.3   86 °F = 30 °C, −17.8 °C = −0.0400 °F
1.4   90 °F = 32 °C, 85 °F = 29 °C
1.5   15 °C
1.6   (a) 1900 (b) 0.08920 (c) 1.82 (d) 14.518
1.7   (a) 1.785 yd (b) 0.001014 mi
1.8   (a) 2.3×10^4 (b) 2.17×10^7 (c) 1.5×10^{-3} (d) 2.7×10^{-5}
1.9   (a) 2700
       (b) 35,000,000,000,000,000,000,000,000,000
       (c) 0.000000000002
1.10 (a) 108 in. (b) 1.25×10^5 cm (c) 0.0107 ft (d) 0.116 liter/km
1.11 2.70 g/ml
1.12 0.272 cm³, 171 g
1.13 (a) 1 Ni, 2 Cl (b) 1 Fe, 1 S, 4 O (c) 3 Ca, 2 P, 8 O
1.14 1 Mg, 2 Cl, 4 H, 2 O
1.51 (a) 0.01 (b) 1000 (c) 1×10^{12} (d) 0.1 (e) 1×10^{-3} (f) 1×10^{-3}
1.53 (a) 1.00×10^{-4} km (b) 5.3×10^3 mg (c) 5.3×10^{-6} kg (d) 3.75×10^{-2} liter (e) 125 ml (f) 3.42×10^{-4} mm
1.55 (a) 91 cm (b) 2.3 kg (c) 2.8×10^3 ml (d) 237 ml (e) 88 km/hr (f) 80.5 km
1.57 3.6×10^2 ml
1.59 2.205×10^3 lb
1.61 (a) 6.7 m² (b) 22 cm² (c) 42 liters
1.63 186 francs
1.65 98.83 °F
1.67 4 K = −269 °C = −452 °F
1.69 (a) 2.45×10^2 (b) 3.1×10^4 (c) 2.87×10^{-3} (d) 4.5×10^7 (e) 4×10^{-8} (f) 3.244×10^5
1.71 (a) 2100 (b) 0.000335 (c) 3,800,000 (d) 0.00000000046 (e) 0.346 (f) 85,000
1.73 (a) 3 (b) 4 (c) 3 (d) 2 (e) 4 (f) 1
1.75 (a) 6.9 (b) 83.14 (c) 0.006 (d) 22.52 (e) 946.5
1.77 62 kg, 1.4×10^2 lb
1.79 12.6 ml

Chapter 2

2.1   Large side down, P = 0.0424 lb/in.²
       When long narrow side down, P = 0.164 lb/in.²
2.2   0.961 atm = 97.3 kPa = 730 mm Hg
2.3   29.9 in. Hg
2.4   14.7 lb/in.²
2.5 1.48×10^3 torr
2.6   190 ml
2.7   159 torr
2.8   732 torr and 299 ml
2.9   It must be cooled to 244 K or −29 °C.
2.10 That would carry the line to negative volumes, which are impossible.
2.11 In $\dfrac{P_1 V_1}{T_1} = \dfrac{P_2 V_2}{T_2}$, if $P_1 = P_2$, these cancel from each side of the equals sign leaving $\dfrac{V_1}{T_1} = \dfrac{V_2}{T_2}$, the temperature-volume law.
2.12 When V is a constant, then equation 2.20 becomes:

$$\frac{P}{T} \times (\text{a constant}) = C''$$

We can then rearrange this into:

$$\frac{P}{T} = \frac{C''}{(\text{a constant})} = \text{a new constant}$$

Or we can express this as: $P = T \times$ (a new constant). This is an expression similar to equation 2.14. But this is the same as saying:

$$P \propto T \text{ (at constant volume and mass)}$$

Therefore, we can state Gay-Lussac's law as follows:

The pressure of a gas is directly proportional to its Kelvin temperature if its volume and mass are constant.

We can write this as the following equation:

$$\frac{P_1}{T_1} = \frac{P_2}{T_2}$$

2.13 2.74×10^3 torr or 3.61 atm

2.14 $P_2 = P_1 \times \left(\dfrac{V_1}{V_2}\right) \times \left(\dfrac{T_2}{T_1}\right)$

$V_2 = V_1 \times \left(\dfrac{P_1}{P_2}\right) \times \left(\dfrac{T_2}{T_1}\right)$

$T_2 = T_1 \times \left(\dfrac{P_2}{P_1}\right) \times \left(\dfrac{V_2}{V_1}\right)$

2.15 688 torr

2.16 Microscopic leaks through which methane can't effuse might be large enough to let hydrogen through. Hydrogen effuses 2.83 times as rapidly as methane through the same leak.

2.17 They stop moving. If KE becomes zero, and the mass of the molecule doesn't disappear, its velocity must vanish; $KE = 1/2 \ mv^2$.

2.18 Put the gas sample in contact with something even colder; or give the molecules negative energies.

2.37 1.39 lb/in.² or 9.46×10^{-2} atm

2.39 744 torr

2.41 75/76 or 0.987 atm

2.43 338 torr

2.45 Nitrogen, 79.293 kPa; Oxygen, 21.333 kPa

2.47 1.14×10^3 torr

2.49 200 torr

2.51

P (torr)	V (ml)	P (torr)	V (ml)
760	100	900	84.5
780	97.4	920	82.7
800	95.0	940	80.9
820	92.7	960	79.2
840	90.5	980	77.6
860	88.4	1000	76.0
880	86.4		

2.53 246 ml

2.55 555 K

2.57 The less dense gas, nitrogen, should effuse more rapidly.

$$\frac{\text{rate nitrogen}}{\text{rate carbon dioxide}} = 1.25$$

Chapter 3

3.1 Mass ratio is 9.01/16.0

3.2 (a) 58.5 (b) 342

3.3 (a) 186 (b) 284

3.4 8.36×10^{24} H_2O molecules

3.5 0.15 mol Fe

3.6 1.0 mol Fe and 1.5 mol O

3.7 (a) 10.5 g (b) 105 g (c) 32.7 g (d) 9.40 g

3.8 (a) 5.551 mol H_2O (b) 0.5549 mol $C_6H_{12}O_6$ (c) 1.791 mol Fe (d) 6.234 mol CH_4

3.9 The molecules of methane have much lower masses than the molecules of glucose, so it takes more of the methane molecules to give a total mass of 100 g.

3.10 20 g/mol

3.11 42.10% C, 6.43% H, 51.45% O

3.12 82.6% C, 17.4% H

3.13 23.63% Cr, 32.74% C, 43.62% O. Yes, analysis consistent with calculated percents for $Cr(CO)_6$.

3.14 HgCl

3.15 Empirical formula is NH_2, molecular formula is N_2H_4.

3.16 15.7 liters

3.17 Formula weight = 2.08 g/mol; likely formula is H_2.

3.18 0.38 liters

3.43 (a) 1.49 g chlorine (b) 5.54 g chlorine (c) 0.596 g Cl/1.00 g Sn (d) 1.19 g Cl/1.00 g Sn (e) $\dfrac{0.596 \text{ g}}{1.19 \text{ g}} = \dfrac{1}{2}$, a whole number ratio.

3.45 (a) 106 (b) 387 (c) 397

3.47 (a) 10.6 g (b) 38.7 g (c) 39.7 g

3.49 0.500 mol H_2O

3.51 4.73×10^3 g $(NH_4)_2SO_4$

3.53 6.18 mol $ZnCl_2$

3.55 3.02×10^{23} atoms N, 7.02 g N

3.57 6.02×10^{23} atoms I, 195 g Ca $(IO_3)_2$

3.59 826 g $(NH_4)_2HPO_4$

3.61 344 mol $(NH_4)_2HPO_4$

3.63 380 g $NaBO_3$

3.65 Actual mol ratio $= \dfrac{9.23 \text{ mol NaOH}}{3.06 \text{ mole tallow}}$

$= 3.02 \approx 3$ (rounded)

3.67 8.5×10^{11} mol NH_3 per year

3.69 (a) CH_3 (b) CH_2O (c) $C_9H_8O_3$

3.71 82.3% N in NH_3, 46.7% N in N_2H_4CO

3.73 83.89% C, 10.35% H, 5.756% N. Data are within specified limits of error and are consistent with formula for phencyclidine.

3.75 (a) 40.00% S, 60.00% O (b) SO_3 (c) SO_3 (molecular weight = empirical formula weight = 80.1)

3.77 (a) oxygen ⓛⓛ, carbon dioxide ⓛ⊕ⓛ, water ⊗ⓛ⊗ (b) CH_4

Chapter 4

4.1 (a) $4P + 5O_2 \rightarrow P_4O_{10}$
(b) $N_2 + 3H_2 \rightarrow 2NH_3$

4.2 (a) $2Mg + O_2 \rightarrow 2MgO$
(b) $CH_4 + 4Cl_2 \rightarrow CCl_4 + 4HCl$
(c) $2NO + O_2 \rightarrow 2NO_2$
(d) $2NaOH + H_2SO_4 \rightarrow Na_2SO_4 + 2H_2O$
(e) $CH_4 + 2O_2 \rightarrow CO_2 + 2H_2O$
(f) $2C_2H_6 + 7O_2 \rightarrow 4CO_2 + 6H_2O$
(g) $2Al(OH)_3 + 3H_2SO_4 \rightarrow Al_2(SO_4)_3 + 6H_2O$

4.3 6 mol SO_3

4.4 45 mol O_2

4.5 (a) $\dfrac{3 \text{ mol } Cl_2}{2 \text{ mol } Fe}$ and $\dfrac{2 \text{ mol } Fe}{3 \text{ mol } Cl_2}$,

$\dfrac{2 \text{ mol } Fe}{2 \text{ mol } FeCl_3}$ and $\dfrac{2 \text{ mol } FeCl_3}{2 \text{ mol } Fe}$,

$\dfrac{3 \text{ mol } Cl_2}{2 \text{ mol } FeCl_3}$ and $\dfrac{2 \text{ mol } FeCl_3}{3 \text{ mol } Cl_2}$

(b) 16 mol $FeCl_3$ (c) 16 mol Fe
(d) 0.750 mol Cl_2, 0.500 mol $FeCl_3$

4.6 9.25×10^{-2} mol Al_2O_3; 9.44 g Al_2O_3

4.7 0.395 mol Al; 10.7 g Al

4.8 0.375 mol O_2; 12.0 g O_2

4.9 0.104 mol O_2; 3.33 g O_2

4.10 5.96×10^{-2} mol CO_2; 2.62 g CO_2

4.11 Total mass reactants = 4.22 g
Total mass products = 4.23 g
They are the same. The slight difference is due to rounding off answers during the calculations.

4.12 3.34 g O_2

4.13 2.62 g CO_2

4.14 For Exercise 4.6: 9.26×10^{-2} mol Al_2O_3, 9.44 g Al_2O_3
For Exercise 4.7: 0.395 mol Al, 10.7 g Al

4.15 12.0 g O_2

4.16 57.8 mol NH_3, 983 g NH_3, 1.29×10^3 liters NH_3

4.17 Sodium

4.18 5.0 ml

4.19 Dissolve 4.20 g $NaHCO_3$ in total volume of 250 ml of solution.

4.20 Dissolve 2.25 g of glucose in sufficient water to give 250 ml of solution.

4.21 17.2 ml KOH solution

4.22 61.4 ml HCl solution

4.23 0.110 M HNO_3

4.24 20 g

4.25 Dissolve 75 g NaOH in sufficient water to make the final mass of the solution 750 g (requires 675 g H_2O).

4.26 Dissolve 7.0 quarts of alcohol in sufficient water to give a final volume of 20 quarts.

4.27 Add sufficient water to 2.0×10^2 ml (200 ml) of 0.50 M NaOH to give a final volume of 5.0×10^2 ml (500 ml).

4.28 31.3 ml

4.29 0.250 M

4.30 2.00×10^{-5} M

4.53 (a) The ratio of the coefficients: 1 mol Mg : 1 mol Cl_2 : 1 mol $MgCl_2$. (b) The actual amounts of reactants, 0.1 mol of each.

4.55 $4Fe + 3O_2 \rightarrow 2Fe_2O_3$

4.57 (a) $Ca(OH)_2 + 2HCl \rightarrow CaCl_2 + 2H_2O$
(b) $2AgNO_3 + CaCl_2 \rightarrow Ca(NO_3)_2 + 2AgCl$
(c) $2Fe_2O_3 + 3C \rightarrow 4Fe + 3CO_2$
(d) $2NaHCO_3 + H_2SO_4 \rightarrow Na_2SO_4 + 2H_2O + 2CO_2$
(e) $2C_4H_{10} + 13O_2 \rightarrow 8CO_2 + 10H_2O$

4.59 (a) $Mg(OH)_2 + 2HBr \rightarrow MgBr_2 + 2H_2O$
(b) $2HCl + Ca(OH)_2 \rightarrow CaCl_2 + 2H_2O$
(c) $Al_2O_3 + 3H_2SO_4 \rightarrow Al_2(SO_4)_3 + 3H_2O$
(d) $2KHCO_3 + H_3PO_4 \rightarrow K_2HPO_4 + 2H_2O + 2CO_2$
(e) $C_9H_{20} + 14O_2 \rightarrow 9CO_2 + 10H_2O$

4.61 (a) 50.0 mol O_2 (b) 8.00 mol CO_2
(c) 54.0 mol H_2O (d) 12.5 mol O_2 and 1.00 mol octane

4.63 (a) 0.137 mol H_2O (b) 2.74×10^{-2} mol C_4H_{10}
(c) 1.59 g C_4H_{10} (d) 0.178 mol O_2, 5.70 g O_2

4.65 (a) 1.45 mol $C_8H_6O_4$, 241 g $C_8H_6O_4$
(b) 1.63×10^{10} mol, 1.73×10^{12} g, 1.73×10^6 metric ton
(c) 91.1%

4.67 (a) 50.0 mol (b) 4.67 mol, 495 g
(c) 1.30×10^{11} mol, 7.61×10^{12} g

4.69 (a) 4.05 mol, 226 g (b) 87.6% (c) 69.8 metric tons, 1.25×10^6 mol

4.71 (a) 1.25 mol, 123 g (b) 2.50 M H_3PO_4
(c) 2.22×10^{10} mol, 6.30×10^6 metric tons P_4O_{10}

4.73 (a) 0.409 mol Zn, 26.7 g Zn (b) 0.818 mol HCl (c) 136 ml

4.75 (a) 6.47×10^{13} liters at STP
(b) 2.89×10^{12} mol
(c) 4.91×10^7 metric tons

4.77 75.2%

4.79 1.0×10^2 g C_6H_6 (2 significant figures)

4.81 (a) 1.46 g NaCl (b) 7.92 g $C_6H_{12}O_6$
(c) 24.5 g H_2SO_4

4.83 (a) 3.48 g NaCl (b) 4.95 g glucose
(c) 225 ml ethylene glycol

4.85 Take 33 ml of 15 M NH_3 and dilute it to 500 ml.

4.87 Dilute 57.5 ml of conc. $HC_2H_3O_2$ to a final volume of 1.00 liter.

4.89 (a) 160 ml K_2CO_3 and 240 ml $CaCl_2$
(b) 174 ml K_2CO_3 and 261 ml $CaCl_2$

4.91 0.1135 M

Chapter 5

5.1 1.20×10^3 cal, 1.20 kcal, 5.02×10^3 J, 5.02 kJ

5.2   9477 cal/g, 2696 kcal/mol

5.3   9.9 kcal/mol NaOH

5.4 $Na(s) + 1/2 H_2(g) + C(s) + 3/2 O_2(g) \rightarrow$ NaHCO₃(s)

5.5 $2C(s) + 4H_2(g) + O_2(g) \rightarrow 2CH_3OH(\ell)$
$$\Delta H° = 2(-57.02) = -114.04 \text{ kcal}$$
$2CH_4(g) \rightarrow 4H_2(g) + 2C(s)$
$$\Delta H° = 2(+17.89) = 35.78 \text{ kcal}$$
net: $2CH_4(g) + O_2(g) \rightarrow 2CH_3OH(\ell)$
$$\Delta H° = -78.26 \text{ kcal}$$

5.6   (a) -21.02 kcal (b) -32.0 kcal

5.7 -531.92 kcal/mol

5.8 -163.4 kcal

5.9 $^{240}_{94}Pu$

5.10   two neutrons

5.77 1.92×10^6 J, 1.92×10^3 kJ

5.79 3.97×10^{-3} BTU

5.81 5.7×10^2 kJ

5.83   3.8 kcal

5.85   (a) 3.5×10^3 kcal (b) 57 mi

5.87   (a) $2Fe(s) + 3/2 O_2(g) \rightarrow Fe_2O_3(s)$
$\Delta H° = -196.5$ kcal
(b) $C(s) + 3/2 H_2(g) + 1/2 Cl_2(g) \rightarrow$
$CH_3Cl(g)$ $\Delta H° = -19.6$ kcal
(c) $C(s) + 1/2 O_2(g) + N_2(g) +$
$2H_2(g) \rightarrow CO(NH_2)_2(s)$ $\Delta H° = -79.634$ kcal

5.89   (a) -46.96 kcal (b) -42.5 kcal (c) -23.8 kcal (d) -31.74 kcal

5.91 $\Delta H_f° = -301$ kcal

5.93 -31.14 kcal

5.95   117.24 kcal

5.97   15.49 kcal, 64.81 kJ

5.99 3.00×10^{18} Hz, X-ray region

5.101

	mass no.	atomic no.	no. neutrons
(a)	235	92	143
(b)	13	6	7
(c)	53	26	27, $^{53}_{26}Fe$

5.103

	protons	electrons	neutrons
(a)	15	15	16
(b)	50	50	69
(c)	98	98	153
(d)	74	74	110

5.105 4.53×10^{-13} J/nucleon

Chapter 6

6.1   (a) K, Ar, Al (b) Cl (c) Ba (d) Ne (e) Li (f) Ce

6.2   (a) NaF (b) Na_2O (c) MgF_2 (d) Al_2S_3

6.3   (a) Na_2CO_3 (b) $(NH_4)_2SO_4$ (c) $KC_2H_3O_2$ (d) $Sr(NO_3)_2$

6.4 $HNO_2 + H_2O \rightleftharpoons H_3O^+ + NO_2^-$

6.5   (a) $HClO_4$ (b) H_2SeO_4

6.6 $HClO_4 > HClO_3 > HClO_2 > HClO$

6.7   (a) HBr (b) H_2Te

Chapter 7

7.1   (a) 3s, 3p, 3d (b) 4s, 4p, 4d, 4f

7.2   nine

7.3   (a) Mg $1s^2\ 2s^2\ 2p^6\ 3s^2$
(b) Ge $1s^2\ 2s^2\ 2p^6\ 3s^2\ 3p^6\ 3d^{10}\ 4s^2\ 4p^2$
(c) Cd $1s^2\ 2s^2\ 2p^6\ 3s^2\ 3p^6\ 3d^{10}\ 4s^2\ 4p^6\ 4d^{10}\ 5s^2$
(d) Gd $1s^2\ 2s^2\ 2p^6\ 3s^2\ 3p^6\ 3d^{10}\ 4s^2\ 4p^6\ 4d^{10}\ 4f^7\ 5s^2\ 5p^6\ 5d^1\ 6s^2$

7.4   (a) Na

(b) S

(c) V

7.5   (a) P [Ne] $3s^2 3p^3$
(b) Sn [Kr] $4d^{10} 5s^2 5p^2$

7.6   (a) N $2s^2 2p^3$
(b) Si $3s^2 3p^2$
(c) Sr $5s^2$

7.7   (a) Ni [Ar] $3d^8 4s^2$
(b) Ru [Kr] $4d^6 5s^2$

7.8   (a) Se $4s^2 4p^4$
(b) Sn $5s^2 5p^2$
(c) I $5s^2 5p^5$

7.9   (a) Sn (b) Ga (c) Fe (d) S^{2-}

7.10 $S(3s^2 3p^4) + 2e^- \rightarrow S^{2-}(3s^2 3p^6)$
$Mg([Ne]3s^2) \rightarrow Mg^{2+}([Ne]) + 2e^-$

7.11   (a) $\cdot \ddot{Se} :$ (b) $\cdot \ddot{I} :$ (c) $\cdot Ca \cdot$

7.12

Chapter 8

8.1   (a)

(b)

8.2

8.3 SO_2, 18; PO_4^{3-}, 32; NO^+, 10

8.4

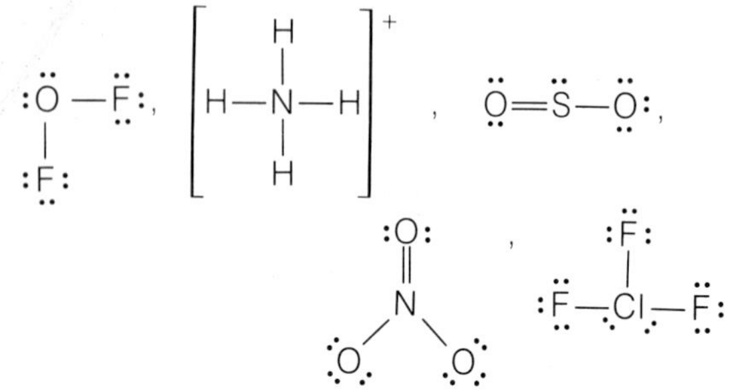

8.5

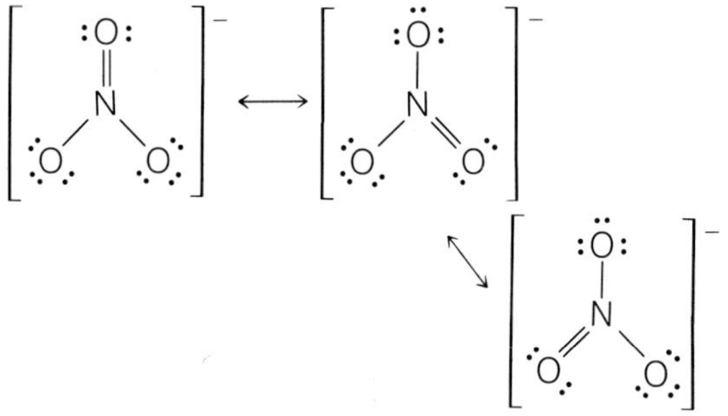

8.6 Trigonal bipyramidal

8.7 ClO_3^-, pyramidal; XeO_4, tetrahedral; OF_2, nonlinear.

8.8 CO_3^{2-}, planar triangular

8.9 Ca is oxidized and is the reducing agent; Cl_2 is reduced and is the oxidizing agent.

$$Ca \rightarrow Ca^{2+} + 2e^- \quad \text{(oxidation)}$$
$$2e^- + Cl_2 \rightarrow 2Cl^- \quad \text{(reduction)}$$

8.10 For $NiCl_2$: Ni, 2+; Cl, 1−. For $MgTiO_4$: Mg, 2+; Ti, 4+; O, 2−. For $K_2Cr_2O_7$: K, 1+; Cr, 6+; O, 2−. For SO_4^{2-}: S, 6+; O, 2−.

8.11 Average oxidation number of C in C_3H_8O is 2−.

8.12 (a) $2KCl + MnO_2 + 2H_2SO_4 \rightarrow K_2SO_4 + MnSO_4 + Cl_2 + 2H_2O$
(b) $2KMnO_4 + 10\ FeSO_4 + 8H_2SO_4 \rightarrow K_2SO_4 + 2MnSO_4 + 5Fe_2(SO_4)_3 + 8H_2O$
(c) $4Zn + 10\ HNO_3 \rightarrow 4Zn(NO_3)_2 + NH_4NO_3 + 3H_2O$

8.13 K_2S, potassium sulfide; Mg_3P_2, magnesium phosphide; $NiCl_2$, nickel(II) chloride; Fe_2O_3, iron(III) oxide.

8.14 (a) Al_2S_3 (b) SrF_2 (c) TiO_2 (d) $CrBr_2$

8.15 PCl_3, phosphorus trichloride; SO_2, sulfur dioxide; Cl_2O_7, dichlorine heptoxide (the *a* of *hepta* omitted for ease of pronunciation).

8.16 HF, hydrofluoric acid; HBr, hydrobromic acid.

8.17 sodium arsenate (*-ate* ending because acid ends in *-ic*)

8.18 $NaHSO_3$

Chapter 9

9.1 The H—Cl bond is formed by the overlap of the half-filled 1*s* orbital of H and the half-filled 3*p* orbital of Cl;

Cl (in HCl) (colored arrow is H electron)

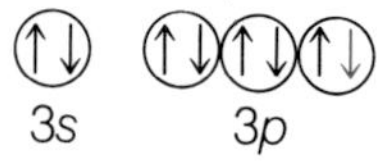

9.2 The half-filled 1*s* orbitals of the H atoms overlap with the half-filled 3*p* orbitals of the P atom. The expected bond angle is 90°.

P (in PH_3) (colored arrows are H electrons)

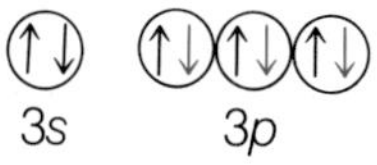

9.3 (a) sp^3 (b) sp^3d

9.4 (a) sp^3 (b) sp^3d

9.5 (a) sp^3d^2

(b) P

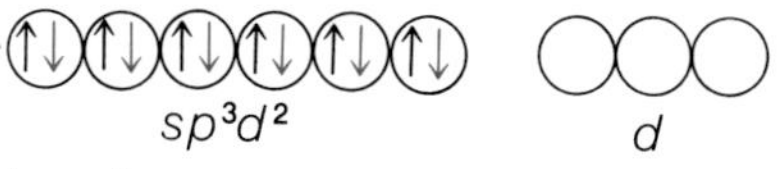

(colored arrows are Cl electrons)

(c) octahedral

9.6 NO has 11 valence electrons

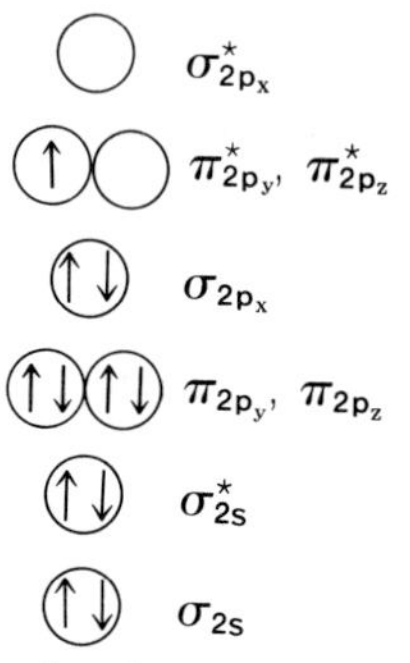

$$\text{bond order} = \frac{8-3}{2} = 2.5$$

9.7 (a) $CH_3—CH_2—CH_3$ [or $CH_3CH_2CH_3$]
(b) $CH_3—CH_2—CH_2—CH_2—CH_3$ [$CH_3CH_2CH_2CH_2CH_3$]
(c) $CH_3—CH_2—CH_2—CH_3$ [or $CH_3CH_2CH_2CH_3$]

(d)

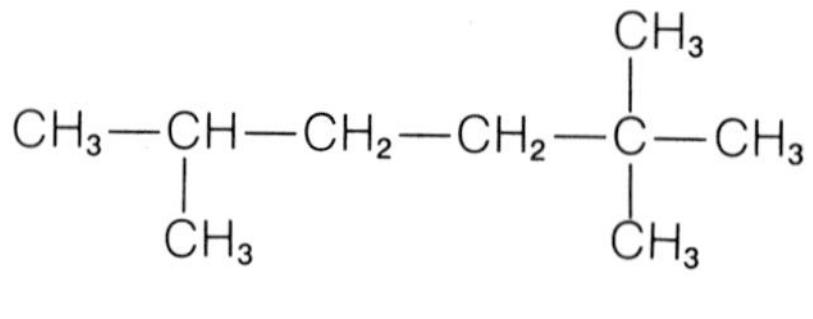

9.8 (a)

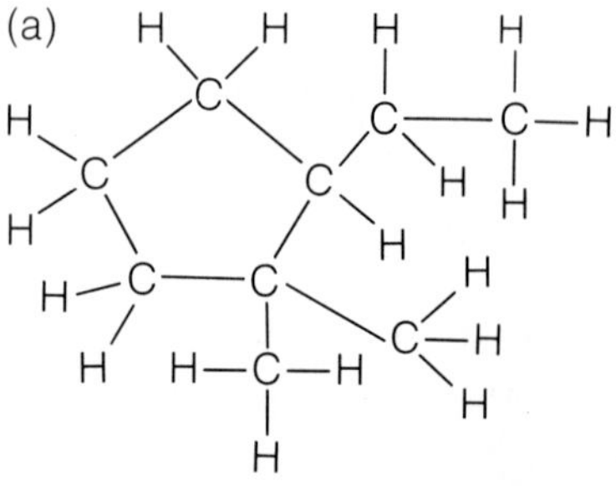

(b)

Chapter 10

10.1 approximately 75 °C
10.2 covalent solid
10.3 molecular solid
10.4 A rise in temperature should increase the vapor pressure.
10.5 Sublimation will occur at −10 °C.
10.6 liquid
10.61 4.07 Å
10.63 0.124 kcal or 124 cal

Chapter 11

11.1 See which solvent dissolves the white solid. If water dissolves it, then it is soap, an ionic compound. If chloroform (a nonpolar solvent) dissolves it, it's candle wax, also nonpolar.
11.2 Moderately polar. The ethyl alcohol molecule must have a waterlike part that helps it dissolve in the polar solvent, and a hydrocarbonlike part that helps it dissolve in the nonpolar benzene.
11.3 $FePO_4 \cdot 2H_2O$
11.4 $Na_3C_6H_5O_7$
11.5 $MgSO_4 \cdot 7H_2O \xrightarrow{150\ °C} MgSO_4 \cdot H_2O + 6H_2O$
11.6 $CaCl_2 + 6H_2O \rightarrow CaCl_2 \cdot 6H_2O$
11.7 8.9×10^{-4} g O_2, 1.5×10^{-3} g N_2
11.8 2.1×10^{-2} g N_2/liter
11.9 16.0 g CH_3OH dissolved in 2000 g H_2O
11.10 0.305 molal
11.11 $X_{CH_3OH} = 0.359$, $X_{H_2O} = 0.641$
       35.9 mol % CH_3OH, 64.1 mol % H_2O
11.12 $X_{NaCl} = 0.0133$, $X_{H_2O} = 0.987$
       1.33 mol % NaCl, 98.7 mol % H_2O
11.13 $X_{O_2} = 0.153$, 15.3 mol % O_2
11.14 $P_{cyclohexane} = 33$ torr, $P_{toluene} = 11$ torr
       $P_{total} = 44$ torr
11.15 For vapor, $X_{cyclohexane} = 0.75$ and $X_{toluene} = 0.25$
       Vapor has greater mole fraction of the more volatile component.
11.16 108.2 °C

11.17 17 g/mol
11.18 −30 °C
11.19 157 g/mol
11.20 168.6 °C
11.21 1.4×10^{-3}
11.22 36 torr
11.23 $Cu^{2+}(aq) + S^{2-}(aq) \rightarrow CuS(s)$
11.24 $3Ca(NO_3)_2(aq) + 2Na_3PO_4(aq) \rightarrow$
       $Ca_3(PO_4)_2(s) + 6NaNO_3(aq)$
11.67 0.035 g/liter
11.69 $X_{toluene} = 0.55$, $X_{chlorobenzene} = 0.45$
11.71 42 mol %
11.73 75 mol % N_2, 14 mol % O_2, 5.3 mol % CO_2, 6.2 mol % H_2O
11.75 1.5×10^2 torr
11.77 (a) 0.100 molal (b) 1.80×10^{-3} (c) 0.998
11.79 23.8 torr
11.81 (a) 22 (b) 1.2×10^3 ml (c) 1.2 qt
11.83 bp = 101.0 °C, fp = −3.72 °C
11.85 152 g/mol
11.87 10% NaCl has a higher osmotic pressure than 10% NaI. Since the formula weight of NaI is greater than that of NaCl, there are more *moles* of solute particles in the NaCl solution.

11.89 (a) $\dfrac{g}{mol} = \dfrac{(g) \times \left(\dfrac{\cancel{liter\ atm}}{mol\ \cancel{K}}\right) \times (\cancel{K})}{(\cancel{atm}) \times (\cancel{liter})} = \dfrac{g}{mol}$

       (b) 1.8×10^6 g/mol

Chapter 12

No exercises or review problems.

Chapter 13

13.1 $CH_2 {=} CH$ with CN above the CH

13.2

13.3

13.4 (a)

(b)

$$
\begin{array}{ccccccc}
& \text{H} & \text{CH}_3 & & \text{H} & \text{CH}_3 & & \text{H} \\
-\text{C}-&\text{C}- & & -\text{C}-&\text{C}- & & -\text{C}- \\
& \text{H} & \text{CO}_2\text{CH}_3 & & \text{H} & \text{CO}_2\text{CH}_3 & & \text{H}
\end{array}
$$

$$
\begin{array}{ccccccc}
\text{CH}_3 & & \text{H} & \text{CH}_3 & & \text{H} & \text{CH}_3 \\
-\text{C}- & & -\text{C}-&\text{C}- & & -\text{C}-&\text{C}- \\
\text{CO}_2\text{CH}_3 & & \text{H} & \text{CO}_2\text{CH}_3 & & \text{H} & \text{CO}_2\text{CH}_3
\end{array}
$$

(c)

$$
\begin{array}{ccccccc}
\text{H} & \text{C}\equiv\text{N} & & \text{H} & \text{C}\equiv\text{N} & & \text{H} \\
-\text{C}-&\text{C}- & & -\text{C}-&\text{C}- & & -\text{C}- \\
\text{H} & \text{CO}_2\text{CH}_3 & & \text{H} & \text{CO}_2\text{CH}_3 & & \text{H}
\end{array}
$$

$$
\begin{array}{ccccccc}
\text{C}\equiv\text{N} & & \text{H} & \text{C}\equiv\text{N} & & \text{H} & \text{C}\equiv\text{N} \\
-\text{C}- & & -\text{C}-&\text{C}- & & -\text{C}-&\text{C}- \\
\text{CO}_2\text{CH}_3 & & \text{H} & \text{CO}_2\text{CH}_3 & & \text{H} & \text{CO}_2\text{CH}_3
\end{array}
$$

Chapter 14

14.1 9.4×10^{-5} mol liter^{-1}s^{-1}

14.2 5.0×10^{-3} mol liter^{-1}s^{-1}

14.3   (a) 8.0×10^{-2} (b) liter mol^{-1}s^{-1}

14.4   Order with respect to HCrO$_4^-$ is 1, H$^+$ is 1, and HSO$_3^-$ is 2. The overall order is 4.

14.5   Each data set gives $k = 2.0 \times 10^2$ liter^{2}mol^{-2}s^{-1}. The values of k are indeed constant.

14.6   First-order

14.7   Rate $= k\,[A]^2[B]^2$, $k = 6.9 \times 10^{-3}$ mol^{-3}liter^{3}s^{-1}

14.8 $t_{1/2} = 18.7$ min. Three-quarters will react in 2 half-lives or 37.4 min.

14.9   The reaction is first-order.

14.10   Rate $= k\,[NO][O_3]$

14.67   At 60 seconds, Rate $= 8.3 \times 10^{-4}$ mol liter^{-1}s^{-1}
At 120 seconds, Rate $= 3.7 \times 10^{-4}$ mol liter^{-1}s^{-1}

14.69   For NO$_2$, Rate $= 5.0 \times 10^{-6}$ mol liter^{-1} s^{-1}
For O$_2$, Rate $= 1.3 \times 10^{-6}$ mol liter^{-1}s^{-1}

14.71 1.0×10^{-8} mol liter^{-1}s^{-1}

14.73   zero order

14.75   Rate $= k\,[ICl][H_2]$, $k = 1.5 \times 10^{-1}$ liter mol^{-1}s^{-1}

14.77 7.5×10^3 s

14.79   From 0.200 M to 0.100 M it takes about 66 seconds, from 0.100 M to 0.050 M it takes

about 128 seconds. The reaction is second-order.

14.81   140 years

Chapter 15

15.1   (a) -54.8 cal/K (b) -28.90 cal/K

15.2 $\Delta G° = -354.3$ kcal

15.3   (a) -16.66 kcal (b) -28.7 kcal

15.4   788 kJ

15.5   614 K or 341 °C

15.6 $\Delta G° = -16.73$ kcal; the reaction should occur spontaneously.

15.7 $\Delta G° = +12.88$ kcal; we should not expect to see any CaCO$_3$(s) formed by this reaction.

15.39   (a) $\Delta H° = -600.2$ kcal; favored
(b) $\Delta H° = -397.5$ kcal; favored
(c) $\Delta H° = -202.6$ kcal; favored
(d) $\Delta H° = +42.24$ kcal; not favored
(e) $\Delta H° = +73.82$ kcal; not favored

15.41   Probability of all 4 in a particular container is 1/16. Probability of an even distribution (2–2) is 6/16. Gases expand spontaneously because the expanded condition has a higher statistical probability than the unexpanded condition.

15.43   (a) -9.5 cal/K (b) -10.60 cal/K
(c) -88.99 cal/K (d) 62.0 cal/K
(e) -68.0 cal/K

15.45   (a) -74.85 cal/K (b) -88.9 cal/K
(c) -71.1 cal/K (d) -89.4 cal/K
(e) -139.3 cal/K

15.47 -32.5 cal/K

15.49 -376.8 kcal/mol

15.51   (a) -44.0 kcal (b) $+0.2$ kcal (c) $+2.04$ kcal
(d) -126.9 kcal (e) -21.98 kcal

15.53 -35.0 kcal

15.55 1.35×10^3 kJ or 323 kcal

15.57   1.51 kcal

15.59 1.06×10^3 K or 787 °C

15.61   97.0 J/mol K or 23.2 cal/mol K

15.63 $\Delta G° = -95.5$ kcal; therefore, the reaction will be observed to be spontaneous.

Chapter 16

16.1 $K_c = \dfrac{[H_2O]^2}{[H_2]^2[O_2]}$, $K_c = \dfrac{[NH_4^+][OH^-]}{[NH_3][H_2O]}$

16.2 $K_p = \dfrac{p_{HI}^2}{p_{H_2}p_{I_2}}$

16.3   Reaction 2 will go farthest to completion.

16.4 $\Delta G° = -7.96$ kcal

16.5 $K_p = 0.27$

16.6   (a) Decrease the amount Cl$_2$
(b) Increase the amount of Cl$_2$
(c) Increase the amount of Cl$_2$
(d) Decrease the amount of Cl$_2$

Only the temperature change will alter K_p. An increase in temperature will lower K_p.

16.7 [CO] decreased by 0.060 mol/liter; $[CO_2]$ increased by 0.060 mol/liter.

16.8 $K_c = 4.06$

16.9 (a) initially, $[PCl_3] = 0.20$ mol/liter, $[Cl_2] = 0.10$ mol/liter, $[PCl_5] = 0.0$ mol/liter
(b) $[PCl_3]$ decreases by 0.08 mol/liter, $[Cl_2]$ decreases by 0.08 mol/liter, $[PCl_5]$ increases by 0.08 mol/liter
(c) Equilibrium concentrations: $[PCl_3] = 0.12\ M$, $[Cl_2] = 0.02\ M$, $[PCl_5] = 0.08\ M$
(d) $K_c = 3 \times 10^1$

16.10 $8.98 \times 10^{-3}\ M$

16.11 $[HI] = 0.312\ M$, $[H_2] = [I_2] = 0.044\ M$

16.12 $1.1 \times 10^{-17}\ M$

16.13 (a) $AgNO_3(aq) + NH_4Cl(aq) \rightarrow AgCl(s) + NH_4NO_3(aq)$
(b) $Na_2S(aq) + Pb(C_2H_3O_2)_2(aq) \rightarrow PbS(s) + 2NaC_2H_3O_2(aq)$
(c) $BaCl_2(aq) + NH_4NO_3(aq) \rightarrow$ no reaction

16.14 $Ba(OH)_2(aq) + H_2SO_4(aq) \rightarrow BaSO_4(s) + 2H_2O$ (molecular)
$Ba^{2+} + 2\ OH^- + 2H^+ + SO_4^{2-} \rightarrow BaSO_4(s) + 2H_2O$ (ionic)
This is also the net ionic equation.

16.15 (a) $K_c = \dfrac{1}{[Cl_2]}$ (b) $K_c = \dfrac{1}{[NH_3][HCl]}$

16.16 (a) $K_{sp} = [Ba^{2+}][CrO_4^{2-}]$
(b) $K_{sp} = [Ag^+]^3[PO_4^{3-}]$

16.17 3.98×10^{-8}

16.18 8.0×10^{-13}

16.19 4.1×10^{-8}

16.20 7.1×10^{-7} mol/liter

16.21 1.2×10^{-4} mol/liter

16.22 7.5×10^{-16} mol/liter

16.23 8.8×10^{-33}

16.24 Ion product $= 7.5 \times 10^{-5} > K_{sp}$. Therefore a precipitate will form.

16.25 Ion product $= 5.9 \times 10^{-50} < K_{sp}$. No precipitate will form.

16.61 $K_p = 8.00 \times 10^8$. Yes, the equilibrium is in favor of the products.

16.63 $K = 1$

16.65 $\Delta G' = 21.1$ kJ $= 5.04$ kcal

16.67 $K_c = 300$

16.69 $[H_2] = 0.53$ mol/liter

16.71 $K_c = 5.0 \times 10^{-3}$

16.73 $[H_2] = [CO_2] = 7.75 \times 10^{-3}\ M$, $[CO] = [H_2O] = 1.23 \times 10^{-2}\ M$

16.75 $[HI] = 6.8 \times 10^{-12}\ M$, $[Cl_2] = 3.4 \times 10^{-12}\ M$, $[HCl] = 1.00\ M$

16.77 (a) 1.43×10^{-4} mol/liter (b) $[Mg^{2+}] = 1.43 \times 10^{-4}$ mol/liter and $[OH^-] = 2.86 \times 10^{-4}$ mol/liter (c) $K_{sp} = 1.17 \times 10^{-11}$

16.79 6.0×10^{-39}

16.81 Molar solubility of $CaCO_3$ is 9.3×10^{-5} mol/liter and 9.3×10^{-4} g $CaCO_3$ will dissolve in 100 ml.

16.83 2×10^{-5} mol/liter

16.85 1.2×10^{-9} mol/liter

16.87 Ion product $= 6.0 \times 10^{-3} > K_{sp}$; a precipitate of $AgC_2H_3O_2$ should form.

Chapter 17

17.1 Brønsted acid: $H_2PO_4^-(aq)$
Brønsted base: $HCO_3^-(aq)$

17.2 (a) OH^- (e) HPO_4^{2-}
(b) I^- (f) PO_4^{3-}
(c) NO_2^- (g) H^-
(d) $H_2PO_4^-$ (h) NH_3

17.3 (a) H_2O (e) NH_3
(b) HSO_4^- (f) NH_4^+
(c) HPO_4^{2-} (g) H_3PO_4
                                 (h) $H_2PO_4^-$
(d) $CH_3 - \overset{\displaystyle O}{\overset{\|}{C}} - O - H$

17.4

$$PO_4^{3-}(aq) + HC_2H_3O_2(aq) \longrightarrow HPO_4^{2-}(aq) + C_2H_3O_2^-(aq)$$

17.5 1.28×10^{-9} mol/liter, basic

17.6 pH $= 7.14$, slightly basic

17.7 (a) $5.6 \times 10^{-3}\ M$ (b) $4.0 \times 10^{-4}\ M$
(c) $3.2 \times 10^{-11}\ M$ (d) $3.2 \times 10^{-4}\ M$
(e) $7.9 \times 10^{-12}\ M$

17.8 (a) acidic
(b) acidic
(c) basic
(d) acidic
(e) basic

17.9 $K_a = \dfrac{[H^+][NO_2^-]}{[HNO_2]}$

17.10 $K_a = \dfrac{[H^+][PO_4^{3-}]}{[HPO_4^{2-}]}$

17.11 $K_a = 1.7 \times 10^{-5}$

17.12 $K_a = 1.1 \times 10^{-4}$

17.13 $[H^+] = 3.7 \times 10^{-4}\ M$, pH $= 3.43$

17.14 (a) $K_b = \dfrac{[HCN][OH^-]}{[CN^-]}$

(b) $K_b = \dfrac{[HC_2H_3O_2][OH^-]}{[C_2H_3O_2^-]}$

(c) $K_b = \dfrac{[C_6H_5NH_3^+][OH^-]}{[C_6H_5NH_2]}$

17.15 1.7×10^{-6}

17.16 $[OH^-] = 6.0 \times 10^{-4}$, pOH $= 3.22$, pH $= 10.78$

17.17 3.34

17.18 Acetic acid is stronger.

17.19 3.19

17.20 $pK_b = 4.69$, $CN^- + H_2O \rightleftharpoons HCN + OH^-$

$$K_b = \frac{[HCN][OH^-]}{[CN^-]}$$

17.21 (a) a Brønsted base
(b) a Brønsted base
(c) a Brønsted base
(d) not a Brønsted base
(e) not a Brønsted base
(f) not a Brønsted base

17.22 It hydrolyzes; the solution is basic.

17.23 It does not hydrolyze

17.24 It hydrolyzes. It tends to lower the pH and make the soil more acidic.

17.25 4.62

17.26 $[H^+] = 6.7 \times 10^{-5} M$
$[S^{2-}] = 1.1 \times 10^{-12} M$
pH = 4.17

17.27 (a) 8.90 (b) 2.89

17.28 (a) $pH = pK_a + 1$
(b) $pH = pK_a - 1$

17.29 3.93

17.30 It is still 4.83.

17.31 Yes; mol ratio, $\frac{[HCO_2^-]}{[HCO_2H]} = 1.4$

17.32 (a) 63.0 g HNO_3/Eq, 56.1 g KOH/Eq
(b) 41.1 g H_2SO_3/Eq, 40.0 g NaOH/Eq
(c) 29.2 g $Mg(OH)_2$/Eq, 120 g $NaHSO_4$/Eq

17.33 (a) 12.0 g H_3PO_4 (b) 6.65×10^{-2} Eq H_2SO_4

17.34 (a) 0.5000 N NaOH
(b) 7.296×10^{-2} N H_2SO_4
(c) 2.70×10^{-1} N Na_2CO_3

17.35 0.560 N

17.36 Lewis acid: CO_2
Lewis base: OH^-

17.37 Al is in the same family as B. Since BF_3 has a sextet of outer shell electrons on the central atom, the trihalide of aluminum, $AlCl_3$, should also have a sextet of outer shell electrons on its central atom.

17.65 $[H^+] = [OH^-] = 1.56 \times 10^{-7} M$,
pH = 6.81 = pOH
$pH + pOH = pK_w = 13.62$. Water is neutral.

17.67 pH = 2

17.69 pOH = 0.82, pH = 13.18

17.71 pH = 4.72

17.73 pH = 2.14, 4.9% ionized

17.75 $K_a = 2 \times 10^{-2}$, $pK_a = 1.7$

17.77 $[H^+] = 6.0 \times 10^{-4} M$, pH = 3.22

17.79 $K_b = 5.6 \times 10^{-4}$; $pK_b = 3.25$; for $CH_3CH_2NH_3^+$, $pK_a = 10.75$

17.81 pH = 10.26

17.83 H-(Mor)$^+$ has $pK_a = 8.21$, pH = 4.46

17.85 $[H^+] = 1.0 \times 10^{-1} M$, pH = 1.0, $[HPO_3^{2-}] = 2.6 \times 10^{-7} M$

17.87 0.27 mol

17.89 1:1 ratio

17.91 pH = 8.23; a good indicator would be cresol red or thymol blue.

17.93 0.192 N

17.95 17.5 N, 17.5 M. The shipment was acceptable.

17.97 (a) 0.2000 N (b) 0.4000 N

17.99 (a) 3.75×10^{-6} mol HCl (b) 3.75×10^{-6} mol morphine (c) 1.07×10^{-3} g morphine
(d) 8.70×10^{-2}%

Chapter 18

18.1 $2Br^-(aq) + H_2O \rightarrow Br_2(aq) + H_2(g) + 2 OH^-(aq)$

18.2 $3Cu + 2NO_3^- + 8H^+ \rightarrow 3Cu^{2+} + 2NO + 4H_2O$

18.3 $2MnO_4^- + 3C_2O_4^{2-} + 4 OH^- \rightarrow 2MnO_2 + 6CO_3^{2-} + 2H_2O$

18.4 8.29×10^{-3} mol OH^-

18.5 7.35 min

18.6 3.67 A

18.7

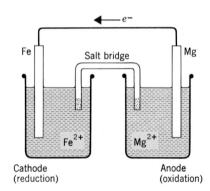

Cathode
(reduction)

Anode
(oxidation)

18.8 $E°_{Mg^{2+}} = -2.37$ V

18.9 $NiO_2(s) + Fe(s) + 2H_2O \rightarrow Ni(OH)_2(s) + Fe(OH)_2(s)$ $E°_{cell} = 1.37$ V

18.10 $3MnO_4^-(aq) + 24H^+(aq) + 5Cr(s) \rightarrow 5Cr^{3+}(aq) + 3Mn^{2+}(aq) + 12 H_2O$
$E°_{cell} = 2.23$ V

18.11 $Br_2(aq) + H_2SO_3(aq) + H_2O \rightarrow 2Br^-(aq) + SO_4^{2-}(aq) + 4H^+(aq)$

18.12 (a) $E°_{cell} = -0.56$ V; nonspontaneous
(b) $E°_{cell} = +0.02$ V; spontaneous

18.13 -264 kJ

18.14 $K_c = 1 \times 10^{-16}$; hardly any products will be formed.

18.15 $E_{cell} = 1.04$ V

18.65 6 $\mathscr{F}$ needed.

18.67 3.36 g $Fe(OH)_2$

18.69 187 ml

18.71 13 hrs (to two significant figures)

18.73 5.1×10^4 A

18.75 $E°_{cell} = +0.01$ V

18.77 $MnO_2(s) + 4H^+(aq) + Pb(s) + 2Cl^-(aq) \rightarrow$
$Mn^{2+}(aq) + 2H_2O + PbCl_2(s)$ $E°_{cell} =$
$+1.50$ V

18.79 $Br_2(aq) + 2I^-(aq) \rightarrow I_2(s) + 2Br^-(aq)$

18.81 $E°_{cell} = +1.52$ V; therefore, the reaction is
spontaneous.

18.83 $\Delta G° = +1.0 \times 10^2$ kJ

18.85 $K_c = 0.31$

18.87 $E_{cell} = 0.105$ V

Chapter 19

19.1 (a) no chiral carbon

(b)
$$H-N-\overset{H}{\underset{\underset{H}{\overset{|}{C}}-H}{\overset{|}{\underset{|}{C^*}}}-\overset{O}{\overset{\|}{C}}-O-H \qquad \begin{matrix} NH_2- \\ H- \\ CH_3- \\ -CO_2H \end{matrix}$$

(c)
$$H-O-\overset{H}{\underset{|}{C}}-\overset{H}{\underset{|}{\underset{OH}{C^*}}}-\overset{O}{\overset{\|}{C}}-H \qquad \begin{matrix} HOCH_2- \\ H- \\ HO- \\ -CHO \end{matrix}$$

19.2
$$\overset{+}{N}H_3-\overset{|}{\underset{CH_2CH_2CH_2CH_2NH_2}{CH}}-CO_2^-$$

19.3
$$\overset{+}{N}H_3-\overset{|}{\underset{CH_3}{CH}}-\overset{O}{\overset{\|}{C}}-NH-\overset{|}{\underset{\underset{CH_3}{CH-CH_3}}{CH}}-\overset{O}{\overset{\|}{C}}-O^-$$

$$\overset{+}{N}H_3-\overset{|}{\underset{\underset{CH_3}{CH-CH_3}}{CH}}-\overset{O}{\overset{\|}{C}}-NH-\overset{|}{\underset{CH_3}{CH}}-\overset{O}{\overset{\|}{C}}-O^-$$

19.4
$$\overset{+}{N}H_3-\overset{|}{\underset{CH_3}{CH}}-\overset{O}{\overset{\|}{C}}-NH-CH_2-\overset{O}{\overset{\|}{C}}-NH-$$

$$-\overset{|}{\underset{CH_2-SH}{CH}}-\overset{O}{\overset{\|}{C}}-O^-$$

$$\overset{+}{N}H_3-\overset{|}{\underset{CH_2-SH}{CH}}-\overset{O}{\overset{\|}{C}}-NH-\overset{|}{\underset{CH_3}{CH}}-\overset{O}{\overset{\|}{C}}-NH-$$

$$-CH_2-\overset{O}{\overset{\|}{C}}-O^-$$

19.5
$$\overset{+}{N}H_3-\overset{|}{\underset{CH_3}{CH}}-\overset{O}{\overset{\|}{C}}-NH-\overset{|}{\underset{CH_3}{CH}}-\overset{O}{\overset{\|}{C}}-NH-\overset{|}{\underset{CH_3}{CH}}-$$

$$-\overset{O}{\overset{\|}{C}}-O^-$$

The only one.

19.6
$$\overset{+}{N}H_3-\overset{|}{\underset{CH_3}{CH}}-\overset{O}{\overset{\|}{C}}-NH-\overset{|}{\underset{\underset{CH_3}{CH-CH_3}}{CH}}-\overset{O}{\overset{\|}{C}}-NH-$$

$$-\overset{|}{\underset{CH_2CH_2CO_2H}{CH}}-\overset{O}{\overset{\|}{C}}-NH-CH_2-\overset{O}{\overset{\|}{C}}-O^-$$

Chapter 20

No exercises or review problems.

Chapter 21

21.1 $^{226}_{88}Ra \rightarrow ^{222}_{86}Rn + ^4_2He + \gamma$

21.2 $^{90}_{38}Sr \rightarrow ^{90}_{39}Y + ^0_{-1}e$

21.3 97 units

21.49 4.69×10^{-2} mg

21.51 1.3×10^9 years old. Assumptions: all ^{40}Ar
came from ^{40}K and no other isotope that
could give ^{40}Ar was present.

21.53 6.6 m

GLOSSARY

Accuracy: Freedom from error. How close a measured quantity is to the true value.

Acid: A substance that produces $H_3O^+(H^+)$ when it is dissolved in water (Arrhenius). A proton donor (Brønsted). An electron-pair acceptor (Lewis).

Acid Ionization Constant, K_a: $K_a = \dfrac{[H^+][A^-]}{[HA]}$ for the equilibrium, $HA \rightleftharpoons H^+ + A^-$.

Acid Rain: Rain made acidic by dissolved sulfur and nitrogen oxides.

Acid Salt: A salt of a partially neutralized polyprotic acid; for example, $NaHSO_4$ and $NaHCO_3$.

Acid-Base Indicator: A dye with one color below a narrow pH range (on the acid side) and a different color above this range (on the basic side).

Acidic Solution: An aqueous solution in which $[H^+] > [OH^-]$.

Actinide Elements (Actinide Series): Elements 90–103.

Activated Complex: The chemical species that exists with partly broken and partly formed bonds in the transition state.

Activation Energy: The minimum kinetic energy that must be possessed by the reactants in order to give an effective collision (one that produces the products).

Addition Polymer: A polymer that forms from monomers by an addition process without the monomers losing any of their structural parts.

Addition Reaction: An organic reaction in which a molecule of one reactant (e.g., H—H) adds to a double (or triple) bond of another reactant.

Alcohol: An organic compound whose molecules have the —O—H group attached to a carbon having only single bonds.

Aldehyde: An organic compound with the $-\overset{\overset{\text{O}}{\|}}{\text{C}}-$H group.

Alkali Metals (Alkalis): Elements of Group IA, except hydrogen.

Alkaline Earth Metals: Elements of Group IIA.

Alkane: A hydrocarbon whose molecules have only single bonds.

Alkene: A hydrocarbon whose molecules have one or more double bonds.

Alkyl group: An organic group of atoms consisting of an alkane less one of its hydrogen atoms (e.g., CH_3—, methyl; CH_3CH_2—, ethyl).

Allotrope: One of the two or more forms of an element.

α-Amino Acid: One of about 20 monomers of polypeptides all having the general structure: $^+NH_3-\overset{\overset{\displaystyle}{|}}{\underset{\underset{\displaystyle G}{|}}{CH}}-CO_2^-$

where G is a sidechain group (or H).

α-Helix: A secondary structure of polypeptides in which the chain is twisted in a right-handed coil that is stabilized by hydrogen bonds.

Alpha Particle: The nucleus of a helium atom—two protons and two neutrons—emitted by radioactive decay. Its symbol is ^4_2He (and it has a charge of $2+$).

Alpha Radiation: A stream of alpha particles produced by radioactive decay.

Alum: A double salt with the general formula: $M^+M^{3+}(SO_4)_2 \cdot 12H_2O$; for example, $KAl(SO_4)_2 \cdot 12H_2O$.

Amalgam: A solution of a metal in mercury.

Amide: An organic compound whose molecules have a carbonyl-nitrogen group: $-\overset{\overset{\text{O}}{\|}}{\text{C}}-\overset{\overset{\displaystyle |}{}}{\text{N}}-$

Amine: An organic compound with one, two, or three hydrocarbon groups joined by single bonds to a nitrogen atom.

Amorphous Solid: A noncrystalline solid. It lacks the long-range order found in crystals. A glass.

Ampere (A): The SI unit for electric current; one coulomb per second.

Amphoteric compound: A compound that can react either as an acid or as a base.

Amphoteric Oxide: An oxide that will neutralize either an acid or a base.

Ångstrom (Å): 1 Å = 10^{-10} m = 100 pm = 0.1 nm.

Anion: A negative ion.

Anode: The electrode where oxidation occurs during an electrochemical change.

Antibonding Molecular Orbital: A MO that removes electron density from between nuclei and destabilizes a molecule.

Antimatter: A particle that is the counterpart of a particle of ordinary matter (as the positron is antimatter to the electron) and with which it undergoes mutual annihilation when the two collide to give gamma radiation.

Aqua Regia: One part concentrated HNO_3 and three parts concentrated HCl by volume.

Aqueous Solution: Any solution having water as the solvent.

Aromatic Compound: Any organic compound with flat rings having delocalized π-electron networks and which give substitution reactions (like benzene) rather than addition reactions.

Atmosphere, Standard: The pressure that supports a column of mercury 760 mm high at 0 °C.

Atomic Mass Unit (amu): 1 amu = 1.6606×10^{-24} g, $\frac{1}{12}$ of the mass of a carbon-12 atom.

Atomic Number: The number of protons in a nucleus.

Atomic Spectrum: The line spectrum produced when energized or excited atoms emit light.

Atomic Weight: The average relative mass of an element's atoms on a scale using atoms of carbon-12 as the reference. Also called the atomic mass.

Avogadro's Law: Equal volumes of gases contain equal numbers of molecules (provided they are compared at identical temperatures and pressures).

Avogadro's Number: 6.02×10^{23}; the number of things in 1 mole of things.

Background Radiation: The radiation to which all people are exposed caused by the presence of natural radionuclides in the soil and other parts of the environment, radioactive pollutants, and by cosmic radiations.

Balance: Apparatus used to measure mass by comparing an unknown mass with known masses.

Balanced Equation: A chemical equation having the same number of atoms and the same net charge on both sides of the arrow.

Band of Stability: An envelope enclosing all stable nuclides when all known nuclides are given locations on a plot of number of neutrons versus number of protons.

Barometer: A device for measuring pressure.

Base: A substance that produces OH^- when it is dissolved in water (Arrhenius). A proton acceptor (Brønsted). An electron-pair donor (Lewis).

Base Ionization Constant, K_b: $K_b = \dfrac{[BH^+][OH^-]}{[B]}$ for the equilibrium: $B + H_2O \rightleftharpoons BH^+ + OH^-$.

Base Units: Fundamental units of the SI.

Basic Oxygen Process: A relatively fast method used to convert pig iron into steel.

Basic Solution: An aqueous solution in which $[H^+] <$ $[OH^-]$.

Battery: One or more galvanic cells arranged to serve as a practical source of electricity.

Becquerel (Bq): The SI unit for activity in nuclear chemistry. 1 Bq = 1 disintegration/sec.

Beta particle: An electron emitted in radioactive decay. Its symbol is $_{-1}^{0}e$.

Beta Radiation: A stream of electrons produced by radioactive decay.

Bidentate Ligand: A ligand that has two atoms that can become simultaneously attached to the same metal ion.

Binary Acid: An acid with the general formula H_nX, where X is a nonmetal.

Binary Compound: A compound composed of two different elements (e.g., HCl and Al_2O_3).

Binding Energy, Nuclear: The energy equivalent of the difference in mass between an atomic nucleus and the sum of the masses of its nucleons.

Blast Furnace: An apparatus that is used to reduce iron ore to metallic iron.

Body-Centered Cubic (bcc) Unit Cell: A cubic unit cell having atoms, molecules, or ions at the corners plus one in the center of the cell.

Boiling Point: The temperature at which the vapor pressure of a liquid is equal to the atmospheric pressure.

Bond Angle: When one atom forms a bond to each of two other atoms, the angle between the two bonds is the bond angle.

Bond Distance: See *Bond Length.*

Bond Energy: The amount of energy needed to separate two bonded atoms to give electrically neutral particles.

Bond Length: The distance between two nuclei that are joined by a chemical bond.

Bond Order: (number of bonding e^- − number antibonding e^-)/2; the *net* number of *pairs* of bonding electrons.

Bonding Molecular Orbital: A MO that gives a buildup of electron density between nuclei and helps stabilize a molecule.

Boyle's Law: See *Pressure-Volume Law.*

Bragg Equation: $n\lambda = 2d \sin \theta$. The equation is used to analyze X-ray diffraction data obtained from crystals.

Branched Chain Compound: An organic compound in whose molecules the carbon atoms do not all occur in a continuous sequence.

Branching Step: A step in a chain reaction that produces more free radicals than it consumes.

Breeder Reactor: An atomic reactor in which a fissile isotope is continuously made ("bred") from a fertile isotope.

Brine: An aqueous sodium chloride solution.

Brønsted Acid: A donor of H^+.

Brønsted Base: An acceptor of H^+.

Brownian movement: The random, erratic motions of colloidally dispersed particles in a fluid medium and that can be observed by using a microscope.

Buffer: (a) A pair of solutes that, together in water, keep the pH of the solution almost constant despite any addition of acids or bases. (b) A solution containing this pair of salts.

Calibration: A check on the accuracy and the precision of the stated capacity of a volumetric flask or pipet, or of the stated mass of some weight. Any check on the accuracy and precision of a stated value associated with any instrument of measurement.

Calorie (cal): The calorie is the energy that will raise the temperature of 1.00 g of water from 14.5 to 15.5 °C. The Calorie is 1000 cal (or 1 kilocalorie), and is the term used to describe the energy content of foods. 1 cal = 4.184 J.

Carbohydrate: A substance in animals or plants consisting of a polyhydroxyaldehyde or polyhydroxyketone or molecules that can be hydrolyzed to these.

Carbon Family: Group IVA in the periodic table; carbon, silicon, germanium, tin, and lead.

Carboxylic Acid: A compound having the carboxyl group,

$$\begin{matrix} & O \\ & \| \\ -C&-O-H \end{matrix}$$

Carbonyl Group: The carbon-oxygen double bond:

$$>C=O$$

Catalysis: The phenomenon of rate-enhancement brought about by a catalyst.

Catalyst: A substance that in a relatively small proportion will accelerate the rate of a reaction without itself being permanently changed, chemically.

Cathode: The electrode where reduction occurs during an electrochemical change.

Cation: A positive ion.

Cell Potential, E_{cell}: The emf that can be produced by any particular galvanic cell when no current is drawn from the cell.

Cell Reaction: The overall chemical change that takes place in an electrolytic cell or a galvanic cell.

Celsius Scale: Temperature scale on which water freezes at 0 °C and boils at 100 °C at a pressure of 1 atm.

Centimeter (cm): 0.01 m = 1 cm; 100 cm = 1 m.

Chain Reaction: A self-sustaining reaction in which the products from one event cause one or more new events.

Change of State: Transformation of matter from one state to another, for example, from liquid to solid.

Charles' law: See *Temperature-Volume Law.*

Chelate: A complex ion containing rings formed by polydentate ligands.

Chelate effect: The extra stability found for complexes that contain chelate rings.

Chemical Bond: Force of attraction that holds atoms together in compounds.

Chemical Equation: A before-and-after description of a chemical reaction.

Chemical Property: Property relating to how a substance interacts with others to form new substances.

Chemical Reaction: The interaction of substances causing them to form new substances.

Chemistry: The study of the composition of substances and the way their properties are related to their composition.

Chiral Carbon: A carbon holding four different atoms or groups.

Chirality: The handedness of a molecular structure such that it cannot pass the test of superimposability with the model of the mirror image of this structure.

Cholesteric Liquid Crystals: Rodlike molecules similar in structure to cholesterol arranged in layers in which the parallel rods in one layer are oriented in a different direction than the parallel rods in an adjoining layer.

Codon: An individual unit of hereditary instruction that consists of three amine sidechains, side by side, on an *m*RNA molecule.

Coefficients: Numbers in front of formulas in a chemical equation.

Coenzyme: A nonprotein, organic substance that forms part of an enzyme.

Coinage Metals: Copper, silver, and gold.

Coke: Coal that has had its volatile components driven off at high temperature. It is mostly carbon.

Colligative Property: A property whose physical value depends only on the ratio of the numbers of solute and solvent particles and not on their chemical identities; vapor pressure lowering, boiling point elevation, freezing point depression, osmotic pressure.

Collision Theory: A theory of reaction rates that postulates that the rate of a reaction is proportional to the number of collisions that occur each second between the reactants.

Colloidal Dispersion: A mixture in which the particles of one (or more) substance are smaller than those in suspensions but larger than those in solutions and that have one dimension in the range of 1 to 10 nm.

Colloidal Osmotic Pressure: That part of the osmotic pressure caused by colloids.

Common Ion: An ion that is common to more than one salt. Na^+ is the common ion in NaCl and $NaNO_3$.

Common Ion Effect: The solubility of a salt is less in a solution containing one of its ions than it is in pure water.

Complex Ion (or simply a *Complex*): A substance formed when one or more anions or neutral molecules become bonded to a metal ion.

Compound: Substance formed from two or more different elements always combined in a fixed ratio.

Compound Nucleus: An atomic nucleus carrying excess energy following the capture of some bombarding particle.

Concentrated Solution: Any solution with a large ratio of the amounts of solute to solvent.

Concentration: The ratio of the quantity of solute to the volume of the solution. See *Molar Concentration; Percent Concentration.*

Condensation Copolymer: A polymer whose formation involves the splitting out of small molecules from the functional groups of the monomers.

Conformations: The different possible relative orientations of the atoms in a molecule.

Conjugate Acid-Base Pair: Two substances, ions or molecules, whose formulas differ only by one H^+. (The acid is the species with this H^+ and the base is the species without this H^+.)

Conservation of Energy, Law of: The energy of the universe is constant; it can be neither created nor destroyed but only transferred and transformed.

Conservation of Mass-Energy, Law of: The sum of all the mass in the universe and of all the energy, expressed as an equivalent in mass, is a constant.

Contact Process: An industrial synthesis of sulfuric acid from sulfur in which one stage, the oxidation of sulfur dioxide to sulfur trioxide, is catalyzed by contact with vanadium pentoxide.

Conversion Factor: Fraction constructed from a relationship between units; for example, 2.54 cm/1.00 in.

Coordinate Covalent Bond: A covalent bond in which both electrons are contributed by only one of the two joined atoms. Once formed, it is no different than any other covalent bond.

Coordination Compound: A complex ion or its salt.

Coordination Number: The number of donor atoms that surround a metal ion.

Copolymer: A polymer made from two or more different monomers.

Coulomb (C): The SI unit for electrical charge; the charge on 6.25×10^{18} electrons, the amount of charge that passes a fixed point along a wire when a current of 1 A flows for 1 s.

Covalent Bond: A chemical bond formed by the sharing of electrons between two atoms.

Covalent Crystal: A crystal in which lattice positions are occupied by atoms covalently bonded to other atoms at neighboring lattice sites.

Critical Pressure, P_c: The vapor pressure of a substance at its critical temperature.

Critical Temperature, T_c: The temperature above which a substance cannot exist as a separate liquid phase, regardless of the pressure.

Crystal field splitting, Δ: The difference in energy between sets of d orbitals in a complex.

Crystal Field Theory: A theory that considers the effects of the polar or ionic ligands of a complex on the energies of the d orbitals of the central metal ion.

Crystal Lattice: The repeating symmetrical pattern of atoms, molecules, or ions that occurs in a crystal.

Curie (Ci): A unit of activity for radioactive substances. 1 Ci = 3.7×10^{10} disintegrations/sec.

Dalton's Atomic Theory: A theory that matter consists of tiny, indestructible particles called atoms; that all atoms of one element are identical; that atoms of different elements have different masses; and that atoms combine in definite ratios *by atoms* when they form compounds.

Dalton's Law of Partial Pressures: See *Partial Pressures, Law of.*

Data: The information (often numbers) obtained in an experiment.

Decimal Multipliers: Factors that modify the size of SI units.

Deliquescence: Absorbtion of moisture by a solid to the extent that the solid dissolves in the absorbed water.

Delocalized Molecular Orbital: A MO that spreads over more than two nuclei.

$\Delta G°$: See *Standard Free Energy Change.*

$\Delta H°$: See *Enthalpy Change, Standard.*

ΔH_{fusion}: See *Molar Heat of Fusion.*

$\Delta H_{sublimation}$: See *Molar Heat of Sublimation.*

$\Delta H_{vaporization}$: See *Molar Heat of Vaporization.*

Denaturation: The loss of biological function by a protein brought about by any action that disorganizes the molecular shape.

Density: An object's mass divided by its volume.

Derived Units: SI units defined in terms of the base units.

Desiccant: A chemical drying agent.

Deuterium: Hydrogen-2; ^2_1H (sometimes ^2_1d); an isotope of hydrogen.

Deuterium Isotope Effect: The reduction in the rate of a reaction caused by the substitution of a deuterium atom for a hydrogen atom at a molecular site undergoing the reaction.

Dialysis: The passage of small molecules and ions through a semipermeable membrane, one that prevents the passage of large molecules or ions in the colloidal state.

Diatomic: A molecule is diatomic if it is composed of *two* atoms (e.g., N_2, HCl).

Diffraction: Constructive and destructive interference by waves.

Diffraction Pattern: The image formed on a screen or on film by diffraction of visible light or X rays.

Dilute Solution: Any solution with a small ratio of the quantities of solute to solvent.

Dipeptide: A dimer of two α-amino acids joined by a peptide bond.

Dipole: A molecule having partial positive and negative charges separated by a distance.

Dipole Moment: The product of the partial charge on either end of a dipole multiplied by the distance between the partial charges. It is a measure of the extent of polarity of a molecule.

Dipole-Dipole Attractions: Attractions between molecules that are dipoles.

Diprotic Acid: An acid that can furnish two H^+ per molecule.

Disaccharide: A carbohydrate whose molecules can be hydrolyzed to two monosaccharides.

Disproportionation: A redox reaction in which a portion of a substance is oxidized while the rest is reduced. The same species undergoes both oxidation and reduction.

DNA: Deoxyribonucleic acid; nucleic acid that, when hydrolyzed, gives deoxyribose, phosphate ion, and chiefly four heterocyclic amines—adenine, thymine, guanine, and cytosine; the carrier of the genetic messages of an organism.

Double Bond: A covalent bond in which two pairs of electrons are shared.

Double Replacement Reaction (Metathesis Reaction): A reaction between two salts in which cations and anions exchange partners, for example, $AgNO_3 + NaCl \rightarrow AgCl + NaNO_3$.

Double Salt: Crystals that contain the components of two different salts in a definite ratio.

Ductility: A metal's ability to be drawn (stretched) into wire.

Dynamic Equilibrium: An equilibrium in which two opposing processes are occurring at equal rates.

Effusion: The movement of a gas through a very tiny opening into a region of lower pressure.

Effusion, Law of (Graham's Law): The rate of effusion of a gas is inversely proportional to the square root of its density when the pressure and temperature are constant.

$$\frac{(\text{effusion rate})_A}{(\text{effusion rate})_B} = \sqrt{\frac{d_B}{d_A}}$$

Einstein Equation: $\Delta E = \Delta mc^2$ where ΔE is the energy obtained when a quantity of mass, Δm, is destroyed, or the energy lost when this quantity of mass is created.

Elastomer: A polymer with elastic properties.

Electrochemical Change: A chemical change that is caused by or that produces electricity.

Electrochemistry: The study of electrochemical changes.

Electrolysis: A chemical change caused by the passage of electricity through a molten ionic compound or through a solution that contains ions.

Electrolysis Cell: An electrolysis apparatus.

Electrolyte: A compound that either in the molten state or in solution conducts electricity.

Electrolytic Cell: An electrolysis apparatus.

Electromagnetic Energy: The energy transmitted by wave-like oscillations in the strengths of electrical and magnetic fields; light energy.

Electromagnetic Radiation: The successive series of oscillations in the strengths of electrical and magnetic fields associated with light, microwaves, gamma rays, and the like.

Electromagnetic Spectrum: The distribution of frequencies of electromagnetic radiation among various classifications such as microwave, infrared, visible, ultraviolet, X-ray, and gamma-ray bands of radiation.

Electromotive Force (emf): The voltage produced by a galvanic cell.

Electron: A subatomic particle with a charge of $1-$ and a mass of 0.0005486 amu (9.109534×10^{-28} g) that occurs outside the nucleus. The particle that moves when an electric current flows. Its symbol is e^- in discussions of chemical changes or $_{-1}^{0}e$ in discussions of nuclear changes.

Electron Affinity (EA): The energy change that occurs when an electron is added to an isolated gaseous atom or ion (usually expressed in kJ/mol).

Electron Capture: A nuclear reaction in which a nucleus captures an orbital electron that changes a nuclear proton into a neutron.

Electron Cloud: Because of its wave properties, the electron is spread out like a cloud around the nucleus.

Electron Configuration: The distribution of electrons in an atom's orbitals.

Electron Density: The concentration of the electron's charge within a given volume.

Electron Spin: A property that the electron appears to have because it behaves like a tiny magnet.

Electronegativity: The relative attraction that an atom has for the electrons in a bond.

Electronic Structure: The distribution of electrons in an atom's orbitals.

Electron-Pair Bond: A covalent bond.

Electron-Volt (eV): The energy an electron receives when it is accelerated under the influence of 1 V 1 eV = 1.6×10^{-19} J.

Electroplating: Depositing a thin metallic coating on an object by electrolysis.

Element: Simplest substance ever obtained in a chemical reaction.

Elementary Process: One of the individual steps in a reaction mechanism.

Empirical Facts: Facts discovered by performing experiments.

Empirical Formula: A chemical formula in which the subscripts are the smallest whole numbers that will express the proportions of the atoms of the different elements present.

Emulsifying Agent: Any substance that stabilizes an emulsion.

Emulsion: A colloidal dispersion of one liquid in another.

Enantiomers: Stereoisomers whose molecular structures are related as an object to its mirror image and that cannot be superimposed.

End Point: The step in a titration when the indicator changes color and the titration is ended.

Endergonic: A change that occurs with a free energy increase.

Endothermic: Descriptive of a change in which energy leaves the surroundings and enters a system.

Energy: Something that matter possesses if it is able to do work.

Energy Level: A particular energy that an electron can have in an atom or molecule.

Energy Source: Any natural change or any material that can be induced to undergo a change that people can use to make their own tasks easier.

Enthalpy, *H*: The heat content of a system.

Enthalpy Change, Standard, $\Delta H°$: The heat of reaction at constant pressure measured under standard conditions (25 °C and 760 torr).

Enthalpy of Solution: The enthalpy change accompanying the formation of a solution when a solute dissolves in a solvent.

Entropy: The thermodynamic quantity that describes the degree of randomness of a system. The greater the disorder or randomness, the higher is the statistical probability of the state and the higher is the entropy.

Enzyme: A catalyst in a living organism; a protein that catalyzes biochemical reactions.

Enzyme-Substrate Complex: A molecular association of enzyme and substrate molecules that forms as the enzyme catalyzes a reaction of the substrate.

Equation, Thermochemical: A balanced chemical equation accompanied by the value of $\Delta H°$ for the mole quantities actually given by the coefficients.

Equilibrium Constant: The value that the mass action expression has when a chemical system is at equilibrium.

Equilibrium Law: An equation that sets the mass action expression equal to the equilibrium constant.

Equilibrium Vapor Pressure: The pressure exerted by a vapor in equilibrium with its liquid.

Equilibrium Vapor Pressure of a Solid: The pressure exerted by a vapor that is in dynamic equilibrium with a solid.

Equivalence Point: The point in a titration when the number of equivalents of one reactant (e.g., an acid) equals the number of equivalents of a second reactant (e.g., a base).

Equivalent (Eq): The grams of an acid that provide 1 mole of H^+ or the grams of a base that neutralize 1 mole of H^+. In a redox reaction, one equivalent of reducing agent loses 1 mole of electrons, and one equivalent of oxidizing agent gains 1 mole of electrons.

Equivalent Weight: If an acid, the mass that provides one mole of H^+. If a base, the mass that neutralizes one mole of H^+. In a redox reaction, the mass of reactant that either gains or loses 1 mole of electrons.

Ester Group: A compound with the carbonyl-oxygen-carbon group:

$$-\overset{\overset{\textstyle O}{\|}}{C}-O-\overset{/}{C}\diagdown$$

Ether: An organic compound with two hydrocarbon groups attached to one oxygen atom.

Exact Numbers: Numbers that come from a direct count or that result from definitions. They have an infinite number of significant figures.

Exergonic: A change that occurs with a free energy decrease.

Exothermic: Descriptive of a change in which energy leaves a system and enters the surroundings.

Exponential Notation: See Scientific notation.

Face-Centered Cubic (fcc) Unit Cell: A cubic unit cell having atoms, molecules, or ions at the corners and in the center of each face.

Factor-Label Method: Problem-solving method that uses the cancellation of units to aid in setting up the proper arithmetic in a problem.

Fahrenheit Scale: Temperature scale on which water freezes at 32 °F and boils at 212 °F.

Family of Elements: See Group.

Faraday ($\mathscr{F}$): One mole of electrons; 96,500 coulombs.

Fatty Acid: One of several long-chain carboxylic acids produced by the hydrolysis of a lipid.

Fertile Isotope: An isotope that can be converted by nuclear reactions into a fissile isotope.

Filler: A substance added during polymerization or casting to prevent shrinkage of the substance as the polymer forms or takes its shape.

First Law of Thermodynamics: A formal statement of the law of conservation of energy—the basis for Hess's law.

Fissile Isotope: An isotope capable of undergoing fission following neutron capture.

Fission: The breaking apart of atomic nuclei and the source of energy in nuclear reactors.

Flame Test: Using the color produced by an ion such as sodium when the ion is introduced into a flame. Sodium gives a yellow color.

Flotation: A method for concentrating sulfide ores of copper and lead. Air is bubbled through a slurry of oil-coated ore particles, which stick to rising air bubbles and collect in the foam at the surface.

Fluid: Any material whose shape adjusts spontaneously and rather quickly to the shape of its container. (Both ordinary liquids and gases are fluids.)

Force: Anything that will cause an object to change its motion or direction.

Formula: Shorthand way of representing the composition of a substance using chemical symbols.

Formula Unit: A particle having the composition given by the chemical formula.

Formula Weight: The sum of the atomic weights of all the atoms represented in the chemical formula of a compound (or element).

Fossil Fuels: Coal, oil, and natural gas.

Free Energy: See Gibbs Free Energy.

Free Radical: An extremely reactive chemical species that contains an unpaired electron.

Frequency, ν: The number of cycles per second of electromagnetic radiation.

Functional Group: A small part of an organic molecule or organic ion that enters into a set of characteristic reactions.

Fusion: The formation of atomic nuclei by the fusing together of the nuclei of lighter isotopes.

Galvanic Cell: An electrochemical cell in which a spontaneous redox reaction produces electricity.

Galvanizing: Coating a steel object with zinc.

Gamma Radiation: A stream of very high energy photons emitted from atomic nuclei when they undergo energy transitions from higher nuclear energy levels to lower ones.

Gangue: The unwanted rock and sand that is separated from an ore.

Gas Constant, Universal (R): $R = 0.0821 \dfrac{\text{liter atm}}{\text{mol K}} = \dfrac{PV}{nT}$

Gas Law, Combined: For a given mass of gas, the product of its pressure and volume divided by its Kelvin temperature is a constant.

$$\frac{P_1 V_1}{T_1} = \frac{P_2 V_2}{T_2}$$

Gay-Lussac's Law: See Pressure-Temperature Law.

Generalization: Summary statement of behavior based on data collected in experiments.

Geometric Isomers: Isomers in which the atoms are arranged in different relative orientations. Among organic compounds, they are isomers with differences in the directions or geometries with which groups point at a double bond or at the carbons of a ring.

Gibbs Free Energy, G: A thermodynamic quantity that relates energy (enthalpy, H) and entropy, S. $G = H - TS$.

Graham's Law: See Effusion, Law of.

Gram (g): $1 \text{ g} = 0.001 \text{ kg}$.

Gray (Gy): The SI unit of radiation-absorbed-dose. $1 \text{ Gy} = 1 \text{ J/kg}$.

Ground State: Lowest energy state of an atom or molecule.

Group: Vertical column of elements in the periodic table.

Haber Process: The industrial synthesis of ammonia from nitrogen and hydrogen under pressure and heat in the presence of a catalyst.

Half-Cell: That part of a galvanic cell in which either oxidation or reduction takes place.

Half-Life, $t_{1/2}$: The time required for a reactant concentration or the mass of a radionuclide to be reduced by half.

Half-Reaction: An individual oxidation or reduction reaction that includes the correct formulas for all species taking part in the reaction; for example, $2H_2O(\ell) \rightarrow O_2(g) + 4H^+(aq) + 4e^-$.

Halogen Family: Group VIIA in the periodic table; fluorine, chlorine, bromine, iodine, and astatine.

Hard Water: Water with dissolved Mg^{2+}, Ca^{2+}, Fe^{2+}, or Fe^{3+} ions at a concentration (above 25 mg/liter) high enough to interfere with the use of soap.

Heat Capacity: The quantity of heat needed to raise the temperature of an object by 1 °C. Its units are cal/°C or J/°C, where °C = the *change* in temperature.

Heat of Combustion, Standard, $\Delta H^\circ_{combustion}$: The enthalpy of combustion for 1 mole of a compound under standard conditions.

Heat of Formation, Standard, ΔH°_f: The enthalpy change for the formation of 1 mole of a compound from its elements, all in their standard states.

Heat of Reaction, Standard, ΔH°: The value of ΔH for a reaction conducted so as to maintain the pressure at 1 atm and the temperature at 25 °C.

Henderson-Hasselbalch Equation:

$$pH = pK_a + \log \frac{[anion]}{[acid]}.$$

Henry's Law: See *Pressure-Solubility Law*.

Hertz (Hz): 1 Hz = 1 cycle/second; the SI unit of frequency.

Hess Law Equation: For the change: $aA + bB + \ldots \rightarrow nN + mM + \ldots$ $\Delta H^\circ = [n\,\Delta H^\circ_f(N) + m\,\Delta H^\circ_f(M) + \ldots] - [a\,\Delta H^\circ_f(A) + b\,\Delta H^\circ_f(B) + \ldots].$

Heterocyclic Compound: An organic ring compound in which one (or more) of the ring atoms is from a nonmetal other than carbon.

Heterogeneous: Two or more phases with different properties.

Heterogeneous Catalyst: A catalyst that is in a different phase than the reactants. The reactants are adsorbed on the catalytic surface where the reaction occurs.

Heterogeneous Reaction: A reaction in which the reactants and/or products are in different phases.

Homogeneous. Uniform properties throughout a sample.

Homogeneous Catalyst: A catalyst that is in the same phase as the reactants.

Homogeneous Reaction: A reaction in which all of the reactants and products are in the same phase.

Hund's Rule: The lowest energy electron configuration results when electrons that occupy orbitals of equal energy are spread out over the orbitals as much as possible with spins unpaired.

Hybrid Atomic Orbitals: Orbitals formed by mixing the basic atomic orbitals of an atom. They are more effective at overlap than ordinary atomic orbitals.

Hydrate: A compound holding molecules of water usually in a stoichiometric ratio.

Hydration: The development of a solvent cage of water molecules about polar molecules or ions dissolved in water.

Hydride: (1) A binary compound of hydrogen. (2) A compound containing the H^- (hydride) ion.

Hydrocarbon: A compound whose molecules consist entirely of carbon and hydrogen.

Hydrogen Bond: An extra strong dipole-dipole attraction that occurs between molecules in which hydrogen is covalently bonded to nitrogen, oxygen, or fluorine.

Hydrogen Electrode: The standard of comparison for reduction potentials, $2H^+(aq) + 2e^- \rightleftharpoons H_2(g)$, $E^\circ_{H^+} = 0.00$ V at 25 °C, 1 atm, and 1 M H^+.

Hydrolysis of Salts: The reaction of a salt with water to affect the pH of the solution.

Hydronium Ion: H_3O^+.

Hydrophilic Group: Any polar organic group capable of participating in forming hydrogen bonds with water molecules.

Hydrophobic Group: Any nonpolar organic group that cannot be involved in hydrogen bonds with water.

Hypertonic Solution: A solution whose osmotic pressure is greater than a reference solution.

Hypothesis: Tentative explanation of the results of experiments.

Hypotonic Solution: A solution whose osmotic pressure is less than a reference solution.

Ideal Gas: A hypothetical gas that obeys the gas laws exactly.

Ideal Gas Law: $PV = nRT$.

Ideal Solution: The hypothetical solution that would obey the vapor pressure—concentration law (Raoult's law) exactly.

Induced Dipole: A dipole created when the electron cloud of an atom or a molecule is distorted by a neighboring dipole or by an ion.

Inert Gases: The noble gases.

Initiation Step: The first step in a chain reaction in which a reactant molecule is converted to one or more free radicals.

Initiator: A substance that, like a catalyst, promotes a reaction, but is used up during the reaction.

Inner Transition Elements: The two long rows of elements below the main body of the periodic table. They include elements 58–71 and 90–103.

Instantaneous Dipole: A momentary dipole caused by the erratic movement of electrons.

Intermolecular Attractions: Attractions *between* neighboring molecules.

Inverse Square Law: In the science of atomic radiations, the relationship between radiation intensity and distance:

radiation intensity $\propto \dfrac{1}{d^2}$.

Ion: An electrically charged particle.

Ion Product: A product of ion concentrations, each of which is raised to a power that is equal to the number of ions of that kind obtained from one formula unit of the salt. The mass action expression for a solubility equilibrium.

Ion-Electron Method: A method for balancing redox reactions that uses half-reactions.

Ionic Bond: The attractions between ions that hold them together in ionic compounds.

Ionic Compound: A compound composed of positive and negative ions (e.g., NaCl).

Ionic Crystal: A crystal that has ions located at the lattice points.

Ionization Energy (IE): The energy needed to remove an electron from an isolated gaseous atom, ion, or molecule (usually expressed in units of kJ/mol).

Ionizing Radiation: Radiation arising from radioactivity or other nuclear reactions that generates ions as it passes through matter.

Ion-Product Constant of Water: $[H^+][OH^-] = K_w$.

Isomer: One of a set of compounds having identical molecular formulas but different structures.

Isomerism: The phenomenon of the existence of sets of compounds having the same molecular formula but different structures.

Isotonic Solution: A solution whose concentration is such that it has the same osmotic pressure as a reference solution.

Isotope Dilution Analysis: An analytic technique for measuring volumes by studying the extent to which the specific activity of a sample of radioactive material is reduced by its being diluted into the larger volume.

Isotopes: Atoms of the same element having different mass numbers caused by differences in their numbers of neutrons.

IUPAC Rules: The formal rules for naming compounds as developed by the International Union of Pure and Applied Chemistry.

Joule (J): The SI unit of energy. $1 \text{ J} = \dfrac{1 \text{ kg m}^2}{s^2}$; $4.184 \text{ J} = 1$ cal.

K_a: See *Acid Ionization Constant*.

K_b: See *Base Ionization Constant*.

K_w: See *Ion-Product Constant of Water*.

Kelvin Scale: Absolute temperature scale having degree units called kelvins (K). Degree size is the same as the Celsius degree. K = °C + 273.15.

Ketone: An organic compound with the $-\overset{|}{\underset{|}{C}}-\overset{O}{\overset{\|}{C}}-\overset{|}{\underset{|}{C}}-$ group.

Kilogram (kg): 1 kg = 1000 g; the base unit for mass in the SI.

Kinetic Energy: Energy of motion: $KE = \frac{1}{2}mv^2$.

Kinetic Theory of Gases: A set of postulates about gases used to explain the gas laws. A gas consists of an extremely large number of very tiny, hard particles in constant, random motion. They have negligible volume and, between collisions, they move in straight lines.

Lanthanide Contraction: The decrease in size experienced by elements 58–71 that causes the elements that follow the lanthanides to have unusually small sizes.

Lanthanide Elements: Elements 58–71.

Lattice Energy: Energy released by the imaginary process in which isolated ions come together to form a crystal of an ionic compound.

Law: Statement of behavior based on the results of many experiments. Laws do not offer *explanations* of behavior.

Law of Conservation of Mass: Mass is neither gained nor lost in a chemical change; mass is conserved.

Law of Constant Heat Summation (Hess's Law): For any reaction that can be written in steps, the standard heat of reaction is the sum of the standard heats of reaction for the steps.

Law of Definite Proportions: In a given chemical compound, the elements are always combined in the same proportion by mass.

Law of Multiple Proportions: Whenever two elements form more than one compound, the different masses of one that combines with the same mass of the other are in the ratio of small whole numbers.

Le Châtelier's Principle: When a system that is in dynamic equilibrium is subjected to a disturbance that upsets the equilibrium, the system undergoes a change that counteracts the disturbance and restores equilibrium.

Lewis Acid: An electron-pair acceptor.

Lewis Base: An electron-pair donor.

Lewis Structure: Also called *Lewis formula*. A structural formula drawn with Lewis symbols that shows the valence electrons using dots and dashes.

Lewis Symbol: The symbol of an element surrounded by dots that represent the valence electrons of an atom of that element.

Ligand: A molecule or anion that can bind to a metal ion to form a complex.

Lipid: Any substance found in plants or animals that can be dissolved in nonpolar solvents.

Liquid Crystal: A substance that is able to flow like a liquid but that has some physical properties normally associated with crystals.

Liter: 1 liter = 1 dm³ = 1000 cm³ = 1000 ml.

London Forces: Weak attractive forces caused by instantaneous dipole-induced dipole attractions.

Lone Pair: A pair of electrons in the valence shell of an atom that is not shared with another atom. An unshared pair of electrons.

Macromolecule: A molecule with hundreds to thousands of atoms such as found in polymers.

Magic Numbers: In nuclear chemistry, the numbers 2, 8, 20, 50, 82, and 126, and whose significance is that nuclides are expected to be particularly stable when the number of protons or neutrons (and especially both) equals one of these numbers.

Magnetic Quantum Number: m, which can have integer values from $-\ell$ to $+\ell$.

Malleability: A metal's ability to be hammered or rolled into thin sheets.

Manganese Nodules: Lumps the size of an orange that contain large concentrations of manganese and iron. They are found on the ocean floor.

Mass: A measure of the amount of matter that there is in a given sample.

Mass Action Expression: A fraction in which the numerator is the product of the molar concentrations of the products raised to powers equal to their coefficients and the denominator is the product of the molar concentrations of the reactants raised to powers equal to their coefficients. For gaseous reactions, partial pressures can be used in place of molar concentrations.

Mass Number: The numerical sum of the protons and neutrons in an atom of a given isotope.

Matter: Anything that has mass and occupies space.

Mechanism: The series of individual steps in a chemical reaction that gives the net, overall change.

Metallic Crystal: A solid having positive ions at the lattice positions that are attracted to a "sea of electrons" that extends throughout the entire crystal.

Metalloids: Elements with properties that lie between those of metals and nonmetals. They are located around the diagonal line running from boron (B) to astatine (At) in the periodic table.

Metallurgy: The science and technology of metals; it is concerned with the procedures and chemical reactions that are used to separate metals from their ores and make them ready for practical uses.

Metathesis Reaction: See *Double Replacement Reaction.*

Meter (m): SI base unit for length.

Metric System: A decimal system of units.

Milliliter (ml): 0.001 liter = 1 ml; 1000 ml = 1 liter

Millimeter (mm): 0.001 m = 1 mm; 1000 mm = 1 m.

Millimeter of Mercury (mm Hg): A unit of pressure equal to $\frac{1}{760}$ atm. 760 mm Hg = 1 atm.

Mineral: A specific solid in the earth's crust having a more or less definite chemical formula.

Mixture: Any matter consisting of two or more substances physically combined in no particular proportion by mass.

Model, Scientific: A picture or a mental construction derived from a set of ideas and assumptions that we imagine to be true because they enable us to explain certain observations and measurements (e.g., the model of an ideal gas; Bohr's atomic model).

Molal Concentration (*m*): The number of moles of solute in 1000 g of solvent; the molality of a solution.

Molality: See *Molal Concentration.*

Molar Concentration (*M*): The number of moles of solute per liter of solution.

Molar Heat of Fusion: Heat absorbed when 1 mol of a solid melts to give 1 mol of liquid at constant temperature and pressure.

Molar Heat of Sublimation: Heat absorbed when 1 mol of solid sublimes to give 1 mol of vapor at constant temperature and pressure.

Molar Heat of Vaporization: Heat absorbed when 1 mol of liquid is converted to 1 mol of vapor at constant temperature and pressure.

Molar Mass: The number of grams per mole of a compound or element.

Molar Solubility: The number of moles of solute required to give 1 liter of a saturated solution of the solute.

Molar Volume, Standard: The volume of 1 mol of a gas at STP; 22.4 liter/mol.

Molarity: See *Molar Concentration.*

Mole (mol): The SI unit for amount of substance; the formula weight in grams; a quantity of chemical substance containing 6.02×10^{23} formula units.

Mole Fraction: The ratio of the number of moles of one component of a mixture to the total number of moles of all components.

Mole Percent: The mole fraction of a component expressed as a percent.

Molecular Crystal: A crystal that has molecules or individual atoms at the lattice points.

Molecular Equation: A chemical equation giving the full formulas of all the reactants and products that will be involved in an actual experiment.

Molecular Formula: A chemical formula that gives the actual composition of one molecule.

Molecular Orbital Theory: Theory of covalent bonding that views a molecule as a collection of positive nuclei surrounded by electrons distributed among a set of bonding and antibonding molecular orbitals of different energies.

Molecular Orbitals: Orbitals that extend over two or more atomic nuclei.

Molecular Weight: See *Formula Weight.*

Molecule: A neutral particle composed of two or more atoms.

Monodentate Ligand: A ligand that can attach itself to a metal ion by only one atom.

Monomer: A substance of low formula weight used to make a polymer.

Monoprotic Acid: An acid that can furnish one H^+ per molecule.

Monosaccharide: A carbohydrate that cannot be hydrolyzed.

Nematic Liquid Crystal: Composed of long, rodlike molecules packed like short pieces of uncooked spaghetti.

Nernst Equation: $E_{cell} = E^\circ_{cell} - \frac{0.0592}{n} \log Q.$

Net Ionic Equation: A chemical equation having no spectator ions or molecules and that satisfies both a material and an electrical balance.

Neutral Solution: A solution in which $[H^+] = [OH^-]$.

Neutralization, Acid-Base: The destruction of an acid by a base or of a base by an acid.

Neutron: A subatomic particle with a charge of zero and a mass of 1.008665 amu (1.674954×10^{-24} g) that exists in all atomic nuclei except those of the hydrogen-1 isotope. Its symbol is n, or ^1_0n in nuclear equations.

Neutron Activation Analysis: A technique for analyzing for trace impurities by rendering them radioactive through neutron bombardment and then studying the frequencies and intensities of the gamma radiations they emit.

Neutron Emission: A nuclear reaction in which a neutron is ejected.

Nitrogen Family: Group VA in the periodic table; nitrogen, phosphorus, arsenic, antimony, and bismuth.

Nitrogen Fixation: The conversion of atmospheric nitrogen into chemical forms used by plants and accomplished by soil microorganisms with the assistance of plants.

Noble Gases: Elements of Group 0 in the periodic table; helium, neon, argon, krypton, xenon, and radon.

Node: A place where the amplitude or intensity of a wave is zero.

Nonelectrolyte: A compound that in its molten state or in solution will not conduct electricity.

Nonstoichiometric Compound: A substance in which the elements are combined in a not quite whole number ratio by moles.

Nonvolatile: Describing a substance with a high boiling point, a low vapor pressure, and that does not evaporate.

Normal Boiling Point: The temperature at which the vapor pressure of a liquid equals 1 atm.

Normality (*N*): The number of equivalents per liter of an acid or a base or other reactant.

Nuclear Equation: A shorthand expression of a nuclear reaction that is balanced when the sums of the atomic numbers on either side of the arrow are equal and the sums of the mass numbers are equal.

Nucleic Acids: Polymers in plants and animals that are involved in the chemistry of heredity and whose molecules, when hydrolyzed, give a sugar unit (ribose or deoxyribose), a phosphate ion, and a set of four or five nitrogen-containing, basic hetercyclic compounds (adenine, thymine, guanine, cytosine, and uracil).

Nucleon: A general name for protons or neutrons.

Nucleus: The hard, dense core at the center of an atom that contains all of the atom's protons and neutrons.

Octet Rule: An atom tends to gain or lose electrons until its outer shell consists of eight electrons.

Odd-Even Rule: In nuclear chemistry, a rule that says when the numbers of protons and neutrons in a nucleus are both even, the isotope is far more likely to be stable than when both numbers are odd.

Open Hearth Furnace: A furnace used to convert pig iron into steel.

Optical Isomers: Stereoisomers other than geometric (cis-trans) isomers.

Orbital: A particular electron waveform with a particular energy. In an atom, each orbital has a specific set of values for n, ℓ, and m.

Orbital Diagram: A diagram showing an atom's orbitals in which the electrons are represented by arrows to indicate paired and unpaired spins.

Order (of Reaction): The sum of the exponents in the rate law is the overall order of the reaction. Each exponent gives the order with respect to a specific reactant.

Ore: A mineral that can be extracted from the earth's crust and processed at a profit.

Organic Compound: Any compound of carbon other than a carbonate, bicarbonate, cyanide, cyanate, carbide, or gaseous oxide.

Osmolality: The molal concentration of all particles of solutes—molecules or separated ions—that contribute to the osmotic pressure of a solution.

Osmolarity: The molar concentration of all particles of solutes—molecules or separated ions—that contribute to the osmotic pressure of a solution.

Osmosis: The passage of solvent molecules, but not those of solutes, through a semipermeable membrane; the limiting case of dialysis.

Osmotic Pressure: The back pressure that would have to be applied to prevent osmosis; a colligative property.

Ostwald Process: An industrial synthesis of nitric acid from ammonia.

Overlap: A portion of two orbitals from different atoms sharing the same space.

Oxidation: Loss of electrons; increase in oxidation number.

Oxidation Number: The charge an atom in a compound would have if all of the electrons in the bonds belonged entirely to the more electronegative atoms.

Oxidation State: Means the same as oxidation number.

Oxidation-Reduction Reaction: A reaction involving the transfer of electrons from one species to another.

Oxidizing Agent: Substance that causes oxidation; it is reduced.

Oxoacid: An acid containing hydrogen, oxygen, and another element. (e.g., HNO_3, H_3PO_4, H_2SO_4).

Oxygen Family: Group VIA in the periodic table: oxygen, sulfur, selenium, tellurium, and polonium.

Paramagnetism: Weak magnetism that a substance has when it contains unpaired electrons.

Partial Charge: Charges at opposite ends of a dipole that are less than full $1+$ or $1-$ charges.

Partial Pressure: The pressure contributed by an individual gas in a mixture of gases.

Partial Pressures, Law of (Dalton's Law of Partial Pressures): The total pressure of a mixture of gases equals the sum of their partial pressures. $P_t = P_a + P_b + P_c + \ldots$

Pascal(Pa): The SI unit of pressure; 133.3224 Pa = 1 torr.

Pauli Exclusion Principle: No two electrons in the same atom can have the same values for all four of their quantum numbers.

Peptide Bond: The amide linkage in protein chemistry:

$$\begin{matrix} & O & \\ & \| & | \\ —& C — N & — \end{matrix}$$

Percent Concentration:

Weight/weight (w/w)	The grams of solute in 100 g of solution.
Volume/volume (v/v)	The number of volumes of solute in 100 v of solution.
Weight/volume (w/v)	The grams of solute in 100 ml of solution.
Milligram percent (mg %)	The milligrams of solute in 100 ml of solution.

Percentage by Weight: The number of grams of an element combined in 100 g of a compound.

Percentage Composition: A complete list of the percentages by weight of a compound's elements.

Period: Horizontal row of elements in the periodic table.

Peroxide: A compound whose molecules have two oxygen atoms joined by a single bond.

pH: $pH = -\log [H^+]$.

Phase: A homogeneous region within a sample.

Phase Diagram: A pressure-temperature graph on which are plotted temperatures and pressures at which equilibrium exists between the states of a substance. It defines $T-P$ regions in which the solid, liquid, and gaseous states of the substance can exist.

Photon: "Packet" of energy in electromagnetic radiation; its energy, $E = h\nu$ where ν is the frequency and h is Planck's constant.

Photosynthesis: The use of solar energy by a plant to synthesize high-energy molecules from simpler, low-energy molecules of carbon dioxide and water. The primary products are carbohydrates. The plant's chlorophyll is the solar energy absorber.

Physical Property: Property that can be specified without reference to another substance, for example, color or volume.

Pi Bond (π Bond): A bond formed by the sideways overlap of a pair of p orbitals. Electron density is concentrated in two separate regions that lie on opposite sides of the imaginary line joining the nuclei.

Pig Iron: The impure iron that comes from the blast furnace.

pK_a: pK_a = −log K_a.

pK_b: pK_b = −log K_b.

Plastic: A finished or semifinished article made by molding, casting, extruding, drawing, or laminating a resin.

Plasticizer: An oily organic material added to a resin to reduce hardness and brittleness in the final plastic article.

pOH: pOH = −log [OH^-].

Polar Covalent Bond (Polar Bond): A covalent bond in which more than half of the bond's negative charge is concentrated around one of the two atoms.

Polyatomic Ion: An ion composed of two or more atoms.

Polydentate Ligand: A ligand that has two or more atoms that can become simultaneously attached to a metal ion.

Polymer: A substance consisting of macromolecules having repeating structural units.

Polymerization: A chemical reaction that converts a monomer into a polymer.

Polyolefin: A polymer of any member of the alkene ("olefin") family.

Polypeptide: A polymer of α-amino acids that makes up all or most of a protein.

Polyprotic Acid: An acid that can furnish more than one H^+ per molecule.

Polysaccharide: A carbohydrate whose molecules can be hydrolyzed to hundreds of monosaccharide molecules.

Positron: A positively charged particle with the mass of an electron. Its symbol is ${}^0_1p$.

Post-transition Metal: A metal that occurs in the periodic table immediately following a row of transition elements.

Potential Energy: Stored energy.

Precipitate: A solid that separates from a solution usually as the result of a chemical reaction.

Precipitation: In chemistry, the formation of a solid that separates from a solution usually as the result of a chemical reaction.

Precision: How reproducible measurements are.

Pressure: Force per unit area.

Pressure-Solubility Law (Henry's Law): The concentration of a gas dissolved in a liquid at any given temperature is directly proportional to the partial pressure of that gas on the solution.

Pressure-Temperature Law (Gay-Lussac's Law): The pressure of a given mass of gas is directly proportional to its Kelvin temperature if the volume is kept constant.

$$\frac{P_1}{T_1} = \frac{P_2}{T_2}$$

Pressure-Volume Law (Boyle's Law): The volume of a given mass of gas is inversely proportional to its pressure (at constant temperature). $P_1V_1 = P_2V_2$.

Principal Quantum Number: n, which can have values of 1, 2, 3, . . . , ∞.

Products: Substances whose formulas appear on the right side of the arrow in a chemical equation. The substances produced in a chemical reaction.

Propagation Step: A step in a chain reaction in which a free radical reacts with a reactant molecule to give a product molecule and another free radical.

Property: A characteristic of matter.

Protein: A macromolecular substance found in nature consisting wholly or mostly of one or more polypeptides often combined with an organic molecule and sometimes also combined with a metal ion.

Proton: (1) A subatomic particle with a charge of 1+ and a mass of 1.007276 amu (1.672649 × 10^{-24} g) that exists in atomic nuclei. (2) The name often used for the hydrogen ion and symbolized as H^+ in acid-base discussions and either ${}^1_1p$ or ${}^1_1H$ in discussions of transmutations.

Pure Substance: Element or compound.

Quantized Energy: Discrete amount of energy.

Quantum: The energy of one photon.

Quantum Mechanics: See *Wave Mechanics.*

Quantum Number: A number related to the energy, shape, or orientation of an orbital. Also, a number related to the spin of the electron.

R: See *Gas Constant, Universal.*

Rad (rd): A unit of radiation-absorbed-dose. 1 rd = 10^{-5} J/g = 10^{-2} Gy.

Radioactive Decay: The change of a nucleus into another nucleus (or a more stable form) by the loss of an alpha or beta particle or a gamma ray photon.

Radioactive Disintegration Series: A sequence of nuclear reactions beginning with an unstable actinide and ending with a stable isotope of lower atomic number.

Radioactivity: The emission of one or more kinds of radiation from an isotope with unstable nuclei.

Radiological Dating: A technique for measuring the age of some geologic formation or ancient artifact by measuring the ratio of the concentrations of two isotopes, one radioactive and the other a stable decay product.

Radionuclide: A radioactive isotope.

Raoult's Law: See *Vapor Pressure—Concentration Law.*

Rare Earth Metals: The lanthanides.

Rate: A ratio in which units of time appear in the denominator, for example, 40 miles/hr or 3.0 mol/liter/s.

Rate Constant: The proportionality constant in the rate law; the rate of reaction when all reactant concentrations are 1 *M.*

Rate of Reaction: How quickly the reactants disappear and the products form, expressed in units of mol liter^{-1} s^{-1}.

Rate Law: An equation that relates the rate of a reaction to the molar concentrations of the reactants raised to powers.

Rate-Determining Step (Rate Limiting Step): The slow step in a reaction mechanism.

Reactant, Limiting: The reactant in a given reaction that is present in a deficient quantity according to the stoichiometry of the reaction.

Reactants: Substances whose formulas appear on the left side of the arrow in a chemical equation. The substances brought together to react.

Reaction Quotient: See *Mass Action Expression.*

Reagent: Any chemical, mixture of chemicals, or solution of chemicals used as part of an experimental reaction.

Redox Reaction: Oxidation-reduction reaction.

Reducing Agent: Substance that causes reduction; it is oxidized.

Reduction: Gain of electrons; decrease in oxidation number.

Reduction Potential: A measure of the tendency of a given half-reaction to occur as a reduction.

Rem: A dose in rads multiplied by a factor that takes into account the variations that different radiations have in their damage-causing abilities in tissue.

Replication: In nucleic acid chemistry, the reproductive duplication of DNA double helices prior to cell division.

Representative Element: An element in one of the A-groups in the periodic table.

Resin: A raw, unfabricated polymer used to make all or most of a plastic article.

Resonance: A concept in which the actual Lewis structure of a molecule or polyatomic ion is represented as a composite or average of two or more resonance structures. The individual resonance structures themselves do not actually exist.

Resonance Hybrid: The actual structure of a molecule or polyatomic ion that is represented by two or more resonance structures.

Reversible Process: A process that occurs by an infinite number of steps during which the driving force for the change is just barely greater than the force that resists the change.

Ring Compound: A compound whose molecules include closed-chain sequences of atoms.

RNA: Ribonucleic acid; nucleic acid that, when hydrolyzed, gives ribose, phosphate ion, and chiefly four heterocyclic amines—adenine, uracil, guanine, and cytosine. Varieties include:

*m*RNA, or messenger RNA made at the direction of DNA, and the carrier of the genetic message (as a series of codons) from the cell nucleus to the assembly site for making polypeptides.

*t*RNA, or transfer RNA—amino acid carrier RNA.

Roasting: Heating an ore in air, which converts sulfides to oxides.

Rock: A complex mixture of minerals.

Salt-Bridge: A tube containing an electrolyte that connects the two half-cells of a galvanic cell.

Saturated Organic Compound: Any organic compound having only single bonds.

Saturated Solution: A solution in which there is an equilibrium between the dissolved and undissolved states of the solute.

Scientific Method: Observation, explanation, and testing of an explanation by additional experiments.

Scientific Notation: Numbers represented as a decimal number between 1 and 10 multiplied by 10 raised to a power; for example, 6.02×10^{23}.

Second law of thermodynamics: Whenever a spontaneous event takes place it is accompanied by an increase in the entropy of the universe.

Secondary Quantum Number: ℓ, which can have values of $0, 1, \ldots, (n - 1)$.

Semiconductor: A substance that conducts electricity weakly.

Shell: All orbitals of a given n.

SI (International System of Units): Modified metric system adopted in 1960 by the General Conference on Weights and Measures.

Sigma Bond (σ Bond): A bond formed by the "head-to-head" overlap of two orbitals. The electron density is concentrated along the imaginary line joining the two nuclei.

Significant Figures: In a measured quantity, the number of digits known for sure plus the first one that is uncertain.

Simple Cubic Unit Cell: A cubic unit cell with atoms, molecules, or ions only at the corners.

Single Bond: A covalent bond in which a single pair of electrons are shared.

Slag: A relatively low melting mixture of impurities that forms in the blast furnace and other furnaces used in refining metals.

Smectic Liquid Crystals: Rodlike molecules in layers of parallel rods.

Sol: The colloidal dispersion of a solid in a fluid.

Solubility: The ratio of solute to solvent in a saturated solution; usually expressed as the number of grams of solute per 100 g of solvent at a specified temperature.

Solubility Product Constant, K_{sp}: The equilibrium constant for the solubility of a salt. For a saturated solution, K_{sp} is equal to the product of the molar concentrations of the ions, each raised to an appropriate exponent.

Solute: Something dissolved in a solvent to make a solution.

Solution: A homogeneous mixture in which all particles are the size of atoms, small molecules or small ions.

Solvation: The development of a cagelike network of solvent molecules about a molecule or ion of the solute.

Solvay Process: Used to prepare Na_2CO_3 from $NaCl$, CO_2, and NH_3.

Solvent: A medium, usually a liquid, into which something (the solute) is dissolved to make a solution.

sp Hybrid Orbitals: Hybrid orbitals formed by mixing one *s* and one *p* atomic orbital. The angle between a pair of *sp* hybrids is 180°.

sp^2 Hybrid Orbitals: Hybrid orbitals formed by mixing one *s* and two *p* atomic orbitals. The angle between any two sp^2 hybrids is 120°.

sp^3 Hybrid Orbitals: Hybrid orbitals formed by mixing one *s* and three *p* atomic orbitals. The angle between any two sp^3 hybrids is 109.5°.

Specific Heat: The quantity of heat that will raise the temperature of 1 g of a substance by 1 °C. (Its units are cal/g °C, where °C = the *change* in temperature.)

Spectrochemical Series: A listing of ligands in order of their ability to produce a large crystal field splitting.

Spin Quantum Number: m_s, which can have values of $+\frac{1}{2}$ and $-\frac{1}{2}$.

Spontaneous Change: A change that occurs by itself without outside assistance.

Standard Cell Potential: The potential of a galvanic cell at 25 °C and 1 atm when all ionic concentrations are exactly 1 *M*.

Standard Conditions of Temperature and Pressure (STP): 273 K (0 °C) and 760 Torr.

Standard Enthropy, $S°$: The entropy 1 mol of a substance has at 25 °C and 1 atm.

Standard Entropy Change, $\Delta S°$: $\Delta S° =$ (Sum of $S°$ of products) − (Sum of $S°$ of reactants).

Standard Free Energy Change, $\Delta G°$: $\Delta G° = \Delta H° - T\Delta S°$.

Standard Reduction Potential: The reduction potential of a half-reaction at 25 °C and 1 atm when all ion concentrations are 1 M.

Standard Solution: Any solution whose concentration is accurately known.

Standard State: The condition in which a substance is in its most stable form at 25 °C and 1 atm.

Standing Wave: A wave whose peaks and nodes do not change position.

State: Solid, liquid, or gas. In thermodynamics, a particular set of conditions of temperature, composition, and pressure.

State Function: A thermodynamic function whose value depends only on the initial and final states of the system and not on the path used to get from the initial to the final state. P, V, T, H, S, and G are all state functions.

State of a System: The specified values of the physical properties of a system—physical form, composition, concentration, temperature, and pressure.

Stereoisomers: Isomers whose structures differ only in spatial orientations (geometric isomers; optical isomers).

Stock System: System of nomenclature that uses Roman numerals to specify oxidation states.

Stoichiometry: A description of the relative quantities by moles of the reactants and products in a reaction as given by the coefficients in the balanced equation.

STP: See *Standard Conditions of Temperature and Pressure.*

Straight-Chain Compound: An organic compound in whose molecules the carbon atoms are joined in one continuous open chain sequence.

Strong Acid: An acid that is essentially 100% ionized in water. A good proton donor. An acid with a very large value of K_a.

Strong Base: Any powerful proton acceptor; a base with a very large value of K_b; a metal hydroxide that ionizes almost 100% in water.

Structural Formula: A chemical formula that shows how the atoms of a molecule are arranged and to which other atoms they are bonded.

Sublimation: Conversion of a solid directly to a gas without passing through the liquid state.

Subshell: All orbitals of a given shell that have the same value of ℓ.

Substrate: In enzyme chemistry, the substance whose reaction is catalyzed by the enzyme.

Supercooled Liquid: A liquid at a temperature below its freezing point. An amorphous solid.

Supercooling: Cooling a liquid to a temperature below its freezing point.

Supercritical Fluid: A substance at a temperature above its critical temperature.

Superimposability: A test of molecular chirality in which a model of a molecule and a model of the mirror image of this molecule are compared to see if the two can be made to blend perfectly, with every part of one coinciding simultaneously with the parts of the other.

Superoxide: An oxide containing the O_2^- ion.

Supersaturated Solution: Any solution containing more solute than a saturated solution would ordinarily hold at the given temperature.

Surface Tension: A measure of the amount of energy needed to expand the surface area of a liquid.

Surfactant: A substance that lowers the surface tension of a liquid and promotes wetting.

Surroundings: That part of the universe other than the system being studied and separated from the system by a real or an imaginary boundary.

Suspension: A homogeneous mixture in which the particles of one (or more) of the substances are relatively large (100 nm in one dimension) and are dispersed uniformly in some medium.

System: That part of the universe under study and separated from the rest of the universe (the surroundings) by a real or an imaginary boundary.

$t_{1/2}$: See *Half-Life.*

Temperature-Volume Law (Charles' Law): The volume of a given mass of gas is directly proportional to its Kelvin temperature if the pressure is kept constant.

$$\frac{V_1}{T_1} = \frac{V_2}{T_2}$$

Theory: Tested explanation of the results of many experiments.

Thermite Reaction: The very exothermic reaction, $2Al + Fe_2O_3 \rightarrow 2Fe + Al_2O_3$.

Thermodynamics: The study of energy changes and the flow of energy from one substance to another.

Thermometer: Device used to measure temperature.

Thermotropic Substance: A substance that behaves like a liquid crystal over a limited temperature range.

Third Law of Thermodynamics: For a pure crystalline substance, $S = 0$ at 0 K.

Torr: A unit of pressure equal to 1 mm Hg.

Trace Element: In enzyme chemistry, any metal or nonmetal ion needed by an enzyme.

Tracer Analysis: The use of radioactivity to trace the movement of something or to locate the site of radioactivity.

Transcription: In nucleic acid chemistry, the synthesis of *m*RNA at the direction of DNA.

Transition Elements: Elements in the periodic table located between Groups IIA and IIIA.

Transition Metals: Transition elements.

Transition State: The brief moment during a reaction when the reactants have collided and are at the high point on the potential energy diagram for the reaction.

Translation: In nucleic acid chemistry, the synthesis of a polypeptide at the direction of a molecule of *m*RNA.

Transmutation: The conversion of one isotope into another.

Transuranium Elements: Elements 93 and higher.

Traveling Wave: A wave whose peaks and nodes move.

Triple Bond: A covalent bond in which three pairs of electrons are shared.

Triple Point: The temperature and pressure at which the liquid, solid, and vapor states of a substance can coexist in equilibrium.

Triprotic Acid: An acid that can furnish three H^+ per molecule.

Tritium: Hydrogen-3; ^3_1H (sometimes ^3_1t); an isotope of hydrogen.

Tyndall Effect: The scattering of light by colloidally dispersed particles that gives a milky appearance to the mixture.

Uncertainty Principle: There are limits in our ability to measure a particle's speed and position simultaneously.

Unit Cell: The smallest portion of a crystal that can be repeated over and over in all directions to give the crystal lattice.

Unsaturated Solution: Any solution with a concentration less than that of a saturated solution of the same solute.

Vacuum: An enclosed space containing no matter whatsoever. A partial vacuum is an enclosed space containing a gas at a very low pressure.

Valence Bond Theory: Theory of covalent bonding that views a bond as being formed by the sharing of one pair of electrons between two overlapping atomic or hybrid orbitals.

Valence Electrons: Electrons in the valence shell.

Valence Shell: In an atom, the shell with highest n.

Valence Shell Electron Pair Repulsion Theory: A theory used to predict molecular structure. It is based on the idea that electron pairs (bonding and lone pairs) in the valence shell of an atom stay as far apart as possible.

Vapor Pressure: The pressure exerted by the vapor above a liquid. If the liquid and vapor are in equilibrium, this pressure is the *equilibrium vapor pressure*.

Vapor Pressure—Concentration Law (Raoult's Law): The vapor pressure of one component above a mixture of molecular compounds equals the product of its vapor pressure when pure (at the same temperature) and its mole fraction.

Vitamin: An organic compound present in trace concentration in various foods that is essential for health but that cannot be made by the body and whose absence leads to a specific vitamin-deficiency disease.

Volatile: Describing a liquid that has a low boiling point, a high vapor pressure and, therefore, evaporates easily.

Volt (V): The SI unit of electrical potential or emf. 1 V = 1 J/C

Water of Hydration: Water molecules trapped in a solid substance, usually in a stoichiometric ratio to the other material.

Wave Mechanics: Theory of atomic structure based on the wave properties of matter.

Wavelength: The distance between crests in the wavelike oscillations of electromagnetic radiations; symbol, λ.

Weak Acid: A poor proton donor; an acid with a value of K_a less than 10^{-3}; any acid with a low percentage ionization in water.

Weak base: Any poor proton acceptor; a base with a low value of K_b; any base with a low percentage ionization in water.

Weight: The force with which a substance is attracted to the earth by gravity.

Wetting: The spreading of a liquid across a surface.

X Ray: A stream of very high-energy photons emitted by substances when they are bombarded by high-energy beams of electrons or are emitted by radionuclides that have undergone K-electron capture.

Yield, Percentage: The ratio, given as a percent, of the quantity of a product actually obtained to the theoretical yield.

Yield, Theoretical: The quantity of a product calculated by the stoichiometry of the reaction.

PHOTO CREDITS

Chapter Eleven

Figure 11.1: *b*, Courtesy Sargent-Welch; *c*, Courtesy Beckman Instruments, Spinco Division. Figure 11.3: U.S. Forest Service.

Chapter Twelve

Figure 12.1: J. Brady & K. Bendo. Figure 12.2: NASA. Figure 12.3: Both, U.S. Forest Service. Figure 12.5: *a*, Runk/Schoenberger/Grant Heilman; *b*, Mario Fantin/Photo Researchers. Figure 12.6: Both, Schmidt-Thomsen, Landesdenkmalamt, Westfalen-Lippe, Muenster, Germany. Figure 12.7: Grant Heilman. Figure 12.8: M.E. Warren/Photo Researchers.

Chapter Thirteen

Figure 13.1: Courtesy of the New York Football Giants. Figure 13.2: Keith Gunnar/The National Audubon Society Collection-Photo Researchers. Figure 13.3: Courtesy of Medtronic, Inc. Figure 13.4: Grant Heilman. Page 431: Lorinda Morris. Figure 13.9: Courtesy of Dow Chemical. Figure 13.10: Courtesy of Sybron Nalge Company, Division of Sybron Corporation. Figure 13.11: Courtesy of Du Pont. Figure 13.12: Courtesy of Permabond International Corporation. Figure 13.13: Nogues/Marlow/Sygma. Figure 13.14: K. Bendo. Figure 13.15: Courtesy of General Electric. Figure 13.18: Courtesy of The American Museum of Natural History. Figure 13:19: Runk/Schoenberger/Grant Heilman.

Chapter Fourteen

Figure 14.1: United Press International. Page 474: NASA. Page 476: Left, An Exxon Photo; right, Courtesy D.J. McCollum, Mobil Research and Development Corporation. Figure 14.13: Courtesy of General Motors. Page 477: Courtesy of The Coleman Company.

Chapter Fifteen

Figure 15.1: Left, Robert A. Isaacs/Photo Researchers; center, Grant Heilman; right, United Press International. Figure 15.3: Left, Belinda Rain/EPA-Documerica; right, Michigan Department of Natural Resources.

Chapter Sixteen

Figures 16.6 and 16.7: J. Brady & K. Bendo. Page 531: Courtesy of The Permutit Company, Division of Sybron Corporation.

Chapter Seventeen

Page 547: K. Bendo & J. Brady. Figure 17.2: Courtesy of Beckman Instruments, Scientific Instruments Division. Pages 567 and 575: K. Bendo & J. Brady.

Chapter Eighteen

Page 599: Courtesy of General Motors and the Department of Energy. Page 607: The Royal Institution, London. Page 611: Courtesy of Kaiser Aluminum & Chemical Corporation. Page 612: Courtesy of Copper Development Association. Figure 18.14: K. Bendo. Page 617: Courtesy of Alyeska Pipeline Service Company.

Chapter Nineteen

Page 638: Top, Lorinda Morris; center, and bottom, K. Bendo. Page 640: Courtesy of Polaroid Corporation.

Chapter Twenty

Page 672: International Salt Company. Figure 20.1: Fred Lyon/Photo Researchers. Figure 20.2: National Science Foundation. Page 673, bottom: Standard Oil Company of New Jersey. Page 674: Bruce Roberts/Rapho-Photo Researchers. Pages 675 and 678: American Iron & Steel Institute. Page 680: J. Brady & K. Bendo. Page 683: Courtesy Kawecki Berylco Industries, Inc. Pages 684 and 688: K. Bendo. Page 685: Lester V. Bergman & Associates. Page 694: National Archives. Page 700: Top, J. Brady & K. Bendo; bottom, K. Bendo.

Chapter Twenty-One

Figure 21.10: Lawrence Radiation Laboratory. Figure 21.11: Courtesy of Nuclear Equipment Chemical. Figure 21.12 *a*: Courtesy of Landauer. Figures 21.13 and 21.14: Courtesy General Electric, Medical Systems Division.

Color Plates

Plate 1 A: K. Bendo. Plate 1 B: G. Tomsich/Photo Researchers. Plate 1 C: Sodium, K. Bendo & J. Brady; chlorine, from LIFE Science Library/Matter, published by Time-Life Books, Inc. Reproduced by permission; salt, K. Bendo. Plate 2 B: Courtesy General Electric, Nela Park Cleveland, Ohio. Plate 3 A: Russ Kinne/Photo Researchers. Plate 3 B: K. Bendo. Plate 3 C: Thomas R. Taylor/Photo Researchers. Plate 3 D: K. Bendo. Plate 3 E and F: K. Bendo & J. Brady. Plate 4 A: *Vein Finder*™ courtesy of Clinitemp; photo by K. Bendo. Plate 4 B: Courtesy of Texasgulf. Plate 4 C: Elizabeth Holum. Plate 5 A and C: K. Bendo. Plate 5 B: Mitchel L. Osborne/Image Bank. Plate 6 A: J. Brady. Plate 6 B: K. Bendo & J. Brady. Plate 7 A: K. Bendo. Plate 7 B: Model courtesy of John Mach, Jr.; photo copyright 1979 by Irving Geis. Plate 7 C: Both, H. Stern/Photo Researchers. Plate 8 A: Tom McHugh/Photo Researchers. Plate 8 B: Left, R. Rowan/Photo Researchers; right, Runk/Schoenberger/Grant Heilman. Plate 8 C: J. Brady & K. Bendo.

INDEX

Page numbers in *italics* indicate tables.